Teacher Edition

Discovering
Advanced
Algebra

THIRD EDITION

Pamela Weber Harris

Ellen Kamischke

Eric Kamischke

Jerald Murdock

Discovering
Mathematics

Kendall Hunt
publishing company

Project Director
Tim Pope

Project Editor
Holly Paige

Developmental Editor
Sandy Berger, Ph.D

Permission Editor
Tammy Hunt

Cover Designer
Tyler Reinhardt

Cover image © Shutterstock, Inc.

Chairman and Chief Executive Officer Mark C. Falb
President and Chief Operating Officer Chad M. Chandlee
Vice President, PreK-12 Division Charley Cook

www.kendallhunt.com
Send all inquiries to:
4050 Westmark Drive
Dubuque, IA 52004-1840
1-800-542-6657

Copyright © 2002, 2007 by Key Curriculum Press

Copyright © 2014, 2017 by Kendall Hunt Publishing Company

ISBN 978-1-4652-9043-4

All rights reserved. No part of this publication may be reproduced, stored in a retrieval system, or transmitted, in any form or by any means, electronic, mechanical, photocopying, recording, or otherwise, without the prior written permission of the copyright owner.

Published in the United States of America

CONTENTS

About the Authors ix

Features of the Student Edition x

Features of the Teacher Edition xv

Student and Teacher ebooks xix

The Common Core State Standards for Mathematics Content and Practice xx

Teaching with *Discovering Advanced Algebra* xxvi
        Differentiating Instruction   xxvi
        Assessing   xxviii
        Planning   xxix

Teaching a *Discovering Advanced Algebra* Lesson xxxii

Materials for Use with *Discovering Advanced Algebra* xxxviii

CHAPTER 0

Developing Mathematical Thinking 1

Chapter 0 Teaching Notes 1A
- **0.1 Modeling and Tools** 2
    Investigation: Polar Bear Crossing the Arctic   3
- **0.2 Reasoning and Explaining** 7
    Investigation: Problems, Problems, Problems   10
- **0.3 Looking for and Generalizing Patterns** 14
    Investigation: Who Owns the Zebra?   17

Chapter 0 Review 21
    Take Another Look   25
    Assessing What You've Learned   26

CHAPTER 1

Linear Modeling 27

Chapter 1 Teaching Notes 27A
- **1.1 Recursively Defined Sequences** 28
    Investigation: Monitoring Inventory   31
- **1.2 Modeling Growth and Decay** 38
    Investigation: Looking for the Rebound   39
- **1.3 Applications and Other Sequences** 46
    Investigation: Doses of Medicine   47
- **1.4 Graphing Sequences** 53
    Investigation: Match Them Up   53
    Performance Task   60

1.5	**Linear Equations and Arithmetic Sequences**	61
	Investigation: Match Point	63
1.6	**Modeling with Intercept and Slope**	68
	Investigation: Balloon Blastoff	69
1.7	**Models and Predictions**	75
	Investigation: The Wave	78
1.8	**Residuals and Fit**	82
	Investigation: Spring Experiment	85
1.9	**Least Squares Line**	90
	Investigation: The Moving Line	90
	Performance Task	94
Chapter 1 Review		95
	Take Another Look	99
	Assessing What You've Learned	100

CHAPTER 2

Systems of Equations and Inequalities — 101

Chapter 2 Teaching Notes		101A
2.1	**Linear Systems**	102
	Investigation: Population Trends	104
2.2	**Substitution and Elimination**	109
	Investigation: What's Your System?	111
2.3	**Linear and Non-linear Systems of Equations**	116
	Investigation: A Designing Dilemma	116
2.4	**Systems of Inequalities**	120
	Investigation: Paying for College	120
2.5	**Linear Programming**	127
	Investigation: Maximizing Profit	127
2.6	**Larger Systems**	133
	Investigation: Three Planes	133
	Performance Task	137
Chapter 2 Review		138
	Take Another Look	140
	Assessing What You've Learned	141

CHAPTER 3

Functions and Relations — 142

Chapter 3 Teaching Notes		142A
3.1	**Interpreting Graphs**	143
	Investigation: Graph Clues	144
	Performance Task	148
3.2	**Function Notation**	149
	Investigation: To Be or Not to Be (a Function)	151
3.3	**Lines in Motion**	158
	Investigation: Movin' Around	158

3.4	**Translations and the Quadratic Family**.................................	165
	Investigation: Make My Graph	166
3.5	**Reflections and the Square Root Family**	172
	Investigation: Take a Moment to Reflect	172
3.6	**Dilations and the Absolute-Value Family**	180
	Investigation: The Pendulum	183
3.7	**Transformations and the Circle Family**................................	188
	Investigation: When Is a Circle Not a Circle?	190
Chapter 3 Review		196
	Take Another Look	198
	Assessing What You've Learned	199

CHAPTER 4 — Exponential, Power, and Logarithmic Functions .. 200

Chapter 4 Teaching Notes 200A

4.1	**Exponential Functions**..	201
	Investigation: Radioactive Decay	201
4.2	**Properties of Exponents and Power Functions**.........................	209
	Investigation: Properties of Exponents	209
4.3	**Rational Exponents and Roots**..	216
	Investigation: Getting to the Root	216
4.4	**Applications of Exponential and Power Equations**.....................	225
4.5	**Building Inverses of Functions**..	230
	Investigation: The Inverse	230
	Performance Task	236
4.6	**Logarithms**..	237
	Investigation: Exponents and Logarithms	237
4.7	**Arithmetic Series**..	243
	Investigation: Arithmetic Series Formula	244
	Performance Task	249
4.8	**Partial Sums of Geometric Series**	250
	Investigation: Geometric Series Formula	251
Chapter 4 Review		255
	Take Another Look	258
	Assessing What You've Learned	258

CHAPTER 5 — Quadratic Functions and Relations 259

Chapter 5 Teaching Notes 259A

5.1	**Equivalent Quadratic Forms**	260
	Investigation: Rolling Along	263
5.2	**Completing the Square**...	268
	Investigation: Complete the Square	269
5.3	**The Quadratic Formula**	277
	Investigation: Hit or Miss	279

v

| 5.4 | **Complex Numbers** | 283 |

Investigation: Complex Arithmetic 285

| 5.5 | **Solving Quadratic Equations** | 288 |

Investigation: The Quadratic Formula Revisited 288
Performance Task 292

| 5.6 | **Solving Radical Equations** | 293 |

Investigation: When is a Solution Not a Solution? 294

| 5.7 | **Using the Distance Formula** | 297 |

Investigation: Bucket Race 297

| 5.8 | **Parabolas** | 305 |

Investigation: Fold a Parabola 309

Chapter 5 Review 313
    Take Another Look 315
    Assessing What You've Learned 316

CHAPTER 6

Polynomial and Rational Functions 317

Chapter 6 Teaching Notes 317A

| 6.1 | **Polynomials** | 318 |

Investigation: Free Fall 321

| 6.2 | **Factoring Polynomials** | 327 |

Investigation: The Box Factory 329

| 6.3 | **Higher-Degree Polynomials** | 334 |

Investigation: The Largest Triangle 335
Performance Task 341

| 6.4 | **More About Finding Solutions** | 342 |
| 6.5 | **Introduction to Rational Functions** | 351 |

Investigation: The Breaking Point 351

| 6.6 | **Graphs of Rational Functions** | 357 |

Investigation: Predicting in the Short Run 358

| 6.7 | **Adding and Subtracting Rational Expressions** | 364 |
| 6.8 | **Multiplying and Dividing Rational Expressions** | 368 |

Investigation: Disappearing Holes or Feeling Restricted 370

| 6.9 | **Solving Rational Equations** | 373 |

Investigation: The (Extraneous) Root of the Problem 374

Chapter 6 Review 381
    Take Another Look 383
    Assessing What You've Learned 383

CHAPTER 7

Trigonometry and Trigonometric Functions 384
Chapter 7 Teaching Notes **384A**

7.1	**Right Triangle Trigonometry**	385
	Investigation: Steep Steps	389
7.2	**Extending Trigonometry**	393
	Investigation: Extending Trigonometric Functions	393
7.3	**Defining the Circular Functions**	399
	Investigation: Paddle Wheel	400
7.4	**Radian Measure**	407
	Investigation: A Radian Protractor	408
	Performance Task	416
7.5	**Graphing Trigonometric Functions**	417
	Investigation: The Pendulum II	420
	Performance Task	428
7.6	**Modeling with Trigonometric Equations**	429
	Investigation: A Bouncing Spring	431
7.7	**Pythagorean Identities**	436
	Investigation: Pythagorean Identities	437
Chapter 7 Review		442
	Take Another Look	445
	Assessing What You've Learned	445

CHAPTER 8

Probability 446
Chapter 8 Teaching Notes **446A**

8.1	**Randomness and Probability**	447
	Investigation: Flip a Coin	448
	Performance Task	456
8.2	**Multiplication Rules of Probability**	457
	Investigation: The Multiplication Rule	458
8.3	**Addition Rules of Probability**	465
	Investigation: Addition Rule	467
8.4	**Bivariant Independence**	473
	Investigation: Are These Things Related?	475
Chapter 8 Review		479
	Take Another Look	481
	Assessing What You've Learned	482

CHAPTER 9

Applications of Statistics 483
Chapter 9 Teaching Notes **483A**

9.1 **Experimental Design** 484
Investigation: Designing a Study   485

9.2 **Normal Distributions** 490
Investigation: Probabilities and Percentiles   493

9.3 *z*-**Values and Confidence Intervals** 499
Investigation: Predicting the True Mean   500
Performance Task   505

9.4 **Bivariate Data and Correlation** 506
Investigation: Looking for Connections   506

Chapter 9 Review 515
Take Another Look   518
Assessing What You've Learned   518

Selected Hints and Answers 519
Glossary 547
Index 561
Photo Credits 570
Common Core State Standards and *Discovering Advanced Algebra* 572

About the Authors

Pamela Weber Harris

Pamela is a former secondary mathematics teacher who currently teaches at The University of Texas, a K-12 mathematics education consultant and a T3 (Teachers Teaching with Technology) National Instructor, an author of various math titles and coauthor of several professional development workshops. She's also a frequent presenter at local and national conferences. Her particular interests include numeracy, technology, assessment and vertical connectivity in curricula in schools K-12.

Ellen Kamischke

Ellen taught high school mathematics for 30 years, covering all levels from Algebra I through AP Calculus. In addition, she mentored advanced students in the independent study of multivariable calculus, cryptography, and other topics. She has been a reader for the AP Calculus exam since 2003, including being a table leader there since 2007. Ellen has led many professional development institutes nationally, focusing on incorporating technology in the teaching of algebra. Since leaving Interlochen Arts Academy, she has earned a masters degree in mathematics with research in partial geometry while teaching first-year mathematics courses at Michigan Technological University. Ellen continues to teach at the college level.

Eric Kamischke

Eric taught high school mathematics for 30 years, covering all levels from Algebra I through AP Calculus and Statistics. He, too, has been a reader for the AP Statistics exam since 2000 and a table leader there since 2009. Eric has led many professional development institutes nationally, focusing on incorporating technology in the teaching of algebra. Since leaving Interlochen Arts Academy, he has earned a masters degree in statistics and has worked as an instructor at Northwestern Michigan College and Michigan Technological University.

Ellen and Eric are also working together to provide professional development assistance to a primary school in rural Haiti.

Jerald Murdock

Jerry, who died in 2013, believed there were two kinds of people in the world, the ones who loved math and the ones who were going to learn to love math. He spent his 40 years in the high school classroom encouraging students to share his enthusiasm for mathematics. Among his many accolades, Jerry received a National Science Foundation grant in 1991 that allowed Jerry and the Kamischkes to create the Graphing Calculator Enhanced Algebra Project. They coauthored *Discovering Advanced Algebra* in 1998 and *Discovering Algebra* in 2002. Jerry retired from the classroom in 2001, but continued his commitment to improving mathematics in the classroom by teaching math workshops nationally through 2010.

Features of the Student Edition

Chapter openers use a trio of images to highlight applications of the mathematics of that chapter. Examples come from a variety of fields that use mathematics in their daily work. A caption connects the opening image to the chapter content.

CHAPTER 4

Exponential, Power, and Logarithmic Functions

Lessons open with a smooth transition from prior learning and a motivational context that keeps students moving forward. Thought-provoking quotations introduce most lessons.

The Richter scale measures the magnitude of an earthquake using the logarithm of the amplitude of waves recorded by a seismograph. Geologists use exponential, power and logarithmic functions to measure distance phenomena that occur in a non-linear manner, including curved exfoliation joints and resultant topographic domes like these in Yosemite National Park. Exponential models have shown that the population growth rate of King penguin colonies, like this one in the South Georgia Islands, have decreased since the 1990s.

OBJECTIVES

In this chapter you will
- review the properties of square roots
- write explicit equations for geometric sequences
- use exponential functions to model real-world growth and decay scenarios
- review the properties of exponents and the meaning of rational exponents
- learn how to find the inverse of a function
- apply logarithms, the inverses of exponential functions

Objectives display the mathematical learning goals the chapter will address in language that students can understand.

LESSON 4.1

Exponential Functions

Life shrinks or expands in proportion to one's courage.
ANAÏS NIN

You have previously used sequences and recursive rules to model geometric growth or decay of money, populations, and other quantities. Recursive formulas generate only discrete values, such as the amount of money after one year or two years, or the population in a certain year. Usually growth and decay happen continuously. In this lesson you will focus on finding explicit formulas for these patterns, which will allow you to model situations involving continuous growth and decay, or to find discrete points without using recursion.

INVESTIGATION

YOU WILL NEED
- one die per person

Radioactive Decay

This investigation is a simulation of radioactive decay. Each person will need a standard six-sided die. Each standing person represents a radioactive atom in a sample. The people who sit down at each stage represent the atoms that underwent radioactive decay.

Procedure Note
1. All members of the class should stand up, except for the recorder. The recorder counts and records the number standing at each stage.
2. Each standing person rolls a die, and anyone who gets a 1 sits down.
3. Wait for the recorder to count and record the number of people standing.
4. Repeat Steps 2 and 3 until fewer than three students are standing.

Step 1 Follow the Procedure Note to collect data in the form (stage, number standing).

Step 2 Graph your data. The graph should remind you of the sequence graphs you have studied. What type of sequence does this resemble?

Step 3 Identify u_0 and the common ratio, r, for your sequence. Complete the table below. Use the values of u_0 and r to help you write an explicit formula for your data.

n	u_n	u_n in terms of u_0 and r	u_n in terms of u_0 and r using exponents
0	u_0		
1	u_1	$u_0 \cdot r$	
2	u_2	$u_0 \cdot r \cdot r$	
3	u_3	$u_0 \cdot r \cdot r \cdot r$	

Step 4 Graph your explicit formula along with your data. Notice where the value of u_0 appears in your equation. Your graph should pass through the original data point $(0, u_0)$. Modify your equation so that it passes through $(1, u_1)$, the second data point. (Think about translating the graph horizontally and also changing the starting value.)

Step 5 Experiment with changing your equation to pass through other data points. Decide on an equation that you think is the best fit for your data. Write a sentence or two explaining why you chose this equation.

Step 6 What equation with ratio r would you write that contains the point $(6, u_6)$?

Investigations provide a careful blend of guidance and discovery that stimulates learning. Some are purely mathematical while others are based in real-world situations and incorporate steps that help guide students through the investigation. Students may work in groups, and each investigation provides rich discussion opportunities for whole-class participation.

Materials lists are provided for each investigation.

LESSON 4.1 Exponential Functions 201

Examples are placed before or after investigations, based on where they will be most helpful. Some are presented in real-world contexts, while others may be provided in purely mathematical form. The examples are worked out computationally and include reasons that justify the process; all notation conventions are also clearly explained.

> **Science**
> **CONNECTION**
>
> Certain atoms are unstable—their nuclei can split apart, emitting radiation and resulting in a more stable atom. This process is called radioactive decay. The time it takes for half the atoms in a radioactive sample to decay is called **half-life**, and the half-life is specific to the element. For instance, the half-life of carbon-14 is 5730 years, whereas the half-life of uranium-238 is 4.5 billion years. Due to the decay time, serious nuclear power plant accidents, which include the Fukushima Daiichi nuclear disaster (2011), Chernobyl disaster (1986), Three Mile Island accident (1979), and the SL-1 accident (1961), leave an area radioactive for decades.

Modern ruins in a high dynamic range (HDR) near Chernobyl (1986).

A table of $y = 2^x$ shows that as x increases, the value of y increases even faster, and as x decreases, the value of y gets closer and closer to zero.

x	−10	−5	−1	−0.5	0	0.5	1	5	10
y	0.001	0.031	0.5	0.707	1	1.41	2	32	1024

EXAMPLE B The functions $g(x) = \frac{1}{2}(2)^x$ and $h(x) = 2^{x-1}$ are transformations of the parent function $f(x) = 2^x$. Describe the graph of the parent function $f(x) = 2^x$. Then describe the transformations of $g(x)$ and $h(x)$ and sketch the graphs.

Solution The graph of the function $f(x) = 2^x$ is an exponential function, where $a = 1$ and $b = 2$. This graph is in both quadrants I and II, the graph is increasing, and the y-intercept is 1.

The graph of $g(x)$ is a vertical dilation of the graph of $f(x)$ by a factor of $\frac{1}{2}$. The marked points on the red graph show how the y-values of the corresponding points on the black graph have been multiplied by $\frac{1}{2}$.

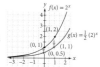

The graph of $h(x)$ is a horizontal translation of the graph of $f(x)$. The marked points on the red graph show how the corresponding points on the black graph have been moved right 1 unit.

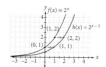

Do you notice anything special about the two red graphs? They appear to be the same. With exponential functions, as with linear and quadratic functions, it is possible to find different transformations that produce the same graphs.

EXAMPLE C Each day, the imaginary caterpillarsaurus eats 25% more leaves than it did the day before. If a 30-day-old caterpillarsaurus has eaten 151,677 leaves in its brief lifetime, how many will it eat the next day?

Solution To solve this problem, you must find u_{31}. The information in the problem tells you that r is $(1 + 0.25)$, or 1.25, and when n equals 30, S_n equals 151,677. Substitute these values into the formula for S_n and solve for the unknown value, u_1.

$$\frac{u_1(1 - 1.25^{30})}{1 - 1.25} = 151,677$$

$$u_1 = 151,677 \cdot \frac{1 - 1.25}{1 - 1.25^{30}}$$

$$u_1 \approx 47$$

Now you can write an explicit formula for the terms of the geometric sequence, $u_n = 47(1.25)^{n-1}$. Substitute 31 for n to find that on the 31st day, the caterpillarsaurus will eat 37,966 leaves.

The explicit formula for the sum of a geometric series can be written in several ways, but they are all equivalent. You probably found these two ways during the investigation:

> **Partial Sum of a Geometric Series**
>
> A partial sum of a geometric series is given by the explicit formula
>
> $$S_n = \left(\frac{u_1}{1-r}\right) - \left(\frac{u_1}{1-r}\right)r^n \text{ or } S_n = \frac{u_1(1-r^n)}{1-r}$$
>
> where n is the number of terms, u_1 is the first term, and r is the common ratio ($r \neq 1$).

Graphing calculator use is modeled in the examples, and calculators are used in many investigations. Separate calculator notes with keystroke-level instructions are provided so that *Discovering Advanced Algebra* can be used with a variety of calculators, including new ones that become available.

4.8 Exercises

You will need a graphing calculator for Exercises **4, 5, 12,** and **15.**

Practice Your Skills

Developmental exercise sets begin with **Practice Your Skills**. The exercises consist of direct applications of what students have learned in the lesson. Additional skills problems are available for each lesson in the student online resources.

1. For each partial sum equation, identify the first term, the ratio, and the number of terms.

    a. $\dfrac{12}{1 - 0.4} - \dfrac{12}{1 - 0.4}(0.4)^8 \approx 19.9869$ @

    b. $\dfrac{75(1 - 1.2^{15})}{1 - 1.2} \approx 5402.633$ @

    c. $\dfrac{40 - 0.46117}{1 - 0.8} \approx 197.69$

    d. $-40 + 40(2.5)^6 = 9725.625$

2. Consider the geometric sequence: 256, 192, 144, 108, . . .

    a. What is the eighth term?
    b. Which term is the first one smaller than 20? @
    c. Find u_7.
    d. Find S_7. @

Reason and Apply provides more challenging exercises that develop reasoning and transfer of knowledge. These exercises often include several steps that build on each other, and they may also require students to combine information from earlier chapters with what they have just learned.

Hints and **answers** can be found in the back of the text.

Review keeps previously used skills from falling out of use, especially those that will be needed in upcoming lessons.

Performance Tasks are tasks that provide less scaffolding than the investigations and require students to determine a solution method and justify their thinking. With at least one performance task highlighted in each chapter, students utilize the concepts from the chapter and previous chapters to solve a more rigorous task.

---

4. Match each function with its inverse. ⓗ

   a. $y = 6 - 2x$
   b. $y = 2 - \frac{6}{x}$ ⓐ
   c. $y = -6(x - 2)$ ⓐ
   d. $y = \frac{-6}{x - 2}$
   e. $y = \frac{-1}{2}(x - 6)$
   f. $y = \frac{2}{x - 6}$
   g. $y = 2 - \frac{1}{6}x$
   h. $y = 6 + \frac{2}{x}$

Reason and Apply

5. Given the functions $f(x) = 4x + 8$ and $g(x) = \frac{1}{4}x - 2$:
   a. Find $f(2)$ and $g(16)$.
   b. What do your answers to 5a imply?
   c. Find $f(1)$.
   d. How can knowing $f(1)$ help you find $g(12)$?

6. Given $f(x) = 4 + (x - 2)^{3/5}$:
   a. Solve for x when $f(x) = 12$.
   b. Find $f^{-1}(x)$ symbolically. ⓗ
   c. How are solving for x and finding an inverse alike? How are they different?

7. Consider the graph of the piecewise function f shown at right.
   a. Find $f(-3), f(-1), f(0)$, and $f(2)$.
   b. Name four points on the graph of the inverse of $f(x)$.
   c. Draw the graph of the inverse. Is the inverse a function? Explain.

8. Write each function using $f(x)$ notation, then find its inverse. If the inverse is a function, write it using $f^{-1}(x)$ notation.
   a. $y = 2x - 3$ ⓐ
   b. $3x + 2y = 4$
   c. $x^2 + 2y = 3$ ⓐ

9. Find the inverse function for each.
   a. $f(x) = 6.34x - 140$
   ... $= 1.8x + 32$ ⓐ

...at the equation in 9b will convert temperatures in °C to temperatures in °F. ...use either this function or its inverse in Exercises 10 and 11.

Consumer CONNECTION

Leaks account for nearly 12% of the average household's annual water consumption. About one in five toilets leaks at any given time, and that can waste more than 50 gallons each day. Dripping sinks add up fast too. A faucet that leaks one drop per second can waste 30 gallons each day.

...story CONNECTION

...lsius (1701–1744) was a Swedish astronomer. He created a thermometric scale using the freezing and boiling ...ures of water as reference points, on which freezing corresponded to 100° and boiling to 0°. His colleagues ...psala Observatory reversed his scale five years later, giving us the current version. Thermometers with this ...e known as "Swedish thermometers" until the 1800s when people began referring to them as "Celsius ...eters."

Review

14. Rewrite the expression $125^{2/3}$ in as many different ways as you can.

15. Find an exponential function that contains the points (2, 12.6) and (5, 42.525). ⓐ

16. Solve by rewriting with the same base.
    a. $4^x = 8^3$
    b. $3^{4x+1} = 9^x$
    c. $2^{x-3} = \left(\frac{1}{4}\right)^x$

17. Give the equations of two different parabolas with vertex (3, 2) passing through the point (4, 5). ⓗ

18. Solve this system of equations.
$$\begin{cases} -x + 3y - z = 4 \\ 2z = x + y \\ 2.2y + 2.2z = 2.2 \end{cases}$$

PERFORMANCE TASK

Here is a paper your friend turned in for a recent quiz in her mathematics class:

If it is a four-point quiz, what is your friend's score? For each incorrect answer, provide the correct answer and explain it so that next time your friend will get it right!

QUIZ

1. Rewrite x^{-1}
   Answer: $\frac{1}{x}$

2. What does $f^{-1}(x)$ mean?
   Answer: $\frac{1}{f(x)}$

3. Rewrite $9^{-1/5}$
   Answer: $\frac{1}{9^5}$

4. What number is 0^0 equal to?
   Answer: 0

INVESTIGATING STRUCTURE

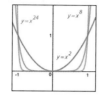

Look at these graphs. Why are the power functions with odd exponents in the first and third quadrants, while the power functions with even exponents are in the first and second quadrants?

How do the graphs on each grid compare? How do they contrast? What points (if any) do they have in common?

Why do the graphs appear to flatten out between −1 and 1 as the power gets larger? Use tables of values to help you explain.

> **Investigating Structure** explorations were created to ensure students were given sufficient opportunity to look for and generalize patterns, focusing specifically on algebraic representations of the mathematics covered in a particular lesson.

You use different methods to solve power equations than to solve exponential equations.

EXAMPLE B | Solve for positive values of x.
a. $x^4 = 3000$
b. $6x^{2.5} = 90$

Solution | To solve a power equation, use the power of a power property and choose an exponent that will undo the exponent on x.

a.
$x^4 = 3000$ Original equation.
$(x^4)^{1/4} = 3000^{1/4}$ Use the power of a power property. Raising both power of $\frac{1}{4}$, the reciprocal of 4, "undoes" the pow
$x \approx 7.40$ Use your calculator to approximate the value of 3

b.
$6x^{2.5} = 90$ Original equation.
$x^{2.5} = 15$ Divide both sides by 6.
$(x^{2.5})^{1/2.5} = 15^{1/2.5}$ Use the power of a power property and choose the
$x \approx 2.95$ Approximate the value of $15^{1/2.5}$.

In this book the properties of exponents are defined only for positive bas there may be negative values that are solutions to power equations, in th you will be asked to find only the positive solutions.

212   CHAPTER 4 Exponential, Power, and Logarithmic Functions

Review

15. Janell starts 10 m from a motion sensor and walks at 2 m/s toward the sensor. When she is 3 m from the sensor, she instantly turns around and walks at the same speed back to the starting point.
 a. Sketch a graph of the function that models Janell's walk.
 b. Give the domain and range of the function.
 c. Write an equation of the function.

16. The graph shows a line and the graph of $y = f(x)$.
 a. Fill in the missing values to make a true statement.
    $f(?) = ?$.
 b. Find the equation of the pictured line.

17. Austin deposits $5,000 into an account that pays 3.5% annual interest. Sami deposits $5,200 into an account that pays 3.2% annual interest.
 a. Write an expression for the amount of money Austin will have in his account after 5 years if he doesn't deposit any more money.
 b. Write an expression for the amount of money Sami will have in her account after 5 years if she doesn't deposit any more money.
 c. How long will it take until Austin has more money than Sami?

18. You can use different techniques to find the product of two binomials, such as $(x − 4)(x + 6)$.
 a. Use a rectangle diagram to find the product.
 b. You can use the distributive property to rewrite the expression $(x − 4)(x + 6)$ as $x(x + 6) − 4(x + 6)$. Use the distributive property again to find all the terms. Combine like terms.
 c. Compare your answers to 18a and b. Are they the same?
 d. Compare the methods in 18a and b. How are they alike?

> **Improving Your ... Skills** is a puzzle-like feature that introduces topics other than those covered in the lesson and provides practice in a particular skill.

IMPROVING YOUR Reasoning SKILLS

Breakfast Is Served

Mr. Higgins told his wife, a mathematics professor, that he would make her breakfast. She handed him this message:

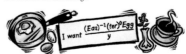

$$I\ want\ \frac{(Eas)^{-1}(ter)^9 Egg}{y}$$

What should Mr. Higgins fix his wife for breakfast?

208   CHAPTER 4 Exponential, Power, and Logarithmic Functions

xiii

The **Chapter Review** begins with a summary of the new mathematical ideas that were addressed in lessons. Review exercises are similar to those in each lesson as well as those that will appear in the Assessment Resources.

Boldface terms are contained in a glossary at the back of the text.

Take Another Look gives students a chance to approach concepts from a different perspective by calling out connections between mathematical ideas.

CHAPTER 4 REVIEW

Exponential functions provide explicit and continuous equations to model geometric sequences. They can be used to model the growth of populations, the decay of radioactive substances, and other phenomena. The general form of an exponential function is $y = ab^x$, where a is the initial amount and b is the base that represents the rate of growth or decay. Because the exponent can take on all real number values, including negative numbers and fractions, it is important that you understand the meaning of these exponents. You also used the **point-ratio form** of this equation, $y = y_1 \cdot b_1^{x - x_1}$.

Until you read this chapter, you had no way to solve an exponential equation, other than guess-and-check. The **logarithmic function** is defined so that $\log_b y = x$ means that $y = b^x$. This means that exponential functions and logarithmic functions are **inverse** functions. The inverse of a function is the relation you get when all values of the independent variable are exchanged with the values of the dependent variable. The graphs of a function and its inverse are reflected across the line $y = x$.

A **series** is the indicated sum of terms of a sequence, defined recursively with the rule $S_1 = u_1$ and $S_n = S_{n-1} + u_n$ where $n \geq 2$. Series can also be defined explicitly. With an explicit formula, you can find any **partial sum** of a sequence without knowing the preceding term(s).

An explicit formula for **arithmetic series** is $S_n = \dfrac{n(u_1 + u_n)}{2}$.

Explicit formulas for **geometric series** are

$$S_n = \left(\dfrac{u_1}{1-r}\right) - \left(\dfrac{u_1}{1-r}\right)r^n \text{ and } S_n = \dfrac{u_1(1 - r^n)}{1 - r} \text{ where } r \neq 1.$$

Exercises

You will need a graphing calculator for Exercises **11** and **13**.

...vers are provided for all exercises in this set.

...luate each expression without using a calculator. Then check your work with a calculator.

...−2 b. $(-3)^{-1}$ c. $\left(\dfrac{1}{5}\right)^{-3}$

...$9^{1/2}$ e. $64^{-1/3}$ f. $\left(\dfrac{9}{16}\right)^{3/2}$

...7^0 h. $(3)(2)^2$ i. $(0.6^{-2})^{-1/2}$

...rite each equation in logarithm form so that the x is no longer an exponent.

...$^x = 7$ b. $5 = 4^x$ c. $10^x = 7^5$

Take Another Look

1. You have learned to use the point-ratio form to find an exponential curve that fits two data points. You could also use the general exponential equation $y = ab^x$ and your knowledge of solving systems of equations to find an appropriate exponential curve. For example, you can use the general form to write two equations for the exponential curve through (4, 40) and (7, 4.7):

   $40 = ab^4$

   $4.7 = ab^7$

   Which constant will be easier to solve for in each equation, a or b? Solve each equation for the constant you have chosen. Use substitution and the properties you have learned in this chapter to solve this system of equations. Substitute the values you find for a and b into the general exponential form to write a general equation for this function.

   Find an exercise or example in this chapter that you solved using the point-ratio form, and solve it again using this new method. Did you find this method easier or more difficult to use than the point-ratio form? Are there situations in which one method might be preferable to the other?

2. The explicit formula for a partial sum of a geometric series is $S_n = \dfrac{u_1(1 - r^n)}{1 - r}$. To find the sum of an infinite geometric series, you can imagine substituting ∞ for n. Explain what happens to the expression $\dfrac{u_1(1 - r^n)}{1 - r}$ when you do this substitution.

Assessing What You've Learned

Assessing What You've Learned helps students capture their learning experiences and identify gaps in understanding. It explains portfolios, journals, presentations, and performance assessments as ways for students to assume responsibility for their own learning.

 GIVE A PRESENTATION Give a presentation about how to fit an exponential curve to data or one of the Take Another Look activities. Prepare a poster or visual aid to explain your topic. Work with a group, or give a presentation on your own.

 ORGANIZE YOUR NOTEBOOK Review your notebook to be sure it's complete and well organized. Make sure your notes include all of the properties of exponents, including the meanings of negative and fractional exponents. Write a one-page chapter summary based on your notes.

 WRITE TEST ITEMS Write a few test questions that explore series. You may use sequences that are arithmetic, geometric, or perhaps neither. You may want to include problems that use sigma notation. Be sure to include detailed solution methods and answers.

Features of the Teacher Edition

An **overview** of the chapter contents is provided on the interleaf pages that open each chapter of the Teacher Edition.

The Mathematics shows the chapter content in context and explains the development of the mathematics within the chapter.

CHAPTER

4

Exponential, Power, and Logarithmic Functions

Overview

Discovering Advanced Algebra is about modeling real-world problems with mathematical functions. This chapter extends students' knowledge of functions to exponential, power, and logarithmic functions. The chapter also uses the function idea of assignment to assign a numerical value to each term of a series.

This chapter begins with an introduction to exponential functions in the form $y = ab^x$ for modeling growth and decay in **Lesson 4.1**. In **Lesson 4.2**, students review properties of exponents while becoming familiar with the power function, $y = ax^n$, and learning about rational exponents. In **Lesson 4.3**, rational exponents ($x^{1/n}$) are related to root notation ($\sqrt[n]{x}$). Students then use the point-ratio form, $y = y_1 b^{x-x_1}$, to find an exponential equation through two points. Real-world applications are presented in **Lesson 4.4. Lesson 4.5** introduces inverses of functions. Students learn about logarithmic functions in **Lesson 4.6**, and they see how these inverses of exponential functions can help solve equations for exponents. **Lesson 4.7** introduces students to series and to the summation notation used to represent them. Students learn to write recursive and explicit formulas for arithmetic series and find partial sums of arithmetic series. In **Lesson 4.8,** students find partial sums of geometric series.

There are two **Performance Tasks** highlighted in this chapter. In Lesson 4.5, students provide explanations for incorrect answers on a quiz. In Lesson 4.7, students create a formula for the sum, S_n, of an arithmetic series.

The Mathematics

Exponents and Exponential Functions

Exponents, if they are positive integers, represent repeated multiplication. The symbol b^n represents multiplying b repeatedly n times. The *exponent* is n, and the *base* is b. The student book calls b^n a *power* of b. (*Power* is a synonym for *exponent*.)

Positive-integer exponents have three major properties: (1) $b^m b^n = b^{m+n}$, (2) $\frac{b^m}{b^n} = b^{m-n}$, and (3) $(b^m)^n = b^{mn}$. The second property leads to negative exponents. If n is a positive integer, then $b^{-n} = \frac{1}{b^n}$. The same three properties apply if the exponents are negative integers.

To extend the third property to fractional exponents, $(b^{1/3})^3$ must equal b^1, or b. Therefore $b^{1/3}$ is defined to be the number whose cube is b, or the cube root of b, which is written $\sqrt[3]{b}$. Just as the symbol $\sqrt{b}$ indicates the positive square root of b, the symbol $b^{1/2}$ represents a positive number. Fractional exponents, however, can lead to contradictions, which can be avoided by limiting fractional exponents to positive bases.

Inverse Functions

If you think of a relation as a two-column table, perhaps infinite, then exchanging the columns gives a relation that's the *inverse* of the original. As with any relation, the inverse may or may not be a function. If both the function and its inverse are functions, it is called a one-to-one function, and the inverse of the function $f(x)$ is written as $f^{-1}(x)$.

Logarithmic Functions

In the case that the function is the exponential function $y = 10^x$, its inverse is also a function, $x = \log y$. A helpful way to think of the logarithmic function is "log y is the exponent put on 10 to get y." Therefore $10^{\log y} = y$ and $\log(10^x) = x$.

While logarithms can have any positive number as a base, two bases are used more than other bases, base 10 and base e. Logarithms with base 10 are called *common logarithms*. Logarithms whose base is the transcendental number e, are called *natural logarithms*.

Logarithmic functions were developed around 1594 as computational tools by Scottish mathematician John Napier (1550–1617). He originally called them "artificial numbers," but changed the name to logarithms, which means "reckoning numbers."

The fact that the inverse of an exponential function is the common logarithmic function helps solve equations in which the unknown appears in an exponent on 10. For example, you can solve the equation $10^{3x-5} = 59$ by realizing that $3x - 5$ is the exponent put on 10 to get 59; that is, $3x - 5 = \log 59$. You can then substitute a calculator value for log 59 to solve for x.

What if the exponent is on some base other than 10? A more general logarithmic function is needed—an inverse of $y = b^x$. This inverse, known as "the logarithm to the base b of y," and notated $x = \log_b y$, gives the exponent put on b to get y. Technology makes it easy to find the logarithm to any base, b, of y.

In addition to solving equations for exponents, logarithms can help find an exponential function that fits a set of data points. You can use the logarithms to transform the data into points that are more linear and then find a line of fit.

200A CHAPTER 4 INTERLEAF Exponential, Power, and Logarithmic Functions

Series

Mathematicians have grappled for centuries with paradoxes that arise from trying to sum infinitely many numbers. If you repeatedly add a nonzero number such as 0.0001 to itself, the sum will eventually get as large as you want. You might say that the sum is infinitely large. But what about the sum $\frac{1}{2} + \frac{1}{4} + \frac{1}{8} + \frac{1}{16} + \cdots$? If you begin adding these numbers, the intermediate results don't seem to be getting infinitely large. In fact, it appears that they'll always be less than 1. The sum $1 - \frac{1}{2} + \frac{1}{3} - \frac{1}{4} + \cdots$ also seems to be finite. But what about the sum $1 - 1 + 1 - 1 + \ldots$? If you start adding, the partial totals flip between 0 and 1. What should we say the complete sum is? Students' difficulties with this topic reflect the historical struggle of mathematicians.

Some standard terminology begins to sort out these questions. Expressions such as $\frac{1}{2} + \frac{1}{4} + \frac{1}{8} + \frac{1}{16} + \cdots$ are called *series*, which is different from a sequence. A series is also not a sum, because a sum is a number. A series is an (infinitely long) expression, a summation, an indicated sum. *Sigma notation* is often used to indicate the terms to be summed. Sigma is a Greek letter, whose capital form is Σ. The series $\frac{1}{2} + \frac{1}{4} + \frac{1}{8} + \frac{1}{16} + \cdots$ is represented by $\sum_{n=1}^{\infty} \frac{1}{2^n}$.

There are many kinds of series. This book emphasizes two of them, based on two types of sequences: *arithmetic series* and *geometric series*. The nth partial sum of an arithmetic series with first term u_1 and nth term u_n is the number of terms times the mean of the first and last terms, or $n\left(\frac{u_1 + u_n}{2}\right)$. The geometric series with initial term u_0 and common ratio r has nth partial sum $u_0\left(\frac{1-r^n}{1-r}\right)$, where $r \neq 1$. If r is close to 0 (actually, less than 1 in absolute value), the powers r^n get closer to 0 as n increases.

Using This Chapter

To save time, you might combine Lessons 4.1 and 4.2 for students familiar with the topics. Using the **Shortened** modifications of the investigations also saves time. It is important, however, to have student presentations of ideas to help them communicate mathematically and justify their answers.

Common Core State Standards (CCSS)

Students are introduced to exponents as early as grade 5 where they use whole-number exponents to denote powers of 10 (5.NBT.2). In the middle grades, students write, evaluate and perform arithmetic operations with numerical expressions involving whole-number exponents, including radicals and integer exponents (6.EE.1, 2c, 8.EE.1, 3). In Algebra 1 students apply solution

logarithms (F.BF.4a, F.LE.4). In Algebra II, students also learn to do a more complete analysis of sequences (F-IF.A.3), especially arithmetic and geometric sequences, and their associated series, developing recursive formulas for sequences by figuring out how you get from one term to the next and then giving a precise description of this process in algebraic symbols (F-BF.A.2). Understanding solving equations as a process of reasoning plays a larger than ever role in Algebra II because the equations students encounter can have extraneous solutions (A-REI.A.2).

Additional Resources For complete references to additional resources for this chapter and others see your digital resources.

Refreshing Your Skills To review the algebra skills needed for this chapter see your digital resources.

Working with Positive Square Roots
Arithmetic and Geometric Sequences

Projects See your digital resources for performance-based problems and challenges.

The Cost of Living
Powers of 10

Explorations See your digital resources for extensions and/or applications of content from this chapter.

Seeing the Sum of a Series

Materials

- dice (1 die per person)
- graph paper
- scissors
- ball, *optional*
- motion sensor, *optional*

CHAPTER 4 INTERLEAF Exponential, Power, and Logarithmic Functions **200B**

Using This Chapter suggests which lessons to emphasize and which may require additional instructional time.

Common Core State Standards illustrates how the mathematical content of this chapter fits within the mathematics progression of the Common Core.

The **Additional Resources** section provides a list of resources available for the chapter.

Materials lists are summarized to support teacher planning.

The wraparound text supports planning by identifying the lesson objectives and mathematical content, presenting new mathematical vocabulary, and listing materials used within the lesson.

Content standards addressed in the lesson are identified in three ways:
- **Applied** standards are those that have been developed previously and are being used to move students forward in the progression of mathematics.
- **Developed** standards are those being taught in the lesson.
- **Introduced** standards are those that the lesson introduces, laying the groundwork for future development.

The **Launch** feature marks the first step of the four-part LISA planning model. It helps spark student interest in the lesson content and connects the lesson to prior learning.

Differentiated Instruction recommendations for each lesson will allow teachers to easily adjust instruction for English language learners, struggling learners, and advanced learners.

The **Investigate** step provides guidance for teachers to facilitate the investigation and provides teachers with suggestions for developing the Standards for Mathematical Practice.

Alerts warn of possible student errors or misunderstanding.

Modifying the Investigation gives three pacing options—**Whole Class, Shortened,** and **One Step**—to help you plan lessons that meet the needs of your students as well as the demands of your school's schedule. The **One Step** version of the lesson provides rich engaging tasks with less scaffolding than the investigation.

LESSON 4.2

Properties of Exponents and Power Functions

When the whole world is silent, even one voice becomes powerful.
MALALA YOUSAFZAI

Frequently, you will need to rewrite a mathematical expression in a different form to make the expression easier to understand or an equation easier to solve. Recall that in an exponential expression, such as 4^3, the number 4 is called the **base** and the number 3 is called the **exponent**. You say that 4 is **raised to the power** of 3. If the exponent is a positive integer, you can write the expression in **expanded form**, for example, $4^3 = 4 \cdot 4 \cdot 4$. Because 4^3 equals 64, you say that 64 is a power of 4.

INVESTIGATION
Properties of Exponents

For expanded forms, see next page.

Use expanded form to review and generalize the properties of exponents.

Step 1 Write each product in expanded form, and then rewrite it in exponential form.
a. $2^3 \cdot 2^4$ 2^7 b. $x^5 \cdot x^{12}$ x^{17} c. $10^2 \cdot 10^5$ 10^7

Step 2 Generalize your results from Step 1. $a^m \cdot a^n = \underline{?}$ a^{m+n}

Step 3 Write the numerator and denominator of each quotient in expanded form. Simplify by reducing the number of common factors, and then rewrite the factors that remain in exponential form.
a. $\dfrac{4^5}{4^2}$ 4^3 b. $\dfrac{x^5}{x^6}$ x^2 c. $\dfrac{(0.94)^{15}}{(0.94)^5}$ $(0.94)^{10}$

Step 4 Generalize your results from Step 3. $\dfrac{a^m}{a^n} = \underline{?}$ a^{m-n}

Step 5 Write each quotient in expanded form, simplify, and rewrite in exponential form.
a. $\dfrac{2^3}{2^4}$ $\dfrac{1}{2^1}$ b. $\dfrac{4^5}{4^7}$ $\dfrac{1}{4^2}$ c. $\dfrac{x^3}{x^8}$ $\dfrac{1}{x^5}$

Step 6 Rewrite each quotient in Step 5 using the property you discovered in Step 4.

Step 7 Generalize your results from Steps 5 and 6. $\dfrac{1}{a^n} = \underline{?}$ a^{-n}

Step 8 Write several expressions in the form $(a^n)^m$. Expand each expression, and then rewrite it in exponential form. Generalize your results. $(a^n)^m = a^{nm}$
sample answer: $(2^3)^4 = (2^3)(2^3)(2^3)(2^3) = 2^{3+3+3+3} = 2^{12}$;

Step 9 Write several expressions in the form $(a \cdot b)^n$. Don't multiply a and b. Expand each expression, and then rewrite it in exponential form. Generalize your results.
sample answer: $(2 \cdot 3)^3 = (2 \cdot 3)(2 \cdot 3)(2 \cdot 3) = 2 \cdot 2 \cdot 2 \cdot 3 \cdot 3 \cdot 3 = 2^3 \cdot 3^3$; $(a \cdot b)^n = a^n b^n$

Step 10 Show that $a^0 = 1$, using the properties you have discovered. Write at least two exponential expressions to support your explanation.
sample answer: $\dfrac{x^3}{x^3} = \dfrac{x \cdot x \cdot x}{x \cdot x \cdot x} = 1$; but, by the property from Step 4, $\dfrac{x^3}{x^3} = x^{3-3} = x^0$

LESSON 4.2 Properties of Exponents and Power Functions 209

In this lesson, students review the properties of exponents and power functions. The properties for exponents are defined only for positive bases. The restriction does not rule out negative bases for integer exponents, so bases in formulas for sequences may be negative.

COMMON CORE STATE STANDARDS

APPLIED	DEVELOPED	INTRODUCED
A.REI.1	N.RN.2	N.RN.1
	A.SSE.3c	F.IF.6
	F.IF.8b	
	F.LE.3	

Objectives
- Introduce algebraic proof
- Review the properties of exponents—specifically, rewriting powers with the same base
- Introduce the parent power function, $y = ax^n$, and distinguish it from the exponential function, $y = ab^x$
- Find some solutions to power equations using the properties of exponents
- Introduce rational exponents as a means of solving equations

Vocabulary
base
exponent
raised to the power
expanded form
power function

Materials
- Calculator Note: Power and Roots

Launch

For each of the following, identify the base and the exponent. Write each in expanded form.
a. 4^5
b. $\left(\dfrac{7}{z}\right)^3$

ELL
Provide numerical examples to support the definitions. This will help students recognize the difference between the letters that represent constants and those that represent variable values in the properties of exponents.

Support
To promote success with the type of exercises shown in Example A, have students

Advanced
ASK "Are power functions and exponential functions inverses of each other?" [No] "Can any

---

Investigate

In place of dice, students might use calculators that are set to find random integers from 1 to 6 or use another number to investigate a different decay rate. There is also a desmos simulation of the investigation in your ebook.
SMP 4, 5

Guiding the Investigation
Step 1 ALERT Make sure that the recorder labels the initial number of people in the class as stage 0. If you have fewer than 15 students, run the procedure several times and analyze the totals, or use the sample data.

Step 2 a decreasing geometric sequence

Step 3 $u_2 = 30$; The ratios between consecutive terms are
$\dfrac{26}{30} \approx 0.867, \dfrac{19}{26} \approx 0.731, \dfrac{17}{19} \approx 0.895,$
$\dfrac{14}{17} \approx 0.824, \dfrac{12}{14} \approx 0.857, \dfrac{11}{13} \approx 0.917,$
$\dfrac{9}{11} \approx 0.818, \dfrac{8}{9} \approx 0.889, \dfrac{8}{8} = 1, \dfrac{5}{8} = 0.625,$
$\dfrac{5}{5} = 1, \dfrac{2}{5} = 0.4.$
Average these 12 ratios to obtain $r \approx 0.8185$.

Step 4

$r1(x) = 30 \cdot (0.8185)^x$

possible answer: $f(x) = 26 \cdot 0.8185^{x-1}$

$r2(x) = 26 \cdot (0.8185)^{x-1}$

You probably recognized the geometric decay model in the investigation. As you learned earlier, geometric decay is nonlinear. At each step, the previous term is multiplied by a common ratio. Because the common ratio appears as a factor more and more times, you can use exponential functions to model the geometric growth. An **exponential function** is a continuous function with a variable in the exponent, and it is used to model growth or decay.

EXAMPLE A Most automobiles depreciate as they get older. Suppose an automobile that originally costs $14,000 depreciates by one-fifth of its value every year.
a. What is the value of this automobile after $2\dfrac{1}{2}$ years?
b. When is this automobile worth half of its initial value?

Solution
a. The recursive formula gives automobile values only after one year, two years, and so on. The value decreases by $\dfrac{1}{5}$, or 0.2, each year, so to find the next term you multiply by another $(1 - 0.2)$. You can use this fact to write an explicit formula.

$14{,}000 \cdot (1 - 0.2)$ Value after 1 year.
$14{,}000 \cdot (1 - 0.2) \cdot (1 - 0.2) = 14{,}000(1 - 0.2)^2$ Value after 2 years.
$14{,}000 \cdot (1 - 0.2) \cdot (1 - 0.2) \cdot (1 - 0.2) = 14{,}000(1 - 0.2)^3$ Value after 3 years.
$14{,}000 \cdot (1 - 0.2)^n$ Value after n years.

So the explicit formula for automobile value is $u_n = 14{,}000(1 - 0.2)^n$. The equation of the continuous function through the points of this sequence is
$$y = 14{,}000(1 - 0.2)^x$$
You can use the continuous function to find the value of the car at any point. To find the value after $2\dfrac{1}{2}$ years, substitute 2.5 for x.
$$y = 14{,}000(1 - 0.2)^{2.5} \approx \$8{,}014.07$$
It makes sense that the automobile's value after $2\dfrac{1}{2}$ years should be between the values of u_2 and u_3, $8,960$ and $7,168$. However, the value of the car after $2\dfrac{1}{2}$ years is not exactly halfway between those values because the function describing the value is not linear.

b. To find when the automobile is worth half of its initial value, substitute 7,000 for y and find x.
$7{,}000 = 14{,}000(1 - 0.2)^x$ Substitute 7,000 for y.
$0.5 = (1 - 0.2)^x$ Divide both sides by 14,000.
$0.5 = (0.8)^x$ Combine like terms.

You don't yet know how to solve for x when x is an exponent, but you can experiment to find an exponent that produces a value close to 0.5. The value of $(0.8)^{3.106}$ is very close to 0.5. This means that the value of the car is about $7,000, or half of its original value, after 3.106 years or about

202 CHAPTER 4 Exponential, Power, and Logarithmic Functions

Modifying the Investigation

Whole Class Follow the procedure note to collect data. For Steps 2 through 6, have students do each step, then discuss answers as a class before moving on to the next step.

Shortened Use the sample data. Discuss Steps 4 through 6.

One Step Follow the Procedure Note of the investigation. Ask students to find recursive formulas and function expressions that describe the data. As needed, remind students that repeated multiplication is represented by exponents. During the discussion, ask what other kinds of growth might be modeled by exponential functions; include depreciation.

202 CHAPTER 4 Exponential, Power, and Logarithmic Functions

Step 2 5 by $(u_1 + u_n)$; or 5 by 20

Step 4 Answers will vary. The rectangle dimensions will be n by $(u_1 + u_n)$; area $= n \cdot (u_1 + u_n)$.

Step 5 Answers will vary, but the formula should be equivalent to $S_n = \frac{n(u_1 + u_n)}{2}$; n is the number of rows or terms, u_1 is the length of the first row, u_n is the length of the last row. The length of the rectangle is $(u_1 + u_n)$, the height or width is n and divide by 2 because the rectangle is twice the partial sum.

Step 2 If you cut out two copies of this figure and slid them together to make a rectangle, what would the dimensions of your rectangle be?

Step 3 Have each member of your group create a different arithmetic sequence. Each of you can choose a starting value and a common difference. (Use positive numbers less than 10 for each of these.) On graph paper, draw two copies of a step-shaped figure representing your sequence.

Step 4 Cut out both copies of the step-shaped figure. Slide the two congruent shapes together to form a rectangle, and then calculate the dimensions and area of the rectangle. Now express this area in terms of the number of rows, n, and the first and last terms of the sequence.

Step 5 Based on what you have discovered, what is a formula for the partial sum, S_n, of an arithmetic series? Describe the relationship between your formula and the dimensions of your rectangle.

In the investigation you found a formula for a partial sum of an arithmetic series. Use your formula to check that when you add $1 + 2 + 3 + \cdots + 100$ you get 5050.

History CONNECTION

According to legend, when German mathematician and astronomer Carl Friedrich Gauss (1777–1855) was 9 years old, his teacher asked the class to find the sum of the integers 1 through 100. The teacher was hoping to keep his students busy, but Gauss quickly wrote the correct answer, 5050. The example shows Gauss's solution method.

EXAMPLE Find the sum of the integers 1 through 100, without using a calculator.

Solution Carl Friedrich Gauss solved this problem by adding the terms in pairs. Consider the series written in ascending and descending order, as shown.

$$\begin{array}{ccccccccccc} 1 & + & 2 & + & 3 & + \ldots + & 98 & + & 99 & + & 100 & = S_{100} \\ 100 & + & 99 & + & 98 & + \ldots + & 3 & + & 2 & + & 1 & = S_{100} \\ \hline 101 & + & 101 & + & 101 & + \ldots + & 101 & + & 101 & + & 101 & = 2S_{100} \end{array}$$

The sum of every column is 101, and there are 100 columns. Thus, the sum of the integers 1 through 100 is

$$\frac{100(101)}{2} = 5050$$

You must divide the product 100(101) by 2 because the series was added twice.

LESSON 4.7 Arithmetic Series **245**

Conceptual/Procedural

Conceptual In this lesson, students further explore the concept of mathematical series and partial sums of arithmetic series.

Procedural Students investigate the formula for finding the partial sum of an arithmetic series.

Summarize

As students present their solutions in the Investigation and example, have them explain their sequence and their strategy. Emphasize the use of correct mathematical terminology. Whether student responses are correct or incorrect, ask other students if they agree and why. SMP 1, 2, 3, 4, 6

Discussing the Investigation

LANGUAGE ASK "Is the common meaning of the word *series* the same thing as the mathematical meaning?" [In common usage, *series* means a sequence rather than a summation.] You might also point out that the word *series* is both plural and singular. SMP 6

As the last students present, ASK "What happens if the common difference or the first term is negative?" Let students experiment to see that the formula still applies. Elicit the fact that the partial sums get farther from 0 as n increases, whether the common difference is positive or negative. SMP 1, 2, 8

If students haven't used non-integer numbers for initial values or common differences, ask if the formula holds for these numbers as well. Students can see that it does from step-shaped figures and rectangles.

The formula for the partial sum of an arithmetic series contains the expression $\frac{u_1 + u_n}{2}$. ASK "Does this expression seem familiar?" Elicit the idea that it's the mean of u_1 and u_n. "How does the mean of the first and last terms relate to the series?" [It's the mean of the terms in the series, because the terms are equally spaced.] "Why does multiplying the mean by the number of terms give you the total of those terms?" Review the often-neglected idea that because a mean is calculated by dividing the total by the number of values, the total is the mean times the number of values. SMP 6, 8

A good opener for new problems is a "what if not" question. ASK "What if it is not an arithmetic series but a geometric series?" Challenge students to think about it overnight.

LESSON 4.7 Arithmetic Series **245**

sequence.

CRITICAL QUESTION How do you find the partial sum of an arithmetic series?

BIG IDEA A basic formula for the sum of the nth partial sum of an arithmetic series is $S_n = \frac{n(u_1 + u_n)}{2}$.

Formative Assessment

Students worked with the idea of summation in their statistical work earlier, so this investigation may be straightforward for them. Check their ability to use previous knowledge in this new context. Their responses in the Investigation, Example, and Exercises provide evidence of meeting the objectives.

Apply

Extra Example

1. Find S_1, S_2, S_3, S_4, and S_5 for the sequence: $-1, 1, 3, 5, 7$. $S_1 = -1, S_2 = 0, S_3 = 3, S_4 = 8, S_5 = 15$

2. Write the expression $\sum_{m=1}^{3}(n^2 + 1)$ as the sum of terms, then calculate the sum. $2 + 5 + 10 = 17$

Closing Question

Find the sum of the first 20 multiples of 3. 630

Assigning Exercises

Suggested: 1–11, 14
Additional Practice: 12, 13, 15–18

Exercises 4, 5 If students have trouble starting these exercises, ask them to find the values of n, u_1, and the common difference.

Exercise 6 Students can use the formula to evaluate 6b and 6c, or they can apply Gauss's method and avoid using a formula. SUPPORT Suggest that students list the first few terms of the series to reinforce the meaning of summation notation.

246 CHAPTER 4 Exponential, Power, and Logarithmic Functions

...the example to any...tinuing, take a moment...e reeds in the original...the expression

...esent in this context?

Partial Sum of an Arithmetic Series

The partial sum of an arithmetic series is given by the explicit formula
$$S_n = \frac{n(u_1 + u_n)}{2}$$
where n is the number of terms, u_1 is the first term, and u_n is the last term.

In the exercises you will use the formula for partial sums to find the sum of consecutive terms of an arithmetic sequence.

Exercises

You will need a graphing calculator for Exercise 15.

Practice Your Skills

1. List the first five terms of this sequence. Identify the first term and the common difference.
   $u_1 = -3 \quad -3, -1.5, 0, 1.5, 3; u_1 = -3, d = 1.5$
   $u_n = u_{n-1} + 1.5$ where $n \geq 2$

2. Find $S_1, S_2, S_3, S_4,$ and S_5 for this sequence: 2, 6, 10, 14, 18. ⓐ $S_1 = 2, S_2 = 8, S_3 = 18, S_4 = 32, S_5 = 50$

3. Write each expression as a sum of terms, then calculate the sum.

   a. $\sum_{n=1}^{3}(n+2)$ $\quad 3 + 4 + 5 + 6; 18$

   b. $\sum_{n=1}^{3}(n^2 - 3)$ $\quad -2 + 1 + 6; 5$

4. Find the sum of the first 50 multiples of 6: $\{6, 12, 18, \ldots, u_{50}\}$. ⓑ $S_{50} = 7650$

5. Find the sum of the first 75 even numbers, starting with 2. ⓐ $S_{75} = 5700$

Reason and Apply

6. Find these values.

   a. Find u_{75} if $u_n = 2n - 1$. $u_{75} = 149$

   b. Find $\sum_{n=1}^{75}(2n - 1)$. ⓐ $S_{75} = 5625$

   c. Find $\sum_{n=20}^{75}(2n - 1)$. $S_{75} - S_{19} = 5264$

246 CHAPTER 4 Exponential, Power, and Logarithmic Functions

The **Chapter Review** revisits key concepts and mathematical vocabulary, and it provides exercises that address the content and procedures covered.

Take Another Look gives students a chance to extend their understanding of concepts by approaching the mathematics from a different perspective, often giving a visual slant or calling out connections between mathematical ideas.

CHAPTER REVIEW

Exponential functions provide explicit and continuous equations to model geometric sequences. They can be used to model the growth of populations, the decay of radioactive substances, and other phenomena. The general form of an exponential function is $y = ab^x$, where a is the initial amount and b is the base that represents the rate of growth or decay. Because the exponent can take on all real number values, including negative numbers and fractions, it is important that you understand the meaning of these exponents. You also used the **point-ratio form** of this equation, $y = y_1 \cdot b_1^{x-x_1}$.

Until you read this chapter, you had no way to solve an exponential equation, other than guess-and-check. The **logarithmic function** is defined so that $\log_b y = x$ means that $y = b^x$. This means that exponential functions and logarithmic functions are **inverse** functions. The inverse of a function is the relation you get when all values of the independent variable are exchanged with the values of the dependent variable. The graphs of a function and its inverse are reflected across the line $y = x$.

A **series** is the indicated sum of terms of a sequence, defined recursively with the rule $S_1 = u_1$ and $S_n = S_{n-1} + u_n$, where $n \geq 2$. Series can also be defined explicitly. With an explicit formula, you can find any **partial sum** of a sequence without knowing the preceding term(s).

An explicit formula for **arithmetic series** is $S_n = \frac{n(u_1 + u_n)}{2}$.

Explicit formulas for **geometric series** are

$$S_n = \left(\frac{u_1}{1-r}\right) - \left(\frac{u_1}{1-r}\right)r^n \text{ and } S_n = \frac{u_1(1-r^n)}{1-r} \text{ where } r \neq 1.$$

Exercises

You will need a graphing calculator for Exercises **11** and **13**.

@ Answers are provided for all exercises in this set.

1. Evaluate each expression without using a calculator. Then check your work with a calculator.

a. 4^{-2} $\frac{1}{16}$ b. $(-3)^{-1}$ $-\frac{1}{3}$ c. $\left(\frac{1}{5}\right)^{-3}$ 125

d. $49^{1/2}$ 7 e. $64^{-1/3}$ $\frac{1}{4}$ f. $\left(\frac{9}{16}\right)^{3/2}$ $\frac{27}{64}$

g. -7^0 -1 h. $(3)(2)^2$ 12 i. $(0.6^{-2})^{-1/2}$ 0.6

2. Rewrite each equation in logarithm form so that the x is no longer an exponent.

a. $3^x = 7$ b. $5 = 4^x$ c. $10^x = 7^5$
$\log_3 7 = x$ $\log_4 5 = x$ $\log 7^5 = x$

CHAPTER 4 REVIEW 255

Reviewing and Assigning Exercises tips support teachers at the end of each chapter.

CHAPTER REVIEW 4

Reviewing
To review the concepts of this chapter, you might use Exercises 4, 7, 11, and 16 as an in-class assignment. Assign the exercises for students to work on in small groups. As they present their solutions, review exponential functions, properties of exponents and logarithms, series, summation notation, and partial sums.

Assigning Exercises
You might ask groups to do parts a and b or half the parts of each exercise and assign the remainder of each problem as homework.

LESSON 4 REVIEW 255

Take Another Look

1. For substitution, students will probably first solve for a: $a = \frac{40}{b^4}$ and $a = \frac{4.7}{b^7}$. $\frac{40}{b^4} = \frac{4.7}{b^7}$, $\frac{40}{4.7} = \frac{b^4}{b^7}$; $8.51 = b^{-3}$; $(8.51)^{-1/3} = (b^{-3})^{-1/3}$; $0.49 \approx b$; $a \approx \frac{40}{0.49^4}$; $a \approx 693.87$; $y \approx 693.87 (0.49)^x$, ($a \approx 695.04$; $y \approx 695.04 (0.49)^x$ if the unrounded value of b is used). Students might also solve the system by elimination, dividing the second equation by the first to get $\frac{4.7}{40} = b^3$; $b = \sqrt[3]{\frac{4.7}{40}} \approx 0.4898$.

2. If $|r| < 1$, then r^n approaches 0 and $\frac{u_1(1-r^n)}{1-r}$ approaches $\frac{u_1(1-0)}{1-r}$, or $\frac{u_1}{(1-r)}$. This behavior supports the explicit formula for the infinite sum of a convergent geometric sequence, $S = \frac{u_1}{(1-r)}$. If $r > 1$, then r^n gets larger without bound, and $1 - r$ will be a fixed negative number; therefore $\frac{u_1(1-r^n)}{1-r}$ also gets larger without bound. If $r < -1$, the expression r^n alternates sign, getting farther from 0, so $\frac{u_1(1-r^n)}{1-r}$ also gets farther from 0.

Assessing the Chapter

The Kendall Hunt Cohesive Assessment System (CAS) provides a powerful tool with which to manage assessment content, create and assign homework and tests, and disseminate assessment data.

You might test the chapter in two separate phases. Allow students to use their notebooks on a test of applications, including at least one constructive assessment item, and then turn in that assessment and take a short test on solving equations using the properties. You might also want to do the individual assessment in two parts: one part without a calculator, including items such as Exercises 1, 2, and 6, and the other part with a calculator.

Take Another Look

1. You have learned to use the point-ratio form to find an exponential curve that fits two data points. You could also use the general exponential equation $y = ab^x$ and your knowledge of solving systems of equations to find an appropriate exponential curve. For example, you can use the general form to write two equations for the exponential curve through (4, 40) and (7, 4.7):

$$40 = ab^4$$
$$4.7 = ab^7$$

Which constant will be easier to solve for in each equation, a or b? Solve each equation for the constant you have chosen. Use substitution and the properties you have learned in this chapter to solve this system of equations. Substitute the values you find for a and b into the general exponential form to write a general equation for this function.

Find an exercise or example in this chapter that you solved using the point-ratio form, and solve it again using this new method. Did you find this method easier or more difficult to use than the point-ratio form? Are there situations in which one method might be preferable to the other?

2. The explicit formula for a partial sum of a geometric series is $S_n = \frac{u_1(1-r^n)}{1-r}$. To find the sum of an infinite geometric series, you can imagine substituting ∞ for n. Explain what happens to the expression $\frac{u_1(1-r^n)}{1-r}$ when you do this substitution.

Assessing What You've Learned

 GIVE A PRESENTATION Give a presentation about how to fit an exponential curve to data or one of the Take Another Look activities. Prepare a poster or visual aid to explain your topic. Work with a group, or give a presentation on your own.

 ORGANIZE YOUR NOTEBOOK Review your notebook to be sure it's complete and well organized. Make sure your notes include all of the properties of exponents, including the meanings of negative and fractional exponents. Write a one-page chapter summary based on your notes.

 WRITE TEST ITEMS Write a few test questions that explore series. You may include sequences that are arithmetic, geometric, or perhaps neither. You may want to include problems that use sigma notation. Be sure to include detailed solution methods and answers.

258 CHAPTER 4 Exponential, Power, and Logarithmic Functions

Facilitating Self-Assessment
Presentations given on topics in this chapter could include desmos demonstrations of an exponential function and its inverse. You might split up the topics covered in this chapter. Each group could do a presentation on a different topic for review. The groups could also be responsible for writing review questions about their topic.

Items from this chapter suitable for student portfolios include Lesson 4.1, Exercise 13; Lesson 4.2, Exercises 7 and 9; Lesson 4.3, Exercises 6 and 7; Lesson 4.4, Exercise 7; Lesson 4.5, Performance Task; Lesson 4.6, Exercise 8; Lesson 4.7, Performance Task; and Lesson 4.8, Exercise 11.

The combination of formal quizzes and tests, informal assessments, and student self-assessment provides feedback about and evidence of what a student knows and is able to do. **Assessments** can be assigned, completed, and graded online through the Kendall Hunt Cohesive Assessment System.

Student and Teacher ebooks

Video clips introduce a truly problematic situation, giving students an intuitive understanding of a particular mathematical phenomenon by showing several, varied examples of the math happening. This vicarious experience creates equity for all students and stimulates intrigue and wonder about the math.

Discovering Advanced Algebra ebooks are compatible with most Internet-enabled devices and are accessible through Flourish, the Kendall Hunt digital learning network.

Color-coded bookmarks support teachers with multiple classes.

Differentiation strategies and tips, CCSS standards, and key concepts and procedures addressed in the lesson can all be accessed at point of use.

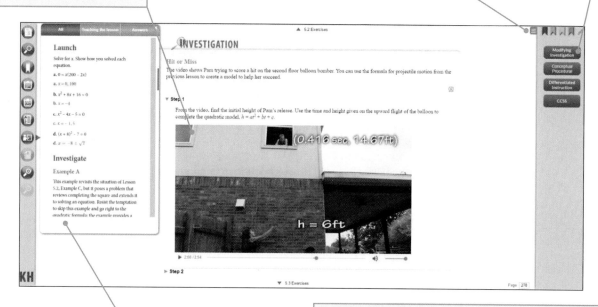

The Teacher ebook offers one-click access to teacher materials including lesson suggestions and problem answers.

desmos objects, including an embedded graphing calculator and simulations for many modeling and data collecting investigations, allows students to create artifacts that are saved to the student's digital book.

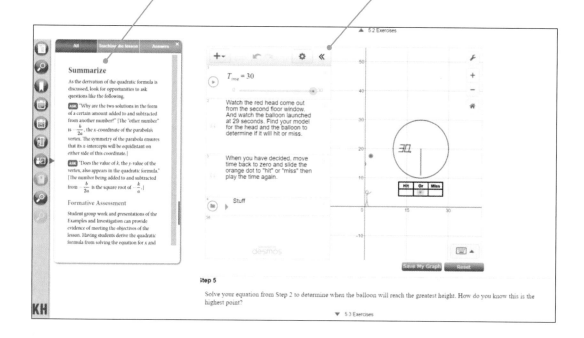

The Common Core State Standards for Mathematics Content and Practice

The Common Core State Standards for Mathematics (CCSS) consist of Content Standards and Standards for Mathematical Practice (SMP). The Content Standards include rigorous content and application of knowledge through high-order skills, requiring both conceptual understanding and procedural fluency. The Standards for Mathematical Practice foster reasoning and sense-making in mathematics

The Standards for Mathematical Practice can be thought of as describing mathematical habits of mind. They are not instructional strategies, or ways of teaching. Instead, they describe ways of thinking about and doing mathematics that mathematics educators at all levels should seek to develop in their students. These eight mathematical practices rest on processes and proficiencies with long standing importance in mathematics education.

The authors of the CCSS stress that the balance of procedure and understanding is achieved by infusing both Content Standards and Standards for Mathematical Practice into every mathematics lesson. This connection of content and practice is a strength of the Discovering Mathematics curriculum. Throughout its history, Kendall Hunt has recognized the value of curriculum that incorporates problem solving, real-world applications, conceptual understanding, and mathematics as sense-making. When students are given the opportunity to be actively involved in their own discovery of mathematics, they become better problem solvers and develop a deeper understanding of the concepts. Scientific research supports these pedagogical approaches, which are central to Discovering Mathematics and the Common Core State Standards.

Balancing Conceptual and Procedural Understanding

Procedural understanding is evidenced by the skill to carry out procedures and processes flexibly, accurately, efficiently, and appropriately. Students demonstrate the ability to select and apply appropriate procedures correctly and to justify or verify the correctness of a procedure using concrete or symbolic methods. Conceptual understanding is evidenced by having the knowledge needed to reason about a concept in a novel situation, not just to replicate a procedure. Students demonstrate the ability to interpret the assumptions and relations involving mathematical concepts. They can compare, contrast, and integrate related concepts and principles.

The CCSS calls for a balanced combination of conceptual understanding and procedural fluency. This balance leads to mathematical understanding that CCSS defines as the ability to justify, in a way appropriate to the student's mathematical maturity, why a particular mathematical statement is true or where a mathematical rule comes from. In other words, students should know the mathematical procedures AND be able to justify why the procedure works.

Discovering Advanced Algebra exemplifies this balance. Investigations and examples provide the balance in developing both procedural and conceptual understanding. The Teacher Edition has a section labeled Conceptual/Procedural to identify where each type of understanding is being cultivated. In addition, the exercises are divided into three categories; Practice Your Skills exercises that provide practice in procedures and processes, Reason and Apply exercises that provide practice in concept application, and Review exercises that provide a blend of procedure and conceptual practice on previously learned concepts.

Discovering Advanced Algebra and the Standards for Mathematics Content

The intent of the CCSS is not to have a checklist and address one standard in one or two days and move to the next standard. The intent is to teach and revisit standards in various contexts that show the connections within the mathematics. At the beginning of all *Discovering Advanced Algebra* lessons you will find a box that lists the CCSS for content under three headings: Developed, Introduced, and Applied. Developed standards are those that are being taught as the focus of that lesson. Applied standards are those that have been developed previously and are being used to move students forward in the progression of mathematics. And Introduced standards are those that the lesson introduces, laying the groundwork for future development.

Discovering Advanced Algebra addresses the Standards for Mathematical Practice deeply

The Standards for Mathematical Practice are an integral part of both the daily investigations and the exercise sets in every lesson of *Discovering Advanced Algebra*. Various investigations in *Discovering Advanced Algebra* encourage students to inquire about relevant ideas and issues beyond the bounds of the course. Students have legitimate opportunities to experiment, hypothesize, measure, analyze, test, talk, write, explain, and justify their ideas. In short, they engage in real mathematics. In addition, Chapter 0 is built around the Mathematical Practices, introducing students to what they are and how they are utilized in problem solving.

A representative lesson is presented to illustrate the correlation to each Standard for Mathematical Practice.

SMP 1: Make sense of problems and persevere in solving them.

Students start by explaining to themselves the meaning of a problem and looking for entry points to its solution. They analyze givens, constraints, relationships, and goals. They make conjectures about the form and meaning of the solution and plan a solution pathway rather than simply jumping into a solution attempt. They consider analogous problems, and try special cases and simpler forms of the original problem in order to gain insight into its solution. They monitor and evaluate their progress and change course if necessary.

While students in advanced algebra learn more sophisticated methods for performing algebraic manipulations and transformations than in earlier courses, sense-making is still paramount. Students who use the *Discovering Advanced Algebra* curriculum gain proficiency in analyzing problems, trying different solution methods, and evaluating their answers. Because rich, nonroutine problems are integral to the textbook, students develop their own understanding and build solid problem-solving skills. For example, in the **Lesson 6.3: Higher-Degree Polynomials** The Largest Triangle Investigation, students fold the upper-left corner of a sheet of paper so that it touches the bottom edge and determine the distance, x, along the bottom that will produce the triangle with the greatest area. They use strategies they have learned in several lessons to create a model for finding that area, surprisingly a cubic function instead of a quadratic function, and write a report explaining both their solution and their strategy. They revisit the Largest Triangle Investigation in Exercise 8, interpreting a graph of a triangle formed from a different size paper and extending to average rate of change for the function.

SMP 2: Reason abstractly and quantitatively.

Students make sense of quantities and their relationships in problem situations. They utilize the ability to decontextualize and/or contextualize when solving problems. Students employ quantitative reasoning as they create a coherent representation of the problem at hand; considering the units involved, attending to the meaning of quantities, and flexibly using different properties of operations and objects.

Discovering Advanced Algebra provides relatable, real-world exercises that deepen students' understanding of concepts. Students become proficient at expressing abstract ideas by graphing relationships, making and verifying conjectures, and solving equations. They become equally adept at drawing larger themes from the lessons. Students move from problem situations to algebraic representations, and then interpret their results in terms of the original context. In **Lesson 4.4: Applications of Exponential and Power Equations**, for example, students solve problems involving interest accrual, the motion of a pendulum, the intensity of light, a simulation of exponential decay, and the orbit of the moons of Saturn. In each exercise in the Reason and Apply section, students use the context to write an algebraic expression, and then use symbolic manipulation to answer questions. Students also make sense of the relationships through tables and graphs.

SMP 3: Construct viable arguments and critique the reasoning of others.

Students understand and use stated assumptions, definitions, and previously established results in constructing arguments. They make conjectures and build a logical progression of statements to explore the truth of their conjectures. They are able to analyze situations by breaking them into cases, and can recognize and use counterexamples. They justify their conclusions, communicate them to others, and respond to the arguments of others. They reason inductively about data, making plausible arguments that take into account the context from which the data arose.

In second year algebra, students move into more generalizable results that pertain to all functions of a certain type, producing results that cannot be checked by checking specific evidence. These require constructing a general argument. The lessons in *Discovering Advanced Algebra* help students develop higher-order thinking and communication skills. Students gather data, make conjectures, explain the reasoning behind their conjectures, and share their ideas verbally and in writing. Students are encouraged to listen carefully to others as they share and, whether correct or not, state whether they agree and why. Through collaboration and group work, they both consider and critique the perspectives of their classmates. Students watch a video in the **Lesson 5.3: The Quadratic Formula** Hit or Miss Investigation and collect evidence of the upward flight of a water balloon to complete a quadratic model that will help them predict whether the "balloon bomber" will score a hit. They make a mathematical argument to support their prediction.

SMP 4: Model with mathematics.

Students can apply the mathematics they know to solve problems arising in everyday life, society, and the workplace and are comfortable making assumptions and approximations to simplify a complicated situation, realizing that these may need revision later. They are able to identify important quantities in a practical situation and map their relationships. They can analyze those relationships mathematically to draw conclusions. They routinely interpret their mathematical results in the context of the situation and reflect on whether the results make sense, possibly improving the model if it has not served its purpose.

Mathematical modeling is the process of using various mathematical structures—graphs, equations, diagrams, scatter plots, statistical displays, and so forth—to represent and understand quantities and their relationships in real-world situations, predict outcomes, make decisions, and develop designs. Models can also explain the mathematical structures themselves—for example, a model of population growth can demonstrate the recursive nature of the exponential function. *Discovering Advanced Algebra* students create models using technology that dynamically represents math concepts, and solve real-world problems using abstract representations such as algebraic equations and graphs. They may also use optional software to simulate and investigate mathematical models, objects, figures, diagrams, and graphs. **Lesson 9.4: Bivariate Data and Correlation** focuses on predicting the associations between parameters of a large populations based in bivariate sampling. Students investigate the correlation of time spent on various after school activities to see if there is a correlation between any pair of variables, and write a report describing their findings.

SMP 5: Use appropriate tools strategically.

Students consider the available tools when solving a mathematical problem. These tools might include pencil and paper, concrete models, a ruler, a protractor, a calculator, a spreadsheet, a computer algebra system, a statistical package, or dynamic geometry software.

Discovering Advanced Algebra fully integrates a wide variety of hands-on and technology tools—more than any other textbook series— including compass-and-straightedge, patty paper, graphing calculators to graph data, observe patterns in tables, and graph functions, motion sensors to gather data, dynamic geometry software to explore transformations, and dynamic data software to investigate probability and statistics concepts. Students become sufficiently familiar with appropriate tools to decide which tool is appropriate, know the benefits and limitations of the tools and detect possible errors as they not only use the tool but explain the mathematics as it relates to the tool. The ebook includes videos for some investigations, as well as embedded desmos simulations for many investigations. Desmos objects are also embedded for students to use when solving examples and exercises. In **Lesson 3.3: Lines in Motion** students use motion sensors to create a function and a translated copy of that function. They then confirm their findings by exploring what happens to the graph of a linear function when you translate the graph of a line. In addition to directions for collecting data with a motion sensor and using a graphing calculator to graph the data, there is a desmos object embedded in the ebook that allows students to collect and graph data from a simulation of the investigation.

SMP 6: Attend to precision.

Students try to communicate precisely to others. They try to use clear definitions in discussion with others and in their own reasoning. They state the meaning of the symbols they choose, including using the equal sign consistently and appropriately. They are careful about specifying units of measure and labeling axes to clarify the correspondence with quantities in a problem. They calculate accurately and efficiently, express numerical answers with a degree of precision appropriate for the problem context.

Discovering Advanced Algebra cultivates students' precision and attention to detail. Students learn the importance of giving accurate definitions by exploring counterexamples. Throughout the textbook, students must be attentive to units of measure, make and test predictions, and clearly communicate their reasoning. They state the meaning of symbols as they write definitions, test their predictions, and check the reasonableness of their answers by asking themselves, "Does this make sense?" They then clearly communicate their reasoning using precise terms. In **Lesson 8.4: Bivariant Independence**, students learn to evaluate the relationship of categorical data. They use marginal frequencies and segmented bar plots to determine if two variables are independent. They examine margin of error intervals for sample proportions.

SMP 7: Look for and make use of structure.

Students look closely to discern a pattern or structure. They also can step back for an overview and shift perspective. They can see complicated things, such as some algebraic expressions, as single objects or as being composed of several objects.

Discovering Advanced Algebra students tackle complex problems by observing similar patterns in problems of varying difficulty. They learn to extend their knowledge of simple properties to more advanced applications, and they apply reasoning strategies to break complex problems into manageable parts—skills they will use throughout their lives. The structure theme in Algebra I centered on seeing and using linear functions, that have a constant difference, and exponential functions, that have a constant ratio. This continues in Algebra II, where students delve deeper into comparing linear and exponential functions in ways that reveal meaning. In the Investigating Structure in **Lesson 6.1: Polynomials,** students synthesize their knowledge of linear and exponential functions to prove that linear functions grow by constant differences over equal intervals, while exponential functions grow by constant ratios over equal intervals.

SMP 8: Look for and express regularity in repeated reasoning.

Students notice if calculations are repeated, and look both for general methods and for shortcuts. As they work to solve a problem, they maintain oversight of the process, while attending to the details. They continually evaluate the reasonableness of their intermediate results.

The investigative approach of *Discovering Advanced Algebra* leads students to look for patterns and structure in mathematics. Students look at repeated calculations to derive formulas and look for generalizable methods or shortcuts. As they investigate mathematics concepts they reflect on the process while attending to details. For example, in **Lesson 4.2: Properties of Exponents and Power Functions,** students move from a concrete example to generalize the properties of exponents. They discover that zooming in or out on the graph of exponential and power functions can give insight on the short and long run behaviors of the two functions, allowing them to determine which grows faster. Students make use of this type of reasoning in the Investigating Structure, where they compare and contrast the graphs of power functions with odd exponents and with even exponents and explain why the graphs appear to "flatten out" between -1 and 1 as the power gets larger.

Incorporating the Mathematical Practices

Not all Standards for Mathematical Practice (SMP) will be embedded in every lesson. The authors of the CCSS have identified two SMP that should be part of every lesson:

>Make sense of problems and persevere in solving them
>Attend to precision.

The other SMP can be grouped in pairs that are connected and at least one pair of SMP should appear in most lessons.

Reasoning and Explaining
>Reason abstractly and quantitatively
>Construct viable arguments and critique the reasoning of others

Modeling and Using Tools
>Model with mathematics
>Use appropriate tools strategically

Seeing Structure and Generalizing
>Look for and make use of structure
>Look for and express regularity in repeated

The Kendall Hunt Common Core Math app

The Kendall Hunt Common Core Math app has been created to support educators and administrators who use *Discovering Advanced Algebra* with the implementation of the CCSS in their classrooms and districts. The Kendall Hunt Common Core Math app provides Look-fors for each program that direct educators to specific Content Standards as they are addressed throughout the curriculum. These classroom observation tools may be used as a non-evaluative assessment resource and a peer-observation tool. The app provides suggested student responses for each of the SMP as observational look-fors. These should be helpful in providing suggestions related to both planning and assessment for teachers, content leaders, and school-based administrators. The app also provides an interface for entering observation data as well as access to all the Content Standards for Mathematics K-12. The collected observation data can be shared via email or with its unique built-in instant meeting feature. Highly interactive graphs allow observers to merge different observations to compare and contrast different teachers or the same teacher over time. Using the Timeline feature, observers can track the observed Look-fors through time. Download the app from the iTunes store.

Teaching with *Discovering Advanced Algebra*

In today's society everyone, not just those who are good at mathematics, must understand algebraic concepts and be able to work with technology. Changes in society and changing expectations of employers have required changes in the algebra curriculum. The Common Core State Standards specify mathematical content for which all students should become proficient in order to be considered college and career ready. All students must have the opportunity to learn and meet the same high standards if they are to access the knowledge and skills necessary in their post-school lives. *Discovering Advanced Algebra* answers that need for change. It teaches a range of skills more applicable to today, it works toward building technological understanding, it connects algebra to other areas of math, and it encourages both conceptual and procedural understanding of algebraic ideas.

The Common Core State Standards stress that mathematical understanding and procedural skill are equally important, and both are accessible using mathematical tasks of sufficient richness. *Discovering Advanced Algebra* uses technology, along with applications, to foster a deeper understanding of algebraic ideas. The investigations emphasize symbol sense, algebraic manipulations, and conceptual understandings. The investigative process encourages the use of multiple representations—numerical, graphical, symbolic, and verbal—to deepen understanding for all students and to serve a variety of learning styles. Explorations from multiple perspectives help students simplify and understand what formerly were difficult algebraic abstractions.

Teaching with *Discovering Advanced Algebra* decreases the time students spend on rote memorization, teacher exposition, and extended periods of paper-and-pencil drill. It changes the rules for what is expected of students and what they should expect of their teacher, and it creates a nontraditional classroom in which students don't just learn procedures, but learn to justify why the procedures work. Once you become comfortable with the program's student-centered teaching approach, and add your own sense of enthusiasm and determination, you will understand why *Discovering Advanced Algebra* has developed such a strong history of success in classrooms worldwide.

Differentiating Instruction

Teachers face the challenge of addressing the needs of all learners in heterogeneous classes that may include struggling students, English Language Learners (ELLs), and advanced learners. With *Discovering Advanced Algebra*, teachers can reach learners of every style and background with educational materials that suit their students' strengths. Many students struggle in math class, not because they "can't do" math, but because traditional methods of instruction do not meet their needs. The "traditional" lesson plan for a math class uses a narrow range of pedagogy—a teacher explains a new topic, demonstrates a specific technique for dealing with a set group of problems, and students then practice solving problems using that technique. This is efficient, but effective for only a limited number of students: those who are independent, auditory learners, who tend to think linearly. In a heterogeneous class, this group describes only a subset of your students. Students who do not fit this description often become frustrated and disengaged. The techniques highlighted in this Teacher Edition will help you reach these students.

Discovering Advanced Algebra provides opportunities for students to work in cooperative groups or with a teacher's direct guidance to master skills. Hands-on investigations give students a chance to connect the mathematics to something they understand (SMP 1). They gather data or construct figures to conduct experiments, explore physical models to solve problems, and use technology that dynamically represents math concepts (SMP 2, 4, 5, 7, 8). They make conjectures, explain the reasoning behind their conjectures, and share their ideas verbally and in writing (SMP 3, 6).

The accessibility of the mathematics in *Discovering Advanced Algebra* is central to the instructional design. Varied forms of pedagogy are required to engage and challenge all learners. *Discovering Advanced Algebra* combines these instructional strategies seamlessly to motivate struggling students, ELLs, and advanced learners alike.

Discovering Advanced Algebra is built on the idea that math should and can be accessible to every student. To help you achieve this goal, the *Discovering Advanced Algebra* Teacher Edition has been written to address issues of the diverse classroom. We provide lesson-planning options and differentiating instruction suggestions for every lesson. The **Differentiating Instruction box** at the beginning of each lesson highlights teaching suggestions for three categories of students: English Language Learners (ELL), struggling students who require extra Support, and Advanced students. Specific alerts and tips for each of these categories can also be found throughout the teacher notes.

English Language Learner Students

The term *English Language Learners* (*ELLs*) describes a wide variety of students. ELL students might have math skills and experience comparable to your highest-achieving native English speakers; or they might have extensive experience reading and writing in English, but minimal experience or confidence speaking; or they might be facing a very different classroom environment and approach to learning mathematics. Whether their math skills are low, middle, or high, ELLs may struggle to understand what is being asked of them. They need hands-on experiences and extra help with the mathematics vocabulary. They benefit significantly from group interactions. In the **Differentiating Instruction** box and in ELL notes in the wraparound text, we focus on addressing the range of language and cultural issues that these diverse students face. Additional practice is available for most lesson in the More Practice Your Skills and Condensed Lessons in your digital resources.

It is important to remember that for students who are learning English, each lesson is actually two lessons—one in math and one in English

Struggling Students

Students struggle in a mathematics classroom for a wide variety of reasons. They are a varied group that includes low-performing students, special needs students, students with behavior issues, or just students who have a different learning style. They are also a group in flux—students who quickly grasp one mathematical concept may struggle with another concept and students who struggle with one concept may quickly grasp the next concept. They may have struggled in earlier mathematics classes and have a low level of understanding of basic mathematics. These students tend to have a high level of math anxiety and may be their own worst enemy in math class. They need additional time to move through the book and gain confidence in each topic. Many students benefit most from starting with ideas that are engaging and relevant and reviewing previous concepts only as they see the need. It is important to not get bogged down in review. Early and ongoing successes with these students begin to overcome the feelings of failure they may have developed over years of schooling. **Refreshing Your Skills** lessons that provide a quick review of the prerequisite skills for each chapter, along with More Practice Your Skills and Condensed Lessons (also available in Spanish) are available in your digital resources.

The **Differentiating Instruction** boxes in this book include **Support** notes that point out places where these students may be challenged and give suggestions for breaking through those tough spots. To help students understand how to apply the concepts they have developed in the investigation and examples, the Teacher Edition provides **Extra Examples** for you to use before students begin work on the exercises.

Advanced Students

Advanced students either have a current understanding of the topic or are able to grasp its fundamentals very quickly, needing more in-depth problems, more challenging work, and extensions. They tend to develop mathematical understanding quickly and are confident in their ability to make sense of new concepts. With this curriculum, they can spend time working with teammates, explaining their thinking, and writing about the mathematics. Their learning will deepen as they exchange ideas and reflect on what they discover from an investigation. The **Advanced** notes in the **Differentiation Instruction** box and wrap around notes reference ideas that take students deeper into a topic, in either more detail or more abstraction, and **Take Another Look** exercises provide engaging extensions for advanced students.

To help you meet the needs of a variety of learning styles, there is **Problem Strings** ancillary. The regular use of the powerful routine called problem strings helps teachers before and after an investigation. Each lesson will have a problem string, a purposefully designed sequence of related problems that helps students mentally construct mathematical relationships and nudges them toward a major, efficient strategy. Increasingly, problem strings are being used by teachers to preview big ideas that will arise in an investigation, solidify the ideas and skills that come up in the investigation, invite students to prove or justify their ideas, describe and solidify strategies, and move towards efficient strategies

The chief objective for a heterogeneous class is to keep all students on the same topics, while also helping students access the lessons in different manners and at different depths. This keeps your preparation time down, allows for inclusive discussions, and, because students are learning together, allows your class to develop a sense of community. Shifting between heterogeneous and homogenous groups of various sizes allows students to experience your class with varying levels of comfort and challenge. Within that context, you can assign different students different aspects of the lesson, as facilitated by the teacher notes provided.

Discovering Advanced Algebra allows you to engage a wider range of students simply through its student-centered and discovery-based approach and that additional accommodation is made easy with the resources provided in this Teacher Edition and in your digital resources.

Assessing

Assessment goes well beyond determining grades. Assessment is the process of gathering evidence of what a student knows and is able to do, while evaluation is the process of interpreting that evidence and making judgments and decisions based on it. When you think of assessment as a way to gather information, rather than just as a way to assign grades, you have many assessment strategies to choose from. Not only do you want to be sure to address all your assessment goals, but you also want to provide ways for students with different learning styles to demonstrate their thinking and learning. The materials accompanying *Discovering Advanced Algebra* support a variety of methods for assessing both processes and results.

While you circulate among groups or watch students' presentations, engage in informal assessment of students' previous learning. Each lesson's **Formative Assessment** section in the *Discovering Advanced Algebra* Teacher Edition provides ideas of what to look for. Also, during class, note how effectively the lesson is proceeding. This allows you to make adjustments as you go. Throughout the Teacher Edition you can find ideas for what to look for. The commentary on the lesson includes **Alerts**, based on teachers' experiences, to difficulties students might have (but not be able to articulate) and suggestions about how you might respond to them. Because *Discovering Advanced Algebra* emphasizes conceptual understanding, journal entries describing how individual students are making sense of the ideas can be especially useful assessment tools. **Critical Questions** and **Big Ideas** point out important topics from each lesson, help you frame discussion, and connect the investigation to the learning objectives. A **Closing Question** helps students synthesize the concepts from the lesson. And **Extra Examples** in each section supplement the lesson objectives and give you further clarifying examples for discussion.

In addition, the Kendall Hunt Cohesive Assessment System (CAS) provides a powerful tool with which to manage assessment content, create and assign homework and tests, and disseminate assessment data. You will also use more intentionally focused activities to evaluate students' learning. *Discovering Advanced Algebra Assessment Resources* includes two versions of quizzes, tests, and Constructive Assessment Options for each chapter, as well as unit exams and a final exam. To evaluate students' understanding in depth, you can use the projects in *Discovering Advanced Algebra* for assessment. You might also have students prepare portfolios of their best work. This Teacher Edition lists exercises you might assign for inclusion in a portfolio. At the end of each chapter of the student book, the section **Assessing What You've Learned** contains suggestions for student self-assessment that will give both you and your students additional feedback on students' learning.

Planning

The *Discovering Advanced Algebra* Teacher Edition includes many ideas for planning. For each chapter and lesson, it lists objectives and materials. But every class is different. As a professional, you'll want to think about each chapter and lesson with your own students and schedule in mind.

Structuring the Class Period

The order of events in a class period will vary with the lesson, with your class's growing experience in independent learning, with the need for variety, and with the class response. If students are lethargic, for example, have them act out the investigation. If they're overly excited, channel the energy into their inquiry. For lessons with investigations, there are helpful suggestions for **Modifying the Investigation**, including guidelines for three possible approaches to the investigation: **Whole Class, Shortened**, and **One Step**.

When equipment or time is in short supply, or early in the year to help students become familiar with this type of class experience, doing the investigation as a **Whole Class** demonstration and discussion allows your class to receive the benefits of the investigative approach. Your challenge is to keep all students engaged and active in the discussion. Create a formal structure for student input so that everyone has a say—discussions like this are challenging for struggling students, ELL students, and quiet students. Using wait time thoughtfully and recognizing that questions can be as important as answers will ease this situation. Monitor Advanced students who want to answer many questions for the class; you can use them as experts to be relied on for reference or as recorders. Asking if students agree with what was said and why, whether the response was correct or not, also involves more students in the discussion while helping students develop the ability to listen critically and defend their thoughts with mathematics as evidence.

The **Shortened** investigation reduces the time needed for the investigation while still retaining many of its qualities and learning objectives.

The **One Step** investigation offers an alternative to the several-step investigation and to some examples. You can use the **One Step** approach for classes that are experienced in investigating. Even if your class needs more guidance at the beginning of the course, read through these alternative investigations to keep in view your goal of helping students become more independent investigators. This practice will help you become better at finding "teachable moments" as you interact with your students. The **One-Step** version of the lesson provides rich engaging tasks with less scaffolding than the investigation.

Planning a Lesson

You also need to think about your students' strengths and weaknesses when you plan individual lessons. Work through the investigation and ask yourself how your students might respond to it and how you can facilitate. If you have time, ask friends and colleagues to think out loud about how they would do the investigation.

While planning lessons, keep in mind that the approach of this course is different from what you might be used to. Instead of telling students everything they need to know before they embark on an investigation (and then retelling them as they work), you should aim to have students first encounter the mathematical ideas during the investigation. Later, during the discussion, those ideas can be formalized with a name and a definition. Remember that your role is to help students discover mathematics—in both the processes and the content of the journey.

Although careful planning is necessary, don't let it get in the way of your teaching. In an inquiry-based classroom, planning does not mean preparing a script you have to follow. What students do with technology-enhanced investigations can be different from what you anticipated. Each day you'll need to adjust your plans for the next few days. The farther you can see ahead, the better you'll be able to decide how to balance allowing students to explore and answer their own questions with getting the class through the course.

The LISA Lesson Plan Model

The LISA lesson plan model is a research-based framework for teaching the content while cultivating the Mathematical Practices. The model serves to organize your thoughts as you consider what you want students to know at the end of the lesson. It also encourages you to think of the kinds of questions you want to address during the classroom discussions. The LISA planning model is a four-step learning cycle model. The four steps are:

Launch Ways to spark interest, assess prior knowledge, make connections, use engaging scenarios, etc.

Considerations for planning include
- *How will I assess students' prior knowledge and experience?*
- *What mathematical understandings do I want students to acquire from this lesson?*
- *How will I introduce students to the activity in a way that engages them and activates their prior knowledge and curiosity?*

Questions for the discussion include
- *What do you know about . . . ?*
- *Have you ever . . . ?*
- *What do you think of when you hear the word . . . ?*

Investigate Ways for students to become actively involved with the problem, skill, or concept

Considerations for planning include
- *What questions will I ask to assess and deepen students' mathematical understanding?*
- *How will I support students with different learning styles or special needs? Do I need to differentiate the lesson? How?*
- *How will students share and discuss their thinking?*
- *How will I help students who are struggling?*

Questions for the discussion include
- *Why did you choose to . . . ?*
- *How did you know that?*
- *Why does that work?*
- *What would happen if . . . ?*

- *What do you predict will happen next?*
- *Do you think that will always happen?*

Summarize Ways for students to articulate mathematical ideas and vocabulary from the lesson, and to compare and contrast ideas and strategies.

Considerations for planning include
- *How will I formatively assess student understanding of the mathematical concepts from the lesson?*
- *What mathematical terms have students learned?*
- *How can I support students who are still struggling with terms or concepts from the lesson?*

Questions for the discussion include
- *What big idea(s) did we learn today?*
- *Can you explain what was just said in another way?*
- *How does what we learned today compare and contrast with what we learned previously?*

Apply Ways for students to practice what they have learned, extend the use of skills and concepts learned, and to make connections to other learning

Considerations for planning include
- *What follow-up tasks or homework will help students apply the core mathematical ideas?*
- *What upcoming lessons build on the mathematical understandings from this lesson?*

Questions for the discussion include
- *Where might you use . . . ?*
- *How does this relate to . . . ?*
- *What would happen if . . . ?*

The Teacher Edition wrap for every lesson in *Discovering Advanced Algebra* incorporates suggestions for utilizing the LISA lesson plan model.

Teaching a *Discovering Advanced Algebra* Lesson

By now, you are probably thinking about how this all comes together for a lesson, wondering what a day in *Discovering Advanced Algebra* classroom really looks like. The focus of the Common Core State Standards for Mathematics on the connection of content and practice adds a new dimension to your daily planning. Let's look at how you might plan and teach a lesson in *Discovering Advanced Algebra*.

As always, start with answering the question, "What's the mathematics of this lesson?" Add to that, "Which of the Mathematical Practices will be cultivated in this lesson?" Remember that the intent of the CCSS is not to have a checklist and address one standard in one or two days, but rather to revisit standards in various contexts that show the connections within the mathematics. Lessons will have a list of standards that are either developed, introduced or revisited.

Here is how **Lesson 4.1 Exponential Functions** might look in your classroom. The content standards and mathematical practices addressed are:

Mathematics Content Standards: (CCSS)

Developed

F.IF.3 Recognize that sequences are functions, sometimes defined recursively, whose domain is a subset of the integers.

F.BF.2 Write arithmetic and geometric sequences both recursively and with an explicit formula, use them to model situations, and translate between the two forms*

F.BF.3 Identify the effect on the graph of replacing $f(x)$ by $f(x) + k$, $k f(x)$, $f(kx)$, and $f(x + k)$ for specific values of k (both positive and negative); find the value of k given the graphs. Experiment with cases and illustrate an explanation of the effects on the graph using technology. Include recognizing even and odd functions from their graphs and algebraic expressions for them.

F.LE.2 Construct linear and exponential functions, including arithmetic and geometric sequences, given a graph, a description of a relationship, or two input-output pairs (include reading these from a table).

Introduced

F.IF.7e Graph exponential and logarithmic functions, showing intercepts and end behavior, and trigonometric functions, showing period, midline, and amplitude.

F.IF.8b Use the properties of exponents to interpret expressions for exponential functions.

F.IF.9 Compare properties of two functions each represented in a different way (algebraically, graphically, numerically in tables, or by verbal descriptions).

Applied

N.Q.2 Define appropriate quantities for the purpose of descriptive modeling.

A.CED.1 Create equations and inequalities in one variable and use them to solve problems.

F.LE.1c Recognize situations in which a quantity grows or decays by a constant percent rate per unit interval relative to another.

F.LE.5 Interpret the parameters in a linear or exponential function in terms of a context.

Standards for Mathematical Practice

Investigations support students in making sense of problems and persevering in solving them (**SMP 1**). Students explain to themselves the meaning of a problem and look for entry points to its solution. They must analyze givens, constraints, relationships, and goals. They should be encouraged to make conjectures about the form and meaning of the solution and plan a solution pathway rather than simply jumping into a solution attempt. Additionally, they always need to use correct mathematical language and be exact with computations (**SMP 6**).

The *Radioactive Decay* Investigation uses a simulation of growth and decay to model the mathematical concept of exponential functions (**SMP 4**). Students experiment with changing their equation to pass through other data points; then they evaluate and adjust their model.

In **Steps 1 and 2**, when students collect data and represent it using a table and graph they are reasoning abstractly and quantitatively (**SMP 2**). Sitting down when their number is not shown on the die models radioactive decay (**SMP 4**).

In **Steps 2–5**, using technology to organize and analyze the data from the simulation is using appropriate tools strategically (**SMP 5**). Additionally, students are reasoning abstractly and quantitatively (**SMP 2**) when they represent the situation it symbolically.

In **Step 3** students are looking for and making use of structure (**SMP 7**) when they draw connections among the recursive formula, the explicit formula, the graph, and the exponential model.

In **Step 5**, as students experiment with changing the equation to pass through data points, they are evaluating and adjusting their model (**SMP 4**).

Students will look for and express regularity (**SMP 8**) in data when they create rules for exponential functions in **Steps 3 and 6**.

PLANNING Lesson 4.1 Exponential Functions

- Read the student lesson, focusing on the key mathematical concepts of the lesson.

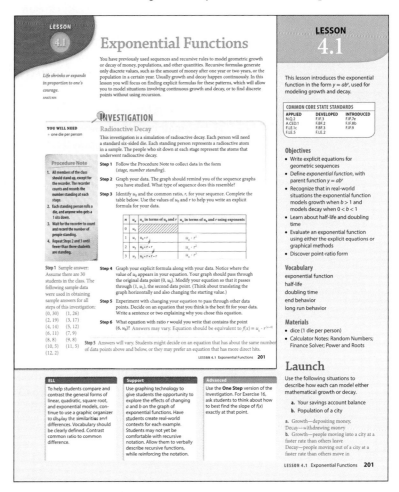

- Write a few sentences encapsulating what you want to students to know at the end of the lesson:

  Interpret equations in intercept form using input and output variables

  Explore the relationships among tables, scatter plots, recursive routines, and equations

- Identify your objectives

  – Write explicit equations for geometric sequences
  – Define *exponential function*, with parent function $y = ab^x$
  – Recognize that in real-world situations the exponential function models growth when $b > 1$ and models decay when $0 < b < 1$
  – Learn about half-life and doubling time
  – Evaluate an exponential function using either the explicit equations or graphical methods
  – Discover point-ratio form

- Read the Summarize section of the Lesson section of the teacher's notes. Instead of giving a list of definitions at the beginning of a class, close the lesson by reaching consensus on definitions, formalizing concepts, or posing thought provoking questions that challenge students to verbalize their understanding of the concepts. Looking ahead to the Critical Questions will help you frame your questions throughout the lesson.

- Think about how this lesson fits in with the chapter (from TE Overview). Previous chapters have addressed some basic statistics for working with real-world problems. They've also taught the powerful concept of function. Chapter 4 extends students' knowledge of functions to exponential, power, and logarithmic functions. Later chapters will address modeling real-word problems with other kinds of functions: polynomial, rational, trigonometric, probabilistic, and statistical. Lesson 4.1 begins with an introduction to exponential functions in the form $y = ab^x$ for modeling growth and decay.

Although you could tell your students the properties of exponential and logarithmic functions very quickly, remember that the goal is for them to understand the relationships deeply by discovering the relationships themselves. To that end, think about how you might use the Problem String for the lesson to help students mentally construct those mathematical relationships. Look over the investigation and the One-step problem and think about which is more appropriate for your class.

The Using this Chapter in the Overview offers suggestions for deciding which lessons are essential or not essential and under what conditions. For example, you might combine Lessons 4.1 and 4.2 for students familiar with the topics. Using the Shortened modification or the desmos simulations of the investigations also saves time. It is important, however, to have student presentations of ideas to help them communicate mathematically and justify their answers.

- Next, decide which investigation to use—the multistep investigation in the student book or an adaptation—whole class, shortened, or the one-step—found in the teacher's notes. Which would work better for your students at this point in the course? Or do you want students to use the embedded desmos object that simulates the investigation? You might even consider whether one or two groups might benefit from doing the one-step problem while other groups work through the steps of the investigation.

- Look closely at the steps of the investigation. Try to predict the various ways students will approach the problem. Look over the margin notes in the Teacher Edition for information about difficulties and ideas students may have. Read through the suggestions in the Investigate section and think about what points you would like to see arise. Think about how you expect your students to explain how they chose the equation to use in Step 4. Think about how you will use the examples: Will you review them in class no matter what or only if students struggle with the investigation?

- Plan the details. The question for the investigation is somewhat complex. Think about whether it would be a good idea to suggest that one person read a step aloud and another paraphrase to help students understand what is being asked of them.

- Look over the exercises and decide which ones to assign. Remember to balance procedural practice (Practice Your Skills) with conceptual practice (Reason and Apply). Some problems lend themselves to group work (Exercises 9 and 11) and might be used in class to get started on homework. Some exercises are excellent portfolio or journal entries (Exercises 7 and 13).

TEACHING LESSON 4.1

The Teacher Edition wrap for each lesson includes suggestions for each step in the LISA learning cycle.

LAUNCH (Introducing the Lesson):

To compare and contrast exponential growth and decay ask students to Think-Pair-Share on how the following situations might model either growth or decay.

- A savings account balance
- The population of a city

After the discussion, ask for examples of other situations that might indicate growth or decay.

INVESTIGATE (Making, Exploring, Finding . . .):

Begin the investigation by completing the simulation with the entire class. Have students return to their groups to complete Steps 2 through 6 on graph paper. Have an individual from each group present their work for one of the steps. Define the answer from Step 4 as an *exponential function* and connect this with what they learned earlier in the course about *geometric sequences*.

Assess students' ability to model data with recursive formulas and with common ratios. They should also be able to make the connection between exponents and repeated multiplication. Note whether the understanding of explicit formulas students developed earlier is deep enough to extend their understanding to the new context in this chapter.

After students begin working, take attendance, then start to circulate among the groups.

If some groups seem to be ahead of the others, have students find real-world examples of growth and decay situations. They might collect data and use transformations to find exponential models. You might also assign The Cost of Living project as an extension.

Decide which groups you will ask to do a presentation and plan the order of their presentations.

As students present their solutions for the investigation, ask probing questions to check for understanding. For example, after Example A, you might ask:

> What might be the equivalent of half-life for increasing functions?
>
> Is the doubling time the same for different interest rates?

And for Example B, you might ask:

> Why are these functions equivalent?

SUMMARIZE (*Closing the Lesson*, Discussing, Writing):

To help students formalize the investigation and examples into mathematical language ask the Critical Questions:

> Can an exponential function model positive growth?
>
> Why is *b* never negative?
>
> How are increasing exponential functions different from decreasing exponential functions?

The discussion portion of the lesson is a great time for students to verbalize their thinking.

The Kendall Hunt Common Core Math App highlights parts of the lesson as mentioned in the Teacher Edition that provide an opportunity to assess one of the mathematical practices. As you discuss ideas with your students, use the Look-Fors to help students develop their skills in each mathematical practice.

APPLY (Solving in a New Context):

You might project the Extra Example while students' books are closed:

> The graph of $y = 2x$ has been translated horizontally 2 and vertically 5 and reflected across the x-axis. What is the equation of the resulting graph?

Then pose and discuss the Closing Question:

> Parker Electronics sold 450 cell phones in January. Since then, their cell phone sales have increased at a rate of 4.75% per month. Create a function that can be used to determine the monthly cell phone sales m months after January, at this rate of growth?

ANALYZING THE LESSON

Assessment of students and analysis of instruction are strongly intertwined, so allow time in your planning to reflect after a lesson. Analyzing the lesson itself to determine how well it accomplished your goals is important to making the adjustments needed to meet the needs of all students. Ask yourself what went well, what could have gone better, and what adjustments need to be made in this lesson. What is your evidence? Did students meet the objectives of the lesson to your satisfaction? Why or why not? Are there other questions you should of asked?

Analyzing the lesson helps you make adjustments immediately in plans for the next lesson as well as for the next time you teach this lesson. To facilitate this reflection, your ebook allows you to make notes that will be readily available when you next plan the lesson. You can then use this information to appraise how well the tasks you used, the discourse you orchestrated, and the environment you fostered cultivated the mathematical understanding in all your students.

Materials for Use with *Discovering Advanced Algebra*

Materials for Lessons	Quantity
algebra tiles, *optional*	1 set per group
balloons	2 per group
books	2 – 4 per group
bowls	1 per group
cardstock paper	3 sheets per group
coins (pennies, nickels, quarters)	1 set per group
coffee can (empty) or poster tubes	1 per group
dice	1 die per person
dynamic data software, *optional*	
film canisters (small)	1 per group
geometry software, *optional*	
graph paper	
graph paper (centimeter)	
graphing calculators, *optional*	1 per student
linguine (uncooked)	
mass holders (such as film canisters)	1 per group
masses of 50–100 grams	1 set per group
measuring cups with milliliters	1 per group
motion sensors	1 per group
mirrors (small)	1 per group
patty paper or tracing paper	
pencils	a few per group
pillows or other soft objects (small)	1 per group
protractors	1 per student
racquetballs, basketballs, or handballs	1 per group
rulers	1 per student
rulers, centimeter	1 per student
scissors	1 per group
springs	weights, small
sticky notes, *optional*	1 pad per class
stopwatches or watches with second hand	1 per group
string	50 meters
support stands	1 per group
tables (long)	
tape	
tape measures or metersticks	1 per group
tinted liquid	
washers	1 per group
waste buckets (or sink)	1 per group
water	1 – 2 gallons
weights, small	1 per group
weights, small (pennies, beans)	

CHAPTER 0: Developing Mathematical Thinking

Overview

Chapter 0 uses problem solving and problem solving strategies to focus on how students engage with mathematics. The problems chosen review basic algebra skills. Starting the year with a review like this allows you to be flexible in assignments during the first few days of a school term, while your class list is stabilizing. It also gives you the opportunity to establish a productive, cooperative classroom environment. Finally, the investigations let you make an initial informal assessment of your students' mathematical strengths and needs, while you observe how they think and do mathematics.

Chapter 0 is built around the eight Mathematical Practices, or ways of thinking about and doing mathematics. While not every mathematical task will utilize all eight practices, two of the practices, making sense and persevering and attending to precision of computation and language, should be evident in every problem. Each lesson highlights two of the remaining six practices through the lens of a particular problem solving strategy.

Lesson 0.1 highlights the mathematical practices of modeling and using appropriate tools strategically, focusing on the problem-solving strategy of visual representations: sketches, graphs, and diagrams. The mathematical practices of reasoning and explaining are highlighted in **Lesson 0.2**, reviewing the strategy of representing English sentences with algebraic symbols. Strategies for organizing information—including making tables and charts and working with measurement units—and direct variation are the focus of **Lesson 0.3**, which highlights the mathematical practices of looking for and generalizing patterns.

The Mathematics

Problem Solving

Problem solving is a critical component of learning mathematics. Students who are engaged in rich problems, or tasks, that can be solved in many ways develop computation skills, higher-order thinking skills, and a repertoire of problem-solving strategies. Traditionally, a lesson where students learned a skill or procedure included "word problems" to give students a chance to practice that skill. Most students could figure out that a lesson on area meant multiply the numbers in the problem, without having to think about it. Problem solving goes far beyond the word problems that were really just an exercise for practicing a skill. A true problem is one in which a solution strategy is not known in advance and may vary among students. Mathematical understanding and procedural skill are equally important, and both are assessable using mathematical tasks of sufficient richness.

George Pólya's 1945 work *How to Solve It* was a seminal discussion of mathematical problem solving. He laid out four facets of solving problems:

- Understand the problem.
- Devise a plan.
- Carry out the plan.
- Look back.

These aspects are too general to provide a method for solving any particular problem. Instead, they describe an art that requires more than following formulas. While this discussion took place around 70 years ago, it is reflected in the current mathematical practices, especially SMP 1:

- Explain the meaning of the problem and analyze givens, constraints, relationships, and goals.
- Make conjectures about the meaning of the solution and plan a pathway rather than simply jumping in to a solution.
- Transform algebraic expressions to get the information you need; explain correspondence between equations, verbal descriptions, and models and search for regularity or trends.
- Monitor and evaluate progress and change course if needed; check answers using a different method; ask "Does this make sense?".

Students may face several challenges in problem solving. One challenge may be discomfort with the problems themselves not being just practice of a learned algorithm. Another challenge for students may be handling confusion—solving problems is an art that can be messy and that requires patience to persist through many attempts. Help students learn from their unsuccessful attempts, and to realize that they're learning. A third challenge comes from students trying to "remember" the "right way" to solve a problem. Encourage them to use strategies that make sense to them, as long as they can defend the mathematics. **ASK** "Could we have solved this problem in a different way?" This helps with a fourth challenge, overcoming students' math anxiety. Positive experiences in math are frequently tempered by past failure. Confidence and success built at the beginning of a course will affect performance positively throughout.

Each lesson in Chapter 0 is based on a key problem-solving technique. The problems in this chapter are not "drill problems," for which it is enough to apply a given method. Rather, these strategies are used for problems that can't be solved by following a predetermined sequence of steps. Problems that students will have to solve in their future careers will require critical thinking and creativity.

Using This Chapter

While the mathematics content of Chapter 0 could be considered optional, the investigations give you an excellent means to review algebraic ideas, get to know students, and introduce cooperative learning into your classroom. It will give you insight into how your students think about mathematics, while setting the tone for the year: students will investigate the mathematics, present their solutions to investigations and exercises, justify those answers, and listen to and critique the reasoning of other students.

With this chapter you will have a chance to assess student strengths and weaknesses. Providing extra practice and support as needed early in the year will help students avoid frustration later. However, don't try to reteach everything from beginning algebra. Students will review those ideas through the daily work of the lessons in this text. And Refreshing Your Skills lessons are provided for skills needed in each chapter for those students who may be weak in a particular skill. In these ways, they'll review in a manner that supports the learning of new topics, rather than being bored by revisiting material they've seen before.

Common Core State Standards (CCSS)

The Common Core State Standards (CCSS) consist of two kinds of standards; Content Standards and Standards for Mathematical Practice (SMP). The Content Standards include rigorous content to ensure that students learn both the "how" and the "why" of the mathematics. The Standards for Mathematical Practice describe ways of thinking about and doing mathematics. The SMPs are, therefore, inherently connected to the balance of procedure and understanding emphasized in the Common Core State Standards. These practices rest on important "processes and proficiencies" with longstanding importance in mathematics education, the National Council of Teachers of Mathematics' Process Standards (NCTM, 2000) and the National Research Council's Strands of Mathematical Proficiency (NRC, 2001).

Pedagogical Practices

The SMP can be thought of as describing mathematical habits of mind. They are not instructional strategies, or ways of teaching. They describe ways of thinking about and doing mathematics that teachers at all levels should seek to develop in their students. There are, however, instructional strategies and pedagogical practices that help cultivate these ways of thinking about mathematics. It is through the tasks and problems used to develop the understanding of the content that students are afforded an opportunity to become flexible in mathematical thinking.

SMP 1 Make sense of problems and persevere in solving them—Give open-ended tasks. Nothing says more about how a teacher defines mathematics than the type of tasks they give students. Give complex contextual problems that students can analyze, try different solution methods, and evaluate their answers in place of focusing on the application of an algorithm in non-contextual problems or a taught solution method. **ASK** "Would you get the same answer if you tried a different strategy?"

SMP 2 Reason abstractly and quantitatively—Give problems that require students to reason quantitatively and to represent real-world problems abstractly, interpreting mathematical answers in terms of the context. **ASK** "What happens if I change this number?"

SMP 3 Construct viable arguments and critique the reasoning of others—ask students to defend their answers both verbally and in writing and whether their answer is correct or incorrect. Only asking when an answer is incorrect tells students to see what they did wrong, instead of justifying their thinking. "Convince us" and "are you convinced" can become the flagship responses to every answer. One hallmark of mathematical understanding is the ability to justify, in a way appropriate to the student's mathematical maturity, why a particular mathematical statement is true or where a mathematical rule comes from. **ASK** "How do you know?"

SMP 4 Model with mathematics—Give open-ended real-world problems where students collect data to investigate a concept, create a mathematical model, and use their model to make predictions. Modeling is not achieved by assigning problems that come with instructions telling students which strategy to use or mimicking a previously demonstrated problem. Choices, assumptions, and approximations should be part of the modeling process. **ASK** "What kind of model is best for this relationship?"

SMP 5 Use appropriate tools strategically—have a variety of tools available and encourage students to consider the advantages and disadvantages of each for the given task. Graphing may not give the exact solution to a problem but it might give a good picture of the end value of a function, a calculator might give a decimal solution when fractions done by hand might reveal a pattern. Allow students to decide on a calculator screen window instead of telling them the parameters. **ASK** "What are we looking for and which tool will get us there?"

SMP 6 Attend to precision—Ask clarifying questions of students. Explicitly model expected behavior—labeling axes, defining variables, using units with significant digits, using precise vocabulary, etc. **ASK** "What do you mean by that?"

SMP 7 Look for and make use of structure—Give tasks and ask questions that require students to shift perspectives. Have students look for patterns beyond the obvious ones to create a rule or state a relationship. **ASK** "What conjecture can you make and how will you test it?"

SMP 8 Look for and express regularity in repeated reasoning—Give tasks that require inductive reasoning. Ask questions that require students to generalize. Have students create rules for manipulating generalities and defend those rules. **ASK** "Under what conditions is that true?"

Additional Resources For complete references to additional resources for this chapter and others see your online resources.

Materials

- graphing calculators and/or access to ebook (assumed for all lessons)
- graph paper
- algebra tiles, *optional*
- sticky notes, *optional*

CHAPTER 0
Developing Mathematical Thinking

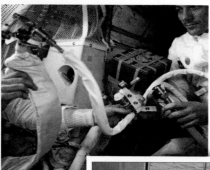

Houston, we've had a problem!
On April 11, 1970, Apollo 13 lifted off for the Moon. Two days into the flight, an oxygen tank exploded, scrubbing the lunar landing and putting the crew in jeopardy. Despite limited power, loss of cabin heat, shortage of potable water, and the critical need for makeshift repairs to the carbon dioxide removal system, the crew returned safely to Earth on April 17, 1970. The splashdown was one of the closest ever to the recovery ship. The mission was classified as a "successful failure" because of the experience gained in rescuing the crew.

OBJECTIVES
In this chapter you will
- make sense of a problem before manipulating the numbers
- be precise in computation and in communication
- solve problems both on your own and as a member of a group, critique and justify mathematical thinking
- use models and other tools in problem-solving
- represent situations with symbolical algebra and understand the meaning of quantities represented in equations
- find and generalize patterns and look for general methods and shortcuts in calculations

CHAPTER 0

Objectives

- Develop proficiency with making sense of problems, choosing a solution strategy, and persevering in solving them
- Communicate precisely to others, stating the meaning of symbols and specifying units
- Calculate accurately and efficiently, expressing numerical answers with an appropriate degree of accuracy
- Use mathematical models, such as pictures, diagrams, and graphs
- Consider available tools when solving a mathematical problem and use appropriate tools strategically
- Construct viable arguments to justify mathematical thinking
- Listen to the arguments of others and critique their mathematical reasoning
- Develop proficiency at looking for and making use of structure
- Develop proficiency at looking for and expressing regularity in repeated reasoning

---

When the oxygen tank blew up on Apollo 13, the ground and flight crew were faced with real world problem solving and failure was not an option. They did not have enough power to return to earth, so the life of the astronauts hung in the balance. They were faced with a new and unusual situation and the best scientific, engineering, and mathematical minds of the country had no ready formula. This scenario can spark a discussion of problem solving and the mathematical practices.

ASK "What do you think they meant by 'successful failure'? How would you describe the problem they were faced with? Why was perseverance in solving the problem so important? **SMP 1** How is this an example of abstract and quantitative reasoning? **SMP 2** How did teamwork impact the success of the mission? How important was communication? **SMP 3** How were models used? **SMP 4** What tools were available? **SMP 5** How precise did their solution have to be? **SMP 6**

What kinds of patterns and structure might they have looked for? Why was it important to evaluate intermediate results? **SMP 7, 8**"

The intent is not to have specific answers, but to generate a discussion of how these ways of thinking are important to problem solving. There are video clips available that you might want to show for students who are unfamiliar with Apollo 13.

LESSON 0.1

This lesson focuses on developing the mathematical practices of modeling and using tools strategically. These problems can be solved in many ways.

COMMON CORE STATE STANDARDS		
APPLIED	DEVELOPED	INTRODUCED
SMP 1	SMP 4	SMP 3
SMP 6	SMP 5	

Objectives

- Use mathematical model, such as pictures, diagrams, and graphs
- Use appropriate mathematical tools strategically

Launch

Find the slope of the line. Explain your strategy.

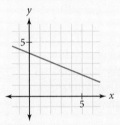

How is the graph a mathematical model? What are some other mathematical models?

$-\frac{2}{5}$ Answers will vary: It is a "picture" of an equation and a relationship. Other models include pictures, diagrams, equations

Investigate

Example A

Consider projecting the example from your ebook and letting students work in pairs. There is also a desmos simulation for the Example in your ebook.

Students may benefit from acting out this solution. **ASK** "What other amounts of water can you get using these two containers?"

A verbal solution to this example is not given. In this case, a diagram is more efficient.

LESSON 0.1

"A whole essay might be written on the danger of thinking without images."

SAMUEL COLERIDGE

Modeling and Tools

In this course, while learning the mathematical content, you will also learn to use different ways of thinking about and doing mathematics. These ways of thinking about and doing mathematics are sometimes called mathematical practices. While different problems will call for different ways of thinking about the mathematics, there are some practices that should be part of your thinking for every problem: always start by thinking of what the problem means, how you want to solve it, and what the expected solution should look like and always be precise with your computations and use clear and mathematically correct definitions and symbols.

In this textbook there are many problems that ask you to look at situations in new and different ways. This chapter offers some strategies to approach these problems. Although some of the problems in this chapter are fictitious, they give you a chance to practice skills that you will use throughout the book and throughout life.

This first lesson focuses on mathematical modeling and tools, such as using a sketch, graph, or diagram to help you find a solution.

EXAMPLE A Allyndreth needs to mix some lawn fertilizer with 7 liters of water. She has two buckets that hold exactly 3 liters and 8 liters, respectively. Describe or illustrate a procedure that will give exactly 7 liters of water in the 8-liter bucket.

Solution There is more than one solution to this problem. The picture sequence below shows one solution.

When faced with a mathematical problem, take a minute to figure out just what the problem means and think about a strategy that will get you where you need to be before you start trying to solve the problem by manipulating numbers.

A written description of the solution to Example A might be complex and hard to understand. Yet the pictures help you make sense of the problem and keep track of the amount of water at each step of the solution. You also see how the water is poured into and out of the buckets. Can you think of a different solution? If so, does your solution take more or fewer steps?

If you get stuck or think you have hit a dead end, don't give up. Review your work for any errors, talk to your classmates to get a different perspective on the problem, or ask your teacher clarifying questions. As you go, make sure the solutions you are getting make sense. Would you get the same answer if you tried a different strategy?

Using pictures is one way to visualize a problem. Another problem-solving strategy is to use objects and act out the problem. For instance, in Example A you could use

ELL
Using pictures, graphs, and diagrams can quickly bridge the understanding gap for ELL students. If they are struggling with the vocabulary, a visual representation of a problem can alleviate confusion. Do not assume that your students have a correct understanding of mathematical terms. Have them make a written vocabulary list that they can refer to regularly.

Support
Now is an excellent time to assess your students' proficiency level with basic algebra skills. The review of concepts in this chapter will allow you to fill in uncertain prior knowledge and to determine the entry level of your students into this advanced algebra course. See **ELL** suggestions for mathematical terms that students may not clearly understand.

Advanced
Although advanced students may work through this material quickly, working cooperatively through this chapter will enhance their problem solving skills. Present the examples and the investigation from your ebook without prior prompting or context; this will challenge them to think for themselves, and generate mathematical debate.

paper cups to represent the buckets and label each cup with the amount of water at each step of the solution. When you act out a problem, it helps to record positions and quantities on paper as you solve the problem so that you can recall your own steps.

Problem solving often requires a group effort. Different people have different approaches to solving problems, so working in a group gives you the opportunity to hear and see different strategies. Sometimes group members can divide the work based on each person's strengths and expertise, and other times it helps if everyone does the same task and then compares results. Each time you work in a group, decide how to share tasks so that each person has a productive role. You might find that working together, your group can solve challenging problems that would have stumped each of you individually. As you communicate within your group and with your presentations, use clear and mathematically correct definitions in your discussions and carefully explain the meaning of the symbols you chose to use.

The following investigation will give you an opportunity to work in a group and an opportunity to practice some problem-solving strategies.

INVESTIGATION

Polar Bear Crossing the Arctic

A polar bear rests by a stack of 2700 pounds of fish he has caught. He plans to travel 1000 miles across the Arctic to bring as many fish as possible to his family. He can pull a sled that holds up to 900 pounds of fish, but he must eat 1 pound of fish at every mile to keep his energy up.

What is the maximum amount of fish (in pounds) the polar bear can transport across the Arctic? How does he do it? Work as a group and prepare a written or visual solution.

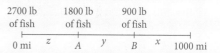

Models are useful problem-solving tools in mathematics but are not limited to diagrams like those in Example A. Coordinate graphs are some of the most important models for problem-solving in mathematics.

EXAMPLE B | A line passes through the point (4, 7) and has slope $\frac{3}{5}$. Find another point on the same line.

Solution | You *could* use a formula for slope and solve for an unknown point. But a graph may be a simpler way to find a solution.

Plot the point (4, 7). Recall that slope is $\frac{\text{change in } y}{\text{change in } x}$ and move from (4, 7) according to the slope, $\frac{3}{5}$. One possible point, (9, 10), is shown. Can you find another point on the line?

possible answers: (14, 13), (−1, 4), (−6, 1)

LESSON 0.1 Modeling and Tools **3**

Modifying the Investigation

Whole Class Have students simulate the situation. Elicit possible solutions and try them out.

Shortened After going over the scenario, demonstrate what happens when the bear travels 1000 miles in one trip. Repeat for 500 miles and 250 miles. Request other suggestions. Assign these suggestions to different groups. Present solutions.

One Step There is no **One Step** alternative for this investigation.

EXTEND If students find the optimal solution, ask, "What if the bear could carry only 450 pounds of fish? What if the distance changed?"

Guiding the Investigation

This Investigation can take a lot of time, so be aware of the time. The process is more important than the student coming up with the correct answer. There is a desmos simulation in your ebook.

If needed, **ASK** "What happens if the bear takes 900 lb of fish 400 miles?" Draw a diagram. **SMP 1, 2, 4, 5**

900 lb of fish on the sled   500 lb of fish left on the sled   no fish left on the sled

0 mi   400 mi   900 mi 1000 mi

ASK "Is there a way around having the bear arriving at 900 miles with no fish?" [Use a drop-off point.] **ASK** "How much fish can the bear leave here?" [100] Students can use this diagram to guess-and-check possible solutions. **SMP 1, 2, 4, 5**

Strategies may vary. One possible strategy is to draw a diagram.

2700 lb of fish   1800 lb of fish   900 lb of fish

0 mi z A y B x 1000 mi

In the diagram, the last part of the polar bear's trip will start at point B with the maximum amount of fish he can carry, 900 lb. He must therefore get to point A with 1800 fish. Using guess-and-check, students can find that z, the greatest distance the bear could take 1800 lb, is 180 miles. They can then follow the same procedure to find that y is 300 miles, so x is 520 miles, and the maximum amount of fish is 533 lb.

A symbolic approach might work backward from the destination. The last part of the polar bear's trip will start at point B with the maximum amount of fish he can carry, 900 lb.

The polar bear will travel x mi and end with $900 - x$ lb of fish. To get 900 lb of fish to point B, the polar bear starts with 1800 lb of fish at point A, makes a round trip and a one-way trip of y mi each way ($3y$ mi), and ends with $1800 - 3y$ lb of fish. $1800 - 3y = 900$, so $y = 300$ miles. To get 1800 lb of fish to point A, the polar bear starts with 2700 lb of fish, makes two round trips and a one-way trip of z mi each way, and ends with $2700 - 5z$ lb of fish. $2700 - 5z = 1800$, so $z = 180$. Total trip is $180 + 300 + x = 1000$ mi, so $x = 520$ miles. He ends with $1000 - 520 = 380$ lb of fish, using drops at 180 mi and 300 mi from the beginning.

LESSON 0.1 Modeling and Tools **3**

Example B

This example reviews coordinate graphing as a problem-solving technique. SMP 4, 5 Consider projecting the example from your ebook and having students work in pairs. Select pairs to share their solutions to reflect different strategies. Encourage students to use appropriate mathematical language, including terminology such as *coordinate graph, Cartesian graph/plane, axis/axes, x-axis/horizontal axis, y-axis/vertical axis,* and *coordinates* (of a point). SMP 1, 3, 6

Example C

This example reviews solving equations in one variable, graphing, and using technology. Consider projecting the example from your ebook and having students work in pairs. As students present their solution, have them justify each step. If necessary, suggest that students set each side of the equation equal to y. SMP 1, 3, 5, 6

Summarize

To encourage critical thinking, select groups that used a variety of strategies to share their solutions, rather than implying one way to do the problem. Have students explain their strategy, not just their answer. To help students develop the practice of listening, whether student responses are correct or incorrect, ask other students if they agree and why. To help students develop the practice of communicating precisely and critiquing the *reasoning* of others, encourage students to use appropriate mathematical language as evidence. SMP 1, 3, 4, 5, 6

ASK "Was there an approach to each example or the investigation that was best? Why?"

[Most mathematics problems can be approached in several legitimate ways—some methods are more efficient, some are better for understanding, some are easier to remember. The best approach for a student is the one that makes sense to them, as long as it is mathematically sound.]

▶ **Mathematics**
CONNECTION

Coordinate graphs are also called Cartesian graphs, named after the French mathematician and philosopher René Descartes (1596–1650). Descartes was not the first to use coordinate graphs, but he was the first to publish his work using two-dimensional graphs with a horizontal axis, a vertical axis, and an origin. Descartes's goal was to apply algebra to geometry, which today is called analytic geometry. Analytic geometry in turn laid the foundations of modern mathematics, including calculus.

René Descartes

Another useful tool in problem solving is technology. Technology is used throughout this course to gather data, observe patterns in tables, graph functions, explore transformations, and investigate probability and statistics concepts. It can also be used to find or verify a solution to an equation or system of equations.

EXAMPLE C Solve the equation. $\frac{x+5}{5} = 2$. Verify your solution by graphing both sides of the equation.

Solution One strategy for solving the equation is by undoing the operations.

$\frac{x+5}{5} = 2$ Original equation.

$x + 5 = 10$ Multiply both sides of the equation by 5 to undo the division.

$x = 5$ Subtract 5 from both sides of the equation to undo the addition.

To verify your solution by graphing, graph $y = \frac{x+5}{5}$ and $y = 2$ on the same set of axes and look for the intersection point. The intersection point is where $x = 5$, so your solution checks.

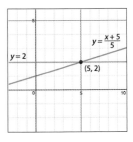

Problem solving requires that you use a variety of strategies. As you work on the exercises, don't limit yourself. Consider all the strategies and tools that are available, what they each do, what the benefits and limitations of each are, and which would be best for solving the problem. And remember that precision in mathematics is extremely important. Precision is not just about calculating accurately and efficiently, although that is certainly an important part of it. It means specifying units of measure and labeling axes. It also involves expressing numerical answers with an appropriate degree of accuracy. Beyond computation, it means communicating precisely to others, using clear and mathematically correct definitions in discussions, and explaining the meaning of the symbols you chose to use.

CRITICAL QUESTION Why do we use models in mathematics?

BIG IDEA Models are representations that link classroom mathematics and statistics to help solve problems that arise in everyday life, society, and the workplace. Models can be physically manipulated and studied. A model can be controlled and adjusted (it is a working model), whereas the real setting may be inaccessible.

CRITICAL QUESTION What tools should you consider using when solving mathematical problems? How do you decide which one to use?

BIG IDEA Paper-and-pencil, graphing calculators, concrete models, rulers, protractors, spreadsheet software, etc. To decide which tool to use, consider the advantages and disadvantages of using a tool for a given task.

0.1 Exercises

You will need a graphing calculator for Exercise **13**.

Practice Your Skills

1. The pictures below show the first and last steps of solutions to bucket problems similar to Example A. Write a statement for each problem.

   a. Begin with a 10-liter bucket and a 7-liter bucket. Find a way to get exactly 4 liters in the 10-liter bucket.

   b. Begin with a 10-liter bucket and a 7-liter bucket. Find a way to get exactly 2 liters in the 10-liter bucket.

2. There are 6 rooms in Eric's house that he wants to wire to an intercom system. Draw a diagram to show how many different lines are needed for a direct connection between each pair of rooms via the intercom system.   15 lines

3. Use a coordinate graph to find the slope of the line that passes through each pair of points.

   a. (2, 5) and (7, 10) @ 1
   b. (3, −1) and (8, 7) $\frac{8}{5}$, or 1.6
   c. (−2, 3) and (2, −6) $-\frac{9}{4}$, or −2.25
   d. (3, 3) and (−5, −2) $\frac{5}{8}$, or 0.625

4. The following problem appears in *Problems for the Quickening of the Mind*, a collection of problems compiled by the Anglo-Saxon scholar Alcuin of York (ca. 735–804 C.E.). Describe a strategy for solving this problem. Do not actually solve the problem. (h)

   A wolf, a goat, and a cabbage must be moved across a river in a boat holding only one besides the ferryman. How can he carry them across so that the goat shall not eat the cabbage, nor the wolf the goat?

   Sample answer: Use three coins to represent the wolf, the goat, and the cabbage. Draw both banks of the river on paper and use an index card as a boat. Act out the problem, trying never to leave the wolf and the goat together or the goat and the cabbage together.

Reason and Apply

5. For each situation, draw and label a diagram. Do not actually solve the problem.

   a. A 25 ft ladder leans against the wall of a building with the foot of the ladder 10 ft from the wall. How high does the ladder reach? @

   b. A cylindrical tank that has diameter 60 cm and length 150 cm rests on its side. The fluid in the tank leaks out from a valve on one base that is 20 cm off the ground. When no more fluid leaks out, what is the volume of the remaining fluid?

   c. Five sales representatives e-mail each of the others exactly once. How many e-mail messages do they send?

6. Tyrese inherited some money from his aunt. He gave half of the money to his wife, a fourth of the money to his kids, an eighth of the money to charity and kept $500 for himself. Use a diagram to determine how much money Tyrese inherited.

7. Use this graph to estimate these conversions between grams (g) and ounces (oz).

   a. 11 oz ≈ _?_ g @ 312 g
   b. 350 g ≈ _?_ oz @ 12.5 oz
   c. 15.5 oz ≈ _?_ g 437 g
   d. 180 g ≈ _?_ oz 6.2 oz
   e. What is the slope of this line? Explain the real-world meaning of the slope.   Using the approximate point (15, 425) and the definite point (0, 0), the slope is $\frac{85}{3}$. There are about 28.3 grams in an ounce.

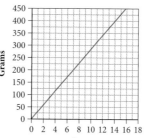

5a.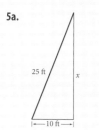

5c.

6. He inherited $4000. $\frac{1}{2} + \frac{1}{4} + \frac{1}{8}$ means there is $\frac{1}{8}$ left for Tyrese, which is $500.

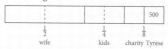

5b.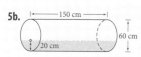

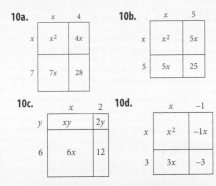

10a. (rectangle diagram: x^2, $4x$, $7x$, 28; sides $x, 4$ and $x, 7$)

10b. (rectangle diagram: x^2, $5x$, $5x$, 25; sides $x, 5$ and $x, 5$)

10c. (rectangle diagram: xy, $2y$, $6x$, 12; sides $x, 2$ and $y, 6$)

10d. (rectangle diagram: x^2, $-1x$, $3x$, -3; sides $x, -1$ and $x, 3$)

Exercise 11 A picture diagram for 11a should show the chlorine levels: 0-0, 0-7, 7-0, 7-7, 10-4, 0-4, 4-0. A picture diagram for 11b should show these levels: 0-0, 10-0, 3-7, 3-0, 0-3, 10-3, 6-7, 6-0, 0-6, 10-6, 9-7, 9-0, 2-7, 2-0.

11a. Sample answer: Fill the 7-liter bucket; pour the contents into the 10-liter bucket; fill the 7-liter bucket again; pour 3 liters from the 7-liter bucket into the 10-liter bucket, completely filling the 10-liter bucket and leaving 4 liters in the 7-liter bucket; empty the 10-liter bucket; pour the 4 liters from the 7-liter bucket into the 10-liter bucket.

11b. Sample answer: Fill the 10-liter bucket; pour the contents into the 7-liter bucket, completely filling the 7-liter bucket and leaving 3 liters in the 10-liter bucket; empty the 7-liter bucket; pour the 3 liters from the 10-liter bucket into the 7-liter bucket. Fill the 10-liter bucket; pour 4 liters into the 7-liter bucket, completely filling the 7-liter bucket and leaving 6 liters in the 10-liter bucket; empty the 7-liter bucket. Pour the 6 liters from the 10-liter bucket into the 7-liter bucket. Fill the 10-liter bucket; pour 1 liter into the 7-liter bucket, completely filling the 7-liter bucket and leaving 9 liters in the 10-liter bucket. Empty the 7-liter bucket; pour 7 liters from the 10-liter bucket into the 7-liter bucket, completely filling the 7-liter bucket and leaving 2 liters in the 10-liter bucket.

12. Take the goat across and come back. Next, take the wolf across and bring the goat back. Next, take the cabbage across and come back. Last, take the goat across.

13a. (graph showing $y = 17$, $y = x + 12$, intersection $(5, 17)$)

---

8. The line has slope $\frac{12}{16}$. The slope from $(0, 0)$ to any point (x, y) on the line will be $\frac{y}{x}$.

8. Explain how this graph helps you solve these proportion problems. Then solve each proportion.

   a. $\frac{12}{16} = \frac{9}{a}$ @ $a = 12$

   b. $\frac{12}{16} = \frac{b}{10}$ $b = 7.5$

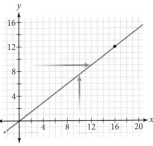

9. You can use diagrams to represent algebraic expressions. Explain how this rectangle diagram demonstrates that $(x + 2)(x + 3)$ is equivalent to $x^2 + 2x + 3x + 6$.

9. The outer rectangle has length $(x + 2)$ and width $(x + 3)$, so the area is length times width, or $(x + 2)(x + 3)$. The areas of the inner rectangles are x^2, $2x$, $3x$, and 6. The sum of the areas of the inner rectangles is equal to the area of the whole outer rectangle, so $(x + 2)(x + 3) = x^2 + 2x + 3x + 6 = x^2 + 5x + 6$.

10. Draw a rectangle diagram like the one in Exercise 9 to represent each product. Use the diagrams to expand each product.

   a. $(x + 4)(x + 7)$ @ $x^2 + 11x + 28$

   b. $(x + 5)^2$ ⓗ $x^2 + 10x + 25$

   c. $(x + 2)(y + 6)$ $xy + 2y + 6x + 12$

   d. $(x + 3)(x - 1)$ $x^2 + 2x - 3$

11. **APPLICATION** Kyle has a summer job cleaning pools. He needs to measure exact amounts of chlorine but has only a 10-liter bucket and a 7-liter bucket. Assuming an unlimited supply of chlorine, describe or illustrate a procedure that will ⓗ

   a. Give exactly 4 liters of chlorine in the 10-liter bucket.

   b. Give exactly 2 liters of chlorine in the 10-liter bucket.

12. Solve the problem in Exercise 4. Explain your solution.

13. Solve each equation. Verify your solution by graphing both sides of the equation.

   a. $x + 12 = 17$ $x = 5$

   b. $2x = 13$ $x = \frac{13}{2}$

   c. $2x + 3 = 15$ $x = 6$

   d. $1.5 = \frac{5 - x}{4}$ $x = -1$

Review

14. Translate each verbal statement into a symbolic expression.

   a. Three more than a number $3 + n$

   b. Four less than twice a number $2n - 4$

   c. Two-thirds of a number $\frac{2}{3} \cdot n$

15. Convert these decimal values to fraction form.

   a. 0.375 $\frac{375}{1000} = \frac{3}{8}$

   b. 1.42 $\frac{142}{100} = \frac{71}{50} = 1\frac{21}{50}$

   c. $0.\overline{2}$ @ $\frac{2}{9}$

   d. $0.\overline{35}$ $\frac{35}{99}$

16. Joel is 16 years old. His cousin Rachel is 12.

   a. What is the difference in their ages?   4 yr

   b. What is the ratio of Joel's age to Rachel's age? $\frac{16}{12} = \frac{4}{3} = 1.\overline{3}$

   c. In eight years, what will be the difference in their ages?   4 yr

   d. In eight years, what will be the ratio of Joel's age to Rachel's age? $\frac{24}{20} = \frac{6}{5} = 1.2$

13b. 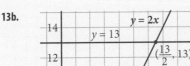 (graph: $y = 2x$, $y = 13$, intersection $(\frac{13}{2}, 13)$)

13c. 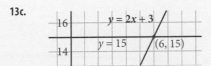 (graph: $y = 2x + 3$, $y = 15$, intersection $(6, 15)$)

13d. 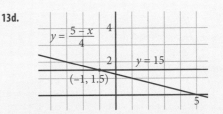 (graph: $y = \frac{5-x}{4}$, $y = 15$, intersection $(-1, 1.5)$)

Exercises 15 ALERT Some students may believe that a repeating decimal cannot be expressed as a fraction.

LESSON 0.2

Reasoning and Explaining

> "Mathematical reasoning may be regarded rather schematically as the exercise of a combination of two facilities, which we may call intuition and ingenuity."
>
> ALAN TURING

On the second day, the customer service team had planned to double the number of calls answered on the first day, but they exceeded that by three dozen. Seventy-five dozen customer service calls in two days set a new record.

You can translate the paragraph above into an algebraic equation. Although you don't know how many calls were answered each day, an equation will help you figure it out.

For many years, problems like this were solved without writing equations. Then, around the 17th century, the development of symbolic algebra made writing equations and finding solutions much simpler. Verbal statements could then be translated into symbols by representing unknown quantities with letters, called variables, and converting the rest of the sentence into numbers and operations. (You will actually translate this telephone call problem into algebraic notation when you do the exercises.)

History CONNECTION

Muhammad ibn Mūsā al-Khwārizmī (ca. 780–850 C.E.), an Iraqi mathematician, wrote the first algebra treatise. The word *algebra* comes from the treatise's title, *Kitāb al-jabr wa'al-muqābalah*, which translates to *The Science of Completion and Balancing*. Al-Khwārizmī, pictured on this postage stamp issued in the former Soviet Union, wanted the algebra in this treatise to address real-world problems that affected the everyday lives of the people, such as measuring land and trading goods.

Language is designed for describing qualities rather than quantities. Mathematical symbols are better for describing quantities. A very helpful problem-solving technique is to use symbols to represent numbers that quantify data such as time, weight, or distance. These numbers might vary or remain constant. Their values might be given in a problem or unknown. Focus on the meaning of the numbers in the context of the solution, not just how to compute them.

EXAMPLE A | When Adam and his sister Megan arrive at a party, they see that there is 1 adult chaperone for every 4 kids. Right behind them come 30 more boys, and Megan notices that the ratio is now 2 boys to 1 girl. However, behind the 30 boys come 30 more girls, and Adam notices that there are now 4 girls for every 3 boys. What is the final ratio of adult chaperones to kids?

Investigate

Example A

This example reviews ratios. Consider projecting the example from your ebook and having students work in groups of 2 – 4. Have students explain their equations and their reasons for each step, not just their answer.
SMP 2, 3

As you circulate, if needed, **ASK** "What do you want to know?" [The number of kids and adults at the end.] **ASK** "What do you need to find out to know the number of adults?" [the number of kids at the beginning, for which we must find the initial numbers of boys and girls] If needed, **ASK** "What equation with a ratio might represent the situation accurately?"
$\left[\text{number of girls} = \frac{4}{3}(\text{number of boys})\right]$

ALERT Students may misrepresent "4 girls for every 3 boys" with the equation 4(number of girls) = 3(number of boys). **ASK** "A set of 3 girls and 4 boys satisfies the equation. Does that give 4 girls for every 3 boys?" Suggest that a good habit to get into is to check equations like this with numbers. **SMP 2, 3**

Example B

This example is good for reviewing how to solve systems of equations by elimination. Consider projecting the example from your ebook and having students work in groups of 2 – 4. To encourage critical thinking, select groups that used a variety of strategies to share their solutions, rather than implying one way to do the problem. If no group solves the system with substitution, you might ask students to do so, for review purposes. The key concept for this example is to write algebraic expressions for the problem.
SMP 1, 2, 3

Guiding the Investigation

You may have each group work on both problems, or assign each group a different problem to present to the whole class. (Problem 2 is easier than Problem 1 for most students.) Students could also work on one problem as a Problem of the Week or use it for their portfolio.

Solution To represent this problem with symbolic algebra, you first need to determine which quantities are unknown. Assign a variable to each unknown.

Let B represent the original number of boys.
Let G represent the original number of girls.
Let A represent the number of adults.

Next, write equations for the information in the problem.

$\frac{A}{B+G} = \frac{1}{4}$, or $A = \frac{1}{4}(B+G)$
When Adam and Megan arrive, there is 1 adult for every 4 kids, so the ratio of adults to kids is 1:4, or $\frac{1}{4}$. The total number of kids is represented by $B + G$. Call this Equation 1.

$\frac{B+30}{G} = \frac{2}{1}$, or $B + 30 = 2G$
After 30 more boys come, the ratio of boys to girls is $\frac{2}{1}$. Call this Equation 2.

$\frac{G+30}{B+30} = \frac{4}{3}$, or $G + 30 = \frac{4}{3}(B+30)$
After 30 more girls come, the ratio of girls to boys is $\frac{4}{3}$. Call this Equation 3.

You want to know the final ratio of adults to kids: $\frac{A}{B+30+G+30} = \frac{?}{}$

Now solve the equations to find values for the variables. Equations 2 and 3 involve only variables B and G, so you might eliminate B and solve for G.

$G + 30 = \frac{4}{3}(2G)$ — Using Equation 2, substitute $2G$ for $B + 30$ in Equation 3.
$3(G + 30) = 4(2G)$ — Multiply both sides by 3.
$3G + 90 = 8G$ — Multiply and distribute.
$90 = 5G$ — Add $-3G$ to both sides of the equation.
$G = 18$ — Divide both sides by 5.
$B + 30 = 2(18)$ — Substitute 18 for G in Equation 2.
$B = 36 - 30 = 6$ — Subtract 30 from both sides.

Now that you have found values for B and G, you can use Equation 1 to find A.

$A = \frac{1}{4}(6 + 18) = 6$ — Substitute for B and G in Equation 1.

Finally, interpret the solution.

$\frac{A}{B+30+G+30} = \frac{6}{6+30+18+30} = \frac{6}{84} = \frac{1}{14}$

The final ratio of adults to kids is 1:14.

You may have noticed that Example A used a four-step solution process. What are the four steps? Let's apply this process to another problem. First, assign a variable to each unknown. Second, write equations for the information in the problem. Third, solve the equations to find values for the variables. Finally, interpret the solution.

8 CHAPTER 0 Developing Mathematical Thinking

Modifying the Investigation

Whole Class Discuss Steps 1–4 for each problem. Note how the approaches to the problems differ.

Shortened Have groups do only Problem 1.

One Step Ask students to work on this problem: A caterer claims that a birthday cake will serve either 20 children or 15 adults. Tina's party presently has 12 children and 7 adults. Is there enough cake? [The current party is equivalent to 16 adults, so, no, there is not enough cake.] As you circulate, encourage the use of symbolic algebra, but don't be too quick to discourage diagrams or fractional reasoning without variables.

EXAMPLE B Three friends went to the gym to work out. None of the friends would tell how much he or she could leg-press, but each hinted at their friends' leg-press amounts.

- Chen said that Juanita and Lou averaged 87 pounds.
- Juanita said that Chen leg-pressed 6 pounds more than Lou.
- Lou said that eight times Juanita's amount equals seven times Chen's amount.

Find how much each friend could leg-press.

Solution First, list the unknown quantities and assign a variable to each.

Let C represent Chen's leg-press weight in pounds.
Let J represent Juanita's leg-press weight in pounds.
Let L represent Lou's leg-press weight in pounds.

Second, write equations from the problem.

$$\begin{cases} \dfrac{J + L}{2} = 87 & \text{Chen's statement translated into an algebraic equation. Call this Equation 1.} \\ C - L = 6 & \text{Juanita's statement as Equation 2.} \\ 8J = 7C & \text{Lou's statement as Equation 3.} \end{cases}$$

Third, solve the equations to find values for the variables.

$J + L = 174$	Multiply both sides of Equation 1 by 2.
$C - L = 6$	Equation 2.
$J + C = 180$	Add the equations.
$7J + 7C = 1260$	Multiply both sides of the previous equation by 7.
$7J + 8J = 1260$	Equation 3 allows you to substitute $8J$ for $7C$.
$15J = 1260$	Combine like terms.
$J = 84$	Divide both sides by 15.
$8(84) = 7C$	Substitute 84 for J in Equation 3.
$C = 96$	Solve for C.
$96 - L = 6$	Substitute 96 for C in Equation 2.
$L = 90$	Solve for L.

Last, interpret your solution. Chen leg-presses 96 pounds, Juanita leg-presses 84 pounds, and Lou leg-presses 90 pounds.

In this course, you will be asked to present your solutions to the exercises and investigations. You will make conjectures and use assumptions, definitions, and previous results to justify your conclusions and explain them to others. You will listen to or read other's arguments, determine if they make sense, and explain why you do or do not agree with them mathematically. You'll ask questions to get clarification of an explanation and use counterexamples in your argument that an answer isn't correct.

Summarize

Have students present their solutions, along with their equations, to the examples and the Investigation. Encourage students to suggest alternative approaches to the problems. To help students develop the practice of listening, whether student responses are correct or incorrect, ask other students if they agree and why. To help students develop the practice of communicating precisely and critiquing the *reasoning* of others, encourage students to use appropriate mathematical language as evidence. **SMP 1, 3, 4, 6**

Students' ideas about Problem 1 may give you insight into different depths of understanding. It's not the case that "either students understand it or they don't." Each student understands concepts in different ways and to different depths. **ASK** "How do your approaches here differ from those you used with the investigation in the previous lesson?" [These problems involve numerical relationships, so you have to use symbolic notation, such as variables, equations, and inequalities.] **SMP 2, 3**

LANGUAGE Some algebra books make a distinction between *unknowns* and *variables*. For example, because x has a definite value in the equation $x + 2 = 3$, it could be called an *unknown*. In the equation $y = x$, however, x could take on any value, so it is a *variable*. This book will use *variable* to mean any letter or word that represents one or more numbers.

CRITICAL QUESTION How does using symbolic representation help in solving some work problems?

BIG IDEA You can organize your information and write it concisely, then use algebra to answer the question.

Formative Assessment

During this lesson you can assess students' level of mastery of the essential reasoning and mechanical processes for solving linear equations and systems.

To assess students' ability to reason abstractly and quantitatively, watch to see that they create a coherent representation of a problem, identifying relationships among quantities and their units. Notice how they translate a problem situation from an Investigation or Example into a mathematical sentence (decontextualize) and how flexibly they use properties of operations when generating equivalent symbolic statements.

To assess students' ability to construct viable arguments and critique the reasoning of others, look for inductive reasoning about data presented in a context. Listen for plausible arguments and critiques which use mathematics as evidence. Listen for how they build a logical progression of statements to justify a conjecture or present a counterexample.

Problem 1

Step 1 Let C represent the hourly wage for clerks. Let S represent the hourly wage for supervisors. Let B represent the daily salary budget.

Step 2
$$\begin{cases} 16C + 8S = B & \text{Mon.} \\ 20C + 4S = B - 10 & \text{Tues.} \\ 24C + 4S = B + 20 & \text{Wed.} \end{cases}$$

Step 3 $C = 7.5$, $S = 10$, $B = 200$

Step 4 Clerks earn $7.50 per hour; supervisors earn $10 per hour. The daily salary budget is $200.

Problem 2

Step 1 Let s represent the number of bags of sand. Let c represent the number of bags of cement.

Step 2
$$\begin{cases} 50s + 40c = 1000 \\ \dfrac{s}{c} = \dfrac{5}{1} \end{cases}$$

The truck holds at most 1000 lb.
The ratio of sand to cement is 5 to 1.

Step 3 $c = 3$. $s = 15$.

Step 4 Sandy can load 3 bags of cement and 15 bags of sand for a total weight of 870 lb.

The investigation will give you a chance to try the four-step solution process on your own or with a group. Practice justifying the thinking of each member of your group, using mathematics as your evidence.

INVESTIGATION
Problems, Problems, Problems

Select one of the problems below, and use these four steps to find a solution.

Step 1 List the unknown quantities, and assign a variable to each quantity.

Step 2 Write one or more equations that relate the unknown quantities to conditions of the problem.

Step 3 Solve the equations to find a value for each variable.

Step 4 Interpret your solution according to the context of the problem.

When you finish, write a paragraph answering this question: Which of the four problem-solving steps was hardest for you? Why?

Problem 1

On Monday the manager scheduled two clerks and one supervisor for 8-hour shifts. She was pleased because she met the daily salary budget.
On Tuesday the manager needed two clerks for 8 hours, a third clerk for 4 hours, and one supervisor for 4 hours. She was very pleased to be $10 under budget.
On Wednesday she needed three clerks for 8 hours and one supervisor for 4 hours. This day she was over budget by $20.
All of the clerks make the same hourly wage. What is the daily salary budget?

Problem 2

Sandy uses a $\frac{1}{2}$-ton pickup (a truck that can carry 1000 pounds) to transport ingredients for mortar. Sand comes in 50-pound bags and cement comes in 40-pound bags. Mortar is made from five bags of sand for every bag of cement. How many bags of each should Sandy load to make the most mortar possible?

History CONNECTION

George Pólya (1887–1985) was a Hungarian-American mathematician often recognized for his contribution to the study of problem solving. In his 1945 book, *How to Solve It,* he describes four steps used in the process of solving a problem:
- Understand the problem
- Devise a plan
- Carry out the plan
- Look back

Solving a challenging problem often involves moving between these steps. For example, a solution plan that fails may lead to deeper understanding of a problem.

The four problem-solving steps in the investigation help you organize information, work through an algebraic solution, and interpret the final answer. As you do the exercises in this lesson, refer back to these four steps and practice using them.

0.2 Exercises

Practice Your Skills

1. Which equation would help you solve the following problem?

   Each member of the student council made three copies of the letter to the principal. Adding these to the 5 original letters, there are now a total of 32 letters. How many members does the student council have? iii

   i. $5 + c = 32$ ii. $3 + 5c = 32$ iii. $5 + 3c = 32$

2. Which equation would help you solve this problem? What does x represent?

   On the second day, the customer service team had planned to double the number of calls answered the first day, but they exceeded that by three dozen. Seventy-five dozen customer service calls in two days set a new record.   ii: x represents the number of dozens of calls made on day one.

   i. $x + 3 = 75$ ii. $x + 2x + 3 = 75$ iii. $2x + 3 = 75$

3. Represent these situations symbolically.

   a. Twice a number, increased by 8, is the same as three times a number, decreased by 11. $2x + 8 = 3x - 11$

   b. Three-fourths of a number, decreased by 12, is greater than or equal to 5. $\frac{3}{4}n - 12 \geq 5$

   c. You must be at least 54 inches tall to go on this ride. @ $h \geq 54$

   d. José rented a car for one day for $60 plus $0.15 per mile. He spent exactly $78. How many miles did he drive the car? $60 + 0.15m = 78$

4. What question is represented by the statement $35 + 8x = 131$? Answers may vary: 35 more than eight times a number is 131.

5. Which of the following results in the expression $4x + 4$? (h) ii, iii, iv

   i. A rectangle has length $2x + 4$ and width 2. Find the area.
   ii. A rectangle has length $2x + 2$ and width 2. Find the area.
   iii. A rectangle has length $2x$ and width 2. Find the perimeter.
   iv. A square has side length $x + 1$. Find the perimeter.

Reason and Apply

6. Julie's math teacher likes to give her students "mathemagical" number tricks. Try the steps of her favorite number trick below.

   Step 1 Write down the first three digits of your seven-digit phone number (don't include the area code).
   Step 2 Multiply by 80. @
   Step 3 Add 1.
   Step 4 Multiply by 250.
   Step 5 Add the last four digits of your phone number to your results from Step 4.
   Step 6 Add the last four digits of your phone number again.
   Step 7 Subtract 250.
   Step 8 Divide your number by 2.

Exercise 6 While it's appropriate for students to use calculators while following the steps with their own phone numbers, you might suggest that it's also a good habit not to do actual calculations until the end. In cases like this, numbers cancel out to avoid calculations; in other cases, roundoff errors are avoided. SMP 5, 6

Apply

Extra Example
The area of a patio is no more than 500 square feet. The total length of the three sides of decorative fence is no more than 50 feet. Write a system of inequalities that could be used to find values of l, the patio's length, and w, the patio's width, that will satisfy these conditions.

$l + 2w \leq 50$
$lw \leq 500$

Closing Question
Mr. Perez assigned this problem:

Let b be the **cost** of a tennis ball. Let c be the **cost** of a can of tennis balls. There are 3 balls in each can. The price per ball is the same whether you buy them separately or in a box. Find an equation linking b and c.

Ada's equation is $c = 3b$.

Vern's equation is $b = 3c$.

Do you agree with either of them? Why?

Ada is correct. Ada's equation says that the cost of a can of balls is the same as the cost of three balls, while Vern's equation says the cost of a ball is the same as the cost of three cans.

Assigning Exercises
Suggested: 2 – 8, 11
Additional Practice: 1, 9, 10, 12 – 15

Exercise 1 If needed, ASK "How do the proposed equations seem to represent the size of the student council? What happens to get from that number up to 32?" SMP 2

Exercise 2 If students are struggling, ASK "What might the x in the proposed equations represent? In which equation does there seem to be doubling and adding 3?" SMP 2

Exercise 3 Students could interpret the problem to mean two different numbers, in which case their answer could be $2x + 8 = 3y - 11$.

6b. Step 1: x
Step 2: $80x$
Step 3: $80x + 1$
Step 4: $20000x + 250$
Step 5: $20000x + 250 + n$
Step 6: $20000x + 250 + 2n$
Step 7: $20000x + 2n$
Step 8: $10000x + n$

Exercises 7, 8 These problems can be solved by reasoning without algebraic symbols: If Ivan had gotten 1.5 times as much as he did, he would have had the same number of large beads as Anita, but he would have had 17.5 more small beads, for which he would have paid $1.575 more. So each small bead must cost 9 cents. From either Anita's or Ivan's total, you can find that each large bead cost 15 cents, making Jill's total $2.64. Commend this reasoning, and point out how it reflects the symbolic solution of a system of equations.

7a. L represents the cost of a single large bead. S represents the cost of a single small bead. J represents the amount that Jill will pay for her purchase.

Exercise 9
Language A *denarius* (plural *denarii*) was the monetary unit of Rome.
ELL *Fivefold* means five times as much.

9a. a represents the number of denarii that A has.

b. b represents the number of denarii that B has.

c. If you take 7 denarii from B and give them to A, then A's new amount equals 5 times B's new amount.

d. If you take 5 denarii from A and give them to B, then B's new amount equals 7 times A's new amount.

---

a. How does your answer to Step 8 compare to your phone number? They are the same.

b. Now you'll represent the number trick to explain why it works. Write an algebraic expression for each step of the process. For each step you should modify your answer from the previous step. Be sure to combine like terms and do the arithmetic for each step. Use x to represent the first three digits of your phone number, and use n to represent the last four digits of your phone number.

c. Will the number trick work for all phone numbers? Use your algebraic expression for Step 8 to explain your answer. The expression $10000x + n$ is the same as the phone number, because you have multiplied the first three digits (x) by 10,000 and added the last four digits (n).

7. Here is a problem and three related equations.

Anita buys 6 large beads and 20 small beads to make a necklace. Ivan buys 4 large and 25 small beads for his necklace. Jill selects 8 large and 16 small beads to make an ankle bracelet. Without tax, Anita pays $2.70, and Ivan pays $2.85. How much will Jill pay?

$$\begin{cases} 6L + 20S = 270 & \text{(Equation 1)} \\ 4L + 25S = 285 & \text{(Equation 2)} \\ 8L + 16S = J & \text{(Equation 3)} \end{cases}$$

a. What do the variables L, S, and J represent?
b. What are the units of L, S, and J? (h) cents
c. What does Equation 1 represent?
Anita's purchase of 6 large beads and 20 small beads for $2.70

8. Follow these steps to solve Exercise 7.
a. Multiply Equation 1 by -2. $-12L - 40S = -540$
b. Multiply Equation 2 by 3. $12L + 75S = 855$
c. Add the resulting equations from 8a and b. $35S = 315$
d. Solve the equation in 8c for S. Interpret the real-world meaning of this solution. $S = 9$. The small beads cost 9¢ each.
e. Use the value of S to find the value of L. Interpret the real-world meaning of the value of L. $L = 15$. The large beads cost 15¢ each.
f. Use the values of S and L to find the value of J. Interpret this solution. $J = 264$. Jill will pay $2.64 for her beads.

An antique, arabic necklace.

9. The following problem appears in *Liber abaci* (1202), or *Book of Calculations*, by the Italian mathematician Leonardo Fibonacci (ca. 1170–1240).

If A gets 7 denarii from B, then A's sum is fivefold B's. If B gets 5 denarii from A, then B's sum is sevenfold A's. How much has each?

$$\begin{cases} a + 7 = 5(b - 7) & \text{(Equation 1)} \\ b + 5 = 7(a - 5) & \text{(Equation 2)} \end{cases}$$

a. What does a represent?
b. What does b represent?
c. Explain Equation 1 with words.
d. Explain Equation 2 with words.

10. Use Equations 1 and 2 from Exercise 9.
a. Explain how to get $b + 5 = 7[(5b - 42) - 5]$ @ Solve Equation 1 for a. Substitute the result, $5b - 42$, for a in Equation 2 to get $b + 5 = 7[(5b - 42) - 5]$.
b. Solve the equation in 10a for b. $b = \frac{167}{17}$
c. Use your answer from 10b to find the value of a. $a = \frac{121}{17}$
d. Use the context of Exercise 9 to interpret the values of a and b. A has $\frac{121}{17}$, or about 7 denarii, and B has $\frac{167}{17}$, or about 10 denarii.

11. A store sells pens at $2 and notebooks at $4.

    n = number of notebooks sold.

    p = number of pens sold.

    The following equations are true:

    $3n = p$

    $4n + 2p = 48$

    Here is what Len and Teri think the equations mean:

    Len: I think the first equation means that the store sells three times as many notebooks as pens.

    Teri: I think the second equation means that the store sold 4 notebooks and 2 pens.

    Is either Len or Teri correct? If not, explain how their thinking is incorrect.

Review

12. **APPLICATION** A 30°-60°-90° triangle and a 45°-45°-90° triangle are two drafting tools used by people in careers such as engineering, architecture, and drafting. The angles in both triangle tools are combined to make a variety of angle measures in hand-drawn technical drawings, such as blueprints. Describe or illustrate a procedure that will give the following angle measures.

    a. 15°
    b. 75°
    c. 105°

13. Find the slope of the line that passes through each pair of points.

    a. (4, 7) and (8, 7)   0
    b. (2, 5) and (−6, 3) $\frac{1}{4}$, or 0.25

14. Rudy is renting a cabin on Lake Tahoe. He wants to make a recipe for hot chocolate that calls for 1 cup of milk. Unfortunately, he has only a 1-gallon container of milk, a 4-cup saucepan, and an empty 12-ounce soda can. Describe or illustrate a process that Rudy can use to measure exactly 1 cup of milk. (There are 8 ounces per cup and 16 cups per gallon.)

15. Draw a rectangle diagram to represent each product. Use the diagrams to expand each product.

    a. $(x + 1)(x + 5)$
    b. $(x + 3)^2$
    c. $(x + 3)(x - 3)$

IMPROVING YOUR Reasoning SKILLS

Fair Share

Abdul, Billy, and Celia agree to meet and can tomatoes from their neighborhood garden. Abdul picks 50 pounds of tomatoes from his plot of land. Billy picks 30 pounds of tomatoes from his plot. Unfortunately, Celia's plants did not get enough sun, and she cannot pick any tomatoes from her plot.

They spend the day canning, and each has 36 quarts of tomatoes to take home. Wanting to pay Abdul and Billy for the tomatoes they gave to her, Celia finds $8 in her wallet. How should Celia divide the money between her two friends?

LESSON 0.3

The focus of this lesson is looking for patterns, whether it is patterns in the structure of algebraic expressions or patterns in repeated calculations. One way to see patterns is by organizing information.

COMMON CORE STATE STANDARDS		
APPLIED	DEVELOPED	INTRODUCED
SMP 1	SMP 7	
SMP 2	SMP 8	
SMP 3		
SMP 4		
SMP 5		
SMP 6		

Objectives

- Develop proficiency at looking for and making use of structure
- Develop proficiency at looking for and expressing regularity in repeated reasoning

Materials

- Sticky notes, *optional*
- Calculator Note: Order of Operations, *optional*

Launch

How is the line through the point (1, 2) with slope 3 related to the equation $\frac{y-2}{x-1} = 3$? Could the point (0, −1) be on the same line? How do you know?

The equation is the slope formula with the coordinates of the point and the slope substituted. The point could be on the line. The slope of a line between the two points is 3, so the point (0, −1) is on the line. You could also substitute (0, −1) into the equation to see if it is still a true statement.

Investigate

Example A

Project Example A from your ebook and have students work in pairs. Have students present their solution, carefully detailing how they used dimensional analysis. Choose presentations that include a variety of

---

LESSON 0.3

Looking for and Generalizing Patterns

"Everybody gets so much information all day long that they lose their common sense."

GERTRUDE STEIN

If one and a half chickens lay one and a half eggs in one and a half days, then how long does it take six monkeys to make nine omelets?

What sort of problem-solving strategy can you apply to the silly problem above? You could draw a picture or make a diagram. You could assign variables to all sorts of unknown quantities. But do you really have enough information to solve the problem? Sometimes the best strategy is to begin by organizing what you know and what you want to know. With the information organized, you may then find a way to get to the solution.

Algebra is the study of patterns, finding a pattern and writing an equation to explain it. In this course, you will look closely to identify a pattern or structure, break things down to an easier problem to find a way to solve the bigger problem, and you'll step back regularly, take another look at your strategy, and change directions if necessary.

EXAMPLE A | How many seconds are in a calendar year?

Solution First, identify what you know and what you need to know.

Know	Need to know
1 year	Number of seconds

It may seem like you don't have enough information, but consider these commonly known facts:

1 year = 365 days (non–leap year)

1 day = 24 hours

1 hour = 60 minutes

1 minute = 60 seconds

This Egyptian calendar is the world's oldest known calendar.

You can write each equality as a fraction and multiply the chain of fractions such that the number of units is reduced to leave seconds.

$$1\,\text{year} \cdot \frac{365\,\text{days}}{1\,\text{year}} \cdot \frac{24\,\text{hours}}{1\,\text{day}} \cdot \frac{60\,\text{minutes}}{1\,\text{hour}} \cdot \frac{60\,\text{seconds}}{1\,\text{minute}} = 31{,}536{,}000\,\text{seconds}$$

There are 31,536,000 seconds in a non–leap-year calendar year.

In Example A, you discovered there are 31,536,000 seconds in 1 year. If you want to know how many seconds there are in 5 years, you can just multiply this number by 5. When quantities are related like this, they are said to be in **direct variation** with each other. Another example of direct variation is the relationship between the number of miles you travel and the time you spend traveling. In each case, the relationship can be expressed in an equation of the form $y = kx$, where k is called the **constant of variation**.

14 CHAPTER 0 Developing Mathematical Thinking

ELL
Some of these problems are text heavy, making them challenging for ELL students. Pair each ELL student with another student who can assist the ELL student in explaining and categorizing terminology. Help students create organizational charts to ease the confusion of differentiating between categories.

Support
Suggest students use flow-charts or diagrams to direct their process when converting units. You might need to review basic units of measure. Students might use sticky notes to demonstrate relationships. For example, students could start with a piece of paper for each color of house, then stick on notes as appropriate.

Advanced
Project Example C from the ebook so that students don't see the solution immediately. Have them work on it with little direction. This will push them to be creative and will increase their independence. Require written explanations of their processes so that when it is time to state their responses, all students will feel more confident in sharing their ideas.

EXAMPLE B To qualify for the Interlochen 470 auto race, each driver must complete two laps of the 5-mile track at an average speed of 100 miles per hour (mi/h). Due to some problems at the start, Naomi averages only 50 mi/h on her first lap. How fast must she go on the second lap to qualify for the race?

Solution Sort the information into two categories: what you know and what you might need to know. Assign variables to the quantities that you don't know. Use a table to organize the information, and include the units for each piece of information.

Know	Might need to know
Speed for first lap: 50 mi/h	Speed for second lap (in mi/h): s
Average speed for both laps: 100 mi/h	Time for first lap (in h): t_1
Length of each lap (in mi): 5	Time for second lap (in h): t_2

Use the units to help you find connections between the pieces of information. Speed is measured in miles per hour and therefore calculated by dividing distance by time. You might also remember the relationship *distance = rate · time*, or $d = rt$. Because you know the distance and rate for the first lap, you can solve for the time: $t = \frac{d}{r}$.

$$t_1 = \frac{5 \text{ miles}}{50 \text{ miles per hour}} = \frac{1}{10} \text{ hour, or 6 minutes}$$

It takes one-tenth of an hour, or 6 minutes, to do the first lap. You know the distance and speed for the first and second laps together, so solve for the time for both laps. Then you can subtract to find what you are looking for, the time for the second lap.

$$(t_1 + t_2) = \frac{10 \text{ miles}}{100 \text{ miles per hour}} = \frac{1}{10} \text{ hour, or 6 minutes}$$

Note that the time for both laps is the same as the time for the first lap. This means the time for the second lap, t_2, must be zero. It's not possible for Naomi to complete the second lap in no time, so she cannot qualify for the race. Would the solution be different if the laps were 10 miles long? d miles long?

Some problems overwhelm you with lots of information. Identifying and categorizing what you know is always a good way to start organizing information.

Career CONNECTION

Event planners make a career organizing information. To plan a New Year's Eve celebration, for example, an event planner considers variables such as location, decorations, food and beverages, number of people, and staff. For the celebration of the year 2000, event planners around the world worked independently and cooperatively to organize unique events for each city or country and to coordinate recording and televising of the events.

Berlin, Germany (left), and Seattle, Washington (right), celebrate the new millennium.

LESSON 0.3 Looking for and Generalizing Patterns 15

strategies, including the intuitive one shown here, or one in which a variable is assigned to the unknown quantity, say x, to the number of seconds in one year, and write the equation 1 year = x seconds. **SMP 1, 6, 7**

ALERT If some students have difficulty with dimensional analysis. **ASK** "What does multiplying a fraction by a fraction equal to 1 do to the the fraction? What does a fraction that equals 1, such as $\frac{1 \text{ yard}}{3 \text{ feet}}$ or $\frac{365 \text{ days}}{1 \text{ year}}$ actually mean?" You might introduce the term *multiplicative identity*. The number 1 is the multiplicative identity because any number times 1 equals the original number. Algebraically, $n \cdot 1 = n$. **SMP 1,6,7,8**

EXTEND "What would change if it were a leap year?" [366 days contain 31,622,400 seconds]

Example B

This example also illustrates organizing information through units, or dimensions. Project Example B from your ebook and have students work in pairs.

If needed, help students with using a variable for the distance. **ASK** "Could Naomi qualify if the length of the lap were different?" [No, she's already used up all available time.] To help distinguish different uses of the word *unit*, some teachers prefer the term *measurement units* or *labels* for dimensional analysis.

ASK "What is the advantage of using subscripts?" [A t can be used for both times.] "Why does miles divided by miles per hour equal hours?" [Ratios of units follow the rules of fraction operations: $\frac{\text{mi}}{\text{mi/h}} = \text{mi} \cdot \frac{\text{h}}{\text{mi}} = \text{h}$.] **SMP 1,6,7,8**

ALERT Students often think that they can find the average speed by averaging the two speeds. In that case, the answer would be 150 mi/h because $\frac{50 + 150}{2} = 100$. If students don't mention this idea, raise it yourself and ask why the answer differs from that of the student book. You might ask for justification of the 150, based on some assumed lap distance, such as 5 miles. Keep pressing until

---

Modifying the Investigation

Whole Class Draw positions 1–5 on the board. Discuss the number of variables in the problem. [5] Have paper labels ready to place on the board as the class goes through each step.

Shortened Give the solutions for the variables to each group.

One Step There is no **One Step** alternative for this investigation.

students see that Naomi took $\frac{1}{10}$ hour for the first lap and, at 150 mi/h, would take $\frac{1}{30}$ hour for the second lap, so her total time for the two laps would be $\frac{2}{15}$ hour, making her average speed over two laps $\frac{10 \text{ mi}}{\frac{2}{15} \text{h}} = 75$ mi/h.

This kind of mistake gives you a chance to mention that doing mathematics often involves the interplay between intuition and logic.

Other solutions may come up in presentations. One way is to assume a convenient lap length (say 5 miles) and look at the units. To find the miles, multiply the rate (speed) by the time: $\frac{\text{mi}}{\text{h}} \cdot \text{h} = \text{mi}$. The hours divide out to make both sides miles. The arrangement could lead to three equations:

First lap $\quad 50t_1 = 5 \quad t_1 = 0.1$ h
Second lap $\quad st_2 = 5 \quad t_2 = \frac{5}{s}$ h
Both laps
$$100(t_1 + t_2) = 10$$
$$(t_1 + t_2) = 0.1 \text{ h}$$

Combining this information into one equation gives $0.1 + \frac{5}{s} = 0.1$ h. But there is no solution to this equation. A different value for the length of a lap would yield the same result.

Example C

LANGUAGE A *flask* is a glass container that is often used in laboratory experiments. An *Erlenmeyer flask* is a flat-bottomed, conical flask.

You could project the example from your ebook and have students work in small groups to organize the data. You might do the example as a class activity, with student input. Students are not asked to solve this problem at this time. They need to organize the information and determine if any information is not needed, a key step in the problem-solving process.

16 CHAPTER 0 Developing Mathematical Thinking

EXAMPLE C Lab assistant Jerry Anderson has just finished cleaning a messy lab table and is putting the equipment back on the table when he reads a note telling him *not* to disturb the positions of three water samples. Not knowing the correct order of the three samples, he finds these facts in the lab notes.

- The water that is highest in sulfur was on one end.
- The water that is highest in iron is in the Erlenmeyer flask.
- The water taken from the spring is not next to the water in the bottle.
- The water that is highest in calcium is left of the water taken from the lake.
- The water in the Erlenmeyer flask, the water taken from the well, and the water that is highest in sulfur are three distinct samples.
- The water in the round flask is not highest in calcium.

Organize the facts into categories. (This is the first step in actually determining which sample goes where.)

Solution Information is given about the types of containers, the sources of the water, the elements found in the samples, and the positions of the samples on the table. You can find three options for each category.

Containers: round flask, Erlenmeyer flask, bottle

Sources: spring, lake, well

Elements: sulfur, iron, calcium

Positions: left, center, right

Now that the information is organized and categorized, you need to see where it leads. You will finish this problem in Exercise 8.

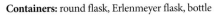

Use this investigation as an opportunity to practice categorizing and organizing information as you did in Example C.

INVESTIGATION

Who Owns the Zebra?

There are five houses along one side of Birch Street, each of a different color. The home-owners each drive a different car, and each has a different pet. The owners all read a different newspaper and plant only one thing in their garden.

- The family with the station wagon lives in the red house.
- The owner of the SUV has a dog.
- The family with the van reads the *Gazette*.
- The green house is immediately to the left of the white house.
- The *Chronicle* is delivered to the green house.
- The man who plants zucchini has birds.
- In the yellow house they plant corn.
- In the middle house they read the *Times*.
- The compact car parks at the first house.
- The family that plants eggplant lives in the house next to the house with cats.
- In the house next to the house where they have a horse, they plant corn.
- The woman who plants beets receives the Daily News.
- The owner of the sports car plants okra.
- The family with the compact car lives next to the blue house.
- They read the *Bulletin* in the house next to the house where they plant eggplant.

Who Owns the Zebra?

Organizing the known information and clarifying what you need to find out is a very useful strategy. To use your time efficiently, you should notice when calculations are repeated and look for both general methods and for short cuts. As you work on exercises and investigations, keep track of where you need to be at the end of the problem while you are doing each step to get there. Continually check to see it the results you get at each step are reasonable, rather than getting to the end and find out you made a mistake in Step 2! Whether it involves simply keeping track of units or sorting out masses of information, organizing your data and making a plan are essential to finding a solution efficiently.

Malaysian-American architect and artist Daniel Castor (b 1966) created this pencil drawing of the Amsterdam Stock Exchange. Castor's work—which he calls "jellyfish" drawings—organizes lots of information and several perspectives into one drawing. He describes his art as "[capturing], in two dimensions, the physical power of the spaces yielded by [the] design process, a power that cannot be adequately described by word or photograph."

Investigation

Position	1 (left)	2	3	4	5 (right)
House color	Yellow	Blue	Red	Green	White
Car	Compact	Van	Station wagon	Sports car	SUV
Newspaper	*Bulletin*	*Gazette*	*Times*	*Chronicle*	*Daily News*
Plant	Corn	Eggplant	Zucchini	Okra	Beets
Pet	Cats	Horse	Birds	Zebra	Dog

The family in the green house owns the zebra.

To assess regularity in repeated reasoning, observe how they reason about different solution strategies and the possible connections within the strategies. Use the Closing Question as evidence of their ability to see and use repeated reasoning in observing patterns.

Apply

Extra Example

1. Your car gets 25 miles to the gallon. If you drive 80 miles, how many quarts of gas have you used? (There are 4 quarts in a gallon.) 12.8 quarts
2. How many inches are there in 5 yards? 180 inches

Closing Question

The first three terms of a sequence are:

$(x - 1)(x + 1)$,
$(x - 1)(x^2 + x + 1)$,
$(x - 1)(x^3 + x^2 + x + 1)$

What's the next term? How would you find the nth term?

$(x - 1)(x^4 + x^3 + x^2 + x + 1)$

$(x - 1)(x^n + x^{n-1} + x^{n-2} + \ldots + x^2 + x + 1)$

Assigning Exercises

Suggested: 1 – 13

2a. Let e represent the average problem solving rate in problems per hour for Emily, and let a represent the average rate in problems per hour for Alejandro.

Exercise 2 In 2d, points i and ii satisfy the first equation. Points ii and iii satisfy the second equation. The last point doesn't work in either equation.

Exercise 5 ALERT Students may not realize that k is the average speed and that its units are mi/min. Students may want to change to more familiar speed units. Although doing so would be a good exercise, it's not necessary in order to answer these questions.

0.3 Exercises

Practice Your Skills

1. Use units to help you find the missing information.
   a. How many seconds would it take to travel 15 feet at 3.5 feet per second? ⓐ approximately 4.3 s
   b. How many centimeters are in 25 feet? (There are 2.54 centimeters per inch and 12 inches per foot.) 762 cm
   c. How many miles could you drive with 15 gallons of gasoline at 32 miles per gallon? 480 mi

2. Emily and Alejandro are part of a math marathon team on which they take turns solving math problems for 4 hours each day. On Monday Emily worked for 3 hours and Alejandro for 1 hour. Then on Tuesday Emily worked for 2 hours and Alejandro for 2 hours. On Monday they collectively solved 139 problems and on Tuesday they solved 130 problems. Find the average problem-solving rate for Emily and for Alejandro.
   a. Identify the unknown quantities and assign variables. What are the units for each variable? ⓐ
   b. What does the equation $3e + 1a = 139$ represent? ⓗ This is the equation for Monday. 3 hours times e problems per hour plus 1 hour times a problems per hour is equal to 139 problems.
   c. Write an equation for Tuesday. $2e + 2a = 130$
   d. Which of these ordered pairs (e, a) is a solution for the problem? ii
      i. (34, 37)   ii. (37, 28)
      iii. (30, 35)   iv. (27, 23)
   e. Interpret the solution from 2d according to the context of the problem.
      Emily averages 37 problems per hour, and Alejandro averages 28 problems per hour.

3. To qualify for the Interlochen 470 auto race, each driver must complete two laps of the track at an average speed of 100 mi/h. Benjamin averages only 75 mi/h on his first lap. How fast must he go on the second lap to qualify for the race? 150 mi/h

4. Solve the equation $24 + 10x - x^2 = p - (x - 5)^2$ to find the value of p. ⓗ $p = 49$

5. Chase rides his bike 5 miles to school in 25 minutes. The relationship between the distance he rides and the time it takes is a direct variation.
   a. Write a direct variation equation between d, the distance Chase rides, and the time it takes, t. Use k for the constant of variation. $d = kt$
   b. Use the information about Chase's ride to school to find the value of k and its units. $k = \frac{5}{25}$, or 0.2; units are mi/min
   c. If Chase always rides at the same rate, how long will it take him to ride 8 miles? 40 min

Reason and Apply

6. **APPLICATION** Alyse earns $15.40 per hour, and she earns time and a half for working past 8:00 P.M. Last week she worked 35 hours and earned $600.60. How many hours did she work past 8:00 P.M.? ⓗ Alyse worked 8 h past 8:00 P.M.

18 CHAPTER 0 Developing Mathematical Thinking

7. The dimensions used to measure length, area, and volume are related by multiplication. Find the information about each rectangular box. Include the units in your solutions.

   a. A box has volume 12,960 cm³, height 18 cm, and length 30 cm. Find the width. ⓐ

   b. A box has volume 486 in³ and height 9 in. Find the area of the base.

   c. A box has base area 3.60 m² and height 0.40 m. Find the volume.

   d. A box has base area 2.40 ft² and volume 2.88 ft³. Find the height.

8. Lab assistant Jerry Anderson has just finished cleaning a messy lab table and is putting the equipment back on the table when he reads a note telling him *not* to disturb the positions of three water samples. Not knowing the correct order of the three samples, he finds these facts in the lab notes.

   - The water that is highest in sulfur was on one end.
   - The water that is highest in iron is in the Erlenmeyer flask.
   - The water taken from the spring is not next to the water in the bottle.
   - The water that is highest in calcium is left of the water taken from the lake.
   - The water in the Erlenmeyer flask, the water taken from the well, and the water that is highest in sulfur are three distinct samples.
   - The water in the round flask is not highest in calcium.

   Determine which water sample goes where. Identify each sample by its container, source, element, and position. ⓗ

9. Kiane has a photograph of her four cats sitting in a row. The cats are different ages, and each cat has its own favorite toy and favorite sleeping spot.

   - Rocky and the 10-year-old cat would never sit next to each other for a photo.
   - The cat that sleeps in the blue chair and the cat that plays with the rubber mouse are the two oldest cats.
   - The cat that plays with the silk rose is the third cat in the photo.
   - Sadie and the cat on Sadie's left in the photo don't sleep on the furniture.
   - The 8-year-old cat sleeps on the floor.
   - The cat that sleeps on the sofa eats the same food as the 13-year-old cat.
   - Pascal likes to chase the 5-year-old cat.
   - If you add the age of the cat that sleeps in a box and the age of the cat that plays with a stuffed toy, you get the age of Winks.
   - The cat that sleeps on the blue chair likes to hide the catnip ball, which belongs to one of the cats sitting next to it in the photo.

   Who plays with the catnip-filled ball?
   Sadie plays with the catnip-filled ball.

Exercise 6 **LANGUAGE** *Time and a half* is the wage that a worker often earns for working overtime; it is equal to one and a half times the worker's normal wage. A nonsymbolic approach might calculate the total that Alyse would earn over 35 hours at the regular rate ($539) and realize that the extra ($61.60) must have come from the extra half pay ($7.70) picked up from late hours. That implies that she worked 8 of the hours after 8:00 P.M.

Exercise 7 As needed, review that the volume of a box equals the area of the base times the height, and wonder aloud if dimensional analysis would help.

7a. $\dfrac{12960 \text{ cm}^3}{(18 \text{ cm} \cdot 30 \text{ cm})} = 24 \text{ cm}$

7b. $\dfrac{486 \text{ in}^3}{9 \text{ in}} = 54 \text{ in}^2$

7c. $3.60 \text{ m}^2 \cdot 0.40 \text{ m} = 1.44 \text{ m}^3$

7d. $\dfrac{2.88 \text{ ft}^3}{2.40 \text{ ft}^2} = 1.20 \text{ ft}$

Exercise 8 **SUPPORT** If students are struggling, you might suggest that they make a chart. Students might number the clues and use those numbers as they fill out the chart. Be open to other approaches, though; some students will be successful at using guess-and-check.

LESSON 0.3 Looking for and Generalizing Patterns

8.

Container	Bottle	Erlenmeyer flask	Round flask
Source	Well	Lake	Spring
Element	Calcium	Iron	Sulfur
Position	Left	Center	Right

9.

Position	1 (left)	2	3	4 (right)
Name	Rocky	Sadie	Winks	Pascal
Age	8	5	13	10
Sleeps	Floor	Box	Chair	Sofa
Toy	Stuffed toy	Catnip ball	Silk rose	Rubber mouse

Exercise 10 Students' intuition may lead them to average the two times. Because 17.5 does not satisfy any of the three equations, students may be confused. This is the good kind of confusion that can lead to deeper understanding. **ASK** "Does the average of 17.5 make sense?" Most students will realize that two people working together can finish the job in less time than either working alone. You might then say that you find it helpful to think of what part of the job each of the workers can do in a minute, and then calculate what part they can do together in a minute.

10a. Equation iii. The area painted is calculated by multiplying painting rate by time. Equation iii represents the sum of Paul's area and China's area, which equals the whole area, or

$$\frac{96 \text{ ft}^2}{15 \text{ min}} \cdot t \text{ min} + \frac{96 \text{ ft}^2}{20 \text{ min}} \cdot t \text{ min} = 96 \text{ ft}^2.$$

Exercise 11 If you need to review the properties of exponents, be sure to go beyond formulas to show why the formulas make sense. For example, $(r^5)(r^2) = (r \cdot r \cdot r \cdot r \cdot r)(r \cdot r) = r^7$. You might review negative exponents here, but they will be taught specifically in later lessons, so don't expect students to master them at this point. **ASK** "How would you write $\frac{1}{x^2}$ without using a fraction?" $[x^{-2}]$

12a.

	3x	1
x	$3x^2$	$1x$
5	$15x$	5

12b.

	x	3
x	x^2	$3x$
3	$3x$	9

12c.

	x	y
2x	$2x^2$	$2xy$
7	$7x$	$7y$

10. **APPLICATION** Paul can paint the area of a 12 ft by 8 ft wall in 15 min. China can paint the same area in 20 min.

   a. In which equation does t represent how long it would take Paul and China to paint the area of a 12 ft by 8 ft wall together? Explain your choice. ⓗ iii

   i. $\frac{15t}{96} + \frac{20t}{96} = \frac{1}{96}$ ii. $(96)15t + (96)20t = 96$ iii. $\frac{96t}{15} + \frac{96t}{20} = 96$

   b. Solve all three equations in 10a. i. $t = \frac{1}{35} \approx 0.03$ ii. $t = \frac{96}{3360} \approx 0.03$ iii. $t = \frac{60}{7} \approx 8.57$

   c. How long would it take Paul and China to paint the wall together?
   It would take approximately 9 min.

Review

11. Rewrite each expression using the properties of exponents so that the variable appears only once.

   a. $(r^5)(r^7)$ ⓐ r^{12} b. $\frac{2s^6}{6s^2} \cdot \frac{s^4}{3}$ 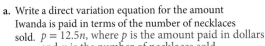   c. $\frac{(t)(t^3)}{t^8}$ t^{-4}, or $\frac{1}{t^4}$ d. $3(2u^2)^4$ $48u^8$

12. Draw a rectangle diagram to represent each product. Use the diagrams to expand each product.

   a. $(3x + 1)(x + 5)$ $3x^2 + 16x + 5$ b. $(x + 3)^2$ $x^2 + 6x + 9$ c. $(x + y)(2x + 7)$
   $2x^2 + 7x + 2xy + 7y$

13. **APPLICATION** Iwanda sells African bead necklaces through a consignment shop. At the end of May, the shop paid her $100 from the sale of 8 necklaces. At the end of June, she was paid $187.50 from the sale of 15 necklaces. Assume that the consignment shop pays Iwanda the same amount of money for each necklace sold.

   a. Write a direct variation equation for the amount Iwanda is paid in terms of the number of necklaces sold. $p = 12.5n$, where p is the amount paid in dollars and n is the number of necklaces sold.

   b. If the shop sells only 6 of her necklaces in July, how much money will Iwanda get? $75

   c. If the materials for each necklace cost $8, how much profit did Iwanda make each month? May profit $36, June profit $67.50, July profit $27

CHAPTER 0 REVIEW

There is no single way to solve a problem. Different people prefer to use different problem-solving strategies, yet not all strategies can be applied to all problems. In this chapter you practiced only a few specific strategies. You may have also used some strategies that you remember from other courses. The next paragraph gives you a longer list of problem-solving strategies to choose from.

Organize the information that is given, or that you figure out in the course of your work, in a list or table. **Draw a picture,** graph, or diagram, and label it to illustrate information you are given *and* what you are trying to find. There are also special types of diagrams that help you represent algebraic expressions, such as rectangle diagrams. **Make a physical representation** of the problem. That is, act it out or make a model. **Look for a pattern** in numbers or units of measure. Be sure all your measures use the same system of units and that you compare quantities with *the same* unit. **Eliminate some possibilities.** If you know what the answer cannot be, you are partway there. **Solve subproblems** that present themselves as part of the problem context, or **solve a simpler problem** by substituting easier numbers or looking at a special case. Don't forget to **use symbolic representation!** Assign variables to unknown quantities and write expressions for related quantities. Translate verbal statements into equations, and solve the equations. **Work backward** from the solution to the problem. For instance, you can solve equations by undoing operations. **Use guess-and-check,** adjusting each successive guess by the result of your previous guess. Finally, but most importantly, **read the problem!** Be sure you know what is being asked.

Exercises

ⓐ **Answers are provided for all exercises in this set.**

1. You are given a 3-liter bucket, a 5-liter bucket, and an unlimited supply of water. Describe or illustrate a procedure that will give exactly 4 liters of water in the 5-liter bucket.

2. Draw a rectangle diagram to represent each product. Use the diagrams to expand each product.
   a. $(x + 3)(x + 4)$
   b. $(2x)(x + 3)$
   c. $(x + 6)(x - 2)$
   d. $(x - 4)(2x - 1)$

3. Use the Pythagorean Theorem to find each missing length.
   a.
   3 cm, x, 3 cm
   $x = \sqrt{18}$ cm $= 3\sqrt{2}$ cm ≈ 4.2 cm

   b. y, 12 in., 13 in.
   $y = \sqrt{25}$ in. $= 5$ in.

CHAPTER 0 REVIEW 21

CHAPTER REVIEW 0

Reviewing

To begin the review, you might pose a single problem that is amenable to a variety of approaches. For example, "If you fit a steel belt tightly around Earth's equator, and then lengthen it by 40 feet and make it stand out equally around the equator, how high is it from Earth?" (The radius of Earth is about 4000 miles.) A diagram can be useful for solving this problem, as can symbolic algebra and generating some data to organize. It's also an example of a problem whose algebraic result $\left(\frac{40}{2\pi}, \text{ or about 6 feet}\right)$ is counterintuitive.

Assigning Exercises

You might use Exercises 1–14 as a pretest of algebra knowledge.

1. Sample answer: Fill the 5-liter bucket; pour 3 liters into the 3-liter bucket, leaving 2 liters in the 5-liter bucket; empty the 3-liter bucket; pour the remaining 2 liters from the 5-liter bucket into the 3-liter bucket. Fill the 5-liter bucket and pour 1 liter into the 3-liter bucket, completely filling the 3-liter bucket and leaving 4 liters in the 5-liter bucket.

Exercise 2 This exercise is the first of this type in which *x* does not have a coefficient of 1.

2a. $x^2 + 7x + 12$

	x	4
x	x^2	$4x$
3	$3x$	12

2b. $2x^2 + 6x$

	x	3
2x	$2x^2$	$6x$

2c. $x^2 + 4x - 12$

	x	-2
x	x^2	$-2x$
6	$6x$	-12

2d. $2x^2 - 9x + 4$

	x	-4
2x	$2x^2$	$-8x$
-1	$-1x$	4

Exercise 3 This exercise reviews the Pythagorean Theorem and side lengths of special triangles. Students may recognize the isosceles right triangle or the 5-12-13 right triangle and not use the Pythagorean Theorem directly.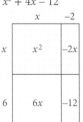

LESSON 0 REVIEW 21

Exercise 4 This exercise reviews ratio and proportion. The exact answers given use the ratios $\frac{120 \text{ mi}}{5 \text{ gal}}$ for Car A and $\frac{180 \text{ mi}}{5 \text{ gal}}$ for Car B. If students approximate each answer from the graph, instead of using proportions, their answers will vary slightly.

Exercise 5 ASK "What is the difference between expressions and equations? Which of these would give an expression and which would give an equation?" [5a and 5b give an expression, 5c an equation] Students who prefer the undoing method of solving equations will find that it doesn't work for equations like this, which contain more than one occurrence of the variable. Encourage them to try other methods. ASK "What other methods might you try?" [guess-and-check, graphing, balancing]
SMP 1, 6

Exercise 6 This exercise requires linear reasoning. If Keisha makes five trips, she will drive the 12-mile distance only nine times (not ten times), because she doesn't need to go back to the old apartment after the fifth time. If students justify an answer to 6b of $65.45 with the argument that she needs to go back to clean or return keys, accept it. Having the discussion with the class encourages students to explain their answer. SMP 8

---

4. This graph shows the relationship between distance driven and gasoline used for two cars going 60 mi/h.

    a. How far can Car A drive on 7 gallons (gal) of gasoline?   168 mi
    b. How much gasoline is needed for Car B to drive 342 mi?   9.5 gal
    c. Which car can drive farther on 8.5 gal of gas? How much farther?
       Car B can go 102 mi farther than Car A.
    d. The graph shows the direct variation between the distance driven and the gasoline used. What is the constant of variation for each car?
       Car A: 24 mi/gal; Car B: 36 mi/gal
    e. What is the slope of each line? Explain the real-world meaning of each slope.   For Car A, the slope is $\frac{120}{5} = 24$, which means the car can drive 24 mi/gal of gasoline. For Car B, the slope is $\frac{180}{5} = 36$, which means the car can drive 36 mi/gal.

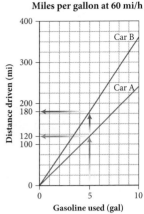

Consumer CONNECTION

The U.S. Environmental Protection Agency (EPA) obtains fuel-economy estimates each year on car manufacturers' new models. In a controlled laboratory setting, professional drivers test the cars on a treadmill-type machine with engines and temperature conditions that simulate highway and city driving. Hybrid cars—a cross between fuel-powered cars and electric cars—are some of the highest-rated vehicles for fuel economy.

This diagram illustrates the components of a parallel hybrid car.

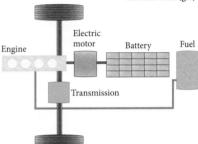

5. In 5a and b, translate each verbal statement to a symbolic expression. Combine the expressions into an equation to solve 5c.

    a. Six more than twice a number
    b. Five times three less than a number
    c. Six more than twice a number is five times three less than the number. Find the number.

    Let n represent the unknown number.
    a. $2n + 6$
    b. $5(n - 3)$
    c. $2n + 6 = 5(n - 3)$; $n = 7$

6. **APPLICATION** Keisha and her family are moving to a new apartment 12 miles from their old one. You-Do-It Truck Rental rents a small truck for $19.95 per day plus $0.35 per mile. Keisha's family hopes they can complete the move with five loads all on the same day. She estimates that she will drive the truck another 10 miles for pickup and return.

    a. Write an expression that represents the cost of a one-day rental with any number of miles.
    b. How much will Keisha pay if she does complete the move with five loads?
    c. How much can Keisha save if she completes the move with four loads?

    Let m represent the number of miles driven, and let c represent the total cost in dollars of renting the truck.
    a. $c = 19.95 + 0.35m$
    b. possible answer: $61.25
    c. $8.40

22   CHAPTER 0   Developing Mathematical Thinking

7. Toby has only a balance scale, a single 40 g mass, and a stack of both white and red blocks. (Assume that all white blocks have the same mass and all red blocks have the same mass.) Toby discovers that four white blocks and one red block balance two white blocks, two red blocks, and the 40 g mass. He also finds that five white blocks and two red blocks balance one white block and five red blocks.

   a. List the unknown quantities and assign a variable to each.

   b. Translate Toby's discoveries into equations.

   c. Solve the equations to find a value for each variable.

   d. Interpret the solution according to the context of the problem.

Science CONNECTION

Calculating how much food a human being needs and planning how much food can be carried are critical factors for space travel. Weight and volume of food must be limited, trash must be minimized, and nutritional value and variety must be provided. Current U.S. space shuttle flights can carry 3.8 pounds of food per person per day, providing 3 meals a day for up to 14 days. In the photo at right, astronaut Linda Godwin handles food supplies aboard the space shuttle *Endeavour* in December 2001.

8. Amy says that it is her birthday and that six times her age five years ago is twice as much as twice her age next year. How old is Amy?   17 years old

9. Scott has 47 coins totaling $5.02. He notices that the number of pennies is the same as the number of quarters and that the sum of the number of pennies and quarters is one more than the sum of the number of nickels and dimes. How many of each coin does Scott have?   12 pennies, 8 nickels, 15 dimes, 12 quarters

10. Jeremy is mixing almonds and dried cranberries to sell at a school fair. He buys almonds in 4-pound bags and dried cranberries in 1-pound bags.

    x = number of bags of almonds he buys.

    y = number of bags of dried cranberries he buys.

    The following equations are true:

    $3x = y$

    $4x + y = 70$

    10a. The number of bags of cranberries is three times the number of bags of almonds. Four times the number of bags of almonds plus the number of cranberries totals 70 (the total weight of all almonds and cranberries is 70 pounds).

    a. Explain in words the meaning of each equation.

    b. Find two pairs of values for x and y that satisfy the first equation.
       Possible values: (4, 12) or (8, 24).
    c. Find two pairs of values for x and y that satisfy the second equation.
       Possible values: (12, 22) or (7, 42).
    d. Find pairs of values for x and y that satisfy both equations simultaneously.
       $x = 10, y = 30$.

11. The height of a golf ball in flight is given by the equation $h = -16t^2 + 48t$, where h represents the height in feet above the ground and t represents the time in seconds since the ball was hit. Find h and interpret the real-world meaning of the result when

    a. $t = 0$ b. $t = 2$ c. $t = 3$

    a. $h = 0$. Before the ball is hit, it is on the ground.
    b. $h = 32$. Two seconds after being hit, the ball is 32 ft above the ground.
    c. $h = 0$. After 3 s, the lands on the ground.

Exercise 7 Instead of using symbolic variables and equations, students may want to manipulate (or think of manipulating) blocks on a balance scale. Have them describe their strategy, then have them represent it in symbols, tracking how it reflects the balancing approach to solving a linear equation. SMP 2

7a. Let w represent the mass in grams of a white block, and let r represent the mass in grams of a red block.

b. $4w + r = 2w + 2r + 40$; $5w + 2r = w + 5r$

c. $w = 60$; $r = 80$

d. The mass of a white block is 60 g, and the mass of a red block is 80 g.

Exercise 8 An answer of 3 indicates that students missed the double doubling in the second condition. Encourage a variety of strategies for solving the problem. SMP 6

Exercise 11 This exercise helps assess students' ability to evaluate and make meaning of expressions, not to solve quadratic equations, which appear in a later chapter. SMP 2

Exercise 12 If necessary, ASK "How would you write x^{-1} with a positive exponent?" $\left[\frac{1}{x}\right]$ "Does the negative sign affect the way you apply the properties of exponents?" [no] SMP 6

Exercise 13 In this exercise, students evaluate expressions and review exponents. Students do not need to know what *exponential function* means at this point. Exponential functions will be studied in depth in a later chapter.

Exercise 15 A chart similar to the one used in Take Another Look activity 2 may be used to solve this problem. SMP 2, 4, 5

15.

Sales Rep	Mr. Mendoza	Mr. Bell	Mrs. Plum
Client	Mr. Green	Ms. Phoung	Ms. Hunt
Location	Conference room	Convention hall	Lunch room
Time	9:00 A.M.	12:00 noon	3:00 P.M.

Exercise 16 This exercise gives a chance to review rational and irrational numbers, as well as converting a repeating decimal to a fraction. Emphasize why students agree or disagree in the problem. SMP 3

Take Another Look

You can use the Take Another Look activities to review and extend the lesson's content. Students might use any one of the Take Another Look activities to deepen their understanding of something that particularly interests them. Or you might assign some of the activities as performance tasks or for a portfolio assessment.

1. A classic model for the multiplication of three binomials is a volume model, in which the length, width, and height each represent a factor, and the volume represents the product. The volume model has 8 different sections. This volume model represents $x^3 + 4x^2 + 3x^2 + 2x^2 + 8x + 6x + 12x + 24$ or $x^3 + 9x^2 + 26x + 24$.

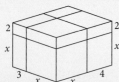

After students have devised a workable diagram (be it a volume model or not), you may want to have them write a few exercises to share with a classmate. The exercises can go both ways: converting an expression into a diagram and converting a diagram into an expression. As an extension, challenge students to think of a model for four factors. How is the fourth dimension represented?

If students have access to geometry software, they can accurately draw volume models with parallel lines and segments in proper ratio, and even create dynamic models that will change size and shape based on parameters or sliders. SMP 5

2. Tables like these are sometimes called *matrix logic tables*. The sample answers in the chart are marked with capital letters when the relationship comes directly from the clues and lowercase when they are deduced from other relationships. Once all bottles are determined, no more cells are filled in. See Exercise 8 on page 19 for a table of answers.

---

12. Rewrite each expression using the properties of exponents.

    a. $(4x^{-2})(x)$ $4x^{-1}$, or $\frac{4}{x}$ b. $\frac{4x^2}{8x^3}$ $\frac{1}{2}x^{-1}$, or $\frac{1}{2x}$ c. $(x^3)^5$ x^{15}

13. Consider the equation $y = 2^x$.

    a. Find y when $x = 0$. $y = 2^0 = 1$
    b. Find y when $x = 3$. $y = 2^3 = 8$
    c. Find y when $x = -2$. $y = 2^{-2} = \frac{1}{2^2} = \frac{1}{4}$
    d. Find x when $y = 32$. $x = 5$

14. Use units to help you find the missing information.

    a. How many ounces are in 5 gallons? (There are 8 ounces per cup, 4 cups per quart, and 4 quarts per gallon.) $5 \text{ gal} \cdot \frac{4 \text{ qt}}{1 \text{ gal}} \cdot \frac{4 \text{ c}}{1 \text{ qt}} \cdot \frac{8 \text{ oz}}{1 \text{ c}} = 640 \text{ oz}$

    b. How many meters are in 1 mile? (There are 2.54 centimeters per inch, 12 inches per foot, 5280 feet per mile, and 100 centimeters per meter.) $1 \text{ mi} \cdot \frac{5280 \text{ ft}}{1 \text{ mi}} \cdot \frac{12 \text{ in}}{1 \text{ ft}} \cdot \frac{2.54 \text{ cm}}{1 \text{ in}} \cdot \frac{1 \text{ m}}{100 \text{ cm}} = 1609.344 \text{ m}$

15. Bethany Rogers just started working as a sales assistant. She learns that all three sales representatives have important meetings with clients, but she uncovers only these clues.

    - Mr. Bell is a sales representative although he is not meeting with Mr. Green.
    - Ms. Hunt is the client who will be meeting in the lunch room.
    - Mr. Green is the client with a 9:00 A.M. appointment.
    - Mr. Mendoza is the sales representative meeting in the conference room.
    - Ms. Phoung is a client, but she will not be in the 3:00 P.M. meeting.
    - Mrs. Plum is a sales representative, but not the one meeting at 12:00 noon.
    - The client with the 9:00 A.M. appointment is not meeting in the convention hall.

    Help Bethany figure out which sales representative is meeting with which client, where, and when.

16. Art, Brandi, ReGi, Koy, and Gina were discussing whether $0.\overline{32}$ is rational or irrational.

    Art said that $0.\overline{32}$ is an irrational number.

    Brandi says it is rational because $0.\overline{32}$ can be written as a fraction.

    ReGi agreed with Brandi, since she says $0.\overline{32} = \frac{32}{100}$.

    Koy pointed out that the bar means $0.\overline{32}$ will go on forever if you tried to write it.

    Gina says that proves $0.\overline{32}$ must be irrational.

    Tell each person why you agree or disagree with their statement. Is $0.\overline{32}$ rational or irrational? Explain why.

    Art: Disagree. repeating decimals are not necessarily irrational. $0.\overline{32}$ $\frac{32}{99}$, so it is a rational number.

    Brandi: Agree. See above.   ReGi: Agree with statement but not with reason.
    Koy: Agree.   Gina: Disagree. Repeating forever does not necessarily mean it is irrational.

Take Another Look

1. You have seen that the multiplication expression $(x + 2)(x + 3)$ can be represented with a rectangle diagram in which the length and width of the rectangle represent the factors and the area represents the product. Find a way to represent the multiplication of three factors, such as $(x + 2)(x + 3)(x + 4)$. Explain how the geometry of the diagram represents the product.

2. Recall Jerry Anderson's problem with the water samples (Exercise 8 in Lesson 0.3). In the table below, each of the six subsections compares two of the characteristics. For example, the upper-left subsection compares the position and the type of container. Each statement given in the problem translates into yeses (Y) or noes (N) in the cells of the table. When you have two noes in the same row or column of one subsection, the third cell must be a yes. And when you have a yes in any cell of a subsection, the other four cells in the same row and column must be noes. The table will eventually show you how the characteristics match up. Use this table to solve Jerry's problem.

	Erlenmeyer flask	Round flask	Bottle	Calcium	Sulfur	Iron	Lake	Well	Spring
Left	n	n	y				N	y	n
Center	y	n	N		N		y		N
Right		y	n	N					y
Lake									
Well	N				N				
Spring									
Calcium	n		N	y					
Sulfur	N		y	n					
Iron		Y	n	n					

3. Find a way to use the distributive property to rewrite $(x + 2)(x + 3)$ without parentheses. Compare and contrast this method to the rectangle diagram method. Look back at Exercise 10 in Lesson 0.1, and explain how you could use the distributive property for each multiplication.

4. Use your graphing calculator or geometry software to explore the slopes of lines. What is the slope of a horizontal line? Of a vertical line? What slopes create a diagonal line at a 45° angle from the x-axis? For which slopes does the line increase from left to right? For which slopes does it decrease? Can you estimate a line's slope simply by looking at a graph? Write a short paper summarizing your findings.

Assessing the Chapter

The Kendall Hunt Cohesive Assessment System (CAS) provides a powerful tool with which to manage assessment content, create and assign homework and tests, and disseminate assessment data. While there are tests for this chapter, you may decide to not give a test or to use Chapter Review Exercises as a pretest that will not affect students' grades. Another good written assessment option is to use one or two constructive assessment items.

You might use the Assessing What You've Learned Presentation as a portfolio assessment.

Facilitating Self-Assessment

It is often helpful to understand students' conceptions of algebra or their views of themselves as problem solvers. Read and comment on some of the writing students do so that they see you are interested in their thinking and that you think their writing is important.

Assessing What You've Learned

 WRITE IN YOUR JOURNAL Recording your thoughts about the mathematics you are learning, including areas of confusion or frustration, helps point out when you should seek assistance from your teacher and what questions you could ask. Keeping a journal is a good way to collect your ideas and questions, and if you write in it regularly, you'll track the progress of your understanding throughout the course. Your journal can also remind you of interesting contributions to make in class or prompt questions to ask during a review period.

Here are some questions you might start writing about.

- How has your idea about what algebra is changed since you finished your first-year course in algebra? Do you have particular expectations about what you will learn in advanced algebra? If so, what are they?

- What are your strengths and weaknesses as a problem-solver? Do you consider yourself well organized? Do you have a systematic approach? Give an everyday example of problem solving that reminds you of work you did in this chapter.

 GIVE A PRESENTATION In the working world, most people will need to give a presentation once in a while or contribute ideas and opinions at meetings. Making a presentation to your class gives you practice in planning, conveying your ideas clearly, and adapting to the needs of an audience.

Choose an investigation or a problem from this chapter, and describe the problem-solving strategies you and your group used to solve it. Here are some suggestions to plan your presentation. (Even planning a presentation requires problem solving!)

- Work with a partner or team. Divide tasks equally. Your role should use skills that are well established, as well as stretch your abilities in new areas.

- Discuss the topic thoroughly. Connect the work you did on the problem with the objectives of the chapter.

- Outline your talk, and decide what details to mention for each point and what diagrams, graphs, or pictures would clarify the presentation.

- Speak clearly and loudly. Reveal your interest in the topic you chose by making eye contact with listeners.

CHAPTER 1: Linear Modeling

Overview

Discovering Advanced Algebra is a study of mathematical functions modeling real-world problems. Some of the most common functions for modeling are linear functions. The first four lessons of this chapter give students a fresh perspective on the beginning algebra content and uses that new approach to bridge the gap between their previous course work and advanced algebra. The treatment of sequences in this chapter reviews concepts from beginning algebra while also laying the groundwork for the later study of linear and exponential functions. Before abstracting to the concept of function, however, this chapter first considers the linear equations with which students may already be familiar. A deep understanding of these equations is encouraged through the approach of making predictions about a two-variable data set.

In **Lesson 1.1**, students learn how to define arithmetic and geometric sequences recursively. **Lesson 1.2** extends recursive formulas to geometric sequences that model growth and decay. **Lesson 1.3** introduces shifted geometric sequences and explores the physical limitations of mathematical models. Graphs are the emphasis of **Lesson 1.4**, and the relationships among sequence representations as graphs, tables, and rules are considered. In **Lesson 1.5**, students move from recursive formulas generating arithmetic sequences to explicit formulas for those sequences and then to the corresponding linear equations and their graphs. The lesson foreshadows the definitions of domain, range, and slope in **Lesson 1.6**, where additional real-world situations are explored. In **Lesson 1.7** we review the point-slope form of the linear equation to graph lines of fit for making predictions about two-variable data. In **Lesson 1.8**, students investigate prediction and accuracy through residuals. And **Lesson 1.9** deepens their understanding of residual plots and least squares.

The **Performance Task** highlighted in this chapter is in Lesson 1.4, where students verbalize the differences in arithmetic and geometric sequences and how each of their graphs may vary.

The Mathematics

Sequences and Formulas

A *sequence* of numbers is an ordered list. The listed numbers are the *terms* of the sequence. Students may have seen sequences in which a formula gives each term. For example, the formula $2n$ describes the sequence $\{2, 4, 6, 8, 10, \ldots\}$. A different approach, called *recursion*, emphasizes how each term is generated from the previous term. For the sequence above, each term is 2 more than the previous one. Sequences like this, in which each term is the same amount more than the previous term, are called *arithmetic sequences*. The amount being added is called the *common difference*. Repeatedly using the recursive formula to find terms is called *iteration*.

Unlike the formula $2n$, a *recursive formula* consists of two equations. One equation gives the starting point, and the other, called the *recursive rule*, tells how to get from one term to the next. For the sequence $\{2, 4, 6, 8, 10, \ldots\}$, for example, the recursive formula has these two equations: $u_1 = 2$ and $u_n = u_{n-1} + 2$, where $n > 1$.

In some sequences a constant is multiplied by, rather than added to, each term to get the next term. These sequences, called *geometric sequences*, are helpful in modeling many real-life situations in which an amount increases, resulting in *growth*, or decreases, resulting in *decay*. The constant by which each term is multiplied is the *common ratio* of the sequence. If the common ratio is between 1 and −1, the terms of the geometric sequence get closer and closer to 0; we say 0 is the *limit* of such a sequence.

Other sequences studied in this chapter have recursive rules that include both multiplication and addition. For example, the rule for the sequence $\{1, 3, 7, 15, 31, \ldots\}$ is $u_n = 2u_{n-1} + 1$. These *shifted geometric sequences* are helpful for modeling many situations. As with geometric sequences, shifted geometric sequences converge to a limit if the multiplier is between −1 and 1. For example, the sequence $\{21, 9, 5, 3.67, 3.22, 3.07, 3.02, 3.01, \ldots\}$ has recursive rule $u_n = \frac{1}{3} u_{n-1} + 2$ and converges to limit 3.00.

Population Models

A *mathematical model* of a situation is a mathematical object, such as a diagram, sequence, or equation, that represents the situation. A problem about a real-life situation can be translated into a mathematical model, such as finding the 50th term of a sequence. If the model is good, a solution to the model translates back into a good solution to the original real-life problem.

Shifted geometric sequences can be good models for the growth of a population, such as a colony of insects. But these models are not particularly accurate, because they don't take into account factors that keep the population in check, such as predators or restraints on resources. It is, therefore, important that all mathematical models of real world situations consider the constraints of that situation.

Linear Models and Residuals

Recursive formulas, which generate sequences, are a powerful vehicle for developing an understanding of rates of change. A more efficient tool for finding terms of a sequence, however, is an *explicit formula*, which expresses term u_n in terms of n rather than u_{n-1}. For example, the sequence $\{3, -4, -11, -18, \ldots\}$, with recursive formula $u_0 = 3, u_n = u_{n-1} - 7$, has explicit formula $u_n = 3 - 7n$.

Recursive and explicit formulas define discrete functions that are evaluated only at integers. (The function values might be non-integers, though.) To model continuous change, we use equations that can be evaluated at any real number. In this book, such equations are usually given in terms of x's and y's, rather than u's and n's. But any variable names may be used. In general, the recursive formula for an arithmetic sequence, $u_0 = a$ and $u_n = u_{n-1} + b$, extends to the continuous linear equation $y = a + bx$.

Like *Discovering Algebra*, this student book emphasizes the *intercept form* of linear equations instead of the more familiar *slope-intercept form*, $y = mx + b$. The intercept form helps students think dynamically of starting at a value a and increasing by b units for each unit that x moves from left to right. This visualization deepens students' understanding of slope as the amount of vertical change for each unit of horizontal change. Note that the variable b has different meanings in the two forms: In the intercept form, b represents the slope; in the slope-intercept form, b represents the y-intercept. You should probably resist any temptation to change everything to slope-intercept form. Emphasize that the two forms are equivalent.

In real life, data rarely fit exactly on a line or any other nice curve. So making predictions requires finding a continuous equation that fits the data as well as possible. In this chapter, students first find *lines of fit* by eyeballing, perhaps picking two data points so that the line through them "looks good." Then they learn a more systematic method called the *least squares line*.

Whereas some books refer to a *best-fit line* or *line of best fit*, the subjective term *best* is not used here. Instead, students evaluate the merits of each line. One way to evaluate the quality of fit is by residuals—the amounts by which the line of fit misses the data points. Each residual is obtained by subtracting from the y-coordinate of a data point the y-coordinate of the point on the line above or below that data point. If the line fits the data perfectly, every residual is 0. The sum of the residuals may be 0 without the line being a good fit, but if the residuals also have a spread close to 0, then the line is a good fit.

Using This Chapter

Much of this chapter reviews basic algebra concepts, but is presented from a fresh perspective. Rather than skipping a topic, you may be able to spend less time on some lessons than on others. Many of the Investigations will allow you to assess prior understanding of familiar topics. In the later lessons of the chapter, students are exposed to the analysis of models that they will need throughout the course.

Common Core State Standards (CCSS)

In a first year algebra course, students view functions as relationships between quantities where one quantity determines another (F.IF.1). They interpret functions represented graphically, numerically, symbolically, and verbally (F.IF.4). They translate between representations, and understand the limitations of various representations (A.CED.3), especially in real world contexts. They recognizing equivalent expressions as representing the same model (F.IF.9) and evaluate a line of fit and adjust their models accordingly (S.ID 6).

Algebra II continues and deepens the themes learned in a basic algebra course. Students do a more complete analysis of sequences (F.IF.3), especially arithmetic and geometric sequences, and their associated series. Developing recursive formulas for sequences is facilitated by the practice of abstracting regularity for how you get from one term to the next and then giving a precise description of this process in algebraic symbols (F.BF.2). Technology can be a useful tool here: most graphing calculators allow one to model recursive function definitions in notation that is close to standard mathematical notation. And spreadsheets make natural the process of taking successive differences and running totals.

Additional Resources For complete references to additional resources for this chapter and others see your digital resources.

Refreshing Your Skills To review the algebra skills needed for this chapter see your digital resources.

Differences and Ratios
Linear Relationships
Measures of Central Tendency and Spread

Projects See your digital resources for performance-based problems and challenges.

Talkin' Trash
Counting Forever

Explorations See your digital resources for extensions and/or applications of content from this chapter.

Graphs of Sequences
Refining the Growth Model
Precision, Accuracy, and Significant Figures
Median-Median Line

Materials

- racquetballs, basketballs, or handballs
- motion sensor (optional)
- meter sticks
- video camera, (optional)
- bowls
- water
- tinted liquid
- measuring cups with milliliters
- sink or waste buckets
- geometry software
- graph paper
- stopwatch or watch with a second hand
- uncooked spaghetti, *optional*
- springs
- mass holders (such as film canisters)
- small unit masses
- centimeter rulers

CHAPTER 1

Linear Modeling

A mathematical model describes a relationship between quantities using a mathematical object, such as an equation, graph, diagram, etc. In a linear model, a graph of the data is a line where the relationship shows a constant rate of change, or slope. Sequences can model recursive processes, such as this sequence of moves by a skier performing a tele-heli. The relationship of distance, rate, and time, as in a race, can be modeled in a linear model. Another linear model can be seen in the ability of a balloon's pilot to heat the air in the balloon at a constant rate, it then becomes less dense and lighter than the surrounding cooler air.

OBJECTIVES

In this chapter you will

- recognize and visualize mathematical patterns called sequences
- write recursive definitions for sequences
- display sequences with graphs
- investigate what happens to sequences in the long run
- recognize the limitations of mathematical models
- explore connections between arithmetic sequences and linear equations
- find lines of fit for data sets that are approximately linear
- assess the fit of a model function by plotting and analyzing residuals

CHAPTER 1

Objectives

- Use recursive formulas for generating arithmetic, geometric, and shifted geometric sequences
- Recognize arithmetic and geometric sequences from their graphs
- Use geometric sequences to model growth and decay
- Explore long-run values of geometric and shifted geometric sequences
- Recognize that mathematical models have real world limitations
- Use graphs to check whether a recursive formula is a good model for data
- Relate recursive formulas for arithmetic sequences to explicit formulas and linear equations
- Deepen understanding of slope in arithmetic sequences and linear equations
- Find lines of fit for data sets and use them to make predictions through interpolation and extrapolation
- Evaluate lines of fit using residuals

Modeling is what links the mathematics we teach to the everyday world in which our students live. The models we choose in class should cultivate the ability to choose and use appropriate mathematics and statistics to analyze and understand a real situation and make informed decisions. While mathematical models such as equations, graphs, diagrams, etc. are used to represent and understand quantities and their relationships, models can also explain the mathematical structures themselves—for example, a model of population growth can demonstrate the recursive nature of an arithmetic sequence (linear function) or a geometric sequence (exponential function). **ASK** "What other things can you think of that are models of mathematical ideas and how do they relate to that mathematics?"

LESSON 1.1

This lesson reviews how arithmetic and geometric sequences can be defined recursively.

COMMON CORE STATE STANDARDS		
APPLIED	**DEVELOPED**	**INTRODUCED**
	F.IF.3	F.LE.2
	F.BF.1a	
	F.BF.2	

Objectives

- Discover recursive formulas for sequences
- Define, explore, and use arithmetic and geometric sequences
- Use recursively defined sequences to model real-life situations

Vocabulary

recursion
sequence
term
general term
recursive formula
arithmetic sequence
common difference
spreadsheet
fractal
geometric sequence
common ratio
Fibonacci sequence

Materials

- Calculator Notes: Reentry; Recursion; Making Spreadsheets Using the CellSheet App

Launch

What is the next term in each sequence?

a. 4, 8, 12, 16, 20, …

b. 1, 0.1, 0.01, 0.001, 0.0001, …

Describe in words how each sequence was generated.

a. 24. Answers will vary. Each term is 4 more than the previous term.

b. 0.00001. Answers will vary. Each term is $\frac{1}{10}$ of the previous term.

LESSON 1.1

Recursively Defined Sequences

For every pattern that appears, a mathematician feels he ought to know why it appears.

W. W. SAWYER

Look around! You are surrounded by patterns and influenced by how you perceive them. You have learned to recognize visual patterns in floor tiles, window panes, tree leaves, and flower petals. In every discipline, people discover, observe, re-create, explain, generalize, and use patterns. Artists and architects use patterns that are attractive or practical. Scientists and manufacturing engineers follow patterns and predictable processes that ensure quality, accuracy, and uniformity. Mathematicians frequently encounter patterns in numbers and shapes.

The arches in the Pershore Abbey in Worcestershire, United Kingdom, show an artistic use of repeated patterns.

Scientists use patterns and repetition to conduct experiments, gather data, and analyze results.

You can discover and explain many mathematical patterns by thinking about recursion. **Recursion** is a process in which each step of a pattern is dependent on the step or steps that come before it. It is often easy to define a pattern recursively, and a recursive definition reveals a lot about the properties of the pattern.

EXAMPLE A A square table seats 4 people. Two square tables pushed together seat 6 people. Three tables pushed together seat 8 people. How many people can sit at 10 tables pushed together? How many tables are needed to seat 32 people? Write a recursive definition to find the number of people who can sit at any linear arrangement of square tables.

Solution Sketch the arrangements of four tables and five tables. Notice that when you add another table, you seat two more people than in the previous arrangement.

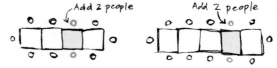

ELL	Support	Advanced
It may help ELL students to develop a number sequence consisting of several terms so that the concepts of common difference and ratio become more evident. Students might be confused by the word *table* being used for both the table to sit at and the data table.	As needed, assign the Refreshing Your Skills: Differences and Ratios lesson. Students could draw several steps of numbers of tables and carefully count the number of tables and students, entering the data into a table.	Use the **One Step** version of the Investigation. This will aid in your assessment of students' prior skills with recursive problems. Focus on clear understanding of the specific notation of recursive rules.

You can put this information into a table, and that reveals a clear pattern. You can continue the pattern to find that 10 tables seat 22 people, and 15 tables are needed for 32 people.

Tables	1	2	3	4	5	6	7	8	9	10	...	15
People	4	6	8	10	12	14	16	18	20	22	...	32

You can also organize the information like this:

number of people at 1 table = 4

number of people at 2 tables = number of people at 1 table + 2

number of people at 3 tables = number of people at 2 tables + 2

If you assume the same pattern continues, then

number of people at 10 tables = number of people at 9 tables + 2.

In general, the pattern is

number of people at n tables = number of people at $(n-1)$ tables + 2.

This rule shows how to use recursion to find the number of people at any number of tables. In recursion, you use the previous value in the pattern to find the next value.

A **sequence** is an ordered list of numbers. The table in Example A represents the sequence

4, 6, 8, 10, 12, ...

Each number in the sequence is called a **term**. The first term, u_1 (pronounced "u sub one"), is 4. The second term, u_2, is 6, and so on.

The nth term, u_n, is called the **general term** of the sequence. A **recursive formula**, the formula that defines a sequence, must specify one (or more) starting terms and a **recursive rule** that defines the nth term in relation to a previous term (or terms).

You generate the sequence 4, 6, 8, 10, 12, ... with this recursive formula:

$u_1 = 4$

$u_n = u_{n-1} + 2$ where $n \geq 2$

> Because the starting value is $u_1 = 4$, the recursive rule $u_n = u_{n-1} + 2$ is first used to find u_2. This is clarified by saying that n must be greater than or equal to 2 to use the recursive rule.

This means *the first term is 4* and *each subsequent term is equal to the previous term plus 2*. Notice that each term, u_n, is defined in relation to the previous term, u_{n-1}. For example, the 10th term relies on the 9th term, or $u_{10} = u_9 + 2$.

Investigate

To introduce the lesson, **ASK** "What's a sequence?" Some students may recall the idea from a previous course. One synonym is *list*. Try to caution against using the word *series* in this context.

Example A

This example reviews the notion of recursion in the context of a sequence of numbers. Consider projecting Example A from your ebook and having students work in pairs. Have students present their strategy as well as their solution. Emphasize the use of correct mathematical terminology, asking probing questions to review the vocabulary of the lesson. For example, if students are using the words *next* and *previous* instead of u_n and u_{n-1}, to ease them into the subscript notation **ASK** "So that would be *u* sub what?" Whether student responses are correct or incorrect, ask other students if they agree and why. **SMP 1, 3, 6**

ASK "What's behind the pattern?" [As a new table is inserted, two new people sitting down, one on each side.] Mathematics isn't only about seeing patterns but also about explaining them. **SMP 3, 6, 8**

ALERT If students are having difficulties with subscript notation, build the connections among the first few terms. **ASK** "What is the next term after u_1?" [u_2] "What is the term right before u_5?" [u_4] "What is the next term after u_n?" [u_{n+1}] "What is the term before u_n?" [u_{n-1}] **SMP 6, 8**

ALERT Students might believe that the terms of a sequence must be in increasing or decreasing order. This is not necessarily true. Another misconception is that there is a simple pattern for each sequence.

ASK "Why is it necessary to include a restriction such as "where $n \geq 2$?" [The goal of the problem is to create a recursive rule that models the given sequence. Thus, in the context of this particular problem, it would not make sense to apply the rule when there are fewer than 2 tables.] **SMP 1, 2, 6, 8**

Modifying the Investigation

Whole Class Draw the table for Step 1 on the board. Then elicit student input and discussion to fill in the information. Discuss Steps 2 and 3. Note that to answer Step 2, you might continue the table to day 18, draw a graph, or use recursion in a spreadsheet or on a calculator. Elicit a variety of approaches.

Shortened Skip Step 2b.

One Step Have students read the investigation problem in the book or project the **One Step** found in your ebook. **ASK** "How many prints will have been delivered to the Little Print Shoppe when FineArt has received twice the number of prints that will still remain to be made?"

Example B

Consider projecting Example B from your ebook and having students work in groups of 2–4. As before, have students present their strategy as well as their solution, continuing to ask questions to emphasize the use of mathematically correct vocabulary. **ASK** "What are the differences between the terms *previous* and u_{n-1} (and, if students worked on calculators, the term *Ans*)?" [They all refer to the same thing.] Whether student responses are correct or incorrect, ask other students if they agree and why. **SMP 1, 3, 6**

Another way to counter notation difficulties is to use home-screen recursion on graphing calculators. See Calculator Note: Recursion. Take some time to explain the connection between stating a starting value and the recursive rule on a graphing calculator— for example, [Ans + 2]. Relate u_{n-1} = (previous answer)(Ans) and u_1 = (starting value). If graphing calculators are new to your students, also give them copies of Calculator Notes: Getting Started, Reentry, and Making Spreadsheets Using the CellSheet App. **SMP 5, 6**

LANGUAGE Even though the term *arithmetic* in *arithmetic sequence* is spelled the same as the noun arithmetic, the adjective is pronounced "ar-ith-'me-tik."

ASK "Why aren't the data points connected with a line?" [You cannot have fractions of a seat.] **SMP 1, 2**

Before moving to the Investigation, ask students to write down their definition of an arithmetic sequence. After several students share their definitions, reach consensus as a class. Project the definition box. Discuss the differences in the definitions from the class and from the book. **SMP 3, 6**

EXAMPLE B

A concert hall has 59 seats in Row 1, 63 seats in Row 2, 67 seats in Row 3, and so on. The concert hall has 35 rows of seats. Write a recursive formula to find the number of seats in each row. How many seats are in Row 4? Which row has 95 seats?

China National Grand Theater (Beijing)

Solution

First, it helps to organize the information in a table.

Row	1	2	3	4	...
Seats	59	63	67		...

Every recursive formula requires a starting term. Here the starting term is 59, the number of seats in Row 1. That is, $u_1 = 59$.

This sequence also appears to have a common difference between successive terms: 63 is 4 more than 59, and 67 is 4 more than 63. Use this information to write the recursive rule for the *n*th term, $u_n = u_{n-1} + 4$.

Therefore, this recursive formula generates the sequence representing the number of seats in each row:

Row	1	2	3	4	...
Seats	59	63	67		...

+4 +4 +4

$$u_1 = 59$$
$$u_n = u_{n-1} + 4 \quad \text{where } n \geq 2$$

You can use this recursive formula to calculate how many seats are in each row.

$u_1 = 59$ — The starting term is 59.
$u_2 = u_1 + 4 = 59 + 4 = 63$ — Substitute 59 for u_1.
$u_3 = u_2 + 4 = 63 + 4 = 67$ — Substitute 63 for u_2.
$u_4 = u_3 + 4 = 67 + 4 = 71$ — Continue using recursion.
⋮

Because $u_4 = 71$, there are 71 seats in Row 4. If you continue the recursion process, you will find that $u_{10} = 95$, or that Row 10 has 95 seats.

Conceptual/Procedural

Conceptual The context of the examples and the investigation help students conceptualize both recursively defined sequences and recursive rules. A comparison of Examples A and C conceptually presents the difference between an arithmetic and a geometric sequence.

Procedural The solutions to the examples and the exercises help students become fluid in the procedure of solving for terms of a sequence.

You can graph the sequence from Example B by plotting (*row, seats*) or, more generally, (n, u_n).

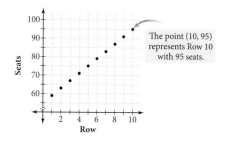

The point (10, 95) represents Row 10 with 95 seats.

In Examples A and B, a constant is added to each term of the sequence to generate the next term. This type of sequence is called an **arithmetic sequence**.

> **Arithmetic Sequence**
>
> An **arithmetic sequence** is a sequence in which each term is equal to the previous term plus a constant. This constant is called the **common difference**. If d is the common difference, the recursive rule for the sequence has the form
>
> $u_n = u_{n-1} + d$

The key to identifying an arithmetic sequence is recognizing the common difference. If you are given a few terms and need to write a recursive formula, first try subtracting consecutive terms. If $u_n - u_{n-1}$ is constant for each pair of terms, then you know your recursive rule must define an arithmetic sequence.

INVESTIGATION

Monitoring Inventory

Art Smith has been providing the prints of an engraving to FineArt Gallery. He plans to make just 2000 more prints. FineArt has already received 470 of Art's prints. The Little Print Shoppe also wishes to order prints. Art agrees to supply FineArt with 40 prints each month and Little Print Shoppe with 10 prints each month until he runs out.

Step 1 As a group, model what happens to the number of unmade prints, the number of prints delivered to FineArt, and the number delivered to Little Print Shoppe in a **spreadsheet** like the one below.

Month	Unmade prints	FineArt	Little Print Shoppe
1	2000	470	0
2	1950	510	10

Step 2 Use your table from Step 1 to answer these questions:

a. How many months will it be until FineArt has an equal number or a greater number of prints than the number of prints left unmade?

b. How many prints will have been delivered to the Little Print Shoppe when FineArt has received twice the number of prints that remain to be made?

Guiding the Investigation

There is a desmos simulation in your ebook. Students do not need any prior knowledge of recursion to do this investigation. To launch the investigation, ASK "What does it mean for a company to *monitor inventory?*"

Step 1 You may want to specify how many months students are to track.

Step 2 Encourage a variety of approaches to these problems.

For example, students might continue the table, use recursion, or draw a graph. Students might use statistics software or a spreadsheet to model the investigation. You could offer students the option of using this onscreen routine to generate the information for the table:

{1,2000,470,0}

{Ans[1] + 1, Ans[2] − 50, Ans[3] + 40, Ans[4] + 10}

See Calculator Note: Recursion for more information about home-screen recursion. SMP 1, 5

Be aware that desmos, which comes with your ebook, does not do recursion.

Choose a few groups to present a variety of approaches to Step 2 of the investigation. Whether student responses are correct or incorrect, ask other students if they agree and why. SMP 1, 3, 6

Step 1

Month	Unmade prints	FineArt	Little Print Shoppe
3	1900	550	20
…	…	…	…
n	$u_{n-1} - 50$	$v_{n-1} + 40$	$w_{n-1} + 10$

The three sequences have terms called u_n, v_n, and w_n.

Step 2a On Month 18, FineArt has 1150 prints, and there are 1150 unmade prints.

Step 2b On Month 27, FineArt has 1510 prints, there are 700 unmade, and Little Print Shoppe has 260. (On Month 26, FineArt had 1470, and there were 750 unmade.)

Example C

LANGUAGE This triangle is sometimes called the *Sierpiński gasket*. If extended for infinitely many steps, it's an example of a fractal, having fractional dimension. There is a desmos simulation for this example in your ebook.

Project the example and have students work in small groups. **ASK** "What would the initial figure, Stage 0, look like? Why?" [Stage 0 would be an equilateral triangle with no triangles inside because no change has occurred.] **SMP 1, 2, 7**

Many different sequences can be studied based on the Sierpiński triangle. **ASK** "If the area of the large triangle is 1 unit, what sequence represents the areas of the red triangles at successive stages?" $\left[\frac{3}{4}, \frac{9}{16}, \frac{27}{64}, \ldots\right]$ "What is the recursive formula for the sequence of areas?" $\left[u_1 = \frac{3}{4}, u_n = \frac{3}{4} u_{n-1} \text{ where } n \geq 2\right]$ "If the length of each side of the large triangle is 1, what sequence represents the perimeter (total distance around all red triangles at a particular stage)?" $\left[\frac{9}{2}, \frac{27}{4}, \frac{81}{8}, \ldots\right]$ "What is the recursive formula for the sequence of perimeters?" $\left[u_1 = \frac{9}{2}, u_n = \frac{3}{2} u_{n-1} \text{ where } n \geq 2\right]$

LANGUAGE *Common ratio* should be defined specifically here. Emphasize that it is the number, which can be a fraction, that each term is multiplied by to get the next. **ASK** "What is the common ratio for the sequence in Example C?" [3] "What would be the common ratio for the sequence if the terms in the series were reversed?" $\left[\frac{1}{3}\right]$

After Example C, ask students to write down their definition of a geometric sequence. After several students share their definitions, project the definition box. Discuss the differences in the definitions from the class and from the book. **SMP 3, 6**

Summarize

Have students present their solutions and explain their work in the Examples and the Investigation. Encourage a variety of approaches, especially making tables by hand or using a calculator or spreadsheet. Students may want to draw graphs or work with formulas. During the discussion, **ASK** "What does the situation in the Investigation have in common with those in the examples?" Bring out the idea of sequences, especially arithmetic versus geometric sequences. Emphasize the use of correct mathematical terminology. Whether student responses are correct or incorrect, ask other students if they agree and why. **SMP 1, 3, 5, 6**

Students may recognize the natural connection between arithmetic sequences and linear functions. The connection between recursive and explicit formulas will be addressed in a later lesson. Students may also see a connection between geometric sequences and exponential functions. If students are impatient with the recursive approach, mention that it is more useful than closed-form formulas for studying growth.

ASK "The arithmetic sequence 5, 2, −1, −4, and so on is decreasing. What's happening?" [We're adding a negative common difference.]

Step 3 Write a short summary of how you modeled the number of prints and how you found the answers to the questions in Step 2. Compare your methods with the methods of other groups. Answers will vary

Career CONNECTION

Economics is the study of how goods and services are produced, distributed, and consumed. Economists in corporations, universities, and government agencies are concerned with the best way to meet human needs with limited resources. Professional economists use mathematics to study and model factors such as supply of resources, manufacturing costs, and selling price.

The sequences in Example A, Example B, and the investigation are arithmetic sequences. Example C introduces a different kind of sequence that is also defined recursively.

EXAMPLE C The geometric pattern below is created recursively. If you continue the pattern endlessly, you create a **fractal** called the Sierpiński triangle. How many red triangles are there at Stage 20?

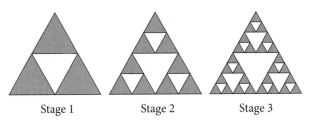

Stage 1   Stage 2   Stage 3

Mathematics CONNECTION

The Sierpiński triangle is named after the Polish mathematician Waclaw Sierpiński (1882–1969). He was most interested in number theory, set theory, and topology, three branches of mathematics that study the relations and properties of numbers, sets, and points, respectively. Sierpiński was highly involved in the development of mathematics in Poland between World War I and World War II. He published 724 papers and 50 books in his lifetime. He introduced his famous triangle pattern in a 1915 paper.

This stamp, part of Poland's 1982 "Mathematicians" series, portrays Waclaw Sierpiński.

Solution

Count the number of red triangles at each stage and write a sequence.

3, 9, 27, ...

The starting term, 3, represents the number of triangles at Stage 1. You can define the starting term as term one, or u_1. In this case, $u_1 = 3$.

Starting with the second term, each term of the sequence is 3 times the previous term, so that 9 is 3 times 3 and 27 is 3 times 9. Use this information to write the recursive rule and complete your recursive formula.

$u_1 = 3$

$u_n = 3 \cdot u_{n-1}$ where $n \geq 2$

Using the recursive rule 19 times, you find that $u_{20} = 3{,}486{,}784{,}401$. There are over 3 billion triangles at Stage 20!

	A	B	
1	n	u_n	
2	1	3	
3	2	9	← B2·3
4	3	27	← B3·3
5	4	81	
6	5	243	
7	6	729	
8	7	2187	
9	8	6561	
10	9	19683	
11	10	59049	
12	11	177147	
13	12	531441	
14	13	1594323	
15	14	4782969	
16	15	14348907	
17	16	43046721	
18	17	129140163	
19	18	387420489	
20	19	1162261467	
21	20	3486784401	← B20·3

In Example C, each term is multiplied by a constant to generate the next term. This type of sequence is called a **geometric sequence**.

> **Geometric Sequence**
>
> A **geometric sequence** is a sequence in which each term is equal to the previous term multiplied by a constant. This constant is called the **common ratio**. If r is the common ratio, the recursive rule for the sequence has the form
>
> $u_n = r \cdot u_{n-1}$

You identify a geometric sequence by dividing consecutive terms. If $\dfrac{u_n}{u_{n-1}}$ has the same value for each pair of terms, then you know the sequence is geometric.

Arithmetic and geometric sequences are the most basic sequences because their recursive rules use only one operation: addition in the case of arithmetic sequences, and multiplication in the case of geometric sequences. Recognizing these basic operations will help you easily identify sequences and write recursive formulas.

Resist the temptation to make a connection to the fact that u_n varies directly with n (with constant of variation u_1), because the emphasis of this lesson is on recursion, with the goal of a deep understanding of rate of change and slope.

CRITICAL QUESTION What are the parts of a recursive formula?

BIG IDEA A recursive formula consists of two parts: the initial value and a recursive rule that tells how to get each term from the previous term.

CRITICAL QUESTION How are arithmetic and geometric sequences and their recursive formulas alike and different?

BIG IDEA They're both sequences; arithmetic sequences are generated recursively by adding a constant, and geometric sequences by multiplying a constant.

CRITICAL QUESTION What generalizations can you make for geometric sequences?

BIG IDEA If the common ratio is between 0 and 1, the sequence is decreasing. If the common ratio is negative, the terms alternate between positive and negative.

CRITICAL QUESTION What is the difference between u_{n-1} and $u_n - 1$? Under what conditions would they be the same?

BIG IDEA u_{n-1} means the previous term, $u_n - 1$ means 1 less than the previous term. They would be the same if the common difference is 1.

Formative Assessment

Student group work and presentations of the Examples, Investigation and Exercises can provide evidence of meeting the objectives of the lesson. Answers and discussion of the Critical Questions also provide anecdotal evidence of understanding. Watch how students understand u_n notation and relate their recursive formulas to the situations and data. Encourage use of appropriate mathematical vocabulary to assess their understanding of the definition of arithmetic sequence (a starting value and a common difference) and a geometric sequence (a starting value and a common and a common ratio).

Apply

Extra Example

1. Write the first five terms of the sequence:
   $u_1 = -3$
   $u_n = u_{n-1} + 4$ where $n \geq 2$
   $-3, 1, 5, 9, 13$

2. Fill in the tables with the missing terms.

   a.

1	2	3	4	...	10	u_n
8	2	-4	-10	...		

   $u_{10} = -46;\ u_n = u_{n-1} - 6$

   b.

0	1	2	3	...	10	u_n
-3	-1	$-\frac{1}{3}$	-10	...		

   $u_{10} = \frac{1}{19{,}683};\ u_n = \frac{1}{3} u_{n-1}$

   c.

1	2	3	4	...	10	u_n
12	18	27		...		

   $u_{10} = \frac{59{,}049}{128};\ u_n = \frac{3}{2} u_{n-1}$

Closing Question

Consider the sequences from the Launch.

a. 4, 8, 12, 16, 20, ...
b. 1, 0.1, 0.01, 0.001, 0.0001

Write a recursive formula to generate each sequence. Then find the eighth term of each. Which sequence is arithmetic and which is geometric? Explain.

a. $u_1 = 4;\ u_n = u_{n-1} + 4;\ u_8 = 32$

b. $u_1 = 1;\ u_n = u_{n-1} \cdot 0.1;\ u_8 = 0.0000001$

Part a is arithmetic, there is a common difference and part b is geometric, there is a common ratio.

Assigning Exercises

Suggested: 1 – 13
Additional Practice: 14 – 17

The exercises emphasize arithmetic sequences. Geometric sequences will receive more attention in later lessons.

2a. (iv) -18, -13.7, -9.4, -5.1; arithmetic; $d = 4.3$
2b. (ii) 47, 44, 41, 38; arithmetic; $d = -3$
2c. (i) 20, 26, 32, 38; arithmetic; $d = 6$
2d. (iii) 32, 48, 72, 108; geometric; $r = 1.5$

1.1 Exercises

Practice Your Skills

1. Match each description of a sequence to its recursive formula.

   a. The first term is –18. Keep adding 4.3. iv
   b. Start with 47. Keep subtracting 3. ii
   c. Start with 20. Keep adding 6. i
   d. The first term is 32. Keep multiplying by 1.5. iii

   i. $u_1 = 20$
      $u_n = u_{n-1} + 6$ where $n \geq 2$

   ii. $u_1 = 47$
       $u_n = u_{n-1} - 3$ where $n \geq 2$

   iii. $u_1 = 32$
        $u_n = 1.5 \cdot u_{n-1}$ where $n \geq 2$

   iv. $u_1 = -18$
       $u_n = u_{n-1} + 4.3$ where $n \geq 2$

2. For each sequence in Exercise 1, write the first 4 terms of the sequence and identify it as arithmetic or geometric. State the common difference or the common ratio for each sequence. @

3. Write a recursive formula and use it to find the missing table values. @

n	1	2	3	4	5	...	9
u_n	40	36.55	33.1	29.65	26.2	...	12.4

   $u_1 = 40$ and $u_n = u_{n-1} - 3.45$ where $n \geq 2$

4. Write a recursive formula to generate an arithmetic sequence with a first term 6 and a common difference 3.2. Find the 10th term. $u_1 = 6$ and $u_n = u_{n-1} + 3.2$ where $n \geq 2$; $u_{10} = 34.8$

5. Write a recursive formula to generate each sequence. Then find the indicated term.

   a. 2, 6, 10, 14, ...   Find the 15th term. $u_1 = 2$ and $u_n = u_{n-1} + 4$ where $n \geq 2$; $u_{15} = 58$
   b. 0.4, 0.04, 0.004, 0.0004, ...   Find the 10th term. $u_1 = 0.4$ and $u_n = 0.1 \cdot u_{n-1}$ where $n \geq 2$; $u_{10} = 0.0000000004$
   c. –2, –8, –14, –20, –26, ...   Find the 30th term. $u_1 = -2$ and $u_n = u_{n-1} - 6$ where $n \geq 2$; $u_{30} = -176$
   d. –6.24, –4.03, –1.82, 0.39, ...   Find the 20th term. @ $u_1 = -6.24$ and $u_n = u_{n-1} + 2.21$ where $n \geq 2$; $u_{20} = 35.75$

History CONNECTION

Hungarian mathematician Rózsa Péter (1905–1977) was the first person to propose the study of recursion in its own right. In an interview she described recursion in this way:

> The Latin technical term "recursion" refers to a certain kind of stepping backwards in the sequence of natural numbers, which necessarily ends after a finite number of steps. With the use of such recursions the values of even the most complicated functions used in number theory can be calculated in a finite number of steps.

In her book *Recursive Functions in Computer Theory*, Péter describes the important connections between recursion and computer languages.

6. Write a recursive formula for the sequence whose first four terms are graphed at right. Find the 46th term. @

   $u_1 = 4$ and $u_n = u_{n-1} + 5$ where $n \geq 2$; $u_{46} = 229$

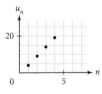

34 CHAPTER 1 Linear Modeling

Exercises 4, 5 Students may use either recursive sequences or home-screen recursion on their calculators to find the terms of these sequences. **SMP 5**

Exercise 6 Students might have difficulty making the transition from single numbers to ordered pairs (and points in a two-dimensional space). To help, **ASK** "What is the first coordinate in each pair?" [1, 2, 3, ...; the term numbers] "The second coordinate?" [4, 9, 14, ...; the terms of the sequence]

Reason and Apply

7. Write a recursive formula that you can use to find the number of segments, u_n, for Figure n of this geometric pattern. Use your formula to complete the table.

$u_1 = 4$ and $u_n = u_{n-1} + 6$ where $n \geq 2$

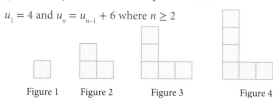

Figure 1 Figure 2 Figure 3 Figure 4

Figure	1	2	3	4	5	...	12	...	32
Segments	4	10	16	22	28	...	70	...	190

8. A 50-gallon (gal) bathtub contains 20 gal of water and is filling at a rate of 2.4 gal/min. You check the tub every minute on the minute.

  a. Suppose that the drain is closed. When will you discover that the water is flowing over the top? @ 13 min

  b. Now suppose that the bathtub contains 20 gal of water and is filling at a rate of 2.4 gal/min, but the drain is open and water drains at a rate of 3.1 gal/min. When will you discover that the tub is empty? 29 min

  c. Write a recursive formula that you can use to find the water level at any minute due to both the rate of filling and the rate of draining.
  $u_0 = 20$ and $u_n = u_{n-1} - 0.7$ where $n \geq 1$

9. A car leaves town heading west at 57 km/h.

  a. How far will the car travel in 7 h? 399 km

  b. A second car leaves town 2 h after the first car, but it is traveling at 72 km/h. To the nearest hour, when will the second car pass the first? 10 h after the first car starts, or 8 h after the second car starts

10. **APPLICATION** Inspector 47 at the Zap battery plant keeps a record of which AA batteries she finds defective. Although the battery numbers at right do not make an exact sequence, she thinks they are close to an arithmetic sequence.

  a. Write a recursive formula for an arithmetic sequence that estimates which batteries are defective. Explain your reasoning.

  b. Predict the numbers of the next five defective batteries.

  c. How many batteries in 100,000 will be defective?

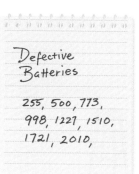

Defective Batteries

255, 500, 773, 998, 1227, 1510, 1721, 2010,

Exercise 9 This exercise does not require that students use recursion. For example, they might use the concept of closing speed. As needed, **ASK** "How can you model this problem recursively?"

10a. The differences are 245, 273, 225, 229, 283, 211, and 289. The average is about 250, so a recursive formula is $u_1 = 250$ and $u_n = u_{n-1} + 250$ where $n \geq 2$.

10b. $u_9 = 2250$, $u_{10} = 2500$, $u_{11} = 2750$, $u_{12} = 3000$, $u_{13} = 3250$

10c. $u_{400} = 100{,}000$. There will be approximately 400 defective batteries in 100,000.

Exercise 11 Students might use the recursive rule $u_n = u_{n-1} - 0.25u_{n-1}$, or, equivalently, $u_n = 0.75u_{n-1}$.

12. David is correct. A common difference of −8 means add −8, or subtract 8, to get the next term. Start with 100, then keep subtracting 8 to get each subsequent term.

Exercise 13 ALERT Students may have difficulty finding the common difference, because they must divide the difference, 16, by 5 (not 4 or 6). As needed, remind students that the difference is *new − previous*.

13a. 3.2; $\frac{51-35}{5} = 3.2$

13b. 25.4, 28.6, 31.8, . . ., 38.2, 41.4, 44.6, 47.8, . . ., 54.2

ADVANCED This is an excellent extension to generating sequences given recursive routines. Students could describe their process for determining the common difference between terms and the starting value of the sequence.

Exercise 14 ELL Using a motion sensor and having students act out exercises like Exercise 14 will help develop a strong, conceptual understanding by modeling time-distance graphs.

14a.

Elapsed time (s)	Distance from motion sensor (m)
0.0	2.0
1.0	3.0
2.0	4.0
3.0	5.0
4.0	4.5
5.0	4.0
6.0	3.5
7.0	3.0

---

11. The week of February 14, the owner of Nickel's Appliances stocks hundreds of red, heart-shaped vacuum cleaners. The next week, he still has hundreds of red, heart-shaped vacuum cleaners. He tells the manager, "Discount the price 25 percent each week until they are gone."

a. On February 14, the vacuums are priced at $80. What is the price of a vacuum during the second week? @ $60

b. What is the price during the fourth week? $33.75

c. When will a vacuum sell for less than $10?
during the ninth week

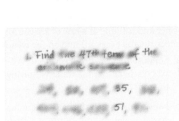

12. Carl and David's teacher asked them to write the first four terms of a sequence that starts with 100 and has a common difference of −8. Carl says the first four terms are 100, 108, 116, 112. David says the first four terms are 100, 92, 86, 78. Who is correct? Write a clear explanation of how to use the common difference to build the sequence.

13. Taoufik picks up his homework paper from the puddle it fell in. Sadly he reads the first problem and finds that the arithmetic sequence is a blur except for two terms.

a. What is the common difference? How did you find it? ⓗ

b. What are the missing terms?

c. What is the answer to Taoufik's homework problem?
172.6

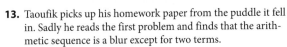

Review

14. Ayaunna starts 2.0 m from a motion sensor. She walks away from the sensor at a rate of 1.0 m/s for 3.0 s and then walks toward the sensor at a rate of 0.5 m/s for 4.0 s.

a. Create a table of values for Ayaunna's distance from the motion sensor at 1-second intervals. @

b. Sketch a time-distance graph of Ayaunna's walk.

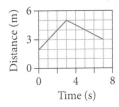

15. Find the area of this triangle using two different strategies. Describe your strategies.

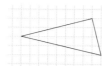

17 square units. Sample answer: Enclose the triangle in a rectangle, and subtract the areas of the right triangles on the outside; or label the triangle ABC with $\angle B$ the largest, and notice that $AB \perp BC$, find their lengths using the distance formula, and use the formula $A = \frac{1}{2}(AB)(BC)$

16. **APPLICATION** Sherez is currently earning $390 per week as a store clerk and part-time manager. She is offered either a 7% increase or an additional $25 per week. Which offer should she accept? the 7% offer sat $417.30 per week

17. The cost of filling your car's gas tank varies directly with the number of gallons of gas you put in. It costs $41.88 to put 12 gallons in the gas tank.

   a. How much does it cost to put 1 gallon of gas in the tank? This is the constant of variation for this relationship. $3.49

   b. Write the equation for the direct variation between c, the cost of filling the tank, and g, the number of gallons put in the tank. $c = 3.49 g$

   c. Use your direct variation equation to find the cost of filling a tank with 22 gallons of gas. $76.78

IMPROVING YOUR Reasoning SKILLS

Fibonacci and the Rabbits

Suppose a pair of newborn rabbits, one male and one female, is put in a field. Assume that rabbits are able to mate at the age of one month, so at the end of its second month a female can produce another pair of rabbits. Suppose that each female who is old enough produces one new pair of rabbits (one male, one female) every month and that none of the rabbits die. Write the first few terms of a sequence that shows how many pairs there will be at the end of each month. Then write a recursive formula for the sequence.

This sequence is called the **Fibonacci sequence** after Italian mathematician Leonardo Fibonacci (ca. 1170–1240), who included a similar problem in his book *Liber abaci* (1202). How is the Fibonacci sequence unique compared with the other sequences you have studied?

In winter, the arctic hare, or polar rabbit, sports a brilliant white coat that provides excellent camouflage in the ice and snow of the North American tundra.

IMPROVING YOUR Reasoning SKILLS

1, 1, 2, 3, 5, 8, 13, 21, 34, 55, 89, 144, . . . ; $u_1 = 1$, $u_2 = 1$, and $u_n = u_{n-1} + u_{n-2}$ where $n \geq 3$. Thus the recursive rule for the Fibonacci sequence is defined by two preceding terms.

Rabbits do actually have a gestation period (from conception, or *breeding*, to birth, or *kindling*) of one month. However, in real life, rabbits usually have to be six to nine months old before breeding, and they usually are not able to breed again immediately after kindling. The Fibonacci sequence, which is related to the golden ratio, has many applications in biology. Fibonacci is credited with informing Europeans about the number system developed by the Hindus and Arabs.

LESSON 1.2

Modeling Growth and Decay

This lesson gives more experience with geometric sequences, both decreasing and increasing.

COMMON CORE STATE STANDARDS
APPLIED	DEVELOPED	INTRODUCED
	F.IF.3	
	F.BF.2	
	F.LE.1c	
	F.LE.2	

Objectives
- Discover applications involving geometric sequences
- Use geometric sequences to model growth and decay situations
- Understand the physical limitations of models

Vocabulary
decay
growth
principal
simple interest
compound interest

Materials
- balls (racquetballs, basketballs, or handballs work well)
- video camera
- paper and colored markers
- motion sensors or meter sticks, *optional*
- Calculator Notes: Looking for the Rebound Using the EasyData App; Entering Data; Plotting Data; Tracing Data Plots; Sharing Data

Launch

Which of the following can be modeled as growth and which can be modeled as decay? Explain.

i. The temperature of hot chocolate
ii. Food left in a locker
iii. Amount of money in a savings account
iv. Value of a car
v. Height of a bouncing ball
vi. Population of a city

Each sequence you generated in the previous lesson was either an arithmetic sequence with a recursive rule in the form $u_n = u_{n-1} + d$ or a geometric sequence with a recursive rule in the form $u_n = r \cdot u_{n-1}$. You compared consecutive terms to decide whether the sequence had a common difference or a common ratio.

In most cases you have used u_1 as the starting term of each sequence. In some situations, it is more meaningful to treat the starting term as a zero term, or u_0. The zero term represents the starting value before any change occurs. You can decide whether it would be better to begin at u_0 or u_1.

EXAMPLE A An automobile depreciates, or loses value, as it gets older. Suppose that a particular automobile loses one-fifth of its value each year. Write a recursive formula to find the value of this car when it is 6 years old, if it cost $23,999 when it was new.

Consumer CONNECTION
The *Kelley Blue Book*, first compiled in 1926 by Les Kelley, annually publishes standard values of every vehicle on the market. Many people who want to know the value of an automobile will ask what its "Blue Book" value is. The *Kelley Blue Book* calculates the value of a car by accounting for its make, model, year, mileage, location, and condition.

Solution Each year, the car will be worth $\frac{4}{5}$ of what it was worth the previous year. Therefore the sequence has a common ratio, which makes it a geometric sequence. It is convenient to start with $u_0 = 23,999$ to represent the value of the car when it was new so that u_1 will represent the value after 1 year, and so on. The recursive formula that generates the sequence of annual values is

$u_0 = 23,999$ Starting value.
$u_n = 0.8 \cdot u_{n-1}$ where $n \geq 1$ $\frac{4}{5}$ is 0.8.

Use this rule to find the 6th term.
After 6 years, the car is worth $6,291.19.

	A	B
	n	u_n
2	0	23999
3	1	19199.2
4	2	15359.36
5	3	12287.488
6	4	9829.9904
7	5	7863.99232
8	6	6291.193856

$fx = 0.8*B7$

In situations like the problem in Example A, it's easier to write a recursive formula than an equation using x and y.

ELL
You might start by asking whether the price of a car should go up, down, or stay the same as it gets older and why. What if the car is a classic sports car in mint condition? Link this to growth and decay. Check reading and verbal comprehension of terms like growth, decay, rebound, etc.

Support
Spend time on the Launch, discussing when each situation might be an example of growth and when it might be an example of decay. Link the specific notation of recursive rules (u_n notation) with the verbal description of a sequence by using starting value and rule. Use the **Whole Class** version of the Investigation.

Advanced
Focus on clear understanding of the specific notation of recursive rules. As an extension or challenge, you might explore finding a rule based on u_0 and u_2 or u_1 and u_4. Use the **One Step** version of the Investigation.

INVESTIGATION

Looking for the Rebound

YOU WILL NEED
- a ball
- a video camera
- paper to make ruler
- color markers

When you drop a ball, the rebound height becomes smaller after each bounce. In this investigation you will write a recursive formula for the height of a real ball as it bounces.

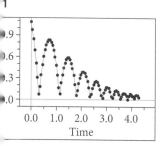

Procedure Note

Collecting Data
1. Create a color ruler by marking each centimeter with rotating colors (Use at least three colors), clearly mark each 10 cm. Attach the ruler to a wall.
2. Sit or kneel with video camera at least 3 meters (10 feet) from the wall.
3. Drop the ball as close to the wall as you can. Record the initial drop and the first few bounces (approximately 5 seconds).

Bounce Number	Rebound Height (m)
0	1.081
1	0.830
2	0.578
3	0.377
4	0.245
5	0.166
6	0.119
7	0.084
8	0.064
9	0.059

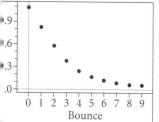

Step 1 Scroll through the video to record the initial height and subsequent heights when the ball reaches the top of each bounce. You will not be able to read the numbers on the ruler but should be able to use the colors to calculate the height.

Step 2 Transfer the data to your calculator in the form (x, y), where x is the time since the ball was dropped, and y is the height of the ball. Trace the data graphed by your calculator to find the starting height and the rebound height after each bounce. Record your data in a table.

Step 3 Graph a scatter plot of points in the form (*bounce number, rebound height*). Record the graphing window you use.

Step 4 Compute the rebound ratio for consecutive bounces.

$$\text{rebound ratio} = \frac{\text{rebound height}}{\text{previous rebound height}}$$

sample answer: 0.77, 0.70, 0.65, …

Step 5 Decide on a single value that best represents the rebound ratio for your ball. Use this ratio to write a recursive formula that models your sequence of *rebound height* data, and use it to generate the first six terms.

Step 6 Compare your experimental data to the terms generated by your recursive formula. Answers will vary.

a. How close are they?

b. Describe some of the factors that might affect this experiment. (For examples how might the formula change if you used a different kind of ball.)

c. According to the recursive formula does the ball ever stop bouncing?

d. Realistically, how many bounces do you think there were before the ball stopped?

1.00 and $u_n = 0.64 \cdot u_{n-1}$ where
; 1.00, 0.64, 0.41, 0.26, 0.17, 0.11

LESSON 1.2 Modeling Growth and Decay **39**

Modifying the Investigation

Whole Class Have two students collect data in front of the class. Complete Steps 3–5 with student input. Discuss Step 6.

Shortened Use the sample data. Have students complete Steps 3–5. Discuss Step 6.

One Step Give the instructions for the investigation and pose this problem: "What is a recursive formula for the height of the ball at the top of each bounce?"

After students gather data, some will look for an additive formula (an arithmetic sequence) and others for a multiplicative formula (a geometric sequence). Have students discuss which models are best, based on differences between predicted and actual data. **ASK** "Does your model have a limit on the number of bounces before the ball stops? How realistic is that?"

Answers could vary. The description will dictate the answer. For example, the population of a city could be either, depending on whether people are moving into or out of the city.

Investigate

Example A

Consider projecting the example from your ebook and having students work in pairs. Have students share their strategies along with their solutions. Emphasize the use of correct mathematical terminology and symbols. Whether student responses are correct or incorrect, ask other students if they agree and why. **SMP 1, 3, 6**

ASK "Why can we find a 20% depreciation by multiplying by $\frac{4}{5}$?" If necessary, suggest that they think about specific numbers, such as $100. Point out that Example A seems to be claiming that $23{,}999 - (0.2)23{,}999 = 23{,}999(0.8)$. **ASK** "Is that reasoning justified? Explain." If needed, suggest they factor the equation. You might generalize to $u_{n-1} - (0.2)u_{n-1} = u_{n-1}(1-0.2) = u_{n-1}(0.8)$ as you review factoring. **SMP 7**

ASK "According to your rule, how low could the value of the car go? Realistically, what is the lowest value of the car?"

Guiding the Investigation

You can use the Investigation Worksheet Looking for the Rebound with Sample Data if you do not wish to conduct the investigation as an activity. There is also a desmos simulation of this investigation in the ebook. If you have motion sensors, students can collect data with the motion sensor. An Investigation Worksheet is available for use with motion sensors.

Introduce the investigation by demonstrating the bounce measurement process. If you are using small balls, balls without seams, such as racquetballs, will work best.

ALERT If the camera is too close to the ball or the ball is too far from the wall you get a bad line of sight any time the ball is above or below the camera.

LESSON 1.2 Modeling Growth and Decay **39**

Step 1 You can't really read measurements marked on the wall but, if your background is distinct, you can measure the peak heights by slowly scrolling through the video. Try to have an initial height of about 2 meters.

To minimize the parallax error from the camera angle you want the camera at the height of the ball when it is at the top of the bounce. As adjusting the camera height is difficult, you want the camera to be far away from the ball and the ball close to the ruler. You will have to translate the colors into numbers as you will not be able to read the ruler in the video.

Step 4 Let students decide what amount of error in the data is acceptable.

Have groups present their work, choosing presentations to include a variety of decimal places in the results, and question which number of decimals is most appropriate. **SMP 3, 5, 6**

ASK "Could you use the first or second rebound height as u_0?" [Yes, either; starting the sequence from the second height would produce an identical set of subsequent bounces.]

Example B

Consider projecting the example from your ebook and having students work in pairs. Have them share their solutions. Emphasize the use of correct mathematical terminology and symbols. Whether student responses are correct or incorrect, ask other students if they agree and why. **SMP 1, 3, 6**

Summarize

Have students present their solutions and explain their work in the Examples and the Investigation. Encourage a variety of approaches. Students may want to draw graphs or work with formulas. During the discussion, **ASK** "What kind of sequences did you see in Examples A and B and in the investigation?" [geometric sequences] Emphasize the use of correct mathematical terminology. Whether student responses are correct or incorrect, ask other students if they agree and why. **SMP 1, 3, 5, 6**

You may find it easier to think of the common ratio as the whole, 1, plus or minus a percent change. In place of r you can write $(1 + p)$ or $(1 - p)$. The car example involved a 20% (one-fifth) loss, so the common ratio could be written as $(1 - 0.20)$. Your bouncing ball may have had a common ratio of 0.75, which you can write as $(1 - 0.25)$ or a 25% loss per bounce.

In Example A, the value of the car decreased each year. Similarly, the rebound height of the ball decreased with each bounce. These and other decreasing geometric sequences are examples of **decay**. These examples of real world decay can be modeled with a geometric recursive rule but every model has error or variation in the application of the model. The value of the car will never be zero and the ball does not keep bouncing forever, yet the mathematical model tells us that the value of the car will keep going down and that the ball will still be bouncing after dozens of bounces. So, we can use models to understand and predict behavior, but every value from the sequence is only an estimate of the true value.

The next example is one of **growth**, or an increasing geometric sequence. Interest is a charge that you pay to a lender for borrowing money or that a bank pays you for letting it invest the money you keep in your bank account. **Simple interest** is a percentage paid on the **principal**, or initial balance, over a period of time. If you leave the interest in the account, then in the next time period you will receive interest on both the principal and the interest that were in your account. This is called **compound interest** because you are receiving interest on the interest.

EXAMPLE B Gloria deposits $2,000 into a bank account that pays 7% annual interest compounded annually. This means the bank pays her 7% of her account balance as interest at the end of each year, and she leaves the original amount and the interest in the account. When will the original deposit double in value?

Solution The balance starts at $2,000 and increases by 7% each year.

$u_0 = 2000$

$u_n = u_{n-1} + 0.07 \cdot u_{n-1}$ where $n \geq 1$ The recursive rule that represents 7% growth.

$u_n = (1 + 0.07) u_{n-1}$ where $n \geq 1$ Factor.

Use technology, such as a spreadsheet or calculator, to compute year-end balances recursively.

Term u_{11} is 4209.70, so the investment balance will more than double in 11 years.

fx	= 1.07*B2
	A	B
1	n	u_n
2	0	$2,000.00
3	1	$2,140.00
4	2	$2,289.80
5	3	$2,450.09

fx	= 1.07*B12
	A	B
1	n	u_n
11	9	$3,676.92
12	10	$3,934.30
13	11	$4,209.70
14	12	$4,504.38

ADVANCED You might discuss the difference between *nominal* and *effective interest rates*. If a 6.5% annual interest rate is compounded monthly, 6.5% is the nominal interest rate and $\left(1 + \frac{0.065}{12}\right)^{12} - 1$ is the effective annual interest rate.

Conceptual/Procedural

Conceptual The real world examples and Investigation situations help students conceptualize increasing and decreasing geometric sequences. Additionally, adding the component of looking at the limits of a mathematical model compared to the physical world makes analyzing and evaluating the model more conceptual.

Procedural Students practice the procedure of writing and evaluating geometric sequences.

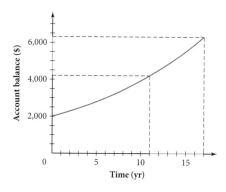

Compound interest has many applications in everyday life. The interest on both savings and loans is almost always compounded, often leading to surprising results. This graph and spreadsheet show the account balance in Example B.

fx	=B18*(1+0.07)	
	A	B
1	n	u_n
16	14	5157.0683
17	15	5518.063081
18	16	5904.327497
19	17	6317.630422

Leaving just $2,000 in the bank at a good interest rate for 11 years can double your money. In another 6 years, the $2,000 will have tripled.

Some banks will compound the interest monthly. You can write the common ratio as $\left(1 + \frac{0.07}{12}\right)$ to represent one-twelfth of the annual interest, compounding monthly. When you do this, n represents months instead of years. How would you change the rule to show that the interest is compounded 52 times per year? What would n represent in this situation?

$\left(1 + \frac{0.07}{52}\right)$; n would represent the number of weeks

1.2 Exercises

You will need a graphing calculator for Exercise 18.

Practice Your Skills

1. Find the common ratio for each sequence, and identify the sequence as growth or decay. Give the percent change for each.
   a. 100, 150, 225, 337.5, 506.25, ... @
      1.5; growth; 50% increase
   b. 73.4375, 29.375, 11.75, 4.7, 1.88, ...
      0.4; decay; 60% decrease
   c. 80.00, 82.40, 84.87, 87.42, 90.04, ...
      1.03; growth; 3% increase
   d. 208.00, 191.36, 176.05, 161.97, ...
      0.92; decay; 8% decrease

2. Write a recursive formula for each sequence in Exercise 1. Use u_0 for the first term and find u_{10}. @

3. Write each sequence or formula as described.
   a. Write the first four terms of the sequence that begins with 2000 and has the common ratio 1.05. @
      2000, 2100, 2205, 2315.25
   b. Write the first four terms of the sequence that begins with 5000 and decays 15% with each term. What is the common ratio?
      5000, 4250, 3612.5, 3070.625; common ratio = 0.85
   c. Write a recursive formula for the sequence that begins 1250, 1350, 1458, 1574.64, ...
      $a_0 = 1250$, $a_n = a_{n-1}(1 + 0.08)$ where $n \geq 1$

Films quickly display a sequence of photographs, creating an illusion of motion

LESSON 1.2 Modeling Growth and Decay 41

CRITICAL QUESTION How do the geometric sequences in the Investigation, Example A, Example B compare with one another?

BIG IDEA The automobile depreciation in Example A and the bouncing ball in the investigation give a decreasing geometric sequence, described as *decay*. The bank account in Example B shows an increasing geometric sequence, described as *growth*.

CRITICAL QUESTION How would you describe the differences between the recursive formulas for geometric growth and decay sequences?

BIG IDEA In growth, the multiplier is greater than 1; in decay, it's less than 1.

Formative Assessment

As students work and present, assess their understanding of geometric sequences and common ratios, as well as their ability to calculate ratios. Observe how students relate situations to notation. Do they have a good understanding of the meaning of u_{n-1} context? Can they represent a loss of 25%?

Apply

Extra Example

1. You buy a pair of limited edition shoes, then immediately sell them on an online auction site. The bidding starts at $100 and each bid pushes the price up by 15%. Make a table. If the 10th bidder purchases the shoes, how much does that person pay? $351.79

2. Rewrite the expression $u_{n-1} + 2u_{n-1}$ so that the variable appears only once. $3u_{n-1}$

Closing Question

Write a recursive formula for the height of a ball that is dropped from 150 cm and has a 60% rebound ratio.
$u_0 = 150$; $u_n = 0.60 \cdot u_{n-1}$ where $n \geq 1$

Assigning Exercises

Suggested: 1 – 11, 14 – 16
Additional Practice: 12, 13, 17 – 21

Exercise 1 **ALERT** Students might offer the common ratio as the percent change.

2a. $u_0 = 100$ and $u_n = 1.5u_{n-1}$ where $n \geq 1$; $u_{10} \approx 5766.5$

2b. $u_0 = 73.4375$ and $u_n = 0.4u_{n-1}$ where $n \geq 1$; $u_{10} \approx 0.0077$

2c. $u_0 = 80.00$ and $u_n = 1.03u_{n-1}$ where $n \geq 1$; $u_{10} \approx 107.513$

2d. $u_0 = 208.00$ and $u_n = .92u_{n-1}$ where $n \geq 1$; $u_{10} \approx 90.35$

LESSON 1.2 Modeling Growth and Decay 41

Exercises 6, 7 These exercises are related to the investigation. Students who missed the investigation may need assistance visualizing the graph of the data.

6d. Yes, when watching a ball bounce from 100 inches the ball stops moving before it has bounced between 20 to 30 times.

Exercises 7, 8 If students have difficulty understanding the recursive formulas because they are printed on one line, suggest that they write them out as they've seen them before. **ELL** Having students describe the real-world meanings for these exercises will give them the chance to practice their vocabulary and will also serve as a checkpoint for comprehension.

7. 100 is the initial height, but the units are unknown. 0.20 is the percent loss, so the ball loses 20% of its height each rebound.

Exercise 8 Because n is related to the year 20, students will probably conclude that the interest rate is annual. Be open to multiple interpretations: 0.025 could be $\frac{1}{12}$ of 30% compounded monthly or $\frac{1}{4}$ of 10% compounded quarterly. With these two options n would need to change to represent months or quarters.

Exercise 9 **ALERT** Discourage students from answering with fractions of people.

---

4. Match each recursive rule to a graph. Explain your reasoning.

a. $u_0 = 10$ ii. decay
   $u_n = (1 - 0.25) \cdot u_{n-1}$ where $n \geq 1$ @

b. $u_0 = 10$ i. growth
   $u_n = (1 + 0.25) \cdot u_{n-1}$ where $n \geq 1$

c. $u_0 = 10$ iii. constant
   $u_n = 1 \cdot u_{n-1}$ where $n \geq 1$

i. 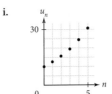   ii. 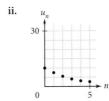   iii.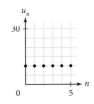

5. Factor these expressions so that the variable appears only once. For example, $x + 0.05x$ factors into $x(1 + 0.05)$.

a. $x + Ax$ @   b. $A - 0.18A$ @   c. $x + 0.08125x$ d. $2u_{n-1} - 0.85u_{n-1}$
   $x(1 + A)$ $(1 - 0.18)A$, or $0.82A$ $(1 + 0.08125)x$, or $1.08125x$ $(2 - 0.85)u_{n-1}$, or $1.15u_{n-1}$

Reason and Apply

6. Suppose the initial height from which a rubber ball drops is 100 in. The rebound heights to the nearest inch are 80, 64, 51, 41, ...

a. What is the rebound ratio for this ball? (h)   0.8

b. What is the height of the tenth rebound?   11 in.

c. After how many bounces will the ball rebound less than 1 in.? Less than 0.1 in.?   21 bounces; 31 bounces

d. Is there reason to suspect that these last two estimates are not correct? Explain.

7. Suppose the recursive formula $u_0 = 100$ and $u_n = (1 - 0.20)u_{n-1}$ where $n \geq 1$ models a bouncing ball. Give real-world meanings for the numbers 100 and 0.20.

8. Suppose the recursive formula $u_{2015} = 250{,}000$ and $u_n = (1 + 0.025)u_{n-1}$ where $n \geq 2016$ describes an investment made in the year 2015. Give real-world meanings for the numbers 250,000 and 0.025, and find u_{2019}. @ $250,000 was invested at 2.5% annual interest in 2015. $u_{2019} = \$27{,}595.32$

9. **APPLICATION** A company with 12 employees is growing at a rate of 20% per year. It will need to hire more employees to keep up with the growth, assuming its business keeps growing at the same rate.

a. How many people should the company plan to hire in each of the next 5 years?
   number of new hires for next 5 years: 2, 3, 3 (or 4), 4, and 5

b. How many employees will it have 5 years from now?
   about 30 employees

10. **APPLICATION** The table below shows investment balances over time.

Elapsed time (yr)	0	1	2	3	. . .
Balance ($)	2,000	2,170	2,354.45	2,554.58	. . .

a. Write a recursive formula that generates the balances in the table. @ $u_0 = 2000$; $u_n = (1 + 0.085)u_{n-1}$ where $n \geq 1$

b. What is the annual interest rate?   8.5%

c. How many years will it take before the original deposit triples in value?   14 years ($6,266.81)

42   **CHAPTER 1** Linear Modeling

11. **APPLICATION** Suppose you deposit $500 into an account that earns 6% annual interest. You don't withdraw or deposit any additional money for 3 years.
   a. If the interest is paid once per year, what will the balance be after 3 years? ⓐ $595.51
   b. If the interest is paid every six months, what will the balance be after 3 years? This is also referred to as 6% compounded semiannually. Divide the annual interest rate by 2 to find the semiannual interest rate. $597.03
   c. What will the balance be if you receive 6% compounded *quarterly* for 3 years? ⓗ $597.81
   d. What will the balance be if you receive 6% compounded *monthly* for 3 years? $598.34

12. **APPLICATION** Suppose $500 is deposited into an account that earns 6.5% annual interest and no more deposits or withdrawals are made.
   a. If the interest is compounded monthly, what is the monthly rate? $\frac{6.5}{12} \approx 0.542\%$
   b. What is the balance after 1 month? $502.71
   c. What is the balance after 1 year? $533.49
   d. What is the balance after 29 months? $584.80

13. Suppose Jill's biological family tree looks like the diagram at right. You can model recursively the number of people in each generation.
   a. Make a table showing the number of Jill's ancestors in each of the past five generations. Use u_0 to represent Jill's generation.
   b. Look in your table at the sequence of the number of ancestors. Describe how to find u_n if you know u_{n-1}. Write a recursive formula.
   c. Find the number of the term of this sequence that is closest to 1 billion. What is the real-world meaning of this answer?
   d. If a new generation is born every 25 years, approximately when did Jill have 1 billion living ancestors in the same generation? 750 years ago
   e. Your answer to 13c assumes there are no duplicates, that is, no common ancestors on Jill's mom's and Jill's dad's sides of the family. Look up Earth's population for the year you found in 13d. Describe any problems you notice with the assumption of no common ancestors.
   The population of the planet at the time was less than 1 billion. Jill must have some common ancestors.

→ Cultural
CONNECTION

Family trees are lists of family descendants and are used in the practice of genealogy. People who research genealogy may want to trace their family's medical history or national origin, discover important dates, or simply enjoy it as a hobby. Author Alex Haley (1921–1992), honored here in this statue in Annapolis, Maryland, told the powerful history of his family's prolonged slavery and decades of discrimination in his 1976 genealogical book, *Roots: The Saga of an American Family*. The book, along with the 1977 television miniseries, inspired many people to trace their family lineage.

13b. Start with 1 and recursively multiply by 2; $u_0 = 1$ and $u_n = 2u_{n-1}$ where $n \geq 1$.
13c. u_{30}; 30 generations ago, Jill had 1 billion living ancestors

13a.

Generations back n	0	1	2	3	4	5
Ancestors in the generation u_n	1	2	4	8	16	32

14. $u_0 = 1$ and $u_n = 0.8855u_{n-1}$ where $n \geq 1$; $u_{25} = 0.048$, or 4.8%. It would take about 25,000 years to reduce to 5%.

Context **Guitar Feedback** Some students will recognize feedback as the loud whine a microphone sometimes makes. Explain that this occurs when the output of the speakers becomes the input to the microphone. The feedback is the result of the system being unable to handle this recursive process.

14. **APPLICATION** Carbon dating is used to find the age of ancient remains of once-living things. Carbon-14 is found naturally in all living things, and it decays slowly after death. About 11.45% of it decays in each 1000-year period of time.

    Let 100%, or 1, be the beginning amount of carbon-14. At what point will less than 5% remain? Write the recursive formula you used.

15. **APPLICATION** Between 1980 and 2010, the population of Grand Traverse County in Michigan grew from 54,899 to 86,999.

    a. Find the actual increase and the percent increase over the 30-year period. @ 32,100 and 58.47%

    b. If you average that total and rate over 30 years, what is the change per year? Find the average rate of change and the average percent rate of change. 1,070 people and 1.95%

    c. Since the growth is not linear, the actual change each year is different. However, overall it averages out to the value you found in 15b. What happens if you use the average percent rate you found in 15b starting in 1980 for 30 years? $u_{1980} = 54,899$ and $u_n = (1 + 0.0195)u_{n-1}$ where $n \geq 1981$ yields 97,990.

    d. Try this again with a rate of 1.55% for 30 years. 87,089 people – much closer

    e. Using 1.55%, find the population estimates for 1990 and 2000. $u_{1990} = 64,027$, $u_{2000} = 74,673$

    f. Using 1980 and the population estimates from 15e, what is the average rate of change in people per year between 1980 and 1990? 1980 and 2000? 1990 and 2000? 912; 989; 1064 people per year

    g. Without computing, which do you think is larger:
       i. The average rate of change from 2000 to 2010, or
       ii. the average rate of change between 1990 and 2010?
       Explain your thinking.

    ii. While the % rate is fixed the number of people added increases each year so the average from the second ten years will be higher than the average over the whole 20 years.

16. Taoufik looks at the second problem of his wet homework that fell in a puddle.

    a. What is the common ratio? How did you find it?
    b. What are the missing terms?
    c. What is the answer he needs to find?

Anthropologist carefully revealing human remains at an ancient burial site.

16a. 3; $\frac{162}{18} = 9$, $3^2 = 9$
16b. 2, 6, …, 54, …, 486, 1458, …, 13122
16c. 118,098

Guitar feedback is a real-world example of recursion. When the amplifier is turned up loud enough, the sound is picked up by the guitar and amplified again and again, creating a feedback loop. Jimi Hendrix (1942–1970), a pioneer in the use of feedback and distortion in rock music, remains one of the most legendary guitar players of the 1960s.

Review

17. The population of the United States grew 9.70% from 2000 to 2010. The population reported in the 2010 census was 308.7 million. What population was reported in 2000? Explain how you found this number. 281.4 million; $(1 + 0.097)x = 308.7$

18. An elevator travels at a nearly constant speed from the ground level to an observation deck at 160 m. This trip takes 40 s. The elevator's trip back down is also at this same constant speed.

  a. What is the elevator's speed in meters per second? 4 m/s

  b. How long does it take the elevator to reach the restaurants, located 40 m above ground level? @ 10 s

  c. Graph the height of the elevator as it moves from ground level to the observation deck.

  d. Graph the height of the elevator as it moves from the restaurant level, at 40 m, to the observation deck.

  e. Graph the height of the elevator as it moves from the deck to ground level.

19. Consider the sequence 180, 173, 166, 159, ...

  a. Write a recursive formula. Use $u_1 = 180$. $u_1 = 180$ and $u_n = u_{n-1} - 7$ where $n \geq 2$

  b. What is u_{10}? $u_{10} = 117$

  c. What is the first term with a negative value? $u_{27} = -2$

20. Solve each equation.

  a. $-151.7 + 3.5x = 0$ @ $x \approx 43.34$

  b. $0.88x + 599.72 = 0$ $x = -681.5$

  c. $18.75x - 16 = 0$ $x \approx 0.853$

  d. $0.5 \cdot 16 + x = 16$ $x = 8$

21. For the equation $y = 47 + 8x$, find the value of y when

  a. $x = 0$ $y = 47$

  b. $x = 1$ @ $y = 55$

  c. $x = 5$ $y = 87$

  d. $x = -8$ $y = -17$

The CN Tower in Toronto is one of Canada's landmark structures and one of the world's tallest buildings. Built in 1976, it has six glass-fronted elevators that allow you to view the landscape as you rise above it at 15 mi/h. At 1136 ft, you can either brace against the wind on the outdoor observation deck or test your nerves by walking across a 256 ft² glass floor with a view straight down.

18c.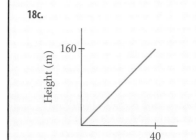

18d.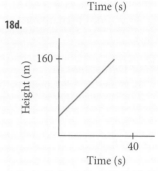

18e.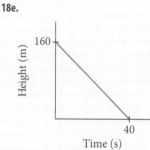

LESSON 1.3

To introduce the lesson, you might
ASK "Do sequences always end?"
Encourage discussion. This lesson
explores infinite sequences, as well as
the physical limitations of mathematical models.

COMMON CORE STATE STANDARDS		
APPLIED	**DEVELOPED**	**INTRODUCED**
F.LE.5	F.IF.3	F.IF.7e
	F.IF.4	
	F.IF.5	

Objectives

- Examine limitations of models
- Investigate shifted geometric sequences

Vocabulary

limit
shifted geometric sequence

Materials

- bowls
- water
- tinted liquid
- measuring cups (with milliliters)
- sink or waste buckets
- Calculator Notes: Creating Sequences; Graphing Sequences, *optional*

Launch

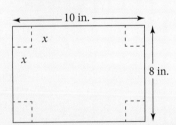

If you were going to make an open-top box by cutting equal sized squares from the corners of an 8 in. by 10 in. piece of cardboard, what are all the sizes of squares that you could cut? Explain.

$0 < x < 4$ in.; If $x \leq 0$ in. or $x \geq 4$ in., one dimension won't exist.

LESSON 1.3

Applications and Other Sequences

A mathematician is a machine for turning coffee into theorems.

PAUL ERDÖS

You have seen applications of arithmetic and geometric sequences, but many real-world changes do not exactly have equal differences or equal ratios. Can a tree continue to grow larger year after year? Can people continue to build taller buildings, run faster, and jump higher, or is there a limit to any of these?

The women's world record for the fastest time in the 100 m dash has decreased by about 3 s in 66 yr. Marie Mejzlíková (Czechoslovakia) set the record at 13.6 s in 1922, and Florence Griffith-Joyner (USA), set the current record at 10.49 s in 1988. In the 1998 article "How Good Can We Get?" Jonas Mureika predicts that the ultimate performance for a woman in the 100 m dash will be 10.15 s. The winning time for the 2012 Olympics was 10.75 s.

How large can a tree grow? It depends partly on environmental factors such as disease and climate. Trees have mechanisms that slow their growth as they age, similar to human growth. (Unlike humans, however, a tree may not reach its full growth until 100 yr after it starts growing.) This giant sequoia, the General Sherman Tree in Sequoia National Park, California, is considered to be the world's largest tree. The volume of its trunk is over 52,500 ft^3.

Decreasing arithmetic sequences are eventually negative and decreasing geometric sequences approach zero. Other sequences can approach different values. For example, the temperature of a cup of hot cocoa as it cools, taken at one-minute intervals, produces a sequence that approaches the temperature of the room. In the long run, the hot cocoa will be at room temperature. Remember that sequences that model can only estimate actual values. What you, the modeler, must keep in mind is your common sense about the reality of things. If calculations yield an unrealistic answer then either you have made an error in your mathematics, chosen the wrong type of model, or the model has limits.

In the next investigation you will explore what happens to a sequence in the long run.

ELL	Support	Advanced
Emphasize the interchangeable terms *limit* and *long-run value*. Many students will be more familiar with determining how something will behave in the long run; students can use this connection to refine their own definition of *limit*. **ASK** "If you are going to the mall, what might impact how long it takes you to get there?"	Students may struggle with making mathematical sense of the competing actions of dosing medicine and the body's filtering system. First demonstrate each process on its own, then discuss how to combine them into one recursive formula. **ASK** "If you are going to the mall, what might impact how long it takes you to get there?"	Students could pick a limit and then create a recursive formula that will meet that limit. Have students define three situations modeled mathematically in this chapter. **ASK** "What factors might impact the physical limitations of the models?"

INVESTIGATION

Doses of Medicine

YOU WILL NEED
- a bowl
- a supply of water
- a supply of tinted liquid
- measuring cups, graduated in milliliters
- a sink or waste bucket

Our kidneys continuously filter our blood, removing impurities. Doctors take this into account when prescribing the dosage and frequency of medicine.

In this investigation you will simulate what happens in the body when a patient takes medicine. To represent the blood in a patient's body, use a bowl containing a total of 1 liter (L) of liquid. Start with 16 milliliters (mL) of tinted liquid to represent a dose of medicine in the blood, and use clear water for the rest.

Step 1 Suppose a patient's kidneys filter out 25% of this medicine each day. To simulate this, remove $\frac{1}{4}$, or 250 mL, of the mixture from the bowl and replace it with 250 mL of clear water to represent filtered blood. Make a table like the one below, and record the amount of medicine in the blood over several days. Repeat the simulation for each day.

Step 2 Write a recursive formula that generates the sequence in your table. $u_0 = 16$ and $u_n = (1 - 0.25)u_{n-1}$ where $n \geq 1$

Step 3 How many days will pass before there is less than 1 mL of medicine in the blood? 10 days

Step 4 According to the model, is the medicine ever completely removed from the blood? In a more practical sense, is the medicine ever completely removed? Explain your reasoning.

Day	Amount of medicine (mL)
0	16
1	12
2	9
3	6.75

Step 5 Sketch a graph and describe what happens in the long run.

A single dose of medicine is often not enough to treat a patient's condition. Doctors prescribe regular doses to produce and maintain a high enough level of medicine in the body. Next you will modify your simulation to look at what happens when a patient takes medicine daily over a period of time.

Step 6 Start over with 1 L of liquid. Again, all of the liquid is clear water, representing the blood, except for 16 mL of tinted liquid to represent the initial dose of medicine. Each day, 250 mL of liquid is removed and replaced with 234 mL of clear water and 16 mL of tinted liquid to

LESSON 1.3 Applications and Other Sequences **47**

Step 9

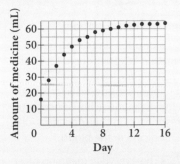

The amount of medicine levels off at 64 mL. Using a calculator to generate the sequence, it gets very close to 64 mL within 10 days and appears to reach 64 mL after 12 days.

Example

This example gives another case of a shifted geometric sequence. Consider projecting the example and having students work in pairs. Have students present their strategy as well as their solution. Encourage the use of appropriate mathematical vocabulary. Whether student responses are correct or incorrect, ask other students if they agree and why. SMP 1, 3, 6

ASK "How is this sequence different from the one in the investigation?" [This sequence is decreasing, whereas the sequence in the investigation was increasing.] SUPPORT Use a recursive routine to create a table of term numbers and their corresponding values before making the graph.

ADVANCED "Will the value ever stop changing? We solve for it, but will the input value ever equal 300?" [In theory, the input value would grow to 300 only in an infinite length of time.]

Summarize

Have students present their solutions and explain their work in the Investigation and the Example. Encourage a variety of approaches and equivalent forms of the equations. Emphasize the use of correct mathematical terminology. Whether student responses are correct or incorrect, ask other students if they agree and why. SMP 1, 3, 6

If you would like to introduce Sequence mode on the calculator, you might use Calculator Notes: Creating Sequences and Graphing Sequences at this point. SMP 5

---

represent a new dose of medicine. Complete another table like the one in Step 1, recording the amount of medicine in the blood over several days.

Step 7 Write a recursive formula that generates this sequence.
$u_0 = 16$ and $u_n = (1 - 0.25)u_{n-1} + 16$ where $n \geq 1$

Step 8 Do the contents of the bowl ever turn into pure medicine? Why or why not?
No, there is always pure water added at each step. The amount of medicine levels off.

Step 9 Sketch a graph and explain what happens to the level of medicine in the blood after many days.

The elimination of medicine from the human body is a real-world example of a dynamic, or changing, system. The cleanup of a contaminated lake is another real-world example. Dynamic systems often reach a point of stability in the long run. The quantity associated with that stability, such as the number of milliliters of medicine, is called a **limit.** Mathematically, we say that the sequence of numbers associated with the system approaches that limit. Being able to predict limits is very important for analyzing real-world situations like these. The long-run value helps you estimate limits.

> **Environmental CONNECTION**
>
> The Cuyahoga River in Cleveland, Ohio, caught fire several times during the 1950s and 1960s because its water was so polluted with volatile chemicals. The events inspired several clean water acts in the 1970s and the creation of the federal Environmental Protection Agency. After testing the toxic chemicals present in the water and locating possible sources of contamination, environmental engineers established pollution control levels and set standards for monitoring waste from local industries. Cleanup efforts are ongoing, and the river has been designated an Area of Concern by the Environmental Protection Agency.

Cleveland Ohio's Warehouse District, Flats, and Cuyahoga river.

Each of the sequences in the investigation approached different long-run values. The first sequence approached zero. The second sequence was shifted and it approached a nonzero value. A **shifted geometric sequence** includes an added term in the recursive rule. Let's look at another example of a shifted geometric sequence.

EXAMPLE Antonio and Deanna are working at the community pool for the summer. They need to provide a "shock" treatment of 450 grams (g) of dry chlorine to prevent the growth of algae in the pool. Then they add 45 g of chlorine each day after the initial treatment. Each day, the sun burns off 15% of the chlorine. Find the amount of chlorine after 1 day, 2 days, and 3 days. Create a graph that shows the chlorine level after several days and in the long run.

Solution The starting value is given as 450. This amount decays by 15% a day, but 45 g is also added each day. The amount remaining after each day is generated by the rule $u_n = (1 - 0.15)u_{n-1} + 45$, or $u_n = 0.85u_{n-1} + 45$. Use this rule to find the chlorine level in the long run.

Conceptual/Procedural

Conceptual The investigation and example explore the difference in mathematical models, which may indicate an activity going on forever, and the physical situation, which has limits on appropriate values. In the example, students explore the concept of a shifted geometric sequence.

Procedural Once students have created a recursive formula that will create a given sequence, they then practice a procedure for solving the sequence.

$u_0 = 450$ The initial shock treatment.
$u_1 = 0.85(450) + 45 = 427.5$ The amount after 1 day.
$u_2 = 0.85(427.5) + 45 \approx 408.4$ The amount after 2 days.
$u_3 = 0.85(408.4) + 45 \approx 392.1$ The amount after 3 days.

To find the long-run value of the amount of chlorine, you can continue evaluating terms until the value stops changing, or see where the graph levels off. From the graph, the long-run value appears to be 300 g of chlorine.

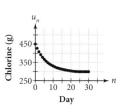

You can also use algebra to find the value of the terms as they level off. The key is to assume that the terms stop changing. Then you can set the value of the next term equal to the value of the previous term and solve the equation.

$u_n = 0.85 u_{n-1} + 45$ Recursive rule.
$c = 0.85c + 45$ Assign the same variable to u_n and u_{n-1}.
$0.15c = 45$ Subtract $0.85c$ from both sides.
$c = 300$ Divide both sides by 0.15.

The amount of chlorine will level off at 300 g, which agrees with the long-run value estimated from the graph.

In the medicine investigation and the pool example you saw that the values approached, but never reached, a limit. In reality, the measure of a drug or of chlorine cannot be made as accurately as the values in the sequence. Under ideal conditions you would find the measured value of chlorine reaches that limit in two to three weeks and then just stays at that value. But on days where there is more sunshine (or less sunshine), the decrease would not be by 15% but by some other value and the sequence would be an approximation of the chlorine level.

LESSON 1.3 Applications and Other Sequences   49

Discussing the Investigation

As students present their ideas about Step 9, **ASK** "Does the sequence actually reach this limit?" [The geometric sequence does not reach its limit of 64, though eventually the calculator doesn't have enough decimal places to represent the difference from 64.] Physically, the concentration of medicine will actually reach the long-run value when it becomes less than one molecule away from it.

ADVANCED "Do all geometric sequences have limits?" [If the common ratio is more than 1, sequences become large without bound. Because infinity isn't considered a number, these sequences have no limits. If the common ratio is between 0 and 1, they approach a limit of 0.] "Does any geometric sequence reach its limit?" [yes; if the common ratio is 1, so that all the terms are the same, or if it is 0, so that all terms after the first term are 0] "If the limit of non-constant geometric sequences can only be 0, how can the sequence of the investigation have a limit of 64?" [It's not a geometric sequence, but rather a shifted geometric sequence.] "Can an arithmetic sequence have a limit?" [only if the common difference is 0, so that all terms are the same]

CRITICAL QUESTION Why do you think this sequence is called a shifted geometric sequence?

BIG IDEA The "geometric" part comes from the multiplier; a graph of the sequence is "shifted" up or down from that of the geometric sequence, so the limit might not be 0.

CRITICAL QUESTION How can you tell from its recursive formula that a sequence is shifted geometric?

BIG IDEA The recursive formula for a shifted geometric sequence involves both multiplication by a constant and addition of a constant.

Formative Assessment

Solutions to the investigation, the example, and the exercises will help you check that students understand how arithmetic sequences and geometric sequences are generated, including shifted geometric sequences. Encourage students to describe sequences in u_n notation and to explain their reasoning. Continue to monitor students' abilities to describe sequences through recursive formulas or words, including the long run values, or limits, of a mathematical model.

Apply

Extra Example

Find the first three terms of the sequence. Is it geometric, arithmetic, or shifted geometric?

$u_0 = 25$

$u_n = (1 - 0.25)u_{n-1} + 4$ where $n \geq 1$

25, 22.75, 21.0625; shifted geometric

Closing Question

What does it mean for a sequence to have a limit, or long run value? What kinds of sequences have limits?

Answers will vary.

Assigning Exercises

Suggested: 1, 3–12

Additional Practice: 2, 13–15

Exercise 2d Have students explain why there is no solution.

4. The correct answer must be b. 70°. Both a. and c. are below freezing so unless this cup is placed in a very, very cold freezer those can't be right. Answer d. is just a few degrees below boiling. It was probably never this hot to begin with and can't be this hot after 30 minutes.

Exercise 6 Students may not realize that this is a shifted geometric sequence, because a constant amount is subtracted, rather than added, each month. *Note:* This is the first time that students have seen a sequence named something other than "u".

1.3 Exercises

Practice Your Skills

1. Find the values of u_1, u_2, and u_3. Identify the type of sequence (arithmetic, geometric, or shifted geometric) and tell whether it is increasing or decreasing.

a. $u_0 = 16$
   $u_n = (1 - 0.05)u_{n-1} + 16$ where $n \geq 1$ @
   31.2, 45.64, 59.358; shifted geometric, increasing

b. $u_0 = 800$
   $u_n = (1 - 0.05)u_{n-1} + 16$ where $n \geq 1$
   776, 753.2, 731.54; shifted geometric, decreasing

c. $u_0 = 50$
   $u_n = (1 - 0.10)u_{n-1}$ where $n \geq 1$
   45, 40.5, 36.45; geometric, decreasing

d. $u_0 = 40$
   $u_n = (1 - 0.50)u_{n-1} + 20$ where $n \geq 1$ @
   40, 40, 40; arithmetic or shifted geometric, neither increasing nor decreasing

2. Solve each equation.

a. $a = 210 + 0.75a$
   $a = 840$

b. $b = 0.75b + 300$ @
   $b = 1200$

c. $d = 0.75d$ @
   $d = 0$

d. $c = 210 + c$
   no solution

3. Find the long-run value for each sequence in Exercise 1. (h)

a. 320   b. 320   c. 0   d. 40

4. An algebra test has a problem about a cooling cup of hot chocolate. It gives specific values for the starting temperature and the temperature every minute for the first ten minutes. Then it asks for the temperature of the liquid after 30 minutes. Which is the best choice for the answer? Without knowing the data values, explain how you know your answer is correct.

a. 2° F   b. 70° F   c. −25° F   d. 200° F

Reason and Apply

5. The Osbornes have a small pool and are doing a chlorine treatment. The recursive formula below gives the pool's daily amount of chlorine in grams.

$u_0 = 300$
$u_n = (1 - 0.15)u_{n-1} + 30$ where $n \geq 1$

a. Explain the real-world meanings of the values 300, 0.15, and 30 in this formula. @
   The first day, 300 g of chlorine were added. Each day, 15% disappears, and 30 g are added.

b. Describe what happens to the chlorine level in the long run.
   It levels off at 200 g.

c. On very sunny days the chlorine will drop by 20%. Explain how you would adjust the daily addition of chlorine if the change was 20% and you still wanted the same long run value.
   You should add 40 g on very sunny days. (200 g · 0.20 = 40 g will be removed)

d. Knowing this sequence, why is it still important that the chlorine level is measured at regular intervals? The recursive rule uses an average of 15% each day but some days are higher and some lower. If you never check the level, you may find the chlorine level drifts too high or too low.

6. APPLICATION On October 2, 2008, Sal invested $24,000 in a bank account earning 3.4% annually, compounded monthly. On November 1, one month's interest was added to his account. The next day Sal withdrew $100. He continued to withdraw $100 on the second day of every month after that. He records his account balance in a table.

a. Find the missing values in the table.

b. Write a recursive formula for the balances on the second day of each month. Start with $a_0 = 24,000$. @

c. What is the value of a_4? What is the meaning of this value? @
   $23,871.45. This is the balance on 2/2/09.

d. What will the balance be on October 2, 2009? On October 2, 2011?
   $23,609.96; $22,789.00

Date	Balance ($)
Oct 2, 2008	24,000.00
Nov 1, 2008	24,068.00
Nov 2, 2008	23,968.00
Dec 1, 2008	24,035.91
Dec 2, 2008	23,935.91
Jan 1, 2009	24,003.73
Jan 2, 2009	23,903.73

6b. $a_0 = 24{,}000$ and $a_n = a_{n-1}\left(1 + \frac{0.034}{12}\right) - 100$, or $a_n = a_{n-1}(1.00283) - 100$

50 CHAPTER 1 Linear Modeling

7. **APPLICATION** Consider the bank account in Exercise 6.
   a. What happens to the balance if the same interest and withdrawal patterns continue for a long time? Does the balance ever level off?
   b. What monthly withdrawal amount would maintain a constant balance of $24,000 in the long run? $68

8. Ronald says "According to my calculations if I deposit $3,000 at 3.4% annual interest and withdraw $100 a month my account will continue to grow". Without using a calculator his friend Hilary tells him that he must be doing something wrong, the amount in the account must be going down. What calculation could Hilary do to know that he is wrong, and what might Ronald have done to get his answer?

9. **APPLICATION** The Forever Green Nursery owns 7000 white pine trees. Each year, the nursery plans to sell 12% of its trees and plant 600 new ones.
   a. Find the number of pine trees owned by the nursery after 10 years. 5557 trees
   b. Find the number of pine trees owned by the nursery after many years, and explain what is occurring.
   c. What equation can you solve to find the number of trees in the long run? @ $c = 0.88c + 600$
   d. Try different starting totals in place of the 7000 trees. Describe any changes to the long-run value.
   e. In the fifth year, a disease destroys many of the nursery's trees. How does the long-run value change?

10. **APPLICATION** Jack takes a capsule containing 20 milligrams (mg) of a prescribed allergy medicine early in the morning. By the same time a day later, 25% of the medicine has been eliminated from his body. Jack doesn't take any more medicine, and his body continues to eliminate 25% of the remaining medicine each day. Write a recursive formula for the daily amount of this medicine in Jack's body. When will there be less than 1 mg of the medicine remaining in his body? @
$u_0 = 20$ and $u_n = (1 - 0.25)u_{n-1}$ where $n \geq 1$; 11 days ($u_{11} \approx 0.84$ mg)

11. Consider the last part of the Investigation Doses of Medicine. If you double the amount of medicine taken each time from 16 mL to 32 mL, but continue to filter only 250 mL of liquid, will the limit of the concentration be doubled? Explain.
Yes, the long-run value was $\frac{15}{0.25}$, or 64 mL, and it is now $\frac{32}{0.25}$, or 128 mL.

12. Suppose square $ABCD$ with side length 8 in. is cut from paper. Another square, $EFGH$, is placed with its corners at the midpoints of $ABCD$, as shown. A third square is placed with its corners at midpoints of $EFGH$, and so on.
    a. What is the perimeter of the ninth square? ⓗ 2 in.
    b. What is the area of the ninth square? 0.25 in²
    c. What happens to the ratio of perimeter to area as the squares get smaller?
       The ratio gets larger and larger without limit.

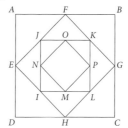

Exercise 14a If graphed, the data look somewhat linear. A good linear formula is $u_0 = 760$, $u_n = u_{n-1} + 83.5$. As needed, point out that multiplicative models are often better for population growth.

14a. $u_{2004} = 760$, $u_n = (1 + 0.1)u_{n-1}$ where $n \geq 2005$

Review

13. Assume two terms of a sequence are $u_3 = 16$ and $u_4 = 128$.
   a. Find u_2 and u_5 if the sequence is arithmetic.
   $u_2 = -96$, $u_5 = 240$
   b. Find u_2 and u_5 if the sequence is geometric.
   $u_2 = 2$, $u_5 = 1024$

14. A park biologist estimated the moose population in a national park over a four-year period of mild winters. She made this table.
   a. Write a recursive formula that approximately models the growth in the moose population for this four-year period.
   b. The winter of 2008 was particularly severe, and the park biologist predicted a decline of 10% to 15% in the moose herd. What is the range of moose population she predicted for 2008? between 859 and 909

Year	Estimated number of moose
2004	760
2005	835
2006	920
2007	1010

15. If a rubber ball rebounds to 97% of its height with each bounce, how many times will it bounce before it rebounds to half its original height? @ 23 times

IMPROVING YOUR Visual Thinking SKILLS

Think Pink

You have two 1-gallon cans. One contains 1 gallon of white paint, and the other contains 3 quarts (qt) of red paint. (There are 4 quarts per gallon.) You pour 1 qt of white paint into the red paint, mix it, and then pour 1 qt of the mixture back into the can of white paint. What is the red-white content of each can now? If you continually repeat the process, when will the two cans be the same shade of pink?

52 CHAPTER 1 Linear Modeling

IMPROVING YOUR Visual Thinking SKILLS

Students might be surprised that at each step the same amount of red paint is in the white as there is white paint in the red. For example, after the first mixing, the "white" can contains 0.75 qt of red paint and 3.25 qt of white paint; the "red" can contains 0.75 qt of white paint and 2.25 qt of red paint. With numerous repetitions, the two cans will get closer and closer to the same shade of pink, but theoretically they will never be exactly the same.

This problem has the nice feature of a nonzero limit. In one can the limit (say, the amount of red paint in the mixture) is approached from below, and in the other it is approached from above.

LESSON 1.4

Graphing Sequences

By looking for numerical patterns, you can write a recursive formula that generates a sequence of numbers quickly and efficiently. You can use graphs to help you identify patterns in a sequence.

INVESTIGATION

Match Them Up

Match each table with a recursive formula and a graph that represent the same sequence. Think about similarities and differences between the sequences and how those similarities and differences affect the tables, formulas, and graphs.

1. B. iv.   **2.** C. vi.   **3.** F. iii.   **4.** D. v.   **5.** A. ii.   **6.** E. i.

n	u_n
0	8
1	4
3	1
6	0.125
9	0.015625

n	u_n
0	0.5
1	1
2	2
3	4
4	8

n	u_n
0	-2
1	1
2	2.5
4	3.625
5	3.8125

n	u_n
0	-2
2	2
5	8
7	12
10	18

n	u_n
0	8
1	6
3	2
5	-2
7	-6

n	u_n
0	-4
1	-4
2	-4
4	-4
8	-4

A. $u_0 = 8$
$u_n = u_{n-1} - 2$ where $n \geq 1$

B. $u_0 = 8$
$u_n = 0.5u_{n-1}$ where $n \geq 1$

C. $u_0 = 0.5$
$u_n = 2u_{n-1}$ where $n \geq 1$

D. $u_0 = -2$
$u_n = u_{n-1} + 2$ where $n \geq 1$

E. $u_0 = -4$
$u_n = u_{n-1}$ where $n \geq 1$

F. $u_0 = -2$
$u_n = 0.5u_{n-1} + 2$ where $n \geq 1$

i.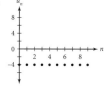

ii.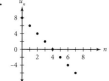

iii.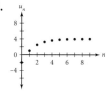

iv.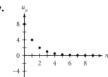

v.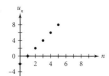

vi.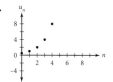

Write a paragraph that summarizes the relationships between different types of sequences, recursive formulas, and graphs. What generalizations can you make? What do you notice about the shapes of the graphs created from arithmetic and geometric sequences?

LESSON 1.4

This lesson focuses on graphs and their connections to tables and recursive formulas. Students will benefit from using a graphing calculator, allowing them to easily move between these representations.

COMMON CORE STATE STANDARDS		
APPLIED	DEVELOPED	INTRODUCED
	F.LE.2	
	F.LE.3	
	F.IF.6	

Objectives
- Develop and explore graphs and sequences
- Recognize and use multiple representations of sequences
- Focus on graphs of recursive models for data

Vocabulary
linear
discrete graphs
model

Materials
- Calculator Notes: Entering Data; Plotting Data; Creating Sequences; Graphing Sequences; Finding Specific Terms

Launch

What do the recursive rule, table and graph below have in common?

$t_0 = 15$ and $t_n = 2.8 + t_{n-1}$ where $n \geq 1$

t	t_n
0	15
1	17.8
2	20.6
3	23.4
4	26.2
5	29

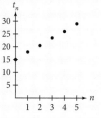

They all represent the same sequence.

ELL
Students may have trouble writing summary paragraphs. You might have students write the paragraph in their primary language first, or explain orally in either language. Alternately, have students create a graphic organizer summarizing the differences.

Support
Students may attempt to match the equations in the investigation, thinking that the n-values are consecutive. Encourage students to create their own table of values for each equation before matching them with the tables and graphs.

Advanced
Have each group make their own set of equations, tables, and graphs. Trade the sets between groups and work through the investigation using that information. You could also use the **One Step** found in your ebook.

Investigate

Guiding the Investigation

This investigation connects the previous three lessons by tying together graphs, equations, and sequences.

ALERT Students may miss the fact that the values given for n are consecutive only in table 2.

If students have trouble writing the paragraph summarizing their findings, suggest that they organize their observations by type of sequence: arithmetic, geometric, or shifted geometric.

Written paragraphs will vary, but may include the following points:

- Arithmetic sequences have linear graphs.
- Geometric sequences have curved graphs.
- When the common difference of an arithmetic sequence is positive, the linear shape slopes upward from left to right.
- When the common difference of an arithmetic sequence is negative, the linear shape slopes downward from left to right.
- When the common ratio of a geometric sequence is greater than 1, the curved shape increases from left to right.
- When the common ratio of a geometric sequence is between 0 and 1, the curved shape decreases from left to right approaching the n-axis.
- Shifted geometric sequences level off to a nonzero value.

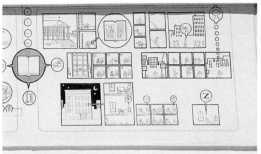 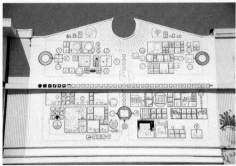

Many cartoons and comic strips show a sequence of events in linear order. The artwork of American artist Chris Ware (b 1967) breaks the convention by showing many sequences intertwined. This complex mural by Ware, at 826 Valencia Street in San Francisco, depicts human development of written and spoken communication.

The general shape of the graph of a sequence's terms gives you an indication of the type of sequence necessary to generate the terms.

The graph at right is a visual representation of the first five terms of the arithmetic sequence generated by the recursive formula

$$u_1 = 1$$

$$u_n = u_{n-1} + 2 \quad \text{where } n \geq 2$$

This graph, in particular, appears to be **linear,** that is, the points appear to lie on a line. The common difference, $d = 2$, makes each new point rise 2 units above the previous point.

Graphs of sequences are examples of **discrete graphs,** or graphs made of isolated points. It is incorrect to connect those isolated points with a continuous line or curve because the term number, n, must be a whole number.

The general shape of the graph of a sequence allows you to recognize whether the sequence is arithmetic or geometric. Even if the graph represents data that are not generated by a sequence, you may be able to find a sequence that is a **model,** or a close fit, for the data. The more details you can identify from the graph, the better you will be at fitting a model.

Weather forecasting is one career that relies on mathematical modeling. Forecasters use computers and sophisticated models to monitor changes in the atmosphere. Trends in the data can help predict the trajectory and severity of an impending storm, such as a hurricane.

54 CHAPTER 1 Linear Modeling

Modifying the Investigation

Whole Class Display the graphs. Complete the matching, with student input. Discuss the questions that follow.

Shortened Have students complete the matching. Don't have them write a paragraph; simply discuss the questions that follow.

One Step Give this assignment: "Each group write two arithmetic sequences, two geometric sequences, and two shifted geometric sequences. Try to make sequences that no other group will make. On one sheet of paper, generate a table of values for each sequence; on another sheet, make a graph for each sequence, but do not identify the sequences. Give your graphs to another group and see how many of the graphs they can label correctly as arithmetic, geometric, or shifted geometric sequences."

EXAMPLE

In deep water, divers find that their surroundings become darker the deeper they go. The data here give the percent of surface light intensity that remains at depth n ft in a particular body of water.

Depth (ft)	0	10	20	30	40	50	60	70
Percent of surface light	100	78	60	47	36	28	22	17

Write a recursive formula for a sequence model that approximately fits these data.

Solution

A graph of the data shows a decreasing, curved pattern. It is not linear, so an arithmetic sequence is not a good model. A geometric sequence with a long-run value of 0 will be a better choice.

The starting value at depth 0 ft is 100% light intensity, so use $u_0 = 100$. The recursive rule should have the form $u_n = (1 - p)u_{n-1}$, but the data are not given for every foot, so you cannot immediately find a common ratio. The ratios between the given values are all approximately 0.77, or $(1 - 0.23)$. Because the light intensity decreases at a rate of 0.23 every 10 feet, it must decrease at a smaller rate every foot. A starting guess of 0.02 gives the model

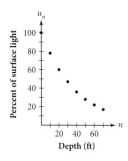

$$u_0 = 100$$
$$u_n = (1 - 0.02)u_{n-1} \quad \text{where } n \geq 1$$

Check this model by graphing the original data and the sequence on your calculator. The graph shows that this model fits only one data point—it does not decay fast enough.

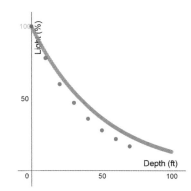

Experiment by increasing the rate of decay. With some trial and error, you can find a model that fits the data better.

$$u_0 = 100$$
$$u_n = (1 - 0.025)u_{n-1} \quad \text{where } n \geq 1$$

ADVANCED "Is a shifted geometric sequence a special kind of geometric sequence?" [Not necessarily; it may not have a common ratio between consecutive terms. On the other hand, every geometric sequence is a shifted geometric sequence, with 0 being added. Every arithmetic sequence is a shifted geometric sequence, with 1 being multiplied.] **SMP 6, 8**

CRITICAL QUESTION What kind of sequence always has a graph that is a straight line?

BIG IDEA The graph of an arithmetic sequence is always a straight line.

CRITICAL QUESTION What kind of sequence has a graph that is a curve approaching the positive x-axis?

BIG IDEA The graph of a geometric sequence that has a limit is a curve that approaches the positive x-axis.

CRITICAL QUESTION What kind of sequence has a graph that is a curve approaching a horizontal line that isn't the x-axis?

BIG IDEA The graph of a shifted geometric sequence that has a limit approaches a horizontal line. If the sequence is shifted, then that horizontal line is not the x-axis.

CRITICAL QUESTION If arithmetic sequences give straight-line graphs, can we tell the slope of the line from the formula? Explain.

BIG IDEA It's the common difference, or the amount of change in y for each unit change in x.

Formative Assessment

Watch for student understanding of relationships among the table, graph, and the recursive rule's initial value and form. Assess students' understanding of recursive formulas, their ability to plot points on a graph, and their skill at reading a table.

Once you have a good sequence model, you can use your calculator to find specific terms. For example, the value of u_{43} means that at depth 43 ft approximately 34% of surface light intensity remains.

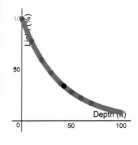

C	D
37	39.1895955
38	38.20985561
39	37.25460922
40	36.32324399
41	35.41516289
42	34.52978382
43	33.66653922
44	32.82487574
45	32.00425385
46	31.2041475

Some calculators use $u(0), u(1), u(2), \ldots, u(n-1)$, and $u(n)$ instead of the subscripted notation $u_0, u_1, u_2, \ldots, u_{n-1}$, and u_n. Be aware that $u(5)$ means u_5, not u multiplied by 5. You may also see other variables, such as a_n or v_n, used for recursive formulas. It is important that you are able to make sense of these equivalent mathematical notations and be flexible in reading other people's work.

Being alert also pays off when working with graphs. Graphs help you understand and explain situations, and visualize the mathematics of a situation. When you make a graph or look at a graph, try to find connections between the graph and the mathematics used to create the graph. Consider what variables and units were used on each axis and what the smallest and largest values were for those variables. Sometimes this will be clear and obvious, but sometimes you will need to look at the graph in a new way to see the connections.

Science CONNECTION

Marine life near the ocean's surface relies on organisms that use the sun for photosynthesis. But lifeforms at the bottom of the ocean, where sunlight is virtually absent, feed on waste material or microorganisms that create energy from chemicals released from Earth's crust, a process called chemosynthesis. Squat lobsters and galatheid crabs are among many deep-sea lifeforms that thrive near hydrothermal vents, where Earth's crust releases chemical compounds.

CONTEXT **Science Connection** **ASK** "What factors might influence surface light intensity?" [Wavelengths of visible light are absorbed as light travels through water; substances in the water, such as waste and vegetation, block light; light intensity decreases with the square of the distance from the source.]

Hydrothermal vents are found 2–5 km below the ocean's surface. Although no sunlight reaches this depth of the sea, the temperature heats up to 300–400°C because of volcanic activity.

 **Exercises**   You will need a graphing calculator for exercises **1, 4, 6, 7–10** and **13**.

Practice Your Skills

1. Suppose you are going to graph the specified terms of these four sequences. For each sequence, what minimum and maximum values of n and u_n would you use on the axes to get a good graph?   0 to 9 for n and 0 to 16 for u_n

   a.

n	0	1	2	3	4	5	6	7	8	9
u_n	2.5	4	5.5	7	8.5	10	11.5	13	14.5	16

   b. The first 20 terms of the sequence generated by

   $u_0 = 400$

   $u_n = (1 - 0.18)u_{n-1}$ where $n \geq 1$ 0 to 19 for n and 0 to 400 for u_n

   c. The first 30 terms of the sequence generated by

   $u_0 = 25$

   $u_n = u_{n-1} - 7$ where $n \geq 1$ 0 to 29 for n and –178 to 25 for u_n

   d. The first 70 terms of the sequence generated by

   $u_0 = 15$

   $u_n = (1 + 0.08)u_{n-1}$ where $n \geq 1$ 0 to 69 for n and 0 to 3037 for u_n

2. Match each graph with a formula and identify the sequence as arithmetic or geometric.

   a. 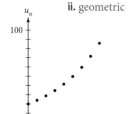   ii. geometric

   b. 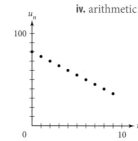   iv. arithmetic

   c. 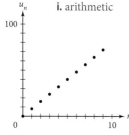   i. arithmetic

   d. 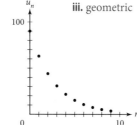   iii. geometric

   i. $u_0 = 0$
   $u_n = u_{n-1} + 8$ where $n \geq 1$

   ii. $u_0 = 20$
   $u_n = 1.2u_{n-1}$ where $n \geq 1$

   iii. $u_0 = 90$
   $u_n = 0.7u_{n-1}$ where $n \geq 1$

   iv. $u_0 = 80$
   $u_n = u_{n-1} - 5$ where $n \geq 1$

LESSON 1.4 Graphing Sequences **57**

Apply

Extra Example

Describe each sequence using exactly three of these terms: *arithmetic, decreasing, geometric, increasing, linear, nonlinear, shifted geometric*.

1a. $u_0 = 200$ and $u_n = u_{n-1} \cdot d + 4$ where $0 < d < 1$ and $n \geq 1$

shifted geometric, decreasing, nonlinear

1b. $u_0 = 150$ and $u_n = u_{n-1} \cdot d + 4$ where $d > 1$ and $n \geq 1$

shifted geometric, increasing, nonlinear

Closing Question

What can you tell about a sequence from its graph?

Answers will vary

Assigning Exercises

Suggested: 1 – 11
Additional Practice: 12, 13

Exercise 4 As needed, suggest that students try several different starting values and look for patterns. They may be surprised to find that the limit is the same for all starting values. **ASK** "Why doesn't the starting value affect the limit?" [Intuitively, multiplying repeatedly by 0.75 tends to move the values toward 0, no matter where they started; adding 210 compensates for this movement.]

4b.

i. sample answer: $u_0 = 200$

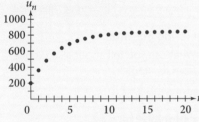

ii. sample answer: $u_0 = 1000$

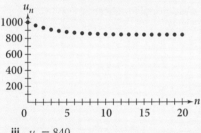

iii. $u_0 = 840$

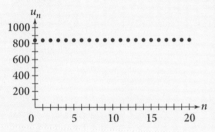

4c. When the starting value is below the long-run value, the graph increases toward the long-run value. When the starting value is above the long-run value, the graph decreases toward the long-run value. When the starting value is the long-run value, the graph is a constant, horizontal, linear pattern at the long-run value.

Exercise 5 Graph i apparently approaches a nonzero limit; the only possible formula is the shifted geometric sequence in 5c, and then only with $r < 1$. Graph ii is a curve that doesn't approach a limit, implying that $r > 1$ in 5b.

3. Imagine the graphs of the sequences generated by these recursive formulas. Describe each graph using exactly three of these terms: arithmetic, decreasing, geometric, increasing, linear, nonlinear, shifted geometric.

   a. $v_0 = 450$ geometric, nonlinear, decreasing
   $v_n = a \cdot v_{n-1}$ where $n \geq 1$ and $0 < a < 1$ @

   b. $v_0 = 450$ arithmetic, linear, decreasing
   $v_n = b + v_{n-1}$ where $n \geq 1$ and $b < 0$

   c. $v_0 = 450$ geometric, nonlinear, increasing
   $v_n = v_{n-1} \cdot c$ where $n \geq 1$ and $c > 1$

   d. $v_0 = 450$ arithmetic, linear, increasing
   $v_n = v_{n-1} + d$ where $n \geq 1$ and $d > 0$

Reason and Apply

4. Consider the recursive rule $u_n = 0.75u_{n-1} + 210$.

   a. What is the long-run value of any shifted geometric sequence that is generated by this recursive rule? Explain.   840

   b. Sketch the graph of a sequence that is generated by this recursive rule and has a starting value

      i. Below the long-run value @
      ii. Above the long-run value
      iii. At the long-run value

   c. Write a short paragraph describing how the long-run value and starting value of each shifted geometric sequence in 4b influence the appearance of the graph.

5. Match each recursive formula with the graph of the same sequence. Give your reason for each choice.

   a. $u_0 = 20$ b. $u_0 = 20$ c. $u_0 = 20$
   $u_n = u_{n-1} + d$ where $n \geq 1$ $u_n = r \cdot u_{n-1}$ where $n \geq 1$ $u_n = r \cdot u_{n-1} + d$ where $n \geq 1$

   i. 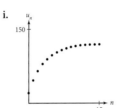   ii. 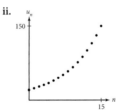   iii.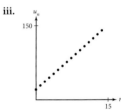

   c. The graph has a nonzero limit, so it must represent a shifted geometric sequence.

   b. The graph of a non-shifted geometric sequence will increase indefinitely or have a limit of 0.

   a. The graph of an arithmetic sequence is linear.

This complex fractal was created by plotting points generated by recursive formulas.

CONTEXT **Computer-Generated Fractals** To generate fractals like the one shown, we can use a recursive formula on the coordinates for each point (on the complex number plane). As the formula is applied over and over again, the value at a point either gets larger without limit (*escapes*) or it reaches a constant value. Each point that escapes is given a color based on how fast it escapes. Points that approach a limit as the recursive process is applied are colored black. Students will work with this computer-generated fractal when they study complex numbers. That project, The Mandelbrot Set, is in your digital resources.

6. Consider the geometric sequence 18, −13.5, 10.125, −7.59375, ... @
   a. Write a recursive formula that generates this sequence. $u_0 = 18$ and $u_n = -0.75u_{n-1}$ where $n \geq 1$
   b. Sketch a graph of the sequence. Describe how the graph is similar to other graphs that you have seen and also how it is different.
   c. What is the long-run value? 0

7. Consider the recursive formula
   $u_0 = 450$
   $u_n = 0.75u_{n-1} + 210$ where $n \geq 1$
   a. Find u_1, u_2, u_3, u_4, and u_5.

   a. 547.5, 620.6, 675.5, 716.6, 747.5
   c. $u_0 = 747.5$ and $u_n = \frac{u_{n-1} - 210}{0.75}$ where $n \geq 1$

   b. How can you calculate backward from the value of u_1 to u_0, or 450? In general, what operations can you perform to any term in order to find the value of the previous term?
   c. Write a recursive formula that generates the values from 7a in reverse order.

8. **APPLICATION** Newtin metal recycling has increased their production of tin by 40% each year for the last six years. At it's beginning, 7 years ago, the company produced 75 pounds of tin.
   a. Plot the production of tin as a function of time over the seven years.
   b. Calculate the average rate of change in tin production between year 1 and year 3. Add a line segment to your plot in part a between these two points to show this slope. $\frac{147 - 75}{2} = 36$ lbs/yr
   c. Calculate the average rate of change in tin production between year 3 and year 6. Add a line segment to your plot in part a between these two points to show this slope. $\frac{403.368 - 147}{3} = 85.456$ lbs/yr
   d. Calculate the average rate of change in tin production between year 6 and year 7. Add a line segment to your plot in part a between these two points to show this slope.
   e. Often you don't need a value for the average rate of change, you just need to know if it is positive or negative, or if is increasing or decreasing as x increases. How could you answer these two questions by just looking at the graph?   positive; increasing

9. **APPLICATION** The Forever Green Nursery has 7000 white pine trees. Each year, the nursery plans to sell 12% of its trees and plant 600 new ones.
   a. Make a graph that shows the number of trees at the nursery over the next 20 years.
   b. Use the graph to estimate the number of trees in the long run. How does your estimate compare to the long-run value you found in Exercise 9b in the previous lesson?

10. Until recently, the African country of Niger has had steady exponential growth.
    a. Experiment with recursive rules to fit the graph of the data through 2005.
    b. Use your recursive rule to predict the population in 2015. How does your prediction compare with the actual population in 2015 of 17.31 million people?

    a. $u_{1965} = 3.836$, $u_n = 1.03\ u_{n-1}$
    b. $17.31 - 16.82 = 0.49$ million

Population of Niger

Year	Population (millions)
1965	3.836
1975	5.071
1985	6.705
1995	9.167
2005	1.180

(The World Bank)

Exercise 6b This sequence can be graphed easily on a calculator. **SMP 5**

6b.
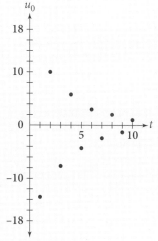

Because every term alternates signs, the graph is novel; each point alternates above or below the n-axis. The points above the n-axis create, however, a familiar geometric pattern, as do the points below the n-axis. If the points below the n-axis were reflected across the n-axis, you would have the graph of $u_0 = 18$ and $u_n = 10.75u_{n-1}$ where $n \geq 1$.

7b. $\frac{547.5 - 210}{0.75} = 450$; subtract 210 and divide the difference by 0.75.

Exercise 7 This exercise uses shifted geometric sequences. It also reviews undoing operations. Students can also visualize the "undoing" sequence as a new sequence in the reverse order.

8a.
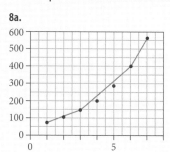

8d.
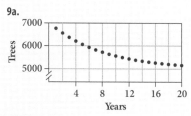
$\frac{564.7152 - 403.368}{1} = 161.3472$ lbs/yr

9a.
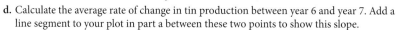

9b. The graph appears to have a long-run value of 5000 trees, which agrees with the long-run value found in Exercise 9b in the last lesson.

Exercise 11 If students write the recursive formula starting with $u_0 = -160$ for a temperature of 0, you might wonder aloud whether crickets can chirp a negative amount. **CONTEXT** The snow tree cricket, *Oecanthus fultoni*, is one of many subspecies of crickets. Every subspecies of cricket has a unique chirp; the snow tree cricket has the highest-pitched chirp.

Exercise 13 It might help students if you are explicit about the patterns elicited in 13a–c. **ASK** "How does the change in the constant term relate to the change in the long-run value?" [In general, the limit of the sequence generated by $u_n = r \cdot u_{n-1} + (m \cdot d)$ will be m times the limit of the sequence generated by $u_n = r \cdot u_{n-1} + d$.]

Extension

Extend Exercise 4 by challenging students to explore whether all shifted geometric sequences behave the way $u_n = 0.75u_{n-1} + 210$ behaves or whether there are exceptions. [When the multiplier is greater than 1, the sequence does not approach a limit.]

---

11. **APPLICATION** As the air temperature gets warmer, snowy tree crickets chirp faster. You can actually use a snowy tree cricket's rate of chirping per minute to determine the approximate temperature in degrees Fahrenheit. Use a graph to find a sequence model that approximately fits these data. ⓗ
possible answer: $u_{50} = 40$ and $u_n = u_{n-1} + 4$ where $n \geq 51$

Snowy Tree Crickets' Rates of Chirping

Temperature (°F)	50	55	60	65	70	75	80
Rate (chirps/min)	40	60	80	100	120	140	160

Review

12. Find the value of a that makes each equation true.
    a. $47,500,000 = 4.75 \times 10^a$ b. $0.0461 = a \times 10^{-2}$ c. $3.48 \times 10^{-1} = a$
    7   4.61   0.348

13. For a–c, find the long-run value of the sequence generated by the recursive formula.
    a. $u_0 = 50$ $33\frac{1}{3}$
    $u_n = (1 - 0.30)u_{n-1} + 10$ where $n \geq 1$

    b. $u_0 = 50$ $66\frac{2}{3}$
    $u_n = (1 - 0.30)u_{n-1} + 20$ where $n \geq 1$ ⓐ

    c. $u_0 = 50$ 100
    $u_n = (1 - 0.30)u_{n-1} + 30$ where $n \geq 1$

    d. Generalize any patterns you notice in your answers to 13a–c. Use your generalizations to find the long-run value of the sequence generated by
    $u_0 = 50$
    $u_n = (1 - 0.30)u_{n-1} + 70$ where $n \geq 1$

    The long-run value grows in proportion to the added constant.
    $7 \cdot 33\frac{1}{3} = 233\frac{1}{3}$

Snowy tree crickets are about 0.7 in. long, are pale green, and live in shrubs and bushes. Only male crickets chirp, and they have different chirps for different activities, such as mating and fighting. All species of crickets chirp by rubbing their wings together.

PERFORMANCE TASK

Your friend is confused about arithmetic and geometric sequences. Help your friend out.

Explain the difference between an arithmetic sequence and a geometric sequence.

Explain how the graphs of arithmetic sequences are similar and how they vary.

Explain how the graphs of geometric sequences are similar and how they vary.

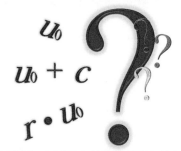

60   CHAPTER 1   Linear Modeling

PERFORMANCE TASK

An arithmetic sequence has a common difference. A geometric sequence has a common ratio (multiplier).

The graph of an arithmetic sequence is always linear. The graph increases when the common difference is positive and decreases when the common difference is negative. The steepness of the graph relates to the common difference.

The graph of a geometric sequence is usually nonlinear and usually approaches the n-axis or recedes more and more quickly from it.

The graph increases when the common ratio is greater than 1 and decreases when the ratio is between 0 and 1. The change in steepness of the curve is related to the common ratio.

LESSON 1.5

Linear Equations and Arithmetic Sequences

And now the sequence of events in no particular order.
DAN RATHER

You can solve many rate problems by using recursion.

Matias is visiting colleges in Chicago. Matias has been watching the meter as he ride cabs around the city. He finds that when you enter the cab the meter starts at $3.25 and that each mile traveled cost $1.80. How much would it cost to travel for 30 miles?

You can calculate the cost of Matias's cab ride with the recursive formula

$u_0 = 3.25$

$u_n = u_{n-1} + 1.80$ where $n \geq 1$

To find the cost of a 30-mile cab ride, calculate the first 30 terms, as shown in the spreadsheet.

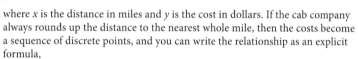

	A	B
26	24	46.45
27	25	48.25
28	26	50.05
29	27	51.85
30	28	53.65
31	29	55.45
32	30	57.25

As you learned in algebra, you or Matias can also find the cost of a 30-mile ride by using the **linear equation**

$y = 3.25 + 1.80x$

where x is the distance in miles and y is the cost in dollars. If the cab company always rounds up the distance to the nearest whole mile, then the costs become a sequence of discrete points, and you can write the relationship as an explicit formula,

$u_n = 3.25 + 1.80n$

where n is the distance traveled in whole miles and u_n is the cost in dollars.

An **explicit formula** gives a direct relationship between two discrete quantities. How does the explicit formula differ from the recursive formula? How would you use each formula to calculate the cost of a 15-mile ride or an n-mile ride?

In this lesson you will write and use explicit formulas for arithmetic sequences. You will also write linear equations for lines through the discrete points of arithmetic sequences.

EXAMPLE A

Consider the recursively defined arithmetic sequence

$u_0 = 2$

$u_n = u_{n-1} + 6$ where $n \geq 1$

a. Find an explicit formula for the sequence.
b. Use the explicit formula to find u_{22}.
c. Find the value of n so that $u_n = 86$.

LESSON 1.5

In this lesson students see how to write an explicit formula for a discrete function defined recursively.

COMMON CORE STATE STANDARDS		
APPLIED	**DEVELOPED**	**INTRODUCED**
A.CED.1	F.IF.3	F.IF.6
A.CED.2	F.BF.2	
F.BF.1a	F.LE.2	
	F.LE.5	

Objectives

- Given a recursive formula, find n for a given u_n
- Graph an arithmetic sequence to locate the u_n-intercept and determine the slope
- Recognize slope as the common difference of an arithmetic sequence
- Use the u_n-intercept and slope to write a linear equation (in x and y)
- Recognize that an arithmetic sequence is always linear
- Introduce explicit formulas for sequences
- Recognize connections between explicit and recursive formulas for arithmetic sequences

Vocabulary

linear equation
explicit formula
slope

Materials

- Calculator Note: Recursion, *optional*

Launch

Sully received a $25 music gift card. He spends $1.35 per song.

a. Find the recursive formula that represents the balance left on the gift card after each song is purchased.

$u_0 = 25, u_n = u_{n-1} - 1.35$

b. How many songs can be purchased with the gift card?

18 songs

ELL
Students should discuss the investigation in groups. Make sure they understand the term *collinear*. Define *explicit* functions by contrasting explicit functions and recursive functions.

Support
If students have difficulty completing Step 1 of the investigation, conduct a focused review of recursive formulas, graphs, and equations. To organize the information provided and needed, suggest they make a table with three columns (recursive formula, graph, equation).

Advanced
Some students will already be familiar with the concepts in this lesson. The **One Step** investigation allows students the freedom to solve the problem using a method of their own choosing. It also allows you to determine the depth of their understanding of these concepts.

Investigate

Consider projecting the paragraph where Matias visits colleges from your ebook. **ASK** "How could we use a recursive formula to calculate the cost of Matias' cab ride?" Work through the problem with student input to review vocabulary and make the connection of a recursive formula and a linear equation.

Example A

Consider projecting Example A and having students work in pairs. To encourage a variety of strategies, have students present their strategy as well as their solution. The equation in part c can be solved in one of several ways, including undoing, arrow diagrams, or graphing. For a graphical approach, draw a horizontal line at $u_n = 86$ and find the point where the plot meets that line. Encourage the use of appropriate mathematical vocabulary, such as term value, initial value, and rate. Whether student responses are correct or incorrect, ask other students if they agree and why. **SMP 1, 3, 6**

ASK "Why is the common difference the rate of change?" [because the change in n is 1]

Example B

Consider projecting Example B and having students work in small groups. Have students present their strategy as well as their solution. Encourage the use of appropriate mathematical vocabulary. Whether student responses are correct or incorrect, ask other students if they agree and why. **SMP 1, 3, 6**

ALERT Students often have difficulty determining whether to use the first term as t_1 or t_0. **ASK** "Why do you think Retta's sequence starts with t_1?" [She has $17 left at the end of Day 1.]

When you download photos from the Internet, sometimes the resolution improves as the percentage of download increases. The relationship between the percentage of download and the resolution could be modeled with a sequence, an explicit formula, or a linear equation.

Solution

a. Look for a pattern in the sequence.

$u_0 \quad 2$
$u_1 \quad 8 = 2 + 6 = 2 + 6 \cdot 1$
$u_2 \quad 14 = 2 + 6 + 6 = 2 + 6 \cdot 2$
$u_3 \quad 20 = 2 + 6 + 6 + 6 = 2 + 6 \cdot 3$

Notice that the common difference (or rate of change) between the terms is 6. You start with 2 and just keep adding 6. That means each term is equivalent to 2 plus 6 times the term number. In general, when you write the formula for a sequence, you use n to represent the number of the term and u_n to represent the term itself.

Term value = Initial value + Rate · Term number

$$u_n = 2 + 6 \cdot n$$

b. You can use the explicit formula to find u_{22} without calculating all of the previous terms. By substituting 22 for n, you get $u_{22} = 2 + 6 \cdot 22$. So, u_{22} equals 134.

c. To find n so that $u_n = 86$, substitute 86 for u_n in the formula $u_n = 2 + 6n$.

$86 = 2 + 6n$ Substitute 86 for u_n.
$14 = n$ Solve for n.

So, 86 is the 14th term.

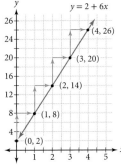

You graphed sequences of points (n, u_n) earlier in Chapter 1. The term number n is a whole number: 0, 1, 2, 3, ... So, using different values for n will produce a set of discrete points. The points on this graph show the arithmetic sequence from Example A.

When n increases by 1, u_n increases by 6, the common difference. In terms of x and y, the number 6 is the change in the y-value that corresponds to a unit change (a change of 1) in the x-value. So the points representing the sequence lie on a line with a slope of 6. In general, the common difference, or rate of change, between consecutive terms of an arithmetic sequence is the **slope** of the line through those points.

The pair (0, 2) names the starting value 2, which is the y-intercept. Using the intercept form of a linear equation, you can now write an equation of the line through the points of the sequence as $y = 2 + 6x$, or $y = 6x + 2$.

In this course you will use n when you are writing a formula for a sequence of discrete points. When you write an equation of the line that contains the points of the sequence, you will use x to indicate that all real numbers can be input values.

In the investigation you will focus on the relationship between the formula for an arithmetic sequence and the equation of the line through the points representing the sequence.

62   CHAPTER 1   Linear Modeling

Modifying the Investigation

Whole Class Display the three graphs. Have students complete Step 1, matching the formulas and equations to the graph. Discuss Steps 2 and 3.

Shortened Have groups complete Step 1. Discuss Steps 2 and 3.

One Step Use the **One Step** found in your ebook. **ASK** "A gardener planted a new variety of ornamental grass and kept a record of its height over the first 2 weeks of growth. What's a recursive formula that will help predict its height after 21 days? What's a quicker way to find that value?" As students work, encourage them to draw graphs, and ask how repeated addition can be represented. During the discussion, introduce the term *explicit formula*.

INVESTIGATION
Match Point

Below are three recursive formulas, three graphs, and three linear equations.

1. $u_0 = 4$
   $u_n = u_{n-1} - 1$ where $n \geq 1$

2. $u_0 = 2$
   $u_n = u_{n-1} + 5$ where $n \geq 1$

3. $u_0 = -4$
   $u_n = u_{n-1} + 3$ where $n \geq 1$

A.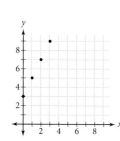

B.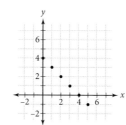

C.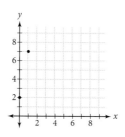

i. $y = -4 + 3x$

ii. $y = 4 + x$

iii. $y = 2 + 5x$

Step 1 Match the recursive formulas, graphs, and linear equations that go together. (Not all of the appropriate matches are listed. If the recursive rule, graph, or equation is missing, you will need to create it.)

Step 2 Write a brief statement relating the starting value and common difference of an arithmetic sequence to the corresponding equation $y = a + bx$.

Step 3 Are points (n, u_n) of an arithmetic sequence always collinear? Write a brief statement supporting your answer.

Step 2
The starting value is a, and the common difference is b.

EXAMPLE B

Retta typically spends $2 a day on lunch. She notices that she has $17 left after today's lunch. She thinks of this sequence to model her daily cash balance.

$t_1 = 17$
$t_n = t_{n-1} - 2$ where $n > 1$

a. Find the explicit formula that represents her daily cash balance and an equation of the line through the points of this sequence.

b. How useful is this formula for predicting how much money Retta will have each day?

Solution

Use the common difference and starting term to write the explicit formula.

a. Each term is 2 less than the previous term, so the common difference of the arithmetic sequence and the slope of the line are both −2.

The term t_1 is 17, so the previous term, t_0, or the y-intercept, is 19.

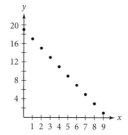

LESSON 1.5 Linear Equations and Arithmetic Sequences 63

Conceptual/Procedural

Conceptual Students explore the relationship between a recursive formula and a linear equation. The concept of an explicit formula is introduced. In the investigation, students see the relationship of sequences represented in recursive formulas, graphs, and linear equations.

Procedural Students practice finding an explicit formula for a recursively defined arithmetic sequence and using it to find a specific term. The investigation explores the process of matching and creating sequences represented in recursive formulas, graphs, and linear equations.

Guiding the Investigation
There is a desmos simulation in your ebook.

Step 1 ALERT Not all the matches are listed. The assignment includes creating the missing pieces. A table with three columns (recursive formula, graph, equation) can help organize the information provided and needed.

Step 1

Recursive formula	Graph	Equation
1	B	$y = 4 - x$
2	C	iii
3	(graph shown)	i
$u_0 = 3$ and $u_n = u_{n-1} + 2$ where $n \geq 1$	A	$y = 3 + 2x$
$u_0 = 4$ $u_n = u_{n-1} + 1$ where $n \geq 1$,	(graph shown)	ii

Step 3 LANGUAGE *Collinear* means lying on the same line.

Step 3 Yes, the points (n, u_n) of an arithmetic sequence are always collinear because the explicit formula for an arithmetic sequence is a linear equation.

Summarize

Have groups present their solutions and explain their work for the Investigation and the examples. Whether student responses are correct or incorrect, ask other students if they agree and why. Encourage students to use appropriate mathematical language as evidence. SMP 1, 3, 6

CRITICAL QUESTION Is $u_n = u_{n-1} + 5$ the same as $u_n = n + 5$?

BIG IDEA The slope of the line through the points on the recursively defined formula is 5, but the slope of the line through the points on the explicitly defined formula is 1.

LESSON 1.5 Linear Equations and Arithmetic Sequences 63

CRITICAL QUESTION What's the difference between discrete and continuous functions?

BIG IDEA There are gaps between the input values of discrete formulas (n) but not between the input values of continuous equations (x), such as linear equations—though there might be gaps in the output values (y).] As students respond, don't stop the thinking by moving on when you hear an answer you like; rather, have the class assess all the responses without prejudice from you.

CRITICAL QUESTION How do explicit formulas relate to linear equations?

BIG IDEA An explicit formula describes a sequence of points. A linear equation describes a line that passes through those points.

Formative Assessment
Assess student understanding of how recursive formulas can define arithmetic sequences, of multiplication as repeated addition, and of linear equations and their graphs. Assess students' understanding of linear recursive sequences and their facility with u_n notation. Also observe students' skills at graphing lines.

Apply
Extra Example
Write a recursive formula for a sequence whose points lie on the line $y = -3 + 1.4x$.

$u_0 = -3$, $u_n = u_{n-1} + 1.4$

Closing Question
Consider the arithmetic sequence 6, 13, 20, 27, 34, ...

Let u_1 represent the first term.

a. Write a recursive formula that describes this sequence.

b. Write an explicit formula for this sequence.

c. What is the slope of your equation in b? What relationship does this have to the arithmetic sequence?

a. $u_1 = 6$ and $u_n = u_{n-1} + 7$ where $n \geq 1$
b. $y = -1 + 7x$
c. The slope is 7. The slope of the line is the same as the common difference of the sequence.

The explicit formula for the arithmetic sequence is $t_n = 19 - 2 \cdot n$, and the equation of the line containing these points is $y = 19 - 2x$.

b. We don't know whether Retta has any other expenses, when she receives her paycheck or allowance, or whether she buys lunch on the weekend. The formula could be valid for eight more days, until she has $1 left (on t_9), as long as she gets no more money and spends only $2 per day.

For both sequences and equations, it is important to consider the conditions for which the relationship is valid. For example, a cab company usually rounds up the distance travelled to the nearest mile to determine the charges, so the relationship between the distance travelled and the cost of the ride is valid only for a distance that is a positive integer. Also, the portion of the line left of the y-axis, where x is negative, is part of the mathematical model but has no relevance for the cab ride scenario.

1.5 Exercises

You will need a graphing calculator for Exercise **1, 7, 8** and **15**.

Practice Your Skills

1. Consider the sequence @

   $a_0 = 18$
   $a_n = a_{n-1} - 3$ where $n \geq 1$

   1a.

   a. Graph the sequence.

   b. What is the slope of the line that contains the points? How is that related to the common difference of the sequence? –3; The common difference is the same as the slope.

   c. What is the y-intercept of the line that contains these points? How is it related to the sequence? 18; The y-intercept is the u_0-term of the sequence.

   d. Write an equation for the line that contains these points. $y = 18 - 3x$

2. Refer to the graph at right.

   a. Write a recursive formula for the sequence. What is the common difference? What is the value of u_0? $u_0 = -1$ and $u_n = u_{n-1} + 5$ where $n \geq 1$; 5; –1

   b. What is the slope of the line through the points? What is the y-intercept? 5; –1

   c. Write an equation for the line that contains these points. $y = -1 + 5x$

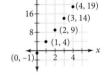

3. Write an equation for the line that passes through the points of an arithmetic sequence with $u_0 = 7$ and a common difference of 3. @ $y = 7 + 3x$

4. Write a recursive formula for a sequence whose points lie on the line $y = 6 - 0.5x$.
$u_0 = 6$ and $u_n = u_{n-1} - 0.5$ where $n \geq 1$

5. Find the slope of each line.

   a. $y = 2 + 1.7x$ @ 1.7   b. $y = x + 5$ 1   c. $y = 12 - 4.5x$ –4.5   d. $y = 12$ ⓗ 0

Assigning Exercises
Suggested: 1–12
Extra Practice: 13–16

Exercises 1–3 It's acceptable for students who are familiar with the intercept form, $y = mx + b$, to use it here. Be sure they use the variable b for the y-intercept instead of for the slope. Encourage students to use u_n and u_0 for discrete formulas and x and y for linear equations.

Exercise 4 If students are having difficulty, suggest that they list the first few terms in the sequence.

Reason and Apply

6. An arithmetic sequence has a starting term, u_0, of 6.3 and a common difference of 2.5.
   a. Write an explicit formula for the sequence. $u_n = 6.3 + 2.5n$
   b. Use the formula to figure out which term is 78.8. 29th term

7. Suppose you drive through Macon, Georgia (which is 82 mi from Atlanta), on your way to Savannah, Georgia, at a steady 54 mi/h.

   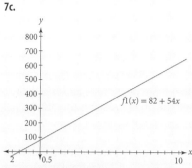

   a. What is your distance from Atlanta two hours after you leave Macon? ⓐ 190 mi
   b. Write an equation that represents your distance, y, from Atlanta x hours after leaving Macon. $y = 82 + 54x$
   c. Graph the equation.
   d. Does this equation model an arithmetic sequence? Why or why not?

   Historic River Street in Savannah, Georgia

8. **APPLICATION** Melissa and Roy both sell cars at the same dealership and have to meet the same profit goal each week. Last week, Roy sold only three cars, and he was below his goal by $2,050. Melissa sold seven cars, and she beat her goal by $1,550. Assume that the profit is approximately the same for each car they sell.
   a. Use a graph to find a few terms, t_n, of the sequence of sales amounts above or below the goal. ⓗ
   b. What is the real-world meaning of the common difference?
   c. Write an explicit formula for this sequence of sales values in relation to the goal. Define variables and write a linear equation.
   d. What is the real-world meaning of the horizontal and vertical intercepts?
   e. What is the profit goal? How many more cars must Roy sell to be within $500 of his goal? $4,750; 2 cars

9. The points on this graph represent the first five terms of an arithmetic sequence. The height of each point is its distance from the x-axis, or the value of the y-coordinate of the point.

   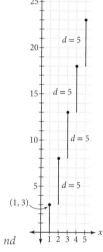

   a. Find u_0, the y-coordinate of the point preceding those given. $u_0 = -2$
   b. How many common differences (d's) do you need to get from the height of $(0, u_0)$ to the height of $(5, u_5)$? 5
   c. How many d's do you need to get from the height of $(0, u_0)$ to the height of $(50, u_{50})$? 50
   d. Explain why you can find the height from the x-axis to $(50, u_{50})$ using the equation $u_{50} = u_0 + 50d$.
   because you need to add 50 d's to the original height of u_0
   e. In general, for an arithmetic sequence, the explicit formula is $u_n = \underline{}$. $u_0 + nd$

LESSON 1.5 Linear Equations and Arithmetic Sequences

8b. The common difference, $900, is the profit per car sold.

8c. $t_n = -4750 + 900n$. Let c represent the number of cars sold, and let p represent the value of the sales in relation to the goal: $p = -4750 + 900c$.

8d. The vertical intercept, -4750, represents how far below the goal Melissa and Roy are if no cars are sold. The horizontal intercept, 5.28, represents the number of cars that must be sold to break even (6 cars must be sold to surpass the goal).

Exercise 10 ALERT Students might not realize that the times in the chart are spaced unequally.

Exercise 12 You may want to mention that free-fall velocity actually has an upper limit, because of air resistance. Until the object reaches that limit, the speed is constantly changing. SUPPORT Use Exercise 12 to assess students' understanding of the difference between a recursive formula and an explicit formula for a single situation. ELL Define *free fall* for students.

Exercise 13 ALERT Students might think that they should add $500 every 6 months for the option with the starting salary of $17,900. Help them realize that in any six-month period only half the current *annual* salary is received (effectively, this means that only $250 is added in each period).

14a. $x = (1 - 0.4)x + 300; x = 750$

14b. No; the long-run value is 750 mL, so eventually the level at the end of the day will approach 750 mL. Once the end-of-the-day level passes 700 mL, when 300 mL is added, it will overflow.

15b. The median and mean prices indicate the midprice and average price, respectively. The standard deviation indicates whether there is price variation (and, therefore, it is worth shopping around for a good deal) or whether all prices are pretty much the same. The median price is the better price to use as a frame of reference. The mean may be low or high if one store is having a sale or is unusually expensive (i.e., is an outlier).

Exercise 16 Students might draw the box before calculating the five-number summary and put the vertical line in the middle of the box. Although this approach gives an appropriate box, challenge students to think of other possible boxes, without the assumption of integer data. For data sets with integer values, it's convenient to start and end bars on 0.5, 1.5, and so on.

---

10. A gardener planted a new variety of ornamental grass and kept a record of its height over the first two weeks of growth.

Time (days)	0	3	7	10	14
Height (cm)	4.2	6.3	9.1	11.2	14

   a. How much does the grass grow each day? 0.7 cm

   b. Write an explicit formula that gives the height of the grass after *n* days. $u_n = 4.2 + 0.7n$

   c. How long will it take for the grass to be 28 cm tall? @ 34 days

11. An arithmetic sequence of six numbers begins with 7 and ends with 27. Follow 11a–c to find the four missing terms.

    a. Name two points on the graph of this sequence: (?, 7) and (?, 27).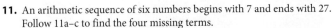
    Possible answer: (0, 7) and (5, 27). The *x*-values should have a difference of 5.

    b. Plot the two points you named in 11a and find the slope of the line connecting the points. Regardless of the points chosen in 11a, the slope should be 4.

    c. Use the slope to find the missing terms. 7, 11, 15, 19, 23, 27

    d. Plot all the points and write the equation of the line that contains them. (h)
    $y = 7 + 4x$. The *y*-intercept will differ if the original points have different *x*-values.

12. **APPLICATION** If an object is dropped, it will fall a distance of about 16 feet during the first second. In each second that follows, the object falls about 32 feet farther than in the previous second.

    a. Write a recursive formula for the distance fallen each second under free fall.
    $u_1 = 16$ and $u_n = u_{n-1} + 32$ where $n \geq 2$

    b. Write an explicit formula for the distance fallen each second under free fall. $u_n = 32n - 16$

    c. How far will the object fall during the 10th second? @ 304 ft

    d. During which second will the object fall 400 feet? 13th second

Leonardo da Vinci (1452–1519) was able to discover the formula for the **velocity** (directional speed) of a freely falling object by looking at a sequence. He let drops of water fall at equally spaced time intervals between two boards covered with blotting paper. When a spring mechanism was disengaged, the boards clapped together. By measuring the distances between successive blots and noting that these distances increased arithmetically, da Vinci discovered the formula $v = gt$, where *v* is the velocity of the object, *t* is the time since it was released, and *g* is a constant that represents any object's downward acceleration due to the force of gravity.

Review

13. **APPLICATION** Suppose a company offers a new employee a starting salary of $18,150 with annual raises of $1,000, or a starting salary of $17,900 with a raise of $500 every six months. At what point is one choice better than the other? Explain. Although the total earnings are different at the end of the odd-numbered 6-month periods, the total yearly income is always the same.

14. Suppose that you add 300 mL of water to an evaporating dish at the start of each day, and each day 40% of the water in the dish evaporates.
    a. Write and solve an equation that computes the long-run water level in the dish. @
    b. Will a 1 L dish be large enough in the long run? Explain why or why not.

15. **APPLICATION** Five stores in Tulsa, Oklahoma, sell the same model of a graphing calculator for $89.95, $93.49, $109.39, $93.49, and $97.69.
    a. What are the median price, the mean price, and the standard deviation? $93.49; $96.80; $7.55
    b. If these stores are representative of all stores in the Tulsa area, of what importance is it to a consumer to know the median, mean, and standard deviation? Which is probably more helpful, the median or the mean?

16. Based on this histogram, create a possible box plot of the data. If you assume all of the data values are integers, how could you find the five-number summary?

    Answers will vary. Given the assumption that all values are integers, the five-number summary is 11, 13, 14, 15, 16.

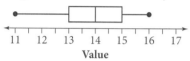

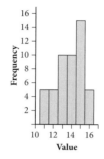

IMPROVING YOUR Reasoning SKILLS

Sequential Slopes

Here's a sequence that generates coordinate points. What is the slope between any two points of this sequence?

$(x_0, y_0) = (0, 0)$
$(x_n, y_n) = (x_{n-1} + 2, y_{n-1} + 3)$ where $n \geq 1$

Now match each of these recursive rules to the slope between points.

a. $(x_n, y_n) = (x_{n-1} + 2, y_{n-1} + 2)$ b. $(x_n, y_n) = (x_{n-1} + 1, y_{n-1} + 3)$
c. $(x_n, y_n) = (x_{n-1} + 3, y_{n-1} - 4)$ d. $(x_n, y_n) = (x_{n-1} - 2, y_{n-1} + 10)$
e. $(x_n, y_n) = (x_{n-1} + 1, y_{n-1})$ f. $(x_n, y_n) = (x_{n-1} + 9, y_{n-1} + 3)$

A. 0   B. 1   C. 3   D. −5   E. $\frac{1}{3}$ F. $\frac{-4}{3}$

In general, how do these recursive rules determine the slope between points?

Extensions

A. Extend Improving Your Reasoning Skills by challenging students to write more than one recursive rule that creates the same slope as rules a–f. Or challenge them to write a generic rule that uses a variable parameter. [For example, rule d could be generalized as any rule in the form $(x_n, y_n) = (x_{n-1} + a, y_{n-1} - 5a)$ where $a \neq 0$.]

B. Pose this problem: "Some cell phone companies now charge their customers in increments of 6 seconds rather than a full minute. Write recursive, explicit, and linear models for a phone call that costs $0.40 for the first minute and $0.02 for each additional 6 seconds. Be sure to give real-world meanings for the variables." This extension could be used as an additional example with the introductory phone call problem.

C. To further extend Improving Your Reasoning Skills, students might explore what happens when x_n or y_n is defined with a geometric or shifted geometric rule. Is it possible to generate collinear points when one or both coordinates are geometric or shifted geometric? If so, how can you recognize the slope from the recursive rules? [If $x_n = rx_{n-1} + h$ and $y_n = sy_{n-1} + k$, the slope $\frac{y_n - y_{n-1}}{x_n - x_{n-1}} = \frac{(s-1)rxn1}{(r-1)x_{n-1} + h}$ can be constant in two ways: if $s = r = 1$, in which case both x_n and y_n are arithmetic sequences and the slope is $\frac{h}{k}$; or if $h = k = 0$ and $r = s \neq 1$, in which case both x_n and y_n are geometric sequences with the same common ratio and the slope is $\frac{y_0}{x_0}$. In this latter case, if $x_0 = 0$, then the line is vertical unless y_0 is also 0, in which case the sequence never leaves the origin.]

IMPROVING YOUR Reasoning SKILLS

Encourage students to plot the first sequence to see that the slope is $\frac{3}{2}$. Students can plot other sequences until they see the pattern that $(x_n, y_n) = (x_{n-1} + h, y_{n-1} + k)$ generates points on a line with slope $\frac{k}{h}$.

Matches: a to B, b to C, c to F, d to D, e to A, and f to E. **ASK** "Why can the slope be determined this way?" [It's the amount of rise per unit of run; by formula, $\frac{y_n - y_{n-1}}{x_n - x_{n-1}} = \frac{y_{n-1} + k - y_{n-1}}{x_{n-1} + h - x_{n-1}} = \frac{k}{h}$.]

ASK "Does changing the starting term affect the slope?" [No; only the recursive rule affects the slope between points.] What effect does the starting term have?" [Changing the starting term would change the y-intercept of a line through the points. However, unless the x-value of the starting term is 0, the starting term does not directly translate into the y-intercept.]

LESSON 1.6

This lesson reviews linear function concepts.

COMMON CORE STATE STANDARDS		
APPLIED	**DEVELOPED**	**INTRODUCED**
N.Q.2	F.IF.5	
A.SSE.1a	F.IF.6	
A.CED.2	F.LE.2	
F.IF.1	F.LE.5	
F.LE.1b		

Objectives
- Use recursion in application contexts
- Deepen understanding of slope
- Define *domain* and *range*
- Define slope formula and intercept form for lines

Vocabulary
dependent variable
independent variable
domain
range
intercept form

Materials
- video, *optional*
- tape
- butcher paper
- balloons
- straws
- string
- graph paper
- Calculator Note: Plotting Data, *optional*

Launch

Line *l* and △LMN and △PQR are graphed below. Which of the following statements is false? Explain.

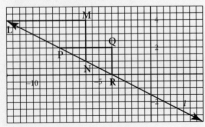

LESSON 1.6

Modeling with Intercept and Slope

There is more to life than increasing its speed.

MOHANDAS GANDHI

While sequences are useful to model values that change discretely (values that jump), many real world measures change continuously (values that flow). Consider a moving object. At any time, the object's position can be measured from a fixed point. So we need more than a model that gives us the position at 1 minute and 2 minutes, we also need a model to give us the position at 1.4 minutes. The explicit model serves better for these situations. When our object is moving at a constant speed, we use a linear model that needs only the starting position and that rate. For these models we often use the **intercept form** of the linear equation.

> **Intercept Form of the Equation of a Line**
>
> You can write the equation of a line as
>
> $y = a + bx$
>
> where the starting value, a, is the y-intercept and the rate, b, is the slope of the line.

Suppose you are taking a long trip in your car. At 5 P.M., you notice that the odometer reads 45,623 miles. At 9 P.M., you notice that it reads 45,831. You find your average speed during that time period by dividing the difference in distance by the difference in time.

$$\text{Average speed} = \frac{45{,}831 \text{ miles} - 45{,}623 \text{ miles}}{9 \text{ hours} - 5 \text{ hours}} = \frac{208 \text{ miles}}{4 \text{ hours}} = 52 \text{ miles per hour}$$

You can also write the rate 52 miles per hour as the ratio $\frac{52 \text{ miles}}{1 \text{ hour}}$. If you graph the information as points of the form (*time, distance*), the slope of the line connecting the two points is $\frac{52}{1}$, which also tells you the average speed.

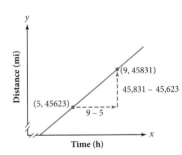

> **Slope**
>
> The formula for the slope between two points, (x_1, y_1) and (x_2, y_2), is
>
> $$\text{slope} = \frac{y_2 - y_1}{x_2 - x_1} \text{ where } x_1 \neq x_2$$

ELL
ELL students often struggle with terms that are used both in the mathematical world and in their daily lives. Terms such as *rate of change*, *domain*, and *range* will probably be more difficult than *independent variable* and *dependent variable*. Also see the Support note.

Support
Using the dynamic graphing capabilities of desmos, built into your ebook, will allow students to explore the concept of slope in a visual and interactive environment. If students have difficulty using the slope formula, relate it to the rate of change by arranging the points in a table format.

Advanced
Advanced students are likely to be comfortable with slope. Assign the **One Step** version of the investigation.

The slope will be the same for any two points selected on the line. In other words, a line has only one slope. Two points on a line can have the same y-value; in that case, the slope of the line is 0. If they had the same x-value, the denominator would be 0 and the slope would be undefined. So the definition of slope specifies that the points cannot have the same x-value. What kinds of lines have a slope of 0? What kinds of lines have undefined slope?

Slope is often represented by the letter m. However, we will use the letter b in linear equations, as in the intercept form $y = a + bx$.

When you are using real-world data, choosing different pairs of points results in choosing lines with slightly different slopes. However, if the data are nearly linear, these slopes should not differ greatly.

INVESTIGATION

Balloon Blastoff

In this investigation you will watch the prepared video for Balloon Blastoff. Then you will write an equation for the rocket's distance as a function of time.

Step 1 The video should provide you with information about time and distance for the rocket. Create a data table and explain which variable is independent and which is dependent for this data.

Step 2 Graph the data. What are the domain and range of your data? Explain.

Step 3 Sketch the graph of the data and select four representative points from the rocket data. Mark the points on your sketch and explain why you chose them.

Step 4 Record the coordinates of the four points and use the points in pairs to calculate slopes. This should give six estimates of the slope.

Step 5 Are all six slope estimates that you calculated in Step 4 the same? Why or why not? Find the mean, median, and mode of your slope estimates. With your group, decide which value best represents the slope of your data. Explain why you chose this value.

Step 6 What is the real-world meaning of the slope, and how is this related to the speed of your rocket?

In many cases, when you try to model the steepness and trend of data points, you may have some difficulty deciding which points to use. In general, select two points that are far apart to minimize the error. They do not need to be data points. Disregard data points that you think might represent measurement errors.

When you analyze a relationship between two variables, you must decide which variable you will express in terms of the other. When one variable depends on the other variable, it is called the **dependent variable**. The other variable is called the **independent variable**. Time is usually considered an independent variable. Why?

LESSON 1.6 Modeling with Intercept and Slope **69**

---

i. The slope of $\overline{LN}$ is equal to the slope of $\overline{PN}$.

ii. The slope of $\overline{LN}$ is equal to the slope of $\overline{PR}$.

iii. The slope of line l is equal to $\frac{QR}{PQ}$.

iv. The slope of line l is equal to $\frac{LM}{MN}$.

iv. should be $\frac{NM}{ML}$

Investigate

This video driven investigation uses water balloons as a model for projectile motion. The ebook and the document Using the Videos found in your online resources offer suggestions for using the video with the investigation. There is also a desmos simulation of this investigation in the ebook. Additionally, there is an Investigation Worksheet that provides sample data. The answers found here are all based on that sample data.

LANGUAGE Independent variable, domain, and range are defined in the paragraphs following the investigation. As needed, define measurement error and trend.

Step 1 See Investigation Worksheet with Sample Data.

Step 2 Answer for sample data:

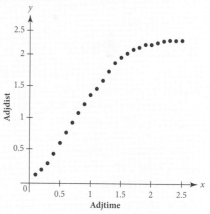

Domain: $0.00 \leq x \leq 2.5$
range: $0.14 \leq y \leq 2.27$; The domain indicates the number of seconds the rocket is in motion. The range indicates the distance from the sensor.

Step 3 Possible answers for sample data:

(0.4, 0.46)
(0.5, 0.64)
(0.6, 0.81)
(1.3, 1.78)

Step 4 Slope of segments:
between A and B: 1.8
between A and C: 1.7
between A and D: 1.39
between B and C: 1.47
between B and D: 1.43
between C and D: 1.75

Step 5 No, the slopes are not the same; the sample data is approximately linear, but not exactly. mean = 1.59; median = 1.59; no mode. The mean or median are the same and would be reasonable choices for a representative slope.

Step 6 The slope is the distance traveled over time. The distance of the balloon rocket from the sensor is increasing by approximately 1.59 m/s.

Modifying the Investigation

Whole Class Have students watch the video. Then show the video again and have a student team collect data in front of the class. Work through Steps 1 through 5 as a class. Discuss Steps 6 and 7.

Shortened Have students watch the video. Use the sample data in the Investigation Worksheet.

One Step Have students watch the video for Balloon Blastoff. Then pose this problem: "What's a linear equation that describes the distance traveled by the rocket after it has been launched?" As students work, be sure they use time as the independent variable, and bring up the ideas of slope, y-intercept, and intercept form as needed.

LESSON 1.6 Modeling with Intercept and Slope **69**

Discussing the Investigation

To review vocabulary, encourage the use of appropriate mathematical vocabulary as students present their solutions from the investigation. Have students specify their independent and dependent variables, as well as their domain and range. **SMP 3, 6**

ASK "Why is time the independent variable? Doesn't the length of time traveled depend on how far the rocket has gone?" [Time is usually taken as the independent variable because the passage of time is unaffected by other events.] **SMP 1, 6**

ASK "In the investigation, what does the slope represent?" [The slope models the average speed of the rocket over its trip.]

As students describe slope, watch to be sure they subtract in the same order in both numerator and denominator. **SMP 6**

Example

This example gives practice with linear equations. For instance, $\frac{1}{20}$ gallons per mile can also be expressed as one gallon every 20 miles or 0.05 gallons every mile. Consider projecting the equation and having students work in small groups. **SMP 1, 3, 6**

Summarize

Have students share their solution and explain their work for both the investigation and the example. Whether student responses are correct or incorrect, ask other students if they agree and why. Encourage students to use appropriate mathematical language as evidence. **SMP 1, 3, 6**

You might remind students of the formula $d = rt$ and rewrite it as $r = \frac{d}{t}$. Then you can relate the ratio of changes to the rate and the slope.

ASK "Is 52 mi/h how fast you were driving most of the time between 5 P.M. and 9 P.M.? [Not necessarily; you might have slowed or stopped and then driven faster than 52 mi/h to compensate.]

Next, you need to think about the domain and range. The set of possible values for the independent variable is called the **domain** and the set of values for the dependent variable is called the **range**.

EXAMPLE Daron's car gets 20 miles per gallon of gasoline. He starts out with a full tank, 16.4 gallons. As Daron drives, he watches the gas gauge to see how much gas he has left.

a. Identify the independent and dependent variables.

b. State a reasonable domain and range for this situation.

c. Write a linear equation in intercept form to model this situation.

d. How much gas will be left in Daron's tank after he drives 175 miles?

e. How far can he travel before he has less than 2 gallons remaining?

Solution The two variables are the distance Daron has driven and the amount of gasoline remaining in his car's tank.

a. The amount of gasoline remaining in the car's tank depends on the number of miles Daron has driven. This means the amount of gasoline is the dependent variable and distance is the independent variable. So use x for the distance (in miles) and y for the amount of gasoline (in gallons).

b. The independent variable is the distance Daron drives. Because his car can go a maximum of 20(16.4), or 328 miles, the domain is $0 \leq x \leq 328$ miles. The dependent variable is the amount of gasoline remaining, so the range is $0 \leq y \leq 16.4$ gallons.

c. Daron starts out with 16.4 gallons. He drives 20 miles per gallon, which means the amount of gasoline decreases $\frac{1}{20}$, or 0.05, gallon per mile. The equation is $y = 16.4 - 0.05x$.

d. You know the x-value is 175 miles. You can substitute 175 for x and solve for y.

$$y = 16.4 - 0.05 \cdot 175$$
$$= 7.65$$

He will have 7.65 gallons remaining.

e. You know the y-value is 2.0 gallons. You can substitute 2.0 for y and solve for x.

$$2.0 = 16.4 - 0.05x$$
$$-14.4 = -0.05x$$
$$288 = x$$

When he has traveled more than 288 miles, he will have less than 2 gallons in his tank.

CRITICAL QUESTION Are the intercept form, $y = a + bx$, and the slope-intercept form, $y = mx + b$, of the equation of a line equivalent? Explain.

BIG IDEA The two forms are equivalent because any variable can be used to represent a quantity. In the slope-intercept form, $y = mx + b$, m represents the slope and b represents the intercept. In the intercept form, $y = a + bx$, b represents the slope and a represents the intercept.

Conceptual/Procedural

Conceptual The investigation helps students attach meaning to slope in modeling a context.

Procedural The example reviews identifying variables, choosing a reasonable domain and range, and writing a linear equation to model a context, and solving the equation to answer questions.

1.6 Exercises

You will need a graphing calculator for exercises **7, 10–12** and **15**.

Practice Your Skills

1. Find the slope of the line containing each pair of points.

   a. (3, −4) and (7, 2)
   $\frac{3}{2}$, or 1.5

   b. (5, 3) and (2, 5)
   $-\frac{2}{3}$, or $-0.\overline{67}$

   c. (−0.02, 3.2) and (0.08, −2.3)
   −55

2. Find the slope of each line.

   a. $y = 3x − 2$ 3
   b. $y = 4.2 − 2.8x$ −2.8
   c. $y = 5(3x − 3) + 2$ 15
   d. $y − 2.4x = 5$ 2.4
   e. $4.7x + 3.2y = 12.9$
      $-\frac{47}{32}$, −1.46875
   f. $\frac{2}{3}y = \frac{2}{3}x + \frac{1}{2}$ 1

3. Solve each equation.

   a. Solve $y = 4.7 + 3.2x$ for y if $x = 3$. $y = 14.3$
   b. Solve $y = −2.5 + 1.6x$ for x if $y = 8$. $x = 6.5625$
   c. Solve $y = a − 0.2x$ for a if $x = 1000$ and $y = −224$. $a = −24$
   d. Solve $y = 250 + bx$ for b if $x = 960$ and $y = 10$. $b = −0.25$

4. Find the equations of both lines in each graph.

   a.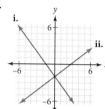

   b.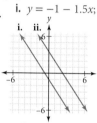

   i. $y = −1 − 1.5x$;   ii. $y = 3.5 − 1.5x$

   a. i. $y = −2 − \frac{4}{3}x$;   ii. $y = −2 + \frac{3}{4}x$

   c. What do the equations in 4a have in common? What do you notice about their graphs?

   c. The equations have the same constant, −2. The lines share the same y-intercept. The lines are perpendicular, and their slopes are reciprocals with opposite signs.

   d. What do the equations in 4b have in common? What do you notice about their graphs?

   d. The equations have the same x-coefficient, −1.5. The lines have the same slope. The lines are parallel.

5. The equation of line m is $y = −4 + 2.5x$. Line l is parallel to line m. Line n is perpendicular to line m. What are the slopes of lines l and n?

   slope of line l: 2.5; slope of line n: $-\frac{1}{2.5} = -0.4$

Reason and Apply

6. Four rocket flights were recorded. Each rocket started at a different distance from the motion sensor and traveled at a different speed. Use the graph to answer the questions.

   a. Which rocket is the fastest? Explain how you know.
   b. Which rocket is the slowest? Explain how you know.
   c. Which rocket is moving at 2 meters per second? Explain how you know.

   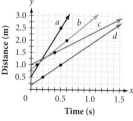

   6a. Rocket a is the fastest. Its graph is the steepest, or has the greatest slope (4 m/s).

   6b. Rocket c is the slowest. Its graph is the flattest, or has the smallest slope (1.25 m/s).

   6c. Rocket b goes through (0.1, 1) and (0.6, 2), so its slope is $\frac{2-1}{0.6 - 0.1} = 2$ m/s.

Exercise 5 If needed, encourage students to sketch and label the lines. To help students see that every line parallel to line m has slope 2.5, and every line perpendicular to line n has slope $-\frac{2}{5}$, have students who graphs different lines l and n present their solutions. After each presentation ask if other students agree and why.

SMP 1, 3, 6

Exercise 7 **LANGUAGE** *Voltage* is electric potential, measured in volts. Emphasize that when describing a real-world meaning, you should use the relevant units. In **7b**, the slope's units are volts per battery.
SMP 2, 4, 6

You may need to remind students that a direct variation relationship is a linear relationship in the form $y = kx$, which includes the value (0, 0).

7a. Possible answer: Using the 2nd and 7th data points gives a slope of approximately 1.47 volts/battery.

7b. Answers will vary. Students should use points that are not too close together. The slope tells you that the voltage increases by about 1.47 volts for every additional battery.

7c. Yes. There is no voltage produced from zero batteries, so the y-intercept should be 0.

8a. Approximately 13. For each additional story, the building increases by about 13 ft.

8b. Approximately 20. This is the additional height in feet that does not depend on the number of stories; the first floor might be taller than the others or its floor could be higher than ground level.

8c. Some of the buildings have more height (or less height) per story.

Context **Engineering Connection**
You might ask students what shapes they see in the bracing in the photo and why those shapes are used. [triangles, parallelograms, x's; the diagonal beams create shear, or sideways support]

Exercise 9 The answers given assume that Anita started her job with no experience 7 years ago. Answers based on different assumptions may also be valid.

---

7. **APPLICATION** Layton measures the voltage across different numbers of batteries placed end to end. He records his data in a table.

Number of batteries	1	2	3	4	5	6	7	8
Voltage (volts)	1.43	2.94	4.32	5.88	7.39	8.82	10.27	11.70

 a. Let x represent the number of batteries, and let y represent the voltage. Find the slope of a line approximating these data. Be sure to include units with your answer. ⓐ

 b. Which points did you use and why? What is the real-world meaning of this slope?

 c. Is this an example of direct variation? Why or why not?

8. This graph shows the relationship between the height of some high-rise buildings in feet and the number of stories in those buildings. A line is drawn to fit the data.

 a. Estimate the slope. What is the meaning of the slope? ⓗ

 b. Estimate the y-intercept. What is the meaning of the y-intercept?

 c. Explain why some of the points lie above the line and some lie below.

 d. According to the graph, what are the domain and range of this relationship? domain: $0 \leq x \leq 80$; range: $20 \leq x \leq 1100$

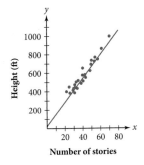

→ *Engineering*
CONNECTION

Many earthquake-prone areas in the United States have strict building codes to ensure that high-rise buildings can withstand earthquakes. Some taller buildings have cross-bracing to make them more rigid and lessen the amount of shaking. The extent of damage that an earthquake can cause depends on several factors: strength of the earthquake, type of underlying soil, and building construction. The shaking increases with height, and the period of shaking (in seconds) is approximately equal to 0.1 times the number of stories in the high-rise. One Maritime Plaza, a 398-foot tall office tower in San Francisco shown here, is one of the first buildings to use seismic bracing in the form of external trusses and X-braces to reduce earthquake damage.

9. This formula models Anita's annual salary for the last seven years: $u_n = 847n + 17{,}109$. The variable n represents the number of years she has worked, and u_n represents her salary in dollars.

 a. What did she earn in her fifth year? What did she earn in her first year? (Think carefully about what n represents.) $20,497; $17,109

 b. What is the rate of change of her salary? $847 per year

 c. What is the first year Anita's salary will be more than $30,000?   in her 17th year

72 CHAPTER 1 Linear Modeling

10. **APPLICATION** This table shows how long it took to lay tile for hallways of different lengths.

Length of hallway (ft)	3.5	9.5	17.5	4.0	12.0	8.0
Time (min)	85	175	295	92	212	153

a. What is the independent variable? Why? Make a graph of the data.

b. Find the slope of the line through these data. What is the real-world meaning of this slope?

c. Which points did you use and why?

d. Find the *y*-intercept of the line. What is the real-world meaning of this value?

11. **APPLICATION** The manager of a concert hall keeps data on the total number of tickets sold and total sales income, or revenue, for each event. Tickets are sold at two different prices.

Total tickets	448	601	297	533	523	493	320
Total revenue ($)	3,357.00	4,495.50	2,011.50	3,784.50	3,334.50	3,604.50	2,353.50

a. Find the slope of a line approximating these data. What is the real-world meaning of this slope?

b. Which points did you use and why?   Answers will vary. Students should use points that are not too close together.

12. **APPLICATION** How much does air weigh? The following table gives the weight of a cubic foot of dry air at the same pressure at various temperatures in degrees Fahrenheit.

Temp. (°F)	0	12	32	52	82	112	152	192	212
Weight (lb)	0.0864	0.0842	0.0807	0.0776	0.0733	0.0694	0.0646	0.0609	0.0591

a. Make a scatter plot of the data.

b. What is the approximate slope of the line through the data?   Using the 3rd and the 7th points gives a slope of −0.000134.

c. Describe the real-world meaning of the slope.
   For each degree the temperature increases, the weight of the cubic foot of dry air decreases by 0.000134 lb.

▶ Recreation
CONNECTION

Hot air rises, cool air sinks. As air is heated, it becomes less dense and lighter than the cooler air surrounding it. This simple law of nature is the principle behind hot-air ballooning. By heating the air in the balloon envelope, maintaining its temperature, or letting it cool, the balloon's pilot is able to climb higher, fly level, or descend. Joseph and Jacques Montgolfier developed the first hot-air balloon in 1783, inspired by the rising of a shirt that was drying above a fire.

12a.
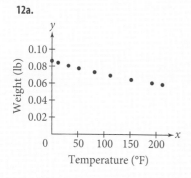

Review

13. Rewrite each expression by eliminating parentheses and then combining like terms.

a. $2 + 3(x - 4)$
   $-10 + 3x$

b. $(11 - 3x) - 2(4x + 5)$
   $1 - 11x$

c. $5.1 - 2.7[1 - (2x + 9.7)]$
   $28.59 + 5.4x$

14. Solve each equation.

a. $12 = 6 + 2(x - 1)$
   $x = 4$

b. $27 = 12 - 2(x + 2)$
   $x = -9.5$

15. Charlotte and Emily measured the pulse rates of everyone in their class in beats per minute and collected this set of data.

{62, 68, 68, 70, 74, 66, 82, 74, 76, 72, 70, 68, 80, 60, 84, 72, 66, 78, 70, 68, 66, 82, 76, 66, 66, 80}

a. What is the mean pulse rate for the class?   ≈71.7 beats/min

b. What is the standard deviation? What does this tell you?   ≈6.47 beats/min. The majority of the data falls within 6.47 beats/min of the mean.

16. Unit conversion: In the activity in this lesson you calculated the speed of the rocket in meters per second. You will now express that same speed in other units. (If you did not do the investigation, then use the value of 3.95 m/s as the speed of the rocket)

a. Find the speed in feet per second using 3.28 feet in a meter. $3.95 \cdot 3.28 = 12.956$ ft/s

b. Find the speed in feet per hour using 3600 seconds in hour. $12.956 \cdot 3600 = 46{,}641.6$ ft/hr

c. Find the speed in miles per hour using 5280 feet in a mile. $46{,}641.6 / 5280 = 8.833635638$ mi/hr

d. Find the speed in miles per century using 876,600 hours in a century. $8.833635638 \cdot 876600 = 7{,}743{,}565$ mi/

e. Find the speed in light-years per century using 5,879,000,000,000 miles in a light-year.
   $7743565 / 5879000000000 \approx 0.000001317$ light-years per century]

17. Each of these graphs was produced by a linear equation in the form $y = a + bx$. For each graph, tell if a and b are greater than zero, equal to zero, or less than zero.

a. 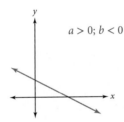 $a > 0; b < 0$

b. 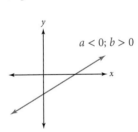 $a < 0; b > 0$

c. 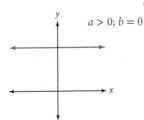 $a > 0; b = 0$

d. 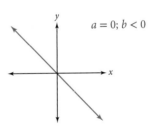 $a = 0; b < 0$

Extension

Have students use a voltage meter, or perhaps a voltage probe with a CBL (calculator-based lab), to collect the data referred to in Exercise 7. Answer for sample data:

LESSON 1.7

Models and Predictions

When you graph points from an arithmetic sequence, they lie on a line. When you collect and graph real-life data, the points may appear to have a linear relationship. However, they will rarely lie exactly on a line. They will usually be scattered, and it is up to you to determine a reasonable location for the line that summarizes or gives the trend of the data set. A line that fits the data reasonably well is called a **line of fit.** A line of fit can be used to make predictions about the data, so it is also called a **prediction line.**

Odd how the creative power at once brings the whole universe to order.
VIRGINIA WOOLF

There is no single list of rules that will give the best line of fit in every instance, but you can use these guidelines to obtain a reasonably good fit.

Finding a Line of Fit

1. Determine the direction of the points. The longer side of the smallest rectangle that contains most of the points shows the general direction of the line.

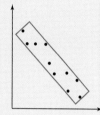

2. The line should divide the points equally. Draw the line so that there are about as many points above the line as below the line. The points above the line should not be concentrated at one end, and neither should the points below the line. The line has nearly the same slope as the longer sides of the rectangle.

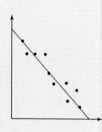

Some computer applications allow you to manually fit a line to data points. You can also graph the data by hand and draw the line of fit.

Once you have drawn a line of fit for your data, you can write an equation that expresses the relationship. To indicate that the line is a prediction line, the variable $\hat{y}$ ("y hat") is used in place of y.

As you learned in algebra, there are several ways to write the equation of a line. You can find the slope and the y-intercept and write the intercept form of the equation. Often, it is easier to choose any two points on the line, use them to calculate the slope, and use the slope and either of the points to write the **point-slope form** of the equation. Either method should give almost exactly the same results. In general, using two points that are farther apart results in a more accurate calculation of the slope than using two points that are closer together.

LESSON 1.7 Models and Predictions 75

LESSON 1.7

In this lesson students review the point-slope form of a linear equation while experimenting with lines that fit data sets.

COMMON CORE STATE STANDARDS		
APPLIED	**DEVELOPED**	**INTRODUCED**
N.Q.2	F.LE.2	S.ID.6a
A.SSE.1a	F.LE.5	
A.CED.2	S.ID.7	
F.LE.1b		

Objectives
- Find a line of fit for data that are approximately linear
- Review point-slope form of a line
- Use interpolation and extrapolation
- Learn the properties of a line of fit

Vocabulary
line of fit
prediction line
point-slope form
interpolation
extrapolation

Materials
- stopwatch (or watch with a second hand)
- uncooked linguine, *optional*
- Calculator Notes: Entering and Graphing Equations; Function Tables, *optional*
- Project: Talkin' Trash

Launch

When asked to draw a line to represent these data points, Kurt drew the red line and Kristin drew the blue line. Who drew the best line for the data? Explain.

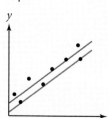

Kristin's line divides the data points more equally. There are no points below Kurt's line.

LESSON 1.7 Models and Predictions 75

ELL
Have students write paragraphs or present their investigation answers in their groups. Have students explain or demonstrate the relationship between the point-slope equation of a line and the equation they wrote in Step 2 of the investigation.

Support
Using the point-slope formula with subscripts (x_1, y_1) might be confusing to students. To increase their recognition of the difference between (x, y) and (x_1, y_1) in the formula, use different colors to indicate the given point and the variables. Emphasize that any point can be used to write an equation in point-slope form.

Advanced
Pose the question from the **One Step** so students can decide the best way to create a linear equation for the investigation. Challenge students to come up with the best method for finding an accurate line of fit and then justify it.

To assess prior knowledge, as students present their answers to the Launch, use questions to review the vocabulary. For example, **ASK** "What do we call a line drawn to represent data?" for **line of fit**. To review the process for finding a line of fit **ASK** "How do you think the way Kurt chose to draw his line differs from Kristin's method?" Whether student responses are correct or incorrect, ask other students if they agree and why. **SMP 3, 6**

A piece of uncooked linguine or the edge of a clear ruler is helpful for seeing lines of fit on paper. Students can input the data into their calculator and use the line of fit. **SMP 5**

ASK "Does the point-slope form work for vertical lines?" [No, because their slope is undefined.]

Investigate

Example

Project the example and let students work on it in small groups. As students present their solutions and explain their work, look for students who have used different strategies. The solution in the book does not show the rectangle from the first step mentioned on page 75, so look for groups who did draw the rectangle. Whether student responses are correct or incorrect, ask other students if they agree and why. **SMP 1, 3, 5, 6**

Point-Slope Form

The formula for the slope is $b = \dfrac{y_2 - y_1}{x_2 - x_1}$, so you can write the equation of a line with slope b and containing point (x_1, y_1) for any general point (x, y) as $b = \dfrac{y - y_1}{x - x_1}$ or, equivalently, as

$$y = y_1 + b(x - x_1)$$

This is called the **point-slope form** for a linear equation.

You can then use this equation to predict points for which data are not available.

EXAMPLE On a barren lava field on top of the Mauna Loa volcano in Hawaii, scientists have been monitoring the concentration of CO_2 (carbon dioxide) in the atmosphere since 1959. This site is favorable because it is relatively isolated from vegetation and human activities that produce CO_2. The average concentrations for 13 different years, measured in parts per million (ppm), are shown here.

Year	CO_2 (ppm)	Year	CO_2 (ppm)	Year	CO_2 (ppm)
1990	354.35	2000	369.52	2010	389.85
1992	356.38	2002	373.22	2012	393.82
1994	358.82	2004	377.49	2014	398.55
1996	362.59	2006	381.90		
1998	366.65	2008	385.59		

(Carbon Dioxide Information Analysis Center)

a. Find a line of fit to summarize the data.
b. Estimate the concentration of CO_2 in the atmosphere in the year 2011.
c. Predict the concentration of CO_2 in the atmosphere in the year 2050.
d. How much confidence do you have in these two predictions? Why?

Science CONNECTION

With a name meaning "long mountain," Mauna Loa has an area of 2035 mi², covers half of the island of Hawaii, and is Earth's largest active volcano. Volcanologists routinely monitor Mauna Loa for signs of eruption and specify hazardous areas of the mountain.

Caldera on Mauna Loa

76   CHAPTER 1   Linear Modeling

Modifying the Investigation

Whole Class Discuss Steps 2 and 3 as a class.

Shortened Use the sample data. Have groups complete Step 2. Discuss Step 3.

One Step Pose this problem: "How can we estimate how long it will take 100 people to do the wave?" Lead the class to design an experiment using different-size groups of students. Agree on a good data set as a class, and then let cooperative groups figure out ways to make the prediction. As students work, you can direct them to use the point-slope form of a linear equation.

Solution | Let x represent the year, and let y represent the concentration of CO_2 in parts per million. A graph of the data shows a linear pattern.

a. You can draw a line that seems to fit the trend of the data.

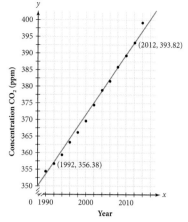

The line shown on the graph is a good fit because many of the points lie on or near the line, the data points above and below the line are roughly equal in number, and they are evenly distributed on both sides along the line. You do not know what the y-intercept should be because you do not know the concentration in the year 0; you have no reason to believe that the line passes through (0, 0).

Next, you need to find the equation of this line of fit. The y-intercept is not easily available, but you can choose two points and use the point-slope form. In this case the line of fit passes through a number of the data points. You might choose the points (1992, 356.4) and (2012, 393.8). They are far enough apart. When you have a data set that is more scattered, you might choose points on the line of fit that are not actually data points.

You can use these points to find the slope.

$$\text{slope} = \frac{393.8 - 356.4}{2012 - 1992} = \frac{37.4}{20} = 1.87$$

This means the concentration of CO_2 in the atmosphere has been increasing at a rate of about 1.87 ppm each year.

Use either of the two points you chose for the slope calculation, and substitute its coordinates for x_1 and y_1 in the point-slope form of a linear equation.

$\hat{y} = 393.8 + 1.87(x - 2012)$ Use point (2012, 393.8), substituting 2012 for x_1 and 393.8 for y_1.

b. You can substitute 2011 for x and evaluate for $\hat{y}$ to estimate the CO_2 concentration in the year 2011. The estimation is 391.93 ppm.

$\hat{y} = 393.8 + 1.87(2011 - 2012) = 391.93$

c. You can substitute 2050 for x and solve for $\hat{y}$ to predict the CO_2 concentration in the year 2050. The prediction is 464.86 ppm.

$\hat{y} = 393.8 + 1.87(2050 - 2012)$
$\phantom{\hat{y}} = 464.86$

d. The estimation for 2011 seems very reasonable (about halfway between the 2010 value and 2012 value), so you trust it. But the 2050 value is some 35 years into the future. What if the data is actually a slight curve instead of a line? This could make a big difference over 35 years. Or what if something changes on earth and the line suddenly moves at a different slope. Lots of things can happen in 35 years. You do not put anywhere near as much confidence in this prediction as the first estimation.

Step 2 answers from the sample data:

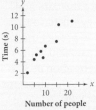

Using the point (5, 4), the equation is $\hat{y} \approx 4 + \frac{6}{13}(x - 5)$. The slope indicates that it takes 0.45 s longer per additional person. The x-intercept is approximately -4.2 and does not have real-world meaning in this context; the y-intercept is 1.9 and indicates the additional time the first person needs to stand and the additional time the last person needs to fully sit down. It might also include reaction time for starting and stopping the stopwatch. The domain could start at 0 people, at 1 person, or at 2 people and increase to the number of students in the entire class.

Summarize

Discussing the Investigation

Having students who have used different strategies present their solutions and then asking other students if they agree and why emphasizes that often more than one line of fit is acceptable, as long as students follow the guidelines and can explain their reasoning.
SMP 1, 3, 5, 6

ASK "What are the units of the slope?" [seconds per person]

ASK "Why does the book replace (x_2, y_2) with (x, y) in the slope formula?" As students struggle to articulate their ideas, you might ask questions such as this: "Are you saying that x and y satisfy an equation if and only if the point with coordinates (x, y) lies on the graph of that equation?" Help the class see that to find a relationship between the coordinates of all points on the line, we let (x, y) represent any point on the line and decide how x and y can be related by an equation.

CRITICAL QUESTION What is the difference between using $y =$ and $\hat{y} =$ in an equation?

BIG IDEA Statistics books customarily use $\hat{y}$ to represent a predicted y-value of a function.

Environment
CONNECTION

Cars and trucks emit carbon dioxide, methane, and nitrous oxide by burning fossil fuels. These gases trap heat from the sun, producing a "greenhouse effect" and causing global warming. Some scientists warn that if accelerated warming continues, higher sea levels will result.

The planet Venus suffers from an extreme greenhouse effect due to constant volcanic activity, which has created a dense atmosphere that is 97% carbon dioxide. As a result, surface temperatures reach 462°C.

Finding a value between other values given in a data set is called **interpolation**. Using a model to extend beyond the first or last data points is called **extrapolation**. How would you use the Mauna Loa data in the example to estimate the CO_2 levels in 1960? In 1991?

INVESTIGATION
The Wave

YOU WILL NEED
- a stopwatch or watch with second hand

Sometimes at sporting events, people in the audience stand up quickly in succession with their arms upraised and then sit down again. The continuous rolling motion that this creates through the crowd is called "the wave." You and your class will investigate how long it takes different-size groups to do the wave.

Step 1 Using different-size groups, determine the time for each group to complete the wave. Collect at least nine pieces of data of the form (*number of people, time*), and record them in a table.

Step 2 Plot the points, and find the equation of a reasonable line of fit. Write a paragraph about your results. Be sure to answer these questions:
- What is the slope of your line, and what is its real-world meaning?
- What are the x- and y-intercepts of your line, and what are their real-world meanings?
- What is a reasonable domain for this equation? Why?

Sample answer: No, for a large group of people, the wave gains momentum and begins to travel faster. So for a large group of people, the data may not be linear.

Step 3 Can you use your line of fit to predict how long it would take to complete the wave if everyone at your school participated? Everyone in a large stadium? Explain why or why not.

Trend lines are often used in the news or in research to project or extrapolate into the future. You may be asked to do that in some of the exercises. You should always carefully consider how much confidence you can place in those projections.

1960: extrapolate $\hat{y} = 296.56$ 1991: interpolate $\hat{y} = 354.53$

CRITICAL QUESTION Why do we want a line of fit for a data set in the first place?

BIG IDEA Lines of fit help us make predictions, using *extrapolation* and *interpolation* for those predictions.

CRITICAL QUESTION Are point-slope equations the same for different data points?

BIG IDEA They are the same if and only if all the points lie exactly on the same line. You might want to produce data points that are truly collinear, have students find point-slope equations using various data points, and point out how those equations are equivalent.

1.7 Exercises

You will need a graphing calculator for exercises **8–10**.

Practice Your Skills

1. Write an equation in point-slope form for each line shown.

   a. 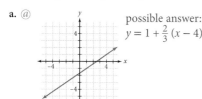 possible answer: $y = 1 + \frac{2}{3}(x - 4)$

   b. 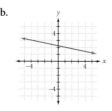 possible answer: $y = 2 - \frac{1}{5}(x - 1)$

2. Write an equation in point-slope form for each line described.
   a. slope $\frac{2}{3}$ passing through $(5, -7)$ $y = -7 + \frac{2}{3}(x - 5)$
   b. slope -4 passing through $(1, 6)$ $y = 6 - 4(x - 1)$
   c. parallel to $y = -2 + 3x$ passing through $(-2, 8)$ $y = 8 + 3(x + 2)$
   d. parallel to $y = -4 - \frac{3}{5}(x + 1)$ passing through $(-4, 11)$ $y = 11 - \frac{3}{5}(x + 4)$

3. Solve each equation.
   a. Solve $u_n = 23 + 2(n - 7)$ for u_n if $n = 11$. $u_n = 31$
   b. Solve $d = -47 - 4(t + 6)$ for t if $d = 95$. $t = -41.5$
   c. Solve $y = 56 - 6(x - 10)$ for x if $y = 107$. $x = 1.5$

4. Consider the line $y = 5$.
   a. Graph this line and identify two points on it. possible answer: $(0, 5), (1, 5)$
   b. What is the slope of this line? 0
   c. Write the equation of the line that contains the points $(3, -4)$ and $(-2, -4)$. $y = -4$
   d. Write three statements about horizontal lines and their equations. Horizontal lines have no x-intercepts (unless the horizontal line is $y = 0$, the x-axis); they have a slope of 0; all points have the same y-coordinate.

5. Consider the line $x = -3$.
   a. Graph it and identify two points on it. possible answer: $(-3, 0), (-3, 2)$
   b. What is the slope of this line? undefined
   c. Write the equation of the line that contains the points $(3, 5)$ and $(3, 1)$. $x = 3$
   d. Write three statements about vertical lines and their equations. Vertical lines have no y-intercepts (unless the vertical line is $x = 0$, the y-axis); they have undefined slope; all points have the same x-coordinate.

Reason and Apply

6. Of the graphs below, choose the *one* with the line that best satisfies the guidelines from the beginning of this lesson. For each of the other graphs, explain which guidelines the line violates.

   a.   b.   c.   d.

LESSON 1.7 Models and Predictions   79

Exercises 7–10 Equations will vary somewhat, depending on which points students choose to be on the lines. Any line that fits the data "by eye" is acceptable if students can explain how they arrived at it. **SMP 1**

Possible answers:

7 i. The y-intercept is about 1.7; (5, 4.6)

ii. The y-intercept is about 7.5; (5, 3.75)

iii. The y-intercept is about 8.6; (5, 3.9)

7a i. $\hat{y} = 1.7 + 0.58x$;

ii. $\hat{y} = 7.5 - 0.75x$;

iii. $\hat{y} = 8.6 - 0.94x$

7b i. $\hat{y} = 4.6 + 0.58(x - 5)$;

ii. $\hat{y} = 3.75 - 0.75(x - 5)$;

iii. $\hat{y} = 3.9 - 0.94(x - 5)$

7c i. for (5, 4.6): $\hat{y} = 4.6 + 0.58x - 2.9$
$= 1.7 + 0.58x$;

ii. for (5, 3.75): $\hat{y} = 3.75 - 0.75x + 3.75$
$= 7.5 - 0.75x$;

iii. for (5, 3.9): $\hat{y} = 3.9 - 0.94x + 4.7$
$= 8.6 - 0.94x$

8a. Possible answer: $\hat{y} = 17 + 0.67(x - 15)$. There is a cost increase of about 67¢ per photo.

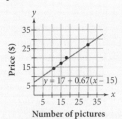

9a. Answers will vary. Possible answer: $145 \le x \le 200$, $40 \le y \le 52$.

9c. On average, a student's forearm length increases by 0.26 cm for each additional 1 cm of height.

9d. The y-intercept is meaningless because a height of 0 cm should not predict a forearm length of 0.71 cm. The domain should be specified.

---

7. For each graph below, lay your ruler along your best estimate of the line of fit. Estimate the y-intercept and the coordinates of one other point on the line.

  a. Write an equation in intercept form for the line of fit for each graph.

  b. Use the other point you identified and write the equation in point slope form.

  c. Use algebra to show that the lines in 7b have the same intercept as those in 7a.

  i. @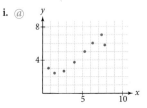

  ii.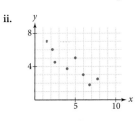

  iii.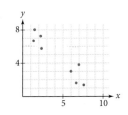

8. APPLICATION A photography studio offers several packages to students posing for yearbook photos. Let x represent the number of pictures, and let y represent the price in dollars.

Number of pictures	30	18	15	11
Price ($)	27.00	20.00	17.00	14.00

  a. Plot the data, and find an equation of a line of fit. Explain the real-world meaning of the slope of this line.

  b. Find the y-intercept of your line of fit. Explain the real-world meaning of the y-intercept.   About 6.95. The sitting fee, or cost just for having your picture taken, is $6.95.

  c. If the studio offers a 50-print package, what do you think it should charge?   about $57

  d. How many prints do you think the studio should include in the package for a $9.99 special?   4 or 5 prints, or perhaps 6 prints if the studio wants it to be a good deal for customers

9. Use height as the independent variable and length of forearm as the dependent variable for the data collected from nine students.

  a. Name a good graphing window for your scatter plot.

  b. Write a linear equation that models the data. @ $\hat{y} = 0.26x + 0.71$

  c. Write a sentence describing the real-world meaning of the slope of your line.

  d. Write a sentence describing the real-world meaning of the y-intercept. Explain why this doesn't make sense and how you might correct it.

  e. Use your equation to estimate the height of a student with a 50 cm forearm and to estimate the length of a forearm of a student 158 cm tall. 189.58 cm; 41.79 cm

Height (cm)	Forearm (cm)
185.9	48.5
172.0	44.5
155.0	41.0
191.5	50.5
162.0	43.0
164.3	42.5
177.5	47.0
180.0	48.0
179.5	47.5

10. **APPLICATION** This data set was collected by a college psychology class to determine the effects of sleep deprivation on students' ability to solve problems. Ten participants went 8, 12, 16, 20, or 24 hours without sleep and then completed a set of simple addition problems. The number of addition errors was recorded.

Hours without sleep	8	8	12	12	16	16	20	20	24	24
Number of errors	8	6	6	10	8	14	14	12	16	12

a. Define your variables and create a scatter plot of the data.

b. Write an equation of a line that approximates the data and sketch it on your graph.

c. Based on your model, how many errors would you predict a person to make if she or he hadn't slept in 22 hours? 14 errors

d. In 10c, did you use interpolation or extrapolation? Explain. interpolation, because you found a value between those given in the table

e. Over what domain would extrapolation make sense in this problem? Explain your answer.

Review

11. The 3rd term of an arithmetic sequence is 54. The 21st term is 81. Find the 35th term. @ 102

12. Given the data set {20, 12, 15, 17, 21, 15, 30, 16, 14}:

    a. Find the mean. 17.778

    b. Add as few elements as possible to the set in order to make 19.5 the mean. Add only the number 35 to the set.

    c. Find the median. 16

    d. Add as few elements as possible to the set in order to make 19.5 the median. Add 19.5 and any three numbers greater than 19.5

13. Earl's science lab group made six measurements of mass and then summarized the results. Someone threw away the measurements. Help the group reconstruct the measurements from these statistics. 2.3 g, 3.0 g, 3.0 g, 3.4 g, 3.6 g, 3.9 g

    • The median and mean are both 3.2 g.
    • The mode is 3.0 g.
    • The IQR is 0.6 g.
    • The largest deviation is −0.9 g.

14. You start 8 meters from a marker and walk toward it at the rate of 0.5 m/s. $u_0 = 8$ and $u_n = u_{n-1} - 0.5$, where $n \geq 1$

    a. Write a recursive rule that gives your distance from the marker after each second.

    b. Write an explicit formula that allows you to find your distance from the marker at any time. $u_n = 8 - 0.5n$

    c. Interpret the real-world meaning of a negative value for u_n in 14a or 14b.
    A negative value means that you have passed the marker.

LESSON 1.8

In this lesson students see how to use residuals to measure how well a line represents a data set.

COMMON CORE STATE STANDARDS
APPLIED	DEVELOPED	INTRODUCED
	S.ID.6	
	S.ID.7	

Objectives
- Define *residual*
- Calculate the residual sum for a small number of data points
- Use residuals to test the fit of a line

Vocabulary
residuals

Materials
- springs
- mass holders
- small unit masses
- centimeter rulers
- Calculator Note: Residuals and the Root Mean Square Error, *optional*

Launch

Project Example A from your ebook. Suggest that students graph the data and find a best fit line to determine how many pizzas should she plan for this November.

After students present their graphs and lines of fit **ASK** "In business, it is important to be as exact as possible with inventory. Is your prediction exact enough? How do you know?"

You might use a line from the Launch to review residual terminology and how to find residuals. Or you can project the graphic in the introduction to the lesson. **SMP 1, 2, 3, 6**

LESSON 1.8

Residuals and Fit

The growth of understanding follows an ascending spiral rather than a straight line.
JOANNA FIELD

You have used different methods of finding linear models to fit a set of data. However, unless your data are perfectly linear, any method may not create the "best" fit. To reach this goal you must first set your criteria for best. You will need a way to evaluate how accurately your model describes the data. One highly valued method is to look at the **residuals**, or differences between the *y*-values of your data and the predicted *y*-values of your model. Graphically, this is the vertical distance between the data points and the line of fit.

residual = *y*-value of data point − *y*-value of point on line

Similar to the deviation from the mean that you learned about in an earlier algebra course, a residual is a signed distance. Here, a positive residual indicates that the point is above the line, and a negative residual indicates that the point is below the line.

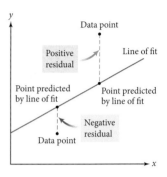

A line of good fit should have about as many points above it as below. This means that the sum of the residuals should be near zero. In other words, if you connect each point to the line with a vertical segment, the sum of the lengths of the segments above the line should be about equal to the sum of those below the line.

EXAMPLE A The manager of Big K Pizza must order supplies for the month of November. The numbers of pizzas sold in November during the past four years were 512, 603, 642, and 775, respectively. How many pizzas should she plan for this November?

Solution Let *x* represent the past four years, 1 through 4, and let *y* represent the number of pizzas. Graph the data. The manager fit this data using the first and last points with the linear model $\hat{y} \approx 417.3 + 87.7x$.

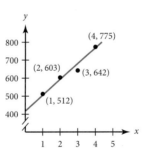

Calculating the residuals is one way to evaluate the accuracy of this model before using it to make a prediction. Evaluate the linear equation when *x* is 1, 2, 3, and 4. When *x* = 1, for example, 417.3 + 87.7(1) = 505.0. A table helps organize the information.

ELL
Using a demonstration from software that allows students to observe the changes in residuals with a movable line will emphasize the visual representation of residuals.

Support
Have students use the regression feature on their calculators to aid with the lengthy process of calculating a line of fit and residuals. Use the technology in the discussions of the examples and the Investigation, reiterating each step to help students become familiar with the process.

Advanced
Challenge students to find a linear model for the data in Example A with a residual sum even closer to 0 than that produced by the best fit line. Use the **One Step** Investigation found in your ebook.

Year (x)	1	2	3	4
Number of pizzas (y)	512	603	642	775
y-value from line $\hat{y} \approx 417.3 + 87.7x$	505.0	592.7	680.3	768.0
Residual ($y - \hat{y}$)	7.0	10.3	−38.3	7.0

The sum of the residuals is $7 + 10.3 + (-38.3) + 7.0$, or -14.0 pizzas, which is fairly close to zero in relation to the large number of pizzas purchased. The linear model is therefore a pretty good fit. The sum is negative, which means that all together the points below the line are a little farther away than those above.

If the manager plans for $417.3 + 87.7(5)$, or approximately 856 pizzas, she will be very close to the linear pattern established over the past four years. Because the residuals range from -38.3 to 7, she may want to adjust her prediction higher or lower depending on factors such as whether or not the supplies are likely to spoil or whether or not she can easily order more supplies.

Making sure that the sum of the residuals is near zero should not be your only criteria. In the next example you will learn another procedure to fit a line to data AND another way to judge the fit using residuals. When a set of data has a large amount of variation it may be difficult to select points to create an equation. The following example uses the statistics of each variable for creating a model.

EXAMPLE B From 2007 to 2014 there were 13 films with a budget of over $250 million. The following data show how well each film has done both in the US and worldwide. How could you predict the worldwide gross based on the US gross?

Release Date	Movie	Production Budget	Domestic Gross	Worldwide Gross
12/18/2009	Avatar	$425,000,000	$760,507,625	$2,783,918,982
5/24/2007	Pirates of the Caribbean: At World's End	$300,000,000	$309,420,425	$963,420,425
7/20/2012	The Dark Knight Rises	$275,000,000	$448,139,099	$1,084,439,099
7/2/2013	The Lone Ranger	$275,000,000	$89,289,910	$259,989,910
3/9/2012	John Carter	$275,000,000	$73,058,679	$282,778,100
11/24/2010	Tangled	$260,000,000	$200,821,936	$586,581,936
5/4/2007	Spider-Man 3	$258,000,000	$336,530,303	$890,875,303
5/1/2015	The Avengers: Age of Ultron	$250,000,000	$458,991,599	$1,404,691,599
12/14/2012	The Hobbit: An Unexpected Journey	$250,000,000	$303,003,568	$1,017,003,568
7/15/2009	Harry Potter and the Half-Blood Prince	$250,000,000	$301,959,197	$935,083,686
12/13/2013	The Hobbit: The Desolation of Smaug	$250,000,000	$258,366,855	$960,366,855
12/17/2014	The Hobbit: The Battle of the Five Armies	$250,000,000	$255,119,788	$955,119,788
5/20/2011	Pirates of the Caribbean: On Stranger Tides	$250,000,000	$241,063,875	$1,045,663,875

http://www.the-numbers.com/movie/budgets/all

ASK "What is the shortest distance from the point to the line?" [The length of a line drawn from the point perpendicular to the line.] "Is it better to calculate distances from the line vertically rather than horizontally?" [Yes, because the line's equation is more likely to be given in $y =$ form than in $x =$ form. The accuracy of predictions must be measured in the same units as y. The residuals are vertical so the units match.] You might draw some similar triangles to show how dividing each residual by the line's slope gives the horizontal distance to the line. **SMP 1, 6**

You will see the term "best fit line" in this lesson. This simply means that a model has been found that "best" satisfies some criteria. While the criteria they are trying to meet may change from one lesson to the next, the student should recognize what the criteria are. Many lines can be best when the criteria are unspecified.

Investigate

Example A

If you did the Launch, ask student groups to find the residuals of their model. Or you can project Example A and work through finding the residuals as a class, with student input. Focus on assessing the model with residuals.

ALERT The order does matter here, so make sure students find the residual by taking the actual data value minus the predicted value. **SMP 6**

ASK "What is the real-world meaning of the residual −38.4?" [In year 3, the number of pizzas actually sold was 38.4 fewer than predicted by the model. By this model, (3, 642) is an outlier because the magnitude of its residual is extreme.]

Modifying the Investigation

Whole Class Collect data with student assistance, or use the desmos simulation in the ebook to collect data. After students have completed Step 2 on their own, discuss Steps 3 and 4.

Shortened Use the Investigation Worksheet Spring Experiment with Sample Data. An alternative approach is to use the same data for all groups and have each group use a different line-fitting method in Step 2. Then in Step 4, students can use the residuals and the residual sums to compare the fit and select the best model.

One Step Use the **One Step** found in your ebook. Challenge students to find a measure of how well a line of fit represents the data set. Suggest, as needed, that they consider how far the line is from each data point.

Example B

Consider projecting Example B and soliciting input from students on "how" you could predict the worldwide gross of the films. Then work through the solution as a class activity, with student input. You could make a handout of the example without the solution and have students work in small groups. Have them present their strategies as well as their solutions.

SMP 1, 3, 5, 6

The point referred to as the center, $(\bar{x}, \bar{y})$, is technically called the centroid of the data, but that vocabulary is not critical at this stage. The process described here for finding a line is actually the beginning steps of a least-squares linear regression. The "regression to the mean," as Francis Galton (1822 – 1911) described, is that values in the center tend to predict values in the center, values in the extremes will predict extremes, BUT not as extreme. To finish the creation of the linear model you need only to multiply your slope by the correlation of your sample data. Thus data with weaker correlation will "regress" more to the mean because your slope is closer to zero than the ratio of the standard deviations.

Guiding the Investigation

There is a desmos simulation of the investigation in your ebook.

Step 1 If you can't borrow spring equipment from the science department, you can use a Slinky Jr. cut in half and place the weights (masses) in film canisters attached to the springs with string. The unit masses can be uniform beans, candies, or pennies. The answers to Steps 2–6 use the sample data from the Investigation Worksheet: Spring Experiment with Sample Data.

Solution The domestic incomes have a mean of $310 million and a standard deviation of $176 million. The worldwide incomes have a mean of $1,013 million and a standard deviation of $620 million.

To create a line, you need a point and a slope. One point you can use is the center of your data. The point $(\bar{x}, \bar{y})$ is the point made from the means of the x and y data values, in this case, (310, 1013). A slope can be found using the ratio of a change in y over a change in x. You can use the standard deviations as average changes in y and x. This gives a slope value of $\frac{s_y}{s_x} = \frac{620}{176} = 3.52$. So the point-slope equation is $y = 1013 + 3.52(x - 310)$.

Now that you have a model, return to your data lists. Using your column for domestic gross, add a new column of data for the amount predicted for worldwide gross by your model and label it Predicted Worldwide Gross. This list can often be created in technology with a single command. These are y-values on the line and lie either directly below or above each data point. Next you calculate the residuals by subtracting your two lists of worldwide grosses. Our "best" line should show no pattern in a scatter plot of these residuals versus the x-values of our points.

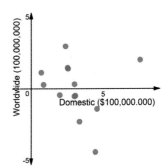

$f(x) = 10.13 + 3.523(x - 3.10)$			
x_1	y_1	$f(x_1)$	$y_1 - f(x_1)$
7.6	27.84	25.9835	1.8565
3.09	9.63	10.09477	−0.46477
4.48	10.84	14.99174	−4.15174
0.89	2.6	2.34417	0.25583
0.73	2.83	1.78049	1.04951
2.01	5.87	6.28993	−0.41993
3.37	8.91	11.08121	−2.17121

Looking at this plot, you may decide that, other than the point on the right, there is a downward trend to the other 12 points. If you twist the line, decreasing the slope, the residuals on the left will go down and those on the right will go up. By experimenting you can find $y = 1013 + 2.55(x - 310)$ is a good model for all but one point.

Conceptual/Procedural

Conceptual This lesson focuses on the concept of looking for patterns in the residual plot to assess the fit of a linear model. Students also look at adjusting their model to be a better fit.

Procedural Students create a linear model for a set of data and learn how to use residuals to assess the fit of the model.

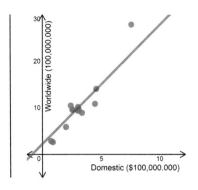 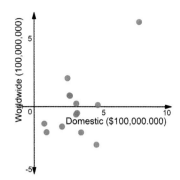

Note that standard deviations are always positive but slopes are sometimes negative. Using the procedure above, you may have to change the sign of your slope when the data has a negative association (downward trend). The investigation gives you another opportunity to explore the meaning of the residuals.

INVESTIGATION

Spring Experiment

YOU WILL NEED
- a spring
- a mass holder
- small unit masses
- a ruler

Procedure Note
1. Attach the mass holder to the spring.
2. Hang the spring from a support, and the mass holder from the spring.
3. Measure the length of the spring (in centimeters) from the first coil to the last coil.

In this investigation you will collect data on how a spring responds to various weights. You will create a model for this relationship and investigate how to judge how well the model fits.

Step 1 Place different amounts of mass on the mass holder, recording the corresponding length of the spring each time. Collect about 10 data points of the form (*mass, spring length*).

Step 2 Plot the data and find a linear model, using the procedure from Example B.

Step 3 Give the real-world meanings of the slope and the *y*-intercept.

Step 4 Use a table to organize your data and calculate the residuals. Make a plot of (*mass, residual*). Comment on the balance of the residuals above and below the line as well as left and right of center.

Step 5 Have each member of the team experiment with your model in a different way (i.e. one person could increase the slope slightly). Then create a new residual plot. Answers will vary.

Step 6 If any team member found that their change improved the balance of the residuals, state your new model and explain what the change did.
Answers will vary.

Step 2 Ask students to use a full sheet of graph paper to give them plenty of space to work. Students can calculate the line of fit by hand or on a calculator. **ASK** "Why aren't all the groups' equations the same?" [Different masses and springs were used. Answers given are for sample data.] **SMP 1, 2, 5**

Step 2 approximately $\hat{y} = 3.173 + 0.038x$

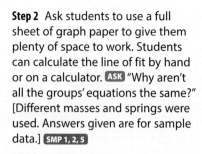

Step 3 The slope of this line is called the *spring constant*.

Step 4 You might want to introduce Calculator Note: Residuals and the Root Mean Square Error so students can calculate residuals on their calculators. Step 6 offers an opportunity to discuss outliers. **SMP 1, 5, 6**

Step 4

Mass x (g)	Spring length y (cm)	Prediction from line $\hat{y} \approx -367.69 + 0.021x$	Residual
50	5.0	5.096	−0.096
60	5.5	3.481	0.019
70	6.0	3.865	0.135
80	6.3	6.250	0.050
90	6.8	6.635	0.165
100	7.1	7.019	0.081
110	7.5	7.404	0.096
120	7.7	7.788	−0.088
130	8.1	8.173	−0.073
140	8.5	8.558	−0.058
150	8.8	8.942	−0.142
160	9.2	9.327	−0.127
170	9.5	9.712	−0.212
180	9.9	10.096	−0.196
190	10.3	10.481	−0.181
200	10.7	10.865	−0.165
210	11.3	11.250	0.050
220	11.4	11.635	−0.235
230	11.9	12.019	−0.119
240	12.4	12.404	−0.004

3 The slope indicates that length of the spring increases at 0.038 cm per gram of ht. The *y*-intercept indicates the spring is 3.173 cm long no weight attached.

The sum of the residuals is which is small relative to the ues, so the linear model is d fit.

Example C
Project Example C and have students work in small groups. Have students explain their strategies as they present their solutions. SMP 1, 3, 5, 6

Summarize

As students present their results from the investigation and examples, include presentations showcasing a variety of ideas. Concentrate on Step 4 – 6 in the investigation. Point out cases in which having a residual sum of 0 does not guarantee a good fit. Emphasize the use of correct mathematical terminology. Whether student responses are correct or incorrect, ask other students if they agree and why. SMP 1, 3, 5, 6

ALERT Students might have difficulty with the ± notation. **ASK** "What is the smallest length you think is reasonable? What is the largest? What number is halfway between the smallest and largest?"

ALERT For Example A, students might be surprised that the units are "pizzas²." The residual is a difference of numbers whose units are pizzas, so the square of each residual (and hence the sum of those squares) has pizzas² as its units.

CRITICAL QUESTION What are the residuals of a line representing a data set?

BIG IDEA The differences between the y-values in the data set and the $\hat{y}$-values on the line right above or below them.

CRITICAL QUESTION What property of the residuals indicates a line of good fit?

BIG IDEA When the sum of the residuals is close to 0; however, some lines with this property are not lines of good fit.

As you continue in this course you will encounter data that needs a non-linear model. Deciding if your model is a good fit to the data takes more than a near zero sum of the residuals. You also need to see that there are no patterns in the residuals. A pattern in the residuals is an indication that you need a different model. This last example will not find a model to fit our data (in this chapter) but will determine that any linear model will make a poor predictor for that data.

EXAMPLE C Data giving the mean distance from our sun and the time for one orbit of the sun is given for 9 residents of the solar system. What is the equation of a linear model and what do the residuals tell us about this model?

Planet	Distance (million km)	Year (earth days)
Mercury	57.9	88
Venus	108.2	224.7
Earth	149.6	365.2
Mars	227.9	687
Jupiter	778.6	4331
Saturn	1433.5	10,747
Uranus	2872.5	30,589
Neptune	4495.1	59,800
Pluto	5906.4	90,560

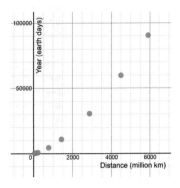

Solution Summary data: Center $(\bar{x}, \bar{y}) = (1781.08$ million km, 21932.4 days$)$

Slope $= \dfrac{s_y}{s_x} = \dfrac{32631.4 \text{ days}}{2166.31 \text{ million km}} = 15.06$ days per million km

Equation: $y = 21932.4 + 15.06(x - 1781.08)$

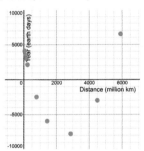

The sum of the residuals is near zero but the nonlinear pattern to the residuals tells us that any prediction made using this line, or any line, is likely to be quite inaccurate.

Residuals have several applications in modeling and statistics. In this lesson you looked at using them to judge the fit of your linear model. They can be used either to justify a good fit to the data or as evidence that the linear fit will not produce good predictions.

1.8 Exercises

You will need a graphing calculator for Exercises **3, 5,** and **8–11.**

Practice Your Skills

1. The best fit line for a set of data is $\hat{y} = 2.4x + 3.6$. Find the residual for each of these data points.

   a. (2, 8.2) @ −0.2 b. (4, 12.8) −0.4 c. (10, 28.2) 0.6

2. The best fit line for a set of data is $\hat{y} = -1.8x + 94$. This table gives the x-value and the residual for each data point. Determine the y-value for each data point. @ 82.3, 82.9, 74.5, 56.9

x-value	5	8	12	20
Residual	−2.7	3.3	2.1	−1.1

3. Sketch a residual plot for the information given in Exercise 2.

4. Match each description to one of the residual plots provided.

   a. The data is linear but the slope of the current model is too low. Graph iii

   b. The data is linear and the model seems a good fit. Graph i

   c. The data is not linear. Graph ii

   i. 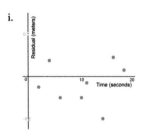 ii. 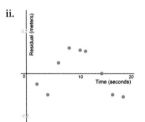 iii.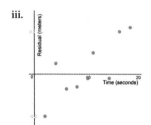

Reason and Apply

5. **APPLICATION** This table gives the mean height in centimeters of boys ages 5 to 13 in the United States.

   a. Find the mean of the ages and the mean of the given heights. @ 9 yrs, 132.44 cm

   b. Find the standard deviation of the ages and of the given heights. 2.739 years, 15.533 cm

   c. Find a line of fit using the mean point and the ratio of standard deviations for the slope.
   $\hat{y} = 132.44 + 5.672(x - 9)$

   d. Find the residuals for the data.
   0.248, 0.576, −0.096, 0.232, −0.44, −1.112, 0.544, 0.872

   e. Create a residual plot for this model.

Age	Height (cm)
5	110
6	116
7	121
8	127
9	132

Age	Height (cm)
10	137
11	143
12	150
13	156

   (www.fpnotebook.com)

6. Consider the residuals from Exercise 5. In place of a single linear model, a type of model called a piecewise model is composed of two line segments joined at a point. Use the residual plot to explain why using two lines would make a better model. At what age would you stop the first line and start the second?
   The residuals below 11 years old line up but the two after that age do not. If we join a line that fits the data from age 5 to 11 then a line that fits from 11 to 13 we would remove that pattern from the plot.

5e.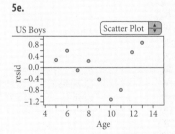

Formative Assessment

Since this is primarily a whole class, teacher centered lesson, student input is essential to provide anecdotal evidence of meeting the objectives. The Critical Questions provide evidence of a student knowing the definition a residual and how they would use residuals to assess the fit of a line. Exercises 1 – 11 provide further evidence of meeting the objectives.

Apply

Extra Example

The best fit line for a set of data is $\hat{y} = -4.1 + 3x$.

a. Find the residual for the point (2, −3). −4.9

b. If the residual value is 2.1 for an x-value of 4, find the actual y-value for the data point. 10

Closing Question

Below is the graph of a set of data. Explain how you would find a best fit line.

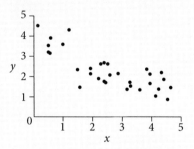

Answers will vary. Sample: Draw a line in which points were evenly divided above and below the line. Check the residuals.

Assigning Exercises

Suggested: 1 – 11
Additional Practice: 12 – 14

3.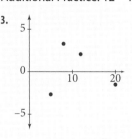

8a.

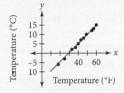

8b.

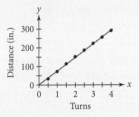

c. 7 positive

d. One solution would be $C = 5 + 4.99(F - 40)$. This has 5 below, 5 above, and one on the line.

e. There is not much difference in the prediction of the two lines. Getting an equal number on both sides of the line is not that important.

9a. $d = 166 + 73.29(t - 2.25)$, This looks to be a good fit.

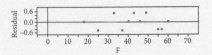

9b.

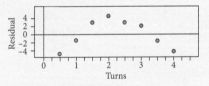

9c. There is a pattern to this residual plot. A non-linear model would make a better predictor.

---

7. With a specific line of fit, the data point (6, 47) has a residual of 2.8. The slope of the line of fit is 2.4. What is the equation of the line of fit? (h) $\hat{y} = 29.8 + 2.4x$

8. The following readings were taken from a display outside the First River Bank. The display alternated between °F and °C. However, there was an error within the system that calculated the temperatures. $C = 6 + 4.99(F - 42.091)$

°F	18	33	37	25	40	46	43	49	55	60	57
°C	−6	2	3	−3	5	8	7	10	12	15	13

a. Plot the data and the line of fit.

b. Calculate the residuals and make a residual plot.

c. Of the 11 values, how many have positive residuals?

d. Modify your equation so that there will be the same number of positive residuals as negative residuals (Some points may have to have 0 residual).

e. Do you think this equation is better than the first equation you found? Why?

9. Leajato experimented by turning the key of a wind-up car different numbers of times and recorded how far it traveled.

Turns	0.5	1	1.5	2	2.5	3	3.5	4
Distance (in.)	33	73	114	152	187	223	256	290

a. Plot the data and a line of fit. Does this look like a good fit to the data?

b. Calculate the residuals and make a residual plot.

c. Interpret this plot. What does it tell you about the model you found in part a.

10. When the data is non-linear there will be a pattern to the points in the linear model residual plot. If you were able to find the correct non-linear equation this pattern would disappear. Using the data from problem 9 and the non-linear model $y = 169.6 + 73.29(x - 2.25) - 2.762(x - 2.25)^2$, calculate the residuals and then create a residual plot. Sketch this plot and comment on this plot.

This plot shows no pattern, this model would make a better predictor.

> **History CONNECTION**
>
> In 1787, Article II, Section 1, of the U.S. Constitution instituted the electoral college as a means of electing the president. The number of electoral votes allotted to each state corresponds to the number of representatives that each state sends to Congress. The distribution of electoral votes among the states can change every 10 years depending on the results of the U.S. census. The actual process of selecting electors is left for each state to decide.

Counting the vote, on November 7th, 1876, at "Elephant Johnnie's".

11. **APPLICATION** Since 1964, the total number of electors in the electoral college has been 538. In order to declare a winner in a presidential election, a majority, or 270 electoral votes, is needed. The table at right shows the number of electoral votes that the Democratic and Republican parties have received in the presidential elections from 1964 to 2004.

   a. Let x represent the electoral votes for the Democratic Party, and let y represent the votes for the Republican Party. Make a scatter plot of the data.

   b. Why are the points nearly linear? What are some factors that make these data not perfectly linear?

   c. Sketch the line $y = 269$ on the scatter plot. How are all the points above the line related?

   d. Find the residuals for the line $y = 269$. What does a negative residual represent?

   e. What does a residual value that is close to 0 represent?   a close election

Electoral Votes

Year	Democrats	Republicans
1964	486	52
1968	191	301
1972	17	520
1976	297	240
1980	49	489
1984	13	525
1988	111	426
1992	370	168
1996	379	159
2000	266	271
2004	251	286
2008	365	173
2012	332	206

Exercise 11 You might discuss the book's choice of x and y. **ASK** "Are these necessarily dependent and independent variables?"

11a.

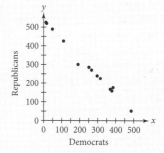

11b. The points are nearly linear because the sum of electoral votes should be 538. The data are not perfectly linear because in a few of the elections, candidates other than the Democrats and Republicans received some electoral votes.

11c.

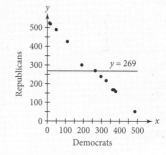

The points above the line are the elections in which the Republican Party's presidential candidate won.

11d. $-217, 32, 251, -29, 220, 256, 157, -101, -110, 2, 17, -96, -63$. A negative residual means that the Democratic Party's presidential candidate won.

Review

12. Write an equation in point-slope form for each of these lines.

   a. The slope is $\frac{3}{5}$ and the line passes through $(2, 4.7)$. $y = 4.7 + 0.6(x - 2)$

   b. The line is a direct variation through $(6, 21)$. $y = 21 + 3.5(x - 6)$

   c. The line passes through $(3, 11)$ and $(-6, -18)$. $y = 11 + \frac{29}{9}(x - 3)$ or $y = -18 + \frac{29}{9}(x + 6)$

13. Travis is riding with his parents on Interstate 15 across Utah. He records the digital speedometer reading in mi/h at 4:00 P.M. and every five minutes for the next hour. The mean of the 13 readings is 58.3 mi/h and the standard deviation is 7.4 mi/h.

   a. If you assume a symmetric mounded set of data, what are your estimates for the fastest and slowest of his 13 readings?   possible answer: $58.3 \pm 2(7.4) = 73.1$ and 43.5

   b. If you assume the data are skewed left, what are your estimates for the fastest and slowest of the readings?   possible answer: $58.3 + 7.4 = 65.7$ for the fastest and $58.3 - 3(7.4) = 36.1$ for the slowest

   c. Which of your assumptions and estimates seem more likely to you and why?

13c. Answers will vary. The standard deviation is fairly large, which may indicate that they ran into traffic or had to navigate more difficult driving conditions for part of the trip. The lower speed readings may be more spread out than the higher readings, which would mean that the data are skewed left.

14. David deposits $30 into his bank account at the end of each month. The bank pays 7% annual interest compounded monthly.

   a. Write a recursive formula to show David's balance at the end of each month.

   b. How much of the balance was deposited and how much interest is earned after

      i. 1 year   ii. 10 years   iii. 25 years   iv. 50 years

   c. What can you conclude about making regular deposits into a bank account that earns compound interest?

14a. $u_0 = 30$ and $u_n = u_{n-1}\left(1 + \frac{0.07}{12}\right) + 30$ where $n \geq 1$

14b.
   i. deposited: $360; interest: $11.78
   ii. deposited: $3,600; interest: $1,592.54
   iii. deposited: $9,000; interest: $15,302.15
   iv. deposited: $18,000; interest: $145,442.13

14c. Sample answer: If the account earns compound interest, the interest earned in the long run will far exceed the total amount deposited.

LESSON 1.9

The least squares line is a line of fit that produces the smallest sum of the distance between the actual value of the dependent variable and the expected value based on the model. This model produces the most accurate prediction across the entire data set but is susceptible to significant influence from outliers.

COMMON CORE STATE STANDARDS		
APPLIED	DEVELOPED	INTRODUCED
S.ID.6a	S.ID.6b	
S.ID.7		
S.ID.8		
F.LE.5		

Objectives

- Find a line of fit using the least squares method
- Identify the strengths and weaknesses of using the least squares method

Vocabulary

least squares line

Materials

- Calculator Program: LEASTSQU

Launch

Sabena collected data on the weight of different brands of an item. She made a scatter plot of her data.

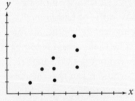

Which line shows the best line of fit for Sabena's data? Explain.

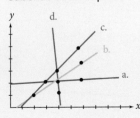

b minimizes the sum of the residuals

LESSON 1.9

Least Squares Line

A line is a dot that went for a walk.
PAUL KLEE

Have you noticed that you and your classmates frequently find different equations to model the same data? Some lines fit data better than others and can be used to make more accurate predictions. In this lesson you will learn a standard method for finding a line of fit that will enable each member of the class to get the same equation for the same set of data.

You can use many different methods to find a statistical line of fit. The **least squares line** makes use of the residuals. In the investigation you will find this line using a trial and error procedure.

INVESTIGATION

The Moving Line

Step 1 Begin by entering the data into lists 1 and 2 in your calculator.

Step 2 Link the program LEASTSQU into your calculator and execute the program. After a page of instructions, you should see a scatter plot of the data you entered and a line through that data. In addition, each residual (vertical segment connecting the points to the line) has been made into a square.

x	y	x	y
4.0	23	8.9	50
5.2	28	9.7	47
5.8	29	10.4	52
6.5	35	11.2	60
7.1	35	11.9	58
7.8	40	12.5	62
8.3	42	13.1	64

Step 3 A criterion for one type of best-fit line is to find the line that has the smallest total area of these squares. This is known as the least squares line. At the top of the graph screen you should see a value labeled SS. This is the total area of the squares on your graph. Record this number.

Step 4 The four arrows control the line on the screen. The up and down arrows move the line without changing the slope. The left and right arrows rotate the line to change the slope. Try each arrow to determine what effect these have on the total sum of squares.

Step 5 When you press WINDOW, you are shown the residual plot for the current model. Looking at this plot is another way to help you determine the fit of the line. Press ENTER to return to the data scatter plot.

Step 6 Why are outliers easy to see in the display made by this program? Explain why making the sum of squares small causes the line to be pulled toward any outlier.

Step 7 Work with these commands until you think that the sum of squares is at its minimum value. This is your least squares line. Press the Y= and record the slope and intercept of your line and record the sum of the squares.

ELL
Provide students with other examples of residual (the remnant at the bottom of a can of soda, remnant of marker left on the whiteboard after wiping).

Support
Rather than focusing on the formula for finding the least squares line, use the desmos simulation in the ebook to allow students to see a visual representation of the squared distance between the expected and actual value of the dependent variable.

Advanced
Ask students to use the formula for the least squares line to explain why the model is influenced by outliers. If not part of your version of the textbook, have students research the median-median line of fit and explain how outliers influence this model.

The actual equation for finding the least squares line is based on the statistics of the data. These statistics can be calculated by hand but they can also be found quickly using a calculator that has the formulas programmed in. There are three statistics you need. The first two you are familiar with, the mean and the standard deviation. The third one is called the correlation and is a measure of how well the points line up. There are several ways to calculate this value, which you will find in the exercises. The letter r is used for this statistic.

> **Least Squares Line**
>
> The least squares line is found using the mean and standard deviation of the x-list, $\bar{x}$ and s_x, the mean and standard deviation of the y-list, $\bar{y}$ and s_y, and the correlation of the two lists, r.
>
> $$y = \bar{y} + r\frac{s_y}{s_x}(x - \bar{x})$$
>
> This can also be written $y = a + bx$, where $b = r\frac{s_y}{s_x}$ and $a = \bar{y} - b\bar{x}$

All these statistics, including a and b, are easily found on your calculator. In the next example the calculator is used to find the equation of a least squares line.

EXAMPLE The data here gives the flight times for morning departures from Detroit, Michigan. Find the equation of the least squares line for the data and interpret the meaning of the slope of your line.

Destination	Flight time (min)	Distance (mi)
Cincinnati, OH	64	229
Houston, TX	189	1092
Los Angeles, CA	288	1979
Memphis, TN	104	610
Denver, CO	180	1129
Phoenix, AZ	248	1671
Louisville, KY	67	306
San Francisco, CA	303	2079
Omaha, NE	120	658
New Orleans, LA	156	938

Solution After entering data into the calculator, use the statistics menu to get the equation of the least-squares line, $y = -240.15 + 7.6146x$

$y = a + bx$
$a = -240.1514946$
$b = 7.614610207$
$r^2 = .9954039241$
$r = .9976993155$

The slope of this line is 7.6146 miles per minute, which is the average speed of the plane in flight. Multiplying this by 60 minutes per hour gives the average speed of about 457 miles per hour.

While the least squares line has many good statistical properties, the use of residuals tends to put more weight on outlier points. Finding an equation for data with one or two influential outliers often gives an equation that does not seem to be the best fit.

ASK "Using the formula, explain how adding the Detroit to Minneapolis data point affects the least squares line model." Adding the Detroit-Minneapolis data point significantly increases the spread of the data, increasing the standard deviation of the dependent variable. Since this is the numerator in the slope of the least squares line and the correlation coefficient is near one, the slope will change significantly.

CRITICAL QUESTION What would the least squares line look like for a data set with a correlation near or at zero?

BIG IDEA A correlation coefficient near zero would produce a least squares line with a slope near zero so the least squares line would be a horizontal line crossing the y-axis at the mean of the y-values of the data set. Conversely, a data set that is highly correlated would produce a least squares line equal to the standard deviation of the y-values divided by the standard deviation of the x-values.

CRITICAL QUESTION Why is the least squares line method most often used as a model to predict the value of a dependent variable?

BIG IDEA The least squares line produces, on average, the most accurate prediction of the dependent variable, or y-value, for a given x-value. Since the model weights all actual y-values equally, the model is also heavily influenced by outliers.

Formative Assessment

Assess students' ability to find not only the line of fit using the least squares model (either graphically or algebraically) but also their ability to interpret the model as a prediction model based on a given data set. Presentations of the Investigation, Example and the Exercises provide evidence of students' understanding of the objectives.

As a performance task, provide the students the location of a nearby city. Ask them to use the model to predict the flight time from Detroit. As an extension, ask them to research flight times and add the data point, predicting how the new data point will affect the least squares line and testing their prediction.

For example, if data was added for a flight in a Piper Cub from Detroit to Minneapolis, a flight of 543 miles in 504 minutes, the line changes radically.

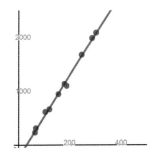 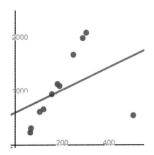

1.9 Exercises

You will need a graphing calculator for for all Exercises

Practice Your Skills

1. Given the lists here, use your calculator to find the mean and standard deviations of each list. @

List x	7	57	1	74	54	81	37	78	29
List y	97	37	64	28	43	3	43	18	48

   mean $x = 46.44$, mean $y = 42.33$, $s_x = 29.86$, $s_y = 27.07$

2. Use the lists from Exercise 1 and your calculator to find the correlation and the equation of the least squares line. @ $r = 0.8998$, $y = 80.23 - 0.8159x$

3. Reverse the two lists and use your calculator to find the correlation and the equation of this least squares line. $r = 0.8998$, $y = 88.46 - 0.9924x$

4. Use your calculator to find the mean and standard deviations of each list.

List x	30	35	89	55	91	32
List y	41	37	99	52	82	42

   mean $x = 55.33$, mean $y = 58.83$, $s_x = 28.30$, $s_y = 25.59$

5. Use the lists from Exercise 4 and your calculator to find the correlation and the equation of the least squares line. $r = 0.9579$, $y = 10.90 + 0.8662x$

Reason and Apply

6. Here is another method to find a least squares line. Use the two lists here.
   List X: 1, 4, 7, 12   List Y: 4, 5, 8, 11

   a. Find the mean and standard deviation of each list. Label the values $\bar{x}$, $\bar{y}$, s_x and s_y. $\bar{x} = 6$, $\bar{y} = 7$, $s_x = 4.6904$, $s_y = 3.1628$

   b. Multiply the two lists together and then add the four products together. Label this value C. $C = 212$

   c. Find the value of $\dfrac{C - n\bar{x}\bar{y}}{n - 1}$ where n is the length of the list. Label this value Cov. $Cov = 14.6667$

   d. Find the value of $\dfrac{Cov}{s_x s_y}$ and label this r. $r = 0.98882$

   e. Use the values you have found to find b and a according to the formulas $b = r\dfrac{s_y}{s_x}$ and $a = \bar{y} - b\bar{x}$. $b = 0.66667$, $a = 3$

   f. Find the equation of the least squares line for the two lists. $a = 3$, $b = 0.66667$
   $y = 3 + 0.66667x$

Conceptual/Procedural

Conceptual This lesson builds student understanding of the relationship between the magnitude of the correlation between two variables, the spread within each of the variables, and a line of fit model (the least squares line). Students will also determine the strengths and weaknesses of using the least squares line to create a model.

Procedural Teachers have the option of focusing on an exclusively technology-driven approach to teaching the procedure for finding the least-squares line or using the formula. Students will examine the formula to determine how the correlation and spread of data affect the least squares model.

7. **APPLICATION** The data shown gives the life expectancy at birth for males in the United States for different years of birth. Let x represent the year of birth, and let y represent the male life expectancy in years.

   a. Find the least squares line. $y = -417.1 + 0.2458x$

   b. Make a scatter plot and add the line.

   c. The year 1998 is missing from you table. Using your model, what would you predict the life expectancy at birth to be for males born in 1998?   74.0 years

   d. Use your model to predict the life expectancy for males born in 2002.   75.0 years

   e. Using this model, when would you expect the life expectancy at birth for males in the United States to exceed 82 years?
   In the year 2030.

8. When the y-intercept is far from the data, any rounding of the slope makes the intercept form of the line seem like a poor fit. You will see this as you look at both forms of the line for the data.

Year	1970	1980	1990	2000	2010
Population (millions)	177.25	204.28	281.36	320.41	352.97

   a. Find the least squares line and round the slope to three digits. $y = -9037.4 + 04.68x$

   b. Find the mean year and population, rounded to three digits. 1990, 267

   c. Create a scatter plot and add your rounded least squares line.

   d. Add the graph of $y = \bar{y} + b(x - \bar{x})$ using the slope from part a and the mean values from part b as your point. $y = 267.2 + 4.68(x - 1990)$

   e. Explain the difference between the two lines on your graph.
   The lines are parallel but the intercept form is 8.8 million higher and looks too high for the data.

9. The number of deaths caused by automobile accidents, D, per hundred thousand population in the United States is given for various years, t.

t	1924	1925	1926	1927	1928	1929	1930	1931	1932	1933
D	15.5	17.1	18.0	19.6	20.8	23.3	24.5	25.2	21.9	23.3

(W. A. Wilson and J. I. Tracey, *Analytic Geometry*, 3rd ed., Boston: D. C. Heath, 1949, p. 246.)

   a. Make a scatter plot of the data.

   b. Would you describe this pattern with a single line? Explain why or why not. @
   No, it appears that there are two patterns here.

   c. Create one or more models and give an appropriate domain and range for each model.

   d. Would you use this model to predict the number of deaths from auto accidents in 2010? Explain why or why not.
   It probably would not be a good idea to extrapolate, because a lot has changed in the automotive industry in the past 75 years. Many safety features are now standard.

Year of birth	Male life expectancy (years)
1910	48.4
1920	53.6
1930	58.1
1940	60.8
1950	65.6
1960	66.6
1965	66.8
1970	67.1
1975	68.8
1980	70.0
1985	71.2
1990	71.8
1995	72.5
2000	74.1
2005	74.9
2010	76.2
2015	76.8

World Bank

Apply

Extra Example
Use the least squares method to determine the equation of the line of best fit for the data. Then plot the line.

x	y
8	3
2	10
11	3
6	6
5	8
4	12
12	1
9	4
6	9
1	14

$y = -1.1x + 14.0$

Closing Question
Explain how you find the least squares line.

Answers will vary.

Assigning Exercises
Suggested: 1–5, 7–8, 10
Additional Practice: 6, 9, 11–13

7b.
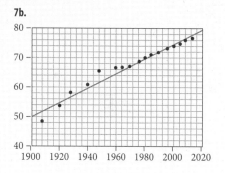

8c.
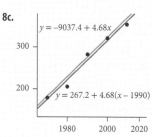

Exercises 9 ADVANCED Have students investigate different methods for determining a line of fit and present to the class the methods, their pros and cons, and a comparison with the median-median line.

Exercise 9 Approach this exercise with sensitivity; some students may know someone who was killed in an auto accident.

9a.

9c. Possible answer: $\hat{y} = -2832.1 - 1.48x$ for $1924 \leq x \leq 1931$, $15.42 \leq y \leq 25.78$ and $\hat{y} = 21.9 + 1.4(x - 1932)$ or $\hat{y} = -2682.9 + 1.4x$ for $1932 \leq x \leq 1933$, $21.9 \leq y \leq 23.3$

Exercise 10 Roger Bannister (UK) was the first person to run a mile in under 4 minutes. **SUPPORT** Remind students to convert the time data into seconds.

10a. $\hat{y} \approx 1075.7 - 0.4276x$

10b. The time for the record decreases about 0.43 seconds each year.

10c. 240.10 s or 4:00.10. This prediction is 0.7 s slower than Roger Bannister actually ran.

10d. 273.89 s or 4:33.89. This prediction is about 9.4 s slower than Walter Slade actually ran.

10e. This suggests that a world record for the mile in the year 0 would have been about 17.9 minutes. This is doubtful because a fast walker can walk a mile in about 15 minutes. The data are only approximately linear and only over a limited domain.

10f. No new record had been set as of 2007.

See your digital resources for the project **Counting Forever** where students collect data and predict how long it would take to count to large numbers, including the amount of the federal debt.

---

10. **APPLICATION** Use these data on world records for the 1-mile run to answer the questions below. Times are in minutes and seconds.

   a. Let x represent the year, and let y represent the time in seconds. What is the equation of the least-squares line?

   b. What is the real-world meaning of the slope? ⓗ

   c. Use the equation to interpolate and predict what new record might have been set in 1954. How does this compare with Roger Bannister's actual 1954 record of 3:59.4? ⓐ

   d. Use the equation to extrapolate and predict what new record might have been set in 1875. How does this compare with Walter Slade's 1875 world record of 4:24.5?

   e. Describe some difficulties you might have with the meaning of 1075.7 as a y-intercept.

   f. Has a new world record been set since 1999? Find more recent information on this subject and compare it with the predictions of your model.

World Records for 1-Mile Run

Year	Runner	Time
1937	Sydney Wooderson, U.K.	4:06.4
1942	Gunder Haegg, Sweden	4:06.2
1945	Gunder Haegg, Sweden	4:01.4
1958	Herb Elliott, Australia	3:54.5
1967	Jim Ryun, U.S.	3:51.1
1979	Sebastian Coe, U.K.	3:49.0
1982	Steve Scott, U.S.	3:47.69
1985	Steve Cram, U.K.	3:46.31
1993	Noureddine Morceli, Algeria	3:44.39
1999	Hicham El Guerrouj, Morocco	3:43.13

(*International Association of Athletics Federations*)

Review

11. What is the equation of the line that passes through a graph of the points of the sequence defined by $y = 7 - 3x$

   $u_1 = 4$

   $u_n = u_{n-1} - 3$ where $n \geq 2$

12. The histogram at right shows the results of a statewide math test given to eleventh graders. If Ramon scored 35, what is the range of his percentile ranking? ⓗ between the 40th and 63rd percentiles

13. Create a set of 7 values with median 47, minimum 28, and interquartile range 12. ⓗ possible answer: 28, 38, 42, 47, 49, 50, 60

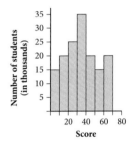

PERFORMANCE TASK

In Example A, you looked at flight times and distances for morning flights from Detroit, Michigan. Use your model to predict the flight time from Detroit, Michigan to Lansing, Michigan, which is 78 miles away. Research the actual flight times and add the data point. Predict how this new data point will affect the least squares line from Example A and testing their prediction. Explain any difference between your prediction and the model you obtained.

94 CHAPTER 1 Linear Modeling

PERFORMANCE TASK

The model from Example A predicts a distance of 132.97 miles from Detroit to Lansing. The flight time from the Detroit Metro Airport to the Lansing Airport is 49 minutes. Since the model without the Lansing data point overestimates the distance, the data point has a negative residual (or falls beneath the least squares line). Minimizing that residual will require some combination of either decreasing the intercept to provide a downward horizontal translation and increasing the rate of change to make the line steeper. The new model with the data point (49, 78) is $y = -256.54 + 7.69x$.

CHAPTER 1 REVIEW

A **sequence** is an ordered list of numbers. In this chapter you used **recursion** to define sequences. A **recursive formula** specifies one or more starting terms and a **recursive rule** that generates the nth term by using the previous term or terms.

There are two special types of sequences—arithmetic and geometric. **Arithmetic sequences** are generated by always adding the same number, called the **common difference**, to get the next term. Your salary for a job on which you are paid by the hour is modeled by an arithmetic sequence. **Geometric sequences** are generated by always multiplying by the same number, called the **common ratio**, to get the next term. The growth of money in a savings account is modeled by a geometric sequence. For some **growth** and **decay** scenarios, it helps to write the common ratio as a percent change, $(1 + p)$ or $(1 - p)$.

Graphs help you recognize whether the data are best modeled by an arithmetic or geometric sequence. The graph of an arithmetic sequence is **linear** whereas the graph of a geometric sequence is curved. For arithmetic sequences you wrote the explicit form for finding a term directly without all the previous values. When the domain of your model includes more than just the points of a sequence, this explicit form is called a linear equation. To write the equation of a line, you can use its slope and y-intercept to write the equation in **intercept form**, or you can use the coordinates of two points to write the equation in **point-slope form**.

You also analyzed sets of two-variable data. A plot of two-variable data may be linear, or it may appear nearly linear over a short domain. If it is linear, you can find a **line of fit** to model the data. This equation can be used to **interpolate** or **extrapolate** points for which data are not available. You learned different methods for finding a line of fit for a set of data. Regardless of the method used, you can examine the **residuals** to determine whether your model is a good fit. With a good model, the residuals should be randomly positive and negative, the sum of the residuals should be near zero, and each residual should be as small as possible. If the graph of the residuals with the x-values is not random but shows a curved pattern, then there is likely a non-linear equation that models the data more accurately than any linear equation.

Reviewing

You might pose this review problem: "Under what conditions is the shifted geometric sequence generated by $u_n = ru_{n-1} + d$ geometric?" [when $d = 0$] "Arithmetic?" [when $r = 1$] Stress the utility of looking at the graph of the sequence when determining what type of recursion is happening. Remind students of the difference between a common ratio and a common difference and their relationships to the graphs.

For review, you might refer students back to Example A in the Residuals and Fit lesson, in which Big K Pizza has sold 512, 603, 642, and 775 pizzas, respectively, in the past four Novembers. Add the information that Big L Pizza sales for those same four months were 223, 341, 486, and 584 pizzas. Have students predict whether Big L will catch up with Big K. Ask if the best model would be a **recursive formula**, an **explicit formula**, or a **linear equation**. The data given are **discrete**, indicating that a recursive or an explicit formula might be best, but the growth in sales is **continuous** (since we can talk about the sales for fractions of a month), and we'll want to be able to **interpolate** or **extrapolate** from our model. (Students thinking critically might point out that the model for the whole year probably won't be linear.) As students find **lines of fit** for the data, review the ideas of **intercept form** and **point-slope form**; have students compare the **residuals** and **root mean square errors** of their lines with those of the **median-median line.** Once the class has decided on good linear equations, students can solve the **system** by graphing, **substitution**, or **elimination** to find the point of intersection. Be sure to include some discussion of the reasonableness of their solutions.

Assigning Exercises

You might assign the exercises as homework and then use them in class to review those concepts in which the class is weakest.

4a. iv. The recursive formula is that of an increasing arithmetic sequence, so the graph must be increasing and linear.

4b. iii. The recursive formula is that of a growing geometric sequence, so the graph must be increasing and curved.

4c. i. The recursive formula is that of a decaying geometric sequence, so the graph must be decreasing and curved.

4d. ii. The recursive formula is that of a decreasing arithmetic sequence, so the graph must be decreasing and linear.

Exercises

You will need a graphing calculator for Exercises **2, 13,** and **15.**

@ **Answers are provided for all exercises in this set.**

1. Consider this sequence:
   256, 192, 144, 108, . . .
   a. Is this sequence arithmetic or geometric? geometric
   b. Write a recursive formula that generates the sequence. Use a_1 for the starting term. $a_1 = 256$ and $a_n = 0.75a_{n-1}$ where $n \geq 2$
   c. What is the 8th term? $a_8 \approx 34.2$
   d. Which term is the first to have a value less than 20? $a_{10} \approx 19.2$, so 10th term
   e. Find a_{17}. $a_{17} \approx 2.57$

2. Consider this sequence:
   3, 7, 11, 15, . . .
   a. Is this sequence arithmetic or geometric? arithmetic
   b. Write a recursive formula that generates the sequence. Use u_1 for the starting term. $u_1 = 3$ and $u_n = u_{n-1} + 4$ where $n \geq 2$
   c. What is the 128th term? $u_{128} = 511$
   d. Which term has the value 159? $u_{40} = 159$, so the 40th term
   e. Find u_{20}. $u_{20} = 79$

3. List the first five terms of each sequence. For each set of terms, what minimum and maximum values of n and a_n would you use on the axes to make a good graph?
   a. $a_1 = -3$
      $a_n = a_{n-1} + 1.5$ where $n \geq 2$
      $-3, -1.5, 0, 1.5, 3$; 0 to 6 for n and -4 to 4 for a_n
   b. $a_1 = 27$ 27, 9, 3, 1, $\frac{1}{3}$; 0 to 6 for n and 0 to 30 for a_n
      $a_n = \frac{1}{3}a_n$ where $n \geq 2$

4. Match each recursive formula with the graph of the same sequence. Give your reason for each choice.
   a. $u_0 = 5$
      $u_n = u_{n-1} + 1$ where $n \geq 1$
   b. $u_0 = 1$
      $u_n = (1 + 0.5)u_{n-1}$ where $n \geq 1$
   c. $u_0 = 5$
      $u_n = (1 - 0.5)u_{n-1}$ where $n \geq 1$
   d. $u_0 = 5$
      $u_n = u_{n-1} - 1$ where $n \geq 1$

i. 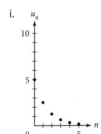 ii. 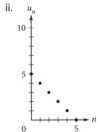 iii. 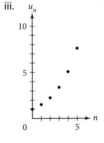 iv.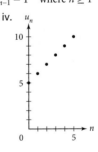

5. A large barrel contains 12.4 gal of oil 18 min after its drain is opened. How many gallons of oil were in the barrel before the drain was opened if it drains at a rate of 4.2 gal/min?   88 gal

6. Find the slope of the line containing the points (16, 1300) and (−22, 3250). $-\frac{975}{19}$

7. Consider the line $y = -5.02 + 23.45x$.
   a. What is the slope of this line?   23.45
   b. Write an equation for a line that is parallel to this line.   possible answer: $y = 23.45x$
   c. Write an equation for a line that is perpendicular to this line.   possible answer: $y = -\frac{20}{469}x \approx -0.0426x$

8. Find the point on each line where y is equal to 740.0.
   a. $y = 16.8x + 405$
      approximately (19.9, 740.0)
   b. $y = -7.4 + 4.3(x - 3.2)$
      approximately (177.0, 740.0)

9. Write an equation in point-slope form for each line described.
   a. slope −1 passing through (−2, 3) $y = 3 - (x + 2)$ or $y = 1 - x$
   b. parallel to $y = 5 + 2(x - 4)$ passing through (0, −8) $y = -8 + 2(x - 0)$ or $y = -8 + 2x$
   c. passing through (1999, 13.2) and (2009, 8.6) $y = 13.2 - 0.46(x - 1999)$ or $y = 8.6 - 0.46(x - 2009)$
   d. horizontal line through (2, 7) $y = 7 + 0(x - 2)$ or $y = 7$

10. Consider the arithmetic sequence 6, 13, 20, 27, 34, ... Let u_1 represent the first term.
    a. Write a recursive formula that describes this sequence. $u_1 = 6$ and $u_n = u_{n-1} + 7$ where $n \geq 1$
    b. Write an explicit formula for this sequence. $y = -1 + 7x$
    c. What is the slope of your equation in 10b? What relationship does this have to the arithmetic sequence?
       The slope is 7. The slope of the line is the same as the common difference of the sequence.
    d. Determine the value of the 32nd term. Is it easier to use your formula from 10a or 10b for this?
       223. It's probably easier to use the equation from 10b.

11. For an arithmetic sequence, $u_1 = 12$ and $u_{10} = 52.5$.
    a. What is the common difference of the sequence?   4.5
    b. Find the equation of the line through the points (1, 12) and (10, 52.5). $y = 7.5 + 4.5x$
    c. What is the relationship between 11a and 11b?
       The slope of the line is equal to the common difference of the sequence.

12. The graphs below show three different lines of fit for the same set of data. For each graph, decide whether the line is a good line of fit or not, and explain why.

    a.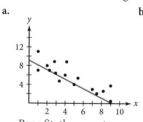
    Poor fit; there are too many points above the line.

    b.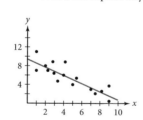

    c.

Exercise 12 As an extension, have students calculate the residuals.

12b. Reasonably good fit; the points are well distributed above and below the line and are not clumped. The line follows the downward trend of the data.

12c. Poor fit; there are an equal number of points above and below the line, but they are clumped to the left and to the right, respectively. The line does not follow the trend of the data.

Exercise 13 As needed, remind students to round final answers but to keep and use all the digits as they work through a problem on their calculators.

13a.

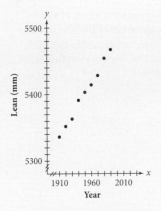

15a. fairly linear (a little logistic)

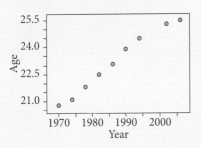

15d.

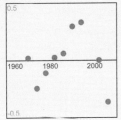

> **Architecture**
> **CONNECTION**
>
> In 1173 C.E., the Tower of Pisa was built on soft ground and, ever since, it has been leaning to one side as it sinks in the soil. This 8-story, 56-meter tower was built with only a 3-meter-deep foundation. The tower was completed in the mid-1300s even though it started to lean after the first 3 stories were completed. The tower's structure consists of a cylindrical body, arches, and columns. Rhombuses and rectangles decorate the surface. In the 1990s, "correction" constructions began on the tower, with tons of lead weights place on one side of the tower so it wouldn't fall while they did repair work. It has now been restored to the lean it had in 1838 and it is no longer moving.

The Tower of Pisa, in Piazza dei Miracoli, Pisa, Italy

13. Read the Architecture Connection. The table lists the amount of lean, measured in millimeters, for nine different years.

    a. Make a scatter plot of the data. Let x represent the year, and let y represent the amount of lean in millimeters.

    b. Find a line of fit for the data. Answers will vary, $y = 5401 + 1.645(x - 1950)$

    c. Interpret the slope from your line of fit in the context of the problem.
       Each year the tower tips another 1.645 millimeters
    d. Find the amount of lean predicted by your equation for 1992 (the year work was started to secure the foundation). 5474.4 mm

    e. Calculate the residuals and create a residual plot.
       0.8, 0.35, −5.1, 6.45, 2, −3.45, −5.9, 3.65, 0.2
    f. What does the residual plot indicate about your model?
       This line is a good fit because there is no pattern to the residuals.

14. The 4th term of an arithmetic sequence is 64. The 54th term is −61. Find the 23rd term.
    $u_{23} = 16.5$

15. Consider these data on the estimated median age of U.S. women who married for the first time in these years between 1970 and 2006. Approximately 0.08% of Americans get married each year.

    a. Create a scatter plot of the data. Do these data seem linear?

    b. Find a line of fit for the data.
       Answers will vary, $y = -274.9 + 0.15x$
    c. Use your line of fit to predict the median age of women at first marriage in 2018. 26.7 years

    d. Calculate the residuals and make a residual plot.
       0.2, −0.1, 0, 0.1, 0.3, 0.3, −0.1, −0.7, −0.5.
    e. What does the residual plot indicate about the linear model?
       There is a slight pattern to the residuals so the best model may not be linear.

Tower of Pisa

Year	Lean
1910	5336
1920	5352
1930	5363
1940	5391
1950	5403
1960	5414
1970	5428
1980	5454
1990	5467

U.S. Women's Median Age at First Marriage

Year	Age	Year	Age
1970	20.8	1990	23.9
1974	21.1	1994	24.5
1978	21.8	2002	25.3
1982	22.5	2006	25.5
1986	23.1		

(www.census.gov)

Take Another Look

1. In Lesson 1.1, Example A, you saw an arithmetic sequence that generates from a geometric pattern: the number of seats around a linear arrangement of square tables is 4, 6, 8, ... Notice that the sequence comes from the increase in the perimeter of the arrangement. What sequence comes from the increase in area?

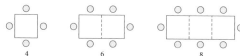

Use these geometric patterns to generate sequences using perimeter, area, height, or other attributes, and give their recursive formulas. Which sequences exhibit arithmetic growth? Geometric growth? Are there sequences you cannot define as arithmetic or geometric?

Invent a geometric pattern of your own. Look for and define different sequences.

2. Imagine a target for a dart game that consists of a bull's-eye and three additional circles. If it is certain that a dart will land within the target but is otherwise random, what sequence of radii gives a set of probabilities that form an arithmetic sequence? A geometric sequence? Sketch what the targets look like.

3. The data at right show the average price of a movie ticket for selected years. Find a line of fit for the years 1935–2001. Does your line seem to fit the data well? Which years are not predicted well by your equation? Consider whether or not two or more line segments would fit the data better. Sketch several connected line segments that fit the data.

Which model, the single line of fit or the connected segments, do you think is more accurate for predicting the current price of a ticket? Is there another line or curve you might draw that you think would be better? Why do you think these data might not be best modeled by a single linear equation?

Year x	Average ticket price ($) y	Year x	Average ticket price ($) y
1935	0.25	1974	1.89
1940	0.28	1978	2.33
1948	0.38	1982	2.93
1954	0.50	1986	3.70
1958	0.69	1990	4.21
1963	0.87	1994	4.10
1967	1.43	1998	4.68
1970	1.56	2001	5.65

(Motion Picture Association of America)

Take Another Look

1. The area of the tables generates a sequence proportional to 1, 2, 3, ...

   The perimeters of the nested squares generate the sequence 4, 8, 12, 16, ... The growth is arithmetic; $u_1 = 4$ and $u_n = u_{n-1} + 4$ where $n \geq 2$. The areas of the nested squares generate the sequence 1, 4, 9, 16, ... The growth is neither arithmetic nor geometric; students probably won't be able to write the recursive rule $u_n = u_{n-1} + 2n - 1$, but they may write $u_n = n^2$ where $n \geq 1$. The heights generate the sequence 1, 2, 3, 4, ... The growth is arithmetic; $u_1 = 1$ and $u_n = u_{n-1} + 1$ where $n \geq 2$.

   The perimeters of the other nested polygons generate the sequence 3, 4, 5, 6, ... The growth is arithmetic; $u_1 = 3$ and $u_n = u_{n-1} + 1$ where $n \geq 2$. Using $s = 1$ and $A = \dfrac{ns^2}{4\tan\left(\frac{360°}{n}\right)}$, their areas generate the sequence $\dfrac{\sqrt{3}}{4}, 1, \dfrac{5}{4\tan 72°}, \dfrac{3}{2\tan 60°}, \ldots$, which generalizes to $u_n = \dfrac{n}{4\tan\left(\frac{360°}{n}\right)}$ where $n \geq 3$. The growth is neither arithmetic nor geometric.

2. Sample arithmetic probability sequence is shown in the table at bottom of page.

   Sample geometric probability sequence is shown in the right table. Begin with areas of bull's-eyes and annuluses as a geometric sequence and work backward through the area formula.

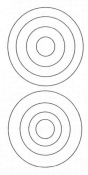

3. The single line of fit, with the equation $\hat{y} = 0.08326x - 162.02993$, does not fit these data well. A better model might be a piecewise function such as
$\hat{y} = 0.01784x - 34.31991$
where $1935 \leq x \leq 1958$,
$\hat{y} = 0.11947x - 233.70649$

Radius	Circle area	Annulus area	Probability
1	π	π	$\frac{1}{16}$
2	4π	3π	$\frac{3}{16}$
3	9π	5π	$\frac{5}{16}$
4	16π	7π	$\frac{7}{16}$
	Totals	16π	1

Radius	Circle area	Annulus area	Probability
1	π	π	$\frac{1}{15}$
$\sqrt{3}$	3π	2π	$\frac{2}{15}$
$\sqrt{7}$	7π	4π	$\frac{4}{15}$
$\sqrt{15}$	15π	8π	$\frac{8}{15}$
	Totals	15π	1

where 1963 ≤ x ≤ 1990, and $\hat{y}$ = 0.22143x − 437.53048 where 1994 ≤ x ≤ 2001. The third segment of the piecewise function using an extended domain would predict the price of a current ticket better than any single line for all the data. Perhaps a curve would fit the data better. A quadratic, logistic, or exponential function might be a better model than a linear or even a piecewise linear model. The data are not modeled best by a single linear equation because inflation is not linear. You might use this activity to introduce the idea of piecewise functions and to discuss the limitations of fitting any mathematical model to a relatively small set of data.

Assessing the Chapter

The Kendall Hunt Cohesive Assessment System (CAS) provides a powerful tool with which to manage assessment content, create and assign homework and tests, and disseminate assessment data.

You might require a presentation or an individual or group performance assessment for the chapter and then give a unit exam. Choose a performance assessment from an exercise you didn't assign that is recommended for that purpose.

Facilitating Self-Assessment

Assessing What You've Learned gives students ideas on organizing their notebooks. To encourage them to keep an orderly notebook, you might allow students to use their notebooks, but not their books, while taking tests.

A good way to check students' understanding is to have them explain a concept to other students. You might encourage this kind of performance assessment by assigning it as part of the chapter review. Allow time for students to present explanations to each other in pairs.

A good sample of student writing in mathematics would be an exercise in which students explain a mathematical concept or relationship. A journal entry in which they also think about their learning can show student understanding and growth. Exercises that require student explanations include Lesson 1.1, Exercise 12; Lesson 1.2, Exercise 16; Lesson 1.3, Exercise 8; Lesson 1.4, Exercise 4; Lesson 1.5, Exercises 7 and 15; Lesson 1.6, Exercises 4 and 8; Lesson 1.7, Exercises 4 and 5; Lesson 1.8, Exercise 6; and Lesson 1.9, Exercise 8.

Good portfolio selections for this chapter include Lesson 1.1, Exercises 10 and 13; Lesson 1.2, Exercise 13; Lesson 1.3, Exercise 5; Lesson 1.4, Performance Task; Lesson 1.5, Exercise 9; Lesson 1.6, Exercise 6; Lesson 1.7, Exercise 6; Lesson 1.8, Exercise 10; and Lesson 1.9, Exercise 9.

Assessing What You've Learned

There are many different ways to assess what you learned—writing in your journal, giving a presentation, organizing your notebook, doing a performance assessment, keeping a portfolio, and writing test items. Assessment is more than just taking tests and more than your teacher giving you a grade.

In the working world, performance in only a few occupations can be measured with tests. All employees, however, must communicate and demonstrate to their employers, coworkers, clients, patients, or customers that they are skilled in their field. Assessing your own understanding and demonstrating your ability to apply what you've learned gives you practice in this important life skill. It also helps you develop good study habits, and that, in turn, will help you advance in school and give you the best possible opportunities in your life.

 ORGANIZE YOUR NOTEBOOK Wouldn't you like to have a record of what you've learned and just what you are expected to know in this algebra course? Your own notebook can be that record if you enter significant information and examples into it on a regular basis. It is a good place for new vocabulary, definitions and distinctions, and worked-out examples that illustrate mathematics that's new to you. On the other hand, if your notebook is simply a stack of returned homework, undated class notes, and scratch-paper computations and graphs, it is too disorganized to perform that service for you.

Before you get far along in the course, take time to go through your notebook. Here are some suggestions:

- Put papers into chronological order. If your work is undated, use the table of contents in this book to help you reconstruct the sequence in which you produced the items in your notebook. You can number pages to help keep them in order.

- Go through the book pages of Chapter 0 and Chapter 1, and see whether you have a good record of how you spent class time—what you learned from investigations and homework. Fill in notes where you need them while information is fresh in your mind. Circle questions that you need to have cleared up before going on.

- Reflect on the main ideas of this chapter. Organize your notes by type of sequence, and make sure you have examples of recursive formulas that produce different types of graphs. Be sure you can identify the starting value and the common difference or common ratio for each type of sequence, and that you know how to use the information to find a later term in the sequence.

 WRITE IN YOUR JOURNAL Use one of these prompts to write a paragraph in your journal.

- Find an exercise from this chapter that you could not fully solve. Write out the problem and as much of the solution as possible. Then clearly explain what is keeping you from solving the problem. Be as specific as you can.

- Compare and contrast arithmetic sequences and linear equations. How do you decide which to use?

 PERFORMANCE ASSESSMENT Show a classmate, family member, or teacher different ways to find a line of fit for a data set. You may want to go back and use one of the data sets presented in an example or exercise, or you may want to research your own data. Discuss how well the line fits and whether you think a linear model is a good choice for the data.

CHAPTER 2
Systems of Equalities and Inequalities

Overview

Discovering Advanced Algebra is about modeling real-world problems with mathematical functions. And most real world situations involve conditions that must be met simultaneously. This chapter looks at mathematical and real world situations where two or more conditions must be met at the same time and has students solve them graphically, by substitution and by elimination. Systems of equations and of inequalities are both represented. The focus is on the meaning of the intersection and how that intersection meets the conditions and constraints of the situation and can possibly be used to maximize or minimize one or more of the conditions.

Lesson 2.1 extends the ideas about lines to systems of linear equations and their graphical representations. In **Lesson 2.2**, students consider substitution and elimination methods for solving a system of equations. They also explore the graphical representations of the equivalent equations obtained through the addition and multiplication properties of equality. **Lesson 2.3** moves from a system of linear equations to a system consisting of a linear and a non-linear equation. After reviewing inequalities in **Lesson 2.4**, students use graphing to represent and solve systems of inequalities and constraints in a real-world situation and to find vertices of feasible regions. In **Lesson 2.5**, students apply these skills to find optimum values in linear programming situations. And in **Lesson 2.6**, students extend from systems of two equations in two unknowns to larger systems of equations, focusing on three equations in three unknowns. They investigate the graph of these systems as being an intersection of planes and solve the systems by substitution and elimination.

The **Performance Task** highlighted in this chapter is in Lesson 2.6. Students solve a system of six equations in six variables.

The Mathematics
Systems of Equations

A linear equation is called *linear* because its graph is a line. A pair of linear equations (or *simultaneous equations*) is graphed as a pair of lines. If these lines intersect at a point, the coordinates of that point make up a *solution* to the system of equations. In this case the system is called *independent*. If the two lines intersect at all points (that is, if they're actually the same line), then there are infinitely many solutions, and the system is *dependent*. A system with one or more solutions is *consistent*. If the two lines are parallel, then there are no solutions, and the system is called *inconsistent*.

A system of equations is not always a pair of lines. A system can be comprised of a combination of linear equations, a combination of linear and nonlinear equations, or even a combination of nonlinear equations. In this chapter, students focus on solving a system of linear equations or a system with a linear and nonlinear equation.

A system of two equations in two variables can be solved using many methods—by graphing, by listing values in tables, and algebraically by substitution or elimination. Regardless of the method used, the solution is one or more ordered pairs that satisfy both equations.

These same methods can be used when solving larger systems of equations. To solve a system of three equations, you use elimination and/or substitution to rewrite the system as a system with two variables. With elimination, you must eliminate one variable from pairs of the three equations, giving you two equations with two variables. With substitution, you solve one equation for one variable in terms of the other two variables, and substitute that expression into the other two equations, giving you two equations with two variables. Graphing is a little more challenging for students. An equation in three variables is three-dimensional, so the solution of three equations in three variables is the intersection of planes. The planes might intersect in a point, a line, not at all, or in all points in the plane.

Systems of Linear Inequalities

An inequality is like an equation except that the equal sign is replaced by < (less than), > (greater than), ≤ (less than or equal to), or ≥ (greater than or equal to). Statements using < or > are called *strict inequalities*. You can solve linear inequalities in one variable as you would solve linear equations, with the exception that when you multiply or divide by a negative number, you must switch the direction of the inequality.

Solutions to an inequality in two variables can be graphed on a plane. First you write an equation by replacing the inequality symbol with an equal sign. Graph that equation. The solutions to the inequality form a half-plane on one side of that line. If the inequality is strict, draw a dashed line, creating an open half-plane. Otherwise, draw a solid line, creating a closed half-plane. Shade in the solution region. The solutions to a system of inequalities in two variables are graphed as the intersection of the half-planes representing the solutions of the individual inequalities in the system.

Linear Programming

Systems of linear inequalities are very useful in the area of mathematics called *linear programming*. In linear programming, a system of linear inequalities represents the constraints of a real-world situation. In a situation with two variables, each constraint can be represented graphically by a region of the plane. A solution to that system is a region of the plane called the feasible region. A linear function of the two variables is maximized or minimized at a vertex or along an edge of the feasible region.

Using This Chapter

While some of this chapter is a review of content from earlier algebra courses, there is new content in the chapter. The study of systems of equations extends beyond two equations in two variables to include three equations in three variables. The chapter also looks at a system consisting of a linear and nonlinear equation. And the study of systems of inequalities extends to studying maximizing or minimizing a vertex or along the edge of a feasible region. Linear programming is introduced to find optimal values. Rather than skip lessons, you can save time by assigning the **Shortened** version of the investigations. The important new lessons are Lesson 2.3, where students solve systems of equations that involve both a linear and a non-linear equation, and Lesson 2.6, which introduces larger systems of equations.

As you teach the chapter, the suggestions for differentiating instruction and modifying the investigations can support you in tailoring the content to your students.

Common Core State Standards (CCSS)

Common Core State Standards holds the expectation that students are able to justify the process used in solving a system of equations. Students should develop fluency writing, interpreting, and translating between various forms of linear equations and inequalities. Students find and interpret their solutions to systems of equations and inequalities.

High school experiences with equations should build on student experiences graphing and solving systems of linear equations in the middle grades. In grade 8, students solve systems of two linear equations in two variables and relate the systems to pairs of lines in the plane; these intersect, are parallel, or are the same line (8.EE.8). In algebra, students further explore systems of equations and inequalities, focusing on justification of the methods used. They find and interpret their solutions, including cases where the two equations describe the same line (yielding infinitely many solutions) and cases where two equations describe parallel lines (yielding no solution) (A.CED.3, A.REI.6, GPE.5). In advanced algebra, these systems are extended to include solving three equations in three variables. Students also algebraically and graphically solve a system consisting of a linear equation and a non-linear equation in two variables (A.REI.7). In addition to solving systems of equations, students are expected to be able to explain why the coordinates of the intersection of the graphs of two functions is the solution to the equation $f(x) = g(x)$ (A.REI.11). Systems of linear inequalities in two variables are graphed as a half-planes and the solution set is the points in the intersection of the corresponding half-planes (A.REI.12).

Additional Resources For complete references to additional resources for this chapter and others see your digital resources.

Refreshing Your Skills To review the algebra skills needed for this chapter see your digital resources.

- Linear Relationships
- Properties of Real Numbers

Projects See your digital resources for performance-based problems and challenges.

- Nutritional Elements

Materials

- Card stock, *optional*
- graph paper

CHAPTER 2 INTERLEAF Systems of Equalities and Inequalities

CHAPTER 2
Systems of Equations and Inequalities

Systems describe situations in which two or more conditions must be met at the same time. These photos present a visual representation of a solution to a system of equations. For a system of equations, there can be one solution (such as when blocking a shot in basketball), more than one solution (such as the cable supports of the Sunshine Skyway Bridge in Florida), or no solution (such as with train tracks used by the steam train in Douro Valley, Portugal).

OBJECTIVES
In this chapter you will
- solve systems of equations and inequalities
- graph systems of equations and inequalities on a coordinate plane
- write and graph inequalities that represent conditions that must be met simultaneously
- solve a simple system consisting of a linear and non-linear equation
- identify a system of equations as consistent and independent, inconsistent, or consistent and dependent

CHAPTER 2

Objectives
- Use systems of linear equations to model problems
- Solve systems of equations and inequalities by graphing, by tables, by substitution, and by elimination
- Write and graph systems of inequalities to describe given real-world constraints
- Learn how to do linear programming—optimizing a function over a feasible region representing real-world situations—with two variables.

Systems describe situations in which two or more conditions must be met simultaneously. **ASK** "How many "functions" do you see?" Students might see intersecting lines in the braces of the train tracks or parallel lines in the tracks themselves. They might recall that the projectile of the basketball is a parabola, while the arm of the player blocking the shot is a "line". Some students might see an inverted absolute value function in the cable supports, or intersecting and parallel lines. **ASK** "How does each photo provide a visual of the solution of a system of equations?" [Each "line" in the braces intersect other "lines" only once. Parallel "lines" of the tracks do not intersect. In the basketball situation, the "line" of the players arm will intersect the "parabola" of the ball one time, if the shot is blocked. The support cables intersect the "line of the bridge in two places, on either side of the main beam.] **ASK** "What conditions are being met in each situation?"

LESSON 2.1

This lesson should be a review for most students.

COMMON CORE STATE STANDARDS

APPLIED	DEVELOPED	INTRODUCED
A.CED.3	A.REI.6	
	A.REI.11	
	F.IF.7b	

Objectives

- Examine problems involving two or more conditions that must be satisfied at the same time
- Use different representations (tables, graphs, equations, models) to create and solve systems of equations
- Solve systems of equations exactly (algebraically) and approximately (graphs)
- Recognize and graph systems of equations represented by step functions

Vocabulary

system of equations
substitution
step function
greatest integer function

Materials

- Calculator Notes: Entering and Graphing Equations; Function Tables; Greatest Integer Function, *optional*

Launch

Project the graph at the beginning of the lesson, or from the Launch found in your ebook.

a. What part of the graph represents profits?
b. What part of the graph represents loss?
c. What is the meaning of the break-even point?

a. expenses < income
b. expenses > income
c. Income = expenses

LESSON 2.1

Linear Systems

The number of tickets sold for a school activity, like a spaghetti dinner, helps determine the financial success of the event. The expenses for the event can be greater than, equal to, or less than the income from the ticket sales. The break-even value is the intersection of the expense function and the income function. This break-even point indicates where the expenses are equal to the income.

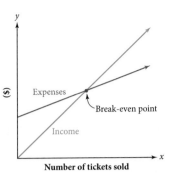

In this lesson you will focus on mathematical situations involving two or more equations or conditions that must be satisfied at the same time. A set of two or more equations that share the same variables and are solved or studied simultaneously is called a **system of equations.**

Geometric ethnic pattern designs are characterized by intersecting lines in mazelike patterns.

EXAMPLE A | Minh and Daniel are starting a business together, and they need to decide between cell phone plans. One company offers the Connected Colleagues Plan for $40 each month for the first phone plus $35 for each additional phone on the plan. A competing company offers the Small Business Plan, which costs $60 a month for the first phone and $25 for each additional phone on the plan. Which plan should Minh and Daniel choose?

Solution | The plan they choose depends on the nature of their business and the most likely number of phones they want to put on the plan. Because the Connected Colleagues Plan has a lower basic monthly rate, it will be cheaper if they only have one phone. However, if they have many phones, the Small Business Plan will cost less, since each additional phone adds less to the total monthly bill. There is a number of phones for which both plans cost the same. To find it, let x represent the number of phones on the plan, and let y represent the cost in dollars.

ELL
Have students start making posters or a graphic organizer that shows all the methods for solving systems of equations.

Support
By demonstrating graphical solutions to check their algebraic work, students are more likely to understand the importance of a "solution" of a system of equations.

Advanced
Ask students to describe the advantages and disadvantages of each method of solving systems of equations. Many times students attempt to learn each method in isolation without recognizing the best time to use each. Have students assess each problem to determine the best method of solution and draw general conclusions from their work.

Because x represents the number of phones, $x - 1$ represents the number of additional phones. Therefore, a cost equation that models the Connected Colleagues Plan is

$$y = 40 + 35(x - 1)$$

A cost equation for the Small Business Plan is

$$y = 60 + 25(x - 1)$$

A graph of these equations shows the Connected Colleagues Plan cost is below the Small Business Plan cost until the lines intersect. You may be able to estimate the coordinates of this point from the graph.

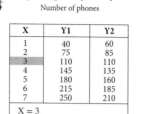

Or you can look in a table to find an answer. The point of intersection at (3, 110) tells you that having three phones on the plan will cost $110 with either plan.

X	Y1	Y2
1	40	60
2	75	85
3	110	110
4	145	135
5	180	160
6	215	185
7	250	210

X = 3

If Minh and Daniel believe that they will not want more than three phones on the plan, they should choose the Connected Colleagues Plan. But if they expect the business to grow and that they will need more than three phones for their employees, the Small Business Plan is the better option.

You can also use intersecting graphs to help you solve a single equation. That's because you can create two equations from one.

EXAMPLE B | Solve the equation $5x + 4(x - 3) = 15 - 3(7 - x)$ by graphing a system of equations.

Solution | Each side of the equation can be used to create its own equation to graph.

$$y = 5x + 4(x - 3) \quad \text{and} \quad y = 15 - 3(7 - x)$$

Graphing the two equations shows that the lines intersect when $x = 1$. This means that $x = 1$ is the solution to the original equation.

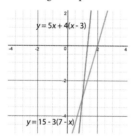

You can use the idea from Example B in reverse. If you have two equations that are each written as $y =$, you can create a single equation and use what you know to solve that single equation for the x-coordinate of the point of intersection. Think about this as you work through the investigation.

You can also use equations to solve for the point of intersection. How would you solve the system of equations symbolically?

LESSON 2.1 Linear Systems   **103**

Investigate

Example A

Project Example A and have students work in small groups. Have them present their strategy as well as their answers. Choose groups that have used different strategies. Have them explain why they chose that particular method. Emphasize the use of correct mathematical terminology. Whether student responses are correct or incorrect, ask other students if they agree and why. SMP 1, 3, 5, 6

ASK "Why do the equations include $(x - 1)$ rather than just x?" [The cost of an additional phones is added on for each phone *after* the first; if $x = 1$, the result should imply no extra cost for more phones.]

On a graphing calculator, you can use the zoom function on the graph and the table to help students find the solution or use the intersect command to check their answers. SMP 5

Pick a point on one of the lines and ASK "What is the real-world meaning of this point?"

Example B

Project Example B and have students work it individually or in pairs. Have students share their graphs. ASK What is the meaning of the intersection point?" [$x = 1$ is the solution to the original equation.] Whether student responses are correct or incorrect, ask other students if they agree and why. SMP 3, 5, 6

Modifying the Investigation

Whole Class Use the desmos image in your ebook to graph the data for the whole class. Elicit each group's best predictions. Discuss the best way to find the intersection. Try two or three methods and have groups choose the answer they feel is best. Each group can defend its preferred method to the class.

Shortened There is no shortened version of this investigation.

One Step There is no **One Step** alternative for this investigation.

Guiding the Investigation

This investigation reviews using a table of values to write a system of equations using a line of fit. There is a desmos tool available in your ebook. Have groups present their strategy as well as their answers. Choose groups that have used different strategies. Most students will choose to solve the system graphically or with substitution, but it is acceptable for them to use elimination or even to consider only the sequence of differences. Emphasize the use of correct mathematical terminology. Whether student responses are correct or incorrect, ask other students if they agree and why. **SMP 1, 3, 5, 6**

Step 1 **ASK** "Are the functions linear? Is it appropriate to use a linear model? Why?" Students might use the residuals to justify their answers.

Step 1. The lines of fit, $\hat{y} \approx 588{,}363 + 14{,}714.96(x-1980)$ for San Jose and $\hat{y} \approx 1{,}275{,}750 - 19{,}172.55(x-1980)$ for Detroit, predict that the populations will be the same between 2000 and 2001; about 886,846.

Example C

Project Example C and have students work in small groups. Have groups present their strategies along with their solutions. Emphasize the use of correct mathematical terminology. Whether student responses are correct or incorrect, ask other students if they agree and why. **SMP 1, 3, 5, 6**

ASK "What allows us to equate the right sides of these equations?" [the substitution property of equality]

ASK "What distance does d represent?" [the distance from Justine's starting point, which is not the same as the distance run by Evan] "Why are the equations combined by setting expressions equal to each other?" [We assume the distances from the starting point are equal, and we want to find the corresponding time.] "Do Justine and Evan run the same distance?" [No; Justine runs 50 ft farther.]

ASK "How long will it take Justine to close the gap?" Justine, running 1.8 ft/s faster than Evan, can close the 50 ft gap in $\frac{50}{1.8}$ s, or about 27.8 s.

INVESTIGATION

Population Trends

The table below gives the populations of San Jose, California, and Detroit, Michigan.

Populations

Year	1950	1960	1970	1980	1990	2000	2010
San Jose	95,280	204,196	459,913	629,400	782,248	894,943	952,560
Detroit	1,849,568	1,670,144	1,514,063	1,203,368	1,027,974	951,270	713,862

(*US Census for 2010*)

Step 1 Estimate the year that the two cities had the same population. What was that population?

Step 2 Show the method you used to make this prediction. Choose a different method to check your answer. Discuss the pros and cons of each method.

Students might solve a system of equations with a graph, a table, or algebraically.

In algebra, you studied different methods for finding the exact coordinates of an intersection point by solving systems of equations. One method is illustrated in the next example.

EXAMPLE C Justine and her little brother Evan are running a race. Because Evan is younger, Justine gives him a 50-foot head start. Evan runs at 12.5 feet per second and Justine runs at 14.3 feet per second. How far will they be from Justine's starting line before Justine passes Evan? What distance should Justine mark for a close race?

Solution You can compare the time and distance that each person runs. Because the distance, d, depends on the time, t, you can write these equations:

$d = rt$ Distance equals the rate, or speed, times time.

$d = 14.3t$ Justine's distance equation.

$d = 50 + 12.5t$ Evan's distance equation.

Graphing these two equations shows that Justine eventually catches up to Evan and passes him if the race is long enough. At that moment, they are at the same distance from the start, at the same time. You can estimate this point from the graph or scroll down until you find the answer in the table. You can also solve the system of equations.

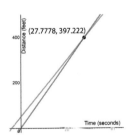

ASK "In this example the distances were set equal. Why? Is there a case in which the *times* would be equal?" As a class, create and solve a problem that reflects this situation.

Conceptual/Procedural

Conceptual This lesson reviews the meaning of the intersection of two lines as solution of both functions. Students look at when and why to use a particular method of solving systems of equations, such as by graphing, a table, or symbolic manipulation.

Procedural Students review writing a system of equations based on data represented verbally, in a table, or graph and how to use different methods of solving systems of equations, such as by graphing, a table, or symbolic manipulation.

Because you want to find out when Justine's distance and Evan's distance are equal, you can substitute Justine's expression for distance, 14.3t, for the distance, d, in Evan's equation. Then you'll solve for t, which will give you the time when the distances are equal.

$14.3t = 50 + 12.5t$ Substitute 14.3t for d in Evan's distance equation.

$1.8t = 50$ Subtract 12.5t from both sides.

$t = \frac{50}{1.8} \approx 27.8$ Divide both sides by 1.8.

So, Justine passes Evan after 27.8 seconds. Now you can substitute this value back into either equation to find their distances from the starting line when Justine passes Evan.

$d = 14.3t = 14.3 \cdot \frac{50}{1.8} \approx 397.2$

If Justine marks a 400 ft distance, she will win, but it will be a close race.

The method of solving a system demonstrated in Example B uses one form of **substitution**. In this case you substituted one expression for distance, 14.3t, for d in the other equation. The resulting equation had only one variable, t. When you have the two equations written in intercept form, substitution is a straightforward method for finding an exact solution. The solution to a system of equations with two variables is a pair of values that satisfies both equations. Sometimes a system will have many solutions or no solution.

Many of the situations that are often modeled with linear equations are more accurately represented by a type of **step function.** Quantities such as numbers of people, costs for rental cars, phone bills, and other things go up only after you reach a certain threshold value. For instance, the price of mailing a letter stays the same until the weight of the letter gets over one ounce and then the cost jumps, rental car costs jump for every additional whole day of rent. Functions of this type look like the one pictured here. This type of function is called a step function because its graph looks like a staircase.

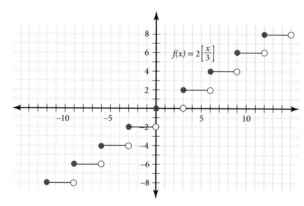

The **greatest integer function** is a parent function for many of these examples. It outputs the greatest integer not greater than its input. For instance, $[3.5] = 3$ and $[4.9] = 4$. You can transform the greatest integer function in the same way as other functions. In the exercises you will have the opportunity to apply this function to a real world situation.

Summarize

Discussing the Lesson

Refer students back to Example A. Ask whether the data are discrete, in which case an explicit formula would be better than a linear equation. Students might think the data are discrete because the y-values can be only integers. Remind them that discreteness depends on the domain (x-values) and that telephone charges can be calculated for non-whole numbers of minutes. Project the graph on this page. **ASK** "How is this graph different from the graph in Example A?" Exercise 9 examines this problem again, using the greatest integer function.

ASK "In Example C, what if Evan ran 14.3 feet per second?" [He would be running at the same rate as Justine and therefore would always be ahead.] "How would that be reflected in the calculations?" [In the calculations, t would disappear and leave the equation 0 = 50.] "What would be the solution of that system?" [There is none.]

EXTEND "When would a system have no solutions? When would it have many solutions?" [graphs of the equations are parallel; the same line]

CRITICAL QUESTION How is the solution of a system of equations represented on a graph, and why is it presented in that way?

BIG IDEA A point lies on a graph if and only if its coordinates satisfy the equation of the graph. Therefore, a pair of values that satisfies both equations represents a point on both lines.

CRITICAL QUESTION How might you solve a system of equations like this? Why?
$$\begin{cases} y = 15 + 8x \\ -10x - 5y = -30 \end{cases}$$

BIG IDEA Students might solve the second equation for y and then set the equations equal to each other. They might use substitution. They might graph. Reasons will vary.

Formative Assessment

Student group work and presentations of the Examples and Investigation can provide evidence of meeting the objectives of the lesson. Confirm that students understand how to solve linear systems algebraically and graphically. The Investigation provides evidence to assess students' understanding of lines of fit and of how equations and their graphs relate. Examples A and C and the Investigation provide evidence of examining problems involving two or more conditions that must be satisfied at the same time. A discussion of the step function and Exercise 9 provide evidence of recognizing and graphing systems of equations represented by step functions.

Apply

Extra Example

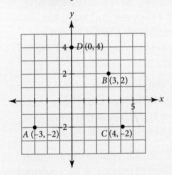

Determine whether $\overleftrightarrow{AB}$ and $\overleftrightarrow{CD}$ are parallel, perpendicular, or neither. Explain.

$\overleftrightarrow{AB} \perp \overleftrightarrow{CD}$; The slopes, $\frac{2}{3}$ and $-\frac{3}{2}$, are negative reciprocals of each other.

Closing Question

Find the point where these two lines intersect, first using a table and then using substitution.

$y = 3 + 5x$
$8x - 2y = -10$
(2, 13)

Assigning Exercises

Suggested: 1 – 11
Additional Practice: 12 – 14

Exercise 2 Students might find it helpful to plot the point and sketch

---

 **Exercises** You will need a graphing calculator for exercises **1, 5,** and **8–10.**

Practice Your Skills

1. Use a table to find the point of intersection for each pair of linear equations.

   a. @ $\begin{cases} y = 3x - 17 \\ y = -2x - 8 \end{cases}$ (1.8, −11.6)
   b. $\begin{cases} y = 28 - 3(x - 5) \\ y = 6 + 7x \end{cases}$ (3.7, 31.9)

2. Write a system of equations that has (2, 7.5) as its solution. Sample answer: $\begin{cases} y = 7.5 + 4(x - 2) \\ y = 7.5 + 1.7(x - 2) \end{cases}$

3. Write the equation of the line perpendicular to $y = 4 - 2.5x$ and passing through the point (1, 5). $y = 5 + 0.4(x - 1)$

4. Solve each equation.

   a. $4 - 2.5(x - 6) = 3 + 7x$ @ $x = \frac{32}{19} \approx 1.684$
   b. $11.5 + 4.1t = 6 + 3.2(t - 4)$ $t = -\frac{61}{3} \approx -20.33$

5. Use substitution to find the point (x, y) where each pair of lines intersect. Use a graph or table to verify your answer.

   a. @ $\begin{cases} y = -2 + 3(x - 7) \\ y = 10 - 5x \end{cases}$ (4.125, −10.625)
   b. $\begin{cases} y = 0.23x + 9 \\ y = 4 - 1.35x \end{cases}$ (−3.16, 8.27)
   c. $\begin{cases} y = -1.5x + 7 \\ 2y = -3x + 14 \end{cases}$ They intersect at every point; they are the same line.

Reason and Apply

6. The equations $s_1 = 18 + 0.4m$ and $s_2 = 11.2 + 0.54m$ give the lengths of two different springs in centimeters, s_1 and s_2, as mass amounts in grams, m, are separately added to each.

   a. When are the springs the same length? a. When you place 48.6 g on each spring, they will be the same length, 37.4 cm.
   b. When is one spring at least 10 cm longer than the other? b. When more than 120 g is placed on the prings, spring 2 will be more than 10 cm longer than spring 1.
   c. Write a statement comparing the two springs.

7. **APPLICATION** This graph shows the Kangaroo Company's production costs and revenue for its pogo sticks. Use the graph to estimate the answers to the questions below.

   a. If 25 pogo sticks are sold, will the company earn a profit? Describe how you can use the graph to answer this question. @ No. At $x = 25$, the cost line is above the income line.
   b. If the company sells 200 pogo sticks, will it earn a profit? If so, approximately how much? Yes. The profit is approximately $120.
   c. How many pogo sticks must the company sell to break even? How do you know?
   About 120 pogo sticks. Look for the point where the cost and income lines intersect.

---

two lines through that point. They would then find the equations of each line. This exercise has numerous solutions. Have students compare their answers and come up with even more equations.

Exercise 5 Make sure your students verify their answers using a graph or a table. It is important that they practice these methods of solving systems also.

6c. Answers will vary but should include comments about spring 1 (s_1) being longer than spring 2 (s_2) in its unstretched condition and the fact that spring 2 stretches more for each added gram of mass than spring 1.

Exercise 7 LANGUAGE Define pogo stick, perhaps with a picture.

8. In many countries of the world, education is not as available as it is in the United States. Men and women also do not necessarily have the same opportunities to go to school. The data below show the average numbers of years of education that adult men and women have historically had in Paraguay.

Years of Education

Year	1980	1984	1988	1992	1996	2000	2004	2008
Men	5.9	6.3	6.6	6.9	7.2	7.5	7.8	8
Women	4	4.4	4.9	5.3	5.7	6.1	6.6	7.1

a. Analyze the data and predict when men and women will have the same average number of years of education in Paraguay, if the current trends continue.
The least-squares lines intersect about the year 2038.

b. How many years of education will that be when both men and women have the same average number of years of education?
In 2038, both men and women will average about 10.4 years of education.

c. Is it appropriate to use a linear model for these data? Why?

9. **APPLICATION** Most wireless phone companies also have charges for data. Suppose the first phone company from Example A charges the same amount for using 1.2 or 1.5 gigabytes of data as it does for using 1 gigabyte of data. There is no increase in charge until you use 2 gigabytes.

a. Use the greatest integer function to write a cost equation for a company that charges $35 for one phone and up to 1 gigabyte of data, with a charge of $18 for each additional gigabyte of data used. $y = 35 + 18[x]$

b. Use the greatest integer function to write a cost equation for a company that charges $40 and up to 1 gigabyte of data, with a charge of $13 for each additional gigabyte of data used. $y = 40 + 13[x]$

c. Graph the two equations representing the phone plans.

d. Determine when each plan is more desirable. Explain your reasoning.

10. **APPLICATION** An anthropologist can use the lengths of certain bones from skeletal remains to estimate the height of the living person. The humerus bone is the single large bone that extends from the elbow to the shoulder socket. The following formulas, attributed to the work of Mildred Trotter and G. C. Gleser, have been used to estimate a male's height, m, or a female's height, f, when the length, h, of the humerus bone is known: $m = 3.08h + 70.45$ and $f = 3.36h + 58.0$. All measurements are in centimeters.

a. Graph the two lines on the same set of axes.

b. If a humerus bone is found and it measures 42 cm, how tall would the person have been according to the model if the bone was determined to come from a male? From a female?

c. At what point do the two equations intersect? What does the point of intersection mean in this context?
(44.46, 207.4). A person with a humerus bone of length 44.46 cm would have been 207.4 cm tall, regardless of the person's gender.

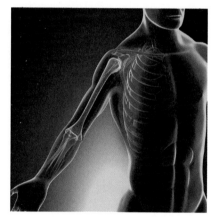

Exercise 8a. SUPPORT Emphasize that being asked to analyze data requires mathematical steps, not just simple observation.

8c. It is reasonable to use a linear model to predict for a while. However, the number of years of education cannot increase indefinitely, so the model is limited in how long it will be appropriate. At some point, the years of education will level out and no longer increase.

9c.

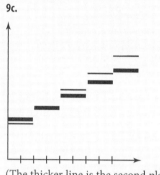

(The thicker line is the second plan)

9d. If you use less than one gigabyte of data the first plan is cheaper. If you use between 1 and 2 gigabytes, the plans cost the same amount. However, if you think you will use 3 or more gigabytes, the second plan is cheaper.

10a. Let x represent the length of the humerus bone in cm, and let y represent the person's height in cm.

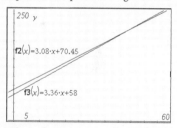

10b. male: approximately 199.81 cm; female: approximately 199.12 cm

LESSON 2.1 Linear Systems

11a. Let l represent length in centimeters, and let w represent width in centimeters; $2l + 2w = 44, l = 2 + 2w$; $w = \frac{20}{3}$ cm, $l = \frac{46}{3}$ cm.

11b. Let l represent length of leg in cm, and let b represent length of base in cm; $2l + b = 40, b = l - 2$; $l = 14$ cm, $b = 12$ cm.

Exercise 11c This is the reading at one time. In general, °F is not equivalent to $3 \cdot °C - 0.4$. The conversion formula is either $C = \frac{5}{9}(F - 32)$ or $F = \frac{9}{5}C + 32$.

11c. Let f represent temperature in °F, and let c represent temperature in °C; $f = 3c - 0.4, f = 1.8c + 32$; $c = 27°C, f = 80.6°F$.

Exercises 12, 13 ELL You might need to define the terms in these exercises.

Exercise 14 ALERT Students might try to solve for x. The solution will involve x, though, and will not be just a single number.

Physical anthropology is a science that deals with the biological evolution of human beings, the study of human ancestors and nonhuman primates, and the in-depth analysis of the human skeleton. By studying bones and bone fragments, physical anthropologists have developed methods that can provide a wealth of information on the age, sex, ancestry, height, and diet of a person who lived in ancient times, just by studying his or her skeleton.

11. Write a system of equations to model each situation, and solve for the values of the appropriate variables.

  a. The perimeter of a rectangle is 44 cm. Its length is 2 cm more than twice its width. ⓐ

  b. The perimeter of an isosceles triangle is 40 cm. The base length is 2 cm less than the length of a leg of the triangle. ⓗ

  c. The Fahrenheit reading on a dual thermometer is 0.4 degree less than three times the Celsius reading. (*Hint:* Your second equation needs to be a conversion formula between degrees Fahrenheit and degrees Celsius.)

Review

12. Use the model $\hat{y} = 393.8 + 1.87(x - 2012)$ that was found in the example in Lesson 1.7. Recall that x represents the year and y represents the concentration of CO_2 in parts per million (ppm) at Mauna Loa.

  a. Predict the concentration of CO_2 in the year 2002. 375.1 ppm

  b. Use the model to predict the concentration of CO_2 in the year 2020. 408.76 ppm

  c. According to this model, when will the level of CO_2 be double the preindustrial level of 280 ppm? approximately the year 2101
  (But we certainly hope this is wrong.)

13. The histogram shows the average annual cost of insuring a motor vehicle in the United States.

  a. How many jurisdictions are included in the histogram? 51

  b. South Carolina is the median jurisdiction. South Carolina is in what bin? 700-800 bin

  c. In what percentage of the jurisdictions is the average cost less than $700? 39%

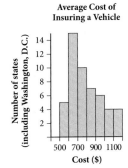

14. Solve these equations for y.

  a. $3x - 8y = 12$ ⓐ
  $y = \frac{3x - 12}{8}$
  $= 0.375x - 1.5$

  b. $5x + 2y = 12$
  $y = \frac{12 - 5x}{2}$
  $= 6 - 2.5x$

  c. $-3x + 4y = 5$
  $y = \frac{5 + 3x}{4}$
  $= 1.25 + 0.75x$

LESSON 2.2

Substitution and Elimination

A place for everything and everything in its place.
ENGLISH PROVERB

A solution to a system of equations in two variables is a pair of values that satisfies both equations and represents the intersection of their graphs. In the previous lesson, you reviewed solving a system of equations using substitution, when both equations are in intercept form. Suppose you want to solve a system and one or both of the equations are not in intercept form. You can rearrange them into intercept form, but sometimes there's an easier method.

If one equation is in intercept form, you can still use substitution.

In *Modern Warrior Series, War Shirt #2* (1998), northern Cheyenne artist Bently Spang (b 1960) reflects on identity. He presents opposing influences as an equivalence of forms—the modern and the traditional, the spiritual and the mundane, the Cheyenne and the non-Cheyenne—all intersecting in an arrangement of family photographs.

EXAMPLE A Solve this system for x and y.
$$\begin{cases} y = 15 + 8x \\ -10x - 5y = -30 \end{cases}$$

Solution You can solve the second equation for y so that both equations will be in intercept form and substitute the right side of one equation for y in the other equation. However, you may find it easier to substitute the right side of the first equation for y in the second equation.

$-10x - 5y = -30$	Original form of the second equation.
$-10x - 5(15 + 8x) = -30$	Substitute the right side of the first equation for y.
$-10x - 75 - 40x = -30$	Distribute -5.
$-50x = 45$	Add 75 to both sides and combine like terms.
$x = -0.9$	Divide both sides by -50.

LESSON 2.2

In this lesson students review the substitution method and the elimination method of solving systems of linear equations.

COMMON CORE STATE STANDARDS
APPLIED	DEVELOPED	INTRODUCED
A.CED.3		
A.REI.5		

Objectives
- Solve systems of equations by substitution
- Use the addition and multiplication properties of equations to solve systems of equations by elimination
- Define *inconsistent, consistent,* and *dependent systems*

Vocabulary
elimination
consistent
inconsistent
dependent
independent

Launch

Consider the following systems. Would it be easier to solve each system by graphing using technology, using a table of values, or by substitution? Why?

a. $\begin{cases} 15x + 3y = 31 \\ x = -\frac{2}{3}y - \frac{3}{5} \end{cases}$

b. $\begin{cases} y = 87 - 0.23x \\ y = -0.05 + 0.23x \end{cases}$

Answers will vary. For part a, substitution might be preferred, since the second equation is already solved in terms of x. Graphing with technology requires solving both equations in terms of y. For part b, graphing or substitution might be the choice, since both equations are in terms of y.

ELL	Support	Advanced
Continue using the graphic organizer to create clear directions on the appropriate times to use each method for solving systems of equations. Give concrete examples of each method to help students connect the vocabulary to the procedures.	Emphasize double–checking solutions by substituting the values for x and y into both of the original equations. Not only does this allow students to verify their work, but it also reinforces what the solution represents.	If students are familiar with the elimination method, have them complete the **One Step**.

Investigate

Example A

Project Example A and have students work it individually or in pairs. Have students present their method and why they chose that method along with their answers. After presentations using different methods, **ASK** "Is there a correct way to solve the system? What are some advantages and disadvantages of each of the methods presented?" **SMP 1, 3, 5, 6**

Students should also check their solutions. **ASK** "How might you use a graph to check that the answer is approximately correct?" For the first equation, students can graph the y-intercept and use the slope to draw an accurate graph. For the second equation, students can use the intercepts. By graphing (0, 6) and (3, 0) and drawing a line through the points, students can quickly graph the second line. Students can use this same technique in the investigation.

Example B

Project Example B and have students work it individually or in pairs. Have students present their method and why they chose that method along with their answers. After presentations using different methods, **ASK** "What are the merits of each of the methods presented?" **SMP 1, 3, 5, 6**

In the system in part b, both expressions equal 6. Students might think of using substitution here. Choose this as one presentation. Ask other students if they agree and why. If it doesn't come up, **ASK** "Is it valid to set the left sides of the equations equal to one another?" [yes] "Would it help us solve the system? Why?" [No; we would have one equation with two variables and no way to solve it.]

Now that you know the value of x, you can substitute it into either equation to find the value of y.

$y = 15 + 8(-0.9)$ Substitute -0.9 for x in the first equation.

$y = 7.8$ Multiply and combine like terms.

Write your solution as an ordered pair. The solution to this system is $(-0.9, 7.8)$.

The substitution method relies on the substitution property, which says that if $a = b$, then a may be replaced by b in an algebraic expression. Substitution is a powerful mathematical tool that allows you to rewrite expressions and equations in forms that are easier to use and solve. Notice that substituting an expression for y, as you did in Example A, eliminates y from the equation, allowing you to solve a single equation for a single variable, x.

A third method for solving a system of equations is the **elimination** method. The elimination method uses the addition property of equality, which says that if $a = b$ and $c = d$, then $a + c = b + d$. In other words, if you add equal quantities to both sides of an equation, the resulting equation is still true. If necessary, you can also use the multiplication property of equality, which says that if $a = b$, then $ac = bc$, or if you multiply both sides of an equation by equal quantities, then the resulting equation is still true.

EXAMPLE B Solve these systems for x and y.

a. $\begin{cases} 4x + 3y = 14 \\ 3x - 3y = 13 \end{cases}$ b. $\begin{cases} -3x + 5y = 6 \\ 2x + y = 6 \end{cases}$

Solution Because neither of these equations is in intercept form, it is probably easier to solve the systems using the elimination method.

a. You can solve the system without changing either equation to intercept form by adding the two equations.

$\begin{array}{ll} 4x + 3y = 14 & \text{Original equations.} \\ 3x - 3y = 13 & \text{Addition property of equality.} \\ \overline{7x = 27} & \text{The variable } y \text{ is eliminated.} \end{array}$

$x = \dfrac{27}{7}$ Multiplication property of equality.

$4\left(\dfrac{27}{7}\right) + 3y = 14$ Substitution property of equality.

$y = -\dfrac{10}{21}$ Addition and multiplication properties of equality.

The solution to this system is $\left(\dfrac{27}{7}, -\dfrac{10}{21}\right)$. You can substitute the coordinates back into both equations to check that the point is a solution for both.

$4\left(\dfrac{27}{7}\right) + 3\left(-\dfrac{10}{21}\right) = \dfrac{108}{7} - \dfrac{30}{21} = \dfrac{294}{21} = 14$

$3\left(\dfrac{27}{7}\right) - 3\left(-\dfrac{10}{21}\right) = \dfrac{81}{7} + \dfrac{30}{21} = \dfrac{273}{21} = 13$

Modifying the Investigation

Whole Class Split the class into six sections. Assign students in each section one of the systems of equations and ask them to do Steps 1 and 2. Have volunteers demonstrate their solutions. Discuss Steps 3 through 6.

Shortened Skip systems d, e, and f.

One Step Challenge the class to solve this system. Solve this system of equations using four different methods. Which method is easier for this system? Why?

$\begin{cases} 4x - 3y = -4 \\ x + 3y = 14 \end{cases}$

Students should come up with graphing, solving both equations for y and setting the two equations equal to each other, solve the second equation for x and substitute into the first equation, and elimination. Have all four methods presented during the discussion and compare them. Reasons will vary.

b. Adding the equations as they are written will not eliminate either of the variables. You need to multiply one or both equations by some value so that if you add the equations, one of the variables will be eliminated.

The easiest choice is to multiply the second equation by −5 and then add it to the first equation.

$$-3x + 5y = 6 \quad \rightarrow \quad -3x + 5y = 6 \quad \text{Original form of the first equation.}$$
$$-5(2x + y) = -5(6) \quad \rightarrow \quad \underline{-10x - 5y = -30} \quad \text{Multiply both sides of the second equation by −5.}$$
$$-13x = -24 \quad \text{Add the equations.}$$

This eliminates the y-variable and gives $x = \frac{24}{13}$. Substituting this x-value back into either of the original equations gives the y-value. Or you can use the same process to eliminate the x-variable.

$$-3x + 5y = 6 \quad \rightarrow \quad -6x + 10y = 12 \quad \text{Multiply both sides by 2.}$$
$$2x + y = 6 \quad \rightarrow \quad \underline{6x + 3y = 18} \quad \text{Multiply both sides by 3.}$$
$$13y = 30$$
$$y = \frac{30}{13}$$

The solution to this system is $\left(\frac{24}{13}, \frac{30}{13}\right)$. You can use your calculator to verify the solution.

It would take a lot of effort to solve this last system using a table on your calculator. If you had used the substitution method to solve the systems in Example B, you would have had to work with fractions to get an accurate answer. To solve these systems, the easiest method to use is the elimination method.

INVESTIGATION

What's Your System?

YOU WILL NEED
- graph paper

In this investigation you will discover different classifications of systems and their properties. You can divide up the work among group members, but make sure each problem is solved by one person and checked by another.

Step 1 Use the method of elimination to solve each system. (Don't be surprised if it doesn't always work.)

a. $\begin{cases} 2x + 5y = 6 \\ 2x - 3y = 22 \end{cases}$ b. $\begin{cases} 3x + 2y = 12 \\ -6x - 4y = -24 \end{cases}$ c. $\begin{cases} 4x - 8y = 5 \\ -3x + 6y = 11 \end{cases}$

d. $\begin{cases} -2x + y = 5 \\ 6x - 3y = -15 \end{cases}$ e. $\begin{cases} x + 3y = 6 \\ 5x - 3y = 6 \end{cases}$ f. $\begin{cases} x + 3y = 8 \\ 3x + 9y = -4 \end{cases}$

Step 2 Graph each system in Step 1.

Step 3 A system that has a solution (a point or points of intersection) is called **consistent**. Which of the six systems in Step 1 are consistent? a, b, d, e

Step 4 A system that has no solution (no point of intersection) is called **inconsistent**. Which of the systems in Step 1 are inconsistent? c, f

Step 5 A system that has infinitely many solutions is called **dependent**. For linear systems, this means the equations are equivalent (though they may not look identical). A system that has a single solution is called **independent**. Which of the systems in Step 1 are dependent? Independent?
dependent: b, d; independent: a, e

LESSON 2.2 Substitution and Elimination 111

Guiding the Investigation

Step 1
a. $(8, -2)$
b. $0 = 0$, or an equivalent equation; infinitely many solutions
c. $0 = 59$, or another false equation; no solution
d. $0 = 0$, or an equivalent equation; infinitely many solutions
e. $\left(2, \frac{4}{3}\right)$
f. $0 = 28$, or another false equation; no solution

2a.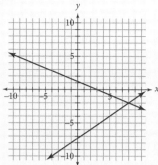

2b.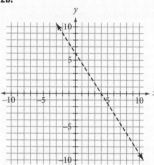

2c.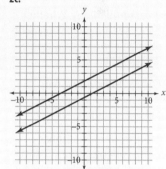

2d.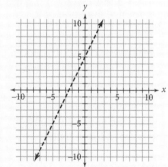

2e.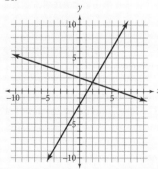

2f.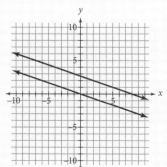

Step 5 ASK "Can inconsistent systems be dependent? Independent?" [No; dependent and independent systems have solutions.]

LESSON 2.2 Substitution and Elimination 111

Summarize

Discussing the Investigation

As students present, **ASK** "How do the lines change as you multiply the entire equation by a number?" [They don't.] **ASK** "Why not? If you start with $x + 3y = 6$, for example (from the investigation), and you multiply by -5 to change it to $-5x - 15y = -30$, shouldn't the line swing around somehow to go through -30 instead of 6?" Elicit the idea that the coordinates of points on the graph are manipulated to get 6; those same coordinates, if manipulated this new way, will yield -30.

Discuss how to write the "solution" for dependent systems and for inconsistent systems. **ASK** "Which points make both equations true?"

As students present their solutions to the investigation, examples, and exercises, emphasize the use of correct mathematical terminology. Whether student responses are correct or incorrect, ask other students if they agree and why. **SMP 1, 3, 5, 6**

CRITICAL QUESTION How would you decide which method to use to solve a system of linear equations?

BIG IDEA If at least one equation is solved for one variable, then substitution is easier; otherwise, elimination is probably better. If both equations are stated in terms of y, then graphing is a good method.

CRITICAL QUESTION Why does the point of intersection not change as the equations are multiplied by constants?

BIG IDEA Multiplying by constants doesn't change the lines, so it doesn't change the point of intersection.

Formative Assessment

Student group work and presentations of the Examples and Investigation can provide evidence of meeting the objectives of the lesson. Pay particular attention to why students choose a particular method. Assess how well students understand graphs of equations, intercepts of lines, and interpretations of intersection points. Presentations of the Investigation will provide evidence of understanding the definition of *inconsistent*, *consistent*, and *dependent systems*.

Step 6 Your graphs from Step 2 helped you classify each system as inconsistent or consistent and as dependent or independent. Now look at your solutions from Step 1. Make a conjecture about how the results of the elimination method can be used to classify a system of equations.

In the elimination method, you combine equations to eliminate one of the variables. Solving for the variable that remains gives you the x- or y-coordinate of the point of intersection, if there is a point of intersection.

Step 6 If the elimination method results in an equation that is always true, like $0 = 0$, then the system is consistent and dependent (the lines are the same). If the elimination method results in an answer for x and y, then the system is consistent and independent (there is a single intersection point). If the final equation is untrue, such as $0 = 24$, then the system is inconsistent (the lines are parallel).

2.2 Exercises

You will need a graphing calculator for Exercises 2, 7, 11, 15, and 16.

Practice Your Skills

1. Solve each equation for the specified variable.
   a. $w - r = 11$, for w $w = 11 + r$
   b. $2p + 3h = 18$, for h @ $h = \frac{18 - 2p}{3} = 6 - \frac{2}{3}p$
   c. $w - r = 11$, for r $r = w - 11$
   d. $2p + 3h = 18$, for p $p = \frac{18 - 3h}{2} = 9 - \frac{3}{2}p$

2. Graph each system and find an approximate solution. Then choose a method and find the exact solution. List each solution as an ordered pair.
   a. $\begin{cases} y = 15 + 3x \\ y = 27 - 3x \end{cases}$ @ $(2, 21)$
   b. $\begin{cases} y = 4x - 5 \\ y = 2x + 1 \end{cases}$ $(3, 7)$
   c. $\begin{cases} y = 5 + \frac{2}{3}x \\ y = 21 - 2x \end{cases}$ $(6, 9)$
   d. $\begin{cases} -2x + y = 5 \\ x - 3y = -30 \end{cases}$ $(3, 11)$
   e. $\begin{cases} x + 3y = 6 \\ 5x - 3y = 6 \end{cases}$ $\left(2, \frac{4}{3}\right)$

3. Solve each system of equations.
   a. $\begin{cases} 4s + 3t = 7 \\ 2s - t = 8 \end{cases}$ @ $(3.1, -1.8)$
   b. $\begin{cases} 5x + 2y = 12 \\ 6x - 4y = -7 \end{cases}$ $(1.0625, 3.34375)$
   c. $\begin{cases} 4x - 3y = 5 \\ -x + 6y = 11 \end{cases}$ $(3, 2.333)$
   d. $\begin{cases} \frac{1}{4}a - \frac{2}{5}b = 3 \\ \frac{3}{8}a + \frac{2}{5}b = 2 \end{cases}$ $\left(8, -\frac{5}{2}\right)$
   e. $\begin{cases} f = 3d + 5 \\ 10d - 4f = 16 \end{cases}$ $(-18, -49)$

4. Classify each system as consistent or inconsistent and as dependent or independent.
   a. $\begin{cases} 3x + 4y = 7 \\ y = -\frac{3}{4}x + 2 \end{cases}$ inconsistent
   b. $\begin{cases} y = 3x - 4 \\ 6x - 2y = 8 \end{cases}$ consistent and dependent
   c. $\begin{cases} y = 3x - 5 \\ 6x + 2y = 10 \end{cases}$ consistent and independent

Reason and Apply

5. Solve each problem.
   a. If $4x + y = 6$, then what is $(4x + y - 3)^2$?   9
   b. If $4x + 3y = 14$ and $3x - 3y = 13$, what is $7x$?   27

6. The formula to convert from Fahrenheit to Celsius is $C = \frac{5}{9}(F - 32)$. What reading on the Fahrenheit scale is three times the equivalent temperature on the Celsius scale? $80°F$

Conceptual/Procedural

Conceptual The Investigation reviews the concept of consistent, inconsistent, dependent, and independent systems of equations.

Procedural The Examples review using the substitution and elimination methods of solving systems of equations.

7. **APPLICATION** Ellen must decide between two cameras. The first camera costs $47.00 and uses two alkaline AA batteries. The second camera costs $59.00 and uses one $4.95 lithium battery. She plans to use the camera frequently enough that she probably would replace the AA batteries six times a year for a total cost of $11.50 per year. The lithium battery, however, will last an entire year.

   a. Let x represent the number of years, and let y represent the cost in dollars. Write an equation to represent the overall expense for each camera. ⓐ

   b. Use a graph to estimate when the overall cost of the less expensive camera will equal that of the more expensive camera. in about 1.8 yr

   c. Use another method to find exactly when the overall cost of the less expensive camera will equal that of the more expensive camera. in ≈ 1.832 yr

   d. Is it important to find an exact answer, as in 7c? Explain.

8. Write a system of two equations that has the solution (−1.4, 3.6).   possible answer: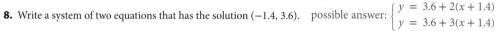

9. The two sequences below have one term that is the same. Determine which term this is and find its value. $u_{31} = v_{31} = 21$

   $u_1 = 12$
   $u_n = u_{n-1} + 0.3$ where $n \geq 2$

   $v_1 = 15$
   $v_n = v_{n-1} + 0.2$ where $n \geq 2$

10. Formulas play an important part in many fields of mathematics and science. You can create a new formula using substitution to combine formulas.

    a. Using the formulas $A = s^2$ and $d = s\sqrt{2}$, write a formula for A in terms of d. $A = \dfrac{d^2}{2}$

    b. Using the formulas $P = IE$ and $E = IR$, write a formula for P in terms of I and R. $P = I^2R$

    c. Using the formulas $A = \pi r^2$ and $C = 2\pi r$, write a formula for A in terms of C. $A = \dfrac{C^2}{4\pi^2}$

Science CONNECTION

In this simple circuit, a battery lights a small light bulb. The power P supplied to the bulb in watts is a product of the voltage E of the battery times the electrical current I in the wire. The electrical current, in turn, can be calculated as the battery voltage E divided by the resistance R of the circuit, which depends on the length and gauge, or diameter, of the wire. It is important to know basic functions and relations when designing a circuit so that there will be enough power supplied to the components, but not excessive power, which would damage the components.

LESSON 2.2 Substitution and Elimination   113

Exercise 11 ASK "Why does the sum of two equations represent a line that passes through their intersection point?" Algebraically, if $ax + by = c$ and $dx + ey = f$, then $(a + d)x + (b + e)y = c + f$ by distributivity.

ADVANCED Geometrically, the reason is easiest to see in the special case that both equations are in intercept form. If $y = a + bx$ and $y = c + dx$, then the sum is equivalent to $y = \frac{a+c}{2} + \frac{b+d}{2}x$. That is, the line representing the sum begins halfway between the other two y-intercepts and has a slope that's halfway between the other two slopes. So it is, in effect, squeezed between them, making it pass through their intersection point.

11c. The graph of the sum passes through the intersection point of the original equations.

12a. Because the first equation contains y with a coefficient of 1, it would be easy to solve this equation for y and then use substitution. So Erris could be correct. From Ryan's perspective it would be easy to use elimination because you only have to multiply the first equation by 4 before adding them together to eliminate the y. While the first equation could be easily solved for y to graph it, the second one is a bit trickier. However, it can still be done and thus graphing is a good option.

12b. Systems that have an equation that is already solved for one variable are good for substitution. Systems that have the equation written in standard form are nice for elimination. Systems that have both equations written in $y =$ form are convenient for either graphing or substitution.

Exercise 13 You might ask students how the fulcrum is related to the centroid (center of gravity) of a triangle. In 13b, you might need to describe a seesaw (teetertotter) as a plank with seats and handles on each end balanced on a fulcrum.

ELL It might be helpful to use a physical model to demonstrate equilibrium for different fulcrum locations.

13a.

$x + y = 40$ and $6x = 9y$; $y = 16$. The fulcrum should be positioned 16 in. from the 9 lb weight.

---

11. Consider the system
$$\begin{cases} 3x - 4y = 7 \\ 2x + 2y = 5 \end{cases}$$

a. Add the two equations. $5x - 2y = 12$

b. Solve the two original equations and their sum for y. Sketch a graph of all three equations on the same axes. $y = -\frac{7}{4} + \frac{3}{4}x$, $y = \frac{5}{2} - x$, $y = -6 + \frac{5}{2}x$

c. What do you observe about the graph of the third equation?

d. Repeat 11a–c for the system
$$\begin{cases} 5x - 7y = 3 \\ -5x + 3y = 5 \end{cases}$$ Sum: $-4y = 8$; $y = -\frac{3}{7} + \frac{5}{7}x$, $y = \frac{5}{3} + \frac{5}{3}x$, $y = -2$; the graphs all intersect at the same point

e. Make a conjecture about the graphs of two linear equations and their sum.
The graph of the sum of two linear equations will intersect the lines at their point of intersection.

12. Erris, Ryan, and ArShawn were discussing solving the system of equations
$$3x + y = 11$$
$$5x - 4y = 7$$
Erris said it would be best to use substitution. Ryan said he preferred to use elimination. ArShawn wondered if graphing would be the best way to solve this system.

a. Who do you agree with and why? What makes that method the best?

b. What would systems look like that you would use each of the other methods on?

13. APPLICATION A support bar will be in equilibrium (balanced) at the fulcrum, O, if $m_1x + m_2y = m_3z$, where m_1, m_2, and m_3 represent masses and x, y, and z represent the distance of the masses to the fulcrum. Draw a diagram for each question and calculate the answer.

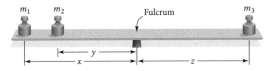

a. A 40 in. bar is in equilibrium when a weight of 6 lb is hung from one end and a weight of 9 lb is hung from the other end. Find the position of the fulcrum.

b. While in the park, Michael and his two sons, Justin and Alden, go on a 16 ft seesaw. Michael, who weighs 150 lb, sits at the edge of one end while Justin and Alden move to the other side and try to balance. The seesaw balances with Justin at the other edge and Alden 3 ft from him. After some additional experimentation, the seesaw balances once again with Alden at the edge and Justin 5.6 ft from the fulcrum. How much does each boy weigh?

Review

14. Classify each statement as true or false. If the statement is false, change the right side to make it true.

a. $x^2 + 8x + 15 = (x + 3)(x + 5)$ true

b. $x^2 - 16 = (x - 4)(x - 4)$ false; $(x - 4)(x + 4)$

c. $(x + 5)^2 = x^2 + 25$ false; $x^2 + 10x + 25$

15. Consider the equation $3x + 2y - 7 = 0$.

a. Solve the equation for y. @ $y = 3.5 - 1.5x$

b. Graph this equation.

c. What is the slope? $-\frac{3}{2}$, or -1.5

d. What is the y-intercept? $\frac{7}{2}$, or 3.5

e. Write an equation for a line perpendicular to this one and having the same y-intercept. Graph this equation.

114 CHAPTER 2 Systems of Equations and Inequalities

13b.

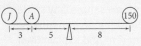

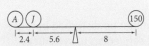

$8J + 5A = 8(150)$ and $8(A) + 5.6(J) = 8(150)$.
Solving the system leads to the fact that Justin weighs 100 lb, and Alden weighs 80 lb.

15b.

15e. $y = \frac{7}{2} + \frac{2}{3}x$

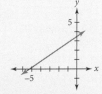

16. **APPLICATION** This table shows the normal monthly precipitation in inches for Pittsburgh, Pennsylvania, and Portland, Oregon.

Month	J	F	M	A	M	J	J	A	S	O	N	D
Pittsburgh	2.7	2.4	3.2	3.0	3.8	4.1	4.0	3.4	3.2	2.3	3.0	2.9
Portland	5.1	4.2	3.7	2.6	2.4	1.6	0.7	0.9	1.7	2.9	5.6	5.7

(The World Almanac and Book of Facts 2007)

a. Display the data in two box plots on the same axis.

b. Give the five-number summary of each data set.

c. Describe the differences in living conditions with respect to precipitation.

d. Which city generally has more rain annually? **Pittsburgh**

Saint John's Bridge in Portland, Oregon.

17. Consider these three sequences.

   i. 243, −324, 432, −576, …

   ii. 22, 26, 31, 37, 44, …

   iii. 24, 25.75, 27.5, 29.25, 31, …

a. Find the next two terms in each sequence.   i. 768, −1024   ii. 52, 61   iii. 32.75, 34.5

b. Identify each sequence as arithmetic, geometric, or other.   i. geometric   ii. other   iii. arithmetic

c. If a sequence is arithmetic or geometric, write a recursive routine to generate the sequence.

d. If a sequence is arithmetic, give an explicit formula that generates the sequence.   iii. $u_n = 1.75n + 22.25$

16a.

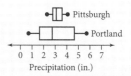

16b. Pittsburgh: 2.3, 2.8, 3.1, 3.6, 4.1;
Portland: 0.7, 1.65, 2.75, 4.65, 5.7

16c. Answers will vary. Pittsburgh has rather constant precipitation all year. Portland has a wet and a dry season.

17c. i $u_1 = 243$ and $u_n = \left(-\frac{4}{3}\right) u_{n-1}$ where $n \geq 2$

iii. $u_1 = 24$ and $u_n = u_{n-1} + 1.75$ where $n \geq 2$

IMPROVING YOUR Reasoning SKILLS

Cartoon Watching Causes Small Feet

Lisa did a study for her health class about the effects of cartoon watching on foot size. Based on a graph of her data, she finds that there was an inverse relationship between foot size and hours spent watching cartoons per week. She concludes that "cartoon watching causes small feet." Is this true? Explain any flaws in Lisa's reasoning.

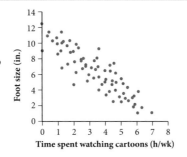

Children have smaller feet and generally watch more cartoons than adults, but this does not mean that one trait causes the other.

The point of this puzzle is that correlation does not necessarily imply causation.

LESSON 2.3

In this lesson, students solve systems of equations in which one equation is linear and the other is non-linear.

COMMON CORE STATE STANDARDS		
APPLIED	DEVELOPED	INTRODUCED
A.CED.3; A.REI.6; A.REI.10	A.REI.7	

Objectives

- Solve a simple system of equations in two variables consisting of a linear equation and a quadratic equation graphically
- Solve a simple system of equations in two variables consisting of a linear equation and a quadratic equation algebraically

Launch

Solve this system of equations:
$$\begin{cases} y = x^2 + 6 \\ y = 5x \end{cases}$$

What does your solution represent?

(2, 10), (3, 15). The values of x and y that makes both equations true at the same time. The intersection of the two graphs.

Investigate

Guiding the Investigation

There is a desmos tool for this investigation in your ebook.

Step 3

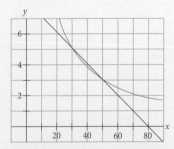

The graphs intersect at (30, 50) and (50, 30), so the dimensions of the building are 30 ft × 50 ft.

Have students share their answers to the Investigation. Spend time discussing Step 4. **SMP 1, 2, 3, 5, 6**

LESSON 2.3

Linear and Non-linear Systems of Equations

One never reaches home, but wherever friendly paths intersect the whole world looks like home for a time.

HERMANN HESSE

You have solved systems of linear equations using graphs and symbolic methods. What if the equations aren't linear? Can you still find solutions? Some of these systems can be solved symbolically while others cannot. In this lesson you will investigate systems that involve non-linear equations.

INVESTIGATION

A Designing Dilemma

A small manufacturing company wants to build a rectangular storage building that has 1500 square feet of floor space to store their product before shipping. There are many possible dimensions for the building. Suppose they also want the building to have a perimeter of 160 feet so they can display their company logo and other advertising on the exterior walls. Is it possible to build a building like this? In this investigation, you'll find a way to answer this and other similar questions.

Step 1 Let x represent the length of the building and y represent the width. $A = lw$, so $1500 = xy$

Step 2 $P = 2l + 2w$, so $160 = 2x + 2y$

Step 4 There is a solution as long as the perimeter at least 4 times the square root of the area. There is no solution if the perimeter is "too small". Graphically, they would intersect if there is a solution and would not intersect if there is no solution.

Step 1 Write an equation for the area of the building. Use x and y to represent the dimensions of the building.

Step 2 Write an equation for the perimeter of the building.

Step 3 You now have a system of equations. Graph the system. Use the trace and/or table features of your technology to find the coordinates of all points of intersection. What is the real world meaning of these points?

Step 4 Is it possible to find a solution to this type of problem for any pair of numbers for area and perimeter? Experiment by changing the values of the area and perimeter to find an answer to this question. What conditions are necessary for this system of equations to have a solution? How would you graphically describe a system of this type that has no solution?

In the investigation you solved a system of equations where only one of the equations was linear. You used technology to find the solution. Technology can help you find solutions when other methods are difficult or impossible.

ELL	Support	Advanced
Spend time discussing a storage building and its area and perimeter. This lesson relies heavily on technology, so make sure students understand the calculator commands. Or you can use the desmos tool found in your ebook.	You may need to review finding area and perimeter. Make sure students know how to use the trace and/or table feature of their calculator. Or you can use the desmos tool found in your ebook. You might also use the Extra Example as an introduction to the intersection of a linear and a non-linear equation.	Use the **One Step**. Have students create a system of two non-linear equations that intersect in one point and a system of two non-linear equations that intersect in two points.

EXAMPLE A | The Community Center is presenting a program. They will charge $10 for each ticket. Their expenses depend on the number of tickets sold according to the equation $y = \dfrac{500}{1 + 2^{-.05x}}$, where x represents the number of tickets sold. There are 200 seats in the auditorium. How many tickets must they sell to break even?

Solution | The income function is $y = 10x$.
The expense function is $y = \dfrac{500}{1 + 2^{-.05x}}$.

Graphing the two equations, you can see that they intersect in one point.

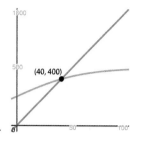

Using a table, you can find that the graphs intersect at the point (40, 400). This means that if they sell 40 tickets both their expenses and income will be $400. If they sell more tickets, then they will make a profit.

Using a graph to solve a system of equations lets you solve many systems that you could not solve in any other way. Some systems, though, are ones that you can solve using your algebra skills.

EXAMPLE B | Solve the system of equations symbolically:

$y = x - 7$

$y = \dfrac{8}{x}$

Solution | You can use substitution to solve this system.

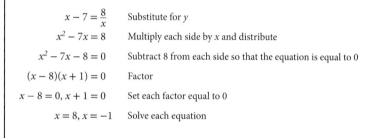

When $x = 8$, $y = 8 - 7 = 1$, so the point (8, 1) is a solution.

When $x = -1$, $y = -1 - 7 = -8$, so the point $(-1, -8)$ is also a solution.

If you graph the system, you might miss one of these solution points. Now that you know the solutions, you can graph the equations with an appropriate window and verify that the solutions are correct.

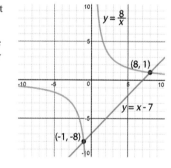

LESSON 2.3 Linear and Non-linear Systems of Equations 117

Modifying the Investigation

Whole Class Do Steps 1 and 2 as a class. Have students do Step 3. Discuss the real world meaning of the intersection. Do Step 4 as a class, with student input.

Shortened Do Steps 1 and 2 as a class. Have students do Step 3. Discuss the real world meaning of the intersection. Have students do Step 4.

One Step Project the **One Step** found in your ebook. Have students present their solutions. Pay particular attention to the conditions necessary for the system of equations to have a solution. [The perimeter must be at least 4 times the square root of the area.]

Example A

Project Example A and have students work in small groups. Have them share their strategy along with their answer. If students are having difficulty with the exponential equation suggest graphing as a way to solve this system. Emphasize the use of correct mathematical terminology. Whether student responses are correct or incorrect, ask other students if they agree and why. SMP 1, 2, 3, 5, 6

Example B

Project Example B and have students work individually or in small pairs. Have them share their strategy, including their reasons for each step, along with their answer. Have students graph the equations to check their answers. Emphasize the use of correct mathematical terminology. Whether student responses are correct or incorrect, ask other students if they agree and why. SMP 1, 2, 3, 5, 6

Summarize

After presentations of Example B showing the graph to check a solution, ASK "What is the solution to this system if we add the line $y = -x$ to the system? Explain. [No solution. While the two lines will intersect, the graph of $y = \dfrac{8}{x}$ will not intersect the line $y = -x$.] What if we add the line $y = x$ to the system? Explain." [$y = x$ is parallel to $y = x - 7$, so they will never intersect. No solution.]

CRITICAL QUESTION What are the advantages and disadvantages of solving a system of a linear and non-linear equation with technology? Algebraically?

BIG IDEA Technology allows you to solve when other methods are difficult or impossible based on your understanding of other methods. Solving algebraically helps insure you don't miss a solution point.

Formative Assessment

Student presentations and responses to Example A and Exercise 5 will provide evidence of solving a simple system of equations in two variables consisting of a linear equation and a quadratic equation graphically. Student presentations and responses to Example B and Exercise 6 will provide evidence of solving a simple system of equations in two variables consisting of a linear equation and a quadratic equation algebraically. The Investigation and Reason and Apply Exercises will provide further insight into whether students are meeting the objectives.

Apply

Extra Example

What is the solution of the system of equations graphed below? How do you know?

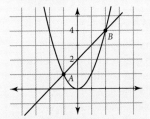

$(-1, 1)$ and $(2, 4)$. Points A and B are the intersection points that make both equations true, so they are the solutions to the system.

Closing Question

Solve this system of equations.
$$\begin{cases} \dfrac{y}{x^2 - 9} = 1 \\ y - 3 = x \end{cases}$$
$(4, 7)$

Students may also include the point $(-3, 0)$ in the solution. Have them check their solution by substituting the values into the equations. **ASK** "What does it mean to have a 0 in the denominator?"

2.3 Exercises

Practice Your Skills

1. Solve each equation for y.

   a. $2x + 3y = 15$ $\quad y = \dfrac{-2x + 15}{3}$ or $y = -\dfrac{2}{3}x + 5$

   b. $4xy = 20$ @ $y = \dfrac{5}{x}$

   c. $xy - 3y + 6 = 0$ $\quad y = \dfrac{-6}{x - 3}$

   d. $7x - 2y = 14$ $\quad y = \dfrac{7x - 14}{2}$ or $y = \dfrac{7}{2}x - 7$

   e. $\dfrac{x}{y - 2} = 3$ $\quad y = \dfrac{x + 6}{3}$ or $y = \dfrac{1}{3}x + 2$

   f. $\dfrac{x + 1}{y - 1} = \dfrac{2}{5}$ $\quad y = \dfrac{5}{2}(x + 1) + 1$ or $y = \dfrac{5}{2}x + \dfrac{7}{2}$

2. Complete each equation so that the point $(2, 5)$ is a solution.

   a. $3x + 4y = ?$ $\quad 26$

   b. $2x - ?y = 12$ $\quad -\dfrac{8}{5}$

   c. $y = \dfrac{?}{x + 2}$ @ 20

3. Determine whether or not the given point is a solution to the given system.

   a. $3x + 5y = 8$
      $\dfrac{6 + 3x}{y + 2} = 3$ $\quad (1, 1)$ @ yes

   b. $xy + 2y = -12$
      $4x - y = 5$ $\quad (2, 3)$ no

   c. $x^2 + 2x + 3 = y$
      $\dfrac{y - 2}{x + 1} = 2$ $\quad (1, 6)$ yes

   d. $2x^2 + y^2 = 5$
      $3xy = -6$ $\quad (2, -1)$ no

4. Solve each system of equations.

   a. $x^2 - 4y = 0$
      $y = x + 3$ $\quad (6, 9), (-2, 1)$

   b. $xy + 10 = 0$
      $y - x = 7$ @ $(-2, 5), (-5, 2)$

   c. $x = \dfrac{14}{y}$
      $y = x + 5$ $\quad (-7, -2), (2, 7)$

5. Solve each system of equations by graphing.

   a. $2x + 6y = 11$
      $xy = 2$ @ $\left(4, \dfrac{1}{2}\right), \left(\dfrac{3}{2}, \dfrac{4}{3}\right)$

   b. $3xy + y = 2$
      $5y = x$ $\quad \left(-2, -\dfrac{2}{5}\right), \left(\dfrac{5}{3}, \dfrac{1}{3}\right)$

   c. $6xy - 5y = -8x^3$
      $y = x$ $\quad (0, 0), \left(\dfrac{1}{2}, \dfrac{1}{2}\right), \left(-\dfrac{5}{4}, -\dfrac{5}{4}\right)$

6. Solve each system of equations symbolically.

   a. $y = 2x$
      $xy - 10y = 48$ @ $(-2, -4), (12, 24)$

   b. $y = x + 1$
      $xy = 18 - 2x$ $\quad (3, 4), (-6, -5)$

   c. $y = x$
      $16y - xy = 7x + 14$ $\quad (2, 2), (7, 7)$

   d. $y = \dfrac{1}{x}$
      $y = 0.25x$ $\quad \left(2, \dfrac{1}{2}\right), \left(-2, -\dfrac{1}{2}\right)$

Reason and Apply

7. Omar solved this system of equations.

   $y = -x^2 + 3x + 18$

   $y - 6 = \dfrac{12}{x + 1}$

   He recorded his answer as $x = -3, 0, 5$ and $y = 0, 8, 18$. Explain what is wrong with his solution and then give a better answer.

   When Omar listed the x and y values separately, you cannot tell what the coordinates of each solution point actually is. You need to pair each x value with its appropriate y value. A better solution is $(-3, 0), (0, 18), (5, 8)$.

Conceptual/Procedural

Conceptual In this lesson students explore the concept that systems of equations do not have to be two linear equations. The system can include both a linear and a non-linear equation.

Procedural Students solve a system of equations consisting a linear equation and a nonlinear equation graphically and algebraically.

8. Any system of linear equations has either one solution, no solution, or an infinite number of solutions. How many solutions are possible for these types of systems?

   a. a system that consists of a line and a parabola @ 0, 1, or 2 solutions

   b. a system that consists of two parabolas 0, 1, 2, or infinitely many solutions

   c. a circle and a line 0, 1, or 2 solutions

   d. a circle and a parabola 0, 1, 2, 3, or 4 solutions

9. Explain how you can tell that this system of equations has no solutions without actually trying to solve it.

   $y = 2^x$
   $y - 1 = -(x-2)^2$

10. Consider the system of equations.

    $y = -2x + 8$
    $y = \dfrac{a}{x}$

    a. Graph this system using $a = 1, 2, 3,$ and 4. Describe how the solutions to the system change as you increase the value of a.

    b. Find a value of a so that there is only one solution to the system. What is this solution? @

    c. Describe what happens to the solutions if a is larger than the value you found in part b.

    d. If a is a negative number, how many solutions will the system have? Explain.

11. The population of Sunnyville, in the southwest, is given by the equation

    $P = \dfrac{850000}{1 + 13(2^{-0.05x})}$, where x is the number of years after the year 2000. Another city in the same state, Cactus City, had a population of 100,000 in the year 2000 and is growing at a steady rate of 2500 people each year.

    a. Write an equation to model the population of Cactus City. @ $P = 100000 + 2500x$

    b. When will the populations of the two cities be the same? In the year 2040, the populations will be the same.

    c. What is the population when the cities have the same population?
       In the year 2040, each city will have 200,000 people.

Review

12. Solve each inequality. Display your solution on a number line.

    a. $2x + 5 < 13$ $x < 4$ b. $3x - 7 \geq 14$ $x \geq 7$ c. $11 - 3x > 5$ $x < 2$

13. Use substitution to solve each system.

    a. $2x - y = 10$ b. $3x + 2y = 5$
       $y = 11 - 5x$ $(3, -4)$ $3 - 4x = 5 + 2y$ $(-7, 13)$

14. Use elimination to solve each system.

    a. $4x + 3y = 11$ b. $2x - 5y = 10$
       $7x + y = 15$ $(2, 1)$ $3x + 7y = 44$ $(10, 2)$

Assigning Exercises
Suggested: 1 – 11
Additional Practice: 12 – 14

5a.
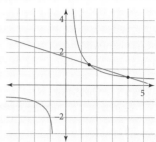

5b.
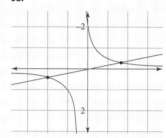

5c.
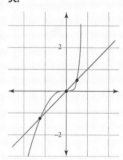

9. The first function has a range of $y > 0$. When $x - 2$, its y-value is 4. The second function is a parabola that is reflected over the x-axis, translated right 2 units and up 1 unit. Its vertex is at $(2, 1)$. Since this is the maximum point for the parabola and it is clearly lower than the exponential at this point, there is no way for these curves to intersect.

10a. As a increases, the two solutions get closer together.

10b. If $a = 8$, there is only one solution at $(2, 4)$.

10c. If a is larger than 8, then there are no solutions to the system.

10d. If a is negative, the two pieces of the second function are in the second and fourth quadrants. The line goes through these quadrants and must always intersect both curves. Thus there are always two solutions.

LESSON 2.4

Systems of Inequalities

This lesson reviews systems of inequalities in two variables as a prelude to linear programming.

COMMON CORE STATE STANDARDS		
APPLIED	DEVELOPED	INTRODUCED
A.CED.3	A.REI.12	

Objectives
- Review linear inequalities and the solution of an inequality
- Write inequalities to describe given real-world constraints
- Graph systems of inequalities
- Interpret the meaning of the points within a feasible region
- Find the vertices of a feasible region

Vocabulary
inequalities
constraints
vertex
half-plane
feasible region

Materials
- Calculator Notes: Graphing Inequalities with the Inequal App

Launch

Project the situations from page 120 or from the Launch found in your ebook. Have students write inequalities to represent the situations.

Investigate

Guiding the Investigation

Step 1 Graphs should be drawn on graph paper, not made on calculators. There is a desmos tool in your ebook that can be used for the graph. The neater and more accurate the graphs, the easier Step 6 will be. As needed, encourage students to graph using intercepts rather than solving for y.

Think left and think right and think low and think high. Oh, the thinks you can think up if only you try!

DR. SEUSS

Frequently, real-world situations involve a range of possible values. Algebraic statements of these situations are called **inequalities**.

Situation	Inequality
Write an essay between two and five pages in length.	$2 \leq E \leq 5$
Practice more than an hour each day.	$P > 1$
The post office is open from nine o'clock until noon.	$9 \leq H \leq 12$
Do not spend more than $10 on candy and popcorn.	$0 \leq c + p \leq 10$
A college fund has $40,000 to invest in stocks and bonds.	$0 \leq s + b \leq 40,000$

Recall that you can perform operations on inequalities very much like you do on equations. You can add or subtract the same quantity on both sides, multiply by the same number or expression on both sides, and so on. The one exception to remember is that when you multiply or divide by a negative quantity or expression, the inequality symbol reverses.

In this lesson you will learn how to graphically show solutions to inequalities with two variables, such as the last two statements in the table above.

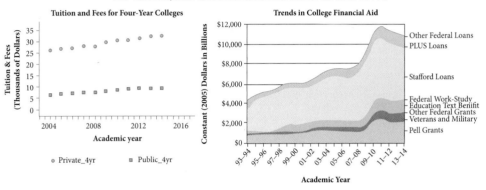

The cost of a college education continues to rise. The good news is that public and private institutions provide billions of dollars of financial aid to students every year. At four-year public colleges, for example, over 65% of students receive some form of financial aid. Financial aid makes college affordable for many students, despite increasing costs.

INVESTIGATION

Paying for College

A total of $40,000 has been donated to a college scholarship fund. The administrators of the fund are considering how much to invest in stocks and how much to invest in bonds. Stocks usually pay more but are often a riskier investment, whereas bonds pay less but are usually safer.

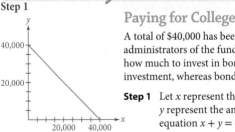

Step 1 Let x represent the amount in dollars invested in stocks, and let y represent the amount in dollars invested in bonds. Graph the equation $x + y = 40,000$.

ELL
Have students create a reference table to identify particular terms and phrases that may be translated by specific inequality symbols. In this lesson, students can add the terms *constraint*, *feasible*, and *vertex* used in new contexts to their table.
ASK "How does *vertex* used here compare with *vertex* from geometry?"

Support
If needed, assign the Refreshing Your Skills: Properties of Real Numbers lesson to review graphing single inequalities. Students may also need practice representing real-world situations with appropriate inequalities. Encourage students to check their answers for the coordinates of each vertex by making sure that the points fall in the appropriate region on the graph.

Advanced
Use the **One Step** found in your ebook. You might also have students write real world situations that could be modeled by the systems of inequalities in Exercises 5 through 8.

Step 2 Name at least five pairs of x- and y-values that satisfy the inequality $x + y < 40{,}000$ and plot them on your graph. In this problem, why can $x + y$ be less than $\$40{,}000$?

Step 3 Describe where all possible solutions to the inequality $x + y < 40{,}000$ are located. Shade this region on your graph.

Step 4 Describe some points that fit the condition $x + y \leq 40{,}000$ but do not make sense for the situation. Negative values for x or y do not make sense.

Assume that each option—stocks or bonds—requires a minimum investment of $\$5{,}000$, and that the fund administrators want to purchase some stocks and some bonds. Based on the advice of their financial advisor, they decide that the amount invested in bonds should be at least twice the amount invested in stocks.

Step 5 Translate all of the limitations, or **constraints**, into a system of inequalities. A table might help you to organize this information.

Step 6 Graph all of the inequalities and determine the region of your graph that will satisfy all the constraints. Find each corner, or **vertex**, of this region.

When there are one or two variables in an inequality, you can represent the solution as a set of ordered pairs by shading the region of the coordinate plane that contains those points. This solution set of ordered pair is known as a **half-plane**, a planar region consisting of all points on one side of a line, and no points on the other side. If the points on the line are included, it is called a *closed* half-plane, whereas if the points on the line are not included, it is called an *open* half-plane.

The solid boundary line indicates that the region *includes* the line.

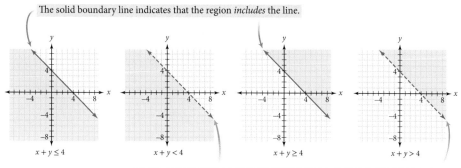

The dashed boundary line indicates that the region does *not* include the line.

When you have several inequalities that must be satisfied simultaneously, you have a system. The solution to a system of inequalities with two variables will be a set of points that is the intersection of the corresponding half-planes. This set of points is called a **feasible region**. The feasible region can be shown graphically as part of a plane, or sometimes it can be described as a geometric shape with its vertices given.

Step 2 Possible answer: (10,000, 10,000), (5000, 5000), (0, 0), (10,000, 29,999), (30,000, 5000). They don't necessarily have to invest all of the money.

Step 2 This inequality is equivalent to one used in the introduction. Here the variables are called x and y to facilitate graphing. Encourage each group member to pick a different pair. Negative values of x and y fit the model through Step 4. **ASK** "Where in your graph does each point lie?"

Step 3 They are all below the line $x + y = 40{,}000$.

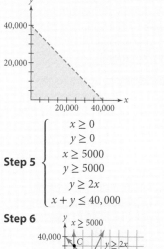

Step 5 $\begin{cases} x \geq 0 \\ y \geq 0 \\ x \geq 5000 \\ y \geq 5000 \\ y \geq 2x \\ x + y \leq 40{,}000 \end{cases}$

Step 6

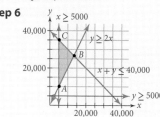

$A(5000, 10{,}000)$, $B(13{,}333.3, 26{,}666.7)$, $C(5000, 35{,}000)$

Modifying the Investigation

Whole Class Have students complete Steps 1 and 2 individually. Discuss Steps 2 through 4. Complete Steps 5 and 6 with student input.

Shortened Have groups complete Steps 1 through 4. Discuss and complete Step 5 with the class. Have groups complete Step 6.

One Step If you do the **One Step** found in your ebook, you will do the first part of the linear programming process that you will complete in Lesson 2.5.

Encourage students to assign variables and set up inequalities. If they're having difficulty deciding how to graph the inequalities, ask whether graphing corresponding equations would help. As students find half-planes, ask whether points below the x–axis or to the left of the y–axis make sense, and help students add commonsense constraints. During the discussion, introduce the terms *constraint, feasible region,* and *vertex*.

Example

Project the example and have students work in small groups. Have students share their graphs. **SMP 1, 3, 6**

As students present their intersection points from the graph, **ASK** "How confident are you in the precision of your answer?" Help them decide to check by substitution. **SMP 1, 2, 6**

ASK "Are all points in the region solutions, or only those with integer coordinates?" [The variables represent times, so every point in the feasible region is a solution.]

ASK "In the example, does it matter whether time spent on mathematics homework is graphed on the *x*–axis or the *y*–axis?" [Unlike with earlier equations relating *x* and *y*, neither variable can be called dependent or independent, so it doesn't matter.] You might have a group work through the example with the meanings of the variables interchanged. If so, have them present their work.

ELL The *commonsense* constraints are those that are not stated in the problem because they are common knowledge. For example, Rachel can't spend fewer than zero hours on homework.

Summarize

Have students share their work in both the Investigation and the Example. Emphasize the use of correct mathematical terminology. Whether student responses are correct or incorrect, ask other students if they agree and why. **SMP 1, 3, 6**

ASK "Why do you need to reverse the direction of an inequality when you multiply by a negative number? Use examples to explain." [Possible answer: The inequality $3 < 5$ is a true statement. Multiplying by -1 gives $-3 < -5$, which is not true. Therefore, the inequality symbol must reverse: $-3 > -5$. Taking the reciprocal of both sides also reverses an inequality symbol. For example, $\frac{1}{2} > \frac{1}{3}$, but $2 < 3$.]

EXAMPLE

Rachel has 3 hours to work on her homework tonight. She wants to spend more time working on math than on chemistry, and she must spend at least a half hour working on chemistry.

a. Let *x* represent time in hours spent on math, and let *y* represent time in hours spent on chemistry. Write inequalities to represent the three constraints of the system.

b. Graph your inequalities and shade the feasible region.

c. Find the coordinates of the vertices of the feasible region.

d. Name two points that are solutions to the system, and describe what they mean in the context of the problem.

Solution

First, read the given information carefully to make sure you understand the constraints.

a. Convert each constraint into an algebraic inequality.

$$\begin{cases} x + y \leq 3, \text{ or } y \leq -x + 3 \\ x > y, \text{ or } y < x \\ y \geq 0.5 \end{cases}$$

Rachel has 3 h to work on homework.

She wants to spend more time working on math than on chemistry.

She must spend at least a half hour working on chemistry.

b. To graph the inequality $y \leq -x + 3$, first graph $y = -x + 3$. This line is part of the solution. Then shade the region that contains the points that satisfy the inequality $y < -x + 3$.

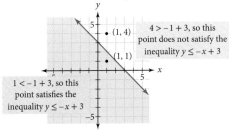

Shade the region in which all points (x, y) have *y*-values that are less than $-x + 3$.

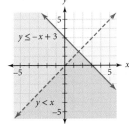

Now add the next inequality, $y < x$. Here the line $y = x$ is not part of the solution, so it is dashed. Shade the region in which each point satisfies the inequality $y < x$.

Then add the last inequality, $y \geq 0.5$. The solution includes $y = 0.5$ and all the points where *y* is greater than 0.5.

The solution to the system, the feasible region, is the set of points common to all three regions.

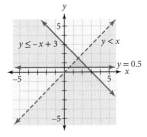

Conceptual/Procedural

Conceptual In this lesson, students investigate the concepts of open and closed half–planes, feasible regions, and constraints on systems of inequalities.

Procedural Students write and solve inequalities in a context.

c. The solution may be identified by giving the coordinates of the vertices of the feasible region. You can find those coordinates by solving systems of equations representing adjacent boundary lines.

Equations	Intersections
$x + y = 3$ and $x = y$	(1.5, 1.5)
$x + y = 3$ and $y = 0.5$	(2.5, 0.5)
$x = y$ and $y = 0.5$	(0.5, 0.5)

Only one of these vertices, (2.5, 0.5), is part of the solution. The other two vertices of the feasible region, (1.5, 1.5) and (0.5, 0.5), are on the dashed line, showing that they do not satisfy the constraint $y < x$.

d. The points (1.5, 1) and (2.5, 0.5) are two solutions to the system. Every point in the feasible region represents a way that Rachel could divide her time. The solution point (1.5, 1) means she could spend 1.5 h on mathematics and 1 h on chemistry and still meet all her constraints. The point (2.5, 0.5) means that Rachel could spend 2.5 h on math and 0.5 h on chemistry. This point represents the boundaries of two constraints: She can't spend less than 0.5 h on chemistry or more than 3 h total on homework.

Sometimes you will need to write nonlinear inequalities to represent constraints.

When you are solving a system of equations based on real-world constraints, it is important to note that sometimes there are constraints that are not specifically stated in the problem. In the example, negative values for x and y would not make sense, because you can't study for a negative number of hours. You could have added the commonsense constraints $x \geq 0$ and $y \geq 0$, although in this case it would not affect the feasible region.

2.4 Exercises

You will need a graphing calculator for Exercises **9**.

Practice Your Skills

1. Solve each inequality for y.
   a. $2x - 5y > 10$ @ $y < \frac{10 - 2x}{-5}$, or $y < -2 + 0.4x$
   b. $4(2 - 3y) + 2x > 14$ $y < \frac{6 - 2x}{-12}$, or $y < -\frac{1}{2} + \frac{1}{6}x$

2. Graph each linear inequality on the coordinate plane.
   a. $y \leq -2x + 5$ @
   b. $2y + 2x > 5$
   c. $x > 5$

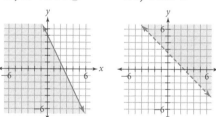
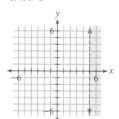

CRITICAL QUESTION What can you say about solutions to a system of inequalities in two unknowns?

BIG IDEA Solutions are ordered pairs. When graphed on a coordinate plane, these points lie in a polygon called the *feasible region*.

CRITICAL QUESTION When should we use a solid line and when should we use a dashed line?

BIG IDEA You should use a solid line when the inequality is inclusive ("greater than or equal to" or "less than or equal to") and a dashed line when the inequality is exclusive ("greater than" or "less than".]

CRITICAL QUESTION Does every system of inequalities have a feasible region? Explain.

BIG IDEA No. The boundary lines could be parallel with the half–planes going in opposite directions.

CRITICAL QUESTION How do you find the vertices of a feasible region?

BIG IDEA By finding the intersections of the sides of the polygon. You often need to solve a system of two equations to find an intersection point.

Formative Assessment

Student presentations of the Launch, Investigation, and Example provide evidence of prior knowledge of how to write and solve inequalities to describe given real–world constraints, graph systems of inequalities, interpret the meaning of the points within a feasible region, and find the vertices of a feasible region. Discussion of the Critical Questions can also provide helpful evidence.

Apply

Extra Example
1. For Steps a through e, use this system of inequalities:
$$\begin{cases} y \geq 3x - 5 \\ y \leq 4 \\ y > -2x \end{cases}$$

a. For each equation, is the line solid or dashed? solid, solid, dashed
b. Graph the feasible region.

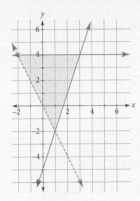

c. Find the vertices of the feasible region.
$(1, -2), (3, 4), (-2, 4)$

d. Which of these vertices is a possible solution? $(3, 4)$

e. Name three integer coordinates that satisfy these inequalities.
possible solutions: $(1, -1), (1, 1), (0, 3)$

Closing Question
Write a system of inequalities such that the graph of its solution forms a square region. How do you know it is a square?

Answers will vary. Possible solution: $x \geq 0$, $y \geq 0$, $x \leq 3$, $y \leq 3$. x and y must have the same bounds (or be the opposite) to be a square.

Assigning Exercises
Suggested: 1–10
Additional Practice: 11–15

5. vertices: $(0, 2), (0, 5)$, $(2.752, 3.596), (3.529, 2.353)$

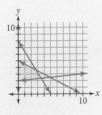

---

3. For 3a–d, write the inequality of each graph.

a. 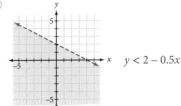 $y < 2 - 0.5x$

b. 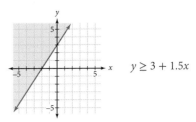 $y \geq 3 + 1.5x$

c. 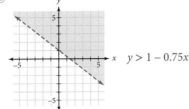 $y > 1 - 0.75x$

d. 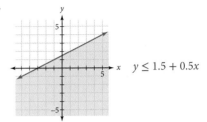 $y \leq 1.5 + 0.5x$

4. The graphs of these equations serve as the boundaries of this feasible region. What three inequalities identify this region?

$y = 2x + 1 \quad y \leq 2x + 1$
$y = 7 - x \quad y \leq 7 - x$
$y = -\frac{1}{4}x + 3 \quad y \geq -\frac{1}{4}x + 3$

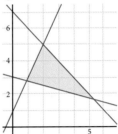

Reason and Apply
For Exercises 5–8, sketch the feasible region of each system of inequalities. Find the coordinates of each vertex.

5. $\begin{cases} y \leq -0.51x + 5 \\ y \leq -16.x + 8 \\ y \geq 0.1x + 2 \\ y \geq 0 \\ x \geq 0 \end{cases}$

6. $\begin{cases} y \geq 1.5x - 6 \\ y \leq \frac{1}{2}(x + 1) \\ y \geq 2 - x \\ y \geq 0 \end{cases}$

7. $\begin{cases} 4x + 3y \leq 12 \\ 1.6x + 2y \leq 8 \\ 2x + y \geq 2 \\ y \geq 0 \\ x \geq 0 \end{cases}$

8. $\begin{cases} y \geq x - 1 \\ y \geq 1 - x \\ y \leq -0.2x + 2.6 \\ y \leq 2.6 \\ y \geq 0 \end{cases}$

124   CHAPTER 2   Systems of Equations and Inequalities

---

6. vertices at $(2, 0), (4, 0), (1, 1)$ and $(6.5, 3.75)$

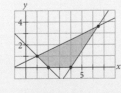

7. vertices: $(0, 4), (3, 0), (1, 0)$, $(0, 2)$

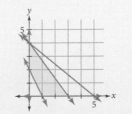

8. vertices at $(1, 0), (3, 2)$, $(-1.6, 2.6), (0, 2.6)$

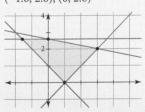

9. In the Lux Art Gallery, rectangular paintings must have a perieter between 66 in. and 80 in. In addition, the length and width must be no less than 10 inches. ⓗ

   a. Write four inequalities involving length and width that represent these constraints. ⓐ
   b. Graph this system of inequalities to identify the feasible region.
   c. Will the gallery accept a painting with the following measures?
      i. 13.4 in. by 18.3 in.   no
      ii. 9.5 in. by 30.6 in.   no
      iii. 16 in. by 17.5 in.   yes

9a. Let x represent the length in inches and y represent the width in inches
$$\begin{cases} x + y \leq 40 \\ x + y \leq 33 \\ x \geq 10 \\ y \geq 10 \end{cases}$$

9b.
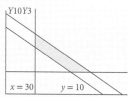

10. Al just got rid of 40 of his dad's old records. He sold each classical record for $5 and each jazz record for $2. He arranged to donate any records that he couldn't sell to a thrift shop. Al knows that he sold fewer than 10 jazz records and that he earned more than $100.

    a. Let x represent the number of classical records sold, and let y represent the number of jazz records sold. Write an inequality expressing that Al earned more than $100. $5x + 2y > 100$
    b. Write an inequality expressing that he sold fewer than 10 jazz records. $y < 10$
    c. Write an inequality expressing that the total number of records sold was no more than 40. $x + y \leq 40$
    d. Graph the solution to the system of inequalities, including any commonsense constraints.
       commonsense constraints: $x \geq 0, y \geq 0$
    e. Name all the vertices of the feasible region.
       (20, 0), (40, 0), (30, 10), (16, 10)

10d.
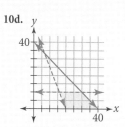

Review

11. Solve the system of equations.
    $3x - y = 5$
    $-4x + 2y = 1$ $\left(\frac{11}{2}, \frac{23}{2}\right)$

12. These data were collected from a bouncing-ball experiment. The heights of successive bounces form a geometric sequence. Complete the table by finding the heights of the bounces.

Bounce number	1	2	3	4	5	6	7
Height (cm)	70	49	34.3	24.0	16.8	11.8	8.2

13. Cezara works at an electronics store where she earns $42 a day and 18% of all purchases as a commission. Adolpho works at a competing store where he earns $48 a day and a 15% commission. At what purchase level do they both earn the same amount?
    Cezara, $y = 42 + 0.18x$, Adolpho, $y = 48 + 0.15x$. The lines intersect at $x = 200$. If they sell $200 worth of merchandise, they earn the same amount.

Exercise 14 ALERT Students may be confused because the unit lb is used for both weight and mass. Weight is a measure of an object's mass under the force of gravity. Although the astronaut's mass is a property of the astronaut and cannot change, the astronaut's weight does change.

14c. According to the equation, they would never be truly weightless. In reality they are weightless whenever they are in freefall.

Science CONNECTION

A typical space shuttle orbits at an altitude of about 400 km. At this height, an astronaut still weighs about 89% of her weight on Earth. You have probably seen pictures in which astronauts in orbit on a space shuttle or space station appear to be weightless. This is actually not due to the absence of gravity but, rather, to an effect called microgravity. In orbit, astronauts and their craft are being pulled toward Earth by gravity, but their speed is such that they are in free fall around Earth, rather than toward Earth. Because the astronauts and their spacecraft are falling through space at the same rate, the astronauts appear to be floating inside the craft. This is similar to the fact that a car's driver can appear to be sitting still, although he is actually traveling at a speed of 60 mi/h. In this NASA picture, astronauts float in space during the 1994 testing of rescue system hardware appear to be weightless. One astronaut floats without being tethered to the spacecraft by using a small control unit.

14. **APPLICATION** As the altitude of a spacecraft increases, an astronaut's weight decreases. The weight of a 180 lb astronaut, w, at a given altitude in kilometers above Earth's sea level, x, is given by the formula

$$w = 180 \cdot \frac{6400^2}{(6400+x)^2}$$

 a. At what altitudes will the astronaut weigh less than 20 lb? above 12,800 km
 b. At an altitude of 400 km, how much will the astronaut weigh? 159.45 lb
 c. Astronauts get the feeling of weightlessness because they are in free fall as they orbit Earth. Could they ever be truly weightless? Why or why not?

15. Find the next three terms in each sequence.
 a. 3, 7, 10, 17, __27__, __44__, __71__
 b. 4, 9, 19, 39, __79__, __159__, __319__

IMPROVING YOUR Visual Thinking SKILLS

Eight square sheets of paper are placed one at a time on a table, each one covering part of some of the sheets below it. In each figure you can see all of the last sheet to be placed on the pile. Determine the order for the other sheets of paper and number them from top to bottom

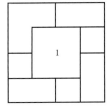 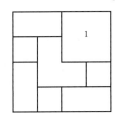

126 CHAPTER 2 Systems of Equations and Inequalities

IMPROVING YOUR Visual Thinking SKILLS

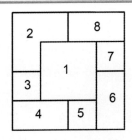 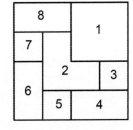

Linear Programming

LESSON 2.5

Love the moment and the energy of that moment will spread beyond all boundaries.
CORITA KENT

Industrial managers often investigate more economical ways of doing business. They must consider physical limitations, standards of quality, customer demand, availability of materials, and manufacturing expenses as restrictions, or constraints, that determine how much of an item they can produce. Then they determine the optimum, or best, amount of goods to produce—usually to minimize production costs or maximize profit. The process of finding a feasible region and determining the point that gives the maximum or minimum value to a specific expression is called **linear programming**.

Problems that can be modeled with linear programming may involve anywhere from two variables to hundreds of variables. Computerized modeling programs that analyze up to 200 constraints and 400 variables are regularly used to help businesses choose their best plan of action. In this lesson you will look at problems that involve two variables because you will be using two-dimensional graphs to help you find the feasible region.

In this investigation you'll explore a linear programming problem and make conjectures about how to find the optimum value in the most efficient way.

INVESTIGATION

Maximizing Profit

The Elite Pottery Shoppe makes two kinds of birdbaths: a fancy glazed and a simple unglazed. An unglazed birdbath requires 0.5 h to make using a pottery wheel and 3 h in the kiln. A glazed birdbath takes 1 h on the wheel and 18 h in the kiln. The company's one pottery wheel is available for at most 8 hours per day (h/d). The three kilns can be used a total of at most 60 h/d, and each kiln can hold only one birdbath. The company has a standing order for 6 unglazed birdbaths per day, so it must produce at least that many. The pottery shop's profit on each unglazed birdbath is $10, and the profit on each glazed birdbath is $40. How many of each kind of birdbath should the company produce each day in order to maximize profit?

LESSON 2.5

Here students see the rest of the linear programming process for two variables.

COMMON CORE STATE STANDARDS		
APPLIED	DEVELOPED	INTRODUCED
	A.CED.3	
	A.REI.12	

OBJECTIVES

- Learn about uses of linear programming and optimization
- Apply linear programming to situations with two variables
- Optimize a function over a feasible region

Vocabulary
linear programming

Launch

Write an inequality for each of the conditions given.

Sami works odd jobs on Saturday. Depending on the job, she earns more than $5 but less than $10 per hour. She can work at most 8 hours. She always works 2 hours for her elderly neighbor.

Let x represent how many hours she works and y represent how much money she earns.

$y \geq 5x$, $y \leq 10x$, $x \geq 2$, $x \leq 8$.

Investigate

Guiding the Investigation

Step 1 ALERT Students may have difficulty combining information from different sentences to fill in the table. Some students might forget the constraint $x \geq 6$. Try to let the groups work out their difficulties before offering assistance.

ELL
Help students create structured tables to list the given information from the questions, and the limiting values of the constraints. The vocabulary in this lesson is on two different levels, relating to both the mathematics and the situation.

Support
Have students divide up the work so they can make better progress on these large problems. You might consider doing a jigsaw, where students look closely at graphing feasible regions, finding vertices, and finding optimal solutions, and then share their work.

Advanced
When students are comfortable with linear programming, have them create a linear programming problem that would model and optimize something in their lives. Have students write optimizing functions for Exercises 6–10 so that each different vertex becomes the best solution. Can they generalize this process?

Step 2 As needed, remind students that they need to be able to identify vertices of the feasible region. Suggest that they draw large graphs on graph paper. **ASK** "Why do you need to graph only the first quadrant?" [Commonsense constraints limit the values of x and y to nonnegative numbers.]

Step 2

$$\begin{cases} 0.5x + y \le 8 & \text{Wheel hrs constraint} \\ 3x + 18y \le 60 & \text{Kiln hrs constraint} \\ x \ge 6 & \text{At least 6 unglazed} \\ x \ge 0 & \text{Common sense} \\ y \ge 0 & \text{Common sense} \end{cases}$$

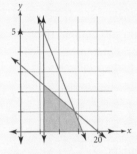

Steps 3, 4 Students may want to record the points in calculator lists so that they can find the profits easily.
SMP 5

Steps 6–8

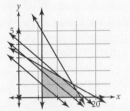

Example

The example applies the linear programming process to finding a minimum value. It uses more constraints than the investigation. Project the example and have students work in small groups. There is a desmos tool available in the ebook. Have students present their inequalities and their graphs along with their answers. Emphasize the use of correct mathematical terminology. Whether student responses are correct or incorrect, ask other students if they agree and why. After presentations, **ASK** "Is the optimum solution reasonable? What might other constraints of concern be?" **SMP 1, 3, 4, 5, 6**

Step 3
$(6, 0), (7, 0), (8, 0), (9, 0),$
$(10, 0), (11, 0), (12, 0),$
$(13, 0), (14, 0), (15, 0),$
$(16, 0), (6, 1), (7, 1), (8, 1),$
$(9, 1), (10, 1), (11, 1), (12, 1),$
$(13, 1), (14, 1), (6, 2), (7, 2),$
$(8, 2)$

Step 4
$\text{profit} = 10x + 40y$
See table on page 129

Step 9
You can graph one profit line and imagine shifting it up until you get the highest profit possible within the feasible region. This should occur at a vertex. If the vertex is not an integer point, you might test the integer points near the optimum vertex. To minimize profit, move the profit line *down* until it leaves the feasible region. This would occur at $(6, 0)$.

Step 1 Organize the information into a table like this one:

	Amount per unglazed birdbath	Amount per glazed birdbath	Constraining value
Wheel hours	0.5	1	≤ 8
Kiln hours	3	18	≤ 60
Profit	$10	$40	Maximize

Step 2 Use your table to help you write inequalities that reflect the constraints given, and be sure to include any commonsense constraints. Let x represent the number of unglazed birdbaths, and let y represent the number of glazed birdbaths. Graph the feasible region to show the combinations of unglazed and glazed birdbaths the shop could produce, and label the coordinates of the vertices. (*Note:* Profit is not a constraint; it is what you are trying to maximize.)

Step 3 It will make sense to produce only whole numbers of birdbaths. List the coordinates of all integer points within the feasible region. (There should be 23.) Remember that the feasible region may include points on the boundary lines.

Step 4 Write the equation that will determine profit based on the number of unglazed and glazed birdbaths produced. Calculate the profit that the company would earn at each of the feasible points you found in Step 3. You may want to divide this task among the members of your group.

Step 5 What number of each kind of birdbath should the Elite Pottery Shoppe produce to maximize profit? What is the maximum profit possible? Plot this point on your feasible region graph. What do you notice about this point? Maximum profit is $180 when 14 unglazed birdbaths and 1 glazed birdbath are produced. This point is a vertex of the feasible region.

Step 6 Suppose that you want profit to be exactly $100. What equation would express this? Carefully graph this line on your feasible region graph. $10x + 40y = 100$

Step 7 Suppose that you want profit to be exactly $140. What equation would express this? Carefully add this line to your graph. $10x + 40y = 140$

Step 8 Suppose that you want profit to be exactly $170. What equation would express this? Carefully add this line to your graph. $10x + 40y = 170$

Step 9 How do your results from Steps 6–8 show you that $(14, 1)$ must be the point that maximizes profit? Generalize your observations to describe a method that you can use with other problems to find the optimum value. What would you do if this vertex point did not have integer coordinates? What if you wanted to *minimize* profit?

Linear programming is a very useful real-world application of systems of inequalities. Its value is not limited to business settings, as the following example shows.

128 CHAPTER 2 Systems of Equations and Inequalities

Modifying the Investigation

Whole Class Complete Step 1 and write inequalities that reflect the given constraints. Have students complete Step 2 individually. Complete Steps 3 through 8 as a class; have students add to their graphs. Discuss Step 9.

Shortened Do a jigsaw in which you divide the work among the groups. Have each group report back to create a classroom graph. Complete Step 1; have groups do Step 2 for wheel hours and kiln hours and report back. [Have groups complete Steps 3 through 5, jigsaw Steps 6 through 8, and discuss Step 9.]

One Step Use the **One Step** found in your ebook. As needed, suggest that students try out various values in the feasible region they found earlier and look for patterns. After each group finds the profit at a particular point, ask whether any other points yield the same profit. Press groups to describe a method to be used for solving similar problems.

EXAMPLE Marco is planning to provide a snack of graham crackers and blueberry yogurt at his school's track practice. He wants to make sure that the snack contains no more than 700 calories and no more than 20 g of fat. He also wants at least 17 g of protein and at least 30% of the daily recommended value of iron. The nutritional content of each food is given below. Each serving of yogurt costs $0.30 and each graham cracker costs $0.06. What combination of servings of graham crackers and blueberry yogurt should Marco provide to minimize cost?

	Serving	Calories	Fat	Protein	Iron (percent of daily recommended value)
Graham crackers	1 cracker	60	2 g	2 g	6%
Blueberry yogurt	4.5 oz	130	2 g	5 g	1%

Solution First organize the constraint information into a table, then write inequalities that reflect the constraints. Be sure to include any commonsense constraints. Let x represent the number of servings of graham crackers, and let y represent the number of servings of yogurt.

	Amount per graham cracker	Amount per serving of yogurt	Limiting value
Calories	60	130	≤ 700
Fat	2 g	2 g	≤ 20 g
Protein	2 g	5 g	≥ 17 g
Iron	6%	1%	≥ 30%
Cost	$0.06	$0.30	Minimize

$$\begin{cases} 60x + 130y \leq 700 & \text{Calories} \\ 2x + 2y \leq 20 & \text{Fat} \\ 2x + 5y \geq 17 & \text{Protein} \\ 6x + 1y \geq 30 & \text{Iron} \\ x \geq 0 & \text{Common sense} \\ y \geq 0 & \text{Common sense} \end{cases}$$

Now graph the feasible region and find the vertices.

Next, write an equation that will determine the cost of a snack based on the number of servings of graham crackers and yogurt.

$$Cost = 0.06x + 0.30y$$

You could try any possible combination of graham crackers and yogurt that is in the feasible region, but recall that in the investigation it appeared that optimum values will occur at vertices. Calculate the cost at each of the vertices to see which vertex provides a minimum value.

LESSON 2.5 Linear Programming **129**

ADVANCED Point out that although neither x nor y is an independent or dependent variable, the profit function seems dependent on both x and y. Wonder aloud whether that means we have a three-dimensional graph. Graphs of functions such as $p = 10x + 40y$ and $c = 0.06x + 0.30y$ are planes in three-dimensional space. These planes are highest and lowest at vertices or along edges of the feasible region.

Summarize

Discussing the Investigation
As students present their ideas about methods, encourage critical analysis by the class. Be open to using students' errors, creative questions, and uncertainty to help deepen understanding. In situations like the one in the investigation, presumably naive students can often raise issues that will lead the class to mathematical insights. For example, a student may ask whether the pottery situation would include constraints on demand. Such constraints are not being considered here, but they would be in real life, where computers handle linear programming with many variables.
SMP 1, 2, 3, 4, 5, 6

ASK "What if those lines were parallel to an edge of the feasible region?" [All feasible points along that edge yield maximum values.] Students will see examples of such regions in the exercises. **ASK** "What if the vertex of the feasible region does not have integer coordinates?" [If all feasible points must have integer coordinates, then the maximum will occur at a nearby point.] **SMP 1, 2, 6**

CRITICAL QUESTION How would you explain to a friend how to solve a linear programming problem?

BIG IDEA Find a feasible region representing the constraints. Then compare values of the optimization function at the vertices of that region.

CRITICAL QUESTION How many half-planes are necessary to have a triangular feasible region?

BIG IDEA Three

Step 4

Point	Profit	Point	Profit	Point	Profit	Point	Profit	Point	Profit	Point	Profit
(6, 0)	$60	(7, 0)	$70	(8, 0)	$80	(9, 0)	$90	(10, 0)	$100	(11, 0)	$110
(12, 0)	$120	(13, 0)	$130	(14, 0)	$140	(15, 0)	$150	(16, 0)	$160	(6, 1)	$100
(7, 1)	$110	(8, 1)	$120	(9, 1)	$130	(10, 1)	$140	(11, 1)	$150	(12, 1)	$160
(13, 1)	$170	(14, 1)	$180	(6, 2)	$140	(7, 2)	$150	(8, 2)	$160		

CRITICAL QUESTION How can you tell if all points on a side of the feasible region give the optimum value?

BIG IDEA Both endpoints of that side give the optimum value.

Formative Assessment

Watch to see that students are able to move beyond the mechanics of creating a feasible region to developing an understanding of profit lines and their use in linear programming problems. Assess students' ability to read and organize information, represent constraints as inequalities, and graph feasible regions. Student presentations and discussions of the investigation, the example, and the exercises provide evidence of meeting the objectives.

Apply

Extra Example

Use this system of inequalities:
$3y + x \geq 3$
$y \leq 2$
$x \leq 3$

a. Find the vertices of the feasible region. $(-3, 2), (3, 0), (3, 2)$

b. What is the maximum value of $P = 3x + y$? $P = 11$ at $(3, 2)$

Closing Question

Why does the maximum or minimum occur at a vertex of the critical region?

Students might explain the situation in terms of lines representing all points with the same profit. These parallel lines shift for higher profits, moving toward a vertex.

Assigning Exercises

Suggested: 1–10
Additional Practice: 11–15

3.
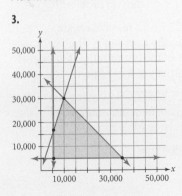

The least expensive combination would be 8.5 crackers and no yogurt. What if Marco wants to serve only whole numbers of servings? The points (8, 1), (9, 1), and (9, 0) are the integer points within the feasible region closest to (8.5, 0), so test which point has a lower cost. The point (8, 1) gives a cost of $0.78, (9, 1) gives a cost of $0.84, and (9, 0) will cost $0.54. Therefore, if Marco wants to serve only whole numbers of servings, he should serve 9 graham crackers and no yogurt.

x	y	Cost
4.444	3.333	$1.27
4.75	1.5	$0.74
8.571	1.429	$0.94
10	0	$0.60
8.5	0	$0.51

The following box summarizes the steps of solving a linear programming problem. Refer to these steps as you do the exercises.

> **Solving a Linear Programming Problem**
> 1. Define your variables, and write constraints using the information given in the problem. Don't forget commonsense constraints.
> 2. Graph the feasible region, and find the coordinates of all vertices.
> 3. Write the equation of the function you want to optimize, and decide whether you need to maximize or minimize it.
> 4. Evaluate your optimization function at each of the vertices of your feasible region, and decide which vertex provides the optimum value.
> 5. If your possible solutions need to be limited to whole-number values, and your optimum vertex does not contain integers, test the whole-number values within the feasible region that are closest to this vertex.

2.5 Exercises

you will need a graphing calculator for exercises **3, 4, 6–10,** and **12.**

Practice Your Skills

1. Carefully graph this system of inequalities and label the vertices. @
$\begin{cases} x + y \leq 10 \\ 5x + 2y \geq 20 \\ -x + 2y \geq 0 \end{cases}$

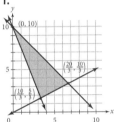

2. For the system in Exercise 1, find the vertex that optimizes these expressions:

  a. maximize: $5x + 2y$ @ $\left(\frac{20}{3}, \frac{10}{3}\right)$
  b. minimize: $x + 3y$ $\left(\frac{10}{3}, \frac{5}{3}\right)$
  c. maximize: $x + 4y$ $(0, 10)$
  d. minimize: $5x + y$ $(0, 10)$
  e. What generalizations can you make about which vertex provides a maximum or minimum value?
  It is not always obvious which point provides a maximum or minimum value.

3. Graph this system of inequalities, label the vertices of the feasible region, and name the integer coordinates that maximize the function $P = 0.08x + 0.10y$. What is this maximum value of P?

$\begin{cases} x \geq 5,500 \\ y \geq 5,000 \\ y \leq 3x \\ x + y \leq 40,000 \end{cases}$

vertices: (5500, 5000), (5500, 16,500), (10,000, 30,000), (35,000, 5000);
maximum: 3800 at (10,000, 30,000)

130 CHAPTER 2 Systems of Equations and Inequalities

Conceptual/Procedural

Conceptual In this lesson, students explore the concept if linear programming.

Procedural Students learn the process of using a feasible region to determine maximum and minimum values.

4. **APPLICATION** During nesting season, two different bird species inhabit a region with area 180,000 m². Dr. Chan estimates that this ecological region can provide 72,000 kg of food during the season. Each nesting pair of species X needs 39.6 kg of food during a specified time period and 120 m² of land. Each nesting pair of species Y needs 69.6 kg of food and 90 m² of land. Let x represent the number of pairs of species X, and let y represent the number of pairs of species Y.

   a. Describe the meaning of the constraints $x \geq 0$ and $y \geq 0$.   a. There are zero or more pairs of each species in the region.

   b. Describe the meaning of the constraint $120x + 90y \leq 180,000$.

   c. Describe the meaning of the constraint $39.6x + 69.6y \leq 72,000$.

   d. Graph the system of inequalities, and identify each vertex of the feasible region.

   e. Maximize the total number of nesting pairs, N, by considering the function $N = x + y$. The maximum number of nesting pairs is 1578. Any of the integer points around (1263.2, 315.8), (1261, 317), (1262, 316), (1263, 315), (1264, 314), or (1265, 313) give this total.

4b. The area required by species X plus the area required by species Y is no more than 180,000 m².

4c. The total food requirement of species X plus the total food requirement of species Y is no more than 72,000 kg.

4d.
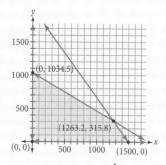

Reason and Apply

5. Use a combination of the four lines shown on the graph along with the axes to create a system of inequalities whose graph satisfies each description.

   a. The feasible region is a triangle. @

   b. The feasible region is a quadrilateral with one side on the y-axis.

   c. The feasible region is a pentagon with sides on both the x-axis and the y-axis.

5a. possible answer: $\begin{cases} y \geq 7 \\ y \leq \frac{7}{5}(x-3) + 6 \\ y \leq -\frac{7}{12}x + 13 \end{cases}$

5b. possible answer: $\begin{cases} x \geq 0 \\ y \geq 7 \\ y \geq \frac{7}{5}(x-3) + 6 \\ y \leq -\frac{7}{12}x + 13 \end{cases}$

5c. possible answer: $\begin{cases} x \geq 0 \\ y \geq 0 \\ x \leq 11 \\ y \leq \frac{7}{5}(x-3) + 6 \\ y \leq 7 \end{cases}$

6. **APPLICATION** The International Canine Academy raises and trains Siberian sled dogs and dancing French poodles. Breeders can supply the academy with at most 20 poodles and 15 Siberian sled dogs each year. Each poodle eats 2 pounds per day (lb/d) of food and each sled dog eats 6 lb/d. Food supplies are restricted to at most 100lb/d. A poodle requires 1,000 h/yr of training, whereas a sled dog requires 250 h/yr. The academy cannot provide more than 15,000 h/yr of training time. If each poodle sells for a profit of $200 and each sled dog sells for a profit of $80, how many of each kind of dog should the academy raise in order to maximize profits? Ⓗ
   12 sled dogs and 12 poodles for a maximum profit of $3,360

7. **APPLICATION** The Elite Pottery Shoppe budgets a maximum of $1,000 per month for newspaper and radio advertising. The newspaper charges $50 per ad and requires at least four ads per month. The radio station charges $100 per minute and requires a minimum of 5 minutes of advertising per month. It is estimated that each newspaper ad reaches 8,000 people and that each minute of radio advertising reaches 15,000 people. What combination of newspaper and radio advertising should the pottery shop use in order to reach the maximum number of people? What assumptions did you make in solving this problem? How realistic do you think they are? @

8. **APPLICATION** A small electric generating plant must decide how much low-sulfur (2%) and high-sulfur (6%) oil to buy. The final mixture must have a sulfur content of no more than 4%. At least 1200 barrels of oil are needed. Low-sulfur oil costs $18.50 per barrel and high-sulfur oil costs $14.70 per barrel. How much of each type of oil should the plant use to keep the cost at a minimum? What is the minimum cost? Ⓗ
   600 barrels each of low-sulfur and high-sulfur oil for a minimum total cost of $19,920

7. 5 radio minutes and 10 newspaper ads to reach a maximum of 155,000 people. This assumes that people who listen to the radio are independent of people who read the newspaper, which is probably not realistic, and that the shop buys both. This also assumes that each newspaper ad reaches a fresh set of readers (instead of the same subscribers day after day who each see the ad 10 times), and that each radio minute reaches a fresh set of listeners (instead of the same group of people who hear the radio spot 5 separate times). 20 newspaper ads and 0 radio minutes would reach 160,000 people.

Exercise 8 ALERT Depending on how much you discussed unbounded feasible regions in class, students may be confused by this exercise. The challenge is to find the minimum cost, which does exist even though the cost function has no maximum value over this feasible region.

9. 6270 acres of coffee and 1230 acres of cocoa for a maximum total income of $327,327.

Exercise 10 If students are having trouble, suggest that they sketch a picture of the package and mark the length and girth.

10a. Let x represent the length in inches, and let y represent the girth in inches. $\begin{cases} x + y \leq 130 \\ x \leq 108 \\ x > 0 \\ y > 0 \end{cases}$

10b.

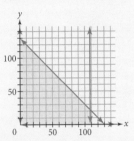

11.

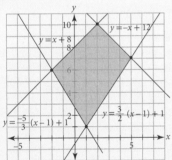

14. $\begin{cases} x \geq 2 \\ y \leq 5 \\ x + y \geq 3 \\ 2x - y \leq 9 \end{cases}$

Extensions

To extend Exercise 10, ask students to research shipping restrictions on package size and to draw feasibility regions for package sizes with different carriers. Students should find out what types of restrictions there are, whether different companies have different restrictions on size, and whether certain things cannot be mailed because of their size.

9. APPLICATION A fair-trade farmers' cooperative in Chiapas, Mexico, is deciding how much coffee and cocoa to recommend to their members to plant. Their 1,000 member families have 7,500 total acres to farm. Because of the geography of the region, 2,450 acres are suitable only for growing coffee and 1,230 acres are suitable only for growing cocoa. A coffee crop produces 30 lb/acre and a crop of cocoa produces 40 lb/acre. The cooperative has the resources to ship a total of 270,000 lb of product to the United States. Fair-trade organizations mandate a minimum price of $1.51 per pound for organic coffee and $0.88 per pound for organic cocoa (note that price is per pound, not per acre). How many acres of each crop should the cooperative recommend planting in order to maximize income?

10. APPLICATION Teo sells a set of DVDs on an online auction. The postal service he prefers puts these restrictions on the size of a package:

Up to 150 lb

Up to 130 in. in length and girth combined

Up to 108 in. in length

Length is defined as the longest side of a package or object. Girth is the distance all the way around the package or object at its widest point perpendicular to the length. Teo is not concerned with the weight because the DVDs weigh only 15 lb.

a. Write a system of inequalities that represents the constraints on the package size. @

b. Graph the feasible region for the dimensions of the package.

c. Teo packages the DVDs in a box whose dimensions are 20 in. by 14 in. by 8 in. Does this box satisfy the restrictions? Yes, the box lies in the feasible region.

These coffee beans are growing in the Mexican state of Chiapas.

Review

11. Sketch a graph of the feasible region described by this system of inequalities:

$$\begin{cases} y \geq -\frac{5}{3}(x - 1) + 1 \\ y \leq -x + 12 \\ y \geq \frac{3}{2}(x - 1) + 1 \\ y \leq x + 8 \end{cases}$$

12. Solve each of these systems in at least two different ways.

a. $\begin{cases} 8x + 3y = 41 \\ 9x + 2y = 25 \end{cases}$ $\left(-\frac{7}{11}, \frac{169}{11}\right)$ b. $\begin{cases} 2x + y - 2z = 5 \\ 6x + 2y - 4z = 3 \\ 4x - y + 3z = 5 \end{cases}$ $(-3.5, 74, 31)$

13. Give the system of inequalities whose solution set is the polygon at right. ⓗ

14. Consider this system of equations:

$\begin{cases} 3x - y = 5 \\ -4x + 2y = 1 \end{cases}$ $(5.5, 11.5)$

a. Solve the system and write the solution as an ordered pair of coordinates.

b. Check your solution values for x and y by substituting them into the original equations. $3(5.5) - 11.5 = 5; -4(5.5) + 2(11.5) = 1$

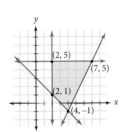

LESSON 2.6

Larger Systems

The time has come, the walrus said, 'to talk of many things: of shoes and ships – and sealing wax – of cabbages and kings.'
LEWIS CARROLL

You have solved systems of two equations using a variety of methods, including the substitution and elimination methods. Can you solve these larger systems using these methods?

A system of two linear equations can have one solution, no solutions, or an infinite number of solutions. These situations correspond to intersecting lines, parallel lines, or two equations for the same line. What happens when you have more equations? What types of situations do they describe?

INVESTIGATION

Three Planes

An equation like $x + 2y + 3z = 6$ cannot be graphed on an x-y grid. For this equation you need three axes, which means your graph must be three dimensional. Instead of a line, you get a plane. What are the possibilities for intersections when you have three planes?

Step 1 Use objects such as the walls, floor, and ceiling of your classroom or the top of a table and pieces of cardboard or paper to show how three planes can have a single point of intersection. Describe, sketch, or take a photo of your model.

Step 2 Now consider how three planes can have no single point of intersection. Make sketches, take photos, or give detailed descriptions of all the different ways three planes can intersect. There are 8 possible ways for three planes to intersect.

Solving a system of three equations using elimination and/or substitution requires organization and care.

EXAMPLE A Solve the system using elimination.
$$x + 2y + z = 3$$
$$3x + y - z = 2$$
$$2x + 3y + 2z = 7$$

Solution The first step is to decide which variable to eliminate first. Looking at the equation, it will be easy to eliminate the z from the first and second equation by simply adding them together.

$$x + 2y + z = 3$$
$$\underline{3x + y - z = 2}$$
$$4x + 3y = 5$$

LESSON 2.6

In this lesson students look at system of three equations and three unknowns.

COMMON CORE STATE STANDARDS		
APPLIED	DEVELOPED	INTRODUCED
	A.REI.6	

Objectives
- Solve a system of three equations in three unknowns by elimination.
- Solve a system of three equations in three unknowns by substitution.
- Visualize the graphical solution of a system of three equations in three unknowns as the intersection of planes.

Materials
- cardstock, *optional*

Launch

Solve this system of equations.

$x + y = 5$

$x + y + z = 7$

$3x + 2y - 2z = 6$

(0, 5, 2)

Investigate

ASK "How would you graph $x + 2y + z = 6$?" [You would need a grid with three axes, x, y, and z.] "What in the room somewhat models the origin of a grid with three axes?" [A corner of the ceiling or floor with one axis being the width of the room, one being the length of the room and one being the height of the room.]

Guiding the Investigation
Step 1

The three planes intersect in a point.

ELL
Review the definition of a plane. Have students identify things in the room that could model a plane.

Support
Use cardstock to demonstrate the intersection of two planes. Then have students use cardstock to build physical models of the intersection of planes. **ASK** "What in the room somewhat models the origin of a grid with three axes?"

Advanced
Assign the **One Step**.
Assign George Pólya's classic problem: Into how many parts will five random planes divide space? [26]

Step 2

The three planes intersect in a line.

Two of the planes are dependent planes that intersect the third plane in a line.

The three planes are dependent planes.

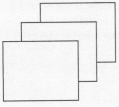
The planes do not intersect. The planes are parallel.

Just two planes are parallel. The third plane intersects each of them in a line. The lines are parallel.

Each of the planes intersect the other two in a line. The lines are pairs of parallel lines.

Two of the planes are coincident planes that that are parallel to the third plane.

Now you need to eliminate the z again, using a different pair of equations. Multiply the second equation by 2 and add it to the third equation.

$$2(3x + y - z) = 2(2) \rightarrow \begin{array}{r} 6x + 2y - 2z = 4 \\ 2x + 3y + 2z = 7 \\ \hline 8x + 5y = 11 \end{array}$$

You now have a system of only two equations

$$4x + 3y = 5$$
$$8x + 5y = 11$$

Multiply the top equation by 5 and the bottom by -3 so that you can eliminate the y.

$$5(4x + 3y) = 5(5) \rightarrow \begin{array}{r} 20x + 15y = 25 \\ -3(8x + 5y) = 11(-3) \rightarrow -24x - 15y = -33 \\ \hline -4x = -8 \\ x = 2 \end{array}$$

Then substitute the value of x back into one of the two variable equations to solve for y.

$$4(2) + 3y = 5$$
$$8 + 3y = 5$$
$$3y = -3$$
$$y = -1$$

Finally, go back to one of the original equations to determine the value of z. Using the first equation:

$$2 + 2(-1) + z = 3$$
$$2 - 2 + z = 3$$
$$z = 3$$

The three planes intersect in the point $(2, -1, 3)$

To check your solution, substitute the three values into each of the equations.

$$(2) + 2(-1) + (3) = 2 - 2 + 3 = 3 \checkmark$$
$$3(2) + (-1) - (3) = 6 - 1 - 3 = 2 \checkmark$$
$$2(2) + 3(-1) + 2(3) = 4 - 3 + 6 = 7 \checkmark$$

The solution works in all three equations so it is correct.

> **Strategy for solving systems of three equations with three variables by elimination**
> 1. Eliminate one variable by using any two of the original equations.
> 2. Eliminate the same variable by using the unused original equation and reusing one of the others.
> 3. Solve the new system of two equations that you created in the first two steps with only two variables.
> 4. Go back to any of the original equations to find the value of the third variable.

The substitution method can also be used in solving larger systems.

EXAMPLE B | Solve the system using substitution.

$$x + y - z = 9$$
$$2x - 3y + z = -1$$
$$x = 5y + 3z + 3$$

Solution | Because the third equation is already solved for x, you can replace the x in each of the other two equations with the expression $(5y + 3z + 3)$.

$$(5y + 3z + 3) + y - z = 9$$
$$2(5y + 3z + 3) - 3y + z = -1$$

Combining the like terms to simplify each equation, they become

$$6y + 2z = 6$$
$$7y + 7z = -7$$

Notice that if you divide the second equation by 7, you can solve it for either y or z quite easily.

$$7y + 7z = -7 \quad \rightarrow \quad y + z = -1 \quad \rightarrow \quad y = -1 - z$$

Substitute for y in the first equation and solve for z.

$$6(-1 - z) + 2z = 6$$
$$-6 - 6z + 2z = 6$$
$$-4z = 12$$
$$z = -3$$

Use this value for z to find y: $y = -1 - (-3) = 2$

And finally go back to the beginning to solve for x.

$$x = 5(2) + 3(-3) + 3 = 4$$

The three planes intersect in the point $(4, 2, -3)$

> **Strategy for solving systems of three equations with three variables by substitution.**
> 1. Solve one equation for one variable.
> 2. Substitute the expression from Step 1 into the other two equations.
> 3. Solve the new system of two equations with only two variables.
> 4. Go back to any of the original equations to find the value of the third variable.

Whether you use substitution or elimination, the big idea is to create a smaller system with fewer variables. In many cases you may use substitution first and then solve the smaller system by elimination. Using a combination of strategies wisely will make the solution easier.

Apply

Extra Example

Solve this system:
$$\begin{cases} 2x + y - 5z = -21 \\ x + 2y - 2z = -15 \\ x - 4y + z = 18 \end{cases}$$
$(-1, -4, 3)$

Closing Question

Solve this system:
$$\begin{cases} 2x - 2y + 4z = 7 \\ -4x + 2y - 3z = 14 \\ x + 4y - 12z = 1 \end{cases}$$
$(-33, -126.5, -45)$

Assigning Exercises

Suggested: 1 – 7
Additional Practice: 8 – 10

Exercise 4 Reasons may vary.

4a. Use elimination because $15x$ and $-15x$ easily add to 0. Solving by substitution will be more difficult because solving for any of the variables will involve fractions.

4b. Either substitution or elimination is a good choice. It is easy to solve for one variable to do substitution. It is also easy to eliminate any one of the three variables.

4c. Substitution may be the easiest method here as it you can solve the second or third equation quickly for any one of the variables. A combination of strategies may also be useful such as adding the second and third equations to eliminate z and then solving either one of these for z and substituting in the first equation.

5a. $4t + 4d + 2p = 79.50,$
$2t + d + p = 35.25,$
$3t + 2d + 2p = 58.50$

2.6 Exercises

Practice Your Skills

1. Solve each system using elimination.
   a. $2x + 3y + z = -2$
      $x - y + 4z = 3$ $(4, -3, -1)$
      $4x + 2y + 3z = 7$
      (*Hint*: Eliminate y first.)

   b. $x + 4y - 2z = -7$
      $3x + 3y + 5z = 19$ $(5, -2, 2)$
      $x - 2y + 3z = 15$
      (*Hint*: Eliminate x first.)

2. Solve each system using substitution.
   a. $2x + 3y - z = -4$ @
      $y = -x + z - 5$ $(-3, 2, 4)$
      $x - 2y + 3z = 5$

   b. $x + y - z = -1$
      $2x + 3y = z - 5$ $(2, -3, 0)$
      $x - y + 4z = 5$

3. Determine if the given point is a solution for the system.
   a. Is the point $(-4, 3, 2)$ a solution for the system: @
      $3x + 2y + 5z = 4$
      $5x - 3y + 8z = -13$ no, the point satisfies the first and second equation, but not the third
      $2x - 3y + 6z = 5$

   b. Is the point $(7, -2, -1)$ a solution for the system:
      $x + 3y + 5z = -4$
      $5x + 8y = 14 - 5z$ yes, the point satisfies all three equations
      $z = 3x - 2y - 26$

Reason and Apply

4. Solve each system of equations. Explain the reason for your choice of method.
   a. $15x + 20y + 3z = -4$ @ $(-2, 1, 2)$
      $-15x - 6y - 14z = -4$
      $7x - 3y + 5z = -7$

   b. $x + 2y + z = 4$ $(-1, 2, 1)$
      $2x + y + z = 1$
      $x + y + 2z = 3$

   c. $2x + y + z = 9$ $(5, 6, -7)$
      $x + y = 4 - z$
      $x - y = y + z$

5. Three groups of people went out to the movies. The first group purchased four tickets, four drinks, and two buckets of popcorn for a total of $79.50. The second group purchased two tickets, one drink, and one bucket of popcorn for a total of $35.25. The third group purchased three tickets, two drinks, and two buckets of popcorn for a total of $58.50.

   a. Write equations to model the spending of each group.

   b. Solve your system of equations using elimination, substitution, or a combination of methods.
   $(12, 4.50, 6.75)$

   c. What would it cost to purchase one ticket and a bucket of popcorn? $18.75

136 CHAPTER 2 Systems of Equations and Inequalities

6. Create a system of three equations that has the point (1, 2, 3) as its unique solution. Solve your system to verify that the solution is unique. Answers will vary. When solving, the system should not produce either an identity or a contradiction. For instance, $x + y + z = 6$, $x + y - z = 0$, $x - y + z = 2$

7. Oreyana and Claire tried to solve the systems shown below. In each case they ran into some difficulties. Based on what happened, try to determine what the graph of the planes represented by the equations would look like.

   a. $3x + 4y + z = 14$
      $5x - y + 2z = 9$ Because each pair of equations in two variables creates an inconsistent system, this
      $2x - 5y + z = 3$ means that the lines of intersection of any pair of planes are parallel.

   In this system, each time they eliminated a variable and created a system of two equations, the new system was inconsistent.

   b. $2x + 3y + 4z = 9$
      $x - y + 2z = 2$ Since each new system was dependent, that indicates that all of the lines of intersection
      $4x + y + 8z = 13$ are the same.

   In this system, each time they eliminated a variable and created a system of two equations, the new system was dependent.

Review

8. Use an appropriate method to solve each system of equations.

   a. $\begin{cases} y = \frac{1}{4}x + 10 \\ 3x - 2y = 0 \end{cases}$ b. $\begin{cases} 3x + 8y = 5 \\ 5x + 4y = 27 \end{cases}$

9. The Lincoln family members are trying to decide which of two catering companies to hire for their anniversary party. Mystery Meals charges $700 plus $35 per person. Peculiar Parties charges $425 plus $50 per person. Both companies are known for their good food and excellent service, so the Lincoln family wants to choose the caterer that will cost less. Which company should they hire? Explain.

10. Solve the equation $2x + 3y = 12$ for y and then graph it.

8a. (8, 12)

8b. (7, −2)

9. It depends on how many people will be at the party. For 18 people or fewer, Peculiar Parties is less expensive. For more than 18 people, Mystery Meals costs less.

10. $y = 4 - \frac{2}{3}x$

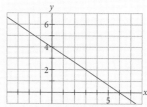

PERFORMANCE TASK

Larger systems can be solved using the same strategies you use to solve a system of three equations. The system below has six variables. Use elimination and substitution carefully to solve the system in an efficient manner.

$$\begin{aligned} x + y + z + u + v + w &= 4 \\ x + z &= 0 \\ y + v &= 0 \\ z + u &= 2 \\ x - y + z &= -2 \\ u - w &= 2 \end{aligned}$$

PERFORMANCE TASK

One strategy is to add the second and third equations and subtract that equation from the first to get $u + w = 4$. Then solve this with the last equation to find the values of u and w. Next use the fourth equation to find z. Then use the second equation to find x. After that use the fifth equation to find the value of y. Finally use either the second or first equation to get the value of v. $x = 1$, $y = 2$, $z = -1$, $u = 3$, $v = -2$, $w = 1$ (1, 2, −1, 3, −2, 1)

CHAPTER REVIEW 2

Reviewing

You might review the chapter by assigning Exercise 1, 3, and 5 for students to work on in class in small groups. Have students share their solutions and discuss why they chose the method they chose. Emphasize the use of correct mathematical terminology. Whether student responses are correct or incorrect, ask other students if they agree and why.

Assigning Exercises

Students might work on the odd Exercises 1, 3, 5, and 9 in groups and do Exercises 2, 4, 6–8, and 10–14 for homework.

1b. every point; same line; consistent, dependent

1e. $\left(\frac{42}{5}, \frac{53}{10}\right)$, or (8.4, 5.3); consistent, independent

CHAPTER 2 REVIEW

You can solve a **system of equations** of two equations with two unknowns to find the intersection of two lines using the methods of **substitution**, **elimination**, or using a graph or table of values. With two linear equations, a system will be **consistent and independent** (intersecting lines, with one solution), **consistent and dependent** (the same line, with infinitely many solutions), or **inconsistent** (parallel lines, with no solution). When finding the solution of a system of three equations with three unknowns, you can think of it as finding the intersection point of three planes. Three planes can intersect in either a point, a line, or not at all. There are more possibilities to consider when the system does not have a unique solution.

When a system is made up of inequalities, the solution usually consists of many points that can be represented by a region in the plane. One important use of systems of linear inequalities is in **linear programming**. In linear programming problems, an equation for a quantity that is to be optimized (maximized or minimized) is evaluated at the vertices of the **feasible region**.

Exercises

@ **Answers are provided for all exercises in this set.**

1. Solve each system (when possible) and classify each system as consistent or inconsistent and as independent or dependent.

   a. $\begin{cases} y = 3.2x - 4 \\ y = 3.1x - 3 \end{cases}$

   b. $\begin{cases} y = \frac{1}{4}(x - 8) + 5 \\ y = 0.25x + 3 \end{cases}$

   c. $\begin{cases} 3x + 2y = 4 \\ -3x + 5y = 3 \end{cases}$

   (10, 28); consistent, independent

   $\left(\frac{2}{3}, 1\right)$, or $(0.\overline{6}, 1)$; consistent, independent

   d. $\begin{cases} 5x - 4y = 5 \\ 2x + 10y = 2 \end{cases}$

   e. $\begin{cases} y = \frac{3}{4}x - 1 \\ \frac{7}{10}x + \frac{2}{5}y = 8 \end{cases}$

   f. $\begin{cases} \frac{3}{5}x - \frac{2}{5}y = 3 \\ 0.6x - 0.4y = -3 \end{cases}$

   (1, 0); consistent, independent

   No intersection; the lines are parallel; inconsistent.

2. Use an appropriate method to solve each system of equations.

   a. $\begin{cases} 2.1x - 3y = -11.43 \\ 5x + 3y = 44.8 \end{cases}$ (4.7, 7.1)

   b. $\begin{cases} y = 1 + \frac{1}{3}x \\ y = \frac{28}{3} - 3x \end{cases}$ $\left(\frac{5}{2}, \frac{11}{6}\right)$

   c. $\begin{cases} 3x + 4y = 23 \\ 2x - 6y = 3.2 \end{cases}$ (5.8, 1.4)

3. Solve each system by any method.

   a. $\begin{cases} 8x - 5y = -15 \\ 6x + 4y = 43 \end{cases}$ $x = 2.5, y = 7$

   b. $\begin{cases} 5x + 3y - 7z = 3 \\ 10x - 4y + 6z = 5 \\ 15x + y - 8z = -2 \end{cases}$ $x = 1.22, y = 6.9, z = 3.4$

138 CHAPTER 2 Systems of Equations and Inequalities

4. Create a system of inequalities that has the feasible region graphed here.

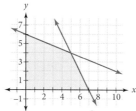

Inequalities should be equivalent to $y \leq 6 - \frac{2}{5}x$, $y < 2(7 - x)$, $y \geq 0$, and $x \geq 0$.

5. Graph the feasible region of each system of inequalities. Find the coordinates of each vertex. Then identify the point that maximizes the given expression.

a. $\begin{cases} 2x + 3y \leq 12 \\ 6x + y \leq 18 \\ x + 2y \geq 4 \\ x \geq 0 \\ y \geq 0 \end{cases}$

maximize: $1.65x + 5.2y$

b. $\begin{cases} x + y \leq 50 \\ 10x + 5y \leq 440 \\ 40x + 60y \leq 2400 \\ x \geq 0 \\ y \geq 0 \end{cases}$

maximize: $6x + 7y$

6. **APPLICATION** Heather's water heater needs repair. The plumber says it will cost $300 to fix the unit, which currently costs $75 per year to operate. Or Heather could buy a new energy-saving water heater for $500, including installation, and the new heater would save 60% on annual operating costs. How long would it take for the new unit to pay for itself? about 4.4 yr

7. Find the values of a, b, and c such that the equation $y = ax^2 + bx + c$ will pass through the points $(2, -15)$, $(-1, 27)$, and $(5, 15)$. $y = 4x^2 - 18x + 5$

8. Solve each inequality and graph the solution.

a. $|2x - 1| \leq 3$ $-1 \leq x \leq 2$
b. $|x + 2| > 3$ $x < -5$ or $x > 1$
c. $-2|x| < -4$ $x < -2$ or $x > 2$

9. **APPLICATION** Yolanda, Myriam, and Xavier have a small business producing handmade shawls and blankets. They spin the yarn, dye it, and weave it. A shawl requires 1 h of spinning, 1 h of dyeing, and 1 h of weaving. A blanket needs 2 h of spinning, 1 h of dyeing, and 4 h of weaving. They make a $16 profit per shawl and a $20 profit per blanket.

Each person has a full-time job, so they have limited time to spend on the business. Xavier does the spinning on Monday, when he can spend at most 8 h spinning. Yolanda dyes the yarn on Wednesday, when she has at most 6 h. Myriam does all the weaving on Friday and Saturday, when she has at most 14 h available. How many of each item should they make each week to maximize their profit?

They should make 4 shawls and 2 blankets to make the maximum profit of $104.

CHAPTER 2 REVIEW **139**

Exercise 5 You might extend this exercise by asking students to write a real–world exercise that would use each set of constraints.

5a.

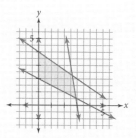

Vertices: $(0, 4)$, $(2.625, 2.25)$, $(2.\overline{90}, 0.\overline{54})$, $(0, 2)$.

The maximum occurs at $(0, 4)$: $1.65(0) + 5.2(4) = 20.8$.

5b.

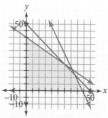

Vertices: $(0, 0)$, $(0, 40)$, $(30, 20)$, $(38, 12)$, $(44, 0)$.

The maximum occurs at $(30, 20)$: $6(30) + 7(20) = 320$.

Exercise 6 LANGUAGE The phrase pay for itself means "reach the point when the additional cost is equal to the amount of savings."

LESSON 2 REVIEW **139**

Exercise 10 Students can do this exercise by graphing, by looking for equations with the same slope or y–intercept, or by finding that the coefficient matrix has no inverse and then looking for reasons why.

Take Another Look

1. Approaches will vary. Some students will prefer to guess-and-check, perhaps randomly but preferably systematically. Using a line of fit is more efficient than using recursion for 60 months many times. The best fit will give a payment of $373.65.

   The recursive formula is $u_n = u_{n-1}\left(1 + \frac{0.09}{12}\right) - payment$. Some sample points: (200, 13097.43), (250, 9326.22), (300, 5555.02), (400, _1987.40). Models will vary. The model of the sample data is $y = 28{,}182 - 75.4241x$, and the intercept is (373.65, 0). When the payment is $373.65, the final balance will be $0 after 60 payments. Testing this value in the formula gives a balance of 2¢ after 5 yr. The slope is -75.4241 and represents the average monthly payment in dollars. The y–intercept is 28,182 which represents the total amount you will spend on the car, including interest, in dollars.

2. This system is different from systems of linear equations because it has two solutions: $(\sqrt{2}, -2)$ and $(-\sqrt{2}, -2)$.

   If students use graphs to check their algebraic results, they might find that they neglected to consider both positive and negative roots.

---

10. Identify each system as consistent and independent (has one solution), inconsistent (has no solution), or consistent and dependent (has infinitely many solutions).

    a. $\begin{cases} y = -1.5x + 7 \\ y = -3x + 14 \end{cases}$ consistent and independent

    b. $\begin{cases} y = \frac{1}{4}(x - 8) + 5 \\ y = 0.25x + 3 \end{cases}$ consistent and dependent

    c. $\begin{cases} 2x + 3y = 4 \\ 1.2x + 1.8y = 2.6 \end{cases}$ inconsistent

    d. $\begin{cases} \frac{3}{5}x - \frac{2}{5}y = 3 \\ 0.6x - 0.4y = -3 \end{cases}$ inconsistent

11. Solve each system of equations

    a. $2x + 3y - 5z = 52$
       $x + 2y + 5z = -17$ (5, 4, −6)
       $3x - 2y + z = 1$

    b. $4x - 5y + z = -16$
       $2x + y + 2z = 8$ (−3, 2, 6)
       $3x - y + 3z = 7$

12. The length of a standard IMAX screen is 34 feet less than twice its width. The perimeter of the screen is 250 feet. What are the dimensions of the screen? 53 feet by 72 feet

13. Suppose the nine justices on the Supreme Court earn a total of $1,934,700 per year. An associate justice earns $9,600 less than the Chief Justice. What is the salary for an Associate Justice? (The court has eight Associate Justices and one Chief Justice.)

    An associate justice earns $213,900 and the chief justice earns $223,500.

Take Another Look

1. You plan to borrow $18,000 to purchase a new car. The lender is advertising an interest rate of 9% compounded monthly. You want to determine how much your payment must be to pay off the loan in 5 years (60 months). Your first step is to guess a payment amount and use the recursive formula from Chapter 1 to determine the balance after exactly 60 payments. The balance may be negative if your guess is too high. In a table, record (*payment, final balance*).

   Make two or three more guesses and record each guess and the resulting final balance in the table. Then plot the ordered pairs you have recorded. Fit a line to the data points and locate the *x*-intercept of the line. What does this point tell you? Check and see if it is correct using the recursive procedure. Find the slope and *y*-intercept of the line and interpret their meanings.

2. In this chapter you learned three methods for solving a system of linear equations—graphing, substitution, and elimination. These methods also can be applied to systems of nonlinear equations. Use all three methods to solve this system.

$$\begin{cases} y = x^2 - 4 \\ y = -2x^2 + 2 \end{cases}$$

Did you find the same solution(s) with all three methods? Describe how the process of solving this system was different from solving a system of linear equations. If you were given another system similar to this one, which method of solution would you choose? What special things would you look out for?

3. You have learned how to do linear programming problems, but how would you do *nonlinear* programming? Carefully graph this system of inequalities and label all the vertices. Then find the point within the feasible region that maximizes the value of P in the equation $P = (x + 2)^2 + (y - 2)^2$. Explain your solution method and how you know that your answer is correct.

$$\begin{cases} y \leq -|x + 2| + 10 \\ 10^{y+2} \geq x + 8 \\ 3x + 8y \leq 50 \\ -3y + x^2 \geq 9 \end{cases}$$

Assessing What You've Learned

 PERFORMANCE ASSESSMENT While a classmate, a friend, a family member, or a teacher observes, show how to solve a system of equations using two different methods. Explain all of your steps and why each method works.

 WRITE IN YOUR JOURNAL You have learned three methods to find a solution to a system of linear equations: graphing, substitution, and elimination. Which method do you prefer? Which one is the most challenging to you? What are the advantages and disadvantages of each method?

 UPDATE YOUR PORTFOLIO Pick a linear programming problem for which you are especially proud of your work, and add it to your portfolio. Describe the steps you followed and how your graph helped you to solve the problem.

3.

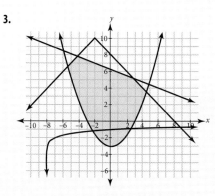

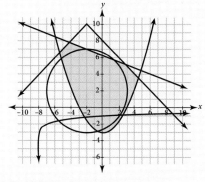

[−10, 10, 1, −5, 10, 1]

Vertices: (2.8, 5.2), (4.437, 3.563), (2.461, −0.980), (−5.373, 6.626), (−4.182, 7.818), (−2.295, −1.244)

Students may find answers by graphing the equation for any value of P (the second graph shows the circle with $P = 25$) and imagine increasing the value of P, enlarging the circle, until the last point within the feasible region is hit. This will show that the vertex on the far right provides the maximum value. The function to be maximized graphs in three dimensions as a paraboloid. The maximum value of P occurs at (4.437, 3.563), where $P = 127 - 7\sqrt{141} \approx 43.879$.

Assessing the Chapter

The Kendall Hunt Cohesive Assessment System (CAS) provides a powerful tool with which to manage assessment content, create and assign homework and tests, and disseminate assessment data. Consider including at least one item from constructive assessment in CAS.

You might require a presentation or an individual or group performance assessment for the chapter and then give a unit exam. Choose a performance assessment from an exercise you didn't assign that is recommended for that purpose.

Facilitating Self–Assessment

A good sample of student writing in mathematics would be an exercise in which students explain a mathematical concept or relationship. A journal entry in which they also think about their learning can show student understanding and growth. Exercises that require student explanations include Lesson 2.1, Exercises 6 and 8; Lesson 2.2, Exercise 7; Lesson 2.3, Exercises 9 and 10; Lesson 2.4, Exercise 14; Lesson 2.5, Exercises 2 and 7; and Lesson 2.6, Exercise 7.

Good portfolio selections for this chapter include Lesson 2.1, Exercise 8; Lesson 2.2, Exercise 7; Lesson 2.3, Exercise 8 and 10; Lesson 2.4, Exercise 10; Lesson 2.5, Exercise 7; and Lesson 2.6, Exercise 7 and the Performance Task.

CHAPTER 3

Functions and Relations

Overview

In *Discovering Advanced Algebra,* students study mathematical functions modeling real-world problems. Chapter 3 is at the core of that study. Here, the abstract idea of a function grows out of students' earlier experiences with linear equations and graphing.

This chapter considers both linear and nonlinear functions and how changing a function's expression transforms its graph. Students also encounter an even further abstraction—the idea of a *relation*— and they study equations and graphs of ellipses.

This chapter begins with exploring graphs as models for data in **Lesson 3.1. Lesson 3.2** makes the distinction between relations and functions as it introduces function notation. Students look at translations of functions in **Lesson 3.3. Lesson 3.4** presents the family of quadratic functions as transformations of the function $y = x^2$ and emphasizes the vertex as a key to writing these equations from a graph or graphing the equations. **Lesson 3.5** uses another transformation, reflection, to examine the square root family, with parent function $y = \sqrt{x}$ and to look at whether a function is even, odd, or neither. Students learn about dilations to help explore the absolute-value family of functions in **Lesson 3.6. Lesson 3.7** considers transformations of the circle and ellipse family of relations.

The **Performance Task** highlighted in this chapter, in Lesson 3.1, is a creative activity in which students sketch a graph from a story and then write a story to reflect a graph.

The Mathematics

Relations and Functions

A *relation* can be thought of as a two-column table of numbers. The items in the first column make up the relation's *domain;* the second column is its *range.* One way you can represent relations is with graphs. You can regard each table row of two numbers as the coordinates of a point on a plane. The relation's *graph* contains all of those points. If a graph consists of disconnected points, the relation is *discrete;* otherwise, it's *continuous.*

The most common kind of relation is a *function,* in which no number appears twice in the first column. A function can therefore be thought of as "taking" each number in the first column to the corresponding number in the second. For functions, the two columns may be called *input* and *output* or *independent variable* and *dependent variable.* In the case of a function, the equation relating the variables is often called the function's *rule.* For example, the equation $y = x^2 + 2$ tells how variable y depends on variable x. To emphasize that y is a function of x, this rule might also be written $f(x) = x^2 + 2$.

Function notation, $f(x)$ [read f of x], is used extensively in mathematics. Any letter may be used to indicate that one variable can be expressed in terms of another. It is important to be able to write, interpret, and evaluate expressions using this notation before exploring new functions and families of functions.

Operations on functions are similar to operations on numbers. Each operation has a definition that produces a new function.

$$f(x) + g(x) = (f + g)(x)$$
$$f(x) - g(x) = (f - g)(x)$$
$$f(x) \cdot g(x) = (f \cdot g)(x)$$
$$\frac{f(x)}{g(x)} = \left(\frac{f}{g}\right)(x)$$

Building Complex Relations

To understand and graph complicated relations, it's often useful to see how they are made up from simpler relations. For example, relations may be *transformations* of simpler relations. Transformations are usually thought of as motions of a graph. This chapter addresses three ways in which graphs might be transformed.

One kind of transformation is the *translation* (shift). A horizontal translation of a graph to the right is like replacing the x in the equation with $(x - h)$. For example, $y = (x - 2)^2$ represents a translation of the graph of $y = x^2$ to the right 2 units. A vertical translation upward is like replacing the y in the equation with $(y - k)$. The graph of the function with equation $y - 3 = x^2$ is a translation of the graph of $y = x^2$ up 3 units.

Another kind of transformation is a *reflection* across an axis. A horizontal reflection (across the vertical axis) corresponds to multiplying x by -1, and a vertical reflection corresponds to multiplying y by -1. For example, $y = (-x)^3$ reflects the graph of $y = x^3$ horizontally, across the y-axis. And $-y = x^3$ reflects the same graph vertically, across the x-axis.

A relation can also be *dilated* (stretched from or shrunk toward an axis). A horizontal stretch corresponds to dividing x by a factor that is greater than 1, and a vertical stretch to dividing y by a factor greater than 1. The equation $\left(\frac{x}{5}\right)^2 + \left(\frac{y}{7}\right)^2 = 1$ represents the ellipse obtained by stretching the unit circle $x^2 + y^2 = 1$ horizontally by a factor of 5 and vertically by a factor of 7. To shrink a graph, the dilation factor is less than 1.

A function, $f(x)$, is an *even* function if $f(-x) = f(x)$. A function $f(x)$ is an *odd* function if $f(-x) = -f(x)$. An even function has the y-axis as a line of symmetry. An odd function is said to have symmetry with respect to the origin. Most functions, however, are neither even nor odd.

Using This Chapter

While much of the content of the early lessons in this chapter should be review, it is presented with a fresh perspective. The ability to interpret graphs is crucial to the rest of the course, so don't skip Lesson 3.1. Defining functions and relations and using function notation is familiar to most students, but they may not have used arithmetic operations on functions before. Transformations of parent functions to get other functions may not be new, but determining whether a function is even, odd, or neither is probably new to most students.

If time is a consideration, rather than skipping a topic, you may be able to spend less time on some lessons than on others. You might also assign the **Shortened** version of the investigations or use the desmos tool to simulate many of the investigations. The suggestions for differentiating instruction and modifying the investigations can support you in tailoring the review of content to meet the needs of your students.

Common Core State Standards (CCSS)

In Algebra I, students formalized the study of functions. They used function notation as they explored many examples of functions, and included more specialized functions—absolute-value, step, and those that are piecewise-defined (F.IF.7, F.BF.3). They interpreted functions given graphically, numerically, symbolically, and verbally, translated between representations, and explored the limitations of various representations. They identified the effect on the graph of $f(x)$ by changing parameters on x (F.BF.3).

In Algebra II, students synthesize and generalize what they learned about that variety of function families in Algebra I. They focus more on the effects of transformations on graphs of functions, including functions in a context, in order to understand that a particular transformation on a graph will always have the same effect regardless of the underlying function. In Algebra II, the emphasis is on the selection of a model function based on the context and the data. (F.IF.4 – 6)

Because we look for dependencies between quantities in our world, functions are important tools in the construction of mathematical models. The ability to flexibly move between verbal descriptions and mathematical representation is an essential skill in understanding those dependencies.

Students compare key features of a situation to a model for that situation, including graphs, to decide if a particular model is appropriate, developing more sophisticated models for more complex situations than before. (F.IF.7 – 9, F.BF.1) Transformations are applied to more mathematically sophisticated graphs and students look for and explain the effect of multiple transformations on a functions or a transformation across function types. (F.BF.3 – 4)

To facilitate this in-depth understanding of functions, *Discovering Advanced Algebra* explores functions through the parent functions and their transformations. In earlier chapters students looked at the parent linear function $y = x$ and transformations on linear functions. In this chapter they will look at other parent functions and transformations on those functions.

Additional Resources For complete references to additional resources for this chapter and others see your digital resources.

Refreshing Your Skills To review the algebra skills needed for this chapter see your digital resources.

Solving Equations

Projects See your digital resources for performance-based problems and challenges.

Step Functions

Explorations See your digital resources for extensions and/or applications of content from this chapter.

Rotation as a Composition of Transformations

Materials

- graph paper
- motion sensors
- geometry software
- string
- small weights
- stopwatches or watches with second hand
- tape measures or metersticks
- small mirrors

CHAPTER 3

Functions and Relations

Objectives

- Describe a graph as discrete or continuous and identify the independent and dependent variables, the intercepts, and the rates of change
- Draw a qualitative graph from a context scenario and create a context scenario given a qualitative graph
- Define *function*, *domain*, and *range*, and use function notation
- Add, subtract, multiply, and divide functions
- Distinguish conceptually and graphically between functions and relations
- Study linear, quadratic, absolute-value, square root, and semicircle ($\sqrt{1-x^2}$) families of functions
- Use $\sqrt{1-x^2}$ and piecewise-constructed functions defined over bounded intervals to explore relationships between transformations and their equations and graphs
- See how translations, reflections, stretches, and compressions of the graphs of these functions and of the unit circle affect their equations
- Determine if a function is even, odd, or neither.

Functions and relations are the rules that define our world. Most of us played with our shadow as kids. The length and size of a shadow is a function of the height and angle of the sun. The age of prehistoric artifacts is predicted by a square root function. And the weight of an object is a function of how hard gravity is pulling on the object. Your weight on Earth is not the same as your weight on the moon. The C-9 jet is one of the tools utilized by NASA to simulate the gravity, or reduced gravity, astronauts feel once they leave Earth.

OBJECTIVES

In this chapter you will
- interpret graphs of functions and relations
- review function notation
- learn to combine functions with arithmetic
- learn about the linear, quadratic, square root, absolute-value, and semicircle families of functions
- apply transformations—translations, reflections, and dilations—to the graphs of functions and relations
- transform functions to model real-world data
- determine if a function is even or odd

Functions and relations are the rules that define our world. The age of prehistoric artifacts is predicted by a square root function similar to $d = \sqrt{5t}$, where *t* is time in thousands of years and *d* is the thickness of the layer of moisture in microns.

Weight is a function of gravity. Using gravity on Earth as 1, gravity on the moon would be .166, while on Jupiter, it would be 2.14. **ASK** "How much would you weigh on the Moon? On Jupiter? (Don't tell!)" In space, without gravity pulling down on them, astronauts are "weightless." The C-9 jet flies a special parabolic pattern that creates several brief periods of reduced gravity. These reduced gravity flights are performed so astronauts, as well as researchers and their experiments, can experience the gravitational forces of the Moon and Mars and the microgravity of space.

This chapter is also about transformations. Most kids have played with shadows. **ASK** "If you held your hand between a lightbulb and the wall, how can you *transform* the size or shape of the shadow?" [different angles, different distances from light source, etc.]

LESSON 3.1

Interpreting Graphs

A picture can be worth a thousand words, if you can interpret the picture. In this lesson you will investigate the relationship between real-world situations and graphs that represent them.

A good picture book should have events that are visually arresting – the pictures should call attention to what is happening in the story.
CHRIS VAN ALLSBURG

What is the real-world meaning of the graph at right, which shows the relationship between the number of customers getting haircuts each week and the price charged for each haircut?

The number of customers depends on the price of the haircut. So the price in dollars is the independent variable and the number of customers is the dependent variable. As the price increases, the number of customers decreases linearly. As you would expect, fewer people are willing to pay a high price; a lower price attracts more customers. The slope indicates the number of customers lost for each dollar increase. The x-intercept represents the haircut price that is too high for anyone. The y-intercept indicates the number of customers when haircuts are free.

EXAMPLE

Students at Central High School are complaining that the juice vending machine is frequently empty. Several student council members decide to study this problem. They record the number of cans in the machine at various times during a typical school day and make a graph.

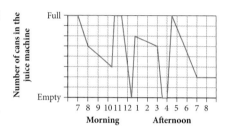

a. Based on the graph, at what times is juice consumed most rapidly?

b. When is the machine refilled? How can you tell?

c. When is the machine empty? How can you tell?

d. What do you think the student council will recommend to solve the problem?

Solution

Each horizontal segment indicates a time interval when juice does not sell. Negative slopes represent when juice is consumed, and positive slopes show when the machine is refilled.

a. The most rapid consumption is pictured by the steep, negative slopes from 11:30 A.M. to 12:30 P.M. and from 3:00 to 3:30 P.M.

b. The machine is completely refilled overnight, again at 10:30 A.M., and again just after school lets out for the day. The machine is also refilled at 12:30 P.M., but only to 75% capacity.

c. The machine is empty from 3:30 to 4:00 P.M., and briefly at about 12:30 P.M.

d. The student council might recommend refilling the machine once more at about 2:00 or 3:00 P.M. in order to solve the problem of its frequently being empty. Refilling the machine completely at 12:30 P.M. may also solve the problem.

LESSON 3.1 Interpreting Graphs   143

LESSON 3.1

This lesson reviews many aspects of representing real-world situations with graphs.

COMMON CORE STATE STANDARDS		
APPLIED	DEVELOPED	INTRODUCED
F.IF.5	F.IF.4	
F.IF.7b		
A.REI.10		

Objectives

- Identify independent and dependent variables
- Interpret features of a qualitative graph, including rates of change and x- and y-intercepts
- Decide whether a graph (or a function) is discrete or continuous when given a description of the variables
- Draw a qualitative graph from a context scenario and create a context scenario given a qualitative graph
- Distinguish between linear change and nonlinear change

Launch

Before showing the Launch found in your ebook, ASK "If you owned a hair salon, how would you determine the cost of a haircut? What would be the independent variable?" Then use the Launch.

Answers are found in the opening paragraphs. It might not be a linear relationship; no matter what the price, it seems that someone is still willing to pay it.

Investigate

Example

Project the example and have students work in pairs. Have students present their answers focusing on the features of the graph. SMP 1, 3, 4, 6

ELL
It may be helpful to relate verbal questions to mathematical expressions. For example, "When is the machine empty?" could be asked as, "When is $y = 0$?" You may also have students tell a story in their primary language.

Support
Encourage students to give detailed descriptions of graphs rather than giving a quick answer. Guide students to break down the graph into segments and to write a brief description for each part of the graph.

Advanced
Have students look ahead to the Polynomials and Rational Functions chapter or the Trigonometry and Trigonometric Functions chapter and write situations for the graphs they find there.

LESSON 3.1 Interpreting Graphs   143

Guiding the Investigation

There is a video to support this investigation. The ebook and the document Using the Videos found in your digital resources offer suggestions for using the video with the investigation. There is also a desmos simulation in your ebook.

Summarize

As students present their solutions, have them be specific as they connect the features of the graph and the shape of the bottle. Emphasize the use of correct mathematical terminology. Whether student responses are correct or incorrect, ask other students if they agree and why. **SMP 1, 3, 4, 6**

After students share their solutions to the Example, **ASK** "Does the graph indicate any other information about the school?" [Apparently students arrive at school at 7:00 in the morning; classes begin at 8:00; lunch begins at 11:30; classes let out at 3:00.] If you have time and your own school has vending machines, suggest that students sketch a graph representing their estimate of the stock in one of these machines.

ASK "How would you describe the slopes of the lines representing refills?" [The slopes are very large.]

CRITICAL QUESTION Do all stories give continuous graphs

BIG IDEA Some of the stories might describe discrete situations; help students see that in those cases continuous graphs are inappropriate.

Formative Assessment

As students work, check their understanding of real-world connections to increasing or decreasing curves and of discrete and continuous phenomena. Watch students' interpretation and understanding of curved lines, which frequently represent acceleration or deceleration, and of step functions.

Health CONNECTION

Many school districts and several states have banned vending machines and the sale of soda pop and junk foods in their schools. Proponents say that schools have a responsibility to promote good health. The U.S. Department of Agriculture already bans the sale of foods with little nutritional value, such as soda, gum, and popsicles, in school cafeterias, but candy bars and potato chips don't fall under the ban because they contain some nutrients.

Although the student council members in the example are interested in solving a problem related to juice consumption, they could also use the graph to answer many other questions about Central High School: When do students begin arriving at school? When are most students gone for the day?

Both the graph of haircut customers and the graph in the example are shown as continuous graphs. In reality, the quantity of juice in the machine can take on only discrete values, because the number of cans must be a whole number. The graph might more accurately be drawn with a series of short horizontal segments, as shown at right. The price of a haircut and the number of customers can also take on only discrete values. This graph might be more accurately drawn with separate points. However, in both cases, a continuous "graph sketch" makes it easier to see the trends and patterns.

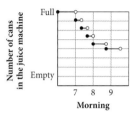

INVESTIGATION
Graph Clues

You have probably experienced filling a thin water bottle and seen the water level rise quickly or filling a wide bucket and seen the level rise slowly. What if the container changes shape? How will the water level change over time? The shape of the bottle gives clues about the graph just as a graph of water height versus time can give clues about the shape of the container.

If this bottle were filled with water at a constant rate over time, a graph of the water height over time would look like this.

144 CHAPTER 3 Functions and Relations

Modifying the Investigation

Whole Class Have students watch the video that supports this investigation. Do Step 1 as a class. Have students read Step 2 and think about it for a minute. Then use student input to sketch the graph. Have students individually draw the bottle for Step 3, and then share and discuss.

Shortened Have students watch the video that supports this investigation. Do Step 1 as a whole class activity. Have students do Steps 2 and 3. Or use the desmos tool to simulate the investigatoin.

One Step Go directly to the investigation, without introduction. Have students explain the difference between a constant rate of change and a varying rate of change, using the bottles in their description. Have them include terms like acceleration or deceleration in their descriptions.

Step 1 Match each graph to the correct vase. Write a short description of what features of the graph make this the right match for that vase.

a. ⓐ iii
b. i
c. ii

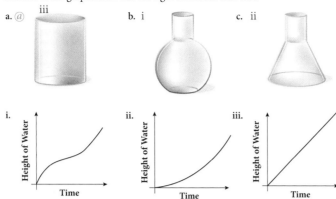

Step 2 Sketch a graph that could represent the water height over time as this crazy bottle is filled with water at a constant rate. Although the water is flowing at a constant rate, the water height does not change at a constant rate. Sometimes the height is changing quickly and sometimes it is changing slowly. Sketch the bottle next to your graph. Put letters at various heights of the bottle and then put the same letters on your graph, so the reader can see when the rate is fast and when it is slow.

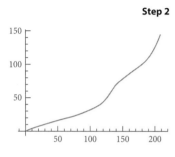

Step 3 Now reverse the process. Here is a graph of water height vs. time as a bottle is filled at a constant rate. Use what you have learned to come up with a possible shape of the bottle. Write a short explanation of what clues in the graph helped you find this shape.

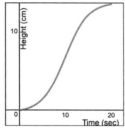

As you interpret data and graphs that show a relationship between two variables, you must always decide which is the independent variable and which is the dependent variable. You should also consider whether the variables are discrete or continuous.

Apply

Extra Example

1. Draw three examples of increasing graphs of real-world situations.

   Answers will vary.

2. Give a real-world example of a decreasing continuous graph.

   possible answer: the temperature of a cup of hot water placed in the freezer

Closing Question

Write a brief story to go with this graph. Your story should convey all of the information in the graph, including when and how the rates of change increase or decrease. Define your variables, including which is the independent variable. Label your axes.

Answers will vary.

Assigning Exercises

Suggested: 1–9
Additional Practice: 10–13

For each of their graphs, ask students to label each axis with a quantity (such as *time* or *distance*); they need not indicate numerical units. **SMP 6** The important factors are which variable is independent, the shape of the graph, and whether the graph is continuous or discrete.

Exercise 1 **ASK** "Graphs a and b are increasing. In which graphs is the rate of growth increasing?" [a and d] The rate itself is given by the slope; the rate is increasing if the slope is getting more positive or less negative. So even when the slope is negative, it can be increasing, as in 1d, from "more negative" to "less negative."

Conceptual/Procedural

Conceptual The example and Investigation emphasize the fact that graphs are representations, or models, for data. Students interpret the relationship of quantities from a graph.

Procedural Students review definitions of increasing and decreasing functions, continuous and discrete data modeled on graphs, and rate of change.

3a. decreasing at a steady rate, suddenly becoming constant, then suddenly increasing at the same rate as it was decreasing

3b. first decreasing, then increasing back to the same level, without any sudden changes in rate

3c. rapidly increasing from 0; suddenly changing to rapidly decreasing, until half the value is reached; constant, then suddenly rapidly decreasing at a constant rate until reaching 0

4a. Possible answer: The curve might describe the relationship between the amount of time the ball is in the air and how far away from the ground it is.

4c. possible answer: domain: $0 \leq t \leq 10$ s; range: $0 \leq h \leq 70$ yd

Exercise 5 Although all of these situations are continuous, it's good to ask students whether the phenomenon is continuous or discrete. **ALERT** The graph of time versus distance would be curved, but this asks for time versus speed. The acceleration due to gravity is constant, so the speed increase is linear.

5a. Time in seconds is the independent variable; the height of the ball in feet is the dependent variable.

5b. The car's speed in miles per hour is the independent variable; the braking distance in feet is the dependent variable.

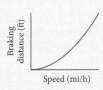

5c. Time in minutes is the independent variable; the drink's temperature in degrees Fahrenheit is the dependent variable

5d. Time in seconds is the independent variable; the acorn's speed in feet per second is the dependent variable.

146 CHAPTER 3 Functions and Relations

3.1 Exercises

You will need a graphing calculator for Exercise **12**.

Practice Your Skills

1. Match a description to each graph.

   a. i b. iii c. iv d. ii

   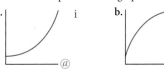 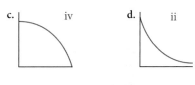

   i. increasing more and more rapidly
   ii. decreasing more and more slowly
   iii. increasing more and more slowly
   iv. decreasing more and more rapidly

   2a.   2b.   2c.

2. Sketch a graph to match each description.
   a. increasing throughout, first slowly and then at a faster rate
   b. decreasing slowly, then more and more rapidly, then suddenly becoming constant @
   c. alternately increasing and decreasing without any sudden changes in rate

3. For each graph, write a description like those in Exercise 2.

   a.   b. 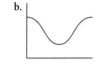 c.

Reason and Apply

4. Harold's concentration often wanders from the game of golf to the mathematics involved in his game. His scorecard frequently contains mathematical doodles and graphs.
   a. What is a real-world meaning for this graph found on one of his recent scorecards?
   b. What units might he be using?   possible answer: seconds and yards
   c. Describe a realistic domain and range for this graph.
   d. Does this graph show how far the ball traveled? Explain.
   No, the horizontal distance traveled is not measured.

5. Sketch what you think is a reasonable graph for each relationship described. In each situation, identify the variables and label your axes appropriately.
   a. the height of a ball during a game of catch with a small child
   b. the distance it takes to brake a car to a full stop, compared to the car's speed when the brakes are first applied @
   c. the temperature of an iced drink as it sits on a table for a long period of time @
   d. the speed of a falling acorn after a squirrel drops it from the top of an oak tree (h)
   e. your height above the ground as you ride a Ferris wheel

146 CHAPTER 3 Functions and Relations

5e. Time in minutes is the independent variable; your height above the ground in feet is the dependent variable.

6. Sample answer: Zeke, the fish, swam slowly, then more rapidly to the bottom of his bowl and stayed there for a while. When Zeke's owner sprinkled fish food into the water, Zeke swam toward the surface to eat. The y-intercept is the fish's depth at the start of the story. The x-intercept represents the time the fish reached the surface of the bowl.

6. Make up a story to go with the graph at right. Be sure to interpret the *x*- and *y*-intercepts. ⓗ

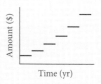

7. Sketch what you think is a reasonable graph for each relationship described. In each situation, identify the variables and label your axes appropriately. In each situation, will the graph be continuous or will it be a collection of discrete points or pieces? Explain why.

   a. the amount of money you have in a savings account that is compounded annually, over a period of several years, assuming no additional deposits are made

   b. the same amount of money that you started with in 7a, hidden under your mattress over the same period of several years

   c. an adult's shoe size compared to the adult's foot length ⓐ

   d. the price of gasoline at the local station every day for a month

   e. the daily maximum temperature of a town for a month

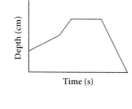

8. Describe a relationship of your own and draw a graph to go with it.

9. Car A and Car B are at the starting line of a race. At the green light, they both accelerate to 60 mi/h in 1 min. The graph at right represents their velocities in relation to time.

   a. Describe the rate of change for each car. ⓗ

   b. After 1 minute, which car will be in the lead? Explain your reasoning.

Review

10. Write an equation for the line that fits each situation.

    a. The length of a rope is 1.70 m, and it decreases by 0.12 m for every knot that is tied in it. Let l represent the length of the rope in meters, and let k represent the number of knots; $l = 1.70 - 0.12k$.

    b. Matt bought an app for his phone for $5.99 that allows for 3 months of free play. He then pays a subscription of $0.99 per month to play the game. Let s represent the amount spent on the game, and let m represent the number of months Matt plays the game; $s = 5.99 + 0.99(m - 3)$ where $m > 3$.

11. **APPLICATION** Albert starts a business reproducing high-quality copies of pictures. It costs $155 to prepare the picture and then $15 to make each print. Albert plans to sell each print for $27. ⓐ

    a. Write a cost equation and graph it.

    b. Write an income equation and graph it on the same set of axes.

    c. How many pictures does Albert need to sell before he makes a profit? 13 pictures

    d. What do the graphs tell you about the income and the cost for eight pictures?

    The income, $216, is less than the cost, $275.

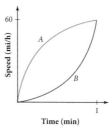

7a. Time in years is the independent variable; the amount of money in dollars is the dependent variable. The graph will be a series of discontinuous segments.

7b. Time in years is the independent variable; the amount of money in dollars is the dependent variable. The graph will be a continuous horizontal segment, because the amount never changes.

7c. Foot length in inches is the independent variable; shoe size is the dependent variable. The graph will be a series of discontinuous horizontal segments, because shoe sizes are discrete.

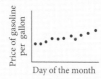

7d. Answers will vary but should be in the form of a discrete graph.

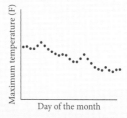

7e. The day of the month is the independent variable; the maximum temperature in degrees Fahrenheit is the dependent variable. The graph will be discrete points, because there is just one temperature reading per day.

9a. Car A speeds up quickly at first and then less quickly until it reaches 60 mi/h. Car B speeds up slowly at first and then quickly until it reaches 60 mi/h.

9b. Car A will be in the lead because it is always going faster than Car B, which means it has covered more distance.

Exercise 9 The goal of 9a is to relate the slopes of the curves to the rates of change. **ALERT** In 9b, students might believe that if the cars reached the same speed in the same amount of time, then the car traveled the same distance. The distance traveled by each car is given by the area of the region between its graph and the horizontal axis.

11a. Let x represent the number of pictures, and let y represent the amount of money (either cost or income) in dollars; $y = 155 + 15x$.

11b. $y = 27x$

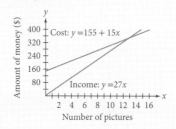

8. sample answer: the cost of parking your car at a lot that charges a certain fixed price for up to an hour and then half as much for each additional hour or fraction thereof

Exercise 13 The choice of using elimination with Equations 1 and 2 and then with Equations 1 and 3 to eliminate z is not the only solution method. You might ask students whether they can think of other approaches. [They can use any two pairs of equations first to eliminate any one variable.] ASK "Why do you need to start by eliminating one variable?" [so you have two equations in two variables that can be solved by either substitution or elimination]

12. **APPLICATION** Suppose you have a $200,000 home loan with an annual interest rate of 6.5%, compounded monthly.

    a. If you pay $1,200 per month, what balance remains after 20 years? @ $142,784.22

    b. If you pay $1,400 per month, what balance remains after 20 years? $44,700.04

    c. If you pay $1,500 per month, what balance remains after 20 years? $0 (You actually pay off the loan after 19 yr 10 mo.)

    d. Make an observation about the answers to 12a–c.

    By making an extra $300 payment per month for 20 yr, or $72,000, you save hundreds of thousands of dollars in the long

13. Follow these steps to solve this system of three equations in three variables.

    $$\begin{cases} 2x + 3y - 4z = -9 & \text{(Equation 1)} \\ x + 2y + 4z = 0 & \text{(Equation 2)} \\ 2x - 3y + 2z = 15 & \text{(Equation 3)} \end{cases}$$

    a. Use the elimination method with Equation 1 and Equation 2 to eliminate z. The result will be an equation in two variables, x and y. @ $3x + 5y = -9$

    b. Use the elimination method with Equation 1 and Equation 3 to eliminate z. @ $6x - 3y = 21$

    c. Use your equations from 13a and b to solve for both x and y. $x = 2, y = -3$

    d. Substitute the values from 13c into one of the original equations and solve for z. What is the solution to the system? $x = 2, y = -3, z = 1$

PERFORMANCE TASK

Every graph tells a story. Make a graph to go with the story in Part 1. Then invent your own story to go with the graph in Part 2.

Part 1

Sketch a graph that reflects all the information given in this story.

"It was a dark and stormy night. Before the torrents of rain came, the bucket was empty. The rain subsided at daybreak. The bucket remained untouched through the morning until Old Dog Trey arrived as thirsty as a dog. The sun shone brightly through the afternoon. Then Billy, the kid next door, arrived. He noticed two plugs in the side of the bucket. One of them was about a quarter of the way up, and the second one was near the bottom. As fast as you could blink an eye, he pulled out the plugs and ran away."

Part 2

Invent your own story to go with this graph that conveys all of this information, including when and how the rates of change increase or decrease.

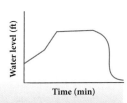

148 CHAPTER 3 Functions and Relations

PERFORMANCE TASK

Part 1 sample answer:

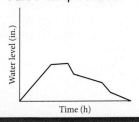

Part 2 Sample answer: Luis and Loretta have a small outdoor swimming pool. The children want to use the pool, but the water level is very low, so Luis turns on the hose and begins filling the pool at a constant rate. The children are restless and persuade Luis to increase the water flow and fill the pool faster. It fills at a faster constant rate than before. When the pool is completely full, he turns off the hose, and the children are careful not to splash water out of the pool. After the children get out of the pool, Luis empties it. The water pours out rapidly at first, then more slowly as there is less and less water left. He leaves just a little water in the bottom, which will slowly evaporate.

LESSON 3.2

Function Notation

> She had not understood mathematics until he had explained to her that it was the symbolic language of relationships. "And relationships," he had told her, "contained the essential meaning of life."
>
> PEARL S. BUCK
> THE GODDESS ABIDES, 1972

Rachel's parents keep track of her height as she gets older. They plot these values on a graph and connect the points with a smooth curve. Why do they connect the points? Do you think you should? For every age you choose on the x-axis, there is only one height that pairs with it on the y-axis. That is, Rachel is only one height at any specific time during her life.

Roger is walking away from a motion detector at a steady rate. His distance from the sensor over time is represented by a continuous line. For every time you choose on the x-axis, is there only one distance from the sensor? Can Roger be in more than one place at a time?

A **relation** is any relationship between two variables. A **function** is a special type of relation such that for every value of the independent variable, there is at most one value of the dependent variable. If x is your independent variable, a function pairs at most one y with each x. You can say that Rachel's height is a function of her age and Roger's distance from the sensor is a function of time.

You may remember the vertical line test from previous mathematics classes. It helps you determine whether or not a graph represents a function. If no vertical line crosses the graph more than once, then the relation is a function.

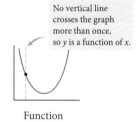
No vertical line crosses the graph more than once, so y is a function of x.

Function

Because a vertical line crosses the graph more than once, y is *not* a function of x.

Not a function

Take a minute to think about how you could apply this technique to the graph of Rachel's height, Roger's walk, and the graph in the next example.

LESSON 3.2

This lesson on function notation, evaluating functions, and the vertical line test may be review for many students.

COMMON CORE STATE STANDARDS		
APPLIED	**DEVELOPED**	**INTRODUCED**
A.REI.10	F.BF.1b	
F.IF.1		
F.IF.2		

Objectives

- Distinguish between functions and relations
- Review definitions of *function*, *domain* and *range*
- Review function notation
- Review the vertical line test for functions
- Learn to combine functions with arithmetic

Vocabulary

relation
function
function notation
zero of the function

Materials

- Calculator Note: Function Notation

Launch

If x is a member of the domain and y is a member of the range, which of the following relations are functions? Explain.

a. x city has y zip code
b. x state has y capital
c. x is friends with y.
d. x is y feet tall.

a. no, large cities have more than one zip code;
b. yes, each state has one capital;
c. no, a person can have many friends;
d. yes, each person has only one height

ELL

Focusing on the vertical line test to determine whether a graph represents a function provides students with a visual connection between the graphs and the definition of *function*. Have students give examples of functional relationships from their own experience. **ASK** "What are some x values for the scenarios? What is the dependent variable?" This will help students understand the connection between the independent variable and function terminology.

Support

Using colors when substituting values of x into the function, as shown in the solution to $f(8)$, may help students understand the process of evaluating functions. Give students multiple examples of functional relationships from their own experience. **ASK** "What are some x values for the scenarios? What is the dependent variable?" This will help students understand the connection between the independent variable and function terminology.

Advanced

Have students create their own function and nonfunction graphs and then ask them to find a mathematical model that will produce the graph.

Investigate

Example A

Project Example A and have students work in pairs. Have pairs present one of a–f, describing their strategy as well as their solution. Emphasize the use of correct mathematical terminology. **SMP 1, 3, 6**

ASK "What is happening to the graph of $f(x)$ when $x = 3$?" [Evaluating $\frac{2x + 5}{x - 3}$ at $x = 3$ would require dividing by 0, so the value is undefined.] This observation can lead to a discussion about the domain of $f(x)$.

For the presentation of part e, if students don't use the terminology *zero of the function*, **ASK** "What is $g(x) = 0$ called?"

ALERT Parts e and f reverse the question and asks for input values instead of output values. This switch may confuse some students.

Function notation emphasizes the dependent relationship between the variables that are used in a function. The notation $y = f(x)$ indicates that values of the dependent variable, y, are explicitly defined in terms of the independent variable, x, by the function f. You read $y = f(x)$ as "y equals f of x."

Graphs of functions and relations can be continuous, such as the graph of Rachel's height or the motion sensor graph of Roger's walk, or they can be made up of discrete points, such as a graph of the maximum temperatures for each day of a month. Although real-world data often have an identifiable pattern, a function does not necessarily need to have a rule that connects the two variables.

EXAMPLE A Function f is defined by the equation $f(x) = \dfrac{2x + 5}{x - 3}$.
Function g is defined by the graph at right.

Find these values.

a. $f(8)$
b. $f(-7)$
c. $g(1)$
d. $g(-2)$
e. Find x when $g(x) = 0$.
f. Find x when $g(x) = 2$

Solution When a function is defined by an equation, you simply replace each x with the x-value and evaluate.

a. $f(x) = \dfrac{2x + 5}{x - 3}$

$f(8) = \dfrac{2 \cdot 8 + 5}{8 - 3} = \dfrac{21}{5} = 4.2$

b. $f(-7) = \dfrac{2 \cdot (-7) + 5}{-7 - 3} = \dfrac{-9}{-10} = 0.9$

c. The notation $y = g(x)$ tells you that the values of y are explicitly defined, in terms of x, by the graph of the function g. To find $g(1)$, locate the value of y when x is 1. The point $(1, 3)$ on the graph means that $g(1) = 3$.

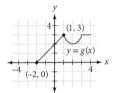

d. The point $(-2, 0)$ on the graph means that $g(-2) = 0$.

e. To find x when $g(x) = 0$, called a **zero of the function**, locate points on the graph with a y-value of 0. There is only one, at $(-2, 0)$, so $x = -2$ when $g(x) = 0$.

f. To find x when $g(x) = 2$, locate points on the graph with a y-value of 2. There are two, at $(0, 2)$ and $(2, 2)$ so $x = 0, 2$ when $g(x) = 2$.

In the investigation you will practice identifying functions and using function notation. As you do so, notice how you can identify functions in different forms.

Modifying the Investigation

Whole Class Display a–i for the class. Classify each as a function or not, with student input. Discuss students' reasoning and Step 2.

Shortened Skip parts c, f, and i.

One Step Pose this problem: "Make a table and a graph of the ages and heights of at least 20 students in this class. Is height a function of age—that is, for every age is there just one height? Is age a function of height?" Be sensitive to students who might be self-conscious about their height. A measurement is not needed from every student. During the discussion, bring out the ideas of the vertical line test and stress that not being one-to-one doesn't mean that a relation isn't a function.

INVESTIGATION

To Be or Not to Be (a Function)

Below are nine representations of relations.

a. 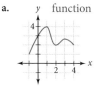 function

b.

c. 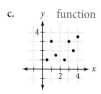 function

d. 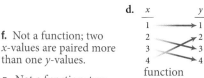 function

e. function

f.

g. independent variable: the age of each student in your class
dependent variable: the height of each student

h. independent variable: an automobile in the state of Kentucky
dependent variable: that automobile's license plate number

i. independent variable: the day of the year
dependent variable: the time of sunset

Step 1 Identify each relation that is also a function. For each relation that is not a function, explain why not.

Step 2 For each graph or table that represents a function in parts a–f, find the y-value when $x = 2$, and find the x-value(s) when $y = 3$. Write each answer in function notation using the letter of the subpart as the function name. For example, if graph a represents a function, $a(2) = \underline{?}$ and $a(\underline{?}) = 3$.

When you use function notation to refer to a function, you can use any letter you like. For example, you might use $y = h(x)$ if the function represents height, or $y = p(x)$ if the function represents population. Often in describing real-world situations, you use a letter that makes sense. However, to avoid confusion, you should avoid using the independent variable as the function name, as in $y = x(x)$. Choose freely but choose wisely.

Example B
This example is a review to highlight what is given in function notation and what is the unknown. Project Example B and ask for student input. **SMP 1, 2, 3, 6**

Example C
This example introduces combining functions using arithmetic operations. Consider projecting Example C and having students work in pairs. Some students may find $(f-g)(5)$ by creating a new function $(f-g)(x) = 2x + 1 - (x - 3) = x + 4$, and then substitute 5 for x, as the book does. Some students may substitute 5 for x in $f(x)$ and in $g(x)$, and then subtract. That leads into the "rules" in the blue box. **SMP 1, 3, 6**

Summarize

As students present their solutions to the investigation and examples, have them explain their strategy along with their answer. Whether student responses are correct or incorrect, ask other students if they agree and why. **SMP 1, 2, 3, 6**

You might introduce the term *one-to-one* to describe a function that has not only one y-value for every x-value but also one x-value for every y-value. This will be addressed formally later. A function might not be expressible as a rule, either mathematically or verbally. Sequence notation for the nth term, u_n, can be thought of as a modified function notation. You could replace u_n with $u(n)$, which is the way many calculators display the notation.

ALERT Students may think that $f(x)$ means f times x and want to divide by x or f to simplify the equation. As needed, point out that $f(x)$ is an expression in itself and cannot be separated into parts.

CRITICAL QUESTION Does the domain of a relation affect whether it's a function? Explain.

BIG IDEA To be a function, a relation's graph must pass the vertical line test. If even one point in the domain corresponds to more than one point in the range, then the relation is not a function.

EXAMPLE B If $h(t)$ is the height in centimeters of water in a vase as it is filled at a constant rate over time in seconds, what is the given and what is the unknown represented in each of the following situations?

a. $h(2)$

b. $h(t) = 5$

Solution

a. In this case, the given is the time of 2 seconds and the unknown is the height of the water after 2 seconds.

b. Here, $h(t) = 5$ tells you the height of the water is 5 centimeters and the unknown is when the water will reach this height.

Functions can be combined in many ways to create new functions. When they are combined using simple arithmetic operations, the notation for these new functions shows how this is to be done.

> **Combining Functions with Arithmetic**
>
> $(f + g)(x) = f(x) + g(x)$ $(f - g)(x) = f(x) - g(x)$
>
> $(f \cdot g)(x) = f(x) \cdot g(x)$ $\left(\dfrac{f}{g}\right)(x) = \dfrac{f(x)}{g(x)}$

EXAMPLE C Suppose $f(x)$ and $g(x)$ are defined as $f(x) = 2x + 1$ and $g(x) = x - 3$. Find the new function and evaluate the following:

a. $(f - g)(5)$

b. $(f \cdot g)(-1)$

c. $\left(\dfrac{f}{g}\right)(5)$

Solution

a. First create the new function $(f - g)(x)$ by subtracting $g(x)$ from $f(x)$ and combining like terms.

$(f - g)(x) = 2x + 1 - (x - 3) = x + 4$.

Then evaluate this new function at $x = 5$.

$(f - g)(5) = 5 + 4 = 9$.

b. First create the new function $(f \cdot g)(x)$ by multiplying $f(x)$ and $g(x)$.

$(f \cdot g)(x) = (2x + 1)(x - 3)$.

You can expand this to get $(f \cdot g)(x) = 2x^2 - 5x - 3$.

Then evaluate this new function at $x = -1$.

$(f \cdot g)(-1) = 2(-1)^2 - 5(-1) - 3 = 2 + 5 - 3 = 4$.

CRITICAL QUESTION What if one point in the range corresponds to more than one point in the domain?

BIG IDEA The relation could be a function, but it's not a one-to-one function. To be one-to-one, a function's graph must pass a horizontal line test.

CRITICAL QUESTION Does the definition of *function* require that there be only one value of x for each value of y? Explain.

BIG IDEA No; the graph need not pass a horizontal line test.

c. First create the new function $\left(\dfrac{f}{g}\right)(x)$ by dividing $f(x)$ by $g(x)$.

$\left(\dfrac{f}{g}\right)(x) = \dfrac{(2x+1)}{(x-3)}$

Then evaluate this new function at $x = 5$.

$\left(\dfrac{f}{g}\right)(5) = \dfrac{(2(5)+1)}{5-3} = \dfrac{11}{2} = 5.5$

When looking at real-world data, it is often hard to decide whether or not there is a functional relationship. For example, if you measure the height of every student in your class and the weight of his or her backpack, you may collect a data set in which each student height is paired with only one backpack weight. But does that mean no two students of the same height could have backpacks of different weights? Does it mean you shouldn't try to model the situation with a function?

Two students of the same height could have different backpack weights, so you could not write that as a function. You could write a function that might predict the average weight as a function of student height, using a line of fit that approximately models the relationship.

3.2 Exercises

You will need a graphing calculator for Exercise 11.

Practice Your Skills

1. Which of these graphs represent functions? Why or why not?

   a. 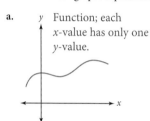 Function; each x-value has only one y-value.

   b. 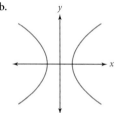 Not a function; there are x-values that are paired with two y-values.

   c. 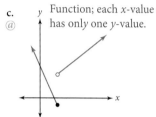 Function; each x-value has only one y-value.

2. Use the functions $f(x) = 3x - 4$ and $g(x) = x^2 + 2$ to find these values.

   a. $f(7)$ 17
   b. $g(5)$ 27
   c. $f(-5)$ -19
   d. $g(-3)$ @ 11
   e. x when $f(x) = 7$ @ $\dfrac{11}{3}$
   f. $(f+g)(x)$ $x^2 + 3x - 2$
   g. $(f-g)(2)$ -4
   h. $(f \cdot g)(-1)$ @ -21
   i. $\left(\dfrac{f}{g}\right)(0)$ @ -2

3. Miguel works at an appliance store. He gets paid $7.25 an hour and works 8 hours a day. In addition, he earns a 3% commission on all items he sells. Let x represent the total dollar value of the appliances that Miguel sells, and let the function m represent Miguel's daily earnings as a function of x. Which function describes how much Miguel earns in a day? ii

   i. $m(x) = 7.25 + 0.03x$
   ii. $m(x) = 58 + 0.03x$
   iii. $m(x) = 7.25 + 3x$
   iv. $m(x) = 58 + 3x$

Exercise 5 Students could logically argue for opposite choices of the independent variable. For example, in 5d, how far you drive might depend on the amount of gas. Most important is students' understanding of the process of choosing an independent variable.

5a. The price of the calculator is the independent variable; function.

5b. The time the money has been in the bank is the independent variable; function.

5c. The amount of time since your last haircut is the independent variable; function.

5d. The distance you have driven since your last fill-up is the independent variable; function.

6a. Let x represent the price of the calculator in dollars, and let y represent the sales tax in dollars.

6b. Let x represent the time in months, and let y represent the account balance in dollars.

6c. Let x represent the time in days, and let y represent the length of your hair.

6d. Let x represent the distance x you have driven in miles, and let y represent the amount of gasoline in your tank in gallons.

Exercise 8 ASK "What is a real-world situation that could be represented by this function?" In 8d, students may need to extend their graphs to show the point where x is negative.

4. Use the graph at right to find each value. Each answer will be an integer from 1 to 26. Relate each answer to a letter of the alphabet (1 = A, 2 = B, and so on), and fill in the name of a famous mathematician.

   a. $f(13)$ $18 = R$ b. $f(25) + f(26)$ $5 = E$

   c. $2f(22)$ $14 = N$ d. $\dfrac{f(3) + 11}{\sqrt{f(3+1)}}$ $5 = E$

   e. $\dfrac{f(1+4)}{f(1)+4} - \dfrac{1}{4}\left(\dfrac{4}{f(1)}\right)$ f. x when $f(x+1) = 26$ $5 = E$

   $\ 4 = D$

   g. $\sqrt[3]{f(21)} + f(14)$ h. x when $2f(x+3) = 52$ $3 = C$

   $\ 19 = S$

   i. x when $f(2x) = 41$ $1 = A$ j. $f(f(2) + f(3))$ $18 = R$

   k. $f(9) - f(25)$ $20 = T$ l. $f(f(5) - f(1))$ $5 = E$

   m. $f(4 \cdot 6) + f(4 \cdot 4)$ $19 = S$

R	e	n	e		D	e	s	c	a	r	t	e	s
a	b	c	d		e	f	g	h	i	j	k	l	m

5. Identify the independent variable for each relation. Is the relation a function?

   a. the price of a graphing calculator and the sales tax you pay

   b. the amount of money in your savings account and the time it has been in the account

   c. the amount your hair has grown since the time of your last haircut

   d. the amount of gasoline in your car's fuel tank and how far you have driven since your last fill-up

6. Sketch a reasonable graph for each relation described in Exercise 5. In each situation, identify the variables and label your axes appropriately. As you create your graph, consider what is an appropriate domain and range for each function.

Reason and Apply

7. Suppose $h(x) = x^2 + 5x$ and $k(x) = 6x + 8$. Find functions $f(x)$ and $g(x)$ for each of the following to make a true statement.   Answers will vary. Samples answers:

   a. $h(x) = (f + g)(x)$ $f(x) = x^2$, $g(x) = 5x$ b. $h(x) = (f \cdot g)(x)$ $f(x) = x$, $g(x) = 5 + x$

   c. $k(x) = (f - g)(x)$ $f(x) = 10x + 10$, $g(x) = 4x + 2$ d. $k(x) = \left(\dfrac{f}{g}\right)(x)$ $f(x) = 6x^2 + 8x$, $g(x) = x$

8. Suppose $f(x) = 25 - 0.6x$.

   a. Draw a graph of this function.   **8a, c, d.**

   b. What is $f(7)$?   20.8

   c. Identify the point $(7, f(7))$ by marking it on your graph.

   d. Find the value of x when $f(x) = 27.4$. Mark this point on your graph. -4

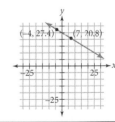

154 CHAPTER 3 Functions and Relations

9. Identify the domain and range of the function g in the graph at right. ⓐ
    domain: $-6 \leq x \leq 5$; range: $-2 \leq y \leq 4$

10. Sketch a graph for each function.
    a. $y = f(x)$ has domain all real numbers and range $f(x) \leq 0$.
    b. $y = g(x)$ has domain $x > 0$ and range all real numbers.
    c. $y = h(x)$ has domain all real numbers and range $h(x) = 3$.

11. Consider the function $f(x) = 3(x + 1)^2 - 4$.
    a. Find $f(5)$. 104
    b. Find $f(n)$. $f(n) = 3(n + 1)^2 - 4$
    c. Find $f(x + 2)$. $f(x + 2) = 3(x + 3)^2 - 4$
    d. Use your calculator to graph $y = f(x)$ and $y = f(x + 2)$ on the same axes. How do the graphs compare? ⓐ

12. Kendall walks toward and then away from a motion sensor.
    a. Is the (*time*, *distance*) graph of his motion a function? Why or why not?
    b. Is the graph continuous or discrete? continuous

13. **APPLICATION** The length of a pendulum in inches, L, is a function of its period, or the length of time it takes to swing back and forth, in seconds, t. The function is defined by the formula $L(t) = 9.73 t^2$.
    a. Find $L(4)$. 155.68 in
    b. What does $L(4)$ mean in terms of pendulum length and period? the length of a pendulum if its period is 4 s.
    c. The Foucault pendulum at the Panthéon in Paris has a 62-pound iron ball suspended on a 220-foot wire. What is its period? approximately 16.5 s

Astronomer Jean Bernard Leon Foucault (1819–1868) displayed this pendulum for the first time in 1851. The floor underneath the swinging pendulum was covered in sand, and a pin attached to the ball traced the pendulum's path. While the ball swung back and forth in straight lines, it changed direction relative to the floor, proving that Earth was rotating underneath it.

14. The number of diagonals of a polygon, d, is a function of the number of sides of the polygon, n, and is given by the formula $d(n) = \dfrac{n(n-3)}{2}$.
    a. Find the number of diagonals in a dodecagon (a 12-sided polygon). 54 diagonals
    b. How many sides would a polygon have if it contained 170 diagonals? ⓗ 20 sides
    c. Write the answers you gave in parts a and b using function notation. $d(12) = \dfrac{12(12-3)}{2} = 54$; $170 = \dfrac{n(n-3)}{2}$, $n = 20$
    d. Find n when $d(n) = 170$. $n = 20$
    e. What does your answer to part c mean in terms of the number of diagonals and the number of sides of a polygon? n represents the number of sides, $d(n)$ represents the number of diagonals; number of diagonals is a function of the number of sides

LESSON 3.2 Function Notation **155**

Exercise 9 Students may wonder whether the graph continues beyond what is drawn. Point out that when a question asks about the domain of a function and only the graph is given, students can assume that the entire graph is showing.

10a. possible answer:

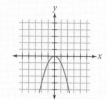

10b. possible answer:

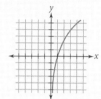

10c.

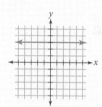

Exercise 11 ALERT Students might be confused by 11b and 11c. They need only replace x with the letter or expression. Expanding or simplifying is unnecessary. In 11d, they can graph on a calculator without squaring $(x + 2)$.

11d.

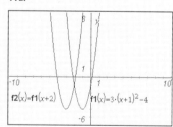

The graphs are the same shape. The graph of $f(x + 2)$ is shifted 2 units to the left of the graph of $f(x)$.

Exercise 12 If students have not used a motion sensor, you may want to give them a brief explanation. If you do have access to a motion sensor, you should demonstrate it here.

12a. Let x represent the time since Kendall started moving, and let y represent his distance from the motion sensor. The graph is a function; Kendall can be at only one position at each moment in time, so there is only one y-value for each x-value.

LESSON 3.2 Function Notation **155**

Exercise 13 Students might think that the period is a function of the length rather than the other way around. Either way is legitimate, because the function is one-to-one if the domain is limited to nonnegative values of t. In 13c, the weight of the ball is unneeded information.

15a.

15b.

15c.

16. Sample answer: Eight students fall into each quartile. Assuming that the mean of each quartile is the midpoint of the quartile, the total will be 8(3.075 + 4.500 + 5.875 + 9.150), or $180.80.

18a. possible answer:

18b. possible answer:
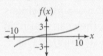

18c. possible answer:

Language CONNECTION

You probably have noticed that some words, like biannual, triplex, and quadrant, have prefixes that indicate a number. Knowing the meaning of a prefix can help you determine the meaning of a word. The word "polygon" comes from the Greek *poly-* (many) and *-gon* (angle). Many mathematical words use the following Greek prefixes.

1 mono	6 hexa	
2 di	7 hepta	
3 tri	8 octa	
4 tetra	9 ennea	
5 penta	10 deca	20 icosa

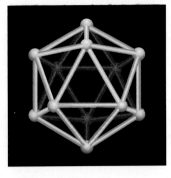

A polyhedron is a three-dimensional shape with many sides. Can you guess what the name of this shape is, using the prefixes given?

Review

15. For each of the following, sketch a graph of the walker's distance from the motion sensor over time.

  a. Started at 2 feet from the sensor, walked away slowly, and gradually sped up for 4 seconds.

  b. Started 8 feet from the sensor, walked quickly toward the sensor, and gradually slowed down over 4 seconds.

  c. Started 1 foot from the sensor, walked away for 2 seconds at a constant rate, and then walked toward the sensor for 2 seconds at a constant rate.

16. APPLICATION The five-number summary of this box plot is $2.10, $4.05, $4.95, $6.80, $11.50. The plot summarizes the amounts of money earned in a recycling fund drive by 32 members of the Oakley High School environmental club. Estimate the total amount of money raised. Explain your reasoning. @

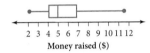

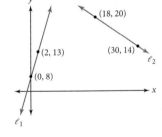

17. Given the graph at right, find the intersection of lines ℓ_1 and ℓ_2. (7, 25.5)

18. Sketch a graph for a function that has the following characteristics.

  a. domain: $x \geq 0$
  range: $f(x) \geq 0$
  linear and increasing

  b. domain: $-10 \leq x \leq 10$
  range: $-3 < f(x) \leq 3$
  nonlinear and increasing

  c. domain: $x \geq 0$
  range: $-2 < f(x) \leq 10$
  increasing, then decreasing, then increasing, and then decreasing

19. You can use rectangle diagrams to represent algebraic expressions. For instance, this diagram demonstrates the equation $(x + 5)(2x + 1) = 2x^2 + 11x + 5$. Fill in the missing values on the edges or in the interior of each rectangle diagram. Complete the equivalence statements for the equations demonstrated.

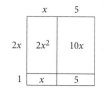

a.

$(x + 7)(x + 3) = \underline{x^2} + \underline{10x} + \underline{21}$

b.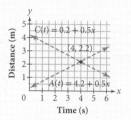

$(\underline{x} + \underline{2})(\underline{x} + \underline{1}) = x^2 + 3x + 2$

c.

$(\underline{x} + \underline{10})(2x + \underline{10}) = 2x^2 + 30x + 100$

d.

$(x + \underline{2})(\underline{3x} + \underline{4}) = 3x^2 + \underline{10x} + \underline{8}$

20. Alice and Carlos are each recording Bao's distance from where they stand. Initially Bao is between Alice and Carlos, standing 0.2 m from Alice and 4.2 m from Carlos. He walks at 0.5 m/s away from Alice and toward Carlos.

   a. On the same axes, sketch graphs of Bao's distance from each student as a function of time.

   b. Write an equation for each graph. $A(t) = 0.2 + 0.5t$; $C(t) = 4.2 - 0.5t$

   c. Find the intersection of the graphs and give the real-world meaning of that point. (4, 2.2); After 4 s, Bao is 2.2 m from both Alice and Carlos.

IMPROVING YOUR Reasoning SKILLS

The Dipper

The group of stars known as the Big Dipper, which is part of the constellation Ursa Major, contains stars at various distances from Earth. Imagine translating the Big Dipper to a new position. Would all of the stars need to be moved the same distance? Why or why not?

Now imagine rotating the Big Dipper around the Earth. Do all the stars need to be moved the same distance? Why or why not?

LESSON 3.3

In this lesson students see how equations of lines change as the lines are translated. Although this lesson can be done with pencil and paper, graphing technology will greatly enhance students' learning.

COMMON CORE STATE STANDARDS		
APPLIED	DEVELOPED	INTRODUCED
	F.IF.7	F.BF.3

Objectives

- Describe translations of a line in terms of horizontal and vertical shifts
- Write the equation of a translated line using h and k
- Understand point-slope form as a translation of the line with its equation written in intercept form
- Apply translations to functions
- Apply and identify translations to piecewise-defined functions

Vocabulary

intercept form
point-slope form
translation
image
standard form

Materials

- motion sensors
- graph paper
- Calculator Note: Movin' Around

Launch

Graph the lines represented by these equations:

$2x + y = 2$

$y - 4 = 2(x - 3)$

Compare the slopes and intercepts of the two lines. Could they be considered translations of each other? Explain.

LESSON 3.3

Lines in Motion

If everyone is moving forward together, then success takes care of itself.

HENRY FORD

In linear modeling you worked with two forms of linear equations:

Intercept form	$y = a + bx$
Point-slope form	$y = y_1 + b(x - x_1)$

In this lesson you will see how these forms are related to each other graphically.

With the exception of vertical lines, lines are graphs of functions. That means you could write the forms above as $f(x) = a + bx$ and $f(x) = f(x_1) + b(x - x_1)$.

The investigation will help you see the effect that moving the graph of a line has on its equation. Moving a graph horizontally or vertically is called a **translation.** The discoveries you make about translations of lines will also apply to the graphs of other functions.

Skateboarding bowls were inspired by the skateboarding culture of the 70s, when empty swimming pools were ridden to emulate the movements of surfing.

INVESTIGATION

Movin' Around

YOU WILL NEED
- two motion sensors
- graph paper

In this investigation you will interpret data from an experiment and then confirm your findings. First, your group will use motion sensors to create a function and a translated copy of that function.

Step 1 Arrange your group as in the photo to collect data.

ELL
Do not skip this investigation. Discuss the Language Connection. Determine whether students can explain the analogy in their own words, in either English or their primary language.

Support
Students might find the $y = y_1 + b(x - x_1)$ horizontal translation of x_1 to be counterintuitive. If so, have students substitute various values for x_1 and observe the effect on the graph. If necessary, remind students that (x_1, y_1) is a point on the line. You could also have students think in terms of $y - y_1 = b(x - x_1)$ as that expression makes the horizontal and vertical translations consistent.

Advanced
Have students examine translations of lines in intercept, point-slope, and standard form so that they can see how the translation differs among the various forms. Have students to find a way to get the slope from an equation in standard form without having to change the equation to intercept form.

Step 4

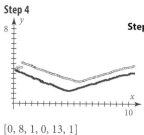

[0, 8, 1, 0, 13, 1]

Step 2 Person D coordinates the collection of data like this:

At 0 seconds:	C begins to walk slowly toward the motion sensors, and A begins to collect data.
About 2 seconds:	B begins to collect data.
About 5 seconds:	C begins to walk backward.
About 10 seconds:	A's sensor stops.
About 12 seconds:	B's sensor stops and C stops walking.

Step 3 After collecting the data, transmit four lists of data to each group member's calculator. Be sure to keep track of which data each list contains.

Step 4 Graph both sets of data on the same screen. Record a sketch of what you see and answer these questions:

a. How are the two graphs related to each other?

b. If A's graph is $y = f(x)$, what equation describes B's graph? Describe how you determined this equation.
$y = f(x + 2) - 1$, because B is delayed by 2 s and sits about 1 ft closer to C.

c. In general, if the graph of $y = f(x)$ is translated horizontally h units and vertically k units, what is the equation of this translated function?
$y = f(x - h) + k$

Your group will now confirm your findings by exploring what happens to the equation of a linear function when you translate the graph of a line. Graph the lines in each step on the same set of axes and look for patterns.

Step 5 On graph paper, graph the line $y = 2x$ and then draw a line parallel to it, but 3 units higher. What is the equation of this new line? If $f(x) = 2x$, what is the equation of the new line in terms of $f(x)$?

Step 6 Draw a line parallel to the line $y = 2x$, but shifted down 4 units. What is the equation of this line? If $f(x) = 2x$, what is the equation of the new line in terms of $f(x)$?

Step 7 Mark the point where the line $y = 2x$ passes through the origin. Plot a point right 3 units from the origin. Draw a line parallel to the original line through this point. Use the point to write an equation in point-slope form for the new line. Then write an equation for the line in terms of $f(x)$.

Step 8 Plot a point left 1 unit and up 2 units from the origin. Draw a line parallel to the original line through this point and use the point to write an equation in point-slope form for the new line. Then write an equation for the line in terms of $f(x)$.

Step 9 If you move every point on the function $y = f(x)$ to a new point up k units and right h units, what is the equation of this translated function?
$y = f(x - h) + k$

Step 5

$y = f(x) + 3$

Step 6

$y = f(x) - 4$

Step 7

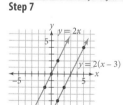

$y = 2(x - 3); y = f(x - 3)$

Step 8

$y = 2 + 2(x + 1); y = 2 + f(x + 1)$

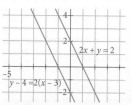

They have the same slope but different intercepts. They could be considered a vertical translation of one line with a shift up or down to get a different y-intercept. It could also be considered a horizontal translation of one line with a shift left or right to get a different x-intercept.

Investigate

If you do not have motion sensors, consider using the Investigation Worksheet Movin' Around without Motion Sensors. There is also a desmos simulation in your ebook.

Guiding the Investigation

Steps 2, 3 It would be most efficient and effective to collect one set of data for the class and distribute it among all calculators. Having four students act out Step 2 allows the rest of the class to focus on the situation and its mathematical meaning. SMP 4, 5

Step 4a A's graph has its vertex farther to the right, indicating A's recorded time is greater when the walker changes direction. The vertex for A is above the vertex for B because A is farther from the walker when the walker changes direction. B's graph is a translation of A's graph left about 2 units and down about 1 unit.

Step 4c If your students are stuck on how to write the equation in Step 4c, have them skip ahead and do Steps 5–9 and then return to Step 4c. Doing an example with a line will help them make the generalization needed to complete Step 4c. SMP 1

Steps 5, 6 Students may be confused by the phrase "in terms of $f(x)$." They are to use $f(x)$ in place of x in the equation.

Step 7 As needed, help students see that $2(x - 3)$ is $f(x - 3)$. ASK "What does function f do to any input?" [multiplies it by 2]

Modifying the Investigation

Whole Class Use graphing calculators to complete Steps 5–9 with student input. Have four students demonstrate and complete Steps 1 and 2. Discuss Steps 3 and 4.

Shortened Use the desmos simulation for Steps 1–4.

One Step Pose this problem in place of the investigation: "What is an equation of the line that results from translating every point on line $y = 2x$ to the right 3 units and up 5 units?" During the discussion, introduce the term *translation* and encourage the class to look for patterns. Elicit the idea that all vertical translations of a line are horizontal translations and vice versa. Investigate together the question of how to determine what translation takes a line to itself.

Example A

Project Example A and have students work in pairs. You might suggest that they use graph paper to draw the line and show the movement. Have pairs present their strategy as well as their solution. **SMP 1, 3, 5, 6**

Example B

Project Example B and have students work in pairs. Have pairs present their strategy as well as their solution. **SMP 1, 3, 4, 6**

ASK "Would you get a different equation if you picked another point to translate?" As an example, translate (2, 4) to (4, 1). Then $y + 3 = f(x - 2)$, just as before.

Summarize

As students present their solutions to the Investigation and the Examples, emphasize the use of correct mathematical terminology. Whether student responses are correct or incorrect, ask other students if they agree and why. **SMP 1, 3, 4, 6**

Discussing the Investigation

As students present their responses and graphs, ask them to clarify confusion about the vertical height of the graph as representative of the walker's horizontal distance from the motion detector. The difference in the heights of the graphs represents the horizontal distance between motion detectors. **ASK** "What does the horizontal axis on the graph represent?" [time]

For Step 4c, **ASK** "What is the real-world meaning of the translated graph?" [The data collection both began and ended 2 seconds later.]

Discussing the Lesson

As students present their solutions to the Investigation and the examples, emphasize the use of correct mathematical terminology, including the terms *map*, *mapped*, and *mapping*. **SMP 1, 3, 4, 6**

If you know the effects of translations, you can write an equation that translates any function on a graph. No matter what the shape of a function $y = f(x)$ is, the graph of $y = f(x - 3) + 2$ will look just the same as $y = f(x)$, but it will be translated up 2 units and right 3 units. Understanding this relationship will enable you to graph functions and write equations for graphs more easily.

> **Translation of a Function**
>
> A **translation** moves a graph horizontally or vertically or both.
>
> Given the graph of $y = f(x)$, the graph of
>
> $y = f(x - h) + k$ or, equivalently, of $y - k = f(x - h)$
>
> is a translation horizontally h units and vertically k units.

Language CONNECTION

The word "translation" can refer to the act of converting between two languages. Similar to its usage in mathematics, *translation* of foreign languages is an attempt to keep meanings parallel. Direct substitution of words often destroys the subtleties of meaning of the original text. The complexity of the art and craft of translation has inspired the formation of Translation Studies programs in universities throughout the world.

United Nations Headquarters in New York City. Speakers deliver their speeches in one of six official languages. U.N. interpreters then translate the lecture into the other five languages and attendees listen to the interpretations on headphones.

In a translation, every point (x_1, y_1) is mapped to a new point, $(x_1 + h, y_1 + k)$. This new point is called an **image** of the original point. If you have difficulty remembering which way to move a function, recall the point-slope form of the equation of a line. In $y = y_1 + b(x - x_1)$, the point at (0, 0) is translated to the new point at (x_1, y_1). In fact, every point is translated horizontally x_1 units and vertically y_1 units.

EXAMPLE A | Describe how the graph of $f(x) = 4 + 2(x - 3)$ is a translation of the graph of $f(x) = 2x$.

Solution | The graph of $f(x) = 4 + 2(x - 3)$ passes through the point (3, 4). Consider this point to be the translated image of (0, 0) on $f(x) = 2x$. The point is translated right 3 units and up 4 units from its original location, so the graph of $f(x) = 4 + 2(x - 3)$ is the graph of $f(x) = 2x$ translated right 3 units and up 4 units.

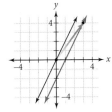

Conceptual/Procedural

Conceptual Students investigate the concept of the point-slope form as a translation of the line with its equation written in intercept form.

Procedural Students learn to describe the graph of a line in terms of a translation. They write the equation of a translated function in terms of $f(x)$ from its graph.

Note that you can distribute and combine like terms in $f(x) = 4 + 2(x - 3)$ to get $f(x) = -2 + 2x$. The fact that these two equations are equivalent means that translating the graph of $f(x) = 2x$ right 3 units and up 4 units is equivalent to translating the line down 2 units. In the graph in the example, this appears to be true.

In the investigation and Example A, you translated a line that passed through the origin. If you are translating a graph of a function that does not pass through the origin, then you will need to identify points on the original function that will match up with points on the translated image.

EXAMPLE B | The red graph is a translation of the graph of function f. Write an equation for the red function in terms of $f(x)$.

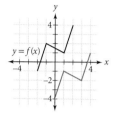

Solution | Any point on $f(x)$ can be matched with a point right 2 units and down 3 units on the red function. For example, the image of $(-1, 2)$ is $(1, -1)$. One notation to show this translation is $(x, y) \rightarrow (x + 2, y - 3)$. The equation of the red graph can be written $y - (-3) = f(x - 2)$, or $y + 3 = f(x - 2)$.

You can describe or graph a transformation of a function graph without knowing the equation of the function. But in the next few lessons, you will find that knowledge of equations for different families of functions can help you learn more about transformations.

A Kuna woman selling molas in an open air market in Panama City. The Kuna, or Guna, people are indigenous people of Panama and Colombia. The Kuna first used translated geometrical patterns for the designs of the mola, which means "shirt" or "clothing."

3.3 Exercises

Practice Your Skills

1. The graph of the line $y = \frac{2}{3}x$ is translated right 5 units and down 3 units. Write an equation of the new line. $y = -3 + \frac{2}{3}(x - 5)$

2. How does the graph of $y = f(x - 3)$ compare with the graph of $y = f(x)$? @ translated right 3 units

3. If $f(x) = -2x$, find

   a. $f(x + 3)$ @
      $-2(x + 3)$, or $-2x - 6$

   b. $-3 + f(x - 2)$ @
      $-3 + (-2)(x - 2)$, or $-2x + 1$

   c. $5 + f(x + 1)$
      $5 + (-2)(x + 1)$, or $-2x + 3$

Apply

Extra Example

1. The graph of the line $y = -2x - 1$ is translated -2 units horizontally and 6 units vertically. Write an equation of the new line.
   $y = -2(x + 2) - 1 + 6$ or $y = -2x + 1$

2. Rewrite $y = f(x)$ as a function that has been translated -3 units vertically and 4 horizontally.
   $y = f(x - 4) - 3$

Closing Question

How might you rewrite $y = 3x - 6$ to show that the equation can represent a vertical and/or horizontal translation of the line with equation $y = 3x$?

vertical: $y + 6 = 3x$; horizontal: $y = 3(x - 2)$

Assigning Exercises

Suggested: 1 – 10
Additional Practice: 11 – 14

Exercise 4 Have students find another equation, and show that the two equations are algebraically equivalent.

Exercise 6 ASK "Earlier, we said every horizontal translation is a vertical translation. Is that true here? Explain." [That property holds only for lines.] SUPPORT If students are confused by these graphs, suggest that they focus on how a single point on the graph moves in order to determine the translation of the entire graph. Then have them verify by checking a second point.

Exercise 7 ALERT Students may miss the point that the ropes have the same thickness because they're cut from the same source. ASK "Why does the rope have to be the same thickness in order to find this equation?" [The equations have the same slope.] "What are the meanings of 102 and 6.3?" [the original length of the rope and the amount it's shortened by each knot]

---

4. Consider the line that passes through the points $(-5.2, 3.18)$ and $(1.4, -4.4)$, as shown.
   a. Find an equation of the line. @ $y = -4.4 - 1.1\overline{48}(x - 1.4)$ or $y = 3.18 - 1.1\overline{48}(x + 5.2)$
   b. Write an equation of the parallel line that is 2 units above this line.
   $y = -2.4 - 1.1\overline{48}(x - 1.4)$ or $y = 5.18 - 1.1\overline{48}(x + 5.2)$

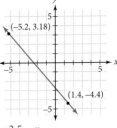

5. Write an equation of each line.
   a. the line $y = 4.7x$ translated down 3 units @ $y = -3 + 4.7x$
   b. the line $y = -2.8x$ translated right 2 units $y = -2.8(x - 2)$
   c. the line $y = -x$ translated up 4 units and left 1.5 units $y = 4 - (x + 1.5)$, or $y = 2.5 - x$

Reason and Apply

6. The graph of $y = f(x)$ is shown in black. Write an equation for each of the red image graphs in terms of $f(x)$.

   a. 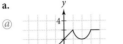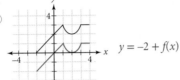 $y = -2 + f(x)$

   b. 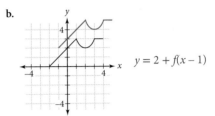 $y = 2 + f(x - 1)$

   c. 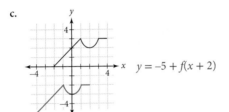 $y = -5 + f(x + 2)$

   d. 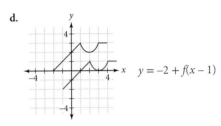 $y = -2 + f(x - 1)$

7. Jeannette and Keegan collect data about the length of a rope as knots are tied in it. The equation that fits their data is $y = 102 - 6.3x$, where x represents the number of knots and y represents the length of the rope in centimeters. Mitch had a piece of rope cut from the same source. Unfortunately he lost his data and can remember only that his rope was 47 cm long after he tied 3 knots. Write an equation that describes Mitch's rope.

$y = 47 - 6.3(x - 3)$

8. Rachel, Pete, and Brian perform Steps 1–4 of the investigation in this lesson. Rachel walks while Pete and Brian hold the motion sensors. A graph of their results is shown at right.

   a. The black curve is made from the data collected by Pete's motion sensor. Where was Brian standing and when did he start his motion sensor to create the red curve? ⓗ Brian stood about 1.5 m behind Pete, and he started his motion sensor 2 s later than Pete started his.
   b. If Pete's curve is the graph of $y = f(x)$, what equation represents Brian's curve? $y = 1.5 + f(x + 2)$

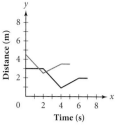

Exercise 9 This is a recursive procedure, because each step depends on the previous one.

9. **APPLICATION** Kari's assignment in her computer programming course is to simulate the motion of an airplane by repeatedly translating it across the screen. The coordinate system in the software program is shown below. In this program, coordinates to the right and down are positive.

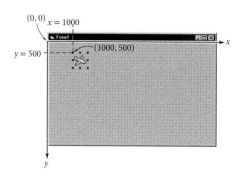

The starting position of the airplane is (1000, 500), and Kari would like the airplane to end at (7000, 4000). She thinks that moving the airplane in 15 equal steps will model the motion well.

   a. What should be the airplane's first position after (1000, 500)? (1400, 733.$\overline{3}$)

   b. If the airplane's position at any time is given by (x, y), what is the next position in terms of x and y? $(x + 400, y + 233.\overline{3})$

   c. If the plane moves down 175 units and right 300 units in each step, how many steps will it take to reach the final position of (7000, 4000)? 20 steps

CONNECTION

Animation simulates movement. An old-fashioned way to animate is to make a book of closely related pictures and flip the pages. Flipbook technique is used in cartooning—a feature-length film might have more than 65,000 images. Today, hand drawing has been largely replaced by computer-generated special effects.

A wall sized Angry Bird poster displayed at IOI Putrajaya Mall in Malaysia. The Angry Birds Movie is a 3D computer-animated action adventure comedy film based on a video game

LESSON 3.3 Lines in Motion 163

Exercise 10 This exercise is a good one to assign to groups. Unlike the coefficients in intercept form or point-slope form, a, b, and c have no direct interpretation as intercepts or slope.

One advantage of standard form is that equations for vertical lines, such as part vi of 10a, can be written. Because b is 0, both the y-intercept and the slope are undefined. You could, however, evaluate $-\frac{c}{a}$ to find that the x-intercept is 2.

Another advantage of the standard form is that it's equally easy to find both intercepts.

The equation in part v of 10a is a horizontal line. Because there is no x-term, the slope is $-\frac{a}{b}$ or $-\frac{0}{2} = 0$.

Because the standard form is not in $y =$ form and the coefficient of y is not necessarily 1, when students just replace y with $(y - k)$, the constant k is multiplied by the original coefficient of y.

	Equation in standard form
$ax + by = c$	
$ax + b(y - k) = c$	Replace y with $(y - k)$
$ax + by - bk = c$	Distribute b
$by = c - ax + bk$	Subtract ax and by from both sides
$\frac{by}{b} = \frac{c - ax + bk}{b}$	Divide both sides by b
$y = \frac{c}{b} - \frac{a}{b}x + \frac{b}{b}k$	Simplify
$y = \frac{c}{b} - \frac{a}{b}x + k$	

This works the same way for horizontal translations.

It is worth pointing out that when you expand the standard form of the equation for the translated line, the constant is the only coefficient that changes. The x- and y-coefficients remain the same as in the equation of the original line.

10c. i. y-intercept: 4; slope: $-\frac{4}{3}$

10c. ii. y-intercept: 5; slope: 1

10c. iii. y-intercept: -1; slope: 7

10c. iv. y-intercept: $-\frac{1}{2}$; slope: $\frac{1}{2}$

10c. v. y-intercept: 5; slope: 0

10c. vi. y-intercept: none; slope: undefined

10d. v. $4x + 3y = 7$

10d. vi. $4x + 3y = 10$

---

10. Linear equations can also be written in standard form.

    Standard form $\quad ax + by = c$

    a. Identify the values of a, b, and c for each of these equations in standard form.
       i. $4x + 3y = 12$ ii. $-x + y = 5$ iii. $7x - y = 1$
       iv. $-2x + 4y = -2$ v. $2y = 10$ vi. $3x = -6$

    b. Solve the standard form, $ax + by = c$, for y. The result should be an equivalent equation in intercept form. What is the y-intercept? What is the slope? ⓐ $y = \frac{c}{b} - \frac{a}{b}x$; y-intercept: $\frac{c}{b}$; slope: $-\frac{a}{b}$

    c. Use what you've learned from 10b to find the y-intercept and slope of each of the equations in 10a.

    d. The graph of $4x + 3y = 12$ is translated as described below. Write an equation in standard form for each of the translated graphs.
       i. a translation right 2 units $\quad 4x + 3y = 20$
       ii. a translation left 5 units ⓐ $\quad 4x + 3y = -8$
       iii. a translation up 4 units $\quad 4x + 3y = 24$
       iv. a translation down 1 unit $\quad 4x + 3y = 9$
       v. a translation right 1 unit and down 3 units
       vi. a translation up 2 units and left 2 units ⓐ

    e. In general, if the graph of $ax + by = c$ is translated horizontally h units and vertically k units, what is the equation of the translated line in standard form? $\quad ax + by = c + ah + bk$

10a. i. $a = 4$, $b = 3$, $c = 12$
10a. ii. $a = -1$, $b = 1$, $c = 5$
10a. iii. $a = 7$, $b = -1$, $c = 1$
10a. iv. $a = -2$, $b = 4$, $c = -2$
10a. v. $a = 0$, $b = 2$, $c = 10$
10a. vi. $a = 3$, $b = 0$, $c = -6$

Review

11. **APPLICATION** The Internal Revenue Service has approved ten-year linear depreciation as one method for determining the value of business property. This means that the value declines to zero over a ten-year period, and you can claim a tax exemption in the amount of the value lost each year. Suppose a piece of business equipment costs $12,500 and is depreciated over a ten-year period. At right is a sketch of the linear function that represents this depreciation.

    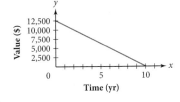

    a. What is the y-intercept? Give the real-world meaning of this value.
       12,500. The original value of the equipment is $12,500.
    b. What is the x-intercept? Give the real-world meaning of this value.
       10. After 10 yr, the equipment has no value.
    c. What is the slope? Give the real-world meaning of the slope.
       -1250. Every year, the value of the equipment decreases by $1,250.
    d. Write an equation that describes the value of the equipment during the ten-year period.
       $y = 12,500 - 1,250x$
    e. When is the equipment worth $6,500?
       after 4.8 yr

12. Suppose that your basketball team's scores in the first four games of the season were 86 points, 73 points, 76 points, and 90 points.

    a. What will be your team's mean score if the fifth-game score is 79 points?
       80.8
    b. Write a function that gives the mean score in terms of the fifth-game score. ⓐ
    c. What score will give a five-game average of 84 points? ⓗ
       95 points

    12b. $y = \frac{1}{5}x + 65$

13. Solve.
    a. $2(x + 4) = 38 \quad x = 15$
    b. $7 + 0.5(x - 3) = 21 \quad x = 31$
    c. $-2 + \frac{3}{4}(x + 1) = -17 \quad x = -21$
    d. $4.7 + 2.8(x - 5.1) = 39.7 \quad x = 17.6$

14. The three summary points for a data set are $M_1(3, 11)$, $M_2(5, 5)$, and $M_3(9, 2)$. Find the line of fit. $\hat{y} = \frac{29}{2} - \frac{3}{2}x$

Exercise 11 ELL Students may need some definitions and context with this exercise.

Exercise 12b The given answer is equivalent to $y = \frac{x + 325}{5}$. The answer equation can be thought of as dividing the sum of the four games (325) by 5 to get 65, the amount each of the 4 games will contribute to the mean for 5 games, then adding $\frac{1}{5}x$, the amount the fifth score will add to the mean.

LESSON 3.4

Translations and the Quadratic Family

I see music as the augmentation of a split second of time.
ERIN CLEARY

Music CONNECTION

When a song is in a key that is difficult to sing or play, it can be translated, or transposed, into an easier key. To transpose music means to change the pitch of each note without changing the relationships between the notes. Jazz saxophonist Ornette Coleman (1930–2015), who grew up with strong interests in mathematics and science, developed "free jazz" by transposing the music, straying from the set standards of harmony and melody.

In the previous lesson, you looked at translations of the graphs of linear functions. Translations can occur in other settings as well. For instance, what will this histogram look like if the teacher decides to add five points to each of the scores? each bin will shift right 5 units

What translation will map the black triangle on the left onto its red image on the right? right 5 units and up 1 unit

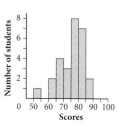
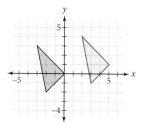

Translations are also a natural feature of the real world, including the world of art. Music can be transposed from one key to another. Melodies are often translated by a certain interval within a composition.

In mathematics, a change in the size or position of a figure or graph is called a **transformation**. Translations are one type of transformation. You may recall other types of transformations, such as reflections, dilations, stretches, shrinks, and rotations, from other mathematics classes.

In this lesson you will experiment with translations of the graph of the function $y = x^2$. The special shape of this graph is called a **parabola**. Parabolas always have a **line of symmetry** that passes through the parabola's **vertex**.

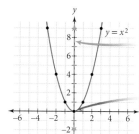

The line of symmetry divides the graph into mirror-image halves. The line of symmetry of $y = x^2$ is $x = 0$.

The vertex is the point where the graph changes from decreasing to increasing. The vertex of $y = x^2$ is $(0, 0)$.

The function $y = x^2$ is a building-block function, or **parent function**. By transforming the graph of a parent function, you can create infinitely many new functions, or a **family of functions**. The function $y = x^2$ and all functions created from transformations of its graph are called **quadratic functions**, because the highest power of x is x-squared.

LESSON 3.4 Translations and the Quadratic Family **165**

LESSON 3.4

This lesson begins a sequence of four lessons that discuss transformations while introducing or reviewing families of relations. This first lesson extends the discussion of translations to parabolic graphs of quadratic equations.

COMMON CORE STATE STANDARDS		
APPLIED	**DEVELOPED**	**INTRODUCED**
F.IF.7a	F.IF.4	A.REI.4b
F.IF.7b	F.IF.9	
	F.BF.3	

Objectives

- Define the parent quadratic function, $y = x^2$
- Determine elements of equations (h and k) that produce translations of the graphs of parent functions
- Introduce the (nonstretched) vertex form of the graph of a parabola, $y = (x - h)^2 + k$
- Define *parabola*, *vertex of a parabola*, and *line of symmetry*
- Determine the graph from an equation and the equation from a graph
- Define and create piecewise functions

Vocabulary

transformation
parabola
line of symmetry
vertex
parent function
family of functions
quadratic functions
piecewise function

Materials

- geometry software
- Calculator Notes: Graphing Transformations, Transformations and Compositions, Entering and Graphing Equations, Setting Windows, *optional*

Launch

Project the Launch found in your ebook. For $y = x^2$, the axis of symmetry goes through the origin; $x = 0$; vertex $(0, 0)$. For $y = x^2 + 3$, the vertex moved up 3 units. For $y = (x + 3)^2$ the vertex moved left 3 units.

ELL
To increase the clarity of the vocabulary surrounding parabolas, such as *vertex* and *line of symmetry*, help students create a visual reminder with the specific vocabulary words labeled on the graph. Students could make, present, and display a poster.

Support
Tie this section closely to the Lines in Motion lesson. Reinforce the use of (h, k) to perform translations. If students still struggle, especially with horizontal translation, continue to emphasize viewing the graph, with (h, k) being substituted into the equation.

Advanced
Piecewise functions open some very interesting avenues of exploration for students. Students can now write equations for some of the very unusual graphs that have been used earlier in this chapter. Explore graphing piecewise functions using Calculator Note: Graphing Piecewise Functions.

LESSON 3.4 Translations and the Quadratic Family **165**

Investigate

Guiding the Investigation

There is a desmos tool for this investigation. You might assign Step 1 for students to "play" with translations individually on their calculators. When students work privately, they tend to be more willing to try equations that might be incorrect and more confident in sharing satisfactory results with the class. Have students share their equations and explain how they got their answers. Then have them do Steps 2 and 3 in pairs and share their responses. You may want to extend the investigation to a class game of Make My Graph, in which you make a graph and students experiment on their calculators to find the equation that will make that graph. **SMP 1, 3, 5, 4, 6**

Step 2 Answers will vary. For a translation right, you subtract from x; for a translation left, you add to x; for a translation up, you add to the entire function (or subtract from y); for a translation down, you subtract from the entire function (or add to y). Students may also notice the coordinates of the vertex are equivalent to (*value of horizontal translation, value of vertical translation*).

The graphs of all quadratic functions are parabolas. **ASK** "Is every parabola the graph of a quadratic function?" [If the line of symmetry of the parabola's graph is vertical, then the parabola is a graph of a function in the family $y = x^2$. If the line of symmetry is horizontal, the parabola has the relation $x = y^2$ as a parent. Here x is a quadratic function of y. Rotations of these graphs through a number of degrees other than a multiple of 90° are parabolas in which neither x nor y is a function of the other, but they still represent quadratic relations.]

Quadratic functions are very useful, as you will discover throughout this book. You can use functions in the quadratic family to model the height of a projectile as a function of time, or the area of a square as a function of the length of its side.

The focus of this lesson is on writing the quadratic equation of a parabola after a translation and graphing a parabola given its equation. You will see that locating the vertex is fundamental to your success with understanding parabolas.

Engineering CONNECTION

Several types of bridge designs involve the use of curves modeled by nonlinear functions. Each main cable of a suspension bridge approximates a parabola. The Mackinac Bridge in Michigan, shown here, was built in 1957.

INVESTIGATION

Make My Graph

Procedure Note
Different calculators have different resolutions. A good graphing window will help you make use of the resolution to better identify points. Enter the parent function $y = x^2$ as the first equation. Enter the equation for the transformation as the second equation. Graph both equations to check your work.

Step 1 Each graph below shows the graph of the parent function $y = x^2$ in black. Find a quadratic equation that produces the congruent, red parabola. Apply what you learned about translations of the graphs of functions in the previous lesson.

a.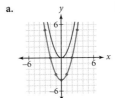
$y = x^2 - 4$

b.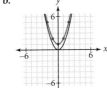
$y = x^2 + 1$

c.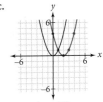
$y = (x - 2)^2$

d.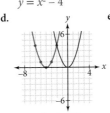
$y = (x + 4)^2$

e.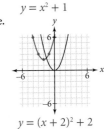
$y = (x + 2)^2 + 2$

f.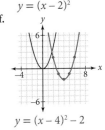
$y = (x - 4)^2 - 2$

Step 2 Write a few sentences describing any connections you discovered between the graphs of the translated parabolas, the equation for the translated parabola, and the equation of the parent function $y = x^2$.

Step 3 In general, what is the equation of the parabola formed when the graph of $y = x^2$ is translated horizontally h units and vertically k units?
$y = (x - h)^2 + k$ or $y - k = (x - h)^2$

Modifying the Investigation

Whole Class Elicit student suggestions for Step 1. Discuss Steps 2 and 3.

Shortened Discuss Steps 2 and 3 as a class.

One Step Use the **One Step** found in your ebook.

Conceptual/Procedural

Conceptual The concept of transformations is introduced through the idea of families of relations, in particular translations to parabolic graphs of quadratic equations.

Procedural Students use transformations to write the equation of a parabola. They interpret a graph of a scenario and adjust it for changes in the scenario. They learn to graph piece-wise defined functions.

The following example shows one simple application involving parabolas and translations of parabolas. In later chapters you will discover many applications of this important mathematical curve.

EXAMPLE A | This graph shows a portion of a parabola. It represents a diver's position (horizontal and vertical distance) from the edge of a pool as he dives from a 5 ft long board 25 ft above the water.

a. Identify points on the graph that represent when the diver leaves the board, when he reaches his maximum height, and when he enters the water.

b. Sketch a graph of the diver's position if he dives from a 10 ft long board 10 ft above the water. (Assume that he leaves the board at the same angle and with the same force.)

c. In the scenario described in part b, what is the diver's position when he reaches his maximum height?

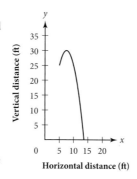

Solution

a. The point (5, 25) represents the moment when the diver leaves the board, which is 5 ft long and 25 ft high. The vertex, (7.5, 30), represents the position where the diver's height is at a maximum, or 30 ft; it is also the point where the diver's motion changes from upward to downward. The x-intercept, approximately (13.6, 0), indicates that the diver hits the water at approximately 13.6 ft from the edge of the pool.

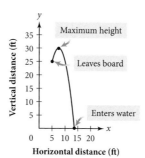

b. If the length of the board increases from 5 ft to 10 ft, then the parabola translates right 5 units. If the height of the board decreases from 25 ft to 10 ft, then the parabola translates down 15 units. If you define the original parabola as the graph of $y = f(x)$, then the function for the new graph is $y = f(x - 5) - 15$.

c. As with every point on the graph, the vertex translates right 5 units and down 15 units. The new vertex is $(7.5 + 5, 30 - 15)$, or $(12.5, 15)$. This means that when the diver's horizontal distance from the edge of the pool is 12.5 ft, he reaches his maximum height of 15 ft.

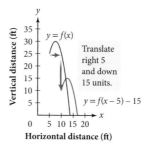

You can extend the ideas you've learned in translating linear and quadratic functions to functions in general. For a linear function, $y = f(x)$, to translate the function horizontally h units, you can replace x in the equation with $(x - h)$. To translate the function vertically k units, replace y in the equation with $(y - k)$. So the translation of $y = f(x)$ horizontally by h and vertically by k is $y - k = f(x - h)$.

When the function is a quadratic function, $f(x) = x^2$, then the translation is $y - k = (x - h)^2$. You may also see this written as $y = (x - h)^2 + k$.

Example A
Whereas the investigation has students translate the graph of the parent function $y = x^2$, this example asks students to relate two parabolas, neither of which is the parent quadratic function. Students may notice that both graphs actually require a reflection of the graph of $y = x^2$ across a horizontal line. The example does not require students to write a function for either graph.

Project Example A and work through it as a class, with student input. **SMP 3, 4, 6**

ASK "Why is the translated vertex (h, k)?" [The vertex of the graph of $y = x^2$ is $(0, 0)$, so a translation horizontally h units and vertically k units puts the translated vertex at (h, k).]

Example B
Project part a of Example B and have students work in pairs to graph it. Have students share and explain their graphs. If needed, **ASK** "What lines did you graph to get each segment?" Look for explanations similar to the Solution. **SMP 3, 6**

Project part b. **ASK** "What equation might this be a graph of?" Work through the example with student input. As needed, relate piecewise functions to the idea to breaking a graph into segments in order to tell a story. Explain that students will now go beyond telling a story to actually finding a mathematical equation that models the entire graph.

Summarize

As students present their solutions to the Investigation and the Examples, emphasize the use of correct mathematical terminology. Whether student responses are correct or incorrect, ask other students if they agree and why.
SMP 1, 3, 4, 6

In their study of quadratic functions, students will see that this is the vertex form of a quadratic equation, with vertical scale factor $a = 1$. **ASK** "What is the line of symmetry of these graphs?"

You might point out that some of the graphs don't really look parallel and question whether they're actually translations. *Corresponding* points of translated parabolas are the same distance apart, but, unlike with lines, the *closest* points may not be.

LANGUAGE The word *quadratic* comes from the Latin root *quadrare*, meaning "to square." The prefix *quad* is usually used in words like *quadrilateral* to mean "four"; its use as "two" in *quadratic* stems from the fact that squared terms were represented as square (four-sided) shapes, as in rectangle diagrams.

LANGUAGE The book uses the term *congruent* to describe parabolas that are translations of each other. In geometry two polygons are congruent if corresponding sides and corresponding angles are congruent. To induce critical thinking, **ASK** "Is the book correct in using the term *congruent*?" Encourage discussion that compares and contrasts parabolas and polygons. Unlike a polygon, a parabola has no angles or sides and is not bounded. But a translation of a polygon is indeed congruent; in fact, figures can be defined to be congruent if one is the image of the other under translations and rotations.

CRITICAL QUESTION In the Investigation, what form of quadratic equations did you use?

BIG IDEA All the translations can be represented by the vertex form of a quadratic equation.

It is important to notice that the vertex of the translated parabola is (h, k). That's why finding the vertex is fundamental to determining translations of parabolas. In every function you study, there will be key points to locate. Finding the relationships between these points and the corresponding points in the parent function enables you to write equations more easily.

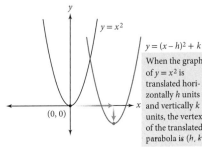

When the graph of $y = x^2$ is translated horizontally h units and vertically k units, the vertex of the translated parabola is (h, k).

EXAMPLE B | A **piecewise function** is a function that consists of two or more standard functions defined on different domains.

a. Graph $f(x) = \begin{cases} 2 + x & -2 \le x \le 1 \\ 3 & 1 < x \le 2 \\ 3 - 2(x - 2) & 2 < x \le 3 \end{cases}$

b. Find the equation for the piecewise function pictured at the right.

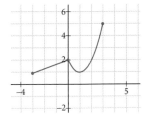

Solution | a. The graph of the first part is a line with a y-intercept of 2 and a slope of 1. It is defined for x-values between -2 and 1, so sketch the line but keep only the segment from $(-2, 0)$ to $(1, 3)$. Now repeat that for the other two parts.

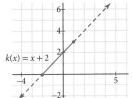 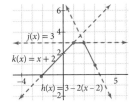 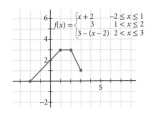

b. The graph has two pieces. The left piece appears to be a line defined between $(-3, 1)$ and $(0, 2)$. This line would have a slope of $\frac{1}{3}$ and a y-intercept of 2, which gives the equation $y = 2 + \frac{1}{3}x$ on the domain $-3 \le x \le 0$. The right piece is a parabola translated one unit up and one unit to the right. The equation of this graph is $y = (x - 1)^2 + 1$ on the domain $0 < x \le 3$. Combining these two pieces, you can represent the piecewise function as

$$g(x) = \begin{cases} 2 + \frac{1}{3}x & -3 \le x \le 0 \\ 1 + (x - 1)^2 & 0 < x \le 3 \end{cases}$$

Notice that even though the two pieces meet at $x = 0$, you include 0 in only one domain piece. It doesn't matter which piece, but it should not be included in both.

You can translate piecewise functions in the same way that you translate a parabola. In this case, the points where the pieces "join" are the key points that you will want to consider when graphing the translated function. The exercises will give you the chance to practice this.

CRITICAL QUESTION When the graph of the quadratic equation $y = x^2$ is translated to put its vertex at (h, k), what is the new equation?

BIG IDEA The equation becomes $y = (x - h)^2 + k$, or, equivalently, $y - k = (x - h)^2$.

Formative Assessment

Check whether students are making the link between translating linear functions and translating quadratic functions. How comfortable are they shifting a parabola in a given direction? Also check to see if they understand how the vertex fits into the new formula. While students investigate, you can begin to see how well they understand the idea of changing a function equation to get a different function with a related graph. Continue to monitor student comfort with function notation and the use of variables in general.

3.4 Exercises

You will need a graphing calculator for Exercise 17.

Practice Your Skills

1. Write an equation for each parabola at right. Each parabola is a translation of the graph of the parent function $y = x^2$.

2. Each parabola described below is congruent to the graph of $y = x^2$. Write an equation for each parabola and sketch its graph.
   a. The parabola is translated vertically −5 units. @
   b. The parabola is translated vertically 3 units.
   c. The parabola is translated horizontally 3 units. @
   d. The parabola is translated horizontally −4 units.

3. If $f(x) = x^2$, then the graph of each equation below is a parabola. Describe the location of the parabola relative to the graph of $f(x) = x^2$.
   a. $y = f(x) - 3$ translated vertically −3 units
   b. $y = f(x) + 4$ translated vertically 4 units
   c. $y = f(x - 2)$ translated horizontally 2 units
   d. $y = f(x + 4)$ @ translated horizontally −4 units

4. Describe what happens to the graph of $y = x^2$ in the following situations.
   a. x is replaced with $(x - 3)$. translated horizontally 3 units
   b. x is replaced with $(x + 3)$. translated horizontally −3 units
   c. y is replaced with $(y - 2)$. translated vertically 2 units
   d. y is replaced with $(y + 2)$. translated vertically −2 units

5. Solve. ⓗ
   a. $x^2 = 4$ $x = 2$ or $x = -2$
   b. $x^2 + 3 = 19$ @ $x = 4$ or $x = -4$
   c. $(x - 2)^2 = 25$ $x = 7$ or $x = -3$

Reason and Apply

6. Write an equation for each parabola.

   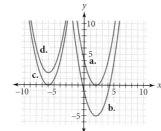

   6a. $y = (x - 2)^2$
   6b. $y = (x - 2)^2 - 5$
   6c. $y = (x + 6)^2$
   6d. $y = (x + 6)^2 + 2$

7. The red parabola at right is the image of the graph of $y = x^2$ after a horizontal translation of 5 units and a vertical translation of −3 units.
   a. Write an equation for the red parabola. $y = (x - 5)^2 - 3$
   b. Where is the vertex of the red parabola? $(5, -3)$
   c. What are the coordinates of the other four points if they are 1 or 2 horizontal units from the vertex? How are the coordinates of each point on the black parabola related to the coordinates of the corresponding point on the red parabola? @
   d. What is the length of blue segment b? Of green segment c? @
      Segment b has length 1 unit, and segment c has length 4 units.

   7c. $(6, -2), (4, -2), (7, 1), (3, 1)$. If (x, y) are the coordinates of any point on the black parabola, then the coordinates of the corresponding point on the red parabola are $(x + 5, y - 3)$.

Apply

Extra Example

1. The parabola $y = x^2$ is shifted to have a vertex of (−2, 4). What is an equation of this new parabola?
   $y = (x + 2)^2 + 4$

2. Where is the vertex of the parabola represented by $y = (x + 0)^2 + 4$? (0, 4)

Closing Question

Describe the location of a parabola $y = f(x - 3) + 4$ relative to $y = f(x)$.
translated horizontally 3 and vertically 4

Assigning Exercises

Suggested: 1 – 9, 11 – 12
Additional Practice: 10, 13 – 17

1a. $y = x^2 + 2$

1b. $y = x^2 - 6$

1c. $y = (x - 4)^2$

1d. $y = (x + 8)^2$

2a. $y = x^2 - 5$

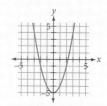

2b. $y = x^2 + 3$

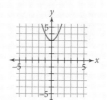

2c. $y = (x - 3)^2$

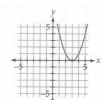

2d. $y = (x + 4)^2$

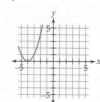

Exercises 6, 7 SUPPORT Due to the restrictions of the graphs' windows, students might mistakenly believe that certain parabolas are "smaller" than others. Emphasize the fact that the parabolas are indeed congruent and that the window limits the full view. SMP 5

Exercise 7d This exercise uses the fact that pairs of corresponding points are the same distance apart. ASK "What is the equation of the line of symmetry?"

8a.
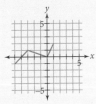

8b.
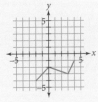

8c.
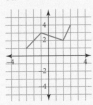

8d.
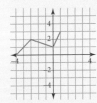

Exercise 9 Students need not graph piecewise function $g(x)$ to complete this exercise, though they should if your standards warrant it. If they choose to, they may be confused by the fact that the function is not continuous. If students aren't sure how to find $g(2)$, you might **ASK** "Which piece of the function contains 2 as part of its domain?" [$g(x) = 3$]

Exercise 10 Because the number of teams and the number of games must be integers, the graph of this function is a collection of points. Its trend can be seen, and predictions made, by drawing a curve through those points. Students can use differences to find the explicit formula; ask whether it makes sense. For each additional team, you will add double the previous number of teams to represent the new team's playing each of the existing teams twice.

10a.

Number of teams (x)	4	5	6	7	8	9	10
Number of games (y)	12	20	30	42	56	72	90

8. Given the graph of $y = f(x)$ at right, draw a graph of each of these related functions.

   a. $y = f(x + 2)$ b. $y = f(x - 1) - 3$
   c. $y = f(x) + 2$ d. $y = f(x + 2) + 1$
   e. Write $f(x)$ as a piece-wise defined function.
   $$f(x) = \begin{cases} 2 + x & -3 \leq x \leq -1 \\ \frac{2}{3} - \frac{1}{3}x & -1 < x \leq 2 \\ -4 + 2x & 2 < x \leq 3 \end{cases}$$

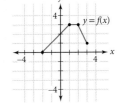

9. Let $f(x)$ be defined as the piecewise function graphed at right, and let $g(x)$ be defined as
   $$g(x) = \begin{cases} 3 & 0 \leq x \leq 2 \\ 2 + 0.5(x - 2) & 2 < x \leq 4 \\ 2 - (x - 4) & 4 < x \leq 6 \\ 1 & 6 < x \leq 7 \end{cases}$$

   Find each value.

   a. $f(0)$ 2   b. x when $f(x) = 0$ -2 c. x when $f(x) = 1$ $-1, 3$
   d. $g(1.8)$ 3   e. $g(2)$ 3   f. $g(4)$ 3
   g. $g(6.999)$ 1

10. **APPLICATION** This table of values compares the number of teams in a pee wee teeball league and the number of games required for each team to play every other team twice (once at home and once away from home).

Number of teams (x)	1	2	3	...
Number of games (y)	0	2	6	...

    a. Continue the table out to 10 teams.
    b. Plot each point and describe the graph produced.
    c. Write an explicit function for this graph. $y = (x - 0.5)^2 - 0.25$
    d. Use your function to find how many games are required if there are 30 teams. 870 games

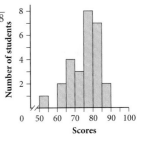

11. Solve.
    a. $3 + (x - 5)^2 = 19$ $x = 9$ or $x = 1$ b. $(x + 3)^2 = 49$ $x = 4$ or $x = -10$
    c. $5 - (x - 1)^2 = -22$ $x = 1 \pm \sqrt{27}$ d. $-15 + (x + 6)^2 = -7$ $x = -6 \pm \sqrt{8}$

12. This histogram shows the students' scores on a recent quiz in Ms. Noah's class. Describe what the histogram will look like if Ms. Noah
    a. adds five points to everyone's score.
       The graph will be translated horizontally 5 points (one bin).
    b. subtracts ten points from everyone's score.
       The graph will be translated horizontally −10 points (two bins).

10b. The points appear to be part of a parabola.

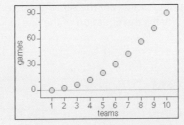

Exercise 11 As needed, remind students to first isolate the quantity in parentheses. **ASK** "What real-world situation might one of these equations represent?" Students may use the "undoing the order of operations" technique suggested for Exercise 5.

Review

13. Match each recursive formula with the equation of the line that contains the sequence of points, (n, u_n), generated by the formula.

   a. $u_0 = -8$
      $u_n = u_{(n-1)} + 3$ where $n \geq 1$

   b. $u_1 = 3$
      $u_n = u_{(n-1)} - 8$ where $n \geq 2$

   a. ii
   b. iii

   i. $y = 3x - 11$
   ii. $y = 3x - 8$
   iii. $y = 11 - 8x$
   iv. $y = -8x + 3$

14. DataForU has a data plan that charges $49 a month for 10 Gb of data downloaded and $10 for each additional 1 Gb downloaded. Saver Data's plan charges $100 a month for unlimited data downloading. BeMobile charges $60 a month for 10 Gb of data downloaded plus $5 for each additional 1 Gb.

   a. Write a cost function for each mobile phone company.
   b. Graph the three functions on the same axes.
   c. Describe which mobile company is the cheapest alternative under various circumstances.

15. A car drives at a constant speed along the road pictured at right from point A to point X. Sketch a graph showing the straight line distance between the car and point X as it travels along the road. Mark points A, B, C, D, E, and X on your graph. @

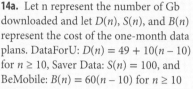

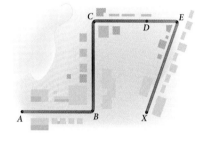

16. The distance between a walker and a stationary observer is shown at right.

   a. Describe the actions of the walker.
   b. What does the equation $3.8 - 0.84(x - 1.2) = 2$ mean in the context of the graph? When is the walker 2 m from the observer?
   c. Solve the equation from 16b and interpret your solution.
      After about 3.34 s, the walker is 2 m from the observer.

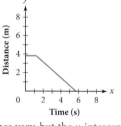

17. Use a graphing calculator to investigate the form $y = ax + b$ of a linear function.

   a. On the same coordinate plane, graph the lines $y = 0.5x + 4$, $y = x + 4$, $y = 2x + 4$, $y = 5x + 4$, $y = -3x + 4$, and $y = -0.25x + 4$. Describe the graphs of the family of lines $y = ax + 4$ as a takes on different values. The slopes vary, but the y-intercept is always 4.

   b. On the same coordinate plane, graph the lines $y = 2x - 7$, $y = 2x - 2$, $y = 2x$, $y = 2x + 3$, and $y = 2x + 8$. Describe the graphs of the family of lines $y = 2x + b$ as b takes on different values. The graphs move up or down, but they all have slope 2.

LESSON 3.4 Translations and the Quadratic Family **171**

Exercise 12 To help students connect this exercise with the lesson, you might suggest that students visualize a parabola that follows the trend of the histogram. Just as lines and parabolas can be translated, so can sets of data. **ASK** "How does this affect the mean and median of the data set? How does it affect the range and IQR?" [it simply shifts them by the amount they were translated; their relationships remain the same]

14a. Let n represent the number of Gb downloaded and let $D(n)$, $S(n)$, and $B(n)$ represent the cost of the one-month data plans. DataForU: $D(n) = 49 + 10(n - 10)$ for $n \geq 10$, Saver Data: $S(n) = 100$, and BeMobile: $B(n) = 60(n - 10)$ for $n \geq 10$

14b.

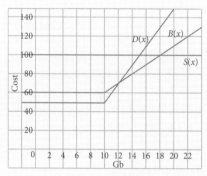

14c. If you plan to use the least amount of data, go with DataforU. If you plan to use between 2 and 8 extra Gb of data, go with BeMobile. If you know you will use more than 8 extra Gb of data, choose Saver Data.

Exercise 16 **ALERT** Despite the labels on the axes, students may consider the graph as a view from above the walker's path.

16a. Possible answer: the walker stayed 3.8 m from the sensor for 1.2 s and then walked at a constant 0.84 m/s toward the sensor.

Exercise 17 Students could use geometry software for this investigation. They could create two sliders, a and b, and use those sliders to manipulate the equation $y = ax + b$. **ELL** Use this exercise to reinforce the effect of varying the values of a and b. It will be beneficial for students to graph an entire family of lines and then create a verbal description of the results.

LESSON 3.5

This lesson discusses reflections (across the axes) and the square root function.

COMMON CORE STATE STANDARDS
APPLIED	DEVELOPED	INTRODUCED
F.IF.2	F.IF.7b	
	F.BF.3	

Objectives

- Define and apply reflections to functions
- Define the parent square root function, $y = \sqrt{x}$
- Define the square root symbol and function as the positive root
- Compare $f(x)$, $-f(x)$, $f(-x)$, and $-f(-x)$
- Apply the square root function in context
- Determine if a function is odd, even, or neither

Vocabulary

square root function
reflection
even function
odd function
radical

Materials

- Calculator Notes: Setting Windows, Graphing Piecewise Functions, Transformations and Compositions

Launch

Find two whole numbers that each of the following is between, without using your calculator.

a. $\sqrt{7}$ 2, 3
b. $\sqrt{19}$ 4, 5
c. $\sqrt{43}$ 6, 7
d. x when $x^2 = 98$ 9, 10

Now graph a–d by setting them equal to y to check your answers.

What happens when you try to find $\sqrt{-68}$ on your calculator?
ERR: NONREAL ANS

LESSON 3.5

Call it a clan, call it a network, call it a tribe, call it a family. Whatever you call it, whoever you are, you need one.
JANE HOWARD

Reflections and the Square Root Family

The graph of the **square root function**, $y = \sqrt{x}$, is another parent function that you can use to illustrate transformations. From the graphs below, what are the domain and range of $f(x) = \sqrt{x}$? If you graph $y = \sqrt{x}$ on your calculator, you can show that $\sqrt{3}$ is approximately 1.732. What is the approximate value of $\sqrt{8}$? How would you use the graph to find $\sqrt{31}$? What happens when you try to find $f(x)$ for values of $x < 0$?

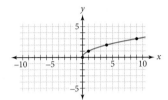
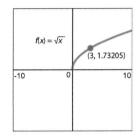

INVESTIGATION

Take a Moment to Reflect

In this investigation you first will work with linear functions to discover how to create a new transformation—a **reflection**. Then you will apply reflections to quadratic functions and square root functions.

Step 1 Graph $f_1(x) = 0.5x + 2$ on your calculator.

  a. Predict what the graph of $-f_1(x)$ will look like. Then check your prediction by graphing $f_2(x) = -f_1(x)$.
  b. Change f_1 to $f_1(x) = -2x - 4$, and repeat the instructions in Step 1a.
  c. Change f_1 to $f_1(x) = x^2 + 1$ and repeat.
  d. In general, how are the graphs of $y = f(x)$ and $y = -f(x)$ related?

Step 1d $y = -f(x)$ is a reflection of $y = f(x)$ across the x-axis.

Step 2 Graph $f_1(x) = 0.5x + 2$ on your calculator.

  a. Predict what the graph of $f_1(-x)$ will look like. Then check your prediction by graphing $f_2(x) = f_1(-x)$.
  b. Change f_1 to $f_1(x) = -2x - 4$, and repeat the instructions in Step 2a.
  c. Change f_1 to $f_1(x) = x^2 + 1$ and repeat. Explain what happens.
  d. Change f_1 to $f_1(x) = (x - 3)^2 + 2$ and repeat.
  e. In general, how are the graphs of $y = f(x)$ and $y = f(-x)$ related?

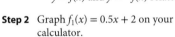

Step 2e $y = f(-x)$ is a reflection of $y = f(x)$ across the y-axis.

ELL
Use a mirror to discuss the idea of a reflection. Draw a diagram of a person and their reflection and link it to the graph of a function and its reflection using a table of values. Define *radical*.

Support
Most students recognize vertical reflections over the x-axis more frequently than they do horizontal reflections over the y-axis. Be sure to emphasize the difference between $y = f(-x)$ and $y = -f(x)$; some students will assume that these are equivalent by using a misinterpretation of the distributive property in function notation. See the ELL note.

Advanced
Ask students how they might change a function's equation to reflect its graph across the line $y = x$. Students can explore extending the idea of translations to cubics and other familiar functions.

Step 3c An entire horizontal parabola wouldn't be the graph of a function. $y = \sqrt{x}$ has a range of $y \geq 0$. $y = \sqrt{-x}$ would complete the bottom half of the parabola.

Step 4 $y = f(-x)$. $y = (x)^2$ and $y = (-x)^2$ both equal $y = x^2$. $f(x) = x^2$, which is a parabola with the axis of symmetry being the y-axis, so substituting $(-x)$ will not change the appearance of the graph.

Step 5 $y = -f(-x)$. $y = (x)^3$ and $y = -(-x)^3$ both equal $y = x^3$. $f(x) = x^3$ is symmetric with respect to the origin, so rotating 180° does not change the appearance of the graph.

Step 3 Graph $f_1(x) = \sqrt{x}$ on your calculator.

a. Predict what the graphs of $f_2 = -f_1(x)$ and $f_3 = f_1(-x)$ will look like. Use your calculator to verify your predictions. Write equations for both of these functions in terms of x.

b. Predict what the graph of $f_4 = -f_1(-x)$ will look like. Use your calculator to verify your prediction.

c. Notice that the graph of the square root function looks like half of a parabola, oriented horizontally. Why isn't it an entire parabola? What function would you graph to complete the bottom half of the parabola?

You may have noticed that some functions are symmetrical. How do reflections affect such functions?

Step 4 Graph $f(x) = x^2$. Which of the following functions, $y = -f(x)$, $y = f(-x)$, or $y = -f(-x)$, has a graph that looks the same? Make a prediction without actually graphing any of the new functions and explain your choice.

Step 5 Graph $f(x) = x^3$. Which of the following functions, $y = -f(x)$, $y = f(-x)$, or $y = -f(-x)$, has a graph that looks the same? Make a prediction without actually graphing any of the new functions and explain your choice.

Reflections over the x- or y-axis are summarized below.

> **Reflection of a Function**
>
> A **reflection** is a transformation that flips a graph across a line, creating a mirror image.
>
> Given the graph of $y = f(x)$, the graph of $y = f(-x)$ is a **horizontal reflection** across the y-axis, and the graph of $-y = f(x)$, or $y = -f(x)$, is a **vertical reflection** across the x-axis.

Because the graph of the square root function looks like half a parabola, it's easy to see the effects of reflections. The square root family has many real-world applications, such as dating prehistoric artifacts.

Science CONNECTION

Obsidian, a natural volcanic glass, was a popular material for tools and weapons in prehistoric times because it makes a very sharp edge. In 1960, scientists Irving Friedman and Robert L. Smith discovered that obsidian absorbs moisture at a slow, predictable rate and that measuring the thickness of the layer of moisture with a high-power microscope helps determine its age. Therefore, obsidian hydration dating can be used on obsidian artifacts, just as carbon dating can be used on organic remains. The age of prehistoric artifacts is predicted by a square root function similar to $d = \sqrt{5t}$, where t is time in thousands of years and d is the thickness of the layer of moisture in microns (millionths of a meter).

Investigate

Most students can complete this investigation and be prepared to work on the exercises with little or no help from you. For assistance in setting up a good window, see Calculator Note: Setting Windows.

Guiding the Investigation

Step 1 Calculator Note: Transformations and Compositions shows how to use f_1 in the equation of f_2. As needed, encourage students to do this instead of entering the first equation with the negative sign distributed, possibly forgetting to negate the second term. If students are neglecting to graph equations, suggest that they reread the instructions carefully. **SMP 1, 5**

Step 1a It is reflected across the x-axis.

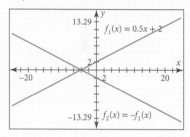

Step 1b It is reflected across the x-axis.

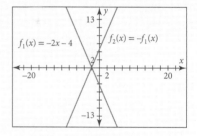

Step 1c It is reflected across the x-axis.

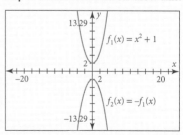

Step 2a It is reflected across the y-axis.

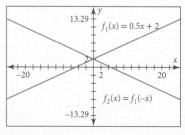

Modifying the Investigation

Whole Class Elicit student predictions for Step 1a. Have students complete Steps 1b-c on calculators. Discuss generalizations. Repeat for Steps 2 and 3.

Shortened Skip Steps 1b, 2d, and 3c.

One Step Ask students to graph the equations $y = \sqrt{x}$, $y = -\sqrt{x}$, and $y = \sqrt{-x}$ and to write down as many observations about the graphs as they can. During the discussion, ask about the domains and ranges of these functions and why inserting a negative sign reflects the graph in various ways.

Conceptual/Procedural

Conceptual Students explore reflections of functions and how they affect the graph and equation of the parent function. They learn the concept of even and odd functions.

Procedural Students learn the procedure for determining if a function is odd, even, or neither.

Step 2b It is reflected across the *y*-axis.

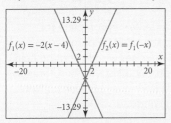

Step 2c It does not change, because the parabola has the *y*-axis as a line of symmetry, so a reflection across the *y*-axis maps the graph onto itself.

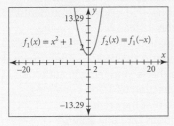

Step 2d It is reflected across the *y*-axis.

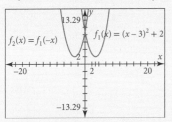

Step 3 **ALERT** Students might struggle with the idea of $f_1(-x)$ when $f_1(x) = \sqrt{x}$ thinking they need to take the square root of a negative. Take time to explain that $-x$ is the opposite of x, so if x itself is negative, then they will actually be taking the square root of its opposite, which is a positive.

Step 3a $f_2 = -f_1(x)$ will be a reflection across the *x*-axis; $f_3 = f_1(-x)$ will be a reflection across the *y*-axis.

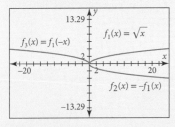

Step 3b $f_4(x) = -f_1(-x)$ will be a reflection across both axes.

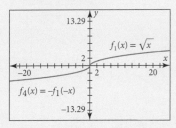

If a function is symmetric with respect to the *y*-axis, then a reflection over the *y*-axis does not change the appearance of the graph. Functions like $y = x^2$ and $y = |x|$ are examples of **even functions**, where $f(-x) = f(x)$ and whose graphs are symmetric with respect to the *y*-axis. If a function is symmetric with respect to the origin, then a 180° rotation does not change the appearance of the graph. Functions like $y = x$, $y = \sqrt[3]{x}$, $y = x^3$, and $y = \frac{1}{x}$ are examples of **odd functions**, where $f(-x) = -f(x)$ and whose graphs are symmetric with respect to the origin. Most functions, however, are neither even nor odd.

> **Even and Odd Functions**
>
> A function, $f(x)$, is an **even function** if $f(-x) = f(x)$. A function $f(x)$ is an **odd function** if $f(-x) = -f(x)$. An even function has the *y*-axis as a line of symmetry. An odd function is said to have symmetry with respect to the origin.

Example Determine if the following functions are even functions, odd functions, or neither.

   a. $f(x) = x^3 - 2x$
   b. $f(x) = x^3 + 2x^2$
   c. $f(x) = x^2 - 4$

Solution Both even and odd functions are defined based on $f(-x)$. If $f(-x) = f(x)$, the function is even and if $f(-x) = -f(x)$, the function is odd. So find $f(-x)$ and compare it to $f(x)$ and to $-f(x)$.

   a. For $f(x) = x^3 - 2x$

   $$f(-x) = (-x)^3 - 2(-x) = -x^3 + 2x$$
   $$-f(x) = -(x^3 - 2x) = -x^3 + 2x$$
   $$f(-x) = -f(x)$$

   The graph of the function $f(x) = x^3 - 2x$ is symmetric with respect to the origin.

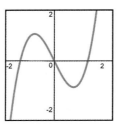

   Because $f(-x)$ is equal to $-f(x)$ and the graph of the function is symmetric with respect to the origin, $f(x)$ is an odd function.

   b. For $f(x) = x^3 + 2x^2$

   $$f(-x) = (-x)^3 + 2(-x)^2 = -x^3 + 2x^2$$
   $$-f(x) = -(x^3 + 2x^2) = -x^3 - 2x^2$$
   $$f(x) \neq f(-x) \neq -f(x)$$

174 CHAPTER 3 Functions and Relations

Example
Project the equation part of the definitions:

If $f(-x) = f(x)$ the function is an even function.

If $f(-x) = -f(x)$ the function is an odd function.

ASK "Is $y = x$, even or odd? [Odd] Is $y = x^2$, even or odd? [Even] Is $y = x^3$, even or odd? [Odd] Is $y = x^2 + x$, even or odd? [Neither]" Have students graph $y = x^2$ and $y = x^3$ to introduce the symmetry of even and odd functions. Then project the example and have students work in pairs. As students share their answers, **ASK** "How do you know?" As needed, remind them that determining the nature of the function includes the symmetry of the function. **SMP 3, 5, 6, 7**

The graph of the function $f(x) = x^3 + 2x^2$ is not symmetric with respect to the y-axis or the origin.

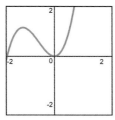

Because $f(x) \neq f(-x) \neq -f(x)$ and the graph of the function is not symmetric with respect to the y-axis or the origin, $f(x)$ is neither an even nor an odd function.

c. For $f(x) = x^2 - 4$

$f(-x) = (-x)^2 - 4 = x^2 - 4$

$f(-x) = f(x)$

The graph of the function $f(x) = x^2 - 4$ is symmetric with respect to the y-axis.

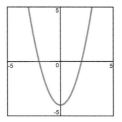

Because $f(-x) = f(x)$ and the graph of the function is symmetric with respect to the y-axis, $f(x)$ is an even function.

Using your knowledge about reflections will help you determine the affects on the equations and graphs of transformed functions.

You will need a graphing calculator for Exercises **8**, **9** and **14**.

Practice Your Skills

1. Each graph below is a transformation of the graph of the parent function $y = \sqrt{x}$. Write an equation for each graph. @

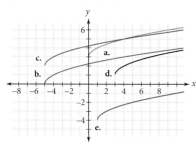

a. $y = \sqrt{x} + 3$
b. $y = \sqrt{x} + 5$
c. $y = \sqrt{x + 5} + 2$
d. $y = \sqrt{x - 3} + 1$
e. $y = \sqrt{x - 1} - 4$

LESSON 3.5 Reflections and the Square Root Family **175**

CRITICAL QUESTION What are the domain of the $f(x) = \sqrt{x}$?

BIG IDEA The domain and range are the nonnegative real numbers.

Formative Assessment

As you circulate while students are doing the investigation, collect anecdotal evidence of understanding reflections. Responses to Critical Questions and Exercises 2–4 will provide further evidence. Class discussion, Critical Questions and Exercises 1, 5, and 8–10 provide evidence of understanding the square root function. Student presentations of the Example and Exercises 3d, 6, 7 and 15 will provide evidence of the ability to determine if a function is odd, even, or neither.

Summarize

As students present their responses, emphasize the use of correct mathematical terminology. Whether student responses are correct or incorrect, ask other students if they agree and why.
SMP 1, 2, 3, 5, 6

Discussing the Investigation

LANGUAGE Mention that in the notation $\sqrt{x}$, the symbol $\sqrt{}$ is called a *radical* and the variable x is called the *radicand*. For example, $\sqrt{3}$ can be read as "radical three." Students may refer to this as "root 3." **SMP 6**

ASK "How can you remember which variable to replace to make a reflection?" When we replace x with $-x$, values horizontally opposite now act the way the original x-values did; so the reflection is horizontal, across the y-axis. Similarly, when y is replaced, values vertically opposite now act as the original y-values did so the reflection is across the x-axis.

As students present their ideas about Step 3c of the investigation, **ASK** "What is the range of the function $f(x) = \sqrt{x}$?" As students look at the graph, they may conjecture that the range omits some positive numbers, because the graph appears to approach a limit. Challenging students to find this limit can get them to explore large values of x and to see that they can get as large a value of y as they want. You might ask them what x-value will result in a y-value of 1000. [1000^2, or 1,000,000]

CRITICAL QUESTION If $x^2 = 5$, then $x = \pm\sqrt{5}$. Why do we use the plus or minus sign in front of the radical? If the radical indicates the square root and there are two of them, isn't the plus or minus sign redundant?

BIG IDEA No; the radical refers only to the positive, or principal, square root.

CRITICAL QUESTION The graph of $y = -f(x)$ is a reflection of $f(x)$ across the which axis?

BIG IDEA The x-axis.

CRITICAL QUESTION The graph of $y = f(-x)$ is a reflection of $f(x)$ across the which axis?

BIG IDEA The y-axis.

LESSON 3.5 Reflections and the Square Root Family **175**

Apply

Extra Example

1. Write an equation for the function $y = \sqrt{x}$ that has been reflected across the y-axis and translated up 3. $y = \sqrt{-x} + 3$

2. Describe what happens to the graph $y = f(x)$ when it is transformed into $y = -f(-x) + 2$.
reflected across x-axis, reflected over y-axis, translated up 2

Closing Question

What equation represents a reflection of the graph $y = \sqrt{x}$ across both axes?
$y = -\sqrt{-x}$ or $-y = \sqrt{-x}$

What equation represents a reflection of the graph of $y = x$ across both axes?
$y = x$

Assigning Exercises

Suggested: 1–10, 12, 15
Additional Practice: 11, 13, 14, 16–21

Exercise 2 As needed, suggest that students graph the equations.

Exercises 3, 4 SUPPORT If students are having difficulty reflecting the entire graph at once, encourage them to reflect each of the four marked points separately before reconnecting the segments. Students could use tracing paper to trace the function $y = f(x)$ and perform the reflection in one step.

4a.

4b.
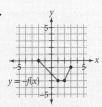

4c.
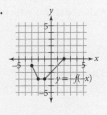

2. Describe what happens to the graph of $y = \sqrt{x}$ in the following situations.
   a. x is replaced with $(x - 3)$ @   translated horizontally 3 units
   b. x is replaced with $(x + 3)$ translated horizontally –3 units
   c. y is replaced with $(y - 2)$ @   translated vertically 2 units
   d. y is replaced with $(y + 2)$ translated vertically –2 units

3. Each graph at right is a transformation of the piecewise function $f(x)$. Match each equation to a graph.
   a. $y = f(-x)$ iii
   b. $y = -f(x)$ i
   c. $y = -f(-x)$ ii
   d. Is $f(x)$ even, odd, or neither? How do you know?
   Neither. It is neither symmetric with respect to the y-axis nor the origin.

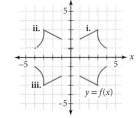

4. Given the graph of $y = f(x)$ at right, draw a graph of each of these related functions.
   a. $y = f(-x)$ @
   b. $y = -f(x)$
   c. $y = -f(-x)$

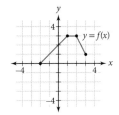

5. Each curve below is a transformation of the graph of the parent function $y = \sqrt{x}$. Write an equation for each curve. @

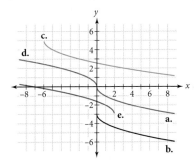

   a. $y = -\sqrt{x}$
   b. $y = -\sqrt{x} - 3$
   c. $y = -\sqrt{x+6} + 5$
   d. $y = \sqrt{-x}$
   e. $y = \sqrt{-(x-2)} - 3$, or $y = \sqrt{-x+2} - 3$

6. Determine if the functions graphed below are even, odd, or neither.

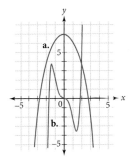
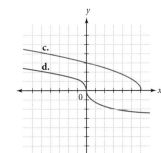

   a. even
   b. odd
   c. neither
   d. odd

176 CHAPTER 3 Functions and Relations

Reason and Apply

7. Determine whether the following functions are even, odd, or neither and justify your answer.

   a. $f(x) = \dfrac{5}{x}$
   b. $f(x) = \dfrac{5}{x^2}$
   c. $f(x) = |x| + 1$
   d. $f(x) = |x + 1|$
   e. $f(x) = 4x - 5$
   f. $f(x) = (x - 2)^2$

8. Consider the parent function $f(x) = \sqrt{x}$.

   a. Name three pairs of integer coordinates that are on the graph of $y = f(x + 4) - 2$. possible answers: $(-4, -2)$, $(-3, -1)$, and $(0, 0)$
   b. Write $y = f(x + 4) - 2$ using a **radical**, or square root symbol, and graph it.
   c. Write $y = -f(x - 2) + 3$ using a radical, and graph it.

9. Consider the parabola below.

   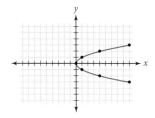

   a. Graph the parabola on your calculator. What two functions did you use?
   b. Combine both functions from 9a using ± notation to create a single relation. Square both sides of the relation. What is the resulting equation? $y = \pm\sqrt{x}$; $y^2 = x$

10. Refer to the two parabolas below.

    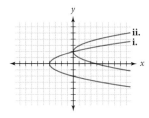

    a. Explain why neither graph represents a function.   a. Neither parabola passes the vertical line test.
    b. Write a single equation for each parabola using ± notation. @   b. i. $y = \pm\sqrt{x + 4}$
    c. Square both sides of each equation in 10b. What is the resulting equation of each parabola?
       ii. $y = \pm\sqrt{x} + 2$
       c. i. $y^2 = x + 4$
       ii. $(y - 2)^2 = x$

7a. Odd because $f(-x) = -f(x)$ and the graph is symmetric with respect to the origin.

7b., c. Even because $f(-x) = f(x)$ and the graph is symmetric with respect to the y-axis.

7d., e., f. Neither because $f(-x) \neq f(x)$ and $f(-x) \neq -f(x)$.

Exercise 8 ALERT In 8b and 8c, students may enclose the entire right side of the equation under the radical. Suggest that they graph $f_1 = \sqrt{x}$, $f_2 = f_1(x + 4) - 2$, and f_3 (the equation they wrote) to see whether the graphs of f_2 and f_3 agree.

8b. $y = \sqrt{x + 4} - 2$

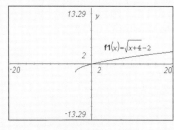

8c. $y = -\sqrt{x - 2} + 3$

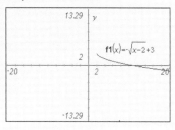

9a. $y = \sqrt{x}$ and $y = -\sqrt{x}$

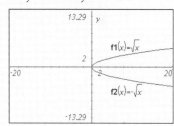

LESSON 3.5 Reflections and the Square Root Family

11a. possible answer

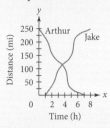

12a. $y = -x^2$

12b. $y = -x^2 + 2$

12c. $y = -(x-6)^2$

12d. $y = -(x-6)^2 - 3$

Exercise 14a **ALERT** Students may not understand that they're being asked simply to substitute 0.7 for f.

14b.

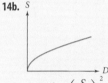

14d. $D = \dfrac{1}{0.7}\left(\dfrac{S}{5.5}\right)^2$; the minimum braking distance, when the speed is known

14e.

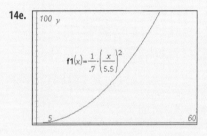

It is a parabola, but the negative half is not used because the distance cannot be negative.

Exercise 15 Suggest that students start by graphing a piece-wise defined function that meets the criteria and then write the function from the graph.

---

11. As Jake and Arthur travel together from Detroit to Chicago, each makes a graph relating time and distance. Jake, who lives in Detroit and keeps his watch on Detroit time, graphs his distance from Detroit. Arthur, who lives in Chicago and keeps his watch on Chicago time (1 hour earlier than Detroit), graphs his distance from Chicago. They both use the time shown on their watches for their x-axes. The distance between Detroit and Chicago is 250 miles.

  a. Sketch what you think each graph might look like. ⓗ

  b. If Jake's graph is described by the function $y = f(x)$, what function describes Arthur's graph?

  c. If Arthur's graph is described by the function $y = g(x)$, what function describes Jake's graph?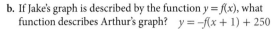

12. Each parabola at right is a transformation of the graph of the parent function $y = x^2$. Write the equation of each parabola. Which parabola(s) are even functions? How do you know? **a.** and **b.** are even because they are symmetric with respect to the y-axis.

13. Write the equation of a parabola that is congruent to the graph of $y = -(x+3)^2 + 4$, but translated right 5 units and down 2 units.

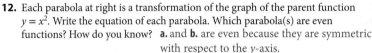

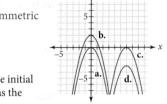

14. **APPLICATION** Police measure the lengths of skid marks to determine the initial speed of a vehicle before the brakes were applied. Many variables, such as the type of road surface and weather conditions, play an important role in determining the speed. The formula used to determine the initial speed is $S = 5.5\sqrt{D \cdot f}$, where S is the speed in miles per hour, D is the average length of the skid marks in feet, and f is a constant called the "drag factor." At a particular accident scene, assume it is known that the road surface has a drag factor of 0.7.

  a. Write an equation that will determine the initial speed on this road as a function of the lengths of skid marks.
  $S = 5.5\sqrt{0.7D}$

  b. Sketch a graph of this function.

  c. If the average length of the skid marks is 60 feet, estimate the initial speed of the car when the brakes were applied.
  approximately 36 mi/h

  d. Solve your equation from 14a for D. What can you determine using this equation?

  e. Graph your equation from 14d. What shape is it? ⓐ

  f. If you traveled on this road at a speed of 65 miles per hour and suddenly slammed on your brakes, how long would your skid marks be? approximately 199.5 ft

15. Write a piece-wise defined function with 2 pieces so that

  a. the function is odd.

  b. the function is even.

  c. the function is neither odd nor even.

178 CHAPTER 3 Functions and Relations

15a. Answers will vary. The pieces together must be symmetric to the origin

15b. Answers will vary. The pieces together must be symmetric to the y-axis

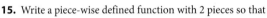

15c. Answers will vary. Any function that is not symmetric to the origin or y-axis.

Review

16. Identify each relation that is also a function. For each relation that is not a function, explain why not.

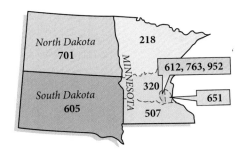

  a. independent variable: state
     dependent variable: area code
  b. independent variable: any pair of whole numbers
     dependent variable: their greatest common factor function
  c. independent variable: any pair of fractions
     dependent variable: their common denominator
  d. independent variable: the day of the year
     dependent variable: the time of sunrise
  Possible answer: Function; the sun rises at only one time on each day of a given year.

17. Solve for x. Solving square root equations often results in answers that don't work in the original equation, so be sure to check your work.

  a. $3 + \sqrt{x-4} = 20$ @ $x = 293$
  b. $\sqrt{x+7} = -3$ no solution
  c. $4 - (x-2)^2 = -21$ $x = 7$ or $x = -3$
  d. $5 - \sqrt{-(x+4)} = 2$ @ $x = -13$

18. Find the equation of the parabola with vertex $(-6, 4)$, a vertical line of symmetry, and containing the point $(-5, 5)$. $y = (x+6)^2 + 4$

19. The graph of the line ℓ_1 is shown at right.

  a. Write the equation of the line ℓ_1. $y = \dfrac{1}{2}x + 5$
  b. The line ℓ_2 is the image of the line ℓ_1 translated right 8 units. Sketch the line ℓ_2 and write its equation in a way that shows the horizontal translation. @
  c. The line ℓ_2 also can be thought of as the image of the line ℓ_1 after a vertical translation. Write the equation of the line ℓ_2 in a way that shows the vertical translation.
  $$y = \left(\dfrac{1}{2}x + 5\right) - 4, \text{ or } y + 4 = \dfrac{1}{2}x + 5$$
  d. Show that the equations in 19b and c are equivalent.
  Both equations are equivalent to $y = \dfrac{1}{2}x + 1$

20. Consider this data set:

  $\{37, 40, 36, 37, 37, 49, 39, 47, 40, 38, 35, 46, 43, 40, 47, 49, 70, 65, 50, 73\}$ @

  a. Give the five-number summary. 35, 37.5, 41.5, 49, 73
  b. Display the data in a box plot.

  20b.

  c. Find the interquartile range. 11.5
  d. Identify any outliers, based on the interquartile range. 70 and 73

21. Find the intersection of the lines $2x + y = 23$ and $3x - y = 17$. $(8, 7)$

LESSON 3.6

Dilations and the Absolute-Value Family

This is the third lesson in the sequence discussing transformations. This lesson focuses on dilations, primarily from absolute-value and square root functions.

COMMON CORE STATE STANDARDS
APPLIED: F.IF.7b
DEVELOPED: F.BF.3
INTRODUCED:

Objectives
- Define absolute value and its notation and use it to model distance
- Define the parent absolute-value function, $y = |x|$, and the absolute-value family, $\frac{y-k}{b} = \left|\frac{x-h}{a}\right|$
- Calculate horizontal and vertical scale factors from points on the image of a graph
- Apply horizontal and vertical dilations to functions in general

Vocabulary
magnitude
dilations
scale factor
vertical dilation
horizontal dilation
rigid transformations
nonrigid transformations

Materials
- string
- small weights
- stopwatches, or watches with second hand
- metersticks or tape measures
- graph paper, *optional*
- Calculator Notes: Graphing Absolute Value Functions, Graphing Transformations

Launch

On a number line, what numbers have a distance of 10 units from 0? 10, −10

What numbers have a distance of −8 units from 0? none

What is the value of x if $(x + 2)$ is 3.675 units from 0? $x = 1.675, -5.675$

A mind that is stretched by a new experience can never go back to its old dimensions.
OLIVER WENDELL HOLMES

Hao and Dayita ride the subway to school each day. They live on the same east-west subway route. Hao lives 7.4 miles west of the school, and Dayita lives 5.2 miles east of the school. This information is shown on the number line below.

The distance between two points is always positive. However, if you calculate Hao's distance from school, or *HS*, by subtracting his starting position from his ending position, you get a negative value:

$$-7.4 - 0 = -7.4$$

In order to make the distance positive, you use the absolute-value function, which gives the **magnitude** of a number, or its distance from zero on a number line. For example, the absolute value of −3 is 3, or $|-3| = 3$. For Hao's distance from school, you use the absolute-value function to calculate

$HS = |-7.4 - 0| = |-7.4| = 7.4$
$DH = |5.2 (-7.4)| = |12.6| = 12.6$ and
$HD = |-7.4 - 5.2| = |-12.6| = 12.6$

What is the distance from *D* to *H*? What is the distance from *H* to *D*?

In this lesson you will explore transformations of the graph of the parent function $y = |x|$. You will write and use equations in the form $\frac{y-k}{b} = \left|\frac{x-h}{a}\right|$. What you have learned about translating and reflecting other graphs will apply to these functions as well. You will also learn about transformations called **dilations** that stretch and shrink a graph.

You may have learned about dilations of geometric figures in an earlier course. Now you will apply dilations to functions.

If you dilate a figure by the same **scale factor** both vertically and horizontally, then the image and the original figure will be similar and perhaps congruent.

If you dilate by different vertical and horizontal scale factors, then the image and the original figure will not be similar.

ELL
There is a lot of new vocabulary in this section. Make connections between the social and mathematical definitions of the terms *transformations, rigid, stretch,* and *shrink*. Use examples and non-examples to help students understand the difference between *rigid* and *nonrigid* transformations, and differentiate the term *scale factor* from other common definitions of the word scale.

Support
Students should use graphing calculators to substitute many different values for *a* and *b* in order to readily recognize the effect of the horizontal and vertical dilations on the graph.

Advanced
Have students experiment with taking the absolute value of some of the functions they have already worked with and discussing their qualities: Why does $y = |\sqrt{x}|$ not change? Why do $y = |x^2 - 3|$ and $y = |x^2| - 3$ differ?

EXAMPLE A | Graph the function $y=|x|$ with each of these functions. How does the graph of each function compare to the original graph?

  a. $\dfrac{y}{2}=|x|$

  b. $y=\left|\dfrac{x}{3}\right|$

  c. $\dfrac{y}{2}=\left|\dfrac{x}{3}\right|$

Solution | In the graph of each function, the vertex remains at the origin. Notice, however, how the points (1, 1) and (−2, 2) on the parent function are mapped to a new location.

a. Replacing y with $\dfrac{y}{2}$ pairs each x-value with twice the corresponding y-value in the parent function. The graph of $\dfrac{y}{2}=|x|$ is a vertical stretch, or a **vertical dilation,** of the graph of $y=|x|$ by a factor of 2.

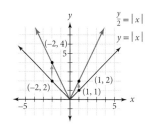

b. Replacing x with $\dfrac{x}{3}$ multiplies the x-coordinates by a factor of 3. The graph of $y=\left|\dfrac{x}{3}\right|$ is a horizontal stretch, or a **horizontal dilation,** of the graph of $y=|x|$ by a factor of 3.

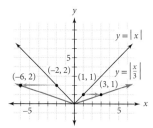

c. The combination of replacing y with $\dfrac{y}{2}$ and replacing x with $\dfrac{x}{3}$ results in a vertical dilation by a factor of 2 *and* a horizontal dilation by a factor of 3.

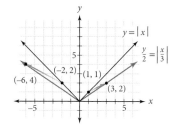

Translations and reflections are **rigid transformations**—they produce an image that is congruent to the original figure. Vertical and horizontal dilations are **nonrigid transformations**—the image is not congruent to the original figure (unless you use a factor of 1 or −1).

Using what you know about translations, reflections, and dilations, you can fit functions to data by locating only a few key points. For quadratic, square root, and absolute-value functions, first locate the vertex of the graph. Then use any other point to find the factors by which to dilate the image horizontally and/or vertically.

Investigate

Example A

Project Example A and have students work in pairs. As they present their graphs, have them compare it to the original function. **SMP 1, 3, 5, 6**

The vertical dilation in part a can also be thought of as a horizontal dilation, because $y=2|x|$ is the same as $y=|2x|$ or $y=\left|\dfrac{x}{\left(\frac{1}{2}\right)}\right|$. Similarly, the horizontal dilation of part b is a vertical dilation, and the combination of part c is equivalent to either a vertical dilation $\left(y=\dfrac{2}{3}|x|\right)$ or a horizontal dilation $\left(y=\left|\dfrac{2}{3}x\right|\right)$.

CONTEXT The French mathematician Augustin-Louis Cauchy ['kō-shē] (1789–1857) first described the absolute-value function in the 1820s. In 1841, the German mathematician Karl Weierstrass (1815–1897) introduced the absolute-value symbol used today.

The equations for the graphs in the **One Step**:

a. $y=\dfrac{1}{2}|x|$
b. $y=2|x|$
c. $y=\dfrac{1}{4}(x-1)^2$
d. $y=3(x-1)^2$
e. $y=2f(x-2)$
f. $y=\dfrac{1}{2}f(x-1)$

Modifying the Investigation

Whole Class Have three students collect data for the whole class. Students can then do Step 2 and discuss Step 3. Have students do Step 4 and discuss Step 5 as a class.

Shortened Use the Investigation Worksheet The Pendulum with Sample Data or the desmos simulation in your ebook.

One Step Use the **One Step** found in your ebook. As needed, remind them of the meaning of the absolute-value function. As groups finish their work, ask them to create (on graph paper) mystery graphs involving transformations of the graph of the square root function and to exchange them with each other as challenges. During the discussion, formalize the rules for dilations and review the rules for translations and reflections.

Example B

You could turn Example B into an investigation by having students collect their own data using a motion sensor and a ball. Or you might project Example B and have students work in small groups. As they present their equation, have them explain their strategy. **SMP 1, 3, 4, 5, 6**

This example is an important illustration of a composition of translations, dilations, and reflections of the quadratic family of equations in a real-world context.

In discussing Example B, **ASK** "How can the book assume that the data point (1.14, 0.18) is the image of (1, 1)? What if some other point on the new curve is the image of (1, 1)? For example, what if we assume that data point (0.54, 0.05) is the image of (1, 1)?" [This data point is 0.54 − 0.86, or −0.32, from the vertex horizontally and 0.05 − 0.60, or −0.55, from the vertex vertically, so the new equation is $\frac{y - 0.6}{-0.55} = \left(\frac{x - 0.86}{-0.32}\right)^2$. This equation is equivalent to $y = -5.37(x - 0.86)^2 + 0.6$, very close to the equation in the example, which can be rewritten as $y = -5.36(x - 0.86)^2 + 0.6$.] Students might see that the equations are close because $\frac{-0.42}{0.28^2} \approx \frac{-0.55}{(-0.32)^2}$. In general, for a parabola, $\frac{b}{a^2}$ is constant (where b and a are the vertical and horizontal scale factors, respectively).

ASK "How would you generalize this for any function?" In general, a vertical dilation of $y = f(x)$ by a factor of b gives $\frac{y}{b} = f(x)$. A horizontal dilation of that function by a factor of a gives $\frac{y}{b} = \frac{f(x)}{a}$. A horizontal translation of h and a vertical translation of k gives $\frac{(y-k)}{b} = \frac{f(x-h)}{a}$.

EXAMPLE B

These data are from one bounce of a ball. Find an equation that fits the data over this domain.

Time (s) x	Height (m) y	Time (s) x	Height (m) y
0.54	0.05	0.90	0.59
0.58	0.18	0.94	0.57
0.62	0.29	0.98	0.52
0.66	0.39	1.02	0.46
0.70	0.46	1.06	0.39
0.74	0.52	1.10	0.29
0.78	0.57	1.14	0.18
0.82	0.59	1.18	0.05
0.86	0.60		

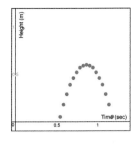

Solution

The graph appears to be a parabola. However, the parent function $y = x^2$ has been reflected, translated, and dilated. Start by determining the translations. The vertex has been translated from (0, 0) to (0.86, 0.60). This is enough information for you to write the equation in the form $y = (x - h)^2 + k$, or $y = (x - 0.86)^2 + 0.60$. If you think of replacing x with $(x - 0.86)$ and replacing y with $(y - 0.60)$, you could also write the equivalent equation, $y - 0.6 = (x - 0.86)^2$.

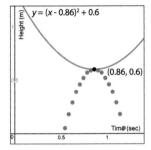

The graph still needs to be reflected and dilated. Select one other data point to determine the horizontal and vertical scale factors. You can use any point, but you will get a better fit if you choose one that is not too close to the vertex. For example, you can choose the data point (1.14, 0.18).

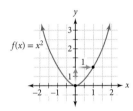
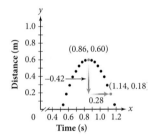

Assume this data point is the image of the point (1, 1) in the parent parabola $y = x^2$. In the graph of $y = x^2$, (1, 1) is 1 unit away from the vertex (0, 0) both horizontally and vertically. The data point we chose in this graph, (1.14, 0.18), is 1.14 − 0.86, or 0.28, unit away from the x-coordinate of the vertex, and 0.18 − 0.60, or −0.42, unit away from the y-coordinate of the vertex.

So the horizontal scale factor is 0.28, and the vertical scale factor is −0.42. The negative vertical scale factor also produces a vertical reflection.

Conceptual/Procedural

Conceptual This lessons reviews the concept of absolute value as a distance. The lesson looks at dilations of functions, primarily the absolute value and square root functions. They review the concept of rigid and nonrigid transformations.

Procedural Students use graphs to compare functions that have been dilated. They find an equation for data that has been reflected, translated, and dilated from the parent function $y = x^2$.

Combine these scale factors with the translations to get the final equation

$$\frac{y - 0.6}{-0.42} = \left(\frac{x - 0.86}{0.28}\right)^2 \text{ or } y = -0.42\left(\frac{x - 0.86}{0.28}\right)^2 + 0.6$$

This model, graphed at right, fits the data nicely.

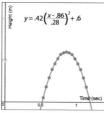

The same procedure works with the other functions you have studied so far. As you continue to add new functions to your mathematical knowledge, you will find that what you have learned about function transformations continues to apply.

INVESTIGATION

The Pendulum

YOU WILL NEED
- string
- a small weight
- a stopwatch or a watch with a second hand

Italian mathematician and astronomer Galileo Galilei (1564–1642) made many contributions to our understanding of gravity, the physics of falling objects, and the orbits of the planets. One of his famous experiments involved the periodic motion of a pendulum. In this investigation you will carry out the same experiment and find a function to model the data.

Procedure Note
1. Tie a weight at one end of a length of string to make a pendulum. Firmly hold the other end of the string, or tie it to something, so that the weight hangs freely.
2. Measure the length of the pendulum, from the center of the weight to the point where the string is held.
3. Pull the weight to one side and release it so that it swings back and forth in a short arc, about 10° to 20°. Time ten complete swings (forward and back is one swing).
4. The **period** of your pendulum is the time for one complete swing (forward and back). Find the period by dividing by 10.

This fresco, painted in 1841, shows Galileo at age 17, contemplating the motion of a swinging lamp in the Cathedral of Pisa. A swinging lamp is an example of a pendulum.

Step 1 Follow the Procedure Note to find the period of your pendulum. Repeat the experiment for several different string lengths and complete a table of values. Use a variety of short, medium, and long string lengths.

Step 2 Graph the data using *length* as the independent variable. What is the shape of the graph? What do you suppose is the parent function?

Step 3 The vertex is at the origin, (0, 0). Why do you suppose it is there?

Step 4 Have each member of your group choose a different data point and use that data point to find the horizontal and vertical dilations. Apply these transformations to find an equation to fit the data.

Step 5 Compare the collection of equations from your group. Which points are the best to use to fit the curve? Why do these points work better than others? Points farther from the vertex work best. These points represent the longer lengths. They are best for fitting a parabola because they are likely to have less measurement error. A parabola that fits the first few points well would probably be quite far from the points farther from the vertex.

Step 4 The equation should be in the form $\frac{y}{b} = \sqrt{\frac{x}{a}}$, or $y = b\sqrt{\frac{x}{a}}$. For the sample data, the function $y = 0.2\sqrt{x}$ is a good fit.

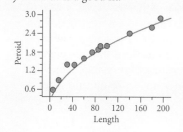

Guiding the Investigation

If you don't want to do the Investigation as an activity, there is an Investigation Worksheet with sample data. Possible answers here are based on that sample data. There is also a desmos simulation for this investigation in your ebook. Give students a string at least 2 m long. Encourage students to use a variety of string lengths, including several very short lengths and at least one very long length. If they don't cut their string, they can collect more data later.

Students may wonder whether the measure of the arc of the swing or the amount of weight will affect the period. (Encourage students to test these parameters if there is time. As long as the horizontal displacement of the weight is small compared to the length of the pendulum, the angle measure does not affect the period.)

Theoretically, the period of a pendulum swinging without resistance is given by $2\pi\sqrt{\frac{L}{g}}$, where L is the length and g is the gravitational constant. If students measure in centimeters, g is about 980 cm/s², so they'll get about $0.2\sqrt{L}$. If they measure in inches, g is about 384 in./s², so they'll get about $0.32\sqrt{L}$.

For some simple functions, assuming that one point is the image of another can determine the two dilation factors for the function. Using a different pair of points gives the same graph and equivalent equations. For more complicated functions and relations, students must check at least two points.

Step 1 See the worksheet for sample data.

Step 2 sample data:

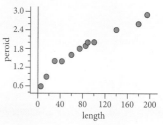

The parent function is the square root function, $y = \sqrt{x}$.

Step 3 The vertex is at the origin because a pendulum of length 0 cm would have no period.

Summarize

As students present their solutions to the Investigation and examples, emphasize the use of correct mathematical terminology. For example, use the term *dilation* rather than *stretch* or *shrink* whenever possible. It's clearer, for example, to say "a vertical dilation by a factor of $\frac{1}{2}$" than "a vertical stretch by a factor of $\frac{1}{2}$," which actually is a shrink. Whether student responses are correct or incorrect, ask other students if they agree and why. **SMP 1, 3, 4, 5, 6**

To the extent possible, choose students for presenting who obtained different results, especially if they measured in different units. Then have the class look for explanations for the differences. **SMP 1, 3, 4, 5, 6**

CRITICAL QUESTION Why do you divide rather than multiply a variable by the scale factor to change an equation?

BIG IDEA When a variable is divided by a constant, the divided value plays the same role in the equation that the original variable did. So if, for example, x is divided by 3, values of x that are 3 times as large will now have the same effect on the equation that the original values of x had.

CRITICAL QUESTION Can lines not through the origin also be thought of as transformations of the parent line, $y = x$?

BIG IDEA A vertical translation of the same line by an amount a gives the familiar equation $y = a + bx$, so every nonvertical line is a dilation followed by a translation of the parent line, $y = x$.

CRITICAL QUESTION In an equation, dividing x by a positive number a produces an equation of what kind of dilation?

BIG IDEA A horizontal dilation by factor a.

CRITICAL QUESTION In an equation, dividing y by positive number b results in an equation of what kind of dilation?

BIG IDEA A vertical dilation by factor b.

In the exercises you will use techniques you discovered in this lesson. Remember that replacing y with $\frac{y}{b}$ dilates a graph by a factor of b vertically. Replacing x with $\frac{x}{a}$ dilates a graph by a factor of a horizontally. When graphing a function, you should do dilations before translations to avoid moving the vertex. When finding the equation for a graph, the process is reversed, so you estimate translations first and dilations second, as shown in Example B.

> **Dilation of a Function**
>
> A **dilation** is a transformation that expands or compresses a graph either horizontally or vertically.
>
> Given the graph of $y = f(x)$, the graph of
>
> $$\frac{y}{b} = f(x) \quad \text{or} \quad y = bf(x)$$
>
> is a vertical dilation by a factor of b. When $|b| > 1$, it is a stretch; when $0 < |b| < 1$, it is a shrink. When $b < 0$, a reflection across the x-axis also occurs.
>
> Given the graph of $y = f(x)$, the graph of
>
> $$y = f\left(\frac{x}{a}\right)$$
>
> is a horizontal dilation by a factor of a. When $|a| > 1$, it is a stretch; when $0 < |a| < 1$, it is a shrink. When $a < 0$, a reflection across the y-axis also occurs.

3.6 Exercises

You will need a graphing calculator for Exercises **5, 6, 12,** and **14**.

Practice Your Skills

1. Each graph is a transformation of the graph of one of the parent functions you've studied. Write an equation for each graph.

   a. $y = |x| + 2$ **b.** $y = |x| - 5$ **c.** $y = |x + 4|$ **d.** $y = |x - 3|$ **e.** $y = |x| - 1$ **f.** $y = |x - 4| + 1$

   g. $y = |x + 5| - 3$ **h.** $\frac{y}{3} = |x - 6|$, or $y = 3|x - 6|$ **j.** $y = (x - 5)^2$ **k.** $-2y = |x + 4|$, or $y = -\frac{1}{2}|x + 4|$ **m.** $y = -(x + 3)^2 + 5$ **n.** $y = \pm\sqrt{x - 4} + 3$

   i. $y = -\left|\frac{x}{4}\right|$

   l. $y = -|x + 4| + 3$

   p. $\frac{y}{-2} = \left|\frac{x - 3}{3}\right|$, or $y = -2\left|\frac{x - 3}{3}\right|$

CRITICAL QUESTION What can you tell about a dilation by looking a and b in the equation $\frac{(y - k)}{b} = \frac{f(x - h)}{a}$?

BIG IDEA The dilation is a stretch if the divisor is more than 1; it is a shrink if it's less than 1. If a or b is negative, the graph is reflected across an axis as well as dilated.

Formative Assessment

Having students work in small groups on the Examples and the Investigation will provide anecdotal evidence of their proficiency with the objectives. As you circulate among the groups, watch to see how flexible students are at transforming the various functions they are working with. See whether students are able to explain the difference between horizontal and vertical dilations.

a. horizontal dilation by a factor of 3

2. Describe what happens to the graph of $y = f(x)$ in these situations.
   a. x is replaced with $\frac{x}{3}$
   b. x is replaced with $-x$ reflection across the y-axis
   c. x is replaced with $3x$ horizontal dilation by a factor of $\frac{1}{3}$
   d. y is replaced with $\frac{y}{2}$ vertical dilation by a factor of 2
   e. y is replaced with $-y$ reflection across the x-axis
   f. y is replaced with $2y$ vertical dilation by a factor of $\frac{1}{2}$

3. Solve each equation for y.
   a. $\frac{y+3}{2} = (x-5)^2$
   $y = 2(x-5)^2 - 3$
   b. $\frac{y+5}{2} = \left|\frac{x+1}{3}\right|$
   $y = 2\left|\frac{x+1}{3}\right| - 5$
   c. $\frac{y+7}{-2} = \sqrt{\frac{x-6}{-3}}$
   $y = -2\sqrt{\frac{x-6}{-3}} - 7$

Reason and Apply

4. Choose a few different values for b. What can you conclude about $y = b|x|$ and $y = |bx|$? Are they the same function? ⓗ

5. The graph at right shows how to solve the equation $|x - 4| = 3$ graphically. The equations $y = |x - 4|$ and $y = 3$ are graphed on the same coordinate axes.
   a. What is the x-coordinate of each point of intersection? What x-values are solutions of the equation $|x - 4| = 3$? 1 and 7; $x = 1$ and $x = 7$
   b. Solve the equation $|x + 3| = 5$ algebraically. Verify your solution with a graph.

6. **APPLICATION** You can use a single radio receiver to find the distance to a transmitter by measuring the strength of the signal. Suppose these approximate distances are measured with a receiver while you drive along a straight road. Find a model that fits the data. Where do you think the transmitter might be located? @

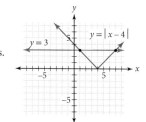

Miles traveled	0	4	8	12	16	20	24	28	32	36
Distance from transmitter (miles)	18.4	14.4	10.5	6.6	2.5	1.8	6.0	9.9	13.8	17.6

7. Assume that the parabola $y = x^2$ is translated so that its vertex is $(5, -4)$.
   a. If the parabola is dilated vertically by a factor of 2, what are the coordinates of the point on the parabola 1 unit to the right of the vertex? @ $(6, -2)$
   b. If the parabola is dilated horizontally instead, by a factor of 3, what are the coordinates of the points on the parabola 1 unit above the vertex? @ $(2, -3)$ and $(8, -3)$
   c. If the parabola is dilated vertically by a factor of 2 *and* horizontally by a factor of 3, name two points on the new parabola that are symmetric with respect to the vertex. $(2, -2)$ and $(8, -2)$

8. A curve with parent function $f(x) = x^2$ has vertex $(7, 3)$ and passes through the point $(11, 11)$. ⓗ
   a. What are the values of h and k in the equation of the curve? $h = 7, k = 3$
   b. Substitute the values for h and k from 8a into $y = k + a \cdot f(x - h)$. Substitute the coordinates of the other point into the equation as values for x and y. $11 = 3 + a(11-7)^2$
   c. Solve for a and write the complete equation of the curve. Confirm that the graph passes through both points. $a = \frac{11-3}{(11-7)^2} = \frac{8}{16} = 0.5$
   d. Write the equation in the form $\frac{y-k}{b} = \left(\frac{x-h}{a}\right)^2$ by considering the horizontal and vertical dilations separately, as in Example B.
   e. Use algebra to show that your answers from 8c and d are equivalent.
   $y - 3 = 8\frac{(x-7)^2}{4^2}$, $y - 3 = 8\frac{(x-7)^2}{16}$, $y = 3 + \frac{8}{16}(x-7)^2$, $y = +0.5(x-7)^2$; the equations are equivalent

LESSON 3.6 Dilations and the Absolute-Value Family **185**

Exercise 6 **ALERT** Students may say that the transmitter is 1.8 mi off the road 20 mi from the starting point. As needed, encourage them to graph the data in order to find the parent function and to write an equation for the transformation.

6. $\hat{y} \approx |x - 18.4|$. The transmitter is located on the road approximately 18.4 mi from where you started.

Exercise 8 Functions like these can also be written so that one of the dilation factors is equal to 1; this makes the equation look less complicated. If $b = 1$, then when you solve for y, you find $y = k + a \cdot f(x - h)$. After finding h and k, you can find the value of a by replacing x and y with some point from the data and solving for a.

8d. $b = \Delta y = 11 - 3 = 8$,
$a = \Delta x = 11 - 7 = 4$, $\frac{y-3}{8} = \left(\frac{x-7}{4}\right)^2$

Apply

Extra Example

1. Describe what happens to the graph of $y = f(x)$ when it is transformed into $3y = f\left(\frac{x}{2}\right)$.
   vertical dilation (shrink) by a factor of $\frac{1}{3}$, horizontal dilation (stretch) by a factor of 2

2. Write an equation for the function that results from translating $y = |x|$ left 3 units and vertically dilating by a factor of 2. $y = 2|x + 3|$

Closing Question

Why does dividing x or y by a number less than 1 result in a dilated image that is smaller than the original, a shrink?

Smaller values of x or y will describe the same points on the graph as larger values did in the original equation.

Assigning Exercises

Suggested: 1, 2, 4 – 9, 11

Additional Practice: 3, 10, 12 – 14

Exercise 4 As needed, point out that these are vertical stretches and horizontal shrinks by the same factor (if $b > 1$). **ASK** "Are there other functions for which a vertical stretch by a factor yields the same graph as a horizontal shrink by the same factor?"

Students can experiment with the parent parabola, the parent square root function, and the parent line.

4. For $b > 0$, the graphs of $y = b|x|$ and $y = |bx|$ are equivalent. For $b < 0$, the graph of $y = |bx|$ is a reflection of $y = |bx|$ across the x-axis.

5b. $x = -8$ and $x = 2$

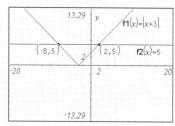

LESSON 3.6 Dilations and the Absolute-Value Family **185**

9. The parabola is dilated vertically by a factor of 3, dilated horizontally by a factor of 4, and translated horizontally −7 units and vertically 2 units.

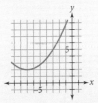

10a.

10b.
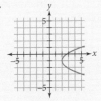

10c.
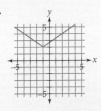

Exercise 12 LANGUAGE μS is the abbreviation for microsiemens. A siemens is equal to 1 ampere per volt (amp/V). CONTEXT The conductivity of the solution is directly related to the concentration of ions, independent of their charge. As the acid is added, the concentration of ions decreases as water molecules are formed, until the solution is neutral; it then increases as the solution becomes more acidic.

12.
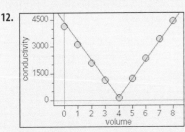

possible equation:
$y = 1050\,|x - 4| + 162$

9. Given the parent function $y = x^2$, describe the transformations represented by the function $\dfrac{y-2}{3} = \left(\dfrac{x+7}{4}\right)^2$. Sketch a graph of the transformed parabola.

10. Sketch a graph of each of these equations.

   a. $\dfrac{y-2}{3} = (x-1)^2$ b. $\left(\dfrac{y+1}{2}\right)^2 = \dfrac{x-2}{3}$ c. $\dfrac{y-2}{2} = \left|\dfrac{x+1}{3}\right|$

11. Given the graph of $y = f(x)$, draw graphs of these related functions.

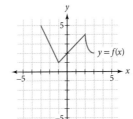   11a. 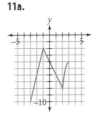   11b. 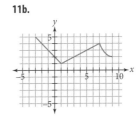   11c.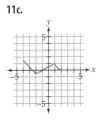

   a. $\dfrac{y}{-2} = f(x)$ b. $y = f\left(\dfrac{x-3}{2}\right)$ c. $\dfrac{y+1}{\tfrac{1}{2}} = f(x+1)$ @

12. **APPLICATION** A chemistry class gathered these data on the conductivity of a base solution as acid is added to it. Graph the data and use transformations to find a model to fit the data.

Acid volume (mL) x	Conductivity ($\mu S/cm^3$) y	Acid volume (mL) x	Conductivity ($\mu S/cm^3$) y
0	4152.95	5	1212.47
1	3140.97	6	2358.11
2	2100.34	7	3417.83
3	1126.55	8	4429.81
4	162.299		

Review

13. A panel of judges rate 20 science fair exhibits as shown. The judges decide that the top rating should be 100, so they add 6 points to each rating.

   a. What are the mean and the standard deviation of the ratings before adding 6 points? @ $\bar{x} = 83.75, s \approx 7.45$

   b. What are the mean and the standard deviation of the ratings after adding 6 points? $\bar{x} = 89.75, s \approx 7.45$

   c. What do you notice about the change in the mean? In the standard deviation?

   By adding 6 points to each rating, the mean increases by 6, but the standard deviation remains the same.

Rank	Rating	Rank	Rating
1	94	11	84
2	92	12	83
3	92	13	83
4	92	14	81
5	90	15	79
6	89	16	79
7	89	17	77
8	88	18	73
9	86	19	71
10	85	20	68

14. **APPLICATION** This table shows the average number of personal computers per household in the United States in various years.

Year	2006	2007	2008	2009	2010	2011
Computers	2.12	2.19	2.27	2.34	2.47	2.75

(http://www.nakono.com/tekcarta/databank/personal-computers-per-household/)

a. Make a scatter plot of these data.
b. Find a line to fit the data.
c. Compare your model's prediction for 2012 with the actual value of 3.00 computers.
d. Is a linear model for this situation good for long-term predictions? Explain your reasoning.

In 1946, inventors J. Presper Eckert and J. W. Mauchly created the first general-purpose electronic calculator, named ENIAC (Electronic Numerical Integrator and Computer). The calculator filled a large room and required a team of engineers and maintenance technicians to operate it.

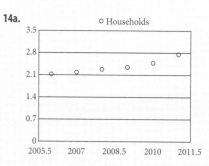

14a.

14b. choose (2007, 2.19) and (2010, 2.47)
$m = \frac{0.28}{3} = 0.093$
$y - 2.47 = 0.093(x - 2010)$
$y = 0.093x - 184.46$

14c. Model predicts $y = 0.093(2012) - 184.46 = 2.656$. The reported 2012 average is 3.00, 0.344 computers per household higher.

14d. It looks like rate is increasing, not staying constant so a linear model may not be the best.

IMPROVING YOUR Geometry SKILLS

Lines in Motion Revisited

Imagine that a line is translated in a direction perpendicular to it, creating a parallel line. What vertical and horizontal translations would be equivalent to the translation along the perpendicular path? Find the slope of each line pictured. How does the ratio of the translations compare to the slope of the lines? Find answers both for the specific lines shown and, more generally, for any pair of parallel lines.

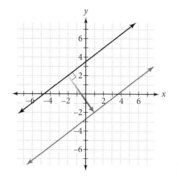

LESSON 3.6 Dilations and the Absolute-Value Family

IMPROVING YOUR Geometry SKILLS

In general, any translation of a line has the same slope as the original, so the translation amounts don't relate to the line's slope. If each point on the line is translated by the same amounts to a point on the perpendicular to the line at that point, however, then there is a relationship. In the example, the vertical translation is −4, and the horizontal translation is 3.

The ratio $-\frac{4}{3}$ is the slope of the perpendicular line, by the definition of slope. The fact that the slope of the original line is $\frac{3}{4}$ gives a clue about the general case: If every point on a line is translated along a perpendicular line horizontally a units and vertically b units, then $-\frac{a}{b}$ is the slope of the line.

LESSON 3.7

Transformations and the Circle Family

This lesson extends to circles the notions of translation and dilation.

COMMON CORE STATE STANDARDS
APPLIED	DEVELOPED	INTRODUCED
A.SSE.1b	F.BF.3	F.TF.2
A.REI.1		

Objectives

- Define *unit circle* and derive the equation $x^2 + y^2 = 1$
- Express a circle equation as two semicircle functions
- Define *ellipse* as "a vertical and/or horizontal dilation of a circle"
- Transform a circle to get an ellipse
- Apply transformations to relations and to a new function expressed in terms of $f(x)$
- Summarize transformations—translations, reflections, rotations, and dilations

Vocabulary

unit circle
ellipse

Materials

- Calculator Note: Drawing Segments, *optional*

Launch

On graph paper, graph these equations. What is the center and radius of each circle?

a. $x^2 + y^2 = 4$ (0, 0), 2
b. $(x - 1)^2 + y^2 = 4$ (1, 0), 2

How could you find the center and radius of each circle from the equation, without graphing?

General form of a circle is $(x - h)^2 + (y - k)^2 = r^2$ where (h, k) is the center and r is the radius.

What is the equation of a circle with center (0, 0) and radius 1?

$x^2 + y^2 = 1$

Many times the best way, in fact the only way, to learn is through mistakes. A fear of making mistakes can bring individuals to a standstill, to a dead center.

GEORGE BROWN

In this lesson you will investigate transformations of a relation that is not a function. A **unit circle** is centered at the origin with a radius of 1 unit. Suppose P is any point on a unit circle with center at the origin. Draw the slope triangle for the radius between the origin and point P.

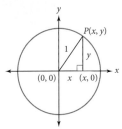

The Aldar headquarters building in Abu Dhabi, UAE is the first circular building of its kind in the Middle East

You can derive the equation of a unit circle from this diagram by using the Pythagorean Theorem. The legs of the right triangle have lengths x and y and the length of the hypotenuse is 1 unit, so its equation is $x^2 + y^2 = 1$. This is true for all points P on the unit circle.

$-1 \leq x \leq 1$ and $-1 \leq y \leq 1$; $y = \pm\frac{\sqrt{3}}{2} \approx 0.866$; it is not a function because there are two y-values for most x-values.

What are the domain and the range of this relation? If a value, such as 0.5, is substituted for x, what are the values of y? Why is the relation not a function?

In order to draw the graph of a circle on your calculator, you need to solve the equation $x^2 + y^2 = 1$ for y. When you do this, you get two equations, $y = +\sqrt{1 - x^2}$ and $y = -\sqrt{1 - x^2}$. Each of these is a function. You have to graph both of them to get the complete circle.

> **Equation of a Unit Circle**
>
> The equation of a **unit circle** is
> $x^2 + y^2 = 1$ or, solved for y, $y = \pm\sqrt{1 - x^2}$

You can apply what you have learned about transformations of functions to find the equations of transformations of the unit circle.

EXAMPLE A Find the equation for each graph.

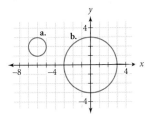

ELL
By this point, all the transformations have been covered. This is a good time to create a graphic organizer, complete with verbal, algebraic, and graphical components of the definition of each transformation.

Support
Students may need extra practice solving a standard circle or elliptical equation for y. Students might want to enter their equations into their calculators to verify the location and shape of their graphs.

Advanced
Urge students to determine how to transform an equation, whether it is in general form or any other form.

Solution

a. Circle *a* is a translation of the unit circle horizontally −6 units and vertically 2 units. Replace x with $(x + 6)$ and y with $(y − 2)$ to get the equation $(x + 6)^2 + (y − 2)^2 = 1$. To check this result on your calculator, solve for y and graph:

$$(y − 2)^2 = 1 − (x + 6)^2$$
$$y − 2 = \pm\sqrt{1 − (x + 6)^2}$$
$$y = 2 \pm \sqrt{1 − (x + 6)^2}$$

You must enter two functions, $y = 2 + \sqrt{1 − (x + 6)^2}$ and $y = 2 − \sqrt{1 − (x + 6)^2}$ into your calculator.

b. Circle *b* is a dilation of the unit circle horizontally and vertically by the same scale factor of 3. Replacing x and y with $\frac{x}{3}$ and $\frac{y}{3}$, you find $\left(\frac{x}{3}\right)^2 + \left(\frac{y}{3}\right)^2 = 1$. This can also be written as $\frac{x^2}{9} + \frac{y^2}{9} = 1$ or $x^2 + y^2 = 9$.

You can transform a circle to get an **ellipse**. An ellipse is a circle where different horizontal and vertical scale factors have been used.

EXAMPLE B | What is the equation of this ellipse?

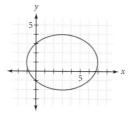

Solution | The original unit circle has been translated and dilated. The new center is at (3, 1). In a unit circle, every radius measures 1 unit. In this ellipse, a horizontal segment from the center to the ellipse measures 4 units, so the horizontal scale factor is 4. Likewise, a vertical segment from the center to the ellipse measures 3 units, so the vertical scale factor is 3. The equation changes like this:

$x^2 + y^2 = 1$ Original unit circle.

$\left(\frac{x}{4}\right)^2 + y^2 = 1$ Dilate horizontally by a factor of 4. (Replace x with $\frac{x}{4}$.)

$\left(\frac{x}{4}\right)^2 + \left(\frac{y}{3}\right)^2 = 1$ Dilate vertically by a factor of 3. (Replace y with $\frac{y}{3}$.)

$\left(\frac{x − 3}{4}\right)^2 + \left(\frac{y − 1}{3}\right)^2 = 1$ Translate to new center at (3, 1). (Replace x with $x − 3$, and replace y with $y − 1$.)

To enter this equation into your calculator to check your answer, you need to solve for y. It takes two equations to graph this on your calculator. By graphing both of these equations, you can draw the complete ellipse and verify your answer.

$$y = 1 \pm 3\sqrt{1 − \left(\frac{x − 3}{4}\right)^2}$$

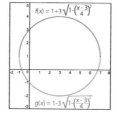

Investigate

Example A

Project Example A and ask for a class volunteer for the equation of each circle. Ask them to explain their answer.
SMP 1, 3, 6

ASK "What would the equation look like if the circle were both dilated and translated?" [If it's dilated first, its equation becomes $\left(\frac{x}{3}\right)^2 + \left(\frac{y}{3}\right)^2 = 1$. The translation then makes its equation $\left(\frac{x + 6}{3}\right)^2 + \left(\frac{y − 2}{3}\right)^2 = 1$, because the translation is represented by replacing the variables themselves. If the circle is translated first, however, its equation becomes $(x + 6)^2 + (y − 2)^2 = 1$, and the dilation turns the equation into $\left(\frac{x}{3} + 6\right)^2 + \left(\frac{y}{3} − 2\right)^2 = 1$. In this case, the variables themselves are replaced.]

Example B

Project Example B and have students work in pairs. As students present their solutions, have them justify each step.
SMP 1, 3, 6

Point out that there is a specific order for any given set of transformations. Suppose students are asked to describe the two transformations that changed $f(x) = x^2$ into $g(x) = \left(\frac{x + 2}{3}\right)^2$. They should start by looking at how the parent function, $f(x)$, has been modified. In function form, this equation would look like $g(x) = f\left(\frac{x + 2}{3}\right)$. First, x has been replaced by $\frac{x}{3}$, representing a horizontal dilation by a factor of 3. Next, x has been replaced by $x + 2$, meaning a horizontal shift 2 units to the left. In this situation, $f(x)$ was dilated horizontally by a factor of 3 and then shifted horizontally −2 units; each new y-value is the result of multiplying an x-value by 3 and then subtracting 2. Note that the order of substitution is not reversible and that this does not follow the logic of the order of operations. If you first replace x with $x + 2$ and then replace x with $\frac{x}{3}$ you get $g(x) = f\left(\frac{x}{3} + 2\right)$.

Modifying the Investigation

Whole Class Use technology to draw the ellipse and axes for the class. Challenge students to create the same ellipse on their own calculator, then discuss.

Shortened There is no shortened version of this investigation.

One Step Use the One Step found in your ebook. As students work, be prepared to refer them to how dilations affect the graphs of the equations of other figures. Remind them as needed how to graph half a unit circle on a calculator.

Conceptual/Procedural

Conceptual Students explore the concept of transforming a circle to get an ellipse. They summarize the transformations of functions and other relations that have been explored in this chapter.

Procedural Students learn the equation of a unit circle and how to transform the equation to get an ellipse.

When a graph is transformed, each variable in the graph's equation is replaced with a variable and a constant representing that aspect of the transformation.

Guiding the Investigation
There is a desmos simulation for this investigation in the ebook. As needed, **ASK** "How does the ellipse compare with the circle from which it's transformed?" [The diameter of the circle is the shortest distance across the ellipse. The centers are the same.]

Example C
Project Example C and work as a class activity, with student input. The coordinates of the right endpoint of the transformed semicircle are now $\left(2\frac{1}{3}, 1\right)$. To describe how the original endpoint was mapped to the new location, track the images of (0, 1) and (1, 0) under various transformations, considering horizontal and vertical dilations first.

ALERT Students may think that
$$g(x) = 2\sqrt{1 - \left(\frac{x-2}{\frac{1}{3}}\right)}$$
indicates a horizontal shift (to get $x - 2$) before a horizontal dilation (by factor $\frac{1}{3}$). Emphasize that a transformation is represented by a replacement of variables in the equation. In this case, x is first divided by $\frac{1}{3}$, then 2 is subtracted.

Start with:	(0, 1)	(1, 0)
Dilate horizontally by a factor of $\frac{1}{3}$.	(0, 1)	$\left(\frac{1}{3}, 0\right)$
Dilate vertically by a factor of 2:	(0, 2)	$\left(\frac{1}{3}, 0\right)$
Translate horizontally 2 units:	(2, 2)	$\left(2\frac{1}{3}, 0\right)$
Translate vertically 1 unit:	(2, 3)	$\left(2\frac{1}{3}, 1\right)$

Summarize

As students present their solutions to the Investigation and examples, Encourage the use of appropriate mathematical vocabulary. Whether student responses are correct or incorrect, ask other students if they agree and why. **SMP 1,3,6**

INVESTIGATION
When Is a Circle Not a Circle?

If you look at a circle, like the top rim of a cup, from an angle, you don't see a circle; you see an ellipse. Choose one of the ellipses from the worksheet. Use your ruler carefully to place axes on the ellipse, and scale your axes in centimeters. Be sure to place the axes so that the longest dimension is parallel to one of the axes. Find the equation to model your ellipse. Graph your equation on your calculator and verify that it creates an ellipse with the same dimensions as on the worksheet.

The tops of these circular oil storage tanks look elliptical when viewed at an angle.

Equations for transformations of relations such as circles and ellipses are sometimes easier to work with in the general form before you solve them for y, but you need to solve for y to enter the equations into your calculator. If you start with a function such as the top half of the unit circle, $f(x) = \sqrt{1-x^2}$, you can transform it in the same way you transformed any other function, but it may be a little messier to deal with.

EXAMPLE C | If $f(x) = \sqrt{1-x^2}$, find $g(x) = 2f(3(x-2)) + 1$. Sketch a graph of this new function.

Solution | In $g(x) = 2f(3(x-2)) + 1$, note that $f(x)$ is the parent function, x has been replaced with $3(x-2)$, and $f(3(x-2))$ is then multiplied by 2 and 1 is added.

You can rewrite the function g as
$$g(x) = 2\sqrt{1 - [3(x-2)]^2} + 1 \quad \text{or} \quad g(x) = 2\sqrt{1 - \left(\frac{x-2}{\frac{1}{3}}\right)^2} + 1$$

This indicates that the graph of $y = f(x)$, a semicircle, has been dilated horizontally by a factor of $\frac{1}{3}$, dilated vertically by a factor of 2, then translated right 2 units and up 1 unit. The transformed semicircle is graphed at right. What are the coordinates of the right endpoint of the graph? Describe how the original semicircle's right endpoint of (1, 0) was mapped to this new location.

You have now learned to translate, reflect, and dilate functions and other relations. These transformations are the same for all equations. $\left(2\frac{1}{3}, 1\right)$. Multiply the x-coordinate by $\frac{1}{3}$ and add 2. Multiply the y-coordinate by 2 and add 1.

Discussing the Investigation
As students present, ask whether they were surprised to find that the images of the circular tanks were not circles. Our brains are very accustomed to receiving elliptical images and interpreting them as circles.

ASK "What are intercepts of the ellipse given by the equation $\frac{x^2}{p} + \frac{y^2}{q} = 1$?"
$[\pm\sqrt{p}, \pm\sqrt{q}]$ To avoid square roots, we often write the equation as $\frac{x^2}{a^2} + \frac{y^2}{b^2} = 1$ so that intercepts are at $\pm a$ and $\pm b$.

Transformations of Functions and Other Relations

Translations

The graph of $y - k = f(x - h)$ translates the graph of $y = f(x)$ horizontally h units and vertically k units.

or

Replacing x with $(x - h)$ translates the graph horizontally h units.
Replacing y with $(y - k)$ translates the graph vertically k units.

Reflections

The graph of $y = f(-x)$ is a reflection of the graph of $y = f(x)$ across the y-axis. The graph of $-y = f(x)$ is a reflection of the graph of $y = f(x)$ across the x-axis.

or

Replacing x with $-x$ reflects the graph across the y-axis. Replacing y with $-y$ reflects the graph across the x-axis.

Dilations

The graph of $\dfrac{y}{b} = f\left(\dfrac{x}{a}\right)$ is a dilation of the graph of $y = f(x)$ by a vertical scale factor of b and by a horizontal scale factor of a.

or

Replacing x with $\dfrac{x}{a}$ dilates the graph by a horizontal scale factor of a.
Replacing y with $\dfrac{y}{b}$ dilates the graph by a vertical scale factor of b.

3.7 Exercises

You will need a graphing calculator for Exercises **9–11**, **14**, **15**, and **16**.

Practice Your Skills

1. Each equation represents a single transformation. Copy and complete this table.

Equation	Transformation (translation, reflection, dilation)	Direction	Amount of scale factor
$y + 3 = x^2$	Translation	Vertical	-3
$-y = |x|$	Reflection	Across x-axis	N/A
$y = \sqrt{\dfrac{x}{4}}$	Dilation	Horizontal	4
$\dfrac{y}{0.4} = x^2$	Dilation	Vertical	0.4
$y = |x - 2|$	Translation	Horizontal	2
$y = \sqrt{-x}$	Reflection	Across y-axis	N/A

LESSON 3.7 Transformations and the Circle Family

Discussing the Lesson

If students are using the standard window on their calculators, **ASK** "Why doesn't the graph look like a circle?" Students can choose Zoom-Square from the Window menu to get a circular graph here. **SMP 5**

CRITICAL QUESTION What do the values of a and b represent in the equation of an ellipse?

BIG IDEA The value of a is the amount the unit circle has been dilated horizontally, because x is divided by a. The value of b is the amount the unit circle has been dilated vertically, because y is divided by b.

CRITICAL QUESTION Is the dilation of a circle always an ellipse?

BIG IDEA It is, unless you dilate both axes the same amount.

CRITICAL QUESTION How do you dilate a circle to get an ellipse whose center is not at the origin?

BIG IDEA Dilate the unit circle and then translate the image.

Formative Assessment

Observe the extent to which students understand that a transformation is represented by a change to individual variables in the equation. Also gauge how well students are extending to circles and ellipses the transformations they studied in previous lessons. Assess students' understanding of domain and range in a new context and monitor the depth of students' knowledge of transformations as a whole. Monitor student facility with solving the equation of a circle for y.

Apply

Extra Example

Describe the transformation of the graph of $y = \sqrt{1 - x^2}$ needed to produce the graph of
$y = \sqrt{1 - (2x)^2} + 4$.

dilated horizontally by a factor of $\dfrac{1}{2}$, translated vertically 4 units

Closing Question

Is a circle an ellipse?

Yes and no; to get an ellipse, the horizontal and vertical dilation factors must differ. If they don't differ, it is still a circle.

Assigning Exercises

Suggested: 1, 4–10
Additional Practice: 2, 3, 11–16

3a.

3b.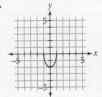

3c.

Exercise 4 **SUPPORT** This exercise provides a good way to assess students' ability to state a transformation algebraically from a given graph. Have students create and proceed through a checklist as they attempt to determine what types of transformations have taken place: Has there been a translation? A dilation? A reflection? Then lead them through making the corresponding algebraic adjustments to the original function.

4d. $\frac{y-1}{2} = \sqrt{1-(x-3)^2}$, or $y = 2\sqrt{1-(x-3)^2} + 1$

4e. $\frac{y-3}{-5} = \sqrt{1-\left(\frac{x+2}{2}\right)^2}$, or $y = -5\sqrt{1-\left(\frac{x+2}{2}\right)^2} + 3$

4f. $\frac{y+2}{4} = \sqrt{1-(x-3)^2}$, or $y = 4\sqrt{1-(x-3)^2} - 2$

5b. $y = \pm\sqrt{1-(x+3)^2} + 2$, or $(x+3)^2 + y^2 = 1$

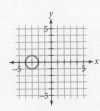

2. The equation $y = \sqrt{1-x^2}$ is the equation of the top half of the unit circle with center (0, 0) shown on the left. What is the equation of the top half of an ellipse shown on the right?

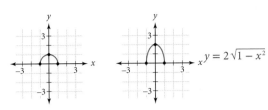

3. Use $f(x) = \sqrt{1-x^2}$ to graph each of the transformations below.

   a. $g(x) = -f(x)$ **b.** $h(x) = -2f(x)$ **c.** $j(x) = -3 + 2f(x)$ @

4. Each curve is a transformation of the graph of $y = \sqrt{1-x^2}$. Write an equation for each curve.

   a. $\frac{y}{3} = \sqrt{1-x^2}$, or $y = 3\sqrt{1-x^2}$ **b.** $\frac{y}{0.5} = \sqrt{1-x^2}$, or $y = 0.5\sqrt{1-x^2}$ **c.** $\frac{y-1}{2} = \sqrt{1-x^2}$, or $y = 2\sqrt{1-x^2} + 1$

   d. @ **e.** @ **f.**

  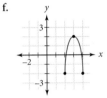

5. Write an equation and draw a graph for each transformation of the unit circle. Use the form $y = \pm\sqrt{1-x^2}$.

   a. Replace y with $(y-2)$. **b.** Replace x with $(x+3)$.
   c. Replace y with $\frac{y}{2}$. @ **d.** Replace x with $\frac{x}{2}$.

5a. $y = \pm\sqrt{1-x^2} + 2$, or $x^2 + (y-2)^2 = 1$

Reason and Apply

6. To create the ellipse at right, the x-coordinate of each point on a unit circle has been multiplied by a factor of 3.

   a. Write the equation of this ellipse. $\left(\frac{x}{3}\right)^2 + y^2 = 1$
   b. What expression did you substitute for x in the parent equation? @ $\frac{x}{3}$
   c. If $y = f(x)$ is the function for the top half of a unit circle, then what is the function for the top half of this ellipse, $y = g(x)$, in terms of f? @ $g(x) = f\left(\frac{x}{3}\right)$

192 **CHAPTER 3** Functions and Relations

5c. $y = \pm 2\sqrt{1-x^2}$, or $x^2 + \left(\frac{y}{2}\right)^2 = 1$

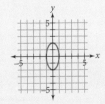

5d. $y = \pm\sqrt{1-\left(\frac{x}{2}\right)^2} + 2$, or $\frac{x^2}{4} + y^2 = 1$

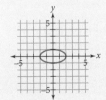

192 **CHAPTER 3** Functions and Relations

7. Given the unit circle at right, write the equation that generates each transformation. Use the form $x^2 + y^2 = 1$.

   a. Each y-value is half the original y-value. $x^2 + \left(\dfrac{y}{0.5}\right)^2 = 1$

   b. Each x-value is half the original x-value. $\left(\dfrac{x}{0.5}\right)^2 + y^2 = 1$

   c. Each y-value is half the original y-value, and each x-value is twice the original x-value. $\left(\dfrac{x}{2}\right)^2 + 2(y)^2 = 1$

8. Consider the ellipse at right. @

   a. Write two functions that you could use to graph this ellipse. $y = 3\sqrt{1 - \left(\dfrac{x}{0.5}\right)^2}$ and $y = -3\sqrt{1 - \left(\dfrac{x}{0.5}\right)^2}$

   b. Use ± to write one equation that combines the two equations in 8a. $y = \pm 3\sqrt{1 - \left(\dfrac{x}{0.5}\right)^2}$

   c. Write another equation for the ellipse by squaring both sides of the equation in 8b. $y^2 = 9\left[1 - \left(\dfrac{x}{0.5}\right)^2\right]$ or $\dfrac{x^2}{0.25} + \dfrac{y^2}{9} = 1$

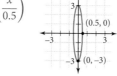

9. Follow these steps to explore a relationship between linear, quadratic, square root, absolute-value, and semicircle functions. Use graphing windows of an appropriate size.

   a. Graph these equations simultaneously on your calculator. The first four functions intersect in the same two points. What are the coordinates of these points? @

   $y = x$ $\quad$ $y = x^2$ $\quad$ $y = \sqrt{x}$ $\quad$ $y = |x|$ $\quad$ $y = \sqrt{1 - x^2}$

   b. Imagine using the intersection points that you found in 9a to draw a rectangle that just encloses the quarter-circle that is on the right half of the fifth function. How do the coordinates of the points relate to the dimensions of the rectangle? ⓗ

   c. Solve these equations for y and graph them simultaneously on your calculator. Where do the first four functions intersect?

   $\dfrac{y}{2} = \dfrac{x}{4}$ $\quad$ $\dfrac{y}{2} = \left(\dfrac{x}{4}\right)^2$ $\quad$ $\dfrac{y}{2} = \sqrt{\dfrac{x}{4}}$ $\quad$ $\dfrac{y}{2} = \left|\dfrac{x}{4}\right|$ $\quad$ $\dfrac{y}{2} = \sqrt{1 - \left(\dfrac{x}{4}\right)^2}$

   d. Imagine using the intersection points that you found in 9c to draw a rectangle that just encloses the right half of the fifth function. How do the coordinates of the points relate to the dimensions of the rectangle? The rectangle has width 4 and height 2. The width is the difference in x-coordinates, and the height is the difference in y-coordinates.

10. Consider the parent function $y = \dfrac{1}{x}$ graphed at right. This function is not defined for $x = 0$. When the graph is translated, the center at (0, 0) is translated as well, so you can describe any translation of the figure by describing how the center is transformed.

    The parent function passes through the point (1, 1). You can describe any dilations of the function by describing how point (1, 1) is transformed. Use what you have learned about transformations to sketch each graph, then check your work with your graphing calculator.

    a. $y = \dfrac{1}{x - 3}$

    b. $y + 1 = \dfrac{1}{x + 4}$

    c. $y = \dfrac{1}{3x}$

    d. $\dfrac{y - 2}{-4} = \dfrac{1}{x}$

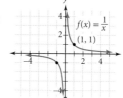

Exercise 7 Equivalent forms of the answers include $x^2 + \left(\dfrac{y}{\frac{1}{2}}\right)^2 = 1$ for 7a, $\left(\dfrac{x}{\frac{1}{2}}\right)^2 + y^2 = 1$ for 7b, and $\left(\dfrac{x}{2}\right)^2 + \left(\dfrac{y}{\frac{1}{2}}\right)^2 = 1$ for 7c.

Exercise 9 This exercise may take a lot of time. Students might benefit from sketching their solutions on graph paper before graphing them on their calculators. The instructions say to imagine drawing a rectangle, but if students want to draw it on their calculators, you can refer them to Calculator Note: Drawing Segments.

9a.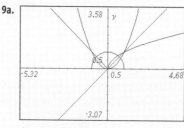

(0, 0) and (1, 1)

9b. The rectangle has width 1 and height 1. The width is the difference in x-coordinates, and the height is the difference in y-coordinates.

9c.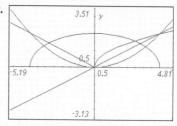

(0, 0) and (4, 2)

Exercise 10c–d Students might legitimately think of these functions as either vertical or horizontal dilations of the parent function. The graphs will be the same.

LESSON 3.7 Transformations and the Circle Family 193

10a.

10b.

10c.

10d.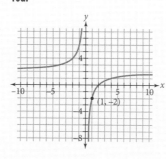

Exercise 11 Students might use statistics software or a spreadsheet for this exercise. SMP 5

11b. Original ratings (from Exercise 13 in Lesson 3.6): $\bar{x} = 83.75$, $s = 7.45$. New ratings: $\bar{x} = 89.10$, $s = 7.92$.

11c.

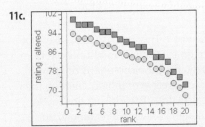

The scores have been stretched by a factor of $\frac{100}{94}$. All scores increased, so the mean increased. The high scores differ from the original by more than the lower ones, so the scores are more spread out, and the standard deviation is increased.

11d. Sample answer: The judge should add 6 points because it does not change the standard deviation. Everyone gets the same amount added instead of those with higher scores getting more.

Science CONNECTION

Satellites are used to aid in navigation, communication, research, and military reconnaissance. The job the satellite is meant to do will determine the type of orbit it is placed in.

A satellite in a geosynchronous orbit revolves west to east above the diameter at the same speed Earth rotates, one revolution every 24 hours. To maintain this velocity, the satellite must have an altitude of about 22,000 miles. In order to stay above the same point on Earth, so that a satellite dish antenna can stay focused on it, the orbit of the satellite must be circular.

Another useful orbit is a north-south elliptical orbit that takes 12 hours to circle the planet. Satellites in these elliptical orbits cover areas of Earth that are not covered by geosynchronous satellites, and are therefore more useful for research and reconnaissance.

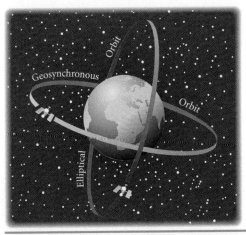

Satellites in a geosynchronous orbit follow a circular path above the equator. Another common orbit is an elliptical orbit in the north-south direction.

Review

11. Refer to Exercise 13 in the previous lesson. The original data are shown at right. Instead of adding the same number to each score, one of the judges suggests that perhaps they should *multiply* the original scores by a factor that makes the highest score equal 100. They decide to try this method.

  a. By what factor should they multiply the highest score, 94, to get 100? $\frac{100}{94}$

  b. What are the mean and the standard deviation of the original ratings? Of the altered ratings?

  c. Let x represent the exhibit number, and let y represent the rating. Plot the original and altered ratings on the same graph. Describe what happened to the ratings visually. How does this explain what happened to the mean and the standard deviation?

  d. Which method do you think the judges should use? Explain your reasoning.

Rank	Rating
1	94
2	92
3	92
4	92
5	90
6	89
7	89
8	88
9	86
10	85

Rank	Rating
11	84
12	83
13	83
14	81
15	79
16	79
17	77
18	73
19	71
20	68

12. Find the next three terms in this sequence: 16, 40, 100, 250, ...   625, 1562.5, 3906.25

13. Solve. Give answers to the nearest 0.01.

  a. $\sqrt{1-(a-3)^2} = 0.5$ $a \approx 2.13$ or 3.87

  b. $1 - |b+2| = -5$ $b = 4$ or -8

  c. $\sqrt{1 - \left(\frac{c-2}{3}\right)^2} = 0.8$ @
  $c = 0.2$ or 3.8

  d. $3 - 5\left(\frac{d+1}{2}\right)^2 = -7$
  $d = -1 \pm 2\sqrt{2}$; $d \approx 1.83$ or $d \approx -3.83$

14. This table shows the distances needed to stop a car on dry pavement in a minimum length of time for various speeds. Reaction time is assumed to be 0.75 s.

Speed (mi/h) x	10	20	30	40	50	60	70
Stopping distance (ft) y	19	42	73	116	173	248	343

a. Construct a scatter plot of these data.

b. Find the equation of a parabola that fits the points and graph it.

c. Find the residuals for this equation and the root mean square error.
For the sample answer: residuals: −5.43, 0.77, 0.97, −0.83, −2.63, −0.43, 7.77; s = 4.45

d. Predict the stopping distance for 56.5 mi/h.
approximately 221 ft

e. How far off might your prediction in 14d be from the *actual* stopping distance?
14d should be correct ± 4.45 ft.

15. This table shows passenger activity in the world's 30 busiest airports in 2015. @

a. Display the data in a histogram.

b. Estimate the total number of passengers who used the 30 airports. Explain any assumptions you make. Using the midpoint value for each histogram bin, there were 1805 million, or 1,805,000,000 passengers.

c. Estimate the mean usage among the 30 airports in 2015. Mark the mean on your histogram.
mean = 60.17 million

d. Sketch a box plot above your histogram. Estimate the five-number summary values. Explain any assumptions you make. Five-number summary: 42.5, 47.5, 57.5, 67.5, 102.5; assume that all data occur at midpoints of bins.

Number of passengers (in millions)	Number of airports
$40 \leq p < 45$	4
$45 \leq p < 50$	5
$50 \leq p < 55$	3
$55 \leq p < 60$	5
$60 \leq p < 65$	4
$65 \leq p < 70$	2
$70 \leq p < 75$	2
$75 \leq p < 80$	3
$85 \leq p < 90$	1
$100 \leq p < 105$	1

(https://en.wikipedia.org/wiki/List_of_the_world%27s_busiest_airports_by_passenger_traffic)

16. Consider the linear function $y = 3x + 1$.

a. Write an equation for the image of the graph of $y = 3x + 1$ after a reflection across the x-axis. Graph both lines on the same axes.

b. Write an equation for the image of the graph of $y = 3x + 1$ after a reflection across the y-axis. Graph both lines on the same axes.

c. Write an equation for the image of the graph of $y = 3x + 1$ after a reflection across the x-axis and then across the y-axis. Graph both lines on the same axes.

d. How does the image in 16c compare to the original line?
The two lines are parallel.

IMPROVING YOUR Visual Thinking SKILLS

4-in-1

Copy this trapezoid. Divide it into four congruent polygons.

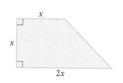

14 a, b.

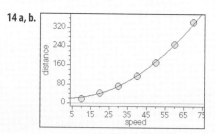

Sample answer: $\hat{y} \approx 0.07(x-3)2 + 21$

Exercise 15 **ALERT** Students might not notice that the table skips a couple of intervals in numbers of passengers. In 15c, students might add up the right-hand column and divide by something. Or they might add up the means of the intervals in the left-hand column and divide. Help them understand that the number in the right-hand column tells how many airports have a number in the corresponding cell of the left-hand column. The estimate is the sum of the products of the interval means and the number of airports in that interval divided by the total number of airports.

15a, c, d.

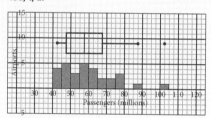

16a. $y = -3x - 1$

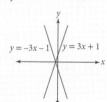

16b. $y = -3x + 1$

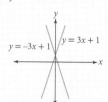

16c. $y = 3x - 1$

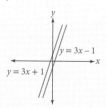

IMPROVING YOUR Visual Thinking SKILLS

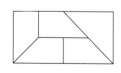

CHAPTER REVIEW 3

Reviewing

To review, present this problem: "For what relations is a vertical translation equivalent to a horizontal translation?" Remind students of relations, functions, and graphs. Consider graphs of the parent functions $y = x$, $y = x^2$, $y = \sqrt{x}$, and $y = |x|$ and of the parent function $x^2 + y^2 = 1$, the unit circle. Take advantage of teachable moments to remind students that $(x + h)^2$ is not the same as $x^2 + h^2$, $\sqrt{x + h}$, is not the same as $\sqrt{x} + \sqrt{h}$, and $|x + h|$ is not the same as $|x| + |h|$. The only function we've seen for which a vertical translation is a horizontal translation is the linear function. Review the laws of exponents and absolute values as students find for each of the above functions that a vertical dilation is equivalent to a horizontal dilation. The unit circle can be thought of as a pair of functions, $y = \pm\sqrt{1 - x^2}$; for neither graph of these two functions is any vertical translation equivalent to a horizontal translation. To review ellipses, ask what the result of a dilation is.

Assigning Exercises

If you are using one day to review this chapter, limit the number of exercises you assign. Several of the exercises have many parts.

1. Sample answer: For a time there are no pops. Then the popping rate slowly increases. When the popping reaches a furious intensity, it seems to level out. Then the number of pops per second drops quickly until the last pop is heard.

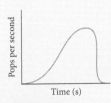

2a. odd; $f(-x) = -f(x)$; y-axis is line of symmetry
2b. neither; $f(-x) \neq f(-x) \neq -f(x)$
2c. even; $f(-x) = f(x)$; symmetric around origin

CHAPTER 3 REVIEW

This chapter introduced the concept of a **function** and reviewed **function notation**. You saw real-world situations represented by rules, sets, functions, graphs, and most importantly, equations. You learned to distinguish between functions and other **relations** by using either the definition of a function—at most one y-value per x-value—or the vertical line test.

This chapter also introduced several **transformations**, including **translations, reflections,** and vertical and horizontal **dilations**. You learned how to transform the graphs of **parent functions** to investigate several families of functions—linear, quadratic, square root, absolute value, and semicircle. For example, if you dilate the graph of the parent function $y = x^2$ vertically by a factor of 3 and horizontally by a factor of 2, and translate it right 1 unit and up 4 units, then you get the graph of the function $y = 3\left(\dfrac{x-1}{2}\right)^2 + 4$.

You learned that some functions can be described as even or odd. An **even function** has the y-axis as a line of symmetry. If the function f is an even function, then $f(-x) = f(x)$ for all values of x in the domain. An **odd function** is symmetrical about the origin. If the function f is an odd function, then $f(-x) = -f(x)$ for all values of x in the domain.

Exercises

You will need a graphing calculator for Exercise **9**.

@ **Answers are provided for all exercises in this set.**

1. Sketch a graph that shows the relationship between the time in seconds after you start microwaving a bag of popcorn and the number of pops per second. Describe in words what your graph shows.

2. Identify each function as even, odd, or neither. Justify your answer.
  a. $f(x) = x^3 - 2x$
  b. $g(x) = 4x + 1$
  c. $h(x) = x^2 + 4$

3. The graph of $y = f(x)$ is shown at right. Sketch the graph of each of these functions:
  a. $y = f(x) - 3$
  b. $y = f(x - 3)$
  c. $y = 3f(x)$
  d. $y = f(-x)$

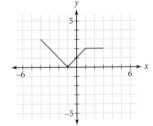

196 **CHAPTER 3** Functions and Relations

3a.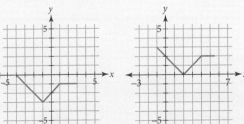

3b.

3c.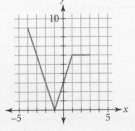

3d.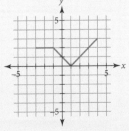

4. Assume you know the graph of $y = f(x)$. Describe the transformations, in order, that would give you the graph of these functions:

   a. $y = f(x + 2) - 3$
   b. $\dfrac{y-1}{-1} = f\left(\dfrac{x}{2}\right) + 1$
   c. $y = 2f\left(\dfrac{x-1}{0.5}\right) + 3$

   Translate horizontally −2 units and vertically −3 units.

5. The graph of $y = f(x)$ is shown at right. Use what you know about transformations to sketch these related functions:

   a. $y - 1 = f(x - 2)$
   b. $\dfrac{y+3}{2} = f(x + 1)$
   c. $y = f(-x) + 1$
   d. $y + 2 = f\left(\dfrac{x}{2}\right)$
   e. $y = -f(x - 3) + 1$
   f. $\dfrac{y+2}{-2} = f\left(\dfrac{x-1}{1.5}\right)$

6. For each graph, name the parent function and write an equation of the graph.

   a.
   $y = \sqrt{1 - x^2}$; $y = 3\sqrt{1 - x^2} - 1$

   b. $y = \sqrt{1 - x^2}$; $y = 2\sqrt{1 - \left(\dfrac{x}{5}\right)^2} + 3$

   c. $y = \sqrt{1 - x^2}$; $y = 4\sqrt{1 - \left(\dfrac{x-3}{4}\right)^2} - 1$

   d.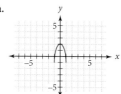
   $y = x^2$; $y = (x - 2)^2 - 4$

   e. $y = x^2$; $y = -2(x + 1)^2$

   f. $y = \sqrt{x}$; $y = -\sqrt{-(x - 2)} - 3$

   g.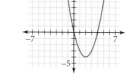
   $y = |x|$; $y = 0.5|x + 2| - 2$

   h. $y = |x|$; $y = -2|x - 3| + 2$

7. Solve for y.

   a. $2x - 3y = 6$ $y = \dfrac{2}{3}x - 2$
   b. $(y + 1)^2 - 3 = x$ $y = \pm\sqrt{x + 3} - 1$
   c. $\sqrt{1 - y^2} + 2 = x$ $y = \pm\sqrt{-(x - 2)^2 + 1}$

8. Solve for x.

   a. $4\sqrt{x - 2} = 10$ $x = 8.25$
   b. $\left(\dfrac{x}{-3}\right)^2 = 5$ $x = \pm\sqrt{45} \approx \pm 6.7$
   c. $\left|\dfrac{x-3}{2}\right| = 4$ $x = 11$ or $x = -5$
   d. $3\sqrt{1 + \left(\dfrac{x}{5}\right)^2} = 2$ no solution

CHAPTER 3 REVIEW **197**

Exercises 4, 5 Remind students that the order of substitution for transformations matters and that they should do dilations before translations.

Exercise 4b ASK "Is the left side the same transformation as $-y + 1$?" [A reflection followed by a translation to the left 1 unit is the same as a translation to the right followed by a reflection.]

4b. Dilate horizontally by a factor of 2, and then reflect across the x-axis.

4c. Dilate horizontally by a factor of $\dfrac{1}{2}$, dilate vertically by a factor of 2, translate horizontally 1 unit and vertically 3 units.

5a.

5b.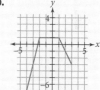

5c.

5d.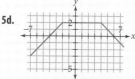

5e.

5f.

LESSON 3 REVIEW **197**

Exercise 9d ASK "Why, in real life, would neither of these fares result in revenue?" [For $0, you would be charging no fare, so you would take in no revenue. For $2.80, the fare is so expensive that no passengers would take the bus.]

9b.

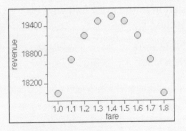

Take Another Look

1. Reflecting the graph across the vertical line $x = a$ is equivalent to translating the graph horizontally by the amount a (to move the line $x = a$ to the y-axis), reflecting it across the y-axis, and then translating it back. This composition of transformations yields the equation $y = f(-(x - a) + a) = f(-x + 2a)$. By a similar composition, a reflection across the horizontal line $y = b$ is given by the equation $y = -(f(x) - b) + b = -f(x) + 2b$.

2. The semicircle function, $y = \sqrt{1 - x^2}$, and the circle relation, $x^2 + y^2 = 1$, are two examples for which a vertical dilation is not equivalent to any horizontal dilation.

---

9. The Acme Bus Company has a daily ridership of 18,000 passengers and charges $1.00 per ride. The company wants to raise the fare yet keep its revenue as large as possible. (The revenue is found by multiplying the number of passengers by the fare charged.) From previous fare increases, the company estimates that for each increase of $0.10 it will lose 1,000 riders.

   a. Complete this table.

Fare ($) x	1.00	1.10	1.20	1.30	1.40	1.50	1.60	1.70	1.80
Number of passengers	18,000	17,000	16,000	15,000	14,000	13,000	12,000	11,000	10,000
Revenue ($) y	18,000	18,700	19,200	19,500	19,600	19,500	19,200	18,700	18,000

   b. Make a graph of the revenue versus fare charged. You should recognize the graph as a parabola.

   c. What are the coordinates of the vertex of the parabola? Explain the meaning of each coordinate of the vertex. (1.40, 19,600). By charging $1.40 per ride, the company achieves the maximum revenue, $19,600.

   d. Find a quadratic function that models these data. Use your model to find $y = -10,000(x - 1.4)^2 + 19,600$

      i. the revenue if the fare is $2.00. $16,000

      ii. the fare(s) that make no revenue ($0). $0 or $2.80

Take Another Look

1. A line of reflection does not have to be the x- or y-axis. Draw the graph of a function and then draw its image when reflected across several different horizontal or vertical lines. Write the equation of each image. Try this with several different functions. In general, if the graph of $y = f(x)$ is reflected across the vertical line $x = a$, what is the equation of the image? If the graph of $y = f(x)$ is reflected across the horizontal line $y = b$, what is the equation of the image?

2. For the graph of the parent function $y = x^2$, you can think of any vertical dilation as an equivalent horizontal dilation. For example, the equations $y = 4x^2$ and $y = (2x)^2$ are equivalent, even though one represents a vertical dilation by a factor of 4 and the other represents a horizontal dilation by a factor of $\frac{1}{2}$. For the graph of any function or relation, is it possible to think of any vertical dilation as an equivalent horizontal dilation? If so, explain your reasoning. If not, give examples of functions and relations for which it is not possible.

Assessing What You've Learned

 ORGANIZE YOUR NOTEBOOK Organize your notes on each type of parent function and each type of transformation you have learned about. Review how each transformation affects the graph of a function or relation and how the equation of the function or relation changes. You might want to create a large chart with rows for each type of transformation and columns for each type of parent function; don't forget to include a column for the general function, $y = f(x)$.

 UPDATE YOUR PORTFOLIO Choose one piece of work that illustrates each transformation you have studied in this chapter. Try to select pieces that illustrate different parent functions. Add these to your portfolio. Describe each piece in a cover sheet, giving the objective, the result, and what you might have done differently.

 WRITE TEST ITEMS Two important skills from this chapter are the ability to use transformations to write and graph equations. Write at least two test items that assess these skills. If you work with a group, identify other key ideas from this chapter and work together to write an entire test.

Assessing the Chapter

The Kendall Hunt Cohesive Assessment System (CAS) provides a powerful tool with which to manage assessment content, create and assign homework and tests, and disseminate assessment data.

You might require a presentation or an individual or group performance assessment for the chapter and then give a unit exam. Choose a performance assessment from an exercise you didn't assign that is recommended for that purpose.

As a good resource for study, refer students to the table on page 191, Lesson 3.7, which includes a summary of all the transformations included in this chapter.

Facilitating Self-Assessment

You might use some student-written items on the chapter assessment. Ask students to specify whether calculators will be allowed in solving the item they write.

Good portfolio items for this chapter include Lesson 3.1, Exercise 7 and the Performance Task; Lesson 3.2, Exercise 18; Lesson 3.3, Exercise 10; Lesson 3.4, Exercises 10 and 17; Lesson 3.5, Exercise 14; Lesson 3.6, Exercise 12; and Lesson 3.7, Exercise 9.

CHAPTER 4

Exponential, Power, and Logarithmic Functions

Overview

Discovering Advanced Algebra is about modeling real-world problems with mathematical functions. This chapter extends students' knowledge of functions to exponential, power, and logarithmic functions. The chapter also uses the function idea of assignment to assign a numerical value to each term of a series.

This chapter begins with an introduction to exponential functions in the form $y = ab^x$ for modeling growth and decay in **Lesson 4.1**. In **Lesson 4.2**, students review properties of exponents while becoming familiar with the power function, $y = ax^n$, and learning about rational exponents. In **Lesson 4.3**, rational exponents ($x^{1/n}$) are related to root notation ($\sqrt[n]{x}$). Students then use the point-ratio form, $y = y_1 b^{x-x_1}$, to find an exponential equation through two points. Real-world applications are presented in **Lesson 4.4**. **Lesson 4.5** introduces inverses of functions. Students learn about logarithmic functions in **Lesson 4.6**, and they see how these inverses of exponential functions can help solve equations for exponents. **Lesson 4.7** introduces students to series and to the summation notation used to represent them. Students learn to write recursive and explicit formulas for arithmetic series and find partial sums of arithmetic series. In **Lesson 4.8**, students find partial sums of geometric series.

There are two **Performance Tasks** highlighted in this chapter. In Lesson 4.5, students provide explanations for incorrect answers on a quiz. In Lesson 4.7, students create a formula for the sum, S_n, of an arithmetic series.

The Mathematics

Exponents and Exponential Functions

Exponents, if they are positive integers, represent repeated multiplication. The symbol b^n represents multiplying b repeatedly n times. The *exponent* is n, and the *base* is b. The student book calls b^n a *power* of b. (*Power* is a synonym for *exponent*.)

Positive-integer exponents have three major properties:
(1) $b^m b^n = b^{m+n}$, (2) $\frac{b^m}{b^n} = b^{m-n}$, and (3) $(b^m)^n = b^{mn}$. The second property leads to negative exponents. If n is a positive integer, then $b^{-n} = \frac{1}{b^n}$. The same three properties apply if the exponents are negative integers.

To extend the third property to fractional exponents, $(b^{1/3})^3$ must equal b^1, or b. Therefore $b^{1/3}$ is defined to be the number whose cube is b, or the cube root of b, which is written $\sqrt[3]{b}$. Just as the symbol $\sqrt{b}$ indicates the positive square root of b, the symbol $b^{1/2}$ represents a positive number. Fractional exponents, however, can lead to contradictions, which can be avoided by limiting fractional exponents to positive bases.

Inverse Functions

If you think of a relation as a two-column table, perhaps infinite, then exchanging the columns gives a relation that's the *inverse* of the original. As with any relation, the inverse may or may not be a function. If both the function and its inverse are functions, it is called a one-to-one function, and the inverse of the function $f(x)$ is written as $f^{-1}(x)$.

Logarithmic Functions

In the case that the function is the exponential function $y = 10^x$, its inverse is also a function, $x = \log y$. A helpful way to think of the logarithmic function is "log y is the exponent put on 10 to get y." Therefore $10^{\log y} = y$ and $\log(10^x) = x$.

While logarithms can have any positive number as a base, two bases are used more than other bases, base 10 and base e. Logarithms with base 10 are called *common logarithms*. Logarithms whose base is the transcendental number e, are called *natural logarithms*.

Logarithmic functions were developed around 1594 as computational tools by Scottish mathematician John Napier (1550–1617). He originally called them "artificial numbers," but changed the name to logarithms, which means "reckoning numbers."

The fact that the inverse of an exponential function is the common logarithmic function helps solve equations in which the unknown appears in an exponent on 10. For example, you can solve the equation $10^{3x-5} = 59$ by realizing that $3x - 5$ is the exponent put on 10 to get 59; that is, $3x - 5 = \log 59$. You can then substitute a calculator value for log 59 to solve for x.

What if the exponent is on some base other than 10? A more general logarithmic function is needed—an inverse of $y = b^x$. This inverse, known as "the logarithm to the base b of y," and notated $x = \log_b y$, gives the exponent put on b to get y. Technology makes it easy to find the logarithm to any base, b, of y.

In addition to solving equations for exponents, logarithms can help find an exponential function that fits a set of data points. You can use the logarithms to transform the data into points that are more linear and then find a line of fit.

Series

Mathematicians have grappled for centuries with paradoxes that arise from trying to sum infinitely many numbers. If you repeatedly add a nonzero number such as 0.0001 to itself, the sum will eventually get as large as you want. You might say that the sum is infinitely large. But what about the sum $\frac{1}{2} + \frac{1}{4} + \frac{1}{8} + \frac{1}{16} + \cdots$? If you begin adding these numbers, the intermediate results don't seem to be getting infinitely large. In fact, it appears that they'll always be less than 1. The sum $1 - \frac{1}{2} + \frac{1}{3} - \frac{1}{4} + \cdots$ also seems to be finite. But what about the sum $1 - 1 + 1 - 1 + \ldots$? If you start adding, the partial totals flip between 0 and 1. What should we say the complete sum is? Students' difficulties with this topic reflect the historical struggle of mathematicians.

Some standard terminology begins to sort out these questions. Expressions such as $\frac{1}{2} + \frac{1}{4} + \frac{1}{8} + \frac{1}{16} + \cdots$ are called *series*, which is different from a sequence. A series is also not a sum, because a sum is a number. A series is an (infinitely long) expression, a summation, an indicated sum. *Sigma notation* is often used to indicate the terms to be summed. Sigma is a Greek letter, whose capital form is Σ. The series $\frac{1}{2} + \frac{1}{4} + \frac{1}{8} + \frac{1}{16} + \cdots$ is represented by $\sum_{n=1}^{\infty} \frac{1}{2^n}$.

There are many kinds of series. This book emphasizes two of them, based on two types of sequences: *arithmetic series* and *geometric series*. The nth partial sum of an arithmetic series with first term u_1 and nth term u_n is the number of terms times the mean of the first and last terms, or $n\left(\frac{u_1 + u_n}{2}\right)$. The geometric series with initial term u_0 and common ratio r has nth partial sum $u_0\left(\frac{1-r^n}{1-r}\right)$, where $r \neq 1$. If r is close to 0 (actually, less than 1 in absolute value), the powers r^n get closer to 0 as n increases.

Using This Chapter

To save time, you might combine Lessons 4.1 and 4.2 for students familiar with the topics. Using the **Shortened** modifications of the investigations also saves time. It is important, however, to have student presentations of ideas to help them communicate mathematically and justify their answers.

Common Core State Standards (CCSS)

Students are introduced to exponents as early as grade 5 where they use whole-number exponents to denote powers of 10 (5.NBT.2). In the middle grades, students write, evaluate and perform arithmetic operations with numerical expressions involving whole-number exponents, including radicals and integer exponents (6.EE.1, 2c, 8.EE.1, 3). In Algebra 1 students apply solution techniques related to solving linear equations and the laws of exponents to the creation and solution of simple exponential equations (N.RN.2, A.REI.2). Students focused on distinguishing between additive and multiplicative change, comparing the intercept form of linear equations, $y = a + bx$ and the exponential form, $y = ab^x$, to emphasize a starting value, a, and repeated addition or multiplication of the constant, b, x times (A.SSE.1b, A.SSE.3). While the language of sequences is not used throughout, students interpret arithmetic sequences as linear functions and geometric sequences as exponential functions (F.F.6).

In Algebra II, students extend their work with exponential functions to include solving exponential equations with logarithms (A.SSE.4, REI.1). Students find inverse functions of simple rational, simple radical, and simple exponential functions and use that idea to connect to the relationship of exponents and logarithms (F.BF.4a, F.LE.4). In Algebra II, students also learn to do a more complete analysis of sequences (F-IF.A.3), especially arithmetic and geometric sequences, and their associated series, developing recursive formulas for sequences by figuring out how you get from one term to the next and then giving a precise description of this process in algebraic symbols (F-BF.A.2). Understanding solving equations as a process of reasoning plays a larger than ever role in Algebra II because the equations students encounter can have extraneous solutions (A-REI.A.2).

Additional Resources For complete references to additional resources for this chapter and others see your digital resources.

Refreshing Your Skills To review the algebra skills needed for this chapter see your digital resources.

Working with Positive Square Roots
Arithmetic and Geometric Sequences

Projects See your digital resources for performance-based problems and challenges.

The Cost of Living
Powers of 10

Explorations See your digital resources for extensions and/or applications of content from this chapter.

Seeing the Sum of a Series

Materials

- dice (1 die per person)
- graph paper
- scissors
- ball, *optional*
- motion sensor, *optional*

CHAPTER 4

Objectives

- Discover point-ratio form
- Review properties of integer exponents and extend them to rational exponents
- Write explicit equations for geometric sequences and generalize them for modeling continuous real-world growth and decay
- Write exponential equations to fit data
- Distinguish a power function from an exponential function and solve power equations
- Define the inverse relation of a function
- Define common logarithms and use them to calculate and solve equations in several ways
- Establish that the inverse of an exponential function is a logarithmic function
- Observe the natural logarithm function and learn about its base, e
- Learn about mathematical series and the summation notation used to represent series
- Write recursive and explicit formulas for the terms of arithmetic and geometric series
- Find formulas for the partial sums of arithmetic and geometric series

CHAPTER 4
Exponential, Power, and Logarithmic Functions

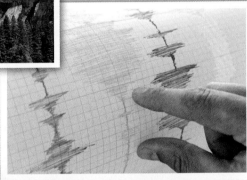

The Richter scale measures the magnitude of an earthquake by taking the logarithm of the amplitude of waves recorded by a seismograph. Geologists use exponential, power and logarithmic functions to measure geoscience phenomena that behave in a non-linear manner, including curved exfoliation joints (and resultant topographic surfaces) like these in Yosemite National Park. Exponential models have shown that the population growth rate of King penguin colonies, like this one in the South Sandwich Islands, have decreased since the 1990s.

OBJECTIVES

In this chapter you will
- review the properties of square roots
- write explicit equations for geometric sequences
- use exponential functions to model real-world growth and decay scenarios
- review the properties of exponents and the meaning of rational exponents
- learn how to find the inverse of a function
- apply logarithms, the inverses of exponential functions

Scientists use exponential, power and logarithmic functions to study the behavior of phenomena in the natural world that is non-linear. Historically, the Richter scale was used to measure the magnitude of an earthquake. It was replaced in 2011 by a new scale called the moment magnitude scale, which measures a variety of seismic waves. Both use a logarithmic scale, meaning each whole-number increase in magnitude represents an increase in amplitude by a power of 10. Another result of seismic activity is the curved exfoliation joints (and resultant topographic surfaces) like these in Yosemite National Park. All living things grow and eventually decay, including populations. Both growth and decay can be modeled using exponential functions. Most King penguin colonies, like this one in the South Sandwich Islands, showed positive growth rates from the 1960s. However, there was evidence for a decrease in the larger colonies since the early 1990s, and for lower growth rates in the smaller colonies during the 1990s. **ASK** "What other examples of phenomena that behave in a non-linear manner can you think of?"

LESSON 4.1

Exponential Functions

You have previously used sequences and recursive rules to model geometric growth or decay of money, populations, and other quantities. Recursive formulas generate only discrete values, such as the amount of money after one year or two years, or the population in a certain year. Usually growth and decay happen continuously. In this lesson you will focus on finding explicit formulas for these patterns, which will allow you to model situations involving continuous growth and decay, or to find discrete points without using recursion.

Life shrinks or expands in proportion to one's courage.
ANAÏS NIN

INVESTIGATION

Radioactive Decay

This investigation is a simulation of radioactive decay. Each person will need a standard six-sided die. Each standing person represents a radioactive atom in a sample. The people who sit down at each stage represent the atoms that underwent radioactive decay.

YOU WILL NEED
- one die per person

Procedure Note
1. All members of the class should stand up, except for the recorder. The recorder counts and records the number standing at each stage.
2. Each standing person rolls a die, and anyone who gets a 1 sits down.
3. Wait for the recorder to count and record the number of people standing.
4. Repeat Steps 2 and 3 until fewer than three students are standing.

Step 1 Follow the Procedure Note to collect data in the form (stage, number standing).

Step 2 Graph your data. The graph should remind you of the sequence graphs you have studied. What type of sequence does this resemble?

Step 3 Identify u_0 and the common ratio, r, for your sequence. Complete the table below. Use the values of u_0 and r to help you write an explicit formula for your data.

n	u_n	u_n in terms of u_0 and r	u_n in terms of u_0 and r using exponents
0	u_0		
1	u_1	$u_0 \cdot r$	$u_0 \cdot r^1$
2	u_2	$u_0 \cdot r \cdot r$	$u_0 \cdot r^2$
3	u_3	$u_0 \cdot r \cdot r \cdot r$	$u_0 \cdot r^3$

Step 4 Graph your explicit formula along with your data. Notice where the value of u_0 appears in your equation. Your graph should pass through the original data point $(0, u_0)$. Modify your equation so that it passes through $(1, u_1)$, the second data point. (Think about translating the graph horizontally and also changing the starting value.)

Step 5 Experiment with changing your equation to pass through other data points. Decide on an equation that you think is the best fit for your data. Write a sentence or two explaining why you chose this equation.

Step 6 What equation with ratio r would you write that contains the point $(6, u_6)$? Answers may vary. Equation should be equivalent to $f(x) = u_6 \cdot r^{(x-6)}$.

Step 1 Sample answer: Assume there are 30 students in the class. The following sample data were used in obtaining sample answers for all steps of this investigation:
(0, 30) (1, 26)
(2, 19) (3, 17)
(4, 14) (5, 12)
(6, 11) (7, 9)
(8, 8)   (9, 8)
(10, 5) (11, 5)
(12, 2)

Step 5 Answers will vary. Students might decide on an equation that has about the same number of data points above and below, or they may prefer an equation that has more direct hits.

LESSON 4.1 Exponential Functions **201**

LESSON 4.1

This lesson introduces the exponential function in the form $y = ab^x$, used for modeling growth and decay.

COMMON CORE STATE STANDARDS		
APPLIED	**DEVELOPED**	**INTRODUCED**
N.Q.2	F.IF.3	F.IF.7e
A.CED.1	F.BF.2	F.IF.8b
F.LE.1c	F.BF.3	F.IF.9
F.LE.5	F.LE.2	

Objectives

- Write explicit equations for geometric sequences
- Define *exponential function*, with parent function $y = ab^x$
- Recognize that in real-world situations the exponential function models growth when $b > 1$ and models decay when $0 < b < 1$
- Learn about half-life and doubling time
- Evaluate an exponential function using either the explicit equations or graphical methods
- Discover point-ratio form

Vocabulary

exponential function
half-life
doubling time
end behavior
long run behavior

Materials

- dice (1 die per person)
- Calculator Notes: Random Numbers; Finance Solver; Power and Roots

Launch

Use the following situations to describe how each can model either mathematical growth or decay.

  a. Your savings account balance
  b. Population of a city

a. Growth—depositing money, Decay—withdrawing money
b. Growth—people moving into a city at a faster rate than others leave
Decay—people moving out of a city at a faster rate than others move in

LESSON 4.1 Exponential Functions **201**

ELL
To help students compare and contrast the general forms of linear, quadratic, square root, and exponential models, continue to use a graphic organizer to display the similarities and differences. Vocabulary should be clearly defined. Contrast common ratio to common difference.

Support
Use graphing technology to give students the opportunity to explore the effects of changing a and b on the graph of exponential functions. Have students create real-world contexts for each example. Students may not yet be comfortable with recursive notation. Allow them to verbally describe recursive functions, while reinforcing the notation.

Advanced
Use the **One Step** version of the Investigation. For Exercise 16, ask students to think about how to best find the slope of $f(x)$ exactly at that point.

Investigate

In place of dice, students might use calculators that are set to find random integers from 1 to 6 or use another number to investigate a different decay rate. There is also a desmos simulation of the investigation in your ebook.
SMP 4, 5

Guiding the Investigation

Step 1 **ALERT** Make sure that the recorder labels the initial number of people in the class as stage 0. If you have fewer than 15 students, run the procedure several times and analyze the totals, or use the sample data.

Step 2 a decreasing geometric sequence

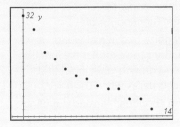

Step 3 $u_0 = 30$; The ratios between consecutive terms are

$\frac{26}{30} \approx 0.867$, $\frac{19}{26} \approx 0.731$, $\frac{17}{19} \approx 0.895$,
$\frac{14}{17} \approx 0.824$, $\frac{12}{14} \approx 0.857$, $\frac{11}{12} \approx 0.917$,
$\frac{9}{11} \approx 0.818$, $\frac{8}{9} \approx 0.889$, $\frac{8}{8} = 1$, $\frac{5}{8} = 0.625$,
$\frac{5}{5} = 1$, $\frac{2}{5} = 0.4$.

Average these 12 ratios to obtain $r \approx 0.8185$.

Step 4

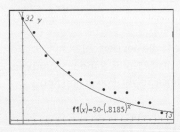

possible answer: $f(x) = 26 \cdot 0.8185^{x-1}$

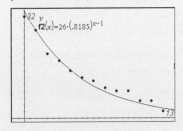

You probably recognized the geometric decay model in the investigation. As you learned earlier, geometric decay is nonlinear. At each step, the previous term is multiplied by a common ratio. Because the common ratio appears as a factor more and more times, you can use exponential functions to model the geometric growth. An **exponential function** is a continuous function with a variable in the exponent, and it is used to model growth or decay.

EXAMPLE A Most automobiles depreciate as they get older. Suppose an automobile that originally costs $14,000 depreciates by one fifth of its value every year.

a. What is the value of this automobile after $2\frac{1}{2}$ years?

b. When is this automobile worth half of its initial value?

Solution a. The recursive formula gives automobile values only after one year, two years, and so on. The value decreases by $\frac{1}{5}$, or 0.2, each year, so to find the next term you multiply by another $(1 - 0.2)$. You can use this fact to write an explicit formula.

$14{,}000 \cdot (1 - 0.2)$ Value after 1 year.

$14{,}000 \cdot (1 - 0.2) \cdot (1 - 0.2) = 14{,}000(1 - 0.2)^2$ Value after 2 years.

$14{,}000 \cdot (1 - 0.2) \cdot (1 - 0.2) \cdot (1 - 0.2) = 14{,}000(1 - 0.2)^3$ Value after 3 years.

$14{,}000 \cdot (1 - 0.2)^n$ Value after n years.

So the explicit formula for automobile value is $u_n = 14{,}000(1 - 0.2)^n$. The equation of the continuous function through the points of this sequence is

$y = 14{,}000(1 - 0.2)^x$

You can use the continuous function to find the value of the car at any point. To find the value after $2\frac{1}{2}$ years, substitute 2.5 for x.

$y = 14{,}000(1 - 0.2)^{2.5} \approx \$8{,}014.07$

It makes sense that the automobile's value after $2\frac{1}{2}$ years should be between the values of u_2 and u_3, $\$8{,}960$ and $\$7{,}168$. However, the value of the car after $2\frac{1}{2}$ years is not exactly halfway between those values because the function describing the value is not linear.

b. To find when the automobile is worth half of its initial value, substitute 7,000 for y and find x.

$7{,}000 = 14{,}000(1 - 0.2)^x$ Substitute 7,000 for y.

$0.5 = (1 - 0.2)^x$ Divide both sides by 14,000.

$0.5 = (0.8)^x$ Combine like terms.

You don't yet know how to solve for x when x is an exponent, but you can experiment to find an exponent that produces a value close to 0.5. The value of $(0.8)^{3.106}$ is very close to 0.5. This means that the value of the car is about $\$7{,}000$, or half of its original value, after 3.106 years, or about

202 CHAPTER 4 Exponential, Power, and Logarithmic Functions

Modifying the Investigation

Whole Class Follow the procedure note to collect data. For Steps 2 through 6, have students do each step, then discuss answers as a class before moving on to the next step.

Shortened Use the sample data. Discuss Steps 4 through 6.

One Step Follow the Procedure Note of the investigation. Ask students to find recursive formulas and function expressions that describe the data. As needed, remind students that repeated multiplication is represented by exponents. During the discussion, ask what other kinds of growth might be modeled by exponential functions; include depreciation.

3 years 39 days. This is the **half-life** of the value of the automobile, or the amount of time needed for the value to decrease to half of the original amount.

You can also find the solution graphically by graphing both sides of the equation $7000 = 14000(0.8)^x$ and finding the intersection point at about (3.11, 7000), so $x \approx 3.11$.

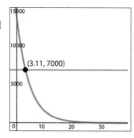

> **Exponential Function**
>
> The general form, or intercept form, of an exponential function is
>
> $y = ab^x \qquad (b > 0)$
>
> where the coefficient a is the y-intercept and the base b is the growth rate.

Exponential growth and decay are both modeled with the general form $y = ab^x$. Growth is modeled by a base that is greater than 1, and decay is modeled by a base that is less than 1, but greater than 0. In general, a larger base models faster growth, and a base closer to 0 models faster decay.

All exponential growth curves have a **doubling time**, just as decay has a half-life. This time depends only on the ratio. For example, if the ratio is constant, it takes just as long to double $1,000 to $2,000 as it takes to double $5,000 to $10,000.

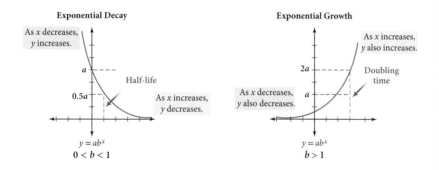

Looking at the graphs for exponential growth and decay, you can see that they have very different *end behaviors* as x values increase and decrease. In the graph of exponential decay, as x increases in value, y values decrease. In the graph of exponential growth, as x increases in value, y values also increase. Looking at these graphs in the opposite direction, as x values decrease, you see the opposite happening. In the graph of exponential decay, as x decreases in value, y values increase. In the graph of exponential growth, as x decreases in value, y values also decrease. In general, the **end behavior**, or **long run behavior**, of the function is the behavior of the function for increasingly larger x-values and for decreasingly smaller x-values.

Example B

Project Example B and have students graph the parent function, $y = 2^x$. Ask for student input describing the graph. Then have them work in pairs to graph $g(x)$ on the same axis and describe the transformation. Repeat with $h(x)$. Have students share their graphs as they describe them. **SMP 3, 5, 6**

ASK "Why are these functions equivalent?" Use the meaning of exponents as repeated multiplication to bring out that $\frac{2^x}{2} = 2^{x-1}$.

Summarize

After Example A, **ASK** "What might be the equivalent of half-life for increasing functions?" [doubling time] You might pose a problem about various amounts of money invested at an annual rate of 4.5%, compounded monthly. "Is the doubling time the same for each initial amount?" **ASK** "Is the doubling time the same for different interest rates?" [no] "Is it the same if the investment is compounded weekly or yearly?" [no] "If the initial investment takes about 185.2 months to double, how long will it take to double again?" [another 185.2 months]

LINK You can also refer to the general form of the exponential as the "intercept form", which ties it to intercept form of a linear equation.

Point out that the same function describes processes as different as automobile depreciation and radioactive decay. Both of these examples are decreasing functions. **ASK** "Can an exponential function model positive growth?" [Yes]

CRITICAL QUESTION How are increasing exponential functions different from decreasing exponential functions?

BIG IDEA The base b, or common ratio, is greater than 1 for increasing functions and between 0 and 1 for decreasing functions.

CRITICAL QUESTION Why is b never negative?

BIG IDEA The exponential function is not defined for negative b.

Certain atoms are unstable—their nuclei can split apart, emitting radiation and resulting in a more stable atom. This process is called radioactive decay. The time it takes for half the atoms in a radioactive sample to decay is called **half-life**, and the half-life is specific to the element. For instance, the half-life of carbon-14 is 5730 years, whereas the half-life of uranium-238 is 4.5 billion years. Due to the decay time, serious nuclear power plant accidents, which include the Fukushima Daiichi nuclear disaster (2011), Chernobyl disaster (1986), Three Mile Island accident (1979), and the SL-1 accident (1961), leave an area radioactive for decades.

Modern ruins in a high dynamic range (HDR) near Chernobyl (1986).

A table of $y = 2^x$ shows that as x increases, the value of y increases even faster, and as x decreases, the value of y gets closer and closer to zero.

x	−10	−5	−1	−0.5	0	0.5	1	5	10
y	0.001	0.031	0.5	0.707	1	1.41	2	32	1024

EXAMPLE B | The functions $g(x) = \frac{1}{2}(2)^x$ and $h(x) = 2^{x-1}$ are transformations of the parent function $f(x) = 2^x$. Describe the graph of the parent function $f(x) = 2^x$. Then describe the transformations of $g(x)$ and $h(x)$ and sketch the graphs.

Solution | The graph of the function $f(x) = 2^x$ is an exponential function, where $a = 1$ and $b = 2$. This graph is in both quadrants I and II, the graph is increasing, and the y-intercept is 1.

The graph of $g(x)$ is a vertical dilation of the graph of $f(x)$ by a factor of $\frac{1}{2}$. The marked points on the red graph show how the y-values of the corresponding points on the black graph have been multiplied by $\frac{1}{2}$.

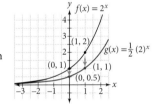

The graph of $h(x)$ is a horizontal translation of the graph of $f(x)$. The marked points on the red graph show how the corresponding points on the black graph have been moved right 1 unit.

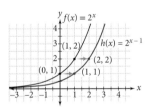

Do you notice anything special about the two red graphs? They appear to be the same. With exponential functions, as with linear and quadratic functions, it is possible to find different transformations that produce the same graphs.

Formative Assessment

Note whether the understanding of explicit formulas students developed earlier is deep enough to extend their concept to the new context in this chapter. Assess students' facility at modeling data with recursive formulas and with common ratios. They should also be able to make the link between exponents and repeated multiplication.

Exercises

You will need a graphing calculator for Exercises **8–14** and **17**.

Practice Your Skills

1. Evaluate each function at the given value.

   a. $f(x) = 4.753(0.9421)^x$, $x = 5$ @ $f(5) \approx 3.52$
   b. $g(h) = 238(1.37)^h$, $h = 14$ $g(14) \approx 19{,}528.32$
   c. $h(t) = 47.3(0.835)^t + 22.3$, $t = 24$ @ $h(24) \approx 22.92$
   d. $j(x) = 225(1.0825)^{x-3}$, $x = 37$ $j(37) \approx 3332.20$

2. Classify each exponential function as exponential growth or decay.

   a. $y = 3.01\,(0.99)^x$ — decay
   b. $y = 0.789\,(2)^x$ — growth
   c. $y = 1.97\,(1.034)^x$ — growth
   d. $y = 0.8\,(0.3)^x$ — decay

3. Record three terms of the sequence and then write an explicit function for the sequence.

   a. $u_0 = 16$ 16, 12, 9; $y = 16(0.75)^x$
      $u_n = 0.75 u_{n-1}$ where $n \geq 1$ @
   b. $u_0 = 24$ 24, 36, 54; $y = 24(1.5)^x$
      $u_n = 1.5 u_{n-1}$ where $n \geq 1$

4. Evaluate each function at $x = 0$, $x = 1$, and $x = 2$ and then write a recursive formula for the pattern.

   a. $f(x) = 125(0.6)^x$ @ $f(0) = 125$, $f(1) = 75$, $f(2) = 45$; $u_0 = 125$ and $u_n = 0.6 u_{n-1}$ where $n \geq 1$
   b. $f(x) = 3(2)^x$ $f(0) = 3$, $f(1) = 6$, $f(2) = 12$; $u_0 = 3$ and $u_n = 2 u_{n-1}$ where $n \geq 1$

5. Describe the end behavior of each function in Exercise 4.

6. Calculate the ratio of the second term to the first term, and express the answer as a decimal value. State the percent increase or decrease.

   a. 48, 36 0.75; 25% decrease
   b. 54, 72 $1.\overline{3}$; $3.\overline{3}$% increase
   c. 50, 47 @ 0.94; 6% decrease
   d. 47, 50 1.0638; 6.38% increase

Reason and Apply

7. In 1995, the population of the People's Republic of China was approximately 1.211 billion, with an annual growth rate of 1.1%. (*www.chinability.com*) @

   a. Write a recursive formula that models this growth. Let u_0 represent the population in 1995. $u_0 = 1.211$, $u_n = u_{n-1} \cdot 1.011$
   b. Complete a table recording the population for the years 1995 to 2002.
   c. Define the variables and write an exponential equation that models this growth. Choose two data points from the table and show that your equation works.
   d. One estimate for the population of China in 2006 was 1.315 billion. How does this compare with the value predicted by your equation?
   e. The reported annual growth rate in 2006 was 0.53%. Write a new exponential equation that models the growth, letting $u_0 = 1.315$.
   f. One estimate for the population of China in 2014 was 1.368 billion. How does this compare with the value predicted by your equation? What does this tell you?

LESSON 4.1 Exponential Functions **205**

7c. y represents the estimated population x years after 1995; $y = 1.211\,(1.011)^x$.

7d. The equation predicts that the population of China in 2006 was 1.37 billion. This is larger than the actual value. This means that the population is growing at a slower rate in 2006 than it was in 1995.

7e. y represents the estimated population x years after 2006: years after 2006: $y = 1.315 \cdot 1.0053^x$

7f. The equation predicts that the population in 2014 is 1.372 billion. This is a little larger than the actual value. This means that the population was growing at a slightly slower rate in 2014 than it was in 2006.

Apply

Extra Example

The graph of $y = 2^x$ has been translated horizontally 2 units right and vertically 5 units down and reflected across the x-axis. What is the equation of the resulting graph?
$y = -2^{x-2} - 5$

Closing Question

Parker Electronics sold 450 cell phones in January. Since then, their cell phone sales have increased at a rate of 4.75% per month. Create a function that can be used to determine the monthly cell phone sales m months after January, at this rate of growth? $p(m) = 450(1.0475)^m$

Assigning Exercises

Suggested: 1–14
Additional Practice: 15–18

Exercise 4 ASK "What is the effect of a 0 exponent?" [The factor always equals 1.]

5a. As x approaches extreme values to the left, y increases without bound. As x approaches extreme values to the right, y approaches 0.

5b. As x approaches extreme values to the left, y approaches 0. As x approaches extreme values to the right, y increases without bound.

Exercise 6 For each part, ASK "Would this be a decay or a growth model?" Point out that 6c and 6d have the same numbers in reverse. ASK "Why are the percentages different?" [because 6% of 50 is not the same as 6% of 47] Suggest that students write the ratio as (1 + *some value*) or (1 − *some value*).

7b.

Year	Estimated population (billions)
1995	1.211
1996	1.224
1997	1.238
1998	1.251
1999	1.265
2000	1.279
2001	1.293
2002	1.307

LESSON 4.1 Exponential Functions **205**

Exercise 8 In this exercise, the data have decimal values in the exponent.

8a. Let x represent the number of the day, and let y represent the height in cm. $y = 2.56(2.5)^x$. For the fifth day, $y = 2.56(2.5)^5 = 250$ cm; for the sixth day, $y = 2.56(2.5)^6 = 625$ cm.

8b. $y = 2.56(2.5)^{3.5} \approx 63.25$

9a–d.

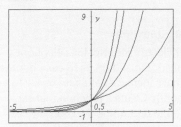

9e. As the base increases, the graph becomes steeper. The curves all intersect the y-axis at $(0, 1)$.

9f. The graph of $y = 6^x$ should be the steepest of all of these. It will contain the points $(0, 1)$ and $(1, 6)$.

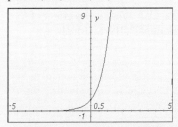

Exercises 10, 12 Students use transformations to find these growth equations. **ALERT** Students may be skeptical of the translations in 10b and 10c because the graphs don't look "parallel." As needed, remind them that translations of parabolas aren't equidistant either. Actually, horizontal translations of graphs of exponential functions are also vertical dilations, because $b^{x+h} = b^x b^h$. **ADVANCED** The graphs in 10c and 10d are dilations. Encourage students to compare the coordinates of the points to find the dilation factors.

---

8. Jack planted a mysterious bean just outside his kitchen window. It immediately sprouted 2.56 cm above the ground. Jack kept a careful log of the plant's growth. He measured the height of the plant each day at 8:00 A.M. and recorded these data:

Day	0	1	2	3	4
Height (cm)	2.56	6.4	16	40	100

   a. Define variables and write an exponential equation for this pattern. If the pattern continues, what will be the heights on the fifth and sixth days?

   b. Jack's younger brother measured the plant at 8:00 P.M. on the evening of the third day and found it to be about 63.25 cm tall. Show how to find this value mathematically. (You may need to experiment with your calculator.) ⓗ

   c. Find the height of the sprout at 12:00 noon on the sixth day. 728 cm

   d. Find the doubling time for this plant. 0.76 day, or 18 hours

   e. Experiment with the equation to find the day and time (to the nearest hour) when the plant reaches a height of 1 km.
   11 days 13 hours, or 9 P.M. on day 11

9. For a–d, graph the equations on your calculator.

   a. $y = 1.5^x$ b. $y = 2^x$ c. $y = 3^x$ d. $y = 4^x$

   e. How do the graphs compare? What points (if any) do they have in common?

   f. Predict what the graph of $y = 6^x$ will look like. Verify your prediction by using your calculator.

10. Each of the red curves is a transformation of the graph of $y = 2^x$, shown in black. Focus on how the two marked points on the black curve are transformed to become the corresponding points on the red curve. Write an equation for each red curve. ⓗ

   a. 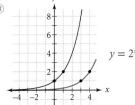 $y = 2^{x-3}$

   b. 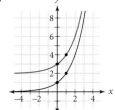 $y - 2 = 2^x$, or $y = 2^x + 2$

   c. 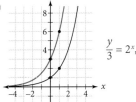 $\frac{y}{3} = 2^x$, or $y = 3 \cdot 2^x$

   d. 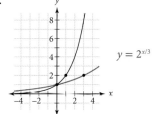 $y = 2^{x/3}$

Kudzu, shown here in Cherokee National Forest, is a fast-growing weed that was brought from Japan and once promoted for its erosion control.

11. For a–d, graph the equations on your calculator.

   a. $y = 0.2^x$ b. $y = 0.3^x$ c. $y = 0.5^x$ d. $y = 0.8^x$

   e. How do the graphs compare? What points (if any) do they have in common?

   f. Predict what the graph of $y = 0.1^x$ will look like. Verify your prediction by using your calculator.

12. Each of the red curves is a transformation of the graph of $y = 0.5^x$, shown in black. Focus on how the two marked points on the black curve are transformed to become the corresponding points on the red curve. Write an equation for each red curve. ⓗ

a.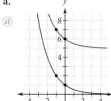

$y - 5 = 0.5^x$, or $y = 5 + 0.5^x$

b.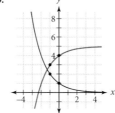

$-(y - 5) = 0.5^x$, or $y = -0.5^x + 5$

c.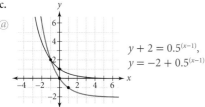

$y + 2 = 0.5^{(x-1)}$, $y = -2 + 0.5^{(x-1)}$

d.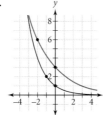

$\dfrac{y}{3} = 0.5^{(x/2)}$, $y = 3(0.5)^{(x/2)}$

13. The general form of an exponential equation, $y = ab^x$, is convenient when you know the y-intercept. Start with $f(0) = 30$ and $f(1) = 27$.

   a. Find the common ratio. ⓐ $\dfrac{27}{30} = 0.9$

   b. Write the exponential function $f(x)$ that passes through the two data points. ⓐ $f(x) = 30(0.9)^x$

   c. Graph $f(x)$ and $g(x) = f(x - 4)$ on the same axes.

   d. What is the value of $g(4)$? $g(4) = 30$

   e. Write an equation for $g(x)$ that does not use its y-intercept. ⓗ possible answer: $g(x) = 30(0.9)^{x-4}$

   f. Explain in your own words why $y = y_1 \cdot b^{x-x_1}$ might be called the point-ratio form. ⓗ

Houston, TX, was the fasttest growing city in US in 2015 Combined, the greater Houston metropolitan area, which includes Houston, The Woodlands and Sugar Land, grew by about 160,000 people between July 2014 and July 2015.

14. Write two equations in the form $y = y_1 \cdot b^{x-x_1}$ that contain the points listed in the table. ⓐ

x	0	1	2	3	4
y	4	7.2	12.96	23.328	41.9904

Answers will vary but will be in the form $y = y_1 \cdot 1.8^{x-x_1}$, with (x_1, y_1) being any point from the table.

11a–d.

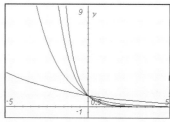

11e. As the base increases, the graph flattens out. The curves all intersect the y-axis at (0, 1).

11f. The graph of $y = 0.1^x$ should be the steepest of all of these. It will contain the points (0, 1) and (−1, 10).

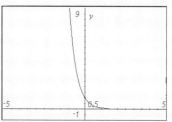

13c.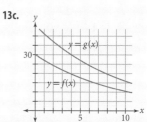

13f. Sample answer: You can use the x- and y-values of any point on the curve and the common ratio to write the equation.

Exercise 14 As needed, suggest that students look for a common ratio and write any equation that fits the data, without worrying about the subtraction in the exponent. If some students write $y = 4(1.8)^{x-0}$, have then share their answer.

LESSON 4.1 Exponential Functions 207

15a. Let x represent time in seconds, and let y represent distance in meters.

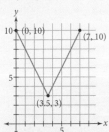

Exercise 17 Exponential functions here are defined with positive bases, b, where $b \geq 1$. If students try to graph these functions, they will see that exponentials with negative bases have real values only at discrete points. In general, when the base of an exponential function is negative, the function produces complex values. Students can get a good estimate by graphing and tracing. Or if they're familiar with properties of exponents, they can simplify the equation $\frac{5200}{5000} = \left(\frac{1.035}{1.032}\right)^x$ to $1.04 = 1.0029^x$. Repeatedly multiplying 1.0029 by itself shows that x is between 13 and 14 years.

Exercise 18 Diagrams for 18a may or may not have dotted lines. Students might say correctly that the first step in 16b could be $x(x-4) + 6(x-4)$.

See your digital resources for the **Project: The Cost of Living**, where students compare the price of an item over the years to find doubling time and predict future cost of the item.

Review

15. Janell starts 10 m from a motion sensor and walks at 2 m/s toward the sensor. When she is 3 m from the sensor, she instantly turns around and walks at the same speed back to the starting point.

  a. Sketch a graph of the function that models Janell's walk.
  b. Give the domain and range of the function. domain: $0 \leq x \leq 7$; range: $3 \leq y \leq 10$
  c. Write an equation of the function. $y = 2|x - 3.5| + 3$

16. The graph shows a line and the graph of $y = f(x)$.

  a. Fill in the missing values to make a true statement.
  $f(\underline{?}) = \underline{?}$. $f(3) = 8.5$
  b. Find the equation of the pictured line.
  $y = 8.5 + 0.5(x - 3)$, $y = 10 + 0.5(x - 6)$, or $y = 7 + 0.5x$

17. Austin deposits \$5,000 into an account that pays 3.5% annual interest. Sami deposits \$5,200 into an account that pays 3.2% annual interest.

  a. Write an expression for the amount of money Austin will have in his account after 5 years if he doesn't deposit any more money. $A = 5000(1 + 0.035)^5$
  b. Write an expression for the amount of money Sami will have in her account after 5 years if she doesn't deposit any more money. $S = 5200(1 + 0.032)^5$
  c. How long will it take until Austin has more money than Sami?
  After 14 years, Austin will have \$8,093.47, and Sami will have \$8,082.

18. You can use different techniques to find the product of two binomials, such as $(x - 4)(x + 6)$.

  a. Use a rectangle diagram to find the product.
  b. You can use the distributive property to rewrite the expression $(x - 4)(x + 6)$ as $x(x + 6) - 4(x + 6)$. Use the distributive property again to find all the terms. Combine like terms. $x^2 + 2x - 24$
  c. Compare your answers to 18a and b. Are they the same? yes
  d. Compare the methods in 18a and b. How are they alike? A rectangle diagram also uses the distributive property. Each term in the first binomial is multiplied by each term in the second binomial.

18a.

	x	6
x	x^2	$6x$
-4	$-4x$	-24

IMPROVING YOUR Reasoning SKILLS

Breakfast Is Served

Mr. Higgins told his wife, a mathematics professor, that he would make her breakfast. She handed him this message:

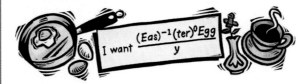

What should Mr. Higgins fix his wife for breakfast?

IMPROVING YOUR Reasoning SKILLS

$\frac{(Eas)^{-1}(ter)^0 Egg}{y} = \frac{1 Egg}{Easy}$, or "one egg over easy"

ELL Students new to English may be unfamiliar with the expression for this way of frying an egg.

LESSON 4.2

Properties of Exponents and Power Functions

When the whole world is silent, even one voice becomes powerful.
MALALA YOUSAFZAI

Frequently, you will need to rewrite a mathematical expression in a different form to make the expression easier to understand or an equation easier to solve. Recall that in an exponential expression, such as 4^3, the number 4 is called the **base** and the number 3 is called the **exponent**. You say that 4 is **raised to the power** of 3. If the exponent is a positive integer, you can write the expression in **expanded form**, for example, $4^3 = 4 \cdot 4 \cdot 4$. Because 4^3 equals 64, you say that 64 is a power of 4.

INVESTIGATION

Properties of Exponents

Use expanded form to review and generalize the properties of exponents.

For expanded forms, see next page.

Step 1 Write each product in expanded form, and then rewrite it in exponential form.

    a. $2^3 \cdot 2^4$ 2^7 **b.** $x^5 \cdot x^{12}$ x^{17} **c.** $10^2 \cdot 10^5$ 10^7

Step 2 Generalize your results from Step 1. $a^m \cdot a^n = \underline{?}$ a^{m+n}

Step 3 Write the numerator and denominator of each quotient in expanded form. Simplify by reducing the number of common factors, and then rewrite the factors that remain in exponential form.

    a. $\dfrac{4^5}{4^2}$ 4^3 **b.** $\dfrac{x^8}{x^6}$ x^2 **c.** $\dfrac{(0.94)^{15}}{(0.94)^5}$ $(0.94)^{10}$

Step 4 Generalize your results from Step 3. $\dfrac{a^m}{a^n} = \underline{?}$ a^{m-n}

Step 5 Write each quotient in expanded form, simplify, and rewrite in exponential form.

    a. $\dfrac{2^3}{2^4}$ $\dfrac{1}{2^1}$ **b.** $\dfrac{4^5}{4^7}$ $\dfrac{1}{4^2}$ **c.** $\dfrac{x^3}{x^8}$ $\dfrac{1}{x^5}$

Step 6
6a $2^{3-4} = 2^{-1}$
6b $4^{5-7} = 4^{-2}$
6c $x^{3-8} = x^{-5}$

Step 6 Rewrite each quotient in Step 5 using the property you discovered in Step 4.

Step 7 Generalize your results from Steps 5 and 6. $\dfrac{1}{a^n} = \underline{?}$ a^{-n}

Step 8 Write several expressions in the form $(a^n)^m$. Expand each expression, and then rewrite it in exponential form. Generalize your results. $(a^n)^m = a^{nm}$
sample answer: $(2^3)^4 = (2^3)(2^3)(2^3)(2^3) = 2^{3+3+3+3} = 2^{12}$;

Step 9 Write several expressions in the form $(a \cdot b)^n$. Don't multiply a and b. Expand each expression, and then rewrite it in exponential form. Generalize your results.
sample answer: $(2 \cdot 3)^3 = (2 \cdot 3)(2 \cdot 3)(2 \cdot 3) = 2 \cdot 2 \cdot 2 \cdot 3 \cdot 3 \cdot 3 = 2^3 \cdot 3^3$; $(a \cdot b)^n = a^n b^n$

Step 10 Show that $a^0 = 1$, using the properties you have discovered. Write at least two exponential expressions to support your explanation.
sample answer: $\dfrac{x^3}{x^3} = \dfrac{x \cdot x \cdot x}{x \cdot x \cdot x} = 1$; but, by the property from Step 4, $\dfrac{x^3}{x^3} = x^{3-3} = x^0$

LESSON 4.2

In this lesson, students review the properties of exponents and power functions. The properties for exponents are defined only for positive bases. The restriction does not rule out negative bases for integer exponents, so bases in formulas for sequences may be negative.

COMMON CORE STATE STANDARDS		
APPLIED	**DEVELOPED**	**INTRODUCED**
A.REI.1	N.RN.2	N.RN.1
	A.SSE.3c	F.IF.6
	F.IF.8b	
	F.LE.3	

Objectives

- Introduce algebraic proof
- Review the properties of exponents—specifically, rewriting powers with the same base
- Introduce the parent power function, $y = ax^n$, and distinguish it from the exponential function, $y = ab^x$
- Find some solutions to power equations using the properties of exponents
- Introduce rational exponents as a means of solving equations

Vocabulary

base
exponent
raised to the power
expanded form
power function

Materials

- Calculator Note: Power and Roots

Launch

For each of the following, identify the base and the exponent. Write each in expanded form.

  a. 4^5
  b. $\left(\dfrac{7}{5}\right)^3$
  c. x^4
  d. $(x-2)^2$

  a. $4; 5; 4 \cdot 4 \cdot 4 \cdot 4 \cdot 4$
  b. $\dfrac{7}{5}; 3; \dfrac{7}{5} \cdot \dfrac{7}{5} \cdot \dfrac{7}{5}$
  c. $x; 4; x \cdot x \cdot x \cdot x$
  d. $(x-2); 2; (x-2) \cdot (x-2)$

ELL
Provide numerical examples to support the definitions. This will help students recognize the difference between the letters that represent constants and those that represent variable values in the properties of exponents.

Support
To promote success with the type of exercises shown in Example A, have students practice rewriting a variety of numbers with common bases and discuss their methods. Students should then find it easier to solve exponential equations by using the power of a power property.

Advanced
ASK "Are power functions and exponential functions inverses of each other?" [No] "Can any of the exponent properties be derived from other properties?" [The power of a quotient property can be derived from the power of a product property and the definition of negative exponents.]

Investigate

Guiding the Investigation

If students use the investigation worksheet rather than the book, they may not be tempted to look up the properties.

Step 1

a. $2^3 \cdot 2^4 = (2\cdot 2\cdot 2)\cdot(2\cdot 2\cdot 2\cdot 2) = 2^7$

b. $x^5 \cdot x^{12} = (x\cdot x\cdot x\cdot x\cdot x)(x\cdot x\cdot x\cdot x\cdot x\cdot x\cdot x\cdot x\cdot x\cdot x\cdot x\cdot x)$
  $= x^{17}$

c. $10^2 \cdot 10^5 = (10\cdot 10)(10\cdot 10\cdot 10\cdot 10\cdot 10) = 10^7$

Step 3

a. $\dfrac{4^5}{4^2} = \dfrac{4\cdot 4\cdot \cancel{4}\cdot \cancel{4}\cdot \cancel{4}}{\cancel{4}\cdot \cancel{4}} = 4^3$

b. $\dfrac{x^8}{x^6} = \dfrac{\cancel{x}\cdot \cancel{x}\cdot \cancel{x}\cdot \cancel{x}\cdot \cancel{x}\cdot \cancel{x}\cdot x\cdot x}{\cancel{x}\cdot \cancel{x}\cdot \cancel{x}\cdot \cancel{x}\cdot \cancel{x}\cdot \cancel{x}} = x^2$

c. See page 211

Step 5

a. $\dfrac{2^3}{x^4} = \dfrac{\cancel{2}\cdot\cancel{2}\cdot\cancel{2}}{2\cdot\cancel{2}\cdot\cancel{2}\cdot\cancel{2}} = \dfrac{1}{2^1}$

b. $\dfrac{4^5}{4^7} = \dfrac{\cancel{4}\cdot\cancel{4}\cdot\cancel{4}\cdot\cancel{4}\cdot\cancel{4}}{4\cdot 4\cdot\cancel{4}\cdot\cancel{4}\cdot\cancel{4}\cdot\cancel{4}\cdot\cancel{4}} = \dfrac{1}{4^2}$

c. $\dfrac{x^3}{x^8} = \dfrac{\cancel{x}\cdot\cancel{x}\cdot\cancel{x}}{x\cdot x\cdot x\cdot x\cdot x\cdot\cancel{x}\cdot\cancel{x}\cdot\cancel{x}} = \dfrac{1}{x^5}$

Step 9 You might have some groups explore $\left(\dfrac{a}{b}\right)^n$ instead of $(a\cdot b)^n$.

Have students present their results, especially their generalizations in Steps 2, 4, 7 and their solutions and explanations for Steps 8 – 9. Emphasize the use of the phrase *dividing out common factors* rather than the word *cancel*. **ASK** "Can terms can be divided out as well? For example, can the expression $\dfrac{x+x+x}{x+x}$ be simplified to x? Why?" [No. Terms are not factors.]
SMP 3, 6, 7, 8

When students present their ideas from Step 7, **ASK** "If $\dfrac{1}{a^n} = a^{-n}$, what does that say about $\left(\dfrac{3}{4}\right)^n$?
$\left[\dfrac{3}{4} \text{ is } \dfrac{1}{\frac{4}{3}}, \text{ so } \left(\dfrac{3}{4}\right)^n = \left(\dfrac{4}{3}\right)^{-n}.\right]$
To generalize the result, $\left(\dfrac{a}{b}\right)^{-n} = \left(\dfrac{b}{a}\right)^n$.

ASK "Why might the book restrict the exponent properties to positive bases?" [The restriction avoids undefined expressions such as 0^0 and dividing by 0. It also avoids taking even roots of negative numbers, but students may not know yet that fractional exponents represent roots.]

Here's a summary of the properties of exponents. You discovered some of these in the investigation. Try to write an example of each property.

> For $a > 0$, $b > 0$, and all values of m and n, these properties are true:
>
> **Product Property of Exponents**
> $a^m \cdot a^n = a^{m+n}$
>
> **Quotient Property of Exponents**
> $\dfrac{a^m}{a^n} = a^{m-n}$
>
> **Definition of Negative Exponents**
> $a^{-n} = \dfrac{1}{a^n}$ or $\left(\dfrac{a}{b}\right)^{-n} = \left(\dfrac{b}{a}\right)^n$
>
> **Zero Exponents**
> $a^0 = 1$
>
> **Power of a Power Property**
> $(a^m)^n = a^{mn}$
>
> **Power of a Product Property**
> $(ab)^m = a^m b^m$
>
> **Power of a Quotient Property**
> $\left(\dfrac{a}{b}\right)^n = \dfrac{a^n}{b^n}$
>
> **Power Property of Equality**
> If $a = b$, then $a^n = b^n$.
>
> **Common Base Property of Equality**
> If $a^n = a^m$, and $a \neq 1$ and $a \neq 0$, then $n = m$.

In the previous lesson, you learned to solve equations that have a variable in the exponent by using a calculator to try various values for x. The properties of exponents allow you to solve these types of equations algebraically. One special case is when you can rewrite both sides of the equation with a common base. This strategy is fundamental to solving some of the equations you'll see later in this chapter.

EXAMPLE A Solve.

a. $8^x = 4$ b. $27^x = \dfrac{1}{81}$ c. $\left(\dfrac{49}{9}\right)^x = \left(\dfrac{3}{7}\right)^{3/2}$

Solution If you use the power of a power property to convert each side of the equation to a common base, then you can solve without a calculator.

a.
$8^x = 4$	Original equation.
$(2^3)^x = 2^2$	$8 = 2^3$ and $4 = 2^2$.
$2^{3x} = 2^2$	Use the power of a power property to rewrite $(2^3)^x$ as 2^{3x}.
$3x = 2$	Use the common base property of equality.
$x = \dfrac{2}{3}$	Divide.

b.
$27^x = \dfrac{1}{81}$	Original equation.
$(3^3)^x = \dfrac{1}{3^4}$	$27 = 3^3$ and $81 = 3^4$.
$3^{3x} = 3^{-4}$	Use the power of a power property and the definition of negative exponents.
$3x = -4$	Use the common base property of equality.
$x = -\dfrac{4}{3}$	Divide.

Modifying the Investigation

Whole Class Complete Steps 1 through 10 with student input. Have a volunteer generalize in each section, then discuss his or her answer.

Shortened Skip every part c.

One Step Use the **One Step** in your ebook. Have the class reach consensus about properties of exponents and about finding exponents by rewriting the same base to different powers.

Conceptual/Procedural

Conceptual Students compare and contrast power functions with odd and even exponents. They compare the long-run behavior of exponential and power functions.

Procedural In this lesson, students define both exponential and power functions. They review the properties of exponents and use them to solve equations involving exponential functions.

c.
$$\left(\frac{49}{9}\right)^x = \left(\frac{3}{7}\right)^{3/2}$$ Original equation.

$$\left(\frac{7^2}{3^2}\right)^x = \left(\frac{3}{7}\right)^{3/2}$$ $49 = 7^2$ and $9 = 3^2$.

$$\left[\left(\frac{7}{3}\right)^2\right]^x = \left[\left(\frac{7}{3}\right)^{-1}\right]^{3/2}$$ Use the power of a quotient property and the definition of negative exponents.

$$\left(\frac{7}{3}\right)^{2x} = \left(\frac{7}{3}\right)^{-3/2}$$ Use the power of a power property.

$$2x = -\frac{3}{2}$$ Use the common base property of equality.

$$x = -\frac{3}{4}$$ Divide.

Remember, it's always a good idea to check your answer with a calculator.

An exponential function has a variable in the exponent. In the exercises you'll use exponent properties to verify the equivalence of exponential equations.

Exponential Function

The general form of an exponential function is

$$y = ab^x$$

where a and b are constants and $b > 0$.

In a **power function**, the variable is in the base. You must learn to distinguish between power functions and exponential functions.

Power Function

The general form of a power function is

$$y = ax^n$$

where a and n are constants.

Who wins? When comparing an exponential to a power function, which function eventually exceeds the other?

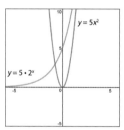
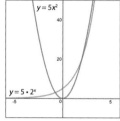
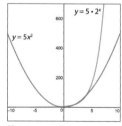

Notice that in the very short run, it appears that $y = 5 \cdot 2^x$ is increasing faster than $y = 5x^2$.

Zooming out shows that $y = 5x^2$ actually intersects with and begins to exceed $y = 5 \cdot 2^x$.

However, zooming out again reveals that in the long run, the end behavior of the exponential exceeds that of the power function.

In the long run, an exponential function will exceed the value of a power function.

Example A

Consider projecting Example A part a and asking for student input on how to solve the equation. If no one suggests getting a common base, you might rewrite the equation as $(2^3)^x = 2^2$. **ASK** "Does that help? How?" If further assistance is needed, **ASK** "What properties could be uses to make the two sides of the equation more alike?" Lead students to $2^{3x} = 2^2$, so $3x = 2$. Have students work parts b and c in pairs. **SMP 1, 8**

Ask students to check the answers on their calculators. **SMP 5**

ALERT This may be the first time students have seen a fractional exponent. **EXTEND** "What might a fractional exponent mean?" **SMP 6**

With books closed, **ASK** "Which will get the largest, $y = 5 \cdot 2^x$ or $y = 5 \cdot x^2$? Explain." Give students a minute to think about it. Ask for volunteers who believe $y = 5 \cdot 2^x$ gets larger to present their rationale. Ask other students if they agree and why. If the rationales did not include graphing, you might suggest students graph both on their calculators to verify. **SMP 1, 3, 5, 6**

Investigating Structure

Project the Investigating Structure and have students work in small groups. Or you might have half of the class graph $y = x^3$, $y = x^5$, $y = x^9$, $y = x^{25}$ and half the class graph $y = x^2$, $y = x^4$, $y = x^8$, $y = x^{24}$. Have students present their graphs and explain why power functions with odd exponents are in the first and third quadrant. [If x-values are positive, y-values are positive (Quadrant 1), if x-values are negative, y-values are negative (Quadrant 3)] power functions with even exponents are in the first and second quadrant. [If x-values are positive, y-values are positive (Quadrant 1), if x-values are negative, y-values are positive (Quadrant 2)] Use the graphs from both groups to compare and contrast power functions with odd and even exponents. **SMP 3, 6, 8**

Step 3c. $\dfrac{(0.94)^{15}}{(0.94)^5} = \dfrac{(0.94)(0.94)(0.94)(0.94)(0.94)(0.94)(0.94)(0.94)(0.94)(0.94)\cancel{(0.94)}\cancel{(0.94)}\cancel{(0.94)}\cancel{(0.94)}\cancel{(0.94)}}{\cancel{(0.94)}\cancel{(0.94)}\cancel{(0.94)}\cancel{(0.94)}\cancel{(0.94)}} = (0.94)^{10}$

Example B

Project Example B and work through part a as you did in Example A. **ASK** "Is −7.40 also a solution to the equation in Example B, part a?" [It is.] "Why wasn't this solution found?" [Because the properties of exponents are defined for positive bases only, using these properties produces only the positive solution to the equation.] You might have students graph the curve $y = x^4$ and the line $y = 3000$ and find their intersections. **SMP 3, 5, 6**

Then have students do part b in pairs. As in Example A, students should use calculators to check answers. You might ask them to check the exact answers rather than the rounded decimal values. **ASK** "What property of exponents is being used?" [the power of a power property] **SMP 3, 5, 6**

Summarize

Have students present their solutions to the investigation, the Investigating Structure, and the examples and how they verified their answers. Whether student responses are correct or incorrect, ask other students if they agree and why. Emphasize the use of correct mathematical terminology. **SMP 3, 5, 6**

CRITICAL QUESTION How can you tell if a function is a power function or an exponential function?

BIG IDEA The variable appears in the exponent of an exponential function and in the base of a power function. Stress having the rules make sense over students' memorizing them.

CRITICAL QUESTION How are power functions of the form $y = ax^n$ different from exponential functions of the form $y = ab^x$?

BIG IDEA In a power function, the independent variable is in the base rather than the exponent. The letters a, b, and n represent constants for any particular function.

▲ INVESTIGATING STRUCTURE

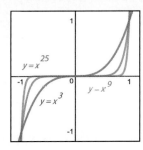 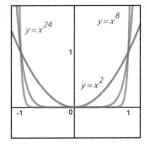

Look at these graphs. Why are the power functions with odd exponents in the first and third quadrants, while the power functions with even exponents are in the first and second quadrants?

How do the graphs on each grid compare? How do they contrast? What points (if any) do they have in common?

Why do the graphs appear to flatten out between −1 and 1 as the power gets larger? Use tables of values to help you explain.

You use different methods to solve power equations than to solve exponential equations.

EXAMPLE B | Solve for positive values of x.

a. $x^4 = 3000$ b. $6x^{2.5} = 90$

Solution | To solve a power equation, use the power of a power property and choose an exponent that will undo the exponent on x.

a.
$x^4 = 3000$ Original equation.
$(x^4)^{1/4} = 3000^{1/4}$ Use the power of a power property. Raising both sides to the power of $\frac{1}{4}$, the reciprocal of 4, "undoes" the power of 4 on x.
$x \approx 7.40$ Use your calculator to approximate the value of $3000^{1/4}$.

b.
$6x^{2.5} = 90$ Original equation.
$x^{2.5} = 15$ Divide both sides by 6.
$(x^{2.5})^{1/2.5} = 15^{1/2.5}$ Use the power of a power property and choose the exponent $\frac{1}{2.5}$.
$x \approx 2.95$ Approximate the value of $15^{1/2.5}$.

In this book the properties of exponents are defined only for positive bases. So while there may be negative values that are solutions to power equations, in this context you will be asked to find only the positive solutions.

4.2 Exercises

You will need a graphing calculator for Exercises **9, 10, 12,** and **17.**

Practice Your Skills

1. Rewrite each expression as a fraction without exponents or as an integer. Verify that your answer is equivalent to the original expression by evaluating each on your calculator.

   a. 5^{-3} @ $\frac{1}{125}$ b. -6^2 -36
   c. -3^{-4} @ $-\frac{1}{81}$ d. $(-12)^{-2}$ $\frac{1}{144}$
   e. $\left(\frac{3}{4}\right)^{-2}$ @ $\frac{16}{9}$ f. $\left(\frac{2}{7}\right)^{-1}$ $\frac{7}{2}$

2. Rewrite each expression in the form a^n.

   a. $a^8 \cdot a^{-3}$ @ a^5 b. $\frac{b^6}{b^2}$ b^4
   c. $(c^4)^5$ c^{20} d. $\frac{d^0}{e^{-3}}$ @ e^3

3. State whether each equation is true or false. If it is false, explain why.

   a. $3^5 \cdot 4^2 = 12^7$ b. $100(1.06)^x = (106)^x$
   c. $\frac{4^x}{4} = 1^x$ d. $\frac{6.6 \cdot 10^{12}}{8.8 \cdot 10^{-4}} = 7.5 \cdot 10^{15}$ true

   One of the authors of this book, Ellen Kamischke, works with two students in Interlochen, Michigan.

4. Solve.

   a. $3^x = \frac{1}{9}$ @ $x = -2$ b. $\left(\frac{5}{3}\right)^x = \frac{27}{125}$ $x = -3$
   c. $\left(\frac{1}{3}\right)^x = 243$ @ $x = -5$ d. $5 \cdot 3^x = 5$ $x = 0$

5. Solve each equation for positive values of x. If answers are not exact, approximate to two decimal places.

   a. $x^7 = 4000$ @ $x \approx 3.27$ b. $x^{0.5} = 28$ $x = 784$
   c. $x^{-3} = 247$ $x \approx 0.16$ d. $5x^{1/4} + 6 = 10.2$ ⓗ $x \approx 0.50$
   e. $3x^{-2} = 2x^4$ $x \approx 1.07$ f. $-3x^{1/2} + (4x)^{1/2} = -1$ @ $x = 1$

Reason and Apply

6. Rewrite each expression in the form ax^n.

   a. $x^6 \cdot x^6$ x^{12} b. $4x^6 \cdot 2x^6$ $8x^{12}$
   c. $(-5x^3) \cdot (-2x^4)$ @ $10x^7$ d. $\frac{72x^7}{6x^2}$ $12x^5$
   e. $\left(\frac{6x^5}{3x}\right)^3$ $8x^{12}$ f. $\left(\frac{20x^7}{4x}\right)^{-2}$ @ $\frac{1}{25}x^{-12}$

7. You've seen that the power of a product property allows you to rewrite $(a \cdot b)^n$ as $a^n \cdot b^n$. Is there a power of a sum property that allows you to rewrite $(a + b)^n$ as $a^n + b^n$? Write some numerical expressions in the form $(a + b)^n$ and evaluate them. Are your answers equivalent to $a^n + b^n$ always, sometimes, or never? Write a short paragraph that summarizes your findings. ⓗ Sample answer: $(a + b)^n$ is not necessarily equivalent to $a^n + b^n$. For example, $(2 + 3)^2 = 25$ but $2^2 + 3^2 = 13$. However, they are equivalent when $n = 1$, or when $a + b = 0$ and n is odd.

LESSON 4.2 Properties of Exponents and Power Functions **213**

8d. Possible answer: Non-integer powers may produce non-integer values. If the exponents form an arithmetic sequence, the decimal powers form a geometric sequence.

9a. Sample answer: The graph of $y = x^6$ will be U-shaped, will be narrower (or steeper) than $y = x^4$, and will pass through $(0, 0), (1, 1), (-1, 1), (2, 64),$ and $(-2, 64)$.

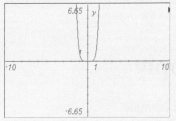

Sample answer: The graph of $y = x^7$ will fall in the first and third quadrants, will be narrower (or steeper) than $y = x^3$ or $y = x^5$, and will pass through $(0, 0), (1, 1), (-1, -1), (2, 128),$ and $(-2, -128)$.

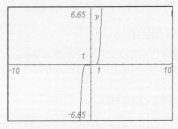

9b. The even-power functions are always in the first and second quadrants. For $x > 0$, as x-values increase, y-values increase. For $x < 0$, as x-values decrease, y-values increase.

9c. The odd-power functions are always in the first and third quadrants. For $x > 0$, as x-values increase, y-values increase. For $x < 0$, as x-values decrease, y-values decrease.

9d. The graph of $y = -x^6$ will be the graph of $y = x^6$ reflected over the x-axis. The graph of $y = -x^7$ will be the graph of $y = x^7$ reflected over the y-axis.

9e. The long run behavior reverses direction with a negative coefficient. With even-power functions, for $x > 0$, as x-values increase, y-values decrease and for $x < 0$, as x-values decrease, y-values decrease. With odd-power functions, for $x > 0$, as x-values increase, y-values decrease and for $x < 0$, as x-values decrease, y-values increase.

Exercise 10 This exercise uses transformations. As needed, remind students that translations of the graphs of functions need not appear parallel.

---

8. Consider this sequence:

$$7^2, 7^{2.25}, 7^{2.5}, 7^{2.75}, 7^3$$

a. Use your calculator to evaluate each term in the sequence. If answers are not exact, approximate to four decimal places. **49; 79.7023; 129.6418; 210.8723; 343**

b. Find the differences between the consecutive terms of the sequence. What do these differences tell you? **30.7023; 49.9396; 81.2305; 132.1277. The sequence is not arithmetic because there is not a common difference.**

c. Find the ratios of the consecutive terms in 8a. What do these values tell you? **1.627; 1.627; 1.627; 1.627. The ratio of consecutive terms is always the same, so the sequence is growing exponen[tially]**

d. What observation can you make about these decimal powers?

9. Use the graphs of $y = x^2, y = x^3, y = x^4$, and $y = x^5$ to help you answer these questions.

a. Predict what the graphs of $y = x^6$ and $y = x^7$ will look like. Verify your predictions by using your calculator.

b. Describe the long run behavior of power functions with even degrees.

c. Describe the long run behavior of power functions with odd degrees.

d. Use your knowledge of transformations to predict what the graphs of $y = -x^6$ and $y = -x^7$ will look like. Verify your predictions by using your calculator.

e. What happens to the long run behavior of a power functions if it has a negative coefficient?

10. Each of the red curves is a transformation of the graph of $y = x^3$, shown in black. Write an equation for each red curve.

a. 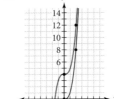 $y - 4 = x^3$, or $y = x^3 + 4$

b. 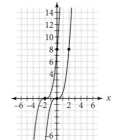 $y = (x + 2)^3$

c. 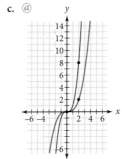 $4y = x^3$, or $y = \frac{1}{4}x^3$

d. 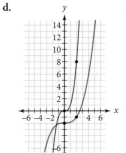 $8(y + 2) = x^3$, or $y + 2 = \frac{1}{8}x^3$, or $y = \frac{1}{8}x^3 - 2$

214 CHAPTER 4 Exponential, Power, and Logarithmic Functions

11. Consider the exponential function $f(x) = 47(0.9)^x$. Several points satisfying the function are shown in the table. Notice that when $x = 0$, $f(x) = 47$.

x	$f(x)$
-1	52.22222
0	47
1	42.3
2	38.07
3	34.263

   a. The expression $47(0.9)^x$ could be rewritten as $47(0.9)(0.9)^{x-1}$. Explain why this is true. Rewrite $47(0.9)(0.9)^{x-1}$ in the form $a \cdot b^{x-1}$.

   b. The expression $47(0.9)^x$ could also be rewritten as $47(0.9)(0.9)(0.9)^{x-2}$. Rewrite $47(0.9)(0.9)(0.9)^{x-2}$ in the form $a \cdot b^{x-2}$. $38.07(0.9)^{x-2}$

   c. Look for a connection between your answers to 11a and b and the values in the table. State a conjecture or write a general equation that summarizes your findings.

12. A ball rebounds to a height of 30.0 cm on the third bounce and to a height of 5.2 cm on the sixth bounce.

   a. Write two different yet equivalent equations in point-ratio form, $y = y_1 \cdot b^{x-x_1}$, using r for the ratio. Let x represent the bounce number, and let y represent the rebound height in centimeters. ⓗ $y = 30.0r^{x-3}$; $y = 5.2r^{x-6}$

   b. Set the two equations equal to each other. Solve for r. ⓐ $30.0r^{x-3} = 5.2r^{x-6}$; $r \approx 0.5576$

   c. What height was the ball dropped from? ⓗ 173 cm

13. Solve.

   a. $(x - 3)^3 = 64$ $x = 7$

   b. $256^x = \frac{1}{16}$ $x = -\frac{1}{2}$

   c. $\frac{(x+5)^3}{(x+5)} = x^2 + 25$ ⓗ $x = 0$

14. **APPLICATION** A radioactive sample was created in 1980. In 2002, a technician measures the radioactivity at 42.0 rads. One year later, the radioactivity is 39.8 rads.

   a. Find the ratio of radioactivity between 2002 and 2003. Approximate your answer to four decimal places. 0.9476

   b. Let x represent the year, and let y represent the radioactivity in rads. Write an equation in point-ratio form, $y = y_1 \cdot b^{x-x_1}$, using the point $(x_1, y_1) = (2002, 42)$. $y = 42(0.9476)^{x-2002}$

   c. Write an equation in point-ratio form using the point (2003, 39.8). $y = 39.8(0.9476)^{x-2003}$

   d. Calculate the radioactivity in 1980 using both equations. ⓐ

   e. Calculate the radioactivity in 2010 using both equations.

   f. Use the properties of exponents to show that the equations in 14b and c are equivalent.
   $y = 42(0.9476)^{x-2002} = 42(0.9476)(0.9476)^{x-2002-1} = 39.8(0.9476)^{x-2003}$

Review

15. Name the x-value that makes each equation true.

   a. $37{,}000{,}000 = 3.7 \cdot 10^x$ $x = 7$

   b. $0.000801 = 8.01 \cdot 10^x$ ⓐ $x = -4$

   c. $47{,}500 = 4.75 \cdot 10^x$ ⓐ $x = 4$

   d. $0.0461 = x \cdot 10^{-2}$ $x = 4.61$

16. Solve this equation for y. Then carefully graph it on your paper. $y = 2(x+4)^2 - 3$

   $\frac{y+3}{2} = (x+4)^2$

17. Paul collects these time-distance data for a remote-controlled car.

Time (s)	5	8	8	10	15	18	22	24	31	32
Distance (m)	0.8	1.7	1.6	1.9	3.3	3.4	4.1	4.6	6.4	6.2

   a. Define variables and make a scatter plot of these data.

   b. Find a line of fit to estimate the car's speed. (*Note:* Don't do more work than necessary.)
   All you need is the slope of the line of fit. Using the points (8, 1.6) and (32, 6.2), the slope is ≈ 0.2. The speed is approximately 0.2 m/s.

Exercise 11 Because $ab^n = \frac{a}{\frac{1}{b}} b^{n-1}$, the graph of every exponential function is a vertical dilation and a translation to the right of itself.

11a. $47(0.9)(0.9)^{x-1} = 47(0.9)^1(0.9)^{x-1} = 47(0.9)^x$ by the product property of exponents; $42.3(0.9)^{x-1}$.

11c. The coefficients are equal to the values of f_1 corresponding to the number subtracted from x in the exponent. If (x_1, y_1) is on the curve, then any equation $y = y_1 \cdot b^{(x-x_1)}$ is an exponential equation for the curve.

Exercise 12 As an alternative solution, students might write $ar^3 = 30$ and $ar^6 = 5.2$, divide the equations to get $r^3 = \frac{5.2}{30}$, and then solve for r.

14d. $42(0.9476)^{1980-2002} \approx 137.2$; $39.8(0.9476)^{1980-2003} \approx 137.2$; both equations give approximately 137.2 rads.

14e. $42(0.9476)^{2010-2002} \approx 27.3$; $39.8(0.9476)^{2010-2003} \approx 27.3$; both equations give 27.3 rads.

16.

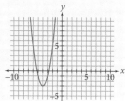

17a. Let x represent time in seconds, and let y represent distance in meters.

LESSON 4.3

This lesson focuses on rational exponents and their use in solving equations.

COMMON CORE STATE STANDARDS
APPLIED	DEVELOPED	INTRODUCED
A.REI.1	N.RN.1	
F.LE.1c	N.RN.2	
	A.SSE.3c	
	F.IF.7b	

Objectives

- Discover that $x^{1/2}$ is equivalent to $\sqrt{x}$
- Introduce the root notation $\sqrt[n]{x}$
- Define rational exponents as equivalent to roots or roots raised to powers
- Formally define the point-ratio form of an exponential function, $y = y_1 b^{x-x_1}$
- Use the point-ratio form to find an exponential equation through two points

Vocabulary

rational
point-ratio form

Materials

- Calculator Notes: Function Tables; Setting Windows; Powers and Roots, *optional*
- Project: Powers of 10

Launch

What does it mean to raise a number, n, to the power of 2?
Square it. n^2. $n \times n$

What does it mean to raise a number to the power of 4?
Multiply it by itself four times. n^4. $n \times n \times n \times n$

What does it mean to raise a number to the power of 4.7?
Multiply it by itself 4.7 times. $n^{4.7}$.

Use students' ideas about this question to lead into definitions of rational exponents.

LESSON 4.3

Rational Exponents and Roots

In a completely rational society, the best of us would be teachers and the rest of us would have to settle for something else.

LEE IACOCCA

In previous lessons you worked mostly with integer exponents and their properties. You saw how the reciprocal of an exponent can be used to undo the exponent. In this lesson you will investigate fractional and other **rational** exponents. Keep in mind that all of the properties you learned earlier apply to this larger class of exponents as well.

The volume and surface area of a cube, such as the Red Cube public art in New York, are related by rational exponents.

INVESTIGATION

Getting to the Root

In this investigation you'll explore the relationship between x and $x^{1/2}$ and learn how to find the values of some expressions with rational exponents.

Step 1 Use your calculator to create a table for $y = x^{1/2}$ at integer values of x. When is $x^{1/2}$ a positive integer? Describe the relationship between x and $x^{1/2}$.

Step 2 Graph $y = x^{1/2}$ in a graphing window with x- and y-values less than 10. This graph should look familiar to you. Make a conjecture about what other function is equivalent to $y = x^{1/2}$, enter your guess as a second equation, and verify that the equations give the same y-value at each x-value.

Step 3 Raising a number to the $\frac{1}{2}$ power is equivalent to taking the square root of the number. For example, $\sqrt{4} = 2$ and $4^{1/2} = 2$.

Step 3 State what you have discovered about raising a number to a power of $\frac{1}{2}$. Include an example with your statement.

Step 4 Clear the previous functions, and make a table for $y = 25^x$ with x incrementing by $\frac{1}{2}$.

Step 5 Each entry is the square root of 25, which is 5, raised to the numerator. $49^{3/2}$ would be the square root of 49 raised to the power of 3, or $7^3 = 343$.

Step 5 Study your table and explain any relationships you see. How could you find the value of $49^{3/2}$ without a calculator? Check your answer using a calculator.

Step 6 $27^{2/3}$ is the cube root of 27, raised to the power of 2. $3^2 = 9$. $8^{5/3} = \left(\sqrt[3]{8}\right)^5 = 2^5 = 32$

Step 6 How could you find the value of $27^{2/3}$ without a calculator? Verify your response and then test your strategy on $8^{5/3}$. Check your answer.

Step 7 Describe what it means to raise a number to a rational exponent, and generalize a procedure for simplifying $a^{m/n}$.
$a^{m/n} = \left(\sqrt[n]{a}\right)^m$; take the nth root of a, then raise it to the mth power.

216 CHAPTER 4 Exponential, Power, and Logarithmic Functions

ELL	Support	Advanced
Make the connection between point-ratio form and point-slope form of a line and have students provide verbal and symbolic definitions in their notes or on posters.	As needed, assign the Refreshing Your Skills: Working With Positive Square Roots lesson. Review writing linear equations in point-slope form. Emphasize the effect of transformations on the graph while in this form. Proceed through the same steps for quadratic and square root functions before progressing to exponential curves.	Students could explore the behavior of exponential functions where $a < 0$ and create a list of facts, problems and contradictions for those a values.

Rational exponents with numerator 1 indicate positive roots. For example, $x^{1/5}$ is the same as $\sqrt[5]{x}$, or the "fifth root of x," and $x^{1/n}$ is the same as $\sqrt[n]{x}$, or the "nth root of x." The fifth root of x is the number that, raised to the power 5, gives x.

For rational exponents with numerators other than 1, such as $9^{3/2}$, the numerator is interpreted as the exponent to which to raise the root. That is, $9^{3/2}$ is the same as $(9^{1/2})^3$, or $(\sqrt{9})^3$, or 27.

> **Definition of Rational Exponents**
> The power of a power property shows that $a^{m/n} = (a^{1/n})^m$ and $a^{m/n} = (a^m)^{1/n}$, so
> $$a^{m/n} = (\sqrt[n]{a})^m \text{ or } \sqrt[n]{a^m} \text{ for } a > 0$$

Properties of rational exponents are useful in solving equations with exponents.

EXAMPLE A | Rewrite each equation with rational exponents, and find the positive solution both algebraically and graphically.
a. $\sqrt[4]{a} = 14$ b. $\sqrt[9]{b^5} = 26$ c. $(\sqrt[3]{c})^8 = 47$

Solution | Rewrite each expression with a rational exponent, then use properties of exponents to find the positive solution algebraically. To solve graphically, graph each side of the equation and find the intersection of the two graphs.

a.
$\sqrt[4]{a} = 14$ Original equation.
$a^{1/4} = 14$ Rewrite $\sqrt[4]{a}$ as $a^{1/4}$.
$(a^{1/4})^4 = 14^4$ Raise both sides to the power of 4.
$a = 38,416$ Evaluate 14^4.

The intersection of the graphs of $y = \sqrt[4]{a}$ and $y = 14$ is approximately $(38,416, 14)$ so $a = 38,416$ checks.

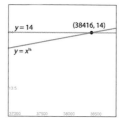

b.
$\sqrt[9]{b^5} = 26$ Original equation.
$b^{5/9} = 26$ Rewrite $\sqrt[9]{b^5}$ as $b^{5/9}$.
$(b^{5/9})^{9/5} = 26^{9/5}$ Raise both sides to the power of $\frac{9}{5}$.
$b \approx 352.33$ Approximate the value of $26^{9/5}$.

The intersection of the graphs of $y = \sqrt[9]{b^5}$ and $y = 26$ is at approximately $(352.33, 26)$ so $b = 352.33$ checks.

LESSON 4.3 Rational Exponents and Roots **217**

Investigate

Guiding the Investigation

Step 1

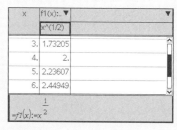

$x^{1/2}$ is a positive integer for values of x that are perfect squares (1, 4, 9, 25, . . .). $x^{1/2}$ is the square root of x.

Step 2

$y = \sqrt{x}$ is equivalent to $y = x^{1/2}$

Step 4

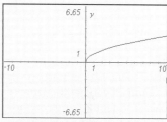

Example A

This example shows how fractional exponents can be used to solve equations. Consider projecting Example A and working through part a as a class activity, with student input. **ASK** "Could we just raise both sides to the power of 4 without using fractional exponents?" Sometimes students can save time if they think about the rational exponent. For example, to solve the equation $\sqrt[10]{x^{20}} = 169$, raising both sides to the 10th power leads to a large number, but rewriting the equation as $x^{20/10} = 169$ helps students see that the equation is simply $x^2 = 169$. Have students work on parts b and c in pairs. Have students present their solutions, including their graphs, not just their answers. **SMP 1, 3, 5, 6, 8**

Modifying the Investigation

Whole Class After students have made tables for Step 1, discuss Steps 1 through 3. Do the same for Steps 4 through 7.

Shortened Students complete Steps 1 through 3. Discuss Steps 4 through 7 with student input.

One Step Pose this problem, adapted from Example B: "Casey hit the bell in the school clock tower. Her pressure reader, held nearby, measured the sound intensity at 40 lb/in.^{2} when 4 seconds had elapsed and at 4.7 lb/in.^{2} after 7 seconds had elapsed. She remembers from her science class that sound decays exponentially. How loud was the bell when it was struck?" Students might solve the problem in a variety of ways; plan to have several approaches presented during the discussion.

Example B

In this example students see how a real-world situation can give rise to an equation that can be solved with rational exponents. Consider projecting Example B and having students work in small groups. Have students present their solutions, not just their answers.
SMP 1, 2, 3, 4, 5, 6

ASK "Why is time the independent variable?" [The sound intensity diminishes with time.] **SMP 6**

ASK "Why can we set the two equations equal to one another?" [We're trying to find a value of b that makes both equations true.] **SMP 6, 7**

ASK "Why does point-ratio form work?" "Can any point on the graph serve as a starting point?" [Yes; one property of an exponential graph is that if you stretch it vertically and shift it horizontally to move one "starting point" to another, you get the same graph back. This feature relies on the product property of exponents, in that the function $f(x) = b^{x-h}$ is equivalent to $f(x) = ab^x$ where $a = b^{-h}$. **SMP 1, 2, 6**

In this discussion of point-ratio form you can reinforce the idea of function transformations. After looking at the solution to Example B, **ASK** "Why can't you simply divide the ratio of the two y-values by 3.2 to find b?" [The ratios aren't linear.] **SMP 8**

Direct students' attention to Example B, part b. **ASK** "What if you used the other equation in the last step?" [The resulting equation is $y = 4.7(0.5121)^{x-7.2}$.] "Are the answers the same?" [The fact that the equations are equivalent can be seen by graphing them on the calculator. The vertical dilation by a factor of $\frac{4.7}{40}$ is canceled by the horizontal translation to the right by 3 units.] "Are there other ways of solving this problem?" There are three possible alternatives. One possibility uses the fact that because the common ratio is $\frac{4.7}{40}$ over 3.2 seconds, it would be $\left(\frac{4.7}{40}\right)^{1/3.2}$, or approximately 0.5121, over each second. (Decay is exponential, so we raise it to the $\frac{1}{3.2}$ power rather than dividing by 3.2.)

---

c. $\left(\sqrt[3]{c}\right)^8 = 47$ Original equation.

$c^{8/3} = 47$ Rewrite $\left(\sqrt[3]{c}\right)^8$ as $c^{8/3}$.

$(c^{8/3})^{3/8} = 47^{3/8}$ Raise both sides to the power of $\frac{3}{8}$.

$c \approx 4.237$ Approximate the value of $47^{3/8}$.

The intersection of the graphs of $y = \left(\sqrt[3]{c}\right)^8$ and $y = 47$ is (4.237, 47) so $c = 4.237$ checks.

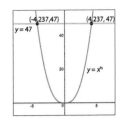

Recall that properties of exponents give only one solution to an equation, because they are defined only for positive bases. Will negative values of a, b, or c satisfy any of the equations in Example A? A negative value will work for c.

In the previous lesson, you learned that functions in the general form $y = ax^n$ are power functions. A rational function, such as $y = \sqrt[9]{x^5}$, is considered to be a power function because it can be rewritten as $y = x^{5/9}$. All the transformations you discovered for parabolas and square root curves also apply to any function that can be written in the general form $y = ax^n$.

Recall that the equation of a line can be written using the point-slope form if you know a point on the line and the slope between points. Similarly, the equation for an exponential curve can be written using **point-ratio form** if you know a point on the curve and the common ratio between points that are 1 horizontal unit apart.

> **Point-Ratio Form**
>
> If an exponential curve passes through the point (x_1, y_1) and the function values have ratio b for values of x that differ by 1, the point-ratio form of the equation is
>
> $y = y_1 \cdot b^{x - x_1}$

You have seen that if $x = 0$, then $y = a$ in the general exponential equation $y = a \cdot b^x$. This means that a is the initial value of the function at time 0 (the y-intercept) and b is the growth or decay ratio. This is consistent with the point-ratio form because when you substitute the point $(0, a)$ into the equation, you get $y = a \cdot b^{x-0}$, or $y = a \cdot b^x$.

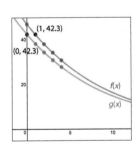

	$f(x) = 47(0.9)^x$	
	$g(x) = 42.3(0.9)^x$	
x	$f(x)$	$g(x)$
0	47	42.3
1	42.3	38.07
2	38.07	34.263
3	34.263	30.8367
4	30.8367	27.75303

Conceptual/Procedural

Conceptual In this lesson, students are introduced to rational exponents and roots. They formalize the power of a power property.

Procedural Students to solve equations with rational exponents and roots both algebraically and graphically.

The graph and table show the exponential functions $f(x) = 47(0.9)^x$ and $g(x) = 42.3(0.9)^x$. Both the graph and table indicate that if the graph of function g is translated right 1 unit, it becomes the same as the graph of function f. So $f(x) = g(x - 1)$, or $f(x) = 42.3(0.9)^{(x-1)} = 47(0.9)^x$. This shows that using the point $(1, 42.3)$ in the point-ratio form gives you an equation equivalent to $y = a \cdot b^x$.

Try substituting another point, (x_1, y_1), along with the ratio $b = 0.9$ into the point-ratio form to convince yourself that any point (x_1, y_1) on the curve can be used to write an equation, $y = y_1 \cdot b^{x-x_1}$, that is equivalent to $y = a \cdot b^x$. You may want to use your graphing calculator or algebraic techniques.

EXAMPLE B

Casey hit the bell in the school clock tower. Her pressure reader, held nearby, measured the sound intensity, or loudness, at 40 lb/in² after 4 s had elapsed and at 4.7 lb/in² after 7.2 s had elapsed. She remembers from her science class that sound decays exponentially.

a. Name two points that the exponential curve must pass through.

b. Find an exponential equation that models these data.

c. How loud was the bell when it was struck (at 0 s)?

The Campanile clock tower at the Iowa State University in Ames, Iowa

Solution

a. Time is the independent variable, x, and loudness is the dependent variable, y, so the two points are $(4, 40)$ and $(7.2, 4.7)$.

b. Start by substituting the coordinates of each of the two points into the point-ratio form, $y = y_1 \cdot b^{x-x_1}$.

$$y = 40b^{x-4} \quad \text{and} \quad y = 4.7b^{x-7.2}$$

Note that you don't yet know what b is. If you were given y-values for two consecutive integer points, you could divide to find the ratio. In this case, however, there are 3.2 horizontal units between the two points you are given, so you'll need to solve for b.

$40b^{x-4} = 4.7b^{x-7.2}$	Use substitution to combine the two equations.
$b^{x-4} = \dfrac{4.7b^{x-7.2}}{40}$	Divide both sides by 40.
$\dfrac{b^{x-4}}{b^{x-7.2}} = \dfrac{4.7}{40}$	Divide both sides by $b^{x-7.2}$.
$b^{(x-4)-(x-7.2)} = 0.1175$	Use the quotient property of exponents.
$b^{3.2} = 0.1175$	Combine like terms in the exponent.
$(b^{3.2})^{1/3.2} = (0.1175)^{1/3.2}$	Raise both sides to the power $\dfrac{1}{3.2}$.
$b \approx 0.5121$	Approximate the value of $0.1175^{1/3.2}$.
$y \approx 40(0.5121)^{x-4}$	Substitute 0.5121 for b in either of the two original equations.

An alternative approach is to use a system of two equations: $40 = ab^4$ and $4.7 = ab^{7.2}$. Divide to eliminate a: $b^{3.2} = \dfrac{4.7}{40}$. Solve for b, and then substitute its value for b in either equation to find the value of a. **SMP 1**

The function can also be found through transformations. Begin with the equation $y = \left(\dfrac{4.7}{40}\right)^x$.

The graph is shifted horizontally by 4, so the exponent becomes $(x - 4)$. The resulting graph goes through point $(4, 1)$; to have it pass through point $(4, 40)$, you must dilate it vertically by 40, giving $y = 40\left(\dfrac{4.7}{40}\right)^{x-4}$. But the ratio $\dfrac{4.7}{40}$ is more than 3 seconds, so the graph must be dilated horizontally by a factor of 3.2; the equation requires dividing $(x - 4)$ by 3.2. The equation

$$y = 40\left(\dfrac{4.7}{40}\right)^{(x-4)/3.2}$$ is equivalent to

$$y = 40\left[\left(\dfrac{4.7}{40}\right)^{1/3.2}\right]^{x-4}$$, or approximately $40(0.5121)^{x-4}$.

Summarize

As students present their solutions to the Investigation and the examples, emphasize the use of correct mathematical terminology. Whether student responses are correct or incorrect, ask other students if they agree and why. **SMP 3, 6**

ADVANCED "What's the domain of $f(x) = x^{2.8}$?" This is an example of a question with no single correct answer. Because $2.8 = \dfrac{14}{5}$ is a rational exponent, it might make sense to say that the function is defined only for positive values of x. Some calculators, however, will graph this function for negative values of x, taking the 5th root of x and then raising it to the 14th power. (Other calculators consider the function equivalent to $f(x) = e^{2.8 \ln x}$, which is not defined for negative values of x, as students will see in a later lesson.)

CRITICAL QUESTION Do the expressions 3^5 and 2^{-4} have rational exponents

BIG IDEA Yes, rational numbers include all integers as well as fractions.

CRITICAL QUESTION Why do we not get some solutions to equations we solve using rational exponents? Explain.

BIG IDEA Rational exponents give only the positive root. For $y = x^a$ to be a function, it must have only one output value for each input value. Therefore, exponential expressions such as $x^{1/2}$ must have only one value.

CRITICAL QUESTION How is the point-ratio form for exponential functions similar to the point-slope form for lines?

BIG IDEA They both use two points to determine an equation of a graph.

Formative Assessment

Watch that students have an understanding of the idea of an nth root and that they are able to convert between radicals and fractional exponents. Student presentations of their solutions to the Investigation, Examples, and Exercises can provide evidence of mastery of the objectives.

Apply

Extra Example

Write two expressions equivalent to $(x^2)^{-1/3}$.

Some possible answers: $x^{-2/3}$, $\left(\sqrt[3]{x}\right)^{-2}$, $\frac{1}{x^{2/3}}$.

Closing Question

Do you evaluate $a^{m/n}$ by calculating $(a^m)^{1/n}$ or by calculating $(a^{1/n})^m$?"

It doesn't matter, though calculating the root of a is generally significantly easier than calculating the root of a^m.

Assigning Exercises

Suggested: 1–13
Additional Practice: 14–17

Exercise 1 As needed, **ASK** "Would it help to rewrite each part using rational exponents." **ALERT** Many students confuse fractions in the base with fractional exponents, rewriting part f as $x^{1/3}$.

Exercise 2 **ALERT** Although x is raised to the 1st power in 2d, students might identify the equation as linear and answer "neither." Students may also have difficulty seeing that the function in 2f is a horizontal and vertical translation of the parent quadratic function.

Exercises 2, 9 **SUPPORT** As needed ask students to include a description of how the graph will be transformed and the reason for their choice.

Exercise 3 **ALERT** The variable b in the stipulated form is not the same as the variables in the various parts.

The exponential equation that passes through the points (4, 40) and (7.2, 4.7) is $y \approx 40(0.5121)^{x-4}$.

c. To find the loudness at 0 s, substitute $x = 0$.

$$y \approx 40(0.5121)^{0-4} \approx 581$$

The sound intensity was approximately 581 lb/in² when the bell was struck.

In Example B, part b, note that the base, 0.5121, was an approximation for b found by dividing 4.7 by 40, then raising that quotient to the power $\frac{1}{3.2}$. You could use $\left(\frac{4.7}{40}\right)^{1/3.2}$ as an exact value of b in the exponential equation.

$$y = 40\left[\left(\frac{4.7}{40}\right)^{1/3.2}\right]^{x-4}$$

Using the power of a power property, you can rewrite this as

$$y = 40\left(\frac{4.7}{40}\right)^{(x-4)/3.2}$$

This equation indicates that the curve passes through the point (4, 40) and that it has a ratio of $\frac{4.7}{40}$ spread over 3.2 units (from $x = 4$ to $x = 7.2$) rather than over 1 unit. Dividing the exponent by 3.2 stretches the graph horizontally by 3.2 units. Using this method, you can write an equation for an exponential curve in only one step.

Science CONNECTION

Sound is usually measured in bels, named after American inventor and educator Alexander Graham Bell (1847–1922). A decibel (dB) is one-tenth of a bel. The decibel scale measures loudness in terms of what an average human can hear. On the decibel scale, 0 dB is inaudible and 130 dB is the threshold of pain. However, sound can also be measured in terms of the pressure that the sound waves exert on a drum. In the metric system, pressure is measured in Pascals (Pa), a unit of force per square meter.

Prolonged or repeated exposure to loud noise can damage the inner ear and lead to noise-induced hearing loss. If you work in a noisy environment or attend loud concerts, you can reduce your risk of hearing damage by wearing protective ear coverings or earplugs.

4.3 Exercises

You will need a graphing calculator for Exercises **6, 7, 10,** and **15**.

Practice Your Skills

1. Match all expressions that are equivalent. a–e–j; b–d–g; c–i; f–h
   a. $\sqrt[5]{x^2}$ ⓐ
   b. $x^{2.5}$
   c. $\sqrt[3]{x}$
   d. $x^{5/2}$
   e. $x^{0.4}$
   f. $\left(\frac{1}{x}\right)^{-3}$
   g. $(\sqrt{x})^5$
   h. x^3
   i. $x^{1/3}$
   j. $x^{2/5}$

2. Identify each function as a power function, an exponential function, or neither of these. (It may be translated, stretched, or reflected.) Give a brief reason for your choice.
   a. $f(x) = 17x^5$
   b. $f(t) = t^3 + 5$
   c. $g(v) = 200(1.03)^v$
   d. $h(x) = 2x - 7$ ⓐ
   e. $g(y) = 3\sqrt{y-2}$
   f. $f(t) = t^2 + 4t + 3$
   g. $h(t) = \frac{12}{3^t}$ ⓐ
   h. $g(w) = \frac{28}{w-5}$
   i. $f(y) = \frac{8}{y^4} + 1$
   j. $g(x) = \frac{x^3 + 2}{1 - x}$ ⓐ
   k. $h(w) = \sqrt[5]{4w^3}$
   l. $p(x) = 5(0.8)^{(x-4)/2}$

3. Rewrite each expression in the form b^n in which n is a rational exponent.
   a. $\sqrt[6]{a}$ $a^{1/6}$
   b. $\sqrt[10]{b^8}$ ⓐ $b^{4/5}, b^{8/10},$ or $b^{0.8}$
   c. $\frac{1}{\sqrt{c}}$ ⓐ $c^{-1/2}$, or $c^{-0.5}$
   d. $\left(\sqrt[5]{d}\right)^7$ $d^{7/5}$, or $d^{1.4}$

4. Solve each equation and show or explain your step(s).
   a. $\sqrt[6]{a} = 4.2$
   b. $\sqrt[10]{b^8} = 14.3$ ⓐ
   c. $\frac{1}{\sqrt{c}} = 0.55$ ⓐ
   d. $\left(\sqrt[5]{d}\right)^7 = 23$

   4a. $a^{1/6} = 4.2$; raise both sides to the power of 6: $a = 4.2^6 = 5489.031744$.
   4b. $b^{4/5} = 14.3$; raise both sides to the power of $\frac{5}{4}$: $b = 14.3^{5/4} \approx 27.808$.
   4c. $c^{-1/2} = 0.55$; raise both sides to the power of -2: $c = 0.55^{-2} \approx 3.306$.
   4d. $d^{7/5} = 23$; raise both sides to the power of $\frac{5}{7}$: $d = 23^{5/7} \approx 9.390$.

Reason and Apply

5. **APPLICATION** Dan placed three colored gels over the main spotlight in the theater so that the intensity of the light on stage was 900 watts per square centimeter (W/cm²). After he added two gels, making a total of five over the spotlight, the intensity on stage dropped to 600 W/cm². What will be the intensity of the light on stage with six gels over the spotlight if you know that the intensity of light decays exponentially with the thickness of material covering it? ⓗ 490 W/cm²

6. For a–d, graph the equations on your calculator.
   a. $y = x^{1/2}$
   b. $y = x^{1/3}$
   c. $y = x^{1/4}$
   d. $y = x^{1/5}$
   e. How do the graphs compare? What points (if any) do they have in common? (0, 0), (1, 0)
   f. Predict what the graph of $y = x^{1/7}$ will look like. Verify your prediction by using your calculator.
   g. What is the domain of each function? Explain why. The domains of $y = x^{1/2}$ and $y = x^{1/4}$ are $x \geq 0$ because you can't take a square root or fourth root of a negative number. The domains of $y = x^{1/3}$ and $x^{1/5}$ are all real numbers.

7. For a–d, graph the equations on your calculator.
   a. $y = x^{1/4}$
   b. $y = x^{2/4}$
   c. $y = x^{3/4}$
   d. $y = x^{4/4}$
   e. How do the graphs compare? What points (if any) do they have in common?
   f. Predict what the graph of $y = x^{5/4}$ will look like. Verify your prediction by using your calculator.

LESSON 4.3 Rational Exponents and Roots **221**

2a. Power; the base is a variable.
2b. Power; the base is a variable.
2c. Exponential; the exponent is a variable.
2d. Power; x is equivalent to x^1.
2e. Power; a square root is equivalent to the exponent $\frac{1}{2}$.
2f. Power; $t^2 + 4t + 3$ is equivalent to $(t+2)^2 - 1$.
2g. Exponential; $\frac{12}{3^t}$ is equivalent to $12(3)^{-t}$.
2h. Power; $\frac{28}{w-5}$ is equivalent to $28(x-5)^{-1}$.
2i. Power; $\frac{8}{y^4}$ is equivalent to $8y^{-4}$.
2j. Neither; the function is not a transformation of either x^a or b^x.
2k. Power; the fifth root of a cube is equivalent to the exponent $\frac{3}{5}$.
2l. Exponential; the exponent contains a variable.

6a–d.
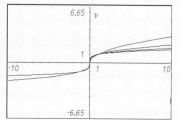

6e. Each curve is less steep than the prior one. The graphs of $y = x^{1/2}$ and $y = x^{1/4}$ are in only the first quadrant, whereas the graphs of $y = x^{1/3}$ and $y = x^{1/5}$ are in the first and third quadrants. All of the functions go through (0, 0) and (1, 1). The graphs of $y = x^{1/3}$ and $y = x^{1/5}$ both go through $(-1, -1)$.

6f. $y = x^{1/7}$ will be less steep than the others graphed and will be in the first and third quadrants. It will pass through (0, 0), (1, 1), and $(-1, -1)$.

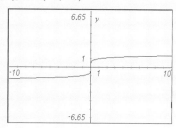

7a–d.
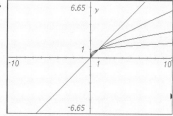

7f. $y = x^{5/4}$ should be steeper and should curve upward.

7e. Each graph is steeper and less curved than the previous one. All of the functions go through (0, 0) and (1, 1). $y = x^{4/4}$ (or $y = x$) is not curved at all.

Exercise 8 ASK "Why is y smaller than x for some of these powers?" [The powers are roots, which are smaller than x for x > 1 and larger than x for x < 1.]

8. Sample answer: Power functions with rational exponents can have limited domain. When the exponent is between 0 and 1, the curve increases slowly with a shape similar to $y = \sqrt{x}$. Exponential curves always have a steadily increasing or decreasing slope, unlike power functions.

8. Compare your observations of the power functions in Exercises 6 and 7 to your previous work with exponential functions and power functions with positive integer exponents. How do the shapes of the curves compare? How do they contrast?

9. Identify each graph as an exponential function, a power function, or neither of these. ⓐ

a.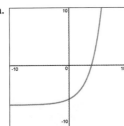
exponential

b.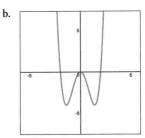
neither

c.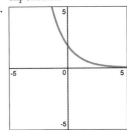
exponential

d.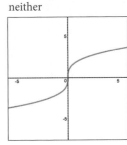
power

10. Each of the red curves is a transformation of the graph of the power function $y = x^{3/4}$, shown in black. Write an equation for each red curve. ⓗ

a.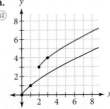
$y = 3 + (x - 2)^{3/4}$

b.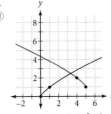
$y = 1 + [-(x - 5)]^{3/4}$

c.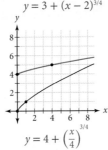
$y = 4 + \left(\dfrac{x}{4}\right)^{3/4}$

d.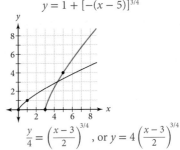
$\dfrac{y}{4} = \left(\dfrac{x-3}{2}\right)^{3/4}$, or $y = 4\left(\dfrac{x-3}{2}\right)^{3/4}$

11. Solve. Approximate answers to the nearest hundredth.

a. $9\sqrt[5]{x} + 4 = 17$

$x \approx \left(\dfrac{13}{9}\right)^5 \approx 6.29$

b. $\sqrt{5x^4} = 30$ ⓐ

$x = 180^{1/4} \approx 3.66$

c. $4\sqrt[3]{x^2} = \sqrt{35}$

$x = \left(\dfrac{\sqrt{35}}{4}\right)^{3/2} \approx 1.80$

222 CHAPTER 4 Exponential, Power, and Logarithmic Functions

12. **APPLICATION** German astronomer Johannes Kepler (1571–1630) discovered in 1619 that the mean orbital radius of a planet, measured in astronomical units (AU), is equal to the time of one complete orbit around the sun, measured in years, raised to a power of $\frac{2}{3}$.

   a. Venus has an orbital time of 0.615. What is its radius?
      0.723 AU
   b. Saturn has a radius of 9.542 AU. How long is its orbital time?
      29.475 yr
   c. Complete this table.

Planet	Mercury	Venus	Earth	Mars
Orbital radius (AU)	0.387	0.7232	1.00	1.523
Orbital time (yr)	0.2408	0.615	1.00	1.8795

Planet	Jupiter	Saturn	Uranus	Neptune
Orbital radius (AU)	5.201	9.542	19.181	30.086
Orbital time (yr)	11.861	29.475	84.008	165.02

13. **APPLICATION** Discovered by Irish chemist Robert Boyle (1627–1691) in 1662, Boyle's law gives the relationship between pressure and volume of gas if temperature and amount remain constant. If the volume in liters, V, of a container is increased, the pressure in millimeters of mercury (mm Hg), P, decreases. If the volume of a container is decreased, the pressure increases. One way to write this rule mathematically is $P = kV^{-1}$, where k is a constant.

   a. Show that this formula is equivalent to $PV = k$. ⓗ $P = kV^{-1}; P = \frac{k}{V}; PV = k$
   b. If a gas occupies 12.3 L at a pressure of 40.0 mm Hg, find the constant, k. $k = (40)(12.3) = 492$
   c. What is the volume of the gas in 13b when the pressure is increased to 60.0 mm Hg? 8.2 L
   d. If the volume of the gas in 13b is 15 L, what would the pressure be? 32.8 mm Hg

Science CONNECTION

Scuba divers are trained in the effects of Boyle's law. As divers ascend, water pressure decreases, and so the air in the lungs expands. It is relatively safe to make an emergency ascent from a depth of 60 ft, but you must exhale as you do so. If you were to hold your breath while ascending, the expanding air in your lungs would cause your air sacs to rupture and your lungs to bleed.

A scuba diver swims with tropical fish at a coral reef in the Red Sea.

Review

14. Use properties of exponents to find an equivalent expression in the form ax^n.

  a. $(3x^3)^3$ @ $27x^9$

  b. $(2x^3)(2x^2)^3$ $16x^9$

  c. $\dfrac{6x^4}{30x^5}$ $0.2x^{-1}$

  d. $(4x^2)(3x^2)^3$ @ $108x^8$

  e. $\dfrac{-72x^5y^5}{-4x^3y}$ (Find an equivalent expression in the form ax^ny^m.) $18x^2y^4$

15. For graphs a–h, write the equation of each graph as a transformation of $y = x^2$ or $y = \sqrt{x}$.

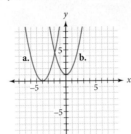
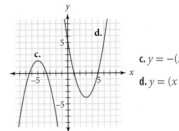

  a. $y = (x + 4)^2$
  b. $y = x^2 + 1$
  c. $y = -(x + 5)^2 + 2$
  d. $y = (x - 3)^2 - 4$

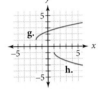

  e. $y = \sqrt{x + 3}$
  f. $y = \sqrt{x} - 1$
  g. $y = \sqrt{x + 2} + 1$
  h. $y = -\sqrt{x - 1} - 1$

16. In order to qualify for the state dart championships, you must be in at least the 98th percentile of all registered dart players in the state. There are about 42,000 registered dart players. How many qualify for the championships? about 840

17. The town of Hamlin has a growing rat population. Eight summers ago, there were 20 rat sightings, and the numbers have been increasing by about 20% each year.

  a. Give a recursive formula that models the increasing rat population. Use the number of rats in the first year as u_1. $u_1 = 20$ and $u_n = 1.2u_{n-1}$ where $n \geq 2$

  b. About how many rat sightings do you predict for this year? $u_9 \approx 86$; about 86 rat sightings

  c. Define variables and write an equation that models the continuous growth of the rat population. Let x represent the year number, and let y represent the number of rats; $y = 20(1.2)^{x-1}$.

The Pied piper of Hamelin is a legend from the town of Hamelin, Lower Saxony, Germany, in the Middle Ages.

224 CHAPTER 4 Exponential, Power, and Logarithmic Functions

LESSON 4.4

Applications of Exponential and Power Equations

I think money is a wonderful thing because it enables you to do things. It enables you to invest in ideas that don't have a short-term payback.
STEVE JOBS

You have seen that many equations can be solved by undoing the order of operations. You learned to apply this strategy to some simple power equations. The strategy also applies for more complex power equations that arise in real-world problems.

EXAMPLE A Rita wants to deposit $500 into a savings account so that its doubling time will be 8 years. What annual percentage rate is necessary for this to happen? (Assume the interest on the account is compounded annually.)

Solution If the doubling time is 8 yr, the initial deposit of $500 will double to $1,000. The interest rate, r, is unknown. Write an equation and solve for r.

$1000 = 500(1 + r)^8$ Original equation.

$2 = (1 + r)^8$ Undo the multiplication by 500 by dividing both sides by 500.

$2^{1/8} = [(1+r)^8]^{1/8}$ Undo the power of 8 by raising both sides to the power of $\frac{1}{8}$.

$2^{1/8} = 1 + r$ Use the properties of exponents.

$2^{1/8} - 1 = r$ Undo the addition of 1 by subtracting 1 from both sides.

$0.0905 \approx r$ Use a calculator to evaluate $2^{1/8} - 1$.

Rita will need to find an account with an annual percentage rate of approximately 9.05%.

You have seen how you can use the point-ratio form of an exponential equation in real-world applications. When an exponential graph models decay, the graph approaches the horizontal axis as x gets very large. When the context is growth, the graph approaches the horizontal axis as x gets increasingly negative. A horizontal line through any long-run value is called a **horizontal asymptote**. The graph approaches the asymptote but never intersects it. In the graphs below, the asymptote is the x-axis.

 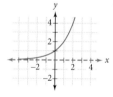

Exponential equations can also be used to model contexts that have long-run values that are not zero. You can still use the point-ratio form in these applications, although it is a bit more complex.

LESSON 4.4 Applications of Exponential and Power Equations **225**

LESSON 4.4

In this lesson, students review the properties of exponents and power functions. The properties for exponents are defined only for positive bases. The restriction does not rule out negative bases for integer exponents, so bases in formulas for sequences may be negative.

COMMON CORE STATE STANDARDS		
APPLIED	**DEVELOPED**	**INTRODUCED**
A.SSE.1a	A.SSE.3c	
F.LE.1c	F.IF.7e	
	F.IF.8b	
	F.LE.2	
	F.LE.5	

Objectives

- Find solutions to real-world applications of rational-exponential, exponential, and power functions
- Understand vertical shifts of exponential functions

Vocabulary

horizontal asymptote

Launch

Project the Launch found in your ebook. (This is Example C).

Investigate

Example A

This example shows an application of a power equation. Project Example A and have students work in small groups. As they present their solutions, have them give reasons for each step. SMP 1, 2, 3, 4, 5, 6

ELL
Because this is an applications lesson, there is a great deal of vocabulary. Support students in making multiple representations of the given information in the form of tables, graphs, and diagrams. You may need to define *pendulum*.

Support
Encourage students to make clear notes for the example solutions, including the mathematical and written explanations, so they can use them for reference.

Advanced
In Example A, ask for other ways of finding this equation. Students might substitute two data points into the equation $y - 1.25 = ab^x$ or the equation $y - 1.25 = (y_1 - 1.25)b^{x-x_1}$ and work with the two equations. The resulting graph will go through both of these data points but may not come very close to other points.

LESSON 4.4 Applications of Exponential and Power Equations **225**

Example B

Project Example B and work through it as a class activity, with student input. Introduce the term *horizontal asymptote*. SMP 1, 2, 3, 4, 5, 6

In this example students see an application of exponential decay in which the observational data don't give an exact value for the base b. To find the values of b in the table, substitute the x- and y-values into the equation for b. You need to do this only six times, because you can't evaluate b for $x = 10$.

Using the point-ratio method, they can come up with a similar b-value (about 0.945), but the equation will not fit the data well. By analyzing this discrepancy, students can see that these data tend toward 1.25, whereas the parent exponential function tends toward 0.

ASK "Does it matter which two points you use in the Example B solution?" [No; the resulting values of b are quite close to each other. In general, though, you should try to find points that are fairly far apart because they are more likely to show trends and less likely to reflect small fit errors.]

Example C

If you did the Launch, you've already done Example C. If not, project the example and have students work in pairs. Have them share their responses. SMP 1, 2, 3, 4, 5, 6

Summarize

As students present their solutions for the examples and exercises, emphasize the use of correct mathematical terminology. Whether student responses are correct or incorrect, ask other students if they agree and why.
SMP 3, 6

CRITICAL QUESTION Why is the solution method of Example B different from the method used earlier with point-ratio form?

BIG IDEA Previously, the ratio of consecutive terms was constant; observed data aren't quite so nice. In addition, the long-run value is 1.25 rather than 0.

EXAMPLE B A motion sensor is used to measure the distance between itself and a swinging pendulum. A table records the greatest distance for every tenth swing. At rest, the pendulum hangs 1.25 m from the motion sensor. Find an equation that models the data in the table below.

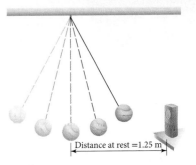

Stage number	0	10	20	30	40	50	60
Greatest distance (m)	2.35	1.97	1.70	1.53	1.42	1.36	1.32

Solution Plot these data. The graph shows a curved shape, so the data are not linear. As the pendulum slows, the greatest distance approaches a long-run value of 1.25 m. The graph appears to have a horizontal asymptote at $y = 1.25$. The pattern looks like a shifted decreasing geometric sequence, so an exponential decay equation is a good choice for the best model.

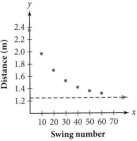

An exponential decay function in point-ratio form, $y = y_1 \cdot b^{x-x_1}$, has the horizontal asymptote $y = 0$. Because these data approach a long-run value of 1.25, the exponential function must be translated up 1.25 units. To find the new function, first create a new data set by shifting the y-values down 1.25 units.

Swing number	0	10	20	30	40	50	60
Greatest distance (m)	2.35	1.97	1.70	1.53	1.42	1.36	1.32
Shifted distances (m)	1.10	0.72	0.45	0.28	0.17	0.11	0.07

Choose a point, such as $(10, 0.72)$, from the new data to be (x_1, y_1). The equation in point-ratio form is $y = 0.72 b^{x-10}$. To find the ratio, select a second point, such as $(50, 0.11)$. Substitute in the equation and solve for b.

$0.11 = 0.72 b^{40}$ Substitute $(50, 0.11)$ for (x, y).

$b^{40} = \dfrac{0.11}{0.72}$ Divide both sides by 0.72.

$b = \left(\dfrac{0.11}{0.72}\right)^{1/40} \approx 0.9541$ Raise both sides to the power $\dfrac{1}{40}$.

The point-ratio equation for the new data is $y = 0.72(0.9541)^{x-10}$. Shifting this up 1.25 gives you the equation for the original data, $y = 1.25 + 0.72(0.9541)^{x-10}$.

Conceptual/Procedural

Conceptual In this application-based lesson, students investigate the concept of horizontal asymptotes.

Procedural Students create and solve equations for real-world applications of rational-exponential, exponential, and power functions.

Using different pairs of points will generate different values for *b*, but all of the values should be relatively close. Graph the model to check whether the points chosen here resulted in a good-fitting model.

You should be able to look at an exponential function and determine the growth or decay rate, which indicates the end behavior.

EXAMPLE C For each function, determine the growth or decay rate and the end, or long run, behavior.

a. $y = 12 + 153.32(0.99)^x$
b. $y = 0.98(1.09)^x$

Solutions

a. The decay rate is 1% ($1 - 0.99 = 0.01$). The long run behavior is as x decreases, the function increases without bound. As x increases, y approaches 12.

b. The growth rate is 9% ($1.09 - 1 = 0.09$). The long run behavior is as x decreases, y approaches 0. As x increases, the function also increases without bound.

 Exercises

You will need a graphing calculator for Exercises **7, 11,** and **14**.

Practice Your Skills

1. Solve.
   a. $x^5 = 50$ @ $x = 50^{1/5} \approx 2.187$
   b. $\sqrt[3]{x} = 3.1$ $x = 29.791$
   c. $x^2 = -121$ @ no real solution

2. Solve.
   a. $x^{1/4} - 2 = 3$ @ $x = 625$
   b. $4x^7 - 6 = -2$ $x = 1$
   c. $3(x^{2/3} + 5) = 207$ $x = 512$
   d. $1450 = 800\left(1 + \dfrac{x}{12}\right)^{7.8}$ @
   $x = 12(-1 + 1.8125^{1/7.8}) \approx 0.951$
   e. $14.2 = 222.1 \cdot x^{3.5}$ $x = \left(\dfrac{14.2}{222.1}\right)^{1/3.5} \approx 0.456$

3. Rewrite each expression in the form ax^n.
   a. $(27x^6)^{2/3}$ @ $9x^4$
   b. $(16x^8)^{3/4}$ $8x^6$
   c. $(36x^{-12})^{3/2}$ $216x^{-18}$

4. Identify the growth or decay rate in each exponential function.
   a. The balance in a bank account after t years is given by the equation $B = 2500(1.015)^t$
      growth rate is 1.5% per year
   b. The temperature of a cup of tea left sitting on a counter after m minutes is given by $T = 68 + 52(0.82)^m$ decay rate is 18% per minute
   c. The value of a car y years after it was purchased is given by the equation $V = 34000(0.76)^y$
      decay rate is 24% per year.

Reason and Apply

5. The population of Austria in millions from 2011 to 2015 can be modeled by $P(t) = 8.39 \cdot 1.08^t$, where t is the number of years after 2011. The growth rate is 8% so the population is growing by 8% a year.
   a. Identify the growth or decay rate in the function. What does this rate mean in this situation?
   b. How can you use the structure of the function to find the population for Austria in 2011 without doing any computation? In the form $y = a \cdot b^x$, the a value is the initial value because any non negative number b raised to the 0 power is 1 so $y = a \cdot 1 = a$.
   c. According to the model, what was the population of Austria for 2014?
      8.59 million

LESSON 4.4 Applications of Exponential and Power Equations **227**

7a. She must replace y with $y - 7$ and y_1 with $y_1 - 7$; $y - 7 = (y_1 - 7) \cdot b^{x - x_1}$.

7c. Possible answers: $x = 0$, $y = 200$, $b = 0.508$; $x = 2$, $y = 57$, $b = 0.510$; $x = 3$, $y = 31$, $b = 0.495$

7d. Possible answer: The mean of the b-values is 0.511. $y = 7 + 98(0.511)^{x-1}$.

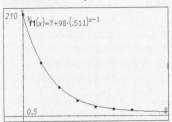

Exercise 7 SUPPORT In 7a, students might need a hint on how the seven unmarked candies relate to a vertical translation of the graph.

Exercise 9 LANGUAGE A *botanist* is a scientist who studies plants. The species name of this tree, *altissima*, means "very tall."

---

6. **APPLICATION** A sheet of translucent glass 1 mm thick is designed to reduce the intensity of light. If six sheets are placed together, then the outgoing light intensity is 50% of the incoming light intensity.
   a. Let r represent the fraction of light that passes through one sheet of glass and let 100 represent the original amount of light. Write an expression for the amount of light passing through six sheets of glass. $100\, r^6$
   b. If 50% of the light passes through six sheets of glass, write an equation that expresses this fact. Solve your equation for r. $100\, r^6 = 50$
   c. What is the reduction rate of one sheet of glass? Write your answer as a percentage. $r = 0.891$; 89.1%

7. Natalie performs a decay simulation using small colored candies with a letter printed on one side. She starts with 200 candies and pours them onto a plate. She removes all the candies with the letter facing up, counts the remaining candies, and then repeats the experiment using the remaining candies. Here are her data for each stage: ⓗ

Stage number x	0	1	2	3	4	5	6
Candies remaining y	200	105	57	31	18	14	12

After stage 6, she checked the remaining candies and found that seven did not have a letter on either side.
   a. Natalie wants to use the point-ratio equation, $y = y_1 \cdot b^{x - x_1}$, to model her data. What must she do to account for the seven unmarked candies? Write the equation. ⓐ
   b. Natalie uses the second data point, (1, 105), as (x_1, y_1). Write her equation with this point and then solve for b in terms of x and y. $y - 7 = (105 - 7)b^{x-1}$; $\left(\dfrac{y-7}{98}\right)^{1/(x-1)} = b$
   c. Select three different data points for (x, y) and use them to find values for b.
   d. How should Natalie choose a value for b? What is her model for the data? Graph the equation with the data, and verify that the model fits reasonably well.

8. **APPLICATION** The relationship between the weight in tons, W, and the length in feet, L, of a sperm whale is given by the formula $W = 0.000137 L^{3.18}$.
   a. A male sperm whale is 52 ft long. What is its weight? 39 tons
   b. How long would a sperm whale be if it weighed 45 tons? ⓐ 54 ft

9. **APPLICATION** In order to estimate the height of an *Ailanthus altissima* tree, botanists have developed the formula $h = \dfrac{5}{3} d^{0.8}$, where h is the height in meters and d is the diameter in centimeters.
   a. If the height of an *Ailanthus altissima* tree is 18 m, find the diameter. 19.58 cm
   b. If the circumference of an *Ailanthus altissima* tree is 87 cm, estimate its height. 23.75 m

10. **APPLICATION** Fat reserves in birds are related to body mass by the formula $F = 0.033 \cdot M^{1.5}$, where F represents the mass in grams of the fat reserves and M represents the total body mass in grams.
    a. How many grams of fat reserves would you expect in a 15 g warbler? 1.9 g
    b. What percentage of this warbler's body mass is fat? 12.8%

228 CHAPTER 4 Exponential, Power, and Logarithmic Functions

11. **APPLICATION** There is a power relationship between the radius of an orbit, x, and the time of one orbit, y, for the moons of Saturn. (The table at right lists 11 of Saturn's many moons.)

   a. Make a scatter plot of these data.

   b. Experiment with different values of a and b in the power equation $y = ax^b$ to find a good fit for the data. Work with a and b one at a time, first adjusting one and then the other until you have a good fit. Write a statement describing how well $y = ax^b$ fits the data. ⓗ

   c. Use your model to find the orbital radius of Titan, which has an orbit time of 15.945 days.   approximately 1,229,200 km

   d. Find the orbital time for Phoebe, which has an orbit radius of 12,952,000 km.   545.390 d

Moons of Saturn

Moon	Radius (100,000 km)	Orbital time (d)
Atlas	1.3767	0.602
Prometheus	1.3935	0.613
Pandora	1.4170	0.629
Epimetheus	1.5142	0.694
Janus	1.5147	0.695
Mimas	1.8552	0.942
Enceladus	2.3802	1.370
Tethys	2.9466	1.888
Dione	3.7740	2.737
Helene	3.7740	2.737
Rhea	5.2704	4.518

(www.solarsystem.nasa.gov)

12. According to the consumer price index, the average cost of a gallon of whole milk was $3.74 in July 2007. If the July 2007 rate of inflation continues, a gallon of whole milk will cost $5.48 in July 2017. What was the rate of inflation in July 2007?   0.319% per month, or 3.9% per year

Review

13. **APPLICATION** A sample of radioactive material has been decaying for 5 years. Three years ago, there were 6.0 g of material left. Now 5.2 g are left.

   a. What is the rate of decay? ⓐ   0.0466, or 4.66% per year

   b. How much radioactive material was initially in the sample? ⓐ   6.6 g

   c. Find an equation to model the decay.
      $y = 6.6(1 - 0.0466)^x = 6.6(0.9534)^x$

   d. How much radioactive material will be left after 50 years (45 years from now)?   0.6 g

   e. What is the half-life of this radioactive material?   14.5 yr

14. Solve this system of three equations with three variables. ⓐ

$$\begin{cases} 2x + y + 4z = 4 \\ x + y + z = \frac{1}{4} \\ -3x - 7y + 2z = 5 \end{cases} \quad (-4.5, 2, 2.75)$$

These images of Saturn's system are a compilation of photos taken by the *Voyager I* spacecraft in 1980.

15. In his geography class, Juan makes a conjecture that more people live in U.S. cities that are warm (above 50°F) in the winter than live in U.S. cities that are cold (below 32°F). In order to test his conjecture, he collects the mean temperatures for January of the 25 largest U.S. cities.

   a. Construct a box plot of these data.

   b. List the five-number summary.   18.9, 29.15, 40.1, 50.35, 57.4

   c. What are the range and the interquartile range for these data?
      range = 38.5, IQR = 21.2

   d. Do the data support Juan's conjecture? Explain your reasoning.
      The data do not support his conjecture. There are approximately the same number of cities in each category.

31.5°, 56.8°, 21.0°, 50.4°, 30.4°, 53.6°, 50.3°, 57.4°, 43.4°, 49.4°, 22.9°, 25.5°, 52.4°, 48.7°, 26.4°, 54.2°, 39.7°, 31.8°, 43.4°, 39.3°, 42.8°, 18.9°, 40.1°, 28.6°, 29.7°

(www.allcountries.org, www.earthday.net)

Exercise 11 **ELL** Define *orbit*.

LANGUAGE *Orbital time* is the time it takes for an object to make one complete revolution around the object it is circling. Here the orbital time is measured in Earth days.

The moons listed in the table are 11 of the moons that have regular orbits.

CONTEXT Earth's own moon is 384,401 km from Earth, and its orbital time is 27.322 d.

11a.

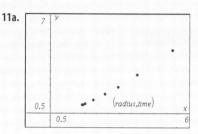

11b. Sample answer: $\hat{y} = 0.37 x^{1.5}$, where x is measured in units of 100,000 km. The graph of the data and the equation appear to be a good fit.

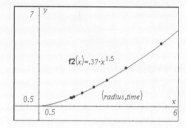

15a

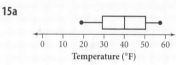

Temperature (°F)

LESSON 4.4 Applications of Exponential and Power Equations   **229**

LESSON 4.5

Building Inverses of Functions

> Success is more a function of consistent common sense than it is of genius.
>
> AN WANG

Gloria and Keith are sharing their graphs for the same set of data.

"I know my graph is right!" exclaims Gloria. "I've checked and rechecked it. Yours must be wrong, Keith."

Keith disagrees. "I've entered these data into my calculator too, and I made sure I entered the correct numbers."

The graphs are pictured below. Can you explain what is happening?

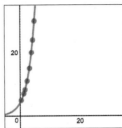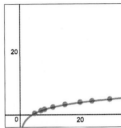

The independent and dependent variables are reversed.

This lesson is about the **inverse** of a function—where the independent variable is exchanged with the dependent variable. Look again at Gloria's and Keith's graphs. If Gloria and Keith labeled the axes, they might see that the only difference is their choice of independent variables. In some real-world situations, it makes equal sense for either of two related variables to be used as the independent variable. In the investigation you will find equations for some inverses and then discover how the inverse equations relate to the original functions.

INVESTIGATION

The Inverse

In this investigation you will use graphs, tables, and equations to explore the inverses of several functions.

Step 1 Graph the equation $f(x) = 6 + 3x$ on your calculator. Complete the table for this function.

x	−1	0	1	2	3
y	3	6	9	12	15

Step 2 Because the inverse is obtained by switching the independent and dependent variables, you can find five points on the inverse of function f by swapping the x- and y-coordinates in the table. Complete the table for the inverse.

x	3	6	9	12	15
y	−1	0	1	2	3

Step 3 Graph the five points you found in Step 2 by creating a scatter plot. Describe the graph and write an equation for it. Graph your equation and verify that it passes through the points in the table from Step 2.

Step 4 Repeat Steps 1–3 for each of these functions. You may need to write more than one equation to describe the inverse.

  i. $g(x) = \sqrt{x+1} - 3$ ii. $h(x) = (x-2)^2 - 5$

Step 5 Study the graphs of functions and their inverses that you made. What observations can you make about the graphs of a function and its inverse? *The graphs are reflections of each other across the line $y = x$.*

Step 6 You create the inverse by switching the x- and y-values of the points. How can you apply this idea to find the equation of the inverse from the original function? Verify that your method works by using it to find the equations for the inverses of functions f, g, and h.
Switch the x- and y-variables and solve for y.

EXAMPLE A A 589 mi flight from Washington, D.C., to Chicago took 118 min. A flight of 1452 mi from Washington, D.C., to Denver took 222 min. Model this relationship both as (*time, distance*) data and as (*distance, time*) data. If a flight from Washington, D.C., to Seattle takes 323 min, what is the distance traveled? If the distance between Washington, D.C., and Miami is 910 mi, how long will it take to fly from one of these two cities to the other?

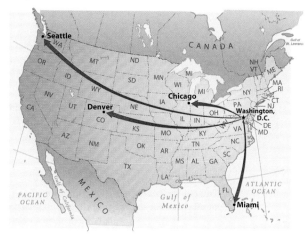

Solution If you know the time traveled and want to find the distance, then time is the independent variable, and the points known are (118, 589) and (222, 1452). The slope is $\frac{1452 - 589}{222 - 118} = \frac{863}{104}$ or approximately 8.3 mi/min. Using the first point to write an equation in point-slope form, you get $d = 589 + \frac{863}{104}(t - 118)$.

To find the distance between Washington, D.C., and Seattle, substitute 323 for the time, t:

$$d = 589 + \frac{863}{104}(323 - 118) \approx 2290.106$$

The distance is approximately 2290 mi.

LESSON 4.5 Building Inverses of Functions

Step 3 a line with the equation $y = \frac{x-6}{3}$

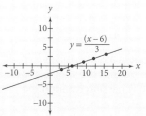

Step 4i Table of values for $g(x)$: $(-1, -3)$, $(0, -2)$, $(1, -1.59)$, $(2, -1.27)$, $(3, -1)$; table of inverse values has these x- and y-coordinates switched. The inverse function has the equation $y = (x+3)^2 - 1$ for $x \geq -3$.

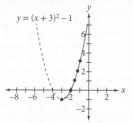

Step 4ii Table of values for $h(x)$: $(-1, 4)$, $(0, -1)$, $(1, -4)$, $(2, -5)$, $(3, -4)$; table of inverse values has these x- and y-coordinates switched. The inverse function has the equations $y = \pm\sqrt{x+5} + 2$.

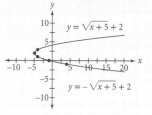

Investigate

Guiding the Investigation

Occasionally asking about the domain and range of various functions and their inverses will help students deepen their understanding of how domain and range are tied to independent and dependent variables, which simply switch.

Step 1 Although students can complete the table without graphing, the graph is important for Step 5.

Step 4 Two equations will be needed to graph the inverse of function ii.

Example A

This example illustrates the use of inverse functions in the many cases in which either variable can be considered to depend on the other.

Modifying the Investigation

Whole Class Demonstrate Steps 1 through 3 with the whole class. Divide the class in half and assign half the students Step 4i and the other half Step 4ii. Have students present and discuss Steps 5 and 6.

Shortened Skip Step 4ii. Discuss Steps 5 and 6.

One Step Use the **One Step** found in your ebook. Encourage creative thinking, but bring out the idea that the independent and dependent variables are switched. During the discussion, elicit the notion of an inverse relation and how the graph of the inverse of a function relates to the graph of the function.

Project Example A and have students work in small groups. As they present their solutions, have them give reasons for each step. **SMP 1, 2, 3, 4, 5, 6**

Students should assume this is a linear relationship but not the simple relationship $d = rt$ passing through the origin. The added constant could be due to various factors.

If all groups use the same point to find the equation for d, **ASK** "What if you hadn't used the first point to find the equation for d?" [Using the second point would have yielded the equivalent equation $d = 1452 + \frac{863}{104}(t - 222)$.]

Example B

This example shows how to find the inverse of a function. Project Example B and have students work in small groups. As they present their solutions, have them give reasons for each step. **SMP 3, 6**

Summarize

As students present their solutions to the investigation and the examples, emphasize the use of correct mathematical terminology. Whether student responses are correct or incorrect, ask other students if they agree and why. **SMP 3, 6**

Discussing the Investigation

If students are having difficulty seeing how the graphs of a function and its inverse are reflections of each other across the line $y = x$, you might demonstrate on wax paper on an overhead projector. Or students can graph a function on paper, darken the graph with a pencil, fold the paper along the line $y = x$, and rub the reflection to create the graph of the inverse. **SMP 5**

ASK "Is there a way to tell whether a function is one-to-one without actually graphing or trying to find an expression for the inverse?" Students may speculate that expressions for one-to-one functions have no even exponents on x. Suggest that they graph some functions such as $y = x^3 + x$ or $y = 2^x$ to check their conjectures. Or point out that the constant in the equation $y = x + 4$ is the same as $4x^0$ and that 0 is an even power. Focus on what happens to the graph as it's reflected across the line $y = x$. Arrive at the idea that a one-to-one function will pass the horizontal line test as well as the vertical line test. **ADVANCED** "If the inverse is not a function, could part of it be a function?" [If the domain of the original function is restricted to an interval over which the function is one-to-one, its inverse will be a function.]

ASK "Is $f^{-1}(x)$ the same as $\frac{1}{f(x)}$?" [No, the first is the inverse, and the second is the reciprocal.] Model good mathematical thinking by asking, "Is there any function $f(x)$ whose inverse equals its reciprocal?" **SMP 1, 3, 6**

If you know distance and want to find time, then distance is the independent variable. The two points then are (589, 118) and (1452, 222). This makes the slope $\frac{222 - 118}{1452 - 589} = \frac{104}{863}$ or approximately 0.12 min/mi. Using the first point again, the equation for time is $t = 118 + \frac{104}{863}(d - 589)$.

To find the time of a flight from Washington, D.C. to Miami, substitute 910 for the distance, d.

$$t = 118 + \frac{104}{863}(910 - 589) \approx 156.684$$

The flight will take approximately 157 min.

You can also rearrange the first equation for distance in terms of time to get the second equation for time in terms of distance.

$$d = 589 + \frac{863}{104}(t - 118) \quad \text{First equation.}$$

$$d - 589 = \frac{863}{104}(t - 118) \quad \text{Subtract 589 from both sides.}$$

$$\frac{104}{863}(d - 589) = (t - 118) \quad \text{Multiply both sides by } \frac{104}{863}.$$

$$118 + \frac{104}{863}(d - 589) = t \quad \text{Add 118 to both sides.}$$

These two equations are inverses of each other. That is, the independent and dependent variables have been switched. Graph the two equations on your calculator. What do you notice? They are reflections across the line $y = x$.

In the investigation you may have noticed that the inverse of a function is not necessarily a function. Recall that any set of points is called a relation. A relation may or may not be a function.

> **Inverse of a Relation**
>
> You get the **inverse** of a relation by exchanging the x- and y-coordinates of all points or exchanging the x- and y-variables in an equation.

When an equation and its inverse are *both* functions, it is called a **one-to-one function**. How can you tell if a function is one-to-one?
It has no more than one value of x for each value of y.
The inverse of a one-to-one function $f(x)$ is written as $f^{-1}(x)$. Note that this notation is similar to the notation for the exponent -1, but $f^{-1}(x)$ refers to the inverse function, not an exponent.

EXAMPLE B | Find the inverse of this function.

$$f(x) = 2 + 5x$$

Solution | To find the inverse, exchange the independent and dependent variables. Then, solve for the new dependent variable.

$x = 2 + 5y$	Exchange x and y.
$x - 2 = 5y$	Subtract 2 from both sides.
$\dfrac{x-2}{5} = y$	Divide by 5.
$f^{-1}(x) = \dfrac{1}{5}x - \dfrac{2}{5}$	Write using function notation.

Look carefully at the graphs below to see the relationship between a function and its inverse. A function and its inverse are reflections across the line $y = x$.

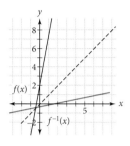 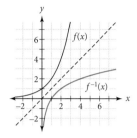

4.5 Exercises

You will need a graphing calculator for Exercise **10**.

Practice Your Skills

1. A function $f(x)$ contains the points $(-2, -3)$, $(0, -1)$, $(2, 2)$, and $(4, 6)$. Give the points known to be in the inverse of $f(x)$. $(-3, -2), (-1, 0), (2, 2), (6, 4)$

2. Given $g(t) = 5 + 2t$, find each value.
   a. $g(2)$ 9
   b. $g^{-1}(9)$ ⓐ 2
   c. $g^{-1}(20)$ ⓗ 15/2, or 7.5

3. Which graph below represents the inverse of the relation shown in the graph at right? Explain how you know.

i.   ii.   iii.

Graph iii is the inverse because the x- and y-coordinates have been switched from the original graph so that the graphs are symmetric across the line $y = x$.

LESSON 4.5 Building Inverses of Functions **233**

CRITICAL QUESTION Does the notation $f^{-1}(x)$ represent all inverses of relations?

BIG IDEA No. It represents an inverse that is a function.

CRITICAL QUESTION Under what conditions is the inverse relation of a function also a function?

BIG IDEA Each y-value corresponds to only one x-value.

CRITICAL QUESTION How can you determine from the graph of a function if its inverse will be a function?

BIG IDEA Use the horizontal line test. If any horizontal line intersects the original function in only one point, then the inverse of that function is also a function.

CRITICAL QUESTION What is the relationship of the graph of a relation and its inverse?

BIG IDEA The graph of a relation and its inverse are reflections of each other across $y = x$.

Formative Assessment

Student responses to the investigation provides evidence of knowing the definition of the inverse relation of a function. Responses to Example A and Exercises 10, 11, and 13 provide evidence of ability to apply inverses in real world situations. The Performance Task can be used as an informal assessment of understanding the difference between x^{-1} and f^{-1}.

Apply

Extra Example

Find the inverse of $y = (x - 3)^2 + 1$.
$y = \pm\sqrt{x - 1} + 3$

Closing Question

Draw a graph of a function that is its own inverse.

The functions $y = x$ and $y = \dfrac{1}{x}$ are both their own inverses.

Assigning Exercises

Suggested: 1–12, Performance Task
Additional Practice: 13–18

Conceptual/Procedural

Conceptual This lesson introduces the concept of inverse functions.

Procedural Students learn the process for finding the inverse of a function.

Exercise 4 Students can use either symbolic or graphical methods. As needed, help them see that a function is the inverse of its inverse; for example, a and e are inverses of each other.

Exercise 6 SUPPORT This exercise points out the similarities between solving for x (a skill most students are now successful at performing) and finding an inverse. This connection can make a big difference in the confidence students feel toward symbolically finding the inverses of function.

6a. $4 + (x-2)^{3/5} = 12$
  $(x-2)^{3/5} = 8$
  $x - 2 = 8^{5/3} = 32$
  $x = 34$

6b. $4 + (y-2)^{3/5} = x$
  $(y-2)^{3/5} = x - 4$
  $y - 2 = (x-4)^{5/3}$
  $f^{-1}(x) = 2 + (x-4)^{5/3}$

7c. Yes, it is a function; it passes the vertical line test.

Exercise 8 You might ASK "How do you know the inverse is a function?" Students may cite the horizontal line test or a table.

8a. $f(x) = 2x - 3$; $f^{-1}(x) = \frac{x+3}{2}$, or
  $f^{-1}(x) = \frac{1}{2}x + \frac{3}{2}$

8b. $f(x) = \frac{-3x+4}{2}$ or $f(x) = -\frac{3}{2}x + 2$;
  $f^{-1}(x) = \frac{-2x+4}{3}$, or $f^{-1}(x) = -\frac{2}{3}x + \frac{4}{3}$

8c. $f(x) = \frac{-x^2+3}{2}$ or $f(x) = -\frac{1}{2}x^2 + \frac{3}{2}$;
  $y = \pm\sqrt{-2x+3}$ (not a function)

---

4. Match each function with its inverse. ⓗ

   a. $y = 6 - 2x$ b. $y = 2 - \frac{6}{x}$ @   c. $y = -6(x-2)$ @   d. $y = \frac{-6}{x-2}$

   e. $y = \frac{-1}{2}(x-6)$ f. $y = \frac{2}{x-6}$ g. $y = 2 - \frac{1}{6}x$ h. $y = 6 + \frac{2}{x}$

   a and e are inverses; b and d are inverses; c and g are inverses; f and h are inverses.

Reason and Apply

5. Given the functions $f(x) = 4x + 8$ and $g(x) = \frac{1}{4}x - 2$:

   a. Find $f(2)$ and $g(16)$. $f(2) = 16$ and $g(16) = 2$

   b. What do your answers to 5a imply? They might be inverse functions

   c. Find $f(1)$. $f(1) = 12$

   d. How can knowing $f(1)$ help you find $g(12)$? If f and g are inverse functions, then $f(1) = 12$ implies $g(12) = 1$. However, if they are not inverses, then knowing $f(1) = 12$ doesn't tell you anything about $g(12)$.

6. Given $f(x) = 4 + (x-2)^{3/5}$:

   a. Solve for x when $f(x) = 12$.

   b. Find $f^{-1}(x)$ symbolically. ⓗ

   c. How are solving for x and finding an inverse alike? How are they different? Sample answer: The steps are the same but you don't have to do the numerical calculations when you find an inverse.

7. Consider the graph of the piecewise function f shown at right.

   a. Find $f(-3), f(-1), f(0),$ and $f(2)$. $-1, 0, 1, 2$

   b. Name four points on the graph of the inverse of $f(x)$. $(-1, -3), (0, -1), (1, 0), (2, 2)$

   c. Draw the graph of the inverse. Is the inverse a function? Explain.

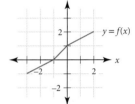

8. Write each function using $f(x)$ notation, then find its inverse. If the inverse is a function, write it using $f^{-1}(x)$ notation.

   a. $y = 2x - 3$ @   b. $3x + 2y = 4$ c. $x^2 + 2y = 3$ @

9. Find the inverse function for each.

   a. $f(x) = 6.34x - 140$ 9a. $f^{-1}(x) = \frac{x + 140}{6.34}$ 9b. $f^{-1}(x) = \frac{x - 32}{1.8}$

   b. $f(x) = 1.8x + 32$ @

   Note that the equation in 9b will convert temperatures in °C to temperatures in °F. You will use either this function or its inverse in Exercises 10 and 11.

> **History CONNECTION**
>
> Anders Celsius (1701–1744) was a Swedish astronomer. He created a thermometric scale using the freezing and boiling temperatures of water as reference points, on which freezing corresponded to 100° and boiling to 0°. His colleagues at the Uppsala Observatory reversed his scale five years later, giving us the current version. Thermometers with this scale were known as "Swedish thermometers" until the 1800s when people began referring to them as "Celsius thermometers."

10. The data in the table describe the relationship between altitude and air temperature.

Feet	Meters	°F	°C
1,000	300	56	13
5,000	1,500	41	5
10,000	3,000	23	−5
15,000	4,500	5	−15
20,000	6,000	−15	−26
30,000	9,000	−47	−44
36,087	10,826	−69	−56

Mount McKinley in Denali National Park, Alaska reaches an altitude of 6194 m above sea level.

a. Write a best-fit equation for $f(x)$ that describes the relationship (*altitude in meters, temperature in °C*). Use at least three decimal places in your answer. @

b. Use your results from 10a to write the equation for $f^{-1}(x)$.

c. Write a best-fit equation for $g(x)$, describing (*altitude in feet, temperature in °F*). @

d. Use your results from 10c to write an equation for $g^{-1}(x)$.

e. What would the temperature in °F be at the summit of Mount McKinley, which is 6194 m high? You will need to use the equation from Exercise 9b. ⓗ

11. **APPLICATION** On Celsius's original scale, freezing corresponded to 100° and boiling corresponded to 0°.

a. Write a formula that converts a temperature given by today's Celsius scale to a temperature on the scale that Celsius invented. $y = 100 - C$

b. Write a formula that converts a temperature in degrees Fahrenheit to a temperature on the original scale that Celsius invented. You will need to use $f(x)$, or its inverse, from Exercise 9b. $C = 100 - \dfrac{F - 32}{1.8}$

12. Match each function with all the statements that apply to it.

a. $f(x) = 2x + 1$ i, ii, iii
b. $f(x) = x^2 + 1$ ii, v
c. $f(x) = \sqrt{x - 1}$ i, iv
d. $f(x) = \dfrac{x - 1}{2}$ i, ii, iii

i. one-to-one
ii. domain is all real numbers
iii. range is all real numbers
iv. domain is $x \geq 1$
v. range is $y \geq 1$

13. In looking over his water utility bills for the past year, Mr. Aviles saw that he was charged a basic monthly fee of $7.18, and $3.98 per thousand gallons (gal) used.

a. Write the monthly cost function in terms of the number of thousands of gallons used.

b. What is his monthly bill if he uses 8000 gal of water? $39.02

c. Write a function for the number of thousands of gallons used in terms of the cost.

d. If his monthly bill is $54.94, how many gallons of water did he use? 12,000 gal

e. Show that the functions from 13a and c are inverses. ⓗ

f. Mr. Aviles decides to fix his leaky faucets. He calculates that he is wasting 50 gal/d. About how much money will he save on his monthly bill? ⓗ about $16

g. A gallon is 231 cubic inches. Find the dimensions of a rectangular container that will hold the contents of the water Mr. Aviles saves in a month by fixing his leaky faucets. ⓗ Answers will vary, but volume should equal $231 \cdot 1500$, or 346,500 in³ or approximately 200 ft³.

Exercises 10, 11 The function from Exercise 9b is required for these exercises.

Answers may vary. Sample answers given.

10a. The equation of a best-fit line is $f(x) \approx -0.006546x + 14.75$.

10b. $f^{-1}(x) \approx \dfrac{x - 14.75}{-0.006546x}$, or $f^{-1}(x) \approx -152.76x + 2252.76$

10c. The equation of a best-fit line is $g(x) \approx -0.003545x + 58.81$.

10d. $g^{-1}(x) \approx \dfrac{x - 58.814}{-0.003545}$, or $g^{-1}(x) \approx -282.1x + 16591$

10e. Use the function in 10a to find the temperature in °C first. $f(6194) \approx -0.006546(6194) + 14.75 \approx -25.80°C$. Then use the function from 9b to change the °C to °F: $y \approx -14.44°F$.

Exercise 11 As needed in 11a, **ASK** "What is the independent variable? The dependent variable?" [independent: today's scale, C; dependent: original scale, y] Students might find it helpful to graph a few points to show that the function is linear. They can then use two of these points to find the point-slope form of the equation. For 11b, students find the composition of the function with one of the functions in Exercise 9.

Exercise 13 As needed, help students see that *and* means "plus." **ALERT** In 13a, students may not realize that the independent variable is the number of thousands of gallons, not the number of gallons. Students may get different answers to 13f and 13g, depending on what assumptions they make about the number of days in a month.

13a. $c(x) = 7.18 + 3.98x$, where c is the cost in dollars and x is the number of thousands of gallons

13d. $g(x) = \dfrac{x - 7.18}{3.98}$, where g is the number of thousands of gallons and x is the cost in dollars

13e. $c(x) = 7.18 + 3.98x$
$x = 7.18 + 3.98y$
$\dfrac{x - 7.18}{3.98} = y = g(x)$

LESSON 4.5 Building Inverses of Functions

Exercise 16 To help motivate the next lesson, on logarithms, **ASK** "What if you can't rewrite the two sides to the same base?"

Exercise 17 There are infinitely many such parabolas; students can easily find at least the one with a horizontal axis and the one with a vertical axis.

Leaks account for nearly 12% of the average household's annual water consumption. About one in five toilets leaks at any given time, and that can waste more than 50 gallons each day. Dripping sinks add up fast too. A faucet that leaks one drop per second can waste 30 gallons each day.

Review

14. Rewrite the expression $125^{2/3}$ in as many different ways as you can.
    possible answers: $\sqrt[3]{125^2}, \left(\sqrt[3]{125}\right)^2, 5^2, \sqrt[3]{15,625}, 25$

15. Find an exponential function that contains the points (2, 12.6) and (5, 42.525). @
    $f(x) = 12.6(1.5)^{x-2}$ or $f(x) = 42.525(1.5)^{x-5}$

16. Solve by rewriting with the same base.
    a. $4^x = 8^3$ $x = \frac{9}{2}$, or 4.5
    b. $3^{4x+1} = 9^x$ $x = -\frac{1}{2}$, or -0.5
    c. $2^{x-3} = \left(\frac{1}{4}\right)^x$ $x = 1$

17. Give the equations of two different parabolas with vertex (3, 2) passing through the point (4, 5). ⓗ $y = 3(x-3)^2 + 2$ and $x = \frac{1}{9}(y-2)^2 + 3$

18. Solve this system of equations. $(-1, 1, 0)$
    $$\begin{cases} -x + 3y - z = 4 \\ 2z = x + y \\ 2.2y + 2.2z = 2.2 \end{cases}$$

PERFORMANCE TASK

Here is a paper your friend turned in for a recent quiz in her mathematics class:

If it is a four-point quiz, what is your friend's score? For each incorrect answer, provide the correct answer and explain it so that next time your friend will get it right!

QUIZ

1. Rewrite x^{-1}
   Answer: $\frac{1}{x}$

2. What does $f^{-1}(x)$ mean?
   Answer: $\frac{1}{f(x)}$

3. Rewrite $9^{-1/5}$
   Answer: $\frac{1}{9^5}$

4. What number is 0^0 equal to?
   Answer: 0

PERFORMANCE TASK

Your friend's score is 1. Sample answers are given for explanations of incorrect answers.
Problem 1 is correct.
Problem 2 is incorrect: The notation $f^{-1}(x)$ indicates the inverse function related to $f(x)$, not the exponent -1.
Problem 3 is incorrect: The expression $9^{-1/5}$ can be rewritten as $\frac{1}{9^{1/5}}$.
Problem 4 is incorrect: The expression 0^0 is not defined.

LESSON 4.6

Logarithms

> If all art aspires to the condition of music, all the sciences aspire to the condition of mathematics.
> GEORGE SANTAYANA

You have used several methods to solve for x when it is contained in an exponent. You've learned that in special cases, it is possible to solve by finding a common base. For example, finding the value of x that makes each of these equations true is straightforward because of your experience with the properties of exponents.

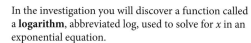

$$10^x = 1000 \qquad 3^x = 81 \qquad 4^x = \frac{1}{16}$$

Solving the equation $10^x = 47$ isn't as straightforward because you may not know how to write 47 as a power of 10. You can, however, solve this equation by graphing $y = 10^x$ and $y = 47$ and finding the intersection—the solution to the system and the solution to $10^x = 47$. Take a minute to verify that $10^{1.672} \approx 47$ is true.

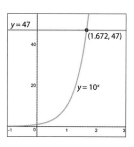

In the investigation you will discover a function called a **logarithm**, abbreviated log, used to solve for x in an exponential equation.

INVESTIGATION

Exponents and Logarithms

In this investigation you'll discover the connection between exponents on the base 10 and logarithms.

Step 1 Graph the function $f(x) = 10^x$ for $-1.5 \leq x \leq 1.5$ on your calculator. Sketch the graph and complete the table of information.

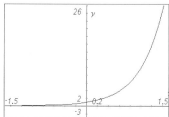

Domain	all real numbers
Range	$y > 0$
x-intercept	none
y-intercept	1
Equation of asymptote	$y = 0$ (x-axis)

Step 2 Complete the table of values for $f(x) = 10^x$ and its inverse.

x	-1.5	-1	-0.5	0	0.5	1	1.5
$f(x)$	0.032	0.1	0.316	1	3.162	10	31.62

x	0.032	0.1	0.316	1	3.162	10	31.62
$f^{-1}(x)$	-1.5	-1	-0.5	0	0.5	1	1.5

Investigate

Guiding the Investigation

Step 1 Review Calculator Note: Function Tables for instructions on using function tables. **SMP 5**

Step 5 **ALERT** Students may not have grasped the idea that they can find the power of 10 by taking the log. For example, in part a, they might not rewrite 300 as $10^{2.4771}$ to check. After students have finished the investigation, you might present part of Calculator Note: Logarithms and Antilogs. The note is not mentioned in the student book because it usurps the discovery of the investigation and contains antilogs, which are not needed yet.

Step 7 This step focuses on seeing (common) logarithms as powers of 10.

Step 3

Step 3 Enter the points for the inverse of $f(x)$ into your calculator and plot them. You will need to adjust the graphing window in order to see these points. Sketch the graph of the inverse function, and complete the table of information about the inverse.

Domain	$x > 0$
Range	all real numbers
x-intercept	1
y-intercept	none
Equation of asymptote	$x = 0$ (y-axis)

Step 4 This inverse function is called the logarithm of x, or $\log(x)$. Enter the equation $y = \log(x)$ into your calculator. Trace your graphs or use tables to find the following values.

a. $10^{1.5} \approx 31.6$ b. $\log(10^{1.5}) = 1.5$ c. $\log 0.32 \approx -0.49$ d. $10^{\log 0.32} = 0.32$

e. $10^{1.2} \approx 15.8$ f. $\log(10^{1.2}) = 1.2$ g. $10^{\log 25} = 25$ h. $\log 10^{2.8} = 2.8$

Step 5 Based on your results from Step 4, what is $\log 10^x$? Explain. x; explanations will vary; the log gives the exponent on base 10.

Step 6 Based on your results from Step 4, what is $10^{\log x}$? Explain. x; explanations will vary; the log undoes the exponent on base 10

Step 7 Using your results from Steps 4–6, complete the following statements:

a. If $100 = 10^2$, then $\log 100 =$? . 2

b. If $400 \approx 10^{2.6021}$, then $\log$? $\approx$? . 400; 2.6021

c. If ? $\approx 10^?$, then $\log 500 \approx$? . 500; 2.6990; 2.6990

Step 8 Complete the following statement: If $y = 10^x$, then $\log$? $=$? . $y; x$

The expression $\log x$ is another way of expressing x as a power of 10. Ten is the commonly used base for logarithms, so $\log x$ is called a **common logarithm** and is shorthand for writing $\log_{10} x$. You read this as "the logarithm base 10 of x." $\log x$ is the exponent you put on 10 to get x.

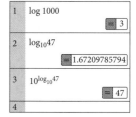

In the investigation you discovered that logarithms and exponents are inverse functions. This means you can use logarithms to "undo" exponents when solving equations.

The general **logarithmic function** is an exponent-producing function. Often you must solve for an exponent that is not on base 10. When bases other than 10 are used, you must specify the base by using a subscript. For example, $\log_2 x$ is an exponent on base 2. Regardless of the base, the logarithm is the exponent you put on the base to get x.

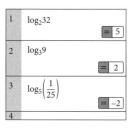

Modifying the Investigation

Whole Class Complete Steps 1 through 3 with student input. Give students time to complete Step 4 and share solutions. Discuss Steps 5 through 8.

Shortened Students complete Steps 1 through 4 in groups. Discuss Steps 5 through 8.

One Step Pose the first part of Exercise 7: "Assume the U.S. national debt can be estimated with the model $y = 0.051517(1.1306727)^x$, where x represents the number of years since 1900 and y represents the debt in billions of dollars. According to the model, when did the debt pass \$1 trillion?" Ask students how to represent 1 trillion.

> **Definition of Logarithm**
> For $a > 0$ and $b > 0$, $\log_b a = x$ is equivalent to $a = b^x$.

EXAMPLE A | Rewrite each exponential equation in logarithmic form using the definition of a logarithm and evaluate using technology.

a. $14.1 = 2^x$
b. $4 \cdot 10^x = 4650$.

Solution

a.
$14.1 = 2^x$	Original equation
$\log_2 14.1 = x$	The logarithm base 2 of 14.1 is the exponent you place on 2 to get 14.1.
$x \approx 3.818$	Evaluate using technology, using log base 2.

b.
$4 \cdot 10^x = 4650$	Original equation.
$10^x = 1162.5$	Divide both sides by 4.
$x = \log_{10} 1162.5$	The logarithm base 10 of 1162.5 is the exponent you place on 10 to get 1162.5.
$x \approx 3.0654$	Use the log key on your calculator to evaluate.

Mathematicians often use a special base when working with logarithms. You've solved problems exploring the amount of interest earned in a savings account when interest is compounded yearly, monthly, or daily. But what if interest is compounded continuously? This means that at every instant, your interest is deposited into your account and the new interest is calculated based on the new balance. This type of continuous growth is related to a number, *e*. This number has a value of approximately 2.718, and like π, it is a **transcendental number**—a special kind of irrational number. The logarithm function with base *e*, $\log_e x$, is also written as $\ln x$, and is called the **natural logarithm** function.

EXAMPLE B | Use the definition of a logarithm to rewrite $e^x = 5.2$ in logarithmic form and evaluate using technology.

Solution

$e^x = 5.2$	Original equation.
$\log_e 5.2 = x$	The logarithm base *e* of 5.2 is the exponent you place on *e* to get 5.2.
$\ln 5.2 = x$	Rewrite $\log_e$ as ln, the abbreviation mathematicians use for $\log_e$.
$x \approx 1.65$	Evaluate using technology, using the ln key.

Conceptual/Procedural

Conceptual In this lesson, students are introduced to the concept of a logarithm, including the natural logarithm *e*. They extend their understanding of inverses to include the inverse relationship of exponential functions and logarithmic functions.

Procedural Students learn to solve logarithmic equations in common bases, including *e*.

Example A

The first part of this example involves solving an equation in base 2. You may want to project the problem and ask for student input to solve it. Have students justify each step. You can use a blank desmos window in your ebook to find the log of a base other than 10 and *e*. **SMP 5**

Consider projecting part b and having students work individually or in pairs. The symbol $\log_{10}$ is the same as the symbol log used in the investigation. The answer, $x \approx 3.0654$, is given to five significant figures because that is the number of decimal places you need so that when you substitute back in for *x* the solution will check. As they present their solutions, have them give reasons for each step. **SMP 3, 5, 6**

Example B

This example provides practice solving exponential equations in base *e* by finding the natural log. Project the example and have students work in pairs. As they present their solutions, have them give reasons for each step. **SMP 3, 5, 6**

Example C

This example gives a real scenario for using logarithms to solve an exponential equation. You might project the example and have students work in pairs. As they present their solutions, have them give reasons for each step. **SMP 1, 2, 3, 4, 5, 6**

The answer is 6 years, not 5.7613 years, because interest is compounded annually.

Summarize

As students present their solutions in the investigation and the examples, have them give reasons for each step. Emphasize the use of correct mathematical terminology. Ask them to try to put into words what *logarithm* means. Don't let them merely say "exponent"; press for what distinguishes logarithms from other exponents. Whether student responses are correct or incorrect, ask other students if they agree and why. **SMP 1, 2, 3, 4, 5, 6**

CRITICAL QUESTION Why does the number you're taking the logarithm of have to be positive?

BIG IDEA Any positive base raised to any power (even a negative power) yields a positive number.

CRITICAL QUESTION What is the difference between a common logarithm and a natural logarithm?

BIG IDEA Common logarithms are in base 10. Natural logarithms are in base e.

Formative Assessment

Assess students' understanding of the concept of logarithm as an exponent. Note how well they grasp the role of the base and see when and how they use logarithms to solve equations. Observe as students work and present that they understand that the logarithm of a number with base b is the exponent put on b to get that number, that as exponents, logarithms can be used to solve equations in which the unknown appears as a power, and that the function $y = \log_b x$ is the inverse of the function $y = b^x$.

Apply

Extra Example

Rewrite the logarithmic equation in exponential form: $\log_6 14 = x$. $6^x = 14$

Closing Question

Use logarithms to write a solution to the equation $3(10)^{2x} = 400$.

possible solutions: $3(10)^{2x} = 400$; $10^{2x} = 133.333$; $\log_{10} 133.333 = 2x$; $2.125 = 2x$; $1.062 = x$

Assigning Exercises

Suggested: 1 – 9

Additional Practice: 10 – 15

EXAMPLE C An initial deposit of $500 is invested at 8.5% interest, compounded annually. How long will it take until the balance grows to $800?

Solution Let x represent the number of years the investment is held. Use the general formula for exponential growth, $y = a(1 + r)^x$.

$500(1 + 0.085)^x = 800$ Growth formula for compounding interest.

$(1.085)^x = 1.6$ Divide both sides by 500.

$x = \log_{1.085} 1.6$ Use the definition of logarithm.

$x \approx 5.7613$ Evaluate using technology, with log base 1.085.

It will take about 6 years for the balance to grow to at least $800.

4.6 Exercises

You will need a graphing calculator for Exercises 4, 11, 12, and 15.

Practice Your Skills

1. Rewrite each logarithmic equation in exponential form using the definition of a logarithm.

   a. $\log 1000 = x$ @ $10^x = 1000$
   b. $\log_5 625 = x$ $5^x = 625$
   c. $\log_7 x = \frac{1}{2}$ @ $7^{1/2} = x$
   d. $\log_x 8 = 3$ $x^3 = 8$
   e. $\ln x = 2.8$ $e^{2.8} = x$
   f. $\ln 1 = x$ $e^x = 1$

2. Solve each equation in Exercise 1 for x. @   2a. $x = 3$ 2b. $x = 4$ 2c. $x = \sqrt{7} \approx 2.65$
   2d. $x = 2$ 2e. $x \approx 16.44$ 2f. $x = 0$

3. Rewrite each exponential equation in logarithmic form using the definition of a logarithm. Then solve for x. (Give your answers rounded to four decimal places.)

   a. $10^x = 0.001$ @   b. $5^x = 100$ c. $35^x = 8$ @ $x = \log_{35} 8$; $x \approx 0.5849$
   $x = \log_{10} 0.001$; $x = -3$ $x = \log_5 100$; $x \approx 2.8614$
   d. $0.4^x = 5$ e. $23 = 2^x$ f. $\ln 0.1 = x$ $\ln 0.1 = x$, $x \approx 2.30$
   $x = \log_{0.4} 5$; $x \approx -1.7565$ $\log_2 23 = x \approx 4.52$

4. Graph each equation. Write a sentence explaining how the graph compares to the graph of either $y = 10^x$ or $y = \log x$.

   a. $y = \log(x + 2)$
   b. $y = 3 \log x$
   c. $y = -\log x - 2$
   d. $y = 10^{x+2}$
   e. $y = 3(10^x)$
   f. $y = -(10^x) - 2$

Reason and Apply

5. Given the functions $f(x) = 3^x$ and $g(x) = \log_3 x$:

   a. Find $f(2)$ and $g(9)$. $f(2) = 9$, $g(9) = 2$
   b. What do your answers to 5a imply?   They might be inverse functions.

6. Given the functions $f(x) = -4 + 0.5(x - 3)^2$ and $g(x) = 3 + \sqrt{2(x + 4)}$:

   a. Find $f(7)$ and $g(4)$. @ $f(7) = 4$; $g(4) = 7$
   b. What do your answers to 6a imply? @   They might be inverse functions.
   c. Find $f(1)$ and $g(-2)$. $f(1) = -2$; $g(-2) = 5$
   d. What do your answers to 6c imply?
   They are *not* inverse functions, at least not over their entire domains and ranges.
   e. Over what domain are f and g inverse functions?
   $f(x)$ for $x \geq 3$ and $g(x)$ for $x \geq -4$ (its entire domain) are inverse functions.

240   CHAPTER 4   Exponential, Power, and Logarithmic Functions

7. Assume the United States' national debt can be estimated with the model $y = 2.07596(1.083415)^x$, where x represents the number of years since 1900 and y represents the debt in billions of dollars.

   a. According to the model, when did the debt pass $1 trillion ($1,000 billion)? @ sometime in 1977

   b. According to the model, what is the annual growth rate of the national debt? 8.3%

   c. What is the doubling time for this growth model? 8.7 yr

8. **APPLICATION** Carbon-14 is an isotope of carbon that is formed when cosmic rays strike nitrogen in the atmosphere. Trees, which get their carbon dioxide from the air, contain small amounts of carbon-14. Once a tree is cut down, it doesn't absorb any more carbon-14, and the amount that is present begins to decay slowly. The half-life of the carbon-14 isotope is 5730 yr.

   a. Find an equation that models the percentage of carbon-14 in a sample of wood. (Consider that at time 0 there is 100% and that at time 5730 yr there is 50%.) @ $y = 100(0.999879)^x$

   b. A piece of wood contains 48.37% of its original carbon-14. According to this information, approximately how long ago did the tree that it came from die? What assumptions are you making, and why is this answer approximate?

Fossilized wood can be found in Petrified Forest National Park, Arizona. Some of the fossils are more than 200 million years old.

9. **APPLICATION** Crystal looks at an old radio dial and notices that the numbers are not evenly spaced. She hypothesizes that there is an exponential relationship involved. She tunes the radio to 88.7 FM. After six "clicks" of the tuning knob, she is listening to 92.9 FM. (h)

   a. Write an exponential model in point-ratio form. Let x represent the number of clicks past 88.7 FM, and let y represent the station number. @ $y \approx 88.7(1.0077)^x$

   b. Use the equation you have found to determine how many clicks Crystal should turn to get from 88.7 FM to 106.3 FM. 23 or 24 clicks

Review

10. Solve.

    a. $(x - 2)^{2/3} = 49$ $x = 345$ b. $3x^{2.4} - 5 = 16$
    $x = 7^{1/2.4} \approx 2.25$

11. The number of airline passengers has been increasing in the United States. The table shows the number of passengers on flights that start or end in the United States.

    a. Plot the passenger data as a function of years since 2000 and find a line of fit to model the data. @

    b. Calculate the residuals.

    c. If the trend continues, what is a good estimate of the number of airline passengers in 2020?

    11a, b. Line of fit equation using years since 2000 is $y = 592 + 5.5x$. See graphs on next page.

    11c. about 702 million passengers

Year	Passengers (millions)	Year	Passengers (millions)
2002	551.9	2009	618.1
2003	583.3	2010	629.5
2004	629.8	2011	638.2
2005	657.3	2012	642.3
2006	658.4	2013	645.7
2007	679.2	2014	662.8
2008	651.7	2015	696.0

(www.transtats.bts.gov)

Exercise 4 Comparisons of graphs should include discussion of transformations. **ELL** Ask students to describe the transformation of each graph when compared with the appropriate parent function.

4a. This is a translation of the graph of $y = \log x$ horizontally -2 units. Note that it actually continues downward indefinitely.

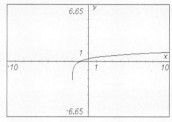

4b. This is a vertical dilation of the graph of $y = \log x$ by a factor of 3.

4c. This is a reflection of the graph of $y = \log x$ across the x-axis and a translation vertically -2 units.

4d. This is a translation of the graph of $y = 10^x$ horizontally -2 units.

4e. This is a vertical dilation of the graph of $y = 10^x$ by a factor of 3.

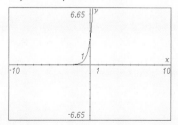

4f. This is a reflection of the graph of $y = 10^x$ across the x-axis and a translation vertically -2 units.

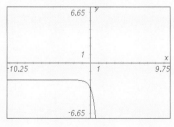

Exercise 8 **LANGUAGE** *Isotopes* of an element are atoms with differing numbers of neutrons. *Radiation*, in this context, is visible light and other forms of energy from the sun.

8b. 6025 yr ago. The technique is approximate and assumes that the carbon-14 concentration in the atmosphere has not changed over the past 6000 yr.

11a, b.

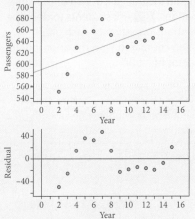

Exercise 12 **LANGUAGE** A musical note sounds when air molecules vibrate back and forth. A single back and forth action that is repeated is called a *cycle*. The number of cycles in a unit of time is called the *frequency* of the vibration and determines what we hear as the pitch of the note. Notes an *octave* apart sound very similar but are at different pitches.

12a. $C_1 = 32.7$, $C_2 = 65.4$, $C_3 = 130.8$, $C_6 = 1046.4$, $C_7 = 2092.8$, $C_8 = 4185.6$

12b. $y = 16.35(2)^x$, where x represents C-note number and y represents frequency in cycles per second

15a.

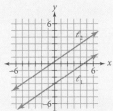

They are parallel.

15b. possible answer: $A(0, -3)$; $P(1, 1)$; $Q(4, 3)$

15c. Possible answer: Translate horizontally 1 unit and vertically 4 units.
$2(x - 1) - 3(y - 4) = 9$.

15d. Possible answer: Translate horizontally 4 units and vertically 6 units.
$2(x - 4) - 3(y - 6) = 9$.

15e. Possible answer:
$2(x - 1) - 3(y - 4) = 9 \rightarrow$
$2x - 2 - 3y + 12 = 9 \rightarrow$
$2x - 3y = -1$, which is l_2
$2(x - 4) - 3(y - 6) = 9 \rightarrow$
$2x - 8 - 3y + 18 = 9 \rightarrow$
$2x - 3y = -1$, which is also l_2

---

12. APPLICATION The C notes on a piano (C_1–C_8) are one octave apart. Their relative frequencies double from one C note to the next.

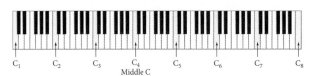
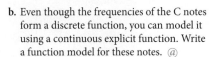

a. If the frequency of middle C (or C_4) is 261.6 cycles per second, and the frequency of C_5 is 523.2 cycles per second, find the frequencies of the other C notes.

b. Even though the frequencies of the C notes form a discrete function, you can model it using a continuous explicit function. Write a function model for these notes. @

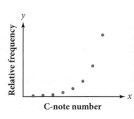

13. In each case below, use the graph and equation of the parent function to write an equation of the transformed image.

a. $y + 1 = x - 3$, or $y = x - 4$

b. $y + 4 = (x + 5)^2$, or $y = (x + 5)^2 - 4$

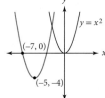

c. $y - 2 = |x + 6|$, or $y = |x + 6| + 2$

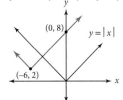

d. $y - 7 = \sqrt{x - 2}$, or $y = \sqrt{x - 2} + 7$

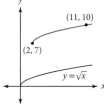

14. A rectangle has perimeter 155 inches. Its length is 7 inches more than twice its width.

a. Write a system of equations using the information given.
$2l + 2w = 155$
$l = 2w + 7$

b. Solve the system and find the rectangle's dimensions.
$l = 54$, $w = 23.5$; length: 54 in., width: 23.5 in.

15. l_1: $2x - 3y = 9$ l_2: $2x - 3y = -1$

a. Graph l_1 and l_2. What is the relationship between these two lines?

b. Give the coordinates of any point A on l_1, and any two points P and Q on l_2.

c. Describe the transformation that maps l_1 onto l_2 and A onto P. Write the equation of the image of l_1 showing the transformation. @

d. Describe the transformation that maps l_1 onto l_2 and A onto Q. Write the equation of the image of l_1 showing the transformation.

e. Algebraically show that the equations in 15c and d are equivalent to l_2.

LESSON 4.7
Arithmetic Series

> Life is a series of collisions with the future; it is not the sum of what we have been, but what we yearn to be.
>
> JOSÉ ORTEGA Y GASSET

According to the U.S. Environmental Protection Agency, each American produced an average of 978 lb of trash in 1960. This increased to 1336 lb in 1980. By 2000, trash production had risen to 1646 lb/yr per person. But by 2011, due to recycling, it had dropped back to 1600 lb/yr per person. You have learned in previous chapters how to write a sequence to describe the amount of trash produced per person each year. If you wanted to find the total amount of trash a person produced in his or her lifetime, you would add the numbers in this sequence.

Environmental CONNECTION

Mount Everest, part of the Himalaya range of southern Asia, reaches an altitude of 29,035 ft and is the world's highest mountain above sea level. It has been nicknamed "the world's highest junkyard" because decades of litter—climbing gear, plastic, glass, and metal—have piled up along Mount Everest's trails and in its camps. Environmental agencies like the World Wildlife Fund have cleared garbage from the mountain's base camp, but removing waste from higher altitudes is more challenging. An estimated 50 tons of junk remain.

Everest Base Camp, with stones, buddhist prayer flags and tents from earlier expeditions.

Series

The indicated sum of terms of a sequence is a **series**.

The sum of the first n terms in a sequence is represented by S_n. You can calculate S_n by adding the terms $u_1 + u_2 + u_3 + \cdots + u_n$.

History CONNECTION

Chu Shih-chieh (ca. 1280–1303) was a celebrated mathematician from Beijing, China, known for his theories on arithmetic series, geometric series, and finite differences. His two mathematical works, *Introduction to Mathematical Studies* and *Precious Mirror of the Four Elements*, were discovered in the 19th century.

Finding the value of a series is a problem that has intrigued mathematicians for centuries. Chinese mathematician Chu Shih-chieh called the sum $1 + 2 + 3 + \cdots + n$ a "pile of reeds" because it can be pictured like the diagram at right. The diagram shows S_9, the sum of the first nine terms of this sequence, $1 + 2 + 3 + \cdots + 9$. The sum of any **finite**, or limited, number of terms is called a **partial sum** of the series.

The expressions S_9 and $\sum_{n=1}^{9} u_n$ are shorthand ways of writing

$$u_1 + u_2 + u_3 + \cdots + u_9.$$

For this series, $u_n = n$, so you can express the partial sum S_9 with sigma notation as $\sum_{n=1}^{9} n$.

The expression $\sum_{n=1}^{9} n$ tells you to substitute the integers 1 through 9 for n in the explicit formula $u_n = n$, and then sum the resulting nine values. You get $1 + 2 + 3 + \cdots + 9 = 45$.

LESSON 4.7 Arithmetic Series **243**

Investigate

Guiding the Investigation
There is a desmos simulation of the investigation in your ebook.

Step 3 These new sequences should be different from each other and different from those in Step 2. To encourage creative thinking (and perhaps strategizing), you might challenge students to choose numbers (positive and less than 10) that nobody else in the class will choose. Non-integers open up the field of possibilities and give practice with adding fractions or decimals.

EXTEND Ask how many sequences have a positive integer less than 10 for both initial term and common difference. [9 possibilities for each number, for a total of 81 sequences]

Step 5 **ALERT** Students may have difficulty connecting the sum of the series to the area of the step-shaped figure.

Example
Project the example from your ebook and have students work in small groups. As they present their solutions, have them explain their strategy. **SMP 1, 2, 3, 4, 6**

If Gauss's method isn't used, tell students that 9-year-old Gauss was able to do this problem in a very short time. Then show his method and ask for student reaction.

ASK "How might we apply Gauss's method to the reed problem of Chu Shih-chieh?" In the expression $\frac{9(1 + 9)}{2}$, the first 9 represents the width of the rectangular pile and the $1 + 9$ represents the height, so $9(1 + 9)$ represents the area (total number of reeds in the double stack). The product is divided by 2 because the pile of reeds now has twice the number of reeds as the original pile. **SMP 4**

Ask students to use the method to find other sums, such as even numbers from 2 to 100 or multiples of 3 from 3 to 99.

How could you find the sum of the integers 1 through 100? The most obvious method is to add the terms, one by one. You can use technology and a recursive formula to do this quickly.

The sum of the first 100 terms is the sum of the first 99 terms plus the 100th term.

	A	B	C
		Sequence (u_n)	Partial Sums (S)
1	n		
2	1	1	1
3	2	2	3
4	3	3	6
5	4	4	10
6	5	5	15

=C2 + B3 — S_2 is the sum of S_1 and u_2.

	A	B	C
	n	Sequence (u_n)	Partial Sums (S)
	96	96	4656
	97	97	4753
	98	98	4851
	99	99	4950
	100	100	5050

=C100 + B101 — S_{100} is the sum of S_{99} and u_{100}.

The recursive definitions for the sequence and series described above are:

Sequence
$u_1 = 1$
$u_n = u_{n-1} + 1$ where $n \geq 2$

Series
$S_1 = 1$
$S_n = S_{n-1} + u_n$ where $n \geq 2$

The graph of S_n appears to form a solid curve, but it is actually a discrete set of 100 points representing each partial sum from S_1 through S_{100}. Each point is in the form (n, S_n) for integer values of n, for $1 \leq n \leq 100$. You can trace to find that the sum of the first 100 terms, S_{100}, is 5050.

When you compute this sum recursively, you or the calculator must compute each of the individual terms. The investigation will give you an opportunity to discover at least one explicit formula for calculating the partial sum of an **arithmetic series** without finding all terms and adding.

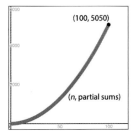
(100, 5050)
(n, partial sums)

INVESTIGATION
Arithmetic Series Formula

YOU WILL NEED
- graph paper
- scissors

Step 1 4, 7, 10, 13, 16; sum = 50

In this investigation you will use a geometric model to help you develop an explicit formula for the partial sum of an arithmetic series.

Step 1 The lengths of the rows of this step-shaped figure represent terms of an arithmetic sequence. Write the sequence u_1, u_2, u_3, u_4, u_5, represented by the figure. What is the sum of the series?

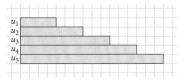

244 CHAPTER 4 Exponential, Power, and Logarithmic Functions

Modifying the Investigation

Whole Class In Step 2, use two transparencies to show the rectangle. Demonstrate the arithmetic sequence for Step 2 and then have students make their own figures and sequences. Guide the group through Step 4. Complete Step 5 with student input.

Shortened Rather than having students create their own figures and sequences, have two additional examples (like Step 2) prepared for groups to explore. Discuss Step 5 with the whole group.

One Step Use the **One Step** in your ebook. Students may approach the problem in a variety of ways. Challenge those who methodically sum 21 terms to find an easier way. Push some groups to find a formula for the sum of the terms of any finite arithmetic sequence.

5 by $(u_1 + u_5)$; or 5 by 20

Answers will vary. The ngle dimensions will be n by u_n); area = $n \cdot (u_1 + u_n)$.

Answers will vary, but the la should be equivalent to $\dfrac{n(u_1 + u_n)}{2}$; n is the number of or terms, u_1 is the length of the ow, u_n is the length of the last The length of the rectangle is u_n), the height or width nd divide by 2 because the ngle is twice the partial sum.

Step 2 If you cut out two copies of this figure and slid them together to make a rectangle, what would the dimensions of your rectangle be?

Step 3 Have each member of your group create a different arithmetic sequence. Each of you can choose a starting value and a common difference. (Use positive numbers less than 10 for each of these.) On graph paper, draw two copies of a step-shaped figure representing your sequence.

Step 4 Cut out both copies of the step-shaped figure. Slide the two congruent shapes together to form a rectangle, and then calculate the dimensions and area of the rectangle. Now express this area in terms of the number of rows, n, and the first and last terms of the sequence.

Step 5 Based on what you have discovered, what is a formula for the partial sum, S_n, of an arithmetic series? Describe the relationship between your formula and the dimensions of your rectangle.

In the investigation you found a formula for a partial sum of an arithmetic series. Use your formula to check that when you add $1 + 2 + 3 + \cdots + 100$ you get 5050.

History CONNECTION

According to legend, when German mathematician and astronomer Carl Friedrich Gauss (1777–1855) was 9 years old, his teacher asked the class to find the sum of the integers 1 through 100. The teacher was hoping to keep his students busy, but Gauss quickly wrote the correct answer, 5050. The example shows Gauss's solution method.

Karl Friedrich Gauß.

EXAMPLE | Find the sum of the integers 1 through 100, without using a calculator.

Solution | Carl Friedrich Gauss solved this problem by adding the terms in pairs. Consider the series written in ascending and descending order, as shown.

$$\begin{array}{ccccccccccccc}
1 & + & 2 & + & 3 & + \ldots + & 98 & + & 99 & + & 100 & = S_{100} \\
100 & + & 99 & + & 98 & + \ldots + & 3 & + & 2 & + & 1 & = S_{100} \\
\hline
101 & + & 101 & + & 101 & + \ldots + & 101 & + & 101 & + & 101 & = 2S_{100}
\end{array}$$

The sum of every column is 101, and there are 100 columns. Thus, the sum of the integers 1 through 100 is

$$\frac{100(101)}{2} = 5050$$

You must divide the product 100(101) by 2 because the series was added twice.

Conceptual/Procedural

Conceptual In this lesson, students further explore the concept of mathematical series and partial sums of arithmetic series.

Procedural Students investigate the formula for finding the partial sum of an arithmetic series.

Summarize

As students present their solutions in the Investigation and example, have them explain their sequence and their strategy. Emphasize the use of correct mathematical terminology. Whether student responses are correct or incorrect, ask other students if they agree and why. **SMP 1, 2, 3, 4, 6**

Discussing the Investigation

LANGUAGE ASK "Is the common meaning of the word *series* the same thing as the mathematical meaning?" [In common usage, *series* means a sequence rather than a summation.] You might also point out that the word *series* is both plural and singular. **SMP 6**

As the last students present, **ASK** "What happens if the common difference or the first term is negative?" Let students experiment to see that the formula still applies. Elicit the fact that the partial sums get farther from 0 as n increases, whether the common difference is positive or negative. **SMP 1, 2, 8**

If students haven't used non-integer numbers for initial values or common differences, ask if the formula holds for these numbers as well. Students can see that it does from step-shaped figures and rectangles.

The formula for the partial sum of an arithmetic series contains the expression $\dfrac{u_1 + u_n}{2}$. **ASK** "Does this expression seem familiar?" Elicit the idea that it's the mean of u_1 and u_n. "How does the mean of the first and last terms relate to the series?" [It's the mean of the terms in the series, because the terms are equally spaced.] "Why does multiplying the mean by the number of terms give you the total of those terms?" Review the often-neglected idea that because a mean is calculated by dividing the total by the number of values, the total is the mean times the number of values. **SMP 6, 8**

A good opener for new problems is a "what if not" question. **ASK** "What if it is not an arithmetic series but a geometric series?" Challenge students to think about it overnight.

CRITICAL QUESTION What is a series?

BIG IDEA A series is a summation of terms of a sequence.

CRITICAL QUESTION What is an arithmetic series?

BIG IDEA An arithmetic series is a summation of terms in an arithmetic sequence.

CRITICAL QUESTION How do you find the partial sum of an arithmetic series?

BIG IDEA A basic formula for the sum of the nth partial sum of an arithmetic series is $S_n = \dfrac{n(u_1 + u_n)}{2}$.

Formative Assessment

Students worked with the idea of summation in their statistical work earlier, so this investigation may be straightforward for them. Check their ability to use previous knowledge in this new context. Their responses in the Investigation, Example, and Exercises provide evidence of meeting the objectives.

Apply

Extra Example

1. Find $S_1, S_2, S_3, S_4,$ and S_5 for the sequence: −1, 1, 3, 5, 7.
$S_1 = -1, S_2 = 0, S_3 = 3, S_4 = 8, S_5 = 15$

2. Write the expression $\sum_{n=1}^{3}(n^2 + 1)$ as the sum of terms, then calculate the sum.
$2 + 5 + 10 = 17$

Closing Question

Find the sum of the first 20 multiples of 3. 630

Assigning Exercises

Suggested: 1–11, 14
Additional Practice: 12, 13, 15–18

Exercises 4, 5 If students have trouble starting these exercises, ask them to find the values of n, u_1, and the common difference.

Exercise 6 Students can use the formula to evaluate 6b and 6c, or they can apply Gauss's method and avoid using a formula. **SUPPORT** Suggest that students list the first few terms of the series to reinforce the meaning of summation notation.

246 CHAPTER 4 Exponential, Power, and Logarithmic Functions

You can extend the method in the example to any arithmetic series. Before continuing, take a moment to consider why the sum of the reeds in the original pile can be calculated using the expression

$$\frac{9(1+9)}{2}$$

What do the 9, 1, 9, and 2 represent in this context?

> **Partial Sum of an Arithmetic Series**
>
> The partial sum of an arithmetic series is given by the explicit formula
>
> $$S_n = \frac{n(u_1 + u_n)}{2}$$
>
> where n is the number of terms, u_1 is the first term, and u_n is the last term.

In the exercises you will use the formula for partial sums to find the sum of consecutive terms of an arithmetic sequence.

4.7 Exercises

You will need a graphing calculator for Exercise **15**.

Practice Your Skills

1. List the first five terms of this sequence. Identify the first term and the common difference.

   $u_1 = -3$
   $u_n = u_{n-1} + 1.5$ where $n \geq 2$

   $-3, -1.5, 0, 1.5, 3$; $u_1 = -3, d = 1.5$

2. Find $S_1, S_2, S_3, S_4,$ and S_5 for this sequence: 2, 6, 10, 14, 18. @ $S_1 = 2, S_2 = 8, S_3 = 18, S_4 = 32, S_5 = 50$

3. Write each expression as a sum of terms, then calculate the sum.

   a. $\sum_{n=1}^{4}(n+2)$ $3 + 4 + 5 + 6; 18$

   b. $\sum_{n=1}^{3}(n^2 - 3)$ @ $-2 + 1 + 6; 5$

4. Find the sum of the first 50 multiples of 6: $\{6, 12, 18, \ldots, u_{50}\}$. ⓗ $S_{50} = 7650$

5. Find the sum of the first 75 even numbers, starting with 2. @ $S_{75} = 5700$

Reason and Apply

6. Find these values.

   a. Find u_{75} if $u_n = 2n - 1$.
   $u_{75} = 149$

   b. Find $\sum_{n=1}^{75}(2n - 1)$. @
   $S_{75} = 5625$

   c. Find $\sum_{n=20}^{75}(2n - 1)$.
   $S_{75} - S_{19} = 5264$

246 CHAPTER 4 Exponential, Power, and Logarithmic Functions

7. Consider the graph of the arithmetic sequence shown at right.

   a. What is the 46th term? @ $u_{46} = 229$

   b. Write a formula for u_n. $u_n = 5n - 1$, or $u_1 = 4$ and $u_n = u_{n-1} + 5$ where $n \geq 2$

   c. Find the sum of the heights from the horizontal axis of the first 46 points of the sequence's graph. $S_{46} = 5359$

8. Suppose you practice the piano 45 min on the first day of the semester and increase your practice time by 5 min each day. How much total time will you devote to practicing during each of the following?

   a. The first 15 days of the semester. 1200 min, or 20 h

   b. The first 35 days of the semester. 4550 min, or 75 h 50 min

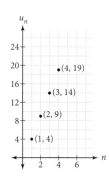

American pianist Van Cliburn (b 1934), shown here arriving in Moscow in 2004, caused a sensation by winning the first International Tchaikovsky Competition in Moscow in 1958, during the height of the Cold War. When asked how many hours he practiced per day, he replied, "Mother always told me that if you knew how long you had practiced, you hadn't done anything. She believed you had to become engrossed … or it just didn't count for anything. Nothing happens if you watch the clock."

9. A concert hall has 59 seats in Row 1, 63 seats in Row 2, 67 seats in Row 3, and so on.

   a. How many seats are there in all 35 rows of the concert hall? 4445 seats

   b. Suppose that Seat 1 is at the left end of Row 1 and that Seat 60 is at the left end of Row 2. Describe the location of Seat 970. row 12, third seat in from the right

10. Jessica arranges a display of soup cans as shown.

    a. List the number of cans in the top layer, the second layer, the third layer, and so on, down to the tenth layer. @ 3, 6, 9, 12, 15, 18, 21, 24, 27, 30

    b. Write a recursive formula for the terms of the sequence in 10a. @

    c. If the cans are to be stacked 47 layers high, how many cans will it take to build the display? 3384 cans

    d. If Jessica uses six cases, with 48 cans in each case, how tall can she make the display? *h* 13 rows with 15 cans left over

    10b. $u_1 = 3$ and $u_n = u_{n-1} + 3$ where $n \geq 2$

LESSON 4.7 Arithmetic Series 247

Exercise 9 ALERT In 9a, students may add 4 a total of 35 times rather than 34 times to 59. The last term of the series is 195. Approaches to 9b may vary. Students may use guess-and-check or add terms until they exceed 970. Or they may generalize, solving the quadratic equation that arises from $n(2n + 57) = 970$ to get n about 11.98; conclude that seat 970 is located after 11 full rows and part of a 12th row; and proceed from there.

Exercise 10 Students can assume that Jessica leaves no gaps in any layer. SUPPORT For 10b, refer students to Refreshing Your Skills: Arithmetic and Geometric Sequences. If students make a diagram for 10d, link to the lesson example. As needed in 10c, ASK "What is n?" [47] If students get stuck in solving 10d, you might ASK "What do you know, and what do you want to find?" [You know that S_n equals 288, and you want to find the number of rows, n.]

12d. The numbers of toothpicks in each row form a sequence, whereas the total numbers of toothpicks used form a series.

Exercise 13 This exercise ignores air resistance. **ALERT** Students may think that 14c and 14d are asking the same question. **EXTENSION** Students could research *terminal velocity* and create a piecewise graph for the actual speed of the quarter.

11. Find each value.
    a. Find the sum of the first 1000 positive integers (the numbers 1 through 1000). ⓗ $S_{1000} = 500{,}500$
    b. Find the sum of the second 1000 positive integers (1001 through 2000). $S_{2000} - S_{1000} = 1{,}500{,}500$
    c. Guess the sum of the third 1000 positive integers (2001 through 3000). Answers will vary.
    d. Now calculate the sum for 11c. $S_{3000} - S_{2000} = 2{,}500{,}500$
    e. Describe a way to find the sum of the third 1000 positive integers if you know the sum of the first 1000 positive integers. @ Each of these 1000 values is 2000 more than the corresponding value between 1 and 1000, so add 1000(2000) to the first sum.

12. It takes 5 toothpicks to build the top trapezoid shown at right. You need 9 toothpicks to build 2 connected trapezoids and 13 toothpicks for 3 trapezoids.
    a. If 1000 toothpicks are available, how many trapezoids will be in the last complete row? 21 trapezoids
    b. How many complete rows will there be? 21 rows
    c. How many toothpicks will you use to construct these rows? 945 toothpicks
    d. Use the context of this exercise to explain the difference between a sequence and a series.

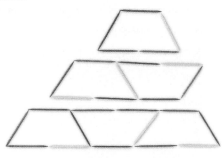

13. **APPLICATION** If an object falls from rest, then the distance it falls during the first second is about 4.9 m. In each subsequent second, the object falls 9.8 m farther than in the preceding second.
    a. Write a recursive formula to describe the distance the object falls during each second of free fall. $u_1 = 4.9$ and $u_n = u_{n-1} + 9.8$ where $n \geq 2$
    b. Find an explicit formula for 13a. @ $u_n = 9.8n - 4.9$
    c. How far will the object fall during the 10th second? 93.1 m
    d. How far does the object fall during the first 10 seconds? @ 490 m
    e. Find an explicit formula for the distance an object falls in n seconds. $S_n = 4.9n^2$
    f. Suppose you drop a quarter from the Royal Gorge Bridge. How long will it take to reach the Arkansas River 321 m below? approximately 8.1 s

14. Consider these two geometric sequences:

    i. 2, 4, 8, 16, 32, ...   ii. $2, 1, \frac{1}{2}, \frac{1}{4}, \frac{1}{8}, \ldots$

    a. What is the long-run value of each sequence?
       i. increases without bound; ii. 0
    b. What is the common ratio of each sequence? i. 2; ii. $\frac{1}{2}$
    c. What will happen if you try to sum all of the terms of each sequence?
       i. The sum will be infinitely large.
       ii. The sum will approach 4.

The Royal Gorge Bridge (built 1929) near Cañon City, Colorado, is the world's highest suspension bridge, with length 384 m (1260 ft) and width 5 m (18 ft).

15. **APPLICATION** There are 650,000 people in a city. Every 15 minutes, the local radio and television stations broadcast a tornado warning. During each 15-minute time period, 42% of the people who had not yet heard the warning become aware of the approaching tornado. How many people have heard the news

   a. after 1 hour? @ 576,443 people
   b. after 2 hours? 641,676 people

In the central United States, an average of 800 to 1000 tornadoes occur each year. Tornado watches, forecasts, and warnings are announced to the public by the National Weather Service.

Review

16. Suppose you invest $500 in a bank that pays 5.5% annual interest compounded quarterly.

    a. How much money will you have after five years? $657.03

    b. Suppose you also deposit an additional $150 at the end of every three months. How much will you have after five years? $4,083.21

17. Consider the explicit formula $u_n = 81 \cdot \left(\frac{1}{3}\right)^{n-1}$. @

    a. List the first six terms, u_1 to u_6. 81, 27, 9, 3, 1, $\frac{1}{3}$
    b. Write a recursive formula for the sequence.
    $u_1 = 81$ and $u_n = \frac{1}{3} u_{n-1}$ where $n \geq 2$

18. Consider the recursive formula

    $u_1 = 0.39$

    $u_n = 0.01 \cdot u_{n-1}$ where $n \geq 2$

    a. List the first six terms.
    0.39, 0.0039, 0.000039, 0.00000039, 0.0000000039, 0.000000000039
    b. Write an explicit formula for the sequence.
    $u_n = 39(0.01)^n$

PERFORMANCE TASK

Create a formula for the sum S_n of an arithmetic series that is based on the number of terms n, the first term u_1, and the common difference d.

PERFORMANCE TASK

Students can substitute $u_1 + d(n - 1)$ for u_n in the known formula and then simplify. **ADVANCED**
Ask students to explain when each formula might be more appropriate **ASK** "Can you also show algebraically that the two formulas for the partial sum of an arithmetic series are equivalent?" [Substitute $u_1 + (n-1)d$ or u_n in the formula you created for this exercise.]

Answers will vary. Possible answers are

$S_n = nu_1 + \frac{n(n-1)}{2}d$, or $S_n = \frac{n[2u_1 + (n-1)d]}{2}$, or $S_n = \left(\frac{d}{n}\right)n^2 + \left(u_1 - \frac{d}{2}\right)n$.

LESSON 4.8

This lesson addresses the question of the nth partial sum of a geometric series. The terms being added to make a partial sum are sometimes called a **finite series**.

COMMON CORE STATE STANDARDS		
APPLIED	DEVELOPED A.SSE.4	INTRODUCED

Objectives

- Find formulas for the partial sums of geometric series
- Given the partial sum of a geometric series and either the common ratio or the initial value, find the other value
- Distinguish between arithmetic series and geometric series
- Investigate other types of series

Vocabulary

geometric series

Materials

- ball, *optional*
- motion sensor, *optional*
- Calculator Note: Partial Sums of Series, *optional*

Launch

If a pair of calculators can be linked and a file transferred from one calculator to the other in 20 s, how long will it be before everyone in a lecture hall of 250 students has the file?

2 minutes and 40 seconds

Investigate

Example A

Project Example A and have students work in pairs. As needed, refer students to Calculator Note: Partial Sums of Series for calculating and graphing the partial sums. You might need to remind students that the calculator might use a symbol other than S_n for the partial sums.

LESSON 4.8

Partial Sums of Geometric Series

We have a problem. 'Congratulations.' But it's a tough problem. 'Then double congratulations.'

W. CLEMENT STONE

If a pair of calculators can be linked and a file transferred from one calculator to the other in 20 s, how long will it be before everyone in a lecture hall of 250 students has the file? During the first time period, the file is transferred to one calculator; during the second time period, to two calculators; during the third time period, to four more calculators; and so on. The number of students who have the file doubles every 20 s.

To solve this problem, you must determine the minimum value of n for which S_n exceeds 250. The solution uses a partial sum of a geometric series. It requires the sum of a finite number of terms of the geometric sequence 1, 2, 4, 8, ...

EXAMPLE A Consider the sequence 2, 6, 18, 54, ...

a. Find u_{15}.

b. Graph the partial sums S_1 through S_{15}, and find the partial sum S_{15}.

Solution The sequence is geometric with $u_1 = 2$ and $r = 3$.

a. A recursive formula for the sequence is $u_1 = 2$ and $u_n = 3u_{n-1}$ where $n \geq 2$. The sequence can also be defined explicitly as $u_n = 2(3)^{n-1}$. Substituting 15 for n into either equation gives $u_{15} = 9{,}565{,}938$.

b. Graph the partial sums. Use a table or trace the graph to find that S_{15} is 14,348,906.

	A	B	C
	n	Sequence	Partial Sums
	11	118098	177146
	12	354294	531440
	13	1062882	1594322
	14	3188646	4782968
	15	9565938	14348906

=C15+B16

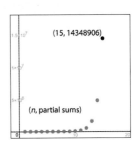

(15, 14348906)

(n, partial sums)

250 **CHAPTER 4** Exponential, Power, and Logarithmic Functions

---

ELL
Students should add to their graphic organizer. If there is time, students could discuss or present their organizers. Review geometric sequences as you discuss geometric series.

Support
The derivation of the formula is not as important as identifying problems that involve partial geometric sums or as being able to apply the formula appropriately. Students should have access to the explicit formula throughout.

Advanced
Understanding a derivation of an explicit formula for partial geometric sums will improve students' facilities with the formula and reduce their need to memorize it. **ASK** "What if the series is not arithmetic or geometric? Are there other kinds of series?" [i.e., harmonic]

In the example you used a recursive method to find the partial sum of a geometric series. For some partial sums, especially those involving a large number of terms, it can be faster and easier to use an explicit formula. The investigation will help you develop an explicit formula.

INVESTIGATION

Geometric Series Formula

A ball is dropped from 180 cm above the floor. With each bounce, it rebounds to 65% of its previous height.

Step 1 heights (to the nearest 0.01): 180, 117, 76.05, 49.43, 32.13, 20.89, 13.58, 8.82, 5.74, 3.73; partial sums: 180, 297, 373.05, 422.48, 454.61, 475.50, 489.07, 497.90, 503.63, 507.36

Step 1 Use your calculator to find the first ten heights, with the initial drop height being the first term of the sequence. Then find the first ten partial sums of this sequence of heights: $S_1, S_2, S_3, \ldots, S_{10}$.

Step 2 Create a scatter plot of points (n, S_n) and find a translated exponential equation to fit the data. The equation will be in the form $S_n = L - a \cdot b^n$. (Hint: L is the long-run value of the partial sums.)

Step 3 Rewrite your equation from Step 2 in terms of n, a or u_1, and r. Use algebraic techniques to write your explicit formula as a single rational expression.

Step 3 Answers may vary; formula should be equivalent to $S_n = \dfrac{u_1(1-r^n)}{1-r}$ for $r \neq 1$.

Step 4 Here is a way to arrive at an equivalent formula. Complete the derivation by filling in the blanks.

$$S_n = u_1 + u_1 \cdot r + u_1 \cdot r^2 + \cdots + u_1 \cdot r^{n-1}$$

$$rS_n = \underline{\ ?\ } \quad u_1 \cdot r + u_1 \cdot r^2 + \cdots + u_1 \cdot r^n$$

$$S_n - rS_n = \underline{\ ?\ } \quad u_1 - u_1 \cdot r^n, \text{ or } u_1(1-r^n)$$

$$S_n = \underline{\ ?\ } \quad \dfrac{u_1(1-r^n)}{1-r} \text{ for } r \neq 1$$

Use algebraic techniques to verify that the formula is equivalent to your formula from Step 3.

Step 5 Use your formula from Step 4 to find S_{10} for the bouncing ball. Then use it to find S_{10} for the geometric sequence 2, 6, 18, 54, ... 507.362; 59,048

In the investigation you found an explicit formula for a partial sum of a geometric series that uses only three pieces of information—the first term, the common ratio, and the number of terms. Now you do not need to write out the terms to find a sum.

EXAMPLE B | Find S_{10} for the series $16 + 24 + 36 + \cdots$.

Solution | The first term, u_1, is 16. The common ratio, r, is 1.5. The number of terms, n, is 10. Use the formula you developed in the investigation to calculate S_{10}.

$$S_{10} = \dfrac{16(1 - 1.5^{10})}{1 - 1.5} = 1813.28125$$

LESSON 4.8 Partial Sums of Geometric Series **251**

Example C

Project Example C and have students work in pairs. As they present their solutions, have them explain their application of the formula (for example, specify what term they are looking for, what they used for u_1, r, and n). **SMP 1, 2, 3, 4, 5, 6**

ASK "Why do we use the value $r = 1.25$ rather than $r = 0.25$?" [The value $r = 1.25$ represents 25% *more than* the number eaten the day before; an r-value of 0.25 would represent 25% *of* the number eaten the day before, and the powers of r would be smaller than r instead of larger.]

Summarize

As students present their solutions to the Investigation and the examples, have them explain their strategy. Emphasize the use of correct mathematical terminology. Whether student responses are correct or incorrect, ask other students if they agree and why. **SMP 1, 2, 3, 4, 5, 6**

CRITICAL QUESTION How do you find the partial sums of a geometric series?

BIG IDEA A standard formula for the nth partial sum of a geometric series with common ratio r is $S_n = \dfrac{u_1(1 - r^n)}{(1 - r)}$.

CRITICAL QUESTION Why does the statement of the formulas specify that r is not 1?

BIG IDEA To avoid dividing by 0

CRITICAL QUESTION Does that mean that the partial sum has no value if $r = 1$?

BIG IDEA The formula is undefined, but the partial sums are defined. When $r = 1$, the terms of the series are all the same, so it becomes a multiplication problem.

Formative Assessment

During the investigation, note students' facility working with the two new formulas, which require some heavy algebraic manipulation. Look for students who are able to apply the new formula and students who can explain why it works.

EXAMPLE C Each day, the imaginary caterpillarsaurus eats 25% more leaves than it did the day before. If a 30-day-old caterpillarsaurus has eaten 151,677 leaves in its brief lifetime, how many will it eat the next day?

Solution To solve this problem, you must find u_{31}. The information in the problem tells you that r is $(1 + 0.25)$, or 1.25, and when n equals 30, S_n equals 151,677. Substitute these values into the formula for S_n and solve for the unknown value, u_1.

$$\frac{u_1(1 - 1.25^{30})}{1 - 1.25} = 151{,}677$$

$$u_1 = 151{,}677 \cdot \frac{1 - 1.25}{1 - 1.25^{30}}$$

$$u_1 \approx 47$$

Now you can write an explicit formula for the terms of the geometric sequence, $u_n = 47(1.25)^{n-1}$. Substitute 31 for n to find that on the 31st day, the caterpillarsaurus will eat 37,966 leaves.

The explicit formula for the sum of a geometric series can be written in several ways, but they are all equivalent. You probably found these two ways during the investigation:

> **Partial Sum of a Geometric Series**
> A partial sum of a geometric series is given by the explicit formula
> $$S_n = \left(\frac{u_1}{1-r}\right) - \left(\frac{u_1}{1-r}\right)r^n \quad \text{or} \quad S_n = \frac{u_1(1 - r^n)}{1 - r}$$
> where n is the number of terms, u_1 is the first term, and r is the common ratio ($r \neq 1$).

4.8 Exercises

You will need a graphing calculator for Exercises **4, 5, 12,** and **15**.

Practice Your Skills

1. For each partial sum equation, identify the first term, the ratio, and the number of terms.

   a. $\dfrac{12}{1 - 0.4} - \dfrac{12}{1 - 0.4}(0.4)^8 \approx 19.9869$ @ $u_1 = 12, r = 0.4, n = 8$

   b. $\dfrac{75(1 - 1.2^{15})}{1 - 1.2} \approx 5402.633$ @ $u_1 = 75, r = 1.2, n = 15$

   c. $\dfrac{40 - 0.46117}{1 - 0.8} \approx 197.69$ $u_1 = 40, r = 0.8, n = 20$

   d. $-40 + 40(2.5)^6 = 9725.625$ $u_1 = 60, r = 2.5, n = 6$

2. Consider the geometric sequence: 256, 192, 144, 108, . . .

   a. What is the eighth term? $u_8 = 34.171875$

   b. Which term is the first one smaller than 20? @ u_{10}

   c. Find u_7. $u_7 = 45.5625$

   d. Find S_7. @ $S_7 = 887.3125$

Conceptual/Procedural

Conceptual In this lesson, students further explore the concept of mathematical series and partial sums of geometric series.

Procedural Students investigate the formula for finding the partial sum of a geometric series.

3. Find each partial sum of this sequence.

   $u_1 = 40$

   $u_n = 0.6u_{n-1}$ where $n \geq 2$

   a. S_5 $S_5 = 92.224$
   b. S_{15} $S_{15} \approx 99.953$
   c. S_{25} $S_{25} \approx 99.9997$

4. Identify the first term and the common ratio or common difference of each series. Then find the partial sum.

   a. $3.2 + 4.25 + 5.3 + 6.35 + 7.4$
      $u_1 = 3.2, d = 1.05, S_5 = 26.5$
   b. $3.2 + 4.8 + 7.2 + \cdots + 36.45$ @
      $u_1 = 3.2, r = 1.5, S_7 = 102.95$
   c. $\sum_{n=1}^{27}(3.2 + 25n)$ @ $u_1 = 5.7, d = 2.5, S_{27} = 1031.4$
   d. $\sum_{n=1}^{10} 3.2(4)^{n-1}$ $u_1 = 3.2, r = 4, S_{10} = 1,118,480$

Reason and Apply

5. Find the missing value in each set of numbers.

   a. $u_1 = 3, r = 2, S_{10} = \underline{?}$ @ 3069
   b. $u_1 = 4, r = 0.6, S_{\underline{?}} \approx 9.999868378$ @ 22
   c. $u_1 = \underline{?}, r = 1.4, S_{15} \approx 1081.976669$ 2.8
   d. $u_1 = 5.5, r = \underline{?}, S_{18} \approx 66.30642497$ 0.95

6. Find the nearest integer value for n if $\dfrac{3.2(1 - 0.8^n)}{1 - 0.8}$ is approximately 15. $n = 12$

7. Consider the sequence $u_1 = 8$ and $u_n = 0.5u_{n-1}$ where $n \geq 2$. Find

   a. $\sum_{n=1}^{10} u_n$ @ $S_{10} = 15.984375$
   b. $\sum_{n=1}^{20} u_n$ $S_{20} \approx 15.99998474$
   c. $\sum_{n=1}^{30} u_n$ $S_{30} \approx 15.99999999$

   d. Explain what is happening to these partial sums as you add more terms.
      They continue to increase, but by a smaller amount each time.

8. Suppose you begin a job with an annual salary of $17,500. Each year, you can expect a 4.2% raise.

   a. What is your salary in the tenth year after you start the job? @ $25,342.39

   b. What is the total amount you earn in ten years? $212,065.89

   c. How long must you work at this job before your total earnings exceed $1 million? 29 yr 272 d, or about 30 yr

9. Consider the geometric series: $5 + 10 + 20 + 40 + \cdots$

   a. Find the first seven partial sums, $S_1, S_2, S_3, \ldots, S_7$. 5, 15, 35, 75, 155, 315, 635

   b. Do the partial sums create a geometric sequence? No, they form a shifted geometric sequence.

   c. If u_1 is 5, find value(s) of r such that the partial sums form a geometric sequence. not possible

10. List terms to find

    a. $\sum_{n=1}^{7} n^2$
       $1 + 4 + 9 + 16 + 25 + 36 + 49 = 140$
    b. $\sum_{n=3}^{7} n^2$ $9 + 16 + 25 + 36 + 49 = 135$

LESSON 4.8 Partial Sums of Geometric Series 253

Apply

Extra Example
Consider the geometric sequence 214, 42.8, 8.56, 1.712, 0.3424, ….

a. What is the 10th term of the sequence? 0.000109568

b. Find S_{10}. 267.499972608

Closing Question
Identify the first term and the common ratio or difference of the series $\sum_{n=1}^{20} 5(0.3)^{n-1}$. $u_1 = 5; r = 0.3$

Assigning Exercises
Suggested: 1–13
Additional Practice: 14–17

Exercise 1d As needed, **ASK** "What is the value of r?" [2.5]

Exercises 10, 12 **ADVANCED** Students could attempt to write explicit formulas for these series.

Exercise 17 As needed, remind students that for polynomial equations with real coefficients, nonreal roots come in conjugate pairs. Therefore the factored form of the equation is $P(x) = (x + 3 - 2i)(x + 3 + 2i)(3x - 2)$.

11. An Indian folktale, recounted by Arab historian and geographer Ahmad al-Yaqubi in the 9th century, begins, "It is related by the wise men of India that when Husiya, the daughter of Balhait, was queen . . . ," and goes on to tell how the game of chess was invented. The queen was so delighted with the game that she told the inventor, "Ask what you will." The inventor asked for one grain of wheat on the first square of the chessboard, two grains on the second, four grains on the third, and so on, so that each square contained twice the number of grains as on the square before. (There are 64 squares on a chessboard.)

   a. How many grains are needed for each of the following?
      i. For the 8th square. 128
      ii. For the 64th square. more than 9×10^{18}
      iii. For the first row. 255
      iv. To fill the board. more than 1.8×10^{19}

   b. In sigma notation, write the series you used to fill the board. $\sum_{n=1}^{64} 2^{n-1}$

12. As a contest winner, you are given the choice of two prizes. The first choice awards $1,000 the first hour, $2,000 the second hour, $3,000 the third hour, and so on. For one entire year, you will be given $1,000 more each hour than you were given during the previous hour. The second choice awards 1¢ the first week, 2¢ the second week, 4¢ the third week, and so on. For one entire year, you will be given double the amount you received during the previous week. Which of the two prizes will be more profitable, and by how much? ⓗ The second choice is more profitable by approximately $45 trillion.

13. Consider the series
$$\sum_{n=1}^{8} \frac{1}{n} = \frac{1}{1} + \frac{1}{2} + \frac{1}{3} + \frac{1}{4} + \cdots + \frac{1}{8}$$ ⓐ

   a. Is this series arithmetic, geometric, or neither? neither
   b. Find the sum of this series. $S_8 \approx 2.717857$

Review

14. The 32 members of the Greeley High chess team are going to have a tournament. They need to decide whether to have a round-robin tournament or an elimination tournament.

   a. If the tournament is round-robin, how many games need to be scheduled? ⓗ 496 games
   b. If it is an elimination tournament, how many games need to be scheduled? 31 games

15. What monthly payment is required to pay off an $80,000 mortgage at 8.9% interest in 30 years? ⓐ $637.95

16. The Magic Garden Seed Catalog advertises a bean with unlimited growth. It guarantees that with proper watering, the bean will grow 6 in. the first week and the height increase each subsequent week will be three-fourths of the previous week's height increase. "Pretty soon," the catalog claims, "Your beanstalk will touch the clouds!" Is this misleading advertising? Yes. The long-run height is only 24 in.

17. Write the polynomial equation of least degree that has integer coefficients and zeros $-3 + 2i$ and $\frac{2}{3}$. ⓐ $f(x) = 3x^3 + 16x^2 + 27x - 26$

CHAPTER 4 REVIEW

Exponential functions provide explicit and continuous equations to model geometric sequences. They can be used to model the growth of populations, the decay of radioactive substances, and other phenomena. The general form of an exponential function is $y = ab^x$, where a is the initial amount and b is the base that represents the rate of growth or decay. Because the exponent can take on all real number values, including negative numbers and fractions, it is important that you understand the meaning of these exponents. You also used the **point-ratio form** of this equation, $y = y_1 \cdot b_1^{x-x_1}$.

Until you read this chapter, you had no way to solve an exponential equation, other than guess-and-check. The **logarithmic function** is defined so that $\log_b y = x$ means that $y = b^x$. This means that exponential functions and logarithmic functions are **inverse** functions. The inverse of a function is the relation you get when all values of the independent variable are exchanged with the values of the dependent variable. The graphs of a function and its inverse are reflected across the line $y = x$.

A **series** is the indicated sum of terms of a sequence, defined recursively with the rule $S_1 = u_1$ and $S_n = S_{n-1} + u_n$ where $n \geq 2$. Series can also be defined explicitly. With an explicit formula, you can find any **partial sum** of a sequence without knowing the preceding term(s).

An explicit formula for **arithmetic series** is $S_n = \dfrac{n(u_1 + u_n)}{2}$.

Explicit formulas for **geometric series** are

$S_n = \left(\dfrac{u_1}{1-r}\right) - \left(\dfrac{u_1}{1-r}\right)r^n$ and $S_n = \dfrac{u_1(1-r^n)}{1-r}$ where $r \neq 1$.

Exercises

You will need a graphing calculator for Exercises **11** and **13**.

@ **Answers are provided for all exercises in this set.**

1. Evaluate each expression without using a calculator. Then check your work with a calculator.

   a. 4^{-2} $\dfrac{1}{16}$
   b. $(-3)^{-1}$ $-\dfrac{1}{3}$
   c. $\left(\dfrac{1}{5}\right)^{-3}$ 125
   d. $49^{1/2}$ 7
   e. $64^{-1/3}$ $\dfrac{1}{4}$
   f. $\left(\dfrac{9}{16}\right)^{3/2}$ $\dfrac{27}{64}$
   g. -7^0 -1
   h. $(3)(2)^2$ 12
   i. $(0.6^{-2})^{-1/2}$ 0.6

2. Rewrite each equation in logarithm form so that the x is no longer an exponent.

   a. $3^x = 7$
      $\log_3 7 = x$
   b. $5 = 4^x$
      $\log_4 5 = x$
   c. $10^x = 7^5$
      $\log 7^5 = x$

4a. $x = \log_{4.7} 28 \approx 2.153$

4b. $x = \pm\sqrt{\log_{4.7} 2209} \approx \pm 2.231$

4c. $x = 2.9^{1/1.25} = 2.9^{0.8} \approx 2.344$

4d. $x = 3.1^{47} \approx 1.242 \times 10^{23}$

4e. $x = \left(\dfrac{101}{7}\right)^{1/2.4} \approx 3.041$

4f. $x = \log_{1.065} 18 \approx 45.897$

Exercise 5 You may want to have students describe each step in their solutions.

5a. $x = 0.5(2432^{1/8} - 1) \approx 0.825$

5c. $x = \log_{1.56} \dfrac{734}{11.2} \approx 9.406$

5d. $x = \left(\dfrac{147}{12.1}\right)^{1/2.3} - 1 \approx 1.962$

Exercise 6 Students may use the point-ratio form to find the solution.

Exercise 6 You might also ask students to graph the equation and its inverse.

Exercise 11b Whether or not students actually use the formula $u_n = u_1 + d(n-1)$ (d is negative in this case), their reasoning will reflect the justification for the formula.

---

3. Rewrite each equation in exponential form.

   a. $\log x = 1.72$ **b.** $2\log x = 4.8$ **c.** $\log_5 x = -1.47$ **d.** $3\log_2 x = 15$
   $10^{1.72} = x$ $10^{2.4} = x$ $5^{-1.47} = x$ $2^5 = x$

4. Use the properties of exponents and logarithms to solve each equation. Confirm your answers by substituting them for x.

   a. $4.7^x = 28$ **b.** $4.7^{x^2} = 2209$ **c.** $\log_x 2.9 = 1.25$ **d.** $\log_{3.1} x = 47$

   e. $7x^{2.4} = 101$ **f.** $9000 = 500(1.065)^x$ **g.** $\log x = 3.771$ **h.** $\sqrt[5]{x^3} = 47$
                                                                                                 $x = 10^{3.771} \approx 5902$ $x = 47^{5/3} \approx 612$

5. Solve for x. Round your answers to the nearest thousandth.

   a. $\sqrt[8]{2432} = 2x + 1$ **b.** $4x^{2.7} = 456$ $x = 114^{1/2.7} \approx 5.779$ **c.** $734 = 11.2(1.56)^x$

   d. $147 = 12.1(1 + x)^{2.3}$ **e.** $2\sqrt{x - 3} + 4.5 = 16$ $x = 5.75^2 + 3 \approx 36.063$

6. Once a certain medicine is in the bloodstream, its half-life is 16 h. How long (to the nearest 0.1 h) will it be before an initial 45 cm³ of the medicine has been reduced to 8 cm³?   about 39.9 h

7. Given $f(x) = (4x - 2)^{1/3} - 1$, find:

   a. $f(2.5)$ 1   **b.** $f^{-1}(x)$ $\dfrac{(x+1)^3 + 2}{4}$ **c.** $f^{-1}(-1)$ $\dfrac{1}{2}$

8. Find the equation of an exponential curve through the points (1, 5) and (7, 32). $y = 5\left(\dfrac{32}{5}\right)^{(x-1)/6}$

9. Draw the inverse of $f(x)$, shown at right.

10. Consider the arithmetic sequence: 3, 7, 11, 15, $\cdots$

   a. What is the 128th term? $u_{128} = 511$

   b. Which term has the value 159? $u_{40} = 159$

   c. Find u_{20}. $u_{20} = 79$

   d. Find S_{20}. $S_{20} = 820$

11. Consider the series: $125.3 + 118.5 + 111.7 + 104.9 + \cdots$

   a. Find S_{67}. ≈ -6639.7

   b. Write an expression for S_{67} using sigma notation. $S_{67} = \sum\limits_{n=1}^{67}[125.3 - 6.8(n-1)]$

12. Consider the geometric sequence: 100, 84, 70, 56, $\cdots$

   a. Which term is the first one smaller than 20? $u_{11} \approx 17.4910$

   b. Find the sum of all the terms that are greater than 20. $S_{10} \approx 515.687$

   c. Find the value of $\sum\limits_{n=1}^{20} u_n$. $S_{20} \approx 605.881$

   d. What happens to S_n as n gets very large? The partial sum approaches 625.

13. A new incentive plan for the Talk Alot long-distance phone company varies the cost of a call according to the formula $cost = a + b \log t$, where t represents time in minutes. When calling long distance, the cost for the first minute is $0.50. The cost for 15 min is $3.44.

    a. Find the a-value in the equation. $a = 0.50$
    b. Find the b-value in the equation. $b = \dfrac{2.94}{\log 15} \approx 2.4998$
    c. What is the x-intercept of the graph of the equation? What is the real-world meaning of the x-intercept?
    d. Use your equation to predict the cost of a 30-minute call. $4.19
    e. If you decide you can afford only to make a $2 call, how long can you talk? about 4 min

14. Your head circumference gets larger as you grow. Most of the growth comes in the first few years of life, and there is very little additional growth after you reach adolescence. The estimated percentage of adult size for your head is given by the formula $y = 100 - 80(0.75)^x$, where x is your age in years and y is the percentage of the average adult size.

    a. Graph this function.
    b. What are the reasonable domain and range of this function? domain: $0 \le x \le 120$; range: $20 \le y \le 100$
    c. Describe the transformations of the graph of $y = (0.75)^x$ that produce the graph in 14a. Vertically dilate by a factor of 80; reflect across the x-axis; vertically shift by 100.
    d. A 2-year-old child's head is what percentage of the average adult-size head? 55%
    e. About how old is a person if his or her head circumference is 75% that of an average adult? about 4 yr old

15. **APPLICATION** A "learning curve" describes the rate at which a task can be learned. Suppose the equation

    $$t = -144 \log\left(1 - \dfrac{N}{90}\right)$$

    predicts the time t (in number of short daily sessions) it will take to achieve a goal of typing N words per minute (wpm).

    a. Using this equation, how long should it take someone to learn to type 40 wpm? approximately 37 sessions
    b. If the typical person had 47 lessons, then what speed would you expect him or her to have achieved? approximately 48 wpm
    c. Interpret the shape of the graph as it relates to learning time. What domain is realistic for this problem?

16. **APPLICATION** All humans start as a single cell. This cell splits into two cells, then each of those two cells splits into two cells, and so on.

    a. Write a recursive formula for cell division starting with a single cell. $u_0 = 1, u_n = (u_{n-1}) \cdot 2, n \ge 1$
    b. Write an explicit formula for cell division. $y = 2x$
    c. Sketch a graph to model the formulas in 16a and b.
    d. Describe some of the features of the graph.
    e. After how many divisions were there more than 1 million cells? after 20 cell divisions
    f. If there are about 1 billion cells after 30 divisions, after how many divisions were there about 500 million cells? after 29 divisions

13c. $\log x = \dfrac{-0.50}{2.4998} \approx -0.2; x \approx 10^{-0.2} \approx 0.63$. The real-world meaning of the x-intercept is that the first 0.63 min of calling is free.

14a.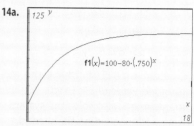

Exercise 14e ALERT Students might use the equation $y = 0.75$ instead of $y = 75$.

15c. Sample answer: It takes much longer to improve your typing speed as you reach higher levels. 60 wpm is a good typing speed, and very few people type more than 90 wpm, so $0 \le x \le 90$ is a reasonable domain.

16c.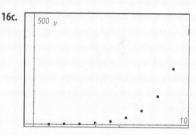

16d. Answers will vary but can include curving upward, increasing, increasing at an increasing rate, discrete.

Take Another Look

1. For substitution, students will probably first solve for a: $a = \dfrac{40}{b^4}$ and $a = \dfrac{4.7}{b^7}$; $\dfrac{40}{b^4} = \dfrac{4.7}{b^7}$; $\dfrac{40}{4.7} = \dfrac{b^4}{b^7}$; $8.51 = b^{-3}$; $(8.51)^{-1/3} = (b^{-3})^{-1/3}$; $0.49 \approx b$; $a \approx \dfrac{40}{0.49^4}$; $a \approx 693.87$; $y \approx 693.87(0.49)^x$, ($a \approx 695.04$; $y \approx 695.04(0.49)^x$ if the unrounded value of b is used). Students might also solve the system by elimination, dividing the second equation by the first to get $\dfrac{4.7}{40} = b^3$; $b = \sqrt[3]{\dfrac{4.7}{40}} \approx 0.4898$.

2. If $|r| < 1$, then r^n approaches 0 and $\dfrac{u_1(1 - r^n)}{1 - r}$ approaches $\dfrac{u_1(1 - 0)}{1 - r}$, or $\dfrac{u_1}{(1 - r)}$. This behavior supports the explicit formula for the infinite sum of a convergent geometric sequence, $S = \dfrac{u_1}{(1 - r)}$. If $r > 1$, then r^n gets larger without bound, and $1 - r$ will be a fixed negative number; therefore $\dfrac{u_1(1 - r^n)}{1 - r}$ also gets larger without bound. If $r < -1$, the expression r^n alternates sign, getting farther from 0, so $\dfrac{u_1(1 - r^n)}{1 - r}$ also gets farther from 0.

Assessing the Chapter

The Kendall Hunt Cohesive Assessment System (CAS) provides a powerful tool with which to manage assessment content, create and assign homework and tests, and disseminate assessment data.

You might test the chapter in two separate phases. Allow students to use their notebooks on a test of applications, including at least one constructive assessment item, and then turn in that assessment and take a short test on solving equations using the properties. You might also want to do the individual assessment in two parts: one part without a calculator, including items such as Exercises 1, 2, and 6, and the other part with a calculator.

Take Another Look

1. You have learned to use the point-ratio form to find an exponential curve that fits two data points. You could also use the general exponential equation $y = ab^x$ and your knowledge of solving systems of equations to find an appropriate exponential curve. For example, you can use the general form to write two equations for the exponential curve through $(4, 40)$ and $(7, 4.7)$:

   $40 = ab^4$

   $4.7 = ab^7$

   Which constant will be easier to solve for in each equation, a or b? Solve each equation for the constant you have chosen. Use substitution and the properties you have learned in this chapter to solve this system of equations. Substitute the values you find for a and b into the general exponential form to write a general equation for this function.

   Find an exercise or example in this chapter that you solved using the point-ratio form, and solve it again using this new method. Did you find this method easier or more difficult to use than the point-ratio form? Are there situations in which one method might be preferable to the other?

2. The explicit formula for a partial sum of a geometric series is $S_n = \dfrac{u_1(1 - r^n)}{1 - r}$. To find the sum of an infinite geometric series, you can imagine substituting ∞ for n. Explain what happens to the expression $\dfrac{u_1(1 - r^n)}{1 - r}$ when you do this substitution.

Assessing What You've Learned

 GIVE A PRESENTATION Give a presentation about how to fit an exponential curve to data or one of the Take Another Look activities. Prepare a poster or visual aid to explain your topic. Work with a group, or give a presentation on your own.

 ORGANIZE YOUR NOTEBOOK Review your notebook to be sure it's complete and well organized. Make sure your notes include all of the properties of exponents, including the meanings of negative and fractional exponents. Write a one-page chapter summary based on your notes.

 WRITE TEST ITEMS Write a few test questions that explore series. You may use sequences that are arithmetic, geometric, or perhaps neither. You may want to include problems that use sigma notation. Be sure to include detailed solution methods and answers.

Facilitating Self-Assessment

Presentations given on topics in this chapter could include desmos demonstrations of an exponential function and its inverse. You might split up the topics covered in this chapter. Each group could do a presentation on a different topic for review. The groups could also be responsible for writing review questions about their topic.

Items from this chapter suitable for student portfolios include Lesson 4.1, Exercise 13; Lesson 4.2, Exercises 7 and 9; Lesson 4.3, Exercises 6 and 7; Lesson 4.4, Exercise 7; Lesson 4.5, Performance Task; Lesson 4.6, Exercise 8; Lesson 4.7, Performance Task; and Lesson 4.8, Exercise 11.

CHAPTER 5
Quadratic Functions and Relations

Overview

In this chapter students learn to model problems with polynomial functions of degree 2, quadratic functions and relations.

The study of quadratic functions (polynomials of degree 2) in **Lesson 5.1** includes the relationships among the general form, the factored form, and the vertex form. In **Lesson 5.2**, students explore projectile motion and use completing the square as one way to convert the general form of a quadratic equation to vertex form to solve problems involving extreme values. Students derive the quadratic formula by completing the square and apply it to real-world problems in **Lesson 5.3**. Complex numbers are introduced in **Lesson 5.4** by including nonreal solutions to polynomial equations. Students also learn to add, subtract, and multiply complex and imaginary numbers. In **Lesson 5.5**, students use the discriminant to determine the nature of the roots of a quadratic equation. They also look at solving equations of degree 4 that can be represented as a quadratic using substitution. In **Lesson 5.6**, students learn to solve equations with radicals and investigate extraneous roots that might be introduced while solving the equation. In **Lesson 5.7**, students review the distance formula as they use it to solve problems involving minimum distances or loci of points, especially in finding the equations for circles. In **Lesson 5.8**, students see the relationship between the locus-of-points definition and focus-directrix definition of a parabola.

The **Performance Task** highlighted in this chapter is in Lesson 5.5, where students explore whether the properties of the sum and product of roots of quadratic equations hold for roots that are not real numbers.

The Mathematics

Polynomials have the form $a_n x^n + a_{n-1} x^{n-1} + \cdots + a_1 x + a_0$, where n is a positive integer (the degree of the polynomial), $a_n \neq 0$, and the coefficients a_j are numbers. In most of this chapter, the coefficients are real numbers. So far in this course, students' work with polynomials has focused on linear polynomials—polynomials of degree 1, which have the general form $a_1 x + a_0$ (or $bx + a$). In this chapter students will study polynomials of degree 2, or quadratic polynomials. In the general form of a quadratic polynomial, n is 2 and $n - 1$ is 1, so $ax^2 + bx + c$.

Polynomials as Functions

Polynomials can be considered either as mathematical expressions or as functions for modeling real-world situations. Like linear functions, polynomial functions can be graphed. Graphs of second-degree polynomial functions, or quadratics, have the shape of parabolas. An input number for a function is a zero of that function if the corresponding output number is 0. For the function's graph, these zeros designate the x-intercepts. If the function is set equal to 0, the solutions to the equation (also called the roots of the equation) are the same numbers as the zeros of the function.

The distinction between zeros, roots, solutions, and x-intercepts is an important one. While they all refer to the same numbers or points, the *graph* has x-intercepts—points where it crosses the x-axis, the *function* has zeros—the x-values the function takes to zero, or the x-coordinates of the x-intercepts, and the equation has roots or solutions, which are the zeros of the function.

Students saw in earlier chapters that quadratic equations of the form $y = ax^2 + bx + c$ represent parabolas with vertical axes. Vertical parabolas give an important piece of information: When the vertical parabola opens up, the vertex is the lowest point on the graph, or the minimum. When the vertical parabola opens down, the vertex is the highest point on the graph, or the maximum. In this chapter, students see that parabolas with horizontal axes have equations of the form $x = ay^2 + by + c$. These horizontal parabolas have no maximum or minimum because they have no limit on how high or how low they can go.

Forms of Quadratics

Several lessons of this chapter focus on quadratic functions. Quadratics have three major forms, each of which has advantages. The expression $ax^2 + bx + c$ is the general form of a polynomial, which makes the y-intercept and the scale factor easy to identify. The factored form of the same polynomial is $a(x - r_1)(x - r_2)$, which readily shows a, r_1 and r_2. a represents the vertical dilation of the parabola given by the function $y = x^2$, and r_1 and r_2 are the zeros of the quadratic function, or roots of the quadratic equation. The vertex form of the polynomial is $a(x - h)^2 + k$, which shows the vertex of a parabola at point (h, k) and the vertical scale factor a. The vertex form gives the best information about how the function's graph is transformed from that of $y = x^2$.

Changing Forms and Solving Equations

To change from vertex form to general form requires squaring the binomial $(x - h)$ and combining like terms. Similarly, to get from the factored form to the general form, you multiply the binomials $(x - r_1)$ and $(x - r_2)$. Students will understand squaring or multiplying binomials more deeply through rectangle diagrams than from memorizing the FOIL (first, outside, inside, last) method. FOIL only works with a binomial times a binomial, but the diagram works for all polynomial multiplication. The diagram is also useful in factoring polynomials.

The primary reason to change quadratic equations from general form is to solve them. In some cases, changing to factored form is easy, especially with rectangle diagrams. Then the roots are apparent. In other cases, completing the square can yield vertex form, which can be solved symbolically. If this process of completing the square is carried out for the general quadratic equation $ax^2 + bx + c = 0$, you derive the quadratic formula

$$x = \frac{-b \pm \sqrt{b^2 - 4ac}}{2a}$$

You can use the quadratic formula to solve all quadratic equations.

Discriminant

When solving a quadratic equation with the quadratic formula, the number under the radical, the discriminant, gives information about the nature of the roots of the equation. If the number under the radical is positive, there will be two real solutions (the square root of a positive number is a real number). If the number under the radical is zero, there will be one real solution (the square root of zero is zero so you are adding and subtracting zero). If the number under the radical is negative, there will be two complex solutions (the square root of a negative number is a nonreal number). Help students to think in terms of why and not just try to memorize the formulas.

Complex Numbers

The quadratic formula includes a square root. If the value inside that square root is negative, then no real number is a root of the equation. These imaginary numbers, when combined with real numbers, form a set of numbers (the complex numbers) that behave like the real numbers. Complex numbers can be written as $a + bi$, where a and b are real numbers and i is shorthand for $\sqrt{-1}$. These numbers can be useful in solving real-world problems. While some people use the terms imaginary number and complex number interchangeably, they have a distinct difference.

Using This Chapter

Much of this chapter may seem like a review of content that was covered in Algebra I, but many of the lessons take the content a step further. Be judicious when deciding what content lessons to skip. If time is a concern, perhaps a quick review at the beginning of a lesson can assess prior knowledge and the lesson time can be spent on the extension from Algebra I. You might also choose to use the Shortened or One Step modifications of the investigations.

Common Core State Standards (CCSS)

The study of quadratic functions in Algebra II adds another level to the understanding of the number systems. In middle grades, students added rational and irrational numbers (6.NS, 7.NS, 8.NS) and worked with radicals and integer exponents (8.EE). In Algebra II, students extended their understanding again to include rational exponents (N-RN) and imaginary numbers (N-CN).

The study of quadratic functions in Algebra I introduced the *existence* of the complex number system (A.REI.4b). Algebra II focuses on a formal study of that system, including arithmetic within the system (N.CN.1, 2, and 7).

The processes of factoring are also introduced early in CCSS, starting as early as Grade 4 students study factors and multiples of whole numbers (4.OA.4) and continuing in the middle grades (6.NS). In the middle grades they extend factoring and multiplying to include expressions (6.EE.3, 7.EE.1). Students are ready in high school to approach the procedures of factoring, completing the square and using the quadratic formula in the context of what you want to show in a quadratic function—zeros, extreme values, and symmetry of the graph and interpret these in terms of a context (F-IF.8.a, F.IF.7c). Students learn to not only graph an equation by plotting data points but to also anticipate the graph of a quadratic function by interpreting various forms of quadratic expressions and through the discriminant. In other words, they become more mindful of why they are manipulating an equation algebraically or graphing a transformation of a family of functions (A.APR.3).

In Algebra II, students work closely with the expressions that define the functions, and expand and refine their abilities to model situations and to solve equations, including solving quadratic equations over the set of complex numbers. They are able to understand and explain the reasons behind the strategies they use to solve equations (A.REI.2). They can analyze and build more sophisticated functions using different representations (F.IF.7c, F.IF.8).

Additional Resources For complete references to additional resources for this chapter and others see your digital resources.

Refreshing Your Skills To review the algebra skills needed for this chapter see your digital resources.

Polynomial Expressions
The Distance Formula

Projects See your digital resources for performance-based problems and challenges.

Calculator Program for the Quadratic Formula
The Mandelbrot Set

Materials

- motion sensors or video cameras
- empty coffee cans or poster tubes
- long tables
- centimeter grid paper
- patty paper or tracing paper
- rulers

CHAPTER 5
Quadratic Functions and Relations

Quadratic functions can describe a variety of things in our world. Nature has carved out many such curves, like those in Utah's Arches National Park. Parabolic reflectors use the characteristics of a parabola to collect or project energy. And architect Frank Gehry incorporated quadratic-like characteristics in the design of the architectural landmark and acoustically sophisticated Walt Disney Concert Hall.

OBJECTIVES

In this chapter you will

- study quadratic functions in general form, vertex form, and factored form
- find roots of a quadratic equation from a graph, by factoring, and by using the quadratic formula
- define complex numbers and operations with them
- use the distance formula to find the distance between two points on a plane and to solve distance and rate problems
- derive the equation of a circle using the Pythagorean Theorem; complete the square to find the center and radius of a circle given by an equation
- determine the location of the parabola's focus, directrix, and vertex from its equation and graph the parabola
- find the equation of a parabola from its graph

CHAPTER 5

Objectives

- Understand the correspondence among the zeros of a polynomial, the *x*-intercepts of its graph, and the roots of an equation
- Find roots of polynomial equations from factored form, using the zero-product property
- Change quadratic polynomials from general form to vertex form by completing the square
- Relate the vertex form of a quadratic equation to its roots, its parent function, its relative extreme points, and the quadratic formula
- Define and use complex and imaginary numbers and conjugate pairs
- Learn to add, subtract, and multiply complex and imaginary numbers
- Apply the distance formula
- Define the circle as a locus of points having fixed distance properties relative to a center
- Define the parabola as a locus of points the same distances from a focus and a directrix
- Derive equations a circle and a parabola.

Quadratic like functions can describe a variety of things in our world. Just look around at natural occurring phenomena, such as the shape of a mountain, a rainbow, or tree limbs on a canopy road. Nature has carved out many such curves, like the arches in Utah's Arches National Park. Delicate Arch is a 65-foot-tall freestanding natural arch. The original sandstone fin was gradually worn away by weathering and erosion, leaving the arch. Artists and architects use curves to create aesthetically pleasing or culturally significant pieces of art or engineering marvels, like the Walt Disney Concert Hall in Los Angeles. This concert hall is one of the most acoustically sophisticated concert halls in the world. Science and technology also makes use of properties of quadratic functions. A parabolic reflector (or dish or mirror) is a reflective surface used to collect or project energy such as light, sound, or radio waves. They collect energy from a distant source (for example sound waves or sunlight) and bring it to a common focal point. Since the principles of reflection are reversible, parabolic reflectors can also be used to project energy of a source at its focus outward in a parallel beam, used in devices such as car headlights. **ASK** "What other representations of mathematical objects can you think of in nature, architecture, engineering or art?" **ASK** "Why do you think a design includes curved shapes?" [Sample answer: They convey a sense of growth, of infinite expansion, or of falling apart.]

LESSON 5.1

Equivalent Quadratic Forms

In this lesson, students explore the three forms of a quadratic equation and what information can be gleaned from each form. They learn to write equations in vertex form from graphs of quadratic functions and they learn how to solve equations written in factored form.

COMMON CORE STATE STANDARDS

APPLIED	DEVELOPED	INTRODUCED
A.SSE.1a	F.IF.7a	F.IF.4
A.SSE.2	F.IF.7c	
A.SSE.3a	F.IF.8a	
A.CED.1	F.IF.9	
A.CED.2	F.BF.1	

Objectives

- Understand the correspondence between the zeros of a polynomial function and the roots of an equation
- Use the zero-product property to find the roots of equations
- Know the relationships among the general form, the factored form, and the vertex form of a quadratic equation and the key information derived from each
- Relate the vertex form of a quadratic equation to the parent function $y = x^2$
- Graph quadratic functions and show intercepts, maxima, and minima

Vocabulary

quadratic functions
general form
factored form
vertex form
maximum
minimum
zeros
zero-product property
roots

Materials

- motion sensors or video cameras
- empty coffee cans or poster tubes
- long tables
- Calculator Notes: Rolling Along; Zero Finding

"I'm very well acquainted, too, with matters mathematical. I understand equations both the simple and quadratical."

WILLIAM S. GILBERT AND ARTHUR SULLIVAN

The Fountains of Bellagio in Las Vegas, Nevada put on a choreographically complex water, music and light show daily. The display of more than 1000 fountains spans more than 1,000 feet with water soaring as high as 460 feet into the air. The arc formed by the spouting water can be described with a quadratic equation.

Second degree polynomial functions are also called **quadratic functions**. The **general form** of a quadratic function is $y = ax^2 + bx + c$. In this lesson you will work with two additional, equivalent forms of quadratic functions.

Every quadratic function can be considered as a transformation of the graph of the parent function $y = x^2$. A quadratic function in the form $\frac{y-k}{b} = \left(\frac{x-h}{a}\right)^2$ or $y = b\left(\frac{x-h}{a}\right)^2 + k$ identifies the location of the vertex, (h, k), and the horizontal and vertical scale factors, a and b. The vertex of a parabola represents either the **maximum** or **minimum** point of the quadratic function.

EXAMPLE A Find the horizontal and vertical scale factors of the parabola at right with vertex $(4, -2)$ and write its equation. Then rewrite the equation with a single scale factor.

Solution If you consider the point $(7, 4)$ to be the image of the point $(1, 1)$ on the graph of $y = x^2$, the horizontal scale factor is 3 and the vertical scale factor is 6. So the quadratic function is

$$\frac{y+2}{6} = \left(\frac{x-4}{3}\right)^2, \text{ or } y = 6\left(\frac{x-4}{3}\right)^2 - 2.$$

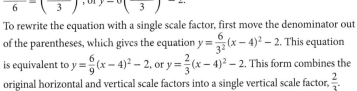

To rewrite the equation with a single scale factor, first move the denominator out of the parentheses, which gives the equation $y = \frac{6}{3^2}(x-4)^2 - 2$. This equation is equivalent to $y = \frac{6}{9}(x-4)^2 - 2$, or $y = \frac{2}{3}(x-4)^2 - 2$. This form combines the original horizontal and vertical scale factors into a single vertical scale factor, $\frac{2}{3}$.

ELL
Ask students for possible definitions of *form*. Provide verbal definitions and symbolic examples of *general*, *vertex*, and *factored forms*. Use graphs to help students understand *vertex* and *factored form*. Discuss the terms *zeros*, *roots*, and *solutions*.

Support
When converting from vertex or factored form to general form, encourage students to use rectangle diagrams to perform the multiplication of the binomials until they become more comfortable with the symbolic manipulation. If necessary, you can use the Refreshing Your Skills: Polynomial Expressions lesson.

Advanced
Keep advanced students focused on the reason for rewriting the equations: that different forms are useful for different purposes. Students might investigate whether *a* means the same thing in all three equations, as well as what relationships exist between the other variables.

In general, the combined scale factor is $\frac{b}{a^2}$, where a and b are the horizontal and vertical scale factors. The coefficient a is often used to represent the combined scale factor. This new form, $y = a(x - h)^2 + k$, is called the **vertex form** of a quadratic function because it identifies the vertex, (h, k), and a single vertical scale factor, a. If you know the vertex of a parabola and one other point, then you can write the quadratic function in vertex form.

Now consider these parabolas. The x-intercepts are marked.

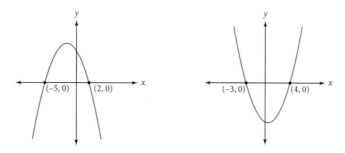

The y-coordinate of any point along the x-axis is 0, so the y-coordinate is 0 at each x-intercept. For this reason, the x-intercepts of the graph of a function are called the **zeros** of the function. You will use this information and the **zero-product property** to find the zeros of a function without graphing.

> **Zero-Product Property**
> For all real numbers a and b, if $ab = 0$, then $a = 0$, or $b = 0$, or $a = 0$ and $b = 0$.

To understand the zero-product property, think of numbers whose product is zero. Whatever numbers you think of will have this characteristic: *At least one of the factors must be zero.* Before moving on, think about numbers that satisfy each equation below.

$\underline{\ ?\ } \cdot 16.2 = 0 \quad 0$

$3(\underline{\ ?\ } - 4)(\underline{\ ?\ } - 9) = 0 \quad 4 \text{ or } 9$

EXAMPLE B | Find the zeros of the function $y = -1.4(x - 5.6)(x + 3.1)$.

Solution | The zeros will be the x-values that make y equal 0. First, set the function equal to zero.

$0 = -1.4(x - 5.6)(x + 3.1)$

Because the product of three factors equals zero, the zero-product property tells you that at least one of the factors must equal zero.

| $-1.4 = 0$ | or | $x - 5.6 = 0$ | or | $x + 3.1 = 0$ |
| not possible | | $x = 5.6$ | | $x = -3.1$ |

So the solutions, or **roots**, of the equation $0 = -1.4(x - 5.6)(x + 3.1)$ are $x = 5.6$ or $x = -3.1$. That means the zeros of the function $y = -1.4(x - 5.6)(x + 3.1)$ are $x = 5.6$ and $x = -3.1$.

LESSON 5.1 Equivalent Quadratic Forms

Launch
Project the two graphs on top of page 261 before Example B.

What do you know about the points that are labeled on these graphs? Be specific.

They are the x-intercepts, the y-coordinate is 0 for each (and all points on the x-axis), and they are the solutions to the function when y is zero.

Investigate

Example A
Project the example and work through it as a whole class, with student input. You may want to review the process of finding scale factors. **ASK** "Why did the book choose the point (7, 4) to be the image of the point (1, 1)? Will other points work?" **SMP 3, 6, 7** [You can consider the point (7, 4) as the image of any nonvertex point on the graph of $y = x^2$.] In the graph of $y = x^2$, the vertex is at (0, 0) and the point (1, 1) is 1 unit right and 1 unit up from the vertex. In the graph of this parabola with vertex (4, −2), the point (7, 4) is 3 units right and 6 units up from the vertex. So the graph is dilated horizontally by a factor of $\frac{3}{1}$ and vertically by a factor of $\frac{6}{1}$. You can also consider the point (7, 4) as the image of the point (3, 9), since both points are 3 units to the right of the vertex of the parabola. Then there's only a *vertical* scale factor of $\frac{6}{9}$, or $\frac{2}{3}$. The resulting equation is $y + 2 = \frac{2}{3}\left(\frac{x-4}{1}\right)^2$, which is equivalent to the equation in the book.

ASK 'What are two nonzero numbers whose product is 0?' Students may suggest pairs of opposites or reciprocals. Let them critique one another's ideas until they reach consensus that there are no such pairs—that the only way the product can be 0 is if at least one of the numbers is 0. Introduce the language *zeros of the function* **SMP 3, 6**.

Example B
Project Example B and let students work on it individually or in pairs. As students present their solutions and explain their work, look for students who have solved the equation using the Zero-Product Property, as well as students who might have graphed the

Modifying the Investigation

Whole Class Have two students demonstrate and collect data for the whole class, either with the motion sensor or the video camera. Have each student complete the graph for Step 4. Discuss remaining questions of Step 4 and Steps 5 through 8 as students continue their work individually.

Shortened Use the sample data. Discuss Step 8 as a class. You could collect one set of data as a class and share the data by linking to one calculator in each group, or use the Rolling Along Sample Data worksheet as well.

One Step Have students gather data as in the investigation, or use the Rolling Along Sample Data worksheet. Ask them to find an equation for the parabola, which they can do using transformations, and to use a rectangle diagram to rewrite the quadratic equation as a product of binomials.

equation. You might ask students to check their answer by graphing the equation. **SMP 1, 3, 5, 6**

ALERT Students may mistakenly believe that they need to expand the quadratic in order to enter it as a calculator function.

ASK "How would you find the coordinates of the vertex of this parabola?" [Evaluate the function at the mean of −3.1 and 5.6 to find the *x*-coordinate, substitute for *x* in the equations to find the *y*-coordinate.]

Example C

Project Example C and have students work on it in groups. You might do each part separately. For example, have students present their solutions to part a before showing part b.

Guiding the Investigation

If you do not have access to a CBR, then lay a meter stick on the table and record the rolling can with a video camera. Download the video to a computer and record the position of the leading edge of the can in every frame of the video by reading the position on the meter stick. You may want to record only every other frame depending on the frequency of the frames.

There is also a desmos simulation for the investigation in your ebook.

Step 1 Students must make sense of the problem and practice for precision of data collection. **SMP 1, 6**

Step 2 If 6 seconds is not long enough, students can adjust the time. The student who is rolling the can should wait until the motion sensor starts clicking before rolling the can. **SMP 5, 6**

Step 3 Sample data can be found in the Investigation worksheet Rolling Along With Sample Data.

ALERT Students may unnecessarily doubt their results when they see negative values for the first and last numbers. **SMP 1, 2**

You can check your work by graphing the equation. You should find that the *x*-intercepts of the graph of $y = -1.4(x - 5.6)(x + 3.1)$ are 5.6 and −3.1.

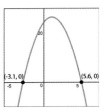

If you know the *x*-intercepts of a parabola, then you can write the quadratic function in **factored form**, $y = a(x - r_1)(x - r_2)$. This form identifies the locations of the *x*-intercepts, r_1 and r_2, and a vertical scale factor, a.

EXAMPLE C Consider the parabola at right.

a. Write an equation of the parabola in vertex form.
b. Write an equation of the parabola in factored form.
c. Show that both equations are equivalent by converting them to general form.

Solution The vertex is (1, −4). From the graph you can see that this vertex is at the minimum. If you consider the point (2, −3) to be the image of the point (1, 1) on the graph of $y = x^2$, then the vertical and horizontal scale factors are both 1. So the single vertical scale factor is $a = \dfrac{1}{1^2} = 1$.

a. The vertex form is $y = 1(x - 1)^2 + (-4)$, or $y = (x - 1)^2 - 4$.

b. The *x*-intercepts are −1 and 3, so the factored form is $y = a(x + 1)(x - 3)$. To verify that the scale factor you found earlier is correct, you can solve for it analytically. Substitute the coordinates of another point on the parabola such as (2, −3) into the equation for *x* and *y*.

$$-3 = a(2 + 1)(2 - 3)$$
$$a = \dfrac{-3}{(3)(-1)} = 1$$

The value of a is 1, so the factored form is $y = 1(x + 1)(x - 3) = (x + 1)(x - 3)$.

c. To convert to general form, multiply the binomials and then combine like terms. The use of rectangle diagrams may help you multiply the binomials.

$y = (x - 1)^2 - 4$ $y = (x + 1)(x - 3)$
$y = (x - 1)(x - 1) - 4$

	x	-1
x	x^2	$-x$
-1	$-x$	1

	x	1
x	x^2	x
-3	$-3x$	-3

$y = (x^2 - x - x + 1) - 4$ $y = x^2 + x - 3x - 3$
$y = x^2 - 2x - 3$ $y = x^2 - 2x - 3$

The vertex form and the factored form are equivalent to each other because they are both equivalent to the same general form.

Conceptual/Procedural

Conceptual In this lesson, students conceptualize the scale factor of a parabola in Example A. In the Investigation, students collect data and use that data to model the three forms of a quadratic equation. They then have to explain when they would use each form to find a model to fit parabolic data.

Procedural In Example B, students apply the zero-product property to an equation already in factored form. In Example C, students write equations from a graph and convert them both to general form to illustrate equivalence.

You now know three different forms of a quadratic function.

> **Three Forms of a Quadratic Function**
> General form $\quad y = ax^2 + bx + c$
> Vertex form $\quad y = a(x - h)^2 + k$
> Factored form $\quad y = a(x - r_1)(x - r_2)$

The investigation will give you practice in using the three forms with real data. You'll find that the form you use guides which features of the data you focus on. Conversely, if you know only a few features of the data, you may need to focus on a particular form of the function.

INVESTIGATION
Rolling Along

YOU WILL NEED
- a motion sensor (or a video camera)
- an empty coffee can
- a long table

Procedure Note
Prop up one end of the table slightly. Place the motion sensor at the low end of the table and aim it toward the high end. With tape or chalk, mark a starting line 0.5 m from the sensor on the table.

The investigation can also be done by laying a meter stick on the table and recording the rolling can with a video camera. Download the video to a computer and record the position of the leading edge of the can in every frame of the video by reading the position on the meter stick.

Step 1 Practice rolling the can up the table directly in front of the motion sensor. Start the can behind the starting line. Give the can a gentle push so that it rolls up the table on its own momentum, stops near the end of the table, and then rolls back. Stop the can after it crosses the line and before it hits the motion sensor.

Step 2 Set up your calculator to collect data for 6 seconds. When the sensor begins, roll the can up the table.

Step 3 The data collected by the sensor will have the form (*time, distance*). Adjust for the position of the starting line by subtracting 0.5 from each value in the distance list.

Step 4 Define your variables. Use your data to graph distance as a function of time. What shape is the graph of the data points? What type of function would model the data?

Step 5 Mark the vertex and another point on your graph. Approximate the coordinates of these points and use them to write the equation of a quadratic model in vertex form. Is the vertex a maximum or minimum point of the quadratic function?

Step 6 From your data, find the distance of the can at 1, 3, and 5 seconds. Use these three data points to find a quadratic model in general form.

Step 7 Mark the x-intercepts on your graph. Approximate the values of these x-intercepts. Use the zeros and the value of a from Step 5 to find a quadratic model in factored form.

Step 8 Verify by graphing that the three equations in Steps 5, 6, and 7 are equivalent, or nearly so. Write a few sentences explaining when you would use each of the three forms to find a quadratic model to fit parabolic data.

Step 5 vertex: (3.2, 4.257); point: (5.2, 0.897); $y = -0.84(x - 3.2)^2 + 4.257$; maximum

Step 6 (1, 0.546), (3, 4.256), and (5, 1.493); $y = -0.81x^2 + 5.09x - 3.74$

Step 7 (0.9, 0) and (5.5, 0); $y = -0.84(x - 0.9)(x - 5.5)$

Step 4 You might have students graph the data on graph paper as well as on calculators. Appropriate technology includes both paper-and-pencil and technology such as graphing calculators or computer programs. SMP 2, 5, 6

Step 4

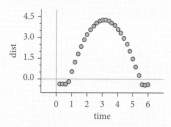

The shape of the graph is parabolic. A quadratic function would model these data.

Step 5 Students can trace the calculator graphs to approximate the vertex and an additional point. SMP 4, 5, 6

Step 6 If students do not have a good data point at 5 seconds, because the can has already stopped, they should use three points that are roughly equally spaced throughout the rolling part of the data. SMP 1, 4

Step 7 The intercepts will likely not be data points. Students should estimate these intercepts to the best of their ability. Students should adjust the values for *a* and the intercepts until they have a good fit to their data. SMP 1, 2, 4

Step 8

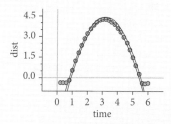

You would use the vertex form when you know the vertex and at least one other point by which to determine the scale factor. You would use the general form when you have any three points. You would use the factored form when you know the x-intercepts and at least one other point by which to determine the scale factor. SMP 1, 2, 3, 7

Summarize

Have students present their solutions and explain their work for the Investigation and the examples. Whether student responses are correct or incorrect, ask other students if they agree and why. Encourage students to use appropriate mathematical language as evidence. **SMP 1, 3, 6**

You might point out that the factored form uses r_1 and r_2 for the x-intercepts because of how the x-intercepts are related to the roots of the equation $f(x) = 0$.

ASK "In the factored form, a is a scale factor. What's being scaled?" [a represents the vertical dilation of the parabola given by the function $y = x^2$.] As needed, remind students that any dilation of a parabola is a vertical dilation (similarly, any shift of a line can be thought of as a vertical shift). In this situation, a is the vertical scale factor that captures both horizontal and vertical scale factors, the vertical over the square of the horizontal. You might also refer to Example C, where both scale factors are 1 and $a = \frac{1}{12^2}$.

ALERT Students are often confused about the difference in the use of the words "and" and "or".

If you're listing zeros of a function, x-intercepts of its graph, or the roots of the corresponding equation, you use the word *and*. If you're describing the values of x that solve an equation, you use the word *or*, because x can't equal more than one number at the same time. For example, you'd say, "The roots are 3 and −4," but "x is 3 or −4." **SMP 6**

CRITICAL QUESTION What are the three forms of a quadratic equation and what can you tell from each form of a quadratic function?

BIG IDEA **General form** is $ax^2 + bx + c$, which gives the vertical scale factor and the intercept. It is the easiest to find from a table of data.

Vertex form is $a(x - h)^2 + k$, which shows the vertex of a parabola at point (h, k) and the vertical scale factor a. The vertex form gives the best information about how the function's graph is transformed from that of $y = x^2$.

Factored form is $a(x - r_1)(x - r_2)$, which readily shows a, r_1 and r_2. a represents the vertical dilation of the parabola given by the function $y = x^2$, and r_1 and r_2 are the **zeros** of the quadratic function, or **roots** of the quadratic equation.

CRITICAL QUESTION What's the difference between zeros, roots, solutions, and x-intercepts? Explain your answers.

BIG IDEA They all refer to the same numbers or points. The *graph* has x-intercepts—points where it crosses the x-axis. The *function* has zeros—the x-values the function takes to zero, or the x-coordinates of the x-intercepts. The *equation* $f(x) = 0$ has roots or solutions, which are the zeros of the function $f(x)$.

You can find a model for data in different ways, depending on the information you have. Conversely, different forms of the same equation give you different kinds of information. Being able to convert one form into another allows you to compare equations written in different forms. In the exercises you will convert both the vertex form and the factored form into the general form. In later lessons you will learn other conversions.

5.1 Exercises

You will need your graphing calculator for Exercise **2–6** and **10–12**.

Practice Your Skills

For the exercises in this lesson, you may find it helpful to show a background grid on your calculator.

1. Identify each quadratic function as being in general form, vertex form, factored form, or none of these forms.

   a. $y = -4.9(x + 4.5)^2 + 2$ vertex form
   b. $y = -4.9(x - 4.5)(x - 0.5)$ factored form
   c. $y = -3.2(x + 4.5)^2$ @ factored form and vertex form
   d. $y = 2.5(x + 1.25)(x - 1.25) + 4$ @ none of these
   e. $y = 2x(3 + x)$ @ factored form
   f. $y = 2x^2 - 4.2x - 10$ general form

When you see an image in a different form, your attention is drawn to different features

2. Each quadratic function below is written in vertex form. What are the coordinates of each vertex? Graph each equation to check your answers.

   a. $y = (x - 2)^2 + 3$ (2, 3)
   b. $y = 0.5(x + 4)^2 - 2$ @ $(-4, -2)$
   c. $y = 4 - 2(x - 5)^2$ (5, 4)

3. Each quadratic function below is written in factored form. What are the zeros of each function? Graph each equation to check your answers.

   a. $y = (x + 1)(x - 2)$ @ -1 and 2
   b. $y = 0.5(x - 2)(x + 3)$ -3 and 2
   c. $y = -2(x - 2)(x - 5)$ 2 and 5

4. Convert each function to general form. Graph both forms to check that the equations are equivalent.

   a. $y = (x - 2)^2 + 3$
      $y = x^2 - 4x + 7$
   d. $y = (x + 1)(x - 2)$
      $y = x^2 - x - 2$

   b. $y = 0.5(x + 4)^2 - 2$
      $y = 0.5x^2 + 4x + 6$
   e. $y = 0.5(x - 2)(x + 3)$
      $y = 0.5x^2 + 0.5x - 3$

   c. $y = 4 - 2(x - 5)^2$ @
      $y = -2x^2 + 20x - 46$
   f. $y = -2(x - 2)(x - 5)$
      $y = -2x^2 + 14x - 20$

5. Find a good viewing window that shows the vertex of the parabola and points where it crosses the x-axis (if they exist). Use your knowledge of each form to help you. Answers will vary. These are practical viewing windows.

   a. $y = 0.05x^2 - 12$
      $[-20, 20]$ by $[-15, 5]$
   b. $y = -(x - 400)(x - 300)$
      $[-50, 450]$ by $[-500, 3000]$
   c. $y = -4.9x^2 + 90x + 250$
      $[-6, 25]$ by $[-50, 700]$

Reason and Apply

6. The graphs of all quadratic functions have a line of symmetry that contains the vertex and divides the parabola into mirror-image halves. Consider this table of values generated by a quadratic function.

x	y
1.5	−8
2.5	7
3.5	16
4.5	19
5.5	16
6.5	7
7.5	−8

   a. What is the line of symmetry for the graph of this quadratic function? $x = 4.5$
   b. Name the vertex of this function and determine whether it is a maximum or minimum. $(4.5, 19)$; maximum
   c. Use the table of values to write the quadratic function in vertex form. @
      $y = -3(x - 4.5)^2 + 19$

7. Write each function in general form.

   a. $y = 4 - 0.5(x + h)^2$
      $y = -0.5x^2 - hx - 0.5h^2 + 4$
   c. $y = a(x - h)^2 + k$
      $y = ax^2 - 2ahx + ah^2 + k$
   e. $y = a(x - 4)(x + 2)$
      $y = ax^2 - 2ax - 8a$

   b. $y = a(x - 4)^2$ @
      $y = ax^2 - 8ax + 16a$
   d. $y = -0.5(x + r)(x + 4)$ @
      $y = -0.5x^2 - (0.5r + 2)x - 2r$
   f. $y = a(x - r)(x - s)$ @
      $y = ax^2 - a(r + s)x + ars$

8. At right is the graph of the quadratic function that passes through $(-2.4, 0)$, $(0.8, 0)$, and $(1.2, -2.592)$.

   a. Use the x-intercepts to write the quadratic function in factored form. For now, leave the scale factor as a.
      $y = a(x + 2.4)(x - 0.8)$
   b. Substitute the coordinates of $(1.2, -2.592)$ into your function from 8a, and solve for a. Write the complete quadratic function in factored form.
      $y = -1.8(x + 2.4)(x - 0.8)$
   c. The line of symmetry for the graph of this quadratic function passes through the vertex and the point on the x-axis halfway between the two x-intercepts. What is the x-coordinate of the vertex? What is the y-coordinate?
      $x = -0.8$, $y = 4.608$
   d. Write this quadratic function in vertex form. $y = -1.8(x + 0.8)^2 + 4.608$

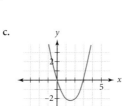

9. Write the factored form for each polynomial function. ⓗ

   a.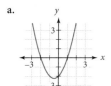
      $y = (x + 2)(x - 1)$
   b. @
      $y = -0.5(x + 2)(x - 3)$
   c.
      $y = x(x - 3)$

CRITICAL QUESTION
What is the role of a and c in the quadratic equation $ax^2 + c$? (Hint: You might graph equations with different a values on one set of axes and equations with different c values on another set of axes.) How does this compare to the slope-intercept form of a linear equation, $y = mx + b$?

BIG IDEA
The greater the absolute value of a is, the narrower the parabola; c is the y-intercept of the parabola. In the slope-intercept form of a linear equation, the greater the absolute value of m, the steeper the line; b is the y-intercept of the line.

CRITICAL QUESTION
How do you know whether the function has a maximum or a minimum?

BIG IDEA
The vertex of a parabola represents the maximum or minimum. If the parabola opens up, the vertex is the lowest point, or minimum point, on the graph. If the parabola opens down, the vertex is the highest point, or maximum point, on the graph.

Formative Assessment
To assess the objectives of the lesson, observe students working in groups and listen carefully to their presentation. As students present their solutions to the examples, Investigation, and exercises, encourage them to use appropriate and precise mathematical language. Whether student responses are correct or incorrect, ask other students if they agree and why. SMP 2, 3, 6

Responses to the Investigation and exercises will provide evidence of meeting all the objectives of the lesson. Allowing students to present their solutions to the examples will provide an opportunity to assess their prior knowledge and evidence of meeting the objectives. Their responses to the Critical Questions and their explanations of the exercises will provide additional evidence of meeting the objectives of the lesson.

Apply

Extra Example

1. Find the zeros of the function $y = -2x(x + 4)$ and rewrite in general form.
   $x = 0$ and $x = -4$, $y = -2x^2 - 8x$

2. Rewrite $y = 2(x - 2)^2 - 18$ in factored form.
   $y = 2(x + 1)(x - 5)$

Closing Question

Use $y = -6 + 2(x + 3)^2$ to answer these questions.

a. What are the coordinates of the vertex? $(-3, -6)$
b. Rewrite in general form. $y = 2x^2 + 12x + 12$
c. What is the line of symmetry of the graph? $x = -3$
d. Is the vertex a minimum or maximum? minimum

Assigning Exercises
Suggested: 1–6, 8, 9–13, 16–18
Additional Practice: 7, 14, 15

Exercises 4 Students may want to use rectangle diagrams to expand the binomials.

Exercise 8 This exercise teaches an important method for finding the vertical scale factor *a* without considering dilations. This method works for either factored form or vertex form.

ALERT Students may be confused because *a* is a constant in the equation but is being treated as a variable until its numeric value is determined.

Exercise 10 Encourage students to ask whether their function expression makes sense. Dimensional analysis might help.

10b.

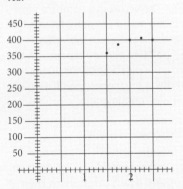

10d.

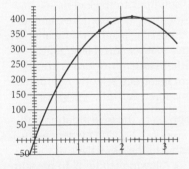

Exercise 11 In 11c, students may recall from geometry that if the perimeter is held constant, the rectangle with the largest area will be a square. If students object to the table, in which a length is less than a width, allow them to use the words *base* and *height* for the dimensions of the rectangles when they create their own tables. **SMP 6**

---

10. **APPLICATION** A local outlet store charges $1.50 for a pack of four AA batteries. On an average day, 240 packs are sold. A survey indicates that the number sold will decrease by 20 packs per day for each $0.25 increase in price.

Selling price ($)	1.50	1.75	2.00	2.25	2.50
Number sold	240	220	200	180	160
Revenue ($)	360	385	400	405	400

a. Complete the table based on the results of the survey.

b. Graph the data (selling price, revenue) and make a conjecture about what type of function will fit this data. *It appears that a parabola will fit the data.*

c. Define your variables. Write a function that describes the relationship between the revenue and the selling price. ⓗ *$x =$ represent the selling price in dollars, $y =$ the revenue in dollars; $y = -80(x - 2.25)^2 + 405$*

d. Graph your function and find the maximum revenue. What selling price provides maximum revenue? Explain. *The maximum revenue is $405 at a selling price of $2.25.*

11. **APPLICATION** Delores has 80 m of fence to surround an area where she is going to plant a vegetable garden. She wants to enclose the largest possible rectangular area.

Width (m)	5	10	15	20	25
Length (m)	35	30	25	20	15
Area (m²)	175	300	375	400	375

a. Copy and complete this table.

b. Define your variables. Write a function that describes the relationship between the area and the width of the garden. *Let x be the width in meters, y the area in square meters; $y = x(40 - x)$, or $y = -x^2 + 40x$*

c. Which width provides the largest possible area? What is that area? *A width of 20 m (and length of 20 m) maximizes the area at 400 m².*

d. Which widths result in an area of 0 m²? Explain. *0 m and 40 m. At these widths, there would be no rectangle.*

12. **APPLICATION** Photosynthesis is the process by which plants use energy from the sun, together with CO_2 (carbon dioxide) and water, to make their own food and produce oxygen. Various factors affect the rate of photosynthesis, such as light intensity, light wavelength, CO_2 concentration, and temperature. Below is a graph of how temperature relates to the rate of photosynthesis for a particular plant. (All other factors are assumed to be held constant.)

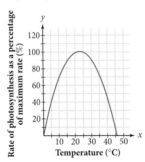

When chlorophyll (a green pigment in plants) breaks down in the winter, leaves change color. When they appear red or purple, it is because glucose trapped in the leaves produces a pigment when exposed to sunlight and cool nights.

a. Describe the general shape of the graph. What does the shape of the graph mean in the context of photosynthesis? *The graph is parabolic. As the temperature increases, the rate of photosynthesis also increases, until a maximum rate is reached; then the rate decreases.*

b. Approximate the optimum temperature for photosynthesis in this plant and the corresponding rate of photosynthesis. ⓐ *At approximately 23°C, the rate of photosynthesis is maximized at 100%.*

c. Temperature has to be kept within a certain range for photosynthesis to occur. If it gets too hot, then the enzymes in chlorophyll are killed and photosynthesis stops. If the temperature is too cold, then the enzymes stop working. At approximately what temperatures does the rate of photosynthesis fall to zero? ⓐ *0° and 46°C*

d. Write a function in at least two forms that will produce this graph. *$y = -0.19x(x - 46)$; $y = -0.19(x - 23)^2 + 100$*

266 CHAPTER 5 Quadratic Functions and Relations

Exercise 12 LANGUAGE *optimum* means "most favorable." The optimum temperature is the temperature that results in the fastest rate of photosynthesis. **ELL** Students may need assistance with the contextual vocabulary in this exercise. Consider bringing in some leaves and drawing diagrams of the process of photosynthesis. Students could research the terminology in their primary language.

13. Some expressions involving polynomials can be used to create special sets of numbers.

   a. Verify the identity $(a^2 + b^2)^2 - (a^2 - b^2)^2 = 4a^2 b^2$ by expanding and simplifying the left side until it matches the right side.

   b. The expressions $a^2 - b^2$, $2ab$, and $a^2 + b^2$ are the lengths of the sides of a right triangle. Find values of a and b so that the sides have lengths 9, 40, and 41. $a = 5, b = 4$

   c. Verify the identity $\dfrac{(a-b)(a^2 - b^2)}{2} + \dfrac{(a+b)(a^2 + b^2)}{2} = a^3 + b^3$ by expanding and simplifying the left side until it matches the right side.

   d. The identity in part c gives you a way to find two numbers that aren't cubes, but that add up to the sum of two cubes. Use the identity to find two noncubic numbers whose sum is $5^3 + 3^3$.   16 and 136

Review

14. Use a rectangle diagram to find each product.

   a. $3x(4x - 5)$ b. $(x + 3)(x - 5)$ c. $(x + 7)(x - 7)$ d. $(3x - 1)^2$ ⓐ

15. Recall that the distributive property allows you to distribute a factor through parentheses. The factor that is distributed doesn't have to be a monomial. Here's an example with a binomial.

   $(x + 1)(x - 3)$
   $x(x - 3) + 1(x - 3)$
   $x^2 - 3x + x - 3$
   $x^2 - 2x - 3$

   Use the distributive property to find each product in Exercise 14. ⓗ

16. You can also use a rectangle diagram to help you factor some trinomials, such as $x^2 - 7x + 12$.

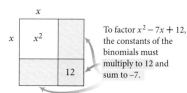
To factor $x^2 - 7x + 12$, the constants of the binomials must multiply to 12 and sum to −7.

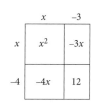
$(x - 3)(x - 4)$
The constants of the binomials need to be −3 and −4.

   Use rectangle diagrams to help you factor these trinomials.

   a. $x^2 + 3x - 10$ b. $x^2 + 8x + 16$ c. $x^2 - 25$

   d. The polynomial $x^2 + 8x + 16$ is called a **perfect-square trinomial**. Use the factored form to help explain why. Give two more examples of perfect-square trinomials. ⓗ

   e. The polynomial $x^2 - 25$ is called a **difference of two squares**. Give two more examples of differences of two squares.   Answers will vary. Possible answers: $x^2 - 16$, $x^2 - 100$.

17. Use the function $f(x) = 3x^3 - 5x^2 + x - 6$ to find these values.

   a. $f(2)$ $\;0$ b. $f(-1)$ $\;-15$ c. $f(0)$ $\;-6$ d. $f\left(\dfrac{1}{2}\right)$ $\;-6\dfrac{3}{8}$ e. $f\left(-\dfrac{4}{3}\right)$ $\;-23\dfrac{1}{3}$

18. Tell whether each statement is always, sometimes, or never true. Give a counterexample if the statement is never true. If the statement is sometimes true, give an example of a true case and an example of a false case.

   a. $x^{-3} = \dfrac{1}{x^3}$ b. $x^3 > x^2$ c. $2^x < 2^{x+1}$

14a. $12x^2 - 15$

14b. $x^2 - 2x - 15$

14c. $x^2 - 49$

14d. $9x^2 - 6x + 1$

15. a. $(3x)4x - (3x)5 = 12x^2 - 15x$
   b. $x(x - 5) + 3(x - 5) = x^2 - 2x - 15$
   c. $x(x - 7) + 7(x - 7) = x^2 - 49$
   d. $3x(3x - 1) - 1(3x - 1) = 9x^2 - 6x + 1$

16a. $(x + 5)(x - 2)$ **16b.** $(x + 4)(x + 4)$

16c. $(x + 5)(x - 5)$

16d. Answers will vary. Possible answer: The factors are the same, so the factored form $(x + 4)(x + 4)$ can be written as the square of a factor: $(x + 4)^2$; $x^2 + 6x + 9$, $x^2 + 2x + 1$.

18a. sometimes true; examples will vary: true for $x = 2$ and other nonzero values for x; false for $x = 0$

18b. sometimes true; examples will vary: true for $x = 2$ and other $x > 1$, false for $x \leq 1$

18c. always true

LESSON 5.2

This lesson focuses on finding the vertex of a graph of a quadratic function. If possible, don't skip this investigation. If you are short on time, consider using the Shortened or Whole Class investigation option.

COMMON CORE STATE STANDARDS

APPLIED	DEVELOPED	INTRODUCED
A.SSE.3	A.REI.4a	
A.REI.1	A.REI.4b	
F.IF.4	F.IF.8a	
F.IF.7a	F.IF.9	
	F.BF.1	

Objectives

- Explore projectile motion
- Understand completing the square as one way to convert the general form of a quadratic equation to vertex form
- Use formulas to convert the general form of a quadratic equation to vertex form
- Use the vertex form of a quadratic equation to solve problems involving maximums or minimums
- Compare properties of two functions

Vocabulary

completing the square
projectile motion

Launch

What value is required in each to make a perfect square?

a. $y = x^2 - 10x + \underline{} = (x - \underline{})^2$ 25; 5
b. $y = x^2 - \underline{}x + 36 = (x - \underline{})^2$
   12 or −12; 6 or −6

Consider bringing a small ball (tennis ball) to class. Start class by tossing or bouncing the ball to a student. **ASK** "How high did the ball go? How long was it in the air?" Have students sketch a graph of the flight of the ball. Project the graph at the beginning of the lesson. Tell students this is the graph of $y = -4.9(x - 0.86)2 + 0.6$, which models one bounce of a ball. **ASK** "What do you know just from the graph?" [It's the graph of a quadratic equation, x is time in s, y is height in m, the

LESSON 5.2

Completing the Square

The graph of $y = -4.9(x - 0.86)^2 + 0.6$ at right models one bounce of a ball, where x is time in seconds and y is height in meters. The maximum height of this ball occurs at the vertex (0.86, 0.6), which means that after 0.86 s the ball reaches its maximum height of 0.6 m.

"Everything should be made as simple as possible, but not simpler."
ALBERT EINSTEIN

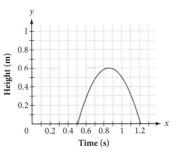

Finding the **maximum** or **minimum** value of a quadratic function is often necessary to answer questions about data. Finding the vertex is straightforward when you are given an equation in vertex form and sometimes when you are using a graph. However, you often have to estimate values on a graph. In this lesson you will learn a procedure called **completing the square** to convert a quadratic equation from general form to vertex form accurately.

The water coming out of a fountain, like this musical fountain in Vinnitsa, Ukraine, follows a path that can be described as projectile motion.

An object that rises and falls under the influence of gravity is called a projectile. You can use quadratic functions to model **projectile motion**, or the height of the object as a function of time.

The height of a projectile depends on three things: the height from which it is thrown, the upward velocity with which it is thrown, and the effect of gravity pulling downward on the object. So the polynomial function that describes projectile motion has three terms. The leading coefficient of the polynomial is based on the acceleration due to gravity, g. On Earth, g has an approximate numerical value of 9.8 m/s^2 when height is measured in meters and 32 ft/s^2 when height is measured in feet. The leading coefficient of a projectile motion function is always $-\frac{1}{2}g$.

> **Projectile Motion Function**
>
> The height of an object rising or falling under the influence of gravity is modeled by the function
>
> $y = ax^2 + v_0 x + s_0$
>
> where x represents time in seconds, y represents the object's height from the ground in meters or feet, a is half the downward acceleration due to gravity (on Earth, a is −4.9 m/s^2 or −16 ft/s^2), v_0 is the initial upward velocity of the object in meters per second or feet per second, and s_0 is the initial height of the object in meters or feet.

268 CHAPTER 5 Quadratic Functions and Relations

ELL	Support	Advanced
Start the lesson with the ball scenario described under the Launch. Use a representation of the bouncing ball to discuss vocabulary. Allow students to represent measures to the more familiar meters.	Help students check their work by graphing the original equation and their answer on the same set of axes. If needed, assign the Refreshing Your Skills: Polynomial Expressions lesson.	Assign the **One-Step**. **ASK** "How high the ball would go if, instead of the times, you knew that the machine gave the ball an initial velocity of 120 ft/s beginning 2 ft above the ground?" Have students rewrite the general form of the projectile motion function $y = ax^2 + v_0 x + s_0$ in vertex form.

EXAMPLE A | A stopwatch records that when Julie jumps in the air, she leaves the ground at 0.25 s and lands at 0.83 s. How high did she jump, in feet?

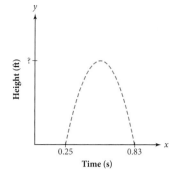

Solution | You don't know the initial velocity, so you can't yet use the projectile motion function. But you do know that height is modeled by a quadratic function and that the leading coefficient must be approximately -16 when using units of feet. Use this information along with 0.25 and 0.83 as the x-intercepts (when Julie's jump height is 0) to write the function

$$y = -16(x - 0.25)(x - 0.83)$$

The vertex of the graph of this equation represents Julie's maximum jump height. The x-coordinate of the vertex will be midway between the two x-intercepts, 0.25 and 0.83. The mean of 0.25 and 0.83 is $\frac{0.25 + 0.83}{2}$, or 0.54.

$y = -16(x - 0.25)(x - 0.83)$ Original function.
$y = -16(0.54 - 0.25)(0.54 - 0.83)$ Substitute the x-coordinate of the vertex.
$y \approx 1.35$ Evaluate.

Julie jumped 1.35 ft.

In Example A, you learned how to average the x-intercepts to find the vertex of a parabola if its equation is in factored form. In the investigation you will convert quadratic functions from general form to the equivalent vertex form by completing the square.

INVESTIGATION

Complete the Square

You can use rectangle diagrams to help convert quadratic functions to other equivalent forms.

Step 1 a. Complete a rectangle diagram to find the product $(x + 5)(x + 5)$, which can be written $(x + 5)^2$. Write out the four-term polynomial, and then combine any like terms you see and express your answer as a trinomial. $x^2 + 5x + 5x + 25 = x^2 + 10x + 25$

b. What binomial expression is being squared, and what is the perfect-square trinomial represented in the rectangle diagram at right? $(x - 8)^2$; $x^2 - 16x + 64$

	x^2	$-8x$
	$-8x$	64

Step 1c.

	x	12
x	x^2	$12x$
12	$12x$	144

c. Use a rectangle diagram to show the binomial factors for the perfect-square trinomial $x^2 + 24x + 144$. $(x + 12)^2$

d. Find the perfect-square trinomial equivalent to $(a + b)^2 = \underline{?} + \underline{?} + \underline{?}$. Describe how you can find the first, second, and third terms of the perfect-square trinomial (written in general form) when squaring a binomial. $a^2 + 2ab + b^2$; a^2 is the square of the first term of the binomial; the middle term, $2ab$, is twice the product of the first and second terms of the binomial; and the last term, b^2, is the square of the second term of the binomial.

ALERT Students may not have thought about the fact that the point halfway between two numbers is the mean.

Guiding the Investigation
The Investigation worksheet provides a blank rectangle, or square in this case, diagram.

ASK "Where do you see symmetry in the square diagrams? Why do you think that is true?" [along a diagonal; you are multiplying the same two factors]

Step 2b ALERT Students may want to add the same thing to both sides of an equation when completing the square, not realizing that here they have only an expression, so the same value must be added *and* subtracted to maintain its value.

Step 2c Students need to set each expression equal to y in order to graph or check the tables on a calculator. **SMP 5**

Step 2c

Step 3b If students are having difficulty, **ASK** "What if the 4 wasn't there? How would you do it then?" "After they have rewritten $x^2 + 6x$ in the form $(x - h)^2 + k$, they simply subtract 4 from k to find their new expression.

Step 3c

Step 6 You might consider splitting the linear term $\left(\frac{b}{a}\right)x$ into two equal parts, such as $a\left(x^2 + \frac{b}{2a}x + \frac{b}{2a}x + \underline{\quad}\right) + c$, before completing the square.

Step 2 Consider the expression $x^2 + 6x$.

a. What could you add to the expression to make it a perfect square? That is, what must you add to complete this rectangle diagram? 9

b. If you add a number to an expression, then you must also subtract the same number in order to preserve the value of the original expression. Fill in the blanks to rewrite $x^2 + 6x$ as the difference between a perfect square and a number. $x^2 + 6x = x^2 + 6x + 9 - 9 = (x+3)^2 - 9$

$$x^2 + 6x = x^2 + 6x + \underline{?} - \underline{?} = (x+3)^2 - \underline{?}$$

c. Use a graph or table to verify that your expression in the form $(x - h)^2 + k$ is equivalent to the original expression, $x^2 + 6x$.

Step 3 Consider the expression $x^2 + 6x - 4$.

a. Focus on the 2nd- and 1st-degree terms of the expression, $x^2 + 6x$. What must be added to and subtracted from these terms to complete a perfect square yet preserve the value of the expression? 9

b. Rewrite the expression $x^2 + 6x - 4$ in the form $(x - h)^2 + k$.
$$x^2 + 6x - 4 = x^2 + 6x + 9 - 9 - 4 = (x+3)^2 - 13$$

c. Use a graph or table to verify that your expression is equivalent to the original expression, $x^2 + 6x - 4$.

Step 4 Rewrite each expression in the form $(x - h)^2 + k$. If you use a rectangle diagram, focus on the 2nd- and 1st-degree terms first. Verify that your expression is equivalent to the original expression.

a. $x^2 - 14x + 3$
$x^2 - 14x + 3 = (x - 7)^2 - 46$

b. $x^2 - bx + 10$
$$x^2 - bx + 10 = \left(x - \frac{b}{2}\right)^2 + \left(-\frac{b^2}{2} + 10\right)$$

When the 2nd-degree term has a coefficient, you can first factor it out of the 2nd- and 1st-degree terms. For example, $3x^2 + 24x + 5$ can be written $3(x^2 + 8x) + 5$. Completing a diagram for $x^2 + 8x$ can help you rewrite the expression in the form $a(x - h)^2 + k$.

	x	4
x	x^2	$4x$
4	$4x$	16

$3x^2 + 24x + 5$	Original expression.
$3(x^2 + 8x) + 5$	Factor the coefficient of the 2nd- and 1st-degree terms.
$3(x^2 + 8x + 16) - 3(16) + 5$	Complete the square. You added $3 \cdot 16$, so you must subtract $3 \cdot 16$.
$3(x + 4)^2 - 43$	An equivalent expression in the form $a(x - h)^2 + k$.

Step 5 Rewrite each expression in the form $a(x - h)^2 + k$. For Step 5a, use a graph or table to verify that your expression is equivalent to the original expression.

a. $2x^2 - 6x + 1$ $2(x - 1.5)^2 - 3.5$

b. $ax^2 + 10x + 7$
$$a\left(x + \frac{5}{a}\right)^2 + \left(7 - \frac{25}{a}\right)$$

Step 6 If you graph the quadratic function $y = ax^2 + bx + c$, what will be the x-coordinate of the vertex in terms of a, b, and c? How can you use this value and the equation to find the y-coordinate?

$-\frac{b}{2a}$. Substitute this value back into the equation to find the y-coordinate of the vertex.

In the investigation you saw how to convert a quadratic expression from the form $ax^2 + bx + c$ to the form $a(x − h)^2 + k$. This process is called **completing the square**.

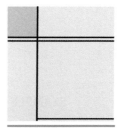

Seamless geometric abstract pattern yellow cube similar to those created by Dutch abstract painter Piet Mondrian (1872–1944).

EXAMPLE B | Convert each quadratic function to vertex form. Identify the vertex.

a. $y = x^2 − 18x + 100$ b. $y = ax^2 + bx + c$

Solution | To convert to vertex form, you must complete the square.

a. First separate the constant from the first two terms.

$$y = (x^2 − 18x) + 100 \quad \text{Original equation.}$$

To find what number you must add to complete the square, use a rectangle diagram.

	x	-9
x	x^2	$-9x$
-9	$-9x$	81

$x^2 − 18x + 100 \rightarrow (x^2 − 18x + ?) + 100 − ? \rightarrow$
$(x^2 − 18x + 81) + 100 − 81 \rightarrow (x − 9)^2 + 19$

When you add 81 to complete the square, you must also subtract 81 to keep the expression equivalent.

$y = (x^2 − 18x + 81) + 100 − 81$ Add 81 to complete the square, and subtract 81 to keep the equation equivalent.

$y = (x − 9)^2 + 19$ Factor into a perfect square, and add the constant terms.

The vertex form is $y = (x − 9)^2 + 19$, and the vertex is $(9, 19)$.

b. This solution will provide a formula for vertex (h, k) and an alternative to completing the square.

$y = ax^2 + bx + c$ Original equation.

$= a\left(x^2 + \dfrac{b}{a}x + \underline{?}\right) + c$ Factor a from the first two terms.

$= a\left[x^2 + \dfrac{b}{a}x + \left(\dfrac{b}{2a}\right)^2\right] + c − a \cdot \left(\dfrac{b}{2a}\right)^2$ Add and subtract the missing term to complete the square.

$y = a\left(x + \dfrac{b}{2a}\right)^2 + c − \dfrac{b^2}{4a}$

The final equation is in vertex form, $y = a(x − h)^2 + k$. This means $h = −\dfrac{b}{2a}$ and $k = c − \dfrac{b^2}{4a}$.

Example B

This example shows how to convert quadratic functions from general form to vertex form.

Consider projecting part a of the example and have the class work on it in pairs. Have groups present their solutions and explain their work. Whether student responses are correct or incorrect, ask other students if they agree and why. **SMP 2, 3, 6** After they present their solutions to part a, project part b and have the class work on it in groups of 2–4. Part b formalizes the nongeometric approach to completing the square that some students will have noticed during the investigation. Have groups present their solutions and explain their work. Whether student responses are correct or incorrect, ask other students if they agree and why. **SMP 2, 3, 6**

After the presentations, have groups compare and contrast their methods, using mathematics as justification. **SMP 2, 3, 6**

Example C

In this example, students are to compare the maximum values of two different quadratic functions. Consider projecting the example and having students work in groups. Have groups present their solutions and explain their work. Whether student responses are correct or incorrect, ask other students if they agree and why. **SMP 2, 3, 6** **LANGUAGE** *Velocity* means "speed in a specific direction." The velocity of 120 ft/s in this example is a speed (rate of change of distance in feet for each second) of 120 ft/s in the upward direction. A negative velocity is downward. For example, a negative initial velocity would indicate that something was being thrown down.

Students may point out that a softball normally doesn't travel straight up and down. Even if the motion has a horizontal component, the projectile motion function describes the vertical component.

Summarize

For Step 6 of the investigation, if you want your students to memorize $\left(-\frac{b}{2a}, c - \frac{b^2}{4a}\right)$, you should have them derive the expression for the *y*-coordinate. If students know that $-\frac{b}{2a}$ represents the *x*-coordinate of the vertex, they can substitute that into the general equation of a parabola to find the corresponding *y*-coordinate.

After students have completed the Investigation **ASK** "Does it make sense that the expression $a\left(x + \frac{b}{2a}\right)^2 + \left(c - \frac{b^2}{4a}\right)$ has an optimum value at $\left(-\frac{b}{2a}, c - \frac{b^2}{4a}\right)$?" One way to make sense of this is to think of the first term as being added to (or subtracted from) the second. The square of the binomial is never negative, so the sum or difference will be optimized when this square is 0, which is when $x = -\frac{b}{2a}$. The optimum value will be the second term, with 0 being added or subtracted.

LANGUAGE The *leading coefficient* of a polynomial is the coefficient of the highest power of *x*.

You may find that using the formulas for *h* and *k* is often simpler than completing the square. Both methods will allow you to find the vertex of a quadratic equation and write the equation in vertex form. However, the method of completing the square will be used again in your work with circles and other geometric shapes, so you should become comfortable using it.

EXAMPLE C Nora hits a softball straight up at a speed of 120 ft/s.

a. If her bat contacts the ball at a height of 3 ft above the ground, how high does the ball travel? When does the ball reach its maximum height?

b. How does this height compare to her previous time at bat, when the ball was in flight for 7 seconds before it was caught at the same height that it left the bat?

2014 World Championship Softball in Haarlem, The Netherlands

Solution This question asks you to compare the maximum values of two different quadratic functions. But first you need to write the equations.

a. Using the projectile motion function, you know that the height of the object at time *x* is represented by the equation $y = ax^2 + v_0 x + s_0$. The initial velocity, v_0, is 120 ft/s, and the initial height, s_0, is 3 ft. Because the distance is measured in feet, the approximate leading coefficient is -16. Thus, the function is $y = -16x^2 + 120x + 3$. To find the maximum height, locate the vertex.

$y = -16x^2 + 120x + 3$ Original equation. Identify the coefficients, $a = -16$, $b = 120$, and $c = 3$.

$h = -\frac{b}{2a} = -\frac{120}{2(-16)} = 3.75$ Use the formula for *h* to find the *x*-coordinate of the vertex.

$y = -16(3.75)^2 + 120(3.75) + 3 = 228$ Substitute 3.75 for *x* into the equation to find the *y*-coordinate of the vertex.

The softball reaches a maximum height of 228 ft at 3.75 s.

b. On Nora's previous time at bat, the ball was at the same height of 3 feet at $x = 0$ and $x = 7$ seconds. So the maximum height occurred at the mean of these two times, or at 3.5 seconds. Because the ball reached its maximum height in less time, it must not have gone as high.

You now have several strategies for finding the vertex of a quadratic function. You can convert from general form to vertex form by completing the square or by using the formula for *h* and substituting to find *k*. Another option is to find the *x*-intercepts and average them. This gives you the *x*-value for the vertex. Of course, some quadratics don't intersect the *x*-axis in two points, so this strategy is not always practical.

CRITICAL QUESTION Why do you complete the square of a quadratic function?

BIG IDEA You complete the square to convert from general to vertex form, so you can identify the vertex of the parabola (and hence the maximum or minimum).

CRITICAL QUESTION Is there an alternative to remembering the formulas for *h* and *k*?

BIG IDEA The values can always be re-derived by completing the square.

CRITICAL QUESTION Why would you want to find the vertex form of a quadratic parabola?

BIG IDEA A parabola's vertex is useful for finding extreme values in real-world situations.

Formative Assessment

To assess the objectives of the lesson, observe students working in groups and listen carefully to their presentation. As students present their solutions to the examples, Investigation, and exercises, encourage them to use appropriate

5.2 Exercises

You will need your graphing calculator for Exercises 11 and 16.

Practice Your Skills

1. Factor each quadratic expression.
   a. $x^2 - 10x + 25$ $(x-5)^2$
   b. $x^2 + 5x + \frac{25}{4}$ @ $\left(x + \frac{5}{2}\right)^2$
   c. $4x^2 - 12x + 9$ $(2x-3)^2$, or $4\left(x - \frac{3}{2}\right)^2$
   d. $x^2 - 2xy + y^2$ @ $(x-y)^2$

2. What value is required to complete the square?
   a. $x^2 + 20x + \underline{?}$ 100
   b. $x^2 - 7x + \underline{?}$ @ 12.25, or $\frac{49}{4}$
   c. $4x^2 - 16x + \underline{?}$ 16
   d. $-3x^2 - 6x + \underline{?}$ @ -3

3. Convert each quadratic function to vertex form by completing the square.
   a. $y = x^2 + 20x + 94$ $y = (x + 10)^2 - 6$
   b. $y = x^2 - 7x + 16$ @ $y = (x - 3.5)^2 + 3.75$
   c. $y = 6x^2 - 24x + 147$ $y = 6(x-2)^2 + 123$
   d. $y = 5x^2 + 8x$ @ $y = 5(x + 0.8)^2 - 3.2$

4. Rewrite each expression in the form $ax^2 + bx + c$, and then identify the coefficients, a, b, and c.
   a. $3x^2 + 2x - 5$ no rewriting necessary; $a = 3, b = 2, c = -5$
   b. $14 + 2x^2$ $2x^2 + 14; a = 2, b = 0, c = 14$
   c. $-3 + 4x^2 - 2x + 8x$ @ $4x^2 + 6x - 3; a = 4, b = 6, c = -3$
   d. $3x - x^2$ $-x^2 + 3x; a = -1, b = 3, c = 0$

Reason and Apply

5. What is the vertex of the graph of the quadratic function $y = -2x^2 - 16x - 20$? Use completing the square to find your solution and explain each step. $(-4, 12)$

6. Complete the solution of the equation $(x-3)^2 - 25 = 0$ by filling in the steps or reasons.

   $(x - 3)^2 - 25 = 0$ Original equation
   $(x - 3)^2 = 25$ Add 25 to each side
   $x - 3 = \pm 5$ Take the square root of each side
   $x = \underline{8}$, $x = \underline{-2}$ Add 3 to each side

7. A rock is thrown upward from the edge of a 50 m cliff overlooking Lake Superior, with an initial velocity of 17.2 m/s. Define variables and write an equation that models the height of the rock. Let x represent time in seconds, and let y represent height in meters; $y = -4.9x^2 + 17.2x + 50$.

8. **APPLICATION** Suppose you are enclosing a rectangular region to create a rabbit cage. You have 80 ft of fence and want to build a pen with the largest possible area for your rabbit, so you build the cage using an existing building as one side.
   a. Make a table showing the areas for some selected values of x, and write a function that gives the area, y, as a function of the width, x.
   b. What width maximizes the area? What is the maximum area? Explain. 20 ft; 800 ft²

   8a. possible table with x representing the sides adjacent to the building:

x	5	10	15	20	25	30	35
y	350	600	750	800	750	600	350

   $y = -2x^2 + 80x$

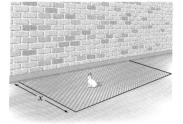

5.
$-2(x^2 + 8x +\quad) - 20$ Factor the -2 from the first two terms

$-2(x^2 + 8x + 16) - 20 - (-32)$ Add 16 to complete the square inside the parentheses, then subtract $-2(16)$ to keep the equation balanced.

$-2(x + 4)^2 + 12$ Factor the square and combine like terms

The equation is now in vertex form and the vertex is at $(-4, 12)$.

---

and precise mathematical language. Whether student responses are correct or incorrect, ask other students if they agree and why. **SMP 2, 3, 6**

Responses to Examples A and C and Exercises 7 and 9 will provide evidence of understanding projectile motion. Example C and Exercise 10 will provide evidence of students' ability to compare properties of two functions represented in different ways. Presentations of Example B, the Investigation and Exercises 3 and 5 will provide evidence of completing the square and converting from general form to vertex form.

Apply

Extra Example
Convert the quadratic function $y = x^2 - 14x + 2$ to vertex form by completing the square.

$y = (x - 7)^2 - 47$

Closing Question
What is the vertex of the graph of the quadratic function $y = 3x^2 - 9x + 5$?

$\left(\frac{3}{2}, -\frac{7}{4}\right)$

Assigning Exercises
Suggested: 1, 2a, 2c, 3a, 3c, 4, 5 – 12

Additional Practice: 2b, 2d, 3b, 3d, 13–16

Exercise 8 **ALERT** Some students will believe that two pens with the same perimeter will have the same area. You might suggest that they sketch several different pens and calculate their areas. In the given answer, x is the length of each side perpendicular to the building. If students let x equal the length of the side parallel to the building, their area function will $y = x\left(\frac{80 - x}{2}\right)$, or $y = -\frac{1}{2}x^2 + 40x$. A value of $x = 40$ ft gives the same area, 800 ft².

ALERT If students remember from geometry that the largest rectangle with a given perimeter is a square, they may be confused that the result here is not a square. That's because one edge of the rectangle is against a wall.

Exercise 9 Encourage students to sketch a picture and label important parts. They can assume that the arrow was shot from the very bottom of the well.

9a. Let x represent time in seconds, and let y represent height in meters; $y = -4.9(x - 1.1)(x - 4.7)$ or $y = -4.9x^2 + 28.42x - 25.333$.

9. Imagine that an arrow is shot from the bottom of a well. It passes ground level at 1.1 s and lands on the ground at 4.7 s. ⓗ

   a. Define variables and write a quadratic function that describes the height of the arrow, in meters, as a function of time. ⓐ
   b. What was the initial velocity of the arrow in meters per second? 28.42 m/s
   c. How deep was the well in meters? 25.333 m

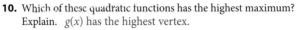

10. Which of these quadratic functions has the highest maximum? Explain. $g(x)$ has the highest vertex.

    i. $f(x) = -2x^2 + 10x + 3$

    ii.
    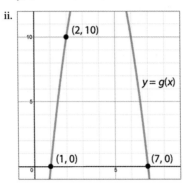

    iii. $h(x)$ is the quadratic function with line of symmetry at $x = 10$, one root at -3 and y-intercept at 6.9

11. An object is projected upward, and these data are collected. ⓐ

Time (s) t	1	2	3	4	5	6
Height (m) h	120.1	205.4	280.9	346.6	402.5	448.6

   a. Write a function that relates time and height for this object. $y = -4.9t^2 + 100t + 25$
   b. What was the initial height? The initial velocity? 25 m; 100 m/s
   c. When does the object reach its maximum height? What is the maximum height? 10.2 s; 535 m

274 **CHAPTER 5** Quadratic Functions and Relations

12. **APPLICATION** The members of the Young Entrepreneurs Club decide to sell T-shirts in their school colors for Spirit Week. In a marketing survey, the members ask students whether or not they would buy a T-shirt for a specific price. In analyzing the data, club members find that at a price of $20 they would sell 60 T-shirts. For each $5 increase in price, they would sell 10 fewer T-shirts.

   a. Find a linear function that relates the price in dollars, p, and the number of T-shirts sold, n. ⓗ

   b. Write a function that gives revenue as a function of price. (Use your function in 11a as a substitute for the number of T-shirts sold.) @

   c. Convert the revenue function to vertex form. What is the real-world meaning of the vertex?

   d. If the club members want to receive at least $1,050 in revenue, what price should they charge for the T-shirts? between $15 and $35

12a. $n = -2p + 100$

12b. $R(p) = -2p^2 + 100p$

12c. Vertex form: $R(p) = -2(p - 25)^2 + 1250$. The vertex is (25, 1250). This means that the maximum revenue is $1,250 when the price is $25.

15b.

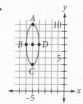

$A(-5, 10)$, $B(-6, 7)$, $C(-5, 4)$, $D(-4, 7)$; center: $(-5, 7)$

→ **Business**
CONNECTION

Entrepreneurs are individuals who take a risk to create a new product or a new business. Some famous American entrepreneurs include Sarah Breedlove ("Madam C. J. Walker") (1867–1919), and Jeff Bezos (b 1964). Sarah Breedlove Walker, commemorated here in a circa 1998 USPS stamp, was a pioneer in the cosmetics industry for African-Americans who expanded her cosmetics company across the United States, the Caribbean, and Europe. Jeff Bezos is an American technology entrepreneur and investor who founded Amazon.com, popularizing a way to sell merchandise without the traditional retail stores.

Review

13. Multiply.

   a. $(x - 3)(2x + 4)$ $2x^2 - 2x - 12$

   b. $(x^2 + 1)(x + 2)$ $x^3 + 2x^2 + x + 2$

14. Solve $(x - 2)(x + 3)(2x - 1) = 0$. @ $x = 2$, $x = -3$, or $x = -\frac{1}{2}$

15. Consider a graph of the unit circle, $x^2 + y^2 = 1$. Dilate it vertically by a scale factor of 3, and translate it left 5 units and up 7 units.

   a. Write the equation of this new shape. What is it called? $(x + 5)^2 + \left(\frac{y-7}{3}\right)^2 = 1$; ellipse

   b. Sketch a graph of this shape. Label the center and at least four points.

LESSON 5.2 Completing the Square   **275**

16a, b. Let x represent the year, and let y represent the number of endangered bird species.

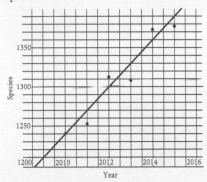

16b. Answers will vary. Approximately $y = 1240 + 30(x - 2010)$

Exercise 16c If students round off differently or begin counting years at some number other than 0, they may get slightly different results.

16c. Answers will vary. Approximately 1150 species in 2007; 2440 species in 2050.

16d. The actual value is about 70 species higher than the predicted number. Because the number of endangered species rose more quickly from 2007 to now, it might be expected to continue to rise more and more quickly. Thus the estimate for 2050 may be too low.

16e. Looking at the data now, it might be better to use a nonlinear model such as a quadratic function or an exponential function since the slopes between the points seem to be increasing.

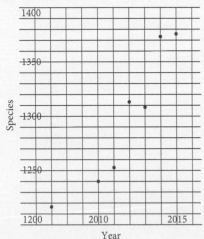

16. **APPLICATION** This table shows the number of endangered bird species in the world for the years from 2010 to 2015.

Endangered Species

Year	2010	2011	2012	2013	2014	2015
Number of endangered species	1240	1253	1313	1308	1373	1375

(www.iucn.org)

a. Define variables and create a scatter plot of these data.

b. Find the equation of a line of fit for these data.

c. Use your equation from 16b to predict the number of endangered species in 2007 and in 2050.

d. The actual number of endangered species in 2007 was 1217. How does this value compare with your prediction from 16c? How might this information change your prediction for 2050?

e. Add the point (2007, 1217) to your scatter plot. How might this information change the type of function you use to model the data? Explain.

The Philippine Eagle is one of the top ten most endangered bird species.

→ Environmental
CONNECTION

Most species become endangered when humans damage their ecosystems through pollution, habitat destruction, and introduction of nonnative species. Over-hunting and over-collecting also threaten animal and plant populations. The current global extinction rate is about 20,000 species a year.

The Karner blue butterfly, native to the Great Lakes region, has become endangered as the availability of its primary food, blue lupine, has become scarce due to development and fire suppression.

The aye-aye is a nocturnal primate native to Madagascar. Their numbers have declined due to habitat destruction and also because, as they are considered to be a bad omen, they are often killed.

LESSON 5.3

The Quadratic Formula

Although you can always use a graph of a quadratic function to approximate the x-intercepts, you are often not able to find exact solutions. This lesson will develop a procedure to find the exact roots of a quadratic equation. Consider again this situation from Example C in the previous lesson.

EXAMPLE A | Nora hits a softball straight up at a speed of 120 ft/s. Her bat contacts the ball at a height of 3 ft above the ground. Recall that the equation relating height in feet, y, and time in seconds, x, is $y = -16x^2 + 120x + 3$. How long will it be until the ball hits the ground?

Solution | The height will be zero when the ball hits the ground, so you want to find the solutions to the equation $-16x^2 + 120x + 3 = 0$. You can approximate the x-intercepts by graphing, but you may not be able to find the exact x-intercept.

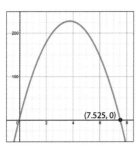

This gives a value for the x-intercept. While it is accurate to three decimal places, it is NOT the EXACT answer.

You will not be able to factor this equation using a rectangle diagram, so you can't use the zero-product property. Instead, to solve this equation symbolically, first write the equation in the form $a(x - h)^2 + k = 0$.

$-16x^2 + 120x + 3 = 0$	Original equation.
$-16x^2 + 120x = -3$	Subtract the constant from both sides.
$-16(x^2 - 7.5x + \underline{\ ?\ }) = -3$	Factor to get the leading coefficient 1.
$-16[x^2 - 7.5x + (\underline{-3.75})^2] = -3 + -16(\underline{-3.75})^2$	Complete the square.
$-16(x - 3.75)^2 = -228$	Factor and combine like terms.
$(x - 3.75)^2 = 14.25$	Divide by -16.
$x - 3.75 = \pm\sqrt{14.25}$	Take the square root of both sides.
$x = 3.75 \pm \sqrt{14.25}$	Add 3.75 to both sides.
$x = 3.75 + \sqrt{14.25}$ or $x = 3.75 - \sqrt{14.25}$	Write the two exact solutions to the equation.
$x \approx 7.525$ or $x \approx -0.025$	Approximate the values of x.

The zeros of the function are $x \approx 7.525$ and $x \approx -0.025$. The negative time, -0.025 s, does not make sense in this situation, so the ball hits the ground after approximately 7.525 s.

LESSON 5.3

This lesson derives the quadratic formula. This may be a familiar topic to students.

COMMON CORE STATE STANDARDS		
APPLIED	**DEVELOPED**	**INTRODUCED**
A.REI.1	A.REI.4a	N.CN.7
F.IF.7a	A.REI.4b	
	F.BF.1	

Objectives
- Use the vertex form of a quadratic equation to find the equation's roots
- Derive the quadratic formula by completing the square
- Use the quadratic formula to solve application problems

Vocabulary
quadratic formula
golden rectangle
golden ratio

Materials
Project: Calculator Program for the Quadratic Formula

Launch

Solve for x. Show how you solved each equation.

a. $0 = x(200 - 2x)$ $x = 0, 100$
b. $x^2 + 8x + 16 = 0$ $x = -4$
c. $x^2 - 4x - 5 = 0$ $x = -1, 5$
d. $(x + 8)^2 - 7 = 0$ $x = -8 \pm \sqrt{7}$

Investigate

Example A

This example revisits the situation of Example C in the previous lesson, but it poses a problem that reviews completing the square and extends it to solving an equation. Resist the temptation to skip this example and go right to the quadratic formula; the example provides a special case for the derivation of that formula. You could use it as a Launch for the lesson. Project the example and have students work in pairs.

ELL
Students should be able to differentiate among quadratic functions (the relationship between two variables), quadratic equations (a specific value of the functions), and the quadratic formula (a method used to solve quadratic equations). Have students provide verbal and symbolic examples of each.

Support
Students should work through Example A as well as another example or two like it before looking at the quadratic formula's derivation. Emphasize the links among the derivation, the formula, and solutions.

Advanced
Ask students to derive the quadratic formula on their own after they see Example A. Have students explain their insights into the relationship between the constant values of a, b, and c and the solutions for x. Assign the **One Step**.

Have students present their solutions and explain their work. Look for groups to present that used completing the square. Emphasize the use of correct mathematical terminology. Whether student responses are correct or incorrect, ask other students if they agree and why. **SMP 1, 3, 4, 5, 6**

In applying the *quadratic formula* to Example A, you might want to simplify to the exact solutions, $x = 3.75 \pm \frac{\sqrt{57}}{2}$.

ALERT Most mistakes made while using a calculator with the quadratic formula can be blamed on not enclosing the radicand within parentheses, not enclosing the entire numerator within parentheses, or not enclosing the denominator within parentheses. **SMP 5, 6**

Since this topic should be familiar to most students, consider projecting the equation $ax^2 + bx + c = 0$ and asking students to solve for *x*. This could be done in groups or as a whole class activity. This can serve to assess prior knowledge and to ask probing questions to lead to a deeper understanding. (See Summarize).

If you follow the same steps with a general quadratic equation, then you can develop the **quadratic formula**. This formula provides solutions to $ax^2 + bx + c = 0$ in terms of *a*, *b*, and *c*.

$ax^2 + bx + c = 0$	Original equation.
$ax^2 + bx = -c$	Subtract *c* from both sides.
$a\left(x^2 + \frac{b}{a}x + \underline{?}\right) = -c$	Factor to get the leading coefficient 1.
$a\left[x + \frac{b}{a}x + \left(\frac{b}{2a}\right)^2\right] = a\left(\frac{b}{2a}\right)^2 - c$	Complete the square.
$a\left(x + \frac{b}{2a}\right)^2 = \frac{b^2}{4a} - c$	Factor the perfect-square trinomial on the left side and multiply on the right side.
$a\left(x + \frac{b}{2a}\right)^2 = \frac{b^2}{4a} - \frac{4ac}{4a}$	Rewrite the right side with a common denominator.
$a\left(x + \frac{b}{2a}\right)^2 = \frac{b^2 - 4ac}{4a}$	Add terms with a common denominator.
$\left(x + \frac{b}{2a}\right)^2 = \frac{b^2 - 4ac}{4a^2}$	Divide both sides by *a*.
$x + \frac{b}{2a} = \pm\sqrt{\frac{b^2 - 4ac}{4a^2}}$	Take the square root of both sides.
$x + \frac{b}{2a} = \pm\frac{\sqrt{b^2 - 4ac}}{2a}$	Use the power of a quotient property to take the square roots of the numerator and denominator.
$x = -\frac{b}{2a} \pm \frac{\sqrt{b^2 - 4ac}}{2a}$	Subtract $\frac{b}{2a}$ from both sides.
$x = \frac{-b \pm \sqrt{b^2 - 4ac}}{2a}$	Add terms with a common denominator.

> **The Quadratic Formula**
>
> Given a quadratic equation written in the form $ax^2 + bx + c = 0$, the solutions are
>
> $$x = \frac{-b \pm \sqrt{b^2 - 4ac}}{2a}$$

To use the quadratic formula on the equation in Example A, $-16x^2 + 120x + 3 = 0$, first identify the coefficients as $a = -16$, $b = 120$, and $c = 3$. The solutions are

$$x = \frac{-120 \pm \sqrt{120^2 - 4(-16)(3)}}{2(-16)}$$

$$x = \frac{-120 + \sqrt{14592}}{-32} \quad \text{or} \quad x = \frac{-120 - \sqrt{14592}}{-32}$$

$$x \approx -0.025 \quad \text{or} \quad x \approx 7.525$$

The quadratic formula gives you a way to find the roots of any equation in the form $ax^2 + bx + c = 0$. The investigation will give you an opportunity to apply the quadratic formula in different situations.

Modifying the Investigation

Whole Class Watch the video as a class, telling students to look for the initial release height of the balloon. Complete Steps 1 through 6 with student input.

Shortened Use the Investigation Worksheet with Sample Data.

One Step In this video, Pam is trying to hit the water balloon bomber on the second floor with a balloon of her own. Since the balloon bomber comes out of the window for only a fraction of a second, do you think Pam will score a hit? Explain your reasoning, using mathematics to justify your answer. You should include the equation you use, when the balloon will pass the window, the greatest height of the balloon, etc.

INVESTIGATION
Hit or Miss

The video shows Pam trying to score a hit on the second floor balloon bomber. You can use the formula for projectile motion from the previous lesson to create a model to help her succeed. Answers will vary. Sample answers provided.

Step 1

$(0.416, 14.67)$ as (time, height)
$h(t) = h_0 + v_0 t + 0.5gt^2$.
Substitute $h_0 = 6$ ft and
$g = -32$ ft/sec to solve for
$v_0 = 27.4973$ ft/s. So the
model for the height of the
balloon over time is
$h(t) = 6 + 27.4973t - 16t^2$.

Step 2

$14.67 = 6 + 27.4973t - 16t^2$
$t = 0.416$ s and 1.3 s.

Step 1 From the video, find the initial height of Pam's release. Use the time and height given on the upward flight of the balloon to complete the quadratic model, $h = at^2 + bt + c$.

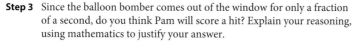

Step 2 Create an equation to determine when the balloon will reach the window height on its downward flight. Solve your equation to find the initial height.

Step 3 Since the balloon bomber comes out of the window for only a fraction of a second, do you think Pam will score a hit? Explain your reasoning, using mathematics to justify your answer.

Step 4 How fast the balloon is going when it passes the window depends on how high up it is. Explain how to use your answers from Step 2 to find the time it takes for the balloon to reach the top. How would you find this maximum height once you know the time.

Step 5 Solve your equation from Step 2 to determine when the balloon will reach the greatest height. How do you know this is the highest point?

Step 6 At what time(s) will the balloon reach 100 feet? Explain your solution.
Graphing the equations $y = 6 + 27.5x - 32x^2$ and $y = 100$ shows that these never intersect. Thus the balloon never reaches this height

It's important to note that a quadratic equation must be in the general form $ax^2 + bx + c = 0$ before you use the quadratic formula.

EXAMPLE B | Solve $3x^2 = 5x + 8$.

Solution | To use the quadratic formula, first write the equation in the form $ax^2 + bx + c = 0$ and identify the coefficients.

$$3x^2 - 5x - 8 = 0$$
$$a = 3, b = -5, c = -8$$

Substitute a, b, and c into the quadratic formula.

$$x = \frac{-b \pm \sqrt{b^2 - 4ac}}{2a}$$
$$= \frac{-(-5) \pm \sqrt{(-5)^2 - 4(3)(-8)}}{2(3)}$$
$$= \frac{5 \pm \sqrt{121}}{6}$$
$$= \frac{5 \pm 11}{6}$$
$$x = \frac{5 + 11}{6} = \frac{8}{3} \quad \text{or} \quad x = \frac{5 - 11}{6} = -1$$

Guiding the Investigation

This video driven investigation uses water balloons as a model for projectile motion. There is an Investigation Worksheet that provides sample data.

There is also a desmos simulation for the investigation in your ebook.

Step 1 Students are to use the projectile motion function.

Step 4 A graph of the function can help students answer this question.

Example B

Consider projecting the equation and asking students to solve it. As they share their answers have them show their work. You might ask students to do the calculations on graphing calculators to give them an opportunity to practice inserting parentheses around the numerator and around the terms under the radical. Emphasize the use of correct mathematical terminology. Whether student responses are correct or incorrect, ask other students if they agree and why. **SMP 3, 5, 6**

Step 3 has no specific answer. If the student thinks the girl has her head out of the window for more than 0.883 seconds, the difference in the times that the balloon is at that height, then the balloon will hit her. If the student believes that the girl has her head out for less time than that, then the balloon will miss.

Step 4 You can find the time when the balloon is at its maximum height by averaging the two times that it is at the window height. This gives you $(0.418 + 1.301)/2 = 0.8595$ seconds. You can then substitute this into the equation for the height to find this maximum height.

Step 5 You know this is the maximum height because the line through the vertex is the axis of symmetry. Thus the points located at equal heights are equidistant from this line. By averaging the x-values of these points, you find the axis of symmetry.

Conceptual/Procedural

Conceptual In the investigation, students collect data and model the projectile motion formula. In Investigating Structure, students derive the sum and product of roots formulas.

Procedural Students derive the quadratic formula symbolically and apply it in the projectile motion formula.

Investigating Structure

In this Investigating Structure, students use the quadratic formula to derive the formulas for finding the sum and product of the roots of a quadratic equation. Consider having students work on it in groups. Have students present their solutions and explain their work. Pay particular attention to their equation created from roots and their explanation of its connection to the general form of a quadratic and the quadratic formula. Emphasize the use of correct mathematical terminology. In particular, discourage the use of the term *quadratic equation* to mean *quadratic formula*. The quadratic formula can be used to solve all quadratic equations. Whether student responses are correct or incorrect, ask other students if they agree and why. SMP 3, 5, 6, 8

Summarize

As the derivation of the quadratic formula is discussed, look for opportunities to ask questions like the following.

ASK "Why are the two solutions in the form of a certain amount added to and subtracted from another number?" [The "other number" is $-\frac{b}{2a}$, the x-coordinate of the parabola's vertex. The symmetry of the parabola ensures that its x-intercepts will be equidistant on either side of this coordinate.]

ASK "Does the value of k, the y-value of the vertex, also appears in the quadratic formula." [The number being added to and subtracted from $-\frac{b}{2a}$ is the square root of $-\frac{k}{a}$.]

Formative Assessment

Student group work and presentations of the Examples and Investigation can provide evidence of meeting the objectives of the lesson. Having students derive the quadratic formula from solving the equation for x and answering questions like those in Summarize can assess how well they understand the derivation of the quadratic formula. Answers to the exercises will provide evidence of being able to use the quadratic formula to solve problems.

The solutions are $x = \frac{8}{3}$ or $x = -1$.

To check your work, substitute these values into the original equation.

Remember, you can find exact solutions to some quadratic equations by factoring. However, most quadratic equations don't factor easily. The quadratic formula can be used to solve any quadratic equation.

▲ INVESTIGATING STRUCTURE

The quadratic formula tells you that the solutions to the equation $ax^2 + bx + c = 0$ are $x = \frac{-b \pm \sqrt{b^2 - 4ac}}{2a}$. Using these expressions for the solutions, you can discover some properties of these two values. Call the two solutions r and s.

What is the sum $r + s$? Simplify this sum as much as possible. $-\frac{b}{a}$

What is the product rs? Simplify this product as much as possible. $\frac{c}{a}$

What equation results if you try to solve the system of equations created from your sum and product? Why is this not surprising? $ax^2 + bx + c$, or $x^2 + \left(\frac{b}{a}\right)x + \frac{c}{a} = 0$

Using what you have discovered, you can find the sum and the product of the solutions to a quadratic equation without actually finding the solutions. What are the sum and product of the solutions to $3x^2 - 5x + 7 = 0$? Sum of $\frac{5}{3}$, product of $\frac{7}{3}$

5.3 Exercises

You will need your graphing calculator for Exercises **8**.

Practice Your Skills

1. Rewrite each equation in general form, $ax^2 + bx + c = 0$. Identify a, b, and c.
   a. $3x^2 - 13x = 10$ $3x^2 - 13x - 10 = 0;$ $a = 3, b = -13, c = -10$
   b. $x^2 - 13 = 5x$ @ $x^2 - 5x - 13 = 0;$ $a = 1, b = -5, c = -13$
   c. $3x^2 + 5x = -1$ @ $3x^2 + 5x + 1 = 0; a = 3, b = 5, c = 1$
   d. $3x^2 - 2 - 3x = 0$ $3x^2 - 3x - 2 = 0;$ $a = 3, b = -3, c = -2$
   e. $14(x - 4) - (x + 2) = (x + 2)(x - 4)$
   $x^2 - 15x + 50 = 0; a = 1, b = -15, c = 50$

2. Evaluate each expression using your calculator. Round your answers to the nearest thousandth.
   a. $\frac{-30 + \sqrt{30^2 - 4(5)(3)}}{2(5)}$  -0.102
   b. $\frac{-30 - \sqrt{30^2 - 4(5)(3)}}{2(5)}$ @ -5.898
   c. $\frac{8 - \sqrt{(-8)^2 - 4(1)(-2)}}{2(1)}$ -0.243
   d. $\frac{8 + \sqrt{(-8)^2 - 4(1)(-2)}}{2(1)}$ @ 8.243

3. Solve by any method.
   a. $x^2 - 6x + 5 = 0$
      $x = 1$ or $x = 5$
   b. $x^2 - 7x - 18 = 0$
      $x = -2$ or $x = 9$
   c. $5x^2 + 12x + 7 = 0$ @
      $x = -1$ or $x = -1.4$

4. Use the roots of the equations in Exercise 3 to write each of these functions in factored form, $y = a(x - r_1)(x - r_2)$.
   a. $y = x^2 - 6x + 5$ $y = (x - 1)(x - 5)$
   b. $y = x^2 - 7x - 18$
      $y = (x + 2)(x - 9)$
   c. $y = 5x^2 + 12x + 7$ @
      $y = 5(x + 1)(x + 1.4)$

280 CHAPTER 5 Quadratic Functions and Relations

CRITICAL QUESTION Why use the quadratic formula?

BIG IDEA The quadratic formula provides a shortcut for finding the zeros of any quadratic function written in general form, without having to complete the square.

CRITICAL QUESTION When the value of a, b, or c is 0, what can be said about the equation, and what happens to the formula?

BIG IDEA When $a = 0$, the equation is linear, not quadratic, because the quadratic formula requires dividing by a, the result when $a = 0$ is not defined; the formula doesn't apply to linear equations. When $b = 0$, the equation's graph is symmetric about the y-axis; the formula gives two solutions equidistant from 0 or a double solution at $x = 0$. When $c = 0$, the equation factors to show a root at $x = 0$, and the formula returns 0 as one of the roots.

5. Use the quadratic formula to find the zeros of each function.

   a. $f(x) = 2x^2 + 7x - 4$ $x = 0.5$ or $x = -4$

   b. $f(x) = x^2 - 6x + 3$ $x = 3 + \sqrt{6}$ or $x = 3 - \sqrt{6}$

   c. $y - 6 = -2x^2$ $x \approx \pm\sqrt{3}$

   d. $5x + 4 + 2x^2 = y$ no real solutions

Reason and Apply

6. Beth uses the quadratic formula to solve an equation and gets

   $$x = \frac{-9 \pm \sqrt{9^2 - 4(1)(10)}}{2(1)}$$

   a. Write the quadratic equation Beth started with. @ $x^2 + 9x + 10 = 0$

   b. Write the simplified forms of the exact answers. $x = \frac{-9 \pm \sqrt{41}}{2}$

   c. What are the x-intercepts of the graph of this quadratic function? $\frac{-9 + \sqrt{41}}{2}$ and $\frac{-9 - \sqrt{41}}{2}$

7. Write a quadratic function whose graph has these x-intercepts.

   a. 3 and -3 @
   $y = a(x-3)(x+3)$ for $a \neq 0$

   b. 4 and $-\frac{2}{5}$ $y = a(x-4)(x+\frac{2}{5})$ or $y = a(x-4)(5x+2)$ for $a \neq 0$

   c. r_1 and r_2 $y = a(x-r_1)(x-r_2)$ for $a \neq 0$

   d. Use the x-intercepts to find the x-coordinate of the vertex for the quadratic functions in parts a, b, and c.

8. Use the quadratic formula to find the zeros of $y = 2x^2 + 2x + 5$. Explain what happens. Graph $y = 2x^2 + 2x + 5$ to confirm your observation. How can you recognize this situation before using the quadratic formula? ⓗ

9. Write a quadratic function that has no x-intercepts. ⓗ   The function can be any quadratic function for which $b^2 - 4ac$ is negative. Sample answer: $y = x^2 + x + 1$.

10. The axis of symmetry passes vertically through the vertex. To locate this axis, you must find the mean of the roots.

    a. Show that the mean of the two solutions provided by the quadratic formula is $-\frac{b}{2a}$. Explain what this tells you about a graph.

    b. What is the equation of the axis of symmetry for the graph of $y = ax^2 + bx + c$? Explain.

11. In the graph below, three quadratics, $y = ax^2 + bx + c$, are graphed along with $y = 5$. Explain how the graphs tell you the number of solutions you would find in each case if you solved $ax^2 + bx + c = 5$.

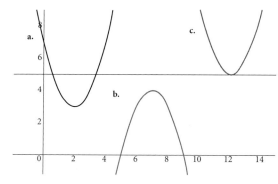

a. two solutions because the parabola crosses the line in two places,

b. no solutions because the parabola and line do not intersect,

c. one solution because the vertex of the parabola is on the line or the parabola and line have only one point in common.

---

8. The solution includes the square root of -36, so there are no real solutions. The graph shows no x-intercepts. Before using the quadratic function, evaluate $b^2 - 4ac$. If $b^2 - 4ac < 0$, then there will be no real solutions.

10a. $\frac{1}{2}\left(\frac{-b + \sqrt{b^2 - 4ac}}{2a} + \frac{-b - \sqrt{b^2 - 4ac}}{2a}\right) = \frac{1}{2}\left(\frac{-2b}{2a}\right)$
$= -\frac{b}{2a}$. The x-coordinate of the vertex, $-\frac{b}{2a}$, is midway between the two x-intercepts.

10b. $x = -\frac{b}{2a}$. The axis of symmetry is a line through the vertex, in this case a vertical line.

Exercise 12 The equation can be solved by the quadratic formula. Only the positive solution, $\frac{1+\sqrt{5}}{2}$, makes sense as a length.

13a. $x^2 + 14x + 49 = (x + 7)^2$ or $x^2 = (-14x) + 49 = [x + (-7)]^2$

13b. $x^2 - 10x - 25 = (x - 5)^2$

13c. $x^2 + 3x + \frac{9}{4} = \left(x + \frac{3}{2}\right)^2$

13d. $2x^2 + 8x + 8 = 2(x^2 + 4x + 4) = 2(x + 2)^2$ or $2x^2 + (-8x) + 8 = 2[x^2 + (-4x) + 4] = 2[x + (-2)]^2$

Exercise 16 Students may need to sketch a diagram to figure out the equation using the Pythagorean Theorem. Assume that the wall is vertical and the ground is horizontal.

See your digital resources for the project **Calculator Program for the Quadratic Formula** where students write a calculator program that uses the quadratic formula to solve problems.

---

12. A **golden rectangle** is a rectangle that can be divided into a square and another smaller rectangle that is similar to the original rectangle. In the figure at right, *ABCD* is a golden rectangle because it can be divided into square *ABFE* and rectangle *FCDE*, and *FCDE* is similar to *ABCD*. Setting up a proportion of the side lengths of the similar rectangles leads to $\frac{a}{a+b} = \frac{b}{a}$. Let $b = 1$ and solve this equation for a.
$\frac{a}{a+1} = \frac{1}{a}, a = \frac{1+\sqrt{5}}{2}$

→ History
CONNECTION

Many people and cultures throughout history have felt that the golden rectangle is one of the most visually pleasing geometric shapes. It was used in the architectural designs of the Cathedral of Notre Dame in Paris, as well as in music and famous works of art. It is believed that the early Egyptians knew the value of the **golden ratio** (the ratio of the length of the golden rectangle to the width) to be $\frac{1+\sqrt{5}}{2}$ and that they used the ratio when building their pyramids, temples, and tombs.

Review

13. Complete each equation.

    a. $x^2 + \underline{?} + 49 = (x + \underline{?})^2$ ⓐ
    b. $x^2 - 10x + \underline{?} = (\underline{?})^2$ ⓐ
    c. $x^2 + 3x + \underline{?} = (\underline{?})^2$
    d. $2x^2 + \underline{?} + 8 = 2(x^2 + \underline{?} + \underline{?}) = \underline{?}(x + \underline{?})^2$

14. Find the inverse of each function. (The inverse does not need to be a function.)

    a. $y = (x + 1)^2$
       $y = \pm\sqrt{x} - 1$
    b. $y = (x + 1)^2 + 4$
       $y = \pm\sqrt{x - 4} - 1$
    c. $y = x^2 + 2x - 5$ ⓐ
       $y = \pm\sqrt{x + 6} - 1$

15. Convert these quadratic functions to general form.

    a. $y = (x - 3)(2x + 5)\ \ y = 2x^2 - x - 15$
    b. $y = -2(x - 1)^2 + 4$
       $y = -2x^2 + 4x + 2$

16. A 20 ft ladder leans against a building. Let *x* represent the distance between the building and the foot of the ladder, and let *y* represent the height the ladder reaches on the building.

    a. Write an equation for *y* in terms of *x*. ⓐ $y = \sqrt{400 - x^2}$

    b. Find the height the ladder reaches on the building if the foot of the ladder is 10 ft from the building. ⓐ approximately 17.32 ft

    c. Find the distance of the foot of the ladder from the building if the ladder must reach 18 ft up the wall. approximately 8.72 ft

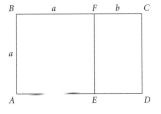

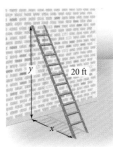

LESSON 5.4

Complex Numbers

"Things don't turn up in the world until somebody turns them up."
JAMES A. GARFIELD

You have explored several ways to solve quadratic equations. You can find the x-intercepts on a graph, you can solve by completing the square, or you can use the quadratic formula. What happens if you try to use the quadratic formula on an equation whose graph has no x-intercepts?

The graph of $y = x^2 + 4x + 5$ at right shows that this function has no x-intercepts. Using the quadratic formula to try to find x-intercepts, you get

$$x = \frac{-4 \pm \sqrt{16 - 4(1)(5)}}{2(1)} = \frac{-4 \pm \sqrt{-4}}{2}$$

How do you take the square root of a negative number? The two numbers $\frac{-4 + \sqrt{-4}}{2}$ and $\frac{-4 - \sqrt{-4}}{2}$ are unlike any of the numbers you have worked with this year—they are nonreal, but they are still numbers.

Throughout the development of mathematics, new sets of numbers have been defined in order to solve problems. Mathematicians have defined fractions and not just whole numbers, negative numbers and not just positive numbers, irrational numbers and not just rational numbers. For the same reasons, we also have square roots of negative numbers, not just square roots of positive numbers. Numbers that include the real numbers as well as the square roots of negative numbers are called **complex numbers**.

History CONNECTION

Since the 1500s, the square root of a negative number has been called an **imaginary number**. In the late 1700s, the Swiss mathematician Leonhard Euler (1707–1783) introduced the symbol i to represent $\sqrt{-1}$. He wrote:

> It is evident that we cannot rank the square root of a negative number amongst possible numbers, and we must therefore say that it is an impossible quantity.... But notwithstanding this these numbers present themselves to the mind; they exist in our imagination, and we still have a sufficient idea of them; since we know that by $\sqrt{-4}$ is meant a number which, multiplied by itself, produces -4; for this reason also, nothing prevents us from making use of these imaginary numbers, and employing them in calculation.

Defining imaginary numbers made it possible to solve previously unsolvable problems.

Leonhard Euler

To express the square root of a negative number, we use an **imaginary unit** called i, defined by $i^2 = -1$ or $i = \sqrt{-1}$. You can rewrite $\sqrt{-4}$ as $\sqrt{4} \cdot \sqrt{-1}$, or $2i$. Therefore, you can write the two solutions to the quadratic equation above as the complex numbers $\frac{-4 + 2i}{2}$ and $\frac{-4 - 2i}{2}$, or $-2 + i$ and $-2 - i$. These two solutions are a **conjugate pair**. That is, one is $a + bi$ and the other is $a - bi$. The two numbers in a conjugate pair are **complex conjugates**. Why will nonreal solutions to the quadratic formula always give answers that are a conjugate pair?

Roots of polynomial equations can be real numbers or nonreal complex numbers, or there may be some of each. If the polynomial has real coefficients, any nonreal roots will come in conjugate pairs such as $2i$ and $-2i$ or $3 + 4i$ and $3 - 4i$.

LESSON 5.4 Complex Numbers **283**

You might project the a copy of the number system tree with only the complex, real, and imaginary numbers boxes filled in and lead the class through filling in the rest of the tree.

Investigate

Example

Consider projecting the example and having students work individually or in pairs. Have students share their solutions and their work. Whether student responses are correct or incorrect, ask other students if they agree and why. Encourage students to use appropriate mathematical language as evidence. Look for students who have used the Quadratic Formula, as well as students who might have graphed the equation, with or without technology. If students use the quadratic formula, emphasize their use of zero coefficients. **SMP 1, 3, 5, 6**

Guiding the Investigation

There is a desmos simulation for the investigation in your ebook.

In this investigation, students will make a conjecture about complex arithmetic. The investigation can be done individually or in groups of 2 – 4. Consider assigning the investigation with little direction, as students will rely on their knowledge of binomial arithmetic. **SMP 1, 3, 5, 6**

Summarize

LANGUAGE An *imaginary number* has only an imaginary component and can be written in the form bi. A *complex number* might have a real part, an imaginary part, or both and can be written in the form $a + bi$ where $b \neq 0$. The set of complex numbers therefore contains the disjoint subsets of real numbers (when $b = 0$) and imaginary numbers (when $a = 0$ and $b \neq 0$), as well as numbers that are neither. You might want to develop a Number System Venn Diagram to review all the subsets of the complex numbers. **SMP 1, 5, 6**

As students share their solutions in the example, remind them that numbers such $\sqrt{-3}$ are usually rewritten as $i\sqrt{3}$ **before** performing any computations. **ALERT** Avoid writing $\sqrt{3}i$, because it's difficult to tell whether i is or isn't under the radical.

Complex Numbers

A **complex number** is a number in the form $a + bi$, where a and b are real numbers and $i = \sqrt{-1}$.

The **conjugate** of the complex number $a + bi$ is $a - bi$.

For any complex number in the form $a + bi$, a is the real part and bi is the imaginary part. The set of complex numbers contains all real numbers and all imaginary numbers. This diagram shows the relationship between these numbers and some other sets you may be familiar with, as well as examples of numbers within each set.

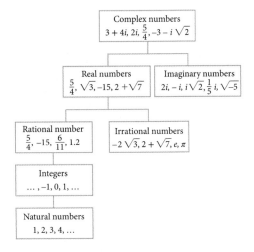

EXAMPLE | Solve $x^2 + 3 = 0$.

Solution You can use the quadratic formula, or you can isolate x^2 and take the square root of both sides.

$$x^2 + 3 = 0$$
$$x^2 = -3$$
$$x = \pm\sqrt{-3}$$
$$x = \pm\sqrt{3} \cdot \sqrt{-1}$$
$$x = \pm\sqrt{3} \cdot i$$
$$x = \pm i\sqrt{3}$$

To check the two solutions, substitute them into the original equation.

$$x^2 + 3 = 0 \qquad\qquad x^2 + 3 = 0$$
$$(i\sqrt{3})^2 + 3 \stackrel{?}{=} 0 \qquad (-i\sqrt{3})^2 + 3 \stackrel{?}{=} 0$$
$$i^2 \cdot 3 + 3 \stackrel{?}{=} 0 \qquad i^2 \cdot 3 + 3 \stackrel{?}{=} 0$$
$$-1 \cdot 3 + 3 \stackrel{?}{=} 0 \qquad -1 \cdot 3 + 3 \stackrel{?}{=} 0$$
$$-3 + 3 \stackrel{?}{=} 0 \qquad -3 + 3 \stackrel{?}{=} 0$$
$$0 = 0 \qquad\qquad 0 = 0$$

The two imaginary numbers $\pm i\sqrt{3}$ are solutions to the original equation, but because they are not real numbers, the graph of $y = x^2 + 3$ shows no x-intercepts.

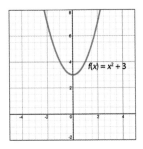

284 CHAPTER 5 Quadratic Functions and Relations

Modifying the Investigation

Whole Class For each part, have students complete the computations and make conjectures individually. Discuss each conjecture as a class before moving on to the next part.

Shortened There is no Shortened version for this investigation.

One Step Have students change the mode on their calculators to one which allows them to enter imaginary numbers. Use the **One Step** in your ebook. As they begin finding results that involve i, challenge them to figure out what i means and to try to find its properties. Have the class develop rules for operating on complex numbers, following the rules for binomials.

Complex numbers have many applications, particularly in science and engineering. To measure the strength of an electromagnetic field, a real number represents the amount of electricity and an imaginary number represents the amount of magnetism. The state of a component in an electronic circuit is also measured by a complex number, where the voltage is a real number and the current is an imaginary number.

The properties of calculations with complex numbers apply to these types of physical states more accurately than calculations with real numbers do. In the investigation you'll explore patterns in arithmetic with complex numbers.

A surreal scene composed of ordinary objects juxtaposed with an imaginary element.

INVESTIGATION

Complex Arithmetic

When computing with complex numbers, there are conventional rules similar to those you use when working with real numbers. In this investigation you will discover these rules. You may use your calculator to check your work or to explore other examples.

Part 1: Addition

Conjecture: You add the real parts, and you add the imaginary parts.

Addition of complex numbers is similar to combining like terms. Use your calculator to add each pair of complex numbers. Make a conjecture about how to add complex numbers without a calculator.

a. $(2 - 4i) + (3 + 5i)$ $5 + i$
b. $(7 + 3i) + (-2 + i)$ $5 + 4i$
c. $(-1 + 5i) + (6 - 11i)$ $5 - 6i$
d. $(-4 + 3i) + (-8 - 9i)$ $-12 - 6i$

Part 2: Subtraction

Conjecture: You subtract the real parts, and you subtract the imaginary parts.

Subtraction of complex numbers is also similar to combining like terms. Use your calculator to subtract each pair of complex numbers. Make a conjecture about how to subtract complex numbers without a calculator.

a. $(2 - 4i) - (3 + 5i)$ $-1 - 9i$
b. $(7 + 3i) - (-2 + i)$ $9 + 2i$
c. $(-1 + 5i) - (6 - 11i)$ $-7 + 16i$
d. $(-4 + 3i) - (-8 - 9i)$ $4 + 12i$

Part 3: Multiplication

Use your knowledge of multiplying binomials to multiply these complex numbers. Express your products in the form $a + bi$. Recall that $i^2 = -1$.

a. $(2 - 4i)(3 + 5i)$ $26 - 2i$
b. $(7 + 2i)(-2 + i)$ $-16 + 3i$
c. $(2 - 4i)^2$ $-12 - 16i$
d. $(4 - 4i)(1 - 3i)$ $-8 - 16i$
e. $(7 + 3i)(-2 + i)$ $-17 + i$

You have seen that arithmetic with complex numbers is not much different than arithmetic with real numbers. When you are simplifying expressions, just remember that $i^2 = -1$.

LESSON 5.4 Complex Numbers **285**

Formative Assessment

Student input in the class discussion of the opening graph and number system tree, as well as responses to Exercises 4, 5, 7, and 10–12 will provide evidence of defining complex numbers and identifying the conjugate of a complex number. Student responses to the Example and Exercises 3, 9, and 12 will provide evidence of finding nonreal solutions as conjugate pairs. Student responses to the Investigation and Exercises 1 and 2 will provide evidence of understanding complex arithmetic with addition, subtraction and multiplication.

Apply

Extra Example
Multiply $(-3 + 2i)(4 - 6i)$. $26i$

Closing Question
Solve: $x^2 + x = -5$.
$x = \dfrac{-1 \pm i\sqrt{19}}{2}$

Assigning Exercises
Suggested: 1–12
Additional Practice: 13–16

4. $(-3 + i)^2 + 6(-3 + i) + 10 =$
$9 - 6i - 1 - 18 + 6i + 10 = 0$,
$(-3 - i)^2 + 6(-3 - i) + 10 =$
$9 + 6i - 1 - 18 - 6i + 10 = 0$

Exercise 5 This exercise reinforces the concepts in the tree diagram earlier in this lesson. As needed, review the properties of Venn diagrams.

LANGUAGE *Taxonomy* means "an orderly classification."

5a.
5b.

5c.
5d.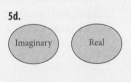
5e.

5.4 Exercises

You will need your graphing calculator for Exercise **15** and **16**.

Practice Your Skills

1. Add or subtract.
   a. $(5 - 1i) + (3 + 5i)$ $8 + 4i$
   b. $(6 + 2i) - (-1 + 2i)$ @ 7
   c. $(2 + 3i) + (2 - 5i)$ @ $4 - 2i$
   d. $(2.35 + 2.71i) - (4.91 + 3.32i)$ $-2.56 - 0.61i$

2. Multiply.
   a. $(5 - 1i)(3 + 5i)$ @ $20 + 22i$
   b. $6(-1 + 2i)$ $-6 + 12i$
   c. $3i(2 - 5i)$ @ $15 + 6i$
   d. $(2.35 + 2.71i)(4.91 + 3.32i)$ $2.5413 + 21.1081i$

3. Solve each equation.
   a. $x^2 + 4 = 0$ $x = 2i, -2i$
   b. $x^2 + 12 = 3$ $x = 3i, -3i$
   c. $x^4 - 16 = 0$ $x = 2, -2, 2i, -2i$

4. Show that $-3 + i$ and $-3 - i$ are both solutions of $x^2 + 6x + 10 = 0$ by substituting them for x and showing that they satisfy the equation.

5. Draw **Venn diagrams** to show the relationships between these sets of numbers.
   a. real numbers and complex numbers @
   b. rational numbers and irrational numbers
   c. imaginary numbers and complex numbers @
   d. imaginary numbers and real numbers
   e. complex numbers, real numbers, and imaginary numbers

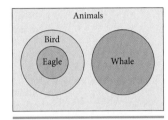

Named after English logician John Venn (1834–1923), Venn diagrams represent logic relationships. For example, this diagram shows all eagles are birds and no whales are birds.

Reason and Apply

6. Rewrite this quadratic equation in general form.
   $[x - (2 + i)][x - (2 - i)] = 0$ $x^2 - 4x + 5 = 0$

7. Use the definitions $i = \sqrt{-1}$ and $i^2 = -1$ to rewrite each power of i as 1, i, -1, or $-i$.
   a. i^3 @ $-i$
   b. i^4 @ 1
   c. i^5 i
   d. i^{10} -1

8. Give examples of each type of number or explain why no such numbers exist.
   a. A number that is both complex and real.   any real number will satisfy this condition
   b. Two complex numbers whose sum is a real number.   any pair of complex numbers whose imaginary parts are equal in magnitude, but opposite in sign
   c. Two imaginary numbers whose product is an imaginary number.   not possible, the product of two imaginary numbers always contains i^2 and thus is real
   d. Two non-real complex numbers, that are not conjugates, whose product is a real number.   any complex numbers such that one is a multiple of the conjugate of the other.

286 CHAPTER 5 Quadratic Functions and Relations

Exercise 7 ASK "In general, how can you find another name for i^n?" [You can simplify any power of i, i^n, by dividing n by 4. If the remainder is 1, $i^n = i$; if the remainder is 2, $i^n = -1$; if the remainder is 3, $i^n = -i$; if the remainder is 0, $i^n = 1$.]

9. Solve each equation. Use substitution to check your solutions. Label each solution as real, imaginary, and/or complex.

   a. $x^2 - 1 = 0$ ± 1; complex and real
   b. $x^2 + 1 = 0$ $\pm i$; complex and imaginary
   c. $x^2 - 4x + 6 = 0$ $2 \pm i\sqrt{2}$; complex
   d. $x^2 + x = -1$ $-\frac{1}{2} \pm \frac{i\sqrt{3}}{2}$; complex
   e. $-2x^2 + 4x = 3$ $1 \pm \frac{i\sqrt{2}}{2}$; complex

10. Write a quadratic function in general form that has the given zeros, and leading coefficient of 1.

    a. $x = -3$ and $x = 5$ $y = x^2 - 2x - 15$
    b. $x = -3.5$ (double root) $y = x^2 + 7x + 12.25$
    c. $x = 5i$ and $x = -5i$ $y = x^2 + 25$
    d. $x = 2 + i$ and $x = 2 - i$ $y = x^2 - 4x + 5$

11. Write a quadratic function with real coefficients in general form that has a zero of $x = 4 + 3i$ and whose graph has y-intercept 50. $y = 2x^2 - 16x + 50$

12. Solve.

    a. $x^2 - 10ix - 9i^2 = 0$ $x = (5 + \sqrt{34})i \approx 10.83i$ or $x = (5 - \sqrt{34})i \approx -0.83i$
    b. $x^2 - 3ix = 2$ $x = 2i$ or $x = i$
    c. Why don't the solutions to 12a and 12b come in conjugate pairs? The coefficients of the quadratic equations are nonreal.

13. The quadratic formula, $x = \dfrac{-b \pm \sqrt{b^2 - 4ac}}{2a}$, provides solutions to $ax^2 + bx + c = 0$. The quantity inside the square root, $b^2 - 4ac$, is called the discriminant of the quadratic equation. How can you use the value of the discriminant to determine each of these solution possibilities?

    a. The solutions are nonreal. $b^2 - 4ac < 0$
    b. The solutions are real. $b^2 - 4ac \geq 0$
    c. There is only one real solution. $b^2 - 4ac = 0$

14. Use these recursive formulas to find the first six terms (z_0 to z_5) of each sequence. Describe what happens in the long run for each sequence.

    a. $z_0 = 0$
       $z_n = z_{n-1}^2 + 0$ where $n \geq 1$
       0, 0, 0, 0, 0, 0; remains constant at 0
    b. $z_0 = 0$
       $z_n = z_{n-1}^2 + i$ where $n \geq 1$
       0, i, $-1 + i$, $-i$, $-1 + i$, $-i$; alternates between $-1 + i$ and $-i$
    c. $z_0 = 0$
       $z_n = z_{n-1}^2 + 1 - i$ where $n \geq 1$
       0, $1 - i$, $1 - 3i$, $-7 - 7i$, $1 + 97i$, $-9407 + 193i$; no recognizable pattern in these six terms
    d. $z_0 = 0$
       $z_n = z_{n-1}^2 + 0.2 + 0.2i$ where $n \geq 1$

Review

15. Consider the function $y = 2x^2 + 6x - 3$.

    15 a. $x = \dfrac{-3 \pm \sqrt{15}}{2}$; $x = -3.44$ or $x = 0.44$

    a. List the zeros in exact radical form and as approximations to the nearest hundredth.
    b. Graph the function and label the exact coordinates of the vertex and points where the graph crosses the x-axis and the y-axis.

16. Consider two positive integers that meet these conditions:
    - three times the first added to four times the second is less than 30
    - twice the first is less than five more than the second

    a. Define variables and write a system of linear inequalities that represents this situation.
    b. Graph the feasible region.
    c. List all integer pairs that satisfy the conditions listed above.

    (1, 1), (2, 1), (1, 2), (2, 2), (3, 2), (1, 3), (2, 3), (3, 3), (1, 4), (2, 4), (3, 4), (4, 4), (1, 5), (2, 5), (3, 5), (1, 6)

16a. Let x represent the first integer, and let y represent the second integer.

$$\begin{cases} x > 0 \\ y > 0 \\ 3x + 4y < 30 \\ 2x < y + 5 \end{cases}$$

16b.
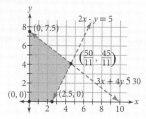

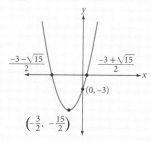

LESSON 5.5

Solving Quadratic Equations

In this lesson students revisit ways to solve quadratic equations, with an emphasis on which strategy is best for a situation. They learn to use the discriminant as a tool for deciding on an appropriate strategy.

COMMON CORE STATE STANDARDS
APPLIED
N.CN.7
A.SSE.1a
A.SSE.1b
A.SSE.2
A.SSE.3a
A.CED.1

DEVELOPED
A.REI.4a
A.REI.4b

INTRODUCED

Objectives
- Solve quadratic equations using a variety of appropriate strategies
- Interpret the discriminant to identify the nature of the roots of a quadratic equation and to identify a solution strategy
- Derive the quadratic formula from an equation of the form $(x - p)^2 = q$.

Vocabulary
discriminant

Launch

Determine a quick way to solve each quadratic equation. Explain why you chose that method. Then find the solution.

a. $-2(x - 9)(x + 7) = 0$
b. $5x^2 - 11 = 0$
c. $x(x + 11) = 12$
d. $2(x - 3)^2 + 3 = 21$
e. $x^2 - 16 = 5x$

a. Solve by looking at it. It is in factored form. $x = 9, -7$
b. The variable appears only once, so solve by undoing. $x = \pm\sqrt{2.2}$
c. Expand the left side and set equation equal to 0. Factor. $x = -12, 1$
d. The variable appears only once, so solve by undoing. $x = 0, 6$
e. Group terms on one side of the equation. Use the quadratic formula. $x = \frac{5 \pm \sqrt{89}}{2}$

LESSON 5.5

Solving Quadratic Equations

"All progress is precarious, and the solution of one problem brings us face to face with another problem."
—MARTIN LUTHER KING, JR.

There are many ways to solve quadratic equations. The one shown here won't get you much credit, but it might make your teacher smile. Sorting out which method to use—factoring, completing the square, or the quadratic formula—can be challenging, especially if there aren't any real solutions.

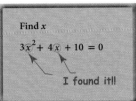

Find x

$3\cancel{x}^2 + 4\cancel{x} + 10 = 0$

I found it!!

Practicing the different methods and thinking about which one is best in each situation is important. In this lesson, you will have the opportunity to revisit these methods and gain confidence in your ability to decide which method is appropriate for a situation.

INVESTIGATION

The Quadratic Formula Revisited

In a previous lesson, you saw a classic derivation of the quadratic formula. In this investigation you will explore a different way to derive this formula.

Step 1
$3x^2 + 5x = 7$
$36x^2 + 60x = 84$

Step 1 Start with the equation $3x^2 + 5x - 7 = 0$. Add 7 to both sides of the equation and then multiply both sides by $4(3)$, or 12.

Step 2 Complete the square for the expression on the left side of the equation by filling in the rectangle diagram below.

	$6x$	5
$6x$	$36x^2$	$30x$
5	$30x$	25

$36x^2 + 60x + 25 = 84 + 25$

Step 3 Write the equation in the form $(\underline{6x} + \underline{5})^2 = \underline{109}$.

Step 4 Solve the equation by taking the square root of both sides. $x = \frac{-5 \pm \sqrt{109}}{6}$

Step 5 Repeat Steps 1–4, using the equation $ax^2 + bx + c = 0$.

Step 6 Why was it important to multiply by $4a$? Would another value have worked just as well? Explain.

We know that $4a$ always works because it makes the first term a perfect square. It turns out you only need to multiply by a, not $4a$, if the b-value is even. If the a-value is already a perfect square, then you only need to multiply by 4.

How can you tell if an equation factors? One clue is found in the quadratic formula. If an equation with integer coefficients can be factored with integers, then the solutions found by the quadratic formula will be rational. That means there won't be any square roots of non-perfect square numbers in the solutions. So, looking at the value of the expression inside the square root in the quadratic formula can give you information about the solutions of the equation.

288 CHAPTER 5 Quadratic Functions and Relations

ELL
The lesson is very mechanical. Consider having students share their work with the class so they can discuss their steps. To help students make a connection with discriminant, explain the term comes from a Latin word meaning to separate or divide out.

Support
Consider having students work together to create a list of criteria for when each solution method is most beneficial. If necessary, you can use the Refreshing Your Skills: Solving Equations lesson.

Advanced
Have students generate examples of quadratic expressions and determine the best method for solving with justification for their selection.

> **Discriminant**
> In a quadratic equation $ax^2 + bx + c = 0$, the value $b^2 - 4ac$ is called the **discriminant**.
> If $b^2 - 4ac < 0$, the equation has two complex solutions.
> If $b^2 - 4ac = 0$, the equation has only one solution.
> If $b^2 - 4ac > 0$, the equation has two real solutions.
> If $b^2 - 4ac$ is a perfect square, then the solutions are rational. That is, the equation can be factored.

The value of the discriminant is closely related to the graph of the quadratic function. When the discriminant is negative, there are no real solutions. This means the graph does not cross the *x*-axis. When it is zero, the two solutions are the same. This corresponds to the graph having the vertex on the *x*-axis. When the discriminant is positive, the graph crosses the *x*-axis in two distinct places. Their *x*-values are the solutions to the equation.

$y = x^2 + x + 1$ $y = x^2 + 2x + 1$ $y = x^2 + 3x + 1$
$b^2 - 4ac = -3$ $b^2 - 4ac = 0$ $b^2 - 4ac = 5$

      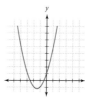

two complex solutions   one real solution   two real solutions

EXAMPLE A | Solve the equation $6x^2 - 7x - 20 = 0$.

Solution | The value of the discriminant is $(-7)^2 - 4(6)(-20) = 49 + 480 = 529$. This is the perfect square, $529 = 23^2$, so the left side of the equation will factor. Now you have the choice of factoring or just using the quadratic formula.

The quadratic formula gives the solutions $x = \frac{(7 \pm 23)}{12}$, which simplifies to $x = \frac{5}{2}$ and $x = -\frac{4}{3}$.

Factoring is more of a challenge, but if you like puzzles then…

To factor this equation, you start by noting that the product of the first and third term is $-120x^2$ and you are looking at values that are $-7x$ apart. After a short search, you find $8x$ and $-15x$ will give the correct sum and product. The common factors can then be found for each row and column of the box.

Step 1

$6x^2$	$?x$
$?x$	-20

Step 2

$6x^2$	$-15x$
$8x$	-20

Step 3

	$2x$	-5
3	$6x^2$	$-15x$
4	$8x$	-20

LESSON 5.5 Solving Quadratic Equations **289**

Investigate

Guiding the Investigation

This investigation is best done individually or in pairs.

Step 5

$$ax^2 + bx = -c$$
$$4a^2x^2 + 4abx = -4ac$$
$$4a^2x^2 + 4abx + b^2 = b^2 - 4ac$$
$$(2ax + b)^2 = b^2 - 4ac$$
$$2ax + b = \pm\sqrt{b^2 - 4ac}$$
$$2ax = -b \pm \sqrt{b^2 - 4ac}$$
$$x = \frac{-b \pm \sqrt{b^2 - 4ac}}{2a}$$

To begin the discussion of the discriminant and its properties, consider projecting three equations and asking students to solve using the quadratic formula and then graph.

$y = x^2 + x + 1$

$y = x^2 + 2x + 1$

$y = x^2 + 3x + 1$

After they present their solutions to the three equations **ASK** "What do you notice about the number under the radical in the quadratic formula and the graph? Why do you think that happens? Do you think that will always be true?" **SMP 1, 8**

Example A

Project Example A and work through it as a class. **ASK** "What's the discriminant of this equation?" [529] What does that tell us about the nature of the roots? [It's positive so there are two real solutions]" Have students solve the equation. Some will use the quadratic formula and some may try to factor. Have students present their solutions, choosing at least one group who used each method. Whether student responses are correct or incorrect, ask other students if they agree and why. **SMP 3, 6 ASK** "In this instance, which strategy, the quadratic formula or factoring, do you prefer? Why?"

Modifying the Investigation

Whole Class Do the Investigation as a class activity, soliciting input from students for each step.

Shortened Do Steps 1 – 4 as a class, have students do Step 5. Discuss Step 6 as a class.

One Step Project the equation $ax^2 + bx + c = 0$. Ask students to solve by completing the square. **ASK** "What did you multiply *a* by and why?"

Conceptual/Procedural

Conceptual Using the discriminant to evaluate the kind of roots a quadratic equation might have helps students conceptualize complex numbers and *x*-intercepts as roots of an equation.

Procedural Students review the procedures for solving quadratic equation. They are introduced to a procedure for using substitution to solve fourth degree equations that can be re-written as a quadratic equation.

Example B

This example asks students to solve a fourth degree equation, not a quadratic. Project the equation and ask for suggestions on how to solve a fourth degree equation. If the idea of substitution does not come up **ASK** "Would it be easier to solve if the equation was $u^2 + 7u - 8 = 0$?" Solicit the response that substituting $u = x^2$, would give that equation. Have students solve the quadratic equation. Have students volunteer to present their solutions and their work. **ALERT** Students may forget to substitute their answer for u back into the equation $u = x^2$ to find the value of x. Whether student responses are correct or incorrect, ask other students if they agree and why. Encourage students to use appropriate mathematical language as evidence. **SMP 3, 6, 8**

Summarize

As students present their responses, have them explain their work. Whether student responses are correct or incorrect, ask other students if they agree and why. Encourage students to use appropriate mathematical language as evidence. **SMP 3, 6**

In the discussion of the discriminant, help students see the connection with the value of the discriminant and the nature of the roots.

CRITICAL QUESTION What purpose does the discriminant serve?

BIG IDEA It tells you the nature of the roots of a quadratic equation.

Formative Assessment

Student responses to the Investigation provide evidence of deriving the quadratic formula from an equation of the form $(x - p)^2 = q$. Student solutions to and discussions of the examples and Exercises 3–5, 7, 10, and 11 provide evidence of solving quadratic equations using a variety of appropriate strategies. Student input in the discussion of the discriminant and solutions to Exercises 2, 6, 8, and 9 provide evidence of interpreting the discriminant to identify the nature of the roots of a quadratic equation and identifying a solution strategy.

So the factored form of $6x^2 - 7x - 20$ is $(2x - 5)(3x + 4)$. Using the zero-product property to solve $(2x - 5)(3x + 4) = 0$, the solutions to $2x - 5 = 0$ and $3x + 4 = 0$ are $x = \frac{5}{2}$ and $x = -\frac{4}{3}$.

Which strategy, the quadratic formula or factoring, do you prefer? Why?

In addition to quadratic equations like the one in Example A, there are other types of equations involving higher degrees of polynomials that can be solved using the same methods.

EXAMPLE B | Solve the equation $x^4 + 7x^2 - 8 = 0$.

Solution | This equation is a fourth degree equation, not a quadratic. However, it resembles a quadratic equation because it can be written as

$$(x^2)^2 + 7(x^2) - 8 = 0$$

If we make the substitution $u = x^2$, the equation becomes the quadratic equation

$$u^2 + 7u - 8 = 0$$

The discriminant, $(49 + 32 = 81)$, is a perfect square, so you can solve this equation by factoring.

$$u^2 + 7u - 8 = 0$$
$$(u + 8)(u - 1) = 0$$

So $u = -8$ or $u = 1$.

Substitute x^2 back into the equation to get $x^2 = -8$ or $x^2 = 1$.

You can solve each of these equations by taking the square root of each side.

$$x = \pm\sqrt{(-8)} = \pm i\sqrt{8} = \pm 2i\sqrt{2} \text{ and } x = \pm 1.$$

There are four solutions to this fourth degree equation.

This example shows that equations that are not quadratic, but are similar in form to quadratics, can be solved using some of the same procedures you have learned. You will continue to use these procedures as you study different areas of mathematics later in the course.

5.5 Exercises

You will need your graphing calculator for Exercises **2** and **9**.

Practice Your Skills

1. Write each equation in standard form $ax^2 + bx + c = 0$. What are the values of a, b, and c in each equation?

   a. $x^2 + 2x = 24$
   $x^2 + 2x - 24 = 0$, $a = 1$, $b = 2$, $c = -24$
   c. $25 - x^2 = 16 - 8x$
   $x^2 - 8x - 9 = 0$, $a = 1$, $b = -8$, $c = -9$

   b. $x^2 + 9 = 6x$ @
   $x^2 - 6x + 9 = 0$, $a = 1$, $b = -6$, $c = 9$
   d. $(x + 4)^2 = 5x$
   $x^2 + 3x + 16 = 0$, $a = 1$, $b = 3$, $c = 16$

2. For each quadratic equation, find the value of the discriminant and classify what type of solutions the equation will have. You do not need to find the solutions. Use a graph to support your answer.

   a. $x^2 + 5x + 14 = 0$ @ −31, solutions are complex, the graph does not intersect the x-axis.
   c. $3x^2 - 100 = 0$ 1200, solutions are real, the graph intersects the x-axis twice.

   b. $x^2 - 12x - 27 = 0$ 252, solutions are real, the graph intersects the x-axis twice.
   d. $2x^2 - 4x + 5 = 0$ −24, solutions are complex, the graph does not intersect the x-axis.

290 CHAPTER 5 Quadratic Functions and Relations

Apply

Extra Example

What is the nature of the roots of $4x^2 - 4x + 1 = 0$?

One real, rational root.

Closing Question

Which of the following could describe the roots of a quadratic equation? How do you know?

i. one complex root
ii. one rational root
iii. one irrational root
iv. all of the above

ii The only way to get one root for a quadratic equation is to have a discriminant that is a perfect square, officially a double root.

3. Solve each equation by factoring.

   a. $x^2 + 5x - 14 = 0$
      $(x + 7)(x - 2) = 0, x = -7, 2$
   b. $x^2 - 12x + 27 = 0$
      $(x - 3)(x - 9) = 0, x = 3, 9$
   c. $x^2 - 100 = 0$
      $(x - 10)(x + 10) = 0, x = 10, -10$
   d. $x^3 + 4x^2 - 5x = 0$ @
      $x(x + 5)(x - 1) = 0\ x = 0, -5, 1$
   e. $2x^2 + 7x + 5 = 0$
      $(2x + 5)(x + 1) = 0, x = -1, -\frac{5}{2}$
   f. $3x^2 - 12 = 0$
      $3(x - 2)(x + 2) = 0, x = 2, -2$

4. Solve each equation by completing the square.

   a. $x^2 + 8x = -1$ $(x + 4)^2 = 15; x = -4 \pm \sqrt{15}$
   b. $x^2 - 10x + 30 = 0$ @ $(x - 5)^2 = -5; x = 5 \pm i\sqrt{5}$
   c. $x^2 + 6x - 2 = 0$ $(x + 3)^2 = 11; x = -3 \pm \sqrt{11}$
   d. $x^2 - 2x = -4$ $(x - 1)^2 = -3; x = 1 \pm i\sqrt{3}$

5. Solve each equation using the quadratic formula.

   a. $3x^2 - 5x + 1 = 0$
      $x = \frac{-(-5) \pm \sqrt{(-5)^2 - 4(3)(1)}}{2(3)} = \frac{5 \pm \sqrt{13}}{6}$
   b. $2x^2 + 6x = 7$ @
      $x = \frac{-(6) \pm \sqrt{(6)^2 - 4(2)(-7)}}{2(2)} = \frac{-6 \pm \sqrt{92}}{4}$
   c. $x^2 + 11x - 5 = 0$
      $x = \frac{-(11) \pm \sqrt{(11)^2 - 4(1)(-5)}}{2(1)} = \frac{-11 \pm \sqrt{141}}{2}$
   d. $5x^2 - x = 8$
      $x = \frac{-(-1) \pm \sqrt{(-1)^2 - 4(5)(-8)}}{2(5)} = \frac{1 \pm \sqrt{161}}{10}$
   e. $4x^2 + 2x = -3$
      $x = \frac{-(2) \pm \sqrt{(2)^2 - 4(4)(3)}}{2(4)} = \frac{-2 \pm i\sqrt{44}}{8}$
   f. $x^2 - 3x + 18 = 7$
      $x = \frac{-(-3) \pm \sqrt{(-3)^2 - 4(1)(11)}}{2(1)} = \frac{3 \pm i\sqrt{35}}{2}$

Reason and Apply

6. Match each quadratic equation on the left with the method on the right that is most appropriate. You may use each method only once. Give a reason for your choice.

   a. $-4.9x^2 + 35.7x + 11.391 = 0$ iii
   b. $x^2 + 4x - 2 = 0$ i
   c. $x^2 - 5x + 6 = 0$ ii
   d. $x^2 + 4x + 7 = 0$ iv

   i. completing the square
   ii. factoring
   iii. graphing
   iv. quadratic formula

7. Suppose the solutions to a quadratic equation are $x = \frac{2}{3}$ and $x = 4$.

   a. Write a quadratic equation in factored form with these solutions. @ possible answer is $(x - \frac{2}{3})(x - 4) = 0$
   b. Write a quadratic equation in standard form with these solutions. possible answer is $x^2 - \frac{14}{3}x + \frac{8}{3} = 0$
   c. Is it possible to write a different quadratic equation in standard form with the same solutions? If so, write one. If not, explain why not.
      any multiple of the previous answer will work, such as $3x^2 - 14x + 8 = 0$.

8. Consider the equation $ax^2 + 5x + 1 = 0$. Use the quadratic formula to help you solve this problem.

   a. For what values of a does the equation have two real solutions? Explain. (h) $a < \frac{25}{4}$
   b. For what values of a does the equation have two complex solutions? Explain. $a > \frac{25}{4}$
   c. Find a value of a so that the equation has a single solution. Explain. $a = \frac{25}{4}$
   d. Write a quadratic equation with the same single solution as in part c, but with no fractions or decimals in the leading coefficient. $25x^2 + 20x + 4 = 0$

9. Consider the equation $3x^2 + bx + 3 = 0$

   a. For what values of b does the equation have two real solutions? Explain. $b < -6$ or $b > 6$
   b. For what values of b does the equation have two complex solutions? Explain. $-6 < b < 6$
   c. For what values of b does the equation have a single solution? Explain. @ $b = 6$ or -6
   d. Graph the equations using the values of b from part c. Describe the relationship between them. The graphs are reflections over the y-axis

LESSON 5.5 Solving Quadratic Equations **291**

Assigning Exercises

Suggested: 1 – 11
Additional Practice: 12 – 15

6a. iii. graphing; the values of a, b, and c will be difficult to use in any method other than graphing with a graphing utility.

6b. i. completing the square; the value of c is a prime number and the factors don't add up to the value of b, so the equation won't factor. Since the value of b is even, completing the square would be a good method to use.

6c. ii. factoring; the values of a, b, and c are small, making it easy to see that the factors of c can be added to get the value of b.

6d. iv. quadratic formula; the value of c is a prime number and the factors don't add up to the value of b, so the equation won't factor. The values of a, b, and c are integers that can easily be used in the quadratic formula.

8a. To get two real solutions, $b^2 - 4ac > 0$, so $25 - 4a > 0; a < \frac{25}{4}$

8b. To get complex solutions, $b^2 - 4ac < 0$, so $25 - 4a < 0; a > \frac{25}{4}$

8c. To get a single solution, $b^2 - 4ac = 0$, so $25 - 4a = 0; a = \frac{25}{4}$

9a. To get two real solutions, $b^2 - 4ac > 0$, so $b^2 - 36 > 0; b < -6$ or $b > 6$

9b. To get complex solutions, $b^2 - 4ac < 0$, so $b^2 - 36 < 0, -6 < b < 6$

9c. To get one real solution, $b^2 - 4ac = 0$ so $b^2 - 36 = 0; b = 6$ or -6

10. Solve the equation $(x + 2)^2 + 7(x + 2) + 12 = 0$. ⓗ $x = -5, -6$

11. Solve the equation $(x^2 + 4)^2 - 3(x^2 + 4) - 10 = 0$. $x = 1, -1, i\sqrt{6}, -i\sqrt{6}$

12. Taoufik has climbed to the top of an apple tree. Finding a good apple, he tosses it up into the air with an initial upward velocity of 60 feet per second. Its height, in feet, is given by the equation $h = -16t^2 + 60t + 16$.

    a. How high up in the tree was Taoufik when he threw the apple? ⓐ
       16 feet
    b. At what time(s) is the apple 52 feet in the air? $\frac{3}{4}$ second and 3 seconds
    c. At what time does the apple hit the ground?
       4 seconds

The apple is a sweet fruit that comes from a flowering tree and tends to fall off at maturity

Review

13. Andrea leaves Littletown at 8 am and drives toward Midville at 65 miles per hour. Brandon leaves Midville at 10 am and drives toward Littletown at 67 miles per hour on the same route. It is 262 miles between the cities. When and where do they meet?
    They meet at 11 am, 195 miles from Littletown

14. Solve the system of equations.

    $3x + 7y = 27$
    $4x + 6y = 21$ $\left(-\frac{3}{2}, \frac{9}{2}\right)$

15. Find the vertex of the parabola $y = x^2 + 12x - 7$ by completing the square.

    $y = (x + 6)^2 - 43$ The vertex is at $(-6, -43)$

PERFORMANCE TASK

In a previous lesson, you learned that the sum of the roots of a quadratic equation was $-\frac{b}{a}$ and the product of the roots was $\frac{c}{a}$. Do these properties still hold if the values of a, b, and c are not all real?

a. Solve the equation $x^2 + 2ix + 3 = 0$. Then find the sum and product of the roots.

b. Solve the equation $x^2 + 2x + 3i = 0$. Then find the sum and product of the roots.

c. For the equations in parts a and b, compare the sums and products with the values of $-\frac{b}{a}$ and $\frac{c}{a}$. Do the properties you found earlier still hold true?

PERFORMANCE TASK

a. $x = i, -3i$; the sum of the roots is $-2i$ and the product is 3

b. $x = \dfrac{-2 \pm \sqrt{4 - 12i}}{2}$, or $-1 \pm \sqrt{1 - 3i}$; the sum of the roots is -2 and the product is $3i$

c. for part a, the sum of the roots is $-2i$, which is $-\frac{b}{a}$, and the product is 3, which is $\frac{c}{a}$, so the properties hold. For part b, the sum of the roots is -2, which is $-\frac{b}{a}$, and the product is $3i$, which is $\frac{c}{a}$, so the properties hold.

LESSON 5.6

Solving Radical Equations

"The roots of education are bitter, but the fruits are sweet"
ARISTOTLE

You have learned to solve quadratic equations in many ways. Sometimes your answers contained square roots. What if the equation starts out with a square root in it? How would you solve it?

Remember that when you square the square root of a positive number, you get that number back. For instance, $(\sqrt{3})^2 = 3$. You can use this idea to solve an equation containing a square root.

History CONNECTION

Aristotle (384 BCE – 322 BCE) was a Greek philosopher and scientist. His writings cover many subjects — including physics, biology, zoology, metaphysics, logic, ethics, poetry, theater, music, rhetoric, linguistics, politics and government — and constitute the first comprehensive system of Western philosophy.

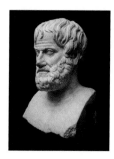

EXAMPLE A | Solve the equation $\sqrt{x+3} + 4 = 10$.

Solution | First, isolate the square root to one side of the equation.

$\sqrt{x+3} = 6$ subtract 4 from each side of the equation
$(\sqrt{x+3})^2 = 6^2$ square each side
$x + 3 = 36$ simplify the equation
$x = 33$ subtract 3 from each side of the equation

Check that the solution works in the original equation.

$\sqrt{(33+3)} + 4 = 10$
$\sqrt{36} + 4 = 10$
$6 + 4 = 10$ the solution works.

To simplify the mathematics and minimize the errors, it is important to always isolate the square root before you square both sides of the equation. In the investigation you'll discover some unusual things that can happen when you solve equations involving square roots.

LESSON 5.6 Solving Radical Equations

LESSON 5.6

In this lesson students learn to solve radical equations. They learn to identify extraneous roots by checking to see if their solutions make the original equation true. They also learn to graph both sides of an equation to see how many roots there are.

COMMON CORE STATE STANDARDS		
APPLIED	**DEVELOPED**	**INTRODUCED**
A.SSE.1b	A.REI.2	
A.SSE.2		
A.REI.1		

Objectives

- Solve radical equations symbolically and graphically
- Identify extraneous roots

Launch

Solve for *x*.

a. $x^2 = 64$ $x = \pm 8$
b. $\sqrt{x} = 9$ $x = 81$
c. $\sqrt{x} - 4 = 6$ $x = 100$
d. $\sqrt{2x^2 - 121} = x$ $x = \pm 11$

Investigate

Example A

Consider projecting the equation and work through it as a class, with student input for each step. Ask for justification of the step. If students suggest squaring both sides of the equation first, follow their instruction. Whether student responses are correct or incorrect, ask other students if they agree and why. Encourage students to use appropriate mathematical language as evidence. **SMP 1, 3, 6** If no one suggests isolating the radical, **ASK** "Can anyone think of a way that might make squaring the left side less "messy?" Lead them to the idea of isolating the radical before you square both sides of the equation. Include checking to see that the solution makes the original equation true.

ELL
Discuss the definition of *extraneous* and ask students to find examples of extraneous objects in their daily lives (e.g., appendix, social conversation during math class)

Support
The Refreshing Your Skills: Working with Positive Square Roots and the Refreshing Your Skills: Solving Equations lessons can be used for students who need to review simplifying and finding the value of positive square roots or solving equations.

Advanced
Ask students to consider how they would solve an equation with a square root and a cube root.

Guiding the Investigation

This investigation might work best if students work individually or in pairs.

Example B

Project the equation. Have students work in pairs. Have students share their solution and explain their work. Whether student responses are correct or incorrect, ask other students if they agree and why. Encourage students to use appropriate mathematical language as evidence. **SMP 1, 3, 6**

Summarize

In solving radical equations, it is easiest to isolate the radical, then square both sides of the equation. This sometimes has to be done more than once.

CRITICAL QUESTION What is an extraneous root?

BIG IDEA An extraneous root is a solution that arises from performing an operation like squaring both sides of an equation that is not a solution to the original equation.

Formative Assessment

Student presentations and justification of their work on the examples and exercises can provide evidence of solving radical equations symbolically and graphically. Student presentations and justification of their work on the investigation and exercises can provide evidence of identifying extraneous roots.

Apply

Extra Example

Solve and check:

$6 - \sqrt{x} = 11$ No solution

Closing Question

What extraneous solutions arise when the equation $\sqrt{x+3} = 2x$ is solved for x by squaring both sides of the equation? How do you know without substituting for x in the equation?

$-\frac{3}{4}$; Since the square root symbol denotes principle square root, $2x$ should be positive. If x is negative, $2x$ is also negative.

INVESTIGATION

When is a Solution Not a Solution?

Step 1 Consider the equation $\sqrt{2x+3} = x$. To find the solution, graph both sides of the equation and determine the x-coordinates of any intersections. 3

Step 2 Square both sides of the equation. Graph both sides of this new equation and determine the x-coordinates of any intersections. 3, −1

Step 3 Substitute each solution to your equation in Step 2 into the original equation. What do you discover?
Substituting $x = 3$ makes the equation true, $x = -1$ does not.

Step 4 When you solve an equation by squaring both sides of an equation, you may cause some extra solutions to appear. These don't work in the original equation and are called *extraneous roots*. It is, therefore, important that you always check solutions to determine if they work in the original equation. Solve the equation $\sqrt{5x+4} = x - 2$ and check for extraneous roots.
Solutions are $x = 0$ and $x = 9$. $x = 0$ is extraneous.

The first step in solving equations with radicals is to isolate the radical, that is, to move everything but the radical to the other side of the equation. But what if there are multiple radicals in the same equation? You still must begin by isolating ONE of the radicals and squaring both sides. This should produce a new equation that has FEWER radicals, so you may have to repeat the process again.

EXAMPLE B | Solve the equation $3 - \sqrt{x} = \sqrt{x - 3}$

Solution When you graph the two sides of the equation you see that there is one solution.

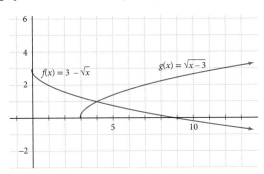

While you can probably read the solution of this equation from the graph, there are times when this is not possible, especially if you need an exact solution. So you need to also be able to solve the equation analytically.

294 CHAPTER 5 Quadratic Functions and Relations

Modifying the Investigation

Whole Class Work on the investigation as a class. Do each step and discuss it before moving to the next step.

Shortened Do Step 1 as a class. Have students do Step 2 and 3. Do Step 4 as a class, introducing the term *extraneous root*.

One Step **ASK** "If $x = 3$, is it true that $x^2 = 9$? If so, if $\sqrt{2x+3} = x$, does $(\sqrt{2x+3})^2 = (x)^2$? Explain."

Have students graph both sides of the equation $\sqrt{2x+3} = x$ to find the intersection of the two graphs. Have them square both sides of the equation and graph both sides of the new equation to find the intersection of the two graphs. **ASK** "How does the intersection of the graphs of the two sides of the equation relate to the roots of the equation? Are they the same for both equations? Explain."

There are two different square roots in this equation. To solve it, you will need to square both sides twice.

$(3 - \sqrt{x})^2 = (\sqrt{x-3})^2$ square each side of the equation

$9 - 6\sqrt{x} + x = x - 3$ expand the left side

$12 = 6\sqrt{x}$ isolate the term containing the square root

$2 = \sqrt{x}$ divide to completely isolate the root

$4 = x$ square both sides

Now check to see that your answer works in the original equation.

$3 - \sqrt{4} \stackrel{?}{=} \sqrt{4 - 3}$

$3 - 2 \stackrel{?}{=} \sqrt{1}$

$1 = 1$ so $x = 4$ is a solution.

In Example B you only found one solution. It was, however, still important to check to see that your solution worked, since sometimes the original equation can have no solution. Graphing the two sides of the original equation will let you know how many, if any, solutions you should expect.

5.6 Exercises

You will need your graphing calculator for Exercises **2** and **6**.

Assigning Exercises
Suggested: 1–12
Additional Practice: 13–15

2a.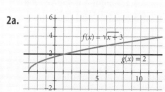

2b.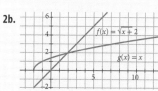

2c.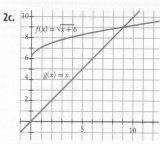

2d.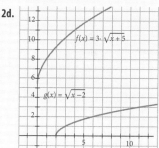

Practice Your Skills

1. Square each expression.
   a. $\sqrt{x+3}$ $x+3$
   b. $\sqrt{x} + 3$ @ $x + 6\sqrt{x} + 9$
   c. $3\sqrt{x+1}$ $9(x+1)$ or $9x + 9$
   d. $\sqrt{x} - \sqrt{5}$ @ $x - 2\sqrt{5x} + 5$

2. Determine the number of solutions for each equation by graphing.
   a. $\sqrt{x+3} = 2$ 1 solution
   b. $\sqrt{x+2} = x$ 1 solution
   c. $\sqrt{x} + 6 = x$ 1 solution
   d. $3\sqrt{x} + 5 = \sqrt{x-2}$ @ no solution

3. Solve each equation.
   a. $\sqrt{x+2} = 5$ $x = 23$
   b. $\sqrt{x-1} + 7 = 3$ @ no solution
   c. $4 + \sqrt{x} = 11$ $x = 49$
   d. $6 - \sqrt{x+3} = 2$ $x = 13$

4. Solve each equation.
   a. $\sqrt{x} = \frac{1}{2}x$ @ $x = 0, 4$
   b. $3\sqrt{3x} = x + 6$ $x = 3, 12$

Reason and Apply

5. Is the equation $\sqrt{a} + \sqrt{b} = \sqrt{a+b}$ ever true?
   a. Experiment with different values for a and b to find a pair that work.   either a or b must be zero
   b. Square both sides of the equation to find the conditions necessary to make the equation true.
      squaring both sides gives the equation $a + 2\sqrt{ab} + b = ab$. This implies that $2\sqrt{ab} = 0$ which means that either or both of a and b must be zero.

Conceptual/Procedural

Procedural Students learn a procedure for solving radical equations algebraically and graphically. They learn to identify extraneous roots by substituting back into the original equation.

11c. Some students may note that extraneous solutions to one equation are actual solutions to another. Others may notice that the right sides of these equations are opposites and then the role of actual and extraneous solutions are switched. Look for any thoughtful consideration of the relationships between the two equations and the solutions.

12. Cal is right. You can just subtract $5 + \sqrt{7}$ from both sides and $x = 3 - \sqrt{7}$. Or you can write $\sqrt{7} = x + 3$, $7 = x^2 + 6x + 9$, $0 = x^2 + 6x + 2$, $x = \frac{-6 \pm \sqrt{36 - 8}}{2} = \frac{-6 \pm \sqrt{28}}{2} = 3 \pm \sqrt{7}$, but only $3 - \sqrt{7}$ works in the original. You only need to square both sides if the variable is in the radical.

---

6. The number of seconds it takes for an object to fall d feet is given by the formula $t = \sqrt{\frac{2d}{32.2}}$.

  a. How far will an object fall in 5 seconds? @ 402.5 feet

  b. The same formula tells how long it takes for a ball to bounce up d feet. How long does it take for the ball to bounce up 3 feet?   0.43 seconds

  c. It takes the same amount of time for a ball to fall a number of feet as it does to bounce up that same number of feet. If it takes 4 seconds for a complete bounce, up and back down, how high did the ball bounce?   64.4 feet

7. The distance between two points (a, b) and (c, d) is given by the formula $\sqrt{(c-a)^2 + (d-b)^2}$. The distance between the points $(x, 5)$ and $(2, 2)$ is $x + 4$. What is the value of x? @ $x = -\frac{1}{4}$

8. The distance between the point $(x, 4)$ and $(2, y)$ is $x + y$.

  a. Find the relationship between x and y. Your answer will be an equation with both variables. $-4x - 8y + 20 = 2xy$

  b. Solve the equation in part a for y. $y = -\frac{2(x-5)}{x+4}$

  c. Are there solutions to the original equation for all values of x? How does the graph of the equation in part b support your answer? @   No, $x \neq -4$ in the original equation. The graph does not have a point at $x = -4$; that would make the denominator 0, which is undefined.

9. The base of a triangle has length 8 inches and the height of the triangle is $\sqrt{3x}$ inches. If the area of the triangle is $2x$ square inches, what is the height?   12 inches

10. Suppose the two legs of a right triangle have lengths $\sqrt{4x}$ and 3 inches. The length of the hypotenuse is $\sqrt{x} + 3$ inches. What is the value of x?   4 inches

11. Solve each equation.

  a. $\sqrt{5-x} = 3x - 1$ @ $x = 1$ ($-\frac{4}{9}$ is extraneous)

  b. $\sqrt{5-x} = 1 - 3x$ $x = -\frac{4}{9}$ (1 is extraneous)

  c. Make a conjecture about extraneous solutions based on the results of these two problems.   Answers will vary.

12. Cal and Alex find the problem $5 - x + \sqrt{7} = 8$. Al says you need to isolate the radical and square both sides, but Cal says that is not necessary in this problem but you can do it if you want to make a lot of work for yourself. Who is right? Explain why.

Review

13. Solve each equation for x.

  a. $3(2^x) = 141$ b. $3x^2 = 141$ $x = \sqrt{47} \approx 6.856$
  $x = \log_2 47 \approx 5.555$

14. Suppose $f(x) = 2x^2 - 5x$ and $g(x) = \frac{3x}{x-1}$.

Find the values of the following expressions.

  a. $f(1) + g(2)$ 3

  b. $\frac{f(3)}{g(3)}$ $\frac{2}{3}$

  c. $(f(-1))^2 + (g(-1))^2$ $\frac{205}{4}$, or 51.25

  d. $(f(-1) + g(-1))^2$ $\frac{289}{4}$, or 72.25

15. Huidobro works at a local bookstore and earns $12.50 per hour. If he works 20 hours each week and pays a total of 18% of his paycheck in taxes, how much money does he actually receive at the end of the week? $205

296   **CHAPTER 5**   Quadratic Functions and Relations

LESSON 5.7

Using the Distance Formula

"It is impossible to be a mathematician without being a poet in soul."
SOPHIA KOVALEVSKAYA

Imagine a race in which you carry an empty bucket from the starting line to the edge of a pool, fill the bucket with water, and then carry the bucket to the finish line. Luck, physical fitness, common sense, a calm attitude, and a little mathematics will make a difference in how you perform. Your performance depends on your speed and on the distance you travel. As you'll see in the investigation, mathematics can help you find the shortest path.

INVESTIGATION

YOU WILL NEED
- centimeter graph paper
- a ruler

Bucket Race

The starting line of a bucket race is 5 m from one end of a pool, the pool is 20 m long, and the finish line is 7 m from the opposite end of the pool, as shown. In this investigation you will find the shortest path from point A to a point C on the edge of the pool to point B. That is, you will find the value of x, the distance in meters from the end of the pool to point C, such that $AC + CB$ is the shortest path possible.

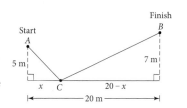

Step 1 Make a scale drawing of the situation on graph paper.

Step 2 Plot several different locations for point C. For each, measure the distance x and find the total length $AC + CB$. Record your data.

Step 3 What is the best location for C such that the length $AC + BC$ is minimized? What is the distance traveled? Is there more than one best location? Describe at least two different methods for finding the best location for C.

Step 4 Make a scale drawing of your solution.

Imagine that the amount of water you empty out at point B is an important factor in winning the race. This means you must move carefully so as not to spill water, and you'll be able to move faster with the empty bucket than you can with the bucket full of water. Assume that you can carry an empty bucket at a rate of 1.2 m/s and that you can carry a full bucket, without spilling, at a rate of 0.4 m/s.

LESSON 5.7 Using the Distance Formula 297

LESSON 5.7

In this lesson students see applications of the Pythagorean Theorem, including a derivation of the distance formula for coordinate geometry.

COMMON CORE STATE STANDARDS		
APPLIED	**DEVELOPED**	**INTRODUCED**
F.BF.1	G.GPE.1	G.SRT.8

Objectives
- Apply the distance formula to problems involving a locus of points
- Derive the equation of a circle of given center and radius using the Pythagorean Theorem

Vocabulary
locus (loci)

Materials
- rulers
- centimeter grid paper
- Calculator Note: Intersections, Maximums, and Minimums

Launch

What is the length of each leg of the triangle? Is it a right triangle? How do you know?

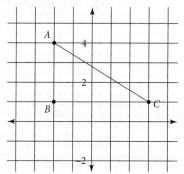

3, 5, $\sqrt{34} \approx 5.8$ Methods will vary.

Students might have counted, used the Pythagorean Theorem, or the distance formula. Try to get presentations of all methods.

ELL
Students are likely to know the Pythagorean Theorem and the distance formula, though not necessarily by these names. Introduce them to new vocabulary. *Locus* may be entirely new. Use numeric, symbolic, and graphical examples to teach and demonstrate a locus.

Support
As needed, encourage students to use rectangle diagrams to expand binomials that arise from applications of the distance formula. Reiterate the fact that the distance formula is derived from the Pythagorean Theorem. If necessary, you can use the Refreshing Your Skills: The Distance Formula lesson.

Advanced
You could extend Example C by asking students to find the locus of points equidistant from 3 points, then 4, 5, and more points, and then have them generalize their results.

LESSON 5.7 Using the Distance Formula 297

Investigate

Guiding the Investigation

There is a desmos simulation for the investigation in your ebook.

Introduce the investigation by describing a bucket race. Have students read the introductory paragraph. **ASK** "Why not just run 5 meters to the pool, 20 meters along the pool, and then 7 meters up to the finish line?" [This is not the shortest path.]

Step 1 LANGUAGE In a *scale drawing*, all the dimensions are proportional to those in the original situation. On centimeter grid paper, students can use 5 cm, 7 cm, and 20 cm for their scale drawings.

Step 2 Students might organize their data in a table with columns for *x*, AC, CB, and AC + CB. Have them leave additional room for other columns, which they will use in Step 5. If students are struggling with finding AC and CB, **ASK** "What kind of figure do you have here?"

Step 2 Sample answer: These values for AC and CB were calculated using the Pythagorean Theorem. Students may also have measured.

x (m)	AC (m)	CB (m)	AC + CB (m)
0	5.00	21.19	26.19
2	5.39	19.31	24.70
4	6.40	17.46	23.87
6	7.81	15.65	23.46
8	9.43	13.89	23.33
10	11.18	12.21	23.39
12	13.00	10.63	23.63
14	14.87	9.22	24.09
16	16.76	8.06	24.83
18	18.68	7.28	25.96
20	20.62	7.00	27.62

Step 3 From the sample data, the best location for point *C* is 8 cm from the end of the pool. The minimum distance by this method is ≈ 23.33 m.

Using the Pythagorean Theorem to write the equation $y = \sqrt{x^2 + 5^2} + \sqrt{(20-x)^2 + 7^2}$ where *y* represents the sum AC + CB and then finding the minimum of the function, the best location for point *C* is ≈ 8.33 m. The minimum distance is ≈ 23.32 m.

Although $y = \sqrt{x^2 + 5^2} + \sqrt{(20-x)^2 + 7^2}$ has a definite minimum, all *x*-values that fall in the domain (approximately) $7.7 \leq x \leq 9.0$ result in paths that differ by less than 0.01 m.

Step 5 Go back to the data collected in Step 2 and find the time needed for each *x*-value.

Step 6 Now find the best location for point *C* so that you minimize the time from point *A* to the pool edge, then to point *B*. What is your minimum time? What is the distance traveled? How does this compare with your answer in Step 3? Describe your solution process.

Refreshing Your Skills: The Distance Formula reviews how to find the distance between points and how to write equations involving distances. You can use expressions created from the distance formula to solve many problems.

In 1790, the U.S. Coast Guard was formed to prevent smuggling and maintain customs laws. Combined with the Life Saving Service in 1915, it now oversees rescue missions, environmental protection, navigation, safety during weather hazards, port security, boat safety, and oil tanker transfers. The U.S. Coast Guard has both military and volunteer divisions.

EXAMPLE A | An injured worker must be rushed from an oil rig 15 mi offshore to a hospital in the nearest town 98 mi down the coast from the oil rig.

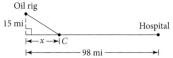

a. Let *x* represent the distance in miles from the point on the shore closest to the oil rig and another point, *C*, on the shore.

How far does the injured worker travel, in terms of *x*, if a boat takes him to *C* and then an ambulance takes him to the hospital?

b. Assume the boat travels at an average rate of 23 mi/h and the ambulance travels at an average rate of 70 mi/h. What value of *x* makes the trip 3 h?

Solution | Use the distance formula, $d = \sqrt{(x_2 - x_1)^2 + (y_2 - y_1)^2}$.

a. The boat must travel $\sqrt{15^2 + x^2}$, and the ambulance must travel $98 - x$.

The total distance in miles is $\sqrt{15^2 + x^2} + 98 - x$.

b. Distance equals rate times time, or $d = rt$. Solving for time, $t = \frac{d}{r}$. So the boat's time is $\frac{\sqrt{15^2 + x^2}}{23}$, and the ambulance's time is $\frac{98 - x}{70}$. The total time in hours, *y*, is represented by

$$y = \frac{\sqrt{15^2 + x^2}}{23} + \frac{98 - x}{70}$$

Modifying the Investigation

Whole Class Complete Step 1. In Step 2, assign different locations of point *C* to each student. Collect the data in a class table. Complete Steps 3 through 6 with student input.

Shortened Skip Steps 5 and 6.

One Step Pose this problem: "A kicked football travels along a path given by the equation $y = -0.005x^2 + x$. What function of *x* describes the distance of the ball from the goalposts, located at the point (200, 10)?" If a group remains stuck for some time, suggest that they draw a right triangle. If necessary, mention the Pythagorean Theorem. Encourage testing of ideas with calculator graphs and tables.

One way to find the value of x that gives a trip of 3 h is to graph the total time equation and $y = 3$, and use the graph to approximate the intersection. The graphs intersect when x is approximately 51.6. For the trip to be 3 h, the boat and the ambulance should meet at the point on the shore 51.6 mi from the point closest to the oil rig.

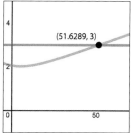

You can use the distance formula to describe the points on a circle. If a circle is centered at the origin and the points on the circle are all r units away from the center, then the equation describing the location of those points is $\sqrt{(x-0)^2 + (y-0)^2} = r$, or $\sqrt{x^2 + y^2} = r$, or $x^2 + y^2 = r^2$.

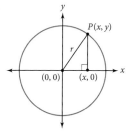

Sometimes the equation of a circle is written in expanded form. You can use your skills in completing the square to change this form into standard form in order to see the center and radius more clearly.

EXAMPLE B | The equation $x^2 - 4x + y^2 + 6y + 10 = 0$ describes a circle. Where is the center of the circle, and what is its radius?

Solution | You need to complete the square in both x and in y.

$(x^2 - 4x + ?) + (y^2 + 6y + ?) = -10$ Subtract 10 from both sides to isolate the variables

$(x^2 - 4x + 4) + (y^2 + 6y + 9) = -10 + 4 + 9$ Add 4 to both sides to complete the square in x and add 9 to both sides to complete the square in y.

$(x - 2)^2 + (y + 3)^2 = 3$ Factor into perfect squares and the equation is in standard form

The center of the circle is at the point $(2, -3)$ and the radius is $\sqrt{3}$.

An equation that describes a set of points that meet a certain condition is called a **locus**. For example, the locus of points that are 3 units from the point $(0, 0)$ is the circle with equation $x^2 + y^2 = 9$. The locus of points midway between $(-4, 0)$ and $(4, 0)$ is the line $x = 0$, the y-axis. In this lesson you will explore equations describing a variety of different **loci** (the plural of locus).

EXAMPLE C | Find the equation of the locus of points that are equidistant from the points $(1, 3)$ and $(5, 6)$.

Solution | First make a sketch of the situation. Plot the two given points and then estimate the location of a few points that are the same distance from the given points. Do they appear to be in any sort of pattern? In this case they appear to lie on a line.

Summarize

Have groups present their solutions and explain their work for the Investigation and the examples. Whether student responses are correct or incorrect, ask other students if they agree and why. Encourage students to use appropriate mathematical language as evidence. SMP 1, 3, 6

As needed, remind students that they need not expand binomials to enter functions into their calculators. They might enter $f_1 = -0.005x^2 + x$ and $f_2 = \sqrt{(200+x)^2 + (10-f_1)^2}$. SMP 5

CRITICAL QUESTION What does the distance formula give you the distance between?

BIG IDEA The distance formula gives the distance between two points whose coordinates in a plane are given.

CRITICAL QUESTION How are the distance formula and the Pythagorean formula related?

BIG IDEA The distance between two points on the coordinate plane can be seen as a right triangle's hypotenuse whose length is determined by the Pythagorean Theorem.

CRITICAL QUESTION Why does the distance formula subtract the coordinates of P_1 from those of P_2 instead of the other way around?

BIG IDEA The two approaches are equivalent, because the square will be positive either way.

Formative Assessment

To assess the objectives of the lesson, observe students working in groups and listen carefully to their presentation. Presentations of their results for the investigation, Examples A and C, and the exercises provide evidence of applying the distance formula to problems involving a locus of points. Example B and Exercise 12 provide evidence of deriving the equation of a circle of given center and radius using the Pythagorean Theorem.

Therefore, you should expect to find the equation of a line as your final answer. Let d_1 represent the distance between $(1, 3)$ and any point, (x, y), on the locus. By the distance formula,

$$d_1 = \sqrt{(x-1)^2 + (y-3)^2}$$

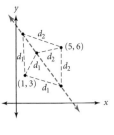

Let d_2 represent the distance between $(5, 6)$ and the same point on the locus, so

$$d_2 = \sqrt{(x-5)^2 + (y-6)^2}$$

The locus of points contains all points whose coordinates satisfy the equation $d_1 = d_2$, or

$$\sqrt{(x-1)^2 + (y-3)^2} = \sqrt{(x-5)^2 + (y-6)^2}$$

Use algebra to transform this equation into something more familiar.

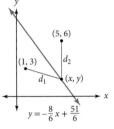

$\sqrt{(x-1)^2 + (y-3)^2} = \sqrt{(x-5)^2 + (y-6)^2}$	Original equation.
$(x-1)^2 + (y-3)^2 = (x-5)^2 + (y-6)^2$	Square both sides.
$x^2 - 2x + 1 + y^2 - 6y + 9 = x^2 - 10x + 25 + y^2 - 12y + 36$	Expand the binomials.
$-2x + 1 - 6y + 9 = -10x + 25 - 12y + 36$	Subtract x^2 and y^2 from both sides.
$8x + 6y = 51.$	Combine like terms and rewrite in the form $ax + by = c$.

As you predicted, the locus is a line. It has equation $8x + 6y = 51$, or $y = -\frac{8}{6}x + \frac{51}{6}$. How could you show that this locus is the perpendicular bisector of the segment joining the two points?

Just as the Pythagorean Theorem and the distance formula are useful for finding the equation of a locus of points, they are helpful for solving real-world problems.

 **Exercises**   You will need your graphing calculator for Exercises **8–11** and **15**.

Practice Your Skills

1. Find the distance between each pair of points.
   a. $(2, 5)$ and $(8, 13)$ ⓐ 10 units
   b. $(0, 3)$ and $(5, 10)$ $\sqrt{74}$ units
   c. $(-4, 6)$ and $(-2, -3)$ $\sqrt{85}$ units
   d. $(3, d)$ and $(-6, 3d)$ $\sqrt{81 + 4d^2}$ units

2. Sketch the locus of points for each situation.
   a. A point is 3 units from the point $(1, -1)$.
   b. A point is 5 units from the point $(-2, 0)$.
   c. A point is the same distance from the point $(1, 2)$ as it is from the point $(5, 7)$.

300 CHAPTER 5 Quadratic Functions and Relations

Apply

Extra Example

The distance between the points $(x, -4)$ and $(-1, 0)$ is 12. Find the possible values of x.

$x \approx 10.314$ or -12.314

Find the center and radius of the circle represented by the equation $x^2 - 6x + y^2 = 16$.

Center $(3, 0)$; radius 5

Closing Question

How does the distance formula apply to two points on a horizontal line or on a vertical line?

For a horizontal line, the y-values are the same so the formula becomes $d = \sqrt{(x_2 - x_1)^2}$, or $d = x_2 - x_1$. For a vertical line, the x-values are the same so the formula becomes $d = \sqrt{(y_2 - y_1)^2}$, or $d = y_2 - y_1$.

3. Write an equation for each locus in Exercise 2.

4. The distance between the points (2, 7) and (5, y) is 5 units. Make a sketch of the situation and find the possible value(s) of y. ⓐ $y = 11$ or $y = 3$

5. The distance between the points (−1, 5) and (x, −2) is 47. Make a sketch of the situation and find the possible value(s) of x. $x = -1 \pm \sqrt{2160} = -1 \pm 12\sqrt{15}$
$x \approx -47.48$ or 45.48

Reason and Apply

6. A point on the line $y = 1$ has coordinates $(x, 1)$. Two other points have coordinates (2, 5) and (4, 9). $\sqrt{(x-2)^2 + (1-5)^2}$, or $\sqrt{(x-2)^2 + 16}$

   a. Write an expression for the distance between the point on the line and the point (2, 5). ⓐ

   b. Write an expression for the distance between the point on the line and the point (4, 9).
   $\sqrt{(x-4)^2 + (1-9)^2}$, or $\sqrt{(x-4)^2 + 64}$

7. In a set of points, each point is twice as far from the point (2, 0) as it is from (5, 0).

   a. Sketch this locus of points. ⓗ

   b. Write an equation for the locus and solve it for y.
   $\sqrt{(x-2)^2 + y^2} = 2\sqrt{(x-5)^2 + y^2}$, or $y = \pm\sqrt{-x^2 + 12x - 32}$

8. If you are too close to a radio tower, you will be unable to pick up its signal. Let the center of a town be represented by the origin of a coordinate plane. Suppose a radio tower is located 2 mi east and 3 mi north of the center of town, or at the point (2, 3). A highway runs north-south 2.5 mi east of the center of town, along the line $x = 2.5$. Where on the highway will you be less than 1 mi from the tower and therefore unable to pick up the signal?
$\frac{6 - \sqrt{3}}{2} < y < \frac{6 + \sqrt{3}}{2}$, or approximately between the points (2.5, 2.134) and (2.5, 3.866)

9. Josh is riding his mountain bike when he realizes that he needs to get home quickly for dinner. He is 2 mi from the road, and home is 3 mi down that road. He can ride 9 mi/h through the field separating him from the road and can ride 22 mi/h on the road.

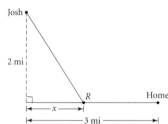
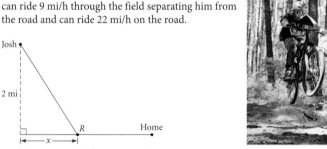

   a. If Josh rides through the field to a point R on the road, and then home along the road, how far will he ride through the field? How far on the road? Let x represent the distance in miles between point R and the point on the road that is closest to his current location. ⓐ in miles, $\sqrt{x^2 + 4}$ through the field and $3 - x$ on the road

   b. How much time will Josh spend riding through the field? How much time on the road? ⓗ in hours, $\frac{\sqrt{x^2+4}}{9}$ through the field and $\frac{3-x}{22}$ on the road

   c. What value of x gets Josh home the fastest? What is the minimum time?
   When x is ≈ 0.897 mi, Josh gets home in ≈ 0.339 h, a little over 20 min.

LESSON 5.7 Using the Distance Formula   301

Assigning Exercises

Suggested: 1–7, 9, 11–13, 15
Additional Practice: 8, 10, 14, 16–21

Students can use geometry software to sketch many of these exercises.
SMP 5

2a.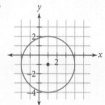

2b.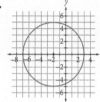

2c.

3a. $\sqrt{(x-1)^2 + (y+1)^2} = 3$

3b. $\sqrt{(x+2)^2 + (y-0)^2} = 5$

3c. $\sqrt{(x-1)^2 + (y-2)^2} = \sqrt{(x-5)^2 + (y-7)^2}$, or $8x + 10y = 69$

4.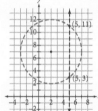

5.

7a.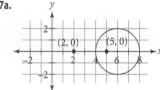

Exercise 7b ALERT Students may write an equation that mimics the words and says, in effect, "2 times the distance of the point from (2, 0) equals its distance from (5, 0)." ASK "Is the point's distance from (2, 0) greater or less than its distance from (5, 0)?" [greater]

Exercise 8 Ask students to make an accurate sketch of this scenario, perhaps using geometry software. Students can more easily see that this is a 30°-60°-90° triangle if the drawing is to scale.

Exercise 9c Students can trace a calculator graph or use the minimum function on a calculator described in Calculator Note: Intersections, Maximums, and Minimums. **SMP 5**

LESSON 5.7 Using the Distance Formula   **301**

10b. domain: $0 \leq x \leq 20$; range: $30 < y < 35$

Exercise 11 Urge students to conjecture before they calculate. They should assume the wall is vertical and the ground is horizontal.

11a.

Time(s)	Height (ft)
0	24.00
1	23.92
2	23.66
3	23.24
4	22.63
5	21.82
6	20.78
7	19.49
8	17.89
9	15.87
10	13.27
11	9.59
12	0

Exercise 13 To use the distance formula, students will consider the point (x, y) on the curve as $(x, 0.5x^2 + 1)$. You can thus check their understanding that an equation of a curve is a relationship that is satisfied by coordinates of exactly those points on the curve.

This cable-stay bridge, from Vladivostok to Russky Island in southeastern Russia, was the longest cable-stayed bridge in the world as of 2013. Taut cables stretch from the tops of two towers to support the roadway.

10. A 10 m pole and a 13 m pole are 20 m apart at their bases. A wire connects the top of each pole with a point on the ground between them.

   a. Let y represent the total length of the wire. Write an equation that relates x and y. @ $y = \sqrt{10^2 + x^2} + \sqrt{(20-x)^2 + 13^2}$

   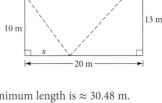

   b. What domain and range make sense in this situation?

   c. Where should the wire be fastened to the ground so that the length of wire is minimized? What is the minimum length?
   When the wire is fastened ≈ 8.696 m from the 10 m pole, the minimum length is ≈ 30.48 m.

11. A 24 ft ladder is placed upright against a wall. Then the top of the ladder slides down the wall while the foot of the ladder is pulled outward along the ground at a steady rate of 2 ft/s.

   a. Find the heights that the ladder reaches at 1 s intervals while the ladder slides down the wall. @

   b. How long will it take before the ladder is lying on the ground? 12 s

   c. Does the top of the ladder also slide down the wall at a steady rate of 2 ft/s? Explain your reasoning.
   No, the distance the ladder slides down increases during each second.

12. Find the center and radius of the circle $x^2 + y^2 + 8x - 10y - 23 = 0$. Then find an equation for another circle with the same radius whose center lies on the first circle.
   $(-4, 5)$, $r = 8$ Answers will vary. Any multiples of the equation are possible answers.

13. Let d represent the distance between the point $(5, -3)$ and any point, (x, y), on the parabola $y = 0.5x^2 + 1$.

   a. Write an equation for d in terms of x. $d = \sqrt{(5-x)^2 + (0.5x^2 + 4)^2}$

   b. What is the minimum distance? What are the coordinates of the point on the parabola that is closest to the point $(5, -3)$?
   ≈ 6.02 units; $\approx (0.92, 1.42)$

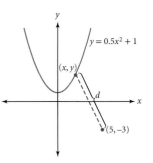

302 CHAPTER 5 Quadratic Functions and Relations

14. Two towns are 6 mi apart. Chanda often bicycles from one town to the other. However, she prefers not to bicycle down the straight highway but rather to explore the region between the towns. She decides that she can bicycle at most 10 mi.

   a. Make a sketch of the situation, locating the two towns at the points (3, 0) and (−3, 0). Estimate the locations of several points such that the straight line distances from each point to the two towns sum to 10 units on your graph.

   b. Suppose Chanda is at the point (x, y). What is her distance from the point (3, 0)? What is her distance from the point (−3, 0)?

   c. Write an equation expressing the fact that the sum of these distances is 10 mi.
   $\sqrt{(x+3)^2 + y^2} + \sqrt{(x-3)^2 + y^2} = 10$

15. **APPLICATION** The city councils of three neighboring towns—Ashton, Bradburg, and Carlville—decide to pool their resources and build a recreation center. To be fair, they decide to locate the recreation center equidistant from all three towns.

   a. When a coordinate plane is placed on a map of the towns, Ashton is at (0, 4), Bradburg is at (3, 0), and Carlville is at (12, 8). At what point on the map should the recreation center be located? ≈ (6.17, 5.50)

   b. If the three towns were collinear (along a line), could the recreation center be located equidistant from all three towns? Explain your reasoning.

   c. What other factors might the three city councils consider while making their decision as to where to locate the recreation center? Answers will vary but could include the population of each town or the arrangement of roads between the towns.

Review

16. Complete the square in each equation so that the left side represents a perfect square or a sum of perfect squares.

   a. $x^2 + 6x = 5$ @
   $(x + 3)^2 = 14$

   b. $y^2 - 4y = -1$
   $(y - 2)^2 = 3$

   c. $x^2 + 6x + y^2 - 4y = 4$ @
   $(x + 3)^2 + (y - 2)^2 = 17$

17. Triangle ABC has vertices $A(8, -2)$, $B(1, 5)$, and $C(4, -5)$.

   a. Find the midpoint of each side.
   midpoint of $\overline{AB}$: (4.5, 1.5); midpoint of $\overline{BC}$: (2.5, 0); midpoint of $\overline{AC}$: (6, −3.5)

   b. Write the equations of the three medians of the triangle. (A median of a triangle is a segment connecting one vertex to the midpoint of the opposite side.) @

   c. Locate the point where the medians meet. $(4.\overline{3}, -0.\overline{6})$, or $\left(4\frac{1}{3}, -\frac{2}{3}\right)$

18. Perform the indicated operation and reduce your answer to lowest terms.

   a. $\frac{10}{x} + \frac{5}{3x}$ @ $\frac{35}{3x}$

   b. $\frac{3x}{2} - \frac{5}{4}$ $\frac{6x-5}{4}$

   c. $\frac{7+x}{2} + \frac{5-3x}{6}$ $\frac{13}{3}$

19. Factor each polynomial.

   a. $x^2 + 7x + 6$ $(x+1)(x+6)$

   b. $x^2 - 8x + 12$ $(x-2)(x-6)$

   c. $x^2 + 4x - 21$ $(x+7)(x-3)$

20. Give the domain and range of the function $f(x) = x^2 + 6x + 7$.
   domain: all real numbers; range: $y \geq -2$

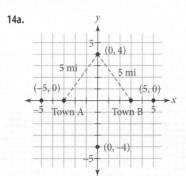

14a.

14b. $\sqrt{(x-3)^2 + (y-0)^2}$, or $\sqrt{(x-3)^2 + y^2}$;

$\sqrt{(x+3)^2 + (y-0)^2}$, or $\sqrt{(x+3)^2 + y^2}$

Exercise 15a The solution is the intersection of the perpendicular bisectors of the triangle's sides. Students might recall from geometry that this point is called the *circumcenter*.

15b. The city councils couldn't build the recreation center equidistant from the towns if they were collinear because the perpendicular bisectors of $\overline{AB}$, $\overline{BC}$, and $\overline{AC}$ would have the same slope, making them parallel. There would be no intersection.

Exercise 17 Students could use geometry software for this exercise. You might want to review from geometry that the intersection of the medians is called the *centroid*. **SMP 5**

17b. median from A to $\overline{BC}$:
$y = -0.\overline{36}x + 0.\overline{90}$ or $y = -\frac{4}{11}x + \frac{10}{11}$;
median from B to $\overline{AC}$:
$y = -1.7x + 6.7$;
median from C to $\overline{AB}$:
$y = 13x - 57$

21. Find w and x.

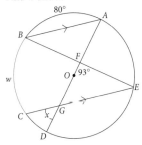

$w = 74°, x = 50°$

The Gateway Arch, a 630-foot monument in St. Louis, Missouri, was started in 1947 and completed in 1965. It is often described as a parabola, but is actually an inverted catenary arch.

IMPROVING YOUR Visual Thinking SKILLS

Miniature Golf

Shannon and Lori are playing miniature golf. The third hole is in a 22-by-14-foot walled rectangular playing area with a large rock in the center. Each player's ball comes to rest as shown. The rock makes a direct shot into the hole impossible.

At what point on the south wall should Shannon aim in order to have the ball bounce off and head directly for the hole? Recall from geometry that the angle of incidence is equal to the angle of reflection. Lori cannot aim at the south wall. Where should she aim?

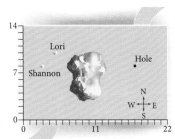

LESSON 5.8

Parabolas

You have previously studied parabolas. Now a locus definition will reveal properties of parabolas that you can use to solve other practical problems.

"There are two ways of spreading light: to be the candle or the mirror that reflects it."
EDITH WHARTON

The designs of telescope lenses, spotlights, satellite dishes, and other parabolic reflecting surfaces are based on a remarkable property of parabolas: A ray that travels parallel to the axis of symmetry will strike the surface of the parabola or paraboloid and reflect toward the **focus**. Likewise, when a ray from the focus strikes the curve, it will reflect in a ray that is parallel to the axis of symmetry. A **paraboloid** is a three-dimensional parabola, formed when a parabola is rotated about its line of symmetry.

Science CONNECTION

Satellite dishes, used for television, radio, and other communications, are always parabolic. A satellite dish is set up to aim directly at a satellite. As the satellite transmits signals to a dish, the signals are reflected off the dish surface and toward the receiver, which is located at the focus of the paraboloid. In this way, every signal that hits a parabolic dish can be directed into the receiver.

This reflective property of parabolas can be proved based on the locus definition of a parabola.

Definition of a Parabola

A **parabola** is a locus of points in a plane whose distance from a fixed point called the **focus** is the same as the distance from a fixed line called the **directrix**. That is, $d_1 = d_2$. In the diagram, F is the focus and l is the directrix.

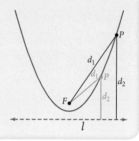

LESSON 5.8

In this lesson students broaden their understanding of parabolas to include the focus-and-directrix locus definition.

COMMON CORE STATE STANDARDS		
APPLIED	DEVELOPED	INTRODUCED
	G.GPE.2	

Objectives

- Introduce the locus definition of *parabola* and new vocabulary associated with a parabola
- Determine the location of the parabola's focus, directrix, and vertex from its equation and graph the parabola
- Determine the equation of a parabola from its graph
- Find the equation of a parabola given a focus and directrix

Vocabulary

parabola
focus
directrix
paraboloid

Materials

- patty paper or tracing paper

Launch

Write the equation of this parabola in factored form, vertex form and general form. Explain how you got each form. Assume $a = 1$.

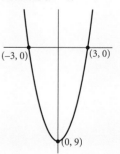

$y = (x - 3)(x + 3)$; $y = (x - 0)^2 - 9$;
$y = x^2 - 9$
Explanations will vary.

ELL
The hands-on investigation can be very effective in helping students understand the new terms *directrix* and *focus* of a parabola. Having students label the focus and directrix of their parabolas in the examples will further reinforce their understanding.

Support
Graph several translated parabolas and ask students to write equations. Once students feel more comfortable with this concept, make the transition into finding the focus and directrix of the parabola. You might project the Investigation from the ebook, with the directions separated in paragraphs.

Advanced
Challenge students to see whether they can create models of circles and the other conic sections by folding paper as in the investigation, and explain why it is or is not possible.

This lesson broadens the definition used earlier for a parabola. Use the Launch to review what students have learned about parabolas up to this point. Then project the blue definition box and introduce the terminology of focus and directrix. SMP 1, 2, 6

Consider projecting the images on the second page of the lesson as you discuss the lesson.

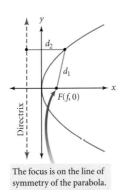

The focus is on the line of symmetry of the parabola.

A parabola is the set of points for which the distances d_1 and d_2 are equal. If the directrix is a horizontal line, the parabola is vertically oriented, like the one in the definition box on the previous page. If the directrix is a vertical line, the parabola is horizontally oriented, like the one at left. The directrix can also be neither horizontal nor vertical, creating a parabola that is rotated at an angle.

How can you locate the focus of a given parabola? Suppose the parabola is horizontally oriented, with vertex (0, 0). It has a focus inside the curve at a point, $(f, 0)$, as shown at left. The vertex is the same distance from the focus as it is from the directrix, so the equation of the directrix is $x = -f$, as shown below at left. You can use this information and the distance formula to find the value of f when the vertex is at the origin, as shown below right.

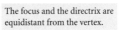

The focus and the directrix are equidistant from the vertex.

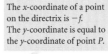

The x-coordinate of a point on the directrix is $-f$. The y-coordinate is equal to the y-coordinate of point P.

First, choose any point P and label the coordinates (x, y).

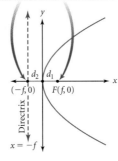

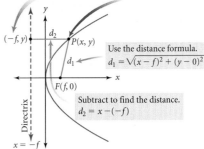

Use the distance formula.
$d_1 = \sqrt{(x - f)^2 + (y - 0)^2}$

Subtract to find the distance.
$d_2 = x - (-f)$

$$\sqrt{(x - f)^2 + (y - 0)^2} = x - (-f)$$ Definition of parabola states that $d_1 = d_2$.

$$\sqrt{(x - f)^2 + y^2} = x + f$$ Subtract.

$$(x - f)^2 + y^2 = (x + f)^2$$ Square both sides.

$$x^2 - 2fx + f^2 + y^2 = x^2 + 2fx + f^2$$ Expand.

$$y^2 = 4fx$$ Combine like terms.

This result means that the coefficient of the variable x is $4f$, where f is the distance from the vertex to the focus. What do you think it means if f is negative?

The parabola opens to the left.

If the parabola is vertically oriented, the x- and y-coordinates are exchanged, for a final equation of $x^2 = 4fy$, or $y = \dfrac{1}{4f}x^2$.

306 CHAPTER 5 Quadratic Functions and Relations

Modifying the Investigation

Whole Class Demonstrate the folding technique using a very large square of paper, or use waxed paper on the overhead. Have students fold their own. Draw coordinate axes on the large paper to support the discussion.

Shortened There is no shortened version of this investigation.

One Step Hand out patty paper and assign the **One Step** from your ebook. During the discussion, explore how to find the focus and the directrix from a parabola's equation.

Designed in 1960, the Theme Building at Los Angeles International Airport in California uses double parabolic arches.

EXAMPLE A | Consider the parent equation of a horizontally oriented parabola, $y^2 = x$.

a. Write the equation of the image of this graph after the following transformations have been performed, in order: a vertical dilation by a factor of 3, a translation right 2 units, and then a translation down 1 unit. Graph the new equation.

b. Where is the focus of $y^2 = x$? Where is the directrix?

c. Where is the focus of the transformed parabola? Where is its directrix?

Solution | Recall the transformations of functions that you studied earlier.

a. Begin with the parent equation, and perform the specified transformations.

$y^2 = x$ Original equation.

$\left(\dfrac{y}{3}\right)^2 = x$ Dilate vertically by a factor of 3.

$\left(\dfrac{y+1}{3}\right)^2 = x - 2$ Translate right 2 units and down 1 unit.

Graph the transformed parabola.

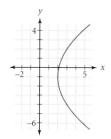

Investigate

Example A

Project the example and work through it as a class, with input from students. Point out that students must write the equation in standard form before they can see the value of 4*f* (and then *f*). **SMP 1, 2, 6**

Projecting the blue box Equation of a Parabola as you discuss the equations can help students who are visual learners or are **ELL**.

Example B

Consider projecting the example and working together as a class, with student input. Have students graph the information they have and label the focus and directrix. Have them add to their graph as you work through the rest of the example. SMP 1, 2, 6

Guiding the Investigation

There is a desmos simulation for the investigation in your ebook.

You might do the folding along with your students and show your result on the overhead, where students can see the folds. The closer together the folds, the more detailed the parabola. Encourage students to lay the focus atop points all the way to the corners of the paper. This construction is an *envelope* of lines.

Once each student has a parabola, groups can work together to find equations of all of them, or students can work individually. As you circulate, try to encourage a mix of equations for both vertically and horizontally oriented parabolas.

ASK "Do you have any distances to which you could apply the distance formula? Which points on the directrix correspond to various points on the curve? Which segment connecting the points is perpendicular to the directrix?"

During the discussion, explore how to find the focus and the directrix from a parabola's equation.

You might also have students measure distances to see how the definition of a parabola applies to their own creations. SMP 1, 2, 6

Sample answer: Suppose this parabola was overlaid on graph paper and traced, with focus $(3, 0)$ and directrix $x = 1$. The general form of the parabola is $y^2 = 4fx$. Because the distance from the focus to the vertex is 1, $f = 1$, and the general equation of the parabola is $y^2 = 4x$. However, the parabola is translated right 2 units, so the final equation is $y^2 = 4(x - 2)$.

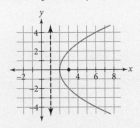

b. Use the standard form, $y^2 = 4fx$, to locate the focus and the directrix of $y^2 = x$. The coefficient of x is $4f$ in the general form, and 1 in the equation $y^2 = x$. So $4f = 1$, or $f = \frac{1}{4}$. Recall that f is the distance from the vertex to the focus and from the vertex to the directrix. The vertex is $(0, 0)$, so the focus is $\left(\frac{1}{4}, 0\right)$ and the directrix is the line $x = -\frac{1}{4}$.

c. To locate the focus and the directrix of $\left(\frac{y+1}{3}\right)^2 = x - 2$, first rewrite the equation as $(y + 1)^2 = 9(x - 2)$. The coefficient of x in this equation is 9, so $4f = 9$, or $f = 2.25$. The focus and the directrix will both be 2.25 units from the vertex in the horizontal direction. The vertex is $(2, -1)$, so the focus is $(4.25, -1)$ and the directrix is the line $x = -0.25$.

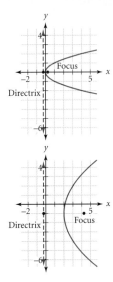

You can now extend what you know about parabolic equations to both horizontally and vertically oriented parabolas.

> **Equation of a Parabola**
>
> The standard form of the equation of a vertically oriented parabola with vertex (h, k), horizontal scale factor of a, and vertical scale factor of b is
> $$\frac{y - k}{b} = \left(\frac{x - h}{a}\right)^2$$
> The focus of a vertically oriented parabola is $(h, k + f)$, where $\frac{a^2}{b} = 4f$, and the directrix is the line $y = k - f$.
>
> The standard form of the equation of a horizontally oriented parabola with vertex (h, k), horizontal scale factor of a, and vertical scale factor of b is
> $$\left(\frac{y - k}{b}\right)^2 = \frac{x - h}{a}$$
> The focus of a horizontally oriented parabola is $(h + f, k)$, where $\frac{b^2}{a} = 4f$, and the directrix is the line $x = h - f$.

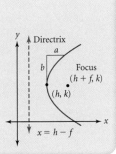

The previous example showed how to graph a parabola and locate the focus and directrix from the equation. Using the locus definition of the parabola, you can find the equation of the parabola when you know the focus and directrix.

Conceptual/Procedural

Conceptual The investigation helps students conceptualize the focus-directrix definition of a parabola.

Procedural The examples and exercises provide opportunities to reinforce the procedure of writing an equation of a parabola from a graph or when given the focus and directrix.

EXAMPLE B | Write the equation of the parabola with directrix $y = 3$ and focus at $(2, 7)$.

Solution | Because the directrix is a horizontal line and the focus lies above the directrix, you know that the parabola opens upward.

The vertex of the parabola lies midway between the focus and directrix. From a graph you can see that the vertex must be at the point $(2, 5)$.

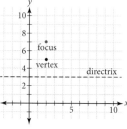

The distance between the focus and vertex is 2. This is half of the distance between the focus and directrix. This means $f = 2$, so $\frac{1}{4f} = \frac{1}{8}$.

You can now write the equation
$y - 5 = \frac{1}{8}(x - 2)^2$ or $y = \frac{1}{8}(x - 2)^2 + 5$.

You can find two more points on the parabola by using the locus definition.

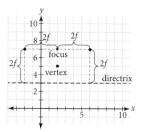

Locating these points can help you draw the graph of the parabola more accurately. You can also use them as a way to verify that your equation is correct.

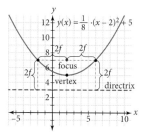

In the investigation you will construct a parabola. As you create your model, think about how your process relates to the locus definition of a parabola.

INVESTIGATION

Fold a Parabola

YOU WILL NEED
- patty paper
- graph paper

Fold the patty paper parallel to one edge to form the directrix for a parabola. Mark a point on the larger portion of the paper to serve as the focus for your parabola. Fold the paper so that the focus lies on the directrix. Unfold, and then fold again, so that the focus is at another point on the directrix. Repeat this many times. The creases from these folds should create a parabola. Think about the definition of the parabola and write an explanation of why these creases create the parabola. Lay the patty paper on top of a sheet of graph paper. Identify the coordinates of the focus and the equation of the directrix, and write an equation for your parabola.

Summarize

ASK "In the equation $y^2 = 4fx$, where will the focus be?" [to the right of the vertex if $f > 0$ and to the left of the vertex if $f < 0$] "Where will the directrix be?" [to the left of the parabola if $f > 0$ and to the right of the parabola if $f < 0$]

ALERT Students might have forgotten the meaning of *axis of symmetry*. **ASK** "Where is the axis of symmetry for a parabola?"

An alternate method for finding d_2 is to subtract: $d_2 = x - (-f)$.

As students present equations, help them see that for either of the two parabolas, $y = kx^2$ and $x = ky^2$, the value of k is $\frac{1}{4f}$ where f is the distance from the vertex to the focus and to the directrix.

CRITICAL QUESTION How would you define a parabola?

BIG IDEA A parabola might be defined by its equation. Or it might be defined as the points in a plane that are the same distance from a given point and line.

CRITICAL QUESTION What's the difference between parabolas in this lesson and the parabolas you were graphing before?

BIG IDEA In this lesson, parabolas may open in directions other than up and down. Therefore they may not be graphs of functions.

CRITICAL QUESTION How do you identify the maximum and minimum of a horizontally oriented parabola?

BIG IDEA There is no maximum or minimum of a horizontally oriented parabola. Horizontally oriented parabolas have no limit on how high or how low they can go.

Formative Assessment

Student input in class discussion and the examples can help you anecdotally assess the objectives of the lesson. Since this is primarily a teacher directed lesson, student responses on the exercises will provide the more objective evidence of meeting the objectives. Since Exercise 1 reviews vocabulary from the lesson, consider projecting the exercise and having students work in pairs.

Have students present their solutions and their work. Whether student responses are correct or incorrect, ask other students if they agree and why. Encourage students to use appropriate mathematical language as evidence. **SMP 1, 3, 6**

Apply

Extra Example

1. Name the vertex, line of symmetry, focus, and directrix of the parabola $y = \frac{(x-2)^2}{16} + 3$.
   vertex: (2, 3); line of symmetry: $x = 2$; focus: (2, 7); directrix: $y = -1$

Closing Question

2. Find the equation of the parabola with focus (3, −1) and vertex (0, −1).
   $(y+1)^2 = 12x$

Assigning Exercises

Suggested: 1–10
Additional Practice: 11–17

Exercise 2 SUPPORT Students may need assistance rewriting the equations in a form from which they can identify the horizontal and vertical shift of the vertex and the direction in which the parabola opens.

2a. line of symmetry: $x = 0$
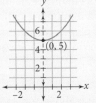

2b. line of symmetry: $y = -2$
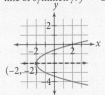

2c. line of symmetry: $x = -3$
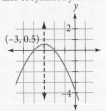

2d. line of symmetry: $y = 0$
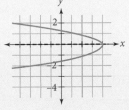

5.8 Exercises

You will need your graphing calculator for Exercises **13** and **14**.

Practice Your Skills

1. For each parabola described, use the information given to find the location of the missing feature. It may help to draw a sketch.
   a. If the focus is (1, 4), and the directrix is $y = -3$, where is the vertex? ⓐ (1, 0.5)
   b. If the vertex is (−2, 2), and the focus is (−2, −4), what is the equation of the directrix? ⓐ $y = 8$
   c. If the directrix is $x = 3$, and the vertex is (6, 2), where is the focus? (9, 2)

2. Sketch each parabola, and label the vertex and line of symmetry.
   a. $\left(\frac{x}{2}\right)^2 + 5 = y$ ⓐ
   b. $(y+2)^2 - 2 = x$ ⓐ
   c. $-(x+3)^2 + 1 = 2y$
   d. $2y^2 = -x + 4$
   e. $\frac{y-3}{2} = \left(\frac{x+1}{4}\right)^2$
   f. $\frac{x-2}{3} = \left(\frac{y}{5}\right)^2$

3. Locate the focus and directrix for each graph in Exercise 2. ⓐ

4. Write an equation in standard form for each parabola.

   a. ⓐ
   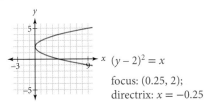
   $(y-2)^2 = x$
   focus: (0.25, 2); directrix: $x = -0.25$

   b. ⓐ
   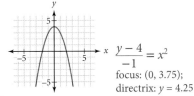
   $\frac{y-4}{-1} = x^2$
   focus: (0, 3.75); directrix: $y = 4.25$

   c.
   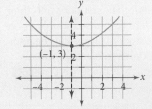
   $y + 1 = \left(\frac{x-3}{2}\right)^2$
   focus: (−3, 0); directrix: $y = -2$

   d.
   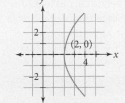
   $\left(\frac{y-2}{3}\right)^2 = \frac{x+6}{-1}$
   focus: (−8.25, 2); directrix: $x = -3.75$

5. Locate the focus and directrix for each graph in Exercise 4. ⓐ See above

Reason and Apply

6. Write the equation of each parabola with the given directrix and vertex.
   a. with directrix $y = -2$, vertex (0, 6) ⓗ
   b. with directrix $x = 3$, vertex (1, 4)
   c. with directrix $y = 4$, vertex (2, −1)
   d. with directrix $x = 1$, vertex (4, 0)

7. Write the equation of each parabola with the given directrix and focus.
   a. with directrix $y = -1$, focus (1, 3) ⓐ
   b. with directrix $x = 2$, focus (5, 1)
   c. with directrix $y = 1$, focus (5, −5)
   d. with directrix $x = 3$, focus (0, 4)

8. Consider the graph at right.

   a. Because $d_1 = d_2$, you can write the equation
   $$\sqrt{(x-0)^2 + (y-3)^2} = y - (-1) \qquad y = \frac{1}{8}x^2 + 1$$
   Rewrite this equation by solving for y. @

   b. Describe the graph represented by your equation from 8a.
   The graph is a parabola with vertex at (0, 1), focus at (0, 3), and directrix $y = -1$.

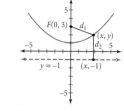

9. The pilot of a small boat charts a course such that the boat will always be equidistant from an upcoming rock and the shoreline.

   a. Make a sketch of the situation, showing the approximate path of the boat.

   b. Describe the path of the boat. The path is parabolic.

   c. If the rock is 2 mi offshore, write an equation for the path of the boat. If you locate the rock at (0, 2) and the shoreline at $y = 0$, the equation is $y = \frac{1}{4}x^2 + 1$

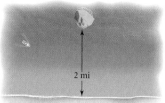

10. **APPLICATION** Sheila is designing a parabolic dish to use for cooking on a camping trip. She plans to make the dish 40 cm wide and 20 cm deep. Where should she locate the cooking grill so that all of the light that enters the parabolic dish will be reflected toward the food? ⓗ   The grill should be located 5 cm from the center of the dish.

Solar cookers focus the heat of the Sun into a single spot, in order to boil water or cook food. A well-designed cooker can create heat up to 400°C. Solar cookers can be created with minimal materials and can help save natural resources, particularly firewood. Inexpensive solar cookers are now being designed and distributed for use in developing countries, like this parabolic reflector in Nepal.

Review

11. The distance between the points (2, 3) and $(x, 7)$ is 5. What are the possible values of x?
    $x = 5, -1$

12. Find the equation of the circle containing the points (4, 9) and (−6, 9).
    $(x + 1)^2 + (y - 9)^2 = 25$

13. Find the minimum distance from the origin to the parabola $y = -x^2 + 1$. What point(s) on the parabola is closest to the origin? $\frac{\sqrt{3}}{2}$, or approximately 0.8660; $\left(\pm\frac{\sqrt{2}}{2}, \frac{1}{2}\right)$, or approximately (±0.7071, 0.5)

14. On a three-dimensional coordinate system with variables x, y, and z, the standard equation of a plane is in the form $ax + by + cz = d$. Find the intersection of the three planes described by $3x + y + 2z = -11$, $-4x + 3y + 3z = -2$, and $x - 2y - z = -3$.
    $x = -1, y = 4, z = -6$

3a. focus: (0, 6); directrix: $y = 4$
3b. focus: (−1.75, −2); directrix: $x = -2.25$
3c. focus: (−3, 0); directrix: $y = 1$
3d. focus: (3.875, 0); directrix: $x = 4.125$
3e. focus: (−1, 5); directrix: $y = 1$
3f. focus: $\left(\frac{49}{12}, 0\right)$; directrix: $x = -\frac{1}{12}$
6a. $y = \frac{1}{32}x^2 + 6$
6b. $x = -\frac{1}{8}(x - 4)^2 + 1$
6c. $y = -\frac{1}{20}(x - 2)^2 - 1$
6d. $x = \frac{1}{12}y^2 + 4$
7a. $y = \frac{1}{8}(x - 1)^2 + 1$
7b. $x = \frac{1}{6}(y - 1)^2 + 3.5$
7c. $y = -\frac{1}{12}(x - 5)^2 - 2$
7d. $x = -\frac{1}{6}(y - 4)^2 + 1.5$

Exercise 8 ALERT The complexity of the algebra in this exercise may be a challenge to some students.

9a. Sketches will vary; the path should be parabolic with the vertex midway between the rock and shoreline.

Exercise 10 Urge students to sketch the parabola to scale on their graph paper.

Exercise 13 ALERT Students probably will not find exact answers to this exercise. Exact answers can be found by completing the square for the polynomial $x^4 - x^2 + 1$, which is the square of the distance to the origin. Note that the minimum value of $x^4 - x^2 + 1 = \left(x^2 - \frac{1}{2}\right)^2 = 0$, at $x = \frac{1}{2}$.

Exercise 14 If students are having trouble getting started, ASK "Is finding the intersection of three planes analogous to finding the intersection of two lines?" [In both cases, students solve a system of equations.]

Exercise 15b ALERT Students might see the cubic power and start trying to factor as a difference of cubes.

Exercise 17 Although there are 11 support cables labeled, their distance apart is $\frac{1}{12}$ of 160 ft, or $13\frac{1}{3}$ ft. Students may correctly reason that a bridge would have two parabolic cables, one on each side of the road. Their answer of 458.3 ft is correct for the assumption they are making.

17a. $a = k = 52.08\overline{3}$ ft;
$b = j = 33.\overline{3}$ ft;
$c = i = 18.75$ ft;
$d = h = 8.\overline{3}$ ft;
$e = g = 2.08\overline{3}$ ft;
$f = 0$;
total length $= 229.1\overline{6}$ ft

15. Completely factor each expression.

   a. $x^2 + 8x + 15$
   $(x + 3)(x + 5)$

   b. $x^3 - 49x$ @
   $x(x - 7)(x + 7)$

   c. $x^2 - 3x - 28$
   $(x - 7)(x + 4)$

16. **APPLICATION** One possible gear ratio on Matthew's mountain bike is 4 to 1. This means that the front gear has four times as many teeth as the gear on the back wheel. So each revolution of the pedal causes the rear wheel to make four revolutions.

    a. If Matthew is pedaling 60 revolutions per minute (r/min), how many revolutions per minute is the rear tire making?
    240 r/min

    b. If the diameter of the rear tire is 26 in., what speed in miles per hour will Matthew attain?
    18.6 mi/h

    c. Matt downshifts to a front gear that has 22 teeth and a rear gear that has 30 teeth. If he keeps pedaling 60 r/min, what will his new speed be?
    6.3 mi/h

17. **APPLICATION** The main cables of a suspension bridge typically hang in the shape of parallel parabolas on both sides of the roadway. The vertical support cables, labeled a–k, are equally spaced, and the center of the parabolic cable touches the roadway at f. If this bridge has a span of 160 ft between towers, and the towers reach a height of 75 ft above the road, what is the length of each support cable, a–k? What is the total length of vertical support cable needed for the portion of the bridge between the two towers?

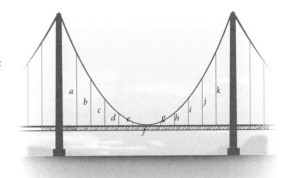

→ **Engineering**
CONNECTION

The roadway of a suspension bridge is suspended, or hangs, from large steel support cables. By itself, a cable hangs in the shape of a *catenary* curve. However, with the weight of a roadway attached, the curvature changes, and the cable hangs in a parabolic curve. It is important for engineers to ensure that cables are the correct lengths to make a level roadway. Just as the silk on a spider's web forms multiple natural catenary curves.

CHAPTER 5 REVIEW

Quadratic equations, which are 2nd-degree polynomial equations, have three major forms. The equation $y = ax^2 + bx + c$ is the **general form** of a quadratic equation. The **factored form** of the equation, $y = a(x - r_1)(x - r_2)$, identifies the x-intercepts and vertical scale factor of the quadratic. And the **vertex form** of the equation, $y = a(x - h) + k$, identifies the vertex and vertical scale factor of the quadratic. In a quadratic equation, you can find the roots by using the **quadratic formula**.

There are the same number of roots as the degree of the polynomial. In some cases these roots may include **imaginary numbers** or **complex numbers**. If the coefficients of a polynomial are real, then any nonreal roots come in **conjugate pairs**.

In this chapter you also saw some special relations. Each relation was described as a set of points, or **locus**, that satisfied some criteria. The **circle** is an example of a relation described by a locus. It is the set of points that are a fixed distance from a fixed point called the **center**. The parabola also has a locus definition. It is the set of points that are equidistant from a fixed point called the focus and a fixed line called the **directrix**.

Exercises

You will need your graphing calculator for Exercises **3** and **5**.

@ Answers are provided for all exercises in this set.

1. Factor each expression completely.

   a. $2x^2 - 10x + 12$
   $2(x - 2)(x - 3)$

   b. $2x^2 + 7x + 3$
   $(2x + 1)(x + 3)$ or $2(x + 0.5)(x + 3)$

   c. $x^3 - 10x^2 - 24x$
   $x(x - 12)(x + 2)$

2. Solve each equation by setting it equal to zero and factoring.

   a. $x^2 - 8x = 9$
   $x = 9$ or $x = -1$

   b. $x^4 + 2x^3 = 15x^2$
   $x = 0, x = 3,$ or $x = -5$

3. Using three noncollinear points as vertices, how many different triangles can you draw? Given a choice of four points, no three of which are collinear, how many different triangles can you draw? Given a choice of five points? n points? 1; 4; 10; $\frac{1}{6}n^3 - \frac{1}{2}n^2 + \frac{1}{3}n$

4. Tell whether each equation is written in general form, vertex form, or factored form. Write each equation in the other two forms, if possible.

   a. $y = 2(x - 2)^2 - 16$

   b. $y = -3(x - 5)(x + 1)$

   c. $y = x^2 + 3x + 2$

   d. $y = (x + 1)(x - 3)(x + 4)$

   e. $y = 2x^2 + 5x - 6$

   f. $y = -2 - (x + 7)^2$

CHAPTER REVIEW

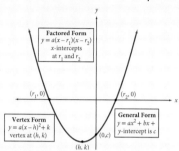

Reviewing

To review the equivalent forms of a quadratic equation, consider projecting the equation $x^2 + 12x = -12$ and having students write the equation in all three forms. This will also review completing the square. To review the importance of each form, project this image and discuss the merits of each form of the equation.

Project the equations
$y = x^2 - 6x + 12$
$y = x^2 - 6x + 9$
$y = x^2 - 6x + 6$

ASK "What is the nature of the roots of these three equations? How do you know?" For the first equation, there are two real roots ($b^2 - 4ac > 0$), for the second equation, there is one (double) real root ($b^2 - 4ac = 0$), and for the third equation, there are two complex roots ($b^2 - 4ac < 0$)

To review complex numbers, consider projecting these:
$(3 + 2i)$ and $(3 - 2i)$
ASK "What is the relationship of these two numbers?" [conjugates]. Have students add, subtract, and multiply the numbers. [6, 0, 13]

To review the distance formula and using it to write the equation of a circle, given a center and radius, **ASK** "What is the equation of the circle with center at $(-3, -6)$ that passes through the point $(-4, 8)$?" [$(x + 3)^2 + (y + 6)^2 = 197$]

You might use Exercise 13 as a class activity to review the focus-directrix definition of a parabola.

Assigning Exercises

If you spend two days on the review to allow some students to give presentations to the class, you can assign all the exercises. Groups might work together on some exercises, such as Exercises 4, 5, 10, and 12. You will need to assign TAL 1 to review the discriminant.

Exercise 3 Encourage students to sketch several diagrams using different numbers of vertices and drawing and counting the triangles.

Exercise 4 You might accept answers for factored form that include only rational or real numbers. For example, 4a could be $y = 2(x^2 - 4x - 4)$ and 4f could be $y = -(x^2 + 14x + 51)$

4a. vertex form; general form: $y = 2x^2 - 8x - 8$; factored form: $y = 2[x - (2 + \sqrt{2})][x - (2 - \sqrt{2})]$

4b. factored form; general form:
$y = -3x^2 + 12x + 15$;
vertex form: $y = -3(x - 2)^2 + 27$

4c. general form; factored form:
$y = (x + 2)(x + 1)$;
vertex form: $y = (x + 1.5)^2 - 0.25$

4d. factored form; general form:
$y = x^3 + 2x^2 - 11x - 12$;
no vertex form for cubic equations

4e. general form; factored form:
$y = 2\left(x - \frac{-5 + \sqrt{73}}{4}\right)\left(x - \frac{-5 - \sqrt{73}}{4}\right)$;
vertex form: $y = 2(x + 1.25)^2 - 9.125$

4f. vertex form; general form:
$y = -x^2 - 14x - 51$; factored form:
$y = -[x - (-7 + i\sqrt{2})][x - (-7 - i\sqrt{2})]$

Exercise 5 For most of these equations, have students graph the function on paper and use calculators to check their work.

5a. zeros: $x = -0.83$ and $x = 4.83$

5b. zeros: $x = -1$ and $x = 5$

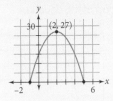

5c. zeros: $x = 1$ and $x = 2$

5d. zeros: $x = -4$, $x = -1$, and $x = 3$

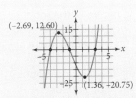

5e. zeros: $x = -5.84$, $x = 1.41$, and $x = 2.43$

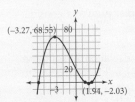

5f. zeros: $x = -2$, $x = -1$, $x = 0.5$, and $x = 2$

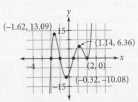

Exercise 8 This exercise uses the projectile motion function from Lesson 5.2.

5. Sketch a graph of each function. Label all zeros and the coordinates of all maximum and minimum points. (Each coordinate should be accurate to the nearest hundredth.)

   a. $y = 2(x - 2)^2 - 16$
   b. $y = -3(x - 5)(x + 1)$
   c. $y = x^2 - 3x + 2$
   d. $y = (x + 1)(x - 3)(x + 4)$
   e. $y = x^3 + 2x^2 - 19x + 20$
   f. $y = 2x^5 - 3x^4 - 11x^3 + 14x^2 + 12x - 8$

6. Write the equation of each graph.

   a.
   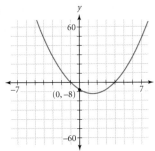
   $y = 2(x + 1)(x - 4)$

   b.
   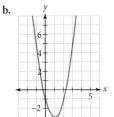
   $y = 2(x - 1)^2 - 3$

7. A farmer has 224 meters of new fencing with which to create three equal sized pens along an already existing fence line. How big should the pens be in order to enclose the maximum total area?

Each pen will be 32m by 32 m for a maximum enclosed space of 3072 m² of pens

8. An object is dropped from the top of a building into a pool of water at ground level. There is a splash 6.8 s after the object is dropped. How high is the building in meters? In feet?
   approximately 227 m, or 740 ft

9. Find the center and radius for each circle.

   a. $x^2 + y^2 - 12x + 8y - 7 = 0$ center $(6, -4)$, radius $\sqrt{59}$
   b. $x^2 + y^2 - 2x + 3y + 1 = 0$ center $(1, -1.5)$, radius 1.5

10. This 26-by-21 cm rectangle has been divided into two regions. The width of the unshaded region is x cm, as shown.

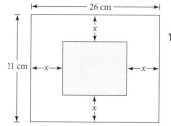

10a.

b. Find the domain and range for this function.

a. Express the area of the shaded part as a function of x, and graph it. $y = (26 - 2x)(21 - 2x)$

The Grande Arche building at La Defense in Paris, France, is in the form of a hollowed-out cube and functions as an office building, conference center, and exhibition gallery.

c. Find the x-value that makes the two regions (shaded and unshaded) equal in area. $x \approx 3.395$ cm

11. Write each expression in the form $a + bi$.

   a. $(4 - 2i)(-3 + 6i)$ $0 + 30i$
   b. $(-3 + 4i) - (3 + 13i)$ $-6 - 9i$
   c. $(2 - i)(3 - 4i)$ $2 - 11i$

12. Write the general quadratic equation $x^2 + y^2 + 8x - 2y - 8 = 0$ in standard form. Identify the shape described by the equation and describe its features. $(x + 4)^2 + (y - 1)^2 = 25$. The graph is a circle with center $(-4, 1)$ and radius 5.

13. Write the general quadratic equation $y^2 - 8y - 4x + 28 = 0$ in standard form. Determine the vertex, focus, and directrix of the parabola defined by this equation. Sketch a graph. $\left(\frac{y-4}{2}\right)^2 = x - 3$; vertex: $(3, 4)$; focus: $(4, 4)$; directrix: $x = 2$

Take Another Look

1. You've already discovered several facts about quadratic polynomials in the form $f(x) = ax^2 + bx + c$. You know the vertex has coordinates $\left(\frac{-b}{2a}, c - \frac{b^2}{4a}\right)$, or $\left(\frac{-b}{2a}, \frac{4ac - b^2}{4a}\right)$, the line of symmetry is $x = -\frac{b}{2a}$, and the discriminant is $b^2 - 4ac$. You also know the quadratic formula for finding roots of the equation $ax^2 + bx + c = 0$: $x = \frac{-b \pm \sqrt{b^2 - 4ac}}{2a}$. Use the quadratic formula to complete parts a and b.

   a. Find a relationship between the coefficients of $ax^2 + bx + c$ and the sum of the two roots.
   b. Find a relationship between the coefficients of $ax^2 + bx + c$ and the product of the two roots.
   c. Use your results for parts a and b to find the sum and product of the roots of these equations.
      i. $2x^2 + x - 3 = 0$
      ii. $14x^2 - 29x - 15 = 0$
      iii. $4(3.5 - x)(x + 5) = 0$
   d. Check your answers for part c by finding the roots.

10b. domain: $0 \leq x \leq 10.5$; range: $0 \leq y \leq 546$

Exercise 10c The area of the unshaded region is $\frac{1}{2}(26)(21)$.

Take Another Look

1a. In finding the sum, the $\pm$ terms in the numerator of the quadratic formula cancel to yield $-\frac{b}{2a}$. Alternatively, students may understand that the mean of the roots is $-\frac{b}{2a}$ and that the sum of two numbers is twice their mean.

1b. The two numerators being multiplied have the form $(x + y)(x - y)$, so their product is $x^2 - y^2$, which is $4ac$. The denominator times itself is $4a^2$, so the result is $\frac{c}{a}$.

1c. i. Substituting $a = 2$, $b = 1$, and $c = -3$ shows that the roots have sum $-\frac{1}{2}$ and product $-\frac{3}{2}$.

   ii. Because $a = 14$, $b = -29$, and $c = -15$, the sum is $\frac{29}{14}$, and the product is $-\frac{15}{14}$

   iii. The expression is equivalent to $-4x^2 - 6x + 70$. The sum is $-\frac{b}{a} = -\frac{-6}{-4} = -1.5$, and the product is $\frac{70}{-4} = -17.5$.

   iv. The roots are $x = 3.5$ and $x = -5$. Their sum and product agree with those found in part iii.

1d. i. Factoring $2x^2 + x - 3$ gives $(2x + 3)(x - 1) = 0$, so $x = \frac{-3}{2}$ and $x = 1$. Then $\frac{-3}{2} + 1 = \frac{-1}{2}$ and $\frac{-3}{2} \cdot 1 = \frac{-3}{2}$. This checks.

   ii. Factoring $14x^2 - 29x - 15$ gives $(7x + 3)(2x - 5) = 0$, so $x = \frac{-3}{7}$ and $x = \frac{5}{2}$. Then $\frac{-3}{7} + \frac{5}{2} = \frac{-6}{14} + \frac{35}{14} = \frac{29}{14}$ and $\frac{-3}{7} \cdot \frac{5}{2} = \frac{-15}{14}$. This checks.

   iii. $(4)(3.5 - x)(x + 5) = 0$, so $x = 3.5$ and $x = -5$. Then $3.5 + -5 = -1.5$ and $3.5 \cdot -5 = -17.5$. This checks.

2. When reflecting the parabola over the vertex, the only change in the equation is the sign of *m*. When finding the roots of the original equation you get

$$x = \frac{2mh \pm \sqrt{4m^2h^2 - 4m(mh^2 + k)}}{2m}$$

$= h \pm i\frac{\sqrt{mk}}{m}$. Note that the real part of this solution is the *x*-coordinate of the vertex.

When finding the roots of the reflection, you get $x =$

$$\frac{-2mh \pm \sqrt{4m^2h^2 + 4m(-mh^2 + k)}}{2m}$$

$= h \pm \frac{\sqrt{mk}}{m}$.

When forming the reflection, the vertex does not change. Thus the two roots of the new equation have the same average as the roots of the original. So the real part of the original solution is the average of the roots. The imaginary part of the original solution is the value that is added or subtracted from the *x*-coordinate. That is, it is the difference between the midpoint and either of the new roots.

Assessing the Chapter

The Kendall Hunt Cohesive Assessment System (CAS) provides a powerful tool with which to manage assessment content, create and assign homework and tests, and disseminate assessment data.

Though the primary purpose of asking students to write test questions is for them to review, you might use some of their questions on your chapter test. You might suggest that students write a question for each lesson as you work through the chapter. Add to the student questions by choosing from constructive assessment or other items in CAS.

Facilitating Self-Assessment

If you can spend two days on review, you might use some time on the second day for students to present some test items they have written and for the class to work through the problems.

Students might consider these exercises for addition to their portfolios:

Lesson 5.1, Exercises 10 and 12; Lesson 5.2, Exercises 9 and 12; Lesson 5.3, Exercise 10 and 11; Lesson 5.4, Exercise 8; Lesson 5.5, Exercises 8 and 9; Lesson 5.6, Exercise 8; Lesson 5.7, Exercise 8 or 10; and Lesson 5.8, Exercise 10.

2. You know that when a quadratic equation $ax^2 + bx + c = 0$ has real roots, these roots are the *x*-intercepts of the graph of the quadratic function $y = ax^2 + bx + c$. When a quadratic equation has complex roots, the graph of the function does not cross the *x*-axis. However, you can still visualize the roots by doing the following procedure. First, reflect the function over the vertex and find the roots of this new quadratic equation. Next, find the midpoint between the two roots. This will be the real part of your complex solutions. Then, determine the distance between this midpoint and either of the roots. This is the complex part of the solutions. Try this procedure with the equation $(x + 1)^2 + 1 = 0$. (when reflecting over the vertex, the vertex stays the same, the parabola simply opens in the opposite direction.) Why does this work? Use the equation $m(x - h)^2 + k = 0$ to explain what is happening.

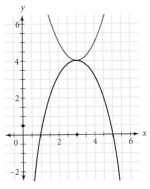

(Note: This is a graph of $y = (x - 3)^2 + 4$ and its reflection over the vertex. The complex solutions are 3 (the midpoint between the solutions of the reflection) $\pm 2i$ (the distance between the midpoint and either of the roots of the reflected graph).

Assessing What You've Learned

WRITE TEST ITEMS In this chapter you learned what complex numbers are, how to do computations with them, and how they relate to polynomial functions. Write at least two test items that assess understanding of complex numbers. Be sure to include complete solutions.

PERFORMANCE ASSESSMENT While a classmate, a friend, a family member, or a teacher observes, show how you would solve a quadratic equation using three different methods, factoring, completing the square, using the quadratic formula, or graphing. Explain how the methods are related to each other and when each method is best to use.

ORGANIZE YOUR NOTEBOOK Update your notebook to include sections on factoring, the quadratic formula, and completing the square. Include examples of each. Be sure to include examples of how to use completing the square to find the center and radius of a circle as well as the vertex of a parabola.

CHAPTER 6

Polynomial and Rational Functions

Overview

Discovering Advanced Algebra is about modeling real-world problems with mathematical functions. Earlier chapters developed linear, exponential, and logarithmic functions, whose domains and ranges are real numbers. In this chapter students learn to model problems with polynomial functions of degree more than 1. This chapter also continues modeling real-world problems with mathematical functions, as students learn to model situations with rational functions.

Lesson 6.1 presents the definition of *polynomial* and the finite differences method to find the degree of a polynomial function. In **Lesson 6.2,** students explore cubic functions and relate the graphs of polynomial functions to the number of zeros. In **Lesson 6.3,** students deepen their understanding of a polynomial function's graph and what it says about degree and extreme values. **Lesson 6.4** shows additional methods for finding zeros of higher-degree polynomial functions with the introduction of the remainder theorem, the rational root theorem, and long division. In **Lesson 6.5,** students explore transformations of the parent function $y = \frac{1}{x}$, a hyperbola whose asymptotes are the coordinate axes. In **Lesson 6.6,** students learn to identify characteristics of the graph of a rational function from the equation and to write the equation of a rational function from the graph. In **Lesson 6.7,** students review factoring as they add and subtract rational expressions. In **Lesson 6.8,** students extend operations with polynomials to include multiplying and dividing rational expressions. And in **Lesson 6.9**, students solve rational equations, with attention to extraneous roots both in the original equation and the ones that may appear in the solution process.

The **Performance Task** highlighted in this chapter is in Lesson 6.3, where students apply all the patterns they have noticed in graphing polynomials to sketch a graph of a 5th-degree function, a 6th-degree function, and a 7th-degree function.

The Mathematics

Polynomials have the form $a_n x^n + a_{n-1} x^{n-1} + \cdots + a_1 x + a_0$, where n is a positive integer (the degree of the polynomial), $a_n \neq 0$, and the coefficients a_i are numbers. In most of this chapter, the coefficients are real numbers. This form, with decreasing powers on x, is the general form of a polynomial. So far in this course, students' work with polynomials has focused on linear polynomials—polynomials of degree 1, which have the general form $a_1 x + a_0$ (or $bx + a$). Polynomials can be considered either as mathematical expressions or as functions for modeling real-world situations.

Polynomials as Expressions

As objects, polynomials have a great deal in common with integers. You can add or subtract them or multiply them. The result of dividing them isn't necessarily a polynomial, just as the quotient of two integers is not necessarily an integer. You can factor many polynomials into "smaller" (lower-degree) polynomials that are basic building blocks, similar to prime numbers. For polynomials with integer coefficients, the Rational Root Theorem helps with this factoring; it says that for $bx + a$ to be a factor of a polynomial, b must be a factor of the coefficient of the highest power, a must be a factor of the constant coefficient and a and b must be relatively prime integers.

Polynomials as Functions

Polynomials can also be thought of as functions. Like linear functions, polynomial functions can be graphed. Graphs of second-degree polynomial functions, or quadratics, have the shape of parabolas. Graphs of higher-degree polynomial functions may be more complicated.

Earlier, students learned how to adjust the coefficients of a linear function to get a good fit to a data set. In this chapter, students find appropriate coefficients for equations representing data sets that can be modeled by quadratic or cubic functions. The finite differences method determines the degree; students then set up and solve a system of linear equations to determine the coefficients.

An input number for a function is a zero of that function if the corresponding output number is 0. For the function's graph, these zeros designate the x-intercepts. If the function is set equal to 0, the solutions to the equation (also called the roots of the equation) are the same numbers as the zeros of the function.

The Remainder Theorem says that, if you divide a polynomial function $P(x)$ by $(x - r)$, then the remainder is $P(r)$ and $(x - r)$ is a factor of $P(x)$ if and only if $P(r) = 0$. So, the number r is a zero of the polynomial function if and only if $(x - r)$ is a factor of the polynomial as an expression. Therefore, if you can write the polynomial in factored form as a product of binomials of the form $(x - r)$, then you can find all the zeros.

Rational Expressions and Functions

A rational expression can be expressed as a ratio of polynomials, just as a rational number can be expressed as a ratio of integers. Like rational numbers, rational expressions can be added, subtracted, multiplied, divided, and simplified by reducing out common factors to get the rational expression in lowest terms.

More useful for solving real-world problems are functions defined by rational expressions. If the polynomials in both the numerator and denominator are linear and neither is a factor of the other, then the rational function is a transformation of the function $y = \frac{1}{x}$. Long or synthetic division can reveal the nature of that transformation.

Graphs of many rational functions have asymptotes to which the graph gets closer and closer. In particular, any zero of the polynomial in the denominator may indicate a vertical asymptote. Any zero of the polynomial in the numerator may show an *x*-intercept. But zeros common to both the numerator and denominator may indicate a one-point gap in the graph.

Using This Chapter
This is an important chapter with a lot of new content. It is, therefore, important not to skip lessons. To save time, you might use the **Shortened** modifications of the investigation. You could also use one of the desmos simulations. It is important, however, to have student presentations of ideas to help them communicate mathematically and justify their answers.

Common Core State Standards (CCSS)
In Algebra I, students worked with linear, quadratic, and exponential functions. In Algebra II, they build on that work and extend it to include polynomial, rational, and radical functions. Students work with the expressions that define the functions, and continue to expand their abilities to model situations and to solve equations, including solving rational equations. Students continue to abstract the general principle that transformations on a graph always have the same effect regardless of the type of the underlying function, which for rational functions is the family of functions defined by $y = \frac{1}{x}$ (F.IF.4, 5, 7, F.BF.3).

Interpreting the structure of expressions (A.SSE.1, 2) and writing expressions in equivalent forms is extended to polynomial and rational expressions. Students perform arithmetic operations on polynomials and understand the relationship between zeroes and factors of polynomials as they apply the Remainder Theorem and use the zeros to sketch a graph (A.APR.1 - 3). They extend solving equations to include rational equations symbolically and graphically (A.REI.2, 11).

Additional Resources For complete references to additional resources for this chapter and others see your digital resources.

Refreshing Your Skills To review the algebra skills needed for this chapter see your digital resources.
   Polynomial Expressions

Materials
- motion sensors
- small pillows or other soft objects
- scissors
- graph paper
- rulers
- tape
- dry linguine
- small film canisters
- string
- small weights (pennies, beans)

CHAPTER 6
Polynomial and Rational Functions

CHAPTER 6

Objectives
- Define types and properties of polynomials
- Use finite differences and systems of equations to find a polynomial function that fits a data set
- Understand the correspondence among the zeros of a polynomial, the *x*-intercepts of its graph, and the roots of an equation
- Find the graph of a rational function from its equation and vice versa
- Learn to add, subtract, multiply, and divide rational expressions
- Learn to solve rational equations

Most mathematical functions commonly used by engineers and scientists can be approximated by polynomials. A precursor to the computer, a difference engine is an automatic mechanical calculator designed to tabulate polynomial functions. Engineers use rational functions in their cantilever engineering–projecting beams that are supported at only one end—like in the skydeck in the Willis Tower in Chicago. The functional response curve is used by ecologist to describe the interrelationship of organisms, like this red panda and the amount of bamboo in a forest.

OBJECTIVES
In this chapter you will
- add, subtract, multiply, and divide polynomial expressions
- find polynomial functions that fit a set of data
- identify features of the graph of a polynomial function
- use division and other strategies to find roots of higher-degree polynomials
- study rational functions and learn special properties of their graphs
- add, subtract, multiply, and divide rational expressions

Polynomial and rational functions are used by scientist and engineers to model many real-world phenomena. Ecologists often look for a mathematical model to describe the interrelationship of organisms, many of which are rational functions. In the case of the endangered red panda, a mammal native to the eastern Himalayas and southwestern China, lack of habitat and food and poaching has caused a decline in the population. **ASK** "What other factors might impact the population of a species?" Engineers use rational function models to design structures like the Skydeck at Willis Tower and to determine the weight that a beam can safely support. **ASK** "What other examples of structures where it might be important to determine weight capacity as they are designed?"

LESSON 6.1

This lesson uses finite differences as a first step toward finding polynomials that fit data.

COMMON CORE STATE STANDARDS		
APPLIED	DEVELOPED	INTRODUCED
	F.LE.1a	S.ID.6a

Objectives

- Define *polynomial, monomial, binomial,* and *trinomial*
- Determine the degree of a polynomial by using finite differences
- Write polynomials in general form
- Use finite differences and systems of equations to find a polynomial function that fits a data set
- Prove that linear functions grow by equal distances and exponential functions grow by equal factors over equal intervals.

Vocabulary

monomial
binomial
trinomial
polynomial term
polynomial function term
degree
general form
finite differences method

Materials

- motion sensors
- small pillows or other soft objects
- Calculator Notes: Looking for the Rebound Using the EasyData App; Movin' Around; Free Fall; Finite Differences

Launch

Use the Launch in your ebook. The Launch will review the vocabulary of polynomials. You can then introduce the finite differences method. Project the three sets of data on the following page and ASK "What patterns do you see?" SMP 7, 8

LESSON 6.1

Differences challenge assumptions.

ANNE WILSON SCHAEF

Polynomials

Earlier, you studied arithmetic sequences, which have a common difference between consecutive terms. This common difference is the slope of a line through the graph of the points. So, if you choose x-values along the line that form an arithmetic sequence, the corresponding y-values will also form an arithmetic sequence.

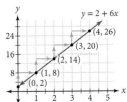

You have also studied several kinds of nonlinear sequences and functions, which do not have a common difference or a constant slope. In this lesson you will discover that even nonlinear sequences sometimes have a special pattern in their differences. These patterns are often described by polynomials.

Definition of a Polynomial

A **polynomial** in one variable is any expression that can be written in the form

$$a_n x^n + a_{n-1} x^{n-1} + \cdots + a_1 x^1 + a_0$$

where x is a variable, the exponents are nonnegative integers, the coefficients are real numbers, and $a_n \neq 0$. Each monomial being added to make the polynomial is a **term**.

When a polynomial is set equal to a second variable, such as y, it defines a **polynomial function**. The **degree** of a polynomial or polynomial function is the power of the term that has the greatest exponent. For example, linear functions are 1st-degree polynomial functions because the largest power of x is 1. The polynomial function below has degree 3. If the degrees of the terms of a polynomial decrease from left to right, the polynomial is in **general form.**

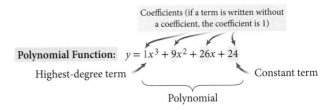

A polynomial that has only one term is called a **monomial.** A polynomial with two terms is a **binomial,** and a polynomial with three terms is a **trinomial.** Polynomials with more than three terms are usually just called "polynomials."

In modeling linear functions, you have already discovered that for x-values that are evenly spaced, the differences between the corresponding y-values must be the same. With 2nd- and 3rd-degree polynomial functions, the differences between the corresponding y-values are not the same. However, finding the differences between those differences produces an interesting pattern.

ELL	Support	Advanced
Start the lesson by having students create a graphic organizer comparing linear and quadratic functions, then introduce the lesson and have students add in the new information and vocabulary on polynomials and finite differences.	As needed, assign the Refreshing Your Skills: Polynomial Expressions lesson to review terms and properties of polynomial expressions.	Have students explain why the finite differences method works. ASK "Why are the nth differences constant for a polynomial of degree n?" [The first differences give the rate of change of a polynomial 1 degree lower than the original. The second differences are the rate of change of the first differences, and so on.]

1st degree		
\multicolumn{3}{c}{$y = 3x + 4$}

x	y	D_1
2	10	
3	13	3
4	16	3
5	19	3
6	22	3
7	25	3

2nd degree
$y = 2x^2 - 5x - 7$

x	y	D_1	D_2
3.7	1.88		
3.8	2.88	1	
3.9	3.92	1.04	0.44
4.0	5.00	1.08	0.44
4.1	6.12	1.12	0.44
4.2	7.28	1.16	0.44

3rd degree
$y = 0.1x^3 - x^2 + 3x - 5$

x	y	D_1	D_2	D_3
−5	−57.5			
0	−5	52.5		
5	−2.5	2.5	−50	
10	25	27.5	25	75
15	152.5	127.5	100	75
20	455	302.5	175	75

The D_4 and D_5 values, respectively, will be constant.

Note that in each case the x-values are spaced equally. You find the first set of differences, D_1, by subtracting each y-value from the one after it. You find the second set of differences, D_2, by finding the differences of consecutive D_1 values in the same way.

For the 2nd-degree polynomial function, the D_2 values are constant, and for the 3rd-degree polynomial function, the D_3 values are constant. What do you think will happen with a 4th- or 5th-degree polynomial function?

Analyzing differences to find a polynomial's degree is called the **finite differences method**. You can use this method to help find the equations of polynomial functions modeling certain sets of data.

Similar types of foods are grouped together at the Toa Payoh Market in Singapore. Polynomials are often grouped into similar types as well.

Number of Terms	Degree	Description
1	0	constant, monomial, polynomial
1	1	linear, monomial, polynomial
2	1	linear, binomial, polynomial
3	2	quadratic, polynomial
4	3	cubic, polynomial
5	4	quartic, polynomial

Investigate

Example

This example illustrates constant second differences in a quadratic function. Project the example from your ebook and have students work in small groups. Have students present their solutions, including their methods of finding the number of diagonals of a dodecagon. Students may notice a pattern in the number of diagonals (add 2, add 3, add 4, and so on); this pattern is described by the first differences.

ASK "What would happen if you continued to find differences after you had a set of constant differences?" [You would obtain only sets of zeros.]
SMP 1, 7, 8

ASK "How did the quadratic equation in one variable become three linear equations in three variables?" "Are a, b, and c constants or variables?" These are constant coefficients of the quadratic polynomial, but they're temporarily being treated as variables because they are unknown. **SMP 4, 6**

ASK "Often in mathematics an answer to a problem suggests that there might be a different way to solve the problem. Is that true here?" Students might realize that each of the x vertices is connected by a diagonal to $x - 3$ vertices (not counting itself or the two adjacent vertices) and that the number of diagonals is half the product of $x(x - 3)$ because each diagonal $\overline{AB}$ is the same as the diagonal $\overline{BA}$. **SMP 1**

EXAMPLE Find a polynomial function that models the relationship between the number of sides and the number of diagonals of a convex polygon. Use the function to find the number of diagonals of a dodecagon (a 12-sided polygon).

Solution You need to create a table of values with evenly spaced x-values. Sketch polygons with increasing numbers of sides. Then draw all of their diagonals.

Let x be the number of sides and y be the number of diagonals. You may notice a pattern in the number of diagonals that will help you extend your table beyond the sketches you make. Calculate the finite differences to determine the degree of

Modifying the Investigation

Whole Class Have two students help collect data for the whole class. Complete Steps 2 and 3 as a class with student input. Discuss Steps 4 and 5. In Step 6, write the system of equations with student input. Have students solve the system individually and then discuss their answers.

Shortened Use the sample data. Skip Steps 3 and 4. Have students complete Steps 2 and 5. Complete Step 6 as a class.

One Step Use the **One Step** found in your ebook. As needed, suggest checking differences of differences, and perhaps differences of those differences and so on, to find the kind of polynomial that fits a set of data.

Guiding the Investigation

You can use the desmos simulation of the investigation in your ebook or the Investigation Worksheet with Sample Data if you do not wish to conduct the investigation as an activity. The answers here are based on that sample data.

You can drop any wieldy objects, not just pillows. A ball or a regular bed pillow would work also. **ALERT** If you use something solid, make sure to protect the sensor—for example, by setting it between two stacks of books. (Or you might use a ball falling *away* from a sensor held 2.5 m above the floor. In this case, subtract each data value from 2.5 m.)

Step 3 Students might complete this step by hand, finding the differences in a table and graphing them on graph paper. **SMP 5**

Step 5 If students want to keep finding finite differences because they don't have differences that are exactly the same, ask them to look for a graph of finite differences that is approximately horizontal, not increasing or decreasing. **SMP 1, 6**

Step 6 **ASK** "How do you decide which three points to choose to represent the data?" [Choose representative points that are not close together.]

You might help students see that the equation in Step 6 shows the distance traveled as the *average speed* times the *time*. The beginning speed is 0; the ending speed is determined by the acceleration due to gravity, 9.8 m/s². Thus the speed of an object x s after it is dropped is $9.8x$ m/s. Because the acceleration is constant, you can calculate the average speed using $\frac{1}{2}(0 + 9.8x)$, or $4.9x$ m/s. The distance fallen is $(4.9x)(x)$, or $4.9x^2$, m. If students measure in feet, their equation will have the coefficient that is half of the acceleration due to gravity, 16.1 ft/s².

---

the polynomial function. (Remember that your x-values must be spaced equally in order to use finite differences.)

Number of sides x	Number of diagonals y	D_1	D_2
3	0		
4	2	2	
5	5	3	1
6	9	4	1
7	14	5	1
8	20	6	1

You can stop finding differences when the values of a set of differences are constant. Because the values of D_2 are constant, you can model the data with a 2nd-degree polynomial function like $y = ax^2 + bx + c$.

To find the values of a, b, and c, you could use a system of three equations. Choose three of the points from your table, say $(4, 2)$, $(6, 9)$, and $(8, 20)$, and substitute the coordinates into $y = ax^2 + bx + c$ to create a system of three equations in three variables. Can you see how these three equations were created?

$$\begin{cases} 16a + 4b + c = 2 \\ 36a + 6b + c = 9 \\ 64a + 8b + c = 20 \end{cases}$$

Solve the system to find $a = 0.5$, $b = -1.5$, and $c = 0$. Use these values to write the function $y = 0.5x^2 - 1.5x$. This equation gives the number of diagonals of any polygon as a function of the number of sides.

Now substitute 12 for x to find that a dodecagon has 54 diagonals.

$y = 0.5x^2 - 1.5x$
$y = 0.5(12)^2 - 1.5(12)$
$y = 54$

With exact function values, you can expect the differences to be equal when you find the right degree. But with experimental or statistical data, as in the investigation, you may have to settle for differences that are nearly constant and that do not show an increasing or decreasing pattern when graphed.

Science CONNECTION

Italian mathematician, physicist, and astronomer Galileo Galilei (1564–1642) performed experiments with free-falling objects. He discovered that the speed of a falling object at any moment is proportional to the amount of time it has been falling. In other words, the longer an object falls, the faster it falls.

Conceptual/Procedural

Conceptual Two concepts are explored in this lesson. The concept of using finite differences to determine the degree of the polynomial that will fit a data set is explored in the Investigation and Example. The Investigating Structure is an informal proof that linear functions grow by equal distances and exponential functions grow by equal factors over equal intervals.

Procedural Students are taught the finite differences method to determine the degree of the polynomial that will fit a data set. They review evaluating polynomials and solving systems of three equations with three unknowns.

INVESTIGATION
Free Fall

YOU WILL NEED
- a motion sensor
- a small pillow or other soft object

Steps 3

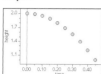
(*time, height*)

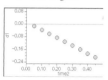
(*time, first difference*)

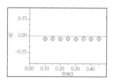
(*time, second difference*)

What function models the height of an object falling due to the force of gravity? Use a motion sensor to collect data, and analyze the data to find a function.

Procedure Note
1. Set the sensor to collect distance data approximately every 0.05 s for 2 to 5 s.
2. Place the sensor on the floor. Hold a small pillow at a height of about 2 m, directly above the sensor.
3. Start the sensor and drop the pillow.

Step 1 Follow the Procedure Note to collect data for a falling object. Let x represent time in seconds, and let y represent height in meters. Select about 10 points from the free-fall portion of your data, with x-values forming an arithmetic sequence. Record this information in a table. Round all heights to the nearest 0.001.

Step 2 Use the finite differences method to find the degree of the polynomial function that models your data. Stop when the differences are nearly constant.

Step 3 Create scatter plots of the original data (*time, height*), then a scatter plot of (*time, first difference*), and finally a scatter plot of (*time, second difference*).

Step 4 Write a description of each graph from Step 3 and what these graphs tell you about the data.

Step 5 Based on your results from using finite differences, what is the degree of the polynomial function that models free fall? Write the general form of this polynomial function.

2nd degree; $y = ax^2 + bx + c$

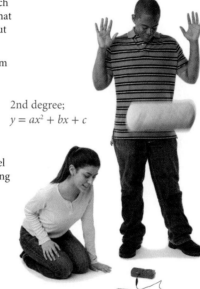

Step 6 Find an equation to model the position of a free-falling object dropped from a height of 2 m.

Systems will vary depending on the points chosen; the function should be approximately equivalent to $y = -4.9x^2 + 2$.

Investigating Structure

This Investigating Structure is an informal proof that linear functions grow by equal distances and exponential functions grow by equal factors over equal intervals. If your students are comfortable with algebraic proofs, you may want to have students work in small groups and present their answers. If they are not comfortable with algebraic proofs, do this Investigating Structure as a whole-class activity, with student input. **SMP 3, 6, 7**

Starting with the linear function, **ASK** "What is the slope of the line in the graph? What is the interval (difference in x-values) between the points in the graph? [m, c]

You might have them check their results with a specific function, say $y = 3x + 4$. Check for the interval length 5. [$m = 3$, $c = 5$ so difference of the y-values should be 15. $3(x + 5) + 4 - (3x + 4) = 3x + 15 + 4 - 3x - 4 = 15$.]

For an exponential function, **ASK** "What is the common factor of the function in the graph? What is the interval between the points in the graph? [b, c]

You might have them check their results with a specific function, say $2 \cdot 5^x$, over the interval length 3. [$b = 5$, $c = 3$, so the ratio of the y-values should be $5^3 = 125$.
$\left[\dfrac{2 \cdot 5^{x+3}}{2 \cdot 5^x} = 5^3 = 125\right]$]

When using experimental data, you must choose your points carefully. When you collect data, as you did in the investigation, your equation will most likely not fit all of the data points exactly due to some errors in measurement and rounding. To minimize the effects of these errors, choose representative points that are not close together, just as you did when fitting a line to data.

Steps 1 and 2 sample data:

Time (s) x	Height (m) y	D_1	D_2
0.00	2.000		
		−0.012	
0.05	1.988		−0.025
		−0.037	
0.10	1.951		−0.024
		−0.061	
0.15	1.890		−0.025
		−0.086	
0.20	1.804		−0.024
		−0.110	
0.25	1.694		−0.025
		−0.135	
0.30	1.559		−0.024
		−0.159	
0.35	1.400		−0.025
		−0.184	
0.40	1.216		−0.024
		−0.208	
0.45	1.008		

Step 4 The graph of (*time, height*) appears parabolic and suggests that the correct model may be a quadratic (2nd-degree) polynomial function. The graph of (*time, first difference*) shows that the first differences are not constant; because they decrease in a linear fashion, the second differences are likely to be constant. The graph of (*time, second difference*) shows that the second differences are nearly constant, so the correct model should be a 2nd-degree polynomial function.

Summarize

As students present their solutions, including explaining their strategy, emphasize the use of correct mathematical terminology and symbols. Be careful not to discourage thinking by talking too quickly or saying too much. Whether student responses are correct or incorrect, ask other students if they agree and why. **SMP 1, 3, 6**

Students are used to seeing an ellipsis to stand for something left out, but they may not have seen one raised, as in the definition of a polynomial. Explain that a raised ellipsis is used when an operation repeats.

ASK "What is the degree of the polynomial x?" [1] "What is the degree of 24?" [0, because $24 = 24x^0$]

ASK "What do common differences represent?" [constant slope] "Are there sequences and functions besides quadratics that do not have common differences?" [yes; exponential functions, for example]

ASK "Is there a way to find the coefficients of the 2nd-degree polynomial other than by solving a system of equations?" For linear functions $f(x)$ with constant difference b, students can find that $f(x) = bx + f(0)$. (This is the intercept form, slightly disguised.)

If students find a constant second difference a, then the quadratic function will have $\frac{a}{2}$ divided by the square of the change in the x-values as a coefficient of the x^2 term. They can then make a table of values for x, $f(x)$, and $f(x) - \frac{a}{2(\Delta x)^2}x^2$ and find the linear expression $bx + c$ for the last column, implying that $f(x) = \frac{a}{2(\Delta x)^2}x^2 + bx + c$.

(In general, the coefficient of the highest power x^n equals the constant nth finite difference divided by $n!$, divided by the nth power of the change in the x-values.)

ASK "Are finite differences ever constant for exponential functions?" For functions of the form $y = ab^x$, the differences will all be geometric sequences with common ratio b.

> **History CONNECTION**
>
> The method of finite differences was used by the Chinese astronomer Li Shun-Fêng in the 7th century to find a quadratic equation to model the Sun's apparent motion across the sky as a function of time. The Persian astronomer Jamshid Masud al-Kashi, who worked at the Samarkand Observatory in the 15th century, also used the finite differences method when calculating the celestial longitudes of planets.

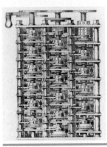

English mathematician and inventor Charles Babbage (1792–1871) designed the first difference engine in the early 1820s, and completed his Difference Engine No. 1, shown here, in 1832.

Note that some functions, such as logarithmic and exponential functions, cannot be expressed as polynomials. The finite differences method will not produce a set of constant differences for functions other than polynomial functions.

🔺 INVESTIGATING STRUCTURE

You have seen examples that linear functions have constant first differences over equal intervals. To generalize, you could use the form $y = mx + b$ and the point $(x_1, mx_1 + b)$. What are the coordinates of a point c units away from x_1? Use these two points to show that the difference in the y-values is constant.

$(x_1 + c, m(x_1 + c) + b)$

$(m(x_1 + c) + b)) - (mx_1 + b)$
$= (mx_1 + mc + b) - (mx_1 + b)$
$= mc$

For any linear function with slope m and interval length c, the function grows by the difference mc.

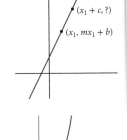

Similarly, you have seen that exponential functions have constant ratios over equal intervals. To generalize, you could use the form $y = a \cdot b^x$ and the point $(x_1, a \cdot b^{x_1})$. What are the coordinates of a point c units away from x_1? Show that the ratios of the y values is constant.

$(x_1 + c, a \cdot b^{x+c})$

$\frac{a \cdot b^{x_1+c}}{a \cdot b^{x_1}} = b^{x_1+c-x_1} = b^c$

For any exponential function with common factor b and interval length c, the function grows by the factor b^c.

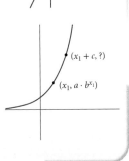

When choosing a model for data, you might begin by finding differences to see if you can find a pattern. If you do not find constant differences, try the other parent functions you have learned. For example, you could find quotients because constant quotients (given an arithmetic sequence in the independent variable) would indicate that an exponential function would be an appropriate model.

CRITICAL QUESTION What is the finite differences method used for?

BIG IDEA You use finite differences to find the degree of a polynomial that represents a set of data.

CRITICAL QUESTION Once you determine the degree, how do you find the polynomial's coefficients?

BIG IDEA One way is to set up and solve a system of linear equations, in which the polynomial's coefficients are the unknowns.

CRITICAL QUESTION Do finite differences need to be exactly the same to determine the nature of the polynomial to fit the data?

BIG IDEA No, if the nth finite differences of a data set are approximately constant, the data set might be fit by a polynomial of degree n.

6.1 Exercises

Practice Your Skills

1. Identify the degree of each polynomial.
   a. $x^3 + 9x^2 + 26x + 24$ 3
   b. $7x^2 - 5x$ 2
   c. $x^7 + 3x^6 - 5x^5 + 24x^4 + 17x^3 - 6x^2 + 2x + 40$ 7
   d. $16 - 5x^2 + 9x^5 + 36x^3 + 44x$ @ 5

2. Determine which of these expressions are polynomials. For each polynomial, state its degree and write it in general form. If it is not a polynomial, explain why not.
   a. $-3 + 4x - 3.5x^2 + \frac{5}{9}x^3$ @
   b. $5p^4 + 3.5p - \frac{4}{p^2} + 16$
   c. $4\sqrt{x^3} + 12$ @
   d. $x^2\sqrt{15} - x - 4^{-2}$
   e. $\frac{1}{x-2} + x$
   f. $3x + 4x^3 - x^4$

3. For each data set, decide whether the last column of differences shows constant values. If it does not, calculate the next set of finite differences.

   a.
x	y
2	4.4
3	6.6
4	9.2
5	11.0
6	10.8
7	7.4

   no; {2.2, 2.6, 1.8, −0.2, −3.4}

   b.
x	y	D_1
3.7	−8.449	
3.8	−8.706	−0.257
3.9	−8.956	−0.250
4.0	−9.200	−0.244
4.1	−9.436	−0.236
4.2	−9.662	−0.226

   no; {0.007, 0.006, 0.008, 0.010}

   c.
x	y	D_1	D_2
−5	−101		
0	−6	95	−100
5	−11	−5	50
10	34	45	200
15	279	245	350
20	874	595	

   no; {150, 150, 150}

4. Find the degree of the polynomial function that models these data.

x	0	2	4	6	8	10	12
y	12	−4	−164	−612	−1492	−2948	−5124

   3

5. Determine which of the data sets can be modeled with a first, second, or third degree polynomial. For each polynomial, state the degree.

   a.
x	y
−1	0.5
0	1
1	2
2	4
3	8

   Does not have constant first, second, or third differences

   b.
x	y
−5	−45
−2	−12
1	3
4	0
7	−21

   Polynomial of degree 2

   c.
x	y
−5	−0.6
−3	−1
−1	−3
1	3
3	1

   Does not have constant first, second, or third differences

   d.
x	y
−3	2.4
−1	1.8
1	1.2
3	0.6
5	0

   Polynomial of degree 1

2a. polynomial; 3; $\frac{5}{9}x^3 - 3.5x^2 + 4x - 3$

2b. not a polynomial because $\frac{4}{p^2} = 4p^{-2}$ has a negative exponent

2c. not a polynomial because $4\sqrt{x^3} = 4x^{3/2}$ has a non-integer exponent

2d. polynomial; 2; already in general form

2e. not a polynomial because of the $(x - 2)$ in the denominator.

2f. polynomial; 4, $-x^4 + 4x^3 + 3x$

Formative Assessment
The Launch will provide evidence of prior knowledge of definitions for polynomials, including the degree of the polynomial. The Investigation and Example provide evidence to assess students' ability to find finite differences, as well as their ability to evaluate polynomials and solve systems of three equations with three unknowns. The Investigating Structure will meet the objective of proving that linear functions grow by equal distances and exponential functions grow by equal factors over equal intervals.

Apply

Extra Example
Find the polynomial function that models this data.

0	−1
1	2
2	7
3	14
4	23
5	34
6	47

$y = x^2 + 2x - 1$

Closing Question
Use the finite differences method to find the degree of the polynomial that models this data:

0	1
1	1
1.5	13
2	73
3	241
4	601
4.5	1261

You can't use the finite differences method because the x-values are not spaced equally.

Assigning Exercises
Suggested: 1–10, 12
Additional Practice: 11, 13–17

LESSON 6.1 Polynomials 323

Exercises 6 SUPPORT Students may need assistance finding a polynomial function that models the data set. Consider discussing Exercise 6 in some depth using students' notes from the Example as a framework.

6a. $D_1 = \{2, 3, 4, 5, 6\}$;
$D_2 = \{1, 1, 1, 1\}$; 2nd degree

6b. The polynomial is 2nd degree, and the D_2 values are constant.

Exercise 6c ASK "In general, how many data points do you need to find the degree of a polynomial? [at least 2 more than the actual degree]

6c. 4 points. You have to find the finite differences twice, so you need at least four data points to calculate two D_2 values that can be compared.

Reason and Apply

6. Consider the data at right.

n	1	2	3	4	5	6
s	1	3	6	10	15	21

   a. Calculate finite differences to find the degree of the polynomial function that models these data.

   b. Describe how the degree of this polynomial function is related to the finite differences you calculated.

   c. What is the minimum number of data points required to determine the degree of this polynomial function? Why? ⓐ

   d. Find the polynomial function that models these data and use it to find s when n is 12. $s = 0.5n^2 + 0.5n$; $s = 78$

   e. The values in the s row are called triangular numbers. Why do you think they are called triangular? ⓗ The pennies can be arranged to form triangles.

7. Find a polynomial to model the data.

   a. $y = x^2 + 1$

x	y
−1	2
0	1
1	2
2	5
3	10

   b. $y = x + 1$

x	y
−5	−4
−2	−1
1	2
4	5
7	8

8. Jennie claims that the following data can be fit exactly with the functions. However, one of the table values is missing. Find the missing values so that she is correct.

   a. *First degree polynomial*

x	y
0	5
1	−15
2	−35
3	−55
4	−75
5	−96

   b. *Second degree polynomial*

x	y
0	5
1	8
2	15
3	26
4	41
5	60

   c. *Third degree polynomial*

x	y
0	5
1	−3
2	−7
3	5
4	45
5	125

9. The data in these tables represent the heights of two objects at different times during free fall.

i.
Time (s) t	0	1	2	3	4	5	6
Height (m) h	80	95.1	100.4	95.9	81.6	57.5	23.6

$D_1 = \{15.1, 5.3, -4.5, -14.3, -24.1, -33.9\}$;
$D_2 = \{-9.8, -9.8, -9.8, -9.8, -9.8\}$

ii.
Time (s) t	0	1	2	3	4	5	6
Height (m) h	4	63.1	112.4	151.9	181.6	201.5	211.6

$D_1 = \{59.1, 49.3, 39.5, 29.7, 19.9, 10.1\}$;
$D_2 = \{-9.8, -9.8, -9.8, -9.8, -9.8\}$

a. Calculate the finite differences for each table.

b. What is the degree of the polynomial function that you would use to model each set of data? **i.** 2; **ii.** 2

c. Write a polynomial function to model each set of data. Check your answer by substituting one of the data points into your function.
i. $h = -4.9t^2 + 20t + 80$; **ii.** $h = -4.9t^2 + 64t + 4$

10. Use the data in the table.

    a. Use the first two table values to find a linear model.

    b. Use the first three table values to find a polynomial model of degree 2 (quadratic).

    c. Use the first three table values to find an exponential model.

    d. Statisticians have methods to determine the best model for data. One way is to consider the predictive value. Which model best fits the data for predicting $x = 4$? Justify your answer.

x	y
0	10
1	11
2	12.1
3	13.31
4	14.641

11. **APPLICATION** In an atom, electrons spin rapidly around a nucleus. An electron can occupy only specific energy levels, and each energy level can hold only a certain number of electrons. This table gives the greatest number of electrons that can theoretically be in any one level (although there are no known atoms that actually have more than 32 electrons in any one level). $D_1 = \{6, 10, 14, 18, 22, 26\}$; $D_2 = \{4, 4, 4, 4, 4\}$. The second differences are constant, so a quadratic function expresses the relationship. Let x represent the energy level, and let y represent the maximum number of electrons. $y = 2x^2$.

Energy level	1	2	3	4	5	6	7
Maximum number of electrons	2	8	18	32	50	72	98

Is it possible to find a polynomial function that expresses the relationship between the energy level and the maximum number of electrons? If so, find the function. If not, explain why not. @

> **Science CONNECTION**
>
> The electrons in an atom exist in various energy levels. When an electron moves from a lower energy level to a higher energy level, the atom absorbs energy. When an electron moves from a higher to a lower energy level, energy is released (often as light). This is the principle behind neon lights. The electricity running through a tube of neon gas makes the electrons in the neon atoms jump to higher energy levels. When they drop back to their original level, they give off light. Times Square in New York City, is a busy tourist intersection of neon art and commerce.

Exercise 9c ASK "What are the similarities and differences between the models for the two data sets? Between their graphs?"

10a. degree 1: $y = 10 + x$

10b. degree 2: (using (0,10) (1, 11), and (2, 12.1)), $y = 0.05x^2 + 0.95x + 10$

10c. exponential: $y = 10 \cdot 1.1^x$

10d. For the linear model, for $x = 4$, $y = 10 + (4) = 14$ which close to the actual value of 10.461, off by 0.461. For the quadratic model, for $x = 4$, $y = 0.05(4)^2 + 0.95(4) + 10 = 14.6$, off by 0.041. The exponential model, for $x = 4$, $y = 10 \cdot 1.1^4 = 14.641$ which matches the data exactly.

LESSON 6.1 Polynomials **325**

Exercise 12 This exercises could be used as a performance task.

12. Using the point $(3, 0)$,
$y = a(x - r_1)(x - 3)$.
Use two other points, say $(5, 5)$ and $(4, 2)$, substitute for x and y to set up a system of equations:
$5 = a(5 - r_1)(5 - 3)$
$2 = a(4 - r_1)(4 - 3)$
Simplifying,
$5 = 2a(5 - r_1)$
$2 = a(4 - r_1)$
Solving, $a = 0.5$, $r_1 = 0$, so
$y = 0.5(x - 0)(x - 3)$,
or $y = 0.5x^2 - 1.5x$

Exercise 13 As needed, ask whether it helps to think about transforming parent functions. Students should use calculators to check their sketches.

13a.   13b.

13c.

15a.

	2x	3
3x	6x²	9x
1	2x	3

15c.

	x	5
x	x²	5x
3	3x	15

Exercise 16 These inequalities will be much easier to write using point-slope form.

16. $y \geq -\dfrac{1}{2}(x+3)+3$ or $y \geq -\dfrac{1}{2}(x-11)-4$

$y \leq \dfrac{1}{2}(x+3)+3$ or $y \leq \dfrac{1}{2}(x-5)+7$

$y \leq -\dfrac{11}{6}(x-5)+7$ or $y \leq -\dfrac{11}{6}(x-11)-4$

---

12. In the example, you used a system of three equations and three unknowns to find a, b, and c in order to write a quadratic function of the form $y = ax^2 + bx + c$. Using what you know about the factored form of a quadratic equation, show how you could have used a system of two equations with two unknowns to write the function. (Hint: Since one of the data points is an x-intercept, find another point to create a system of two linear equations with two unknowns for $y = a(x - r_1)(x - 3)$.)

Review

13. Sketch a graph of each function without using your calculator.

    a. $y = (x - 2)^2$ b. $y = x^2 - 4$ c. $y = (x + 4)^2 + 1$

14. Solve.

    a. $12x - 17 = 13$ $x = 2.5$ b. $2(x - 1)^2 + 3 = 11$ c. $3(5^x) = 48$ ⓐ $x = \log_5 16 \approx 1.7227$

    $x = 3$ or $x = -1$

15. The rectangle diagram at right represents the product $(x + 2)(x + 3)$, which you can write as the trinomial $x^2 + 5x + 6$.

    a. Draw a rectangle diagram that represents the product $(2x + 3)(3x + 1)$.
    b. Express the area in 15a as a polynomial in general form. $6x^2 + 11x + 3$
    c. Draw a rectangle diagram whose area represents the polynomial $x^2 + 8x + 15$. ⓗ
    d. Express the area in 15c as a product of two binomials. $(x + 3)(x + 5)$

16. Write a system of inequalities that describes the feasible region graphed at right. ⓐ

17. Find the product $(x + 3)(x + 4)(x + 2)$. $x^3 + 9x^2 + 26x + 24$

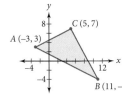

LESSON 6.2

Factoring Polynomials

Imagine a cube with any side length. Imagine increasing the height by 2 cm, the width by 3 cm, and the length by 4 cm.

Ideas are the factors that lift civilization.
JOHN H. VINCENT

Imagine a cube.

Increase its height 2 cm.

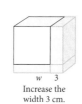
Increase the width 3 cm.

Increase the length 4 cm.

The starting figure is a cube, so you can let x be the length of each of its sides. So, $l = w = h = x$. The volume of the starting figure is x^3. To find the volume of the expanded box, you can see it as the sum of the volumes of eight different boxes. You find the volume of each piece by multiplying length by width by height.

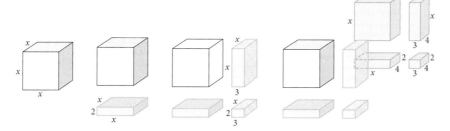

The total expanded volume is this sum:

$$V = x^3 + 2x^2 + 3x^2 + 6x + 4x^2 + 8x + 12x + 24 = x^3 + 9x^2 + 26x + 24$$

You can also think of the expanded volume as the product of the new height, width, and length.

$$V = (x+2)(x+3)(x+4)$$

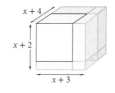

A 3rd-degree polynomial function is called a **cubic function.** This cubic function in factored form is equivalent to the polynomial function in general form. (Try graphing both functions on your calculator.)

You already know that there is a relationship between the factored form of a quadratic equation, and the roots and x-intercepts of that quadratic equation. In this lesson you will learn how to write higher-degree polynomial equations in factored form when you know the roots of the equation. You'll also discover useful techniques for converting a polynomial in general form to factored form.

The factored form, the roots, and the x-intercepts of a polynomial all inform you about the short run behavior of polynomials. The **short run behavior** is what happens for x-values near zero. Later we will learn to analyze the long run behavior for polynomials, the behavior for x-values far from zero.

LESSON 6.2 Factoring Polynomials **327**

LESSON 6.2

This lesson concerns the relationship between the zeros and the factors of cubic functions.

COMMON CORE STATE STANDARDS		
APPLIED	DEVELOPED	INTRODUCED
A.SSE.1a	A.APR.3	
A.SSE.3a	F.IF.7c	
	F.BF.3	

Objectives
- Explore functions defined by 3rd-degree polynomials (cubic functions)
- Use graphs of polynomial equations to find the roots and write the equations in factored form
- Relate the graphs of polynomial equations to the number and types of roots

Vocabulary
cubic function
short run behavior

Materials
- graph paper
- scissors
- tape, *optional*

Launch

Project the "growing" cube from your ebook. Have students find the volume, stated in general form, of the new cube using two different strategies.

$V = x^3 + 9x^2 + 26x + 24$

Students could find the volume using the dimensions of the final cube.

$V = (x + 3)(x + 2)(x + 4)$

They could find the volume of the original cube and add the volumes of each additional rectangular prism.

$V = x^3 + 2x^2 + (3x^2 + 6x) + (4x^2 + 8x + 12x + 24)$
$= x^3 + 9x^2 + 26x + 24$

ELL
Emphasize the physical models as a means of exploring a polynomial. Students will be able to refer back to the models as they find roots of cubic equations. Some students may confuse the physical representation of a cube with the graphical one. Compare the graph of $y = x^3$ and a cube with the graph of the $y = x^2$ and a square.

Support
Students could use calculators to find x- and y-intercepts in the graphs of cubic functions. They may benefit from adding the graphs and information showing how quadratic functions might intersect the x-axis to their graphic organizer.

Advanced
Students could research and apply tools such as the Cubic Formula. Assign the **One Step** version of the investigation.

LESSON 6.2 Factoring Polynomials **327**

Investigate

Example A

This example builds on the relationship between the zeros of a cubic function and the *x*-intercepts of its graph. Project Example A from your ebook and ASK "What is alike and what is different between the graphs? [They both have the same intercepts but different vertical scale factors, *a*.] Ask for students input for how to find the vertical scale factor for each graph.

What does *a* mean geometrically? [It's the vertical dilation factor.] SMP 2, 4, 6

EXAMPLE A

Write cubic functions for the graphs below.

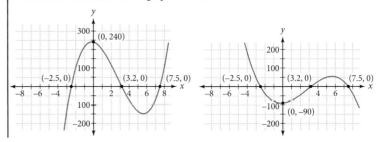

Solution

Both graphs have the same *x*-intercepts: -2.5, 3.2, and 7.5. So both functions have the factored form $y = a(x + 2.5)(x - 7.5)(x - 3.2)$. But the vertical scale factor, a, is different for each function.

One way to find a is to substitute the coordinates of one other point, such as the *y*-intercept, into the function. Why is the *y*-intercept a nice choice?

The curve on the left has *y*-intercept $(0, 240)$. Substituting this point into the equation gives $240 = a(2.5)(-7.5)(-3.2)$. Solving for a, you get $a = 4$. So the equation of the cubic function on the left is

$$y = 4(x + 2.5)(x - 7.5)(x - 3.2)$$

The curve on the right has *y*-intercept $(0, -90)$. Substituting this point into the equation gives $-90 = a(2.5)(-7.5)(-3.2)$. So $a = -1.5$, and the equation of the cubic function on the right is

$$y = -1.5(x + 2.5)(x - 7.5)(x - 3.2)$$

The factored form of a polynomial function tells you the zeros of the function and the *x*-intercepts of the graph of the function. Recall that zeros are solutions to the equation $f(x) = 0$. Factoring, if a polynomial can be factored, is one strategy for finding the real solutions of a polynomial equation. In the investigation you will practice writing a higher-degree polynomial function in factored form.

English sculptor Cornelia Parker (b 1956) creates art from damaged objects that have cultural or historical meaning. His Transitional Object (Psycho Barn), a sculpture commissioned by the Metropolitan Museum of Art in New York and inspired by Alfred Hitchcock's 1960 film Psycho, sits atop the roof of the Museum.

Modifying the Investigation

Whole Class Demonstrate the procedure by making a box. Then have each student cut and fold a box. Collect data from students to complete the table in Step 1. Complete the remaining steps as students follow along.

Shortened Demonstrate the procedure with one rectangle. Use sample data or the desmos simulation. Complete Step 2 as a class. Have groups complete Steps 3 through 6. Discuss Step 7 as a class.

One Step Challenge groups to find as many different shapes as they can of cubic functions. (You may need to define cubic functions as 3rd-degree polynomial functions.) Geometry software is useful but not necessary. During the discussion, try to distill how the various coefficients affect the shapes and see how the zeros relate to the *x*-intercepts.

INVESTIGATION

The Box Factory

YOU WILL NEED
- graph paper
- scissors
- tape (optional)

What are the different ways to construct an open-top box from a 16-by-20–unit sheet of material? What is the maximum volume this box can have? What is the minimum volume? Your group will investigate this problem by constructing open-top boxes using several possible integer values for x.

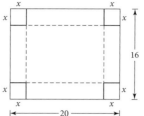

Procedure Note
1. Cut several 16-by-20–unit rectangles out of graph paper.
2. Choose several different values for x.
3. For each value of x, construct a box by cutting squares with side length x from each corner and folding up the sides.

Step 1 Follow the Procedure Note to construct several different-size boxes from 16-by-20–unit sheets of paper. Record the dimensions of each box and calculate its volume. Make a table to record the x-values and volumes of the boxes.

Step 2 For each box, what are the length, width, and height, in terms of x? Use these expressions to write a function that gives the volume of a box as a function of x.

Step 3 Graph your volume function from Step 2. Plot your data points on the same graph. How do the points relate to the function?

Step 4 What is the degree of this function? Give some reasons to support your answer. 3rd-degree. If the factors are multiplied, the highest-degree term is x^3. The shape is that of a cubic function.

Step 5 Locate the x-intercepts of your graph. (There should be three.) Call these three values r_1, r_2, and r_3. Use these values to write the function in the form $y = (x - r_1)(x - r_2)(x - r_3)$. 0, 8, 10; $y = (x - 0)(x - 8)(x - 10)$

Step 6 Graph the function from Step 5 with your function from Step 2. What are the similarities and differences between the graphs? How can you alter the function from Step 5 to make both functions equivalent?

Step 2 $20 - 2x$, $16 - 2x$, and x; $y = (20 - 2x)(16 - 2x)(x)$

Step 7 What happens if you try to make boxes by using the values r_1, r_2, and r_3 as x? What domain of x-values makes sense in this context? What x-value maximizes the volume of the box?

For $x = r_1$, $x = r_2$, or $x = r_3$, the box will be flat. Only a domain of $0 < x < 8$ makes sense in this context. An x-value of about 2.94 maximizes the volume.

Step 3

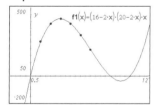

The points lie on the graph of the function.

Guiding the Investigation

This video driven investigation uses creating boxes as a model for maximizing volume. The ebook and the document Using the Videos found in your digital resources offer suggestions for using the video with the investigation. In place of doing the investigation as an activity, you could use the desmos simulation in your ebook. Additionally, there is an Investigation Worksheet that provides sample data. The answers found here are all based on that sample data.

Step 1 Students can use graph paper. **ASK** "What is the domain of x? Explain." $[0 < x < 8; x$ need not be an integer. For $x \geq 8$, each corner would be more than half the length of the short side of the paper.] In Step 7, students will rethink the domain of x. **SMP 1, 2, 6**

Step 4 If students use the finite differences method, they will find a constant after three iterations.

Step 6 **ASK** "Why doesn't this graph fit your data points?" [The value of the leading coefficient, a, has been left out.] **SMP 1**

Step 7 Students may say there is no maximum; the graph of the function continues upward as x increases. Although the function has no maximum, only the values of x between 0 and 8 model boxes, and over this domain the function does have a maximum. **SMP 2, 6**

Step 6 They have the same x-intercepts and general shape but different vertical scale factors. A vertical scale factor of 4 makes them equivalent: $y = 4x(x - 8)(x - 10)$.

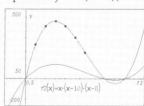

LESSON 6.2 Factoring Polynomials **329**

Conceptual/Procedural

Conceptual This lesson explores functions defined by 3rd-degree polynomials through volume. They investigate the concept of short run behavior of polynomials. Students relate the graphs of polynomial equations to the number and types of roots.

Procedural Use graphs of polynomial equations to find the roots and write the equations in factored form

Example B

Project Example B from your ebook. Have students work in pairs. Have them share their strategy along with their solution.

ALERT Some students will be confused by the changing role of *a* from constant to unknown variable and back again.

ASK "In the solution to part b, why does the leading coefficient need to be 4?" [In the original equation, which is in general form, the leading coefficient is 4.] **SMP 6**

Summarize

As students present their solutions to the Launch, the Investigation and the Examples, including explaining their strategy, emphasize the use of correct mathematical terminology and symbols. Be careful not to discourage thinking by talking too quickly or saying too much. Whether student responses are correct or incorrect, ask other students if they agree and why. **SMP 1, 2, 3, 6**

Take time to make sure students can visualize the expanded cube. **ASK** "Which boxes share a face with the cube?" [$2x^2$, $3x^2$, and $4x^2$—the 2nd-degree terms] "Which share an edge with the cube?" [$8x$, $6x$, and $12x$—the linear terms] "Which share only a vertex?" [24, which is 2(3)(4)]

ASK "Why is x^3 called x cubed?" [It can be represented by a physical cube, just as x^2 can be represented by a square.] **SMP 2, 6**

LANGUAGE When the parabola has exactly one point of intersection with the *x*-axis, this root of the corresponding quadratic equation is known as a *double root*. This follows through for multiple roots with an even power; $y = (x - r)^{2n}$ has n coincidental roots.

The statements following Example B about conjugate pairs and factoring do not apply if the polynomial has complex coefficients but instead relate only to polynomials with real-number coefficients.

The connection between the roots of a polynomial equation and the *x*-intercepts of a polynomial function helps you factor any polynomial that has real roots.

EXAMPLE B | Use the graph of each function to determine its factored form.
a. $y = x^2 - x - 2$
b. $y = 4x^3 + 8x^2 - 36x - 72$

Solution | You can find the *x*-intercepts of each function by graphing. The *x*-intercepts tell you the real roots, which help you factor the function.

a. The graph shows that the *x*-intercepts are -1 and 2. Because the coefficient of the highest-degree term, x^2, is 1, the vertical scale factor is 1. The factored form is $y = (x + 1)(x - 2)$.

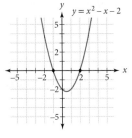

You can verify that the expressions $x^2 - x - 2$ and $(x + 1)(x - 2)$ are equivalent by graphing $y = x^2 - x - 2$ and $y = (x + 1)(x - 2)$. You can also check your work algebraically by finding the product $(x + 1)(x - 2)$. This rectangle diagram confirms that the product is $x^2 - x - 2$.

b. The *x*-intercepts are -3, -2, and 3. So, you can write the function as

$$y = a(x + 3)(x + 2)(x - 3)$$

Because the leading coefficient needs to be 4, the vertical scale factor is also 4.

$$y = 4(x + 3)(x + 2)(x - 3)$$

To check your answer, you can compare graphs or find the product of the factors algebraically.

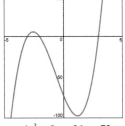

$y = 4x^3 + 8x^2 - 36x - 72$

In Example B, you converted a function from general form to factored form by using a graph and looking for the *x*-intercepts. This method works especially well when the zeros are integer values. Once you know the zeros of a polynomial function, r_1, r_2, r_3, and so on, you can write the factored form,

$$y = a(x - r_1)(x - r_2)(x - r_3) \cdots$$

You can also write a polynomial function in factored form when the zeros are not integers, or even when they are nonreal.

330 CHAPTER 6 Polynomial and Rational Functions

As you discuss the three quadratic functions that end the lesson, you might ask students to find the factored form.

$$\left[y = \left[x - \left(\frac{1}{2} + i\frac{\sqrt{3}}{2} \right) \right] \left[x - \left(\frac{1}{2} - i\frac{\sqrt{3}}{2} \right) \right]; \right.$$

$$y = \left(x - \frac{1 + \sqrt{5}}{2} \right) \left(x - \frac{1 - \sqrt{5}}{2} \right);$$

$$\left. y = (x + 1)(x - 2) \right]$$

Polynomials with real coefficients can be separated into three types: polynomials that can't be completely factored with real numbers; polynomials that can be factored with real numbers, but some of the roots are not "nice" integer or fractional values; and polynomials that can be factored and have all integer or fractional roots. For example, consider these cases of quadratic functions:

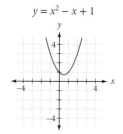
$y = x^2 - x + 1$

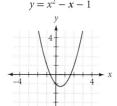
$y = x^2 - x - 1$

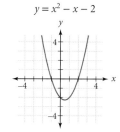
$y = x^2 - x - 2$

If the graph of a quadratic function does not intersect the x-axis, then you cannot factor the polynomial using real numbers. However, you can use the quadratic formula to find the complex zeros, which will be a conjugate pair.

If the graph of a quadratic function intersects the x-axis, but not at integer or common fractional values, then you can use the quadratic formula to find the exact real zeros or technology to find the approximate real zeros.

If the graph of a quadratic function intersects the x-axis at integer or common fractional values, then you can use the x-intercepts to factor the polynomial. This is often quicker and easier than using the quadratic formula or a rectangle diagram.

What happens when the graph of a quadratic function has exactly one point of intersection with the x-axis? It has one root that occurs twice, so it is a perfect square. Its factored form is $y = a(x - r_1)^2$.

 Exercises

You will need a graphing calculator for Exercises **1, 4, 5,** and **17**.

Practice Your Skills

1. Without graphing, find the x-intercepts and the y-intercept for the graph of each equation. Check each answer by graphing. **1a.** x-intercepts: $-1.5, -6$; y-intercept: -2.25

   a. $y = -0.25(x + 1.5)(x + 6)$

   b. $y = 3(x - 4)(x - 4)$ x-intercept: 4; y-intercept: 48

   c. $y = -2(x - 3)(x + 2)(x + 5)$
   x-intercepts: 3, -2, -5; y-intercept: 60

   d. $y = 5(x + 3)(x + 3)(x - 3)$
   x-intercepts: $-3, 3$; y-intercept: -135

2. Write the factored form of the quadratic function for each graph. Don't forget the vertical scale factor.

   a.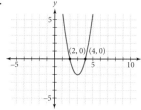
   $y = 2(x - 2)(x - 4)$

   b.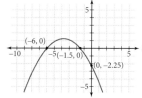
   $y = -0.25(x + 1.5)(x + 6)$

   c.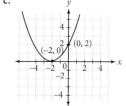
   $y = 0.5(x + 2)^2$

LESSON 6.2 Factoring Polynomials **331**

CRITICAL QUESTION Why would you want to factor a polynomial?

BIG IDEA From the factored form of a polynomial function, you can determine the zeros of the function, and the x-intercepts of its graph.

CRITICAL QUESTION What shape is the graph of a 1st-degree function? Of a 2nd-degree function? Of higher-degree functions?

BIG IDEA The graph of a 1st-degree function is a line. The graph of a 2nd-degree function is a parabola. The graph of a higher-degree functions vary depending on the degree.

Formative Assessment

Use the Investigation to assess how well students connect the ideas of polynomial factors with concepts related to volume. Example A and B provide evidence of students understanding the relationship of a graph and the roots of a cubic equation.

Apply

Extra Example

1. Without graphing, find the x-intercepts and the y-intercept for the graph of $y = -4(x - 2)(x + 1)(x - 3)$.
   x-intercepts: $(2, 0), (-1, 0), (3, 0)$; y-intercept: $(0, -24)$

2. Convert the polynomial function $y = -x(x + 1)(x - 5)$ to general form.
   $y = -x^3 + 4x^2 + 5x$

3. Expand $(x - 1)(x - 2)(x - 3)$ in the form $ax^3 + bx^2 + cx + d$.
   $x^3 - 6x^2 + 11x - 6$

Closing Question

How many roots does the equation $y = (x - 1)(x + 2)^2$ have?

Cubic equations have three roots. This one has two distinct roots, at 1 and -2. The root at -2 is considered a double root so the three roots are $1, -2, -2$.

Assigning Exercises

Suggested: 1–6, 8–11
Additional Practice: 7, 12–17

LESSON 6.2 Factoring Polynomials **331**

Exercises 3, 4 To expand these expressions, students can either draw rectangle diagrams or distribute systematically. **SUPPORT** To multiply three or more factors, suggest that students use the problem-solving technique of breaking down the problem into smaller problems. **ASK** "What do you already know how to do?" They can use a rectangle diagram to multiply two of the binomials. Then suggest that they use a 2 × 3 rectangle diagram to multiply the remaining binomial by the trinomial resulting from the first diagram. They can finish by multiplying the result by the constant, if one is given.

 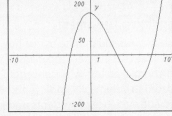

Exercise 5 In 5a, students can find the maximum either by examining the graph or by substituting into the function expression the value of x they found in the investigation.

5c. The function doesn't model volumes of real boxes outside the interval $0 < x < 8$. For $x > 10$, two dimensions of the box are negative, so the product is positive. This means that this impossible box has a possible volume.

5d. The function doesn't model volumes of real boxes outside the interval $0 < x < 8$. For $8 < x < 10$, one of the dimensions of the box would be negative, so the product is negative.

8a. $4(x - 12)(x - 10)$

8b. $6\left(x - \dfrac{5}{3}\right)\left(x + \dfrac{1}{2}\right)$, or $(3x - 5)(2x + 1)$

8c. $(x + 2)(x - 2)(x + 5)$

8d. $2(x + 1)(x + 3)(x + 4)$

8e. $(a + b)(a + b)$ or $(a + b)^2$

Exercise 9 As needed, remind students to look for x-intercepts in the graphs.

---

3. Convert each polynomial function to general form.
    a. $y = (x - 4)(x - 6)$ $y = x^2 - 10x + 24$
    b. $y = (x - 3)(x - 3)$ $y = x^2 - 6x + 9$
    c. $y = x(x + 8)(x - 8)$ @ $y = x^3 - 64x$
    d. $y = 3(x + 2)(x - 2)(x + 5)$ $y = 3x^3 + 15x^2 - 12x - 60$

4. Given the function $y = 2.5(x - 7.5)(x + 2.5)(x - 3.2)$:
    a. Find the x-intercepts without graphing. 7.5, −2.5, 3.2
    b. Find the y-intercept without graphing. 150
    c. Write the function in general form. $y = 2.5x^3 - 20.5x^2 - 6.875x + 150$
    d. Graph both the factored form and the general form of the function to check your work.

Reason and Apply

5. Use your work from the investigation to answer these questions.
    a. What x-value maximizes the volume of your box? What is the maximum volume possible? approximately 2.94 units; approximately 420 cubic units
    b. What x-value or values give a volume of 300 cubic units? 5 and approximately 1.28
    c. The portion of the graph with domain $x > 10$ shows positive volume. What does this mean in the context of the problem?
    d. Explain the meaning of the parts of the graph showing negative volume.

6. Sandy's group did The Box Factory Investigation using a different size sheet of material. They found the following polynomial to represent the volumes of the possible boxes from their sheet of material $V(x) = (11 - 2x)(8.5 - 2x)x$.
    a. What was the size of their sheet of material? 8.5 units by 11 units
    b. What x-value maximizes the volume of their box? What is the maximum volume possible? Maximum volume of about 66.15 cubic units at $x \approx 1.59$ units.
    c. What is the practical domain of x in the box volume context? The height, x, must be between 0 and 4.25 for all dimensions to be positive.

7. Ron's pen is leaking. Fill in the blanks to make true statements.
    a. $3(x + 1)(x + 3)(x - 2) = \underline{3}\, x^3 + 6x^2 - 15x - \underline{18}$
    b. $x^{\underline{3}}\, \text{_____} + 60 = (x - 5)(x + 3)(x - \underline{4})$
    c. $2x^3 + \text{_____} x = 2x(x + \underline{1})(x + \text{\_\_})$

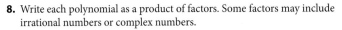

8. Write each polynomial as a product of factors. Some factors may include irrational numbers or complex numbers.
    a. $4x^2 - 88x + 480$
    b. $6x^2 - 7x - 5$ @
    c. $x^3 + 5x^2 - 4x - 20$
    d. $2x^3 + 16x^2 + 38x + 24$
    e. $a^2 + 2ab + b^2$
    f. $x^2 - 64$ @ $(x - 8)(x + 8)$
    g. $x^2 + 64$ $(x + 8i)(x - 8i)$
    h. $x^2 - 7$ @ $(x - \sqrt{7})(x + \sqrt{7})$
    i. $x^2 - 3x$ $x(x - 3)$

9. Sketch a graph for each situation if possible.
    a. a quadratic function with only one real zero
    b. a quadratic function with no real zeros
    c. a quadratic function with three real zeros
    d. a cubic function with only one real zero
    e. a cubic function with two real zeros
    f. a cubic function with no real zeros not possible

10. Write the equation for a cubic function with x-intercepts −4.5, −1, and 2 that contains the point $(-4, -2.7)$. $y = -0.3(x + 4.5)(x + 1)(x - 2)$

332 CHAPTER 6 Polynomial and Rational Functions

9a. sample answer: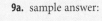

9b. sample answer:

9c. not possible

9d. sample answer:
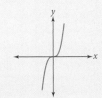

9e. sample answer:

11. Consider the function in this graph.
   a. Write the equation of a polynomial function that has the x-intercepts shown. Use a for the scale factor.
   $y = a(x + 5)(x - 3)(x - 6)$
   b. Use the y-intercept to determine the vertical scale factor. Write the function from 11a, replacing a with the value of the scale factor. ⓐ $y = 2(x + 5)(x - 3)(x - 6)$
   c. Imagine that this graph is translated up 100 units. Write the equation of the image. ⓐ $y = 2(x + 5)(x - 3)(x - 6) + 100$
   d. Imagine that this graph is translated left 4 units. Write the equation of the image. $y = 2(x + 9)(x + 1)(x - 2)$

12. **APPLICATION** The way you taste certain flavors is a genetic trait. For instance, the ability to taste the bitter compound phenyltheocarbamide (PTC) is inherited as a dominant trait in humans. In the United States, approximately 70% of the population can taste PTC, whereas 30% cannot.

   Every person inherits a pair of genes, one from each parent. A person who inherits the taster gene from *either* parent will be able to taste PTC. Let T represent the percentage of taster genes found in the population, and t represent the percentage of nontaster genes.

   a. Complete the rectangle diagram of possible gene-pair combinations. What algebraic expression does this diagram represent? ⓐ $(T + t)^2$, or $T^2 + 2Tt + t^2$
   b. The sum of all possible gene-pair combinations must equal 1, or 100%, for the entire population. Write an equation to express this relationship. ⓐ $(T + t)^2 = 1$, or $T^2 + 2Tt + t^2 = 1$
   c. Use the fact that 70% of the U.S. population are tasters to write an equation that you can use to solve for t. ⓗ $0.70 + t^2 = 1$
   d. Solve for t, the frequency of the recessive gene in the population. $t \approx 0.548$
   e. What is the frequency of the dominant gene in the population? $T \approx 0.452$
   f. What percentage of the U.S. population has two taster genes? $TT \approx 0.205$, or about 20% of the population

Review

13. Is it possible to find a quadratic function that contains the points $(-4, -2)$, $(-1, 7)$, and $(2, 16)$? Explain why or why not. ⓐ No. These points are collinear.

14. Find the quadratic function whose graph has vertex $(-2, 3)$ and contains the point $(4, 12)$. $y = 0.25(x + 2)^2 + 3$

15. Find all real solutions.
   a. $x^2 = 50.4$ ⓐ $x = \pm\sqrt{50.4} \approx \pm 7.1$
   b. $x^4 = 169$ ⓐ $x = \pm\sqrt{13} \approx \pm 3.6$
   c. $(x - 2.4)^2 = 40.2$
   $x = 2.4 \pm \sqrt{40.2} \approx 2.4 \pm 6.3$; $x \approx 8.7$ or $x \approx -3.9$
   d. $x^3 = -64$ $x = -4$

16. Find the inverse of each function algebraically. Then choose a value of x and check your answer.
   a. $f(x) = \frac{2}{3}(x + 5)$
   b. $g(x) = -6 + (x + 3)^{2/3}$
   c. $h(x) = 7 - 2^x$

17. Use the finite differences method to find the function that generates this table of values. Explain your reasoning. $f(x) = -2.5x + 1$
   The differences are all -1, so the function is linear. Find an equation for the line through any two of the points.

x	2.2	2.6	3.0	3.4
f(x)	-4.5	-5.5	-6.5	-7.5

Exercise 12 The rectangle diagram might show the gene inherited from the mother in the rows and the gene from the father in the columns. If the dominant characteristic, *T*, is inherited from either parent, the child is a taster. Students need to realize that the set of tasters includes both *TT* and *Tt*, because *T* is dominant. Therefore $T^2 + 2Tt = 0.70$. **ASK** "What proportion of the population will, if they have children, have only children who are tasters?" [The answer to 10f, about 20%.]

Exercise 13 Instead of stating that the points are collinear, students might answer that any attempt to find a 2nd-degree equation gives an *a*-value of 0.

Exercise 15 **ASK** "Are there other solutions?" [In 15b, there are two nonreal roots; in 15d, there are two nonreal roots.] You might challenge some students to look for nonreal solutions. For example, in 15d, students can factor $x^3 + 64$ into $(x + 4)(x^2 - 4x + 16)$ and then apply the quadratic formula to factor the quadratic factor.

16a. $f^{-1}(x) = \frac{3}{2}x - 5$
sample check: $f(4) = 6$; $f^{-1}(6) = 4$

16b. $g^{-1}(x) = -3 + (x + 6)^{3/2}$
sample check: $g(5) = -2$; $g^{-1}(-2) = 5$

16c. $h^{-1}(x) = \log_2(7 - x)$
sample check: $h(2) = 3$; $h^{-1}(3) = 2$

LESSON 6.3

In this lesson students extend their knowledge of functions of degree 3 or more. They explore the extreme values and end behavior of the graphs of these functions.

COMMON CORE STATE STANDARDS

APPLIED	DEVELOPED	INTRODUCED
A.SSE.1b	A.APR.3	
	F.IF.4	
	F.IF.6	
	F.IF.7c	

Objectives

- Investigate extreme values and end behavior of polynomial functions
- Identify possible degrees of a polynomial function by looking at its graph
- Identify and find the lowest-degree polynomial that has given roots
- Find additional roots when given one or more complex roots
- Relate the graphs of polynomial equations to the number and types of roots

Vocabulary

local maximum
local minimum
extreme values
end behavior
long run behavior
double root

Materials

- rulers

Launch

$f(x) = 2(x - 3)(x - 5)(x - 4)^2$

a. What are the zeros of $f(x)$?
b. What is the degree of $f(x)$?
c. What are the roots of $f(x)$?
d. How many roots should $f(x)$ have? Explain.

a. $x = 3, 5, 4$
b. 4
c. 3, 5, 4
d. 4; Functions of degree four should have four roots. 4 is a double root.

LESSON 6.3

Higher-Degree Polynomials

It is good to have an end to journey towards, but it is the journey that matters in the end.

URSULA K. LEGUIN

Polynomials with degree 3 or higher are called higher-degree polynomials. Frequently, 3rd-degree polynomials are associated with volume measures.

If you create a box by removing small squares of side length x from each corner of a square piece of cardboard that is 30 inches on each side, the volume of the box in cubic inches is modeled by the function $y = x(30 - 2x)^2$, or $y = 4x^3 - 120x^2 + 900x$. The zero-product property tells you that the zeros are $x = 0$ or $x = 15$, the two values of x for which the volume is 0. The x-intercepts on the graph below confirm this.

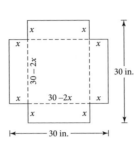
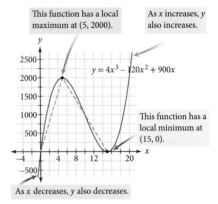

This function has a local maximum at (5, 2000).
As x increases, y also increases.
This function has a local minimum at (15, 0).
As x decreases, y also decreases.

Notice that the volume is changing faster for x values from 0 to 5 than it is from 5 to 15. You can confirm that by finding the average rate of change in these intervals. The average rate of change of the volume is $\frac{2000}{5} = 400$ in³ per inch of side length removed when $0 \leq x \leq 5$, but for $5 \leq x \leq 15$, the volume is changing on average $\frac{-2000}{10} = -200$ in³ per inch of side length removed. What does this mean about the symmetry of the polynomial between $0 \leq x \leq 15$?

The shape of this graph is typical of the odd-degree polynomial graphs you will work with in this lesson. Note that this graph has one **local maximum** at (5, 2000) and one **local minimum** at (15, 0). These are the points that are higher or lower than all other points near them. Maximums and minimums, collectively, are called **extreme values**.

Graphs of all polynomials with real coefficients have a y-intercept and possibly one or more x-intercepts. You can also describe other features of polynomial graphs, such as local maximums or minimums and end behavior. You can describe the **end behavior**, or **long run behavior**, of a polynomial—what happens to $f(x)$ as x takes on extreme positive and negative values. You can think about how the graph looks as you look to the left for extreme negative values of x and when you look to the right for very large positive values of x. To describe the long run behavior, you describe what happens as x increases and what happens as x decreases. In the case of

334 CHAPTER 6 Polynomial and Rational Functions

ELL
Provide several examples of graphs of higher-degree polynomials. Ask students to identify local extreme values and to predict end behavior; discuss both.

Support
Students may continue to have difficulty determining the complex zeros of a polynomial function. Have students convert several examples of higher-order polynomial functions from factored form to general form so they can observe how the complex conjugates behave.

Advanced
Challenge students to come up with paper investigations that yield 1st-, 2nd-, and 4th-degree functions or have them explain why it is not possible to do so.

ASK "Do any functions have absolute extreme values?" [Polynomial functions of even degrees have either an absolute maximum or an absolute minimum but not both.]

this cubic function, you would say as x increases, y also increases. As x decreases, y also decreases. The behavior of a polynomial in the long run (left or right) is dominated by the term of highest degree, which for this cubic function is the $4x^3$ term. Each of the graphs below is the graph of $y = 4x^3$ (in blue) and $y = 4x^3 - 120x^2 + 900x$ (in red). You can see how your choice of viewing window determines what behavior is visible.

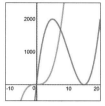
In a small viewing window, you see the short run behavior.

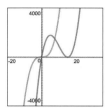
Zooming out, you can see more end behavior.

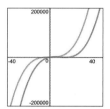
Zooming out more, the short run behavior isn't visible, the long run behavior dominates.

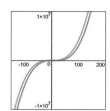
In a very large window, the two functions look almost the same.

Long-run Behavior

In general, when viewed on a large enough scale, the **long-run behavior** of a polynomial $f(x) = a_n x^n + a_{n-1} x^{n-1} + \cdots + a_1 x^1 + a_0$ looks like the graph of the power function, $p(x) = a_n x^n$. Therefore, the long-run behavior is determined by the long-run behavior of the polynomial's leading term.

INVESTIGATION
The Largest Triangle

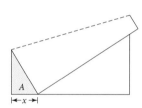
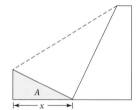

Take a sheet of notebook paper and orient it so that the longest edge is closest to you. Fold the upper-left corner so that it touches some point on the bottom edge. Find the area, A, of the triangle formed in the lower-left corner of the paper. What distance, x, along the bottom edge of the paper produces the triangle with the greatest area?

Work with your group to find a solution. You may want to use strategies you've learned in several lessons in this chapter. Write a report that explains your solution and your group's strategy for finding the largest triangle. Include any diagrams, tables, or graphs that you used.

Modifying the Investigation

Whole Class Demonstrate how to fold the paper and find the area of a triangle. Assign different values of x to individual students. Collect and analyze the data to find the largest triangle.

Shortened Use the desmos simulation or sample data. Suggest possible strategies.

One Step Illustrate the paper folding of the investigation and pose the problem from the investigation.

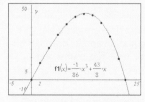

Maximum: approximately (12.4, 44.5); when x is approximately 12.4 cm, the area is maximized at approximately 44.5 cm^2.

Investigate

Guiding the Investigation

It is not necessary to discuss the opening remarks before doing the investigation.

There is a desmos simulation of this investigation in your ebook. You can also use the Investigation Worksheet with Sample Data. **SMP 5**

This investigation allows for rich discovery. Students often expect a quadratic model for area, so they may be surprised to find that the model is a cubic function. **ASK** "Why is the model a cubic function, rather than a quadratic function? [The model is ((length (cm), area (cm^2))]. If it is graphed only in Quadrant I, though, it looks quadratic. **SMP 1, 2, 6**

For presentations, select groups that have used a variety of methods to find the area of the triangle: counting squares on graph paper, measuring base and height to calculate the area, using the Pythagorean Theorem, etc. For some approaches, students might find the area function by using finite differences and creating a system of four equations whose four unknowns are the coefficients of a cubic function. **SMP 1**

You can use a general solution to check student work; you might have students work through it themselves. If the corner is being folded to cut a height b of the paper down to a height y of the triangle, then the Pythagorean Theorem gives the equation $x^2 + y^2 = (b - y)^2$, or $y = \dfrac{b^2 - x^2}{2b}$. This gives the area function $A = \dfrac{1}{2} xy = -\dfrac{1}{4b} x^3 + \dfrac{b}{4} x$.

These answers assume that the piece of notebook paper is 21.5 × 28 cm or $8\dfrac{1}{2} \times 11$ in. Answers will vary if paper with different dimensions is used. Answers are from the sample data found in the Investigation Worksheet with Data.

polynomial function:
$$y = \frac{1}{2} x \left(\frac{462.25 - x^2}{43} \right), \text{ or}$$
$$y = \frac{1}{86} x (21.5 - x)(21.5 + x), \text{ or}$$
$$y = -\frac{1}{86} x^3 + \frac{43}{8} x$$

Example A

This example could be solved with a higher-degree polynomial function, but the simplest solution is a cubic function. If you discussed the introductory content before the investigation, you might project Example A and have students work in pairs. Otherwise, you can use the example, with student input, as a means of reviewing the introductory content. Point out that neither (4, −24) nor (−1.4, 220) is an absolute extreme value because the end behavior of the graph goes to plus and minus infinity. (The symbols, ∞ and −∞, will be introduced later.)

ASK "Is there a 4th-degree polynomial with the same three roots and an additional real root?" [Yes; multiply the factors by a fourth factor, $(x - r)$.] "Is there a 4th-degree polynomial with real coefficients and the same three roots and an additional imaginary root?" [No; nonreal roots come in conjugate pairs.] **SMP 1**

A small window may show features of a graph but might not indicate end behavior. A large viewing screen can show end behavior, but individual features might be hard to see. **SMP 5**

Example B

This example asks students to find any polynomial function that satisfies the criteria. The solution given is one possibility. Multiplying this solution by a constant or by any other factor $(x - r)$ will also give a solution. **SMP 2, 7**

Project Example B and have students work in small groups. As needed, **ASK** "How many zeros does the function have?" [4. Complex zeros occur in conjugate pairs] Suggest that students use a graph to check their solution.

Cubic functions have local extreme values but no absolute extreme values, because their end behavior exceeds all bounds.

In the remainder of this lesson you will explore the connections between a polynomial equation and its graph, which will allow you to predict when certain features will occur in the graph.

EXAMPLE A Find a polynomial function whose graph has x-intercepts 3, 5, and −4, and y-intercept 180. Describe the features of its graph.

Solution A polynomial function with three x-intercepts has too many x-intercepts to be a quadratic function. It could be a 3rd-, 4th-, 5th-, or higher-degree polynomial function. Consider a 3rd-degree polynomial function, because that is the lowest degree that has three x-intercepts. Use the x-intercepts to write the equation $y = a(x - 3)(x - 5)(x + 4)$, where $a \neq 0$.

Substitute the coordinates of the y-intercept, (0, 180), into this function to find the vertical scale factor.

$$180 = a(0 - 3)(0 - 5)(0 + 4)$$
$$180 = a(60)$$
$$a = 3$$

The polynomial function of the lowest degree through the given intercepts is

$$y = 3(x - 3)(x - 5)(x + 4)$$

Graph this function to confirm your answer and look for features.

This graph shows a local minimum at about (4, −24) because that is the lowest point in its immediate neighborhood of x-values. There is also a local maximum at about (−1.4, 220) because that is the highest point in its immediate neighborhood of x-values.

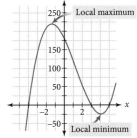

Even the small part of the domain shown in the graph suggests the end behavior. As x increases, y also increases. As x decreases, y also decreases. If you increase the domain of this graph to include more x-values at the right and left extremes of the x-axis, you'll see that the graph does continue this long run behavior.

You can identify the degree of many polynomial functions by looking at the shapes of their graphs. Every 3rd-degree polynomial function has essentially one of the shapes shown here. Graph A shows the graph of $y = x^3$. It can be translated, stretched, or reflected. Graph B shows one possible transformation of Graph A. Graphs C and D show the graphs of general cubic functions in the form $y = ax^3 + bx^2 + cx + d$. In Graph C, a is positive, and in Graph D, a is negative.

Conceptual/Procedural

Conceptual Students explore the connections between a polynomial equation and its graph, including the concepts of local maximums and minimums, extreme values, long run behavior, and rate of change.

Procedural Students describe features of the graph of a polynomial, given the x and y intercepts. They write a polynomial function with real coefficients, given the zeros of the function.

Graph A	Graph B	Graph C	Graph D
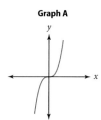	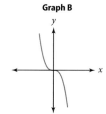	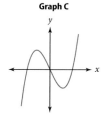	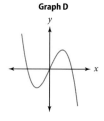

You'll explore the general shapes and characteristics of other higher-degree polynomials in the exercises.

EXAMPLE B | Write a polynomial function with real coefficients and zeros $x = 2$, $x = -5$, and $x = 3 + 4i$.

Solution | For a polynomial function with real coefficients, complex zeros occur in conjugate pairs, so $x = 3 - 4i$ must also be a zero. In factored form the polynomial function of the lowest degree is

$$y = (x - 2)(x + 5)[x - (3 + 4i)][x - (3 - 4i)]$$

Multiplying the last two factors to eliminate imaginary numbers gives

$$y = (x - 2)(x + 5)(x^2 - 6x + 25)$$

Multiplying all factors gives the polynomial function in general form:

$$y = x^4 - 3x^3 - 3x^2 + 135x - 250$$

Graph this function to check your solution. You can't see the complex zeros, but you can see x-intercepts 2 and -5.

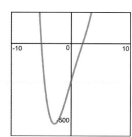

Note that in Example B, the solution was a 4th-degree polynomial function. It had four complex zeros, but the graph had only two x-intercepts, corresponding to the two real zeros.

INVESTIGATING STRUCTURE

A factor raised to the power of 1 results in an x-intercept that crosses the x-axis. A factor raised to the power of 2 results in an x-intercept that touches, but does not cross, the x-axis. A factor raised to the power of 3 results in an x-intercept that crosses the x-axis in a curved fashion.

The graph of $f(x) = 2(x - 3)(x - 5)(x - 4)^2$ has x-intercepts 3, 5, and 4 because they are the only possible x values that make $y = 0$. This is a 4th-degree polynomial, but it has only three x-intercepts. The root $x = 4$ is called a **double root** because the factor $(x - 4)$ occurs twice.

What if one of the other factors was squared? For example, what does $y = 2(x - 3)^2(x - 5)(x - 4)$ look like? Experiment by graphing functions with the same factors as f but where the factors have different powers.

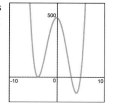

Describe a connection between the power of a factor and what happens to the graph at that x-intercept. It appears that If the power of the factor is odd, the graph crosses the x-axis, whereas if the power of the factor is even, it touches, but does not cross, the x-axis at that value.

Investigating Structure

This Investigating Structure explores the connection between the power of a factor and the graph at that x-intercept. Project the problem from your ebook and have students work in small groups. Or you could assign it for homework. Have students present their conclusion and their graphs.

Summarize

As students present their solutions to the Launch, the Investigation and the Examples, including explaining their strategy, emphasize the use of correct mathematical terminology and symbols. Be careful not to discourage thinking by talking too quickly or saying too much. Whether student responses are correct or incorrect, ask other students if they agree and why.
SMP 1, 2, 3, 6

CRITICAL QUESTION How are greatest and least values represented on a graph?

BIG IDEA The extreme values of a function are represented by maximums or minimums on a graph. The maximum or minimum may be local—that is, relative only to nearby points.

CRITICAL QUESTION Does a cubic polynomial always have a relative maximum between every two zeros?

BIG IDEA It has a relative maximum or a relative minimum.

CRITICAL QUESTION How could you switch the end behaviors of this cubic function and still have a cubic function?

BIG IDEA Change the sign of the leading coefficient.

CRITICAL QUESTION Can you always tell the degree of a polynomial from the shape of its graph?

BIG IDEA No. If students believe so, suggest that they graph the functions $y = x^4$ and $y = x^9$. A polynomial curve with n local extreme values has degree at least $n + 1$ and perhaps much higher.

Formative Assessment

Assess how well students integrate the ideas of roots of polynomials as being representative of physical events. The Investigation will allow you to assess students' concepts of area and their ability to find the area of a right triangle and to apply the Pythagorean Theorem. Example A will provide evidence of understanding how to write a function from the intercepts and how to determine the degree and features of the graph. Example B can provide evidence of understanding how to write a function when given the zeros. The Exercises provide further evidence of meeting the objectives.

Apply

Extra Example

Name the missing zero for a polynomial function with real coefficients and zeros $x = -3$, $x = 0$, and $x = 1 + i$. $1 - i$

Closing Question

Write the polynomial function for the polynomial function with real coefficients and zeros $x = -3$, $x = 0$, and $x = 1 + i$.

$y = x^4 + x^3 - 4x^2 + 6x$

Assigning Exercises

Suggested: 1–12, Performance Task
Additional Practice: 13–18

3a. 3; as x decreases, y decreases and as x increases, y increases

3b. 4; as x decreases, y increases and as x increases, y increases

3c. 2; as x decreases, y increases and as x increases, y increases

3d. 5; as x decreases, y decreases and as x increases, y increases

4a. $y = (x + 5)(x - 3)(x - 7)$

4b. $y = 0.5(x + 6)(x + 3)(x - 2)(x - 6)$

4c. $y = 10(x + 5)(x - 2)$

4d. $y = 0.25(x + 5)(x + 3)(x - 1)(x - 4)(x - 6)$

6.3 Exercises

You will need a graphing calculator for Exercises **4, 10, 11,** and **14.**

Practice Your Skills

For Exercises 1–4, use these four graphs.

a.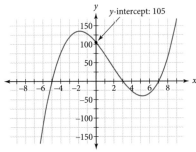
$x = -5, x = 3,$ and $x = 7$

b.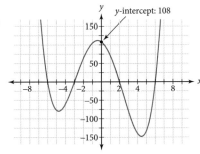
$x = -6, x = -3, x = 2,$ and $x = 6$

c.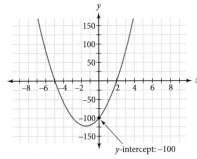
$x = -5$ and $x = 2$

d.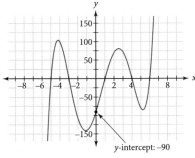
$x = -5, x = -3, x = 1, x = 4,$ and $x = 6$

1. Identify the zeros of each function.

2. Give the coordinates of the y-intercept of each graph.
   2a. $(0, 105)$; **2b.** $(0, 108)$ **2c.** $(0, -100)$; **2d.** $(0, -90)$

3. Identify the lowest possible degree of each polynomial function and describe its long run behavior.

4. Write the factored form for each polynomial function. Check your work by graphing on your calculator.

Reason and Apply

5. Write polynomial functions with these features.

   a. A linear function whose graph has x-intercept 4. $y = a(x - 4)$ where $a \neq 0$

   b. A quadratic function whose graph has only one x-intercept, 4. $y = a(x - 4)^2$ where $a \neq 0$

   c. A cubic function whose graph has only one x-intercept, 4.
      $y = a(x - 4)^3$ where $a \neq 0$; or $y = a(x - 4)(x - r_1)(x - r_2)$ where $a \neq 0$ and r_1 and r_2 are complex conjugates

338 CHAPTER 6 Polynomial and Rational Functions

6. The graph at right is a complete graph of a polynomial function. @

   a. How many x-intercepts are there? 4

   b. What is the lowest possible degree of this polynomial function? 5

   c. Write the factored form of this function if the graph includes the points (0, 0), (−5, 0), (4, 0), (−1, 0), and (1, 216). $y = -x(x + 5)^2(x + 1)(x - 4)$

   d. Find the average rate of change from $x = -5$ to $x = -3$, and from $x = 3$ to $x = 4$. In which interval is the graph the most steep? How can you tell by the average rates?

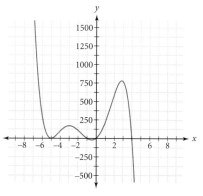

7. Write the lowest-degree polynomial function that has the given set of zeros and whose graph has the given y-intercept.

   a. zeros: $x = -4$, $x = 5$, $x = -2$ (double root); y-intercept: −80 $y = (x + 4)(x - 5)(x + 2)^2$

   b. zeros: $x = -4$, $x = 5$, $x = -2$ (double root); y-intercept: 160 @ $y = -2(x + 4)(x - 5)(x + 2)^2$

   c. zeros: $x = \frac{1}{3}$, $x = -\frac{2}{5}$, $x = 0$; y-intercept: 0 $y = ax\left(x - \frac{1}{3}\right)\left(x + \frac{2}{5}\right)$, or $y = ax(3x - 1)(5x + 2)$ where $a \neq 0$

   d. zeros: $x = -5i$, $x = -1$ (triple root), $x = 4$; y-intercept: −100 ⓗ
      $y = (x + 5i)(x - 5i)(x + 1)^3(x - 4)$, or $y = (x^2 + 25)(x + 1)^3(x - 4)$

8. The polynomial and graph represent the area of the triangle formed by folding a piece of paper, like in The Largest Triangle Investigation. This paper is 30 in. by 30 in.

   a. What is the domain that makes sense for the area of the triangle? Why? @
      $0 \leq x \leq 30$

   b. What is the x value that gives the maximum area? $x \approx 17.3$ inches

   c. What is the maximum area? How do you know? $A(17.3) \approx 86.6$ inches²

   d. What is the average rate of change of the area from $x = 0$ to $x \approx 17$? From $x \approx 17$ to $x = 30$? @

   e. What does your answer to part d mean in terms of the area of the triangle?

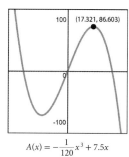

$A(x) = -\frac{1}{120}x^3 + 7.5x$

9. Look back at Exercises 1–4. Find the products of the zeros in Exercise 1. How does the value of the leading coefficient, a, relate to the y-intercept, the product of the zeros, and the degree of the function? @

10. A 4th-degree polynomial function has the general form $y = ax^4 + bx^3 + cx^2 + dx + e$ for real values of a, b, c, d, and e, where $a \neq 0$. Graph several 4th-degree polynomial functions by trying different values for each coefficient. Be sure to include positive, negative, and zero values. Make a sketch of each different type of curve you get. Concentrate on the shape of the curve. You do not need to include axes in your sketches. Compare your graphs with the graphs of your classmates, and come up with six or more different shapes that describe all 4th-degree polynomial functions.

LESSON 6.3 Higher-Degree Polynomials 339

Exercise 10 You might suggest that students use computer software, such as the desmos object in your ebook, in which sliders can make it easy to see how changing coefficient values affects the graph.

10. Graphs will be a basic W shape, pointed either up or down. In some cases the W shape is less apparent because the local maximums and minimums and/or points of inflection are close together or concurrent.

Sample graphs:

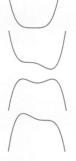

Exercise 11 Students might write each of the multiple roots as a separate factor or use powers to show multiplicity.

11. Each of these is the graph of a polynomial function with leading coefficient $a = 1$ or $a = -1$.

i.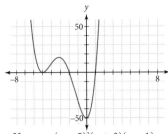

11a. $y = (x + 5)^2(x + 2)(x - 1)$
11b. $x = -5, x = -5, x = -2,$ and $x = 1$

ii.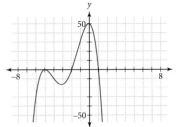

11a. $y = -(x + 5)^2(x + 2)(x - 1)$
11b. $x = -5, x = -5, x = -2,$ and $x = 1$

iii.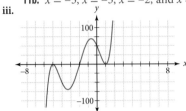

11a. $y = (x + 5)^2(x + 2)(x - 1)^2$
11b. $x = -5, x = -5, x = -2, x = 1,$ and $x = 1$

iv.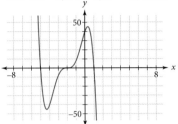

11a. $y = -(x + 5)(x + 2)^3(x - 1)$
11b. $x = -5, x = -2, x = -2, x = -2,$ and $x = 1$

   a. Write a function in factored form that will produce each graph.

   b. Name the zeros of each polynomial function in 11a. If a factor is raised to the power of n, list the zero n times.

12. Consider the polynomial functions in Exercise 11.

   a. What is the degree of each polynomial function? **i.** 4; **ii.** 4; **iii.** 5; **iv.** 5

   b. How many extreme values does each graph have? **i.** 3; **ii.** 3; **iii.** 4; **iv.** 2

   c. What is the relationship between the degree of the polynomial function and the number of extreme values? The number of extreme values of a polynomial function of degree n is at most $n - 1$.

   d. Complete these statements:

      i. The graph of a polynomial curve of degree n has at most _?_ x-intercepts. n

      ii. A polynomial function of degree n has at most _?_ real zeros. n

      iii. A polynomial function of degree n has at most _?_ complex zeros. n

      iv. The graph of a polynomial function of degree n has at most _?_ extreme values. $n - 1$

13. In this chapter, you have seen the appearance of many polynomials. Consider the possible end behaviors for even and odd degree polynomials. Graph a variety of polynomials to help you complete these statements.

   a. The end behavior of a polynomial with an odd degree with positive first term is that as x increases, y _____ and as x decreases, y _____. increases, decreases

   b. The end behavior of a polynomial with an even degree with positive first term is that as x increases, y _____ and as x decreases, y _____. increases, increases

Review

14. Find the roots of these quadratic equations. Express them as fractions.
   a. $0 = 3x^2 - 13x - 10$ $x = -\frac{2}{3}$ or $x = 5$
   b. $0 = 6x^2 - 11x + 3$ $x = \frac{3}{2}$ or $x = \frac{1}{3}$
   c. List all the factors of the constant term, c, and the leading coefficient, a, for 15a and b. What do you notice about the relationship between the factors of a and c, and the roots of the functions?

15. If $3 + 5\sqrt{2}$ is a solution of a quadratic equation with rational coefficients, then what other number must also be a solution? Write an equation in general form that has these solutions. @
   $3 - 5\sqrt{2}$; $0 = a(x^2 - 6x - 41)$ where $a \neq 0$

16. Given the function $Q(x) = x^2 + 2x + 10$, find these values.
   a. $Q(-3)$ 13
   b. $Q\left(-\frac{1}{5}\right)$ $\frac{241}{25}$
   c. $Q(2 - 3\sqrt{2})$ @ $36 - 18\sqrt{2}$
   d. $Q(-1 + 3i)$ 0

17. **APPLICATION** According to Froude's Law, the speed at which an aquatic animal can swim is proportional to the square root of its length. If a 75-foot blue whale can swim at a maximum speed of 20 knots, write a function that relates its speed to its length. How fast would a similar 60-foot-long blue whale be able to swim? @
   $S = \frac{20}{\sqrt{75}}\sqrt{l}$; approximately 17.9 knots

A blue whale surfaces for air in the Indian Ocean. The blue whale is the world's largest mammal.

18. Write an equation of the image of the absolute-value function, $y = |x|$, after performing each of the following transformations in order. Sketch a graph of your final equation.
   a. Dilate vertically by a factor of 2. $y = 2|x|$
   b. Then translate horizontally 4 units. $y = 2|x - 4|$
   c. Then translate vertically -3 units. $y = 2|x - 4| - 3$

PERFORMANCE TASK

In the lesson you saw various possible appearances of the graph of a 3rd-degree polynomial function, and in Exercise 10 you explored possible appearances of the graph of a 4th-degree polynomial function. In Exercises 11 and 12, you found a relationship between the degree of a polynomial function and the number of zeros and extreme values. Use all the patterns you have noticed in these problems to sketch one possible graph of

a. A 5th-degree function.
b. A 6th-degree function.
c. A 7th-degree function.

PERFORMANCE TASK

a. sample answer: b. sample answer: c. sample answer:

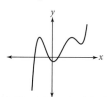 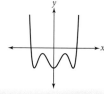 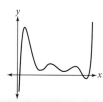

LESSON 6.4

In this lesson students see ways to find zeros of higher-degree polynomial functions.

COMMON CORE STATE STANDARDS

APPLIED	DEVELOPED	INTRODUCED
A.SSE.2	A.APR.2	
	A.APR.4	

Objectives

- Learn to find zeros of higher-degree polynomial functions algebraically
- Use the Rational Root Theorem to identify rational numbers that might be zeros
- Use the Remainder Theorem and long division to find zeros
- Understand and use synthetic division

Vocabulary

Remainder Theorem
Rational Root Theorem
synthetic division

Materials

- Calculator Notes: SYNDIV program; Zero Finding

Launch

Use long division to divide 4089 by 28.

146 r 1 or $146\frac{1}{28}$

```
      146
   _____
28)4089
    -28
    ____
    128
   -112
    ____
     169
    -168
    ____
       1
```

LESSON 6.4

More About Finding Solutions

It isn't that they can't see the solution. It is that they can't see the problem.

G. K. CHESTERTON

You can find zeros of a quadratic function by factoring or by using the quadratic formula. How can you find the zeros of a higher-degree polynomial?

Sometimes a graph will show you zeros in the form of x-intercepts, but only if they are real. And this method is often accurate only if the zeros have integer values. Fortunately, there is a method for finding exact zeros of many higher-degree polynomial functions. It is based on the multiplicative relationship of the factors. When you have one factor, you can divide to find the other factors.

Blood is often separated into two of its factors, plasma and red blood cells.

You first need to find one or more zeros. Then you divide your polynomial function by the factors associated with those zeros. Repeat this process until you have a polynomial function that you can find the zeros of by factoring or using the quadratic formula. Let's start with an example in which we already know several zeros. Then you'll learn a technique for finding some zeros when they're not obvious.

EXAMPLE A | What are the zeros of $P(x) = x^5 - 6x^4 + 20x^3 - 60x^2 + 99x - 54$?

Solution | The graph appears to have x-intercepts at 1, 2, and 3. You can confirm that these values are zeros of the function by substituting them into $P(x)$.

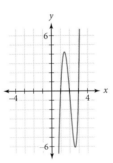

$P(1) = (1)^5 - 6(1)^4 + 20(1)^3 - 60(1)^2 + 99(1) - 54 = 0$

$P(2) = (2)^5 - 6(2)^4 + 20(2)^3 - 60(2)^2 + 99(2) - 54 = 0$

$P(3) = (3)^5 - 6(3)^4 + 20(3)^3 - 60(3)^2 + 99(3) - 54 = 0$

ELL

When students first encounter the terms *rational, factor,* and *root,* provide several examples with clear labels.

Support

Students who are struggling with polynomial long division may benefit from working on several basic long division problems and comparing the algorithms.

Advanced

Project the function from Example B and challenge students to try to factor it by finding its zeros exactly. Encourage them to think about how the factors of 3 and 25 might relate to a rational root. After they find the rational root, draw analogies to the quadratic case to encourage dividing by the binomial.

For all three values, you get $P(x) = 0$, which shows that $x = 1$, $x = 2$, and $x = 3$ are zeros. This also means that $(x - 1)$, $(x - 2)$, and $(x - 3)$ are factors of $x^5 - 6x^4 + 20x^3 - 60x^2 + 99x - 54$. None of the x-intercepts has the appearance of a repeated root, and you know that a 5th-degree polynomial function has five complex zeros, so this function must have two additional nonreal zeros.

You know that $(x - 1)$, $(x - 2)$, and $(x - 3)$ are all factors of the polynomial, so the product of these three factors, $x^3 - 6x^2 + 11x - 6$, must be a factor also. Your task is to find another factor such that

$$(x^3 - 6x^2 + 11x - 6)(factor) = x^5 - 6x^4 + 20x^3 - 60x^2 + 99x - 54$$

$$(factor) = \frac{x^5 - 6x^4 + 20x^3 - 60x^2 + 99x - 54}{x^3 - 6x^2 + 11x - 6}$$

You can find this factor by dividing, using a rectangle diagram or long division.

$$(x^5 - 6x^4 + 20x^3 - 60x^2 + 99x - 54) \div (x^3 - 6x^2 + 11x - 6)$$

	x^2		
x^3	x^5		
$-6x^2$	$-6x^4$		
$11x$	$11x^3$		
-6	$-6x^2$		

First, divide x^5 by x^3 to get x^2
$(x^3)(?) = x^5$
$(x^3)(\underline{x^2}) = x^5$

Then, multiply x^2 by the rest of the divisor.

$$(x^5 - 6x^4 + 20x^3 - 60x^2 + 99x - 54) \div (x^3 - 6x^2 + 11x - 6)$$

	$x^2 + 0x$		
x^3	x^5	$0x^4$	
$-6x^2$	$-6x^4$	$0x^3$	
$11x$	$11x^3$	$0x^2$	
-6	$-6x^2$	$0x$	

The like terms in the diagonals must add to the terms in the dividend.
$-6x^4 + \underline{} = -6x^4$
$-6x^4 + \underline{0x^4} = -6x^4$

Then divide $0x^4$ by x^3 to get $0x$.
$(x^3)(?) = 0x^4$
$(x^3)(\underline{0x}) = 0x^4$

Then, multiply $0x$ by the rest of the divisor.

$$(x^5 - 6x^4 + 20x^3 - 60x^2 + 99x - 54) \div (x^3 - 6x^2 + 11x - 6)$$

	$x^2 - 0x + 9$		
x^3	x^5	$0x^4$	$9x^3$
$-6x^2$	$-6x^4$	$0x^3$	$-54x^2$
$11x$	$11x^3$	$0x^2$	$99x$
-6	$-6x^2$	$0x$	-54

The like terms in the diagonals must add to the terms in the dividend.
$11x^3 + 0x^3 + \underline{} = 20x^3$
$11x^3 + \underline{9x^3} = 20x^3$

Then divide $9x^3$ by x^3 to get 9.
$(x^3)(?) = 9x^3$
$(x^3)(\underline{9}) = 9x^3$

Then, multiply 9 by the rest of the divisor.

LESSON 6.4 More About Finding Solutions

Investigate

Example A

Project Example A from your ebook. **ASK** "How many zeroes should $P(x)$ have?" [5] Ask for suggestions on how to find the zeros. If no one suggests graphing, **ASK** "What do the zeros represent? [x-intercepts] How can we find the x-intercepts?" Graphing gives you three of the zeroes. Connect the zeros to the factors. Continue with the discussion at the top of page, either with or without books open. Work through using the rectangle diagram to find other factors, since those are more familiar to students. Then go over the long division. You might compare it to the long division of the Launch.

As you talk about representing missing terms for some powers in a polynomial, you can make an analogy to the difference between 10,358 and 1358. The zero is necessary as a placeholder and must be used in both long division and synthetic division.

Polynomial long division is tedious and can be frustrating for students. Additional examples, especially shorter ones, or discussion might relieve some of the frustration.

Formalize the discussion with a discussion of the Remainder Theorem.

LANGUAGE In a division problem, the *dividend* is the expression being divided, and the *divisor* is the expression that is doing the dividing.

Conceptual/Procedural

Conceptual In this lesson, students explore the Remainder Theorem and the Rational Root Theorem.

Procedural Students learn to find the zeros of a higher-degree polynomial algebraically, using long division, rectangle diagrams, and synthetic division.

$$(x^5 - 6x^4 + 20x^3 - 60x^2 + 99x - 54) \div (x^3 - 6x^2 + 11x - 6)$$

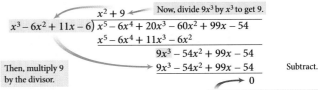

The like terms in the diagonals must add to the terms in the dividend. They do! Because they do, the remainder is 0 and the division is finished. The quotient is $x^2 + 9$.

You can also use long division.

First, divide x^5 by x^3 to get x^2.

$$\begin{array}{r} x^2 \\ x^3 - 6x^2 + 11x - 6 \overline{) x^5 - 6x^4 + 20x^3 - 60x^2 + 99x - 54} \\ x^5 - 6x^4 + 11x^3 - 6x^2 \\ \hline 9x^3 - 54x^2 + 99x - 54 \end{array}$$

Then, multiply x^2 by the divisor. Subtract.

Now, divide $9x^3$ by x^3 to get 9.

$$\begin{array}{r} x^2 + 9 \\ x^3 - 6x^2 + 11x - 6 \overline{) x^5 - 6x^4 + 20x^3 - 60x^2 + 99x - 54} \\ x^5 - 6x^4 + 11x^3 - 6x^2 \\ \hline 9x^3 - 54x^2 + 99x - 54 \\ 9x^3 - 54x^2 + 99x - 54 \\ \hline 0 \end{array}$$

Then, multiply 9 by the divisor. Subtract.

The remainder is zero, so the division is finished, resulting in two factors.

Now you can rewrite the original polynomial as a product of factors:

$$x^5 - 6x^4 + 20x^3 - 60x^2 + 99x - 54 = (x^3 - 6x^2 + 11x - 6)(x^2 + 9)$$
$$= (x - 1)(x - 2)(x - 3)(x^2 + 9)$$

Now that the polynomial is in factored form, you can find the zeros. You knew three of them from the graph. The two additional zeros are contained in the factor $x^2 + 9$. What values of x make $x^2 + 9$ equal zero? If you solve the equation $x^2 + 9 = 0$, you get $x = \pm 3i$.

Therefore, the five zeros are $x = 1$, $x = 2$, $x = 3$, $x = 3i$, and $x = -3i$.

To confirm that a number is a zero, you can use the **Remainder Theorem**.

> **Remainder Theorem**
> If you divide a polynomial function $P(x)$ by $(x - r)$, then the remainder is $P(r)$. $(x - r)$ is a factor of a polynomial function $P(x)$ if and only if $P(r) = 0$.

In the example you showed that $P(1)$, $P(2)$, and $P(3)$ equal zero. You can check that $P(3i)$ and $P(-3i)$ will also equal zero.

$P(3i) = (3i)^5 - 6(3i)^4 + 20(3i)^3 - 60(3i)^2 + 99(3i) - 54 = 243i^5 - 486i^4 + 540i^3 - 540i^2 + 297i - 54 = 243i - 486 - 540i + 540 + 297i - 54 = 0$

$P(-3i) = (-3i)^5 - 6(-3i)^4 + 20(-3i)^3 - 60(-3i)^2 + 99(-3i) - 54 = -243i^5 - 486i^4 - 540i^3 - 540i^2 - 297i - 54 = -243i - 486 + 540i + 540 - 297i - 54 = 0$

When you divide, either using a rectangular diagram or using long division, both the original polynomial and the divisor are written in descending order of the powers of x. If any degree is missing, insert a term with coefficient 0 as a placeholder. For example, you can write the polynomial $x^4 + 3x^2 - 5x + 8$ as

$$x^4 + 0x^3 + 3x^2 - 5x + 8$$

Insert a zero placeholder because the polynomial did not have a 3rd-degree term.

Often you won't be able to find any zeros for certain by looking at a graph. However, there is a pattern to rational numbers that might be zeros. The **Rational Root Theorem** helps you narrow down the values that might be zeros of a polynomial function.

> **Rational Root Theorem**
>
> If the polynomial equation $P(x) = 0$ has rational roots, they are in the form $\frac{p}{q}$, where p is a factor of the constant term and q is a factor of the leading coefficient.

Notice that this theorem will identify only possible *rational* roots. It won't find roots that are irrational or contain imaginary numbers.

EXAMPLE B | Find the roots of this polynomial equation:

$$3x^3 + 5x^2 - 15x - 25 = 0$$

Solution | First, graph the function $y = 3x^3 + 5x^2 - 15x - 25$ to see if there are any identifiable integer x-intercepts.

There are no integer x-intercepts, but the graph shows x-intercepts between -3 and -2, -2 and -1, and 2 and 3. Any rational root of this polynomial will be a factor of -25, the constant term, divided by a factor of 3, the leading coefficient. The factors of -25 are ± 1, ± 5, and ± 25, and the factors of 3 are ± 1 and ± 3, so the possible rational roots are ± 1, ± 5, ± 25, $\pm\frac{1}{3}$, $\pm\frac{5}{3}$, or $\pm\frac{25}{3}$. The only one of these that looks like a possibility on the graph is $-\frac{5}{3}$. Try substituting $-\frac{5}{3}$ into the original polynomial.

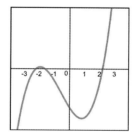

$$3\left(-\frac{5}{3}\right)^3 + 5\left(-\frac{5}{3}\right)^2 - 15\left(-\frac{5}{3}\right) - 25 = 0$$

Because the result is 0, you know that $-\frac{5}{3}$ is a root of the equation. If $-\frac{5}{3}$ is a root of the equation, then $\left(x + \frac{5}{3}\right)$ is a factor. Use a rectangle diagram or long division to divide out this factor.

Investigating Structure

This Investigating Structure explores the sum and difference of cubes. You might project it and have students work in small groups or you might assign it as homework. Have students present their solutions.

Summarize

As students present their solutions to the Examples, Investigating Structure, and Exercises, including explaining their strategy, emphasize the use of correct mathematical terminology and symbols. Be careful not to discourage thinking by talking too quickly or saying too much. Whether student responses are correct or incorrect, ask other students if they agree and why.
SMP 1, 2, 3, 5, 6

During the discussion, formalize students' observations with the terms *Remainder Theorem* and *Rational Root Theorem* and have students develop synthetic division as a shortcut for long division of polynomials.

Synthetic division simplifies long division, working with just the coefficients and not having to write variables repeatedly. As you look at the graphic, guide students to follow the arrows. You might also give students Calculator Note: SYNDIV Program, which includes a calculator program that performs synthetic division. **SMP 5**

Students may wonder whether formulas such as the quadratic formula exist for solving higher-degree equations. **CONTEXT** Formulas for solving general cubic and quartic (4th-degree polynomial) equations were found beginning in the 1500s, but no one was able to find a formula for solving all polynomial equations of higher degree in terms of sums, differences, products, quotients, and roots. In the mid-1820s, the young Norwegian mathematician Niels Abel proved that there can be no such formula for equations of any degree higher than 4.

$$(3x^3 + 5x^2 - 15x - 25) \div \left(x + \frac{5}{3}\right) = 3x^2 - 15$$

	$3x^2 + 0x - 15$		
x	$3x^3$	$0x^2$	$-15x$
$\frac{5}{3}$	$5x^2$	$0x$	-25

$$x + \frac{5}{3} \overline{\smash{)}\begin{array}{l} 3x^2 - 15 \\ 3x^3 + 5x^2 - 15x - 25 \\ \underline{3x^3 + 5x^2} \\ 0 - 15x - 25 \\ \underline{-15x - 25} \\ 0 \end{array}}$$

So $3x^3 + 5x^2 - 15x - 25 = 0$ is equivalent to $\left(x + \frac{5}{3}\right)(3x^2 - 15) = 0$. You already knew $-\frac{5}{3}$ was a root. Now solve $3x^2 - 15 = 0$.

$$3x^2 = 15$$
$$x^2 = 5$$
$$x = \pm\sqrt{5}$$

The three roots are $x = -\frac{5}{3}$, $x = \sqrt{5}$, and $x = -\sqrt{5}$. As decimal approximations, $-\frac{5}{3}$ is about -1.7, $\sqrt{5}$ is about 2.2, and $-\sqrt{5}$ is about -2.2. These values appear to be correct based on the graph.

Now that you know the roots, you could write the equation in factored form as $3\left(x + \frac{5}{3}\right)(x - \sqrt{5})(x + \sqrt{5}) = 0$, or $(3x + 5)(x - \sqrt{5})(x + \sqrt{5}) = 0$. You need the coefficient of 3 to make sure you have the correct leading coefficient in general form.

When you divide a polynomial by a linear factor, such as $\left(x + \frac{5}{3}\right)$, you can use a shortcut method called **synthetic division**. Synthetic division is simply an abbreviated form of long division that only works when the divisor is a linear factor $(x - a)$.

Consider this division of a cubic polynomial by a linear factor:

$$\frac{6x^3 + 11x^2 - 17x - 30}{x + 2}$$

Here are the procedures using a rectangle diagram, long division and synthetic division:

Rectangle Diagram

	$6x^2$	$-x$	-15
x	$6x^3$	$-1x^2$	$-15x$
2	$12x^2$	$-2x$	-30

Long Division

$$x + 2 \overline{\smash{)}\begin{array}{l} 6x^2 - 1x - 15 \\ 6x^3 + 11x^2 - 17x - 30 \\ (-)\ \underline{6x^3 + 12x^2} \\ -1x^2 - 17x \\ (-)\ \underline{-1x^2 - 2x} \\ -15x - 30 \\ (-)\ \underline{-15x - 30} \\ 0 \end{array}}$$

Synthetic Division

$$\begin{array}{r|rrrr} -2 & 6 & 11 & -17 & -30 \\ & & -12 & 2 & 30 \\ \hline & 6 & -1 & -15 & 0 \end{array}$$

$$6x^2 - 1x - 15$$

All of these methods give a quotient of $6x^2 - 1x - 15$. The corresponding numbers in long division and synthetic division are shaded. Notice that synthetic division contains all of the same information, but in a condensed form.

Here's how to do synthetic division:

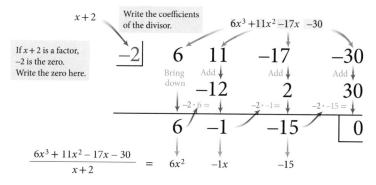

$$\frac{6x^3 + 11x^2 - 17x - 30}{x+2} = 6x^2 \quad -1x \quad -15$$

The number farthest to the right in the last row of a synthetic division problem is the remainder, which in this case is 0. When the remainder in a division problem is 0, you know that the divisor is a factor. This means -2 is a zero and the polynomial $6x^3 + 11x^2 - 17x - 30$ factors into the product of the divisor and the quotient, or $(x+2)(6x^2 - 1x - 15)$. You could now use any of the methods you've learned—simple factoring, the quadratic formula, synthetic division, or perhaps graphing—to factor the quotient even further. Synthetic division can **only** be used when the divisor is a linear factor. Both the rectangle diagram and long division are more general and can help you find the quotient using any polynomial divisor.

▲ INVESTIGATING STRUCTURE

The difference of two squares, $a^2 - b^2$, can be factored over the set of real numbers, $a^2 - b^2 = (a - b)(a + b)$. Note that you can use a rectangle diagram or polynomial long division to find one factor if you know the other factor.

	$a - b$	
a	a^2	$-ab$
b	ab	b^2

a. Complete this polynomial division to find the missing factor for the difference of two cubes, $a^3 - b^3 = (a - b)(\underline{?})$. $a^2 + ab + b^2$

	a	
a	a^3	
$-b$		$-b^3$

Hint: think about $a^3 - b^3$ as $a^3 + 0a^2b + 0ab^2 - b^3$

b. Find the missing factor for the sum of two cubes, $a^3 + b^3 = (a + b)(\underline{?})$. $a^2 - ab + b^2$

c. Use what you have learned to factor these polynomials.

  i. $x^3 - 8$ ii. $x^3 + 8y^3$ iii. $a^2 - b^2i^2$ iv. $27x^3 - 8y^3$

d. Summarize how to factor the difference and sum of two cubes using

  i. $x^3 - y^3$ ii. $x^3 + y^3$

  i. $x^3 - y^3 = (x - y)(x^2 + xy + y^2)$
  ii. $x^3 + y^3 = (x + y)(x^2 - xy + y^2)$

c. i. $(x - 2)(x^2 + 2x + 4)$
   ii. $(x + 2y)(x^2 - 2xy + 4y^2)$
   iii. $(a + bi)(a - bi)$
   iv. $(3x - 2y)(9x^2 + 6xy + 4y^2)$

CRITICAL QUESTION What use is the Rational Root Theorem?

BIG IDEA It allows you to guess the factors of a polynomial expression before checking your guess.

CRITICAL QUESTION How does synthetic division relate to a function's zeros and factors?

BIG IDEA If you've used the Rational Root Theorem to find one zero of the function, then you can use synthetic division to divide by the factor corresponding to that zero.

CRITICAL QUESTION Where have you seen examples of the Remainder Theorem before?

BIG IDEA The Remainder Theorem describes a correspondence between a function's zeros and its factors. Students saw that correspondence for quadratic functions.

Formative Assessment

With no investigation, this is a teacher-centered lesson. Listen for students' responses for anecdotal evidence of meeting the objectives. Consider having students present their solutions to the Exercises to get more insight in to their understanding.

Apply

Extra Example

The number 3 is a zero of the polynomial function
$y = x^4 - 7x^3 + 16x^2 - 12x$.

a. Using division and factoring, find the other zeros.
$x = 2, x = 2, x = 0$

b. Write the equation in factored form.
$y = x(x-3)(x-2)^2$

Closing Question

What is the remainder when $P(x) = 8x^5 + 3x^3 + 5x + 7$ is divided by $x + 4$? Is $x + 4$ a factor of $P(x)$? -8397; no

Assigning Exercises

Suggested: 1–13
Additional Practice: 14–18

2a. $3x^3 + 22x^2 + 38x + 15 = (x+5)(3x^2 + 7x + 3)$

2b. $6x^3 + 11x^2 - 19x + 6 = (3x-2)(2x^2 + 5x - 3)$

2c. $2x^4 - 7x^3 + 7x^2 - x - 2 = (x-2)(2x^3 - 3x^2 + x + 1)$

2d. $-3x^4 - 8x^3 + 7x^2 + 11x - 3 = (x+3)(-3x^3 + x^2 + 4x - 1)$

4a. $3x^3 - 11x^2 + 7x - 44 = (x-4)(3x^2 + x + 11)$

4b. $x^3 + 5x^2 - x - 21 = (x+3)(x^2 + 2x - 7)$

4c. $4x^3 - 8x^2 + 7x - 6 = (x-1.5)(4x^2 - 2x + 4)$

4d. $x^3 + 7x^2 + 11x - 4 = (x+4)(x^2 + 3x - 1)$

6a. The zeros so far are $2i$, $-2i$, and 2, and there are 2 more complex zeros.

6b. The factored form so far is
$p(x) = (x-2)(x-2i)(x+2i)(?)(?)$.

6c. The polynomial has a term of x^5, the graph has an x-intercept at $(2,0)$ and the graph will have at most 3 x-intercepts because of the 2 complex roots.

6d. The long run behavior is as x decreases, y increases and as x increases, y decreases.

Exercises

You will need a graphing calculator for Exercises **7, 8, 11,** and **15.**

Practice Your Skills

1. Find the missing polynomial in each division problem.

    a.
    $$\begin{array}{r} a3x^2 + 7x + 3 \\ x+5\overline{)3x^3 + 22x^2 + 38x + 15} \\ \underline{3x^3 + 15x^2} \\ 7x^2 + 38x \\ \underline{7x^2 + 35x} \\ 3x + 15 \\ \underline{3x + 15} \\ 0 \end{array}$$

    b.
    $$\begin{array}{r} 2x^2 + 5x - 3 \\ 3x-2\overline{)6x^3 + 11x^2 - 19x + 6} \\ \underline{b6x^3 - 4x^2} \\ 15x^2 - 19x \\ \underline{15x^2 - 10x} \\ -9x + 6 \\ \underline{-9x + 6} \\ 0 \end{array}$$

    c. $(2x^4 - 7x^3 + 7x^2 - x - 2) \div (x-2)$

	$2x^3$	$-3x^2$	x	1
x	$2x^4$	$-3x^3$	x^2	x
-2	$-4x^3$	$6x^2$	$-2x$	-2

    d. $(-3x^4 - 8x^3 + 7x^2 + 11x - 3) \div (x+3)$

	$-3x^3$	x^2	$4x$	-1
x	$-3x^4$	x^3	$4x^2$	$-x$
3	$-9x^3$	$3x^2$	$12x$	-3

2. Use the dividend, divisor, and quotient to rewrite each long-division problem in Exercise 1 as a factored product in the form $P(x) = D(x) \cdot Q(x)$. For example, $x^3 + 2x^2 + 3x - 6 = (x-1)(x^2 + 3x + 6)$. @

3. Find the missing value in each synthetic-division problem.

    a. $\begin{array}{r|rrrr} 4 & 3 & -11 & 7 & -44 \\ & & a & 4 & 44 \\ \hline & 3 & 1 & 11 & 0 \end{array}$ @
    $a = 12$

    b. $\begin{array}{r|rrrr} -3 & 1 & 5 & -1 & -21 \\ & & -3 & -6 & 21 \\ \hline & 1 & b & -7 & 0 \end{array}$
    $b = 2$

    c. $\begin{array}{r|rrrr} -1.5 & 4 & -8 & c & -6 \\ & & 6 & -3 & 6 \\ \hline & 4 & -2 & 4 & 0 \end{array}$ @
    $c = 7$

    d. $\begin{array}{r|rrrr} d & 1 & 7 & 11 & -4 \\ & & -4 & -12 & 4 \\ \hline & 1 & 3 & -1 & 0 \end{array}$
    $d = -4$

4. Use the dividend, divisor, and quotient to rewrite each synthetic-division problem in Exercise 3 as a factored product in the form $P(x) = D(x) \cdot Q(x)$. @

5. Make a list of the possible rational roots of each of the polynomial equations.

    a. $0 = 2x^3 + 3x^2 - 32x + 15$.
    $\pm 15, \pm 5, \pm 3, \pm 1, \pm\frac{15}{2}, \pm\frac{5}{2}, \pm\frac{3}{2}, \pm\frac{1}{2}$

    b. $0 = 2x^3 + 9x^2 + x - 12$
    $\pm 12, \pm 6, \pm 4, \pm 3, \pm 2, \pm 1, \pm 1.5, \pm 0.5$

Reason and Apply

6. Given $p(x)$ is a 5th degree polynomial with $p(2) = 0$ and $p(2i) = 0$.

    a. What can you say about the zeros of $p(x)$?

    b. What can you say about the factored form of $p(x)$?

    c. What can you say about the number of x-intercepts of $y = p(x)$?

    d. If the highest degree term of the expanded form is negative then what can you say about the long run behavior?

7. Division often results in a remainder. In each of these problems, use the polynomial that defines P as the dividend and the polynomial that defines D as the divisor. Write the result of the division in the form $P(x) = D(x) \cdot Q(x) + R$, where R is an integer remainder. For example, $x^3 + 2x^2 + 3x - 4 = (x - 1)(x^2 + 3x + 6) + 2$. ⓗ

   a. $P(x) = 47, D(x) = 11$ ⓐ $47 = 11 \cdot 4 + 3$
   b. $P(x) = 6x^4 - 5x^3 + 7x^2 - 12x + 15, D(x) = x - 1$ ⓐ $P(x) = (x - 1)(6x^3 + x^2 + 8x - 4) + 11$
   c. $P(x) = x^3 - x^2 - 10x + 16, D(x) = x - 2$ $P(x) = (x - 2)(x^2 + x - 8) + 0$

8. Consider the function $P(x) = 2x^3 - x^2 + 18x - 9$.

   a. Verify that $3i$ is a zero.
   $2(3i)^3 - (3i)^2 + 18(3i) - 9 = -54i + 9 + 54i - 9 = 0$
   b. Find the remaining zeros of the function P.
   $x = -3i$ and $x = \dfrac{1}{2}$

9. Consider the function $y = x^4 + 3x^3 - 11x^2 - 3x + 10$. ⓐ

   a. How many zeros does this function have?   4
   b. Name the zeros. $x = 1, x = 2, x = -5,$ and $x = -1$
   c. Write the polynomial function in factored form. $y = (x + 5)(x + 1)(x - 1)(x - 2)$

10. Factor to find the zeros of these polynomial functions.

    a. $y = x^3 - 5x^2 - 14x$
    $y = x(x - 7)(x + 2)$; zeros are 0, 7, and -2.
    b. $y = x^3 + 3x^2 - 28x - 60$ if one zero is 5
    $y = (x - 5)(x + 6)(x + 2)$; zeros are 5, -6, -2.

11. Use your list of possible rational roots from Exercise 5 to write the functions in factored form.
    a. $y = 2(x - 3)(x + 5)\left(x - \dfrac{1}{2}\right)$
    b. $y = 2(x - 1)(x + 1.5)(x + 4)$

12. Sketch a possible graph of the functions using what you know about the zeros and the end behavior of the functions.

    a. $y = (x - 2i)(x + 2i)(x - a)$, where $a > 0$
    b. $y = (x - 2)(x + 1)(x - 1)(x + b)$, where $b > 4$
    c. $y = x(x + 2)(x - c)$, where $c > 6$
    d. $y = (x + 6)(x + 2)(x - d)$, where $d > 4$
    e. $y = 2\left(x - \dfrac{1}{2}\right)(x - 3)(x + e)$, where $e > 4$

13. When you trace the graph of a function on your calculator to find the value of an x-intercept, you often see the y-value jump from positive to negative when you pass over the zero. By using smaller windows, you can find increasingly accurate approximations for x. Use your calculator to find good approximations for the zeros of these functions. Then use synthetic or long division to find any nonreal zeros.

    a. $y = x^5 - x^4 - 16x + 16$ ⓐ $x = \pm 2, x = 1,$ and $x = \pm 2i$
    b. $y = 2x^3 + 15x^2 + 6x - 6$ ⓐ    $x \approx -7.01, x \approx -0.943,$ and $x \approx 0.454$
    c. $y = 0.2(x - 12)^5 - 6(x - 12)^3 - (x - 12)^2 + 1$
    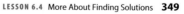 $x \approx 6.605, x \approx 12.501, x \approx 17.556,$ and $x \approx 11.669 \pm 0.472i$
    d. $y = 2x^4 + 2x^3 - 14x^2 - 9x - 12$
    $x \approx -3.033, x \approx 2.634,$ and $x \approx -0.3 \pm 0.813i$

LESSON 6.4 More About Finding Solutions   **349**

Exercise 7 [Alert] If students are having difficulty, you may want to work through the given example with the class. The degree of remainder R must be less than the degree of D, just as the remainder is less than the divisor when we divide numbers.

Exercise 8b [Ask] "Do you know more than one zero of this function?" [The conjugate of $3i$, $-3i$, is also a zero.] "Dividing through by an imaginary number can be difficult; is there an easier method?" [Dividing by $(x - 3i)(x + 3i)$, or $x^2 + 9$, is simpler because its coefficients are real.]

Exercise 10a [Alert] Students might be so focused on the coefficients that they neglect to realize that the polynomial is cubic with an x in each term.

Exercise 11 A graph might help students pick which rational roots are reasonable.

12. Answers will vary. These should be sketches that include the zeros and the long run behavior. Sample answers:

12a.

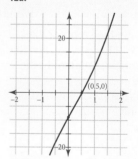

12b.

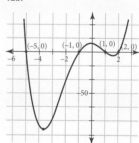

12c.

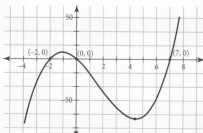

12d.

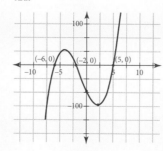

12e.

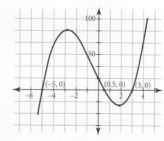

Exercise 13 As needed, impress upon students how difficult these problems would be without calculators.

LESSON 6.4 More About Finding Solutions   **349**

Exercise 16 Challenge students to rephrase each inequality as a statement about the real-world situation.

16a.

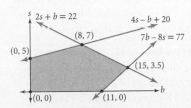

17a. $y = x^2 - 4x - 12$,
$y = (x - 6)(x + 2)$; vertex: $(2, -16)$;
y-intercept: -12; x-intercepts: $6, -2$

17b. $y = 3x^2 + 6x - 24$,
$y = 3(x - 2)(x + 4)$; vertex: $(-1, -27)$;
y-intercept: -24; x-intercepts: $2, -4$

17c. $y = -\frac{1}{2}x^2 + 5x + 12$,
$y = -\frac{1}{2}(x - 12)(x + 2)$; vertex: $\left(5, \frac{49}{2}\right)$;
y-intercept: 12; x-intercepts: $12, -2$

17d. $y = 2x^2 - 12x + 21$;
$y = 2\left(x - \frac{6 + i\sqrt{6}}{2}\right)\left(x - \frac{6 - i\sqrt{6}}{2}\right)$;
vertex: $(3, 3)$; y-intercept: 21;
x-intercept: none

Extension

Have students find the roots of the functions $y = x^3 - 1$, $y = x^4 - 1$, and $y = x^6 - 1$; graph the solutions to these functions on the complex plane; and make a conjecture about the roots of the general function $y = x^n - 1$.

Review

14. APPLICATION The relationship between the height and the diameter of a tree is approximately determined by the equation $f(x) = kx^{3/2}$, where x is the height in feet, $f(x)$ is the diameter in inches, and k is a constant that depends on the kind of tree you are measuring. @

   a. A 221 ft British Columbian pine is about 21 in. in diameter. Find the value of k, and use it to express diameter as a function of height. $f(x) = 0.00639x^{3/2}$

   b. Give the inverse function. $f^{-1}(x) \approx (156x)^{2/3}$

   c. Find the diameter of a 300 ft British Columbian pine. 33 in.

   d. What would be the height of a similar pine that is 15 in. in diameter? about 177 ft

15. Find a polynomial function of lowest possible degree whose graph passes through the points $(-2, -8.2)$, $(-1, 6.8)$, $(0, 5)$, $(1, -1)$, $(2, 1.4)$, and $(3, 24.8)$. $y = 2.1x^3 - 2.1x^2 - 6x + 5$

16. APPLICATION Sam and Beth have started a hat business in their basement. They make baseball caps and sun hats. Let b represent the number of baseball caps, and let s represent the number of sun hats. Manufacturing demands and machinery constraints confine the production per day to the feasible region defined by

$$\begin{cases} b \geq 0 \\ s \geq 0 \\ 4s - b \leq 20 \\ 2s + b \leq 22 \\ 7b - 8s \leq 77 \end{cases}$$

They make a profit of $2 per baseball cap and $1 per sun hat. @

   a. Graph the feasible region. Give the coordinates of the vertices.

   b. How many of each type of hat should they produce per day for maximum profit? (*Note:* At the end of each day, all partially made hats are recycled at no profit.) What is the maximum daily profit? 14 baseball caps and 4 sun hats; $33

17. Write each quadratic function in general form and in factored form. Identify the vertex, y-intercept, and x-intercepts of each parabola.

   a. $y = (x - 2)^2 - 16$ @

   b. $y = 3(x + 1)^2 - 27$ @

   c. $y = -\frac{1}{2}(x - 5)^2 + \frac{49}{2}$

   d. $y = 2(x - 3)^2 + 3$

18. Solve.

   a. $6x + x^2 + 5 = -4 + 4(x + 3)$ $x = -3$ or $x = 1$

   b. $7 = x(x + 3)$ $x = \dfrac{-3 \pm \sqrt{37}}{2}$

   c. $2x^2 - 3x + 1 = x^2 - x - 4$ $x = 1 \pm 2i$

LESSON 6.5

Introduction to Rational Functions

There's no point breaking a lot of crockery unnecessarily.
J. CARTER BROWN

You probably know that a lighter tree climber can crawl farther out on a branch than a heavier climber can, before the branch is in danger of breaking. What do you think the graph of (*length, mass*) data will look like when mass is added to a length of pole until it breaks? Is the relationship linear, like line A, or does it resemble one of the curves, B or C?

Engineers study problems like this because they need to know the weight that a beam can safely support. In the next investigation you will collect data and experiment with this relationship.

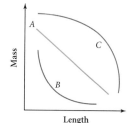

The Louise M. Davies Symphony Hall, built in 1980, is part of the Civic Center in San Francisco, California. The design of both the balcony and the covered entrance rely upon cantilevers—projecting beams that are supported at only one end.

INVESTIGATION
The Breaking Point

YOU WILL NEED
- several pieces of dry linguine
- a small film canister
- string
- some weights (pennies, beans, or other small units of mass)
- a ruler
- tape

Procedure Note

1. Lay a piece of linguine on a table so that its length is perpendicular to one side of the table and the end extends over the edge of the table.
2. Tie the string to the film canister so that you can hang it from the end of the linguine. (You may need to use tape to hold the string in place.)
3. Measure the length of the linguine between the edge of the table and the string. (See the photo on the next page.) Record this information in a table of (*length, mass*) data.
4. Place mass units into the container one at a time until the linguine breaks. Record the maximum number of weights that the length of linguine was able to support.

LESSON 6.5

In this lesson, students explore the parent rational function $y = \dfrac{1}{x}$.

COMMON CORE STATE STANDARDS

APPLIED	DEVELOPED	INTRODUCED
A.SSE.1a	A.APR.6	F.IF.7d
A.SSE.1b	A.APR.7	
	A.CED.1	
	A.CED.2	
	A.REI.2	

Objectives
- Investigate inverse variations
- Define *rational function*
- Graph and find the equations of transformations of the parent function $y = \dfrac{1}{x}$
- Use rational functions to solve problems

Vocabulary
inverse variation
rational function
hyperbola

Materials
- uncooked linguine
- small film canisters
- string
- small weights
- rulers
- tape
- Calculator Note: Asymptotes, Holes, and Drag Lines

Launch

Graph the parent function $y = \dfrac{1}{x}$ on your calculator. Use what you have learned about transformations to predict what the graph of these functions will look like.

a. $y = \dfrac{1}{x+1} - 2$

b. $y = \dfrac{3}{x}$

a. A translation of the graph of $y = \dfrac{1}{x}$ left 1 unit and down 2 units

b. A vertical stretch of the graph of $y = \dfrac{1}{x}$ by a factor of 3

ELL
Distinguish between *inverse variation* $\left(y = \dfrac{1}{x}\right)$ and the more general *rational function*. Use graphical examples. This is a new context for asymptotes, so relate asymptotes in hyperbolas to asymptotes in rational functions by stressing the mathematical definition.

Support
To help students better understand asymptotes, have them look at $y = \dfrac{1}{x}$ and think about 1 divided by larger and larger numbers, getting closer to 0, and 1 divided by smaller and smaller fractions, getting larger without bound. All other rational functions are transformations of $y = \dfrac{1}{x}$, so students generalize from there.

Advanced
Have students see if they can determine end behavior based on the degree of the polynomials in the numerator and denominator.

Investigate

Guiding the Investigation

In place of doing the investigation as an activity, you could use the desmos simulation found in your ebook. Or you can use the Investigation Worksheet with Sample Data. The answers given here are based on that sample data.

Test the procedure first. The type of weights and the size canister you choose will depend on the strength of the pasta. Flat pastas, such as linguine, work best. Pennies or equivalent masses are about the right weight. To prevent spilling of the mass units, you could cut a slot in the lid of the film canister, like a piggy bank, and then attach the lid.

Step 1 Caution students to make sure they hang the film canister the same distance from the end of each piece of pasta. Students should use a new piece of pasta, rather than part of the broken pasta, for each data point. Encourage students to collect data from a wide domain of lengths.

As they work, students may suggest that the amount of pasta on the table, not just the amount extending beyond the table, affects the breaking point. Let them check this hypothesis by breaking some of the pasta into shorter pieces and recording a third variable. **ASK** "What do you observe about the influence of the length of pasta on the table?"

Step 3 Students can use guess-and-check to try different models. The function $f(x) = 1 - x$ decreases but is linear. Students may also try an exponential function.

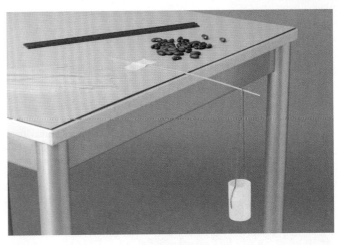

Step 2

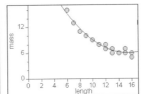

The graph does not appear to be linear. It is a curve that decreases quickly at first and then more slowly.

$f(0)$ is undefined.

Step 1 Work with a partner. Follow the Procedure Note to record at least five data points, and then compile your results with those of other group members.

Step 2 Make a graph of your data with length as the independent variable, x, and mass as the dependent variable, y. Does the relationship appear to be linear? If not, describe the appearance of the graph.

The relationship between length and mass is an **inverse variation**. The parent function for an inverse variation curve, $f(x) = \frac{1}{x}$, is the simplest **rational function**.

Step 3 Your data should fit a dilated version of the parent function $f(x) = \frac{1}{x}$. Write an equation that is a good fit for the plotted data.

$$y = \frac{90}{x}$$

> **Rational Function**
>
> A **rational function** is one that can be written as a quotient, $f(x) = \frac{p(x)}{q(x)}$, where $p(x)$ and $q(x)$ are both polynomial expressions. The denominator polynomial cannot equal the constant 0.

Rational functions can be transformed just like all the other functions you have previously studied. In the investigation you created a dilation of $f(x) = \frac{1}{x}$. This is an important rational function where $p(x) = 1$.

Graph the function $f(x) = \frac{1}{x}$ on your calculator and observe some of its special characteristics.

The graph is made up of two branches. One part occurs where x is negative and the other where x is positive. There is no value for this function when $x = 0$. What happens when you try to evaluate $f(0)$?

This graph is a **hyperbola**. It has vertices $(1, 1)$ and $(-1, 1)$, and its asymptotes are the x- and y-axes.

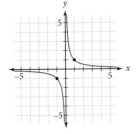

Modifying the Investigation

Whole Class Have two students demonstrate and collect data for the class. Complete Steps 2 and 3 with student input.

Shortened Use the sample data.

One Step There is no **One Step** for this lesson.

Conceptual/Procedural

Conceptual This lesson introduces the concept of a rational function.

Procedural In this lesson, students look at graphs of rational functions and transformations of the graphs.

To understand the behavior of the graph close to the axes, make a table with values of x very close to zero and very far from zero and examine the corresponding y-values. Consider these values of the function $f(x) = \frac{1}{x}$.

x	-1	-0.1	-0.01	-0.001	0	0.001	0.01	0.1	1
y	-1	-10	-100	-1000	undefined	1000	100	10	1

As x approaches zero from the negative side, the y-values have an increasingly large absolute value. Why?

So $x = 0$ is a vertical asymptote.

As x approaches zero from the positive side, the y-values have an increasingly large absolute value. Why?

The behavior of the y-values as x gets closer to zero shows that the y-axis is a vertical asymptote for this function.

x	-10000	-1000	-100	-10	0	10	100	1000	10000
y	-0.0001	-0.001	-0.01	-0.1	undefined	0.1	0.01	0.001	0.0001

As x takes on negative values farther from zero, the y-values approach zero.

So $y = 0$ is the long run behavior.

As x takes on larger positive values, the y-values approach zero.

As x approaches the extreme values at the left and right ends of the x-axis, the curve approaches the x-axis. The horizontal line $y = 0$, then, is called a horizontal asymptote.

If you think of $y = \frac{1}{x}$ as a parent function, then $y = \frac{1}{x} + 1$, $y = \frac{1}{x-2}$, and $y = 3\left(\frac{1}{x}\right)$ are examples of transformed rational functions. What happens to a function when x is replaced with $(x - 2)$? The function $y = \frac{1}{x-2}$ is shown below.

The graph is translated right 2 units.

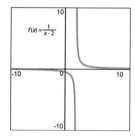

EXAMPLE Describe the function $f(x) = \frac{2x-5}{x-1}$ as a transformation of the parent function, $f(x) = \frac{1}{x}$. Then sketch a graph.

LESSON 6.5 Introduction to Rational Functions   **353**

ALERT Calculators often misrepresent discontinuous graphs of rational functions. They may display drag lines to fill in jumps, and they often can't display single-point gaps. Students will need to use reasoning to verify features of graphs.

ASK "Is it the numerator or denominator of a fraction that can't be zero?" As needed, remind students that a fraction with denominator 0 is undefined. "Why is this so?" [As the denominator gets closer to 0, you're dividing by smaller and smaller pieces, so the result gets infinitely large in absolute value.] Emphasize that a fraction with 0 numerator but nonzero denominator has value 0.

CRITICAL QUESTION What is a rational function?

BIG IDEA The expression for a rational function can be written as a ratio of polynomials, but it might also be written in other forms.

CRITICAL QUESTION What's the shape of the graph of the rational function $y = \dfrac{1}{x}$?

BIG IDEA The graph of $y = \dfrac{1}{x}$ is a hyperbola. In the general quadratic equation $Ax^2 + Bxy + Cy^2 + Dx + Ey + F = 0$, all coefficients are 0 except $B = 1$ and $F = -1$.

CRITICAL QUESTION $y = \dfrac{1}{x}$ is a parent function for what rational functions?

BIG IDEA Any rational function that is the ratio of linear functions is a transformation of $y = \dfrac{1}{x}$.

CRITICAL QUESTION If you translate $y = \dfrac{1}{x}$ horizontally by h and vertically by k and dilate it horizontally by a and vertically by b, do you get a rational function?

BIG IDEA Yes. The general equation would be $y = \dfrac{1}{b}\left(\dfrac{1}{a(x-h)} + k\right)$.

To verify, you might have students start with graphing in the desmos app in your ebook. They can add a new function with a value for h and compare it with the parent function. They can then repeat with a new function that adds a value for k to their second function, and continue until they have a graph of each transformation. This will review the effects of the transformation on the parent function and show that the resulting function is rational.

Solution You can change the form of the equation so that the transformations are more obvious. Because the numerator and denominator both have degree 1, you can use division to rewrite the expression.

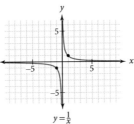 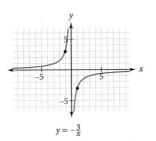

The remainder, -3, means the function is the same as $f(x) = 2 + \dfrac{-3}{x-1}$, or $f(x) = 2 - \dfrac{3}{x-1}$. In this form, you can see that the parent function has been dilated vertically by a factor of -3, then translated right 1 unit and up 2 units.

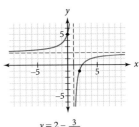

$y = \dfrac{1}{x}$
The parent rational function, $y = \dfrac{1}{x}$, has vertices $(1, 1)$ and $(-1, -1)$.

$y = -\dfrac{3}{x}$
A vertical dilation of -3 moves the vertices of the original hyperbola to $(1, -3)$ and $(-1, 3)$. The points $(3, -1)$ and $(-3, 1)$ are also on the curve. Notice that this graph looks more "spread out" than the graph of $y = \dfrac{1}{x}$.

$y = 2 - \dfrac{3}{x-1}$
A translation right 1 unit and up 2 units moves the vertices of the original hyperbola to $(2, -1)$ and $(0, 5)$. The asymptotes are also translated to $x = 1$ and $y = 2$.

Notice that the asymptotes have been translated. How are the equations of the asymptotes related to your final equation above?
The asymptotes are translated right 1 unit and up 2 units.

To identify an equation that will produce a given graph, reverse the procedure you used to graph Example A. You can identify translations by looking at the translations of the asymptotes. To identify scale factors, pick a point, such as a vertex, whose coordinates you would know after the translation of $f(x) = \dfrac{1}{x}$. Then find a point on the dilated graph that has the same x-coordinate. The ratio of the vertical distances from the horizontal asymptote to those two points is the vertical scale factor.

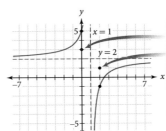

An undilated (but reflected) inverse variation function with these translations would have vertices $(2, 1)$ and $(0, 3)$, 1 horizontal unit and 1 vertical unit away from the center. Because the distance is now 3 vertical units from the center, include a vertical scale factor of 3 to get the equation $\dfrac{y-2}{3} = \dfrac{-1}{x-1}$, or $y = 2 - \dfrac{3}{x-1}$.

354 CHAPTER 6 Polynomial and Rational Functions

Formative Assessment

This curve in the investigation is a new one for students, so pay attention to their interpretations of the data they collect. If students have seen rational functions in earlier courses, watch to see what they remember. This section also deals with transformations and asymptotes, so pay attention to students' developing understanding of those ideas.

6.5 Exercises

You will need a graphing calculator for Exercises **1, 8,** and **9**.

Practice Your Skills

1. Write an equation and graph each transformation of the parent function $f(x) = \frac{1}{x}$.
   a. Translate the graph up 2 units. @
   b. Translate the graph right 3 units.
   c. Translate the graph down 1 unit and left 4 units.
   d. Vertically dilate the graph by a scale factor of 2.
   e. Horizontally dilate the graph by a factor of 3, and translate it up 1 unit. @

2. Which transformations in Exercise 1 resulted in:
   a. a shift in the vertical asymptote? b, c
   b. a shift in the horizontal asymptote? ⓗ a, c, d, e
   c. no shift in either the vertical or horizontal asymptote? none

3. Describe how each function was transformed from the parent function $f(x) = \frac{1}{x}$.
   a. $g(x) = \frac{1}{x+3} - 2$ @   b. $h(x) = \frac{5}{x-7}$ c. $j(x) = 3 - \frac{4}{x-1}$
   translated left 3 and down 2   translated right 7 and vertically dilated by a factor of 5

   3c. vertically dilated by a factor of -4 (or vertically dilated by a factor of 4 and vertically reflected), translated right 1 and up 3

4. As the rational function $y = \frac{1}{x}$ is translated, its asymptotes are translated also. Write an equation for the translation of $y = \frac{1}{x}$ that has the asymptotes described.
   a. horizontal asymptote $y = 2$ and vertical asymptote $x = 0$ @ $y = \frac{1}{x} + 2$
   b. horizontal asymptote $y = -4$ and vertical asymptote $x = 2$ $y = \frac{1}{x-2} - 4$
   c. horizontal asymptote $y = 3$ and vertical asymptote $x = -4$ @ $y = \frac{1}{x+4} + 3$

5. What are the equations of the asymptotes for each hyperbola?
   a. $y = \frac{2}{x} + 1$ @   b. $y = \frac{3}{x-4}$ c. $y = \frac{4}{x+2} - 1$ @   d. $y = \frac{-2}{x+3} - 4$
   $y = 1, x = 0$ $y = 0, x = 4$ $y = -1, x = -2$ $y = -4, x = -3$

Reason and Apply

6. Match the graphs of the rational functions with their equations.

   a. 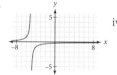 iv
   b.  ii
   c. 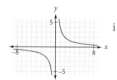 i
   d. 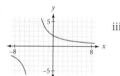 iii

   i. $f(x) = \frac{4}{x}$ ii. $f(x) = \frac{1}{x-6} - 2$ iii. $f(x) = \frac{9}{x+4}$ iv. $f(x) = \frac{-1}{x+5}$

LESSON 6.5 Introduction to Rational Functions **355**

1c. $f(x) = \frac{1}{x+4} - 1$

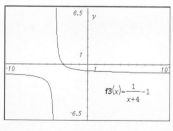

1d. $f(x) = 2\left(\frac{1}{x}\right)$, or $f(x) = \frac{2}{x}$

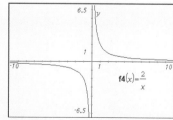

1e. $f(x) = 3\left(\frac{1}{x}\right) + 1$, or $f(x) = \frac{3}{x} + 1$

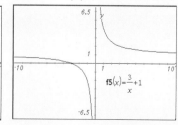

Apply

Extra Example
Write an equation for the translation of $y = \frac{1}{x}$ that has a horizontal asymptote $y = 1$ and vertical asymptote $x = -3$.
$y = \frac{1}{x+3} + 1$

Closing Question
Describe the function $f(x) = \frac{2}{x+1} + 4$ as a transformation of the parent function $f(x) = \frac{1}{x}$.
vertical dilation of 2, translated horizontally -1 unit and vertically 4 units

Assigning Exercises
Suggested: 1–9
Additional Practice: 10–15

1a. $f(x) = \frac{1}{x} + 2$

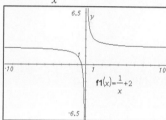

1b. $f(x) = \frac{1}{x-3}$

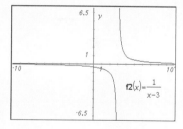

Exercise 4 It may be useful for students to sketch graphs of these situations.

Exercise 6 SUPPORT Provide students with a copy of the graphs so they can mark information such as the asymptotes and the points on the curve that may be used to determine the scale factor of the function.

7a.
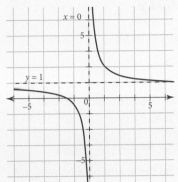

7b.
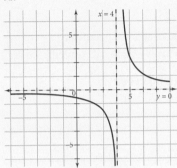

7c.
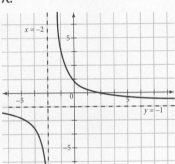

7d.
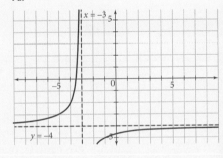

7. Sketch a graph of the functions in Exercise 5. Label the asymptotes.

8. Consider these functions.
   i. $y = \dfrac{2x - 13}{x - 5}$ $y = 2 + \dfrac{-3}{x - 5}$
   ii. $y = \dfrac{3x + 11}{x + 3}$ $y = 3 + \dfrac{2}{x + 3}$

   a. Rewrite each rational function to show how it is a transformation of $y = \dfrac{1}{x}$. (h)
   b. Describe the transformations of the graph of $y = \dfrac{1}{x}$ that will produce graphs of the equations in 8a.
   c. Graph each equation on your calculator to confirm your answers to 8b.

9. Graph the function $g(x) = \dfrac{2}{x - 3} + 1$ on your calculator.
   a. Describe the transformations of the graph of $f(x) = \dfrac{1}{x}$ that will produce the graph of $g(x)$. Dilate vertically by a factor of 2; translate right 3 and up 1.
   b. What are the domain and range of $f(x) = \dfrac{1}{x}$? domain: $x \neq 0$; range: $y \neq 0$
   c. What are the domain and range of $g(x) = \dfrac{2}{x - 3} + 1$? domain: $x \neq 3$; range: $y \neq 1$
   d. How does identifying the transformations help you determine the domain and range? The features of the domain and range (asymptotes) are also translated. The dilation has no effect on the domain and range.

Review

10. Factor each expression completely.
    a. $x^2 - 7x + 10$ $(x - 5)(x - 2)$
    b. $x^3 - 9x$ $x(x - 3)(x + 3)$

11. Write the equation of the circle with center $(2, -3)$ and radius 4. $(x - 2)^2 + (y + 3)^2 = 16$

12. A 2 m rod and a 5 m rod are mounted vertically 10 m apart. One end of a 15 m wire is attached to the top of each rod. Suppose the wire is stretched taut and fastened to the ground between the two rods. How far from the base of the 2 m rod is the wire fastened? 9.8 m

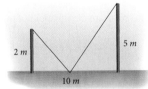

13. Write the general quadratic equations of two concentric circles with center $(6, -4)$ and radii 5 and 8. (h)
    $x^2 + y^2 - 12x + 8y + 27 = 0$, $x^2 + y^2 - 12x + 8y - 12 = 0$

14. Find the length of each labeled segment.

    a.   b.   c.
    $a = \sqrt{3}$ $b = 3\sqrt{3}$ $c = \dfrac{6}{\sqrt{3}} = 2\sqrt{3}$

15. The radius of the circle is 1. Find the coordinates of the labeled point.
    $\left(\dfrac{1}{\sqrt{2}}, \dfrac{1}{\sqrt{2}}\right)$, or $\left(\dfrac{\sqrt{2}}{2}, \dfrac{\sqrt{2}}{2}\right)$

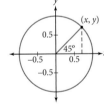

8b.
i. For $y = 2 + \dfrac{-3}{x - 5}$, dilate vertically by a factor of -3 and translate right 5 units and up 2 units.

ii. For $y = 3 + \dfrac{2}{x + 3}$, dilate vertically by a factor of 2 and translate left 3 units and up 3 units.

8c.
i. 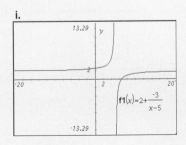   ii.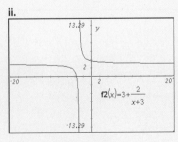

9.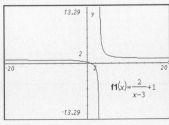

LESSON 6.6
Graphs of Rational Functions

> Besides learning to see, there is another art to be learned— to see what is not.
> — MARIA MITCHELL

Rational functions are ratios of polynomials so the long run behavior of the polynomials determines the long run behavior of the rational function.

$$r(x) = \frac{a_n x^n + a_{n-1} x^{n-1} + \cdots + a_1 x^1 + a_0}{b_m x^m + b_{m-1} x^{m-1} + \cdots + b_1 x^1 + b_0}.$$

Just as the long run behavior of a polynomial is based on its term of highest degree, the long run behavior of the rational function is based on the quotient of its polynomials highest degree terms.

Consider the rational function $r(x) = \frac{x^4 + 3x^3 - 11x^2 - 3x + 10}{2x^2 + 1}$. The numerator behaves like x^4 and the denominator behaves like $2x^2$, so $r(x)$ behaves like $\frac{x^4}{2x^2} = \frac{1}{2}x^2$.

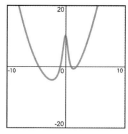
The graph of $r(x)$ in a window that is suitable to see the short run behavior.

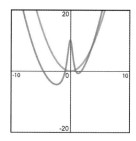
The graph of $r(x)$ and the power function $y = 0.5x^2$.

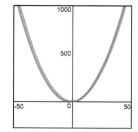
The graph of $r(x)$ and $y = 0.5x^2$ in a large scaled window where you can see that the graphs have very similar long run behavior.

Even though the long run behavior is much like power functions, the short run behavior is quite different.

Some rational functions create very different kinds of graphs from those you have studied previously. The graphs of these functions can have two or more parts. This is because the denominator, a polynomial function, may be equal to zero at some point, so the function will be undefined at that point.

Sometimes it's difficult to see the different parts of the graph because they may be separated by only one missing point, called a **hole**. At other times you will see two parts that look very similar—one part may look like a reflection or rotation of the other part. Or you might get multiple parts that look totally different from each other. Look for these features in the graphs shown here.

$$y = \frac{x^3 + 2x^2 - 5x - 6}{x - 2}$$

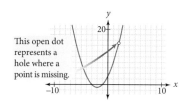
This open dot represents a hole where a point is missing.

LESSON 6.6

This lesson extends the study of rational functions beyond the family of functions with parent function $y = \frac{1}{x}$.

COMMON CORE STATE STANDARDS
APPLIED	DEVELOPED	INTRODUCED
	A.APR.6	F.IF.7d
	A.REI.2	
	F.IF.4	
	F.BF.3	

Objectives
- Identify characteristics of the graph of a rational function from its equation
- Learn to write the equation of a rational function from its graph
- Rewrite a function as a rational function
- Identify holes, vertical asymptotes, x-intercepts, y-intercepts, and horizontal asymptotes of rational functions

Vocabulary
hole

Launch
What are the restrictions on the variables in the following?

a. $y = \frac{x^2 - 9}{x(x + 3)}$ $x \neq -3, 0$

b. $y = \frac{3(x - 1)}{x - 1}$ $x \neq 1$

c. $y = \frac{x^3 + 2x^2 - 5x - 6}{x - 2}$ $x \neq 2$

What do you think might happen at that point on the graphs?

Answers may vary.

Don't expect the answer here. This is a lead-in to the lesson.

ELL
Start the lesson by discussing the functions covered so far in the book and students' strategies for graphing them. Encourage students to build techniques for graphing asymptotes and holes from those strategies.

Support
If you think your students will have difficulty with the factoring, you might do a few factoring exercises for review. Create and discuss a graphic organizer covering graphing rational functions and including factoring quadratics.

Advanced
Rather than having students proceed through the investigation step-by-step, use the **One Step** in your ebook.

Investigate

Guiding the Investigation

Geometry software is not appropriate for this investigation because it may not leave visible holes in the graphs. In place of doing the investigation as an activity, you could use the desmos simulation found in your ebook.

Step 4 ALERT Students may think that because the value $x = 2$ makes both the numerator and denominator equal 0, it indicates a hole. Point out that $y = \dfrac{x-2}{(x-2)^2}$ simplifies to $y = \dfrac{1}{x-2}$. When you simplify a rational expression, you may lose information about a value for which the denominator is undefined. In this case, the factor $(x-2)$ is still represented in the denominator, so $y = \dfrac{x-2}{(x-2)^2}$ is equivalent to $y = \dfrac{1}{x-2}$. This function has a vertical asymptote, not a hole.

Step 1a B; vertical asymptote $x = 2$. $x = 2$ makes the denominator 0 and this is a translation of the parent hyperbola $\dfrac{1}{x}$ horizontally 2 units.

Step 1b D; hole at $x = 2$. $x = 2$ makes both the numerator and denominator 0 and y simplifies to 1 for all $x \neq 2$.

Step 1c A; vertical asymptote $x = 2$. $x = 2$ makes the denominator 0, and the denominator is squared so that all values become positive.

Step 1d C; hole at $x = 2$. $x = 2$ makes both the numerator and denominator 0. If you simplify the equation, you get $y = x - 2$, which produces the graph shown, except you get a hole at $x = 2$.

Step 2a $y = \dfrac{1}{x+1}$. The graph has asymptote $x = -1$. Other than that, the equation looks like $y = \dfrac{1}{x}$.

Step 2b $y = \dfrac{2(x+1)}{x+1}$. There is a hole at $x = -1$, so the denominator is $x + 1$. However, when you simplify, you get the line $y = 2$, so there must be factors of 2 and an $x + 1$ in the numerator.

$y = \dfrac{1}{x-2}$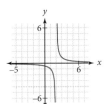

$y = \dfrac{x^3 - x^2 - 8x + 12}{6x^2 + 6x - 12}$
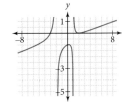

$y = \dfrac{1}{x^2 + 1}$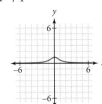

$y = \dfrac{x-1}{x^2 + 4}$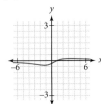

In this lesson you will explore local and end behavior of rational functions, and you will learn how to predict some of the features of a rational function's graph by studying its equation. When examining a rational function, you will often find it helpful to look at the equation in factored form.

INVESTIGATION
Predicting in the Short Run

In this investigation you will consider the graphs of four rational functions.

Step 1 Match each rational function with a graph. Investigate each graph by dragging a point on the graph, tracing, or looking at a table of values. Describe the unusual occurrences at exactly $x = 2$ and other values nearby. What features in the equation cause the different types of graph behavior? (You will not see an actual "hole," as pictured in graphs b and d.)

a. 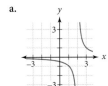 b. 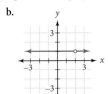 c. 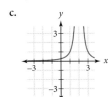 d.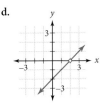

i. $y = \dfrac{1}{(x-2)^2}$ ii. $y = \dfrac{1}{x-2}$ iii. $y = \dfrac{(x-2)^2}{x-2}$ iv. $y = \dfrac{x-2}{x-2}$

358 CHAPTER 6 Polynomial and Rational Functions

Modifying the Investigation

Whole Class Display the four graphs. Use an overhead graphing calculator to graph and match the functions. Discuss each graph's behavior. Have students individually predict functions for the graphs in Step 2. Discuss Steps 2 through 4.

Shortened Discuss Steps 3 and 4 as a class.

One Step Display the function $y = \dfrac{x^3 + 3x^2 - x - 3}{x^2 + x - 6}$ and ask students to find intercepts and asymptotes exactly, using a calculator only to check their work approximately. Press faster groups to state a general method for finding intercepts and asymptotes of rational functions.

Step 2 Have each group member choose one of the graphs below. Find a rational function equation for your graph, and write a few sentences that explain the appearance of your graph. Share your answers with your group.

a.   b. 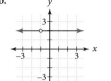   c.   d.

Step 3 Write a paragraph explaining how you can use an equation to predict where holes and asymptotes will occur, and how you can use these features in a graph to write an equation.

Step 4 Consider the graph of $y = \dfrac{x-2}{(x-2)^2}$. What features does it have? What can you generalize about the graph of a function that has a factor that occurs more times in the denominator than in the numerator?

In the investigation you saw that the graph of a rational function contains many clues about the equation. The next example shows how you can put that information to use.

EXAMPLE A | Write an equation for each graph.

a.   b.   c.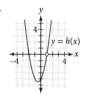

Solution | a. This graph looks like the line $y = 3$, but there is a hole when $x = 2$. This means the function is rational with the factor $(x - 2)$ in both the numerator and denominator. So its equation is $f(x) = \dfrac{3(x-2)}{(x-2)}$.

b. This graph has vertical asymptotes $x = -2$ and $x = 1$. So the denominator contains the factors $(x + 2)$ and $(x - 1)$. Also, the graph has x-intercept -1, which means the numerator contains the factor $(x + 1)$. Write the equation $g(x) = \dfrac{a(x+1)}{(x+2)(x-1)}$. To find a, check the value of the y-intercept by evaluating $g(0)$. It is $-\dfrac{1}{2}a$, and the graph shows y-intercept $-\dfrac{1}{2}$. So $a = 1$ and the final equation is $g(x) = \dfrac{(x+1)}{(x+2)(x-1)}$. It appears that the function approaches the line $y = 0$ as x increases without bound.

LESSON 6.6 Graphs of Rational Functions **359**

Conceptual/Procedural

Conceptual This lesson explores asymptotes and discontinuities in graphs of rational functions. Students describe features of graphs from the equation of a rational function.

Procedural Students write equations for graphs of rational functions. They rewrite equations of rational functions in fractional form.

Step 2c $y = \dfrac{1}{(x+1)^2}$. There is a vertical asymptote at $x = -1$, so the denominator includes $x + 1$. Because all y-values end up positive, the equation must have an even number of factors of $x + 1$ in the denominator.

Step 2d $y = \dfrac{(x+1)^2}{x+1}$. There is a hole at $x = -1$, so the denominator includes $x + 1$. When the equation is simplified, it must be $y = x + 1$ to produce the given graph, so the equation must be $y = \dfrac{(x+1)^2}{x+1}$.

Step 3 Paragraph should include the following:

- A vertical asymptote occurs at zeros that appear only in the denominator or at zeros that appear more times in the denominator than in the numerator.

- A hole occurs at values that make both the denominator and the numerator 0, provided there are no vertical asymptotes at these values.

Step 4 The graph has vertical asymptote $x = 2$. When a factor occurs in both the denominator and the numerator, but occurs more times in the denominator, it indicates a vertical asymptote rather than a hole.

Example A

You might project Example A from your ebook and assign each graph to a part of the class. Have students present their solutions, including explaining their strategy. **SMP 3, 5, 6**

As needed, **ASK** "How do you know there is a hole there?" or "How do you know that is the asymptote?" Emphasize that a zero in the numerator and denominator means a hole, unless there's a vertical asymptote there. Also emphasize that, after the common factors have been factored out, a 0 in the numerator indicates an x-intercept, whereas a 0 in the denominator indicates a vertical asymptote. As needed, tie this discussion to fractions. Bring out the difference between $\dfrac{0}{0}$, which is undefined because it can equal any real number, and $\dfrac{3}{0}$, which is undefined because it can equal no real number.

LESSON 6.6 Graphs of Rational Functions **359**

Example B

Project Example B from your ebook. Have students work in pairs. Ask students to conjecture about the features of the graph of the function's factored version before graphing. Ask about holes, asymptotes, and x- and y-intercepts. To help students organize their thinking, you might project a table with columns labeled graph feature, location, and equation feature. Horizontal asymptotes are a little more difficult to find. Students might substitute values of x far from 0 to investigate end behavior. Have students present their solutions, including explaining their strategy.
SMP 3, 5, 6

Example C

Project Example C and ask for input as to how to write the function in factored form. As needed, you might want to ask how to calculate a sum of rational numbers, such as $\frac{3}{2} + 5$. You could have students work the two expressions side by side to show the similarities in the solutions.

Summarize

As students present their solutions to the Examples, Investigation, and Exercises, including explaining their strategy, emphasize the use of correct mathematical terminology and symbols. Be careful not to discourage thinking by talking too quickly or saying too much. Whether student responses are correct or incorrect, ask other students if they agree and why.
SMP 1, 2, 3, 5, 6

---

c. This graph looks like a parabola with x-intercepts -2 and 1, so its basic equation is $h(x) = a(x+2)(x-1)$. But the graph has a hole at $x = 1$, so its equation is a rational function with the factor $(x - 1)$ in both the numerator and the denominator. Write the equation $h(x) = \frac{a(x+2)(x-1)^2}{(x-1)}$. To find a, check the value of the y-intercept. You find that $h(0) = -2a$. The graph has y-intercept -4, so the function needs to be dilated vertically by the scale factor $a = 2$. The final equation is $h(x) = \frac{2(x+2)(x-1)^2}{(x-1)}$.

Similar thinking can help you determine the features of a rational function graph before you actually graph it. Values that make the denominator or numerator equal to zero, the short run behavior of the polynomials, give you important clues about the appearance of the graph.

EXAMPLE B | Describe the features of the graph of $y = \frac{x^2 + 2x - 3}{x^2 - 2x - 8}$

Solution | To confirm the horizontal asymptotes, consider what happens to y-values as x-values get increasingly large in absolute value.

x	$-10{,}000$	$-1{,}000$	-100	100	$1{,}000$	$10{,}000$
y	0.9996001	0.9960129	0.9612441	1.0413603	1.0040131	1.0004001

A table shows that the y-values get closer and closer to 1 as x gets farther from 0. So, $y = 1$ is a horizontal asymptote. Note that for extreme positive values of x, y-values decrease to 1, whereas for extreme negative values of x, y-values increase to 1.

Short run features of rational functions are apparent when the numerator and denominator are factored.

$$y = \frac{x^2 + 2x - 3}{x^2 - 2x - 8} = \frac{(x+3)(x-1)}{(x-4)(x+2)}$$

This table summarizes how the equation can help you identify features in the graph.

Graph feature	Location	Equation feature
Holes	none	Common factors in numerator and denominator
x-intercepts	-3 and 1	Numerator (but not denominator) is 0 at these values
Vertical asymptotes	$x = 4$ and $x = -2$	Denominator (but not numerator) is 0 at these values
y-intercepts	0.375	y-value when $x = 0$
Horizontal asymptotes	$y = 1$	Long-run value of y (see table above)

A graph of the function confirms these features.

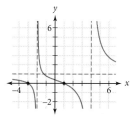

Rational functions can be written in different forms. The factored form is convenient for locating vertical asymptotes and intercepts. And you saw in the previous lesson how some rational functions can be written in a form that shows you clearly how the parent function has been transformed. You can use properties of arithmetic with fractions to convert from one form to another.

EXAMPLE C | Rewrite $f(x) = \dfrac{3}{x-2} - 4$ in fractional form.

Solution | The original form shows that this function is related to the parent function, $f(x) = \dfrac{1}{x}$. It has been vertically dilated by a factor of 3 and translated right 2 units and down 4 units. To change to fractional form, you must add the two parts to form a single fraction.

$$f(x) = \dfrac{3}{x-2} - 4 \quad \text{Original equation.}$$

$$= \dfrac{3}{x-2} - \dfrac{4}{1} \cdot \dfrac{x-2}{x-2} \quad \text{Create a common denominator of } (x-2).$$

$$= \dfrac{3}{x-2} - \dfrac{4x-8}{x-2} \quad \text{Rewrite second fraction.}$$

$$= \dfrac{3 - 4x + 8}{x-2} \quad \text{Combine the two fractions.}$$

$$f(x) = \dfrac{-4x + 11}{x-2} \quad \text{Add like terms.}$$

You can check this result by using a rectangle diagram or long division. The expression is indeed equivalent to the original form, which shows that the horizontal asymptote is $y = -4$.

	-4	
x	$-4x$	3
-2	8	

$$\begin{array}{r} -4 \\ x-2 \overline{\smash{)}-4x + 11} \\ \underline{-4x + 8} \\ 3 \end{array}$$

LESSON 6.6 Graphs of Rational Functions **361**

Discussing the Investigation

After students have presented ideas, ask, **ASK** "How would you describe how to predict, from a rational function expression, the holes, vertical asymptotes, and x-intercepts in its graph?" Students may conjecture that vertical asymptotes occur where the denominator is 0 and x-intercepts occur where the numerator is 0. You might put forth the example $y = \dfrac{(x+3)(x-1)}{(x-4)(x+3)}$ for testing the conjecture. Points at which both the numerator and the denominator are 0 indicate potential holes, or possibly vertical asymptotes. If the numerator and denominator share a common factor, there's a hole where that factor equals 0, unless there's a vertical asymptote there. Divide out the lower power of that common factor. Now examine factors in the numerator for x-intercepts and factors in the denominator for vertical asymptotes.

CRITICAL QUESTION How are the asymptotes and intercepts related to the graph of the parent function, $y = \dfrac{1}{x}$?

BIG IDEA The parent function $y = \dfrac{1}{x}$ has asymptotes $x = 0$ and $y = 0$ and no x-intercept. A translation right 2 units and down 4 units will move the asymptotes to $x = 2$ and $y = -4$. This translation to the right and down will introduce intercepts, but their exact values can't be seen easily from the translation.

CRITICAL QUESTION How would you describe the graph of the rational function whose parent function is $y = \dfrac{1}{x}$?

BIG IDEA It is a hyperbola.

Formative Assessment

As students present their solutions, watch to see that students are able to predict from an equation the holes and asymptotes in its graph and how the graph will appear as it approaches one of those discontinuities. Observe students' understanding of the graphs of the parent family $y = \dfrac{1}{x}$ and their ideas about asymptotes.

Apply

Extra Example

1. Write $y = \dfrac{x^2 + x - 2}{x^2 - 2x + 1}$ in factored form. Find the vertical asymptotes and the domain.

   $y = \dfrac{(x+2)(x-1)}{(x-1)^2}$; vertical asymptote: $x = 1$; domain: $x \neq 1$

2. Write a rational function with a hole at $x = 3$.
   Answers will vary. One possible solution:
   $y = \dfrac{x-3}{3x-9}$

Closing Question

Compare the graphs of $y = \dfrac{x+1}{x-2}$ and $y = \dfrac{(x+1)(x-1)}{(x-2)(x-1)}$.

They are identical, except the latter has a hole at $x = 1$.

Assigning Exercises

Suggested: 1–12
Additional Practice: 13–15

Exercise 2 It's important for students to make sure zeros of the denominator are not also zeros of the numerator before deciding that they mark vertical asymptotes.

Exercise 6 ASK "Can you simply divide out the common factors from the numerator and the denominator?" [The original function and the simplified function may not be equivalent if the common factor can equal 0; the graph of the function with common factors has a hole.]

Exercise 8 Encourage thinking rather than blind algebraic manipulation. For example, in 8a, encourage students to sketch by hand the graphs of the individual terms and to think about how the graphs might be added. The first term ensures that the horizontal asymptote is at $y = 3$. To the right of the vertical asymptote $x = 1$, the middle term contributes little, and to the left of the line $x = -2$, the last term contributes little. In between the vertical asymptotes, these two terms interact to make a U shape. The function in 8b is very similar to $y = \dfrac{2}{x+3} - 2$, which students can graph, but they need to add a hole at $x = 3$.

6.6 Exercises

You will need a graphing calculator for Exercises 6, and 8–11.

Practice Your Skills

1. Rewrite each rational expression in factored form.

   a. $\dfrac{x^2 + 7x + 12}{x^2 - 4}$ @ $\dfrac{(x+3)(x+4)}{(x+2)(x-2)}$

   b. $\dfrac{x^3 - 5x^2 - 14x}{x^2 + 2x + 1}$ $\dfrac{x(x-7)(x+2)}{(x+1)(x+1)}$

2. Identify the vertical asymptotes for each equation.

   a. $y = \dfrac{3}{x+1}$ $\quad x = -1$

   b. $y = \dfrac{x^2 + 7x + 12}{x^2 - 4}$ @ $x = 2$ and $x = -2$

   c. $y = \dfrac{x^3 - 5x^2 - 14x}{x^2 + 2x + 1}$ $\quad x = -1$

   d. $y = \dfrac{x^4}{x^2 - 1}$ $\quad x = 1$ and $x = -1$

3. What is the domain for each equation in Exercise 2? **3a.** $x \neq -1$ **3b.** $x \neq 2, x \neq -2$ **3c.** $x \neq -1$ **3d.** $x \neq 1, x \neq -1$

4. Add by finding a common denominator.

   a. $3 + \dfrac{2}{x-4}$ @ $\dfrac{3x-10}{x-4}$

   b. $2x + \dfrac{x-1}{x+3}$ $\dfrac{2x^2 + 7x - 1}{x+3}$

   c. $5 + 2x - \dfrac{3}{x+1}$ $\dfrac{2x^2 + 7x + 2}{x+1}$

5. Rewrite each expression in rational form (as the quotient of two polynomials). ⓗ

   a. $3 + \dfrac{4x-1}{x-2}$ @ $\dfrac{7x-7}{x-2}$

   b. $\dfrac{3x+7}{2x-1} - 5$ $\dfrac{-7x+12}{2x-1}$

6. Graph each equation on your calculator, and make a sketch of the graph on your paper. Indicate any holes on your sketches.

   a. $y = \dfrac{5-x}{x-5}$

   b. $y = \dfrac{3x+6}{x+2}$

   c. $y = \dfrac{(x+3)(x-4)}{x-4}$

   d. What causes a hole to appear in the graph? ⓗ A hole occurs when both the numerator and the denominator are 0 for a value of x and the zero appears the same or more times in the numerator than in the denominator.

Reason and Apply

7. Write an equation for each graph.

   a. @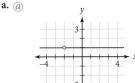
   $y = \dfrac{x+2}{x+2}$

   b.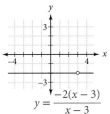
   $y = \dfrac{-2(x-3)}{x-3}$

   c.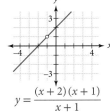
   $y = \dfrac{(x+2)(x+1)}{x+1}$

8. Graph each function on your calculator. List all holes and asymptotes.

   a. $y = 3 + \dfrac{1}{x+2} - \dfrac{1}{x-1}$

   b. $y = \dfrac{2}{x+3} + \dfrac{6-2x}{x-3}$

   c. $y = \dfrac{x}{x+2} - \dfrac{x}{x-2}$

362 CHAPTER 6 Polynomial and Rational Functions

6a.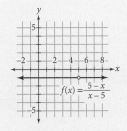
hole at $x = 5$

6b.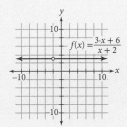
hole at $x = -2$

6c.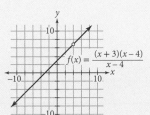
hole at $x = 4$

9. Write an equation for each graph.

a. $y = \dfrac{(x-1)(x+4)}{x(x+2)}$

b. $y = \dfrac{x+2}{(x+4)(x-2)}$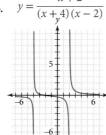

c. $y = \dfrac{(x+1)(x+2)(x-2)}{2(x-1)}$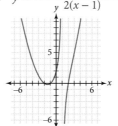

10. Write an equation that fits each description.

   a. vertical asymptotes $x = 2$ and $x = 5$, x-intercept -1, y intercept 1 $y = \dfrac{10(x+1)}{(x-2)(x-5)}$

   b. vertical asymptote $x = 4$, no x-intercept, horizontal asymptote $y = 2$, y-intercept 3 $y = \dfrac{16}{(x-4)^2} + 2$

   c. vertical asymptote $x = 1$, hole at $x = 3$, y-intercept 4 $y = \dfrac{-4(x-3)}{(x-3)(x-1)}$

11. Consider the equation $y = \dfrac{(x-1)(x+4)}{(x-2)(x+3)}$.

   a. Describe the features of the graph of this function.
   b. Describe the end behavior of the graph. The graph approaches horizontal asymptote $y = 1$.
   c. Sketch the graph.

12. Without technology, sketch a graph of the functions. Label asymptotes and holes.

   a. $f(x) = \dfrac{4x}{(x-4)}$
   b. $f(x) = \dfrac{(x-4)}{(x-4)(x+1)}$
   c. $f(x) = \dfrac{(x-4)}{x(x-4)}$
   d. $f(x) = \dfrac{x^2(x-4)}{(x-4)}$

Review

13. Find the points of intersection, if any, of the circle with center $(2, 1)$ and radius 5 and the line $x - 7y + 30 = 0$. $(-2, 4)$ and $(5, 5)$

14. A 500 g jar of mixed nuts contains 30% cashews, 20% almonds, and 50% peanuts.

   a. How many grams of cashews must you add to the mixture to increase the percentage of cashews to 40%? What is the new percentage of almonds and peanuts? $83\tfrac{1}{3}$ g; approximately 17% almonds and 43% peanuts

   b. How many grams of almonds must you add to the original mixture to make the percentage of almonds the same as the percentage of cashews? Now what is the percentage of each type of nut? 50 g; approximately 27.3% almonds, 27.3% cashews, and 45.5% peanuts

15. Solve each quadratic equation.

   a. $2x^2 - 5x - 3 = 0$ $x = -\tfrac{1}{2}$ or $x = 3$
   b. $x^2 + 4x - 4 = 0$ $x = -2 \pm 2\sqrt{2}$
   c. $x^2 + 4x + 1 = 0$ $x = -2 \pm \sqrt{3}$

8a.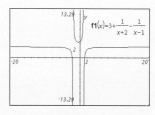
$x = -2, 1;\ y = 3$

8b.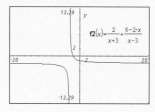
$x = -3,\ y = -2$, hole at $x = 3$

8c.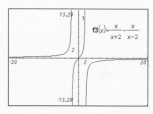
$x = -2, 2;\ y = 0$

10. Answers will vary. The sample answers use the lowest-degree numerator and denominator possible given the constraints.

11a. The graph has x-intercepts 1 and -4, vertical asymptotes $x = 2$ and $x = -3$, and y-intercept $\dfrac{2}{3}$.

11c.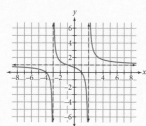

12a.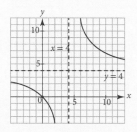

12b.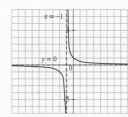

12c.

12d.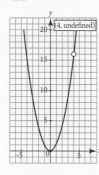

LESSON 6.7

In this lesson students extend the arithmetic operations of addition and subtraction to rational functions.

COMMON CORE STATE STANDARDS

APPLIED	DEVELOPED	INTRODUCED
	A.APR.2	
	A.APR.6	

Objectives

- Learn to add and subtract rational expressions
- Learn about slant asymptotes
- Simplify rational expressions in order to find x-intercepts, vertical asymptotes and holes

Vocabulary
slant asymptote

Launch

Add the following. Explain your process.

a. $\dfrac{2}{9} + \dfrac{4}{9} \quad \dfrac{2}{3}$

b. $\dfrac{3}{10} + \dfrac{7}{15} \quad \dfrac{23}{30}$

c. $\dfrac{4x+2}{5x^2} + \dfrac{6x+3}{5x^2} \quad \dfrac{2x+1}{x^2}$

d. $\dfrac{3}{x+1} + \dfrac{2}{3x} \quad \dfrac{11x+2}{3x(x+1)}$

How would you subtract rational expressions?

Rewrite the expressions with a common denominator and subtract numerators.

Investigate

Example A

Project Example A and have students work in pairs. Have pairs present their answer, explaining each step in their solution. **SMP 3, 6**

ASK "Why is it important to factor the end result?" [From the factored form, you can identify holes, x-intercepts, and vertical asymptotes.]

LESSON 6.7

Adding and Subtracting Rational Expressions

Images/split the truth/in fractions.
DENISE LEVERTOV

In this lesson you will learn to add and subtract rational expressions. In the previous lesson you combined a rational expression with a single constant or variable by finding a common denominator. That process was much like adding a fraction to a whole number. Likewise, all of the other arithmetic operations you will do with rational expressions have their counterparts in working with fractions. Keeping the operations with fractions in mind will help you understand the procedures.

Recall that to add $\dfrac{3}{10} + \dfrac{7}{15}$, you need a common denominator. The smallest number that has both 10 and 15 as factors is 30. So use 30 as the common denominator.

$\dfrac{3}{10} + \dfrac{7}{15}$ Original expression.

$\dfrac{3}{10} \cdot \dfrac{3}{3} + \dfrac{7}{15} \cdot \dfrac{2}{2}$ Multiply each fraction by an equivalent of 1 to get a denominator of 30.

$\dfrac{9}{30} + \dfrac{14}{30}$ Multiply.

$\dfrac{23}{30}$ Add.

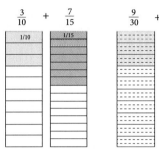

You could use other numbers, such as 150, as a common denominator, but using the least common denominator keeps the numbers as small as possible and eliminates some of the reducing the number of common factors, or simplifying, afterward. Recall that you can find the least common denominator by factoring the denominators to see which factors they share and which factors are unique to each one. In the example above, 10 factors to 2 · 5 and 15 factors to 3 · 5. The least common denominator must include factors that multiply to give each denominator, with no extras. So in this case you need a 2, a 3, and a 5. You can use this same process to add two rational expressions.

EXAMPLE A | Add the rational expressions and write the sum as a single rational expression.

$\dfrac{x-1}{(x-2)(x+1)} + \dfrac{x}{(x+1)(x-1)}$

ELL
Define the term *least common denominator*. Break the term into individual words so that students understand the overall term. Reinforce students' understanding of *numerator* and *denominator* throughout the discussion.

Support
Practice operations using basic fractions before beginning this lesson. Stress the need for a common denominator when adding and subtracting. Emphasize the instances when students can simplify factored numerators and denominators before introducing rational expressions with polynomial terms.

Advanced
Have students make up some rational expressions that have holes, vertical asymptotes, and/or x-intercepts. Have them experiment with adding and subtracting them, and describe where holes, vertical asymptotes, and x-intercepts will be in the results. **ASK** "Is it true that if a rational expression has a hole at $x = a$, then its sum with any other rational expression has a hole where $x = a$?"

Solution | The common denominator contains the factors of each denominator. Here the common denominator will contain $(x-2)$, $(x+1)$, and $(x-1)$, so it can be written $(x-2)(x+1)(x-1)$.

$\dfrac{x-1}{(x-2)(x+1)} + \dfrac{x}{(x+1)(x-1)}$	Original expression.
$\dfrac{(x-1)}{(x-2)(x+1)} \cdot \dfrac{(x-1)}{(x-1)} + \dfrac{x}{(x+1)(x-1)} \cdot \dfrac{(x-2)}{(x-2)}$	Multiply by an equivalent of 1 to get a common denominator.
$\dfrac{x^2-2x+1}{(x-2)(x+1)(x-1)} + \dfrac{x^2-2x}{(x-2)(x+1)(x-1)}$	Multiply factors in the numerators.
$\dfrac{2x^2-4x+1}{(x-2)(x+1)(x-1)}$	Since you have a common denominator, add the numerators and combine like terms.

The numerator cannot be factored, so the expression cannot be factored further.

To verify your results, try substituting a value for x into the original expressions and the final expression. We'll try $x = 4$.

Original expressions: $\dfrac{4-1}{(4-2)(4+1)} + \dfrac{4}{(4+1)(4-1)} = \dfrac{3}{2 \cdot 5} + \dfrac{4}{5 \cdot 3} = \dfrac{3}{10} + \dfrac{4}{15}$

Final expression: $\dfrac{2(4)^2 - 4(4) + 1}{(4-2)(4+1)(4-1)} = \dfrac{32 - 16 + 1}{2 \cdot 5 \cdot 3} = \dfrac{17}{30}$

So this problem is a generalization of the problem with fractions that opened this lesson.

Earlier you learned what to look for to graph a single rational function. If a function equation contains more than one rational expression, you must combine the terms into a single expression in order to use those graphing clues you learned.

EXAMPLE B | Write the difference as a single rational expression.

$$\dfrac{x+2}{(x-3)(x+4)} - \dfrac{5}{x+1}$$

Solution | Begin by finding a common denominator so that you can write the expression as a single rational expression in factored form. The common denominator is $(x-3)(x+4)(x+1)$.

$\dfrac{(x+2)}{(x-3)(x+4)} \cdot \dfrac{(x+1)}{(x+1)} - \dfrac{5}{(x+1)} \cdot \dfrac{(x-3)(x+4)}{(x-3)(x+4)}$	Multiply each fraction by an equivalent of 1 to get a common denominator.
$\dfrac{x^2+3x+2}{(x-3)(x+4)(x+1)} - \dfrac{5(x^2+x-12)}{(x+1)(x-3)(x+4)}$	Multiply the factors in the numerators.
$\dfrac{x^2+3x+2-5(x^2+x-12)}{(x-3)(x+4)(x+1)}$	Since you have the same denominator, subtract the numerators.
$\dfrac{x^2+3x+2-5x^2-5x+60}{(x-3)(x+4)(x+1)}$	Use the distributive property.
$\dfrac{-4x^2-2x+62}{(x-3)(x+4)(x+1)}$	Combine like terms.

Example B

Project Example B and have students work in pairs. Have pairs present their answer, explaining each step in their solution. **SMP 3, 6**

In the next-to-last step, **ASK** "Why is the expression in the numerator $-5x + 60$ instead of $+5x - 60$?" [The negative sign has been distributed.]

ASK "Why are there no holes?" [The numerator and denominator have no polynomial factors in common.]

Conceptual/Procedural

Conceptual Students review the concept that, like numerical fractions, it is necessary to find a common denominator to add or subtract rational expressions. They also investigate slant asymptotes.

Procedural Students learn to add and subtract rational expressions.

Investigating Structure

You might use this as an investigation for this lesson. Project the Investigating Structure from your ebook and have students work in small groups. As students present their solutions, have them explain their strategy. **SMP 1, 3, 5, 6, 7**

ELL Define *slant*. **ASK** "Can you tell from its fractional form when a rational function might have a slant asymptote?" [If the numerator has degree 1 more than the degree of the denominator, then when long division is applied, the result will have the form indicating a slant asymptote.]

Summarize

As students present their solutions to the Examples, Investigating Structure, and Exercises, including explaining their strategy, emphasize the use of correct mathematical terminology and symbols. Be careful not to discourage thinking by talking too quickly or saying too much. Whether student responses are correct or incorrect, ask other students if they agree and why. **SMP 1, 2, 3, 5, 6**

CRITICAL QUESTION Why do the rules for adding and subtracting fractions apply to rational algebraic expressions?

BIG IDEA The variables represent numbers.

CRITICAL QUESTION What is $\frac{a}{c} + \frac{b}{c}$? What is $\frac{a}{c} - \frac{b}{d}$?

BIG IDEA $\frac{a}{c} + \frac{b}{c} = \frac{a+b}{c}$

$\frac{a}{c} - \frac{b}{d} = \frac{ad - bc}{cd}$

Formative Assessment

Assess students' skills at manipulating fractions and factoring polynomials. Student answers for Examples A and B will provide evidence of the ability to add and subtract rational expressions. The Investigating Structure provides evidence of understanding slant asymptote.

Check to see whether the numerator factors. There is a common factor of -2, so you can rewrite the numerator.

$$\frac{-2(2x^2 + x - 31)}{(x-3)(x+4)(x+1)}$$ Factor the numerator.

No further factoring is possible.

▲ INVESTIGATING STRUCTURE

The graph of $y = x + \frac{1}{x}$ has a vertical asymptote at $x = 0$ and a slant asymptote, $y = x$.

The graph of $y = x - 1 + \frac{2}{x+1}$ has a slant asymptote, $y = x - 1$.

The graph of $y = 2x + 1 - \frac{1}{x-1}$ has a slant asymptote, $y = 2x + 1$.

In previous lessons, you saw many examples of graphs with vertical and horizontal asymptotes, including the parent function $f(x) = \frac{1}{x}$. Are there other types of asymptotes?

Investigate by graphing the functions $y = x + \frac{1}{x}$, $y = x - 1 + \frac{2}{x+1}$, and $y = 2x + 1 - \frac{1}{x-1}$.

These functions have a **slant asymptote**. Find the equations of the slant asymptotes and look for a pattern to help you write the equation for the slant asymptote for the equation $y = ax + b + \frac{c}{x-d}$. $y = ax + b$

The rules for the addition and subtraction of rational expressions are the same as for regular fraction arithmetic. First, rewrite the rational expressions as equivalent expressions with a common denominator. Then you can add or subtract the numerators. Simplify your answer by dividing out the common factors.

6.7 Exercises

You will need a graphing calculator for Exercises **5–7**.

1a. $\frac{x(x+2)}{(x-2)(x+2)} = \frac{x}{x-2}$

1b. $\frac{(x-1)(x-4)}{(x+1)(x-1)} = \frac{x-4}{x+1}$

1c. $\frac{3x(x-2)}{(x-4)(x-2)} = \frac{3x}{x-4}$

1d. $\frac{(x+5)(x-2)}{(x+5)(x-5)} = \frac{x-2}{x-5}$

Practice Your Skills

1. Factor the numerator and denominator of each expression completely, and simplify by reducing the number of common factors.

   a. $\frac{x^2 + 2x}{x^2 - 4}$ @   b. $\frac{x^2 - 5x + 4}{x^2 - 1}$ c. $\frac{3x^2 - 6x}{x^2 - 6x + 8}$ @   d. $\frac{x^2 + 3x - 10}{x^2 - 25}$

2. What is the least common denominator for each pair of rational expressions? Don't forget to factor the denominators completely first.

   a. $\frac{x}{(x+3)(x-2)}, \frac{x-1}{(x-3)(x-2)}$ @   b. $\frac{x^2}{(2x+1)(x-4)}, \frac{x}{(x+1)(x-2)}$

   c. $\frac{2}{x^2 - 4}, \frac{x}{(x+3)(x-2)}$ @   d. $\frac{x+1}{(x-3)(x+2)}, \frac{x-2}{x^2 + 5x + 6}$ $(2x+1)(x-4)(x+1)(x-2)$

   $(x+2)(x-2)(x+3)$ $(x-3)(x+2)(x+3)$

3. Add or subtract as indicated. Simplify by dividing any common factors.

   a. $\frac{x}{x+1} + \frac{x+2}{x+1}$ @   2   b. $\frac{x}{x+1} - \frac{x+2}{x+1}$ @ $\frac{-2}{x+1}$

   c. $\frac{x^2 - 6}{x^2 - 4} - \frac{x}{x^2 - 4}$ $\frac{x-3}{x-2}$ d. $\frac{x^2 - 6}{x^2 - 4} + \frac{x}{x^2 - 4}$ $\frac{x+3}{x+2}$

4. Add or subtract as indicated. **4a.** $\dfrac{(2x-3)(x+1)}{(x+3)(x-2)(x-3)}$

   a. $\dfrac{x}{(x+3)(x-2)} + \dfrac{x-1}{(x-3)(x-2)}$ @   b. $\dfrac{2}{x^2-4} - \dfrac{x}{(x+3)(x-2)}$ @ $\dfrac{-x^2+6}{(x+2)(x+3)(x-2)}$

   c. $\dfrac{x+1}{(x-3)(x+2)} + \dfrac{x-2}{x^2+5x+6}$ d. $\dfrac{2x}{(x+1)(x-2)} - \dfrac{3}{x^2-1}$ $\dfrac{2x^2-5x+6}{(x+1)(x-2)(x-1)}$

   4c. $\dfrac{2x^2-x+9}{(x-3)(x+2)(x+3)}$

Reason and Apply

5. Graph $y = \dfrac{x+1}{x^2-7x-8} - \dfrac{x}{2(x-8)}$ on your calculator.   **5a.** vertical asymptote: $x = 8$; horizontal asymptote: $y = -0.5$; hole at $x = -1$; y-intercept: -0.125; x-intercept: 2

   a. List all asymptotes, holes, and intercepts based on your calculator's graph.
   b. Rewrite the right side of the equation as a single rational expression. $\dfrac{2-x}{x(x-8)}$
   c. Use your answer from 5b to verify your observations in 5a. Explain.

6. Consider the equation $y = \dfrac{x-3}{x^2-4}$.

   a. Without graphing, identify the zeros and asymptotes of the graph of the equation. Explain your methods. ⓗ
   b. Verify your answers by graphing the function.

7. Consider the equation $y = 1 + \dfrac{1}{x-1} + \dfrac{2}{x-2}$.

   a. Without graphing, name the asymptotes of the function. ⓗ $x=1, x=2, y=1$
   b. Rewrite the equation as a single rational function. @ $\dfrac{x^2-2}{(x-1)(x-2)}$
   c. Sketch a graph of the function without using your calculator.
   d. Confirm your work by graphing the function with your calculator.

Review

8. This graph is the image of $y = \dfrac{1}{x}$ after a transformation.

   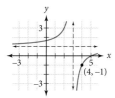

   a. Write an equation for each asymptote. $x=3, y=1$
   b. What translations are involved in transforming $y = \dfrac{1}{x}$ into its image? @   translation right 3 units and up 1 unit
   c. The point $(4, -1)$ is on the image. What is the vertical scale factor in the transformation? -2
   d. Write an equation of the image. $y = 1 - \dfrac{2}{x-3}$ or $y = \dfrac{x-5}{x-3}$
   e. Name the intercepts. x-intercept: 5; y-intercept: $\dfrac{5}{3}$

9. Factor each expression completely.

   a. $x^2 + 9x + 8$ b. $x^2 - 9$ c. $x^2 - 7x - 18$ d. $3x^3 - 15x^2 + 18x$
   $(x+1)(x+8)$ $(x-3)(x+3)$ $(x-9)(x+2)$ $3x(x-2)(x-3)$

LESSON 6.7 Adding and Subtracting Rational Expressions **367**

6b. 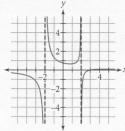 **7c.** 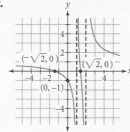 **7d.**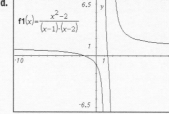

Apply

Extra Example

Rewrite as a single rational expression:
$$\dfrac{x+3}{x^2+2x} - \dfrac{x}{(x+2)(x+3)}$$
$$\dfrac{3(2x+3)}{x(x+2)(x+3)}$$

Closing Question

Solve these equations.

a. $\dfrac{3}{x+2} + \dfrac{4}{x+5} = \dfrac{9}{x^2+7x+10}$

b. $\dfrac{4x^2}{2x^2-x} - \dfrac{2}{2x-1} = \dfrac{1}{x}$

No solution. Standard manipulation of each equation produces a result that doesn't check because it makes at least one denominator 0.

Assigning Exercises

Suggested: 1–7
Additional Practice: 8, 9

5.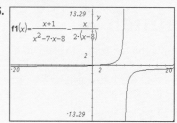

5c. The expression is undefined at $x = 8$, and the numerator is nonzero at $x = 8$, causing the vertical asymptote. The factor $x + 1$ was divided out, and there are no more factors of $x + 1$ in the denominator after simplifing; this causes a hole. When $x = 0$, the expression has the value $-\dfrac{1}{8}$, or -0.125, the y-intercept. The numerator is 0 when $x = 2$; this is the x-intercept. For large values of x, the expression approaches the value -0.5, the horizontal asymptote.

6a. $x = 3$ is a zero, because that value causes the numerator to be 0 and doesn't make the denominator 0. The vertical asymptotes are $x = 2$ and $x = -2$, because these values make the denominator 0 and do not also make the numerator 0. The horizontal asymptote is $x = 0$, because this is the value that y approaches when $|x|$ is large.

LESSON 6.8

Multiplying and Dividing Rational Expressions

In this lesson students extend the arithmetic operations of multiplication and division to rational functions.

COMMON CORE STATE STANDARDS

APPLIED	DEVELOPED	INTRODUCED
	A.APR.2	
	A.APR.6	

Objectives

- Learn to multiply and divide rational expressions
- Understand the "invert and multiply" algorithm
- Determine if a simplified rational expression is always equivalent to the original expression.

Launch

Multiply the following. Explain your process.

a. $\dfrac{2}{3} \cdot \dfrac{1}{3} \quad \dfrac{2}{9}$

b. $\dfrac{3}{8} \cdot \dfrac{1}{2} \quad \dfrac{3}{16}$

c. $\dfrac{3}{5} \cdot \dfrac{1}{15} \quad \dfrac{1}{25}$

d. $\dfrac{3x^2}{2} \cdot \dfrac{4}{x} \quad 6x$

How would you divide rational expressions?

Answers may vary: invert and multiply, multiply by the reciprocal, etc.

To help students understand how and why the commutative property can be used to simplify when multiplying rational expressions, spend time discussing the area models showing that $\dfrac{3}{5} \cdot \dfrac{2}{9} = \dfrac{2}{5} \cdot \dfrac{3}{9}$.

Compare it to the area model showing that $\dfrac{x+1}{x+4} \cdot \dfrac{x+2}{2x+1} = \dfrac{x+2}{x+4} \cdot \dfrac{x+1}{2x+1}$.

Nothing is particularly hard if you divide it into small jobs.

— HENRY FORD

In the previous section, you learned to add and subtract rational expressions by finding a common denominator. In this section, you will learn to multiply and divide rational expressions. Rational expressions are like numerical fractions, so multiplication and division of rational expressions is done much the same as multiplication and division of other fractions.

Like numerical fractions, you don't need a common denominator to multiply rational expressions. The rectangle below shows the product of $\dfrac{3}{5}$ and $\dfrac{2}{9}$. The rectangle is divided into $5 \cdot 9$, or 45 parts. The shaded area shows $3 \cdot 2$, or 6, of those parts. Thus $\dfrac{3}{5} \cdot \dfrac{2}{9} = \dfrac{6}{45}$.

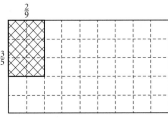

The denominators form a 5×9 outer array. The numerators form a 3×2 inner array. The area of the shaded $\dfrac{3}{5} \times \dfrac{2}{9}$ array is $\dfrac{6}{45}$, or $\dfrac{2}{15}$. You can factor and use the commutative property to simplify.

$$\dfrac{3}{5} \cdot \dfrac{2}{3 \cdot 3} = \dfrac{3}{3} \cdot \dfrac{2}{5 \cdot 3} = 1 \cdot \dfrac{2}{15} = \dfrac{2}{15}$$

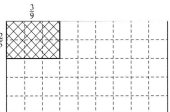

Another way to look at this is to rotate the inner array. Notice that

$$\underbrace{\dfrac{3}{5} \cdot \dfrac{2}{9}}_{\text{the original}} = \underbrace{\dfrac{2}{5} \cdot \dfrac{3}{9}}_{\text{the rotated}} = \dfrac{2}{5} \cdot \dfrac{1}{3} = \dfrac{2}{15}$$

Can you find the equivalent $\dfrac{2}{15}$ and $\dfrac{6}{45}$?

Rotating the inner array to obtain an equivalent expression is actually using the commutative property with the numerators, that is, $a \cdot b = b \cdot a$.

With rational expressions, a similar open array model can be used. The pictures below of a rectangle shows the product of $\dfrac{x+2}{2x+1}$ and $\dfrac{x+1}{x+4}$.

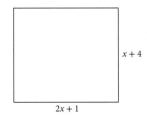

Start with an open array that is $(x+4)$ by $(2x+1)$.

368 CHAPTER 6 Polynomial and Rational Functions

ELL

Reinforce the meaning of restrictions on the variable. Review the definition of equivalent expressions, focusing on equivalent rational expressions.

Support

Practice operations using basic fractions before beginning this lesson. Stress that there is no need for a common denominator when multiplying and dividing. Complex rational expressions may be confusing for some students. Work numerical examples before the rational expressions.

Advanced

Use the **One Step** version of the Investigation. Ask students to provide descriptions and sketches of the graphs of functions once they have been written in factored form.

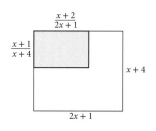

The ratio of $\frac{x+1}{x+4}$ can be represented by the part out of the whole vertical side. The ratio of $\frac{x+2}{2x+1}$ can be represented by the part out of the whole horizontal side. The area of the shaded region is
$$\frac{x+1}{x+4} \cdot \frac{x+2}{2x+1} = \frac{(x+1)(x+2)}{(x+4)(2x+1)}$$

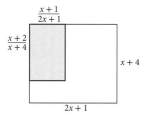

Rotating the inner array maintains an equivalent product. The area of the shaded region is
$$\frac{x+2}{x+4} \cdot \frac{x+1}{2x+1} = \frac{(x+2)(x+1)}{(x+4)(2x+1)}.$$

By the commutative property,
$$\frac{(x+2)(x+1)}{(2x+1)(x+4)} = \frac{(x+1)(x+2)}{(2x+1)(x+4)}$$

The commutative property can also be used to help you simplify before you even begin multiplying. For example:

$\frac{x-2}{x+3} \cdot \frac{x+3}{x-4}$ Notice the common factor of $(x+3)$

$\frac{x+3}{x+3} \cdot \frac{x-2}{x-4}$ Use the commutative property of multiplication to rearrange the factors.

$1 \cdot \frac{x-2}{x-4} = \frac{x-2}{x-4}$ Simplify, since $\frac{x+3}{x+3} = 1$

In multiplication and division problems with rational expressions, it is best to factor all expressions first. This will make it easy to simplify common factors and identify x-intercepts, holes, and vertical asymptotes.

EXAMPLE A | Multiply $\frac{(x+2)}{(x-3)} \cdot \frac{(x+1)(x-3)}{(x-1)(x^2-4)}$

Solution |

$\frac{(x+2)}{(x-3)} \cdot \frac{(x+1)(x-3)}{(x-1)(x^2-4)}$

$\frac{(x+2)}{(x-3)} \cdot \frac{(x+1)(x-3)}{(x-1)(x-2)(x+2)}$ Factor any expressions that you can. (x^2-4) factors to $(x-2)(x+2)$

$\frac{(x+2)(x+1)(x-3)}{(x-3)(x-1)(x-2)(x+2)}$ Combine the two expressions using multiplication.

$\frac{(x+2)}{(x+2)} \cdot \frac{(x-3)}{(x-3)} \cdot \frac{(x+1)}{(x-1)(x-2)}$ Use the commutative property of multiplication to rearrange factors in a useful way.

$1 \cdot 1 \cdot \frac{(x+1)}{(x-1)(x-2)}$ Divide common factors to 1.

$\frac{(x+1)}{(x-1)(x-2)}$ Rewrite the expression.

LESSON 6.8 Multiplying and Dividing Rational Expressions **369**

Investigate

Example A
Project Example A from your ebook. Have students work in pairs. As they share their response, have them explain their steps. **SMP 1, 2, 3, 5, 6**

Have students graph both the original expression and the simplified expression. **ASK** "Are the expressions equivalent, even though the graph of one has a hole and the graph of the other doesn't?" [The expressions are equivalent; the graphs are of functions, which aren't equivalent because the functions have different domains. **ALERT** The function $y = \frac{(x+1)}{(x-1)(x-2)}$ has domain all real numbers except 1 and 2, whereas the function from the original expression, even after it is simplified, has domain all real numbers except −2, 1, 2, and 3.]

Guiding the Investigation
Step 3 $f(x)$

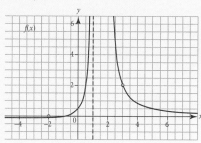

$g(x)$

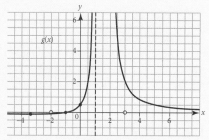

Step 4 $f(-2)$ is undefined and $g(-2) = -0.08\overline{3}$ so $f(-2) \neq g(-2)$.
$f(1)$ and $g(1)$ are both vertical asymptotes.
$f(2)$ and $g(2)$ are both vertical asymptotes.
$f(3)$ is undefined and $g(3) = 2$ so $f(3) \neq g(3)$.

Modifying the Investigation

Whole Class Do Steps 1 – 4 as a class, with student input. Discuss Step 6.

Shortened Do Steps 1 – 4 with half the class using the expressions in Step 1 and half the class using the expressions in Step 5. Discuss Step 6 as a class.

One Step Use the **One Step** in your ebook.

Conceptual/Procedural

Conceptual Students review the concept that, like numerical fractions, it is not necessary to find a common denominator to multiply or divide rational expressions. They also explain why you "invert and multiply" when dividing rational expressions. They explore whether the simplified rational expression is always equivalent to the original expression.

Procedural Students learn to multiply and divide rational expressions.

Step 5 $h(x) = \dfrac{1}{(x^2+1)} \cdot \dfrac{(x^2+1)(x-3)}{(x^2-4)}$

Domain: $x \neq -2, 2$

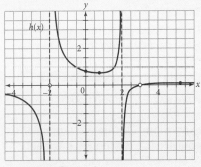

$k(x) = \dfrac{(x-3)}{(x^2-4)}$

Domain: $x \neq -2, 2$

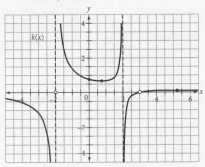

$h(2)$ and $k(2)$ are both vertical asymptotes.

$h(-2)$ and $k(-2)$ are both vertical asymptotes.

$h(1) = k(1) = \dfrac{2}{3}, h(3) = k(3) = 0$

Step 6 Even though you can use properties of equality to simplify a rational expression, the simplified expression does not always have the same domain as the original function. In the first example, g is a simplified version of f but they are not equivalent expressions because their domains are not equivalent and their graphs are not the same because of the undefined values when there is a value that produces a hole or a vertical asymptote. In the second example, the domains are the same so the expressions are equivalent and the graphs are the same.

Example B

Example B is best done as a whole-class activity. Emphasize the connection of the numerical fraction above the example and the rational expression in the example. The goal is for students to see that, when denominators are the same, you are actually dividing the numerators.

Is the simplified form found by reducing the common factors of an expression always equivalent to the original form? The investigation will explore the answer to that question.

INVESTIGATION
Disappearing Holes or Feeling Restricted

In Example A, we found that $\dfrac{(x+2)}{(x-3)} \cdot \dfrac{(x+1)(x-3)}{(x-1)(x^2-4)} = \dfrac{(x+1)}{(x-1)(x-2)}$.

Is this equation true for all values of x?

Step 1 Label the original expression $f(x) = \dfrac{(x+2)}{(x-3)} \cdot \dfrac{(x+1)(x-3)}{(x-1)(x^2-4)}$ and the simplified form $g(x) = \dfrac{(x+1)}{(x-1)(x-2)}$.

Step 2 Using the expressions for $f(x)$ and $g(x)$, identify their domains. How do their domains compare?   domain of $f(x)$: $x \neq -2, 1, 2, 3$
domain of $g(x)$: $x \neq 1, 2$

Step 3 Graph f and g.

Step 4 If the domain is undefined at an x-value, the graph has a hole or a vertical asymptote at that x-value.

Step 4 Determine if the following are true: $f(-2) = g(-2), f(1) = g(1), f(2) = g(2), f(3) = g(3)$. How do the domains of the functions affect the graphs of f and g?

Step 5 Repeat Steps 1 – 3 with $\dfrac{1}{(x^2+1)} \cdot \dfrac{(x^2+1)(x-3)}{(x^2-4)} = \dfrac{x-3}{x^2-4}$

Step 6 When you simplify a rational expression, is the simplified form always equivalent to the original? Why or why not? Give examples to justify your answer.

Sometimes, division of fractions is pretty straightforward, sometimes the divisor is the quotient, like $20 \div 5$ can be thought of as how many 5's in 20? Since 5×4 is 20, $20 \div 5$ is 4. Similarly, you can think about $\dfrac{5}{6} \div \dfrac{1}{6}$ as how many $\dfrac{1}{6}$'s are in $\dfrac{5}{6}$? Since there are five $\dfrac{1}{6}$'s in $\dfrac{5}{6}$, $\dfrac{1}{6} \cdot 5 = \dfrac{5}{6}$, then $\dfrac{5}{6} \div \dfrac{1}{6} = 5$.

EXAMPLE B | Divide $\dfrac{6x^2}{x+1} \div \dfrac{x}{x+1}$

Solution | $\dfrac{6x^2}{x+1} \div \dfrac{x}{x+1}$ can be thought about as $x \cdot \underline{\quad} = 6x^2$, and since $x \cdot 6x = 6x^2$ then $\dfrac{6x^2}{x+1} \div \dfrac{x}{x+1} = 6x^2 \div x = 6x$. This only works if you have a common denominator.

Notice that in both $\dfrac{5}{6} \div \dfrac{1}{6}$ and $\dfrac{6x^2}{x+1} \div \dfrac{x}{x+1}$ you have common denominators. What happens if the denominators are not the same?

You can also think about division as multiplying by the reciprocal of the divisor: $20 \div 5$ as $20 \cdot \dfrac{1}{5}$ and $\dfrac{5}{6} \div \dfrac{1}{6}$ as $\dfrac{5}{6} \cdot 6$. When reasoning isn't quite as easy as the problem in Example B, you can change the division problem into a multiplication problem and then use what you know about multiplication to solve the problem.

INVESTIGATING STRUCTURE

You learned to divide fractions in middle school. You were probably told to just invert and multiply. But did you ever really wonder why? Let's investigate.

The algorithm you were probably shown was

$$\frac{a}{b} \div \frac{c}{d} = \frac{a}{b} \times \frac{d}{c} = \frac{ad}{bc}$$

$$\frac{a}{b} \div \frac{c}{d} = \frac{\frac{a}{b}}{\frac{c}{d}} = \frac{\frac{a}{b} \cdot \frac{d}{c}}{\frac{c}{d} \cdot \frac{d}{c}} = \frac{ad}{bc}$$

Think of changing the problem into an equivalent expression that has a divisor of 1. Anything divided by 1 is easy, right? You know from past experience that $1 \div 6$ can be written as $\frac{1}{6}$. Use that to show that $\frac{a}{b} \div \frac{c}{d} = \frac{ad}{bc}$. Justify each step.

EXAMPLE C Divide $\dfrac{\frac{x^2-1}{x^2+5x+6}}{\frac{x^2-3x+2}{x+3}}$.

Solution

$\dfrac{x^2-1}{x^2+5x+6} \cdot \dfrac{x+3}{x^2-3x+2}$ Change the division problem into a multiplication problem by multiplying by the reciprocal

$\dfrac{(x+1)(x-1)}{(x+2)(x+3)} \cdot \dfrac{(x+3)}{(x-2)(x-1)}$ Factor all expressions.

$\dfrac{(x+1)(x-1)(x+3)}{(x+2)(x+3)(x-2)(x-1)}$ Multiply.

$\dfrac{(x+1)}{(x+2)(x-2)}$ Simplify by reducing all common factors.

Rational expressions like those in Example C can look intimidating. However, the rules are the same as for regular fraction arithmetic. Work carefully and stay organized.

6.8 Exercises

1. Use the commutative property of multiplication to rewrite the expressions and simplify.
   a. $\dfrac{(x-2)}{(x+4)} \cdot \dfrac{(x+4)}{(x-3)} \quad \dfrac{x-2}{x-3}$
   b. $\dfrac{(x+55)}{x} \cdot \dfrac{x^3}{(x-22)} \cdot \dfrac{(x-22)}{x}$ @ $x(x+55) = x^2 + 55x$

2. What is the domain of each expression in Exercise 1? a. $x \neq -4, 3$ b. $x \neq 0, 22$

3. Divide.
   a. $\dfrac{8x^3}{x-4} \div \dfrac{2}{x-4} \quad 4x^3$
   b. $\dfrac{9x^5}{x^2-4x+15} \div \dfrac{3x}{x^2-4x+15}$ @ $3x^4$

4. Multiply or divide as indicated. Simplify by dividing any common factors.
   a. $\dfrac{6x^2}{2x} \cdot \dfrac{x^3}{x^2} \quad 3x^2$
   b. $\dfrac{6x^2}{2x} \div \dfrac{x^3}{x^2} \quad 3$
   c. $(x^2 - x - 12) \div (x^2 - 3x - 4)$ @ $\dfrac{x+3}{x+1}$
   d. $\dfrac{x+1}{x-4} \cdot \dfrac{x+1}{x-4}$ @ $\dfrac{x^2+2x+1}{x^2-8x+16}$ or $\dfrac{(x+1)^2}{(x-4)^2}$
   e. $\dfrac{x+1}{x-4} \div \dfrac{x+1}{x-4} \quad 1$
   f. $\dfrac{x^2}{x+13} \div \dfrac{x}{x+13} \quad x$

Formative Assessment

Student presentations of their solutions for the example will provide evidence of students' skills at multiplying and dividing rational expressions. Presentations of solutions for the investigation provides evidence of understanding that a simplified rational expression is not always equivalent to the original expression. The Investigating Structure provides evidence that students understand and can explain the "invert and multiply" algorithm.

Apply

Extra Example

Rewrite as a single rational expression:

a. $\dfrac{x^2 + x - 2}{x(x-3)^2} \cdot \dfrac{x^2 - 2x - 3}{x - 1}$ $\dfrac{(x+2)(x+1)}{x(x-3)}$

b. $\dfrac{x+1}{x^2+3x-10} \div \dfrac{x^2-2x-3}{x^2-x-2}$ $\dfrac{(x+1)}{(x+5)(x-3)}$

Closing Question

Rewrite $\dfrac{\dfrac{9x+9}{15x+15} + \dfrac{15x-6}{75x-30}}{\dfrac{4x^2-8x+12}{10x^2-20x+30}}$ as simply as possible. 2

Assigning Exercises

Suggested: 1–8
Additional Practice: 9–12

Exercise 6 As needed, suggest that students work in stages, first combining the terms in the numerators and denominators separately and then simplifying the quotient.

Reason and Apply

5. Multiply or divide as indicated. Reduce any common factors to simplify.

   a. $\dfrac{x+1}{(x+2)(x-3)} \cdot \dfrac{x^2-4}{x^2-x-2}$ @ $\dfrac{1}{x-3}$

   b. $\dfrac{x^2-16}{x+5} \div \dfrac{x^2+8x+16}{x^2+3x-10}$ @ $\dfrac{(x-4)(x-2)}{x+4}$

   c. $\dfrac{x^2+7x+6}{x^2+5x-6} \cdot \dfrac{2x^2-2x}{x+1}$ $2x$

   d. $\dfrac{\dfrac{x+3}{x^2-8x+15}}{\dfrac{x^2-9}{x^2-4x-5}} \cdot \dfrac{x+1}{(x-3)^2}$

6. Rewrite as a single rational expression.

   a. $\dfrac{1-\dfrac{x}{x+2}}{\dfrac{x+1}{x^2-4}}$ @ $\dfrac{2(x-2)}{x+1}$

   b. $\dfrac{\dfrac{1}{x-1}+\dfrac{1}{x+1}}{\dfrac{x}{x-1}-\dfrac{x}{x+1}}$ 1

   c. $\left(\dfrac{1}{x-2}-\dfrac{1}{x+1}\right) \div \left(\dfrac{2}{x-1}+\dfrac{1}{x-2}\right)$ $\dfrac{3(x-1)}{(x+1)(3x-5)}$

   d. $\dfrac{\dfrac{1}{x^2+5x+6}+\dfrac{2}{x+3}}{\dfrac{x^2}{x^2+3x}}$ $\dfrac{2x+5}{x(x+2)}$

7. Matthew says that he can divide much quicker. Do you agree? If not, how would you help him? *You cannot divide out common "numbers", you divide out common "factors".*

   $\dfrac{x^2+2x+15}{2x^2+5} = \dfrac{x^2+2x+15^3}{2x^2+5} = \dfrac{x^2+x+3}{x^2+1}$

8. Sometimes Abby likes to divide after finding a common denominator and other times to divide by multiplying to get a divisor of 1. She spilled some ketchup on her work. What belongs under the ketchup blobs?

   a. $\dfrac{x+1}{6x} \div \dfrac{x}{3} = \dfrac{x+1}{6x} \div \dfrac{\blacksquare}{6x} = \dfrac{x+1}{\blacksquare}$ $2x^2, 2x^2$

   b. $\dfrac{12x^3}{2x-8} \div \dfrac{x}{x-4} = \left(\dfrac{12x^3}{2x-8} \cdot \blacksquare\right) \div \left(\dfrac{x}{x-4} \cdot \dfrac{x-4}{x}\right) = 6x^2 \div 1 - \blacksquare$ $\dfrac{(x-4)}{x}, 6x^2$

Review Exercises

9. Add or subtract as indicated.

   a. $\dfrac{1}{x+1} + \dfrac{2}{x+2}$ $\dfrac{3x+4}{(x+1)(x+2)}$

   b. $\dfrac{1}{x+1} - \dfrac{2}{x+2}$ $\dfrac{-x}{(x+1)(x+2)}$

10. Solve for x.

    a. $x^2 + 3x + 1 = -1$ $x = -2, -1$

    b. $x^2 + 3x + 1 = 0$ $x = \dfrac{-3 \pm \sqrt{5}}{2}$

    c. $\dfrac{1}{x+1} = \dfrac{2}{x+2}$ $x = 0$

    d. $\sqrt{x+1} = 2\sqrt{x+2}$ $x = -\dfrac{3}{7}$

11. Find either a polynomial or exponential model for the data. Justify your choice.

    a.
x	y
-1	0.5
0	2
1	8
2	32

    $y = 2 \cdot 4^x$, because there is an arithmetic sequence for x and a constant multiplier of 4 for y.

    b.
x	y
-1	0.5
0	2
1	3.5
2	5

    $y = 2 + 1.5x$, because there is an arithmetic sequence for x and a constant difference of 1.5 for y.

12. Sketch a graph of the functions.

    a. $y = \dfrac{2}{x+3}$

    b. $y = 3^x$

    c. $y = (x-3)^2 + 2$

372 CHAPTER 6 Polynomial and Rational Functions

12a.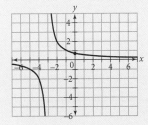

12b.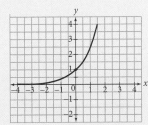

12c.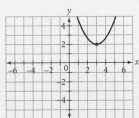

LESSON 6.9

Solving Rational Equations

Our whole life is solving puzzles.
ERNO RUBIK

A rational equation is the quotient of polynomials, $r(x) = \frac{p(x)}{q(x)}$. When you are solving an equation that involves a rational expression, you can undo the division by multiplying.

EXAMPLE A | Solve $\frac{15}{x+1} = 3$.

Solution | To solve, you can rewrite the equation and then undo the operations in reverse order.

$$(x+1)\left(\frac{15}{x+1}\right) = (x+1) \cdot 3 \quad \text{Multiply by } (x+1) \text{ to undo the division}$$
$$15 = 3x + 3 \quad \text{Distribute}$$
$$12 = 3x \quad \text{Subtract 3 to undo adding 3}$$
$$4 = x \quad \text{Divide by 3 to undo multiplying by 3}$$

In some cases, there is a different strategy that can be helpful.

You could remember from proportional reasoning that if $\frac{a}{b} = \frac{c}{d}$ then $\frac{b}{a} = \frac{d}{c}$ as long as $a, b, c, d \neq 0$.

As long as $x + 1 \neq 0$, or $x \neq -1$, then $\frac{15}{x+1} = 3$ means that $\frac{x+1}{15} = \frac{1}{3}$.

$$\frac{x+1}{15}(15) = \frac{1}{3}(15) \quad \text{Multiply by 15 to undo the division}$$
$$x + 1 = 5 \quad \text{Simplify}$$
$$x = 4 \quad \text{Subtract 1 to undo adding 1}$$

Notice that the graph of $y = \frac{15}{x+1}$ intersects the graph of $y = 3$ at the solution $x = 4$. What do you think would be true of solutions you found algebraically if the graphs did not intersect?

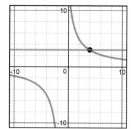

Check the answer in the original equation. If the solution you found produces a 0 in the denominator of the original equation, then the solution is not in the domain of the equation and is extraneous. Can you explain why? Therefore, it is good practice to consider first what values cannot be solutions to a rational equation. If the algebraic moves you make while solving the equation produce an extraneous solution, you can disregard that solution, since it makes the equation undefined for that x.

LESSON 6.9

In this lesson students learn to solve rational equations. They create and solve rational equations in context.

COMMON CORE STATE STANDARDS
APPLIED	DEVELOPED	INTRODUCED
A.SSE.1a	A.REI.2	
A.SSE.1b	A.CED.1	
A.SSE.2		

Objectives
- Solve rational equations in one variable
- Determine when extraneous solutions may arise

Vocabulary
joint variation
combined variation

Launch

Solve for x.

a. $\frac{3}{8} + \frac{4}{5} = \frac{x}{20} \quad x = \frac{47}{2}$

b. $\frac{x}{3} - \frac{3}{2} = 1 \quad x = \frac{15}{2}$

c. $\frac{5}{3x} + \frac{3}{x} = 1 \quad x = \frac{14}{3}$

d. $x + \frac{5}{x} = 6 \quad x = 1, 5$

Investigate

Example A

Project Example A. **ASK** "Are there any restrictions on the variable in this equation?" [$x \neq -1$] Have students solve the equation individually. Ask for volunteers to share their solution and have them describe what they did in each step of their process. **ASK** "Did anyone use a different strategy?" If not, **ASK** "Could I just invert both sides of the equation? Why?" [If $\frac{a}{b} = \frac{c}{d}$ then $\frac{b}{a} = \frac{d}{c}$] Have students find the solutions to the new equation. Again, ask for volunteers to share their solution and have them describe what they did in each step of their process. **ASK** "How could we verify our answer by graphing?" Have students graph each side of the equation on the same axes to find the intersection point.

ELL
The lesson is very mechanical. Consider having students share their work with the class so they can discuss their steps. Revisit the definition of *extraneous*. Some of the terms of Example C will need to be defined: *focal point, image, magnification*, etc.

Support
Practice solving equations using numerical fractions and the variable in the numerator only before the rational expressions. If necessary, you can use the Refreshing Your Skills: Solving Equations lesson.

Advanced
Have students look for other context for solving rational equations in the real-world.

Example B

Project Example B. **ASK** "Can we use proportional reasoning and just invert both sides of the equation? Why? [No. There is a sum on one side.] Are there any restrictions on the variable in this equation?" [$x \neq 5, 6$] Solve the equation as a class activity, with student direction for each step.

Guiding the Investigation

As needed, remind students that the process of solving an equation with a quadratic expression often introduces an extraneous solution.

Step 1

a. $f(x) = \dfrac{x}{x-2} + \dfrac{1}{x-3}$; $g(x) = \dfrac{1}{x^2-5x+6}$

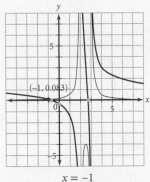

$x = -1$

b. $h(x) = \dfrac{5}{x^2-x} + \dfrac{5}{x-1}$; $k(x) = \dfrac{4}{x}$

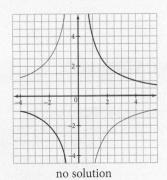

no solution

EXAMPLE B

Solve for x.
$$\dfrac{1}{x-6} + \dfrac{x}{x-5} = \dfrac{1}{x^2-11x+30}$$

You can check for extraneous solutions at the beginning or the end. Which do you think you will remember better? If you check for extraneous solutions at the beginning, make sure you check them again when you are verifying your solution. Let's find the possible extraneous values and then check them against the solution(s) we find.

Looking at the rational expressions in the equation, you can see that the domain of the first term cannot include 6. If you use $x = 6$ in this expression, the denominator becomes zero. In the same way, the value 5 must be excluded since it will make the denominator of the second expression zero. To find the excluded values from the expression on the right side of the equation, you need to factor, $(x^2 - 11x + 30) = (x - 5)(x - 6)$. You already know that 5 and 6 are excluded as possible values of x, so this right side does not add any more excluded values.

$$x - 6 = 0 \qquad x - 5 = 0 \qquad x^2 - 11x + 30 = 0$$
$$x = 6 \qquad x = 5 \qquad (x-5)(x-6) = 0$$
$$x = 5 \text{ or } x = 6$$

Therefore, $x \neq 5, 6$.

$$\dfrac{1}{x-6} + \dfrac{x}{x-5} = \dfrac{1}{x^2-11x+30}$$

$\dfrac{1}{x-6} + \dfrac{x}{x-5} = \dfrac{1}{(x-5)(x-6)}$ Factor the denominator

$\dfrac{1}{x-6} \cdot (x-5)(x-6) + \dfrac{x}{x-5} \cdot (x-5)(x-6)$ Multiply each term by the common denominator.
$= \dfrac{1}{(x-5)(x-6)} \cdot (x-5)(x-6)$

$(x - 5) + x(x - 6) = 1$ Simplify each rational expression.

$x^2 - 5x - 6 = 0$ Expand and then subtract 1 so the equation is equal to zero.

$(x + 1)(x - 6) = 0$ Factor and use zero product property.

$x = -1, 6$

But we had already determined that $x \neq 6$, so $x = -1$ is the **only** solution.

Why do extraneous roots show up in the solution process of solving some rational equations? You'll explore this question in the investigation.

INVESTIGATION

The (Extraneous) Root of the Problem

Step 1 Solve each equation graphically by graphing the left and right side of the equation and finding the intersection point(s).

a. $\dfrac{x}{x-2} + \dfrac{1}{x-3} = \dfrac{1}{x^2-5x+6}$ b. $\dfrac{5}{x^2-x} - \dfrac{5}{x-1} = \dfrac{4}{x}$

Modifying the Investigation

Whole Class Do the Investigation as a class activity, soliciting input from students for each step. Have students follow along with their own graphs.

Shortened Do Steps 1–3 as a class. Have students do Steps 4 and 5.

One Step There is no **One Step** for this investigation.

Conceptual/Procedural

Conceptual The focus of the investigation is to conceptualize when and how extraneous roots are introduced when solving rational equations.

Procedural Students learn to solve rational equations by undoing, by graphing, and by proportionality.

Step 2 Solve each equation in Step 1 algebraically by multiplying the equation by the common denominator of the rational expressions.

Step 3 Compare the graphic and algebraic solutions.

Step 4 For both equations in Step 1, graph the left and right side of the equivalent equations you got in Step 2 when you multiplied the original equation by the common denominator. What are the solutions to those equations? Are they the same as the solutions to the original equations?

Step 5 Describe what happened in the graphs when the extraneous solutions appear. Why do you algebraically get an extraneous solution to each equation?

Step 2a $x = -1, 3$

Step 2b $x = 1$

Step 3 Algebraically, all values of x, including the extraneous ones, are found. Graphically, the functions do not intersect at the extraneous roots.

Step 4a

$f(x) = x(x - 3) + (x - 2); g(x) = 1$

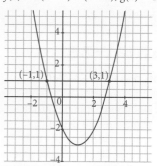

$x = -1$ and $x = 3$

Step 4b

$h(x) = 5 - 5x; + k(x) = 4(x - 1)$

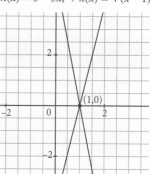

$x = 1$

The solutions are the same as the algebraic solutions.

Step 5 Algebraically, there are no restrictions on the domain until you put them in.

The side mirrors on many cars say "Objects in mirror are closer than they appear." Objects appear closer because the objects are made smaller by the mirror, so they appear farther away!

Concave lenses are used in reflecting telescopes and in mirrors that provide magnification such as for applying makeup or shaving. When you look at an image of a light bulb in a concave mirror, depending on the position of the bulb, the image may be the same size as the object or it could appear larger or smaller than the object. It might appear inverted or it may disappear entirely. The image can also appear in front of or behind the mirror.

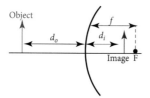
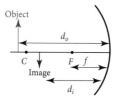

Rational equations can be used to find information about the images of objects in mirrors. Two such equations are the mirror equation and the magnification equation.

The mirror equation is the relationship between the object distance (d_o), the image distance (d_i), and the focal length (f):

$$\frac{1}{f} = \frac{1}{d_o} + \frac{1}{d_i}$$

The magnification equation is the relationship between the ratio of the image distance and object distance to the ratio of the image height (h_i) and object height (h_o):

$$M = \frac{h_i}{h_o} = -\frac{d_i}{d_o}$$

Example C

This example uses rational equations to find information about the images of objects in a concave mirror. You might project the two graphics and use those to define the terms *object*, *image*, *object distance*, *image distance*, and *focal length*. If possible, you might use a concave mirror and light bulb to demonstrate.

Summarize

As students present their solutions to the Examples, Investigation, and Exercises, including explaining their strategy, emphasize the use of correct mathematical terminology and symbols. Be careful not to discourage thinking by talking too quickly or saying too much. Whether student responses are correct or incorrect, ask other students if they agree and why.
SMP 1, 2, 3, 5, 6

In solving rational equations, you must get a common denominator and multiply both sides of the equation by that common denominator. It is important to check for restrictions on the variable, either at the beginning or end (or both) of the solution process.

CRITICAL QUESTION If an equation has no variables in the denominator, how many extraneous roots are there?

BIG IDEA None

Formative Assessment

Student presentations and justification of their work on the examples and exercises can provide evidence of meeting the objective of solving rational equations symbolically and graphically. Student presentations and justification of their work on the investigation and exercises can provide evidence of meeting the objective of identifying extraneous roots.

EXAMPLE C | The focal length of a mirror is 37 cm. An 8.00-cm tall light bulb is sitting 20 cm in front of the mirror. What is the image distance and image size?

Solution | Use the mirror equation to find the image distance. In this case, the focal length, f, is 37 and the object distance, d_o, is 20 cm.

$$\frac{1}{37} = \frac{1}{20} + \frac{1}{d_i} \quad \text{Substitute known values.}$$

$$\frac{1}{37} - \frac{1}{20} = \frac{1}{d_i} \quad \text{Isolate } \frac{1}{d_i}.$$

$$d_i = \left(\frac{1}{37} - \frac{1}{20}\right)^{-1} \quad \text{Take the reciprocal of both sides of the equation.}$$

$$d_i = \left(\frac{1}{37} - \frac{1}{20}\right)^{-1} = -43.53 \quad \text{Find the reciprocal of } \left(\frac{1}{37} - \frac{1}{20}\right).$$

The negative value for the image distance means that the image appears behind the mirror.

Now use the image distance in the magnification equation to find the image size.

$$\frac{h_i}{8} = -\frac{-43.53}{20} \quad \text{Substitute known values.}$$

$$h_i = -\frac{-43.53 \cdot 8}{20} = 17.41 \quad \text{Simplify.}$$

This positive value for the height of the image means that image appears to be 17.41 cm high, larger than the original 8 cm tall light bulb.

If we know the focal length of a mirror, we can look at the image distance, d_i, as a function of the object distance, d_o. Using a focal length of 37 from Example C, $f(x)$ for d_i, and x for d_o in the mirror equation, the equation becomes

$$\frac{1}{37} = \frac{1}{x} + \frac{1}{f(x)}.$$

$$\frac{1}{f(x)} = \frac{1}{37} - \frac{1}{x} \quad \text{Isolate } \frac{1}{f(x)}.$$

$$f(x) = \left(-\frac{1}{x} + \frac{1}{37}\right)^{-1} \quad \text{Take the reciprocal of both sides of the equation.}$$

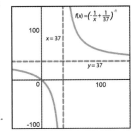

A graph of both sides of the equation reveals both a horizontal and vertical asymptote. Notice anything special about the asymptotes?

If we know the focal length and the object height, we can look at the image height as a function of the object distance. The graph uses the 8 cm object and the image distance of −43.53 cm from Example C.

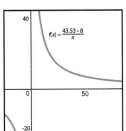

Exercises

Practice Your Skills

1. What values are excluded from the domains of the rational expressions in these equations? You do not need to solve.

   a. $3 = \dfrac{9}{x-2}$ b. $\dfrac{-17}{x+20} = -1$ c. $\dfrac{2x-4}{x-2} = x+4$ d. $x+2 = \dfrac{25}{x+2}$ @
   $x = 2$ $x = -20$ $x = 2$ $x = -2$

2. Solve the equations in Exercise 1.
   a. $x = 5$ b. $x = -3$ c. $x = -2$ d. $x = -7, 3$

3. Solve.

   a. $12 = \dfrac{x-8}{x+3}$ b. $21 = \dfrac{3x+8}{x-5}$ c. $3 = \dfrac{2x+5}{4x-7}$ d. $-4 = \dfrac{-6x+5}{2x+3}$
   $x = -4$ $x = \dfrac{113}{18}$, or $6.2\overline{7}$ $x = 2.6$ $x = -8.5$

Reason and Apply

4. Solve.

   a. $\dfrac{2}{x+1} = -2x$ @ no solutions because the resulting quadratic has no real solutions

   b. $\dfrac{18}{x+4} - 1 = \dfrac{14}{4(x+1)}$ $x = 0, 9.5$

   c. $\dfrac{1}{x-6} + \dfrac{x}{x-3} = \dfrac{3}{x^2 - 9x + 18}$ $x = -1$

   d. $\dfrac{3}{x^2 - 5x + 4} = \dfrac{2}{x-4} - \dfrac{x}{x-1}$ $x = 5$

5. If a basketball team's present record is 42 wins and 36 losses, how many consecutive games must the team win so that its winning record reaches 60%? @   12 games

6. **APPLICATION** Ohm's law states that $I = \dfrac{V}{R}$. This law can be used to determine the amount of current I, in amps, flowing in the circuit when a voltage V, in volts, is applied to a resistance R, in ohms.

   a. If a hairdryer set on high is using a maximum of 8.33 amps on a 120 volt line, what is the resistance in the heating coils? @   14.4 ohms

   b. In the United Kingdom, power lines use 240 volts. If a traveler were to plug in a hairdryer, and the resistance in the hairdryer were the same as in 6a, what would be the flow of current?   16.7 amps

   c. The additional current flowing through the hairdryer would cause a meltdown of the coils and the motor wires. In order to reduce the current flow in 6b back to the value in 6a, how much resistance would be needed?   28.8 ohms

7. The cost of a square piece of glass varies jointly with the thickness of the glass and the square of the side length. A **joint variation** means that the cost varies directly with each of the other variables taken one at a time. Suppose a $\dfrac{1}{4}$ in. thick piece of glass that measures 15 in. on a side costs $2.40. Find the cost of a $\dfrac{1}{2}$ in. thick piece of glass whose sides measure 18 in. $6.91

8. The equation $P = \dfrac{kT}{V}$ is an example of a **combined variation** because P varies directly with T and inversely with V. Suppose that when $P = 3$, $V = 8$ and $T = 300$.

   a. Find the value of the constant of variation, k. $k = 0.08$

   b. When $T = 350$ and $V = 14$, what is P? $P = 2$

Apply

Extra Example
$\dfrac{4}{x-3} + \dfrac{2x}{x^2-9} = \dfrac{1}{x+3}$

No solution

Closing Question
Solve for x.

$\dfrac{x}{x - \dfrac{4}{9}} - \dfrac{x}{x + \dfrac{4}{9}} = \dfrac{1}{x}$

$x = \dfrac{4}{3}, -\dfrac{4}{3}$

Assigning Exercises
Suggested: 1–10, 12, 13
Additional Practice: 11, 14–18

Exercise 3 To help students deepen their understanding of rational functions, you might encourage them to solve these equations by graphing.

Exercise 4a EXTEND ASK "What if we allow complex roots?" [Answers would be $\dfrac{-1 \pm i\sqrt{3}}{2}$]

Exercise 5 Students can let P represent the portion of wins and let x represent the number of consecutive wins.

LESSON 6.9 Solving Rational Equations   **377**

10b.

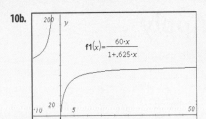

10d. The vertical asymptote is meaningless because there cannot be a negative number of moose. The horizontal asymptote means that as the moose density increases, more moose will be attacked—up to a maximum of 96 during 100 days. This is the maximum number that the wolves can eat.

CONTEXT Environmental Connection
Mathematical modeling of population change has helped ecologists better understand ecological *niches*. Niches are the unique roles that various species play in an *ecosystem*, or natural community. These roles depend on a variety of factors, such as an animal's eating habits, its predators, and the amount of heat and moisture it needs to survive. The niches of the various species in an ecosystem help maintain a fragile population balance that may be disturbed by human activity.

Exercise 11 ALERT Students might miss the word *any* in *any cylinder*. No matter what cylinder is used, the same amount of material is to remain after drilling.

11a.

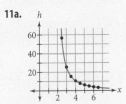

11b. The height gets larger as the radius gets smaller. The radius, x, must be greater than 2.

---

9. Solve. Give exact solutions.

    a. $\dfrac{2}{x-1} + x = 5 \quad x = 3 \pm \sqrt{2}$

    b. $\dfrac{2}{x-1} + x = 2 \quad$ no real solutions; $x = \dfrac{3 \pm i\sqrt{7}}{2}$

10. **APPLICATION** The functional response curve given by the function $y = \dfrac{60x}{1 + 0.625x}$ models the number of moose attacked by wolves as the density of moose in an area increases. In this model, x represents the number of moose per 1000 km^2, and y represents the number of moose attacked every 100 days.

    a. How many moose are attacked every 100 days if there is a herd containing 260 moose in a land preserve with area 1000 km^2? 95

    b. Graph the function.

    c. What are the asymptotes for this function? vertical asymptote $x = -1.6$ and horizontal asymptote $y = 96$

    d. Describe the significance of the asymptotes in this problem.

Environmental CONNECTION

Ecologists often look for a mathematical model to describe the interrelationship of organisms. C. S. Holling, a Canadian researcher, came up with an equation in the late 1950s for what he called a Type II functional response curve. The equation describes the relationship between the number of prey attacked by a predator and the density of the prey. For example, the wolf population increases through reproduction as moose density increases. Eventually, wolf populations stabilize at about 40 per 1000 km^2, which is the optimum size of their range based on defense of their territories.

A moose rises from the surface of a pond in Baxter State Park, Maine.

11. A machine drill removes a 2 inch core from any cylinder. Suppose you want the amount of material left after the core is removed to remain constant. The table below compares the height and radius needed if the volume of the hollow cylinder is to remain constant.

Radius x	2.5	3.0	3.5	4.0	4.5	5.0	5.5	6.0	6.5
Height h	56.6	25.5	15.4	10.6	7.8	6.1	4.9	4.0	3.3

a. Plot the data points, (x, h), and draw a smooth curve through them.

b. Explain what happens to the height of the figure as the radius gets smaller. How small can x be?

c. Write a formula for the volume of the hollow cylinder, V, in terms of x and h. $V = \pi x^2 h - 4\pi h$

d. Solve the formula in 11c for h to get a function that describes the height as a function of the radius. $h = \dfrac{V}{\pi(x^2 - 4)}$

e. What is the constant volume? approximately 400 units^3

12. **APPLICATION** How long should a traffic light stay yellow before turning red? One study suggests that for a car approaching a 40 ft wide intersection under normal driving conditions, the length of time, y, that a light should stay yellow, in seconds, is given by the equation $y = 1 + \frac{v}{25} + \frac{50}{v}$, where v is the velocity of an approaching car in feet per second.

   a. Rewrite the equation in rational function form. @ $y = \frac{v^2 + 25v + 1250}{25v}$

   b. Enter both the original equation and your simplified equation into your calculator. Check using the table feature that the values of both functions are the same. What does this tell you?

   c. If the speed limit at a particular intersection is 45 mi/h, how long should the light stay yellow?   4.4 s

   d. If cars typically travel at speeds ranging from 25 mi/h to 55 mi/h at that intersection, what is the possible range of times that a light should stay yellow?   3.8 s to 4.8 s

Career CONNECTION

Traffic engineers time traffic signals to minimize the "dilemma zone." The dilemma zone occurs when a driver has to decide to brake hard to stop or to accelerate to get through the intersection. Either decision can be risky. It is estimated that 22% of traffic accidents occur when a driver runs a red light.

Exercise 12 This formula is valid only for speeds over approximately 25 mi/h. Students must convert mi/h to ft/s to complete 12c and 12d.

12b.

The values are the same, so it was simplified correctly.

CONTEXT Career Connection States use a number of variables to determine the length of the yellow light. In California, for example, the yellow light interval is determined by the equation $T = r + \frac{V}{d}$ where T is the yellow light interval in seconds, V is the speed limit in meters per second, d is the deceleration rate (presumed to be 3.05 m/s²), and r is the driver's reaction time (presumed to be 1 s).

13. An 8-cm tall lightbulb is placed in front of a mirror with focal length 37 cm.

   a. Find the image distance and image height if the bulb is placed 40 cm in front of a mirror.

   b. Find the image distance and image height if the bulb is placed 10 cm in front of a mirror.

   c. If the image height is positive, the image is upright. If the image height is negative, the image is inverted. If the image distance is positive, the image appears *in front* of the mirror and is considered a *real image*—if you put a piece of paper at the image distance, the light rays converge and the image could be projected on a sheet of paper. If the image distance is negative, the image is considered *virtual*—the viewer sees the image *behind* the mirror. How would you describe the images in 13a and 13b.

13a. image distance is ≈ 493.33 cm; image height is ≈ 98.67 cm

13b. image distance is ≈ −13.7 cm; image height is ≈ 10.96 cm

13c. In part a, the image is upright and real. In part b, the image is upright and virtual.

LESSON 6.9 Solving Rational Equations

14.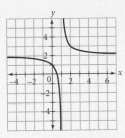

17a. $y = (x - 2)^2(x - 1)(x + 1)$
$= x^4 - 4x^3 + 3x^2 + 4x - 4$

17b. $y = (x - 2)^2 (x + 1) = x^3 - 3x^2 + 4$

Review

14. Sketch a graph $f(x) = 2 + \dfrac{1}{x - 1}$.

15. If $(x - 2i)$ and $(x - 5)$ are factors of a cubic function, write the equation of the function. $y = (x - 2i)(x + 2i)(x - 5)$

16. Solve $2 \cdot 3^x = 162$ $x = 4$

17. Write a possible polynomial function for each graph.

a.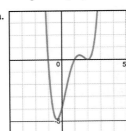
b.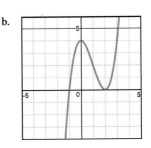

18. If you invest $1,000 at 6.5% interest for 5 years, how much interest do you earn in each of these situations?
    a. The interest is compounded annually. @ $370.09
    b. The interest is compounded monthly. $382.82
    c. The interest is compounded weekly. $383.75
    d. The interest is compounded daily. $383.99

Digital Orca is a 2009 sculpture by Douglas Coupland, located adjacent to the Vancouver Convention Centre, is commonly referred to as Lego Orca or Pixel Whale.

CHAPTER 6 REVIEW

Polynomials can be used to represent the motion of projectiles, the areas of regions, and the volumes of boxes. When examining a set of data whose x-values form an arithmetic sequence, you can calculate the **finite differences** to find the **degree** of a polynomial that will fit the data. When you know the degree of the polynomial, you can define a system of equations to solve for the coefficients. Polynomial equations can be written in several forms. The form $a_n x^n + a_{n-1} x^{n-1} + \cdots + a_1 x^1 + a_0$ is called **general form**.

The degree of a polynomial function and the values of its coefficients determine the shape of its graph. The graphs may have hills and valleys where you will find **local minimums** and **maximums**. The **Remainder Theorem** helps you identify zeros, and the **Rational Root Theorem** helps you identify possible rational roots. You can use **rectangle diagrams**, **long division**, or **synthetic division** to find factors of complicated polynomial expressions.

You were also introduced to **rational functions**, which can be written in the form $y = \dfrac{p(x)}{q(x)}$, where $p(x)$ and $q(x)$ are polynomial expressions. The graphs of rational functions contain **vertical asymptotes** or **holes** when the denominator is undefined. You also learned how to do arithmetic with rational expressions, which can help solve rational equations.

Exercises

You will need a graphing calculator for Exercise **2**.

@ **Answers are provided for all exercises in this set.**

1. Write the equation of each graph in factored form.

a.
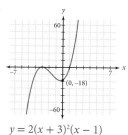
$y = 2(x + 3)^2(x - 1)$

b.
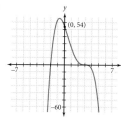
$y = -(x + 2)(x - 3)^3$

c.
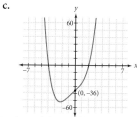
$y = 0.5(x + 4)(x - 2)(x - 3i)(x + 3i)$
(*Hint:* One of the zeros occurs at $x = 3i$.)

Exercise 3 Students can use finite differences to find a model, but they probably will need to generate more data first.

Exercise 4b Students might first graph this polynomial to find any integer roots.

7.

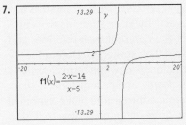

horizontal asymptote: $y = 2$; vertical asymptote: $x = 5$

8. Multiply the numerator and denominator by the factor $(x + 3)$.
$$y = \frac{(2x - 14)(x + 3)}{(x - 5)(x + 3)}$$

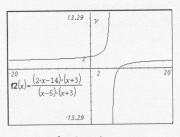

9a. $\dfrac{3x^2 + 8x + 3}{(x - 2)(x + 1)(x + 2)}$

---

2. **APPLICATION** By postal regulations, the maximum combined girth and length of a rectangular package sent by Priority Mail is 108 in. The length is the longest dimension, and the girth is the perimeter of the cross section. Find the dimensions of the package with maximum volume that can be sent through the mail. (Assume the cross section is always a square with side length x.) Making a table might be helpful. 18 in. × 18 in. × 36 in.

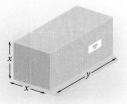

3. Consider this puzzle:

   a. Write a formula relating the greatest number of pieces of a circle, y, you can obtain from x cuts. $y = 0.5x^2 + 0.5x + 1$

   b. Use the formula to find the maximum number of pieces with five cuts and with ten cuts. 16 pieces; 56 pieces

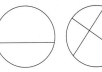

One cut   Two cuts   Three cuts

4. Consider the polynomial equation $0 = 3x^4 - 20x^3 + 68x^2 - 92x - 39$.

   a. List all possible rational roots.

   b. Find the four roots of the equation.

   $\pm 1, \pm 3, \pm 13, \pm 39, \pm\frac{1}{3}, \pm\frac{13}{3}$

   $x = -\frac{1}{3}, x = 3, x = 2 + 3i,$ and $x = 2 - 3i$

5. Divide. $\dfrac{6x^3 + 8x^2 + x - 6}{3x - 2}$ $2x^2 + 4x + 3$

6. Write an equation of each rational function described as a translation of the graph of $y = \dfrac{1}{x}$.

   a. The rational function has asymptotes $x = -2$ and $y = 1$.

   a. $y = 1 + \dfrac{1}{x + 2}$, or $y = \dfrac{x + 3}{x + 2}$

   b. The rational function has asymptotes $x = 0$ and $y = -4$.

   b. $y = -4 + \dfrac{1}{x}$, or $y = \dfrac{-4x + 1}{x}$

7. Graph $y = \dfrac{2x - 14}{x - 5}$. Write equations for the horizontal and vertical asymptotes.

8. How can you modify the equation $y = \dfrac{2x - 14}{x - 5}$ so that the graph of the new equation is the same as the original graph except for a hole at $x = -3$? Verify your new equation by graphing it on your calculator.

9. Rewrite each expression as a single rational expression in factored form.

   a. $\dfrac{2x}{(x - 2)(x + 1)} + \dfrac{x + 3}{x^2 - 4}$

   b. $\dfrac{x^2}{x + 1} \cdot \dfrac{3x - 6}{x^2 - 2x}$ $\dfrac{3x}{x + 1}$

   c. $\dfrac{x^2 - 5x - 6}{x} \div \dfrac{x^2 - 8x + 12}{x^2 - 1}$ $\dfrac{(x + 1)^2(x - 1)}{x(x - 2)}$

10. Solve for x: $\dfrac{x}{x + 2} = \dfrac{x + 1}{x - 1}$ $-\dfrac{1}{2}$

Take Another Look

1. Use the method of finite differences to find the degree of a polynomial function that fits the data at right. What do you notice? Plot the points. What type of curve do you think might best fit the data?

2. How can you find the horizontal asymptote of a rational function without graphing or using a table? Use a calculator to explore these functions, and look for patterns. Some equations may not have horizontal asymptotes. Make a conjecture about how to determine the equation of a horizontal asymptote just by looking at the equation.

x	y
0	60
1	42
2	28
3	20
4	14
5	10
6	7

$$y = \frac{3x^2 + 4x - 5}{2x^4 + 2} \qquad y = \frac{4x^4 - 2x^3 - 2}{x^5 - 5x} \qquad y = \frac{3x^2 + 4x - 5}{2x^2}$$

$$y = \frac{4x^5 - 2x^3 - 2}{x^5 - 5x} \qquad y = \frac{-x^3}{x^2 - 2x + 5} \qquad y = \frac{3x^4 + 2x}{5x^2 - 1}$$

Assessing What You've Learned

 PERFORMANCE ASSESSMENT While a classmate, a friend, a family member, or a teacher observes, show how you would find all zeros of a polynomial equation given in general form, or how you would find an equation in general form given the zeros. Explain the relationship between the zeros and the graph of a function, including what happens when a particular zero occurs multiple times.

 GIVE A PRESENTATION By yourself or with a group, demonstrate how to factor the equation of a rational function. Then describe how to find asymptotes, holes, and intercepts, and graph the equation. Or present your solution of a project or Take Another Look activity from this chapter.

Take Another Look

1. Using the method of finite differences, you run out of differences before reaching a constant difference, so these data can't be modeled by a polynomial function of degree 5 or less. There aren't enough points to determine whether it can't be modeled by a 6th-degree polynomial function.

Based on the graph, it appears that an exponential decay curve may fit these data.

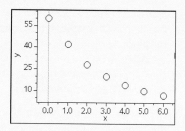

2. When the degree of the numerator is less than the degree of the denominator, the horizontal asymptote is the line $y = 0$. When the degree of the numerator is greater than the degree of the denominator, there is no horizontal asymptote. When the degree of the numerator equals the degree of the denominator, the horizontal asymptote is determined by the ratio of the leading coefficients. When the degree of the numerator is one more than the degree of the denominator, there is a slant asymptote.

Assessing the Chapter

The Kendall Hunt Cohesive Assessment System (CAS) provides a powerful tool with which to manage assessment content, create and assign homework and tests, and disseminate assessment data.

Though the primary purpose of asking students to write test questions is for them to review, you might use some of their questions on your chapter test. You might suggest that students write a question for each lesson as you work through the chapter. Add to the student questions by choosing from constructive assessment or other items in CAS.

Facilitating Self-Assessment

If you can spend two days on review, you might use some time on the second day for students to present some test items they have written and for the class to work through the problems. If you would like your students to give presentations, you might want to divide the topics covered in the chapter and spend a couple of days reviewing as each group gives its presentation.

Students might consider these exercises for addition to their portfolios: Lesson 6.1, Exercise 11; Lesson 6.2, Exercise 9; Lesson 6.3, Exercises 10 and Performance Task; Lesson 6.4, Exercise 12; Lesson 6.5, Exercise 8; Lesson 6.6, Exercise 11; Lesson 6.7, Exercise 7; Lesson 6.8, Exercise 6; and Lesson 6.9, Exercise 10.

CHAPTER 7
Trigonometry and Trigonometric Functions

Overview

Discovering Advanced Algebra is about modeling real-world problems with mathematical functions. Earlier chapters have developed polynomial, exponential, logarithmic, and rational functions. This chapter revisits trigonometric ratios and some of their applications and develops trigonometric functions for modeling periodic phenomena.

In **Lesson 7.1,** the trigonometric ratios sine, cosine, and tangent are reviewed, and the inverse ratios are used for finding angles. Trigonometric ideas are extended to non-acute angles in **Lesson 7.2. Lesson 7.3** expands students' understanding of trigonometric ratios by defining the trigonometric functions for angle measures greater than 90°. Students are introduced to radian measure in **Lesson 7.4,** where they explore angular speed. In **Lesson 7.5,** students apply their knowledge of transformations to the graphs of trigonometric functions and learn vocabulary associated with sinusoidal graphs, including *amplitude* and *phase shift*. In **Lesson 7.6,** students fit real-world data to trigonometric functions. **Lesson 7.7** introduces students to the cotangent, cosecant, and secant functions and to trigonometric identities and their proofs.

The **Performance Tasks** highlighted in this chapter are in Lesson 7.4 and Lesson 7.5. In Lesson 7.4, students consider their motion relative to the center of Earth to find their speed. In Lesson 7.5, students are given a graph with identified maximums and minimums and write equations expressing the curves as a transformation of both the sine and cosine functions.

The Mathematics

Trigonometry

Trigonometry is based on two properties of similar triangles: The angles of similar triangles have the same measure, and their corresponding side lengths have the same ratio. When applied to similar right triangles, these facts imply that ratios of side lengths depend only on the size of the angle. For example, if one angle of a right triangle measures 30°, then the ratio of the lengths of the leg opposite that angle to the hypotenuse is the same, no matter what the lengths themselves are. This ratio is called the *sine* of the angle. Similarly, the ratio of the lengths of the leg adjacent to the angle to the hypotenuse is the *cosine* of the angle, and the ratio of the lengths of the leg opposite to the side adjacent is the *tangent* of the angle.

Trigonometric ratios are useful for indirectly measuring angles and sides of right triangles, as in surveying. For example, if you know that the hypotenuse of a right triangle measures 100 feet and the angle between the hypotenuse and one leg of length x measures 72°, then the ratio $\frac{x}{100}$ is the cosine of 72, written cos 72. So $x = 100 \cos 72$. A calculator says that cos 72 ≈ 0.309, so the unknown length is about 30.9 feet.

On the other hand, if you know that one leg of a triangle has length 30 and the other leg length 60, then trigonometry can help find the angle between the longer leg and the hypotenuse. The tangent of that angle is the ratio of opposite to adjacent legs, which is $\frac{30}{60} = 0.5$. The angle itself is called the *inverse* tangent. A calculator will give the value of $\tan^{-1}(0.5)$ as about 26.6°.

Cyclic Functions

Many phenomena, both natural and humanmade, are cyclic, or *periodic*. For example, average monthly high temperatures at any location change cyclically, with a *period* of one year. What kind of function can model this kind of repetition in order to make predictions?

The term *cyclic* refers to a circle, and circles provide a way in which the ideas of trigonometric ratios can be extended to functions. As a point goes around the unit circle repeatedly, the reference angles defined by the point repeat those of the first cycle, and the sines take on the same values again. So extending the sine ratios beyond angles of 360° gives a function that can be a good model for periodic phenomena. The cosine ratios can also be extended to a periodic function. The value of the sine function is the y-coordinate of the moving point, and the value of the cosine function is the x-coordinate of that point. Because of their close connection to the unit circle, these extensions of the sine and cosine ratios are called *circular functions*. In general, extensions of trigonometric ratios are the *trigonometric functions*.

Radians

When used for modeling, trigonometric functions often have independent variables that are not angles (time, for instance). Thus the extension of trigonometric ratios to trigonometric functions should also move beyond angles and degrees. The independent variable is thought of as the distance that the point has moved around the circle, or a real-life cycle. The fraction of the circle's circumference it travels is given by the number of radius lengths that the point has traveled. To indicate that the distance is measured this way, it's said to be in *radians*.

Some ancient mathematicians considered degree angle measures to be arbitrary, since 360° for a complete circle was an arbitrary choice, like 24 hours in a day. Mathematicians prefer to measure angles in radians, a unit derived naturally from the geometry of the circle. The circumference of a circle of radius r

is $2\pi r$; so a point that makes a full cycle around a circle travels a distance of 2π radius lengths, or radians. In general, if the point travels over an arc intercepted by an angle measuring θ, then the actual distance it travels (the arc length) is $\frac{\theta}{360}2\pi r$ if θ is expressed in degrees, or θr if θ is expressed in radians. Similarly, the area of the circle's sector based on that arc is $\frac{\theta}{360}\pi r^2$ if θ is in degrees, but it is $\frac{1}{2}\theta r^2$ if θ is in radians. Thus some formulas are simpler in radians than in degrees.

Sinusoidal Functions

Transformations of the circular functions $y = \sin x$ and $y = \cos x$ are called the *sinusoidal functions,* because their wavelike graphs resemble that of the sine function. To find a sinusoidal function that models a set of periodic data, you can use either the sine or the cosine function as the parent function. The parent function $y = \cos x$, for example, has a maximum at point $(0, 1)$, period 2π, and amplitude 1. You can find the *period* of the data, T, by calculating the distance between peaks. The *amplitude* of the data, b, is half the difference between the maximum and the minimum. The dilation factors are $a = \frac{T}{2\pi}$ and b, so the function $y = \cos x$ is dilated to $\frac{y}{b} = \cos\frac{x}{a}$, or $y = b\cos\frac{x}{a}$. Next you consider translations: The x-coordinate of the first data maximum to the right of the y-axis is the amount h of the horizontal translation. The average of the data's maximum and minimum values is the vertical shift k. The resulting sinusoidal model is $y = b\cos\left(\frac{x-h}{a}\right) + k$.

Other Trigonometric Functions

The tangent ratio can also be extended to a trigonometric function. Its graph is quite different from those of the sinusoidal functions; it has vertical asymptotes at $x = \frac{\pi}{2} \pm \pi n$. Three other trigonometric functions can be useful in modeling periodic phenomena: $\sec x = \frac{1}{\cos x}$ and the cofunctions, $\cot x = \frac{1}{\tan x}$ and $\csc x = \frac{1}{\sin x}$. Again, their graphs are quite different from those of the sinusoidal functions.

Pythagorean Identities

Numerous equations relate trigonometric functions. An equation that is true for all values of x for which it's defined is called an *identity*. The most fundamental of the identities are the Pythagorean identities, or the three identities that come from the Pythagorean theorem. For example, for any value of x where the functions are defined, $\sin^2 x + \cos^2 x = 1$. Proving trigonometric identities is a good exercise in logical reasoning.

Using This Chapter

Students may have been introduced to the basic trigonometric ratios and inverse ratios in previous courses. As always, the content is seen here from a fresh perspective. Lessons with familiar content may require less time than lessons with new content. This chapter is replete with desmos simulations that can be used in place of Investigations, saving class time. Rather than skipping lessons, or not having student presentations of ideas, consider using the **Shortened** modifications of the investigations.

Common Core State Standards (CCSS)

The Algebra I work with linear, quadratic, and exponential functions and the Algebra II work with polynomial, rational, and radical functions is once again extended to include trigonometric functions. In addition to building on students' previous experience with functions, trigonometry builds on their work with trigonometric ratios and circles in Geometry (G.SRT.6 – 8). The definitions of sine, cosine, and tangent for acute angles are founded on right triangles and similarity, and, with the Pythagorean Theorem, are fundamental in many real-world and theoretical situations. Students in Grade 8 applied the Pythagorean Theorem to determine unknown side lengths in right triangles (8.G.7). In high school, students apply their earlier experience with dilations and proportional reasoning to apply similarity in right triangles to understand right triangle trigonometry, with attention to special right triangles and the Pythagorean theorem to solve problems and find missing side lengths or angle measures of right triangles (G.SRT.6 – 8). In advanced algebra, students build on the trigonometry from their geometry course, to explore more with the trigonometric ratios, extending the domain of trigonometric functions from acute angles of right triangles to all real numbers using the unit circle (T.FT.1, 2). Students now use the coordinate plane to extend trigonometry to model periodic phenomena (T.TF.5). They extend their understanding of proof (G.CO.9 – 11) to prove and apply trigonometric identities (F.TF.8).

Additional Resources For complete references to additional resources for this chapter and others see your digital resources.

Refreshing Your Skills To review the algebra skills needed for this chapter see your digital resources.

Special Right Triangles

Projects See your digital resources for performance-based problems and challenges.

Design a Picnic Table
Sunrise, Sunset

Explorations See your digital resources for extensions and/or applications of content from this chapter.

Circular Functions

Materials

- protractors
- graph paper
- string
- washers
- motion sensors
- springs
- masses of 50–100 g
- support stands, *optional*

CHAPTER 7

Objectives

- Define and use the trigonometric ratios sine, cosine, tangent and their inverse ratios
- Use the definitions of trigonometric ratios and inverse trigonometric ratios to find the unknown side lengths and angle measures of a right triangle
- Expand the definition of trigonometric functions to angles of any measure using the unit circle
- Use the unit circle to extend trigonometric ratios to the circular, or sinusoidal, functions
- Evaluate trigonometric functions through reference angles and reference triangles
- Introduce angle measure in radians and convert between radians and degrees
- Apply knowledge of transformations to find the period, amplitude, phase shift, vertical translation, and average value of a data set or sinusoidal function from an equation or graph
- Use graphs to help find solutions to inverse trigonometric functions other than the principal value
- Use trigonometric functions to model real-world data
- Define the other trigonometric functions through identities
- Discover, prove, and apply the Pythagorean identities

CHAPTER 7 Trigonometry and Trigonometric Functions

Throughout history, circles and triangles have played an important role in design and architecture. The circular terraces created by the Incas, like these in Moray in Peru, may have been used for agricultural experimentation. Today, traffic circles, or roundabouts, direct traffic onto a one-way circular roadway for an almost continuous traffic flow. Meanwhile, triangles have played an important role in bridge structure from the past until today, as seen is this historic bridge in Budapest. Architects continue to discover new ways to use these simple shapes in modern design.

OBJECTIVES

In this chapter you will

- review properties of similar triangles and special right triangles
- review the trigonometric ratios—sine, cosine, and tangent
- use the trigonometric ratios and their inverses to solve application problems
- identify the relationship between circular motion and the sine and cosine functions
- use a unit circle to find values of sine and cosine for various angles
- learn a new unit of measurement for angles, called radians
- study Pythagorean identities

As with all functions, trigonometric functions define our world and give us tools for understanding and, in some situations, controlling our environment. Trigonometry's earliest applications involved circles and spheres. **ASK** "Why do you think Peruvians chose circular terraces for agriculture?" [maximizes area] Today's traffic circles are used to control traffic. **ASK** "How does a traffic circle help with traffic congestion?" Architects, engineers, and surveyors use trigonometry for many things, from calculating structural load and roof slopes, to measuring property boundaries and heights of mountains. **ASK** "Why do you think the triangle is so popular in bridge design?"

Trigonometric functions are *periodic*, or repetitive, like tides. **ASK** "What are some periodic phenomena?" [pendulums, amounts of daylight, distances of Earth from the Sun, the bounce of a spring, musical sounds, gear-driven mechanisms]. **ASK** "How has trigonometry impacted your life?" Have students research how technology depends on trigonometry.

Right Triangle Trigonometry

LESSON 7.1

Happiness is not a destination. It is a method of life.
BURTON HILLIS

The steepest paved road in the world is Baldwin Street in Dunedin, New Zealand. It runs up a hill for a quarter of a mile at an angle of 20.8° with the valley floor. Can you imagine riding a bicycle up this hill?

Houses built along Baldwin Street can have one end of the house completely above street level while the other end is nearly hidden behind the street.

How much higher are you at the top of the hill than at the bottom? You can make a diagram of this situation using a right triangle. But this isn't one of the special right triangles.

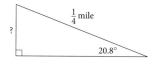

How can you find the height of this triangle? One way is to use the fact that corresponding side lengths of similar triangles have the same ratio. For example, in the right triangles shown below, all corresponding angles are congruent, so the triangles are similar. The ratio of the length of the shorter leg to the length of the longer leg is always 0.75, and the ratios of the lengths of other pairs of corresponding sides are also equal.

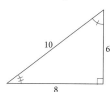 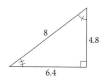

These ratios are called **trigonometric ratios**. The word **trigonometry** comes from the Greek words for "triangle" and "measure."

For the New Zealand road problem, the ratio of the length of the leg opposite the 20.8° to the length of the hypotenuse will be the same in every similar triangle. If you know this ratio, you can solve for the height.

LESSON 7.1

In this lesson, students look at right triangle trigonometry as a tool for calculating distances and angles. Much of this lesson should be a review for most students.

COMMON CORE STATE STANDARDS
APPLIED	DEVELOPED	INTRODUCED
	G.SRT.6	F.TF.7
	G.SRT.7	
	G.SRT.8	

Objectives
- Define the trigonometric ratios sine, cosine, and tangent
- Use trigonometric ratios to find the unknown lengths of the sides of a right triangle
- Use inverse trigonometric ratios to find the unknown angle measures of a right triangle

Vocabulary
trigonometric ratios
trigonometry
sine
cosine
tangent

Materials
- Calculator Notes: Changing Settings; Trigonometric and Inverse Trigonometric Functions

Launch
Project the Launch in your eBook.

$\sin A = \dfrac{\text{opposite leg}}{\text{hypotenuse}} = \dfrac{a}{c}$

$\cos A = \dfrac{\text{adjacent leg}}{\text{hypotenuse}} = \dfrac{b}{c}$

$\tan A = \dfrac{\text{opposite leg}}{\text{adjacent leg}} = \dfrac{a}{b}$

$\sin B = \dfrac{\text{opposite leg}}{\text{hypotenuse}} = \dfrac{b}{c}$

$\cos B = \dfrac{\text{adjacent leg}}{\text{hypotenuse}} = \dfrac{a}{c}$

$\tan B = \dfrac{\text{opposite leg}}{\text{adjacent leg}} = \dfrac{b}{a}$

ELL
Write careful definitions of triangle vocabulary, especially *hypotenuse* and *opposite* and *adjacent* legs. Help students read aloud the abbreviated trigonometric ratios. You could start class by finding a set of stairs and measuring them, or two sets of stairs and comparing them.

Support
As needed, assign the Refreshing Your Skills: Special Right Triangles lesson. Have students practice locating opposite and adjacent legs, regardless of the orientation of the triangle. Point out that symbols such as tan 38° represent numbers, which can be thought of as the values of ratios.

Advanced
If students already have experience with right triangle trigonometry, make this a one-day review. **ASK** "Are other ratios of the side lengths possible?" Elicit the idea that there are three more possible such ratios—the reciprocals of the sine, cosine, and tangent ratios.

Investigate

Example A

Project Example A from your eBook. Have students work in pairs. As they present their solution, have them state the trigonometric function and ratio and justify their steps. **SMP 1, 3, 5, 6**

ALERT If students get an answer of 36.03, their calculator is in Radian mode. Calculators should be in Degree mode for this lesson. Radian mode will be used in a later lesson. See Calculator Note: Changing Settings to learn about changing modes.

Warn students that rounding trigonometric values too much can lead to inaccurate answers. Students should simply enter $\frac{20}{\cos 43°}$ into the calculator.

ASK "What is the measure of $\angle A$?" [47°] "What is the length of the side opposite $\angle B$? How can we find it?" [about 18.7 units. To find b, students can use the Pythagorean Theorem or another trigonometric ratio, such as $b = 20 \tan 43°$.]

Trigonometric Ratios

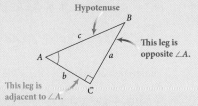

For any acute angle A in a right triangle, the **sine** of $\angle A$ is the ratio of the length of the leg opposite $\angle A$ to the length of the hypotenuse.

$$\sin A = \frac{\text{opposite leg}}{\text{hypotenuse}} = \frac{a}{c}$$

The **cosine** of $\angle A$ is the ratio of the length of the leg adjacent to $\angle A$ to the length of the hypotenuse.

$$\cos A = \frac{\text{adjacent leg}}{\text{hypotenuse}} = \frac{b}{c}$$

The **tangent** of $\angle A$ is the ratio of the length of the opposite leg to the length of the adjacent leg.

$$\tan A = \frac{\text{opposite leg}}{\text{adjacent leg}} = \frac{a}{b}$$

➤ Language CONNECTION

The word *sine* has a curious history. The Sanskrit term for sine was *jya-ardha* ("half-chord"), later abbreviated to *jya*. Islamic scholars, who learned about sine from India, called it *jiba*. In Segovia, around 1140, Robert of Chester read *jiba* as *jaib* when he was translating al-Khwārizmī's book, *Kitāb al-jabr wa'al-muqābalah*, from Arabic into Latin. One meaning of *jaib* is "indentation" or "gulf." So *jiba* was translated into Latin as *sinus*, meaning "fold" or "indentation," and from that we get the word *sine*.

You can use the trigonometric ratios to find unknown side lengths of a right triangle when you know the measure of one acute angle and the length of one of the sides.

EXAMPLE A | Find the unknown length, x.

Solution | In this problem, you know the length of one leg and the measure of one acute angle. You want to find the length of side $\overline{AB}$, which is the hypotenuse. The known leg is adjacent to the known angle. The cosine ratio relates the lengths of the adjacent leg and hypotenuse, so you can use it to find the length of the hypotenuse.

Modifying the Investigation

Whole Class Work through Step 1 with student input. Design one set of stairs in Step 2 with the entire class. Have students individually design other sets of stairs and discuss their designs as a class. Complete Steps 3 and 4 with student input, or have students find their solutions individually and then discuss.

Shortened Assign half the groups Step 3 and the other half Step 4. Or skip Step 3.

One Step Pose this problem: "Wheelchair ramps are supposed to have a slope between $\frac{1}{16}$ and $\frac{1}{20}$. Suppose you construct the shortest legal ramp up to a porch that is 24 inches above the surrounding ground. What angle will the ramp make with the ground?" As needed, urge students to draw diagrams.

$$\cos 43° = \frac{20}{x}$$ Write the cosine ratio and substitute the known values.

$$x \cos 43° = 20$$ Multiply both sides by x.

$$x = \frac{20}{\cos 43°}$$ Divide both sides by $\cos 43°$.

$$x \approx 27.35$$ Find the cosine of 43° on a calculator and divide.

The length, x, of hypotenuse $\overline{AB}$ is approximately 27.35 cm.

Note that the trigonometric ratios apply to both acute angles in a right triangle. In the triangle below, b is the leg adjacent to A and the leg opposite B, whereas a is the leg opposite A and the leg adjacent to B. The hypotenuse, labeled c in this triangle, is always the side opposite the right angle.

$$\sin A = \frac{a}{c} \quad \sin B = \frac{b}{c}$$
$$\cos A = \frac{b}{c} \quad \cos B = \frac{a}{c}$$
$$\tan A = \frac{a}{b} \quad \tan B = \frac{b}{a}$$

INVESTIGATING STRUCTURE

In a right triangle, the measures of the two acute angles add up to 90°. You may recall angle pairs like these are called complementary. If you represent the measure of an acute angle with the variable x, then its complement is represented by $90 - x$. Looking back at the definitions of the sine and cosine of angles A and B above, what can you say about the values of $\sin x$, $\cos x$, $\sin(90 - x)$, and $\cos(90 - x)$?

Each trigonometric ratio is a function of an acute angle measure because each angle measure has a unique ratio associated with it. The inverse of each trigonometric function gives the measure of the angle. For example, $\sin^{-1}\left(\frac{3}{5}\right) \approx 36.87°$. (*Note:* This can be read as "the angle whose sine is $\frac{3}{5}$ measures 36.87 degrees.")

EXAMPLE B Two hikers leave their campsite. Emily walks east 2.85 km and Savannah walks south 6.03 km.

a. After Savannah gets to her destination, she looks directly toward Emily's destination. What is the measure of the angle between the path Savannah walked and her line of sight to Emily's destination?

b. How far apart are Emily and Savannah?

Investigating Structure

Project the Investigating Structure. Have students work in pairs. As students present their solution, have them state the trigonometric function and ratio and justify their steps.
SMP 1, 3, 6

Possible answer:

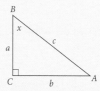

Because the angles of the triangle sum to 180° and $m\angle C = 90°$, $m\angle A + m\angle B = 90°$. If $m\angle B = x$, $m\angle A = 90° - x$. Using the diagram and the definitions of sine and cosine, $\sin B = \frac{b}{c}$ and $\cos A = \frac{b}{c}$. Because $m\angle A = 90° - x$, this can be written $\cos A = \sin(90° - x) = \frac{b}{c}$.

Example B

Project Example B from your eBook. Work through it as a class, with student input. As they make their suggestions, have them state the trigonometric function and ratio and justify their steps. **SMP 1, 2, 3, 5, 6**

Ask all students to draw diagrams; some students are reluctant to sketch pictures. **SMP 4, 5**

LANGUAGE Angles measured clockwise from north are called *bearings*.

ASK "What do you think $\tan^{-1} x$ means? In particular, does it mean $\frac{1}{\tan x}$?" Try to delay approving any answer to this question until you know what all students think.

LANGUAGE Reading and saying $\tan^{-1} x$ as "the angle whose tangent is x" will help reinforce the meaning of the inverse function notation for trigonometric ratios.

LESSON 7.1 Right Triangle Trigonometry **387**

Conceptual/Procedural

Conceptual Students review the concept that for similar right triangles, the ratios of the corresponding sides are equal.

Procedural Students review the definitions of the trigonometric functions and use them and their inverses to find unknown side lengths and angle measures of a right triangle.

Guiding the Investigation

There is a desmos simulation for this investigation in the eBook.

Step 1 ALERT Students may not see that the limiting dimensions indicate a maximum but not a minimum slope. The minimum slope of 0 occurs when the unit rise is 0.

Step 3 Students might refer to the photo of Baldwin Street. Encourage students to use variables and equations.

The assumption of both the angle and the rule of thumb determines either the equation $\frac{17.5 - run}{run} = \tan 20.8°$ or a similar equation involving the rise. Students may be surprised that the problem has only one solution.

Step 4 ASK "Is slope related to the rise and run of the stairs? How?" The ratio of rise to run gives the slope of a line along the staircase, such as that formed by a board lying on the steps. As needed, ASK "What does it mean to design a ramp?" [determine the run required] Encourage students to translate their design into feet or meters to get a better feel for the amount of space needed.

Summarize

As students present their solutions to the Investigation, Investigating Structure, examples, and exercises, have them discuss their strategy. Emphasize the use of correct mathematical terminology. Whether student responses are correct or incorrect, ask other students if they agree and why.
SMP 1, 2, 3, 5, 6

Discussing the Investigation

ASK "Do you think all the stairs that satisfy the building code would be comfortable to climb?" Students may have experienced the awkwardness of climbing stairs with a very small ratio of rise to run.

EXTEND "How long is the wheelchair ramp?" [32 to 40 ft]

Solution Draw a diagram using the information given.

a. Angle S is the angle between Savannah's path and her line of sight to Emily's destination. Triangle SCE is a right triangle and you know the lengths of both legs, so you can use the tangent ratio to find S.

$$\tan S = \frac{2.85}{6.03}$$

Take the inverse tangent of both sides to find the angle measure. Because this is an inverse function, it undoes the tangent and gives the value of S.

$$S = \tan^{-1}\left(\frac{2.85}{6.03}\right) \approx 25°$$

The angle between Savannah's path and her line of sight to Emily's destination measures about 25°.

b. You can use the Pythagorean Theorem to find the distance from S to E.

$$6.03^2 + 2.85^2 = (SE)^2$$
$$SE = \sqrt{6.03^2 + 2.85^2}$$
$$SE \approx 6.67$$

The distance between the two hikers is approximately 6.67 km.

Recreation CONNECTION

A compass is a useful tool to have if you're hiking, camping, or sailing in an unfamiliar setting. The circular edge of a compass has 360 marks representing degrees of direction. North is 0° or 360°, east is 90°, south is 180°, and west is 270°. Each degree on a compass shows the direction of travel, or bearing. The needle inside a compass always points north. Magnetic poles on Earth guide the direction of the needle and allow you to navigate along your desired course.

Your calculator will display each trigonometric ratio to many digits, so your final answer could be displayed to several decimal places. However, it is usually appropriate to round your final answer to the nearest degree or 0.1 unit of length, as in the solution to part a in Example B. If the problem gives more precise measurements, you can use that same amount of precision in your answer. In the example, distances were given to the nearest 0.01 km, so the answer in part b was rounded to the nearest 0.01 also.

INVESTIGATION

Steep Steps

Have you ever noticed that some sets of steps are steeper than others? Building codes and regulations place restrictions on how steep steps can be. Over time these codes change, so stairs built in different locations and at different times may vary quite a bit in their steepness.

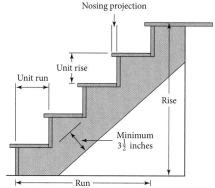

(www.sizes.com/home/stairs.htm)

Step 1 Refer to the diagram of stairs. According to the 1996 Council of American Building Officials and the 2000 International Code Council, the unit run should be not less than 10 inches, and the unit rise should not be more than 7.75 inches. With these limiting dimensions, what is the angle of inclination for the stairs? about 37.8°

Step 2 A rule of thumb for designing stairs is that the sum of the unit rise and unit run should be about 17.5 inches. Design three different sets of stairs that meet this condition. Make two of your designs within the approved building code given in Step 1. The third design should not meet the building code. Find the angle of inclination for each set of stairs. Answers will vary.

Step 3 Consider designing steps to be built alongside Baldwin Street, Dunedin, at an angle of 20.8°.

a. How many designs are possible? Do all possible designs meet the code given in Step 1? Infinitely many designs; not all meet the code.

b. Create a design for the steps that meets the code. Does your design meet the rule of thumb in Step 2? If not, create a new design in which the sum of the rise and run is 17.5 in.

Step 4 Wheelchair ramps are supposed to have a slope between $\frac{1}{16}$ and $\frac{1}{20}$. For each of these slopes, design a ramp to get up to a door 24 inches above the surrounding ground. What is the angle of each ramp? (www.mobility-advisor.com/wheelchair-ramp-specs.html)

Step 3b. Answers will vary; rise: 10 in. $\leq x \leq$ 20.4 in.; run: 3.8 in. $\leq x \tan 20.8°$ $\leq$ 7.75 in. To meet both the code and the rule of thumb, run = 12.68 in., rise = 4.82 in.

Step 4. For slope $\frac{1}{16}$, a ramp with run = 32 ft and angle 3.58°. For slope $\frac{1}{20}$, a ramp with run = 40 ft and angle 2.86°.

Discussing the Lesson

ALERT Students may be confused about the distinctions among the hypotenuse and adjacent and opposite legs when calculating trigonometric ratios. **ASK** "Is the hypotenuse ever the opposite or adjacent leg?" [Although it may be adjacent to an angle, it's not the side that is called the leg adjacent to the angle.] Help students understand that the *opposite leg* is the side across from the angle and the *adjacent leg* is the nonhypotenuse side that forms one side of the angle.

If students do not have a firm understanding of trigonometric ratios, have them complete the Similar Right Triangles supplemental Investigation found in your digital resources.

Context **Language Connection** As the Sanskrit word indicates, trigonometric ratios originally were associated with chords of a circle rather than with angles. Trigonometry was first developed by the Greeks to aid in the study of astronomy.

CRITICAL QUESTION What are trigonometric ratios?

BIG IDEA Trigonometric ratios are ratios of side lengths of right triangles. The ratios depend only on the angle measures of the triangles.

CRITICAL QUESTION What ratios would give you the sine, cosine and tangent of an angle?

BIG IDEA The sine of an angle is the ratio of the opposite side to the hypotenuse. The cosine is the ratio of the adjacent leg to the hypotenuse. The tangent is the ratio of the opposite side to the adjacent leg.

CRITICAL QUESTION Does $\frac{1}{\tan x}$ ever equal $\tan^{-1} x$?

BIG IDEA No; $\frac{1}{\tan x}$ is a ratio of the lengths of sides of a triangle, but $\tan^{-1} x$ refers to an angle measure—specifically, the measure of the angle whose tangent is x. Even if they had the same numerical value at some point, their units would be different.

Formative Assessment

Assess students' skill at finding trigonometric ratios and applying those ratios to find lengths of triangles' sides. Check that students can determine angles by evaluating inverse trigonometric ratios. Check students' ability to think about parts of a triangle (its angles and sides) independently of its shape, their familiarity with ratios and right triangles, and their skill at setting up and solving a linear equation in which the unknown appears in the numerator and denominator of a fraction.

Apply

Extra Example

Draw a right triangle for each problem. Label the sides and angle, then solve to find the unknown measure.

a. $\sin 27° = \dfrac{3}{x}$ $x \approx 6.6$

b. $\tan^{-1} \dfrac{5}{6} = A$ $A \approx 40°$

Closing Question

What is the result of dividing the sine of an angle by the cosine of the same angle?

the tangent of that angle

Assigning Exercises

Suggested: 1–12
Additional Practice: 13–17

Exercises 3, 4 ASK "Is your answer realistic?" Help students make sure that the hypotenuse is always the longest side, for instance, and that the shortest side is always opposite the smallest angle.

Exercise 5 If students are not sure what calculations to make, suggest that they use measurement units. ELL Help students create an appropriate diagram. Discuss the terminology of "angle 20.8° to the floor of the valley."

As you've seen in this lesson, trigonometry can be used to solve problems relating to roads, navigation, and construction. You'll get more practice using the trigonometric functions and their inverses in the exercises.

7.1 Exercises

Practice Your Skills

1. Write all the trigonometric formulas (including inverses) relating the sides and angles in this triangle. There should be a total of 12. @

   $\sin A = \dfrac{k}{j}$; $\sin B = \dfrac{h}{j}$; $\sin^{-1}\left(\dfrac{k}{j}\right) = A$; $\sin^{-1}\left(\dfrac{h}{j}\right) = B$; $\cos B = \dfrac{k}{j}$; $\cos A = \dfrac{h}{j}$;

   $\cos^{-1}\left(\dfrac{k}{j}\right) = B$; $\cos^{-1}\left(\dfrac{h}{j}\right) = A$; $\tan A = \dfrac{k}{h}$; $\tan B = \dfrac{h}{k}$; $\tan^{-1}\left(\dfrac{k}{h}\right) = A$; $\tan^{-1}\left(\dfrac{h}{k}\right) = B$

2. Draw a right triangle for each problem. Label the sides and angle, then solve to find the unknown measure.

   a. $\sin 20° = \dfrac{a}{12}$ @
   b. $\cos 80° = \dfrac{25}{b}$
   c. $\tan 55° = \dfrac{c+4}{c}$ @
   d. $\sin^{-1}\left(\dfrac{17}{30}\right) = D$

Reason and Apply

3. For each triangle, find the length of the labeled side.

   a.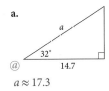
   $a \approx 17.3$

   b.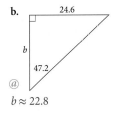
   $b \approx 22.8$

   c.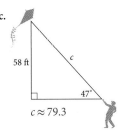
   $c \approx 79.3$

4. For each triangle, find the measure of the labeled angle.

   a.
   $A \approx 17°$

   b.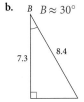
   $B \approx 30°$

   c.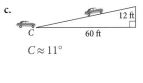
   $C \approx 11°$

5. The steepest paved road in the world runs for a quarter of a mile at an angle of 20.8° to the floor of the valley. (*Note:* 1 mile = 5280 feet.)

   a. How many feet higher are you at the top of the road than at the bottom?   about 468.7 ft

   b. If stair steps average 7.75 inches of rise for each step, how many steps would you have to climb to go up the same height as the hill? @   726 steps

---

390 CHAPTER 7 Trigonometry and Trigonometric Functions

2a. $a \approx 4.1$

2b. $b \approx 144.0$
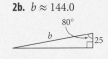

2c. $c \approx 9.3$
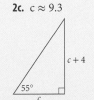

2d. $D \approx 34.5°$

6. Use the properties of the special right triangles to complete the table using exact values. *(h)*

Angle	Sine	Cosine	Tangent
30°	$\frac{1}{2}$	$\frac{\sqrt{3}}{2}$	$\frac{1}{\sqrt{3}}$, or $\frac{\sqrt{3}}{3}$
45°	$\frac{1}{\sqrt{2}}$, or $\frac{\sqrt{2}}{2}$	$\frac{1}{\sqrt{2}}$, or $\frac{\sqrt{2}}{2}$	1
60°	$\frac{\sqrt{3}}{2}$	$\frac{1}{2}$	$\sqrt{3}$

7. In the figure below, $AB = 75$ cm, $AC = 90$ cm, and $\angle A$ measures 20°.
   a. Find BD, the height of $\triangle ABC$. *(@)* $BD \approx 25.7$ cm
   b. Find the area of $\triangle ABC$. area ≈ 1154.32 cm²
   c. Find AD and DC. *(@)* $AD \approx 70.5$ cm, $DC \approx 19.5$ cm
   d. Find the measure of $\angle C$. $m\angle C \approx 53°$

8. Woody owns a triangular shaped piece of land with a pond on it. He makes the measurements shown below. What are the area and perimeter of his property? *(h)*

   area ≈ 4496 ft², perimeter ≈ 353 ft

9. **APPLICATION** Civil engineers generally bank, or angle, a curve on a road so that a car going around the curve at the recommended speed does not skid off the road. Engineers use this formula to calculate the proper banking angle, θ, where v represents the velocity in meters per second, r represents the radius of the curve in meters, and g represents the gravitational constant, 9.8 m/s².

$$\tan \theta = \frac{v^2}{rg}$$

   a. If the radius of an exit ramp is 60 m and the recommended speed is 40 km/h, at what angle should the curve be banked? *(h)* 12°
   b. A curve on a racetrack is banked at 36°. The radius of the curve is about 1.7 km. What speed is this curve designed for? Express your answer in km/h. *(@)* 396 km/h

→ Science
CONNECTION

When a car rounds a curve, the driver must rely on the friction between the car's tires and the road surface to stay on the road. Unfortunately, this does not always work—especially if the road surface is wet!

In car racing, where cars travel at high speeds, tracks banked steeply allow cars to go faster, especially around the corners. Banking on NASCAR tracks ranges from 36° in the corners to just a slight degree of banking in the straighter portions.

LESSON 7.1 Right Triangle Trigonometry 391

Exercise 11 ADVANCED Help students translate the question into one about graphs: To what extent does the graph of $y = \sin x$ resemble that of $y = \sin 2x$?

11.

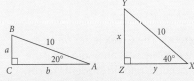

Let $m\angle A = 20°$ and $m\angle X = 2A = 40°$ with hypotenuse of 10.
$\sin 20° = \dfrac{a}{10}$ $\sin 40° = \dfrac{x}{10}$
$a = 10 \sin 20° \approx 3.4$ and $x = 10 \sin 40° \approx 6.4$; $x \neq 2a$, so Shane's statement is false.

Exercise 13 As needed, ASK "How can you tell which angles measure 30° and which measure 60°?" [You must rely on how the diagram appears. Encourage students to reproduce the diagram and label the angles correctly on it.] ADVANCED "Can you tell from the diagram alone that all the triangles are 30°–60°–90° triangles?" [no] "Would it be enough to know that the angle at C measures 30°?" [no] "Is there one additional condition that would tell you?" [yes; for example, that $AC = BC$]

Exercise 17 This exercise reviews the volume of a cone. Remind students to justify their answers so they don't just stop with the intuitive answer "half." In fact, they may find the result very counterintuitive. Students might solve the problem in several different ways.

If the linear dimensions of a solid are multiplied by k, the volume is multiplied by k^3. The volume of the top (empty) cone inside the original is therefore $\dfrac{1}{8}$ of the original, so the frustum below it has $\dfrac{7}{8}$ of the volume of the original cone. When the cone is turned over, $\dfrac{7}{8}$ of its volume is taken up by a cone having $\sqrt[3]{\dfrac{7}{8}}$, or approximately 0.956 of its height.

---

10. There is an interesting connection between the tangent ratio and the equation of a line.
    a. Write an equation for the line containing the points (3, 1) and (5, 7). $y = 3x - 8$
    b. The angle of inclination of a line is the angle it makes with the x-axis. Which trigonometric ratio of this angle is part of the equation of the line? Why?

10 b. The tangent of the angle is the same as the slope. This is because in both cases it is a ratio of the vertical change over the horizontal change.

11. Shane says, "If you double the size of an acute angle that is less than 45° in a right triangle and keep the hypotenuse the same length, then the length of the leg opposite that angle also doubles." Is he correct? Use specific examples either to refute or to justify Shane's statement.

12. Graph the line $y = \dfrac{3}{4}x$, and plot a point on the line. What is the angle between the line and the x-axis? ⓗ about 37°

Review

13. All of the right triangles in the figure below are 30°-60°-90° triangles. What is the length labeled x? $x = 4$

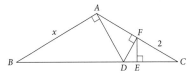

14. Solve this system. ⓐ $(7, -3)$
    $\begin{cases} 3x + 5y = 6 \\ 4y = x - 19 \end{cases}$

15. Solve the equation $x^2 - 3x + 2 = 0$ for x. $x = 1, 2$

16. Consider this sequence.
    $-6, -4, 3, 15, 32, \ldots$
    a. Find the next three terms of the sequence. 54, 81, 113
    b. An explicit formula for the terms of the sequence is $2.5n^2 - bn - c$. What are the values of b and c, if $u_1 = -6$? $b = 5.5, c = 3$

17. A sealed 10 cm tall cone resting on its base is filled to half its height with liquid. It is then turned upside down. To what height, to the nearest hundredth of a centimeter, does the liquid reach? 9.56 cm

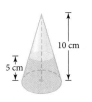

392 CHAPTER 7 Trigonometry and Trigonometric Functions

LESSON 7.2

Extending Trigonometry

In an earlier lesson, the definitions given for the sine, cosine, and tangent ratios applied to acute angles in right triangles. In this lesson you'll extend the definitions of trigonometric ratios to apply to any size angle.

The wheel is come full circle.
WILLIAM SHAKESPEARE

Start by placing the angles on the coordinate plane. When measuring angles on a clock or a compass, you begin at the top and move around the circle to the right (clockwise). For measuring angles on the coordinate plane, however, you begin at the right and move in the opposite direction. You might think of beginning in Quadrant I and moving in order through Quadrants II, III, and IV.

  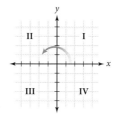

INVESTIGATION

Extending Trigonometric Functions

YOU WILL NEED
- graph paper
- a protractor

In this investigation you'll learn how to calculate the sine, cosine, and tangent of non-acute angles on the coordinate plane.

Step 1 Follow the Procedure Note for each angle measure given below. An example is shown for 120°.

  a. 135°
  b. 210°
  c. 270°
  d. 320°
  e. −100°

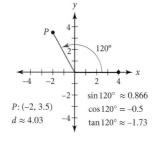

$P: (-2, 3.5)$
$d \approx 4.03$
$\sin 120° \approx 0.866$
$\cos 120° = -0.5$
$\tan 120° \approx -1.73$

Procedure Note

1. Draw a point on the positive *x*-axis. Rotate the point counterclockwise about the origin by the given angle measure and draw the image point. (If the angle measure is negative, rotate clockwise.) Then connect the image point to the origin. The angle between the segment and the positive *x*-axis, in the direction of rotation, represents the amount of rotation.
2. Use your calculator to find the sine, cosine, and tangent of this angle.
3. Estimate the coordinates of the rotated point.
4. Use the distance formula to find the length of the segment.

Step 2 Experiment with the estimated *x*- and *y*-coordinates and the segment length to find a way to calculate the sine, cosine, and tangent. (Your values are all estimates, so just try to get close.)
Answers will depend on the points selected.

Step 3 Plot the point (−3, 1) and draw a segment from it to the origin. Label as *A* the angle between the segment and the positive *x*-axis. Use your method from Step 2 to find the values of sin *A*, cos *A*, and tan *A*. What happens when you try using the inverse, $\sin^{-1}$, to find the value of *A*? What happens for $\cos^{-1}$ and $\tan^{-1}$?

Step 4 Now consider the general case of the angle θ between the positive *x*-axis and a segment connecting the origin to the point (*x*, *y*). Give definitions for the values of sin θ, cos θ, and tan θ. See box after Example A.

LESSON 7.2 Extending Trigonometry 393

Investigate

Guiding the Investigation

There is a desmos simulation for this investigation in your eBook.

Step 1 The image point is the result of the rotation. Students may be more accustomed to thinking of angles in a static way than as rotations. **ELL** *Counterclockwise* is called *anticlockwise* in other parts of the world. **SUPPORT** Students may be confused about which direction to rotate and may be uncomfortable with angles larger than 180°.

Investigation

Sample answers to investigation; the point (4, 0) is used as the starting point for the sample answers.

Step 1

a.
$\sin 135° \approx 0.707$, $\cos 135° \approx -0.707$, $\tan 135° = -1$; $\approx (-2.8, 2.8)$; ≈ 3.96 units

b.
$\sin 210° = -0.5$, $\cos 210° \approx -0.866$, $\tan 210° \approx 0.577$; $\approx (-3.5, -2)$; ≈ 4.03 units

c.
$\sin 270° = -1$, $\cos 270° = 0$, $\tan 270°$ is undefined; $(0, -4)$; 4 units

d.
$\sin 320° \approx -0.643$, $\cos 320° \approx 0.766$, $\tan 320° \approx -0.839$; $\approx (3, -2.6)$; ≈ 3.97 units

e.
$\sin(-100°) \approx -0.985$, $\cos(-100°) \approx -0.174$, $\tan(-100°) \approx 5.671$; $\approx (-0.7, -3.9)$; ≈ 3.96 units

---

Greek letters like θ (theta) and α (alpha) are frequently used to represent the measures of unknown angles. You'll see both of these variables used in the exercises.

Suppose you know that $\sin B \approx 0.47$. That information isn't enough to determine whether B is about 28°, 151°, or even −208°. You need additional information to determine the measure of angle B.

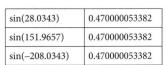

In Step 3 of the investigation, the point at $(-3, 1)$ had a sine of $\dfrac{1}{\sqrt{10}}$, a cosine of $\dfrac{-3}{\sqrt{10}}$ and a tangent of $\dfrac{1}{-3}$.

The calculator gives three different angles for the three inverses. Only one corresponds to the angle from Step 3.

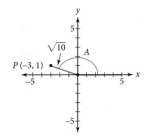

$\sin^{-1}\left(\dfrac{1}{\sqrt{10}}\right)$	18.4349488229
$\cos^{-1}\left(\dfrac{-3}{\sqrt{10}}\right)$	161.565051177
$\tan^{-1}\left(\dfrac{1}{-3}\right)$	−18.4349488229

You can use a graph to find the angle. Create a right triangle by drawing a vertical line from the end of the segment to the x-axis. Use right triangle trigonometry to find the measure of the angle with its vertex at the origin. The acute angle in this **reference triangle**, labeled B, is called the **reference angle.** Use the measure of the reference angle to find the angle you are looking for. In this case, angle B has measure 18.435°. Angle A measures 180° − 18.435°, or 161.565°.

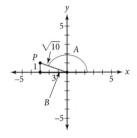

EXAMPLE A | Find $\sin 300°$ without a calculator.

Solution | Rotate a point counterclockwise 300° from the positive x-axis. The image point is in Quadrant IV, 60° below the x-axis. The reference angle is 60°. The sine of a 60° angle is $\dfrac{\sqrt{3}}{2}$. The sine of an angle is the y-value divided by the distance from the origin to the point. Because the y-value is negative in Quadrant IV, $\sin 300° = -\dfrac{\sqrt{3}}{2}$.

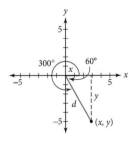

394 CHAPTER 7 Trigonometry and Trigonometric Functions

Modifying the Investigation

Whole Class Demonstrate Step 1 using the 120° angle shown. Repeat Step 1 with the angles given as students follow along. If students seem to understand the process, have them individually complete Step 1 with an assigned angle. Discuss Step 2. Complete Steps 3 and 4 with student input.

Shortened Assign one angle measure from Step 1 to each group. Discuss results and Step 2 as a class. Have groups complete Steps 3 and 4.

One Step Pose this problem: "Your small plane flies for 52 miles. Then it turns 30° to the left and flies an additional 75 miles. How far is it from the starting point?" They may not hesitate to have calculators find the cosine of 150° and solve the problem, getting about 123 miles. During the discussion, ask what the cosine of a non-acute angle means. Can it be thought of as a ratio of adjacent to hypotenuse? Is it the sine of the complement of some angle?

You can generalize this process to find the sine, cosine, and tangent of any angle.

> **Trigonometric Functions for Real Angle Values**
>
> Suppose angle θ is formed by rotating a ray along the positive x-axis about the origin. Point (x, y) on the ray is d units from the origin, where $d = \sqrt{x^2 + y^2}$. Then $\sin\theta = \frac{y}{d}$, $\cos\theta = \frac{x}{d}$, and $\tan\theta = \frac{y}{x}$.

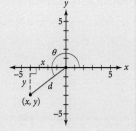

EXAMPLE B What measure describes an angle, measured counterclockwise, from the positive x-axis to the ray from the origin through $(-4, -3)$?

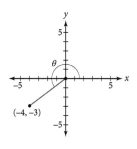

Solution A reference triangle constructed with a perpendicular to the x-axis has sides 3-4-5. The reference angle has measure $\tan^{-1}\left(\frac{3}{4}\right) \approx 36.9°$. The graph shows that θ is more than 180°, so $\theta \approx 180° + 36.9°$, or 216.9°.

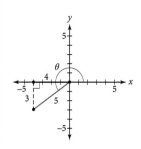

You've seen in this lesson that when you need to find the measure of an angle that is larger than 90° or negative, you will have to decide whether to add to or subtract from 180° or 360°. Your choice will depend on the quadrant of the rotated ray or point and whether the angle measure is positive or negative.

Summarize

As students present their solution to the Investigation and examples, have them describe their strategy. Help them formalize the trigonometric ratios as ratios of coordinates and distances. Emphasize the use of correct mathematical terminology. Whether student responses are correct or incorrect, ask other students if they agree and why. SMP 1, 3, 5, 6

CRITICAL QUESTION Can the point (2, −2) represent two angles with different directions or amounts of rotation?

BIG IDEA A point might represent angles with negative measure (rotated clockwise) or angles generated by rotating (in either direction) by more than 360°.

CRITICAL QUESTION How are the trigonometric ratios related to the coordinates?

BIG IDEA The horizontal coordinate is the cosine of the related angle multiplied by the distance to the origin, and the vertical coordinate is the sine multiplied by the distance to the origin.

Formative Assessment

Watch that students are able to use the basic trigonometric functions and their inverses in this new setting. At this point students should be able to create an acute reference angle for any angle. Watch their understanding of the appropriate sign of the ratio for each angle and check to see that they have a basic understanding of why a given sign is appropriate.

You will need a graphing calculator for Exercise **15**.

Practice Your Skills

1. Find the trigonometric value requested for each angle. Points have integer coordinates.

   a. $\sin \theta$ @   b. $\tan \alpha$ c. $\cos \beta$

   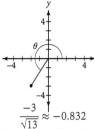 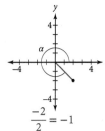 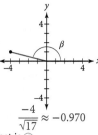

   $\dfrac{-3}{\sqrt{13}} \approx -0.832$ $\dfrac{-2}{2} = -1$ $\dfrac{-4}{\sqrt{17}} \approx -0.970$

2. For each graph in Exercise 1, determine whether it represents an angle that is @

   i. between 0° and 90°
   ii. between 90° and 180°   **2a.** iii.
   iii. between 180° and 270°   **2b.** iv.
   iv. between 270° and 360°   **2c.** ii.

3. Sketch each angle on the coordinate plane.

   a. 280° @   b. 101°   c. 222°   d. −135°

4. Find the measure of the reference angle for each angle.

   a. 280° @ 80°   b. 101° 79°   c. 222° 42°   d. −135° 45°

5. Without using a calculator, determine whether each value is positive or negative.

   a. sin 280° @ negative   b. cos 280° positive   c. sin 101° positive
   d. tan 101° negative   e. cos 222° negative   f. sin −135° negative

> **Recreation CONNECTION**
>
> Skateboarders, snowboarders, and stunt bicyclists all do tricks that involve rotations of greater than 180°. The conservation of angular momentum, the force of gravity, Newton's laws of motions, and other physics concepts can be used to explain how the athletes are able to propel themselves and do aerial flips and twists.
>
> To celebrate the launch of Einstein Year at the Science Museum of London, Cambridge University physicist Helen Czerski worked with Professional BMX rider Ben Wallace to design a new stunt, which they named the "Einstein Flip." Wallace performed the stunt, which involved a 360° backward rotation, for the first time in January 2005.

396 CHAPTER 7 Trigonometry and Trigonometric Functions

3a. 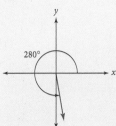   3b. 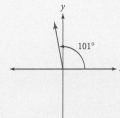   3c. 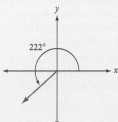   3d.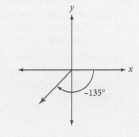

Reason and Apply

6. Use trigonometry to find the measure of each angle in Exercise 1. (*Note:* Assume the measures are between 0° and 360°.) ⓐ **6a.** about 236.3° **6b.** 315° **6c.** about 166°

7. Robyn stands 80 m east of Nathan. Keshon stands directly north of Robyn. Nathan measures the angle between Robyn and Keshon as 28°. How far is Nathan from Keshon? ⓗ about 90.6 m

8. Amnah stands 200 yd east of Warren, and Jana stands 100 yd north and 30 yd west of Warren. From Warren, what is the angle between Amnah and Jana? ⓐ about 106.7°

9. Ryan stands 50 m west and 20 m north of Candice, and Kyle stands 40 m east and 90 m south of Candice. From Candice, what is the angle between Ryan and Kyle? about 135.8°

10. A point starts at the coordinates (6, 0) and rotates counterclockwise 140° about the origin. What are its new coordinates? ⓐ about (−4.60, 3.86)

11. A point starts at the coordinates (−8, 2) and rotates counterclockwise 100° about the origin. What are its new coordinates? ⓗ about (−0.58, −8.23)

12. A point has a *y*-coordinate of −4, and the cosine of the angle from the positive *x*-axis is −0.7. What is the *x*-coordinate? about −3.92

13. Follow these steps to explore the relationship between the tangent ratio and the slope of a line. For 13a and b, find tan θ and the slope of the line.

   a. $\dfrac{3}{2}; \dfrac{3}{2}$
   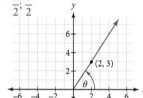

   b. $-2; -2$
   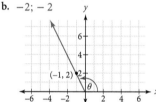

   c. What is the relationship between the tangent of an angle in standard position, and the slope of its nonhorizontal side? They are equal.

   d. Find the slope of this line. approximately 2.414
   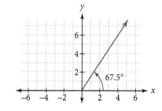

Apply

Extra Example

1. Find the measure of the reference angle for 300°. 60°

2. Find the measure of angle α in the diagram.

$\alpha \approx 116.57°$

Closing Question

What steps do you follow to find the sine of a non-acute angle?

Sketch a graph, determine the reference angle, find the sine, decide whether it's negative or positive.

Assigning Exercises

Suggested: 1–13
Additional Practice: 14–18

Exercise 1b Ask for explanations. Students can get the same result in a variety of mistaken ways.

Exercise 9 To solve this problem, students will need to find the angle between Kyle and due east, and Ryan and due east. For Ryan, remind students that an angle measure from due east is the same as 180° minus an angle from due west. Students might also find approximately 224.2°, which is also correct.

15c. Substitute each y-value found in 15a for x into the inverse. Check that the output is equivalent to the original x-value.

15d.

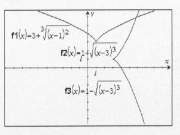

Exercise 16 If students are struggling, ask them what another root must be.

Exercise 18 ASK "Does the solution $x = -1.021$ have any meaning?" [The signal will also reach 1.021 mi north of Pleasant Beach.]

Context Technology Connection Most Americans now carry cellular phones, a situation that has caused a growing need for cellular phone towers. One problem is that although everyone wants good service for his or her phone, many people don't want a cellular tower in their backyards. Cellular companies have begun camouflaging their towers so they won't be obtrusive in people's communities. About 25% of cellular towers now reside in flagpoles, artificial trees, fake chimneys, water towers, silos, large signs and billboards, or windmills.

Review

14. Find the equation of the line through (–3, 10) and (6, –5). $y = 5 - \frac{5}{3}x$

15. Consider the function $f(x) = 3 + \sqrt[3]{(x-1)^2}$. @
   a. Find
      i. $f(9)$ 7
      ii. $f(1)$ 3
      iii. $f(0)$ 4
      iv. $f(-7)$ 7
   b. Find the equation(s) of the inverse of $f(x)$. Is the inverse a function? $y = 1 \pm \sqrt{(x-3)^3}$; not a function
   c. Describe how you can use your calculations in 15a to check your inverse in 15b.
   d. Use your calculator to graph $f(x)$ and its inverse on the same axes.

16. Find a polynomial equation of least degree with integer coefficients that has roots –3 and $\left(\frac{1}{2} - \sqrt{3}\right)$. $a(4x^3 + 8x^2 - 23x - 33) = 0$, where a is an integer, $a \neq 0$

17. Find the length of the minor arc AC.
   a. $\frac{44\pi}{9}$ cm, or ≈ 15.359 cm

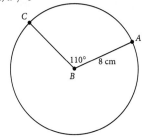

18. APPLICATION The Hear Me Now Phone Company plans to build a cell tower to serve the needs of Pleasant Beach and the beachfront. It decides to locate the cell tower so that Pleasant Beach is 1 mi away at an angle of 60° clockwise from north. The range of the signal from the cell tower is 1.75 mi. The beachfront runs north to south. How far south of Pleasant Beach will customers be able to use their cell phones? 2.02 mi

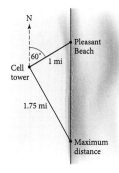

> **Technology**
> **CONNECTION**
>
> The placement of cell towers is crucial to providing a variety of cellular services. Mathematical models are used to analyze possible sites. Cell towers are located by individual companies based on their own business plan for the market they serve. For greatest cost efficiency, cellular companies consider the physical design, topography, population density, environmental impact, engineering, and aesthetics of their towers.

LESSON 7.3

Defining the Circular Functions

> It is by will alone I set my mind in motion.
> MENTAT CHANT

Many phenomena are predictable because they are repetitive, or cyclical. The water depths caused by the tides, the motion of a person on a swing, your height above ground as you ride a Ferris wheel, and the number of hours of daylight each day throughout the year are all examples of cyclical patterns. In this lesson you will discover how to model phenomena like these with the sine and cosine functions.

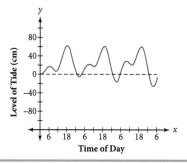

The water levels in the canals of Venice rise and fall periodically, as shown in the graph. During very high tides, the lower-lying areas of Venice can be flooded. This phenomenon is called *Acqua Alta* (high water). Acqua Alta occurs more frequently when the moon is full or new and when there are high winds.

The Rialto Bridge is reflected in the Grand Canal in Venice, Italy.

Science CONNECTION

High and low ocean tides repeat in a continuous cycle, with two high tides and two low tides every 24.84 hours. The highest tide is an effect of the gravitational pull of the Moon, which causes a dome of water to travel under the Moon as it orbits Earth. The entire planet Earth experiences the Moon's gravitational pull, but because water is less rigid than land, it flows more easily in response to this force. The lowest tide occurs when this dome of water moves away from the shoreline.

The circle at right has radius r and center at the origin. A central angle of t degrees is shown. You can use your knowledge of right triangle trigonometry to write the equations $\sin t = \frac{y}{r}$ and $\cos t = \frac{x}{r}$. Sometimes it is useful to write these equations as $y = r \sin t$ and $x = r \cos t$.

In this investigation you'll see how to use the sine and cosine functions to model circular motion.

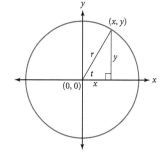

LESSON 7.3

In this lesson, students connect the graphs of sine and cosine to the basic trigonometric functions.

COMMON CORE STATE STANDARDS		
APPLIED	**DEVELOPED**	**INTRODUCED**
F.IF.5	F.IF.4	F.TF.4
	F.TF.2	

Objectives
- Identify periodic functions and their periods
- Find reference angles and draw corresponding reference triangles
- Identify relationships between trigonometric ratios and circular functions
- Use the unit circle and function graphs to find values of sine and cosine
- Understand that coterminal angles have the same trigonometric value

Vocabulary
periodic
period
standard position
terminal side
reference angle
reference triangle
coterminal

Materials
- graph paper
- protractor
- compass
- Calculator Notes: Function Tables; Setting Windows; Unit Circle

Launch

Project the Launch from your ebook.

a. $\sin t = \frac{y}{r}$; $\cos t = \frac{x}{r}$

b. $y = r \sin t$ and $x = r \cos t$, so $(r \cos t, r \sin t)$

ELL
Have students create posters of the unit circle and of sine and cosine waves (for angles between -2π and 2π) to help discuss the vocabulary from this lesson. You could use a bicycle wheel to demonstrate how circular functions extend trigonometric ratios on a unit circle.

Support
Focus the discussion of reference angles on the similarities between the x- and y-values of points in the first quadrant and those in other quadrants. Students can think of these in terms of transformations.

Advanced
Students can explore creating piecewise functions that describe the movement of the frog in the investigation if it jumps from the wheel.

Investigate

Although paddle wheels are used to propel boats, the movement of the boat is ignored in the investigation.

Guiding the Investigation

There is a desmos simulation for the investigation in your eBook.

Step 2

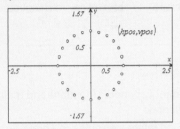

Step 3 The symbol *hpos* refers to horizontal position, and *vpos* means vertical position. Students are working with a unit circle in this investigation, so *hpos* and *vpos* could be renamed cos and sin. In the general case, $r\cos$ and $r\sin$ would be more appropriate.

Step 3 The x-coordinate for every point is cos(*time*), and the y-coordinate is sin (*time*). The coordinates can be found by tracing on the graph or by calculating the cosine and sine. More appropriate names for *hpos* and *vpos* might be *cos* and *sin*.

Step 4 After 360°, the table values repeat. The x- and y-values are cyclical. For degree values in Quadrant I, the x- and y-values are positive; in Quadrant II, the x-values are negative and y-values are positive; in Quadrant III, the x- and y-values are negative; and in Quadrant IV, the x-values are positive and y-values are negative. The pairing of x- and y-values is always the same. That is, ± 0.866 is always paired with ± 0.5, and so on.

As students present the results of Step 6 of the investigation, ask why the graphs of the sine and cosine functions are horizontal translations of each other. The fact that *cosine* means *complement's sine* leads to the equation $\cos t = \sin(90° - t) = \sin[-(t - 90°)]$, which is technically a reflection as well as a horizontal translation by 90°.

INVESTIGATION

Paddle Wheel

While swimming along, a frog reaches out and grabs onto the rim of a paddle wheel with radius 1 m. The center of the wheel is at water level. The frog, clinging tightly to the wheel, is immediately lifted from the surface of the river.

Step 1 The wheel spins counterclockwise at a rate of one rotation every 6 minutes. Through how many degrees does the frog rotate each minute? Each second?
60° each minute, 1° each second

Step 2 Create three lists on your calculator. Name the first list *time* and fill it with the values $\{0, 15, 30, 45, \ldots, 900\}$. Name the second list *hpos* and define it as $hpos = \cos(time)$. Name the third list *vpos* and define it as $vpos = \sin(time)$. Make a scatter plot of (*hpos*, *vpos*) in a square window and trace the path of the frog.

Step 3 Explain how to find the x- and y-coordinates of any point on the circle. In this context, what is the meaning of those points? What might be more appropriate names for lists *hpos* and *vpos*?

Step 4 Scroll down your lists and describe any patterns you see.

Step 5 Use your lists to answer these questions.

a. What is the frog's location after 1215°, or 1215 s? When, during the first three rotations of the wheel, is the frog at that same location?
(−0.707, 0.707); 135 s, 495 s, 855 s

b. When is the frog at a height of −0.5 m during the first three rotations? 210 s, 330 s, 570 s, 690 s, 930 s, 1050 s

c. What are the maximum and minimum x- and y-values?
$-1 \leq y \leq 1, -1 \leq x \leq 1$

Step 6 Make scatter plots of (*time*, *hpos*) and (*time*, *vpos*) on the same screen, using a different symbol for each plot. Use the domain $0 \text{ s} \leq time \leq 360 \text{ s}$. How do the graphs compare? How can you use the graphs to find the frog's position at any time? Why do you think the sine and cosine functions are sometimes called circular functions?

Step 6

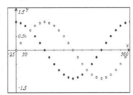

The graphs both show cyclical motion, in the same pattern. However, cos t starts at a height of 1 and sin t starts at a height of 0. They both pass through one full cycle and return to the starting location after 360°. In this graph, x represents t, and y represents *hpos* or *vpos*. To find the frog's position at time t, subtract multiples of 360° from t until the result is between 0° and 360°, and then find the corresponding x-value on the first graph and y-value on the second graph.

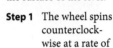
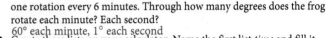

Recall that a circle with radius 1 unit centered at the origin is called the unit circle. While using the unit circle in the investigation, you discovered that values for sine and cosine repeat in a regular pattern. When output values of a function repeat at regular intervals, the function is **periodic**. The **period** of a function is the smallest distance between values of the independent variable before the cycle begins to repeat.

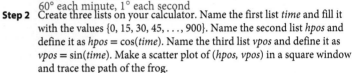

Modifying the Investigation

Whole Class Work through each step with student input.

Shortened Complete Steps 1 and 2 as a class. Have groups complete the remaining steps.

One Step Pose this problem: "A huge Ferris wheel has a 100-meter diameter, takes 15 minutes to complete a full rotation, and reaches a maximum height of 112.5 meters. Sketch a graph showing the height of a rider as a function of time. Exactly how high will the rider be after 10 minutes? When will the rider be at $\frac{2}{3}$ the maximum height above the water?"

EXAMPLE A | Find the period of the cosine function.

Solution | In the investigation the frog returned to the same position each time the paddle wheel made one complete rotation, or every 360°. It seems reasonable to say that the function cos x has a period of 360°. You can verify this by looking at a graph of $y = \cos x$.

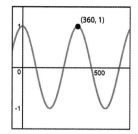

The graph shows that from $x = 0°$ to $x = 360°$, the function completes one full cycle.

Because the period of the cosine function is 360°, you can translate the graph of the function $y = \cos x$ left or right by 360° and it will look the same. Verify this by graphing $y = \cos(x + 360°)$ and $y = \cos(x - 360°)$ on your calculator on the same axes as the parent function $y = \cos x$.

The graph at right of $y = \sin x$, with one cycle highlighted between 0 and 360°, confirms that the sine function also has a period of 360°.

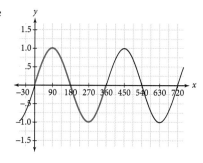

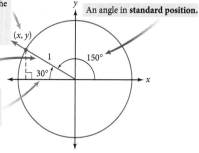

The cosine of an angle is the x-coordinate of the point where the terminal side and the unit circle intersect. The sine of an angle is the y-coordinate of this point.

The **terminal side**.

An angle of 150° has a **reference angle** of 30°.

An angle in **standard position**.

Notice that the domains of the sine and cosine functions include all positive and negative numbers. In a unit circle, angles are measured from the positive x-axis. Positive angles are measured in a counter-clockwise direction, and negative angles are measured in a clockwise direction. An angle in **standard position** has one side on the positive x-axis, and the other side is called the **terminal side**.

The x- and y-coordinates of the point where the terminal side touches the unit circle determine the values of the cosine and sine of the angle. Identifying a **reference angle**, the acute angle between the terminal side and the x-axis, and drawing a **reference triangle** can help you find these values.

LESSON 7.3 Defining the Circular Functions **401**

Conceptual/Procedural

Conceptual Students learn the concept of cyclic, or periodic, phenomena.

Procedural Students use the unit circle and function graphs to find values of sine and cosine.

Example A

Project Example A from your eBook. Ask for student input. Have them share their thinking along with their answer. **ASK** "What three cosine functions could we graph to verify?" [$y = \cos x$, $y = \cos(x + 360°)$, and $y = \cos(x - 360°)$]

ALERT Students may not understand why the coordinates on the unit circle determine the cosine and the sine. Point out that $\cos x = \frac{y}{r}$; $r = 1$, therefore $\cos x = y$.

Example B

Project Example B from your eBook. Have students work in pairs. Encourage students to sketch each angle on a small set of coordinate axes so that they can draw a right triangle between the given segment and the x-axis. Students do not have to use the quadrant numbers, but they do need to know the location of the triangles and the appropriate signs of the values. **SMP 1, 3, 6**

You can also show students how to trace a sine or cosine graph on a calculator to find approximate values of the functions and to check that $\sin 30° = \sin 150°$ and that $\cos 150° = -\cos 30°$. You may want to discuss the difference between exact values and values estimated on the calculator. **SMP 5**

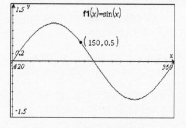

Summarize

As students present their solutions to the Investigation and examples, have them describe their strategy. Emphasize the use of correct mathematical terminology. Whether student responses are correct or incorrect, ask other students if they agree and why. **SMP 1, 2, 3, 4, 5, 6, 7, 8**

ASK "Does the period of a periodic function have anything to do with the period of a pendulum?" As needed, refer students back to where they saw the period of a pendulum—the time for one complete swing (forward and back). The location of a swinging pendulum bob (without dampening) is indeed described by a periodic function.

ASK "Are sin x and cos x ratios of side lengths of a triangle, or are they x- and y-values of points on the unit circle?" [They can be seen from either perspective. For measuring distances, they're most useful as ratios; for modeling periodic phenomena, the extension to functions is more appropriate.]

CRITICAL QUESTION What are the circular functions?

BIG IDEA Circular functions are the sine and cosine functions that describe the coordinates of a point traveling around and around a unit circle.

CRITICAL QUESTION How do these functions relate to trigonometric ratios?

BIG IDEA The x-coordinate of the moving point is the cosine of the vertex angle at the origin, of the right triangle having its hypotenuse along the radius to that moving point, its horizontal leg along the horizontal axis and its vertical leg intercepting the circle at that moving point. Similarly, the y-coordinate of the moving point is the sine of that angle.

CRITICAL QUESTION What does it mean to say that these functions are periodic?

BIG IDEA Periodic functions repeat over each interval of a certain minimum length, called their *period*.

Formative Assessment

Note students' ability to graph and investigate a circle using two separate functions. Students should be able to use reference angles to find coordinates on the unit circle for angles greater than 360°. Assess students' understanding of the basic trigonometric functions and reference angles. Watch students' comfort when working with and plotting functions.

EXAMPLE B Find the value of the sine or cosine for each angle. Explain your process.

a. sin 150°   b. cos 150°   c. sin 210°   d. cos 320°

Solution For each angle in a–d, rotate the point (1, 0) counterclockwise about the origin. Draw a ray from the origin through the image point to be the terminal side. Drop a perpendicular line from the image point to the x-axis to create a reference triangle and then identify the reference angle.

a. The y coordinate of the rotated point on the unit circle is the sine of the angle. For 150°, the reference angle measures 30°. Using your knowledge of 30°-60°-90° triangle side relationships and the fact that the hypotenuse of the triangle is 1, you can find that the lengths of the legs of the reference triangle are $\frac{1}{2}$ and $\frac{\sqrt{3}}{2}$. Thus the coordinates of the image point are $\left(-\frac{\sqrt{3}}{2}, \frac{1}{2}\right)$. Therefore, $y = \sin 150° = \frac{1}{2}$.

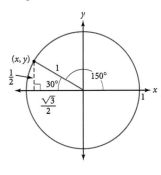

Notice that y remains positive in Quadrant II. Use your calculator to verify that sin 150° equals 0.5.

b. You can use the same unit circle diagram as above. The cosine is the x-coordinate of the rotated point. The length of the adjacent leg in the reference triangle is $\frac{\sqrt{3}}{2}$. Because the reference triangle is in Quadrant II, the x-coordinate is negative. Therefore, $x = \cos 150° = -\frac{\sqrt{3}}{2}$.

The calculator gives −0.866, which is approximately equal to $-\frac{\sqrt{3}}{2}$, as a decimal approximation of cos 150°.

c. Rotate the point counterclockwise 210°, then draw the reference triangle. The reference angle in this triangle again measures 30°. Because this angle is in Quadrant III, the y-value is negative, so $\sin 210° = -\frac{1}{2}$.

The point (210, −0.5) on the graph of $f(x) = \sin x$ verifies this result.

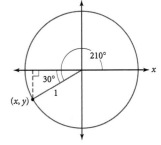

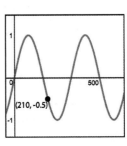

402 CHAPTER 7 Trigonometry and Trigonometric Functions

d. If you draw a reference triangle for 320°, you will find a reference angle of 40° in Quadrant IV. In this quadrant, *x*-values are positive. So, cos 320° = cos 40°. An angle measuring 40° is not a special angle, so you don't know its exact trigonometric values. According to the calculator, either cos 320° or cos 40° is approximately 0.766.

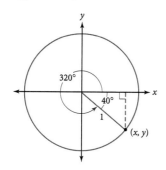
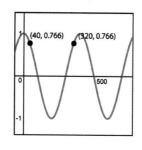

Angles in standard position are **coterminal** if they share the same terminal side. For example, the angles measuring −145°, 215°, and 575° are coterminal, as shown. Coterminal angles have the same trigonometric values.

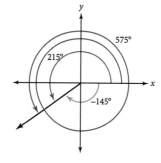

Apply

Extra Example

1. Create a sine graph and trace to find the value of the expression sin 95°. 0.9962
2. Identify an angle θ that is coterminal with the angle 132°. possible answer: −228°

Closing Question

How are the circular functions different from the sine and cosine ratios? Circular functions allow the possibility of a point moving around the unit circle more than once.

Assigning Exercises

Suggested: 1–15
Additional Practice: 16–20

1a.
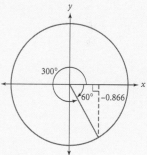

approximately −0.866 m

1b.
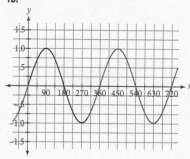

 **Exercises**

You will need a graphing calculator for Exercises **3, 11, 13, 14,** and **20**.

Practice Your Skills

1. After 300 s, the paddle wheel in the investigation has rotated 300°. @
   a. Draw a reference triangle and find the frog's height at this time.
   b. Draw a graph of $y = \sin x$ and highlight the portion of that graph that pictures the frog's height during the first 300 s.

2. Use your calculator to find each value, approximated to four decimal places. Then draw a diagram in a unit circle to represent the value. Name the reference angle.
   a. sin (−175°) @   b. cos 147°   c. sin 280° @   d. cos 310°   e. sin (−47°)

3. Create a sine or cosine graph, and trace to find the value of each expression.
   a. sin 120°   b. sin (−120°)   c. cos (−150°)   d. cos 150°

LESSON 7.3 Defining the Circular Functions **403**

2a. −0.0872; sin(−175°) = −sin 5°; reference angle 5°
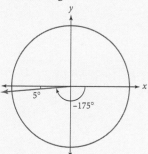

2b. −0.8387; cos 147° = −cos 33°; reference angle 33°
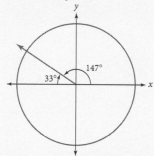

2c. −0.9848; sin 280° = −sin 80°; reference angle 80°
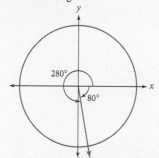

2d. 0.6428; cos 310° = cos 50°; reference angle 50°
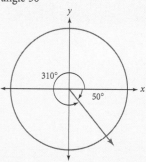

2e. −0.7314; sin −47° = −sin 47°; reference angle 47°

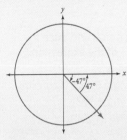

Exercise 3 You might ask students to draw the reference triangles. Be tolerant if a student interprets the instructions as accepting a sine or cosine graph for any of the functions. The property cos x = sin (90° − x) allows you to find a value of either the sine or the cosine function from the other function.

3a.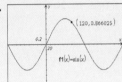

3b.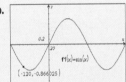

3c.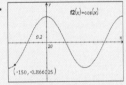

3d.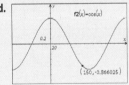

7. Quadrant I: cos θ and sin θ are positive; Quadrant II: cos θ is negative, and sin θ is positive; Quadrant III: cos θ and sin θ are negative; Quadrant IV: cos θ is positive, and sin θ is negative.

Exercise 8 ASK "What is the range of the function y = sin x?" [−1 ≤ y ≤ 1]

8.
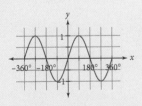

$x = \{-360°, -180°, 0°, 180°, 360°\}$

---

4. The functions y = sin x and y = cos x are periodic. How many cycles of each function are pictured?

a. 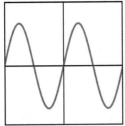   2

b. 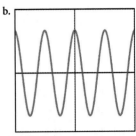   4

5. Which of the following functions are periodic? For each periodic function, identify the period.

a. 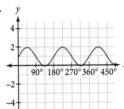 periodic, 180°

b. 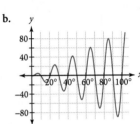 not periodic

c. 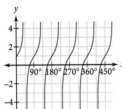 periodic, 90°

d. 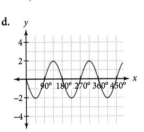 periodic, 180°

Reason and Apply

6. Identify an angle θ that is coterminal with the given angle. Use domain 0° ≤ θ ≤ 360°.

  a. −25° θ = 335°   **b.** −430° θ = 290°   **c.** 435° θ = 75°   **d.** 1195° θ = 115°

7. For each Quadrant, I–IV, shown at right, identify whether the values of cos θ and sin θ are positive or negative.

8. Carefully sketch a graph of the function y = sin x over the domain −360° ≤ x ≤ 360°. Identify all values of x in this interval for which sin x = 0.

9. Carefully sketch a graph of the function y = cos x over the domain −360° ≤ x ≤ 360°. Identify all values of x in this interval for which cos x = 0.

10. Suppose sin θ = −0.7314 and 180° ≤ θ ≤ 270°.

  a. Locate the point where the terminal side of θ intersects the unit circle.

  b. Find θ and cos θ. $\theta \approx$ 227°; cos θ = −0.6820

  c. What other angle α has the same cosine value? Use domain 0° ≤ α ≤ 360°. α = 360° − θ = 133°

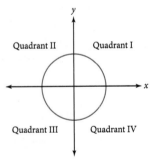

Exercise 9 ASK "What is the range of the function y = cos x?" [−1 ≤ y ≤ 1]

9.
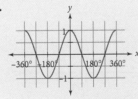

$x = \{-270°, -90°, 90°, 270°\}$

10a.
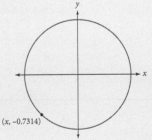

11. Graph $y = \sin x$ using the domain $0 \leq x \leq 720°$.

   a. What are the x-coordinates of all points on the graph such that $\sin x = -0.7314$? ⓗ
   approximately 227°, 313°, 587°, and 673°
   b. Consider the frog from the Investigation Paddle Wheel. How do your answers from 11a relate to the frog's location and the amount of time passed?
   The frog was 0.7314 m below the surface at approximately 227 s, 313 s, 587 s, and 673 s.

12. Let θ represent the angle between the x-axis and the ray with endpoint (0, 0) passing through (2, 3).

   a. Find $\sin \theta$ and $\cos \theta$. ⓗ $\sin \theta = \dfrac{3}{\sqrt{13}}$ or $\dfrac{3\sqrt{13}}{13}$, $\cos \theta = \dfrac{2}{\sqrt{13}}$ or $\dfrac{2\sqrt{13}}{13}$

   b. Find the coordinates of the point where the terminal side of θ intersects the unit circle. ⓗ $\left(\dfrac{2\sqrt{13}}{13}, \dfrac{3\sqrt{13}}{13}\right)$ or about (0.55, 0.83)

13. **APPLICATION** For the past several hundred years, astronomers have kept track of the number of sunspots. This table shows the average number of sunspots each day for the years from 1980 to 2014.

Year	Number of sunspots	Year	Number of sunspots	Year	Number of sunspots	Year	Number of sunspots
1980	154.6	1989	157.6	1998	64.3	2007	7.6
1981	140.4	1990	142.6	1999	93.3	2008	2.9
1982	115.9	1991	145.7	2000	119.6	2009	3.1
1983	66.6	1992	94.3	2001	111.0	2010	16.5
1984	45.9	1993	54.6	2002	104.0	2011	55.7
1985	17.9	1994	29.9	2003	63.7	2012	57.6
1986	3.4	1995	17.5	2004	40.4	2013	64.7
1987	29.4	1996	8.6	2005	29.8	2014	79.3
1988	100.2	1997	21.5	2006	15.2		

(www.sws.bom.gov.au/Educational/2/3/6)

   a. Make a scatter plot of the data and describe any patterns that you notice. ⓐ
   b. Estimate the length of a cycle. ⓐ   10–11 yr
   c. Predict the next period of maximum solar activity (years with larger numbers of sunspots) after 2014.   about 2022

Science CONNECTION

Sunspots are dark regions on the Sun's surface that are cooler than the surrounding areas. They are caused by magnetic fields on the Sun and seem to follow short-term and long-term cycles.

Sunspot activity affects conditions on Earth. The particles emitted by the Sun during periods of high sunspot activity disrupt radio communications and have an impact on Earth's magnetic field and climate. Some scientists theorize that ice ages are caused by relatively low solar activity over a period of time.

LESSON 7.3 Defining the Circular Functions **405**

11.

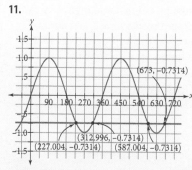

Exercise 12 Students may be surprised that the two questions in this exercise are asking the same thing.

Exercise 13 The cyclical nature of the graph in 13a is much easier to see with a connected scatter plot.

Because it's not possible to get an accurate count of sunspots from a single location, daily counts are calculated by averaging the counts from international observatories. The yearly values given in the table are the average daily means. **LANGUAGE** The phrase *maximum solar activity* refers to a period in the solar cycle with a large number of sunspots, such as occurred in 1979–1982 and 1988–1991.

13a.

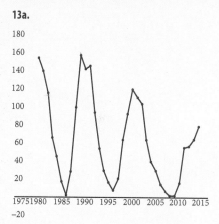

CONTEXT Science Connection The number and position of sunspots varies according to the approximately 11-year solar cycle. At the point in the cycle with the minimum number of sunspots, sunspots appear close to the solar equator. At the high point in the cycle, sunspots appear closer to the poles. The magnetic activity in the sunspots, however, has a 22-year cycle. Most sunspots have one magnetic polarity for a period of 11 years and then switch to the opposite polarity for 11 years. Scientists do not yet know why sunspots follow this cycle.

Exercise 14 These angle measures can be found by drawing the reference triangle in the appropriate quadrant, and finding the angle with respect to the positive x-axis.

20. $f(x) = \dfrac{2(x+2)(x-5)}{(x+4)(x-1)}$

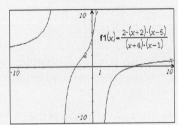

14. Find each angle θ with the given trigonometric value. Use the domain $0 \leq \theta \leq 360°$.

   a. $\cos\theta = -\dfrac{\sqrt{3}}{2}$ $\theta = 150°$ and $\theta = 210°$

   b. $\cos\theta = -\dfrac{\sqrt{2}}{2}$ $\theta = 135°$ and $\theta = 225°$

   c. $\sin\theta = -\dfrac{3}{5}$ $\theta \approx 217°$ and $\theta \approx 323°$

   d. $\sin\theta = 1$ $\theta = 90°$

15. In a previous lesson you defined $\sin\theta = \dfrac{y}{d}$, $\cos\theta = \dfrac{x}{d}$, and $\tan\theta = \dfrac{y}{x}$, where x and y were the coordinates of a point and d was the distance of the point from the origin. Solve the equations for sine and cosine for y and x and substitute these into the tangent equation to find a new definition for the tangent of an angle. Check that your formula works for a randomly chosen angle. $\tan\theta = \dfrac{\sin\theta}{\cos\theta}$

Review

16. Annie is standing on a canyon floor 20 m from the base of a cliff. Looking through her binoculars, she sees the remains of ancient cliff dwellings in the cliff face. Annie holds her binoculars at eye level, 1.5 m above the ground.

   a. Write an equation that relates the angle at which she holds the binoculars to the height above ground of the object she sees. @
   $h = 1.5 + 20\tan A$

   b. The top of the cliff is at an angle of 58° above horizontal when viewed from where Annie is standing. How high is the cliff, to the nearest tenth of a meter? @ 33.5 m

   c. Lower on the cliff, Annie sees ruins at angles of 36° and 40° from the horizontal. How high are the ruins, to the nearest tenth of a meter?
   16.0 m and 18.3 m

   d. There is a nest of cliff swallows in the cliff face, 10 m above the canyon floor. At what angle should Annie point her binoculars to observe the nest? Round your answer to the nearest degree. 23°

The Cliff Palace in Mesa Verde National Park, Colorado, was constructed by Ancestral Puebloan people around 1200 c.e.

17. Convert to the specified units using ratios. For example, to convert 0.17 meter to inches:

   $0.17\,\text{m} \cdot \dfrac{100\,\text{cm}}{1\,\text{m}} \cdot \dfrac{1\,\text{in.}}{2.54\,\text{cm}} \approx 6.7\,\text{in.}$

   a. 0.500 day to seconds 43,200 s

   b. 3.0 mi/h to ft/s (*Note:* 1 mile = 5280 feet) 4.4 ft/s

18. Find the circumference and area of the circle with equation
   $2x^2 + 2y^2 - 2x + 7y - 38 = 0$. $C \approx 29.7$ units, $A \approx 70.1$ square units

19. Rewrite each expression as a single rational expression in factored form. @

   a. $\dfrac{x+1}{3x-4} - \dfrac{x+2}{x+4} + \dfrac{4x}{16-x^2}$ $\dfrac{3}{x-4}$, $x \neq \pm 4$

   b. $\dfrac{2x^2-2}{x^2+3x+2} \cdot \dfrac{x^2-x-6}{x^2-4x+3}$ 2, $x \neq -2, -1, 1, 3$

   c. $\dfrac{1+\dfrac{a}{3}}{1-\dfrac{a}{6}}$ $\dfrac{2(3+a)}{6-a}$, $a \neq 6$

20. Write an equation for a rational function, $f(x)$, that has vertical asymptotes $x = -4$ and $x = 1$, horizontal asymptote $y = 2$, and zeros $x = -2$ and $x = 5$. Check your answer by graphing the equation on your calculator.

LESSON 7.4

Radian Measure

The measure of our intellectual capacity is the capacity to feel less and less satisfied with our answers to better and better problems.

C. WEST CHURCHMAN

You've learned a number of things about circles in the past. For instance you should recall the following facts about circles:

The **circumference** of a circle is the distance around the circle. For a circle with radius r, $C = 2\pi r$.

A **central angle** is an angle formed by two radii.

An **arc** is a piece of the circle.

The **measure of an arc** of a circle is the same as the measure of the central angle that intercepts the arc.

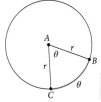

$\angle BAC$ is a central angle. $m\angle BAC = m\overset{\frown}{BC}$

The **length of an arc** is given in the same units as the radius or circumference.

You can use ratios and proportions to find the lengths and measures of arcs.

In this way, you can see that the measure of an arc in degrees, m, the length of the arc, s, and the circumference, C, of the circle are related by the proportion $\dfrac{m}{360°} = \dfrac{s}{C}$.

This formula applies when you are measuring the arc in degrees. However, the choice to divide a circle into 360° is rooted in the history of mathematics and is not connected to any fundamental properties of circles. In this lesson you will learn about a different angle measure that is based on an essential property of circles.

The circular gears of a clock remind us that the fact that hours are divided into 60 minutes and a circle is divided into 360 degrees both have their origins in the same Babylonian number system.

History CONNECTION

More than 5000 years ago, the Sumerians in Mesopotamia used a base-60 number system. They may have chosen this system because numbers like 30, 60, and 360 can be evenly divided by many numbers. The Babylonians and Egyptians then borrowed this system and divided the circle into 360 degrees. The Egyptians also devised the symbol for degrees and went on to divide both Earth's equator and north-south great circles into 360 degrees, inventing latitude and longitude lines. The Greek astronomer and mathematican Hipparchus of Rhodes (ca. 190–120 B.C.E.) is credited with introducing the Babylonian division of the circle to Greece and producing a table of chords, the earliest known trigonometric table. Hipparchus is often called the "founder of trigonometry."

LESSON 7.4

In this lesson, students learn the concept of radians and discover formulas for arc length and the area of a sector.

COMMON CORE STATE STANDARDS		
APPLIED	**DEVELOPED**	**INTRODUCED**
	F.TF.1	F.TF.3
	F.TF.2	
	G.C.5	

Objectives

- Introduce angle measure in radians and convert between radians and degrees
- Understand a 1-radian angle as the angle whose intercepted arc is the length of the radius
- Find angular speed, the amount of rotation in radians (or degrees) per unit of time
- Learn formulas for arc length and the area of a sector in terms of radians

Vocabulary

circumference
central angle
arc
measure f an arc
length of an arc
radian

Materials

- string
- Calculator Notes: Radians

Launch

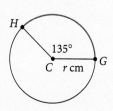

a. What is the measure of $\overset{\frown}{HG}$? 135°

b. What fraction of the circumference is the length of $\overset{\frown}{HG}$? $\dfrac{135°}{360°} = \dfrac{3}{8}$

c. What is the length of $\overset{\frown}{HG}$?
$\dfrac{3}{8} \cdot 2\pi r = \dfrac{3\pi r}{4} \approx 2.36r$

ELL
Students can overlay the frog metaphor from the last lesson with radians and the new vocabulary.

Support
Remind students that radians can be thought of as another way to describe the size of angles. Use the Radian Protractor worksheet included in the Investigation Worksheet so students can mark the lengths of the radius on their radian protractor.

Advanced
Have students develop their own methods for calculating arc length and sector area. Use the **One Step** version of the Investigation.

Investigate

Guiding the Investigation

There is a desmos simulation for this investigation in your eBook.

Steps 3–4 Students may benefit from lightly drawing the segments to make some of the angles.

Step 5 Have students write the radian measures in terms of π rather than as decimal approximations.

Step 7 The completed protractor should have angles of 0, 0.5, $\frac{\pi}{6}$, $\frac{\pi}{4}$, 1, $\frac{\pi}{3}$, 1.5, $\frac{\pi}{2}$, 2, $\frac{2\pi}{3}$, $\frac{3\pi}{4}$, 2.5, $\frac{5\pi}{6}$, 3, and π radians labeled.

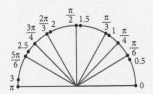

INVESTIGATION

A Radian Protractor

YOU WILL NEED
- string

Use the semicircle on the worksheet to complete the investigation.

Step 1 Mark the length of the radius of the semicircle on your string.

Step 2 Starting from the right-hand base, use the radius length of string to measure off an arc whose length is the same as the radius.

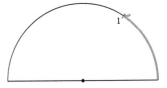

Step 3 The central angle intercepting this arc has a measure of 1 **radian**. Use your radius length of string to mark points that correspond to angles with measures of 2 and 3 radians.

Step 4 Fold your radius length of string in half and use this length to locate points corresponding to angles with measures 0.5, 1.5, and 2.5 radians.

Step 5 If the radius of your circle is r, calculate the length of a semicircular arc. How many radians is this? Mark this radian value on your protractor. πr units, or π radians

Step 6 What is the length of the arc intercepted by a right angle? How many radians are associated with a right angle? Mark this radian value on your protractor. $\frac{\pi r}{2}$ units; a right angle corresponds to $\frac{\pi}{2}$ radians.

Step 7 Find radian values that correspond with these common angle measurements and mark them on your protractor: 30°, 45°, 60°, 120°, 135°, and 150°.

Definition of a Radian

One **radian** is the measure of the angle that cuts off an arc whose length is equal to the radius of the circle. There are 2π radians in a circle.

Like degrees, radians are used to measure angles and rotations. The radian measure of an angle can be found by constructing *any* arc intercepted by that angle, with its center at the vertex of the angle, and dividing the length of that arc by its radius.

$$\text{radian measure} = \frac{\text{arc length}}{\text{radius}}$$

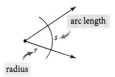

You can solve this equation for the arc length to get

arc length = radian measure · radius.

This equation says that the arc length is directly proportional to the radius and the constant of proportionality is the radian measure of the angle. You can also use this relationship to find a way to convert between radian measure and degrees.

Modifying the Investigation

Whole Class Complete Steps 1 through 5 as students follow along with their own protractor. Discuss Steps 5 through 8.

Shortened Use the desmos simulation found in your ebook.

One Step Use the **One Step** in your eBook. As students decide that the number of radii in the arcs' lengths is the same for all the circles, ask for that number ($\frac{2\pi}{9}$) and encourage them to generalize to angles of all measures. During the discussion, define *radian measure* and point out that it counts the number of radii (as opposed to the number of degrees).

The circumference of a circle depends on the length of its radius. Using the radian measure definition and the formula $C = 2\pi r$, you will find that the radian measure of a full circle, or 360° rotation, is $\frac{arc\,length}{radius} = \frac{2\pi r}{r} = 2\pi$ radians.

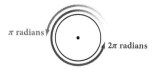

The conversion formula

$$\frac{angle\,in\,degrees}{360} = \frac{angle\,in\,radians}{2\pi}$$

is based on the full circle, or one rotation. You can also use an equivalent formula based on a half rotation.

$$\frac{angle\,in\,degrees}{180} = \frac{angle\,in\,radians}{\pi}$$

If you solve the last proportion for either the angle in degrees or the angle in radians, you get a formula that you can use to convert from one system to the other.

$$angle\,in\,degrees = \frac{180}{\pi}(angle\,in\,radians)$$

$$angle\,in\,radians = \frac{\pi}{180}(angle\,in\,degrees)$$

Radian measure is based on a fundamental property of a circle: The circumference of any circle is 2π times its radius. For this reason, radian measure is often preferred in advanced mathematics and in physics. You can learn to recognize, compare, and use radians and degrees, just as in the past you have worked with inches and centimeters or Fahrenheit and Celsius.

Degree measures are always labeled with the symbol °. Radian measures are not usually labeled, but they can be labeled for clarity.

Radian measures of most common angles are irrational numbers, but you can express them as exact values in terms of π.

EXAMPLE A Convert degrees to radians or radians to degrees.

a. $\frac{2\pi}{3}$ b. n radians

c. 225°   d. n°

Solution You can use either conversion formula. Here we will use the equivalence 180° = π radians.

a. You can think of $\frac{2\pi}{3}$ as $\frac{2}{3}(\pi)$, or $\frac{(2\pi)}{3}$. Although no units are given, it is clear that these are radians.

$$\frac{x}{180} = \frac{\frac{2\pi}{3}}{\pi} \text{ or } \frac{x}{180} = \frac{2}{3}$$

So $x = 120°$. Therefore, $\frac{2\pi}{3} = 120°$.

b. $\frac{x}{180} = \frac{n}{\pi}$ or $x = \frac{n}{\pi} \cdot 180$

So n radians $= \left(\frac{180n}{\pi}\right)°$.

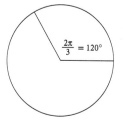

$\frac{2\pi}{3} = 120°$

c. $\frac{225}{180} = \frac{x}{\pi}$ or $x = \frac{225}{180} \cdot \pi = \frac{5\pi}{4}$

So $225° = \frac{5\pi}{4}$ radians.

d. $\frac{n}{180} = \frac{x}{\pi}$ or $x = \frac{n}{180} \cdot \pi$

So $n° = \frac{n\pi}{180}$ radians.

You can use the relations you found in parts b and d in Example A to convert easily between radians and degrees.

You can also use dimensional analysis to convert between degrees and radians. Dimensional analysis is a procedure based on multiplying by fractions formed of conversion factors to change units. For example, to convert 140° to radian measure, you write it as a fraction over 1 and multiply it by $\frac{\pi \text{ radians}}{180°}$, which is a fraction equal to 1.

$$\frac{140 \text{ degrees}}{1} \cdot \frac{\pi \text{ radians}}{180 \text{ degrees}} = \frac{140\pi \text{ radians}}{180} = \frac{7\pi}{9} \text{ radians}$$

You can use your calculator to check or approximate conversions between radians and degrees.

Earlier you reviewed the proportion $\frac{m}{360°} = \frac{s}{C}$, which relates the arc length, s, and arc measure in degrees, m. Using radians, you can write the relationship $\frac{s}{2\pi r} = \frac{\theta}{2\pi}$, where s is the arc length, r is the radius, and θ is the central angle measured in radians. If you solve this equation for s, you'll get $s = r\theta$ as a formula for arc length. You can also use radian measure to write a simple formula for the area of a sector.

EXAMPLE B Circle O has diameter 10 m. The measure of central angle BOC is 1.4 radians. What is the length of its intercepted arc, $\widehat{BC}$? What is the area of the shaded sector?

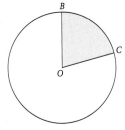

Solution In the formula $s = r\theta$, substitute 5 for r and 1.4 for θ, and solve for s. The length of $\widehat{BC}$ is $5 \cdot 1.4$, or 7 m.

The ratio of the area of the sector to the total area of the circle is the same as the ratio of the measure of $\angle BOC$ to 2π radians.

$$\frac{A_{\text{sector}}}{A_{\text{circle}}} = \frac{1.4}{2\pi}$$

The area of the circle is πr^2, or 25π m².

$$\frac{A_{\text{sector}}}{25\pi} = \frac{1.4}{2\pi}$$

Solving the equation, you find that the area of the sector is 17.5 m².

As you saw in the example, the area of a sector with radius r and central angle θ can be found using the proportion

$$\frac{A_{\text{sector}}}{\pi r^2} = \frac{\theta}{2\pi}$$

You can rewrite this relationship as

$$A_{\text{sector}} = \frac{\pi r^2 \theta}{2\pi} = \frac{r^2 \theta}{2} = \frac{1}{2}r^2\theta$$

> **Length of an Arc and Area of a Sector**
>
> When a central angle, θ, of a circle with radius r is measured in radians, the length of the intercepted arc, s, is given by the equation
>
> $$s = r\theta$$
>
> and the area of the intercepted sector, A, is given by the equation
>
> $$A = \frac{1}{2}r^2\theta$$

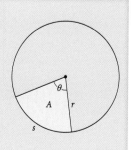

When an object follows a circular path, the distance it travels is the arc length. You can calculate its speed as distance traveled per unit of time. The amount of rotation, or angle traveled per unit of time, is called the **angular speed**.

EXAMPLE C

The Cosmo Clock 21 Ferris wheel at the Cosmo World amusement park in Yokohama, Japan, has a 100 m diameter. This giant Ferris wheel, with 60 gondolas and 8 people per gondola, makes one complete rotation every 15 minutes. The wheel reaches a maximum height of 112.5 m from the ground.

a. Find the speed of a person on this Ferris wheel as it is turning.

b. Find the angular speed of this person.

The Cosmo Clock 21 Ferris wheel, which opened in 1999, can hold as many as 480 passengers at a time.

Solution

The speed is the distance traveled per unit of time, measured in units such as meters per second. The angular speed is the rate of rotation, measured in units like radians per second or degrees per second.

a. The person travels one complete circumference every 15 min, or

$$\frac{2\pi r \,\text{m}}{15 \,\text{min}} = \frac{2\pi \cdot 50 \,\text{m}}{15 \,\text{min}} \approx 20.94 \,\frac{\text{m}}{\text{min}}$$

CRITICAL QUESTION Why do we use radians instead of degrees for trigonometric functions?

BIG IDEA Some ancient mathematicians considered degree angle measures to be arbitrary, since 360° for a complete circle was an arbitrary choice, like 24 hours in a day. Mathematicians prefer to measure angles in radians, a unit derived naturally from the geometry of the circle. The point moving around the circle can go around more than once, so angles in degrees are less relevant.

Formative Assessment

As students work and present, assess their understanding of arc length and their comfort level in working with ratios and proportions. Watch to see whether students make effective use of radians and the conversion proportion to measure angles in multiples of 90°. Ideally, by the end of this lesson, students will extend this process to angles in multiples of 30° and 45°.

Apply

Extra Example

1. Find the length of the intercepted arc for the central angle with $r = 5$ and $\theta = \frac{\pi}{3}$. $\frac{5\pi}{3}$
2. 180° is equivalent to how many radians? π

Closing Question

How is the number of radians in an angle related to the number of radii that can be laid out along the arc that the angle intercepts?

The number of radians in an angle is the number of radii that can be laid out along the arc that the angle intercepts, where the radii are those of a circle centered at the angle's vertex and containing the arc. Because a 360° angle intercepts a full circle, which has arc length 2π radii, 2π radians corresponds to 360°.

Assigning Exercises

Suggested: 1–13
Additional Practice: 14–19

Exercise 1 Encourage students to think about ratios instead of applying formulas.

Exercise 2 Suggest that students sketch these angles and arcs. Point out that 2c uses d instead of r.

So the person travels at about 21 m/min. You can also use dimensional analysis to express this speed as approximately 1.26 km/h.

$$\frac{20.94 \text{ m}}{1 \text{ min}} \cdot \frac{60 \text{ min}}{1 \text{ h}} \cdot \frac{1 \text{ km}}{1000 \text{ m}} \approx \frac{1.26 \text{ km}}{1 \text{ h}}$$

b. The person completes one rotation, or 2π radians, every 15 min.

$$\frac{2\pi \text{ radians}}{15 \text{ min}} \approx 0.42 \text{ radian/min}$$

So the person's angular speed is 0.42 radian/min.

Using dimensional analysis, you can convert between units to express answers in any form.

$$\frac{2\pi \text{ radians}}{15 \text{ min}} \cdot \frac{360°}{2\pi \text{ radians}} = \frac{24°}{1 \text{ min}}$$

So you can also express the angular speed as 24°/min.

A central angle has a measure of 1 radian when its intercepted arc is the same length as the radius. Similarly, the number of radians in an angle measure is the number of radii in the arc length of its intercepted arc.

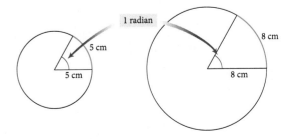

7.4 Exercises

You will need a graphing calculator for Exercises 7–9.

Practice Your Skills

1. Convert between radians and degrees. Give exact answers. (Remember that degree measures are always labeled °, and radians generally are not labeled.)

   a. 80° @ $\frac{4\pi}{9}$
   b. 570° @ $\frac{19\pi}{6}$
   c. $-\frac{4\pi}{3}$ −240°
   d. $\frac{11\pi}{9}$ 220°
   e. $-\frac{3\pi}{4}$ −135°
   f. 3π 540°
   g. −900° @ -5π
   h. $\frac{5\pi}{6}$ 150°

2. Find the length of the intercepted arc for each central angle.

   a. $r = 3$ and $\theta = \frac{2\pi}{3}$ @ 2π
   b. $r = 2$ and $\theta = 2$ 4
   c. $d = 5$ and $\theta = \frac{\pi}{6}$ $\frac{5\pi}{12}$

412 CHAPTER 7 Trigonometry and Trigonometric Functions

3. Draw a large copy of this diagram on your paper. Each angle shown has a reference angle of 0°, 30°, 45°, 60°, or 90°.

   a. Find the counterclockwise degree rotation of each segment from the positive x-axis. Write your answers in both degrees and radians.

   b. Find the exact coordinates of points A–N.

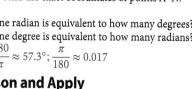

4. One radian is equivalent to how many degrees? One degree is equivalent to how many radians? $\frac{180}{\pi} \approx 57.3°$; $\frac{\pi}{180} \approx 0.017$

Reason and Apply

5. Are 6 radians more than, less than, or the same as one rotation about a circle? Explain. Less than; one rotation is 2π, which is more than 6.

6. The minute hand of a clock is 15.2 cm long.

   a. What is the distance the tip of the minute hand travels during 40 minutes? 63.7 cm

   b. At what speed is the tip moving, in cm/min? @ 1.6 cm/min

   c. What is the angular speed of the tip in radians/minute? 0.105 radian/min

Republic Square in the capital city of Yerevan, Armenia

7. Without using a graphing calculator, graph $y = \sin x$ over the domain $0 \leq x \leq 2\pi$.

   a. On the x-axis, label all the x-values that are multiples of $\frac{\pi}{6}$.

   b. On the x-axis, label all the x-values that are multiples of $\frac{\pi}{4}$.

   c. What x-values in this domain correspond to a maximum value of $\sin x$? A minimum value of $\sin x$? $\sin x = 0$? $\frac{\pi}{2}$; $\frac{3\pi}{2}$; 0, π, and 2π

   d. Use your graphing calculator to check your graph. ⓗ Students may use a graph or table to check their answers.

8. Without using a graphing calculator, graph $y = \cos x$ over the domain $0 \leq x \leq 2\pi$.

   a. On the x-axis, label all the x-values that are multiples of $\frac{\pi}{6}$.

   b. On the x-axis, label all the x-values that are multiples of $\frac{\pi}{4}$.

   c. What x-values in this domain correspond to a maximum value of $\cos x$? A minimum value of $\cos x$? $\cos x = 0$? 0 and 2π; π; $\frac{\pi}{2}$ and $\frac{3\pi}{2}$

   d. Use your graphing calculator to check your graph. ⓗ Students may use a graph or table to check their answers.

9. Without using a graphing calculator, graph $y = \frac{\sin x}{\cos x}$ (that is, $y = \tan x$) over the domain $-\pi \leq x \leq \pi$.

   a. On the x-axis, label all the x-values that are multiples of $\frac{\pi}{6}$ or $\frac{\pi}{4}$.

   b. Name the x-values in this domain where the denominator, $\cos x$, is 0. What can you say about the function at these values? Odd multiples of $\frac{\pi}{2}$; the function is undefined.

   c. Which x-values in the domain $-\pi \leq x \leq \pi$ produce vertical asymptotes? $-\frac{\pi}{2}$ and $\frac{\pi}{2}$

   d. Use your graphing calculator to verify your graph of $y = \frac{\sin x}{\cos x}$ over the domain $-\pi \leq x \leq \pi$. Confirm that the graph is the same as the graph of $y = \tan x$. ⓗ

LESSON 7.4 Radian Measure 413

Exercise 3a SUPPORT Start students off by helping them find the measures of angles A–C as fractions of 90° and point out that all other angles are multiples of A–C.

3a, b.

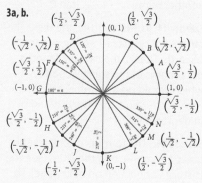

Exercises 7–9 SUPPORT If students have difficulty recognizing multiples of $\frac{\pi}{6}$ and $\frac{\pi}{4}$, suggest that they divide the domain into increments equal to $\frac{\pi}{12}$, marking them off as $\frac{2\pi}{12}, \frac{3\pi}{12}, \frac{4\pi}{12}$, and so on, all the way to $\frac{24\pi}{12}$. At this point, have students simplify the fractions in order to notice patterns.

Exercise 7 You might suggest that students first mark off π and 2π and then divide each section into six sections for $\frac{\pi}{6}$ and into four sections for $\frac{\pi}{4}$.

ASK "Do any of these values coincide?" [Yes; $\frac{3\pi}{6} = \frac{2\pi}{4}$, and $\frac{9\pi}{6} = \frac{6\pi}{4}$.]

9a.

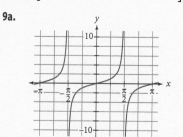

9d.

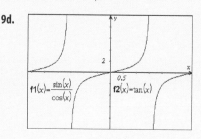

7a, b.

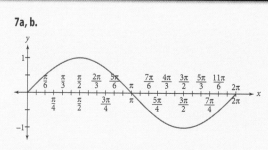

8a, b.

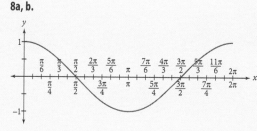

LESSON 7.4 Radian Measure 413

Exercise 11 Suggest that students sketch these reference triangles in the appropriate quadrant. If students are stuck on 11c, ask whether they know another name for $\frac{\sin \theta}{\cos \theta}$. [tan θ]

Exercise 14 In 14a and 14c, the distances are measured on the surface of Earth. That is, they are arc lengths, not lengths of the shortest paths, which would go through the planet. Remind students that they should be in degree mode.

Exercise 14c The formula calculates the great arc connecting the cities (see the History Connection on next page). To demonstrate the idea of a great arc, use a globe, ball, or balloon and pull a string tight between two points on the surface.

10. Find the value of *r* in the circle at right. *r* = 48

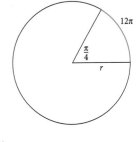

11. Solve for θ. Express your answers in radians.
    a. $\cos \theta = -\frac{1}{2}$ and $\pi \leq \theta \leq \frac{3\pi}{2}$ @ $\frac{4\pi}{3}$
    b. $\cos \theta = \frac{\sqrt{2}}{2}$ and $\frac{3\pi}{2} \leq \theta \leq 2\pi$ $\frac{7\pi}{4}$
    c. $\frac{\sin \theta}{\cos \theta} = \sqrt{3}$ and $0 \leq \theta \leq \frac{\pi}{2}$ $\frac{\pi}{3}$

12. Suppose you are biking down a hill at 24 mi/h. What is the angular speed, in radians per second, of your 27-inch-diameter bicycle wheel? @ 31.3 radians/s

> **Recreation**
> **CONNECTION**
>
> Fred Rompelberg (b 1945) of the Netherlands broke the world record in 1995 for the fastest recorded bicycle speed, 167.043 mi/h. He rode a specially-designed bicycle behind a race car, which propelled him forward by slipstream, an airstream that reduces air pressure. The record-breaking event took place at Bonneville Salt Flats in Utah, the venue for numerous land speed records and home of the annual World of Speed event.

Coleman Brother Racing team pushing their yellow super bike during the World of Speed 2012

13. A sector of a circle with radius 8 cm has central angle $\frac{4\pi}{7}$.
    a. Use the formula $A = \frac{1}{2}r^2\theta$ to find the area of the sector. $A \approx 57.45$ cm²
    b. Set up a proportion of the area of the sector to the total area of the circle, and a proportion of the central angle of the sector to the total central angle measure. $\frac{A}{64\pi} = \frac{\frac{4\pi}{7}}{2\pi}$
    c. Solve your proportion from 13b to show that the formula you used in 13a is correct. $\frac{4\pi}{7}$

    13c. $A = 64\pi \cdot \frac{\frac{4\pi}{7}}{2\pi} \approx 57.45$ cm²

14. **APPLICATION** The two cities Minneapolis, Minnesota, and Lake Charles, Louisiana, lie on the 93° W longitudinal line. The latitude of Minneapolis is 45° N (45° north of the equator), whereas the latitude of Lake Charles is 30° N. The radius of Earth is approximately 3960 mi. ⓗ
    a. Calculate the distance between the two cities. 1037 mi
    b. Like hours, degrees can be divided into minutes and seconds for more precision. (There are 60 seconds in a minute and 60 minutes in a degree.) For example, 61°10′ means 61 degrees 10 minutes. Write this measurement as a decimal. 61.17°

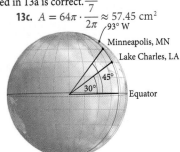

414 **CHAPTER 7** Trigonometry and Trigonometric Functions

c. If you know the latitudes and longitudes of two cities, you can find the distance in miles between them, D, using this formula:

$$D = \frac{\pi \cdot r}{180} \cos^{-1}\left[\sin \phi_A \sin \phi_B + \cos \phi_A \cos \phi_B \cos(\theta_A - \theta_B)\right]$$

In the formula, r is the radius of Earth in miles, ϕ_A and θ_A are the latitude and longitude of city A in degrees, and ϕ_B and θ_B are the latitude and longitude of city B in degrees. North and east are considered positive angles, and south and west are considered negative.

Using this formula, find the distance between Anchorage, Alaska (61°10′ N, 150°1′ W), and Tucson, Arizona (32°7′ N, 110°56′ W).   2660 mi

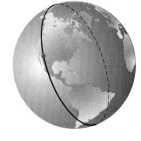

History CONNECTION

To find the shortest path between two points on a sphere, you connect the two points with a great arc, an arc of a circle that has the same center as the center of Earth. Pilots fly along great circle routes to save time and fuel. Charles Lindbergh's carefully planned 1927 flight across the Atlantic was along a great circle route that saved about 473 miles compared with flying due east.

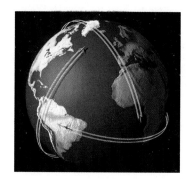

Review

15. List the transformations of each graph from its parent function.

a. $y = 2 + (x + 4)^2$
b. $\frac{y}{3} = \left(\frac{x-5}{4}\right)^2$ @
c. $y + 1 = |x - 3|$
d. $y = 3 - 2|x + 1|$ @

16. Write an equation for each graph. @

a.
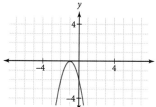
$y = -2(x + 1)^2$

b.
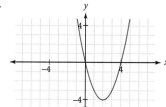
$y + 4 = (x - 2)^2$, or $y = (x - 2)^2 - 4$

c.
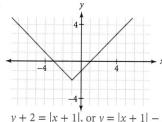
$y + 2 = |x + 1|$, or $y = |x + 1| - 2$

d.
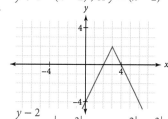
$-\frac{y-2}{2} = |x - 3|$, or $y = -2|x - 3| + 2$

Context History Connection Two-dimensional maps are deceptive for purposes of navigation. Drawing a straight line between two points, for example, appears to be shorter than following a plane's curved flight path. Flat maps also present a distorted representation of the rounded surface of Earth. This distortion is especially significant on most world maps, where distances are greatly exaggerated near the poles. Globes are useful when an accurate representation of Earth is needed. In addition to helping plot great-circle routes, globes are useful for tracing earthquake shocks, tidal-wave paths, and satellite movement.

15a. translated vertically 2 units and horizontally 4 units

15b. dilated vertically by a factor of 3 and horizontally by a factor of 4 and translated horizontally 5 units

15c. translated vertically 1 unit and horizontally 3 units

15d. dilated vertically by a factor of 2, reflected across the x-axis, and translated vertically 3 units and horizontally −1 unit

17. Find a second value of θ that gives the same trigonometric value as the angle given. Use domain 0° ≤ θ ≤ 360°.

   a. sin 23° = sin θ   157°
   b. sin 216° = sin θ   324°
   c. cos 342° = cos θ   18°
   d. cos 246° = cos θ   114°

18. While Yolanda was parked at her back steps, a slug climbed onto the wheel of her go-cart just above where the wheel makes contact with the ground. When Yolanda hopped in and started to pull away, she did not notice the slug until her wheels had rotated $2\frac{1}{3}$ times. Yolanda's wheels have a 12 cm radius.

    a. How far off the ground is the slug when Yolanda notices it?   18 cm
    b. How far horizontally has the slug moved?   166 cm

19. Circles O and P, shown below, are tangent at A. Explain why AB = BC. (Hint: One method uses congruence.)

    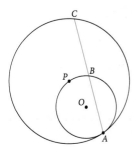

    Possible answer: Construct $\overline{AP}$, $\overline{BP}$, and $\overline{CP}$. △APC is isosceles because $\overline{AP}$ and $\overline{CP}$ are radii of the same circle. ∠A ≅ ∠C because they are base angles of an isosceles triangle. ∠ABP measures 90° because the angle is inscribed in a semicircle. ∠PBC = 90° because it is supplementary to ∠ABP. △APB ≅ △CPB by AAS. Therefore, AB = BC because corresponding parts of congruent triangles are congruent.

PERFORMANCE TASK

Sitting at your desk, you are approximately 6380 km from the center of Earth. Consider your motion relative to the center of Earth as Earth rotates on its axis. What is your angular speed? What is your speed in km/h? What is your speed in mi/h?

PERFORMANCE TASK

Students might convert kilometers to miles, they can use the fact that 1 mi = 1.61 km to perform a dimensional analysis.

Angular speed - 2π radians/d, or 0.26 radian/h

Speed in km/h - approximately 1670 km/h

Speed in mi/h - approximately 1040 mi/h

Exercise 17 As needed, suggest that students use the unit circle and reference angles to find these values.

LESSON 7.5

Graphing Trigonometric Functions

The best way to predict the future is to invent it.
BENJAMIN PIMENTEL

Graphs of the functions $y = \sin x$ and $y = \cos x$ and transformations of these graphs are collectively called **sine waves** or **sinusoids**. You will find that reflecting, translating, and dilating sinusoidal functions and other **trigonometric functions** is very much like transforming any other function. In this lesson you will explore many real-world situations in which two variables can be modeled by a sinusoidal function in the form $y = k + b \sin\left(\frac{x-h}{a}\right)$, or $\frac{y-k}{b} = \sin\left(\frac{x-h}{a}\right)$.

The seven wavy terraces of the Hsinbyume Pagoda in Mingum, Myanmar, may represent the seven surrounding hills or the seven seas of the universe.

EXAMPLE A | The graph of one cycle ($0 \leq x \leq 2\pi$) of $y = \sin x$ is shown at right. Sketch the graph of one cycle of
a. $y = 2 + \sin x$
b. $y = \sin(x - \pi)$
c. $y = 3 + 2 \sin(x + \pi)$

Solution | Use your knowledge of transformations of functions.

a. In relation to the graph of any parent function, $y = f(x)$, the graph of $y = 2 + f(x)$ is a translation up 2 units. The graph of one cycle of $y = 2 + \sin x$ is therefore a translation up 2 units from the graph of $y = \sin x$.

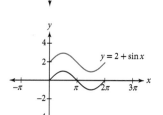

LESSON 7.5

In this lesson, students see how the period, amplitude, and phase shift of sinusoidal and tangent functions relate to transformations.

COMMON CORE STATE STANDARDS		
APPLIED	**DEVELOPED**	**INTRODUCED**
	F.IF.4	F.TF.3
	F.IF.7e	
	F.BF.3	

Objectives
- Apply knowledge of transformations to the graphs of trigonometric functions
- Learn vocabulary associated with sinusoidal graphs, including *amplitude* and *phase shift*
- Model real-world data with sine and cosine functions

Vocabulary
sine waves
trigonometric functions
amplitude
phase shift
midline
period

Materials
- string
- washer or other small weight
- motion sensor
- Calculator Notes: Looking for the Rebound Using the Easy Data App; Asymptotes, Holes and Drag Lines, *optional*

LANGUAGE A *sinusoid* is a curve shaped like the graph of the sine (or cosine) function.

Launch

Project the Launch from your eBook. This is Example A.

ELL
For Example B, review the separation of position components. Continue to use the frog metaphor.

Support
Students may need to review translations, reflections, and vertical dilations of previously discussed functions (absolute value, quadratic, and exponential) in order to make a smooth transition into transforming trigonometric functions.

Advanced
Suggest that students graph the reciprocals of the sine, cosine, and tangent functions and explore transformations of their graphs.

Investigate

Example A

If you didn't do the Launch, project Example A and ask students to predict, without using graphing calculators, how the graphs will differ from the graph of the parent function. Then have students work in pairs. Have them share their solutions, not just their answers. **SMP 1, 3, 6**

b. When x is replaced by $(x - \pi)$ in any function, the graph translates right π units.

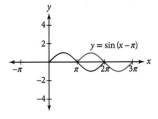

c. The coefficient 2 means the graph $y = \sin x$ must be dilated vertically by a factor of 2. The graph must also be translated left π units and up 3 units.

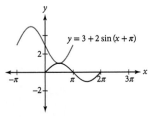

Recall that the period of a function is the smallest distance between values of the independent variable before the cycle begins to repeat. You've discovered that the period of both $y = \sin x$ and $y = \cos x$ is 2π, or 360°. The period of each of the three functions in Example A is also 2π, because they were not dilated horizontally.

When working with the graphs of the sine and cosine, the standard transformations that you have used in the past have special names. The **amplitude** of a sinusoid is the absolute value of the vertical scale factor, or $|b|$. To find this value you can calculate half the difference of the maximum and minimum function values, or $\frac{maximum - minimum}{2}$. The amplitude of $y = \sin x$ and $y = \cos x$ is 1 unit. The horizontal translation value is called the **phase shift**. Look carefully at the figures below to see how this relates to both the sine and cosine functions. The vertical translation value, k, gives you the equation of the **midline**, or average value. This is the line that corresponds to the x-axis in the basic sine or cosine function. The horizontal dilation factor, a, must be multiplied by 2π to determine the **period**. This is because the period of either $y = \sin x$ or $y = \cos x$ is 2π.

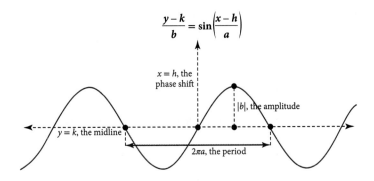

418 CHAPTER 7 Trigonometry and Trigonometric Functions

Modifying the Investigation

Whole Class Have three students collect data for the whole class. Find a model for the data and real-world meanings with student input.

Shortened Use the sample data or the Pendulum II Sample Data worksheet.

One Step Ask students to do the investigation. As they work to find functions to model the data, encourage them to consider horizontal and vertical dilations and horizontal translations of the parent functions. During the discussion, introduce the tangent function and the terms *sinusoid, sinusoidal function, amplitude,* and *phase shift.*

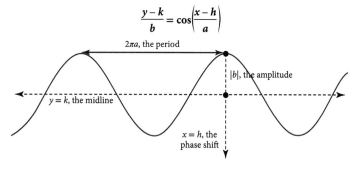

Consider the graph of the function $\dfrac{y-4}{3} = \cos\left(\dfrac{x - \frac{\pi}{3}}{2}\right)$.

The parent cosine function is shown in black. First perform the dilations. The red image graph is a dilation of the parent graph vertically by a factor of 3 and horizontally by a factor of 2.

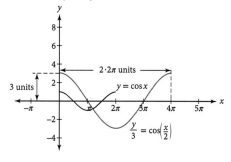

Now apply the translations. Translate the red image graph right $\dfrac{\pi}{3}$ units and up 4 units.

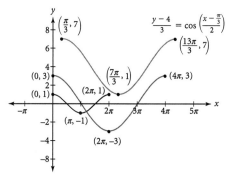

Values of the function $\dfrac{y-4}{3} = \cos\left(\dfrac{x - \frac{\pi}{3}}{2}\right)$, graphed in blue, vary from a minimum of 1 to a maximum of 7, so it has amplitude 3. The horizontal scale factor 2 means its period is dilated to 4π. The horizontal translation makes the phase shift $\dfrac{\pi}{3}$ so that the cycle of this graph starts at $\dfrac{\pi}{3}$, has a minimum at $2\pi + \dfrac{\pi}{3}$, or $\dfrac{7\pi}{3}$, and ends at $4\pi + \dfrac{\pi}{3}$, or $\dfrac{13\pi}{3}$.

LESSON 7.5 Graphing Trigonometric Functions

Conceptual/Procedural

Conceptual Students see how trigonometric functions are transformed as they explore the period, amplitude, and phase shift of sinusoidal and tangent functions.

Procedural Students graph trigonometric functions with transformations of the period, amplitude, and/or phase shift.

Guiding the Investigation

You can use the Investigation Worksheet with Sample Data if you do not wish to conduct the investigation as an activity. The answers here are based on that sample data. There is also a desmos simulation of the investigation in your eBook.

Students investigated the pendulum earlier, where they related period to string length. This investigation looks for a periodic function that describes the location of the swinging bob. The bob might be a metal washer, a ring, a spool of tape, or a water bottle partly filled with water.

As students saw earlier, the length of the pendulum's string determines the period. Different lengths will thus yield different sinusoidal function models.

As needed, encourage students to think about how to transform the parent function to fit the data. Students may find it easier to model the data with a cosine function, because the cosine function has a peak at $x = 0$. The sine model can be derived through the relationship $\sin x = \cos\left(\frac{\pi}{2} - x\right)$.

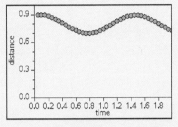

$y = 0.103 \cos\left[\frac{2\pi}{1.375}(x - 0.1)\right] + 0.8$;

$y = 0.103 \sin\left\{\frac{2\pi}{1.375}\left[x - \left(0.1 - \frac{1.375}{4}\right)\right]\right\} + 0.8$

For the sample data, 0.8 is the average distance from the motion sensor; 0.103 is the distance from the average value to the minimum or maximum value of the swing; 0.1 is the number of seconds before the first maximum value occurs; 1.375 is the number of seconds it takes to complete one cycle; x is measured in radians.

In the investigation you will transform sinusoidal functions to fit real-world, periodic data.

INVESTIGATION

The Pendulum II

YOU WILL NEED
- a washer
- string
- a motion sensor

Suspend a washer from two strings so that it hangs 10 to 15 cm from the floor between two tables or desks. Place the motion sensor on the floor about 1 m in front of the washer hanging at rest. Pull the washer back about 20 to 30 cm and let it swing. Collect data points for 2 s of time. Model your data with both a sine function and a cosine function. Give real-world meanings for all numerical values in each equation.

You can use sinusoidal functions to model many kinds of situations.

EXAMPLE B   After flying 300 mi west from Detroit to Chicago, a plane is put in a circular holding pattern above Chicago's O'Hare International Airport. The plane flies an additional 10 mi west past the airport and then starts flying in a circle with diameter 20 mi. The plane completes one circle every 15 min. Model the east-west component of the plane's distance from Detroit as a function of time.

Solution To help understand the situation, first sketch a diagram. The plane flies 300 mi to Chicago and then 10 mi past Chicago. At this time, call it 0 min, the plane begins to make a circle every 15 minutes. The diagram helps you see at least five data points, recorded in the table.

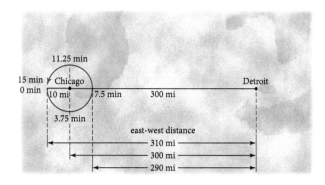

Time from beginning of circle (min) x	East-west distance from Detroit (mi) y
0	310
3.75	300
7.5	290
11.25	300
15	310

A quick plot of these points suggests that a cosine function might be a good model.

The period of $y = \cos x$ is 2π (from 0 to 2π). The period of this model should be 15 min (from 0 to 15). You know that the period = $2\pi a$. So $15 = 2\pi a$ or $a = \dfrac{15}{2\pi}$.

The amplitude of the model should be 10 mi, so use the vertical scale factor $b = 10$.

The plane's initial flight of 300 mi means that the function must be translated up 300 units, so use $k = 300$.

One possible model is

$$y = 300 + 10 \cos\left(\dfrac{x}{\frac{15}{2\pi}}\right), \text{ or } y = 300 + 10 \cos\left(\dfrac{2\pi x}{15}\right)$$

Example C

Project Example C and have students work in small groups. Have them share their solution, not just their answer.
SMP 1, 3, 6

LANGUAGE You may want to define the points at which the graph of the tangent function changes the direction of curving (at $x = 0$, $x = \pm n\pi$, and so on) to be *inflection points*. Students saw this term earlier as the inflection points on the normal curve.

Students will see the identity $\tan A = \dfrac{\sin A}{\cos A}$ in a later lesson. You may prefer to introduce it here.

You might **ASK** "Will the denominator of the tangent function ever be 0, making the function undefined?" [The value of x (or $\cos A$) is 0 when A is $\dfrac{\pi}{2} \pm n\pi$. At these points the graph of the tangent function has a vertical asymptote.]

Investigating Structure

Project the Investigating Structure and have student work in small groups. Have them present their answer and their justification. **SMP 1, 3, 6, 7**

Summarize

As students present, have them share their solution, not just their answer. Emphasize the use of correct mathematical terminology. Whether student responses are correct or incorrect, ask other students if they agree and why. **SMP 1, 2, 3, 6**

ADVANCED "Is a sinusoid a graph or a function?" [A sinusoid is a graph; a sinusoidal function is a transformation of the parent function $y = \sin x$.] Terms that are defined for graphs, such as *amplitude* and *phase shift,* are often also applied to functions. Likewise, terms such as *period,* defined for functions in earlier, are also applied to graphs.

ASK "How do you think the values of h, k, a, and b will affect the graphs of the sine and cosine functions?" You might need to point out that a represents not amplitude but rather the horizontal dilation factor, which is associated with the period. You may want to ask students to write equations relating the period to a and 2π, the period

---

Although the sine and cosine functions describe many periodic phenomena, such as the motion of a pendulum or the number of hours of daylight each day, there are other periodic functions that you can create from the unit circle.

In right triangle trigonometry, the tangent of angle A is the ratio of the length of the opposite leg to the length of the adjacent leg.

$$\tan A = \frac{a}{b}$$

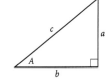

The definition of tangent can be extended to apply to any angle. Here, the tangent of angle A is the ratio of the y coordinate to the x coordinate of a point rotated $A°$ (or radians) counterclockwise about the origin from the positive ray of the x axis.

$$\tan A = \frac{y\text{-coordinate}}{x\text{-coordinate}}$$

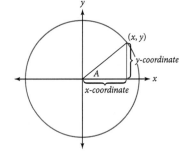

You can transform the graph of the tangent function in the same way that you transform sinusoidal functions. The asymptotes are changed only by horizontal transformations. Use the x-intercepts to help you determine translations. You can use the fact that $\tan \dfrac{\pi}{4} = 1$ to help you with vertical dilations.

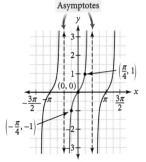

EXAMPLE C | The graph of $y = \tan x$ from $-\pi$ to 2π is shown in black. Find an equation for the red curve, which is a transformation of the graph of $y = \tan x$.

Solution | Notice that the parent tangent curve, $y = \tan x$, has a period of π. So the period of any transformed tangent curve is πa. The red curve appears to run from $-\pi$ to 2π, a distance of 3π, so the horizontal scale factor is 3.

Notice how the parent curve bends as it passes through the origin, $(0, 0)$. The red curve appears to bend in the same way at $\left(\dfrac{\pi}{2}, 1\right)$. That is a translation right $\dfrac{\pi}{2}$ units and up 1 unit. So an equation for the red curve is

$$y = 1 + \tan\left(\frac{x - \dfrac{\pi}{2}}{3}\right)$$

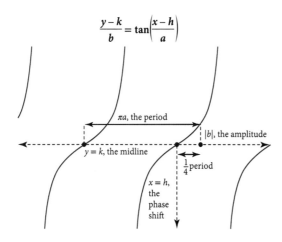

$$\frac{y-k}{b} = \tan\left(\frac{x-h}{a}\right)$$

In Example C, the point $\left(\frac{\pi}{2}, 1\right)$, on the red curve could have been considered the image of $(\pi, 0)$, rather than $(0, 0)$. This would indicate a translation *left* $\frac{\pi}{2}$ units and up 1 unit. So the equation

$$y = 1 + \tan\left(\frac{x + \frac{\pi}{2}}{3}\right)$$

will also model the red curve. Periodic graphs can always be modeled with many different, but equivalent, equations.

INVESTIGATING STRUCTURE

Do you remember the definitions of even and odd functions? A function is even if $f(-x) = f(x)$, and a function is odd if $f(-x) = -f(x)$. So far you have seen the graphs of sin x, cos x, and tan x. Which of these three functions are even? Which are odd?

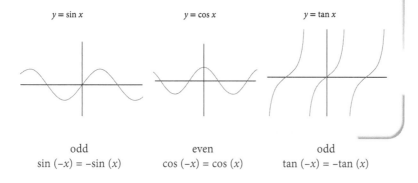

of the parent sinusoidal functions. For example, they might write the equation period = $2\pi a$ or $a = \frac{period}{2\pi}$. The period of any periodic function is the period of the parent function times the horizontal dilation factor.

Now that students are working primarily with radians, you might ask them how the labels (scale) on the x-axis differ for radians and degrees. Elicit the idea that the graph doesn't need to change if the scale changes appropriately, with 360° replacing 2π.

ASK "Is the tangent function sinusoidal?" [It's periodic with period π, but its graph doesn't have the same shape as that of the sine function, so it's not sinusoidal.]

CRITICAL QUESTION Why is the period of the tangent function π instead of 2π, like the sine and cosine functions?

BIG IDEA The tangent function is given by a ratio of x- and y-coordinates of a point rotating around the unit circle. A translation of the graphs of $y = \sin x$ and $y = \cos x$ horizontally by π gives a reflection of those same graphs across the x-axis. So the ratio $\tan x = \frac{\sin x}{\cos x}$ remains the same when translated by π.

CRITICAL QUESTION Is it true that the functions $y = \sin x$ and $y = \cos x$ are related to each other by phase shifts? Explain.

BIG IDEA Yes; the relationship $y = \cos x = \cos(-x)$ is equivalent to
$y = \sin\left[\frac{\pi}{2} - (-x)\right] = \sin\left(x + \frac{\pi}{2}\right)$, a phase shift of $-\frac{\pi}{2}$ from the graph of $y = \sin x$.

7.5 Exercises

You will need a graphing calculator for Exercises **4, 6, 7, 11, 13,** and **18**.

Practice Your Skills

1. For 1a–f, write an equation for each sinusoid as a transformation of the graph of either $y = \sin x$ or $y = \cos x$. More than one answer is possible. Describe the amplitude, period, phase shift, and vertical shift of each graph.

a.

$y = \sin x + 1$;
amplitude = 1,
period = 2π,
phase shift = 0,
vertical shift = 1

b.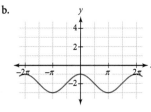

$y = \cos x - 2$;
amplitude = 1,
period = 2π,
phase shift = 0,
vertical shift = -2

c.

$y = \sin x - 0.5$;
amplitude = 1,
period = 2π,
phase shift = 0,
vertical shift = -0.5

d.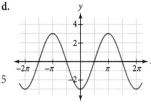

$y = -3 \cos x$;
amplitude = 3,
period = 2π,
phase shift = 0,
vertical shift = 0

e.

$y = -2 \sin x$;
amplitude = 2,
period = 2π,
phase shift = 0,
vertical shift = 0

f.

$y = 2 \cos x + 1$;
amplitude = 2,
period = 2π,
phase shift = 0,
vertical shift = 1

2. For 2a–f, write an equation for each sinusoid as a transformation of the graph of either $y = \sin x$ or $y = \cos x$. More than one answer is possible. Describe the amplitude, period, phase shift, and vertical shift of each graph.

a.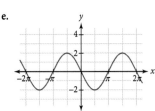

$y = \cos \dfrac{x}{0.5}$, or
$y = \cos 2x$;
amplitude = 1,
period = π,
phase shift = 0,
vertical shift = 0

b.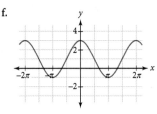

$y = \sin \dfrac{x}{2} - 1$;
amplitude = 1,
period = 4π,
phase shift = 0,
vertical shift = -1

c.

$y = -2 \sin 3x$;
amplitude = 2,
period = $\dfrac{2\pi}{3}$,
phase shift = 0,
vertical shift = 0

d.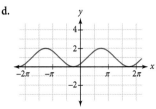

$y = \sin\left(x - \dfrac{\pi}{4}\right) + 1$;
amplitude = 1,
period = 2π,
phase shift = $\dfrac{\pi}{4}$,
vertical shift = 1

424 CHAPTER 7 Trigonometry and Trigonometric Functions

e. 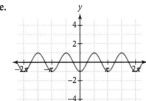 $y = -\cos\dfrac{x}{0.5}$, or
$y = -\cos 2x$;
amplitude $= 1$,
period $= \pi$,
phase shift $= 0$,
vertical shift $= 0$

f. 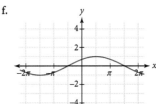 $y = \sin\dfrac{x + \dfrac{\pi}{2}}{2}$;
amplitude $= 1$,
period $= 4\pi$,
phase shift $= -\dfrac{\pi}{2}$,
vertical shift $= 0$

Reason and Apply

3. Consider the graph of $y = k + b \sin\left(\dfrac{x-h}{a}\right)$.

   a. What effect does k have on the graph of $y = k + \sin x$?

   b. What effect does b have on the graph of $y = b \sin x$? What is the effect if b is negative?

   c. What effect does a have on the graph of $y = \sin\left(\dfrac{x}{a}\right)$?

   d. What effect does h have on the graph of $y = \sin(x - h)$?

4. Sketch the graph of $y = 2 \sin\left(\dfrac{x}{3}\right) - 4$. Use a calculator to check your sketch.

5. Describe a transformation of the graph of $y = \sin x$ to obtain an image that is equivalent to the graph of $y = \cos x$.

6. Write three different equations for the graph at right. Use a calculator to verify your answers. ⓐ possible answer: $y = \sin\left(x - \dfrac{\pi}{2}\right)$, $y = -\sin\left(x + \dfrac{\pi}{2}\right)$, or $y = -\cos x$

7. **APPLICATION** The percentage of the lighted surface of the Moon that is visible from Earth can be modeled with a sinusoid. Assume that tonight the Moon is full (100%) and in 14 days it will be a new moon (0%).

   a. Define variables and find a sinusoidal function that models this situation. ⓗ

   b. What percentage will be visible 23 days after the full moon? 72%

   c. What is the first day after a full moon that shows less than 75% of the lit surface? day 5

Science CONNECTION

Half of the Moon's surface is always lit by the Sun. The phases of the Moon, as visible from Earth, result from the Moon's orientation changing as it orbits Earth. A full moon occurs when the Moon is farther away from the Sun than Earth, and the lighted side faces Earth. A new moon is mostly dark because its far side is receiving the sunlight, and only a thin crescent of the lighted side is visible from Earth.

The clock tower in Zug, Switzerland was built in the 13th century. The hands on the clocks revolve with periods equal to one solar year, one cycle of the moon, one week, and four years to keep track of leap years. The longest hand makes one turn in 365.24 days. The second longest hand makes one turn in 29 days, 12 hours, 44 minutes, 2.8 seconds. Another hand makes a complete turn in 7 days, while the smallest hand makes one turn every 4 years. By looking at the clock it is possible to tell the day, month, and year.

3a. The k-value vertically translates the graph of the function.

3b. The b-value vertically dilates the graph of the function. The absolute value of b represents the amplitude. When b is negative, the curve is reflected across the x-axis.

3c. The a-value horizontally dilates the graph of the function. It also determines the period with the relationship $2\pi a = \text{period}$.

3d. The h-value horizontally translates the graph of the function. It represents the phase shift.

4.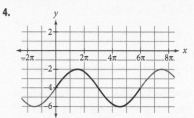

5. horizontal translation left $\dfrac{\pi}{2}$ units (phase shift of $-\dfrac{\pi}{2}$ units); $y = \sin\left(x + \dfrac{\pi}{2}\right)$

Exercise 7 ASK "What are the independent and dependent variables?" [The independent variable is time; the dependent variable is the percentage of the surface visible from Earth that is lit.] "What are the domain and range?" [The domain is unbounded; the range is $0 \leq y \leq 100$.]

7a. Let x represent the number of days after a full moon (today), and let y represent the percentage of lighted surface that is visible.

$$y = 0.5 + 0.5 \cos \dfrac{2\pi x}{28}$$

LESSON 7.5 Graphing Trigonometric Functions **425**

Exercise 8 If students are having difficulty considering the graphs separately, suggest that they trace each one in turn with their finger as they examine its properties.

Exercise 9c As needed remind students that *continuous* means having no breaks or gaps. Students may note a kind of symmetry to the graph of the sine function; in some sense it is symmetric across the *x*- and *y*-axes together.

Exercise 10 Encourage students to come up with several equations. An equation of the form
$y = \cos\left(x - \frac{\pi}{2} + 2n\pi\right)$ for any integer value of *n* will accomplish the desired translation.

Exercise 11 ASK "What are the independent and dependent variables?" [The independent variable is time; the dependent variable is brightness.] "What are the domain and range?" [The domain is unbounded; the range is $20 \leq y \leq 50$.]

11b.
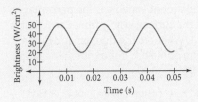

---

8. Estimate the amplitudes and periods of the three graphs pictured.

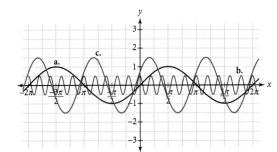

8a. Amplitude is 1 and period is about 2π.
8b. Amplitude is 0.5 and period is about 0.6.
8c. Amplitude is 1.5 and period is about π.

9. For each property, tell whether it applies to sine, cosine, tangent, or none of these functions.
   a. The graph is symmetric over the *y*-axis. cosine
   b. The graph is symmetric over the *x*-axis. none
   c. The graph is continuous. sine and cosine
   d. The domain is the set of real numbers. sine and cosine
   e. The range is the set of real numbers. tangent
   f. The function contains the point $\left(\frac{\pi}{4}, 1\right)$. tangent

10. Give an equation showing the transformation of the graph of $y = \cos x$ to $y = 1 + 3 \sin\left(\frac{x}{2}\right)$. Answers will vary; possible answer: $y = 1 + 3 \cos\left(x - \frac{\pi}{2}\right)$.

11. Fluorescent lights do not produce constant illumination as incandescent lights do. Ideally, fluorescent light cycles sinusoidally from dim to bright 60 times per second. At a certain distance from the light, the maximum brightness is measured at 50 watts per square centimeter (W/cm²) and the minimum brightness at 20 W/cm².
    a. To collect data on light brightness from three complete cycles, for how much total time should you record data? 0.05 s
    b. Sketch a graph of the sinusoidal model for the data collected in 11a if the light climbed to its mean value of 35 W/cm² at 0.003 s.
    c. Write an equation for the sinusoidal model. $y = 35 + 15 \sin 120\pi(x - 0.003)$

12. Imagine a unit circle in which the point (1, 0) is rotated *A* radians counterclockwise about the origin from the positive *x* axis. Copy this table and record the *x*-coordinate and *y*-coordinate for each angle. Then use the definition of tangent to find the slope of the segment that connects the origin to each point.

Angle *A*	0	$\frac{\pi}{6}$	$\frac{\pi}{4}$	$\frac{\pi}{2}$	$\frac{3\pi}{4}$	π	$\frac{4\pi}{3}$	$\frac{5\pi}{3}$	$\frac{11\pi}{6}$
x-coordinate	1	$\frac{\sqrt{3}}{2}$	$\frac{\sqrt{2}}{2}$	0	$-\frac{\sqrt{2}}{2}$	-1	$-\frac{1}{2}$	$\frac{1}{2}$	$\frac{\sqrt{3}}{2}$
y-coordinate	0	$\frac{1}{2}$	$\frac{\sqrt{2}}{2}$	1	$\frac{\sqrt{2}}{2}$	0	$-\frac{\sqrt{3}}{2}$	$-\frac{\sqrt{3}}{2}$	$-\frac{1}{2}$
Slope or tan *A*	0	$\frac{1}{\sqrt{3}}$	1	undefined	-1	0	$\sqrt{3}$	$-\sqrt{3}$	$-\frac{1}{\sqrt{3}}$

13. Graph $y = \tan x$ on your calculator. Set your graphing window to $-2\pi \leq x \leq 2\pi$ and $-5 \leq y \leq 5$.
    a. What happens at $x = \frac{\pi}{2}$? Explain why this is so, and name other values when this occurs.
    b. What is the period of $y = \tan x$? π
    c. Explain, in terms of the definition of tangent, why the values of $\tan \frac{\pi}{5}$ and $\tan \frac{6\pi}{5}$ are the same.
    d. Carefully graph two cycles of $y = \tan x$ on paper. Include vertical asymptotes at x values where the graph is undefined.

14. Write equations for sinusoids with these characteristics:
    a. a cosine function with amplitude 1.5, period π, and phase shift $-\frac{\pi}{2}$ $y = 1.5 \cos 2\left(x + \frac{\pi}{2}\right)$
    b. a sine function with minimum value -5, maximum value -1, and one cycle starting at $x = \frac{\pi}{4}$ and ending at $x = \frac{3\pi}{4}$ ⓐ $y = -3 + 2 \sin 4\left(x - \frac{\pi}{4}\right)$
    c. a cosine function with period 6π, phase shift π, vertical translation 3, and amplitude 2 $y = 3 + 2 \cos \frac{x - \pi}{3}$

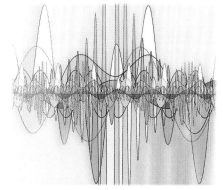

Review

15. Find the measure of the labeled side in each triangle.
    a. $a \approx 17.79$
    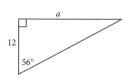
    b. $b \approx 17.75$

16. Make a table of angle measures from 0° to 360° by 15° increments. Then find the radian measure of each angle. Express the radian measure as a multiple of π.

17. The second hand of a wristwatch is 0.5 cm long. ⓐ
    a. What is the speed, in meters per hour, of the tip of the second hand?   1.885 m/h
    b. How long would the minute hand of the same watch have to be for its tip to have the same speed as the second hand?   30 cm
    c. How long would the hour hand of the same watch have to be for its tip to have the same speed as the second hand?   360 cm
    d. What is the angular speed, in radians per hour, of the three hands in 17a–c?
       second hand: 377 radians/h; minute hand: 6.28 radians/h; hour hand: 0.52 radian/h
    e. Make an observation about the speeds you found.
       The speed of the tip of a clock hand varies directly with the length of the hand. Angular speed is independent of the length of the hand.

13.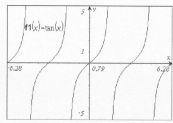

13a. The value of $\tan \frac{\pi}{2}$ is undefined, and $x = \frac{\pi}{2}$ is a vertical asymptote. The vertical asymptotes are $x = \cdots -\frac{3\pi}{2}, -\frac{\pi}{2}, \frac{\pi}{2}, \frac{3\pi}{2}, \cdots$, or, in general, $x = \frac{\pi}{2} + \pi n$ where n is an integer.

13c. On a unit circle, the point $(1, 0)$ rotated $\frac{\pi}{5}$, the origin, and the point $(1, 0)$ rotated $\frac{6\pi}{5}$ are collinear. So the segment from the origin to the point rotated $\frac{\pi}{5}$ has the same slope as the segment from the origin to the point rotated $\frac{6\pi}{5}$.

13d.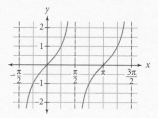

Exercise 17c The hour hand completes one revolution in 12 hours. It takes 12 times as many hours as the minute hand; so to have the same speed, it would have to be 12 times as long.

16.

Degrees	0°	15°	30°	45°	60°	75°	90°	105°	120°	135°	150°	165°	180°
Radians	0	$\frac{\pi}{12}$	$\frac{\pi}{6}$	$\frac{\pi}{4}$	$\frac{\pi}{3}$	$\frac{5\pi}{12}$	$\frac{\pi}{2}$	$\frac{7\pi}{12}$	$\frac{2\pi}{3}$	$\frac{3\pi}{4}$	$\frac{5\pi}{6}$	$\frac{11\pi}{12}$	π
Degrees	195°	210°	225°	240°	255°	270°	285°	300°	315°	330°	345°	360°	
Radians	$\frac{13\pi}{12}$	$\frac{7\pi}{6}$	$\frac{5\pi}{4}$	$\frac{4\pi}{3}$	$\frac{17\pi}{12}$	$\frac{3\pi}{2}$	$\frac{19\pi}{12}$	$\frac{5\pi}{3}$	$\frac{7\pi}{4}$	$\frac{11\pi}{6}$	$\frac{23\pi}{12}$	2π	

18b. i.

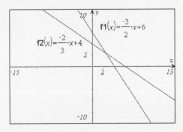

18b. ii.

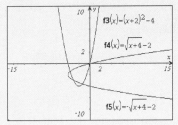

Exercise 20 Students can create right triangles by dropping perpendicular segments from each circle center to the horizontal line. You may want to help students see that the length of the hypotenuse of each right triangle is twice the radius, so the distance from the angle's vertex to the closest point on any circle equals the radius of that circle.

20a. The radius of C_2 is 18. The radius of C_3 is 54.

See your digital resources for the project **Designing a Picnic Table** where students design a picnic table and draw the plans for it, including scale drawings with all the lengths and all the angle measures labeled.

---

18. Consider these three functions:

    i. $f(x) = -\frac{3}{2}x + 6$ @ $y = -\frac{2}{3}x + 4$

    ii. $g(x) = (x+2)^2 - 4$ $\quad y = \pm\sqrt{x+4} - 2$

    a. Find the inverse of each function.
    b. Graph each function and its inverse.
    c. Which of the inverses, if any, are functions?   The inverse of i is a function.

19. Use what you know about the unit circle to find possible values of θ in each equation. Use domain $0° \leq \theta \leq 360°$ or $0 \leq \theta < 2\pi$.

    a. $\cos \theta = \sin 86°$
    $\theta = 4°$ and $\theta = 356°$

    b. $\sin \theta = \cos \frac{19\pi}{12}$ $\theta = \frac{\pi}{12}$ and $\theta = \frac{11\pi}{12}$

    c. $\sin \theta = \cos 123°$
    $\theta = 213°$ and $\theta = 327°$

    d. $\cos \theta = \sin \frac{7\pi}{6}$ $\theta = \frac{2\pi}{3}$ and $\theta = \frac{4\pi}{3}$

20. Circles $C_1, C_2, \ldots$ are tangent to the sides of $\angle P$ and to the adjacent circle(s). The radius of circle C_1 is 6. The measure of $\angle P$ is 60°.

    a. What are the radii of C_2 and C_3? (*Hint:* One method uses similar right triangles.)
    b. What is the radius of C_n?
    The radius of C_n is $6(3^{n-1})$.

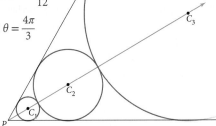

PERFORMANCE TASK

In this graph, the first minimum to the left of the origin is $(-37.5°, -1.5)$, and the second maximum to the right of the origin is $(97.5°, 4.5)$.

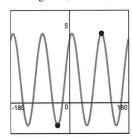

Write an equation expressing the curve as a transformation of:

a. $y = \sin x$
b. $y = \cos x$ in which the coefficient of the cosine function is positive
c. $y = \cos x$ in which the coefficient of the cosine function is negative

428 **CHAPTER 7** Trigonometry and Trigonometric Functions

PERFORMANCE TASK

a. Possible answer: $y = 1.5 + 3 \sin 4(\theta + 15°)$
b. Possible answer: $y = 1.5 + 3 \cos 4(\theta - 7.5°)$
c. Possible answer: $y = 1.5 + 3 \cos 4(\theta + 37.5°)$

LESSON 7.6
Modeling with Trigonometric Equations

Language exerts hidden power, like the moon on the tides.
RITA MAE BROWN

Tides are caused by gravitational forces, or attractions, between the Moon, Earth, and the Sun as the Moon circles Earth. You can model the height of the ocean level in a seaport with a combination of sinusoidal functions of different phase shifts, periods, and amplitudes. Because tides and other real world periodic phenomena have periods that don't involve π, it can be easier to work with a parent function that has a period of one. The functions $f(x) = \sin(2\pi x)$ and $g(x) = \cos(2\pi x)$ both have a period of one. You'll see how this works in Example A.

Like tides, the motion of waves can be modeled with trigonometric functions.

Life in a tide pool is challenging due to the waves and also the constantly changing levels of sunlight and salinity. However, the constant action of the tides also brings food and oxygen into the pool, enabling a wide variety of creatures to live there.

Science CONNECTION

The Moon's gravitational pull causes ocean tides and also tidal bores in rivers. Tidal bores occur during new and full moons when ocean waves rush upstream into a river passage, sometimes at speeds of more than 40 mi/h. The highest recorded river bore was 15 ft high, in China's Fu-ch'un River.

EXAMPLE A The height of water at the mouth of a certain river varies during the tide cycle. The time in hours since midnight, t, and the height in feet, h, are related by the equation

$$h = 15 + 7.5 \cos\left[\frac{2\pi(t-3)}{12}\right]$$

a. Identify the transformations in the height equation in the context of tides.
b. When is the first time the height of the water is 11.5 feet?
c. When will the water be at that height again?

LESSON 7.6

In this lesson, students apply the characteristics of sinusoidal functions to modeling real-world situations.

COMMON CORE STATE STANDARDS		
APPLIED	DEVELOPED	INTRODUCED
	F.IF.4	
	F.IF.7e	
	F.BF.5	

Objectives
- Use trigonometric equations to model real-world phenomena
- Define *frequency* and calculate it
- Find the average value of a sine or cosine function

Vocabulary
frequency

Materials
- spring
- mass of 50 to 100 g
- support stand
- motion sensor
- Calculator Notes: Reentry; Intersections, Maximums, and Minimums *optional*

LANGUAGE A *sinusoid* is a curve shaped like the graph of the sine (or cosine) function.

Launch

How many complete cycles do each of the graphs below make, $0 < x < 2\pi$?

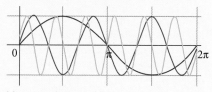

blue, 1; red, 3; green, 6

Investigate

Example A
This example illustrates the interpretation of data that can be modeled by a trigonometric function. Notice that the functions highlighted are

---

ELL	Support	Advanced
Discuss tides and their importance to coastal communities. Discuss other cyclical events. Consider temporarily adding labels relating to these situations to the class graphs.	Help students list the critical values in the situation and relate these to designing an appropriate functional model.	Challenge students to solve the examples on their own, before reading the text.

$f(x) = \sin(2\pi x)$ and $g(x) = \cos(2\pi x)$, both of which have a period of 1. **ELL**

The *mouth of a river* is the place where the river empties into another body of water, such as a lake or an ocean.

Project Example A from your eBook. Ask for student input. Have them share their thinking along with their answer.

A slightly different approach to part c is to think about the symmetry of the function. If the period of this function is 12 h and the cycle pictured is considered an example, the middle time will be 6 h; so the maximum will occur at 3 h and the minimum at 9 h. Therefore, if the height were 11.5 at 6.9 h, which is 2.1 h before the minimum, the height would be 11.5, again 2.1 h after the minimum, or at 11.1 h.

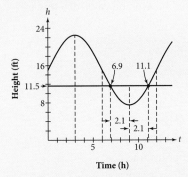

Time (h)

Guiding the Investigation

You can use the Investigation Worksheet with Sample Data if you do not wish to conduct the investigation as an activity. Sample answers given here are based on that data. There is also a desmos simulation of this investigation in your eBook.

Have Calculator Note: Looking for the Rebound Using the EasyData App available to help students gather data from motion sensors.

A Slinky Jr. cut in half works well as the spring, and a heavy washer works well as the mass. Students might use a ruler between two desks as the support stand.

Step 1 How far students pull down the mass does not matter as long as they are consistent. Different motion sensors have different ranges. Students will need to adjust their set-ups accordingly.

Step 2 Students can report average values of the numbers characterizing the function.

430 CHAPTER 7 Trigonometry and Trigonometric Functions

Solution

In this equation, consider the parent function to be $y = \cos(2\pi x)$.

a. Rewrite the equation as $\dfrac{h-15}{7.5} = \cos\dfrac{2\pi(t-3)}{12}$ to help you see the transformations. The vertical translation, or midline, is 15 feet and the vertical dilation factor, or amplitude, is 7.5 feet. The horizontal translation, or phase shift, is 3 hours and the horizontal dilation factor, or period, is 12 hours. So the cycle of high and low tides takes 12 hours. The tide is high 3 hours after midnight. The water level goes 7.5 feet above and below the 15 foot level.

b. To find when the height of the water is 11.5 feet, substitute 11.5 for h.

$11.5 = 15 + 7.5 \cos\left[\dfrac{2\pi(t-3)}{12}\right]$ Substitute 11.5 for h.

$\dfrac{-3.5}{7.5} = \cos\left[\dfrac{\pi(t-3)}{6}\right]$ Subtract 15, and divide by 7.5.

$\cos^{-1}\left(\dfrac{-3.5}{7.5}\right) = \dfrac{\pi(t-3)}{6}$ Take the inverse cosine of both sides.

$\dfrac{6}{\pi} \cdot \cos^{-1}\left(\dfrac{-3.5}{7.5}\right) + 3 = t$ Multiply by $\dfrac{6}{\pi}$, and add 3.

$t \approx 6.93$ hours

So the water height will be 11.5 feet after approximately 6.93 hours, or at 6:56 A.M.

You can also find this answer by graphing the height function and the line $h = 11.5$ and by approximating the point of intersection. The graph verifies your solution. (*Note*: On your calculator, use x in place of t and y in place of h.)

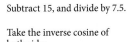

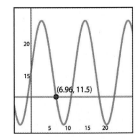

The cycle begins at (3, 22.5). The water is at its highest point, 22.5 feet, at 3 hours after midnight.

c. The graph in part b shows several different times when the water depth is 11.5 feet. The third occurrence is 12 hours after the first, because each cycle is 12 hours in length. So the third occurrence is at 6:56 P.M. But how can you find the second occurrence? First, consider that the graph is shifted right 3 units, so the cycle begins at $t = 3$. It has a period of 12, so it ends at $t = 15$.

The first height of 11.5 feet occurred when $t \approx 6.93$, or about 3.93 hours after the cycle's start. The next solution will be about 3.93 hours before the end of the cycle, that is, at $15 - 3.93 \approx 11.07$ hours, or 11:04 A.M. This value will also repeat every 12 hours. A graph confirms this approximation.

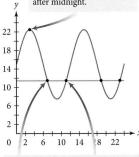

At 6.9 hours after midnight, the height has fallen to 11.5 feet and is still falling. At 11.1 hours after midnight, the height has risen back to 11.5 feet.

You can use both sine and cosine functions to model sinusoidal patterns because they are just horizontal translations of each other. Often it is easier to use a cosine function, because you can identify the maximum value as the start value of a cycle.

430 CHAPTER 7 Trigonometry and Trigonometric Functions

Modifying the Investigation

Whole Class Have three students collect data for the whole class, or demonstrate the setup and use sample data. Discuss Steps 3 and 4.

Shortened Collect a set of class data and link it to a calculator in each group. Or students can use the Bouncing Spring Sample Data worksheet. Or use the desmos simulation.

One Step Pose this problem: "The time between high and low tides in a river harbor is approximately 7 h. The high-tide depth of 16 ft has just occurred at noon today, and the average river depth is 11 ft. What are the next two time periods when a boat requiring at least 9 ft of water will not be able to enter the harbor?"

When a function has a short cycle, it can be more useful to talk about how many cycles are completed in one unit of time. This value, called the **frequency** of the function, is the reciprocal of the period. For example, if a wave has a period of 0.01 second, then it has a frequency of 100 cycles per second. In Example A, the period is $\frac{1}{2}$ day, so the frequency is 2 cycles per day.

$$\text{frequency} = \frac{1}{\text{period}}$$

INVESTIGATION

A Bouncing Spring

YOU WILL NEED
- a motion sensor
- a spring
- a mass of 50 to 100 g
- a support stand

In this experiment you will suspend a mass from a spring. When you pull down on the mass slightly, and release, the mass will move up and down. In reality, the amount of motion gradually decreases, and eventually the mass returns to rest. However, if the initial motion is small, then the decrease in the motion occurs more slowly and can be ignored during the first few seconds.

Procedure Note
1. Attach a mass to the bottom of a spring. Position the motion sensor directly below the spring, leaving space to pull down on the mass.
2. Set the motion sensor to collect about 5 s of data. Pull the mass down slightly, and release at the same moment as you begin gathering data.

Step 1 Follow the Procedure Note to collect data on the height of the bouncing spring for a few seconds.

Step 2 Delete values from your lists to limit your data to about four cycles. Identify the phase shift, amplitude, period, frequency, and vertical shift of your function.

Step 3 Write a sine or cosine function that models the data.

Step 4 Answer these questions, based on your equation and your observations.

Step 4a 0.652 m is the average height of the spring. 0.076 m is the amount up and down from the average value the spring moves. 0.8064 s is the time it takes to complete one cycle. 0.43 s is the time when the first maximum occurs.

a. How does each of the numbers in your equation from Step 3 correspond to the motion of the spring?

b. How would your equation change if you moved the motion sensor 1 m farther away? The vertical translation would increase by 1. All other values would remain the same.

c. How would your equation change if you pulled the spring slightly lower when you started? The amplitude would increase. The period might change too.

In the investigation, you modeled cyclical motion with a sine or cosine function. As you read the next example, think about how the process of finding an equation based on given measurements compares to your process in the investigation.

EXAMPLE B Earlier, you learned that the Cosmo Clock 21 Ferris wheel has a 100 m diameter, takes 15 min to rotate, and reaches a maximum height of 112.5 m.

Find an equation that models the height in meters, h, of a seat on the perimeter of the wheel as a function of time in minutes, t. Then determine when, in the first 30 min, a given seat is 47 m from the ground if the seat is at its maximum height 10 min after the wheel begins rotating.

Step 2 answers for sample data:

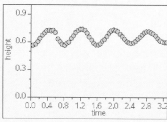

The average maximum is 0.728, and the average minimum is 0.575, so the amplitude is $\frac{0.153}{2} \approx 0.076$. The average value, the vertical translation, is 0.652. The minima occur at 0, 0.80640, 1.61280, 2.41920, and 3.22560, at regular intervals of 0.80640, so the period equals 0.8064 s. The frequency is $\frac{1}{0.8064}$, or 1.240 cycles per second. The first maximum occurs at $t = 0.43$; so if you choose a cosine function, it has been translated right 0.43.

Step 3 $h = 0.652 + 0.076 \cos\left[\frac{2\pi(t - 0.43)}{0.8064}\right]$

A graph of the curve with the data shows a good fit.

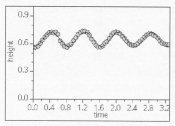

Step 4 If needed, help students see that the length of one cycle is the same as the period. In 4b, students can assume that t starts at 0.

Example B

This example shows how to set up a trigonometric function model. Project Example B and have students work in small groups. Have them share their solution, not just their answer.
SMP 1, 3, 4, 5, 6

ASK "Why does the average height of the parent sinusoid curve equal zero?" [Each positive value is matched with a negative value, so the average is zero.] "Why is the average height of a point on the Ferris wheel 62.5 m?" [The maximum height is 112.5 m and the wheel is 100 m in diameter, so the minimum height is 12.5 m and the average height is 62.5 m.]

ASK "Why will the function have a value of 47 m before the maximum point?" [The symmetry of the cosine function guarantees that if the function reaches 47 m after the maximum, it must also reach that value before the maximum.]

Conceptual/Procedural

Conceptual Students explore the concept of frequency.

Procedural Students use trigonometric functions to model real-world phenomena.

Summarize

As students present their solutions to the investigation and the examples, have them share their solution, not just their answer. Emphasize the use of correct mathematical terminology. Whether student responses are correct or incorrect, ask other students if they agree and why. **SMP 1, 3, 4, 5, 6**

CRITICAL QUESTION How would you describe to a friend how to use a sinusoidal function to model a data set that appears to be periodic?

BIG IDEA Steps should include (not necessarily in this order):

- Subtract the minimum from the maximum and divide by 2 to get the amplitude, or vertical dilation, b.
- Find the average of the maximum and minimum values to determine the amount of vertical translation, k.
- Subtract the x-values of consecutive maximums (or minimums) to calculate the period, a.
- Find the x-value of the maximum with the smallest positive x-coordinate. For a cosine model, that number is the phase shift, h; for a sine model, the phase shift h is that number minus $\frac{1}{4}$ of the period.
- Write the function as $y = b \cos\left[\frac{2\pi(x-h)}{a}\right] + k$ or $y = b \sin\left[\frac{2\pi(x-h)}{a}\right] + k$ depending on how you calculated h.

Students need not memorize these steps, as long as they can find the function. To check the algebra, they should continually refer back to the physical characteristics represented by the various numbers. **SMP 1, 2**

CRITICAL QUESTION What are the units of frequency?

BIG IDEA They are the reciprocal of the units of the period. For example, if the period is 15 minutes per cycle, then the frequency is $\frac{1}{15}$ cycle per minute.

Solution The parent sinusoid curve has an amplitude of 1. The diameter of the Ferris wheel is 100 m, so the vertical scale factor, or amplitude, is 50. The period is 1 for either parent function, $\sin(2\pi x)$ or $\cos(2\pi x)$. Since the period of the Ferris wheel is 15 min, the horizontal scale factor is 15. The midline of the parent sinusoid is 0, but the midline of the wheel is 62.5 m, so the vertical translation is 62.5.

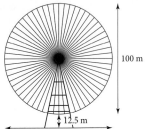

The top of a sinusoid corresponds to the maximum height of a seat. The cosine curve is at its maximum point when x is zero, so it will be easiest to use a cosine function. Because the first maximum of the Ferris wheel occurs after 10 min, the horizontal translation, or phase shift, of its equation is 10 to the right. Incorporate these values into a cosine function $h = \cos(2\pi x)$ to get

$$h = 50 \cos\left[\frac{2\pi(t-10)}{15}\right] + 62.5.$$

Now, to find when the height is 47 m, substitute 47 for h.

$50 \cos\left[\frac{2\pi(t-10)}{15}\right] + 62.5 = 47$	Substitute 47 for h.
$\cos\left[\frac{2\pi(t-10)}{15}\right] = \frac{47 - 62.5}{50} = -0.31$	Subtract 62.5 and divide by 50. Evaluate.
$\frac{2\pi(t-10)}{15} = \cos^{-1}(-0.31)$	Take the inverse cosine of both sides.
$t - 10 = \frac{15}{2\pi} \cdot \cos^{-1}(-0.31)$	Multiply by $\frac{15}{2\pi}$ on both sides.
$t = \frac{15}{2\pi} \cdot \cos^{-1}(-0.31) + 10$	Add 10.
$t \approx 14.5$	Approximate the principal value of t.

So the given seat is at a height of 47 m approximately 14.5 min after the wheel starts rotating. But this is not the only time. The period is 15 min, so the seat will also reach 47 m on the second rotation, after 29.5 min. Also, the height is 47 m once on the way up and once on the way down. The seat is at 47 m 4.5 min after its maximum point, which is at 10 min. It will also be at the same height 4.5 min before its maximum point, at 5.5 min, and 15 min after that, at 20.5 min.

So in the first 30 min, the seat reaches a height of 47 m after 5.5, 14.5, 20.5, and 29.5 min. A graph of the function $h = 50 \cos\left[\frac{2\pi(t-10)}{15}\right] + 62.5$ and $y = 47$ confirms this.

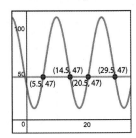

Using sinusoidal models, you can easily find y when given an x-value. As you saw in the examples, the more difficult task is finding x when given a y-value. There will be multiple answers because the graphs are periodic. You should always check the values and number of solutions with a calculator graph.

7.6 Exercises

You will need a graphing calculator for Exercises **2**, **5–15** and **18**.

Practice Your Skills

1. Find the first four positive solutions. Give exact values in radians.
   a. $\cos x = 0.5$ @ $x = \left\{\dfrac{\pi}{3}, \dfrac{5\pi}{3}, \dfrac{7\pi}{3}, \dfrac{11\pi}{3}\right\}$
   b. $\sin x = -0.5$ $x = \left\{\dfrac{7\pi}{6}, \dfrac{11\pi}{6}, \dfrac{19\pi}{6}, \dfrac{23\pi}{6}\right\}$

2. Find all solutions for $0 \leq x < 2\pi$, rounded to the nearest thousandth.
   a. $2\sin(x + 1.2) - 4.22 = -4$ @ $x = \{1.831, 5.193\}$
   b. $7.4\cos(x - 0.8) + 12.3 = 16.4$ $x = \{1.784, 6.100\}$

3. Compare the graph of the function $h = 5 + 7\sin\left[\dfrac{2\pi(t-9)}{11}\right]$ with the parent function $h = \sin(2\pi x)$.
   a. What are the vertical translation and average value? ⓗ 5; 5
   b. What are the vertical scale factor, minimum and maximum values, and amplitude? @ 7; −2; 12; 7
   c. What are the horizontal scale factor and period? 11; 11
   d. What are the horizontal translation and phase shift? 9; 9

4. Compare the graph of the function $h = 18 - 17\cos\left[\dfrac{2\pi(t+16)}{15}\right]$ with the parent function $h = \cos(2\pi x)$.
   a. What are the vertical translation and average value? 18; 18
   b. What are the vertical scale factor, minimum and maximum values, and amplitude? −17; 1; 35; 17
   c. What are the horizontal scale factor and period? 15; 15
   d. What are the horizontal translation and phase shift? −16; −16

5. Find the frequency of each sinusoidal function.
   a. $y = 4.7\cos\left(\dfrac{2\pi x}{0.075}\right)$ @ $\dfrac{1}{0.075}$ or ≈ 13.33
   b. $y = 7.8 + 26.2\sin\left(\dfrac{2\pi(x-3.75)}{24.59}\right)$ $\dfrac{1}{24.59}$ or ≈ 0.04

Reason and Apply

6. A walker moves counterclockwise around a circle with center (1.5, 2) and radius 1.2 m and completes a cycle in 8 s. A recorder walks back and forth along the x-axis, staying even with the walker, with a motion sensor pointed toward the walker. What equation describes the distance as a function of time? Assume the walker starts at point (2.7, 2). $y = 1.2\sin\dfrac{2\pi t}{8} + 2$, or $y = 1.2\sin\dfrac{\pi t}{4} + 2$

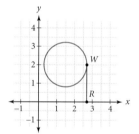

7. A mass attached to a spring is pulled down 3 cm from its resting position and then released. It makes ten complete bounces in 8 s. At what times during the first 2 s was the mass 1.5 cm above its resting position? 0.267 s, 0.533 s, 1.067 s, 1.333 s, 1.867 s

LESSON 7.6 Modeling with Trigonometric Equations

Formative Assessment
Assess students' understanding of and ability to work with the phase shift, amplitude, period, and vertical translation of a periodic function. Assess how clearly students are able to link the attributes of the weight oscillating on the spring to the various parts of the equation for sine or cosine.

Apply

Extra Example
1. Find the first four positive solutions in degrees of $\sin x = \dfrac{1}{\sqrt{2}}$.
   45°, 135°, 405°, 495°
2. Find all solutions for $0 \leq x \leq 2\pi$, rounded to the nearest thousandth, of equation $3\cos(x + 2.1) + 5.1 = 7.9$.
   3.816, 4.550

Closing Question
What steps should you take to model periodic data with a trigonometric function?
Do any dilations first and then any translations.

Assigning Exercises
Suggested: 1–14
Additional Practice: 15–18

Exercise 1 An equation such as $\cos x = 0.5$ has infinitely many solutions. The first positive solution is $\dfrac{\pi}{3}$. The equation $y = 5\cos x$ is periodic with a period of 2π, so $\dfrac{\pi}{3} + 2\pi$, $\dfrac{\pi}{3} + 4\pi$, $\dfrac{\pi}{3} + 6\pi$, and so on are all solutions. These solutions can be generated by $\dfrac{\pi}{3} + 2\pi n$, where n is any integer. The second positive solution is $\dfrac{5\pi}{3}$, so values generated by $\dfrac{5\pi}{3} + 2\pi n$ are also solutions.

Exercise 2 SUPPORT In addition to finding the solutions, ask students to describe the amplitude, period, phase shift, frequency, and vertical shift of each function.

Exercises 3a, 4a LANGUAGE For sinusoidal functions, the *average value* is the mean of the maximum and minimum values.

Exercise 6 Students may want to walk through this exercise before starting to solve it; it could be done as a class activity as well. ASK "What is the real-world meaning of each part of the equation?" Although y-coordinates usually are given by sines, students might choose to model the phenomenon with a cosine function, leading to the equation $y = 1.2\cos\left[\dfrac{2\pi(x-2)}{8}\right] 2$.

LESSON 7.6 Modeling with Trigonometric Equations

Exercise 8 Students can assume that the function has a phase shift of 0.

8b.

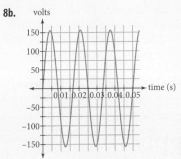

10a. $y_1 = -3\cos\left[\dfrac{2\pi(t+0.17)}{\frac{2}{3}}\right]$,

$y_2 = -4\cos\left(\dfrac{2\pi t}{\frac{2}{3}}\right)$, or $y_1 = 3\sin 3\pi t$,

$y_2 = -4\cos 3\pi t$

10b. at 0.2 s, 0.6 s, 0.9 s, 1.2 s, 1.6 s, and 1.9 s

Exercise 12 **ALERT** Many answers are possible, depending on where students choose to place the point's initial location. To encourage creativity, you might suggest that each student choose an initial point that they're pretty sure will differ from every other student's choice (and give the coordinates of that initial point).

12.

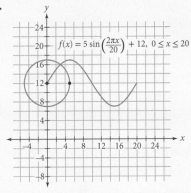

8. Household appliances are typically powered by electricity through wall outlets. The voltage provided varies sinusoidally between $-110\sqrt{2}$ volts and $110\sqrt{2}$ volts, with a frequency of 60 cycles per second. **8a.** possible answer: $v = 110\sqrt{2}\sin 120\pi t$

   a. Use a sine or cosine function to write an equation for voltage as a function of time. ⓗ
   b. Sketch and label a graph picturing three complete cycles.

9. The time between high and low tide in a river harbor is approximately 7 h. The high-tide depth of 16 ft occurs at noon and the average harbor depth is 11 ft.

   a. Write an equation modeling this relationship. possible answer: $d = 11 + 5\cos\dfrac{\pi t}{7}$
   b. If a boat requires a harbor depth of at least 9 ft, find the next two time periods when the boat will not be able to enter the harbor.
   between 4:25 P.M. and 9:35 P.M., and between 6:25 A.M. and 11:35 A.M. the next day

10. Two masses are suspended from springs, as shown. The first mass is pulled down 3 cm from its resting position and released. A second mass is pulled down 4 cm from its resting position. It is released just as the first mass passes its resting position on its way up. When released, each mass makes 12 bounces in 8 s.

    a. Write a function for the height of each mass. Use the moment the second mass is released as $t = 0$. ⓐ
    b. At what times during the first 2 s will the two masses be at the same height? Solve graphically, and state your answers to the nearest 0.1 s.

11. **APPLICATION** An AM radio transmitter generates a radio wave given by a function in the form $f(t) = A\sin 2000\pi nt$. The variable n represents the location on the broadcast dial, $550 \le n \le 1600$, and t is the time in seconds.

    a. For radio station WINS, located at 1010 on the AM radio dial, what is the period of the function that models its radio waves? $\dfrac{1}{1,010,000}$ s
    b. What is the frequency of your function from 11a? 1,010,000 cycles/s
    c. Find a function that models the radio waves of an AM radio station near you. Find the period and frequency. Find the dial number of an AM radio station and substitute it for n. Period will be $\dfrac{1}{1000n}$. Frequency will be $1000n$. Answers will vary.

12. A point rotates at 3 rev/min counterclockwise around a circle with center $(0, 12)$ and radius 5 m, as shown at right. Write and graph a function of height versus time showing one complete revolution.

13. The number of hours of daylight, y, on any day of the year in Philadelphia can be modeled using the equation
    $y = 12 + 2.4\sin\left[\dfrac{2\pi(x-80)}{365}\right]$, where x represents the day number (with January 1 as day 1).

    a. Find the amount of sunlight in Philadelphia on day 354, one of the shortest days of the year. ⓐ about 9.6 h
    b. Find the dates for which the model predicts exactly 12 hours of daylight.
    day 80, March 21 (or March 20 in leap years); and day 262.5, September 19 or 20 (or September 18 or 19 in leap years)

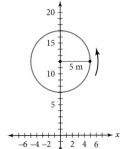

434 **CHAPTER 7** Trigonometry and Trigonometric Functions

14. A popular amusement park ride is the double Ferris wheel. Each small wheel takes 20 s to make a single rotation. The two-wheel set takes 30 s to rotate once. The dimensions of the ride are given in the diagram.

The Sky Wheel, a double Ferris wheel, operated at the Cedar Point theme park in Sandusky, Ohio, from 1962 to 1981.

a. Sandra gets on at the foot of the bottom wheel. Write an equation that will model her height above the center of this wheel as the wheels rotate. possible answer: $h = -10 \cos \frac{2\pi t}{20}$

b. The entire ride (the two-wheel set) starts revolving at the same time that the two small wheels begin to rotate. Write an equation that models the height of the center of Sandra's wheel as the entire ride rotates. possible answer: $h = 23 - 11 \cos \frac{2\pi t}{30}$

c. Because the two motions occur simultaneously, you can add the two equations to write a final equation for Sandra's position. Write this equation.

d. In a 5 min ride, during how many distinct intervals of time is Sandra within 6 ft of the ground? 6

Review

15. Solve $\tan \theta = -1.111$ graphically. Use domain $-180° \leq \theta \leq 360°$. Round answers to the nearest degree. @

16. Which has the larger area, an equilateral triangle with side 5 cm or a sector of a circle with radius 5 cm and arc length 5 cm? Give the area of each shape to the nearest 0.1 cm.
The sector has the larger area. The triangle's area is 10.8 cm²; the sector's area is 12.5 cm².

17. Find the equation of the circle with center $(-2, 4)$ and tangent line $2x - 3y - 6 = 0$.
$(x + 2)^2 + (y - 4)^2 = \frac{484}{13}$

18. Consider the equation $P(x) = 2x^3 - x^2 - 10x + 5$.
  a. Graph $P(x)$ and find the value of any rational roots.
  b. Graph $\frac{P(x)}{2x - 1}$ and write an equation for this parabola.
  c. Find exact values for the roots of the parabola. $x = -\sqrt{5}, \sqrt{5}$
  d. Write $P(x)$ in factored form.
  $P(x) = 2\left(x - \frac{1}{2}\right)(x + \sqrt{5})(x - \sqrt{5})$,
  or $P(x) = (2x - 1)(x + \sqrt{5})(x - \sqrt{5})$

Exercise 14 This exercise illustrates how the sum of two periodic functions is also a periodic function. Sound waves provide other examples of sums of sinusoidal functions.

14c. possible answer:
$h = 23 - 11 \cos \frac{2\pi t}{30} - 10 \cos \frac{2\pi t}{20}$

15.

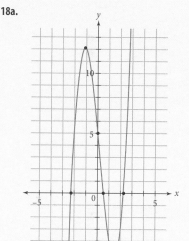

$\theta \approx \{-48°, 132°, 312°\}$

Exercise 17 LANGUAGE A *tangent line* to a circle has exactly one point in common with the circle. Contrast this with the trigonometric definition of the word.

18a.

$x = \frac{1}{2}$

18b.

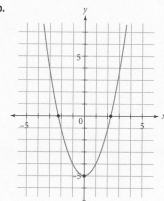

$y = x^2 - 5$

See your digital resources for the project **Sunrise, Sunset** where students choose a city and create a model for the number of hours of daylight between sunrise and sunset.

LESSON 7.7

Pythagorean Identities

In this lesson you will discover several equations that express relationships among trigonometric functions. When an equation is true for all values of the variables for which the expressions are defined, the equation is called an **identity**.

First say to yourself what you would be; and then do what you have to do.
EPICTETUS

You've already learned that $\tan A = \dfrac{y\text{-coordinate}}{x\text{-coordinate}}$ of the image point when a point $P(x, y)$ is rotated counterclockwise through $\angle A$ about the origin from the positive x-axis. You've also seen that the y-coordinate is equivalent to $r \sin A$, where r is the distance between the point and the origin, and the x-coordinate is equivalent to $r \cos A$. You can use these relationships to derive an identity that relates $\tan A$, $\sin A$, and $\cos A$.

$\tan A = \dfrac{y}{x}$ Definition of tangent.

$\tan A = \dfrac{r \sin A}{r \cos A}$ Substitute $r \sin A$ for y and $r \cos A$ for x.

$\tan A = \dfrac{\sin A}{\cos A}$ Simplify.

The reciprocals of the tangent, sine, and cosine functions are also trigonometric functions. The reciprocal of tangent is called **cotangent**, abbreviated cot. The reciprocal of sine is called **cosecant**, abbreviated csc. The reciprocal of cosine is called **secant**, abbreviated sec. These definitions lead to six more identities.

History CONNECTION

The reciprocal trigonometric functions were introduced by Muslim astronomers in the 9th and 10th centuries C.E. Before there were calculating machines, these astronomers developed remarkably precise trigonometric tables based on earlier Greek and Indian findings. They used these tables to record planetary motion, to keep time, and to locate their religious center of Mecca. Western Europeans began studying trigonometry when Arabic astronomy handbooks were translated in the 12th century.

Reciprocal Identities

$\csc A = \dfrac{1}{\sin A}$ or $\sin A = \dfrac{1}{\csc A}$

$\sec A = \dfrac{1}{\cos A}$ or $\cos A = \dfrac{1}{\sec A}$

$\cot A = \dfrac{1}{\tan A}$ or $\tan A = \dfrac{1}{\cot A}$

Your calculator probably does not have special keys for secant, cosecant, and cotangent. However, you can use the reciprocal identities to enter these functions into your calculator. For example, to graph $y = \csc x$, you use $y = \dfrac{1}{\sin x}$. The graphs at right show the principal cycles of the parent sine and cosecant functions.

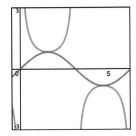

Once you know a few identities, you can use them to prove other identities. One strategy for proving that an equation is an identity is to verify that both sides of the equation are always equivalent. You can do this by writing equivalent expressions for one side of the equation until it is the same as the other side.

436 CHAPTER 7 Trigonometry and Trigonometric Functions

EXAMPLE Is the statement $\cot A = \dfrac{\cos A}{\sin A}$ an identity? Investigate with a graph or table and then write an algebraic proof.

Solution First, look at the tables for $y = \cot A$ and $y = \dfrac{\cos A}{\sin A}$. The table shows that the function values appear to be the same. Now use an algebraic proof to verify this conjecture.

x	$y = \cot(x)$	$y = \dfrac{\cos(x)}{\sin(x)}$
0	UNDEF	UNDEF
$\dfrac{\pi}{12}$	3.73205	3.73205
$\dfrac{\pi}{6}$	1.73205	1.73205
$\dfrac{\pi}{4}$	1	1
$\dfrac{\pi}{3}$	0.57735	0.57735

Use definitions and identities that you know in order to show that both sides of the equation are equivalent. Be sure to work on only one side of the equation.

$\cot A \stackrel{?}{=} \dfrac{\cos A}{\sin A}$ Original statement.

$\dfrac{1}{\tan A} \stackrel{?}{=} \dfrac{\cos A}{\sin A}$ Use the reciprocal identity to replace $\cot A$ with $\dfrac{1}{\tan A}$.

$\dfrac{1}{\frac{\sin A}{\cos A}} \stackrel{?}{=} \dfrac{\cos A}{\sin A}$ Replace $\tan A$ with $\dfrac{\sin A}{\cos A}$.

$\dfrac{\cos A}{\sin A} = \dfrac{\cos A}{\sin A}$ $1 \div \dfrac{\sin A}{\cos A}$ is equivalent to $1 \cdot \dfrac{\cos A}{\sin A}$, or simply $\dfrac{\cos A}{\sin A}$.

Therefore, $\cot A = \dfrac{\cos A}{\sin A}$ is an identity. Now that you have proved this identity, you can use it to prove other identities.

In the investigation you will discover a set of trigonometric identities that are collectively called the Pythagorean identities. To prove a new identity, you can use any previously proved identity.

INVESTIGATION
Pythagorean Identities

Step 1 Use your calculator to graph the equation $y = \sin^2 x + \cos^2 x$. (You'll probably have to enter this as $y = (\sin x)^2 + (\cos x)^2$.) Does this graph look familiar? Use your graph to write an identity.

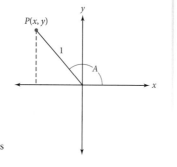

Step 2 Use the definitions for $\sin A$, $\cos A$, and the diagram at right to prove your identity.

Step 3 Explain why you think this identity is called a Pythagorean identity.

Step 2 By the Pythagorean Theorem, the sum of the squares of the right triangle's legs' lengths equals the square of the hypotenuse's length. For this right triangle, that means $(\sin A)^2 + (\cos A)^2 = 1$, or $\sin^2 A + \cos^2 A = 1$.

Step 3 In a unit circle with a reference triangle, $\sin x$ and $\cos x$ are the lengths of the legs of the right triangle, and 1 is the length of the hypotenuse. So, $\sin^2 x + \cos^2 x = 1$ is similar to $a^2 + b^2 = c^2$, the Pythagorean Theorem.

LESSON 7.7 Pythagorean Identities **437**

Investigate

Example
Project the example and have students work in small groups. Have them share their solution, not just their answer. **SMP 1, 2, 3, 4, 5, 6**

Emphasize that working on only one side of an equation is essential for a valid proof. A proof must go from what is known to what is being proved. Working with only one side of the equation helps students avoid irreversible steps.

Guiding the Investigation

Step 1 A squared trigonometric function can be written as $\sin^2 x$ or $(\sin x)^2$.

Step 1

$\sin^2 x + \cos^2 x = 1$

Step 5 $\dfrac{\sin^2 x}{\cos^2 x} + 1 = \dfrac{1}{\cos^2 x}$, or

$\tan^2 x + 1 = \sec^2 x$

Proof:

$\tan^2 x + 1 \stackrel{?}{=} \sec^2 x$ Original equation.

$(\tan x)^2 + 1 \stackrel{?}{=} \sec^2 x$ $\tan^2 x = (\tan x)^2$.

$\left(\dfrac{\sin x}{\cos x}\right)^2 + 1 \stackrel{?}{=} \sec^2 x$ Identity for $\tan x$.

$\dfrac{(\sin x)^2}{(\cos x)^2} + \dfrac{(\cos x)^2}{(\cos x)^2} \stackrel{?}{=} \sec^2 x$ Find a common denominator.

$\dfrac{(\sin x)^2 + (\cos x)^2}{(\cos x)^2} \stackrel{?}{=} \sec^2 x$ Add.

$\dfrac{1}{(\cos x)^2} \stackrel{?}{=} \sec^2 x$ Pythagorean identity

$\left(\dfrac{1}{\cos x}\right)^2 \stackrel{?}{=} \sec^2 x$ $\dfrac{1^2}{(\cos x)^2} = \left(\dfrac{1}{\cos x}\right)^2$

$(\sec x)^2 \stackrel{?}{=} \sec^2 x$ Reciprocal identity

$\sec^2 x = \sec^2 x$ $\sec^2 x = (\sec x)^2$

Modifying the Investigation

Whole Class Work through Steps 1 through 8 with student input as students follow along from their own workspace.

Shortened Complete Steps 1 through 3 with the class.

One Step Challenge students to write a convincing argument that the equation $1 + \dfrac{\cos^2 A}{\sin^2 A} = \dfrac{1}{\sin^2 A}$ is valid. Then challenge students to try dividing the identity $\sin^2 A + \cos^2 A = 1$ by $\cos^2 A$. [$\tan^2 A + 1 = \sec^2 A$]

Conceptual/Procedural

Conceptual Students are introduced to the concept of an identity. They extend their experience with proof to proving trigonometric identities.

Procedural Students use a graph or table to determine if a statement is an identity. They then write an algebraic proof.

LESSON 7.7 Pythagorean Identities **437**

Step 6

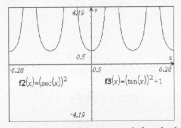

The domain values are undefined when $\cos x = 0$, or when $x = \frac{\pi}{2} + \pi n$ or $x = 90° + 180°n$ where n is an integer.

Step 7 In the standard identity, the quotient $\frac{\cos^2 x}{\sin^2 x}$ is simplified to $\cot^2 x$. A good first step is to rewrite $\cot^2 x$ as $\frac{\cos^2 x}{\sin^2 x}$.

Step 7 $1 + \frac{\cos^2 x}{\sin^2 x} = \frac{1}{\sin^2 x}$, or $1 + \cot^2 x = \csc^2 x$

Proof:

$1 + \cot^2 x \stackrel{?}{=} \csc^2 x$ Original equation.

$1 + (\cot x)^2 \stackrel{?}{=} \csc^2 x \quad \cot^2 x = (\cot x)^2$.

$1 + \left(\frac{\cos x}{\sin x}\right)^2 \stackrel{?}{=} \csc^2 x$ Identity for $\cot x$.

$\frac{(\sin x)^2}{(\sin x)^2} + \frac{(\cos x)^2}{(\sin x)^2} \stackrel{?}{=} \csc^2 x$ Find a common denominator.

$\frac{(\sin x)^2 + (\cos x)^2}{(\sin x)^2} \stackrel{?}{=} \csc^2 x$ Add.

$\frac{1}{(\sin x)^2} \stackrel{?}{=} \csc^2 x$ Pythagorean identity.

$\left(\frac{1}{\sin x}\right)^2 \stackrel{?}{=} \csc^2 x \quad \frac{1^2}{(\sin x)^2} = \left(\frac{1}{\sin x}\right)^2$

$(\csc x)^2 \stackrel{?}{=} \csc^2 x$ Reciprocal identity.

$\csc^2 x = \csc^2 x \quad \csc^2 x = (\csc x)^2$.

Summarize

As students present, have them share their solution, not just their answer. Emphasize the use of correct mathematical terminology. Whether student responses are correct or incorrect, ask other students if they agree and why.

SMP 1, 2, 3, 4, 5, 6

Step 4 Solve the identity from Step 1 for $\cos^2 x$ to get another identity. Then solve for $\sin^2 x$ to get another variation.

$\cos^2 x = \underline{\ ?\ } \quad 1 - \sin^2 x$

$\sin^2 x = \underline{\ ?\ } \quad 1 - \cos^2 x$

Step 5 Divide both sides of the identity from Step 1 by $\cos^2 x$ to develop a new identity. Simplify so that there are no trigonometric functions in the denominator.

Step 6 Verify your identity from Step 5 with a graph or table. Name any domain values for which the identity is undefined.

Step 7 Divide both sides of the identity from Step 1 by $\sin^2 x$ to develop a new identity. Simplify so that there are no trigonometric functions in the denominator.

Step 8

Step 8 Verify the identity from Step 7 with a graph. Name any domain values for which the identity is undefined. The domain values are undefined when $\sin x = 0$, or when $x = \pi n$ or $x = 180°n$ where n is an integer.

You have verified identities by setting each side of the equation equal to y and graphing. If the graphs and table values match, you may have an identity. You may have used your calculator in this way to verify the Pythagorean identities in the investigation. If not, try it now on one of the three identities below.

> **Pythagorean Identities**
> $\sin^2 A + \cos^2 A = 1$
> $1 + \tan^2 A = \sec^2 A$
> $1 + \cot^2 A = \csc^2 A$

You should always use an algebraic proof to be certain that you have an actual identity.

7.7 Exercises

You will need a graphing calculator for Exercises 2, 9–12, 15, and 17.

Practice Your Skills

1. Explain how you can graph $y = \cot x$ on your calculator without using a built-in cotangent function. @ Graph $y = \frac{1}{\tan x}$.

2. Use graphs or tables to determine which of these equations may be identities.

   a. $\cos x = \sin\left(\frac{\pi}{2} - x\right)$ yes
   b. $\cos x = \sin\left(x - \frac{\pi}{2}\right)$ @ no
   c. $(\csc x - \cot x)(\sec x + 1) = 1$ no
   d. $\tan x (\cot x + \tan x) = \sec^2 x$ @ yes

3. Prove algebraically that the equation in Exercise 2d is an identity. @

4. Evaluate.

   a. $\sec \frac{\pi}{6} \quad \frac{2}{\sqrt{3}}$, or $\frac{2\sqrt{3}}{3}$
   b. $\csc \frac{5\pi}{6} \quad 2$
   c. $\csc \frac{2\pi}{3} \quad \frac{2}{\sqrt{3}}$, or $\frac{2\sqrt{3}}{3}$
   d. $\sec \frac{3\pi}{2}$ undefined
   e. $\cot \frac{5\pi}{3} \quad -\frac{1}{\sqrt{3}}$, or $-\frac{\sqrt{3}}{3}$
   f. $\csc \frac{4\pi}{3} \quad -\frac{2}{\sqrt{3}}$, or $-\frac{2\sqrt{3}}{3}$

3. Proof:

$\tan x(\cot x + \tan x) \stackrel{?}{=} \sec^2 x$ Original equation.

$\tan x \cot x + (\tan x)^2 \stackrel{?}{=} \sec^2 x$ Distribute.

$\tan x \cdot \frac{1}{\tan x} + \tan^2 x \stackrel{?}{=} \sec^2 x$ Reciprocal identity.

$1 + \tan^2 x \stackrel{?}{=} \sec^2 x$ Simplify.

$\sec^2 x = \sec^2 x$ Pythagorean identity.

Reason and Apply

5. In your own words, explain the difference between a trigonometric equation and a trigonometric identity.

6. Use the reciprocal identities and the definition of tangent to rewrite each expression in a simpler form.
   a. $\csc A \tan A$ $\sec A$
   b. $\dfrac{\cot A}{\csc A}$ $\cos A$
   c. $\sec A \cot A \csc A$
   d. $\dfrac{\tan A}{\sec A}$ $\sin A$

7. Use the Pythagorean Identities to rewrite each expression in a simpler form.
   a. $\sin^2 A + 1 - \cos^2 A$ $2\sin^2 A$
   b. $(1 + \tan^2 A)\cos^2 A$ 1
   c. $(\cot^2 A + 1)(1 - \sin^2 A)$ $\cot^2 A$
   d. $\sec^2 A (1 - \sin^2 A)$ 1

8. Sketch the graph of $y = \dfrac{1}{f(x)}$ for each function.

   a.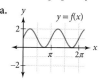

   b. (graph of y = f(x))

9. Find another function that has the same graph as each function named below. More than one answer is possible.
   a. $y = \cos\left(\dfrac{\pi}{2} - x\right)$ @ $y = \sin x$
   b. $y = \sin\left(\dfrac{\pi}{2} - x\right)$ $y = \cos x$
   c. $y = \tan\left(\dfrac{\pi}{2} - x\right)$ $y = \cot x$
   d. $y = \cos(-x)$ @ $y = \cos x$
   e. $y = \sin(-x)$ @ $y = -\sin x$
   f. $y = \tan(-x)$ $y = -\tan x$
   g. $y = \sin(x + 2\pi)$ $y = \sin x$
   h. $y = \cos\left(\dfrac{\pi}{2} + x\right)$ $y = -\sin x$
   i. $y = \tan(x + \pi)$ $y = \tan x$

10. Find the first three positive x-values that make each equation true.
    a. $\sec x = -2.5$ 1.98, 4.30, 8.27
    b. $\csc x = 0.4$ no such values

11. Sketch a graph for each equation with domain $0 \le x < 4\pi$. Include any asymptotes and state the x-values at which the asymptotes occur. ⓗ
    a. $y = \csc x$
    b. $y = \sec x$
    c. $y = \cot x$

12. Write an equation for each graph. More than one answer is possible. Use your calculator to check your work.

    a. 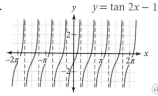 $y = \tan 2x - 1$ @

    b. 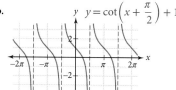 $y = \cot\left(x + \dfrac{\pi}{2}\right) + 1$

    c. 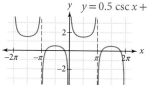 $y = 0.5 \csc x + 1$ @

    d. 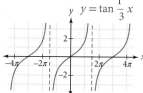 $y = \tan \dfrac{1}{3} x$

LESSON 7.7 Pythagorean Identities **439**

8a.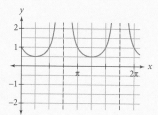

8b.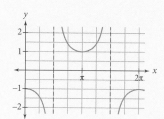

Exercise 9 Encourage students to graph the functions. You might have them find two other functions that share each graph.

Exercise 10 Encourage students to use a combination of graphing and inverse trigonometric functions rather than the calculator's Solve function.

11a.

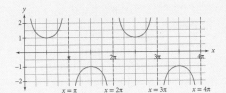

11b.

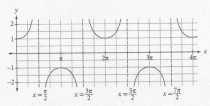

11c.

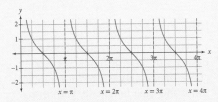

13a.
$$\frac{\sin x}{1+\cos x} \stackrel{?}{=} \frac{1-\cos x}{\sin x}$$
$$\frac{\sin x}{1+\cos x} \cdot \frac{1-\cos x}{1-\cos x} \stackrel{?}{=} \frac{1-\cos x}{\sin x}$$
$$\frac{\sin x(1-\cos x)}{1-\cos^2 x} \stackrel{?}{=} \frac{1-\cos x}{\sin x}$$
$$\frac{\sin x(1-\cos x)}{\sin^2 x} \stackrel{?}{=} \frac{1-\cos x}{\sin x}$$
$$\frac{\sin x(1-\cos x)}{\sin x \cdot \sin x} \stackrel{?}{=} \frac{1-\cos x}{\sin x}$$
$$\frac{1-\cos x}{\sin x} = \frac{1-\cos x}{\sin x}$$

13b.
$$\frac{1-\cos^2 x}{\tan x} \stackrel{?}{=} \sin x \cos x$$
$$\frac{\sin^2 x}{\frac{\sin x}{\cos x}} \stackrel{?}{=} \sin x \cos x$$
$$\frac{\sin^2 x \cos x}{\sin x} \stackrel{?}{=} \sin x \cos x$$
$$\sin x \cos x = \sin x \cos x$$

---

13. Use definitions and identities that you know to prove that the two sides of each equation are equivalent. Be sure to change only one side of the equation.

a. $\dfrac{\sin x}{1+\cos x} = \dfrac{1-\cos x}{\sin x}$ ⓗ

b. $\dfrac{1-\cos^2 x}{\tan x} = \sin x \cos x$

c. $\dfrac{\sec^2 x - 1}{\sin x} = \tan x \sec x$

14. Use the trigonometric identities to rewrite each expression in a simplified form that contains only one type of trigonometric function. For each expression, give values of θ for which the expression is undefined. Use domain $0 \leq \theta < 2\pi$.

a. $2\cos^2\theta + \dfrac{\sin\theta}{\csc\theta} + \sin^2\theta$ 2; undefined when θ equals 0 or π

b. $(\sec\theta)(2\cos^2\theta) + (\cot\theta)(\sin\theta)$ 3 $\cos\theta$; undefined when θ equals 0, $\dfrac{\pi}{2}$, π, or $\dfrac{3\pi}{2}$

c. $\sec^2\theta + \dfrac{1}{\cot\theta} + \dfrac{\sin^2\theta - 1}{\cos^2\theta}$ $\tan^2\theta + \tan\theta$; undefined when θ equals 0, $\dfrac{\pi}{2}$, π, or $\dfrac{3\pi}{2}$

d. $\sin\theta(\cot\theta + \tan\theta)$ $\sec\theta$; undefined when θ equals 0, $\dfrac{\pi}{2}$, π, or $\dfrac{3\pi}{2}$

Review

15. **APPLICATION** Tidal changes can be modeled with a sinusoidal curve. This table gives the time and height of the low and high tides for Burntcoat Head, Nova Scotia, on two consecutive days. (AST stands for Atlantic Standard Time.)

High and Low Tides for Burntcoat Head, Nova Scotia

	Low	High	Low	High
	Time (AST)/Height (m)	Time (AST)/Height (m)	Time (AST)/Height (m)	Time (AST)/Height (m)
Day 1	04:13/1.60	10:22/13.61	16:40/1.42	22:49/14.15
Day 2	05:12/1.01	11:19/14.09	17:38/0.94	23:45/14.71

(www.ldeo.columbia.edu)

a. Select one time value as the starting time and assign it time 0. Express the other time values in minutes relative to the starting time.

b. Let x represent time in minutes relative to the starting time, and let y represent the tide height in meters. Make a scatter plot of the eight data points.

c. Find mean values for
   i. Height of the low tide. 1.24 m
   ii. Height of the high tide. 14.14 m
   iii. Height of "no tide," or the mean water level. 7.69 m
   iv. Time in minutes of a tide change between high and low tide. 373 min

d. Write an equation to model these data.

e. Graph your equation with the scatter plot in order to check the fit.

f. Predict the water height at 12:00 on Day 1. 12.07 m

g. Predict when the high tide(s) occurred on Day 3. only one high tide, at 12.07

The Bay of Fundy borders Maine and the Canadian provinces of New Brunswick and Nova Scotia. This bay experiences some of the most dramatic tides on Earth, with water depths fluctuating up to 18 m.

16. **APPLICATION** Juan's parents bought a $500 savings bond for him when he was born. Interest has compounded monthly at an annual fixed rate of 6.5%.

    a. Juan just turned 17, and he is considering using the bond to pay for college. How much is his bond currently worth? @ $1,505.12

    b. Juan also considers saving the bond and using it to buy a used car after he graduates. If he would need about $4,000, how long would he have to wait?   another 15 yr, or until he is 32

17. **APPLICATION** A pharmacist has 100 mL of a liquid medication that is 60% concentrated. This means that in 100 mL of the medication, 60 mL is pure medicine and 40 mL is water. She alters the concentration when filling a specific prescription. Suppose she alters the medication by adding water.

    a. Write a function that gives the concentration of the medication, $d(x)$, as a function of the amount of water added in milliliters. $d(x) = \dfrac{60}{100 + x}$

    b. What is the concentration if the pharmacist adds 20 mL of water?   50%

    c. How much water should she add if she needs a 30% concentration?   100 mL

    d. Graph $y = d(x)$. Explain the meaning of the asymptote.

IMPROVING YOUR Visual Thinking SKILLS

An Equation Is Worth a Thousand Words

You have learned to model real-world data with a variety of equations. You've also seen that many types of equations have real-world manifestations. For example, the path of a fountain of water is parabolic because its projectile-motion equation is quadratic.

Look for a photo of a phenomenon that suggests the graph of an equation. Impose coordinate axes, and find the equation that models the photo. If you have geometry software, you can import an electronic version of your photo and graph your function over it.

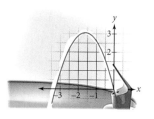

13c.
$$\dfrac{\sec^2 x - 1}{\sin x} \stackrel{?}{=} \tan x \sec x$$
$$\dfrac{\tan^2 x}{\sin x} \stackrel{?}{=} \tan x \sec x$$
$$\dfrac{\dfrac{\sin^2 x}{\cos^2 x}}{\sin x} \cdot \dfrac{1}{\dfrac{\sin x}{1}} \stackrel{?}{=} \tan x \sec x$$
$$\dfrac{\sin x}{\cos x \cos x} \stackrel{?}{=} \tan x \sec x$$
$$\tan x \sec x = \tan x \sec x$$

Exercise 15 ALERT If students are confused by times such as 16:40, explain that these are times on a 24-hour clock; 16:40 is 4:40 P.M.

15a. Sample answer: Select the second time value, 10:22, as time 0. The new time values are −369, 0, 378, 747, 1130, 1497, 1876, 2243.

15b.
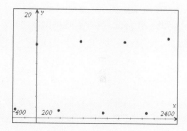

15d. possible answer:
$$y = 6.45 \cos\left(\dfrac{2\pi}{746}x\right) + 7.69$$

15e.
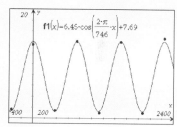

17d.
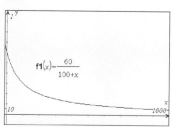

The asymptote is the x-axis, $y = 0$. The more water that is added, the closer the concentration will get to 0%, but it will never actually become 0%.

IMPROVING VISUAL Thinking Skills

Students could find photos that represent linear, quadratic, polynomial, exponential, logarithmic, conic section, or trigonometric equations. For the illustration in the book, a sample equation that models the water fountain is $y = -1.15x(x + 3.25)$.

CHAPTER 7 REVIEW

Reviewing

Ask students to revisit lesson on Defining the Circular Functions, Exercise 13, about sunspots. Pose the problem of predicting the next two years (after 2006) during which the average monthly number of sunspots will be about 30. Model the data with both sine and cosine functions, reminding students of the meanings of radians, sinusoids, amplitude, phase shift, vertical translation, period, frequency, and inverse functions. Then remind students of the other trigonometric functions, the inverse trigonometric functions, and how to prove some trigonometric identities.

CHAPTER 7 REVIEW

The **trigonometric ratios**—sine, cosine, and tangent—relate the side lengths of a triangle to the measure of an angle. The **sine** of an angle is the ratio of the opposite leg to the hypotenuse in a right triangle. The **cosine** of an angle is the ratio of the adjacent leg to the hypotenuse, and the **tangent** of an angle is the ratio of the opposite leg to the adjacent leg. You can use these ratios to find missing side lengths and angles in right triangles.

In this chapter you learned to measure angles in **radians**, and you then identified relationships among radian measure, arc length, speed, and **angular speed**. You studied circular functions, their graphs, and their applications, and you learned to think of the sine and cosine of an angle as the y- and x-coordinates of a point on the unit circle. This allowed you to identify angles in **standard position** that are **coterminal**. That is, they share the same **terminal side**. Coterminal angles have the same trigonometric values. The remaining **trigonometric functions**—tangent, **cotangent**, **secant**, and **cosecant**—are also defined either as a ratio involving an x-coordinate and a y-coordinate, or as a ratio of one of these coordinates and the distance from the origin to the point.

Graphs of the trigonometric functions model periodic behavior and have domains that extend in both the positive and negative directions. You worked with many relationships that can be modeled with **sinusoids**. You studied transformations of sinusoidal functions and defined their **amplitude, period, phase shift, vertical translation,** and **frequency**.

You also learned the fundamental reciprocal and Pythagorean identities and used them to rewrite expressions in alternate forms.

Exercises

You will need a graphing calculator for Exercises **4, 9, 10, 11** and **13**.

@ Answers are provided for all exercises in this set.

1. For each triangle, find the measure of the labeled angle or the length of the labeled side.

   a. $A \approx 43°$, sides 15, 11

   b. sides 13, 7, $B \approx 28°$

   c. 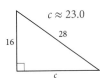 $c \approx 23.0$, sides 16, 28

   d. 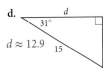 $d \approx 12.9$, $31°$, side 15

   e. 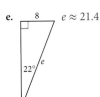 side 8, $e \approx 21.4$, $22°$

   f. 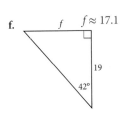 $f \approx 17.1$, side 19, $42°$

2. When looking up at an object, the angle above the horizontal is called the angle of elevation. When looking down at an object, the angle below the horizontal is called the angle of depression. about 73°

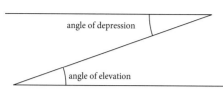

Aliya sees a coconut 66 ft up in a coconut palm tree that is 20 ft away from her. What is the angle of elevation from her position to the nut?

3. The point (6, 0) is rotated 150° counterclockwise about the origin. What are the coordinates of its image point? $(-3\sqrt{3}, 3)$

4. Use the parametric equations $x = -3t + 1$ and $y = \dfrac{2}{t+1}$ to answer each question.
   a. Find the x- and y-coordinates of the points that correspond to the values of $t = 3$, $t = 0$, and $t = -3$. $t = 3$: $x = -8$, $y = 0.5$; $t = 0$: $x = 1$, $y = 2$; $t = -3$: $x = 10$, $y = -1$
   b. Find the y-value that corresponds to an x-value of -7. $y = \dfrac{6}{11}$
   c. Find the x-value that corresponds to a y-value of 4. $x = \dfrac{5}{2}$
   d. Sketch the curve for $-3 \le t \le 3$, showing the direction of movement. Trace the graph and explain what happens when $t = -1$. When $t = -1$, the y-value is undefined.

5. The circle at right has radius 4 cm, and the measure of central angle ACB is 55°.
   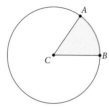
   a. What is the measure of $\angle ACB$ in radians? $\dfrac{11\pi}{36}$
   b. What is the length of $\widehat{AB}$? approximately 3.84 cm
   c. What is the area of sector ACB? approximately 7.68 cm²

6. For each angle in standard position given, identify the quadrant that the angle's terminal side lies in, and name a coterminal angle. Then convert each angle measure from radians to degrees or vice versa.
   a. 60° I; 420°; $\dfrac{\pi}{3}$
   b. $\dfrac{4\pi}{3}$ III; $\dfrac{10\pi}{3}$; 240°
   c. 330° IV; $-30°$; $\dfrac{11\pi}{6}$
   d. $-\dfrac{\pi}{4}$ IV; $\dfrac{7\pi}{4}$; $-45°$

7. Find exact values of the sine and cosine of each angle in Exercise 6.

8. State the period of the graph of each equation, and write one other equation that has the same graph. Other equations are possible.
   a. $y = 2\sin\left[3\left(x - \dfrac{\pi}{6}\right)\right]$
   b. $y = -3\cos 4x$
   c. $y = \sec 2x$
   d. $y = \tan(-2x) + 1$

9. For the sinusoidal equations in 8a and b, state the amplitude, phase shift, vertical translation, and frequency. Then sketch a graph of one complete cycle.

CHAPTER 7 REVIEW **443**

4d.
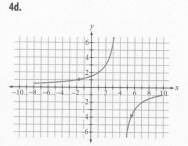

Exercise 7 As needed, suggest that students draw reference triangles.

7a. $\sin 60° = \dfrac{\sqrt{3}}{2}$; $\cos 60° = \dfrac{1}{2}$

7b. $\sin\dfrac{4\pi}{3} = -\dfrac{\sqrt{3}}{2}$; $\cos\dfrac{4\pi}{3} = -\dfrac{1}{2}$

7c. $\sin 330° = -\dfrac{1}{2}$; $\cos 330° = \dfrac{\sqrt{3}}{2}$

7d. $\sin\left(-\dfrac{\pi}{4}\right) = -\dfrac{\sqrt{2}}{2}$; $\cos\left(-\dfrac{\pi}{4}\right) = \dfrac{\sqrt{2}}{2}$

Exercise 8 Looking at the graphs of these equations might help students find other equations.

8a. period $= \dfrac{2\pi}{3}$; $y = -2\cos\left[3\left(x - \dfrac{2\pi}{3}\right)\right]$

8b. period $= \dfrac{\pi}{2}$; $y = 3\sin\left[4\left(x - \dfrac{\pi}{8}\right)\right]$

8c. period $= \pi$; $y = \csc\left[2\left(x + \dfrac{\pi}{4}\right)\right]$

8d. period $= \dfrac{\pi}{2}$; $y = \cot\left[2\left(x - \dfrac{\pi}{4}\right)\right] + 1$

9a. 2; $\dfrac{\pi}{6}$; 0; $\dfrac{3}{2\pi}$

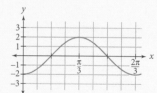

9b. 3; 0; 0; $\dfrac{2}{\pi}$

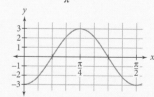

LESSON 7 REVIEW **443**

11c.

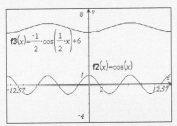

Exercise 13 Remind students that 00:00 is midnight on a 24-hour clock.

13a. Let *x* represent time in hours, and let *y* represent tide height in meters.

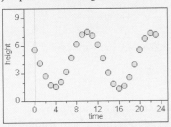

13b. possible answer:
$$y = 2.985 \cos\left[\frac{\pi}{6}(x-10)\right] + 4.398$$

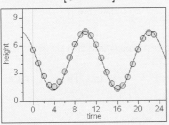

Take Another Look

1. For a supplementary pair, ∠A and ∠B, sin A and sin B, cos A = −cos B, and tan A = −tan B.

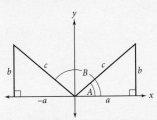

For a complementary pair, ∠C and ∠D, sin C = cos D, cos C = sin D, and tac C = $\frac{1}{\tan D}$. (In fact, the word *cosine* means "complement's sine.")

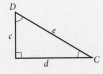

---

10. Write an equation for each graph.

a.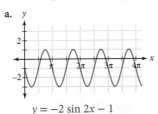
$y = -2 \sin 2x - 1$

b.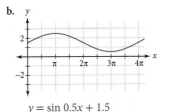
$y = \sin 0.5x + 1.5$

11. Consider the function $y = \cos x$.

a. Write the equation of the image after the function is reflected across the *x*-axis, dilated by a vertical scale factor of $\frac{1}{2}$ and a horizontal scale factor of 2, and translated up 6 units. $y = -\frac{1}{2}\cos\left(\frac{1}{2}x\right) + 6$

b. What is the period of the image, in radians? What are the amplitude and phase shift? period 4π, amplitude $\frac{1}{2}$; phase shift: none

c. Graph the function and its image on the same graph.

12. Find the area and arc length of a sector of a circle that has radius 3 cm and central angle $\frac{\pi}{4}$. Give exact answers. area = $\frac{9\pi}{8}$ cm²; arc length = $\frac{3\pi}{4}$ cm

13. APPLICATION These data give the ocean tide heights each hour on November 17, 2002, at Saint John, New Brunswick, Canada.

a. Create a scatter plot of the data.

b. Write a function to model the data, and graph this function on the scatter plot from 13a.

c. What would you estimate the tide height to have been at 3:00 P.M. on November 19, 2002? approximately 1.84 m

d. A ship was due to arrive at Saint John on November 20, 2002. The water had to be at least 5 m for the ship to enter the harbor safely. Between what times on November 20 could the ship have entered the harbor safely?

between 00:00 and 00:37, between 7:22 and 12:38, and between 19:22 and 23:59

Time	Tide height (m)	Time	Tide height (m)
00:00	5.56	12:00	6.09
01:00	4.12	13:00	4.65
02:00	2.71	14:00	3.08
03:00	1.77	15:00	1.85
04:00	1.56	16:00	1.33
05:00	2.09	17:00	1.60
06:00	3.21	18:00	2.54
07:00	4.70	19:00	3.96
08:00	6.18	20:00	5.51
09:00	7.20	21:00	6.75
10:00	7.50	22:00	7.33
11:00	7.09	23:00	7.17

Take Another Look

1. Select a pair of noncongruent angles that are supplementary (whose measures sum to 180°). Use your calculator to find the sine, cosine, and tangent of each angle measure. What relationships do you notice? Try other supplementary pairs to verify your relationships. Then select a pair of complementary angles (whose measures sum to 90°), and find the sine, cosine, and tangent of each angle measure. What relationships do you notice? Verify these relationships with other complementary pairs. Draw geometric diagrams that prove the relationships you find.

2. In an earlier math course, you probably learned the formula $Area = bh$ for the area of a parallelogram, where b represents the length of the base, and h represents the height of the parallelogram, drawn perpendicular from the base to the opposite side. Find a formula for the area of a parallelogram given two adjacent side lengths and the measure of one angle.

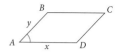

3. You are now familiar with angle measures in degrees and radians, but have you ever heard of gradians? Research the gradian angle measure. Explain how it compares with radians and degrees, when and where it was used, and any advantages it might have. Can you find any other units that measure angle or slope?

Assessing What You've Learned

 WRITE IN YOUR JOURNAL Are your understandings of the sine, cosine, and tangent functions different now than they were when you started this chapter? Write a journal entry that describes how your understanding of the trigonometric functions has changed over time.

 ORGANIZE YOUR NOTEBOOK Make sure your notebook is complete and well organized. Be sure to include all of the definitions and formulas that you have learned in this chapter. You may want to include at least one example of each of the different applications of trigonometry.

 WRITE TEST ITEMS Write at least four test items for this chapter. Include items that cover different applications of trigonometry. You may want to include the use of trigonometry for right triangles and oblique triangles, as well as applications of vectors and parametric equations.

CHAPTER 8 Probability

Overview

Discovering Advanced Algebra is about modeling real-world problems with mathematical functions. Earlier chapters developed polynomial, exponential, logarithmic, and rational functions. Underlying the probability theory of this chapter are functions called *random variables* that assign numbers to events.

In **Lesson 8.1,** students simulate random processes to compare experimental and theoretical probabilities. Tree diagrams in **Lesson 8.2** help students break down dependent compound events into independent events and explore conditional probability. In **Lesson 8.3,** students break down events into mutually exclusive events with the aid of Venn diagrams. In **Lesson 8.4,** students construct and interpret two-way frequency tables and use the two-way table as a sample space to decide if events are independent and to approximate conditional probabilities.

The **Performance Task** highlighted in this chapter are in Lesson 8.1 where students rank scenarios according to the method that will best produce a random integer from 0 to 9.

The Mathematics

What's the chance of rain? What's the probability that the basketball player will make a free throw? Probability is a very applicable topic, because it can allow making predictions despite the uncertainty around us.

Much uncertainty is due to randomness; a study of probability can challenge students' ideas of randomness. Some students might believe that there really is no randomness; perhaps they believe events are determined by divine intervention or events can be explained completely by science. You might explain to them that randomness means there is uncertainty in any prediction, and probability gives us the means of measuring that uncertainty.

Mathematical probability theory is based on *outcomes*. For example, one outcome might be "rain." If you flip a coin twice, an outcome is getting exactly one head. Each outcome is assigned a *probability,* which is a number between 0 and 1.

Probabilities of outcomes such as rain are *experimental probabilities.* They are calculated from data gathered over time. Experimental probabilities can be very useful if they arise from many careful observations. When the experimentation is less systematic, however, these probabilities can lead to superstition. For example, people might realize that they were wearing certain socks when they had two very good performances, so they plan to wear those "lucky" socks for future performances.

To formalize probabilistic ideas, it helps to study *theoretical probabilities,* which occur in the most random situations. For example, what's the probability of getting a 3 when you roll a die? What's the chance of choosing a red playing card from a deck? The *Law of Large Numbers* says that as the number of trials increases, the experimental probability approaches the theoretical probability. For example, the fraction of coin flips that come up heads will get closer and closer to $\frac{1}{2}$ as you keep flipping.

Suppose you're asked, "What's the theoretical probability of getting exactly one head when flipping a coin twice?" To answer this question, you need to figure out which of the possible outcomes are *equally likely*. The outcomes "exactly one head," "two heads," and "no heads" are not equally likely, but the four outcomes HH, HT, TH, and TT are equally likely. Because "exactly one head" consists of the two equally likely outcomes HT and TH, its probability is $\frac{2}{4} = \frac{1}{2}$. In general, probability is defined to be P(outcome of interest) = $\frac{\text{number of equally likely outcomes of interest}}{\text{total number of equally likely outcomes}}$.

Another common situation requires calculating the probability of one event under the condition that another event has occurred. For example, suppose you have a group of 6 female and 3 male students. One male and one female have red hair (at the moment). The conditional probability of red hair knowing that the person is male, symbolized P(red | male), is $\frac{1}{3}$. Similarly, P(red | female) = $\frac{1}{6}$. In general, $P(A | B) = \frac{P(A \text{ and } B)}{P(B)}$.

Using This Chapter

This chapter builds on the probability content from a first year algebra course to focus on more decision making. It is important that you not skip this chapter as statistically sound data production, the focus of the next chapter, is based on the probability concept of random sampling. If time is a concern, you can use the modifications listed for **Shortened** versions of the investigation, the desmos technology simulations, or the Investigation Worksheets with Sample Data.

Common Core State Standards (CCSS)

In the middle grades, primarily in Grade 7, students were introduced to the concepts of basic probability, including chance processes, probability models, and sample spaces. They saw probabilities of chance events as long-run relative frequencies of their occurrence, examining the proportion of "successes" in a chance process—one involving repeated observations of random outcomes of a given event, such as a series of coin tosses. Students saw, and created, probability models to find probabilities of events, including tree diagrams and tables, and used them as a tool for organized counting of possible outcomes from chance processes (7.SP.5 - 8)

In high school, students understand that the probabilities of real world events are often approximated by data about those events. They use two-way frequency tables of data to decide if events are independent and to approximate conditional probability (S.ID.5). They look at rules of probability and their use in finding probabilities of compound events. They use the language of "and" or "or" to compute and interpret theoretical and experimental probabilities for compound events, attending to mutually exclusive events, independent events, and conditional probability. In developing their understanding of conditional probability and independence, students should see problems in which the uniform probabilities of the outcomes leads to independence and problems in which it does not (S.CP.1 – 3). Students should make use of geometric probability models wherever possible, understanding the role that randomness and careful design play in the conclusions that can be drawn and use probability to make informed decisions (S.IC.2, S.CP.4).

Because of its strong connection with modeling, all the standards pertaining to probability are starred, meaning they are modeling standards.

Additional Resources For complete references to additional resources for this chapter and others see your digital resources.

Refreshing Your Skills To review the algebra skills needed for this chapter see your digital resources.

Basic Probability

Explorations See your digital resources for extensions and/or applications of content from this chapter.

Geometric Probability
The Law of Large Numbers

Materials
- coins (pennies, nickels, dimes, quarters)

CHAPTER 8

Objectives

- Learn about randomness
- See how experimental probabilities relate to theoretical probabilities
- Define and calculate geometric probability
- Use tree diagrams and conditional probabilities to break down compound events into independent events
- Use the multiplication rule for independent events
- Use Venn diagrams to break down compound events into mutually exclusive events
- Understand the addition rule for mutually exclusive events
- Construct and interpret two-way frequency tables of data when two categories are associated with each object being classified.
- Use the two-way table as a sample space to decide if events are independent and to approximate conditional probabilities.
- Develop a margin of error through the use of simulation models for random sampling.

CHAPTER 8 Probability

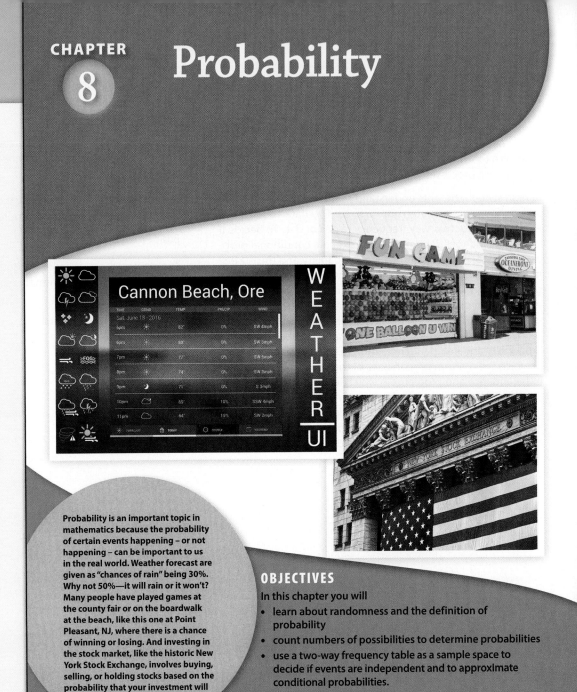

Probability is an important topic in mathematics because the probability of certain events happening – or not happening – can be important to us in the real world. Weather forecast are given as "chances of rain" being 30%. Why not 50%—it will rain or it won't? Many people have played games at the county fair or on the boardwalk at the beach, like this one at Point Pleasant, NJ, where there is a chance of winning or losing. And investing in the stock market, like the historic New York Stock Exchange, involves buying, selling, or holding stocks based on the probability that your investment will make, not lose, money.

OBJECTIVES

In this chapter you will
- learn about randomness and the definition of probability
- count numbers of possibilities to determine probabilities
- use a two-way frequency table as a sample space to decide if events are independent and to approximate conditional probabilities.

Probabilities of outcomes such as rain or choosing a stock to buy are *experimental probabilities*. They are calculated from data gathered over time. Experimental probabilities can be very useful if they are the result of systematic, accurate data collection. Sometimes decisions based on inaccurate data collection or interpretation, leads to superstition. For example, in the 2016 World Series, the Chicago Cubs broke the legendary "curse of the billy goat," a curse stemming from losing seasons after a goat was evicted from Wrigley Field in 1945. **ASK** "What are some "superstitions" resulting from interpretation of observations have you heard of?" [lucky shirts, special color, lucky numbers, etc.]

LESSON 8.1

Randomness and Probability

It's choice—not chance—that determines your destiny.
JEAN NIDETCH

"It isn't fair," complains Noah. "My car insurance rates are much higher than yours." Rita replies, "Well, Noah, that's because insurance companies know the chances are good that it will cost them less to insure me."

How much do you think Demolition Derby drivers pay for auto insurance?

Insurance companies can't know for sure what kind of driving record you will have. So they use the driving records for people of your age group, gender, and prior driving experience to determine your chances of an accident and, therefore, your insurance rates. This is just one example of how probability theory and the concept of randomness affect your life.

Career CONNECTION

An actuary uses mathematics to solve financial problems. Actuaries often work for insurance, consulting, and investment companies. They use probability, statistics, and risk theory to decide the cost of a company's employee benefit plan, the cost of a welfare plan, and how much funding an insurance company will need to pay for expected claims.

Probability theory was developed in the 17th century as a means of determining the fairness of games, and it is still used to make sure that casino customers lose more money than they win. Probability is also important in the study of sociological and natural phenomena.

At the heart of probability theory is randomness. Rolling a die, flipping a coin, drawing a card, and spinning a game-board spinner are examples of **random processes**. In a random process, no individual outcome is predictable, even though the long-range pattern of many individual outcomes often is predictable.

LESSON 8.1

In this lesson, students see the basic concepts of randomness and experimental and theoretical probability.

COMMON CORE STATE STANDARDS		
APPLIED	DEVELOPED	INTRODUCED
	S.IC.2	S.MD.6

Objectives
- Learn about randomness
- Define experimental and theoretical probability
- Simulate experimental probability using technology
- Define and calculate geometric probability

Vocabulary
random processes
fair
random
simulation
outcomes
event
simple event
compound
experimental probabilities
theoretical probability
geometric probability

Materials
- Calculator Notes: Random Numbers; Histograms; Dice Simulation; Dice Simulation with the Prob Sim App

Launch

1. If you flip a fair coin 50 times, how many times do you expect heads to occur? 25

2. A fair coin is tossed twice. One possible outcome is HH. What are the other possible outcomes?
   HT, TH, TT

3. What is a *fair* coin?
   One that it is equally likely to land on heads or tails.

ELL
Working through the investigation will make the definitions of *experimental* and *theoretical* probabilities much clearer. Introduce the two terms before the investigation and clarify them during the discussion.

Support
Use the Refreshing Your Skills: Basic Probability lesson to scaffold the lesson. Students may need to list outcomes to provide tangible data. Help them solve the problem of making the lists complete. See also the ELL note.

Advanced
Have students extend the simulations from the investigation to experiments with dice or playing cards.

Investigate

Guiding the Investigation

Students can use the Flip a Coin worksheet included with the investigation worksheet to record their outcomes in this investigation.

There is also a desmos simulation of this investigation in your ebook.

Steps 1, 2 In order to collect as much data as possible for comparison in Step 4, each student should go through these steps individually.

Step 1 When students invent a list of H's and T's, they are likely to think about what they have already listed in order to decide what to list next. They may try to avoid having many of the same letters in a row or having an unbalanced number of H's and T's.

Although the "random" numbers generated by calculators and computers are called *pseudo-random,* the numbers that students generate in this step are not pseudo-random. The term is generally reserved for numbers generated by computer methods, which, though deterministic, produce results that appear random. The methods embedded in calculators and computers usually generate pseudo-random numbers using previous numbers as seed values.

Step 4 The longest strings in Sequence A will, in general, be shorter than the longest strings in Sequence B.

Step 5 The number of H's in sequence A will probably be between 4 and 6 for most students. The results for Sequence B will be less consistent.

As students present, **ASK** "Can you imagine someone thinking that your sequence was more random than the sequence generated randomly?"

INVESTIGATION

YOU WILL NEED
- a coin

Step 1. Sample Sequence A: H, T, T, H, T, T, H, H, T, H

Step 2. Sample Sequence B: T, T, T, T, T, H, T, T, T, H

Step 3. For the sample data sequence B has more uneven results, with more T's than H's. Sequence B has 5 T's in a row, whereas the most Sequence A has in a row is 2.

Flip a Coin

In this investigation you will explore the predictability of random outcomes. You will use a familiar random process, the flip of a coin.

Step 1 Imagine you are flipping a **fair** coin, one that is equally likely to land heads or tails. Without flipping a coin, record a random arrangement of H's and T's, as though you were flipping a coin ten times. Label this Sequence A.

Step 2 Now flip a coin ten times and record the results on a second line. Label this Sequence B.

Step 3 How is Sequence A different from the result of your coin flips? Make at least two observations.

Step 4 Find the longest string of consecutive H's or T's in Sequence A. Do the same for Sequence B. Then find the second-longest string. Record these lengths for each person in the class as tally marks in a table like the one below. How do the lengths of the longest strings in Sequence A compare with the lengths of the longest strings in Sequence B?
Answers will vary.

Longest string	Sequence A	Sequence B	2nd-longest string	Sequence A	Sequence B
1			1		
2			2		
3			3		
4			4		
5 or more			5 or more		

Step 5 Count the number of H's in each set. Record the results of the entire class in a table like the one at right. What do you notice about the numbers of H's in Sequence A compared with Sequence B?
Answers will vary.

Step 6 If you were asked to write a new random sequence of H's and T's, how would it be different from what you recorded in Sequence A?

Sample answer: It would include longer runs of H's or T's.

Number of H's	Sequence A	Sequence B
0		
1		
2		
3		
4		
5		
6		
7		
8		
9		
10		

Modifying the Investigation

Whole Class Have each student write down a different sequence for Step 1. Give each student a coin and then have students complete Step 2 individually. Discuss Step 3. Compile class data in Steps 4 and 5. Discuss Step 6 with the whole group.

Shortened Complete Steps 4, 5, and 6 as a whole group.

One Step Pose this problem: "What is the probability that when two dice are rolled, their sum will be 6?" Let students discuss the question in groups. Ask groups to experiment with pairs of dice or calculator lists (refer them to Lesson Example B as needed). During the discussion, ask why the experimental results differ from all the conjectured results.

History CONNECTION

Many games are based on random outcomes, or chance. Paintings and excavated material from Egyptian tombs show that games using *astragali* were established by the time of the First Dynasty, around 3500 B.C.E. An *astragalus* is a small bone in the foot and was used in games resembling modern dice games. Senet, shown here, is one of the oldest board games, dating to around 3100 B.C.E. In the Ptolemaic dynasty (323–30 B.C.E.), games with 6-sided dice became common in Egypt. People in ancient Greece made icosahedral (20-sided) and other polyhedral dice. The Romans were such enthusiastic dice players that laws were passed forbidding gambling except in certain seasons.

You often use a random process to generate **random numbers**. Over the long run, each number is equally likely to occur, and there is no pattern in any sequence of numbers generated.

EXAMPLE A | Use a random-number generator to find the probability of rolling a sum of 6 with a pair of dice.

Solution | As you study this solution, follow along on your own calculator. Your results will be slightly different.

To find the probability of the event "the sum is 6," also written as $P(\text{sum is } 6)$, simulate a large number of rolls of a pair of dice. You can use three spreadsheet columns or calculator lists, one for the first die, one for the second die, and one for the sum. Create a list of 300 random integers from 1 to 6 to simulate 300 tosses of the first die. Store the results in the first column or list. Then create a second list of 300 outcomes and store it in the second column. Add the two columns to get a list of 300 random sums of two dice, and store the results in the third column.

fx =RANDBETWEEN(1, 6)

	A	B	C
1	Die #1	Die #2	Die Sum
2	1	3	4
3	2	6	8
4	6	4	10
5	4	5	9
6	4	6	10
7	5	3	8
8	6	2	8
9	6	6	12
10	2	6	8
11	5	3	8
12	2	6	8
13	4	3	7
14	3	1	4
15	2	1	3
16	1	5	6
17	6	6	12
18	3	1	4
19	4	5	9
20	6	6	12
21	2	6	8
22	6	4	10

Create a histogram of the 300 entries in the sum list. The table shows the number of each of the sums from 2 to 12. (Your lists and histogram will show different entries.) The bin height of the "6" bin is 30. So out of 300 simulated rolls,

$P(\text{sum is } 6) = \dfrac{30}{300} = 0.10$.

Repeating the entire process five times gives slightly different results: $\dfrac{249}{1800} \approx 0.138$.

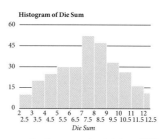

LESSON 8.1 Randomness and Probability **449**

Example A

ELL The singular of *dice* is *die*. Compare this to *mouse* and *mice*, and contrast it with *house* and *houses*.

Consider projecting Example A and working through it with student input. If you are using a calculator, have students follow along on their own calculator. **SMP 5**

Calculator Note: Dice Simulation with the Prob Sim App explains how to use Texas Instruments' application called Probability Simulation. Students can use the App as they follow the example. Some technology tools like a spreadsheet may have sufficient memory and speed to quickly simulate many more than 300 random toss sums. **SMP 5**

ASK "Could you generate these numbers without using a calculator?" [yes, by rolling a pair of dice or using a random number table] Students can find random number tables online. "What advantages are there to using the calculator simulation?" [You can quickly simulate 300 rolls.]

You might refer to Calculator Note: Histograms for creating the histogram.

LANGUAGE As needed, remind students that a *bin* is one bar on the histogram and that the *bin height* tells how many results fall within the bin. Here each bin represents a single digit, but in many histograms, the bin contains numbers over an interval.

ASK "Which probability, 0.16 or 0.138, is the actual probability?" [Neither; the theoretical probability is $\dfrac{5}{36}$. The value 0.138 is closer to this, which is to be expected because it's the result of more trials.]

Conceptual/Procedural

Conceptual Students learn the difference between experimental and theoretical probability. They learn about randomness.

Procedural Students define and calculate geometric probability. They simulate experimental probability with technology.

Example B

Consider projecting Example B and having students work in small groups. Have students present their strategy, not just their answers. SMP 1, 3, 5, 6

If no group uses a grid like the one in the book, suggest it as a way to organize the outcomes. As needed, help students understand how the grid represents all the possible sums. For example: sum of 2: (1, 1); sums of 3: (1, 2), (2, 1); and so on.

ASK "What is your interpretation of the upper-rightmost grid point?" [6 on both dice] "Why do you need to consider both point (1, 5) and point (5, 1)?" [All grid points represent equally likely events; point (1, 5) represents 1 on the green die and 5 on the white die and is different from point (5, 1), which represents 5 on the green die and 1 on the white die.]

Ask students to critique the argument that the probability is $\frac{3}{21}$ because there are 21 outcomes—(1, 1), (1, 2), (1, 3), (1, 4), (1, 5), (1, 6), (2, 2), (2, 3), (2, 4), (2, 5), (2, 6), (3, 3), (3, 4), (3, 5), (3, 6), (4, 4), (4, 5), (4, 6), (5, 5), (5, 6), (6, 6)—three of which have a sum of 6. Students may or may not realize that these outcomes are not equally likely. Indeed, the doubles, such as (3, 3), are half as likely as the other pairings, each of which represents two different pairs. For example, (2, 4) represents both itself and (4, 2). If each double counted as only half a possibility here, making all outcomes equally likely, then the probability would be $\frac{2\frac{1}{2}}{18} = \frac{5}{36}$, as found in Example B. SMP 1, 3, 8

Encourage students to pose "why" and "what if not" problems for the class to discuss.

Example C

Project Example C and have students work in small groups. Have students present their strategy, not just their answers. SMP 1, 3, 5, 6

ASK "Can you use inequalities to describe the region representing a sum less than or equal to 5?" [yes; $n_1 \geq 0$, $n_2 \geq 0$, $n_2 \leq 5 - n_1$]

Rather than actually rolling a pair of dice 1800 times in Example A, you performed a **simulation,** representing the random process electronically. You can use dice, coins, spinners, or random-number generators to simulate trials and explore the probabilities of different **outcomes,** or results.

An **event** consists of one or more outcomes. A **simple event** consists of only one outcome. Events that aren't simple are called **compound.** You might recall that the probability of an event, such as "the sum of two dice is 6," must be a number between 0 and 1. The probability of an event that is certain to happen is 1. The probability of an impossible event is 0. The solution for Example A showed that $P(\text{sum is 6})$ is approximately 0.14, or 14%.

Probabilities that are based on trials and observations like this are called **experimental probabilities.** A pattern often does not become clear until you observe a large number of trials. Find your own results for 300 or 1800 simulations of a sum of two dice. How do they compare with the outcomes in Example A?

Sometimes you can determine the **theoretical probability** of an event without conducting an experiment. To find a theoretical probability, you count the number of ways a desired event can happen and compare this number to the total number of equally likely possible outcomes. Outcomes that are "equally likely" have the same chance of occurring. For example, you are equally likely to flip a head or a tail with a fair coin.

> **Experimental Probability**
> If $P(E)$ represents the probability of an event, then
> $$P(E) = \frac{\text{number of occurrences of an event}}{\text{total number of trials}}$$
>
> **Theoretical Probability**
> If $P(E)$ represents the probability of an event, then
> $$P(E) = \frac{\text{number of different ways the event can occur}}{\text{total number of equally likely outcomes possible}}$$

Refreshing Your Skills: Basic Probability reviews one method for determining the theoretical probability of rolling a specific sum with two dice, organizing the possible outcomes in a table. Another way to find the theoretical probability is to use a graph.

EXAMPLE B | Find the theoretical probability of rolling a sum of 6 with a pair of dice.

Solution | The possible equally likely outcomes, or sums, when you roll two dice are represented by the 36 grid points in this diagram. The point in the upper-left corner represents a roll of 1 on the green die and 6 on the white die, for a total of 7.

The five possible outcomes with a sum of 6 are labeled A–E in the diagram. Point D, for example, represents an outcome of 4 on the green die and 2 on the white die. What outcome does point A represent? 1 on green die and 5 on white die

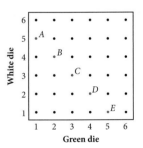

The theoretical probability is the number of ways the event can occur, divided by the number of equally likely events possible. So $P(\text{sum is 6}) = \frac{5}{36} = 0.13\overline{8}$ or about 13.89%.

Before moving on, compare the experimental and theoretical results of this event. Do you think the experimental probability of an event can vary? How about its theoretical probability? yes; no

When you roll dice, the outcomes are whole numbers. What if outcomes could be other kinds of numbers? In those cases, you can often use an area model to find probabilities.

EXAMPLE C | What is the probability that any two real numbers you select at random between 0 and 6 have a sum that is less than or equal to 5?

Solution | Because the two values are no longer limited to integers, counting would be impossible. The possible outcomes are represented by all points within a 6-by-6 square.

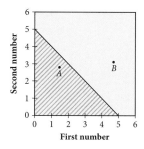

In the diagram, point A represents the outcome $1.47 + 2.8 = 4.27$, and point B represents $4.7 + 3.11 = 7.81$. The points in the triangular shaded region are all those with a sum less than or equal to 5. They satisfy the inequality $n_1 + n_2 \leq 5$, where n_1 is the first number and n_2 is the second number. The area of this triangle is $(0.5)(5)(5) = 12.5$. The area of all possible outcomes is $(6)(6) = 36$.

The probability is therefore $\frac{12.5}{36} \approx 0.347$, or 34.7%.

A probability that is found by calculating a ratio of lengths or areas is called a **geometric probability**.

Experimental probabilities can help you estimate a trend if you have enough cases. But obtaining enough data to observe what happens in the long run is not always feasible. Calculating theoretical probabilities can help you predict these trends. In the rest of this chapter, you'll explore different ways to calculate numbers of outcomes in order to find theoretical probabilities.

In skee ball the probability of getting different point values is based on the geometric area of the regions and their distance from the player.

Summarize

As students present their solutions to the investigation and examples, whether student responses are correct or incorrect, ask other students if they agree and why. Emphasize the use of correct mathematical terminology.
SMP 3, 6

ASK "How do the results of Examples A and B differ." Reiterate the difference between experimental and theoretical probabilities, and emphasize that the former approaches the latter as the number of trials increases. Ponder that this idea sounds familiar, and elicit students' memory of long-run values of sequences.

ASK "What if the two numbers in Example C did have to be integers? Would the result be the same?" [It would be $\frac{21}{49}$, or about 0.429] "Why are the results different?" [There are infinitely many ways to get a sum of 5 in Example C, but only finitely many ways if two non-negative integers are being added.]

EXTEND "What if the sum does not have to be less than or equal to 5 in Example C? What if it must be exactly 5?" [The probability is the area of a line, which is 0.]

CRITICAL QUESTION What's the difference between theoretical or experimental probability?

BIG IDEA Experimental probability is calculated from counting outcomes and trials. Theoretical probability is calculated from equally likely outcomes in an ideal world.

CRITICAL QUESTION Why do we represent probabilities sometimes as fractions and sometimes as decimals?

BIG IDEA In discrete cases, it's easier to come up with fractions, and often fractions are more exact; decimals are usually easier to compare.

Formative Assessment

Student presentations to the Investigation, Examples, and Exercises will provide evidence of meeting the objectives. Class discussion of the differences between will theoretical and experimental probability will help you assess understanding. Discussion of strategies as opposed to just answers will provide anecdotal evidence of understanding randomness, simulation with technology, and ability to calculate probabilities.

Apply

Extra Example

1. This table shows the numbers of students at Kittridge High who download music, buy CDs, or do both.

Grade	Download	Buy	Both	Total
9th	53	85	130	268
10th	70	75	100	245
11th	85	49	120	254
Total	208	209	350	767

   a. What is the probability that a randomly chosen student is a tenth-grader? $\frac{245}{767}$

   b. What is the probability that a randomly chosen student only buys CDs? $\frac{209}{767}$

   c. What is the probability that a randomly chosen student is a ninth-grader who only downloads music? $\frac{53}{767}$

2. If 1.1% of the products produced were defective, what is the probability that a randomly selected item would not be defective? 98.9

Closing Question

When you flip two coins, you might get two heads, two tails, or one of each. Is the probability of two heads therefore $\frac{1}{3}$? Why or why not?

The probability of heads is $\frac{1}{4}$; these three outcomes are not equally likely.

Assigning Exercises

Suggested: 1–7, 9–13
Additional Practice: 8, 14–18

Exercises 1, 2 ELL SUPPORT You might have a class discussion of students' categorization of the probabilities in these exercises.

Exercise 2 ALERT Students who don't easily think about parts of a sentence relative to one another may find it difficult to determine which cell of the table represents the total number of outcomes. You might ASK "What part of the questions in 2a and 2b refers to the total number of outcomes?" [a randomly chosen student from a total of 1424 students] "Is the same true for the questions in 2c and 2d?" [No; the total is no longer 1424.]

8.1 Exercises

You will need a graphing calculator for Exercises **6, 8,** and **19.**

Practice Your Skills

1. Nina has observed that her coach does not coordinate the color of his socks to anything else that he wears. Guessing that the color is a random selection, she records these data during three weeks of observation: @

   black, white, black, white, black, white, black, red, white, red, white, white, white, black, black

   a. What is the probability that he will wear black socks the next day? White socks? Red socks? $\frac{6}{15} = 0.4$; $\frac{7}{15} = 0.4\overline{6}$; $\frac{2}{15} = 0.1\overline{3}$

   b. Are the probabilities you found in 1a experimental or theoretical? experimental

2. This table shows numbers of students in several categories at Ridgeway High. Find the probabilities described below. Express each answer to the nearest thousandth.

	Male	Female	Total
10th grade	263	249	512
11th grade	235	242	477
12th grade	228	207	435
Total	726	698	1424

   a. What is the probability that a randomly chosen student is female? $\frac{698}{1424} \approx 0.490$

   b. What is the probability that a randomly chosen student is an 11th grader? $\frac{477}{1424} \approx 0.335$

   c. What is the probability that a randomly chosen 12th grader is male? @ $\frac{228}{435} \approx 0.524$

   d. What is the probability that a randomly chosen male is a 10th grader? $\frac{263}{726} \approx 0.362$

   e. Are the probabilities in 2a–d experimental or theoretical? theoretical

3. Consider number pairs (x, y) selected from the shaded region of the graph at right. Use the graph and basic area formulas to answer each question. Express each answer to the nearest thousandth.

   a. What is the probability that x is between 0 and 2? @ $\frac{4}{14} \approx 0.286$

   b. What is the probability that y is between 0 and 2? $\frac{10}{14} \approx 0.714$

   c. What is the probability that x is greater than 3? $\frac{7.5}{14} \approx 0.536$

   d. What is the probability that y is greater than 3? $\frac{1.5}{14} \approx 0.107$

   e. What is the probability that $x + y$ is less than 2? (h) $\frac{2}{14} \approx 0.143$

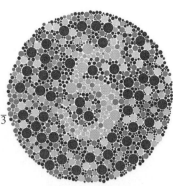

This is a visual test for color blindness, a condition that affects about 8% of men and 0.5% of women. A person with red-green color blindness (the most common type) will see the number 2 more clearly than the number 5 in this image. Dr. Shinobu Ishihara developed this test in 1917.

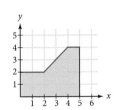

Exercise 3 SUPPORT Provide students with a photocopy of this diagram so they can more easily define the intervals in question.

4. Find each probability.

   a. Each day, your teacher randomly calls on 5 students in your class of 30. What is the probability you will be called on today? $\frac{5}{30} = 0.1\overline{6}$

   b. If 2.5% of the items produced by a particular machine are defective, then what is the probability that a randomly selected item will *not* be defective? $\frac{97.5}{100} = 0.975$

   c. What is the probability that the sum of two tossed dice will not be 6? (h) $\frac{31}{36} = 0.86\overline{1}$

Reason and Apply

5. To prepare necklace-making kits, three camp counselors pull beads out of a box, one at a time. They discuss the probability that the next bead pulled out of the box will be red. Describe each probability as theoretical or experimental.

   a. Claire said that $P(\text{red}) = \frac{1}{2}$, because 15 of the last 30 beads she pulled out were red.   experimental

   b. Sydney said that $P(\text{red}) = \frac{1}{2}$, because the box label says that 1000 of the 2000 beads are red.   theoretical

   c. Mavis says $P(\text{red}) = \frac{1}{3}$, because 200 of the 600 beads the three of them have pulled out so far have been red.   experimental

6. Suppose you are playing a board game for which you need to roll a 6 on a die before you can start playing.

   a. Predict the average number of rolls a player should expect to wait before starting to play.

   b. Describe a simulation, using random numbers, that you could use to model this problem.

   c. Do the simulation ten times and record the number of rolls you need to start playing in each game. (For example, the sequence of rolls 4, 3, 3, 1, 6 means you start playing on the fifth roll.)

   d. Find the average number of rolls needed to start during these ten games.

   e. Combine your results from 6d with those of three classmates, and approximate the average number of rolls a player should expect to wait.

A stamp, printed in 1983 by Greece, shows Achilles throwing dice with Ajax

7. Find the number of equally likely outcomes of each event described for a roll of two dice. Then write the probability of each event. @

   a. The dice sum to 9. $4; \frac{4}{36} = 0.\overline{1}$

   b. The dice sum to 6. $5; \frac{5}{36} = 0.13\overline{8}$

   c. The dice have a difference of 1. $10; \frac{10}{36} = 0.2\overline{7}$

   d. The sum of the dice is 6, and their difference is 2. $2; \frac{2}{36} = 0.0\overline{5}$

   e. The sum of the dice is at most 5. $10; \frac{10}{36} = 0.2\overline{7}$

6a. Answers will vary.

6b. Possible answer: Use the random integer command on the calculator to simulate rolling a die.

6c. Answers will vary.

6d. Answers will vary. Sum your answers from 6c and divide the answer by 10.

6e. Answers will vary. Long-run averages should tend toward 6 turns in order to roll a 6.

8b. The long-run experimental probability should show that $\frac{1}{6}$ of all rolls are a 3.

8c. Answers will vary. The points should level out to a straight line at $y = 0.1\overline{6}$. If you considered 5's instead of 3's, the data should level out to the same value.

8d. Answers will vary but should be close to $\frac{1}{6}$.

8e. $P(3) = \frac{1}{6} = 0.1\overline{6}$. There are six equally likely outcomes, and 3 is one of them, so the theoretical probability is $\frac{1}{6}$.

Exercise 9f The probability of hitting a line or a point is 0 because the area of a point or a line is 0. If students are uncomfortable with this, you can describe the probability as *infinitesimally small*, but make it clear that no real numbers other than 0 are that small.

---

8. Simulate rolling a fair die 100 times with your calculator's random-number generator. Display the results in a histogram to see the number of 1's, 2's, 3's, and so on. Do the simulation 12 times.

   a. Make a table storing the results of each simulation. After each 100 rolls, calculate the experimental probability, so far, of rolling a 3.   Answers will vary.

Simulation number	1's	2's	3's	4's	5's	6's	Ratio of 3's	Cumulative ratio of 3's
1								$\frac{?}{100} = ?$
2								$\frac{?}{200} = ?$
3								$\frac{?}{300} = ?$

   b. What do you think the long-run experimental probability will be?

   c. Make a graph of the cumulative ratio of 3's versus the number of tosses. Plot the points (cumulative number of tosses, cumulative ratio of 3's). Then plot three more points as you extend the domain of the graph to 2400, 3600, and 4800 trials by adding the data from three classmates. Would it make any difference if you were considering 5's instead of 3's? Explain.

   d. What is $P(3)$ for this experiment?

   e. What do you think the theoretical probability, $P(3)$, should be? Explain.

9. Consider the diagram at right.

   a. What is the total area of the square?   144 square units

   b. What is the area of the shaded region?   44 square units

   c. Suppose the horizontal and vertical coordinates are randomly chosen numbers between 0 and 12, inclusive. Over the long run, what ratio of these points will be in the shaded area? ⓐ  $\frac{44}{144}$

   d. What is the probability that any randomly chosen point within the square will be in the shaded area? $\frac{44}{144} = 0.30\overline{5}$

   e. What is the probability that the randomly chosen point will *not* land in the shaded area? $\frac{100}{144} = 0.69\overline{4}$

   f. What is the probability that any point randomly selected within the square will land on a specific point? On a specific line? ⓗ   0; 0

10. A sportscaster makes the following statement about a table-tennis game between two rivals: "The odds are 3 to 2 that May will win the first game." A statement like this means that in general, if the rivals play five games, May will likely win three and lose two.

    a. What is the probability that May will win the first game? $\frac{3}{5}$, or 0.6

    b. The probability that May will win a second game against a different player is $\frac{2}{7}$. What are her odds of winning the second game?   2 to 5

454   CHAPTER 8 Probability

11. Suppose x and y are both randomly chosen numbers between 0 and 8. (The numbers are not necessarily integers.)

   a. Write a symbolic statement describing the event that the sum of the two numbers is at most 6. $x + y \leq 6$

   b. Draw a two-dimensional picture of all possible outcomes, and shade the region described in 11a.

   c. Determine the probability of the event described in 11a. $\dfrac{18}{64} \approx 0.281$

12. Use the histogram below for 12a–d.

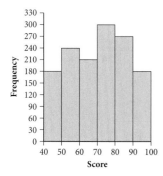

   a. Approximate the frequency of scores between 80 and 90. @ 270

   b. Approximate the sum of all the frequencies. 1380

   c. Find P(a score between 80 and 90). @ $\dfrac{270}{1380} \approx 0.196$

   d. Find P(a score that is not between 80 and 90). $\dfrac{1110}{1380} \approx 0.804$

13. A 6 in. cube painted on the outside is cut into 27 smaller congruent cubes. Find the probability that one of the smaller cubes, picked at random, will have the specified number of painted faces. ⓗ

   a. exactly one $\dfrac{6}{27} = 0.\overline{2}$ b. exactly two $\dfrac{12}{27} = 0.\overline{4}$

   c. exactly three $\dfrac{8}{27} = 0.\overline{296}$ d. no painted face $\dfrac{1}{27} = 0.\overline{037}$

14. **APPLICATION** Where, in atoms, do electrons reside? The graph at right shows the probability distribution of the distance from the nucleus of a hydrogen atom for an electron at any moment. The total area between the curve and the x-axis is 1. The probability that the electron might be between any two values, such as 10 to 20 pm, is the area shown in the graph. This probability is about 0.28. We can estimate this using the width of the slice $(20 - 10 = 10)$ and the average height of the slice $\left(\dfrac{.02 + .036}{2} = 0.028\right)$ and multiply these together.

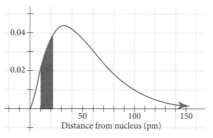

Use this trick to estimate probabilities.

   a. What is the probability that the electron would be between 40 and 50 pm? about 0.4

   b. What is the probability that the electron would be between 90 and 100 pm? about 0.11

   c. Why is it important that the total area is exactly 1?

   The probability that the electron is somewhere has to be 1.

11b.

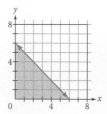

Exercise 13 ALERT Visualization of the situation will be difficult for some students. You may want to display a model cube that is marked off with the dimensions 3 by 3 by 3 or have students build a 3 by 3 by 3 cube out of sugar cubes. You might suggest that students check that their answers to 13a–d sum to 1.

16a.

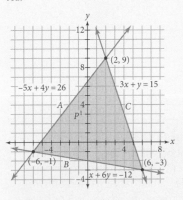

Exercise 16c You might suggest that students inscribe the triangle in a square with horizontal and vertical sides.

18b.
i. $\bar{x} = 35$, $s \approx 22.3$
ii. $\bar{x} = 117$, $s \approx 3.5$

18c. The original values of $\bar{x}$ and s are multiplied by 10.
i. $\bar{x} = 350$, $s \approx 223.5$
ii. $\bar{x} = 1170$, $s \approx 35.4$

18d. The original values of $\bar{x}$ are increased by 10, and the original values of s are unchanged.
i. $\bar{x} = 45$, $s \approx 22.3$
ii. $\bar{x} = 127$, $s \approx 3.5$

Science Connection The atomic models proposed by Niels Bohr and Erwin Schrödinger contributed to the development of quantum theory, which uses probability models to describe the behavior of subatomic particles. The properties of these particles are not accurately described by classical physics, because they follow rules that have a negligible effect in the perceivable world. One such rule is the Heisenberg uncertainty principle, which states that it is impossible to simultaneously determine a subatomic particle's position and its velocity. Probability models describe the possible positions and velocities of the particle. In addition, quantum theory treats particles both as waves and as matter. Among Schrödinger's contributions to quantum theory was his theory of wave mechanics, which uses the wavelike behavior of electrons to determine the particles' probable location in an atom.

Review

15. Expand $(x - y)^4$. @ $x^4 - 4x^3y + 6x^2y^2 - 4xy^3 + y^4$

16. Consider this system of inequalities: @
$$\begin{cases} 3x + y \leq 15 \\ x + 6y \geq -12 \\ -5x + 4y \leq 26 \end{cases}$$

a. Graph the triangle defined by this system.

b. Give the coordinates of the vertices of the triangle in 16a. $(2, 9), (-6, -1), (6, -3)$

c. Find the area of the triangle in 16a. 68 units²

17. Describe the locus of points equidistant from the line $y = 6$ and the point $(3, 0)$. Then write the polynomial equation in general form. @ a parabola with focus $(3, 0)$ and directrix $y = 6$; $y = -\frac{1}{12}x^2 + \frac{1}{2}x -$

18. Consider these two sets of data. @

i. $\{5, 23, 36, 48, 63\}$ ii. $\{112, 115, 118, 119, 121\}$

a. Which set would you expect to have the larger standard deviation? Explain your reasoning. Set i should have a larger standard deviation because the values are more spread out.

b. Calculate the mean and the standard deviation of each set.

c. Predict how the mean and the standard deviation of each set will be affected if you multiply every data value by 10. Then do calculations to verify your answer. How do these measures compare with those you found in 18b?

d. Predict how the mean and the standard deviation of each set will be affected if you add 10 to every data value. Then do calculations to verify your answer. How do these measures compare with those you found in 18b?

PERFORMANCE TASK

Rank i–iii according to the method that will best produce a random integer from 0 to 9. Support your reasoning with complete statements.

i. The number of heads when you drop nine pennies

ii. The length, to the nearest inch, of a standard 9 in. pencil belonging to the next person you meet who has a pencil

iii. The last digit of the page number closest to you after you open a book to a random page and spin the book flat on the table so that a random corner is pointing to you

PERFORMANCE TASK

Answers will vary. Each of these methods has shortcomings.

i. Middle numbers (3–7) are more common than getting only 1 or 2 or 8 or 9 heads in one trial of dropping pennies.

ii. Very few pencils will be at 0 or 1 in.; students throw away their pencils long before that.

iii. This is the best method, although books tend to open to pages that are used more than others.

LESSON 8.2

Multiplication Rules of Probability

> When people have too many choices, they make bad choices.
> — THOM BROWNE

In the previous lesson, you determined the probability of single or **simple** events, like the probability you would find a particular toy in a box of cereal when one of several toys are randomly placed in each box. A **compound** event is where you have multiple outcomes joined by words like "and" and "or". As these problems become more complicated you will explore ways to organize and visualize both the possible and the desired outcomes.

EXAMPLE A A national advertisement says that every puffed-barley cereal box contains a toy and that the toys are distributed equally. Talya wants to collect a complete set of the different toys from cereal boxes.

  a. If there are two different toys, what is the probability that she will find both of them in her first two boxes?

  b. If there are three different toys, what is the probability that she will have them all after buying her first three boxes?

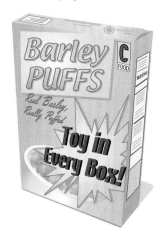

Solution A figure called a tree diagram is a useful tool to find the list of outcomes.

  a. In this tree diagram, the first branching represents the possibilities for the first box and the second branching represents the possibilities for the second box. Thus, the four paths from left to right represent all outcomes for two boxes and two toys. Path 2 and Path 3 contain both toys. If the advertisement is accurate about equal distribution of toys, then the paths are equally likely. So the probability of getting both toys is $\frac{2}{4}$, or 0.5.

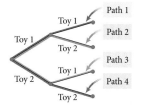

LESSON 8.2

Students learn to use tree diagrams to arrive at the multiplication rule for finding probabilities of independent events.

COMMON CORE STATE STANDARDS		
APPLIED	DEVELOPED	INTRODUCED
	S.CP.2	
	S.CP.3	
	S.CP.5	
	S.CP.6	
	S.CP.8	

Objectives

- Use tree diagrams as an aid to counting possibilities for compound events
- Use the multiplication rule for independent events
- Explore conditional probability

Vocabulary

simple event
compound event
independent
dependent
conditional probability

Launch

What is the probability of rolling a 3 on a fair die? $\frac{1}{6}$

What is the probability of not rolling a 3 on a fair die? $\frac{5}{6}$

A fair die is rolled three times. What is the probability of rolling a 3 on the second roll? $\frac{1}{6}$

Investigate

Example A

Project Example A from your ebook. Work through part a with student input. As needed, suggest tree diagrams as a tool for counting possibilities. Have students work on part b in pairs. Have groups present their solutions and explain their work.
SMP 2, 3, 6

ELL
The difference between *independent* and *dependent* events is crucial for applying probability. Provide several examples of both types of events and have students act them out, if time and space allow.

Support
Tree diagrams provide a concrete approach to conditional probability. Encourage them to search for shortcuts as they work.

Advanced
Before teaching Example B, challenge students to come up with a process or formula for calculating probabilities of dependent events. Challenge students to write a formula for conditional probability.

$$\left[P(n_2|n_1) = \frac{P(n_1 \text{ and } n_2)}{P(n_1)}\right]$$

LANGUAGE In this situation *distributed equally* means that each type of toy is put into the same number of boxes.

ASK "Will Talya collect the toys in a particular order?" [No; there are two possibilities: Toy 1 then Toy 2, or Toy 2 then Toy 1.]

You might point out that each of the four outcomes occurs in two stages. Tree diagrams help analyze multistage outcomes.

Guiding the Investigation

There is a desmos simulation of this investigation in your ebook.

Step 1a Students can check their work against Example A.

As students present their work on the investigation, **ASK** "Are the events independent?" **LANGUAGE** Two events are *independent* when neither event's happening affects the probability of the other event. Theoretically each event has a slight effect on the probability of the next. For example, if the first box contains Toy 1, then the number of Toy 1's in the rest of the boxes decreases by 1. Because the number of boxes is so large and other boxes are being purchased, the effect is minimal, and we can ignore it. **SMP 2, 3, 6**

b. This tree diagram shows all the toy possibilities for three boxes. There are 27 possible paths. You can determine this quickly by counting the number of branches on the far right. Six of the 27 paths contain all three toys, as shown. Because the paths are equally likely, the probability of having all three toys is $\frac{6}{27}$, or $0.\overline{2}$.

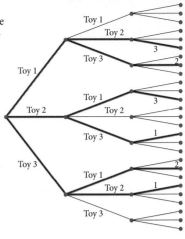

Any piece of the path is an outcome to one stage of the problem and any path from left to right through the entire tree is an outcome of the complete problem. There are probabilities for each stage as well as for each complete path. In the investigation you will explore how those probabilities are related.

INVESTIGATION
The Multiplication Rule

Step 1 In Example A, we assumed that the company placed equal numbers of each toy into the boxes. What if the company produced three times as many Toy 1 as they did Toy 2?

a. On your paper, redraw the tree diagram for Example A, part a. This time, write the probability on each branch.

Step 1a
1st Box: 0.75; 0.25
2nd Box: 0.75; 0.25; 0.75; 0.25

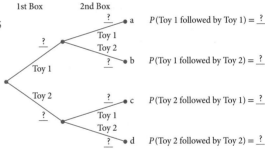

Step 1b
0.5625; 0.1875;
0.1875; 0.0625
or $\frac{9}{16}$; $\frac{3}{16}$; $\frac{3}{16}$; $\frac{1}{16}$

b. In Example A, each path (outcome) had the same probability of $\frac{1}{4}$. The probabilities on the path were also equal, each being $\frac{1}{2}$ and if you multiply those two values together, you found the product to be $\frac{1}{4}$. Multiply the probabilities along each path in your tree diagram and use that to label the probability of the path.

c. Find the sum of your four path probabilities. Why should you always expect this sum? $\frac{16}{16} = 1$, These are all the possible outcomes so it is a sure that one of these must happen. The probability of a sure thing is one.

458 CHAPTER 8 Probability

Modifying the Investigation

Whole Class Work through each step while students write their solutions individually.

Shortened There is no shortened version of this investigation.

One Step Pose this problem: "Each week, one of the four officers of the math club is randomly chosen to write a problem for a club competition. For each of the first three weeks of the competition, Angela's name is drawn. Angela claims that this is not very likely, but Becca argues, 'It is just as likely as any other three names.' Your group needs to defend either Angela's or Becca's point of view. Be sure to use diagrams as you prepare your case."

Step 2
The probability of this sequence of steps is the product of the probabilities of each step
$\left(\frac{3}{4}\right)\left(\frac{3}{4}\right)\left(\frac{1}{4}\right) = \frac{9}{64} = 0.140625$

Step 2 Now suppose you buy three boxes but there are still only two toys with probabilities of $\frac{3}{4}$ and $\frac{1}{4}$. Without drawing a tree diagram, explain how to find the probability of getting Toy 1, then Toy 1, then Toy 2.

Step 3 Imagine there are several outcomes to each stage and three of those outcomes (A, B, and C) always have the same probabilities, $P(A) = e$, $P(B) = f$, and $P(C) = g$. What is the probability of the outcome P(A and then B and then C)?
$P(A \text{ and then } B \text{ and then } C) = P(A) \cdot P(B) \cdot P(C) = efg$.

In Example A and the investigation, you used probabilities for outcomes that never changed. But in many cases the probability of an event at stage 2 or 3 might change, depending on what happens in earlier stages.

EXAMPLE B Devon is going to draw three cards, one after the other, from a standard deck. What is the probability that she will draw exactly two hearts?

Solution The outcome of each draw can be represented by branches on a tree diagram. Rather than list all of the cards in the deck, classify the result of each draw as a heart (H) or nonheart (NH).

Study the tree diagram. There are 52 cards in a standard deck, with 13 cards in each suit. Notice that the denominator of each probability for the second draw is 51 because there are only 51 cards left to choose from.

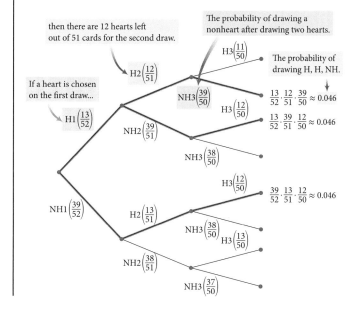

LESSON 8.2 Multiplication Rules of Probability 459

Summarize

As groups present their solutions to the investigation and exercises, have them also explain their work. Whether student responses are correct or incorrect, ask other students if they agree and why. **SMP 1, 2, 3, 6**

CRITICAL QUESTION What would you say is the purpose of tree diagrams?

BIG IDEA Tree diagrams are useful for representing outcomes that occur in stages.

CRITICAL QUESTION What allows you to multiply probabilities along the branches of a path?

BIG IDEA The probabilities are of independent events.

Formative Assessment

Students should be able to interpret travel along each path of the tree diagram as a sequence of decisions and see that the total probability of the decisions at each stage is 1. Watch students' understanding of the multiplication rule; although this rule may seem straightforward, a deep understanding is not simple.

The three highlighted paths show the ways to get exactly two hearts. For example, the top highlighted path shows getting a heart on the first draw, a heart on the second draw, and a nonheart on the third draw.

The probability of any event that can occur along multiple paths is the sum of the probabilities of those paths. Thus, the probability that Devon gets exactly two hearts is the sum of the probabilities of the highlighted paths, which is about $0.046 + 0.046 + 0.046 = 0.138$.

When the probability of event B does not change for an outcome of a stage whether event A has occurred or not, then we say that events A and B are **independent**. In Example B the events were **dependent**. That is to say, the probability of drawing a heart depended on what was drawn before. Imagine if, after drawing a card, the card is mixed back into the deck before the next draw. Then there is only one probability of drawing a heart.

> **The Multiplication Rule for Independent Events**
>
> If n_1, n_2, n_3, and so on, represent independent events, then the probability that this sequence of events will occur can be found by multiplying the probabilities of the events.
> $P(n_1 \text{ and } n_2 \text{ and } n_3 \text{ and} \ldots) = P(n_1) \cdot P(n_2) \cdot P(n_3) \cdots$

You can use **conditional probability** notation to describe both independent and dependent events. The probability of event A given event B is denoted with a vertical line:

$P(A|B)$

For example, to denote the probability of drawing a heart on the second draw given that you have already drawn a nonheart, you write $P(H_2|NH_1)$. And $P(H_2|H_1)$ denotes the probability of drawing a heart on the second draw given that you have already drawn a heart. Using the tree diagram from Example B, you can see that $P(H_2|NH_1) = \frac{13}{51}$ and $P(H_2|H_1) = \frac{12}{51}$.

> **The General Multiplication Rule**
>
> If n_1, n_2, n_3, and so on, represent events, then the probability that this sequence of events will occur can be found by multiplying the conditional probabilities of the events.
> $P(n_1 \text{ and } n_2 \text{ and } n_3 \text{ and} \ldots) = P(n_1) \cdot P(n_2|n_1) \cdot P(n_3|(n_1 \text{ and } n_2)) \cdots$

If events A and B are independent, then the probability of A is the same whether B happens or not. In this case, $P(A|B) = P(A)$. Can you explain the difference in the two multiplication rules?

This tree could represent any two-stage event with two options at each stage. To find the probability of event C, you must add all paths, or outcomes, that contain C. $P(C) = P(A \text{ and } C) + P(B \text{ and } C) = P(A) \cdot P(C|A) + P(B) \cdot P(C|B)$.

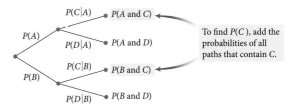

To find $P(C)$, add the probabilities of all paths that contain C.

Exercises

Practice Your Skills

1. Create a tree diagram showing the different outcomes if the cafeteria has three main entrée choices, two vegetable choices, and two dessert choices. @

2. Find the probability of each path, a–d, in the tree diagram at right. What is the sum of the values of a, b, c, and d? @
$P(a) = 0.675$; $P(b) = 0.075$; $P(c) = 0.05$; $P(d) = 0.2$; 1

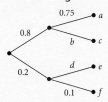

3. Find the probability of each branch or path, a–g, in the tree diagram below. (h)

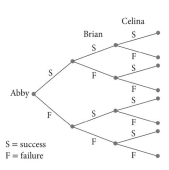

$P(a) = 0.7$
$P(b) = 0.3$
$P(c) = 0.18$
$P(d) = 0.4$;
$P(e) = 0.8$
$P(f) = 0.2$;
$P(g) = 0.08$

4. Three friends are auditioning for different parts in a comedy show. Each student has a 50% chance of success. Use the tree diagram at right to answer 4a–c.

   a. Find the probability that all three students will be successful. @ $\frac{1}{8} = 0.125$

   b. Find the probability that exactly two students will be successful. $\frac{3}{8} = 0.375$

   c. If you know that exactly two students have been successful, but do not know which pair, what is the probability that Celina was successful? (h) $\frac{2}{3} = 0.\overline{6}$

S = success
F = failure

LESSON 8.2 Multiplication Rules of Probability **461**

Apply

Extra Example

1. Fill in the missing probabilities in the tree diagram.

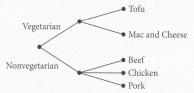

$a = 0.6$, $b = 0.25$, $c = 0.2$, $d = 0.9$,
$e = 0.18$, $f = 0.02$

2. Create a tree diagram showing the outcomes of choosing a vegetarian or a nonvegetarian meal. If a vegetarian meal is chosen, there is a choice of tofu stir-fry or macaroni and cheese. If a nonvegetarian meal is chosen, there is a choice of chicken, beef, or pork.

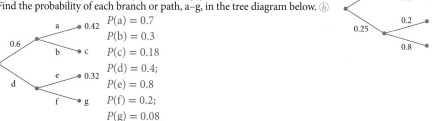

Closing Question

Why are events A and B independent if $P(A|B) = P(A)$?

The probability of A is the same whether or not B occurs.

Assigning Exercises

Suggested: 1–7,
Additional Practice: 8, 18–21

Exercise 1 ASK "How can you find the number of outcomes without using a tree diagram?"
[3(2)(2) = 12]

1.

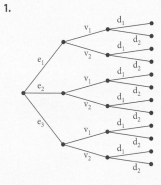

Exercise 4b You may want to discuss the importance of the word *exactly* here, contrasting it with the terms *at least* and *at most*.

LESSON 8.2 Multiplication Rules of Probability **461**

Exercise 5 ASK "Are the two choices independent events?" [No; the probabilities at the second branch point depend on which branch was chosen first.] SUPPORT Students may benefit from writing the numbers of students instead of only probabilities along each branch of the tree diagram.

5a. The probability of selecting a junior given that a sophomore has already been selected; $P(J2|S1)$

8a.

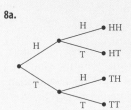

8b.

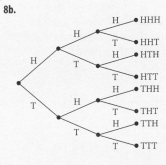

8c.

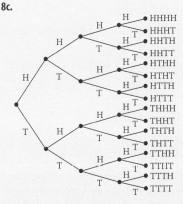

Exercises 8, 12 Help students with tree diagrams only to the extent they need to. Some students will want to use them extensively; others will be able to solve these problems without drawing any diagrams. In either case, make sure students understand what they are doing. Tree diagrams help students think through the situation rather than just working with an expression.

Exercise 10 [Math] Dawn's "you're due" argument is sometimes called the *gambler's fallacy*, and Sue's "hot streak" argument is also fallacious. Both arguments assume that the coin somehow

462 CHAPTER 8 Probability

Reason and Apply

5. This tree diagram models the outcomes of selecting two different students from a class of 7 juniors and 14 sophomores.
   a. Locate the probability $\frac{7}{20}$ on the diagram and explain what it represents.
   b. What is $P(S2|S1)$? $\frac{13}{20} = 0.65$
   c. What is $P(S2|J1)$? $\frac{7}{10} = 0.7$

   S1 $\left(\frac{14}{21} \text{ or } \frac{2}{3}\right)$ — S2 $\left(\frac{13}{20}\right)$
   — J2 $\left(\frac{7}{20}\right)$
   J1 $\left(\frac{7}{21} \text{ or } \frac{1}{3}\right)$ — S2 $\left(\frac{14}{20} \text{ or } \frac{7}{10}\right)$
   — J2 $\left(\frac{6}{20} \text{ or } \frac{3}{10}\right)$

6. Use the diagram from Exercise 5 to answer each question. @
   a. Use the multiplication rule to find the probability of each path. $\frac{182}{420} = 0.4\overline{3}$; $\frac{98}{420} = 0.2\overline{3}$; $\frac{98}{420} = 0.2\overline{3}$; $\frac{42}{420} = 0.1$
   b. Are the paths equally likely? Explain. No, because the probabilities of the four paths are not all the same
   c. What is the sum of the four answers in 6a? $\frac{420}{420} = 1$

7. A recipe calls for four ingredients: flour, baking powder, shortening, and milk (F, B, S, M). But there are no directions stating the order in which they should be combined. Chris has never followed a recipe like this before and has no idea which order is best, so he chooses the order at random.
   a. How many different possible orders are there? @ 24
   b. What is the probability that Chris will choose milk first? .25
   c. What is the probability that Chris will choose flour first and shortening second? $\frac{2}{24} = 0.08\overline{3}$
   d. What is the probability that Chris will choose FBSM? $\frac{1}{24} \approx 0.042$
   e. What is the probability that Chris will not choose FBSM? $\frac{23}{24} \approx 0.958$
   f. What is the probability that Chris will choose flour and milk next to each other? @ $\frac{12}{24} = 0.5$

8. Draw a tree diagram that pictures all possible equally likely outcomes if a coin is flipped as specified.
   a. two times   b. three times   c. four times

9. How many different equally likely outcomes are possible if a coin is flipped as specified?
   a. two times 4   b. three times 8   c. four times @ 16
   d. five times 32   e. ten times @ 1024   f. n times 2^n

10. Consider these questions about flipping a coin multiple times.
    a. Janny flipped a coin and it came up heads. What is the probability it will come up heads if she flips it again? $\frac{1}{2}$
    b. Kevin has flipped a coin three times and it has come up heads each time. What is the probability it will come up heads the next time he flips it? $\frac{1}{2}$
    c. Are multiple flips of a coin independent or dependent events? independent
    d. Jeremy flipped a coin five times and it came up heads each time. Sue says, "You're on a hot streak, it will be heads the next time you flip it." Dawn says, "It's come up heads so many times, you're due for tails next time." Evaluate each of their statements. Both statements reveal misconceptions about the probability of independent events. The probability of heads is always $\frac{1}{2}$; the coin does not remember how it landed previously.

462 CHAPTER 8 Probability

remembers its previous behavior. But flips of a coin are independent. Hot streaks are limited to events that are dependent, such as a basketball player making free throws or investors picking stocks; in those cases, feelings of confidence or clarity inspired by early successes affect the probability of a later success. ELL Define the concept of "hot streak" in part d as a long string of successes.

11. You are totally unprepared for a true-false quiz, so you decide to guess randomly at the answers. There are four questions. Find the probabilities described in 11a–e. ⓗ
   a. P(none correct) $\dfrac{1}{16} = 0.0625$
   b. P(exactly one correct) $\dfrac{4}{16} = 0.25$
   c. P(exactly two correct) $\dfrac{6}{16} = 0.375$
   d. P(exactly three correct) $\dfrac{4}{16} = 0.25$
   e. P(all four correct) $\dfrac{1}{16} = 0.0625$
   f. What should be the sum of the five probabilities in 11a–e? 1
   g. If a passing grade means you get at least three correct answers, what is the probability that you passed the quiz? @ 0.3125

12. **APPLICATION** The ratios of the number of phones manufactured at three sites, M1, M2, and M3, are 20%, 35%, and 45%, respectively. The diagram at right shows some of the ratios of the numbers of defective (D) and good (G) phones manufactured at each site. The top branch indicates a 0.20 probability that a phone made by this manufacturer was manufactured at site M1. The ratio of these phones that are defective is 0.05. Therefore, 0.95 of these phones are good. The probability that a randomly selected phone is both from site M1 and defective is (0.20)(0.05), or 0.01.

   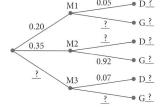

   a. Copy the diagram and fill in the missing probabilities. @
   b. Find P(a phone from site M2 is defective). 0.08
   c. Find P(a randomly chosen phone is defective). 0.0695
   d. Find P(a phone was manufactured at site M2 if you already know it is defective). @ ≈ 0.403

13. The Pistons and the Bulls are tied, and time has run out in the game. However, the Pistons have a player at the free throw line, and he has two shots to make. He generally makes 83% of the free throw shots he attempts. The shots are independent events, so each one has the same probability. Find these probabilities:
   a. P(he misses both shots) 0.0289
   b. P(he makes at least one of the shots) 0.9711
   c. P(he makes both shots) 0.6889
   d. P(the Pistons win the game) 0.9711

14. What is the probability that there are exactly two girls in a family with four children? Assume that girls and boys are equally likely. ⓗ $\dfrac{6}{16} = 0.375$

15. Consider one roll of a six-sided die. Let A represent the event "odd" and let B represent the event "3 or 5." Express each probability in words, and then give its value.
   a. $P(A \text{ and } B)$
   b. $P(B|A)$
   c. $P(A|B)$

16. The table at right gives numbers of students in several categories. @
   a. Find P(10th grade|female) $\dfrac{349}{798} \approx 0.437$
   b. Find P(10th grade). $\dfrac{512}{1424} \approx 0.360$
   c. Are the events "10th grade" and "female" dependent or independent? Explain your reasoning.

	Male	Female	Total
10th grade	163	349	512
11th grade	243	234	477
12th grade	220	215	435
Total	626	798	1424

   The events are dependent, because P(10th grade|female) $\neq P$(10th grade). The probability of choosing a 10th grader from the female students is greater than the probability of choosing a 10th grade student.

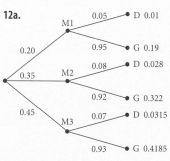

12a.

Exercise 13 Students might be concerned that if the player makes or misses the first shot, the probability of making the second will change. Point out that the exercise states that each shot has the same probability. The probability is based on so many trials that the outcome of one trial won't affect it significantly.

15a. the probability that the roll is odd and it is a 3 or a 5; $\dfrac{1}{3} = 0.\overline{3}$

15b. the probability that the roll is a 3 or a 5 given that the roll is odd; $\dfrac{2}{3} = 0.\overline{6}$

15c. the probability that the roll is odd given that it is a 3 or a 5; 1

Exercise 16 Remind students that two events, A and B, are independent if $P(A|B) = P(A)$; it is dependent in all other cases. **ALERT** Many students find tables like this difficult to deal with because the tables require thinking about ratios.

LESSON 8.2 Multiplication Rules of Probability **463**

Exercise 17 If students assume that the possibility of no raised dots is not a character, the answer is 63.

17. Braille is a form of writing for sight-impaired people. Each Braille character consists of a cell containing six positions that can have a raised dot or not have one. How many different Braille characters are possible?  64

> History
> **CONNECTION**
>
> At age 12, French innovator and teacher Louis Braille (1809–1852) invented a code that enables sight-impaired people to read and write. He got the idea from a former soldier who used a code consisting of up to 12 raised dots that allowed soldiers to share information on a dark battlefield without having to speak. Braille modified the number of dots to 6, and added symbols for math and music. In 1829, he published the first book in Braille. Many aids are available today, including computers with braille displays and keyboards.

Review

18. Write each expression in the form $a + bi$. @
    a. $(2 + 4i) - (5 + 2i)$ $-3 + 2i$
    b. $(2 + 4i)(5 + 2i)$ $2 + 24i$
    c. $(2 + 4i) + (5 + 2i)$ $7 + 6i$

19. Refer to the diagram at right. What is the probability that a randomly selected point within the rectangle is in the orange region? The blue region? $P(\text{orange}) \approx 0.152$; $P(\text{blue}) = 0.45\overline{6}$

20. A sample of 230 students is categorized as shown.

	Male	Female
Junior	60	50
Senior	70	50

    a. What is the probability that a junior is female? $\frac{50}{110} = 0.\overline{45}$

    b. What is the probability that a student is a senior? $\frac{120}{230} \approx 0.522$

21. The side of the largest square in the diagram at right is 4. Each new square has side length equal to half of the previous one. If the pattern continues infinitely, what is the long-run length of the spiral made by the diagonals? $8\sqrt{2}$

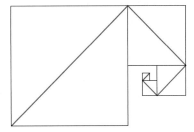

Addition Rules of Probability

LESSON 8.3

Of course there is no formula for success except perhaps an unconditional acceptance of life and what it brings.
—ARTHUR RUBENSTEIN

In the previous lesson you learned how to find the probability of compound events joined by the word "and", such as $P(A$ nd $B)$. In this lesson you will find the probability of compound events joined by the word "or", such as $P(A$ or $B)$. Two outcomes or events that cannot both happen are **mutually exclusive**. You already worked with theoretical probabilities and mutually exclusive events when you added probabilities of different paths.

For example, this tree diagram represents the possibilities of success in auditions by Abby, Bonita, and Chih-Lin. One path from left to right represents the outcome that Abby and Chih-Lin are successful but Bonita is not (outcome AC). Another outcome is success by all three (outcome ABC). These two outcomes cannot both take place, so they're mutually exclusive.

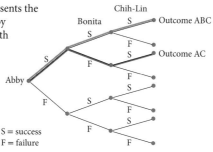

S = success
F = failure

Suppose that each student has a 0.5 probability of success. Then the probability of any single path in the tree is $(0.5)(0.5)(0.5) = 0.125$. So the probability that either AC or ABC occurs is the sum of the probabilities on two particular paths, $0.125 + 0.125$, or 0.25.

Another way to visualize these eight outcomes (paths) is using a Venn diagram. Each circle represents one of the path segments. Being within the circle is success and outside the circle is failure. Each area in the diagram is one full path. Can you identify the area where Abby and Chih-Lin are successful but Bonita is not? Can you find one area of the diagram for EACH of the eight paths in the tree?

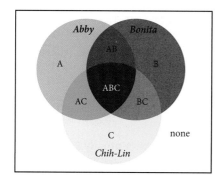

Venn diagrams help break down compound events into mutually exclusive events and help us to see unions and intersections of events more clearly.

COMMON CORE STATE STANDARDS		
APPLIED	**DEVELOPED**	**INTRODUCED**
S.CP.5	S.CP.1	
S.CP.6	S.CP.3	
	S.CP.7	

Objectives

- Explore mutually exclusive events
- Use Venn diagrams as a tool for breaking down compound events into mutually exclusive events
- Understand the addition rule for finding the probabilities of events described by a Venn diagram
- Differentiate between mutually exclusive events and independent events

Vocabulary

mutually exclusive
intersection
union
complements

Launch

Suppose that H is the set of students taking History and E is the set of students taking English. Describe the students in each region of this Venn diagram.

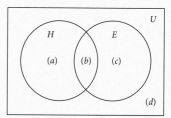

a is students taking history, but not English

b is students taking both history and English

c is students taking English, but not history

d is students taking neither history nor English

ELL
Define and discuss the new vocabulary, *mutually exclusive*, *union*, and *complements*, by comparing them with their colloquial meanings.

Support
Spend extra time discussing the differences between independent and mutually exclusive events using the examples from the book and further concrete examples.

Advanced
Challenge students to create a Venn diagram from a situation where as little information as possible is given. What is the least amount of information required to complete a Venn diagram with two, three, or four events?

Investigate

Example A

Project Example A and have students work in pairs. Have groups present their solutions and explain their thinking. **SMP 2, 3, 6**

ALERT Students may confuse the event "The violin string breaks" with "The only one of the three events is that the violin string breaks." You might want students to write about or discuss the meaning of each of the eight probabilities.

ASK "What do the probabilities add up to?" [1] "What is the probability that at least one of these three events will occur?" [1 − 0.57, or 0.43] "Can you find the probability that either there is a math quiz or the violin string breaks?" [Yes; add all the probabilities of outcomes within those two circles.]

EXAMPLE A Melissa has been keeping a record of probabilities of events involving

  i. Her violin string breaking during orchestra rehearsal (Event B).

 ii. A pop quiz in math (Event Q).

iii. Her team losing in gym class (Event L).

Although the three events are not mutually exclusive, they can be broken into eight mutually exclusive events. These events and their probabilities are shown in the Venn diagram.

a. What is the meaning of the region labeled 0.01?

b. What is the meaning of the region labeled 0.03?

c. What is the probability of either a pop quiz or Melissa's team losing today?

d. Find the probability of a pretty good day, P(not B and not Q and not L). This means no string breaks and no quiz and no loss.

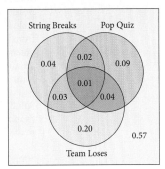

Solution The meaning of each region is determined by the circles that contain it.

a. The region labeled 0.01 represents the probability of a really bad day. In this **intersection**, or overlap, of all three circles, Melissa's string breaks, she gets a pop quiz, and her team loses. The word *and* is often associated with the intersection.

b. The region labeled 0.03 represents the probability that Melissa's string will break and her team will lose, but there will be no pop quiz in math.

c. You can find the probability of either a pop quiz or the team losing by adding the probabilities in regions inside the two circles: 0.02 + 0.09 + 0.01 + 0.04 + 0.03 + 0.20 = 0.39. This is the same as finding the **union** of the two circles. The word *or* is often associated with the union.

d. The probability of a pretty good day, P(not B and not Q and not L), is pictured by the region outside the circles and is 0.57.

In general, the probability that one of a set of mutually exclusive events will occur is the sum of the probabilities of the individual events.

> **The Addition Rule for Mutually Exclusive Events**
>
> If n_1, n_2, n_3, and so on, represent mutually exclusive events, then the probability that any event in this collection of mutually exclusive events will occur is the sum of the probabilities of the individual events.
>
> $$P(n_1 \text{ or } n_2 \text{ or } n_3 \text{ or } \ldots) = P(n_1) + P(n_2) + P(n_3) + \cdots$$

466 CHAPTER 8 Probability

Modifying the Investigation

Whole Class Complete Steps 1 through 8 with student input.

Shortened Skip Step 7

One Step Project the **One Step** found in your ebook. As students work, encourage the use of diagrams (Venn or other) to see how the intersection is being double-counted when adding 0.6 and 0.7 (or 60 and 70). Be open to a variety of approaches. During the discussion, bring up Venn diagrams, mutually exclusive events, the general addition rule, and complements.

But what if you don't know all the probabilities that Melissa knew? In the investigation you'll discover one way to figure out probabilities of mutually exclusive events when you know probabilities of non–mutually exclusive events.

INVESTIGATION

Addition Rule

Of the 100 students in 12th grade, 70 are enrolled in mathematics, 50 are in science, 30 are in both subjects, and 10 are in neither subject.

Step 1 "A student takes mathematics" and "a student takes science" are two events. Are these events mutually exclusive? Explain.
No; a student can take both.

Step 2 Sketch a Venn diagram, similar to the one below. In each of the four regions of the diagram enter the enrollment of students. Draw a square around each value.

Step 2 (□) and Step 3 (◇)

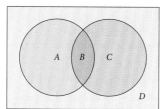
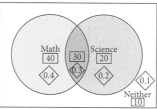

Step 3 Use the numbers of students to calculate probabilities and place a number in each region. Draw a diamond around the probability.

Step 4 Find $P(M \text{ and } S)$, $P(M)$, and $P(M \text{ only})$. Explain why $P(M) \neq P(M \text{ only})$.

Step 4 $P(M \text{ and } S) = (0.4)(0.2) = 0.08$
$P(M) = 0.4$
$P(M \text{ only}) = 0.4 - 0.3 = 0.1$
$P(M) \neq P(M \text{ only})$ because $P(M)$ includes students that take both math and science, while $P(M \text{ only})$ includes students who only take math.

Step 5 Explain why the probability that a randomly chosen student takes mathematics or science, $P(M \text{ or } S)$, does not equal $P(M) + P(S)$.
$P(M) + P(S)$ counts the intersection twice.

Step 6 Create a formula for calculating $P(M \text{ or } S)$ that includes the expressions $P(M)$, $P(S)$, and $P(M \text{ and } S)$. $P(M \text{ or } S) = P(M) + P(S) - P(M \text{ and } S)$

Step 7 Sketch the grid (with labels) pictured here. Enclose all the points where the sum of the two die are 7 and label this region A. Now enclose all the points where both die are > 2 and label this B.

Find the probabilities in parts a–e by counting dots:

a. $P(A)$
b. $P(B)$
c. $P(A \text{ and } B)$
d. $P(A \text{ or } B)$
e. $P(\text{not } A \text{ and not } B)$
f. Find $P(A \text{ or } B)$ by using a rule or formula similar to your response in Step 6.

Step 8 Complete the statement: For any two events A and B, $P(A \text{ or } B) = \underline{?}$.
$P(A \text{ or } B) = P(A) + P(B) - P(A \text{ and } B)$

Guiding the Investigation

Step 2 ASK "Can you know when you have accounted for all 100 students?" [Yes; the sum of the numbers of all four events equals 100.] ALERT Students may put 50 into the region representing science only rather than making it the sum of the numbers in the entire science circle.

Step 7

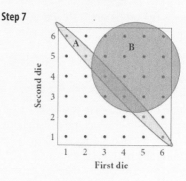

Step 7a $\frac{6}{36} = 0.1\overline{6}$

Step 7b $\frac{16}{36} = 0.\overline{4}$

Step 7c $\frac{2}{36} = 0.0\overline{5}$

Step 7d $\frac{20}{36} = 0.\overline{5}$

Step 7e $\frac{16}{36} = 0.\overline{4}$

Step 7f $\frac{6}{36} + \frac{16}{36} - \frac{2}{36} = \frac{20}{36} = 0.\overline{5}$

Discussing the Investigation

After students present the results of their investigation, ASK "If the number of students who are enrolled in mathematics had not been given, could you have found it?" [Yes. Subtract the other values from 100 to get 40, the number of students taking only mathematics. Then add 40 to the number of students taking both mathematics and science, 30, to get 70 students.]

Suggest that students open questions with "In general . . ." or "Is it possible . . .?" Listen carefully to students' questions; you can learn a great deal from your students.

Conceptual/Procedural

Conceptual Students explore the concept of subsets of a sample space and the concept of conditional probability.

Procedural Students apply the Addition Rule and interpret the answer in terms of the model.

Example B

Project Example B from your ebook and have students work in pairs. Have groups present their solutions and explain their work. SMP 2, 3, 6

To help deepen students' understanding of the ideas, ASK "What are all the possibilities of having a 3 or a 6 on the first roll and the second roll? What is the probability? How does this probability fit into the Venn diagram?" [There are four possibilities: (6, 3), (3, 6), (3, 3), and (6, 6). The probability is $\frac{4}{36}$. Put the number 4 in the intersection of the circles in the Venn diagram.] "What are all the possibilities of having a 3 or a 6 on the first or second roll but not on both rolls? What is the probability? How does this probability fit into the Venn diagram?" [There are two possibilities of having either a 3 or a 6 on the first roll and four possibilities of having neither on the second roll, for a total of eight possibilities. The same is true if you reverse the first and second rolls. Thus the probability of having a 3 or a 6 come up on only one roll is $\frac{2}{6} \cdot \frac{4}{6} + \frac{2}{6} \cdot \frac{4}{6} = \frac{16}{36}$. You can put the number 8 in each circle (outside the intersection) to represent this probability.] "What is the sum of the probability that 3 and 6 will come up and the probability that 3 or 6 will come up?" $\left[\frac{4}{36} + \frac{16}{36} = \frac{20}{36} = \frac{5}{9}\right]$

As needed, point out that this is the probability calculated in the example, expressed as *and/or*.

Example C

Project Example C and have students work in small groups. Have groups present their solutions and explain their work. SMP 2, 3, 6

A general form of the addition rule allows you to find the probability of an "or" statement even when two events are not mutually exclusive.

> **The General Addition Rule**
>
> If n_1 and n_2 represent event 1 and event 2, then the probability that at least one of the events will occur can be found by adding the probabilities of the events and subtracting the probability that both will occur.
>
> $$P(n_1 \text{ or } n_2) = P(n_1) + P(n_2) - P(n_1 \text{ and } n_2)$$

You might wonder whether independent events and mutually exclusive events are the same. Independent events don't affect the probabilities of each other. Mutually exclusive events affect each other dramatically: If one occurs, the probability of the other occurring is 0.

But there is a connection between independent and mutually exclusive events. In calculating probabilities of non–mutually exclusive events, you use the probability that they both will occur. In the case of independent events, you can calculate this probability using the probability of each event.

EXAMPLE B The probability that a rolled die comes up 3 or 6 is $\frac{1}{3}$. What's the probability that a die will come up 3 or 6 on the first or second roll?

Solution Let F represent getting a 3 or 6 on the first roll, regardless of what happens on the second roll. Let S represent getting a 3 or 6 on the second roll, regardless of what happens on the first roll.

Notice that both F and S include the possibility of getting a 3 or 6 on both rolls, as shown by the overlap in the Venn diagram. To compensate for the overlap, the general addition rule subtracts $P(F \text{ and } S)$ once.

$P(F \text{ or } S) = P(F) + P(S) - P(F \text{ and } S)$	The general addition rule.
$P(F \text{ or } S) = P(F) + P(S) - P(F) \cdot P(S)$	F and S are independent, so $P(F \text{ and } S) = P(F) \cdot P(S)$.
$P(F \text{ or } S) = \frac{1}{3} + \frac{1}{3} - \left(\frac{1}{3}\right)\left(\frac{1}{3}\right)$	Substitute the probability.
$P(F \text{ or } S) = \frac{5}{9}$	Multiply and add.

The probability of getting a 3 or 6 on the first or second roll is $\frac{5}{9}$.

When two events are mutually exclusive and make up all possible outcomes, they are referred to as **complements**. The complement of "a 1 or a 3 on the first roll" is "not a 1 and not a 3 on the first roll," an outcome that is represented by the regions outside the F circle. Because $P(F)$ is $\frac{1}{3}$, the probability of the complement, $P(\text{not } F)$, is $1 - \frac{1}{3}$, or $\frac{2}{3}$.

EXAMPLE C | In the backstage area prior to the annual school concert there are band members, choir members and the sound and lighting students who are in neither group. Use C to represent the event that a student is in the choir, and B to represent the event that a student is in the band. A reporter who approaches a student at random backstage knows these probabilities:

  i. $P(B \text{ or } C) = 0.88$
  ii. $P(\text{not } B) = 0.30$
  iii. B and C are independent

a. Turn each of these statements into a statement about percentage in plain English.

b. Create a Venn diagram of probabilities describing this situation.

Solution | a. Convert each probability statement to a percentage statement.

  i. $P(B \text{ or } C) = 0.88$ means that 88% of the students are in band or in choir.
  ii. $P(\text{not } B) = 0.30$ means that 30% of the students are not in band.
  iii. B and C are independent can be stated in several ways. The percent of those backstage in band is the same as the percent of choir members in band. Being in choir does not change this percentage, as choir membership does not increase or decrease the probability of band membership.

b. Start with a general Venn diagram showing two circles overlapping and identify the four areas with the letters W, X, Y, and Z. Use these letters with the given information (and the sum of all areas) to make four statements. Then use those statements to find the values for the four letters.

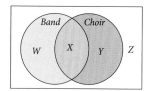

Using i: $W + X + Y = 0.88$ (Equation 1)

Using ii: $Y + Z = 0.3$ (Equation 2)

Using iii: $X = (W + X)(X + Y)$ (Equation 3)

Using total probability: $W + X + Y + Z = 1$ (Equation 4)

Subtract Equation 1 from Equation 4: $Z = 0.12$

Substitute for Z in Equation 2 and solve for Y: $Y = 0.18$

Subtract Equation 2 from Equation 4: $W + X = 0.7$

Substitute for $(W + X)$ and Y in Equation 3 and solve for X
$$X = 0.7(X + 0.18)$$
$$X = 0.7X + 0.126$$
$$0.3X = 0.126$$
$$X = 0.42$$

Finally substitute for X in $W + X = 0.7$ and solve for W: $W = 0.28$

Summarize

As groups present their solutions and explain their thinking in the investigation and examples, emphasize the use of appropriate mathematical vocabulary. Whether student responses are correct or incorrect, ask other students if they agree and why. **SMP 1, 2, 3, 6**

ASK "In general, are mutually exclusive events the same as independent events?" Help students rephrase the explanation in the book in their own words. Emphasize the idea that mutually exclusive events are dependent unless one event is always impossible. The occurrence of one event definitely affects the probability of the other (making it 0). Algebraically, if events A and B are mutually exclusive, then $P(A \text{ and } B) = 0$; this probability can't equal $P(A) \cdot P(B)$ unless either $P(A)$ or $P(B)$ is 0; so except in this special case, the events aren't independent.

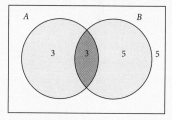

CRITICAL QUESTION How are Venn diagrams useful in studying probability?

BIG IDEA Venn diagrams allow breaking up events into mutually exclusive components, whose probabilities can be added to get the probability of a larger event.

CRITICAL QUESTION Is it possible to tell from a Venn diagram whether events are independent?

BIG IDEA Not directly, but by calculation; if the ratio of the number of outcomes in the intersection to the number of outcomes in one event is the same as the ratio of the number of outcomes in that same event to the total number of outcomes, then the events are independent. In this diagram, for example, $P(B|A) = \frac{3}{6}$ and $P(B) = \frac{8}{16}$; because these probabilities are equal, the events are independent.

Formative Assessment

Check students' comfort level with seeing parts of wholes, such as seeing students who take only mathematics as a part of students who take mathematics, as in the investigation. Assess students' ability to express probabilities as fractions and to generalize from data to formulas. Students should have a basic understanding of mutually exclusive events and how they differ from independent events. Students should also become comfortable with Venn diagrams in the context of probability.

Apply

Extra Example
Find the missing probabilities in this Venn diagram if P(wearing shorts) = 0.6 and P(wearing sandals) = 0.5.
$a = 0, b = 0.2, c = 0.25, d = 0$

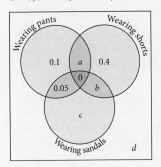

a. What is P(wearing shorts or pants)? 0.75
b. Are any of the three events mutually exclusive? shorts and pants
c. Are any of the three events independent? no

Closing Question
In the General Addition Rule, why is $P(n_1$ and $n_2)$ subtracted off?

Otherwise, the elements in the intersection would be counted twice—once with $P(n_1)$ and once with $P(n_2)$.

Assigning Exercises
Suggested: 1–11
Additional Practice: 12–17

1. 10% of the students are sophomores and not in advanced algebra. 15% are sophomores in advanced algebra. 12% are in advanced algebra but are not sophomores. 63% are neither sophomores nor in advanced algebra.

Exercise 3 ALERT Students may neglect to convert one or more of the four probabilities to a frequency.

3.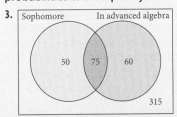

Exercise 5 ALERT Students may think that this is not a Venn diagram because it doesn't look like those they've seen before. Circles in a Venn diagram can be drawn as nonintersecting to represent that it is not possible for both events to occur. They

470 CHAPTER 8 Probability

You should take the time to check that the sum of the probabilities in all the areas is 1. Solving this kind of puzzle exercises your understanding of probabilities and the properties of *and*, *or*, *independent*, and *mutually exclusive*.

8.3 Exercises

Practice Your Skills

Exercises 1–4 refer to the diagram at right. Let S represent the event that a student is a sophomore, and A represent the event that a student takes advanced algebra.

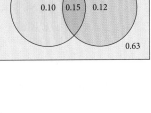

1. Describe the events represented by each of the four regions.

2. Find the probability of each event. @
   a. P(S) 0.25
   b. P(A and not S) 0.12
   c. $P(S \mid A)$ $\dfrac{0.15}{0.15 + 0.12} = 0.\overline{5}$
   d. P(S or A) 0.37

3. Suppose the diagram refers to a high school with 500 students. Change each probability into a frequency (number of students). @

4. Are the two events, S and A, independent? Show mathematically that you are correct. (h) No. $P(S) \cdot P(A) = 0.25 \cdot 0.27 = 0.0675$, $P(S$ and $A) = 0.15$. These must be equal if the events are independent.

5. Events A and B are pictured in the Venn diagram at right. @
   a. Are the two events mutually exclusive? Explain.
      yes, because they do not overlap
   b. Are the two events independent? Assume $P(A) \neq 0$ and $P(B) \neq 0$. Explain.
      No. $P(A$ and $B) = 0$. This would be the same as $P(A) \cdot P(B)$ if they were independent.

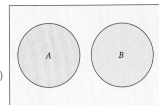

Reason and Apply

6. Of the 420 students at Middletown High School, 126 study French and 275 study music. Twenty percent of the music students take French. @
   a. Create a Venn diagram of this situation.
   b. What percentage of the students take both French and music? ≈ 13%
   c. How many students take neither French nor music? 74

7. Two events, A and B, have probabilities $P(A) = 0.2$, $P(B) = 0.4$, $P(A \mid B) = 0.2$.
   a. Create a Venn diagram of this situation. (h)
   b. Find the value of each probability indicated.
      i. P(A and B) 0.08
      ii. P(not B) 0.60
      iii. P(not (A or B)) 0.48

7a.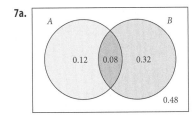

470 CHAPTER 8 Probability

---

can also be drawn as concentric to indicate that if one occurs, the other must occur. Alternatively, they can be drawn overlapping, with impossible regions labeled with 0.

Exercise 7 ASK "What meaning would $P(A \mid B)$ have in this situation?" [It is the portion of B that is also in A.]

6a.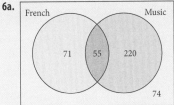

8. If $P(A) = 0.4$ and $P(B) = 0.5$, what range of values is possible for $P(A \text{ and } B)$? What range of values is possible for $P(A \text{ or } B)$? Use a Venn diagram to help explain how each range is possible. @

9. Assume that the diagram for Exercises 1–4 refers to a high school with 800 students. Draw a new diagram showing the probabilities if 20 sophomores moved from geometry to advanced algebra. (h)

10. At right are two color wheels. Figure A represents the mixing of light, and Figure B represents the mixing of pigments. (h)

    Using Figure A, what color is produced when equal amounts of

    a. Red and green light are mixed? @ yellow
    b. Blue and green light are mixed? cyan
    c. Red, green, and blue light are mixed? white

    Figure A   Figure B

    Using Figure B, what color is produced when equal amounts of

    d. Magenta and cyan pigments are mixed? blue
    e. Yellow and cyan pigments are mixed? green
    f. Magenta, cyan, and yellow pigments are mixed? black

Science CONNECTION

The three primary colors of light are red, green, and blue, each having its own range of wavelengths. When these waves reach our eyes, we see the color associated with the wave. These colors are also used to project images on TV screens and computer monitors and in lighting performances on stage. In the mixing of colors involving paint, ink, or dyes, the primary colors are cyan (greenish-blue), magenta (purplish-red), and yellow. Mixing other pigment colors cannot duplicate these three colors. Pigment color mixing is important in the textile industry as well as in art, design, and printing.

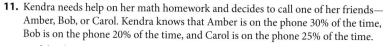

This magnified image shows how the color in a printed photograph appears on paper. The colors are combinations of cyan, magenta, yellow, and black. These four colors can be blended in varying amounts to create any color.

11. Kendra needs help on her math homework and decides to call one of her friends—Amber, Bob, or Carol. Kendra knows that Amber is on the phone 30% of the time, Bob is on the phone 20% of the time, and Carol is on the phone 25% of the time.

    a. If the three friends' phone usage is independent, make a Venn diagram of the situation.
    b. What is the probability that all three of her friends will be on the phone when she calls? (h) 0.015
    c. What is the probability that none of her friends will be on the phone when she calls? 0.42

8. $0 \leq P(A \text{ and } B) \leq 0.4$, $0.5 \leq P(A \text{ or } B) \leq 0.9$. The first diagram shows $P(A \text{ and } B) = 0$ and $P(A \text{ or } B) = 0.9$. The second diagram shows $P(A \text{ and } B) = 0.4$ and $P(A \text{ or } B) = 0.5$.

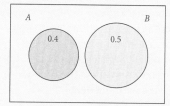

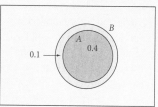

9.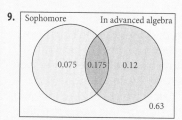

Exercise 10 It might help students to read the Science Connection before completing this exercise. You may need to point out to students that this is not a probability question. **LANGUAGE** *Cyan* is pronounced "'sī-an."

11a.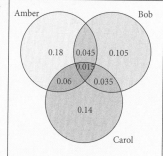

Exercise 12 If students are having difficulty, ask them what information they are missing in order to help them see that they need to find row and column totals.

Exercise 14 ASK "Are the results of the thumbtack tosses independent events?" [Yes; even though the probabilities are not the same, as they would be with a fair coin, the result of one toss doesn't affect the probability associated with the next.]

Exercise 16 ADVANCED This exercise is a good challenge for advanced students. Their intuition may well be that the probability is 0.5 because one of the two coins has a head on the other side. These outcomes are not equally likely, however. If students make a tree diagram, with a choice of coins at the first stage and the result of the flip at the second, they can see that three of the four paths have a head as the outcome, and two of these three cases came from the double-headed coin. ASK "Can you make sense of this nonintuitive result?" One way is to realize that if a head comes up, it's more likely that you've chosen the two-headed coin than the standard coin.

Review

12. The registered voters represented in the table have been interviewed and rated. Assume that this sample is representative of the voting public in a particular town. Find each probability. @

	Liberal	Conservative
Age under 30	210	145
Age 30–45	235	220
Age over 45	280	410

a. P(a randomly chosen voter is over 45 yr old and liberal) $\frac{280}{1500} = 0.18\overline{6}$

b. P(a randomly chosen voter is conservative) $\frac{775}{1500} = 0.51\overline{6}$

c. P(a randomly chosen voter is conservative if under 30 yr old) $\frac{145}{355} \approx 0.408$

d. P(a randomly chosen voter is under 30 yr old if conservative) $\frac{145}{775} \approx 0.187$

13. The most recent test scores in a chemistry class were
{74, 71, 87, 89, 73, 82, 55, 78, 80, 83, 72} ≈ 77
What was the average (mean) score?

14. Claire and Sydney toss a thumbtack and keep track of whether it lands point up (U) or point down (D). If $P(U) = \frac{2}{5}$ and $P(D) = \frac{3}{5}$, then what is the probability that six tosses come up with U, D, D, D, U, D in exactly this order? $\frac{324}{15625} \approx 0.021$

15. Rewrite each expression in the form $a\sqrt{b}$, such that b contains no factors that are perfect squares.
  a. $\sqrt{18}$ $3\sqrt{2}$
  b. $\sqrt{54}$ $3\sqrt{6}$
  c. $\sqrt{60\,x^3y^5}$ @ $2xy^2\sqrt{15xy}$

16. A bag contains two coins. One is a regular coin, with heads and tails. The second coin is a trick coin with two heads. If you choose one of the coins at random and flip it and it lands a head, what is the probability that the other side of the coin will also be a head? $\frac{2}{3} = 0.\overline{6}$

17. Janie keeps track of her free throw percentage in basketball. Over the last year, she has made 50% of her free throw attempts. However, in the last three games, she made 5 out of 6 free throw attempts. Her teammate says, "You're on a hot streak, Janie!" What do you think? Explain your answer completely.

Answers will vary. Making a free throw is not random as flipping a coin is, so Janie may be improving. However, even if her overall accuracy is still 50%, she could have a sequence of 5 successes in a row.

Bivariant Independence

LESSON 8.4

The overwhelming majority of Americans are possessed of two great qualities—a sense of humor and a sense of proportion.

FRANKLIN D. ROOSEVELT

In this course you have looked at data in the form of two numeric variables and worked to discover if (or how) the variables were related. Usually this meant that when the variables were dependant you found a function or equation that explained how they were dependant. But the study of two variable data is not limited to numeric data. Consider the following survey results. Each high school student in the survey was asked to classify themselves as an athlete or not and if they thought they were over, under or about the right weight.

	Body Image		
	Overweight	About Right	Underweight
Athlete	74	295	71
Non-athlete	125	509	126

Here we have two variables, athlete and body image. The first variable has two values, Yes and No, and the second variable has three values. What may interest you is whether these two variables are independent or not. In other words, you want to know if being an athlete or not is related to the proportions of students who feel they are overweight, underweight, or about the right weight. Of course, athletic status and body image are not numeric variables and you can't produce a scatter plot of ordered pairs to look for a pattern, but there are other graphs to be considered.

The values in the table are called joint frequencies because they count cases of two values jointly. The totals of the rows and columns are called **marginal frequencies**. If each row or each column had the same total then we might compare these values but that is not the case. Here the marginal frequencies are different.

	Body Image			
	Overweight	About Right	Underweight	Marginal
Athlete	74	295	71	440
Non-athlete	125	509	126	760
Marginal	199	804	197	1200

Choosing one of the sets of marginal frequencies, you can compute the relative frequencies and the conditional relative frequencies. How were each of the values in this new table found? The numbers inside the table are called the conditional relative frequencies or conditional proportions, while those on the outer edges are the relative frequencies.

LESSON 8.4 Bivariant Independence 473

LESSON 8.4

In this lesson, students construct and interpret two-way frequency tables and use them to decide if events are independent and to approximate conditional probabilities.

COMMON CORE STATE STANDARDS
APPLIED	DEVELOPED	INTRODUCED
S.ID.5	S.IC.4	
	S.CP.4	

Objectives
- Construct and interpret two-way frequency tabless
- Use the two-way table as a sample space to decide if events are independent and to approximate conditional probabilities.

Vocabulary
marginal frequencies

Launch

Use the Launch found in your ebook. Complete the two-way frequency table for this transportation survey of Ms. Neal's math students.

	Male	Female	Total
Walk	22	15	37
Car	46	46	92
Bus	47	38	85
Bike	17	4	21
Total	132	103	235

Investigate

Have students examine the data table at the top of the lesson. **ASK** "Given this data, does the fact that someone is an athlete affect their body image perception? How could you use the data to justify your hypothesis?"
SMP 1, 3 Use student responses to discuss filling in the second table on the page. After the chart is filled in, you can then use the table to introduce marginal frequency, relative frequency, and conditional relative frequency.

ELL
After going through the example, create a new version of the completed table, adding in the appropriate term along with the value in each cell. Have students do the same as they complete the Investigation.

Support
Use an obvious example of non-independence (e.g., age and grade level) to provide students a contrast to the data presented in the Example.

Advanced
Have students examine the data in the Example, the formula for margin of error, and their Investigation data. Ask them to explore methods someone might use to manipulate the data to draw a preferred conclusion.

Example

Consider showing the intervals on a number line. Invite students to discuss how close the intervals need to be for independence.

To make sure students are ready for the Investigation, **ASK** "Assuming the proportions did not change, would asking more students, both athletes and non-athletes, make it more or less likely that you can conclude the variables are independent." [Increasing the sample size will make the intervals smaller, thus the overlap in intervals will also be smaller.]

Guiding the Investigation

There is a desmos simulation for creating a segmented relative bar chart for this investigation in your ebook.

When approving student data topics, avoid private information (grades, weight, athletic ability, etc…). In setting the parameters for the amount of data having students collect, options include having the entire class collect data on the same topics, having different groups collect different sample sizes, and limiting data collection to students in the class. Further discussion of appropriate data collection is in the statistics chapter.

	Body Image			
	Overweight	About Right	Underweight	Marginal
Athlete	0.168	0.670	0.161	0.367
Non-athlete	0.164	0.670	0.166	0.633
Marginal	0.166	0.670	0.164	1.000

One of many ways to plots these relative frequencies is in a segmented bar plot like this.

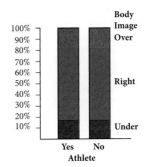

Based on this graph, would you think that self-described body image and athlete status are independent or not? This is math class so in the end it is about the numbers.

EXAMPLE Determine if the two variables in this table are independent.

	Body Image		
	Overweight	About Right	Underweight
Athlete	74	295	71
Non-athlete	125	509	126

Solution The proportion of those who claim to be athletes and feel they are overweight is $\frac{74}{440} \approx 0.168$ and the proportion that do not claim to be athletes and feel they are overweight is $\frac{125}{760} \approx 0.164$. These are not exactly equal, but this is a sample, so we can expect some error. The approximate error in each of these estimates is $2\sqrt{\frac{0.168(1-0.168)}{440}} \approx 0.036$ and $2\sqrt{\frac{0.164(1-0.164)}{760}} \approx 0.027$. This means that the real proportion of those who claim to be athletes and feel they are overweight is likely to be between 0.168 ± 0.036, or between 0.132 and 0.204. The proportion for non-athletes is between 0.137 and 0.191. These two intervals have much overlap, so we conclude that it is possible the true values could be equal. You can repeat this calculation for other pairs of numbers in the table and discover that it is possible that the conditional proportions could be equal. This now means that these two variables might be independent, and your answer is a qualified maybe.

You can say that in the sample the variables were NOT independent (but close). But if we want to talk about the population this sample represents, there is always some uncertainty. It turns out that it is easier to say when two population values can't be equal than to say when they can.

Modifying the Investigation

Whole Class The day before doing this lesson, choose the subjects for Step 1 as a class and have students collect data for homework. Then use the class set of data to do Steps 3 – 6 as a class activity.

Shortened The day before doing this lesson, choose the subjects for Step 1 as a class and have students collect data for homework. Have groups use the class set of data to do Steps 3 – 6.

One Step There is no **One Step** for this investigation.

Margin of Error for Sample Proportion

Given a sample proportion, $p = \frac{x}{n}$, the interval likely to contain the true proportion is $p \pm 2\sqrt{\frac{p(1-p)}{n}}$. Note that the larger your sample, the smaller the interval will be.

INVESTIGATION

Are These Things Related?

You will sample a population, create a graph (or two), and determine if the two variables are independent.   Answers will vary.

Step 1 Choose your subjects. These could be people, book bags, phones, shopping websites, or whatever. Then choose two variables. To keep things simple, make your variables have values of Yes/No or True/False. (Such as "Are you 16 or older?") Once your group has decided on the subjects and the two variables get them approved by your instructor.

Step 2 Collect some data. Your instructor will set parameters on how long you have or how much data you should collect.

Step 3 Create a two-way table of frequencies for your two values from the data you have collected. Then create one or more tables of conditional proportions.

Step 4 Create one or more graphical displays to show the conditional proportions you have calculated.

Step 5 Determine intervals for all the true conditional proportions based on your sample values.

Step 6 Decide if your two variables are independent and write a short argument to convince the reader of what you believe. Your instructor may ask your team to make a short presentation where you show your graph and explain your calculations and your final conclusion.

Summarize

Discussing the Investigation

Have students share their arguments, along with their graphical displays, for the investigation. Whether student responses are correct or incorrect, ask other students if they agree and why.
SMP 2, 3, 6

Depending on time, every student can present their data to the class or you may select certain groups. Try to select one group with data that is close to independent and one group with data that is clearly not independent.

ADVANCED You can ask students where the true proportion most probably lies within the interval as a lead-in to normal distribution.

CRITICAL QUESTION Why does increasing the group size when collecting data decrease the size of the margin of error?

BIG IDEA Increasing the value of the denominator in the formula for margin of error will decrease the value of the margin in both the positive and negative direction.

CRITICAL QUESTION Do the marginal frequencies always have to be identical for the variables to be independent?

BIG IDEA As long as the marginal frequencies are proportional, the conditional proportions will be identical and the variables are independent.

Formative Assessment

Student responses to the Investigation will indicate their ability to compute the margin of error. Be attentive during student presentations for correct use of the vocabulary of the lesson. The Critical Questions and Closing Question will indicate understanding of the concept of using two-way tables to test for independence.

Conceptual/Procedural

Conceptual Students learn to recognize and explain the concepts of conditional probability and independence.

Procedural Students construct two-way frequency tables of data and use it as a sample space to decide if events are independent and to approximate conditional probabilities.

Apply

Extra Example
Determine the conditional proportions for the data in the Launch. Are student gender and method of transportation to school independent? Use a segmented bar plot to justify your response.

The proportions indicate that gender and transportation to school are not independent.

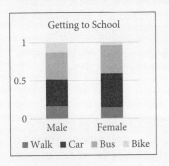

Closing Question
Describe the marginal frequencies of a two-way table when the two variables are independent.

The marginal frequencies are equal or otherwise proportional.

Assigning Exercises
Suggested: 1–10
Additional Practice: 11–14

3a.
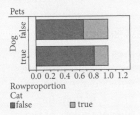

3b.
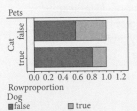

6a. They are not independent, the proportions for Dog-Yes conditional on cats are not equal to 0.22 for yes and 0.39 for no.

Remember that within the data, two variables are independent only if the conditional proportions or probabilities are equal. To claim some relationship for the population about the two variables your need to see if the likely intervals for the proportion overlap.

8.4 Exercises

Practice Your Skills

1. Find the marginal totals for each row and column. @

	Live with a Dog	
Live with a cat	Yes	No
Yes	20	70
No	85	135

   Cat, Yes = 90, Cat, No = 220, Dog, Yes = 105, Dog, No = 205

2. Given the two-way table in Exercise 1, find the requested conditional proportions.

   a. Proportion who live with a dog given they live with a cat. $\frac{20}{90} = 0.22$

   b. Proportion who live with a dog given they do not live with a cat. $\frac{85}{220} \approx 0.39$

   c. Proportion who live with a cat given they live with a dog. $\frac{20}{105} \approx 0.19$

   d. Proportion who live with a cat given they do not live with a dog. @ $\frac{70}{205} \approx 0.34$

3. Use your proportions from Exercise 2 to create a segmented bar plot for each of the following.

   a. Proportions based on the conditions of cat ownership.

   b. Proportions based on the conditions of dog ownership.

4. Find the sample proportion and estimate the true proportion given each of the following.

   a. $x = 20, n = 105$
   0.19 and <0.114, 0.267>

   b. $x = 44, n = 231$ @ 0.19 and <0.139, 0.242>

   c. $x = 188, n = 987$
   0.19 and <0.165, 0.215>

   d. $x = 372, n = 1953$ 0.19 and <0.173, 0.208>

5. Do the interval estimates for the true proportions of these two samples overlap?

   Sample A: $x_1 = 40, n_1 = 150$ and Sample B: $x_2 = 60, n_2 = 750$
   <0.19, 0.34> and <0.036, 0.124> do NOT overlap

Reason and Apply

6. Return to the data about cats and dogs in Exercise 1.

   a. If this were the whole population, then would you say the two variables are independent? Why or why not? @

   b. If this were just a sample of the population, then would you say the two variables are independent? Why or why not?

6b. They are still not independent. The interval for yes cat is <0.15, 0.29> and for no it is <0.32, 0.45>, these two intervals do not overlap. Having one type of pet appears to make it less likely you will have the other.

7. There are two hospitals near where you live and you must choose one for an upcoming procedure. Here is some data you found on the survival of patients at each.

Patient Survival Data

Hospital	Patient's Status	
	Died	Survived
Regional Hospital (RH)	63	2037
Local Hospital (LH)	16	784

a. What is the proportion of those in RH that died? $p = 0.03$ or 3%
b. What is the proportion of those in LH that died? $p = 0.02$ or 2%
c. Would you consider death and hospital to be independent variables? Why? @
d. Which hospital would you pick and why?
   The lowest probability is with the local hospital.

7c. Maybe not independent. These proportions are not equal, but do the intervals <0.022, 0.037> and <0.01, 0.03> overlap?

8. More information comes to light when you also include the type of illness into the information.

Severely Ill Patients Survival Data

Hospital	Patient's Status	
	Died	Survived
RH	57	1443
LH	8	192

Not Severely Ill Patients Survival Data

Hospital	Patient's Status	
	Died	Survived
RH	6	594
LH	8	592

a. In the case of severely ill patients, what is the proportion of those in RH that died? What about those that were not severely ill? @ $p = 0.038$ or 3.8% for severely ill in RH and 0.01 or 1% for those who are not severely ill.
b. In the case of severely ill patients, what is the proportion of those in LH that died? What about those that were not severely ill? $p = 0.04$ or 4% for severely ill in LH and 0.013 or 1.3% for those who are not severely ill. $p = 0.04$ or 4% for LH
c. Would you consider death and hospital to be independent variables for those that were severely ill? What about those that were not severely ill? Maybe not independent. These proportions are not equal.
d. Which hospital would you pick and why?
   The lowest probability for both groups is with the regional hospital.

9. Imagine a survey done on 200 people. Twenty were left-handed and 180 were not. Eighty subjects said they prefer Mac computers and the rest named another make.

a. Create a two-way table where the sample has these two variables as independent.

b. Create a two-way table where these two variables are clearly not independent and explain how you know this.

9a. Left-Mac = 8, Right-Mac = 72, Left-other = 12, Right-Other = 108

9b. Answers will vary. The most extreme answer would be Left-Mac = 20, Right-Mac = 60, Left-other = 0, Right-Other = 120

LESSON 8.4 Bivariant Independence   477

14b.

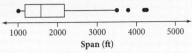

The data are skewed right.

10. Use this two-way table to answer the questions.

		Var Y	
		Yes	No
Var X	Yes	40	60
	No	50	50

a. Find the proportions when variable Y is Yes for each condition of variable X. $p = 0.4$ or 40% for X, Yes and 0.5 or 50% for X, No.

b. Find the interval estimates for variable Y, Yes for each condition of the variable X. <0.30, 0.50> and <0.4, 0.6>

c. You should have found the intervals in part b overlap. Now imagine that the proportions for the sample stay at 0.4 and 0.5 but the size of the sample increase. Would the intervals still overlap if the there were 400 sampled with variable X, having 200 Yes and 200 No? <0.331, 0.469> and <0.429, 0.571>, yes, they still overlap.

d. How many samples would it take until you could say the variables are independent? If each n were 400 (a total or 800 surveyed) then the interval <0.351, 0.449> and <0.45, 0.55> would not o

Review

11. In a group of mice, 30% of the mice are left-handed (pawed?) and 12% are both left-handed and squeak with a French accent.

a. If these two events are independent, then what percent of mice squeak with a French ccent? $P(L \text{ and } F) = P(L) \cdot P(F), P(F) = 40\%$

b. If they are not independent, then what is smallest value that is possible for the percent of mice squeak with a French accent? Explain. If no right-handed mouse speaks with an accent then $P(F) = 12\%$

c. If they are not independent, then what is largest value that is possible for the percent of mice squeak with a French accent? Explain. If every right-handed mouse speaks with an accent then $P(F) = 8$

12. By rounding 4 of the 8 corners on a 6 sided die the probabilities are changed to $P(1) = \frac{1}{12}$, $P(2) = P(3) = P(4) = P(5)$ and $P(6) = \frac{1}{4}$

a. Find $P(2)$. $P(2) = \frac{1}{6}$

b. Find $P(\text{Even})$. $P(\text{Even}) = \frac{7}{12}$

c. Find $P(\text{Prime})$. $P(\text{Prime}) = \frac{1}{2}$

13. Expand each expression.

a. $(x+y)^2$ $x^2 + 2xy + y^2$

b. $(x+y)^3$ $x^3 + 3x^2y + 3xy^2 + y^3$

c. $\left(\frac{1}{2} + \frac{1}{2}\right)^3$ $\frac{1}{8} + \frac{3}{8} + \frac{3}{8} + \frac{1}{8} = 1$

14. The lengths in feet of the main spans of 40 notable suspension bridges in North America are

{4260, 4200, 3800, 3500, 2800, 2800, 2388, 2310, 2300, 2190, 2150, 2000, 1850, 1801, 1750, 1632, 1600, 1600, 1600, 1596, 1550, 1500, 1495, 1470, 1447, 1400, 1380, 1263, 1207, 1200, 1150, 1108, 1105, 1080, 1060, 1059, 1057, 1050, 1030, 1010}

(The World Almanac and Book of Facts 2007)

a. What are the mean, median, and mode of these data? mean: 1818.7 ft; median: 1573 ft; mode: 1600 ft

b. Make a box plot of these data. Describe the shape.

c. What is the standard deviation? approximately 865.3 ft

New York's Manhattan Bridge was constructed from 1901 to 1909 over the East River.

CHAPTER 8 REVIEW

In this chapter you were introduced to the concept of randomness and you learned how to use technology to generate random numbers. Random numbers should all have an equal chance of occurring and should, in the long run, occur equally frequently. You can perform an experiment or use random-number procedures to simulate the experiment and determine the **experimental probability** of an **event**. You can determine **theoretical probability** by comparing the number of successful **outcomes** to the total number of possible outcomes. You represented situations involving probability with Venn diagrams and learned the meaning of events that are **dependent, independent,** and **mutually exclusive.** You used **tree diagrams** to help you count possibilities and learned that sometimes probability situations can be represented geometrically. You also used a **two-way frequency table** as a sample space to decide if events are independent and to approximate **conditional probabilities**.

Exercises

@ Answers are provided for all exercises in this set.

1. Name two different ways to generate random numbers from 0 to 10.

2. Suppose you roll two octahedral (eight-sided) dice, numbered 1–8.
   a. Draw a diagram that shows all possible outcomes of this experiment.
   b. Indicate on your diagram all the possible outcomes for which the sum of the dice is less than 6.
   c. What is the probability that the sum is less than 6? $\frac{10}{64} = 0.15625$
   d. What is the probability that the sum is more than 6? $\frac{49}{64} \approx 0.766$

3. Answer each geometric probability problem.
   a. What is the probability that a randomly plotted point will land in the shaded region at right? 0.5
   b. One thousand points are randomly plotted in the rectangular region shown below. Suppose that 374 of the points land in the shaded portion of the rectangle. What is an approximation of the area of the shaded portion? 17.765 units²

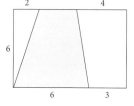

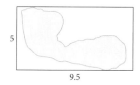

Reviewing

To review the chapter, assign Exercises 2, 5, and 6 in class. Have students share their responses and their thinking. Emphasize the use of appropriate mathematical vocabulary to review the vocabulary of the chapter.

1. Answers will vary depending upon whether you interpret the problem to imply random decimal numbers (between 0 and 10, non-inclusive) or random integers (0 to 10, inclusive). To generate random decimal numbers, you might look at a randomnumber table and place the decimal point after the first digit in each group of numbers. Alternatively, you could use a calculator command, such as 10*rand(numTrials) on the TI-Nspire, or 10*rand on the TI-84 Plus. To generate random integers, you might number 11 chips or slips of paper and randomly select one. Alternatively, you could use a calculator command, such as randInt (0, 10) on the TI-Nspire or the TI-84 Plus.

Exercise 2 An 8-by-8 table is preferable to a tree diagram.

2a, b.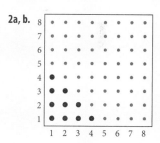

CHAPTER 8 REVIEW 479

LESSON 8 REVIEW 479

4a.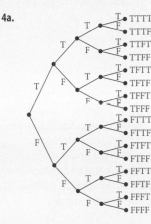

5a.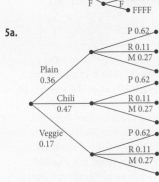

4. A true-false test has four questions.

   a. Draw a tree diagram representing all of the possible results. (Assume all four questions are answered.)

   b. How many possible ways are there of getting one false and three true answers? 4

   c. Suppose you are sure that the answers to the first two questions on the test are true, and you write these answers down. Then you guess the answers to the remaining two questions at random. What is the probability that there will be one false and three true answers on the test? $\frac{2}{4} = 0.5$

5. The local outlet of Frankfurter Franchise sells three types of hot dogs: plain, with chili, and veggie. The owners know that 47% of their sales are chili dogs, 36% are plain, and the rest are veggie. They also offer three types of buns: plain, rye, and multigrain. Sixty-two percent of their sales are plain buns, 27% are multigrain, and the rest are rye. Assume that the choice of bun is independent of the choice of type of hot dog.

   a. Make a tree diagram showing this information.

   b. What is the probability that the next customer will order a chili dog on rye? 0.0517

   c. What is the probability that the next customer will *not* order a veggie dog on a plain bun? 0.8946

   d. What is the probability that the next customer will order either a plain hot dog on a plain bun or a chili dog on a multigrain bun? 0.3501

6. All students in a school were surveyed regarding their preference for whipped cream or ice cream to be served with chocolate cake. The results, tabulated by grade level, are reported in the table.

	9th grade	10th grade	11th grade	12th grade	Total
Ice cream	18	37	85	114	254
Whipped cream	5	18	37	58	118
Total	23	55	122	172	372

   a. Copy and complete the table.

   b. What is the probability that a randomly chosen 10th grader will prefer ice cream? $\frac{37}{55} = 0.6\overline{72}$

   c. What is the probability that a randomly chosen 11th grader will prefer whipped cream? $\frac{37}{122} \approx 0.303$

   d. What is the probability that someone who prefers ice cream is a 9th grader? $\frac{18}{254} \approx 0.071$

   e. What is the probability that a randomly chosen student will prefer whipped cream? $\frac{118}{372} \approx 0.317$

7. Misty polls residents of her neighborhood about the types of pets they have: cat, dog, other, or none. She determines these facts:

   Ownership of cats and dogs is mutually exclusive.

   32% of homes have dogs.

   54% of homes have a dog or a cat.

   16% of homes have only a cat.

   42% of homes have no pets.

   22% of homes have pets that are not cats or dogs.

   Draw a Venn diagram of these data. Label each region with the probability of each outcome.

Take Another Look

1. In a group of 1000 people, only 30 have a particular attribute. If you choose 4 people at random, what is the probability that none will have the attribute? It is typically easier to calculate probabilities when events are independent because you can use the same probability regardless of what happened before. In statistics, probabilities are calculated using independence even when it is not true because of this. The rule is that if the sample is very large then the probability will be close even if they are not independent. The question above would be different if you were allowed to pick the same person more than once so that each time you chose a person from the list of 1000 rather than removing the name from the list after it was selected. How different are these two probabilities?

2. Geometric probability uses the concepts of area or length to find probability when there are an infinite number of events. This can work with any measured value. Explain how you would use geometric probability to solve this problem.

   Each day Mr. A goes to the lunch truck for 5 minutes at some point between 12:00 and 1:00. Also each day Ms. B goes to the lunch truck for 7 minutes between 12:15 and 1:15. What it the probability that they would meet on a random day?

7.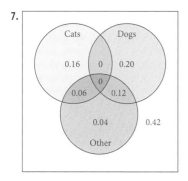

CHAPTER 8 REVIEW 481

Take Another Look

1. Without independence
$$\left(\frac{970}{1000}\right)\left(\frac{969}{999}\right)\left(\frac{968}{998}\right)\left(\frac{967}{997}\right) \approx 0.885128,$$
with independence $\left(\frac{970}{1000}\right)^4 \approx 0.885293$.

The difference between these is only about 0.000165 while the standard error is $\sqrt{\frac{0.97 \cdot 0.03}{4}} = 0.085294$. So using the independent value is only off by about two one-hundredths of a percent of the value or 0.2 percent of the standard error. As long as the sample size is small (less than 10%) of the population, the probability calculation using independence will always be very close to the more difficult calculation without independence.

2. The "A" axis of my picture goes from 0 to 55 to represent the arrival time or Mr A in minutes after 12:00. The "B" axis goes from 15 to 78 to represent the arrival time of Ms. B. Every point in this rectangle now represents two possible arrival times. Some of these, like (1, 16), belong to the points where they will not meet, and some, like (16, 17), where they will meet. Now if we shade all the points where there is some overlap time we have a geometric representation of the event out of all possible events. The area of the shaded region is 468 min² and the total area is 2915 min² so the probability is 468/2915 $\frac{468}{2915} \approx 0.1605$.

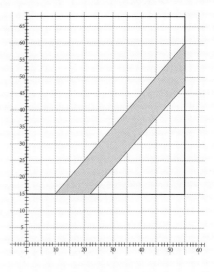

LESSON 8 REVIEW 481

Assessing the Chapter

The Kendall Hunt Cohesive Assessment System (CAS) provides a powerful tool with which to manage assessment content, create and assign homework and tests, and disseminate assessment data.

You can use items from any of the assessment resources for a chapter test. Student portfolios might count as part of the chapter assessment grade.

Facilitating Self-Assessment

If you ask students to create and solve their own probability problem, be aware that many difficult probability problems are easy to state. If the problem involves conditional probability or if there is a question about whether events are dependent or independent, student solutions may not be correct. If students can't solve their problems with the techniques they've learned, encourage them to think critically about what information or methods they would need to solve the problem.

Exercises with prompts that require students to explain the mathematics and thus make good journal entries include Lesson 8.1, Exercise 8; Lesson 8.2, Exercise 16; Lesson 8.3, Exercises 5 and 8; and Lesson 8.4, Exercise 9.

Exercises that would make good portfolio entries include Lesson 8.1, Exercise 6; Lesson 8.2, Exercise 11; Lesson 8.3, Exercise 8; and Lesson 8.4, Exercises 7 and 8.

Assessing What You've Learned

 GIVE A PRESENTATION By yourself or with a group, write and present a probability exercise. Detail the outcome you want to find the probability of, and tell how to solve the problem. If possible, explain how to find both experimental and theoretical probabilities of the outcome occurring. You may even want to do a simulation with the class. Or you could present your work on a Take Another Look from this chapter.

 WRITE IN YOUR JOURNAL You have seen how to represent situations involving probabilities using tree diagrams and Venn diagrams. Describe situations in which each of these approaches would be appropriate, and explain how each method can help you solve problems. Specify the kinds of questions that can be answered with each type of diagram.

 UPDATE YOUR PORTFOLIO Choose a couple of investigations or exercises from this chapter that you are particularly proud of. Write a paragraph about each piece of work. Describe the objective of the problem, how you demonstrated understanding in your solution, and anything you might have done differently.

CHAPTER 9
Application of Statistics

Overview

Discovering Advanced Algebra models real-world problems with mathematical functions. Earlier chapters have developed a variety of functions. This chapter addresses the *normal distribution,* which is a kind of exponential function.

Lesson 9.1 gives the perspective of various kinds of studies in which statistics are applied and discusses the importance of using random samples to reduce bias. In **Lesson 9.2,** students are introduced to normal distributions, learn to find the equation of a normal curve whose mean and standard deviation are known, and learn the 68-95-99.7 rule. As students continue to study the normal curve in **Lesson 9.3,** they learn about *z*-values and find confidence intervals. **Lesson 9.4** introduces correlation coefficients.

The **Performance Task** highlighted in this chapter is in Lesson 9.3, where students look at the change in size of confidence levels.

The Mathematics

One purpose of a mathematical model is to make a prediction. For example, a function modeling the location of a car can predict where the car will be in the future. But functions are about certainty: If you evaluate the function at a point, you get a certain result. How do you make a prediction in an uncertain world?

That question has become increasingly important to mathematics over the past 150 years, and it has given rise to the study of probability. A probability function assigns to an event a number indicating how certain the occurrence of that event is. A similar approach to uncertainty is through inference.

Inference

The word *prediction* indicates telling something about the future. Telling something about large numbers of people or objects is *inference*. Scientists who want to infer properties of a chemical compound study a few examples of the compound extensively and assume that all other examples of that compound have the same properties. But what do you do if you want to infer, for instance, the voting preferences of a large group of people? You can't assume that the people you might be able to study extensively are just like the whole group of voters.

The mathematical tools for studying a large population by way of a smaller sample make up the field of *inferential statistics*. The *descriptive statistics* that students used earlier in this book— means, medians, standard deviations, and so on—can describe the sample. But inferential statistics goes on to describe the population and, very importantly, to quantify how accurate the description of the population is.

Normal distributions

Probability distributions are based on relative frequencies. Probabilities can be calculated by examining areas of regions over intervals and under the graph of the function. Inference often uses the probability distribution called a *normal distribution,* whose graph is mound shaped and whose parent function has equation $y = \frac{1}{\sqrt{2\pi}} \sqrt{e}^{-x^2}$. Helpful in analyzing normally distributed data is the *68-95-99.7 rule of thumb,* which says that 68% of such data are within 1 standard deviation of the mean; 95% fall within 2 standard deviations of the mean; and 99.7% fall within 3 standard deviations of the mean. Here's one example: Suppose a sample of 36 randomly selected female 11th graders were found to have a mean height of 64.2 inches with a standard deviation of 1.8 inches. The margin of error of the mean is found by dividing the standard deviation by the square root of the sample size, so for this example, $\frac{1.8}{\sqrt{36}} = 0.3$ inches. If you were to take many such random samples of 36 11th grade females, the means of these samples would have a normal distribution with a standard deviation of 0.3 inches. Therefore you could be 95% confident that the population mean (the mean height of all 11th grade females) is within 2 margins of error of the sample mean. So the population mean lies between $64.2 - 2 \cdot 0.3 = 63.6$ inches and $64.2 + 2 \cdot 0.3 = 64.8$ inches.

Correlation

Suppose you want to see if taller 11th-grade females tend to weigh more. Comparing mean weight with the mean height doesn't help; instead, you want to see how the spread of heights around the mean height compares with the spread of weights around the mean weight. You can begin the comparison by measuring the number of standard deviations each data value lies away from the mean. For example, suppose that the mean height is 64.2 inches with a standard deviation of 1.8 inches. Then a height of 59.7 inches is 4.5 inches, or -2.5 standard deviations, from the mean. The number -2.5 is the *z-value* of 59.7.

To see how well heights and weights correlate, you can use the *correlation coefficient,* called r, which is the mean of products of corresponding *z*-values. For example, if a particular student's weight has a *z*-value of 1.3 and the student's height has a *z*-value of -0.4, then the product for that student will be $(-0.4)(1.3) = -0.52$. Sum all of these products and divide by the number of students to find the value of r. If heights and weights match up perfectly, r will have value 1. If they're exactly opposite, r will have value -1. The sign of r indicates the direction of lines of fit through the data points.

Using This Chapter

Statistics is the study of data. Data pervades our conversation about most aspects of our life: income, sports, health, politics, weather, prices, etc. In this information technology age, data is coming at a rapid rate. Being able to understand and analyze that data is an important skill in making informed decisions, making this an important chapter. If time is a concern, you can use the modifications listed for **Shortened** versions of the investigations, the desmos technology simulations, or the Investigation Worksheets with Sample Data.

Common Core State Standards (CCSS)

Until Grade 8, almost all of students' statistical topics and investigations have dealt with univariate data, e.g., collections of counts or measurements of one characteristic. By grade 8, students have enough experience with linear functions to study bivariate data (8.F.4-5). They construct scatterplots and look for patterns in the relationship of the two variables, especially a linear association (8.SP.1-2).

In high school, students see how the visual displays and summary statistics from the middle grades relates to different types of data. Students interpret categorical and quantitative data, allowing them to give more precise answers to deeper questions. They learn about the different ways of collecting data, and how to make it unbiased. (S.ID.1- 6) They quickly learn that not all the data in the world has a linear association. Students understand that the process of fitting and interpreting models for discovering possible relationships between variables requires more tools, like finding residuals or the correlation coefficient. They use their analysis of the data to make inferences and justify conclusion (S.IC.1 – 6).

Because of its strong connection with modeling, all of the Statistics standards are starred, indicating that all the standards are modeling standards.

Additional Resources For complete references to additional resources for this chapter and others see your digital resources.

Refreshing Your Skills To review the algebra skills needed for this chapter see your digital resources.

Statistics

Projects See your digital resources for performance-based problems and challenges.

Simpson's Paradox
Correlation vs. Causation
Making It Fit

Explorations See your digital resources for extensions and/or applications of content from this chapter.

Normally Distributed Data
Nonlinear Regression

Materials

- centimeter rulers
- pencils (a few per group)
- sticky notes, *optional*
- different-size coins
- stopwatches or watches with second hand

CHAPTER 9
Applications of Statistics

In this age of information technology, data is everywhere, and statistics is the way we analyze that data to better understand a situation, and to make decisions based on the data. The Centers for Disease Control and Prevention (CDC) collects and analyzes data to protect America from health, safety, and security threats. The Census Bureau collects population data to accurately apportion congressional districts and to provide the federal government with data to determine the allocation of federal funding for education programs in states and communities, among other things. Sports statistics give important information on how teams and individual players are doing.

Objectives

- Learn about, design, and critique studies
- Learn the difference between association and causation
- Investigate the graph and equation of the normal curve
- Learn the 68-95-99.7 rule and its relationship to the area under the normal curve
- Find confidence intervals of the population mean
- Find and understand z-values
- Find and interpret the correlation coefficient

OBJECTIVES

In this chapter you will
- learn different ways to collect data as part of designing a study
- apply different methods for making predictions about a population based on a random sample
- discover the association between the statistics of a sample and the parameters of an entire population
- study population distributions, including normal distributions

Statistics is the study of data. Data pervades our conversation about most aspects of our life: from the latest influenza outbreak to the statistics on your favorite sports team to how many people have moved into or out of your community and what services are needed for that change in population. Ask students to brainstorm all the ways data impacts their life. **ASK** "What data have you used recently to make a decision? How did you analyze the data in order to make that decision?"

LESSON 9.1

This lesson begins a study of statistical inference, in which statistics about samples are used to make predictions about larger populations. Students learn three major types of studies and the strengths and weaknesses of each. The lesson's approach intentionally involves more critical thinking than calculation; methods of analyzing and interpreting data will follow in later lessons.

COMMON CORE STATE STANDARDS		
APPLIED	DEVELOPED	INTRODUCED
	S.IC.1	S.ID.9
	S.IC.3	

Objectives

- Learn the strengths and weaknesses of three major types of studies
- Know the difference between association and causation
- Design a study appropriate for testing a hypothesis
- Critique a study design behind reported statistics

Vocabulary

anecdotal
experiment
observational study
survey
causation
association

Launch

In an advertisement for a brand of toothpaste, your favorite movie star claims that 4 out of 5 dentists recommend that brand. Based on the ad, would you buy that brand? Why?

Answers will vary. Reasons not to buy the brand include not knowing how the dentists were chosen, the total number of dentists surveyed, and whether the star had any expertise in dentistry or in chemistry of toothpaste.

LESSON 9.1

Experimental Design

Rebecca and Tova have math class right after lunch. Rebecca always eats a hot lunch on days when she has an exam, because she has a theory that people score a little higher on math exams when they have eaten a hot meal. Tova is very skeptical of this theory. How can Rebecca and Tova test whether the theory is valid?

Rebecca and Tova write down as many of their own exam scores as they can remember, and they record whether they had a cold lunch or hot lunch on the days of those exams. They ask a few of their friends to give them the same information. Then they find the average of all the scores associated with a cold lunch and all the scores associated with a hot lunch.

This type of study, in which data are collected from a sample convenient to the investigators, is called **anecdotal**. Though it is not an accurate or scientific kind of data collection, it is used frequently to create a working hypothesis. Then proper data collection is designed to produce accurate and meaningful data. Good data collection begins with a proper sample. This means that you must use randomization, but each type of study has its own type of randomization.

Type of study	Randomization and treatment selection	Example
Experimental	Treatments (assignments) are assigned randomly to subjects by the researcher.	A number of students agree to help. The students (subjects) are assigned to Group 1 or 2 by a coin flip. Group 1 is provided a hot lunch and Group 2 is provided a cold lunch.
Observational	Time and/or locations of observations are selected randomly. Subjects choose the treatment without interference of the researcher, who merely records their choice.	A random number is used to locate a seat in the lunchroom. The person closest to that location is asked for their name and if they are about to take a test. An observation of what kind of lunch they ate is recorded.
Survey	Subjects are selected randomly from the population of interest. Subjects identify the treatments they choose to the researcher.	Class lists of all students that took a test after lunch are numbered and then, using random numbers, the researcher selects a group of students from this list (frame) to complete a survey about what type of lunch they had that day.

ELL
Students get a lot of vocabulary quickly here. Define the terminology in relation to the test scores situation. At the end of the section, have students make a mind map of the terminology as it relates to the Practice Your Skills exercises.

Support
Start the lesson with a discussion of where students see bias in society and how it clouds peoples' judgment. Then ask how students would alter those situations to remove that bias. As needed, assign the Refreshing Your Skills: Statistics lesson for review.

Advanced
Have students carry out the investigation on a hypothesis they find in a magazine, newspaper, or textbook.

The diagram shows the stages of an experimental study.

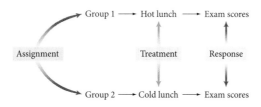

After using any of these methods to collect information about the type of lunch students ate, Rebecca and Tova would need to learn the exam scores for the students who participated in the study. For this study it would be best if they gave the names of the students in each lunch-type group to the teacher, who could provide the mean and standard deviation of the exam scores of each group. Each type of study has advantages and disadvantages. In the investigation you will use two of the methods to conduct a study.

INVESTIGATION
Designing a Study

Consider this hypothesis: Listening to music while doing math homework will shorten the time it takes to complete the assignment.

Your group will design two studies that could test this hypothesis. Choose two of the types of study described on the previous page, and describe how you would collect sample data for the study. Be sure to explain how randomization enters into your design. Provide enough detail so that another group could use your description to collect data in the same way. Keep these points in mind as you design your study:

Answers will vary quite a bit. Possible answers are given for each type of study.

- If you design an **experiment**, you must describe what treatments will be used and how subjects will be assigned to a treatment.
- To design an **observational study**, you must describe what is to be observed and measured and how you'll decide which subjects to observe.
- A **survey** plan must address who will be surveyed, the method of survey (interview, paper questionnaire, e-mail, and so on), and the exact questions to be asked.

When the assignment of treatments in an experiment is done so that the groups are not biased, then the conclusions of the experiment may indicate **causation**. If Rebecca implements a carefully designed experimental study, she might find that students having a hot lunch did much better on exams than those eating a cold lunch. If so, she would have evidence that eating the hot lunch "caused" the students to perform better on the exam. However, if Rebecca used either an observational study or a survey, she could conclude only that there's an **association** between the type of lunch and exam scores.

Investigate

Guiding the Investigation

Students may try to take into consideration practical details, such as how to measure the amount of time spent on homework. Although these details would be important in designing a full study, they are not required for this investigation.

Experiment: Ask for volunteers from several different math classes. In each class, assign half the volunteers to Group 1 and half to Group 2. Ask all the volunteers to write down how long it takes them to complete their math assignment each night for two weeks. For the first week, ask Group 1 to listen to music while they do their math homework. Ask Group 2 not to listen to music. During the second week, have the groups switch roles.

Observational: It will be challenging to do an observational study that does not have a biased sample. Students might observe students doing math homework in the library or in a tutoring center. They can record whether the students are listening to music and note how much time it takes students to complete an assignment. However, it may be that students who work at home are more likely to listen to music while they do their homework. A possibility would be to invite students to participate in a study on multitasking. Give students a form with 10-minute time intervals down the side and different activities, such as Do Homework, Watch TV, Listen to Music, Talk on Phone, Chat Online, and Other across the top. Ask them to keep a nightly record for the next two weeks.

Survey: Distribute surveys to students in several math classes. Ask them to record the length of time it takes them to complete their math assignments for one week. Ask them whether they listen to music while they do their math homework.

As groups present, emphasize how they will choose the subjects. No matter what the method, the subjects should be chosen as randomly as possible. As groups present, emphasize the use of appropriate mathematical terminology. Ask others if they agree and why. **ASK** "How might bias creep into the process?"

Modifying the Investigation

Whole Class Elicit suggestions as the class designs the three different types of study.

Shortened Assign one type of study, rather than two, to each group.

One Step There is no **One Step** alternative for this lesson.

Conceptual/Procedural

Conceptual In this lesson, students explore the strengths and weaknesses of three experimental designs for studies. They also explore the difference between association and causation.

Procedural In this lesson, students design a study appropriate for testing a hypothesis and critique a study design behind reported statistics

Example A

Project Example A from your ebook and have students discuss it in pairs, before having a class discussion. Look for a variety of reasons for the discrepancy. If suggestions don't include that Brandon's results might have been more accurate than those of the later study, **ASK** "How might the later, "more thorough" study have been flawed?" [The study might have been "more thorough" because it used a larger sample, but that sample might have been more biased than Brandon's.]
SMP 1, 3, 6

Example B

Project Example B from your ebook and have students work in small groups. You might have groups discuss only one of the scenarios and present their evaluation to the class.
SMP 1, 3, 6

For part b, **ASK** "Why might the spending pattern not be uniform for a month?" [Students with jobs might get paid every two weeks and might spend more right after a pay day. Others might argue that pay days may differ among these students enough to make the spending pattern uniform.

In part c, **ASK** "Was the increase really 67%?" Some may claim that if the study body numbered 600, the percent increased from $\frac{3}{600} = 0.5\%$ to $\frac{5}{600} \approx 0.8\overline{3}\%$, so the increase was about 0.33%. Bring out the difference between percent increase and percentage point increase.

Summarize

As students present their solutions, including explaining their reasoning, emphasize the use of correct mathematical terminology and symbols. Whether student responses are correct or incorrect, ask other students if they agree and why. **SMP 1, 3, 6**

In all types of study, the legitimacy of the results depends on how the subjects are chosen.

Careful experimental studies are the primary way to learn about causation. Only a huge observational study over a long period of time can indicate causation. On the other hand, observational studies and surveys are often more practical than experimental studies, and they do not require subjects to give up control of their own treatment.

In all types of studies, a major concern is to minimize bias that will invalidate the results. In experimental studies, the way subjects are assigned to a treatment may affect the data. Observational studies may omit consideration of some subjects. Surveys may be biased by the selection of the participants, by the truthfulness of the subjects, and by how the questions are phrased.

Health CONNECTION

Studies have found that insufficient sleep is associated with many chronic diseases, including diabetes, obesity, heart disease, and depression. Many motor vehicle accidents and industrial accidents are also linked to insufficient sleep. More than 25% of adults report not getting enough sleep on occasion, and almost 10% have chronic insomnia. The National Sleep Foundation recommends that teenagers get 8.5 to 9.5 hours of sleep a night.

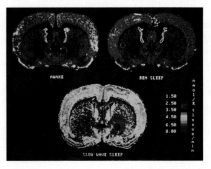

These brain images were used in a study of sleep at La Villette Science Project in Paris, France. During an extended period of sleep, we cycle through five phases: stages 1, 2, 3, and 4, and REM (rapid eye movement) sleep. Most dreams occur during REM sleep, and researchers believe that REM sleep stimulates areas of the brain involved in learning. *Slow wave* sleep describes the deep sleep of stages 3 and 4, during which there is no eye movement. (www.ninds.nih.gov)

EXAMPLE A | Brandon surveyed some students and found that the average amount of sleep for female students was much higher than that for male students. Later, a thorough follow-up study concluded that the average amount of sleep was the same for males and females. Give some possible reasons for the discrepancy in the results. How could Brandon design a better survey?

Solution | The description given does not say that Brandon selected a random sample from his population of interest. So, the original survey may have been biased by how Brandon selected the students to be sampled. For example, the only males he sampled might have been friends who had similar sleeping patterns to his own. Or students who were reluctant to talk about their sleep habits may have declined to participate. Or the participants may have misunderstood the questions.

One way Brandon could minimize bias is to choose subjects randomly. In addition, Brandon should ask some other people to review the questions in his survey to make sure the phrasing of the questions does not influence the answers given or the likelihood that a person would choose not to respond.

No matter how the data are collected, a researcher must be concerned about over- or underrepresenting subgroups of the population. For instance, Brandon might have overrepresented his male friends.

As another example, suppose you want to design a study about the effectiveness of a new fertilizer. If all of your evidence comes from using the fertilizer on tomato plants, then your conclusion can be only about tomato plants, not about all plants. If your data come from several types of plants, but most of your data are from

---

ASK "What are some examples of studies that might be experimental?" Encourage lots of ideas. For example, researchers may be testing medications, new-to-market health drinks, results of sleep deprivation, or the educational value of an online learning service.

A researcher may not always choose groups completely randomly—in a study involving twins, the groups may be chosen specifically to split each pair of randomly chosen twins between two groups as a comparison.

tomato plants, then the data will be biased because one segment of the population is overrepresented. A study based on biased data will not necessarily have incorrect conclusions, but the methods of data collection should be described so that those who read about the study know of the sampling bias.

Newspapers or other sources do not always include enough information about a study to determine whether the statistics are reliable. As an educated consumer, before accepting a conclusion from a study you need to know how the data were collected and whether any bias or other errors could be present.

EXAMPLE B | Evaluate each of these statistics.

a. Arthur says that 71.43% of students hate the new school dress code. He interviewed seven people in the parking lot after school and asked them, "Do you like the way the school is forcing us to dress?"

b. Betty claims that the typical student spends $19.93 on fast food every month. She passed out a survey to the 30 students in her algebra class, and 20 of the surveys were returned. The survey question read, "How much did you spend last week at fast-food restaurants?"

c. Carl states that the number of students with multiple suspensions has increased by 67% in the last 10 years. He found out from school records that last year there were 5 students with multiple suspensions and 11 years ago there were only 3 students.

Solution | Although none of these students have lied about their results, all of them are using biased data or misrepresenting a statistic to make a point.

a. Arthur's conclusion does not match his question. A participant may object to being forced to dress in a certain way, but find nothing objectionable in the present dress code. Or subjects may have one small issue with the current code but overall think that it is fair.

Arthur's sample is biased because it didn't include students who weren't in the parking lot (such as those involved in after-school activities). Therefore, the students that he sampled may not share the same view about uniforms as the other students in the school.

b. Betty multiplied her one-week figures to obtain a monthly value. This assumes that spending is uniform every week. You do not know whether the value is a mean or a median value. You do not know why $\frac{1}{3}$ of the people did not return the survey. If participants had similar reasons for not returning the survey, it may be they had similar spending habits also. Because Betty chose to sample only students from algebra class, her coverage of students is biased.

c. Carl's claim may sound as if he has many years of data instead of only two years. Also, the population of the school needs to be considered. Suppose there are 600 students at the school. Although it is true that there was an increase of 67% in the actual number of students with multiple suspensions, the *ratio* of students with multiple suspensions in the student body increased by only $0.\overline{3}\%$ from $\frac{3}{600}$ to $\frac{5}{600}$. Also, if the population of the school has increased in the last ten years, the suspension rate may have actually decreased.

LESSON 9.1 Experimental Design   487

Types of study	Advantages	Disadvantages
Experimental	Have greatest control over factors that might interfere with the research. Might indicate a causal relationship without a huge sample.	Subjects give up choice of treatment, so it may be unethical. Usually difficult to implement, especially to keep subjects unaware of what group they're in. Subjects may behave differently knowing that they're being observed.
Observational	Observation is after the fact, so it can't influence behavior. Data are potentially very reliable.	It's difficult to avoid experimenter bias. Cannot infer causation except with very large samples, and some statisticians doubt it even then.
Survey	Usually easier to implement than the other types.	Questions are difficult to design. Consciously or unconsciously, subjects may not tell the truth.

Exercise 1–4 Encourage students to answer the questions in complete sentences that use the vocabulary of the lesson. "Type of data collection" means "type of study." SMP 3, 6

1a. This is a survey (more specifically, an interview survey). Nick chose the subjects. They choose whether to respond.

1b. This is an experiment. Elise chose the subjects and applied the treatment.

1c. This is an observational study. The subjects were not aware of the measurements made.

1d. This is anecdotal evidence. There is no structure or control to the data. It was just what was available on the websites that Jill found in her search.

2a. The treatment is that the person has experienced the movie in question. The measure is how much he or she liked the movie.

2b. The treatment is eating a cookie from one of the recipes. The measure is a count of the students who returned for a second cookie.

2c. Assuming this class has been meeting for a while, the treatment is sitting in the front or back of the room and the measure is the arrival time. On the other hand, if this is the first time the class has met, then the treatment is the arrival time and the measure is where the students chose to sit.

2d. The treatment is owning a particular type of cell phone. The response is the owner's evaluation of the phone.

3a. The subjects who choose this movie are more likely to enjoy this type of movie than if you had selected a random group of people to watch the movie.

3b. There could be any number of reasons for getting a second cookie or not getting one other than the taste of the cookie.

3c. The time a person arrives to class is not always under that person's control. A person who typically arrives early may be held up on the day of the study. A person who typically arrives late may have a special need to arrive early that day.

3d. Typically only people who feel strongly will make a comment, so most of the comments will be from those who really like or dislike the phone. So, Jill only has the extreme values in her measures.

4a. As it is a survey, Nick needed a random process to determine which movie

 Exercises

You will need a graphing calculator for Exercise **10**.

Practice Your Skills

For Exercises 1–4, use each of the four scenarios to answer the questions.

a. Nick stood outside the theater and asked many of the patrons if the movie was good. @

b. Elise gave each member of the class a cookie, some from recipe A and some from recipe B. Then she observed which students returned for a second cookie.

c. Steven recorded whether the first half of the students who arrived for class chose to sit in the front half of the room.

d. Jill searched the Internet to find comments from people who owned the type of cell phone she was thinking of buying.

1. Identify which type of data collection was used. Explain your reasoning.
2. Identify the treatments and what is being measured in each scenario.
3. Identify at least one source of bias in each study design in each scenario.
4. None of the descriptions gave details of randomization. Explain how this part of data collection should have been done.

1–4. Answers will vary. Sample answers are given.

Reason and Apply

5. Darita's government class plans to conduct a poll to determine which political candidate has the most support in an upcoming election. They decide to place phone calls to the fourth name in the second column of every page of the local phone book. What problems, if any, do you see in their plan?

6. For each scenario, decide which of the three types of data collection you would use, give a reason for the choice, and give your design for the study. (Refer to the investigation for a description of each type of study.) Answers will vary. These may be considered optimal answers.

    a. You wish to find which brand of microwave popcorn has the fewest unpopped kernels. @
    b. You wish to find if there is a connection between sunny days and people's moods.
    c. You wish to find out the proportion of students at your school who are planning to attend the school play two weekends from now.

7. Eighty-one students have signed up to participate in your experiment. Using the signup sheet we can refer to them as students 1 through 81.

    a. You want to separate them into testing groups A, B, and C. Explain why there might be a bias if you just put the first 27 in group A, then 28 through 54 in group B, and the last 27 in group C.
    b. Put all 81 into your three groups. Explain what process you used to do this.
    c. Your friend wants to interview 5 of the students to find out why they signed up. Select 5 numbers for them and explain your process.
    d. Because of an error 25% of the students at the school never heard about the experiment or the signup sheets. Will this fact change or invalidate either the experiment or the survey?

488 CHAPTER 9 Applications of Statistics

patrons he should ask. In an exit poll (like this) you could roll a die to see how many people to skip before you ask the next person.

4b. As this is an experiment, Elise should randomly assign each person to cookie A or B. Any systematic or patterned dispersal will bias your data.

4c. As an observational study, Steven must do this for randomly selected classes on randomly selected days.

4d. As this is anecdotal evidence, there is no randomization here.

5. Answers will vary. The answer should mention that using the phone book will bias the sample because not all voters have a phone number listed in the book. This is not a random sample. It is unlikely that people would have a similar political bias because of the location of their names in the phone book but it still goes against practice to take a non-random sample.

6a. This is a cross between an observational study and an experiment. You are not assigning kernels of corn to particular brands but you are controlling or randomizing many of the factors, such as microwaves

Review

8. Find the area of each shape.

a. 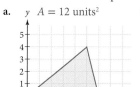 $A = 12$ units^2

b. 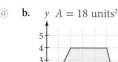 $A = 18$ units^2

c. 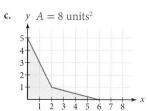 $A = 8$ units^2

9. In your class, if three students are picked at random, what is the probability that

a. All were born on a Wednesday? $\frac{1}{343} \approx 0.003$

b. At least one was born on a Wednesday? $\frac{127}{343} \approx 0.370$

c. Each was born on a different day of the week? $\frac{30}{49} \approx 0.612$

10. At a maple tree nursery, a grower selects a random sample of 5-year-old trees. He measures their heights to the nearest inch.

Height (in.)	60	61	62	63	64	65	66	67	68	69	70	71	72	73	74	75
Frequency	1	1	4	6	8	10	9	8	13	10	5	6	6	3	6	3

a. Find the mean and the standard deviation of the heights. $\bar{x} \approx 67.8; s \approx 3.6$

b. Make a histogram of these data.

c. Find the percentile rank of 65 inches. 20th percentile

History CONNECTION

Polls before the 1948 presidential election predicted that Republican Thomas Dewey would beat Democrat Harry Truman by 5 to 15 percentage points. The polls missed a late swing in favor of Truman, who won the election with a margin of more than 4 percentage points. Some analysts believe that reports of Dewey's lead may have encouraged more Democrats to come out and vote for Truman. In response to this embarrassing mistake, pollsters changed some of their methods. They extended polling until closer to the election day, and used probability sampling to poll voters chosen by chance throughout the country.

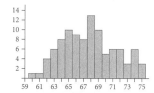

President Truman holds up the newspaper erroneously predicting his defeat in the 1948 presidential election.

---

used, time and temperature settings. It is designed like an experiment but you will need to test a good number of bags to be confident that one brand is better and it is not just the freshness of the box you got off the shelf. Use an experiment to control as many of the factors as you can. Purchase the same size bag of popcorn for each brand, and assume that there are approximately the same number of kernels in each bag. Using one microwave, pop one bag of each brand and count the unpopped kernels in each. Record the deviation from the mean. In other words, if you have five brands and you found 20, 35, 15, 21, and 24 unpopped kernels, then you calculate the mean (23) and record −3, 12, −8, −2, 1. Now do the same with nine different microwaves. The brand with the lowest sum of all these deviations is the winner.

6b. Use an observational study so that your subjects are not influenced by the measurement. Make a random selection of the people you come in contact with every day at school. The sample should include 20 to 30 people. Record the names in a logbook. Each day record the level of sunshine, and shortly after your contact with each person, rate their mood on a scale of 0 to 5. Continue this process until you have data for at least 20 sunny days and 20 overcast days.

6c. Use a survey to get an answer quickly. Number the chairs in the lunch room and use a random-number generator to pick 20 chairs. At each lunch period, go to the designated chair and ask the person there if he or she is hoping to attend the play. If the chair is empty, then ask the person who is seated closest to that chair.

7a. People who signed up together may be friends and have other similarities. Putting groups of similar people into the same group may bias your result.

7b. Answers will vary. You could put all 81 numbers in a hat and just draw out 27 to be in a group. You could use a random number generator to pick a number between 1 and 81 and if that person is not assigned then add them to the next group. You could roll a die for each person and put them in group A if it is a 1 or 2, in B if it is a 3 or 4, and in group C if it is a 5 or 6. (Note that the last method may not give 27 in each group but it is still a valid method)

7c. Answers will vary. You could put all 81 numbers in a hat and just draw out 5 to be in a group. You could use a random number generator to pick 5 unique numbers between 1 and 81.

7d. It changes the interpretation of the results because the sample was not drawn from the entire student population and there may be a bias in who had never heard of the experiment. It does not affect the validity of the experiment.

LESSON 9.2

Normal Distributions

Many continuous variables follow a distribution that is symmetric and mounded in a "bell" shaped distribution. Graphs of distributions for large populations, such as heights, clothing sizes, and test scores, often have this shape. The bell-shaped distribution is so common that it is called the **normal distribution,** and its graph is called the **normal curve.**

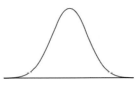

This 10-foot-tall machine drops balls through a grid of pins. The balls land in a bell-shaped curve—a visual representation of their probability distribution.

The formulas you have learned for the sample mean, $\bar{x}$, and sample standard deviation, s, are estimates for values in the population. When you find the mean and standard deviation for an entire population, they are called the **population mean,** μ, and **population standard deviation,** σ. These are the Greek letters mu (pronounced "mew") and sigma.

In this lesson you'll see some properties of normal distributions. The general equation for a normal distribution curve is in the form $y = ab^{-x^2}$. If you graph a function like $y = 3^{-x^2}$, you'll get a bell-shaped curve that is symmetric about the vertical axis.

To describe a particular distribution of data, you translate the curve horizontally to be centered at the mean of the data, and you dilate it horizontally to match the standard deviation of the data. Then you dilate it vertically so that the area is 1. These steps are shown graphically below. You'll want to begin with a parent function that has standard deviation 1.

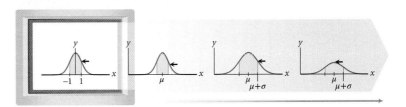

Lesson 9.2 (Teacher notes)

In this lesson students see what it means for a continuous distribution representing a population to be a normal distribution.

COMMON CORE STATE STANDARDS
APPLIED	DEVELOPED	INTRODUCED
	S.ID.4	

Objectives
- Learn the 68-95-99.7 rule and its relationship to the area under the normal curve
- Investigate the graph of the normal curve
- Use the equation of a normal distribution to determine characteristics of normally distributed populations

Vocabulary
normal distribution
normal curve
population mean, μ
population standard deviation, σ
standard normal distribution
inflection points

Material
- Calculator Notes: Entering e; Normal Graphs; Probabilities of Normal Distributions; Creating Random Probability Distributions

Launch
What does it mean when you want your teacher to "curve" the grades on a test? Explain.

Answers will vary. A "curve" presumably puts the grades on a more normal curve, to offset a lack of representative grades on the test.

ELL
Before starting on the exercises, revisit the new vocabulary as it relates to the curve students have derived in the lesson. See also the Support suggestions.

Support
At a minimum, students should understand how changing the values of μ and σ in the normal distribution formula translates or dilates the graph.

Advanced
Show students the normal curve and parent function and have them use a calculator to find values of the base and multiplier, knowing that the total area and standard deviation are 1. **ASK** "Is it really possible for the unbounded region under this graph to have a finite area?"

History CONNECTION

In 1733 French mathematician Abraham de Moivre (1667–1754) developed the normal distribution as a means to approximate the binomial distribution. Given a binomial sample of n trials with a probability of success of p for each trial, the probabilities can be estimated using a normal curve with mean $\mu = n \cdot p$ and standard deviation $\sigma = \sqrt{n \cdot p \cdot (1-p)}$.

The parent function of a normal distribution has standard deviation 1, and is called the **standard normal distribution**. To meet the conditions for the standard normal distribution, statisticians have used advanced mathematics to determine the values of a and b in the equation $y = ab^{-x^2}$. The value of a is related to the number π.

$$a = \sqrt{\frac{1}{2\pi}} \approx 0.399$$

The value of b is related to another common mathematical constant, the transcendental number e.

$$b = \sqrt{e} \approx 1.649$$

Calculators allow you to work with these numbers fairly easily.

EXAMPLE A The general equation for a normal curve is in the form $y = ab^{-x^2}$.

a. Write the equation for a standard normal curve, using $a = \sqrt{\frac{1}{2\pi}}$ and $b = \sqrt{e}$.
   Find a good graphing window for this equation and describe the graph.

b. Write the equation for a normal distribution with mean μ and standard deviation σ.

c. Another bell shaped curve is $f(x) = \frac{0.32}{1+x^2}$. How does this curve compare to the standard normal curve?

Solution Substitute the values given for the constants a and b into the general equation for a normal curve.

a. The equation for a standard normal curve is

$$y = \sqrt{\frac{1}{2\pi}} (\sqrt{e})^{-x^2}$$

Note that $\sqrt{\frac{1}{2\pi}}$ is the same as $\frac{1}{\sqrt{2\pi}}$.

A good window for the graph is $-3.5 \leq x \leq 3.5$, $-0.1 \leq y \leq 0.5$. The graph is bell-shaped and symmetric about $x = 0$. So the mean, median, and mode are all 0. Almost all of the data are in the interval $-3 \leq x \leq 3$.

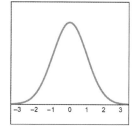

b. Translate the curve horizontally to shift the mean from 0 to μ. And dilate the curve horizontally to change the standard deviation from 1 to σ. The area under a probability distribution must be 1. The horizontal dilation will increase the area, so it must be accompanied by a vertical dilation.

Investigate

You might project the progression of graphs at the bottom of the page and use it to discuss the new vocabulary.

A graphing window of $0 \leq y \leq \frac{0.4}{\sigma}$ will always contain the full range of the normal curve. When $\sigma = 1$, the maximum y-value is less than 0.4. For larger values of σ, the graph is more spread out, and therefore the maximum y-value is even smaller. **SMP 5**

CONTEXT History Connection Although Carl Gauss developed the idea of the normal curve, the curve was first invented by French mathematician Abraham de Moivre (1667–1754) in his 1718 book *Doctrine of Chances*. De Moivre also introduced the use of mortality statistics to calculate the maturation of insurance policies and successfully predicted the exact date of his own death.

Example A

Project Example A and work through it with student input as a class. As needed, make it clear that the standard normal distribution is the parent function. **ASK** "Why is the range of the function so small?" [Actually, the range is all positive y. As σ gets closer to 0, the height of the curve becomes infinitely large. This can be seen arithmetically by inserting smaller and smaller values of σ into the function for the normal curve; as the denominator of the fraction gets smaller, the value of the whole fraction increases, thus increasing the value of the function. This can also be seen geometrically through the areas of triangles. Imagine that the area of a triangle is 1. The smaller its base is, the larger the height has to be for the area to remain 1.]

In part b, **ASK** "Why are we transforming the graph here?" [The parent graph is translated horizontally to move the mean from 0, is dilated horizontally to adjust the standard deviation, and is dilated vertically to keep the area 1.]

After discussing inflection points, you might ask students to sketch by hand normal curves with given means and standard deviations.

Modifying the Investigation

Whole Class Choose a class variable for Step 1. Work Steps 2 – 5 as a class, with student input.

Shortened Use the desmos simulation of the investigation.

One Step Use the **One Step** found in your ebook.

Conceptual/Procedural

Conceptual Students investigate the graph of the normal curve. They explore the relationship of the 68-95-99.7 rule to the area under the normal curve.

Procedural Students learn the 68-95-99.7 rule and use the equation of a normal distribution to determine characteristics of normally distributed populations

$$y = \sqrt{\frac{1}{2\pi}} (\sqrt{e})^{-x^2}$$ Start with the parent function.

$$y = \sqrt{\frac{1}{2\pi}} (\sqrt{e})^{-(x-\mu)^2}$$ Substitute $(x - \mu)$ for x to translate the mean horizontally to μ.

$$y = \sqrt{\frac{1}{2\pi}} (\sqrt{e})^{-[(x-\mu)/\sigma]^2}$$ Divide $(x - \mu)$ by the horizontal scale factor, σ, so the curve reflects the correct standard deviation.

$$y = \frac{1}{\sigma\sqrt{2\pi}} (\sqrt{e})^{-[(x-\mu)/\sigma]^2}$$ Divide the right side of the equation by the vertical scale factor, σ, to keep the area under the curve equal to 1.

c. The two curves have a maximum at $x = 0$ and are asymptotic to the x-axis in both directions. They also appear to have about the same total area between the curve and the x-axis. But the normal curve peaks higher and decays faster with the new function drops to zero at a slower rate. Using these two curves to estimate probability would give very different answers.

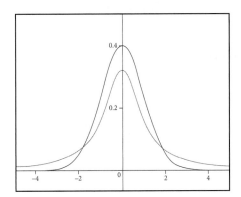

The Normal Distribution

The equation for a **normal distribution** with mean μ and standard deviation σ is

$$y = \frac{1}{\sigma\sqrt{2\pi}} (\sqrt{e})^{-[(x-\mu)/\sigma]^2}$$

You can view the graph of a normal distribution well in the window

$$\mu - 3\sigma \leq x \leq \mu + 3\sigma \text{ and } 0 \leq y \leq \frac{0.4}{\sigma}$$

This window will show three standard deviations to the left and the right of the mean, and the minimum and maximum frequencies of the distribution.

In the equation for a normal distribution, the data values are represented by x and their relative frequencies by y. The area under a section of the curve gives the probability that a data value will fall in that interval.

Most graphing calculators provide the normal distribution equation as a built-in function, and you have to provide only the mean and standard deviation. In this chapter we will use the notation $n(x)$ to indicate a standard normal distribution function with mean 0 and standard deviation 1. Using this notation, a nonstandard normal distribution is written

$n(x, mean, standard\ deviation)$

For example, a normal distribution with mean 3.1 and standard deviation 0.14 is written $n(x, 3.1, 0.14)$. The area under a portion of a normal curve is written

$N(lower, upper, mean, standard\ deviation)$

This notation, with a capital N, indicates the lower and upper endpoints of the interval, and the mean and standard deviation of the distribution. This area determines the probability that a value in a normal distribution will fall within a particular range.

INVESTIGATION

Probabilities and Percentiles

A percentile is a value of a random variable that has a certain percent of values below it. This percent also represents the probability that a randomly chosen value will be below your percentile. If your dog's weight is the 75%ile for adults of that breed then you are saying that 75% of all adult dogs of this breed weigh less than your dog. You are also saying there is a 75% chance that a randomly selected dog of that breed weighs less than your dog. *IF* values follow a normal distribution, it is possible to know the percentile of every value by knowing only the mean and standard deviation of the values.

Step 1 First you must choose a variable that you think has a bell shaped distribution. (Like heights of professional football players, or weights of 4 door sedans). Then find 30 values in the distribution of this variable. (You might measure these items or find values already measured.)

Step 2 Find the mean ($\bar{x}$) and standard deviation (s) of these values. Plot the values with a dot plot or a relative histogram. Add the graph of a normal curve using the same mean and standard deviation to your graph.

Step 3 Assuming a normal distribution, use technology to find the percentile of the value of $\bar{x} + s$ (using your values from the sample). Now find the percent of your data values less than this number.

Step 4 Use technology to find the 16th percentile of a normal distribution based on $\bar{x}$ and s. How is this value similar to the value used in Step 3?

Step 5 What values would bracket the middle 95% of your normal distribution? How do these value relate to your values for $\bar{x}$ and s?

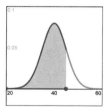 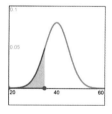 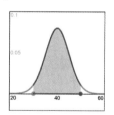

Guiding the Investigation

There is a desmos simulation for this investigation in your ebook.

Answers will vary.

Step 1 In this sample solution we have used the play time of the top thirty hits in 2015 according to Triple J's annual countdown.

Step 2 Here is a histogram of the sample

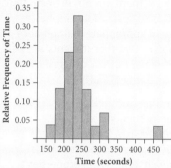

The mean of the sample is 240.7 sec with a standard deviation of 56 sec. But there is one outlier and if you recalculate mean and standard deviation without it you get 232.9 sec and 36.8 sec. Here is a plot of the normal curve (multiplied by the bin width of 25) using the actual x-bar and s on the left and the x-bar and s from the trimmed data on the right.

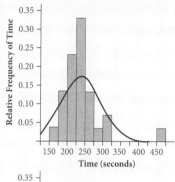

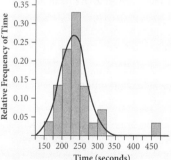

You can see that the first one is a better fit to the data.

Step 3 The value of $\bar{x} + s$ is 296.7 sec (or 269.7). In a normal distribution this would be about the 84%ile and in this distribution there are 27 values below 296.7 or 90% and 26 values below 269.7 or 87%.

Step 4 The 16%ile of the normal distribution would be at 185 sec (or 196.3 sec). These values are about one standard deviation below the mean (240.7 − 56 = 284.7 sec) and 232.9 − 36.8 = 196.1 sec)

Step 5 The middle 95% is bracketed by the 2.5%ile is at 130.9 (or 160.8) sec and the 97.5%ile is at 350.5 (or 305). These values are within two seconds of x-bar plus and minus two standard deviations. 128.7 (or 159.3) and 352.7 (or 306.5)

Example B

Project Example B from your ebook and have students work in small groups. Have them present their strategy as well as their solution.

Summarize

As students present their solutions to the investigation, examples and exercises, have them present their strategy as well as their solution. Emphasize the use of appropriate mathematical vocabulary. Whether students' answers are correct or not, ask others if they agree and why. **SMP 1, 3, 5, 6**

Rather than giving the formula for the parent function, symmetric about the *y*-axis, for the *standard normal distribution,* lead students to derive it. **ASK** "What properties does the graph of the parent function, symmetric about the *y*-axis, have?" [$y(0) = 1$; the function is even; the *x*-axis is an asymptote.] "What functions have some of these properties?" [Exponential functions have *y*-intercept 1 and the *x*-axis as an asymptote.] "Can you adjust the exponential function to be symmetric about the *y*-axis?" [The graph of the function $y = b^{x^2}$ doesn't have the asymptote, but the graph of $y = b^{-x^2}$ does, and this is the parent function of the standard normal distribution.]

To check student understanding **ASK** "What does *x* represent?" [data values] "What does *y* represent?" [frequencies of the data values] "What does the area under the curve represent?" [the probability] "What is the total area under the curve? Why?" [1; because the sum of the probabilities for all possible outcomes must be 1.]

CRITICAL QUESTION What might be a good way to find inflection points on a graph?

BIG IDEA Be open to a variety of ideas. One way is to cover the right (or left) part of the graph until the visible curve is the largest part that appears concave up. A line drawn to the *x*-axis will locate the inflection point.

Look at the curvature of a normal curve. At the points that are exactly one standard deviation from the mean, the curve changes between curving downward (the part of the curve with decreasing slope) and curving upward (the parts of the curve with increasing slope). These points are called **inflection points.** You can estimate the standard deviation of any normal distribution by locating the inflection points of its graph.

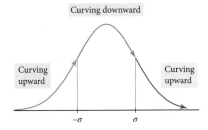

The 68 – 95 – 99.7 Rule

In a normal distribution …
Approximately 68% of values will fall within one standard deviation of the mean.
Approximately 95% of values will fall within two standard deviations of the mean.
Approximately 99.7% of values will fall within three standard deviations of the mean.

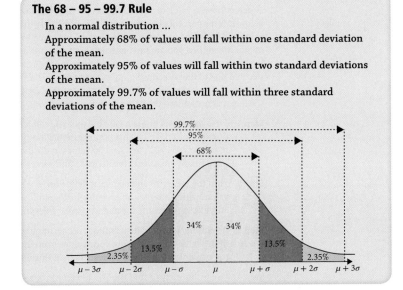

While calculators and tables often provide values for area/probability for any interval of the normal distribution, it is good to keep these three numbers in mind to check how practical the values provided are. For example, if asked what is the probability that a random value is higher than 2.1 standard deviations above the mean, you know the answer should be about 0.02 or 2% even before you start your calculations.

EXAMPLE B Piarell Cones sells a single dip cone with a mean weight of 4 ounces and a standard deviation of 0.3 ounces. The distribution of weights follows a normal distribution. Find the probability of a random cone that weights more than 4.5 ounces.

Solution | Since the standard deviation is 0.3 ounces, then the value of 4.5 is more than one standard deviation and close to two standard deviations above the mean, so the probability might be estimated to be around 5%. The calculation of the area under the normal curve (centered at 4 with a 0.3 horizontal dilation) from 4.5 on up is about 0.0478.

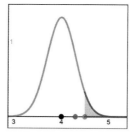

Not every distribution of values is normal but many are mounded and symmetric. Using the normal function will give pretty good estimates for those types of values. In other cases where the distribution is not a single symmetric mound, the area under the normal curve is not as valuable. It is best not to quote too many decimal places of accuracy in your answers when you are unsure of the distribution.

History CONNECTION

English nurse Florence Nightingale (1820–1910) contributed to the field of applied statistics by collecting and analyzing data during the Crimean War (1853–1856). While stationed in Turkey, she systematized data collection and record keeping at military hospitals and created a new type of graph, the polar-area diagram. Nightingale used statistics to show that improved sanitation in hospitals resulted in fewer deaths. In this 1930 lithograph by Robert Riggs, Nightingale ministers to soldiers near Istanbul during the Crimean War.

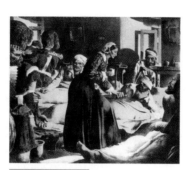

Florence Nightingale

CRITICAL QUESTION Can the curve be increasing but concave down?

BIG IDEA Yes, between $\mu - \sigma$ and the mean, the curve is increasing at a decreasing rate.

CRITICAL QUESTION Where is the curve decreasing at an increasing rate? Where is the curve decreasing but curving up?

BIG IDEA The curve is decreasing at an increasing rate from the mean to $\mu + \sigma$. It is decreasing but curving up above $\mu + \sigma$.

Formative Assessment

Student answers and presentation will provide evidence of meeting the objectives of the lesson. Assess students' understanding of probability distributions and their means and standard deviations, as well as the 68-95-97.7 rule.

Apply

Extra Example

1. Find the mean and standard deviation of the equation $y = \frac{1}{2\sqrt{2\pi}}\sqrt{e}^{-\left(\frac{x-31}{2}\right)^2}$.
   $\mu = 31, \sigma = 2$

2. Where is the mean located on a graph of a standard normal distribution?
   the x value directly below the maximum point

Closing Question

Carefully graph the parent function of the normal distribution and label the inflection points and approximate locations for integer values of sigma.

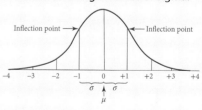

Assigning Exercises

Suggested: 1–12
Additional Practice: 13, 14

Exercises

You will need a graphing calculator for Exercises **1, 5–8,** and **10–12**.

Practice Your Skills

1. The standard normal distribution equation, $y = ab^{-x^2}$, where $a = \sqrt{\frac{1}{2\pi}}$ and $b = \sqrt{e}$, is equivalent to the calculator's built-in function $n(x, 0, 1)$.
   a. Use a table or graph to verify that these functions are equivalent.
   b. Evaluate each function at $x = 1$. $y \approx 0.242$ and $n(x, 0, 1) \approx 0.242$

1a. The tables and graphs of both functions agree.

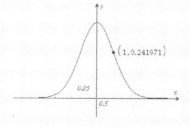

Exercise 2 On graphs, the first standard deviation is estimated by finding inflection points.

4a. $y = \dfrac{1}{2.5\sqrt{2\pi}} (\sqrt{e})^{-[(x-18)/2.5]^2}$

4b. $y = \dfrac{1}{0.8\sqrt{2\pi}} (\sqrt{e})^{-[(x-10)/0.8]^2}$

4c. $y = \dfrac{1}{6\sqrt{2\pi}} (\sqrt{e})^{-[(x-68)/6]^2}$

4d. $y = \dfrac{1}{0.12\sqrt{2\pi}} (\sqrt{e})^{-[(x-0.47)/0.12]^2}$

Exercise 5 As needed, remind students to put the maximum at the mean and inflection points 1 standard deviation above and below the mean. Ask them to label the mean and standard deviations. **ALERT** Some students will think that the mean is at the maximum point on the curve rather than on the x-axis.

5.
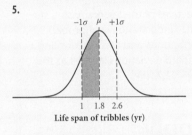
Life span of tribbles (yr)

Exercise 6 **ALERT** Students may try to determine the inflection points by taking 1 standard deviation above and below 0 rather than above and below the mean. Remind them to find the values of the points 1 and 2 standard deviations away from the mean.

6a, c.
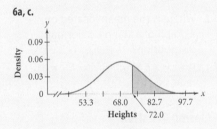

6b, c.
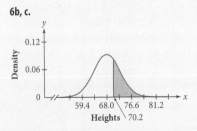

2. From each graph, estimate the mean and standard deviation. Estimates will vary.

a.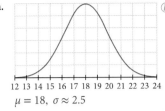
$\mu = 18, \sigma \approx 2.5$

b.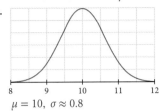
$\mu = 10, \sigma \approx 0.8$

c.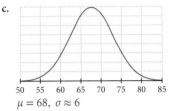
$\mu = 68, \sigma \approx 6$

d.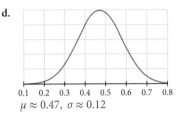
$\mu \approx 0.47, \sigma \approx 0.12$

3. Use the 68-95-99.7 rule to estimate the probability for these events.

   a. The probability of a value between 10 and 22 in a distribution with a mean of 16 and a standard deviation of 3. 0.95

   b. The probability of a value between 84 and 90 in a distribution with a mean of 90 and a standard deviation of 6. 0.34

   c. The probability of a value below 30 or above 54 in a distribution with a mean of 42 and a standard deviation of 4. 0.3

4. Estimate the equation of each graph in Exercise 2. @

Reason and Apply

5. The life spans of wild tribbles are normally distributed with a mean value of 1.8 years and a standard deviation of 0.8 year. Sketch the normal curve, and shade the portion of the graph showing tribble life spans of 1.0 to 1.8 years.

6. Assume that the mean height of an adult male gorilla is 5 ft 8 in., with a standard deviation of 7.2 in.

   a. Sketch the graph of the normal distribution of gorilla heights. @

   b. Sketch the graph if, instead, the standard deviation were 4.3 in.

   c. Shade the portion of each graph representing heights greater than 6 ft. Compare your sketches and explain your reasoning.

6c. The first graph is sketched in the domain $m - 3s \le x \le m + 3s$. The second graph displays less spread because the standard deviation is smaller; because there is more area under the central portion of the curve, there is less area under the tail where $x > 72$ in.

William Shatner (b 1931) and Leonard Nimoy (1931–2015) played Captain James T. Kirk and Dr. Spock in the original *Star Trek* television series (1966–1969). The series was the first of seven *Star Trek* televison series and thirteen films. Captain Kirk and Dr. Spock toys and two fictional alien creatures called tribbles are shown here.

7a.

Exercise 8 Area calculations will vary based on the technology used. Students should confirm that their method is accurate and that their calculations are reasonable.

7. Frosted Sugar Squishies are packaged in boxes labeled "Net weight: 16 oz." The filling machine is set to put 16.8 oz in the box, with a standard deviation of 0.7 oz.

   a. Sketch a graph of the normal distribution of box weights. Shade the portion of the graph representing boxes that are below the labeled weight.

   b. What percentage of boxes does the shading represent? Is this acceptable? Why or why not? 12.7%. Sample answer: Not acceptable; more than 10% of boxes do not meet minimum weight requirements.

8. Makers of Sweet Sips 100% fruit drink have found that their filling machine will fill a bottle with a standard deviation of 0.75 oz. The control on the machine will change the mean value but will not affect the standard deviation.

   a. Where should they set the mean so that 90% of the bottles have at least 12 oz of fruit drink in them? *ⓗ*

   b. If a fruit drink bottle can hold 13.5 oz before overflowing, what percentage of the bottles will overflow at the setting suggested in 8a?

9. The pH scale measures the acidity or alkalinity of a solution. Water samples from different locations and depths of a lake usually have normally distributed pH values. The mean of those pH values, plus or minus one standard deviation, is defined to be the pH range of the lake.

   Lake Fishbegon has a pH range of 5.8 to 7.2. Sketch the normal curve, and shade those portions that are outside the pH range of the lake. *@*

10. Data collected from 493 women are summarized in the table.

Height (cm)	148–50	150–52	152–54	154–56	156–58	158–60	160–62	162–64	164–66
Frequency	2	5	9	15	27	40	52	63	66

Height (cm)	166–68	168–70	170–72	172–74	174–76	176–78	178–80	180–82	182–84
Frequency	64	53	39	28	15	8	4	1	2

   a. Find the mean and standard deviation of the heights, and sketch a histogram of the data. *@*

   b. Write an equation based on the model $y = ab^{-x^2}$ that approximates the histogram. $y = 66(0.99)^{(x-165)^2}$

   c. Find the equation for a normal curve using the height data. $y = \dfrac{1}{5.91\sqrt{2\pi}}(\sqrt{e})^{-[(x-165)/5.91]^2}$

11. These data are the pulse rates of 50 people.

   a. Find the mean and standard deviation of the data and sketch a histogram.

   b. Sketch a distribution that approximates the histogram.

   c. Find the equation of a normal curve using the pulse-rate data. $y = \dfrac{1}{7.49\sqrt{2\pi}}(\sqrt{e})^{-[(x-79.1)/7.49]^2}$

   d. Are these pulse rates normally distributed? Why or why not?
   Answers will vary. The data do not appear to be normally distributed. They seem to be approximately symmetrically distributed with several peaks.

66	75	83	73	87	94	79	93	87	64
80	72	84	82	80	73	74	80	83	68
86	70	73	62	77	90	82	85	84	80
80	79	81	82	76	95	76	82	79	91
82	66	78	73	72	77	71	79	82	88

8a. Guess-and-check will provide a mean of about 12.96 oz so that the region with leftmost endpoint of 12 provides an area of 0.90.

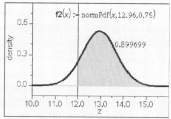

8b.

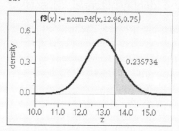

$N(13.5, 20, 12.96, 0.75) \approx 0.24 = 24\%$

9.

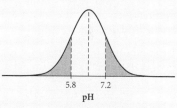

Shade the area farther out than 1 standard deviation away from the mean.

Exercise 10 Because the area of the histogram is not 1, the continuous distribution approximating the histogram in 10b will also have an area other than 1, so this distribution is not a probability distribution and hence not the normal distribution requested in 10c. Even if the heights of the histogram bins are adjusted to give portions of the total, the area is still not 1, because the bins have width 2. The normal curve of 10c includes the bottom half of this relative frequency histogram.

10a. $\mu \approx 165$; $\sigma \approx 5.91$

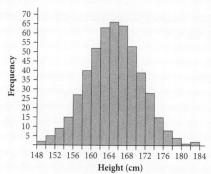

11a. $\mu \approx 79.1$; $\sigma \approx 7.49$

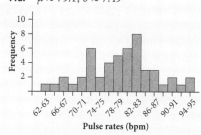

11b. sample answer:

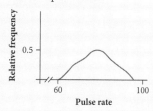

CONTEXT **Science Connection** For law enforcement officers, fingerprint identification is vital for the investigation of crimes. At a crime scene, investigators examine smooth, hard surfaces for finger prints that are made by the oily secretions of the skin. Powder or chemicals are used to make the prints visible so that they can be photographed or preserved. Law enforcement agencies keep databases of fingerprints that they compare with the fingerprints found at the crime scene to identify potential suspects.

13a. Possible answers: The means for both the French and Spanish tests are lower than that for the Mandarin test, which may indicate more difficult tests. Also, the larger standard deviation on the French exam indicates that there is a larger variation in scores, meaning that more students did worse. However, it is difficult to compare three completely different exams.

13b. the French exam, because it has the greatest standard deviation

13c. Determine how many standard deviations each student's score is from the mean.

Paul: 1.88σ above

Kenyatta: 2.07σ above

Rosanna: 0.98σ above

Kenyatta's score is better than Paul's and Rosanna's because her score is 2.07 standard deviations above the mean.

12. Ridge counts in fingerprints are approximately normally distributed with a mean of about 150 and a standard deviation of about 50. Find the probability that a randomly chosen individual has a ridge count
    a. Between 100 and 200.
       approximately 0.68
    b. Of more than 200.
       approximately 0.16
    c. Of less than 100.
       approximately 0.16

Science CONNECTION

Dactyloscopy is the comparison of fingerprints for identification. Francis Galton (1822–1911), an English anthropologist, demonstrated that fingerprints do not change over the course of an individual's lifetime and that no two fingerprints are exactly the same. Even identical twins, triplets, and quadruplets have completely different prints. According to Galton's calculations, the odds of two individual fingerprints being the same are 1 in 64 billion. Researchers in Atsugi, Japan, developed a microchip that scans a finger to identify a fingerprint with 99% accuracy in half a second. Fingerprint scans are now used for security on personal devices like phones, computers, and bank accounts

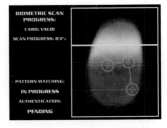

Review

13. Paul, Kenyatta, and Rosanna each took one national language exam. Paul took the French exam and scored 88. Kenyatta took the Spanish exam and scored 84. Rosanna took the Mandarin exam and scored 91. The national means and standard deviations for the tests are as follows:

    French: $\mu = 72, \sigma = 8.5$ Spanish: $\mu = 72, \sigma = 5.8$ Mandarin: $\mu = 85, \sigma = 6.1$

    a. Can you determine which test is most difficult? Why or why not?
    b. Which test had the widest range of scores nationally? Explain your reasoning.
    c. Which of the three friends did best when compared with the national norms? Explain your reasoning.

14. Mr. Hamilton gave his history class an exam in which a student must choose 3 out of 6 parts and complete 2 out of 4 questions in each part selected. How many different ways are there to complete the exam? 120

IMPROVING YOUR Visual Thinking SKILLS

Acorns

Acorns fall around the base of a particular tree in an approximately normal distribution with standard deviation 20 ft away from the base.

1. What is the probability that any 1 acorn will land more than 10 ft from the tree?
2. What is the probability that any 4 randomly chosen acorns will land an average of more than 10 ft from the tree?
3. What is the probability that any 16 randomly chosen acorns will land an average of more than 10 ft from the tree?

498 CHAPTER 9 Applications of Statistics

IMPROVING YOUR Visual Thinking SKILLS

ALERT Students might forget that "more than 10 ft from the tree" can be on either tail of the normal distribution.

1. 0.6171
2. 0.3178
3. 0.0455

LESSON 9.3

z-Values and Confidence Intervals

The way to do research is to attack the facts at the point of greatest astonishment.
CELIA GREEN

In the previous lesson, you learned about normally distributed populations and the probability that a randomly chosen item from such a population will fall in various intervals. That is, you knew something about the population and you saw how to find information about a sample.

In most real-life situations, however, you have **statistics** from one or more **samples** and want to estimate **parameters** of the **population,** which can be quite large. For example, suppose you know the mean height and standard deviation of 50 students that you survey, and you want to know the mean height and standard deviation of the entire population of students in your school. In this lesson you'll see how to describe some population parameters based on sample statistics.

First, it will be useful to learn a new method for describing a data value in a normal distribution. Knowing how a value relates to the mean value is important, but it does not tell you how typical the value is. For instance, to say that a penny's weight is 0.4 g less than the mean does not tell you whether this measurement is a rare event or a common event. In this first example you will see a method for describing how typical or unusual a particular data value is.

EXAMPLE A Andres and his cousin Imani have always been very competitive. This year they are both taking French at their respective schools. Andres boasted that he scored 86 on a recent exam. Imani had scored only 84 on her latest exam. But she countered by asking Andres about the mean and standard deviation on his test. The mean score on Andres's exam was 74 and the standard deviation was 9. The mean score on Imani's exam was 75 and the standard deviation was 6. Which of the two scores was actually better relative to the population of scores on the same exam?

Solution First we can determine how many points from the mean each of the scores are. Recall that this value is called the deviation.

Andres: $86 - 74 = 12$ points

Imani: $84 - 75 = 9$ points

Both of the deviations are positive because they are above the mean. Since these distributions have different spreads, we need to find where the values fall in the distributions, that is "How many standard deviations from the mean are these values?" So the next step of the calculation is to divide by the standard deviation.

Andres: $\frac{12}{9} \approx 1.33$ standard deviations above the mean

Imani: $\frac{9}{6} = 1.5$ standard deviation above the mean

Thus if we were to scale these two exams onto the same distribution, we would see that Imani has done a little better than Andres.

Investigate

Example A

Project Example A from your ebook. Have students work in small groups and present their answers along with their approach. As students present, **ASK** "What assumptions are you making?" One assumption is that the test scores are normally distributed. If no group tried using percentile ranks, bring it up yourself. Elicit the idea that percentile rank is another way of looking at the portion of the area lying to the left of the given score in the probability distribution. **SMP 1, 2, 3, 6**

LANGUAGE Other names for *z*-values are *z*-scores, standard scores, and *z*-statistics.

Guiding the Investigation

Answers may vary. Sample answers are given.

Step 6 The original deck is a uniform (flat) histogram. The first sample should be mounded with peak near 7. The second sample should be like that but with even less variation than the first.

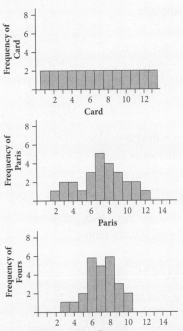

Step 1 The mean is 7, $1 + 2 + \ldots 13 = 91$ and $\frac{91}{13} = 7$ or $4 \cdot \frac{91}{52} = 7$

Step 2 For example, $\frac{6+4}{2} = 5$

Step 3 This should always be 7 unless students made a mistake in their averages.

Step 4 For example $(4 + 13 + 7 + 1) = 6.25$. There is a good probability that this answer will be closer to 7 than Step 2.

Step 5 Again this should be exactly 7

The number of standard deviations that a variable *x* is from the mean is called the *z*-value, or *z*-score. To say a cent coin's weight is 0.4 g less than the mean doesn't indicate whether this is unusual. But to say the weight is 2.86 standard deviations below the mean tells you that this measurement is quite rare. You can think of the *z*-value of *x* as the image of *x* under the transformation that translates and dilates the normal distribution to the standard normal distribution $n(x)$, with a mean $\mu = 0$ and standard deviation $\sigma = 1$.

> **Standardize the Variable**
> The equation $z = \frac{x - \mu}{\sigma}$ is the transformation of *x* to a *z*-value.

In the real world you do not know the true mean or standard deviation of the population like the two students in Example A because populations are typically very large. The values calculated from statistics such as $\bar{x}$ and *s* are estimates for the matching parameters in the population μ and σ. Thus, we can estimate a *z*-value using $z = \frac{x - \bar{x}}{s}$. But how does the sample mean relate to the population mean? In the investigation we will look at this relationship.

INVESTIGATION
Predicting the True Mean

In this investigation you will look at predicting from a sample. Start with a standard deck of 52 cards. This deck is your population and each card represents a value in a uniform distribution. Ace = 1, Two = 2, Three = 3, … Jack = 11, Queen = 12, and King = 13.

Step 1 What is the mean of your population? Explain how you found it.

Step 2 Shuffle the cards well and take the top two cards. This is your random sample of two cards. What is the average (mean) of the two cards? If you didn't know the true mean, would the mean of these two cards give you an accurate prediction?

Step 3 Repeat taking the top two cards and recording the average of each pair until you reach the bottom of the deck. You should have 26 averages now. Find the average of all the averages.

Step 4 Shuffle the deck again, this time take the top four cards and find the mean. Is this a better estimate of the true mean?

Step 5 Repeat taking the top four cards and recording the average until you reach the bottom of the deck. Now reshuffle the deck and do it all again. You should have 26 averages now. Find the average of all the averages.

Step 6 Create histograms of the original deck, of the pair averages, and of the four card averages. Describe any similarities or differences in the three histograms.

Step 7 Make a plot of what you think would happen if you took averages of 8 cards and created a histogram of 26 of these 8 card averages. Explain why you think the plot would look this way.

Modifying the Investigation

Whole Class Work through the investigation using the desmos simulation. Have student's work as you demonstrate and solicit their input.

Shortened Use the desmos simulation of the investigation found in your ebook.

One Step There is no **One Step** for this investigation. You might have students use the desmos simulation.

Every sample can be used to estimate the population mean, but from the investigation you have seen that small samples don't always give a very good estimate. However, the best estimate for the population mean is the sample mean. This is not true for every parameter or for all types of sampling. For a simple random sample, the estimate for the population standard deviation is the sample standard deviation and the estimate for the population median is the sample median.

EXAMPLE B Mrs. Corto sent her Algebra students to collect data. Each student was assigned to get the height of four different groups of 5 students. When they returned, they gave the four means of size five and one mean of size 20.

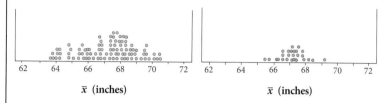

a. Which plot is the mean of size 20?

b. What value would you estimate is the mean of all students in the school?

c. Give two values that you are pretty sure the mean height of all students is between.

Solution

a. The second plot comes from samples of size 20. You know this because there are four times as many values in the first plot and the values in the second plot are much closer together.

b. The mean height of all students is about 67 inches. It is near the center of both plots.

c. The actual mean height is probably between 66 inches and 68 inches. Most of the values in the 20-person plot are between these two numbers.

The answer to part b is called a **point estimate**. It is a single value that you have reason to believe is close to being correct. But it doesn't tell much about how accurate that predictions is. The answer to part c of the example is called a **confidence interval**. It includes both the point estimate (at the center of the interval) and **margin of error**. This margin of error tells you about how accurate you believe the estimate to be. You don't know for sure the value is in the interval, but based on the data, you are pretty confident that this is true. In real situations, though, you don't get 27 mean values to look at; you just get one mean from your data. Using this mean, as well as the spread of the data, the size of the sample, and the confidence level you want, an interval estimate for the population mean can be created.

> **Confidence Interval**
>
> Suppose a sample from a normally distributed population has size n and mean $\bar{x}$, and the population standard deviation is σ. Then the $p\%$ **confidence interval** is an interval about $\bar{x}$ in which you can be $p\%$ confident that the population mean, μ, lies. If z is the number of standard deviations from the mean within which $p\%$ of normally distributed data lie, the $p\%$ confidence interval is
>
> $$\bar{x} - \frac{z\sigma}{\sqrt{n}} < \mu < \bar{x} + \frac{z\sigma}{\sqrt{n}}$$
>
> The **margin of error** is the expression $\frac{z\sigma}{\sqrt{n}}$.

Step 7 This should have even less variation that the sample by 4, but still centered at 7. The average of the averages should still be the same as the actual average because a random sample should still be a good estimate of the population. But the spread should be less. There are only so many cards in the extremes, and it is unlikely that any one sample would have way more than its share. For example, to get an average of 10 you will need four or more of your cards to be larger than ten and likely all the cards larger than six. This could happen, but being unlikely it would happen often.

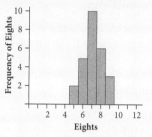

Example B
Project Example B from your ebook and work through it with student input. As students make suggestions, ask other students if they agree and why. Emphasize the use of correct mathematical vocabulary and the use of mathematics as evidence. SMP 3, 6

Example C
Project Example A from your ebook. Have students work in small groups and present their answers, including their formulas with substituted values. Whether students' responses are correct or not, ask other students if they agree and why. Emphasize the use of correct mathematical vocabulary and the use of mathematics as evidence. For example, students may make statements such as, "The probability that the population mean is within 1 standard deviation of the sample mean is 68%." Actually, the population mean either is or isn't in that interval, so the probability that it lies there is either 0 or 1. Not knowing which, we say that we're 68% confident that it's there, because it would be there for 68% of samples if we took many samples. SMP 3, 6

The assumption is that 0.28 s is the standard deviation of this population.

Conceptual/Procedural

Conceptual Students explore confidence intervals and z-values.

Procedural Students find z-values and confidence intervals of the population mean given the population standard deviation.

Summarize

Encourage students to pose problems and ask questions, such as those here.

ASK "What does confidence really mean?" [Intervals created from 95% of the samples will actually capture the true mean of the population. Your sample might be one of those that does.]

ASK "If the population is not normally distributed, can you still calculate confidence intervals?" [Yes.]

ASK "Is the z-value associated with 95% of the data exactly 2?" [It's really closer to 1.96, but the 68-95-99.7 rule is easy to remember and good enough for most purposes.]

ASK "What's the sum of the z-values?" [The sum of the n values in any data set is n times the mean, and the mean of the z-values is 0, so their sum is 0.]

ASK "Is the z in the confidence interval definition a z-value?" [It's not the z-value of any particular data point, but rather the number of standard deviations from the mean within which $p\%$ of the data lie. Statistics books often call it by another name, such as z^* or $z_{\alpha/2}$, to emphasize the difference.]

CRITICAL QUESTION What do these symbols mean in the confidence interval definition: $n, \bar{x}, \mu, s, \sigma$?

BIG IDEA n is the number of data points in the sample; $\bar{x}$ and μ are the sample and population means, respectively; s and σ are the sample and population standard deviations, respectively. English letters refer to the sample statistics, and Greek letters refer to the population parameters.

CRITICAL QUESTION Why do we use $\frac{\sigma}{\sqrt{n}}$ rather than just s?

BIG IDEA Point out as the sample size increases so will the precision of the estimate for the mean.

CRITICAL QUESTION What's the 68-95-99.7 rule in terms of z-values?

BIG IDEA In a normal distribution, about 68% of the z-values lie between −1 and 1, about 95% between −2 and 2, and about 99.7 between −3 and 3. That is, the mean of the z-values is 0, and their standard deviation is 1.

Suppose a science class is asked to estimate the mean pH of the water in a lake. The students take 30 samples of water. When they analyze their results, they find a mean pH of 5.7 and a standard deviation of 0.8. It is very unlikely that any sample will give the true mean pH of the lake water, so they express the estimate as an interval of possible values of the mean pH. The 68-95-99.7 rule tells the students that 95% of normally distributed values have a z-value between −2 and 2. Using this z-value, the class can predict with 95% confidence that the mean pH of the lake lies between $5.7 - \frac{2(0.8)}{\sqrt{30}}$ and $5.7 + \frac{2(0.8)}{\sqrt{30}}$, or between about 5.41 and 5.99. This interval can also be expressed as 5.7 ± 0.29, where 0.29 is the margin of error.

What if you want a confidence interval that is not 68%, 95%, or 99.7%? You will need a z-value other than 1, 2, or 3. You have used your technology to find probabilities (areas) between z-values. You can use an inverse function to find the z-values for a given probability. A 97% confidence interval has 1.5% of the area below the interval and 1.5% above the interval.

invNorm(0.015)	−2.17009037517
invNorm(0.985)	2.17009037517
normCdf(−2.1700903517, 2.1700903517)	0.97000010708

The table shows that a z-value of about 2.17 standard deviations corresponds to a 97% confidence interval.

EXAMPLE C Jackson is training for the 100 m race. His coach timed his last run at 11.47 s. Experience in previous training sessions indicates that the standard deviation for timing this race is 0.28 s.

a. Find the 95% confidence interval.

b. What confidence interval corresponds to ± 2.3 standard deviations?

c. Find the 90% confidence interval.

Start of Men's 100 meter sprint where Usain Bolt wins and sets a new world record at the 2008 Summer Olympic Games in Beijing, China

Solution Because only one run time is known, the value for n is 1. All confidence intervals are given as $\left(11.47 - \frac{z(0.28)}{\sqrt{1}}, 11.47 + \frac{z(0.28)}{\sqrt{1}}\right)$, where the value of z is determined by the level of confidence. This is the interval with endpoints $11.47 \pm 0.28z$.

a. The center region under the normal curve is 95% of the total area, so the "tail" of the curve on each side represents 2.5% of the data. Use technology to find that $P(z < -1.96) \approx 0.025$. So the margin of error, with 95% confidence, corresponds to a z-score of approximately 1.96, and therefore is about $0.28(1.96) = 0.55$ s. The 95% confidence interval has endpoints about 11.47 ± 0.55 s, or 10.92 s and 12.02 s.

Using the 68-95-99.7 rule to find the 95% confidence interval gives a result that differs only by the amount of round-off error. By the 68-95-99.7 rule, the z-value for 95% is about two standard deviations. The endpoints of the confidence interval, then, are $11.47 \pm 0.28(2)$. The coach is 95% confident that the actual time is between approximately 10.91 and 12.03 s.

Formative Assessment

Assess students' understanding of the mean and standard deviation of a normal distribution and their ability to find probabilities related to normal distributions. Student input and presentations will provide evidence of meeting the objectives of the lesson.

b. The area of the region within 2.3 standard deviations of the mean is $P(-2.3 < z < 2.3) \approx 0.979$, so it represents a 98% confidence interval. Alternatively, because $11.47 \pm 0.28(2.3)$ gives the interval between 10.826 and 12.114, you can get the same confidence level by finding $P(-2.3 < x < 2.3 | \mu = 11.47, \sigma = 0.28)$.

```
normCdf(-2.3,2.3)        .9785518379
normCdf(10.826,12.114,11.47,.28)
                         .9785518379
```

c. If the center region under the normal curve has area 90%, then the tail on each side must have area 5%. Use the calculator to find that $P(z < -1.645) \approx 0.05$. The margin of error here, with 90% confidence, is $0.28(1.645) = 0.46$ s. Therefore, the time interval has endpoints 11.47 ± 0.46, so the coach can be 90% confident that the actual mean time lies between 11.01 s and 11.93 s.

In Example C, the coach had to rely on only one time measurement. If four people, instead of only one, had timed the run, and the mean of those times had been 11.47 s, then the 95% interval would have had endpoints $11.47 \pm \dfrac{2(0.28)}{\sqrt{4}}$, making it between 11.19 and 11.75 s. In general, the larger the sample size, the narrower the interval in which you can be confident that the population mean lies.

9.3 Exercises

You will need a graphing calculator for Exercises **5–8, 11**, and **12**.

Practice Your Skills

1. A set of normally distributed data has mean 63 and standard deviation 1.4. Find the z-value for each of these data values.

   a. 64.4 @ $z = 1$
   b. 58.8 $z = -3$
   c. 65.2 @ $z \approx 1.57$
   d. 62 $z \approx -0.71$

2. A set of normally distributed data has mean 125 and standard deviation 2.4. Find the data value for each of these z-values.

   a. $z = -1$ @ 122.6
   b. $z = 2$ 129.8
   c. $z = 2.9$ @ 131.96
   d. $z = -0.5$ 123.8

3. Given the sample set of {1, 3, 4, 7}.
   a. Find mean and standard deviation of this data. mean = 3.75, standard deviation = 2.165
   b. Using the sample standard deviation as σ, find the 95% margin of error in estimating the true mean. margin of error = 2.12
   c. Give a 95% confidence interval for the mean. <1.63, 5.87>

4. Given the sample set of {3, 5, 6, 8, 9}.
   a. Find mean and standard deviation of this data. mean = 6.2, standard deviation = 2.135
   b. Using the sample standard deviation as σ, find the 90% margin of error estimating the true mean. margin of error = 1.57
   c. Give a 90% confidence interval for the mean. <4.63, 7.77>

Apply

Extra Example

1. Shade the region that represents the area greater than 1 standard deviation away from the mean.

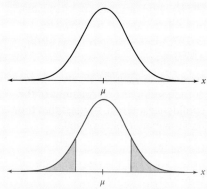

2. A set of normally distributed data has a mean of 25 and a standard deviation of 0.8. Find the z-value of 26. 1.25

Describe where this value is located.
The data value is 1.25 standard deviations above the mean.

Closing Question

What z-value corresponds to 80% of the data? about 1.28

Assigning Exercises

Suggested: 1–11
Additional Practice: 12–14

Exercise 3 The answer to 3b and 3c are precisely correct (meaning, if you use technology these are the answers you will get). If you use the formulas in the book, there are rounding errors (e.g., a z-score of ± 2 is not exactly 95% but that is the common rule used so using the common rule introduces some error in computation)

LESSON 9.3 z-Values and Confidence Intervals

Exercises 7, 8 ADVANCED Together these exercises ask students to investigate the difference that sample size plays in determining levels of confidence. Extend this idea by asking students to find the appropriate sample size to meet a specified confidence level.
ALERT If students use $s = 1.96$ instead of $s = 2$, the answer to 7b will be (3.050, 3.150) and 8b will be (3.073, 3.127).

Exercises 9, 10 SUPPORT If students are getting lost in the calculations of the confidence interval, suggest that they find the margin of error first.

Exercise 10a If students use $s = 1.96$ instead of $s = 2$, the range will be between 204.5 and 210.5.

Reason and Apply

5. The mean travel time between two bus stops is 58 min with standard deviation 4.5 min.
   a. Find the z-value for a trip that takes 66.1 min. ⓐ
      $z = 1.8$
   b. Find the z-value for a trip that takes 55 min.
      $z = -0.\overline{6}$
   c. Find the probability that the bus trip takes between 55 and 66.1 min. approximately 0.71

6. A set of normally distributed data has mean 47 s and standard deviation 0.6 s. Find the percentage of data within these intervals: ⓗ
   a. Between 45 and 47 s approximately 50%

San Marco water bus stop sign in Venice, Italy

   b. Greater than 1.5 s above or below the mean approximately 1.2%

7. A sample has mean 3.1 and standard deviation 0.14. Find each confidence interval, and round values to the nearest 0.001. Assume $n = 30$ in each case.
   a. 90% confidence interval ⓐ   b. 95% confidence interval   c. 99% confidence interval
      (3.058, 3.142)   (3.049, 3.151)   (3.034, 3.166)

8. Repeat Exercise 7 assuming $n = 100$.
   a. (3.077, 3.123)   b. (3.072, 3.128)   c. (3.064, 3.136)

9. **APPLICATION** Fifty recent tests of an automobile's mileage indicate it averages 31 mi/gal with standard deviation 2.6 mi/gal. Assuming the distribution is normal, find the 95% confidence interval.
   between 30.3 and 31.7 mi/gal

10. **APPLICATION** A commercial airline finds that over the last 60 days a mean of 207.5 ticketed passengers actually show up for a particular 7:24 A.M. flight. The standard deviation of their data is 12 passengers.

    a. Assuming this distribution is normal, find the 95% confidence interval.
       between 204.4 and 210.6 passengers
    b. If the plane seats 225 passengers, what is the probability the plane will be overbooked?
       0.07

11. A random sample of 250 residents in town were asked if they were voting for or against annexation. The sample showed that 100 were for annexation.

    a. What is the proportion of voters that favor annexation? 0.4

    b. The standard deviation of a proportion, p is $s = p\sqrt{(1-p)}$ What is the standard deviation of this sample proportion? 0.49

    c. Give a 95% confidence interval for the true proportion. (0.34, 0.46)

    d. Do you believe there is any chance that more than 50% of all voters might support annexation? No, if is very unlikely that the true proportion would be greater than 50%.

Review

12. Given a linear model $y = 13.47 + 0.94x$, find the residual for each value and explain what the residual tells you about the point in relation to the model.

    a. (16.2, 31.0)   **b.** (19.5, 31.8)   **c.** (14.7, 28.1)

13. Five hundred integer values, −3 through 3, are randomly selected (and replaced) from a hat containing tens of thousands of integers. The frequency table at right lists the results. Find these values:

    a. $P(-3)$ 0.064

    b. P(less than 0)   0.49

    c. P(not 2)   0.82

    d. the expected sum of the next ten selected values   −1.2

Value	Frequency
−3	32
−2	60
−1	153
0	92
1	45
2	90
3	28

14. You are about to sign a long-term rental agreement for an apartment. You are given two options:

Plan 1: Pay $400 the first month with a $4 increase each month.

Plan 2: Pay $75 the first month with a 2.5% increase each month.

    a. Write a function to model the accumulated total you will pay over time for each plan. ⓗ

    b. Use your calculator to graph both functions on the same screen.

    c. Which rental plan would you choose? Explain your reasoning.
If you stay 11 yr 9 mo or less, choose Plan 2. If you stay longer, choose Plan 1.

12a. 2.302, the y-value when $x = 16.2$ is 2.302 higher than predicted by the model

12b. 0, the y-value when $x = 19.2$ is exactly what is predicted by the model

12c. 0.812, the y-value when $x = 14.7$ is 0.812 higher than predicted by the model

14a. Let n represent the number of months, and let S_n represent the accumulated total.

Plan 1: $S_n = 398n + 2n^2$;

Plan 2: $S_n = \dfrac{75(1 - 1.025^n)}{1 - 1.025}$

14b.

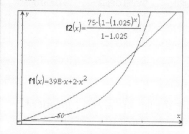

Performance Task

Make a statement about the change in size of each new confidence interval.

a. If you increase the size of your sample, then the confidence interval will __?__ .

b. If you increase your confidence from 90% to 99%, then the interval will __?__ .

c. If your sample has a larger mean, then the interval will __?__ .

d. If your sample has a larger standard deviation, then the interval will __?__ .

Performance Task

If students are having difficulty, you might suggest that they use sample data and change the variables for each step.

a. decrease

b. increase

c. stay the same size

d. increase

LESSON 9.4

Bivariate Data and Correlation

In previous lessons students saw how to make predictions about a population from a sample. Now they turn to making predictions about one characteristic from another in a sample or population.

COMMON CORE STATE STANDARDS		
APPLIED	DEVELOPED	INTRODUCED
	S.ID.5	
	S.ID.6	
	S.IC.6	
	S.ID.8	

Objectives
- Learn the language and concept of bivariate sampling
- Find and interpret the coefficient of correlation to measure association
- Understand the difference between correlation and causation

Vocabulary
bivariate sampling
correlation
correlation coefficient
explanatory variables
response variables
lurking variables

Materials
- Calculator Note: Correlation Coefficient

Launch

If you take a random sample of public school students in grades K – 12 and measure weekly allowance and size of vocabulary, you will find a strong relationship. Which of the following might be a cause of the strong correlation? iii

i. Getting a higher allowance causes an increase in vocabulary.

ii. A larger vocabulary causes a higher allowance.

iii. Age could cause the relationship between allowance and vocabulary for public school students.

iv. Getting a larger allowance causes a decrease in vocabulary.

It is a capital mistake to theorize before one has data. Insensibly one begins to twist facts to suit theories, instead of theories to suit facts.

SIR ARTHUR CONAN DOYLE

Dr. Aviles and Dr. Scott collected data on tree diameters and heights. Dr. Aviles thought that the height was closely associated with the diameter, but Dr. Scott claimed that the height was more closely associated with the square of the diameter. Which model is better? Each researcher plotted data and found a good line of fit. Their graphs are shown here.

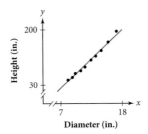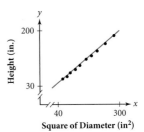

Does it appear that a line is the appropriate model? If so, is there a better linear relationship in Dr. Aviles's data or in Dr. Scott's data? How good is the fit for each line? In this lesson you'll learn how to answer questions like these.

A giant Sequoia tree in Sequoia National Park, California.

In the last lesson you learned how to make predictions about population parameters from sample statistics. You can also predict associations between parameters, such as height and diameter or height and the squares of diameter, for a large population, such as all trees. You can even make predictions about populations that are infinitely large.

The process of collecting data on two possibly related variables is called **bivariate sampling.** How can you measure the strength of the association in the sample?

An association between two variables is called **correlation.** A commonly used statistical measure of linear association is called the **correlation coefficient.** In the investigation you'll use a calculator to explore properties of the correlation coefficient for a bivariate sample.

INVESTIGATION

Looking for Connections

Step 1 Work with your group to create a survey with five questions that are all answered with a number, or use the sample survey here.

Step 1 Sample data provided are for the survey described in the text. The answer to question 4 is recorded in minutes past 8 P.M.

1. How many minutes of homework did you do last night?

2. How many minutes did you spend talking, calling, e-mailing, or writing to friends?

ELL	Support	Advanced
Start the lesson with a discussion about causality and correlation in students' lives between time spent reading and time spent watching TV, number of siblings and GPA, or the distance students live from school and the number of tardies in the morning.	Students may attempt to connect the correlation coefficient with the slope of a line. The correlation coefficient for a set of truly linear data is either +1 or −1; the only thing it has in common with the slope of the line is the sign. Stress that a correlation coefficient may be found for any set of data, no matter how nonlinear it appears.	Students could try to create data sets with specific *r*-values. Focus the discussion on the impact of outliers.

3. How many minutes did you spend just watching TV or listening to music?

4. At what time did you go to bed?

5. How many academic classes do you have?

Step 2 possible answers:

i. positive: 1 and 5; 3 and 4

ii. negative: 1 and 2; 2 and 5

iii. weak: 1 and 3; 1 and 4; 2 and 3; 2 and 4; 3 and 5; 4 and 5

Step 3

The graphs that increase have positive correlation coefficients, and the graphs that decrease have negative correlation coefficients. When the relationship is strong, the correlation coefficient is closer to ±1, and when the relationship is weak, the correlation coefficient is close to 0.

Step 2 Conjecture with your group about the strengths of correlations between pairs of variables. For example, you may decide that the number of minutes of homework is strongly correlated with the number of academic classes. Consider each of the ten pairs of variables and identify which combinations you believe will have

i. A positive correlation (as one increases, the other tends to increase).

ii. A negative correlation (as one increases, the other tends to decrease).

iii. A weak correlation.

Step 3 Gather data from each student in your class. Then enter the data into five calculator lists. Plot points for each pair of lists, and find the correlation coefficients. You may want to divide this work among members of your group. Describe the relationship between the appearance of the graph and the value of the correlation coefficient.

Step 4 Write a paragraph describing the correlations you discover. Include any pairs that are not correlated that you find surprising. You have collected a small and not very random sample; do you think these relationships would still be present if you collected answers from a random sample of your entire school population?

In 1896, English mathematician Karl Pearson (1857–1936) proposed the correlation coefficient, now abbreviated r. To compute the correlation coefficient, Pearson replaced each x- and y-value in a data set with its corresponding z-value. If a particular x- or y-value is larger than the mean value for that variable, then its z-value is positive. And if a particular x- or y-value is smaller than the mean value for that variable, then its z-value is negative.

Pearson then found the product of z_x and z_y for each data point and summed these products. In a data set that is generally increasing, the products of z_x and z_y are positive. This is because for every point, usually either both x and y are above the mean or both x and y are below the mean. In a data set that is generally decreasing, usually either z_x is positive and z_y is negative or z_x is negative and z_y is positive. Therefore, the products will be negative. After summing the products, Pearson divided by $(n-1)$ to get a number between −1 and 1. So he defined the correlation coefficient as $\frac{\sum z_x z_y}{n-1}$. But what do values of this coefficient mean?

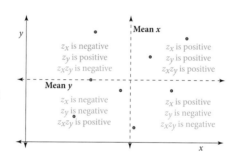

LESSON 9.4 Bivariate Data and Correlation 507

Investigate

Guiding the Investigation

You might use the Investigation Worksheet with Sample Data if you do not wish to conduct the investigation as an activity. The sample data was used for these answers.

Step 1 One way to enter data from Question 4 numerically into the calculator is to enter the number of minutes past 8 P.M.

Step 2 ⓔⓛⓛ To *conjecture* means to make a statement that you think is true but you have not yet proved. Some students will include 1 and 3, 1 and 4, 3 and 5, or 4 and 5 as negative rather than weak.

Step 3 You may want to have the class critique various methods of data collection. Students can poll one another or anonymously answer the questions on a piece of paper. If each student in a group asks a separate question and makes one list, remind group members to make sure they enter data for students in the same order. Putting answers to Question n into spreadsheet column n or into list Ln might also help avoid confusion.

Step 3

questions 1 and 2: $r \approx -0.521$

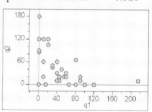

questions 1 and 3: $r \approx -0.320$

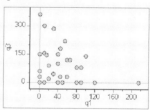

questions 1 and 4: $r \approx -0.206$

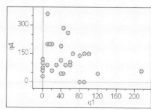

Modifying the Investigation

Whole Class Pick five questions to use for the whole class. Discuss student predictions in Step 2. Compile data for Step 3. Test the conjectures in Step 3. Have students write paragraphs in Step 4 individually.

Shortened Use sample data survey questions for Step 1. Have groups make conjectures for Step 2. Use the sample data to complete Steps 3 and 4. Assign each group a pair of data to examine.

One Step Use the **One Step** in your ebook. As needed, emphasize that the values displayed are z-values. As students realize that they can't make the prediction from the z-scores alone, ask what other information they would need [means and standard deviations]. Even if students don't come up with the Pearson correlation coefficient, they'll be open to understanding it when you describe it during the discussion.

LESSON 9.4 Bivariate Data and Correlation **507**

questions 1 and 5: $r \approx 0.771$

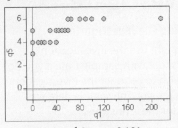

questions 2 and 3: $r \approx -0.191$

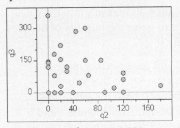

questions 2 and 4: $r \approx -0.059$

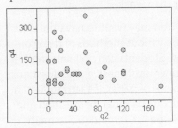

questions 2 and 5: $r \approx -0.632$

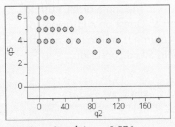

questions 3 and 4: $r \approx 0.576$

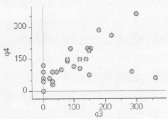

questions 3 and 5: $r \approx -0.207$

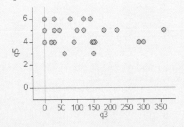

You may have noticed in the investigation that values of r can range from -1 to 1. A value of 1 means the x-values are positively correlated with the y-values in the strongest possible way. That is, as x-values increase, y-values increase linearly. A value of -1 means the x-values are negatively correlated with the y-values in the strongest possible way. That is, as x-values increase, y-values decrease linearly. A value of 0 means there's no linear correlation between the values of x and y.

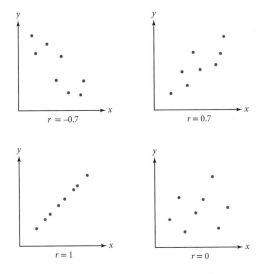

If data are highly correlated, a straight line will model the data points well.

For Dr. Aviles's data in the tree example at the beginning of this lesson, the correlation coefficient is 0.99. So his data are very close to linear. For Dr. Scott's data, the correlation coefficient is 0.999. That's even better. This means that for the trees measured, the squares of diameters are better predictors of the heights than are the diameters themselves.

> **The Correlation Coefficient**
>
> The correlation coefficient, r, can be calculated with the formula
> $$r = \frac{\sum z_x z_y}{n-1} = \frac{\sum(x-\bar{x})(y-\bar{y})}{s_x s_y (n-1)}$$
> A value of r close to ± 1 indicates a strong correlation, whereas a value of r close to 0 indicates little or no correlation.

Note that the definition of the correlation coefficient includes no reference to any particular line, though it describes how linear a bivariate data set is. In contrast, the residuals you studied in Chapter 1 can describe how well a *particular* line fits a data set.

Often, bivariate data are collected from a study or an experiment in which one variable represents some condition and the other represents measurements based on that condition, as shown in the next example. In statistics, the x- and y-variables are often called the **explanatory** and **response variables** instead of the independent and dependent variables.

508 CHAPTER 9 Applications of Statistics

questions 4 and 5: $r \approx -0.185$

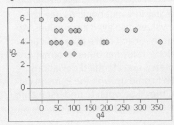

Step 4 The correlations between time spent on homework and number of academic classes and time watching TV and bedtime are relatively strong and positive. The correlations between time talking to friends and number of academic classes and time spent on homework and time talking to friends are relatively strong and negative. The other pairs of variables do not appear to be correlated. These results do not necessarily represent the school, because people in this class are not a random sample of the school.

EXAMPLE A Kiane belongs to many committees and notices that different groups take different amounts of time to make decisions. She wonders if the time it takes to make a decision is linearly related to the size of the committee. So she collects some data. Find the correlation coefficient of this data set and interpret your result.

Size (people)	4	6	7	9	9	11	15	15	18	20	21	24	25
Time (min)	5.2	3.8	8.2	8.5	12.0	10.8	14.7	15.5	22.0	19.1	35.3	29.2	32.1

Solution In this instance, it makes sense to let the explanatory variable, x, represent the committee size and the response variable, y, represent the time it takes to make a decision. When plotted, the data show an approximately linear pattern.

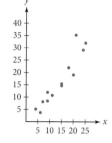

You can find the value of r directly using a calculator, but you should also try to use the formula—assisted by a calculator, of course. First, enter the data into lists and verify these statistics:

$\bar{x} \approx 14.15$ people, $s_x \approx 7.0456$ people

$\bar{y} \approx 16.65$ minutes, $s_y \approx 10.302$ minutes

Then use the lists in the formula

$$\frac{\Sigma(x - 14.15)(y - 16.65)}{(7.0456)(10.302)(13 - 1)} \approx 0.9380$$

A correlation coefficient $r \approx 0.9380$ means there is a strong positive correlation between the size of a committee and the time it takes to reach a decision. This means that as the size of a committee increases, the change in time it takes to reach a decision increases proportionately.

Be careful that you don't confuse the ideas of correlation and causation. A strong correlation may exist between two sets of data, but this does not necessarily imply a causal relationship. For instance, in Example A, Kiane found a strong correlation between committee size and decision-making time. But this does not necessarily mean that the size of the committee *caused* the decision to take longer. Whether it did or did not can be proved only by a carefully controlled experiment.

EXAMPLE B Tigo is a runner and he wonders if there is any truth to the idea that outfits made from natural fibers are better for runners. He designs an experiment and recruits a fellow runner, Daniel, to help, but doesn't tell him what the objective is. Each day he provides Daniel with a randomly chosen outfit of shorts, shirt, and socks made with a fixed ratio of natural and synthetic fibers and then times him in a two kilometer run. What conclusion, if any, can you make?

Natural (%)	20	15	90	70	75	0	50
Time (min)	5.42	5.39	5.08	5.11	5.14	5.41	5.24

Natural (%)	45	35	30	25	100	40	60
Time (min)	5.26	5.34	5.31	5.37	5.02	5.32	5.22

LESSON 9.4 Bivariate Data and Correlation

Summarize

Discussing the Investigation

As students present their results, ask them to make sense of the correlation coefficient. (You might want to remind students of the meaning of z-values as the number of standard deviations from the mean. The sum of the z-values of a data set is 0, and their standard deviation is 1.) Encourage students to make sense of the idea of the correlation coefficient by asking questions such as these.

ADVANCED "Is the correlation coefficient r like variance, dividing by $(n - 1)$?" [Indeed, the correlation coefficient is a generalization of the variance; if only one sample is involved, corresponding perfectly to itself, then r equals the variance of the sample's z-values, which is 1. In fact, as with variance, to find the correlation coefficient of a population, you use denominator N (the population size) rather than $(n - 1)$, where n is the sample size.]

CRITICAL QUESTION Is correlation about samples or about populations?

BIG IDEA In this lesson the correlation coefficient is being treated as a sample statistic; it can be used to predict a corresponding population parameter, usually indicated by the Greek letter ρ (rho).

CRITICAL QUESTION How does r relate to the straight line through the data?

BIG IDEA The sign of r will be the same as the sign of the slope of a line through the data, but otherwise has nothing to do with the slope.

CRITICAL QUESTION Which indicates more correlation, a correlation coefficient of 0.94 or one of −0.94?

BIG IDEA The correlations are equally strong, though one is negative. You might be prepared to illustrate this idea with various graphs and correlation coefficients.

Solution Graph the data and calculate the correlation. The graph of the data shows a clear downward trend. The correlation, $r = -0.976$ indicates a strong association between the percent of natural fiber and the time of the 2K run.

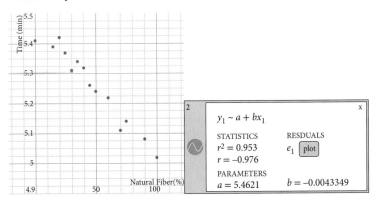

This was a well-designed single blind experiment including the randomized assignment of treatment each day. We can say that there is evidence of causation in this case. But since the subject was only Daniel we can apply this conclusion only to a limited population (until more testing is done.) Conclusion: For people like Daniel we believe that higher percent of natural fibers in clothing worn will decrease the time of a 2K run.

Drawing conclusions from data requires such things as properly designed and executed experiments. Even then it is easy to overstate your conclusion too broadly. We might suspect that all physical exercise is better for all people when they wear clothing with less synthetic fiber, but that is not what the experiment tells us. The purpose of the randomization is to rule out other **lurking variables,** such as the runner just getting faster each day, but there is always a concern that some other variable, like the heat or humidity, might by some chance be correlated with the random choice of outfits. The purpose of the blinding is so that Daniel's performance is not altered by what Daniel believes. It might be even better if it were a double blind experiment and the person timing the runner also did not know the level of natural fiber for that day.

Athletes competing in the Boston Marathon 2015 faced a cold and raining day on April 20, 2015

Practice Your Skills

1. Approximate the correlation coefficient for each data set.

   a. 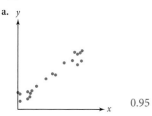 0.95

   b. 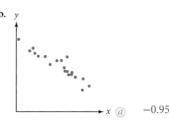 −0.95

   c. 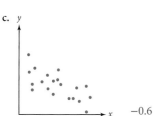 −0.6

   d. 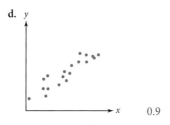 0.9

2. Sketch a graph of a data set with approximately these correlation coefficients.

   a. $r = -0.8$ b. $r = -0.4$

   c. $r = 0.4$ d. $r = 0.8$

3. Copy and complete the table at right. Then answer 3a–d to calculate the correlation coefficient.

x	y	$x-\bar{x}$	$y-\bar{y}$	$(x-\bar{x})(y-\bar{y})$
12	8	−5	2	−10
14	8	−3	2	−6
18	6	1	0	0
19	5	2	−1	−2
22	3	5	−3	−15

   a. What is the sum of the values for $(x-\bar{x})(y-\bar{y})$? −33

   b. What are $\bar{x}$ and s_x? 17; 4

   c. What are $\bar{y}$ and s_y? 6; 2.1213

   d. Calculate $r = \dfrac{\Sigma(x-\bar{x})(y-\bar{y})}{s_x s_y (n-1)}$. −0.9723

   e. What does this value of r tell you about the data? There is a strong negative correlation in the data.

   f. Draw a scatter plot to confirm your conclusion in 3e.

4. In each study described, identify the explanatory and response variables.

   a. Doctors measured how well students learned finger-tapping patterns after various amounts of sleep.   explanatory variable: amount of sleep; response variable: ability to learn the finger-tapping pattern

   b. Scientists investigated the relationship between the weight of a mammal and the weight of its brain.   explanatory variable: weight of mammal; response variable: weight of brain

   c. A university mathematics department collected data on the number of students enrolled each year in the school and the number of students who signed up for a basic algebra class.   explanatory variable: university enrollment; response variable: algebra class enrollment

2. possible answers:

2a. 2b. 2c. 2d.

3f.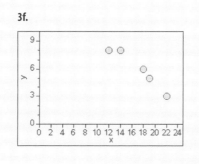

Formative Assessment

Watch that students understand how z-values relate to the probability of an event. Also confirm their understanding of the differences and similarities between samples and populations.

Watch that students begin to use the new vocabulary of correlation. Assess whether they are able to find correlation coefficients on calculators. Check that they have developed a basic idea of the relationship between that number and their data.

Apply

Extra Example

1. A group of students completed an experiment about the relationship between student behavior in the afternoon and the amount of sugar consumed in their lunches. Identify the explanatory and response variables. explanatory: amount of sugar; response: behavior

2. Sketch a graph of a data set with a correlation coefficient of approximately $r = -1$.

Closing Question

Create two data points with $x_1 \neq x_2$ and $y_1 \neq y_2$, and find the correlation coefficient. Explain your result.

The correlation coefficient is 1 because there's a line that goes exactly through those two points.

Assigning Exercises

Suggested: 1–10
Additional Practice: 11–15

Exercise 1 The answers given are only estimates. A range of answers is acceptable.

Exercise 5 **ELL** Students may need help interpreting these situations.

Exercise 6 **LANGUAGE** Dieback is the decay that moves incrementally from the tips of a plant's leaves toward its center.

6a.

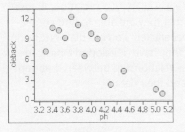

Exercise 6c **ELL** To *refute* an answer means to show it is wrong.

CONTEXT **Environmental Connection**
Scientists, environmentalists, and some economists estimate that conserving rain forests has a far greater economic value than deforestation for lumber or agriculture. The rain forests provide numerous environmental services. They absorb carbon dioxide, a greenhouse gas, thus helping to fend off global warming. Rain forests cycle nutrients that are essential to life, such as phosphorus and nitrogen. Thanks to the forests' high biodiversity, researchers often discover the ingredients for new pharmaceuticals in rain forest plants.

5. For each research finding, decide whether there is evidence of causation, correlation, or both. If it is only a correlation, name a possible lurking variable that may be the cause of the results.

   a. As the sales of television sets have increased, so has the number of overweight adults. Does owning a television cause weight gain? ⓗ Correlation; weight gain probably has more to do with amount of physical activity than with television ownership.
   b. A study in an elementary school found that children with larger shoe sizes were better readers than those with smaller shoe sizes. Do big feet make children read better? Correlation; the age of the children may be the variable controlling both size of feet and reading ability.
   c. The more firefighters sent to a fire, the longer it takes to put out the fire. Does sending more firefighters cause a fire to burn longer? Correlation; the size of a fire may be the variable controlling both the number of firefighters and the length of time the fire burns.

Reason and Apply

6. An environmental science class conducted a research project to determine whether there was a relationship between the soil pH, x, and the percent dieback of new growth, y, for a particular type of tree. The table contains the data the class collected.

x	3.3	3.4	3.5	3.6	3.7	3.8	3.9	4.0	4.1	4.2	4.3	4.5	5.0	5.1
y	7.3	10.8	10.4	9.3	12.4	11.2	6.6	10.0	9.2	12.4	2.3	4.3	1.6	1.0

   a. Make a scatter plot of the data.
   b. Describe the relationship between the two variables. Possible answer: The variables appear to have a negative correlation. As the soil pH increases, the percent of dieback decreases.
   c. Find the correlation coefficient. Does this confirm or refute your answer to 6b? How can you tell? $r \approx -0.740$. Answers will vary. A value of -0.740 confirms a fairly strong negative correlation.
   d. Can you conclude that higher soil pH causes less dieback of new growth? Explain. There is a correlation, but you cannot assume it is causation. Other variables, such as temperature, location, and amount of sunlight, would also have to be considered.

→ Environmental
CONNECTION

Trees take in carbon dioxide from the air, which helps offset carbon dioxide emissions caused by cars and industry. Destruction of forests has contributed to an increase in the amount of carbon dioxide in the atmosphere, leading to global warming and other problems. Although nature can eventually reforest most areas on its own, we can speed the process by improving soil quality and selecting the best species of trees for a particular area.

Deforestation from logging and pasture clearing have reduced the size of the Amazon rain forest in Brazil by more than 15% since 1970.

7. The table contains information from a selection of four-year colleges and universities for the 2005–2006 school year. Describe the correlation between the number of students and the number of faculty. $r \approx 0.938$. There is a strong positive correlation between the number of students and the number of faculty.

Four-Year Colleges

College	Number of students	Number of faculty	College	Number of students	Number of faculty
Alfred University	2,235	205	Mills College	1,372	184
Brandeis University	5,189	472	Morehouse College	3,029	223
Brown University	8,261	888	Mt. Holyoke College	2,127	241
Bryn Mawr College	1,799	185	Princeton University	6,916	1,060
California College of the Arts	1,616	370	Rhode Island School of Design	2,258	494
Carleton College	1,959	216	Rhodes College	1,692	184
College of William & Mary	7,544	763	Saint John's University	1,196	176
DePauw University	2,397	253	St. Olaf College	3,058	332
Drake University	5,277	388	Spelman College	2,318	245
Duquesne University	9,916	908	Swarthmore College	1,479	195
Gallaudet University	1,834	230	Syracuse University	17,266	1,391
Hampshire College	1,376	137	Tufts University	9,780	1,194
Illinois Wesleyan University	2,146	222	Tuskegee University	2,880	265
Lehigh University	6,748	621	University of Tulsa	4,084	422
Maryland Institute, College of Art	1,717	267	Wesleyan University	3,207	368
Miami University of Ohio	16,338	1,198	Wheaton College	1,568	161

(*The World Almanac and Book of Facts 2007*)

8. For each data set in 8a–c, draw a scatter plot and find the correlation coefficient, r. State what this value of r implies about the data, and note any surprising results you find.

   a. {(0.5, 1), (0.6, 0.9), (0.7, 0.8), (0.8, 0.7), (0.9, 0.6)} @
   b. {(0.5, 1), (0.6, 0.9), (0.7, 0.8), (0.8, 0.7), (0.9, 0.6), (1.9, 1.9)}
   c. {(0.5, 1), (0.6, 0.9), (0.7, 0.8), (0.8, 0.7), (0.9, 0.6), (1.9, 0.9)}
   d. Based on your answers to 8a–c, do you think the correlation coefficient is strongly affected by outliers? Yes, one outlier can drastically affect the value of r.

9. For each data set in 9a and b, draw a scatter plot and find the correlation coefficient, r. State what this value of r implies about the data. ⓗ

   a. {(0.3, 0.9), (0.4, 0.6), (0.6, 0.4), (1, 0.3), (1.4, 0.4), (1.6, 0.6), (1.7, 0.9), (0.3, 1.1), (0.4, 1.4), (0.6, 1.6), (1, 1.7), (1.4, 1.6), (1.6, 1.4), (1.7, 1.1)}
   b. {(0.4, 1.4), (0.6, 1.1), (0.8, 0.8), (1, 0.5), (1.2, 0.8), (1.4, 1.1), (1.6, 1.4)}
   c. For the data sets in 9a and b, does there appear to be a relationship between x- and y-values? Is this reflected in the values you found for r?

   In 9a, the data are circular, and in 9b, the data appear to fit an absolute-value function. However, the values of r are 0, implying that there is no correlation. This is because there is no *linear* correlation, although there is in fact a strong relationship in the data.

8a.
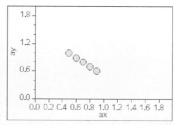
$r = -1$. This value of r implies perfect negative correlation consistent with the plot.

8b.
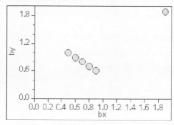
$r \approx 0.833$. This value of r implies strong positive correlation, but the plot suggests negative correlation with one outlier.

8c.
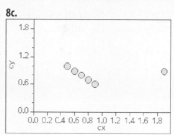
$r = 0$. This value of r implies no correlation, but the plot suggests negative correlation with one outlier.

9a.
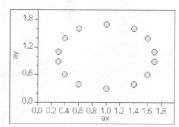
$r = 0$; no correlation

9b.
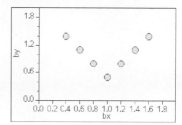
$r = 0$; no correlation

13. $y = 0 - 1.5(x - 4)$, or
$y = -3 - 1.5(x - 6)$, or
$y = -1.5x + 6$

Exercise 14 ALERT Students may still want to average the averages—that is, find the mean of 50 and 75. As needed, remind them that these are average speeds over different amounts of time and thus can't be averaged to find the mean speed.

Exercise 15 Students should also assume that the fishing pole was not bent.

If students need help you might suggest that the new package has a larger volume but a shorter length.

See your digital resources for the **Project: Correlation and Causation** where students make a claim, research data related to the claim and determine whether or not the data seem to show a correlation between the two variables, as well as whether or not one event really causes the other.

10. These data show numbers of country radio stations and numbers of oldies radio stations for several years.

Year	1994	1995	1996	1997	1998	1999	2001	2002	2003	2004	2005	2006
Country	2642	2613	2525	2491	2368	2306	2190	2131	2088	2047	2019	2097
Oldies	714	710	738	755	799	766	785	813	807	816	773	755

(*The World Almanac and Book of Facts 2007*)

a. What is the correlation coefficient for year and number of country stations, and for year and number of oldies stations? @ The correlation coefficients for (*year, country*) and (*year, oldies*) are −0.972 and 0.659, respectively.

b. Is the number of country stations or the number of oldies stations more strongly correlated with the year? Explain your reasoning. The number of country stations is more closely correlated with the year, because | −0.972 | is closer to 1 than 0.659.

Willie Nelson, shown here performing in 2014, is one of the most recognized artists in country music. He recorded his first album in 1962.

Since winning the 2005 season of American Idol, Carrie Underwood has become one of the most successful artists in any musical genre.

Review

11. What is the slope of the line that passes through the point (4, 7) and is parallel to the line $y = 12(x - 5) + 21$? @ 12

12. The data at right give the reaction times of ten people who were administered different dosages of a drug. Find the best fit line for these data. $y \approx 0.045x - 3.497$; 0.480

13. Find an equation of the line passing through (4, 0) and (6, −3).

14. On a car trip, your speed averages 50 km/h as you drive to your destination. You return by the same route and average 75 km/h. What's your average speed for the entire trip? (h) 60 km/h

15. David wants to send his nephew a new 5-foot fishing pole. David wraps up the pole and takes it to Bob's Courier Service. But Bob's has a policy of not accepting any parcels longer than 4 feet. David returns an hour later with the fishing pole wrapped and sends it with no problem. How did he do it? (Assume the pole is not broken into pieces.)

David got a thin box 4 ft long and 3 ft wide. He fit the fishing rod into the box along the diagonal.

Dosage (mg)	Reaction time (s)
85	0.5
89	0.6
90	0.2
95	1.2
95	1.6
103	0.6
107	1.0
110	1.8
111	1.0
115	1.5

CHAPTER 9 REVIEW

In this chapter you learned about three different methods for collecting data: **experimental studies**, **observational studies**, and **surveys**. You saw some statistical tools for estimating **parameters** of a very large (perhaps infinite) population from **statistics** of samples taken from that population. Many large populations can be described by **probability distributions** of continuous random variables. The area under the graph of a probability distribution is always 1. When using any probability distribution, you do not find the probability of a single exact value; you find the probability of an interval of values.

The probability distributions of many sets of data are **normal**. Their graphs, called **normal curves**, are bell-shaped. To write an equation of a normal distribution curve, you need to know the mean and standard deviation of the data set. To make predictions about a population based on a sample, you can make a nonstandard normal distribution standard by using **z-values**, and you can predict things about the population mean with a **confidence interval**.

You can also make predictions by collecting **bivariate data** and analyzing the relationship between two variables. The **correlation coefficient**, r, tells to what extent the variables have a linear relationship.

Exercises

You will need a graphing calculator for Exercises **5**, **6**, and **8**.

@ **Answers are provided for all exercises in this set.**

1. Classify each method for collecting data. Then describe a different method for collecting the data. *Answers will vary. Sample answers are given.*

   a. A random group of dogs was allowed to choose one of two types of dog food. Once the dogs made the choice, they were fed only that food for a week. Data were recorded on the number of minutes each dog spent actively playing each day for a week.

   b. A number of students were asked if they would return money to a cashier if they were given back the incorrect amount of change.

2. The heights of all adults in Bigtown are normally distributed with a mean of 167 cm and a standard deviation of 8.5 cm.

   a. Sketch a graph of the normal distribution of these heights.

   b. Shade the portion of that graph showing the percentage of people who are shorter than 155 cm.

   c. What percentage of people are shorter than 155 cm? *approximately 7.9%*

CHAPTER REVIEW 9

Reviewing

To review the chapter, assign Exercises 5, 6, and 8 for students to work on in class in groups. Have groups present their solutions. Use the presentations as a way of reviewing the vocabulary and content of the chapter.

As part of the review, you might discuss the idea that the process of inferring facts (estimates or parameters) about a population involves four steps:

1. Select a sample that represents the population. A randomly selected sample from the entire population will be unbiased.

2. Collect measures from the sample. Your method of data collection affects the conclusions that can be drawn about the population.

3. Create estimates about the population or relationships using statistics from the sample.

4. Determine the precision or likelihood of the estimates.

1a. Observational study; for an experimental study, the dogs would be randomly assigned to one of the two types of food.

1b. Survey; for an observational study, the researcher could have a cashier give out the incorrect amount of change and observe the behavior of a randomly selected group of customers.

2a and 2b.

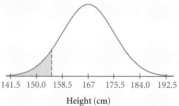

Height (cm)

Exercise 4 If students are having difficulty, remind them that the mean is the midpoint of the confidence interval. They can also apply the 68-95-99.7 rule.

5a.

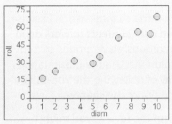

5b. yes; $r \approx 0.965$, indicating a relationship that is close to linear

5c. The value at 9.5 in. is out of line and would improve the r value if it lined up better. This is also true of the value at 5 in., but this value is close to the mean so a change here would have less effect on the correlation.

---

3. A sample of 40 data values is taken from a normally distributed population. The mean of the sample is 117 and the standard deviation is 13. Find the 68% and 95% confidence intervals.
   (114.94, 119.06); (112.89, 121.11)

4. The weights of house cats in northern Michigan can be modeled by a normal distribution, with the middle 68% of cats weighing between 8.4 and 12.7 lb.

   a. What are the population mean and standard deviation for the weight of a house cat in northern Michigan? $\bar{x} = 10.55$ lb; $s = 2.15$ lb

   b. What interval of weights would include the middle 95% of house cats in northern Michigan? 6.25 lb to 14.85 lb

5. Recently, Adam put larger wheels on his skateboard and noticed it would coast farther. He decided to test this relationship. He collected some skateboards, attached wheels of various sizes to them, and rolled them to see how far they would roll on their own. He collected the data shown in the table.

Wheel diameter (in.) x	Rolling distance (in.) y
1	17
2	23
3.5	32
5	30
5.5	36
7	52
8.5	57
9.5	55
10	70

Dirtboards use wheels up to 10 inches in diameter.

   a. Make a scatter plot of these data.

   b. Does there appear to be a linear relationship between the variables? Find the correlation coefficient, and use this value to justify your answer.

   c. Wanting a better value for correlation, Adam decides to re-check his distance for one of the wheel diameters. Which diameter should he check to possibly give him a better correlation? Explain why you picked that value.

6. Suppose the weights of all male baseball players who are 6 ft tall and between the ages of 18 and 24 are normally distributed. The mean is 175 lb, and the standard deviation is 14 lb.

   a. What percentage of these males weigh between 180 and 200 lb? approximately 32.3%

   b. What percentage of these males weigh less than 160 lb? approximately 14.2%

   c. Find an interval that should include the weights of the middle 90% of these males. between 152 and 198 lb

   d. Find the equation of a normal curve that provides a probability distribution for this information. $y = \dfrac{1}{14\sqrt{2\pi}}(\sqrt{e})^{-[(x-175)/14]^2}$

7. The length of a book is measured by many people. The measurements have mean 284 mm and standard deviation 1.3 mm. If four students measure the book, what is the probability that the mean of their measurements will be less than 283 mm? approximately 0.062

8. Consider the monthly rainfall totals for three cities in northern Florida.
   a. Which two of the cities show the strongest correlation?
   b. Which two cities have the weakest correlation?
   c. If Panama City had above average rainfall for a period of time, then which city is more likely to also have above average rainfall in that same period? Explain why you believe that.

	Rain per month (in.), 2006		
Month	Crestview	Chipley	Panama City
Jan	3.01	3.12	3.69
Feb	5.09	5.36	4.31
Mar	0.29	0.50	0.21
Apr	2.42	2.54	1.56
May	3.66	5.18	4.03
Jun	1.51	2.57	4.15
Jul	3.77	2.17	3.19
Aug	4.35	6.72	6.71
Sep	3.42	2.94	2.95
Oct	3.71	4.40	4.98
Nov	4.91	3.43	2.64
Dec	2.42	3.63	4.12

(Northwest Florida Water Management)

Sand Dunes at Panama City Beach, FL

8a. Chipley and Panama City ($r \approx 0.859$)
8b. Panama City and Crestview have the weakest correlation.
8c. A high positive correlation mean that when one variable has a large value then you can expect the other to also be large. So if Panama City is above average then it is most likely that Chipley will also be above average.

Take Another Look

1. If r is multiply by both s_x and s_y you will get the same value as the covariance. The covariance has units. If x and y had units seconds and meters then the units of the covariance are meter-seconds, though it is tough to interpret this value. Correlation has no units. Covariance can be positive or negative, same as correlation and is close to zero when the data is not linear like correlation. But this value can be very large where the correlation can be no greater than one. When there is a large spread in x or y then covariance may also be large. It will never be larger than the product of the two standard deviations. Thus to interpret the value you should always compare it to this product.

2. Summaries should emphasize that the width and height of the normal curve vary inversely, so the area under the curve always remains 1. That is, as σ increases, the curve widens and gets shorter; as σ decreases, the curve narrows and gets taller.

Suggest that students using graphing calculators graph in the same window several equations of the form $y = \frac{1}{\sigma\sqrt{2\pi}}(\sqrt{e})^{-(x/\sigma)^2}$ and change the value of σ slightly in each (such as $\sigma = \{0.5, 1.0, 1.5, 2.0\}$). Suggest that students using geometry software create a slider whose length represents the value of σ and then define and graph the function $f(x) = \frac{1}{\sigma\sqrt{2\pi}}(\sqrt{e})^{-(x/\sigma)^2}$.

Changing the length of the segment will dynamically change the curve. Students using geometry software should use a rectangular coordinate grid in which the scale of the x-axis can be adjusted independent of the y-axis.

3. Students may find that the equation $y = \frac{1}{\sigma\sqrt{2\pi}}\left(1 - \frac{1}{2\sigma^2}\right)^{x^2}$ is a particularly good approximation when σ is greater than 2. For the interval $\sqrt{0.5} < \sigma < 2$, the curve is narrower than the actual normal curve. For σ less than or equal to $\sqrt{0.5}$ (or $\frac{\sqrt{2}}{2}$, or approximately 0.7), the function is undefined.

Suggest that students using graphing calculators graph the equations $y = \frac{1}{\sigma\sqrt{2\pi}}(\sqrt{e})^{-(x/\sigma)^2}$ and $y = \frac{1}{\sigma\sqrt{2\pi}}\left(1 - \frac{1}{2\sigma^2}\right)^{x^2}$ in the same window and change the value of σ in both equations to see how they compare (such as $\sigma = \{0.4, 0.5, 0.6, 0.7, 0.8, 0.9, 1.0, 1.1, \ldots\}$). Suggest that students using geometry software create a slider whose length represents the value of σ and then define and graph the functions $f(x) = \frac{1}{\sigma\sqrt{2\pi}}(\sqrt{e})^{-(x/\sigma)^2}$ and $g(x) = \frac{1}{\sigma\sqrt{2\pi}}\left(1 - \frac{1}{2\sigma^2}\right)^{x^2}$.

Changing the length of the segment will dynamically change the curves. Students using geometry software should use a rectangular coordinate grid in which the scale of the *x*-axis can be adjusted independent of the *y*-axis.

Assessing the Chapter

Test items can include problems students wrote, along with items from any of the assessment resources. The exams in Assessment Resources cover Chapters 10–13. Consider grading students' portfolios as part of the unit grade.

Facilitating Self-Assessment

As you encourage students to organize their notebooks to include the statistical measures, emphasize the importance of including descriptions of how the statistics are used.

See your digital resources for the **Project: Making it Fit**, where students find any bivariate data that they think might be related and use techniques from this chapter to find a function that fits the data well.

Take Another Look

1. Another measure of correlation is called the covariance. Recall the variance (the square of the standard deviation) is found by the formula $s^2 = \frac{1}{n-1}\sum_{i=1}^{n}(x_i - \bar{x})^2$. The formula for covariance is $s_{xy} = \frac{1}{n-1}\sum_{i=1}^{n}(x_i - \bar{x})(y_i - \bar{y})$. Explain how this value is both different and similar to the correlation coefficient you used in this chapter. Take any set of data and find both the covariance and the correlation coefficient. If the variance is a measure of spread then what spread is the covariance measuring?

2. Consider a normal distribution with mean 0. Use your graphing calculator or geometry or statistics software to explore how the standard deviation, σ, affects the graph of the normal curve,
$$y = \frac{1}{\sigma\sqrt{2\pi}}(\sqrt{e})^{-(x/\sigma)^2}$$
Summarize how the normal curve changes as the value of σ changes.

3. Consider a normal distribution with mean 0. You've already seen that the equation of the normal curve is
$$y = \frac{1}{\sigma\sqrt{2\pi}}(\sqrt{e})^{-(x/\sigma)^2}$$
To avoid using *e*, you can use another equation that approximates the normal curve:
$$y = \frac{1}{\sigma\sqrt{2\pi}}\left(1 - \frac{1}{2\sigma^2}\right)^{x^2}$$
Use your graphing calculator or geometry or statistics software to explore how the graphs of these two equations compare for different values of σ. For which values of σ is the second equation a good approximation, a poor approximation, or even undefined?

Assessing What You've Learned

WRITING TEST QUESTIONS Write a few test questions that reflect the topics of this chapter. You may want to include questions on experimental design, probability distributions, confidence intervals, and bivariate data and correlation. Include detailed solutions.

ORGANIZE YOUR NOTEBOOK Make sure that your notebook has complete notes on all of the statistical tools and formulas that you have learned. Specify which statistical measures apply to populations and which apply to samples, and explain which tools allow you to make conclusions about a population based on a sample and vice versa. Be sure you know when to use the various statistical measures and exactly what each one allows you to predict.

PERFORMANCE ASSESSMENT As a friend, family member, or teacher watches, solve a problem from this chapter that deals with fitting a line to data and analyzing how well the function fits. Explain the various tools for analyzing how well a function fits data.

Selected Hints and Answers

This section contains hints and answers for exercises marked with ⓗ or ⓐ in each set of Exercises.

LESSON 0.1

1a. Begin with a 10-liter bucket and a 7-liter bucket. Find a way to get exactly 4 liters in the 10-liter bucket.

3a. 1

4. *Hint:* Your strategy could include using objects to act out the problem and/or using pictures to show a sequence of steps leading to a solution.

5a.

7a. 312 g **7b.** 12.5 oz

8a. $a = 12$

10a. $x^2 + 11x + 28$

	x	4
x	x^2	$4x$
7	$7x$	28

10b. *Hint:* First rewrite the expression as $(x + 5)(x + 5)$.

11. *Hint:* Try using a sequence of pictures similar to those on page 2.

15c. $\frac{2}{9}$

LESSON 0.2

3c. $h \geq 54$

5. *Hint:* Find the expression for the area or perimeter of each rectangle first.

6b. If x represents the first three digits of your phone number, the expression for Step 2 is $80x$.

7b. *Hint:* L represents the price of each large bead. What units are the prices in?

10a. Solve Equation 1 for a to get $a = 5b - 42$. Substitute $5b - 42$ for a in Equation 2 to get $b + 5 = 7[(5b - 42) - 5]$.

12a. *Hint:* $45° - 30° = 15°$. How can you show this with the triangle tools?

14. *Hint:* Try using a sequence of pictures similar to those on page 2. Also be sure to convert all measurements to cups.

LESSON 0.3

1a. approximately 4.3 s

2a. Let e represent the average problem-solving rate in problems per hour for Emily, and let a represent the average rate in problems per hour for Alejandro.

2b. *Hint:* The variable e represents the number of problems that Emily solves in 1 hour.

4. *Hint:* Expand the right side of the equation and simplify by undoing.

6. *Hint:* How much does Alyse earn when she works overtime? Let r represent the hours worked at the regular rate and t represent the hours worked at the time-and-a-half rate. Write two equations. One equation is $r + t = 35$.

7a. $\dfrac{12960 \text{ cm}^3}{(18 \text{ cm} \cdot 30 \text{ cm})} = 24 \text{ cm}$

8. *Hint:* The well water is in the bottle.

10a. *Hint:* The area painted is calculated by multiplying *painting rate · time*. What fraction expresses Paul's painting rate in $\dfrac{\text{ft}^2}{\text{min}}$?

11a. r^{12}

CHAPTER 0 REVIEW

1. Sample answer: Fill the 5-liter bucket; pour 3 liters into the 3-liter bucket, leaving 2 liters in the 5-liter bucket; empty the 3-liter bucket; pour the remaining 2 liters from the 5-liter bucket into the 3-liter bucket. Fill the 5-liter bucket and pour 1 liter into the 3-liter bucket, completely filling the 3-liter bucket and leaving 4 liters in the 5-liter bucket.

2a. $x^2 + 7x + 12$

	x	4
x	x^2	$4x$
3	$3x$	12

2b. $2x^2 + 6x$

	x	3
$2x$	$2x^2$	$6x$

2c. $x^2 + 4x - 12$

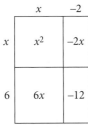

2d. $2x^2 - 9x + 4$

3a. $x = \sqrt{18}$ cm $= 3\sqrt{2}$ cm ≈ 4.2 cm

3b. $y = \sqrt{25}$ in. $= 5$ in.

4a. 168 mi **4b.** 9.5 gal

4c. Car B can go 102 mi farther than Car A.

4d. Car A: 24 mi/gal; Car B: 36 mi/gal

4e. For Car A, the slope is $\frac{120}{5} = 24$, which means the car can drive 24 mi/gal of gasoline. For Car B, the slope is $\frac{180}{5} = 36$, which means the car can drive 36 mi/gal.

5. Let n represent the unknown number.

5a. $2n + 6$ **5b.** $5(n - 3)$

5c. $2n + 6 = 5(n - 3)$; $n = 7$

6. Let m represent the number of miles driven, and let c represent the total cost in dollars of renting the truck.

6a. $c = 19.95 + 0.35m$

6b. possible answer: $61.25 **6c.** $8.40

7a. Let w represent the mass in grams of a white block, and let r represent the mass in grams of a red block.

7b. $4w + r = 2w + 2r + 40$; $5w + 2r = w + 5r$

7c. $w = 60$; $r = 80$

7d. The mass of a white block is 60 g, and the mass of a red block is 80 g.

8. 17 years old

9. 12 pennies, 8 nickels, 15 dimes, 12 quarters

10a. The number of bags of cranberries is three times the number of bags of almonds.

Four times the number of bags of almonds plus the number of cranberries totals 70 (the total weight of all almonds and cranberries is 70 pounds).

10b. Possible values: (4, 12) or (8, 24).

10c. Possible values: (12, 22) or (7, 42).

10d. $x = 10$, $y = 30$.

11a. $h = 0$. Before the ball is hit, it is on the ground.

11b. $h = 32$. Two seconds after being hit, the ball is 32 ft above the ground.

11c. $h = 0$. After 3 s, the ball lands on the ground.

12a. $4x^{-1}$, or $\frac{4}{x}$ **12b.** $\frac{1}{2}x^{-1}$, or $\frac{1}{2x}$ **12c.** x^{15}

13a. $y = 2^0 - 1$ **13b.** $y = 2^3 = 8$

13c. $y = 2^{-2} = \frac{1}{2^2} = \frac{1}{4}$ **13d.** $x = 5$

14a. 5 gal $\cdot \frac{4 \text{ qt}}{1 \text{ gal}} \cdot \frac{4 \text{ c}}{1 \text{ qt}} \cdot \frac{8 \text{ oz}}{1 \text{ c}} = 640$ oz

14b. 1 mi $\cdot \frac{5280 \text{ ft}}{1 \text{ mi}} \cdot \frac{12 \text{ in.}}{1 \text{ ft}} \cdot \frac{2.54 \text{ cm}}{1 \text{ in.}} \cdot \frac{1 \text{ m}}{100 \text{ cm}}$
$= 1609.344$ m

15.

Sales Rep	Mr. Mendoza	Mr. Bell	Mrs. Plum
Client	Mr. Green	Ms. Phoung	Ms. Hunt
Location	Conference room	Convention hall	Lunch room
Time	9:00 A.M.	12:00 noon	3:00 P.M.

16. Art: Disagree. repeating decimals are not necessarily irrational. $0.\overline{32} = \frac{32}{99}$, so it is a rational number.

Brandi: Agree. See above.

ReGi: Agree with statement but not with reason.

Koy: Agree.

Gina: Disagree. Repeating forever does not necessarily mean it is irrational.

LESSON 1.1

2a. iv. $-18, -13.7, -9.4, -5.1$; arithmetic; $d = 4.3$

2b. ii. 47, 44, 41, 38; arithmetic; $d = -3$

2c. i. 20, 26, 32, 38; arithmetic; $d = 6$

2d. iii. 32, 48, 72, 108; geometric; $r = 1.5$

3.

n	1	2	3	4	5	…	9
u_n	40	36.55	33.1	29.65	26.2	…	12.4

$u_1 = 40$ and $u_n = u_{n-1} - 3.45$ where $n \geq 2$

5d. $u_1 = -6.24$ and $u_n = u_{n-1} + 2.21$ where $n \geq 2$; $u_{20} = 35.75$

6. $u_1 = 4$ and $u_n = u_{n-1} + 5$ where $n \geq 2$; $u_{46} = 229$

8a. 13 min

11a. $60

13a. *Hint:* How many times do you add the common difference to 35 to get to 51?

14a.

Elapsed time (s)	Distance from motion sensor (m)
0.0	2.0
1.0	3.0
2.0	4.0
3.0	5.0
4.0	4.5
5.0	4.0
6.0	3.5
7.0	3.0

LESSON 1.2

1a. 1.5; growth; 50% increase

2a. $u_0 = 100$ and $u_n = 1.5u_{n-1}$ where $n \geq 1$; $u_{10} \approx 5766.5$

3a. 2000, 2100, 2205, 2315.25

4a. ii. The graph and rule both indicate decay.

5a. $x(1 + A)$

5b. $(1 - 0.18)A$, or $0.82A$

6a. *Hint:* rebound ratio $= \dfrac{\text{rebound height}}{\text{previous rebound height}}$

8. $250,000 was invested at 2.5% annual interest in 2015; $U_{2019} = \$27{,}595.32$

10a. $u_0 = 2000$
$u_n = (1 + 0.085)u_{n-1}$ where $n \geq 1$

11a. $595.51

11c. *Hint:* You will receive $\dfrac{6\%}{4}$ of the balance 12 times.

15a. 32,100 and 58.47%

18b. 10 s

20a. $x \approx 43.34$

21b. $y = 55$

LESSON 1.3

1a. 31.2, 45.64, 59.358; shifted geometric, increasing

1d. 40, 40, 40; arithmetic or shifted geometric, neither increasing nor decreasing

2b. $b = 1200$

2c. $d = 0$

3a. *Hint:* Solve the equation $a = 0.95a + 16$.

5a. The first day, 300 g of chlorine were added. Each day, 15% disappears, and 30 g are added.

6b. $a_0 = 24{,}000$ and $a_n = a_{n-1}\left(1 + \dfrac{0.034}{12}\right)a - 100$, or $a_n = a_{n-1}(1.00283) - 100$

6c. $23,871.45. This is the balance on 2/2/09.

9c. $c = 0.88c + 600$

10. $u_0 = 20$ and $u_n = (1 - 0.25)u_{n-1}$ where $n \geq 1$; 11 days ($u_{11} \approx 0.84$ mg)

12a. *Hint:* You might use the Pythagorean Theorem or use the properties of special right triangles. Look for patterns.

15. 23 times

LESSON 1.4

1b. 0 to 19 for n and 0 to 400 for u_n

1d. 0 to 69 for n and 0 to 3037 for u_n

2b. iv. arithmetic

3a. geometric, nonlinear, decreasing

4b. i. sample answer: $u_0 = 200$

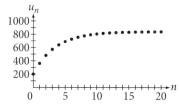

6a. $u_0 = 18$ and $u_n = -0.75u_{n-1}$ where $n \geq 1$

6b.

Because every term alternates signs, the graph is different from other graphs you have seen; each point alternates above or below the n-axis. The points above the n-axis, however, create a familiar geometric pattern, as do the points below the n-axis. If the points below the n-axis were reflected across the n-axis, you would have the graph of $u_0 = 18$ and $u_n = 10.75u_{n-1}$ where $n \geq 1$.

6c. 0

11. *Hint:* If the rate of chirping increases by 20 chirps/min per 5° increase in temperature, how much does the rate of chirping increase per 1° increase in temperature?

13b. $66\dfrac{2}{3}$

LESSON 1.5

1a.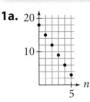

1b. −3; The common difference is the same as the slope.

1c. 18; The y-intercept is the u_0-term of the sequence.

1d. $y = 18 − 3x$

3. $y = 7 + 3x$

5a. 1.7

5d. *Hint:* Plot some points on the line: (0, 12), (1, 12), (2, 12). What is the change in the y-value for each unit change in the x-value?

7a. 190 mi

8a. *Hint:* The point (3 cars, −$2050), corresponds to the term $t_3 = −2050$.

10c. 34 days

11a. *Hint:* The x-values must have a difference of 5.

11d. *Hint:* The linear equation will have a slope of 4. Find the y-intercept.

12c. 304 ft

14a. $x = (1 − 0.4)x + 300; x = 750$

LESSON 1.6

1a. $\frac{3}{2} = 1.5$

2c. 15

3a. $y = 14.3$

4c. The equations have the same constant, −2. The lines share the same y-intercept. The lines are perpendicular, and their slopes are reciprocals with opposite signs.

5. *Hint:* Graph the original line and then graph several parallel lines and a perpendicular line. Also, refer to your answers to 4c and d.

7a. Possible answer: Using the 2nd and 7th data points gives a slope of approximately 1.47 volts/battery.

8a. *Hint:* Think about the units of y and x and the units of the slope.

13a. $−10 + 3x$

LESSON 1.7

1a. possible answer: $y = 1 + \frac{2}{3}(x − 4)$

2a. $y = −7 + \frac{2}{3}(x − 5)$

3a. $u_n = 31$

5d. *Hint:* What can you say about the slope and the x- and y-intercepts of a vertical line?

7a. i. Possible answer: The y-intercept is about 1.7; (5, 4.6); $\hat{y} = 1.7 + 0.58x$

9b. possible answer: $\hat{y} = 0.26x + 0.71$

11. 102

LESSON 1.8

1a. −0.2

2. 82.3, 82.9, 74.5, 56.9

5a. 9 yrs, 132.44 cm.

7. *Hint:* Using the data point and its residual, find the coordinates of a point on the line.

12a. $y = 4.7 + 0.6(x − 2)$

12b. *Hint:* You might need to review the meaning of *direct variation*.

14b. i. deposited: $360; interest: $11.78

LESSON 1.9

1. mean $x = 46.44$, mean $y = 42.33$, $s_x = 29.86$, $s_y = 27.07$

2. $r = 0.8998, y = 80.23 − 0.8159x$

9b. No, it appears that there are two patterns here.

10b. *Hint:* Think about the units for x and y.

10c. 240.10 s, or 4:00.10. This prediction is 0.7 s slower than Roger Bannister actually ran.

12. *Hint:* What percentage of the students may have scored lower than Ramon? Consider the students that definitely scored lower than 35 and the students that *may* have scored lower than 35 according to the histogram.

13. *Hint:* The difference between the 2nd and 6th values is 12.

CHAPTER 1 REVIEW

1a. geometric

1b. $a_1 = 256$ and $a_n = 0.75a_{n−1}$ where $n \geq 2$

1c. $a_8 \approx 34.2$ **1d.** $a_{10} \approx 19.2$; 10th

1e. $a_{17} \approx 2.57$

2a. arithmetic

2b. $u_1 = 3$ and $u_n = u_{n−1} + 4$ where $n \geq 2$

2c. $u_{128} = 511$ **2d.** $u_{40} = 159$; 40th

2e. $u_{20} = 79$

3a. −3, −1.5, 0, 1.5, 3; 0 to 6 for n and −4 to 4 for a_n

3b. 27, 9, 3, 1, $\frac{1}{3}$; 0 to 6 for n and 0 to 30 for a_n

4a. iv. The recursive formula is that of an increasing arithmetic sequence, so the graph must be increasing and linear.

4b. iii. The recursive formula is that of a growing geometric sequence, so the graph must be increasing and curved.

4c. i. The recursive formula is that of a decaying geometric sequence, so the graph must be decreasing and curved.

4d. ii. The recursive formula is that of a decreasing arithmetic sequence, so the graph must be decreasing and linear.

5. 88 gal

6. $-\dfrac{975}{19}$

7a. 23.45

7b. possible answer: $y = 23.45x$

7c. possible answer: $y = -\dfrac{20}{469}x$, or $y \approx -0.0426x$

8a. approximately (19.9, 740.0)

8b. approximately (177.0, 740.0)

9a. $y = 3 - (x + 2)$, or $y = 1 - x$

9b. $y = -8 + 2(x - 0)$, or $y = -8 + 2x$

9c. $y = 13.2 - 0.46(x - 1999)$, or $y = 8.6 - 0.46(x - 2009)$

9d. $y = 7 + 0(x - 2)$, or $y = 7$

10a. $u_1 = 6$ and $u_n = u_{n-1} + 7$ where $n \geq 1$

10b. $y = -1 + 7x$

10c. The slope is 7. The slope of the line is the same as the common difference of the sequence.

10d. 223. It's probably easier to use the equation from 10b.

11a. 4.5 **11b.** $y = 7.5 + 4.5x$

11c. The slope of the line is equal to the common difference of the sequence.

12a. Poor fit; there are too many points above the line.

12b. Reasonably good fit; the points are well distributed above and below the line and are not clumped. The line follows the downward trend of the data.

12c. Poor fit; there are an equal number of points above and below the line, but they are clumped to the left and to the right, respectively. The line does not follow the trend of the data.

13a.

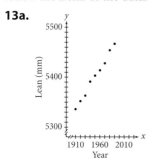

13b. Answers will vary, $y = 5401 + 1.645(x - 1950)$

13c. Each year the tower tips another 1.645 millimeters.

13d. 5474.4 mm

13e. 0.8, 0.35, −5.1, 6.45, 2, −3.45, −5.9, 3.65, 0.2

13f. This line is a good fit because there is no pattern to the residuals.

14. $u_{23} = 16.5$

15a. fairly linear (a little logistic)

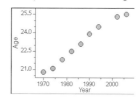

15b. Answers will vary, $y = -274.9 + 0.15x$

15c. 26.7 years

15d. 0.02, −0.25, −0.11, 0.02, 0.06, 0.300.033, 0, −0.36

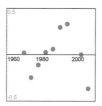

15e. There is a slight pattern to the residuals so the best model may not be linear.

LESSON 2.1

1a. (1.8, −11.6)

4a. $x = \dfrac{32}{19} \approx 1.684$

5a. (4.125, −10.625)

7a. No. At $x = 25$, the cost line is above the income line.

9d. *Hint:* Two step functions may not have any intersection or they may intersect over a short interval rather than a single point.

11a. Let l represent length in centimeters, and let w represent width in centimeters; $2l + 2w = 44$, $l = 2 + 2w$; $w = \dfrac{20}{3}$ cm, $l = \dfrac{46}{3}$ cm.

11b. *Hint:* An isosceles triangle has two or more sides that are equal in length.

14a. $y = \dfrac{3x - 12}{8} = 0.375x - 1.5$

LESSON 2.2

1b. $h = \dfrac{18 - 2p}{3} = 6 - \dfrac{2}{3}p$

2a. (2, 21)

3a. (3.1, −1.8)

5a. *Hint:* Look for a way to use substitution.

6. *Hint:* The second equation in this system is $F = 3C$.

7a. cost for first camera: $y = 47 + 11.5x$; cost for second camera: $y = 59 + 4.95x$

15a. $y = \dfrac{7 - 3x}{2} = \dfrac{7}{2} - \dfrac{3}{2}x = 3.5 - 1.5x$

LESSON 2.3

1b. $y = \frac{5}{x}$

2c. 20

3a. yes

4b. $(-2, 5), (-5, 2)$

5a. $\left(4, \frac{1}{2}\right), \left(\frac{3}{2}, \frac{4}{3}\right)$

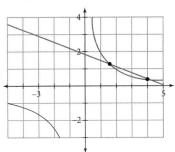

6a. $(-2, -4), (12, 24)$

8a. 0, 1, or 2 solutions

10b. If $a = 8$, there is only one solution at $x = 2, y = 4$.

11a. $P = 100000 + 2500x$

LESSON 2.4

1a. $y < \frac{10 - 2x}{-5}$, or $y < -2 + 0.4x$

2a.

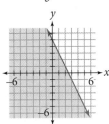

3a. $y < 2 - 0.5x$ **3c.** $y > 1 - 0.75x$

5.

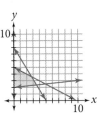

vertices: $(0, 2), (0, 5), (2.752, 3.596), (3.529, 2.353)$

7.

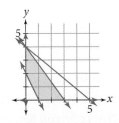

vertices: $(0, 4), (3, 0), (1, 0), (0, 2)$

9. *Hint:* If the length is x and the height is y, then what are the formulas for area and perimeter?

9a. Let x represent length in inches, and let y represent width in inches.
$$\begin{cases} x + y \leq 40 \\ x + y \leq 33 \\ x \geq 10 \\ y > 10 \end{cases}$$

LESSON 2.5

1.

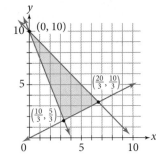

2a. $\left(\frac{20}{3}, \frac{10}{3}\right)$

5a. possible answer:
$$\begin{cases} y \geq 7 \\ y \leq \frac{7}{5}(x - 3) + 6 \\ y \leq -\frac{7}{12}x + 13 \end{cases}$$

6. *Hint:* One constraint (concerning food) is $2p + 6s \leq 100$.

7. 5 radio minutes and 10 newspaper ads to reach a maximum of 155,000 people. This assumes that people who listen to the radio are independent of people who read the newspaper, which is probably not realistic, and that the shop buys both. This also assumes that each newspaper ad reaches a fresh set of readers (instead of the same subscribers day after day who each see the ad 10 times), and that each radio minute reaches a fresh set of listeners (instead of the same group of people who hear the radio spot 5 separate times). 20 newspaper ads and 0 radio minutes would reach 160,000 people.

8. *Hint:* One constraint (concerning sulfur) is $0.02L + 0.06H \leq 0.04(L + H)$.

10a. Let x represent the length in inches, and let y represent the girth in inches.
$$\begin{cases} x + y \leq 130 \\ x \leq 108 \\ x > 0 \\ y > 0 \end{cases}$$

13. *Hint:* One inequality is $y \leq 5$.

LESSON 2.6

1a. $(4, -3, -1)$

2a. $(-3, 2, 4)$

3a. no, the point satisfies the first and second equation, but not the third

4a. (−2, 1, 2) Use elimination because 15*x* and −15*x* easily add to 0. Solving by substitution will be more difficult because solving for any of the variables will involve fractions.

CHAPTER 2 REVIEW

1a. (10, 28); consistent, independent

1b. every point; same line; consistent, dependent

1c. $\left(\frac{2}{3}, 1\right)$, or $(0.\overline{6}, 1)$; consistent, independent

1d. (1, 0); consistent, independent

1e. $\left(\frac{42}{5}, \frac{53}{10}\right)$, or (8.4, 5.3); consistent, independent

1f. No intersection; the lines are parallel; inconsistent.

2a. (4.7, 7.1)   **2b.** $\left(\frac{5}{2}, \frac{11}{6}\right)$

2c. (5.8, 1.4)

3a. $x = 2.5, y = 7$

3b. $x = 1.22, y = 6.9, z = 3.4$

4. Inequalities should be equivalent to $y \leq 6 - \frac{2}{5}x$, $y \leq 2(7 - x)$, $y \geq 0$, and $x \geq 0$.

5a.

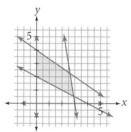

Vertices: (0, 4), (2.625, 2.25), (2.$\overline{90}$, 0.$\overline{54}$), (0, 2). The maximum occurs at (0, 4): 1.65(0) + 5.2(4) = 20.8.

5b.

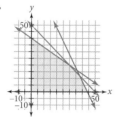

Vertices: (0, 0), (0, 40), (30, 20), (38, 12), (44, 0). The maximum occurs at (30, 20): 6(30) + 7(20) = 320.

6. about 4.4 yr

7. $y = 4x^2 - 18x + 5$

8a. $-1 \leq x \leq 2$

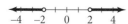

8b. $x < -5$ or $x > 1$

8c. $x < -2$ or $x > 2$

9. They should make 4 shawls and 2 blankets to make the maximum profit of $104.

10a. consistent and independent

10b. consistent and dependent

10c. inconsistent   **10d.** inconsistent

11a. (5, 4, −6)   **11b.** (−3, 2, 6)

12. 53 feet by 72 feet

13. An associate justice earns $213,900 and the chief justice earns $223,500.

LESSON 3.1

1a. i   **1d.** ii

2b.

5b. The car's speed in miles per hour is the independent variable; the braking distance in feet is the dependent variable.

5c. Time in minutes is the independent variable; the drink's temperature in degrees Fahrenheit is the dependent variable.

5d. *Hint:* Remember, you are graphing the speed of the acorn, not its height.

6. *Hint:* Consider the rate and direction of change (increasing, decreasing, constant) of the various segments of the graph.

7c. Foot length in inches is the independent variable; shoe size is the dependent variable. The graph will be a series of discontinuous horizontal segments, because shoe sizes are discrete.

9a. *Hint:* This is not a (*time, distance*) graph.

11a. Let x represent the number of pictures, and let y represent the amount of money (either cost or income) in dollars; $y = 155 + 15x$.

11b. $y = 27x$

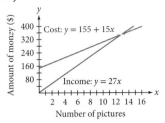

11c. 13 pictures

11d. The income, $216, is less than the cost, $275.

12a. $142,784.22

13a. $3x + 5y = -9$ **13b.** $6x - 3y = 21$

LESSON 3.2

1c. Function; each x-value has only one y-value.

2d. 11   **2e.** $\dfrac{11}{3}$

2h. -21 **2i.** -2

4d. $5 = E$

4g. *Hint:* $\sqrt[3]{\ }$ means the cube root. The cube root of x is the number that you cube to get x. For example, $\sqrt[3]{8} = 2$ because $2^3 = 8$.

4j. $18 = R$

9. domain: $-6 \leq x \leq 5$; range: $-2 \leq y \leq 4$

11d.

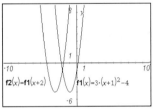

The graphs are the same shape. The graph of $f(x + 2)$ is shifted 2 units to the left of the graph of $f(x)$.

14b. *Hint:* You might use guess-and-check or a graphing calculator.

16. Sample answer: Eight students fall into each quartile. Assuming that the mean of each quartile is the midpoint of the quartile, the total will be $8(3.075 + 4.500 + 5.875 + 9.150)$, or $180.80.

19a.

$x^2 + 10x + 21$

LESSON 3.3

2. translated right 3 units

3a. $-2(x + 3)$, or $-2x - 6$

3b. $-3 + (-2)(x - 2)$, or $-2x + 1$

4a. $y = -4.4 - 1.1\overline{48}(x - 1.4)$ or $y = 3.18 - 1.1\overline{48}(x + 5.2)$

5a. $y = -3 + 4.7x$

6a. $y = -2 + f(x)$

8a. *Hint:* Which axis pictures differences in time and which pictures differences in distance?

10b. $y = \dfrac{c}{b} - \dfrac{a}{b}x$; y-intercept: $\dfrac{c}{b}$; slope: $-\dfrac{a}{b}$

10d. ii. $4x + 3y = -8$

10d. vi. $4x + 3y = 10$

12b. $y = \dfrac{1}{5}x + 65$

12c. *Hint:* Substitute 84 for y in the function from 12b and solve for x.

LESSON 3.4

2a. $y = x^2 - 5$

2c. $y = (x - 3)^2$

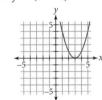

3d. translated horizontally -4 units

5. *Hint:* You can verify your answers with a table or a graph.

5b. $x = 4$ or $x = -4$

7c. $(6, -2), (4, -2), (7, 1), (3, 1)$. If (x, y) are the coordinates of any point on the black parabola, then the coordinates of the corresponding point on the red parabola are $(x + 5, y - 3)$.

7d. Segment b has length 1 unit, and segment c has length 4 units.

9. *Hint:* Which piece of the graph is defined for $x = 2$?

10. *Hint:* You might model this situation with points on a circle to represent the teams and segments that show which teams have played each other.

11a. $x = 9$ or $x = 1$

15.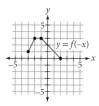

LESSON 3.5

1c. $y = \sqrt{x+5} + 2$ **1d.** $y = \sqrt{x-3} + 1$

2a. translated horizontally 3 units

2c. translated vertically 2 units

4a.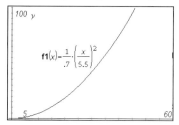

5a. $y = -\sqrt{x}$ **5b.** $y = -\sqrt{x} - 3$

5c. $y = -\sqrt{x+6} + 5$ **5d.** $y = \sqrt{-x}$

5e. $y = \sqrt{-(x-2)} - 3$, or $y = \sqrt{-x+2} - 3$

10b. i. $y = \pm\sqrt{x+4}$ **10b. ii.** $y = \pm\sqrt{x} + 2$

11a. *Hint:* Use Chicago time on the horizontal axis and distance on the vertical axis. What is Arthur's distance at the beginning of the trip?

14e.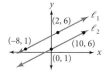

It is a parabola, but the negative half is not used because the distance cannot be negative.

17a. $x = 293$ **17d.** $x = -13$

19b. $y = \frac{1}{2}(x-8) + 5$

20a. 35, 37.5, 41.5, 49, 73

20b.

20c. 11.5 **20d.** 70 and 73

LESSON 3.6

1e. $y = |x| - 1$ **1f.** $y = |x-4| + 1$

1m. $y = -(x+3)^2 + 5$ **1n.** $y = \pm\sqrt{x-4} + 4$

1p. $\frac{y}{-2} = \left|\frac{x-3}{3}\right|$, or $y = -2\left|\frac{x-3}{3}\right|$

4. *Hint:* Remember to choose some negative values and some values between -1 and 1.

6. $\hat{y} \approx |x - 18.4|$. The transmitter is located on the road approximately 18.4 mi from where you started.

7a. $(6, -2)$ **7b.** $(2, -3)$ and $(8, -3)$

8. *Hint:* Refer to Example B of this lesson for help.

11c.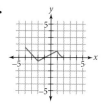

13a. $\bar{x} = 83.75$, $s \approx 7.45$

LESSON 3.7

1.

Reflection	Across x-axis	N/A
Dilation	Horizontal	4
Dilation	Vertical	0.4
Translation	Horizontal	2
Reflection	Across y-axis	N/A

3c.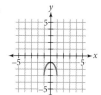

4d. $\frac{y-1}{2} = \sqrt{1 - (x-3)^2}$, or
$y = 2\sqrt{1 - (x-3)^2} + 1$

4e. $\frac{y-3}{-5} = \sqrt{1 - \left(\frac{x+2}{2}\right)^2}$, or
$y = -5\sqrt{1 - \left(\frac{x+2}{2}\right)^2} + 3$

5c. $y = \pm 2\sqrt{1-x^2} + 2$, or $x^2 + \left(\frac{y}{2}\right)^2 = 1$

6b. $\frac{x}{3}$ **6c.** $g(x) = f\left(\frac{x}{3}\right)$

8a. $y = 3\sqrt{1 - \left(\frac{x}{0.5}\right)^2}$ and $y = -3\sqrt{1 - \left(\frac{x}{0.5}\right)^2}$

8b. $y = \pm 3\sqrt{1 - \left(\frac{x}{0.5}\right)^2}$

8c. $y^2 = 9\left[1-\left(\dfrac{x}{0.5}\right)^2\right]$, or $\dfrac{x^2}{0.25}+\dfrac{y^2}{9}=1$

9a.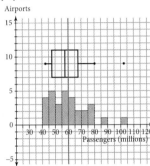

(0, 0) and (1, 1)

9b. *Hint:* Compare the coordinates of the points to the radius of the quarter-circle.

13c. $c = 0.2$ or 3.8

15a, c, d.

15b. Using the midpoint value for each histogram bin, there were 1805 million, or 1,805,000,000 passengers.

15c. mean = 60.17 million

15d. Five-number summary: 42.5, 47.5, 57.5, 67.5, 102.5; assume that all data occur at midpoints of bins.

CHAPTER 3 REVIEW

1. Sample answer: For a time there are no pops. Then the popping rate slowly increases. When the popping reaches a furious intensity, it seems to level out. Then the number of pops per second drops quickly until the last pop is heard.

2a. odd **2b.** neither **2c.** even

3a.

3b.

3c.

3d.

4a. Translate horizontally −2 units and vertically −3 units.

4b. Dilate horizontally by a factor of 2, and then reflect across the x-axis.

4c. Dilate horizontally by a factor of $\dfrac{1}{2}$, dilate vertically by a factor of 2, translate horizontally 1 unit and vertically units.

5a.

5b.

5c.

5d.

5e.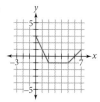

5f.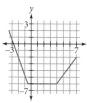

6a. $y = \sqrt{1-x^2}$; $y = 3\sqrt{1-x^2} - 1$

6b. $y = \sqrt{1-x^2}$; $y = 2\sqrt{1-\left(\frac{x}{5}\right)^2} + 3$

6c. $y = \sqrt{1-x^2}$; $y = 4\sqrt{1-\left(\frac{x-3}{4}\right)^2} - 1$

6d. $y = x^2$; $y = (x-2)^2 - 4$

6e. $y = x^2$; $y = -2(x+1)^2$

6f. $y = \sqrt{x}$; $y = -\sqrt{-(x-2)} - 3$

6g. $y = |x|$; $y = 0.5|x+2| - 2$

6h. $y = |x|$; $y = -2|x-3| + 2$

7a. $y = \frac{2}{3}x - 2$ **7b.** $y = \pm\sqrt{x+3} - 1$

7c. $y = \pm\sqrt{-(x-2)^2} + 1$

8a. $x = 8.25$ **8b.** $x = \pm\sqrt{45} \approx \pm 6.7$

8c. $x = 11$ or $x = -5$ **8d.** no solution

9a.

17,000	16,000	15,000	14,000	13,000	12,000	11,000	10,000
18,700	19,200	19,500	19,600	19,500	19,200	18,700	18,000

9b.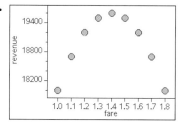

9c. (1.40, 19,600). By charging $1.40 per ride, the company achieves the maximum revenue, $19,600.

9d. $\hat{y} = -10,000(x - 1.4)^2 + 19,600$

9d. i. $16,000   **9d. ii.** $0 or $2.80

LESSON 4.1

1a. $f(5) \approx 3.52$ **1c.** $h(24) \approx 22.92$

3a. 16, 12, 9; $y = 16(0.75)^x$

4a. $f(0) = 125, f(1) = 75, f(2) = 45$; $u_0 = 125$ and $u_n = 0.6u_{n-1}$ where $n \geq 1$

6c. 0.94; 6% decrease

7a. $u_0 = 1.211$, $u_n = u_{n-1} \cdot 1.011$

7b.

Year	Estimated population (billions)
1995	1.211
1996	1.224
1997	1.238
1998	1.251
1999	1.265
2000	1.279
2001	1.293
2002	1.307

7c. y represents the estimated population x years after 1995; $y = 1.211(1.011)^x$.

7d. The equation predicts that the population of China in 2006 was 1.37 billion. This is larger than the actual value. This means that the population is growing at a slower rate in 2006 than it was in 1995.

7e. y represents the estimated population x years after 2006: years after 2006: $y = 1.315 \cdot 1.0053^x$

7f. The equation predicts that the population in 2014 is 1.372 billion. This is a little larger than the actual value. This means that the population was growing at a slightly slower rate in 2014 than it was in 2006.

8b. *Hint:* Because you have an explicit equation for the height, you can use decimal values for x.

10. *Hint:* You will need to experiment with transformations of the parent function $\frac{y-k}{a} = 2^{[(x-h)/b]}$, or $y = k + a \cdot 2^{[(x-h)/b]}$. Compare the marked points on the black curve and the red curve to identify translations and stretches.

10a. $y = 2^{x-3}$ **10c.** $\frac{y}{3} = 2^x$, or $y = 3 \cdot 2^x$

12. *Hint:* You will need to experiment with transformations of the parent function, $\frac{y-k}{a} = 0.5^{[(x-h)/b]}$, or $y = k + a \cdot 0.5^{[(x-h)/b]}$.

Use what you learned in Exercise 8.

12a. $y - 5 = 0.5^x$, or $y = 5 + 0.5^x$

12c. $y + 2 = 0.5^{(x-1)}$, $y = -2 + 0.5^{(x-1)}$

13a. $\frac{27}{30} = 0.9$ **13b.** $f(x) = 30(0.9)^x$

13e. *Hint:* Your equation should contain the number 30.

13f. *Hint:* Think about what x_1, y_1, and b represent.

14. Answers will vary but will be in the form $y = y_1 \cdot 1.8^{x-x_1}$, with (x_1, y_1) being any point from the table.

LESSON 4.2

1a. $\frac{1}{125}$ **1c.** $-\frac{1}{81}$ **1e.** $\frac{16}{9}$

2a. a^5 **2d.** e^3

4a. $x = -2$ **4c.** $x = -5$

5a. $x \approx 3.27$

5d. *Hint:* Think about the order of operations to determine the first step to undo.

5f. $x = 1$

6c. $10x^7$ **6f.** $\frac{1}{25}x^{-12}$

7. *Hint:* Is $(2+3)^2$ equivalent to $2^2 + 3^3$? Is $(2+3)^1$ equivalent to $2^1 + 3^1$? Is $(2-2)^3$ equivalent to $2^3 + (-2)^3$? Is $(2-2)^2$ equivalent to $2^2 + (-2)^2$?

10c. $4y = x^3$, or $y = \frac{1}{4}x^3$

12a. *Hint:* Write the original information as two ordered pairs, $(3, 30)$ and $(6, 5.2)$.

12b. $30.0r^{x-3} = 5.2r^{x-6}$; $r \approx 0.5576$

12c. *Hint:* The height the ball was dropped from is the same as the height of bounce zero.

13c. *Hint:* First simplify the left side of the equation. Then expand what is left.

14d. $42(0.9476)^{1980-2002} \approx 137.2$; $39.8(0.9476)^{1980-2003} \approx 137.2$; both equations give approximately 137.2 rads.

15b. $x = -4$ **15c.** $x = 4$

LESSON 4.3

1a. One set of equivalent expressions is a, e, and j.

2d. Power; x is equivalent to x^1.

2g. Exponential; $\frac{12}{3^t}$ is equivalent to $12(3)^{-t}$.

2j. Neither; the function is not a transformation of either x^a or b^x.

3b. $b^{4/5}, b^{8/10}$, or $b^{0.8}$ **3c.** $c^{-1/2}$, or $c^{-0.5}$

4b. $b^{4/5} = 14.3$; raise both sides to the power of $\frac{5}{4}$: $b = 14.3^{5/4} \approx 27.808$.

4c. $c^{-1/2} = 0.55$; raise both sides to the power of -2: $c = 0.55^{-2} \approx 3.306$.

5. *Hint:* Write the given information as two ordered pairs in the form (*number of gels, intensity of light*).

9a. exponential **9b.** neither

9c. exponential **9d.** power

10. *Hint:* Compare the two points on the black curve with the points on the red curve to identify the translations and stretches.

10a. $y = 3 + (x-2)^{3/4}$

10b. $y = 1 + [-(x-5)]^{3/4}$

11b. $x = 180^{1/4} \approx 3.66$

13a. *Hint:* Solve for k.

14a. $27x^9$ **14d.** $108x^8$

LESSON 4.4

1a. $x = 50^{1/5} \approx 2.187$ **1c.** no real solution

2a. $x = 625$

2d. $x = 12(-1 + 1.8125^{1/7.8}) \approx 0.951$

3a. $9x^4$

7. *Hint:* Refer to Example B of this lesson.

7a. She must replace y with $y - 7$ and y_1 with $y_1 - 7$; $y - 7 = (y_1 - 7) \cdot b^{x-x_1}$.

8b. 54 ft

11b. *Hint:* Start with a value for a that is less than 1 and a value for b that is a little larger than 1.

13a. 0.0466, or 4.66% per year

13b. 6.6 g

14. $x = -4.5$, $y = 2$, $z = 2.75$

LESSON 4.5

2b. 2

2c. *Hint:* If $g^{-1}(20) = v$, then $g(v) = g(v) = \underline{?}$.

4. *Hint:* Graph the four linear functions and identify the function-inverse pairs. Then do the same for the nonlinear functions.

4b. b and d are inverses **4c.** c and g are inverses

6b. *Hint:* Find an equation for $f^{-1}(x)$.

8a. $f(x) = 2x - 3$; $f^{-1}(x) = \frac{x+3}{2}$, or $f^{-1}(x) = \frac{1}{2}x + \frac{3}{2}$

8c. $f(x) = \frac{-x^2 + 3}{2}$ or $f(x) = -\frac{1}{2}x^2 + \frac{3}{2}$; $y = \pm\sqrt{-2x+3}$ (not a function)

9b. $f^{-1}(x) = \frac{x - 32}{1.8}$

10a. A best-fit equation is $f(x) \approx -0.006546x + 14.75$.

10c. A best-fit equation is $g(x) \approx -0.003545x + 58.81$.

10e. *Hint:* Think carefully about the units and what each equation uses as its independent variable.

13e. *Hint:* Find the inverse of both functions.

13f. *Hint:* Use an "average" month of 30 days.

13g. *Hint:* The product of length, width, and height should be equivalent to the volume of water, in cubic inches, saved in a month.

15. $f(x) = 12.6(1.5)^{x-2}$, or $f(x) = 42.525(1.5)^{x-5}$

17. *Hint:* Consider a vertically oriented parabola and a horizontally oriented parabola.

LESSON 4.6

1a. $10^x = 1000$ **1c.** $7^{1/2} = x$

2a. $x = 3$ **2b.** $x = 4$

2c. $x = \sqrt{7} \approx 2.65$ **2d.** $x = 2$

2e. $x \approx 16.44$ **2f.** $x = 0$

3a. $x = \log_{10} 0.001; x = -3$

3c. $x = \log_{35} 8; x \approx 0.5849$

6a. $f(7) = 4; g(4) = 7$

6b. They might be inverse functions.

7a. sometime in 1977

8a. $y = 100(0.999879)^x$

9. *Hint:* Write the given information as two points: (0, 88.7) and (6, 92.9).

9a. $y \approx 88.7(1.0077)^x$

11a. Line of fit equation using years since 2000 is $y = 592 + 5.5x$.

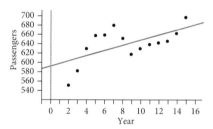

12b. $y = 16.35(2)^x$, where x represents C-note number and y represents frequency in cycles per second

15c. Possible answer: Translate horizontally 1 unit and vertically 4 units. $2(x-1) - 3(y-4) = 9$.

LESSON 4.7

2. $S_1 = 2, S_2 = 8, S_3 = 18, S_4 = 32, S_5 = 50$

3b. $-2 + 1 + 6; 5$

4. *Hint:* 6 divides each multiple without a remainder. Find the value of u_{50}.

5. $S_{75} = 5700$

6b. $S_{75} = 5625$

7a. $u_{46} = 229$

10a. 3, 6, 9, 12, 15, 18, 21, 24, 27, 30

10b. $u_1 = 3$ and $u_n = u_{n-1} + 3$ where $n \geq 2$

10d. *Hint:* Write an equation that shows the sum of the cans in the display is 288.

11a. *Hint:* Find the partial sums recursively until you have a total as close to 1000 as possible without going over.

11e. Each of these 1000 values is 2000 more than the corresponding value between 1 and 1000, so add 1000(2000) to the first sum.

13b. $u_n = 9.8n - 4.9$ **13d.** 490 m

15a. 576,443 people

17a. 81, 27, 9, 3, 1, $\frac{1}{3}$

17b. $u_1 = 81$ and $u_n = \frac{1}{3}u_{n-1}$ where $n \geq 2$

LESSON 4.8

1a. $u_1 = 12, r = 0.4, n = 8$

1b. $u_1 = 75, r = 1.2, n = 15$

2b. u_{10} **2d.** $S_7 = 887.3125$

4b. $u_1 = 3.2, r = 1.5, S_7 = 102.95$

4c. $u_1 = 5.7, d = 2.5, S_{27} = 1031.4$

5a. 3069 **5b.** 22

7a. $S_{10} = 15.984375$

8a. $25,342.39

12. *Hint:* What type of sequence does each prize represent?

13a. neither **13b.** $S_8 \approx 2.717857$

14a. *Hint:* How many games will each team member play?

15. $637.95

17. $f(x) = 3x^3 + 16x^2 + 27x - 26$

CHAPTER 4 REVIEW

1a. $\frac{1}{16}$ **1b.** $-\frac{1}{3}$ **1c.** 125

1d. 7 **1e.** $\frac{1}{4}$ **1f.** $\frac{27}{64}$

1g. -1 **1h.** 12 **1i.** 0.6

2a. $\log_3 7 = x$ **2b.** $\log_4 5 = x$

2c. $\log 7^5 = x$

3a. $10^{1.72} = x$ **3b.** $10^{2.4} = x$

3c. $5^{-1.47} = x$ **3d.** $2^5 = x$

4a. $x = \log_{4.7} 28 \approx 2.153$
4b. $x = \pm\sqrt{\log_{4.7} 2209} \approx \pm 2.231$
4c. $x = 2.9^{1/1.25} = 2.9^{0.8} \approx 2.344$
4d. $x = 3.1^{47} \approx 1.242 \times 10^{23}$
4e. $x = \left(\frac{101}{7}\right)^{1/2.4} \approx 3.041$
4f. $x = \log_{1.065} 18 \approx 45.897$
4g. $x = 10^{3.771} \approx 5902$ **4h.** $x = 47^{5/3} \approx 612$
5a. $x = 0.5(2432^{1/8} - 1) \approx 0.825$
5b. $x = 114^{1/2.7} \approx 5.779$
5c. $x = \log_{1.56}\frac{734}{11.2} \approx 9.406$
5d. $x = \left(\frac{147}{12.1}\right)^{1/2.3} - 1 \approx 1.962$
5e. $x = 5.75^2 + 3 \approx 36.063$
6. $x = 16\log_{0.5}\frac{8}{45} \approx 39.9$; about 39.9 h
7a. 1 **7b.** $\frac{(x+1)^3 + 2}{4}$
7c. $\frac{1}{2}$
8. $y = 5\left(\frac{32}{5}\right)^{(x-1)/6}$
9.

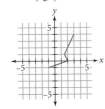

10a. $u_{128} = 511$ **10b.** $u_{40} = 159$
10c. $u_{20} = 79$ **10d.** $S_{20} = 820$
11a. -6639.7
11b. $S_{67} = \sum_{n=1}^{67}[125.3 - 6.8(n-1)]$
12a. $u_{11} \approx 17.490$ **12b.** $S_{10} \approx 515.687$
12c. $S_{20} \approx 605.881$
12d. The partial sum approaches 625.
13a. $a = 0.50$ **13b.** $b = \frac{2.94}{\log 15} \approx 2.4998$
13c. $\log x = \frac{-0.50}{2.4998} \approx -0.2; x \approx 10^{-0.2} \approx 0.63$. The real-world meaning of the x-intercept is that the first 0.63 min of calling is free.
13d. $4.19 **13e.** about 4 min
14a.

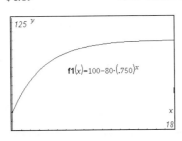

14b. domain: $0 \leq x \leq 120$; range: $20 \leq y \leq 100$
14c. Vertically dilate by a factor of 80; reflect across the x-axis; vertically shift by 100.
14d. 55% **14e.** about 4 yr old
15a. approximately 37 sessions
15b. approximately 48 wpm
15c. Sample answer: It takes much longer to improve your typing speed as you reach higher levels. 60 wpm is a good typing speed, and very few people type more than 90 wpm, so $0 \leq x \leq 90$ is a reasonable domain.
16a. $u_0 = 1, u_n = (u_{n-1}) \cdot 2, n \geq 1$
16b. $y = 2x$
16c.

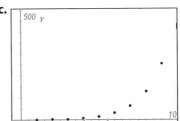

16d. Answers will vary but can include curving upward, increasing, increasing at an increasing rate, discrete.
16e. after 20 cell divisions
16f. after 29 divisions

LESSON 5.1

1c. factored form and vertex form
1d. none of these forms
1e. factored form
2b. $(-4, -2)$
3a. -1 and 2
4c. $y = -2x^2 + 20x - 46$
6c. $y = -3(x - 4.5)^2 + 19$
7b. $y = ax^2 - 8ax + 16a$
7d. $y = -0.5x^2 - (0.5r + 2)x - 2r$
7f. $y = ax^2 - a(r + s)x + ars$
9. *Hint:* Substitute the coordinates of the y-intercept into each function to solve for the scale factor, a.
9b. $y = -0.5(x + 2)(x - 3)$
10c. *Hint:* You are writing an equation that describes the ordered pairs (selling price, revenue).

12b. At approximately 23°C, the rate of photosynthesis is maximized at 100%.

12c. 0°C and 46°C

14d. $9x^2 - 6x + 1$

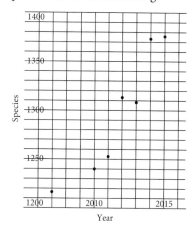

15. *Hint:* Distribute the second factor to each term in the first factor.

16d. *Hint:* Another perfect-square trinomial is $x^2 + 6x + 9$.

LESSON 5.2

1b. $\left(x + \dfrac{5}{2}\right)^2$ **1d.** $(x - y)^2$

2b. 12.25, or $\dfrac{49}{4}$ **2d.** -3

3b. $y = (x - 3.5)^2 + 3.75$

3d. $y = 5(x + 0.8)^2 - 3.2$

4c. $4x^2 + 6x - 3$; $a = 4, b = 6, c = -3$

9. *Hint:* Sketch a graph with time on the horizontal axis and height above ground on the vertical axis. Plot all the points you know.

9a. Let x represent time in seconds, and let y represent height in meters; $y = -4.9(x - 1.1)(x - 4.7)$, or $y = -4.9x^2 + 28.42x - 25.333$.

11a. $y = -4.9t^2 + 100t + 25$

11b. 25 m; 100 m/s   **11c.** 10.2 s; 535 m

12a. *Hint:* Make a table showing price and number sold to help you find this function.

12b. $R(p) = -2p^2 + 100p$

14. $x = 2, x = -3$, or $x = \dfrac{1}{2}$

16a. Let x represent the year, and let y represent the number of endangered bird species.

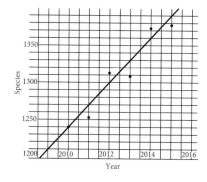

16b. Answers will vary. A line of fit is $y = 1240 + 30(x - 2010)$.

16c. Answers will vary. Using the equation from 16b, approximately 1150 species in 2007; 2440 species in 2050.

16d. The actual value is about 70 species higher than the predicted number. Because the number of endangered species rose more quickly from 2007 to now, it might be expected to continue to rise more and more quickly. Thus the estimate for 2050 may be too low.

16e. Looking at the data now, it might be better to use a nonlinear model such as a quadratic function or an exponential function since the slopes between the points seem to be increasing.

LESSON 5.3

1b. $x^2 - 5x - 13 = 0$; $a = 1, b = -5, c = -13$

1c. $3x^2 + 5x + 1 = 0$; $a = 3, b = 5, c = 1$

2b. -5.898 **2d.** 8.243

3c. $x = -1$ or $x = -1.4$

4c. $y = 5(x + 1)(x + 1.4)$

6a. $x^2 + 9x + 10 = 0$

7a. $y = a(x - 3)(x + 3)$ for $a \neq 0$

8. *Hint:* Think carefully about the value inside the square root, $b^2 - 4ac$.

9. *Hint:* When will the quadratic formula result in no real solutions?

13a. $x^2 + 14x + 49 = (x + 7)^2$ or $x^2 + (-14x) + 49 = [x + (-7)]^2$

13b. $x^2 - 10x + 25 = (x - 5)^2$

14c. $y = \pm\sqrt{x + 6} - 1$

16a. $y = \sqrt{400 - x^2}$

16b. approximately 17.32 ft

LESSON 5.4

1b. 7

1c. $4 - 2i$

2a. $20 + 22i$

2c. $15 + 6i$

5a.

5c.

7a. $-i$

7b. 1

10. *Hint:* Write each function in factored form first.

11. *Hint:* Assume complex zeros occur in conjugate pairs.

12c. The coefficients of the quadratic equations are nonreal.

14b. $0, i, -1+i, -i, -1+i, -i$; alternates between $-1+i$ and $-i$

14d. $0, 0.2+0.2i, 0.2+0.28i, 0.162+0.312i, 0.129+0.301i, 0.126+0.277i$; approaches $0.142+0.279i$

LESSON 5.5

1a. $x^2 - 6x + 9 = 0, a = 1, b = -6, c = 9$

2a. -31, solutions are complex, the graph does not intersect the x-axis.

3d. $x(x+5)(x-1) = 0$ $x = 0, -5, 1$

4b. $(x-5)^2 = -5; x = 5 \pm i\sqrt{5}$

5b. $x = \dfrac{-(6) \pm \sqrt{(6)^2 - 4(2)(-7)}}{2(2)} = \dfrac{-6 \pm \sqrt{92}}{4}$

7a. possible answer is $\left(x - \dfrac{2}{3}\right)(x-4) = 0$

8a. *Hint:* To get two real solutions, $b^2 - 4ac > 0$.

9c. To get one real solution, $b^2 - 4ac = 0$ so $b^2 - 36 = 0; b = 6$ or -6

10. *Hint:* Let $u = (x+2)$.

12a. 16 ft

LESSON 5.6

1b. $x + 6\sqrt{x} + 9$

1d. $x - 2\sqrt{5x} + 5$

2d. no solution

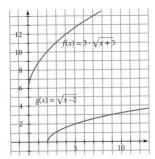

3b. no solution

4a. $x = 0, 4$

6a. 402.5 feet

7. $x = -\dfrac{1}{4}$

8c. No, $x \neq -4$ in the original equation. The graph does not have a point at $x = -4$, that would make the denominator 0, which is undefined.

11a. $x = 1 \left(-\dfrac{4}{9} \text{ is extraneous}\right)$

LESSON 5.7

1a. 10 units

4. $y = 11$ or $y = 3$

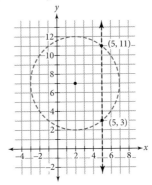

6a. $\sqrt{(x-2)^2 + (1-5)^2}$, or $\sqrt{(x-2)^2 + 16}$

7a. *Hint:* The point $(8, 0)$ satisfies the condition because it is 6 units from $(2, 0)$ and 3 units from $(5, 0)$. What other points can you find?

9a. $\sqrt{x^2 + 4}$ miles through the field and $3 - x$ miles on the road

9b. *Hint:* Refer to Example A of this lesson to see how to express the time.

10a. $y = \sqrt{10^2 + x^2} + \sqrt{(20-x)^2 + 13^2}$

11a.

Time (s)	Height (ft)
0	24.00
1	23.92
2	23.66
3	23.24
4	22.63
5	21.82
6	20.78
7	19.49
8	17.89
9	15.87
10	13.27
11	9.59
12	0

16a. $(x+3)^2 = 14$

16c. $(x+3)^2 + (y-2)^2 = 17$

17b. median from A to $\overline{BC}$: $y = -0.\overline{36}x + 0.\overline{90}$, or $y = -\frac{4}{11}x + \frac{10}{11}$; median from B to $\overline{AC}$: $y = -1.7x + 6.7$; median from C to $\overline{AB}$: $y = 13x - 57$

18a. $\frac{35}{3x}$

LESSON 5.8

1a. $(1, 0.5)$ **1b.** $y = 8$

2a. line of symmetry: $x = 0$

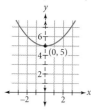

2b. line of symmetry: $y = -2$

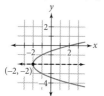

3a. focus: $(0, 6)$; directrix: $y = 4$

3b. focus: $(-1.75, -2)$; directrix: $x = -2.25$

4a. $(y-2)^2 = x$ **4b.** $\frac{y-4}{-1} = x^2$

5a. focus: $(0.25, 2)$; directrix: $x = -0.25$

5b. focus: $(0, 3.75)$; directrix: $y = 4.25$

6a. *Hint:* Make a sketch and locate the focus.

7a. $y = \frac{1}{8}(x-1)^2 + 1$

8a. $y = \frac{1}{8}x^2 + 1$

10. *Hint:* Make a sketch to scale. The food should be placed at the focus of the parabola.

15b. $x(x-7)(x+7)$

CHAPTER 5 REVIEW

1a. $2(x-2)(x-3)$

1b. $(2x+1)(x+3)$ or $2(x+0.5)(x+3)$

1c. $x(x-12)(x+2)$

2a. $x = 9$ or $x = -1$

2b. $x = 0$, $x = 3$, or $x = -5$

3. $1; 4; 10; \frac{1}{6}n^3 - \frac{1}{2}n^2 + \frac{1}{3}n$

4a. vertex form; general form: $y = 2x^2 - 8x - 8$; factored form:
$y = 2[x - (2 + 2\sqrt{2})][x - (2 - 2\sqrt{2})]$

4b. factored form; general form: $y = -3x^2 + 12x + 15$; vertex form: $y = -3(x-2)^2 + 27$

4c. general form; factored form: $y = (x+2)(x+1)$; vertex form: $y = (x+1.5)^2 - 0.25$

4d. factored form; general form: $y = x^3 + 2x^2 - 11x - 12$; no vertex form for cubic equations

4e. general form; factored form:
$y = 2\left(x - \frac{-5+\sqrt{73}}{4}\right)\left(x - \frac{-5-\sqrt{73}}{4}\right)$;
vertex form: $y = 2(x+1.25)^2 - 9.125$

4f. vertex form; general form:
$y = -x^2 - 14x - 51$; factored form:
$y = -[x - (-7 + i\sqrt{2})][x - (-7 - i\sqrt{2})]$

5a.

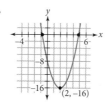

zeros: $x = -0.83$ and $x = 4.83$

5b.

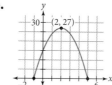

zeros: $x = -1$ and $x = 5$

5c.

zeros: $x = 1$ and $x = 2$

5d.

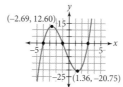

zeros: $x = -4$, $x = -1$, and $x = 3$

5e.

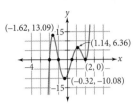

zeros: $x = -5.84$, $x = 1.41$, and $x = 2.43$

5f.

zeros: $x = -2$, $x = -1$, $x = 0.5$, and $x = 2$

6a. $y = 2(x + 1)(x - 4)$

6b. $y = 2(x - 1)^2 - 3$

7. Each pen will be 32 m by 32 m for a maximum enclosed space of 3072 m².

8. approximately 227 m, or 740 ft

9a. center $(6, -4)$, radius $\sqrt{59}$

9b. center $(1, -1.5)$, radius 1.5

10a. $y = (26 - 2x)(21 - 2x)$

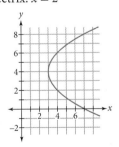

10b. domain: $0 \leq x \leq 10.5$; range: $0 \leq y \leq 546$

10c. $x \approx 3.395$ cm

11a. $0 + 30i$ **11b.** $-6 - 9i$

11c. $2 - 11i$

12. $(x + 4)^2 + (y - 1)^2 = 25$. The graph is a circle with center $(-4, 1)$ and radius 5.

13. $\left(\dfrac{y - 4}{2}\right)^2 = x - 3$; vertex: $(3, 4)$; focus: $(4, 4)$; directrix: $x = 2$

LESSON 6.1

1d. 5

2a. polynomial; 3; $\dfrac{5}{9}x^3 - 3.5x^2 + 4x - 3$

2c. not a polynomial because $4\sqrt{x^3} = 4x^{3/2}$ has a non-integer exponent

3b. no; {0.007, 0.006, 0.008, 0.010}

6c. 4 points. You have to find the finite differences twice, so you need at least four data points to calculate two D_2 values that can be compared.

6e. *Hint:* Find some pennies and try to arrange each number of pennies into a triangle.

11. $D_1 = \{6, 10, 14, 18, 22, 26\}$; $D_2 = \{4, 4, 4, 4, 4\}$. The second differences are constant, so a quadratic function expresses the relationship. Let x represent the energy level, and let y represent the maximum number of electrons. $y = 2x^2$.

14c. $x = \log_5 16 \approx 1.7227$

15c. *Hint:* You need to break $8x$ into two terms.

16. $y \geq -\dfrac{1}{2}(x + 3) + 3$ or $y \geq -\dfrac{1}{2}(x - 11) - 4$
$y \leq \dfrac{1}{2}(x + 3) + 3$ or $y \leq \dfrac{1}{2}(x - 5) + 7$
$y \leq -\dfrac{11}{6}(x - 5) + 7$ or $y \leq -\dfrac{11}{6}(x - 11) - 4$

LESSON 6.2

1c. x-intercepts: 3, -2, -5; y-intercept: 60

2b. $y = -0.25(x + 1.5)(x + 6)$

3c. $y = x^3 - 64x$

8b. $6\left(x - \dfrac{5}{3}\right)\left(x + \dfrac{1}{2}\right)$, or $(3x - 5)(2x + 1)$

8f. $(x - 8)(x + 8)$

8h. $(x - \sqrt{7})(x + \sqrt{7})$

11b. $y = 2(x + 5)(x - 3)(x - 6)$

11c. $y = 2(x + 5)(x - 3)(x - 6) + 100$

12a. $(T + t)^2$, or $T^2 + 2Tt + t^2$

	T	t
T	TT	Tt
t	Tt	tt

12b. $(T + t)^2 = 1$, or $T^2 + 2Tt + t^2 = 1$

12c. *Hint:* Tasters include those who have at least one taster gene.

13. No. These points are collinear.

15a. $x = \pm\sqrt{50.4} \approx \pm 7.1$

15b. $x = \pm\sqrt{13} \approx \pm 3.6$

LESSON 6.3

1b. $x = -6, x = -3, x = 2,$ and $x = 6$

2b. $(0, 108)$

3b. 4; As x decreases, y increases and as x increases, y increases

4b. $y = 0.5(x + 6)(x + 3)(x - 2)(x - 6)$

5b. $y = a(x - 4)^2$ where $a \neq 0$

6a. 4   **6b.** 5

6c. $y = -x(x + 5)^2(x + 1)(x - 4)$

6d. The average rate of change from $x = -5$ to $x = -3$ is approximately $150/2 = 75$. The average rate of change from $x = 3$ to $x = 4$ is approximately $750/1 = 750$. The graph is steeper from $x = 3$ to $x = 4$. The absolute value of the rate from $x = 3$ to $x = 4$ is higher.

7b. $y = -2(x + 4)(x - 5)(x + 2)^2$

7d. *Hint:* Remember that imaginary and complex roots come in conjugate pairs.

8a. $0 \leq x \leq 30$, lengths of the paper

8d. $86.66/17 \approx 5.09$ in.^{2} per in. and $-86.6/13 \approx -6.66$ in.^{2} per in.

9. The leading coefficient is equal to the y-intercept divided by the product of the zeros if the degree of the function is even or the y-intercept divided by -1 times the product of the zeros if the degree of the function is odd.

11a. ii. $y = -(x + 5)^2(x + 2)(x - 1)$

11b. ii. $x = -5, x = -5, x = -2,$ and $x = 1$

15. $3 - 5\sqrt{2}; 0 = a(x^2 - 6x - 41)$ where $a \neq 0$

16c. $36 - 18\sqrt{2}$

17. $S = \dfrac{20}{\sqrt{75}} \sqrt{\ell}$; approximately 17.9 knots

LESSON 6.4

1a. $3x^2 + 7x + 3$

2a. $3x^3 + 22x^2 + 38x + 15 = (x + 5)(3x^2 + 7x + 3)$

2b. $6x^3 + 11x^2 - 19x + 6 = (3x - 2)(2x^2 + 5x - 3)$

2c. $2x^4 - 7x^3 + 7x^2 - x - 2 = (x - 2)(2x^3 - 3x^2 + x + 1)$

2d. $-3x^4 - 8x^3 + 7x^2 + 11x - 3 = (x + 3)(-3x^3 + x^2 + 4x - 1)$

3a. $a = 12$ **3c.** $c = 7$

4a. $3x^3 - 11x^2 + 7x - 44 = (x - 4)(3x^2 + x + 11)$

4c. $4x^3 - 8x^2 + 7x - 6 = (x - 1.5)(4x^2 - 2x + 4)$

7. *Hint:* Try working through the procedure shown on pages 433–434.

7a. $47 = 11 \cdot 4 + 3$

7b. $P(x) = (x - 1)(6x^3 + x^2 + 8x - 4) + 11$

9a. 4

9b. $x = 1, x = 2, x = -5,$ and $x = -1$

9c. $y = (x + 5)(x + 1)(x - 1)(x - 2)$

13a. $x = \pm 2, x = 1,$ and $x = \pm 2i$

13b. $x \approx -7.01, x \approx -0.943,$ and $x \approx 0.454$

14a. $f(x) = 0.00639x^{3/2}$

14b. $f^{-1}(x) \approx (156x)^{2/3}$

14c. 33 in.   **14d.** about 177 ft

16a.

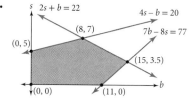

16b. 14 baseball caps and 4 sun hats; $33

17a. $y = x^2 - 4x - 12, y = (x - 6)(x + 2)$; vertex: $(2, -16)$; y-intercept: -12; x-intercepts: $6, -2$

17b. $y = 3x^2 + 6x - 24, y = 3(x - 2)(x + 4)$; vertex: $(-1, -27)$; y-intercept: -24; x-intercepts: $2, -4$

LESSON 6.5

1a. $f(x) = \dfrac{1}{x} + 2$

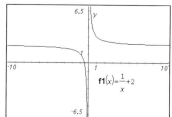

1e. $f(x) = 3\left(\dfrac{1}{x}\right) + 1,$ or $f(x) = \dfrac{3}{x} + 1$

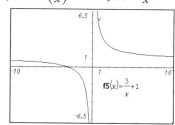

2b. *Hint:* What changed extreme x values?

3a. translated left 3 and down 2

4a. $y = \dfrac{1}{x} + 2$ **4c.** $y = \dfrac{1}{x + 4} + 3$

5a. $y = 1, x = 0$ **5c.** $y = -1, x = -2$

8a. *Hint:* Divide the expression as in Example A of this lesson.

13. *Hint:* Concentric circles have the same center but different radii.

LESSON 6.6

1a. $\dfrac{(x+3)(x+4)}{(x+2)(x-2)}$

2b. $x = 2$ and $x = -2$

4a. $\dfrac{3x-10}{x-4}$

5. *Hint:* See Example C of this lesson.

5a. $\dfrac{7x-7}{x-2}$

6d. *Hint:* Look at a table of values for each graph if your graphing calculator doesn't display the holes.

7a. $y = \dfrac{x+2}{x+2}$

LESSON 6.7

1a. $\dfrac{x(x+2)}{(x-2)(x+2)} = \dfrac{x}{x-2}$

1c. $\dfrac{3x(x-2)}{(x-4)(x-2)} = \dfrac{3x}{x-4}$

2a. $(x+3)(x-3)(x-2)$

2c. $(x+2)(x-2)(x+3)$

3a. 2

3b. $\dfrac{-2}{x+1}$

4a. $\dfrac{(2x-3)(x+1)}{(x+3)(x-2)(x-3)}$

4b. $\dfrac{-x^2+6}{(x+2)(x+3)(x-2)}$

6a. *Hint:* Factor the denominator.

7a. *Hint:* Look at the denominators.

7b. $\dfrac{x^2-2}{(x-1)(x-2)}$

8b. translation right 3 units and up 1 unit

LESSON 6.8

1b. $x^2 + 55x$

3b. $3x^4$

4c. $\dfrac{x+3}{x+1}$

4d. $\dfrac{x^2+2x+1}{x^2-8x+16}$, or $\dfrac{(x+1)^2}{(x-4)^2}$

5a. $\dfrac{1}{x-3}$

5b. $\dfrac{(x-4)(x-2)}{x+4}$

6a. $\dfrac{2(x-2)}{x+1}$

LESSON 6.9

1d. $x = -2$

4a. no real solution

5. 12 games

6a. 14.4 ohms

12a. $y = \dfrac{v^2 + 25v + 1250}{25v}$

18a. $370.09

CHAPTER 6 REVIEW

1a. $y = 2(x+3)^2(x-1)$

1b. $y = -(x+2)(x-3)^3$

1c. $y = 0.5(x+4)(x-2)(x-3i)(x+3i)$

2. 18 in. by 18 in. by 36 in.

3a. $y = 0.5x^2 + 0.5x + 1$

3b. 16 pieces; 56 pieces

4a. $\pm 1, \pm 3, \pm 13, \pm 39, \pm \dfrac{1}{3}, \pm \dfrac{13}{3}$

4b. $x = -\dfrac{1}{3}, x = 3, x = 2 + 3i,$ and $x = 2 - 3i$

5. $2x^2 + 4x + 3$

6a. $y = 1 + \dfrac{1}{x+2}$, or $y = \dfrac{x+3}{x+2}$

6b. $y = -4 + \dfrac{1}{x}$, or $y = \dfrac{-4x+1}{x}$

7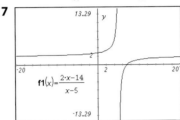

horizontal asymptote: $y = 2$; vertical asymptote: $x = 5$

8. Multiply the numerator and denominator by the factor $(x+3)$.

$y = \dfrac{(2x-14)(x+3)}{(x-5)(x+3)}$

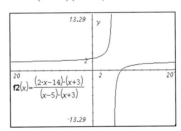

9a. $\dfrac{3x^2 + 8x + 3}{(x-2)(x+1)(x+2)}$

9b. $\dfrac{3x}{x+1}$

9c. $\dfrac{(x+1)^2(x-1)}{x(x-2)}$

10. $x = \dfrac{1}{2}$

LESSON 7.1

1. $\sin A = \frac{k}{j}$; $\sin B = \frac{h}{j}$; $\sin^{-1}\left(\frac{k}{j}\right) = A$; $\sin^{-1}\left(\frac{h}{j}\right) = B$;
$\cos B = \frac{k}{j}$; $\cos A = \frac{h}{j}$; $\cos^{-1}\left(\frac{k}{j}\right) = B$; $\cos^{-1}\left(\frac{h}{j}\right) = A$;
$\tan A = \frac{k}{h}$; $\tan B = \frac{h}{k}$; $\tan^{-1}\left(\frac{k}{h}\right) = A$; $\tan^{-1}\left(\frac{h}{k}\right) = B$;

2a.

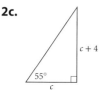

$a \approx 4.1$

2c.

$c \approx 9.3$

3a. $a \approx 17.3$ **3b.** $b \approx 22.8$

4a. $A \approx 17°$

5b. 726 steps

6. *Hint:* Refer to Refreshing Your Skills: Special Right Triangles to review the properties of special right triangles.

7a. $BD \approx 25.7$ cm

7c. $AD \approx 70.5$ cm, $DC \approx 19.5$ cm

8. *Hint:* First find the height of the triangle.

9a. *Hint:* Pay careful attention to units. You will need to convert km/h to m/s.

9b. 396 km/h

12. *Hint:* Make a right triangle by drawing a segment from the point perpendicular to the x-axis.

14. $(7, -3)$

LESSON 7.2

1a. $\frac{-3}{\sqrt{13}} \approx -0.832$

2a. iii.

3a.

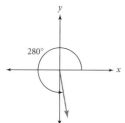

4a. $80°$

5a. negative

6a. about $236.3°$

7. *Hint:* Place Robyn at the origin and plot the positions of Nathan and Keshon.

8. about $106.7°$

10. about $(-4.60, 3.86)$

11. *Hint:* First calculate the angle of the original point. Add $100°$ to find the angle of the point after it rotates.

15a. i. 7   **15a. ii.** 3   **15a. iii.** 4   **15a. iv.** 7

15b. $y = 1 \pm \sqrt{(x-3)^3}$; not a function

15c. Substitute each y-value found in 15a for x into the inverse. Check that the output is equivalent to the original x-value.

15d.

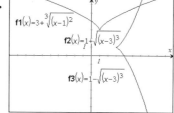

LESSON 7.3

1a.

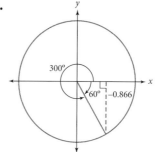

approximately -0.866 m

1b.

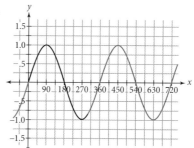

2a. -0.0872; $\sin(-175°) = -\sin 5°$; reference angle $5°$

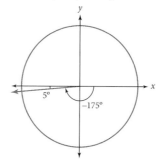

2c. -0.9848; $\sin 280° = -\sin 80°$; reference angle $80°$

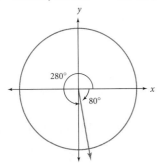

5b. not periodic **5c.** periodic, $90°$
6b. $\theta = 290°$ **6c.** $\theta = 75°$
10a.

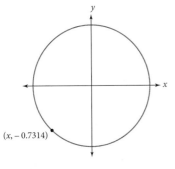

10b. $\theta \approx 227°$; $\cos \theta = -0.682$
10c. $\alpha = 360° - \theta = 133°$
11. *Hint:* Graph a constant function and find the intersection points with $y = \sin(x)$.
12a. *Hint:* Use the Pythagorean Theorem.
12b. *Hint:* Draw similar right triangles.
13a.

13b. 10–11 yr
16a. $h = 1.5 + 20 \tan A$ **16b.** 33.5 m
19a. $\dfrac{3}{x-4}$, $x \neq \pm 4$
19b. 2, $x \neq -2, -1, 1, 3$
19c. $\dfrac{2(3+a)}{6-a}$, $a \neq 6$

LESSON 7.4

1a. $\dfrac{4\pi}{9}$ **1b.** $\dfrac{19\pi}{6}$ **1g.** -5π
2a. 2π
6b. 1.6 cm/min
7d, 8d, 9d. *Hint:* Make sure your calculator is set to radian mode.
11a. $\dfrac{4\pi}{3}$
12. 31.3 radians/s
14. *Hint:* Draw an arc (a great arc) on a ball, balloon, or globe.
15b. dilated vertically by a factor of 3 and horizontally by a factor of 4 and translated right 5 units
15d. dilated vertically by a factor of 2, reflected across the *x*-axis, and translated up 3 units and left 1 unit
16a. $y = -2(x+1)^2$
16b. $y + 4 = (x-2)^2$, or $y = (x-2)^2 - 4$
16c. $y + 2 = |x+1|$, or $y = |x+1| - 2$
16d. $-\dfrac{y-2}{2} = |x-3|$, or $y = -2|x-3| + 2$
19. *Hint:* Construct $\overline{AP}$, $\overline{BP}$, and $\overline{CP}$. $\triangle APC$ is isosceles because $\overline{AP}$ and $\overline{CP}$ are radii of the same circle. $\angle ABP$ measures $90°$ because the angle is inscribed in a semicircle. Use these facts to prove that $\triangle ABP \cong \triangle CBP$.

LESSON 7.5

1a. $y = \sin x + 1$; amplitude = 1, period = 2π, phase shift = 0, vertical shift = 1
1b. $y = \cos x - 2$; amplitude = 1, period = 2π, phase shift = 0, vertical shift = -2
1c. $y = \sin x - 0.5$; amplitude = 1, period = 2π, phase shift = 0, vertical shift = -0.5
2b. $y = \sin\dfrac{x}{2} - 1$; amplitude = 1, period = 4π, phase shift = 0, vertical shift = -1
2c. $y = -2 \sin 3x$; amplitude = 2, period = $\dfrac{2\pi}{3}$, phase shift = 0, vertical shift = 0
6. possible answer: $y = \sin\left(x - \dfrac{\pi}{2}\right)$, $y = -\sin\left(x + \dfrac{\pi}{2}\right)$, or $y = -\cos x$
7a. *Hint:* What are the independent and dependent variables?
14b. $y = -3 + 2 \sin 4\left(x - \dfrac{\pi}{4}\right)$
17a. 1.885 m/h **17b.** 30 cm **17c.** 360 cm
17d. second hand: 377 radians/h; minute hand: 6.28 radians/h; hour hand: 0.52 radian/h
17e. The speed of the tip of a clock hand varies directly with the length of the hand. Angular speed is independent of the length of the hand.
18a. i. $y = -\dfrac{2}{3}x + 4$

LESSON 7.6

1a. $x = \left\{\dfrac{\pi}{3}, \dfrac{5\pi}{3}, \dfrac{7\pi}{3}, \dfrac{11\pi}{3}\right\}$

2a. $x = \{1.831, 5.193\}$

3a. *Hint:* The average value of a sinusoid is the center line.

3b. 7; −2; 12; 7

5a. $\dfrac{1}{0.075}$, or 13.33

8a. *Hint:* You can assume a phase shift of 0.

10a. $y_1 = -3\cos\left[\dfrac{2\pi(t + 0.17)}{\frac{2}{3}}\right]$, $y_2 = -4\cos\left(\dfrac{2\pi t}{\frac{2}{3}}\right)$ or $y_1 = 3\sin 3\pi t$, $y_2 = -4\cos 3\pi t$

13a. about 9.6 h

15.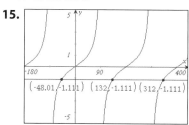

$\theta \approx \{-48°, 132°, 312°\}$

LESSON 7.7

1. Graph $y = \dfrac{1}{\tan x}$.

2b. not an identity

2d. may be an identity

3. Proof:

$\tan x(\cot x + \tan x) \stackrel{?}{=} \sec^2 x$	Original equation.
$\tan x \cot x + (\tan x)^2 \stackrel{?}{=} \sec^2 x$	Distribute.
$\tan x \cdot \dfrac{1}{\tan x} + \tan^2 x \stackrel{?}{=} \sec^2 x$	Reciprocal identity.
$1 + \tan^2 x \stackrel{?}{=} \sec^2 x$	Simplify.
$\sec^2 x = \sec^2 x$	Pythagorean identity.

8a.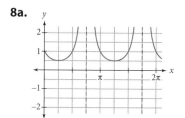

9a. $y = \sin x$ **9d.** $y = \cos x$ **9e.** $y = -\sin x$

11a–c. *Hint:* Use the reciprocal trigonometric identities to graph each equation on your calculator, with window $0 \le x \le 4\pi$, $-2 \le y \le 2$.

12a. $y = \tan 2x - 1$ **12c.** $y = 0.5 \csc x + 1$

13a. *Hint:* Multiply the left side by a fraction equivalent to 1. You will be using the identity $1 - \cos^2 x = \sin^2 x$.

16a. $1,505.12

CHAPTER 7 REVIEW

1a. $A \approx 43°$ **1b.** $B \approx 28°$ **1c.** $c \approx 23.0$
1d. $d \approx 12.9$ **1e.** $e \approx 21.4$ **1f.** $f \approx 17.1$

2. about 73°

3. $(-3\sqrt{3}, 3)$

4a. $t = 3$: $x = -8$, $y = 0.5$; $t = 0$: $x = 1$, $y = 2$; $t = -3$: $x = 10$, $y = -1$

4b. $y = \dfrac{6}{11}$

4c. $x = \dfrac{5}{2}$

4d. When $t = -1$, the y-value is undefined.

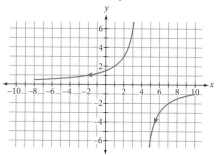

5a. $\dfrac{11\pi}{36}$

5b. approximately 3.84 cm

5c. approximately 7.68 cm²

6a. I; 420°; $\dfrac{\pi}{3}$ **6b.** III; $\dfrac{10\pi}{3}$; 240°

6c. IV; −30°; $\dfrac{11\pi}{6}$ **6d.** IV; $\dfrac{7\pi}{4}$; −45°

7a. $\sin 60° = \dfrac{\sqrt{3}}{2}$; $\cos 60° = \dfrac{1}{2}$

7b. $\sin \dfrac{4\pi}{3} = -\dfrac{\sqrt{3}}{2}$; $\cos \dfrac{4\pi}{3} = -\dfrac{1}{2}$

7c. $\sin 330° = -\dfrac{1}{2}$; $\cos 330° = \dfrac{\sqrt{3}}{2}$

7d. $\sin\left(-\dfrac{\pi}{4}\right) = -\dfrac{\sqrt{2}}{2}$; $\cos\left(-\dfrac{\pi}{4}\right) = \dfrac{\sqrt{2}}{2}$

8. Other equations are possible.

8a. period = $\dfrac{2\pi}{3}$; $y = -2\cos\left[3\left(x - \dfrac{2\pi}{3}\right)\right]$

8b. period = $\dfrac{\pi}{2}$; $y = 3\sin\left[4\left(x - \dfrac{\pi}{8}\right)\right]$

8c. period = π; $y = \csc\left[2\left(x + \dfrac{\pi}{4}\right)\right]$

8d. period = $\dfrac{\pi}{2}$; $y = \cot\left[2\left(x - \dfrac{\pi}{4}\right)\right] + 1$

9a. $2; \frac{\pi}{6}; 0; \frac{3}{2\pi}$

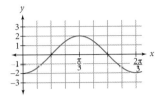

9b. $3; 0; 0; \frac{2}{\pi}$

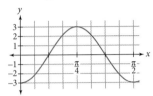

10a. $y = -2\sin 2x - 1$

10b. $y = \sin 0.5x + 1.5$

11a. $y = -\frac{1}{2}\cos\left(\frac{1}{2}x\right) + 6$

11b. period: 4π, amplitude: $\frac{1}{2}$; phase shift: none

11c.

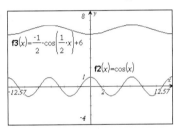

12. area $= \frac{9\pi}{8}$ cm^2; arc length $= \frac{3\pi}{4}$ cm

13a. Let x represent time in hours, and let y represent tide height in meters.

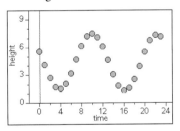

13b. possible answer:
$y = 2.985 \cos\left[\frac{\pi}{6}(x - 10)\right] + 4.398$

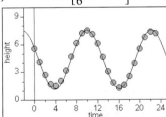

13c. approximately 1.84 m

13d. between 00:00 and 00:37, between 07:22 and 14:10, and between 19:22 and 23:59

LESSON 8.1

1a. $\frac{6}{15} = 0.4; \frac{7}{15} \approx 0.4\overline{6}; \frac{2}{15} \approx 0.1\overline{3}$

1b. experimental

2c. $\frac{228}{435} \approx 0.524$

3a. $\frac{4}{14} \approx 0.286$

3e. *Hint:* Think about graphing the line $x + y = 2$.

4c. *Hint:* Use the graph in Example B of this lesson to help you.

7a. $4; \frac{4}{36} \approx 0.\overline{1}$ **7b.** $5; \frac{5}{36} \approx 0.13\overline{8}$

7c. $10; \frac{10}{36} \approx 0.2\overline{7}$ **7d.** $2; \frac{2}{36} \approx 0.0\overline{5}$

7e. $10; \frac{10}{36} \approx 0.2\overline{7}$

9c. $\frac{44}{144}$

9f. *Hint:* How much area does the line itself occupy?

12a. 270

12c. $\frac{270}{1380} \approx 0.196$

13. *Hint:* Where is each type of face located on the original cube?

15. $x^4 - 4x^3y + 6x^2y^2 - 4xy^3 + y^4$

16a.

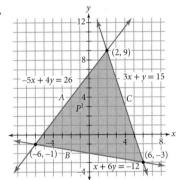

16b. $(2, 9), (-6, -1), (6, -3)$

16c. 68 units^2

17. a parabola with focus $(3, 0)$ and directrix $y = 6$;
$y = -\frac{1}{12}x^2 + \frac{1}{2}x + \frac{9}{4}$

18a. Set i should have a larger standard deviation because the values are more spread out.

18b. i. $\bar{x} = 35, s \approx 22.3$

18b. ii. $\bar{x} = 117, s \approx 3.5$

18c. The original values of $\bar{x}$ and s are multiplied by 10.

18c. i. $\bar{x} = 350, s \approx 223.5$

18c. ii. $\bar{x} = 1170, s \approx 35.4$

18d. The original values of $\bar{x}$ are increased by 10, and the original values of s are unchanged.

18d. i. $\bar{x} = 45, s \approx 22.3$

18d. ii. $\bar{x} = 127, s \approx 3.5$

LESSON 8.2

1.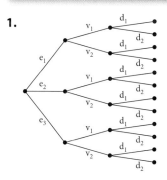

2. $P(a) = 0.675; P(b) = 0.075; P(c) = 0.05; P(d) = 0.2; 1$

3. *Hint:* 0.42 is the product of 0.6 and the probability of path a.

4a. $\frac{1}{8} = 0.125$

4c. *Hint:* Which paths show Celina and one other student were successful? Which show two successful students?

6a. $\frac{182}{420} = 0.4\overline{3}; \frac{98}{420} = 0.2\overline{3}; \frac{98}{420} = 0.2\overline{3}; \frac{42}{420} = 0.1$

6b. no, because the probabilities of the four paths are not all the same

6c. $\frac{420}{420} = 1$

7a. 24

7f. $\frac{12}{24} = 0.5$

9c. 16

9e. 1024

11. *Hint:* Guessing randomly is like flipping a coin for each question on the quiz.

11g. 0.3125

12a.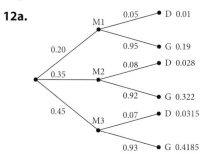

12d. ≈ 0.403

14. *Hint:* This is like flipping a coin with sides boy and girl instead of heads and tails.

16a. $\frac{349}{798} \approx 0.437$

16b. $\frac{512}{1424} \approx 0.360$

16c. The events are dependent, because $P(10\text{th grade} \mid \text{female}) \neq P(10\text{th grade})$. The probability of choosing a 10th grader from the female students is greater than the probability of choosing a 10th grader from all students.

17. 64

18a. $-3 + 2i$

18b. $2 + 24i$

18c. $7 + 6i$

LESSON 8.3

2a. 0.25

2b. 0.12

2c. $\frac{0.15}{0.15 + 0.12} \approx 0.\overline{5}$

2d. 0.37

3.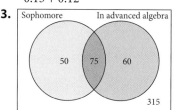

4. *Hint:* Does $P(S) \cdot P(A) = P(S \text{ and } A)$?

5a. yes, because they do not overlap

5b. No. $P(A \text{ and } B) = 0$. This would be the same as $P(A) \cdot P(B)$ if they were independent.

6a.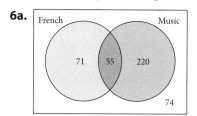

6b. approximately 13%

6c. 74

7a. *Hint:* $P(A|B) = \dfrac{P(A \text{ and } B)}{P(B)}$

8. $0 \leq P(A \text{ and } B) \leq 0.4, 0.5 \leq P(A \text{ or } B) \leq 0.9$. The first diagram shows $P(A \text{ and } B) = 0$ and $P(A \text{ or } B) = 0.9$. The second diagram shows $P(A \text{ and } B) = 0.4$ and $P(A \text{ or } B) = 0.5$.

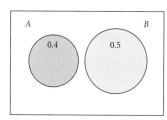

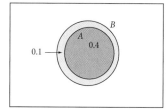

9. *Hint:* Find the new numbers of students in each section before the 20 sophomores move.

10. *Hint:* See the related Science Connection.

10a. yellow

11b. *Hint:* Because all three friends' phone usage is independent, $P(A) \cdot P(B) \cdot P(C) = P(A \text{ and } B \text{ and } C)$.

12a. $\frac{280}{1500} = 0.18\overline{6}$ **12b.** $\frac{775}{1500} = 0.51\overline{6}$

12c. $\frac{145}{355} \approx 0.408$ **12d.** $\frac{145}{775} \approx 0.187$

15c. $2xy^2\sqrt{15xy}$

LESSON 8.4

1. Cat, Yes = 90, Cat, No = 220, Dog, Yes = 105, Dog, No = 205

2d. $70/205 \approx 0.34$

4b. 0.19 and <0.139, 0.242>

6a. They are not independent, the proportions for Dog-Yes conditional on cats are not to equal 0.22 for yes and 0.39 for no.

7c. Maybe not independent. These proportions are not equal, but do the intervals <0.022, 0.037> and <0.01, 0.03> overlap?

8a. p = 0.038, or 3.8%, for severely ill in RH and 0.01, or 1% for those who are not severely ill.

CHAPTER 8 REVIEW

1. Answers will vary depending upon whether you are generating random decimal numbers or random integers. To generate random decimal numbers, you might look at a random-number table and place the decimal point after the first digit in each group of numbers. Alternatively, you could use a calculator command, such as **10*rand(numTrials)** on the TI-Nspire, or **10*rand** on the TI-84 Plus. To generate random integers, you might number 11 chips or slips of paper and randomly select one. Alternatively, you could use a calculator command, such as **randInt(0, 10)**.

2a, b.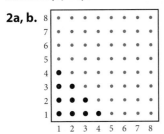

2c. $\frac{10}{64} = 0.15625$ **2c.** $\frac{49}{64} \approx 0.766$

3a. 0.5 **3b.** 17.765 units^2

4a.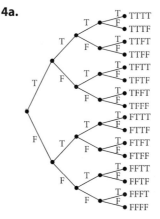

4b. 4

4c. $\frac{1}{2} = 0.5$

5a.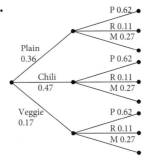

5b. 0.0517

5c. 0.8946 **5d.** 0.3501

6a. Ice cream: 254; Whipped cream: 118; Total: 23, 55, 122, 172, 372

6b. $\frac{37}{55} = 0.6\overline{72}$ **6c.** $\frac{37}{122} \approx 0.303$

6d. $\frac{18}{254} \approx 0.071$ **6e.** $\frac{118}{372} \approx 0.317$

7.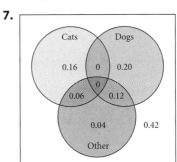

LESSON 9.1

1a–4a. Answers will vary. Sample answers are given.

1a. This is a survey (more specifically, an interview survey). Nick chose the subjects. They choose whether to respond.

2a. The treatment is that the person has experienced the movie in question. The measure is how much he or she liked the movie.

3a. The subjects who choose this movie are more likely to enjoy this type of movie than if you had selected a random group of people to watch the movie.

4a. As it is a survey, Nick needed a random process to determine which movie patrons he should ask. In an exit poll (like this) you could roll a die to see how many people to skip before you ask the next person.

6a. Use an experiment to control as many of the factors as you can. Purchase the same size bag of popcorn for each brand, and assume that there are approximately the same number of kernels in each bag. Using one microwave, pop one bag of each brand and count the unpopped kernels in each. Record the deviation from the mean. In other words, if you have five brands and you found 20, 35, 15, 21, and 24 unpopped kernels, then you calculate the mean (23) and record −3, 12, −8, −2, 1. Now do the same with nine different microwaves. The brand with the lowest sum of all these deviations is the winner.

8a. $A = 12$ units^2

10c. 20th percentile

LESSON 9.2

1a. *Hint:* Enter the functions as f_1 and f_2. Then graph or create a table of values, and confirm that they are the same.

1b. $y \approx 0.242$ and $n(x, 0, 1) \approx 0.242$

2a. *Hint:* To estimate the standard deviation, look for the inflection points.

4a. $y = \dfrac{1}{2.5\sqrt{2\pi}} (\sqrt{e})^{-1[(x-18)/2.5]^2}$

6a, c.

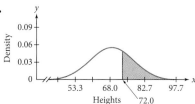

8a. *Hint:* Ninety percent of the area must be above the value 12.

9.

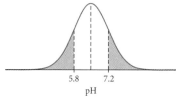

Shade the area greater than or less than 1 standard deviation away from the mean.

10a. $\mu \approx 165$; $\sigma \approx 5.91$

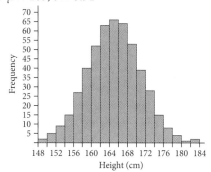

13a. *Hint:* Consider what the mean and standard deviation tell you about the distribution of test scores. Can you be sure which test is more difficult?

13c. *Hint:* Determine how many standard deviations each student's score is from the mean. This will tell you how each student scored relative to other test-takers.

LESSON 9.3

1a. $z = 1$
1c. $z \approx 1.57$
2a. 122.6
2c. 131.96
5a. $z = 1.8$
6. *Hint:* Refer to Example C of this lesson.
7a. (3.058, 3.142)
14a. *Hint:* Plan 1 can be represented by an arithmetic series.

LESSON 9.4

1b. −0.95
2a. Sample answer:

3d. −0.9723
3f. *Hint:* Do the points seem to decrease linearly?

4a. *Hint:* The explanatory variable corresponds to the independent variable.

5a. *Hint:* What lurking variable might be correlated with television ownership?

8a.
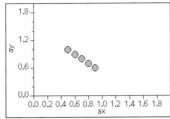

$r = -1$. This value of r implies perfect negative correlation consistent with the plot.

9. *Hint:* The correlation coefficient gives you information about the linearity of the data.

10a. The correlation coefficients for (*year, country*) and (*year, oldies*) are -0.972 and 0.659, respectively.

11. 12

14. *Hint:* The average speed is $\frac{total\ distance}{total\ time}$. If the distance is d, then the total distance is $2d$. To find the total time, use the relationship $t = \frac{d}{r}$.

CHAPTER 9 REVIEW

1. Answers will vary. Sample answers are given.

1a. Observational study; for an experimental study, the dogs would be randomly assigned to one of the two types of food.

1b. Survey; for an observational study, the researcher could have a cashier give out the incorrect amount of change and observe the behavior of a randomly selected group of customers.

2a and b.
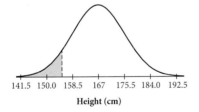

2c. approximately 7.9%

3. (114.94, 119.06); (112.89, 121.11)

4a. $\bar{x} = 10.55$ lb; $s = 2.15$ lb

4b. 6.25 lb to 14.85 lb

5a.
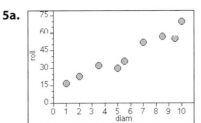

5b. yes; $r \approx 0.965$, indicating a relationship that is close to linear

5c. The value at 9.5 in. is out of line and would improve the r value if it lined up better. This is also true of the value at 5 in., but this value is close to the mean so a change here would have less effect on the correlation.

6a. approximately 32.3%

6b. approximately 14.2%

6c. between 152 lb and 198 lb

6d. $y = \dfrac{1}{14\sqrt{2\pi}} (\sqrt{e})^{-[(x-175)/14]^2}$

7. approximately 0.062

8a. Chipley and Panama City ($r \approx 0.859$)

8b. Panama City and Crestview have the weakest correlation.

8c. A high positive correlation means that when one variable has a large value then you can expect the other to also be large. So if Panama City is above average then it is most likely that Chipley will also be above average.

Glossary

absolute value A number's distance from zero on the number line. The absolute value of a number gives its size, or magnitude, whether the number is positive or negative. The absolute value of a number x is shown as $|x|$. For example, $|-9| = 9$ and $|4| = 4$.

ambiguous case A situation in which more than one possible solution exists.

amplitude Half the difference of the maximum and minimum values of a periodic function.

anecdotal A type of study in which data are collected from informational reports or observations.

angular speed The amount of rotation, or angle traveled, per unit of time.

antilog The inverse function of a logarithm.

arc (of a circle) Two points on a circle and the continuous part of the circle between them.

arc length The portion of the circumference of the circle described by an arc, measured in units of length.

arc measure The measure of the central angle that intercepts an arc, measured in degrees.

arithmetic mean See **mean**.

arithmetic sequence A sequence in which each term after the starting term is equal to the sum of the previous term and a common difference.

arithmetic series A sum of terms of an arithmetic sequence.

association (between variables) A relationship between two variables. In a positive association, an increase in one variable correlates with an increase in the other variable. See also **correlation**.

associative property of addition For any values of a, b, and c, $a + (b + c) = (a + b) + c$.

associative property of multiplication For any values of a, b, and c, $a(bc) = (ab)c$.

asymptote A line that a graph approaches, but does not reach, as the magnitude of the x- or y-values increases without bound.

augmented matrix A matrix that represents a system of equations. The entries include columns for the coefficients of each variable and a final column for the constant terms.

base The base of an exponential expression, b^x, is b. The base of a logarithmic expression, $\log_b x$, is b.

bearing An angle measured clockwise from north.

bias A statistical sample of a population may be biased if some of the members of the population are less likely to be included in the sample than others.

bin A column in a histogram that represents a certain interval of possible data values.

binomial A polynomial with two terms.

binomial expansion See **Binomial Theorem** and **expansion**.

Binomial Theorem For any binomial $(p + q)$ and any positive integer n, the binomial expansion is
$(p + q)^n = {}_nC_n p^n q^0 + {}_nC_{(n-1)} p^{n-1} q^1 + {}_nC_{(n-2)} p^{n-2} q^2 + \cdots + {}_nC_0 p^0 q^n$.

bivariate sampling The process of collecting data on two variables per case.

box plot A one-variable data display that shows the five-number summary of a data set.

box-and-whisker plot See **box plot**.

causation A relationship in which changes in one variable, called the *explanatory* or *predictor* variable, cause changes in another variable, called the *response* variable. Causation is difficult to prove.

center (of a circle) See **circle**.

center (of an ellipse) The point midway between the foci of an ellipse.

center (of a hyperbola) The point midway between the vertices of a hyperbola.

central angle An angle whose vertex is the center of a circle and whose sides pass through the endpoints of an arc.

circle A locus of points in a plane that are located a constant distance, called the *radius*, from a fixed point, called the *center*.

circumference The perimeter of a circle, which is the distance around the circle. Also, the curved path of the circle itself.

combination An arrangement of choices in which the order is unimportant.

combined variation An equation in which one variable varies directly with one variable and inversely with another variable. For example, in the equation $P = \frac{kT}{V}$, P varies directly with T and inversely with V, and k is the constant of variation.

common base property of equality For $a \neq 1$ and all real values of m and n, if $a^n = a^m$, then $n = m$.

common difference The constant difference between consecutive terms in an arithmetic sequence.

common logarithm A logarithm with base 10, written $\log x$, which is shorthand for $\log_{10} x$.

common ratio The constant ratio between consecutive terms in a geometric sequence.

commutative property of addition For any values of a and b, $a + b = b + a$.

commutative property of multiplication For any values of a and b, $ab = ba$.

compass rose A diagram that shows the orientation of the directions north, south, east, and west, and often intermediate directions such as northeast, southeast, northwest, and southwest.

complements Two events that are mutually exclusive and make up all possible outcomes.

completing the square A method of converting a quadratic equation from general form to vertex form.

complex conjugate A number whose product with a complex number produces a nonzero real number. The complex conjugate of $a + bi$ is $a - bi$.

complex number A number with a real part and an imaginary part. A complex number can be written in the form $a + bi$, where a and b are real numbers and i is the imaginary unit, $\sqrt{-1}$.

complex plane A coordinate plane used for graphing complex numbers, where the horizontal axis is the real axis and the vertical axis is the imaginary axis.

composition of functions The process of using the output of one function as the input of another function. The composition of f and g is written $f(g(x))$.

composition of transformations The resulting transformation when a transformation is applied to a figure and a second transformation is applied to the image of the first.

compound event An event consisting of more than one outcome.

compound interest Interest charged or received based on the sum of the principal and the accrued interest.

conditional probability The probability of a particular dependent event, given the outcome of the event on which it depends.

confidence interval A $p\%$ confidence interval is an interval about $\bar{x}$ in which you can be $p\%$ confident that the population mean, μ, lies.

congruent Two polygons or other geometric figures are congruent if they are identical in shape and size.

conic section Any curve that can be formed by the intersection of a plane and an infinite double cone. Circles, ellipses, parabolas, and hyperbolas are conic sections.

conjugate pair A pair of expressions in which one is the sum of two terms and the other is the difference of the same terms, such as $x + 3$ and $x - 3$ or $2 + \sqrt{5}$ and $2 - \sqrt{5}$. The complex numbers $a + bi$ and $a - bi$ form a complex conjugate pair.

consistent (system) A system of equations that has at least one solution.

constant of variation (k) The constant ratio in a direct variation or the constant product in an inverse variation. The value of k in the direct variation equation $y = kx$ or the inverse variation equation $y = \frac{k}{x}$.

constraint A limitation in a linear programming problem, represented by an inequality.

continuous random variable A quantitative variable that can take on any value in an interval of real numbers.

convergent series A series in which the terms of the sequence approach zero and the partial sums of the series approach a long-run value as the number of terms increases.

correlation A relationship between two variables.

correlation coefficient (r) A value between -1 and 1 that measures the strength and direction of a linear relationship between two variables.

cosecant The reciprocal of the sine ratio. If A is an acute angle in a right triangle, then the cosecant of angle A is the ratio of the length of the hypotenuse to the length of the opposite leg, or $\csc A = \frac{hyp}{opp}$. See **trigonometric function.**

cosine If A is an acute angle in a right triangle, then the cosine of angle A is the ratio of the length of the adjacent leg to the length of the hypotenuse, or $\cos A = \frac{adj}{hyp}$. See **trigonometric function.**

cotangent The reciprocal of the tangent ratio. If A is an acute angle in a right triangle, then the cotangent of angle A is the ratio of the length of the adjacent leg to the length of the opposite leg, or $\cot A = \frac{adj}{opp}$. See **trigonometric function.**

coterminal Describes angles in standard position that share the same terminal side.

counterexample An example that shows that a given conjecture is not true.

counting principle When there are n_1 ways to make a first choice, n_2 ways to make a second choice, n_3 ways to make a third choice, and so on, the product $n_1 \cdot n_2 \cdot n_3 \cdot \cdots$ represents the total number of different ways in which the entire sequence of choices can be made.

Cramer's rule An algorithm involving determinants for solving systems of equations.

cubic function A polynomial function of degree 3.

cumulative distribution function (CDF) The function (formula) that gives the probability that a random variable is between two values.

curve straightening A technique used to determine whether a relationship is logarithmic, exponential, power, or none of these. See **linearization.**

cycloid The path traced by a fixed point on a circle as the circle rolls along a straight line.

decay See **geometric decay.**

degree In a one-variable polynomial, the power of the term that has the greatest exponent. In a multivariable polynomial, the greatest sum of the powers in a single term.

dependent (events) Events are dependent when the probability of occurrence of one event depends on the occurrence of the other.

dependent (system) A system with infinitely many solutions.

dependent variable A variable whose values depend on the values of another variable.

determinant The difference of the products of the entries along the diagonals of a square matrix. For any 2×2 matrix $\begin{bmatrix} a & b \\ c & d \end{bmatrix}$ the determinant is $ad - bc$.

deviation For a one-variable data set, the signed difference between a data value and some standard value, usually the mean.

difference of two squares An expression in the form $a^2 - b^2$, in which one squared number is subtracted from another. A difference of two squares can be factored as $(a + b)(a - b)$.

dilation A transformation that stretches or shrinks a function or graph both horizontally and vertically by the same scale factor. See also **vertical dilation** and **horizontal dilation.**

dimensions (of a matrix) The number of rows and columns in a matrix. A matrix with m rows and n columns has dimensions $m \times n$.

direct variation The relationship between two variables such that the ratio of their values is always the same, or constant. This relationship is written as $y = kx$, where k is the constant of variation.

direction (vector) The orientation of a vector.

directrix See **parabola.**

discontinuity A jump, break, or hole in the graph of a function.

discrete graph A graph made of distinct, nonconnected points.

discrete random variable A random variable that can take on only distinct (not continuous) values.

discriminant The expression under the square root symbol in the quadratic formula. If a quadratic equation is written in the form $ax^2 + bx + c = 0$, then the discriminant is $b^2 - 4ac$. If the discriminant is greater than 0, the quadratic equation has two solutions. If the discriminant equals 0, the equation has one real solution. If the discriminant is less than 0, the equation has no real solutions.

distance formula The distance, d, between points (x_1, y_1) and (x_2, y_2) is given by the formula $d = \sqrt{(x_2 - x_1)^2 + (y_2 - y_1)^2}$.

distributive property For any values of a, b, and c, $a(b + c) = a(b) + a(c)$.

domain The set of input values for a relation.

dot plot A one-variable data display in which each data value is represented by a dot above that value on a horizontal number line.

double root A value r is a double root of an equation $f(x) = 0$ if $(x - r)^2$ is a factor of $f(x)$.

doubling time The time needed for an amount of a substance to double.

e A transcendental number related to continuous growth, with a value of approximately 2.718.

eccentricity A measure of how elongated an ellipse is. The ratio of the focal length to the length of the semi-major axis.

element (matrix) See **entry**.

elimination A method for solving a system of equations that involves adding or subtracting multiples of the equations to eliminate a variable.

ellipse A shape produced by dilating a circle horizontally and/or vertically. The shape can be described as a locus of points in a plane for which the sum of the distances to two fixed points, called the foci, is constant.

ellipsoid A three-dimensional shape formed by rotating an ellipse about one of its axes.

end behavior The behavior of a function $y = f(x)$ for x-values that are large in absolute value.

entry Each number in a matrix. The entry identified as a_{ij} is in row i and column j.

equivalent vectors Vectors with the same magnitude and direction.

even function A function that has the y-axis as a line of symmetry. For all values of x in the domain of an even function, $f(-x) = f(x)$.

event A specified outcome or set of outcomes.

expanded form The form of a repeated multiplication expression in which every occurrence of each factor is shown. For example, $4^3 \cdot 5^2 = 4 \cdot 4 \cdot 4 \cdot 5 \cdot 5$.

expansion An expression that is rewritten as a single polynomial.

expected value An average value found by multiplying the value of each possible outcome by its probability, then summing all the products.

experiment A type of study in which a researcher tests a hypothesis by assigning subjects, or experimental units, to specific treatments.

experimental probability A probability calculated based on trials and observations, given by the ratio of the number of occurrences of an event to the total number of trials.

explanatory variable In statistics, the variable used to predict (or explain) the value of the response variable.

explicit formula A formula that gives a direct relationship between two discrete quantities. A formula for a sequence that defines the nth term in relation to n, rather than the previous term(s).

exponent The exponent of an exponential expression, b^x, is x. The exponent tells how many times the base, b, is a factor.

exponential function A function with a variable in the exponent, typically used to model growth or decay. The general form of an exponential function is $y = ab^x$, where the coefficient, a, is the y-intercept and the base, b, is the ratio.

extraneous solution An invalid solution to an equation. Extraneous solutions are sometimes found when both sides of an equation are raised to a power.

extrapolation Estimating a value that is outside the range of all other values given in a data set.

extreme values Maximums and minimums.

factored form The form $y = a(x - r_1)(x - r_2) \cdots (x - r_n)$ of a polynomial function, where $a \neq 0$. The values $r_1, r_2, \ldots, r_n$ are the zeros of the function, and a is the vertical scale factor.

factorial For any integer n greater than 1, n factorial, written $n!$, is the product of all the consecutive integers from n decreasing to 1.

Factor Theorem If $P(r) = 0$, then r is a zero and $(x - r)$ is a factor of the polynomial function $y = P(x)$. This theorem is used to confirm that a number is a zero of a function.

fair Describes a coin that is equally likely to land heads or tails. Can also apply to dice and other objects.

fair game An unbiased contest in which each player is equally likely to win.

family of functions A group of functions with the same parent function.

feasible region The set of points that is the solution to a system of inequalities.

Fibonacci sequence The sequence of numbers 1, 1, 2, 3, 5, 8, . . ., each of which is the sum of the two previous terms.

finite A limited quantity.

finite differences method A method of finding the degree of a polynomial that will model a set of data, by analyzing differences between data values corresponding to equally spaced values of the independent variable.

first quartile (Q_1) The median of the values less than the median of a data set.

five-number summary The minimum, first quartile, median, third quartile, and maximum of a one-variable data set.

focus (plural **foci**) A fixed point or points used to define a conic section. See **ellipse, hyperbola,** and **parabola.**

fractal The geometric result of infinitely many applications of a recursive procedure or calculation.

frequency (of a data set) The number of times a value appears in a data set, or the number of values that fall in a particular interval.

frequency (of a sinusoid) The number of cycles of a periodic function that can be completed in one unit of time.

function A relation for which every value of the independent variable has at most one value of the dependent variable.

function notation A notation that emphasizes the dependent relationship between the variables used in a function. The notation $y = f(x)$ indicates that values of the dependent variable, y, are explicitly defined in terms of the independent variable, x, by the function f.

general form (of a polynomial) The form of a polynomial in which the terms are ordered such that the degrees of the terms decrease from left to right.

general form (of a quadratic function) The form $y = ax^2 + bx + c$, where $a \neq 0$.

general quadratic equation An equation in the form $Ax^2 + Bxy + Cy^2 + Dx + Ey + F = 0$, where A, B, and C do not all equal zero.

general term The nth term, u_n, of a sequence.

geometric decay A decay pattern in which amounts decrease by a constant ratio, or percent. In a geometric sequence modeling decay, the common ratio can be represented by $(1 - p)$, where p is the percent change.

geometric growth A growth pattern in which amounts increase by a constant ratio, or percent. In a geometric sequence modeling growth, the common ratio can be represented by $(1 + p)$, where p is the percent change.

geometric probability A probability that is found by calculating a ratio of geometric characteristics, such as lengths or areas.

geometric random variable A random variable that represents the number of trials needed to get the first success in a series of independent trials. The probabilities form a geometric sequence.

geometric sequence A sequence in which each term is equal to the product of the previous term and a common ratio.

geometric series A sum of terms of a geometric sequence.

golden ratio The ratio of two numbers (larger to smaller) whose ratio to each other equals the ratio of their sum to the larger number. Or, the positive number whose square equals the sum of itself and 1. The number $\frac{1 + \sqrt{5}}{2}$, or approximately 1.618, often represented with the lowercase Greek letter phi, ϕ.

golden rectangle A rectangle in which the ratio of the length to the width is the golden ratio.

greatest integer function The function $f(x) = [x]$ that returns the largest integer that is less than or equal to a real number, x.

growth See **geometric growth**.

half-life The time needed for an amount of a substance to decrease by one-half.

head (or **tip**) The end of a vector with the arrowhead.

histogram A one-variable data display that uses bins to show the distribution of values in a data set. Each bin corresponds to an interval of data values; the height of a bin indicates the number, or frequency, of values in that interval.

hole A missing point in the graph of a relation.

horizontal asymptote See **asymptote**.

horizontal dilation A transformation that increases or decreases the width of a figure or graph. A horizontal dilation by a factor of a multiplies the x-coordinate of every point on a graph by a.

horizontal reflection A reflection of a function across the y-axis.

hyperbola A locus of points in a plane for which the difference of the distances to two fixed points, called the foci, is constant.

hyperboloid A three-dimensional shape formed by rotating a hyperbola about the line through its foci or about the perpendicular bisector of the segment connecting the foci.

identity An equation that is true for all values of the variables for which the expressions are defined.

identity matrix The square matrix, symbolized by $[I]$, that does not alter the entries of a square matrix $[A]$ under multiplication. Matrix $[I]$ must have the same dimensions as matrix $[A]$, and it has entries of 1's along the main diagonal (from top left to bottom right) and 0's in all other entries.

image A graph of a function or point(s) that is the result of a transformation of an original function or point(s).

imaginary axis See **complex plane**.

imaginary number A number that is the square root of a negative number. An imaginary number can be written in the form bi, where b is a real number ($b \neq 0$) and i is the imaginary unit, $\sqrt{-1}$.

imaginary unit The imaginary unit, i, is defined by $i^2 = -1$ or $i = \sqrt{-1}$.

inconsistent (system) A system of equations that has no solution.

independent (events) Events are independent when the occurrence of one has no influence on the probability of the other.

independent (system) A system of equations that has exactly one solution.

independent variable A variable whose values are not based on the values of another variable.

incquality A statement that one quantity is less than, less than or equal to, greater than, greater than or equal to, or not equal to another quantity.

infinite A quantity that is unending, or without bound.

infinite geometric series A sum of infinitely many terms of a geometric sequence.

inflection point A point where a curve changes between curving downward and curving upward.

intercept form The form $y = a + bx$ of a linear equation, where a is the y-intercept and b is the slope.

interpolation Estimating a value that is within the range of all other values given in a data set.

interquartile range (*IQR*) A measure of spread for a one-variable data set that is the difference between the third quartile and the first quartile.

intersection A set of elements common to two or more sets. In a Venn diagram, it is the overlapping region between two or more sets.

inverse The relationship that reverses the independent and dependent variables of a relation.

inverse matrix The matrix, symbolized by $[A]^{-1}$, that produces an identity matrix when multiplied by $[A]$.

inverse variation A relation in which the product of the independent and dependent variables is constant. An inverse variation relationship can be written in the form $xy = k$, or $y = \dfrac{k}{x}$.

joint variation An equation in which one variable varies directly with the other variables taken one at a time. In the formula $A = \dfrac{1}{2}bh$, A varies directly with both b and h, and $\dfrac{1}{2}$ is the constant of variation.

Law of Cosines For any triangle with angles A, B, and C, and sides of lengths a, b, and c (a is opposite $\angle A$, b is opposite $\angle B$, and c is opposite $\angle C$), $c^2 = a^2 + b^2 - 2ab \cos C$.

Law of Sines For any triangle with angles A, B, and C, and sides of lengths a, b, and c (a is opposite $\angle A$, b is opposite $\angle B$, and c is opposite $\angle C$), these equalities are true: $\dfrac{\sin A}{a} = \dfrac{\sin B}{b} = \dfrac{\sin C}{c}$.

least squares line A line of fit for which the sum of the squares of the residuals is as small as possible.

like terms Monomials with the same base and the same exponents. For example, $3x^2$ and $5x^2$ are like terms, and $2xy^2z$ and $-5xy^2z$ are like terms.

limit A long-run value that a sequence or function approaches. The quantity associated with the point of stability in dynamic systems.

line of fit A line used to model a set of two-variable data.

line of symmetry A line that divides a figure or graph into mirror-image halves.

linear In the shape of a line or represented by a line, or an algebraic expression or equation of degree 1.

linear equation An equation characterized by a constant rate of change. The graph of a linear equation in two variables is a straight line.

linear programming A method of modeling and solving a problem involving constraints that are represented by linear inequalities.

linearization A method of finding an equation to fit data. The x- and/or y-values are transformed until the relation appears linear. Then inverse transformations are applied to the linear model to produce an equation that models the original data.

local maximum A value of a function or graph that is greater than other nearby values.

local minimum A value of a function or graph that is less than other nearby values.

locus A set of points that fit a given condition.

logarithm A value of a logarithmic function, abbreviated log. For $a > 0$ and $b > 0$, $\log_b a = x$ means that $a = b^x$.

logarithm change-of-base property For $a > 0$ and $b > 0$, $\log_a x$ can be rewritten as $\dfrac{\log_b x}{\log_b a}$.

logarithmic function The logarithmic function $y = \log_b x$ is the inverse of $y = b^x$, where $b > 0$ and $b \neq 1$.

logistic function A function used to model a population that grows and eventually levels off at the maximum capacity supported by the environment. A logistic function has a variable growth rate that changes based on the size of the population.

long run behavior What happens to the values of $f(x)$ as x takes on extreme positive and negative values.

lurking variable An unmeasured variable that affects the value of the response variable (the variable being studied).

magnitude The distance of a number from zero on a number line. Also, the length of a vector.

magnitude of a complex number The distance from the origin (pole) to a complex number $a + bi$ in the complex plane. It is equal to $\sqrt{a^2 + b^2}$.

major axis The longer dimension of an ellipse. Or the line segment with endpoints on the ellipse that has this dimension.

margin of error A measure of accuracy for estimates of a population mean, given by $\frac{z\sigma}{\sqrt{n}}$, where n is the sample size, z is the number of standard deviations from the mean, and σ is the standard deviation or population proportions, $z\sqrt{\frac{p(1-p)}{n}}$ where p is the sample proportion.

marginal frequencies The totals of each row and column in a frequency table.

marginal totals The sum of values in each row and each column of a two-way table.

mathematical induction A type of mathematical proof used to show that a given statement is true for all natural numbers n. A proof by mathematical induction involves first proving the statement is true for $n = 1$, then assuming it is true for $n = k$, and finally proving that it is true for $n = k + 1$. This establishes the truth of the statement for all values of n.

matrix A rectangular array of numbers or expressions, enclosed in brackets.

matrix addition The process of adding two or more matrices. To add matrices, you add corresponding entries.

matrix multiplication The process of multiplying two matrices. The entry c_{ij} in the matrix $[C]$ that is the product of two matrices, $[A]$ and $[B]$, is the sum of the products of corresponding entries in row i of matrix $[A]$ and column j of matrix $[B]$.

maximum The greatest value in a data set or the greatest value of a function or graph.

mean ($\bar{x}$) A measure of central tendency for a one-variable data set, found by dividing the sum of all values by the number of values. See also **population mean.**

measure of central tendency A single number used to summarize a one-variable data set, commonly the mean, median, or mode.

median A measure of central tendency for a one-variable data set that is the middle value, or the mean of the two middle values, when the values are listed in order. See also **population median.**

median-median line A line of fit found by ordering a data set by its x-values, dividing it into three groups, finding three points (M_1, M_2, and M_3) based on the median x-value and the median y-value for each group, and writing the equation that best fits these three points.

midline A horizontal line, where half of the function is above it, and half of the function is below it.

minimum The least value in a data set or the least value of a function or graph.

minor axis The shorter dimension of an ellipse. Or the line segment with endpoints on the ellipse that has this dimension.

mode A measure of central tendency for a one-variable data set that is the value(s) that occur most often. See also **population mode.**

model A mathematical representation (sequence, expression, equation, or graph) that closely fits a set of data.

modified box plot A box plot in which any values that are more than 1.5 times the IQR from the ends of the box are plotted as separate points.

modulus of a complex number See magnitude of a complex number.

monomial A polynomial with one term.

multiplicative identity The number 1 is the multiplicative identity because any number multiplied by 1 remains unchanged.

multiplicative inverse Two numbers are multiplicative inverses, or reciprocals, if their product is 1.

mutually exclusive (events) Two outcomes or events are mutually exclusive when they cannot both occur simultaneously.

natural logarithm A logarithm with base e, written $\ln x$, which is shorthand for $\log_e x$.

negative exponents For $a > 0$, and all real values of n, the expression a^{-n} is equivalent to $\frac{1}{a^n}$ and $\left(\frac{a}{b}\right)^{-n} = \left(\frac{b}{a}\right)^n$.

nonrigid transformation A transformation that produces an image that is not congruent to the original figure. Dilations are nonrigid transformations (unless the scale factor is 1 or -1).

normal curve The graph of a normal distribution.

normal distribution A symmetric bell-shaped distribution. The equation for a normal distribution with mean μ and standard deviation σ is $y = \frac{1}{\sigma\sqrt{2\pi}}(\sqrt{e})^{-[(x-\mu)/\sigma]^2}$.

oblique (triangle) A triangle that does not contain a right angle.

observational study A type of study in which the researcher measures variables of interest, but does not assign a treatment to the subjects of the study, and the subjects are not aware of what is being measured.

odd function A function that is symmetric about the origin. For all values of x in the domain of an odd function, $f(-x) = -f(x)$.

one-to-one function A function whose inverse is also a function.

outcome A possible result of one trial of an experiment.

outlier A value that stands apart from the bulk of the data.

parabola A locus of points in a plane that are equidistant from a fixed point, called the focus, and a fixed line, called the directrix.

paraboloid A three-dimensional shape formed by rotating a parabola about its line of symmetry.

parameter (in parametric equations) See **parametric equations.**

parameter (statistical) A number, such as the mean or standard deviation, that describes an entire population.

parametric equations A pair of equations used to separately describe the x- and y-coordinates of a point as functions of a third variable, called the parameter.

parent function The most basic form of a function. A parent function can be transformed to create a family of functions.

partial sum A sum of a finite number of terms of a series.

Pascal's triangle A triangular arrangement of numbers containing the coefficients of binomial expansions. The first and last numbers in each row are 1's, and each other number is the sum of the two numbers above it.

percentile rank The percentage of values in a data set that are below a given value.

perfect square A number that is equal to the square of an integer, or a polynomial that is equal to the square of another polynomial.

perfect-square trinomial Any trinomial in the form $a^2 + 2ab + b^2$, whose factored form is a binomial squared, $(a + b)^2$.

period The time it takes for one complete cycle of a cyclical motion to take place. Also, the minimum amount of change of the independent variable needed for a pattern in a periodic function to repeat.

periodic function A function whose graph repeats at regular intervals.

permutation An arrangement of choices in which the order is important.

phase shift The horizontal translation of a periodic graph.

piecewise function A function that consists of two or more functions defined on different intervals.

point estimate a single value given as an estimate of a parameter of a population.

point-ratio form The form $y = y_1 \cdot b^{x-x_1}$ of an exponential function equation, where the curve passes through the point (x_1, y_1) and has ratio b.

point-slope form The form $y = y_1 + b(x - x_1)$ of a linear equation, where (x_1, y_1) is a point on the line and b is the slope.

polar coordinates A method of representing points in a plane with ordered pairs in the form (r, θ), where r is the distance of the point from the origin and θ is the angle of rotation of the point from the positive x-axis.

polar form (of a vector) A way to represent a vector that describes the magnitude of the vector and the angle it makes with the positive x-axis.

polar form of a complex number A complex number written in the form $r(\cos(\theta) + i\sin(\theta))$ where $r = \sqrt{a^2 + b^2}$ and θ is the angle between the positive x-axis and the ray from the origin through $a + bi$. An abbreviated version of this is written as $r\operatorname{cis}\theta$.

polar form of a vector A vector written in the form $\vec{v} = \langle r, -\theta \rangle$ where r is the magnitude of the vector and θ is the angle the vector makes with the positive x-axis.

polynomial A sum of terms containing a variable raised to different powers, often written in the form $a_n x^n + a_{n-1} x^{n-1} + \cdots + a_1 x^1 + a_0$, where x is a variable, the exponents are nonnegative integers, and the coefficients are real numbers, and $a_n \neq 0$.

polynomial function A function in which a polynomial expression is set equal to a second variable, such as y or $f(x)$.

population A complete set of people or things being studied.

population mean (μ) The mean of an entire population or a distribution of a random variable.

population median For a probability distribution, the median is the number d such that the line $x = d$ divides the area into two parts of equal area.

population mode For a probability distribution, the mode is the value(s) of x at which the graph reaches its maximum value.

population standard deviation (σ) The standard deviation of an entire population or of a distribution of a random variable.

power function A function that has a variable as the base. The general form of a power function is $y = ax^n$, where a and n are constants.

power of a power property For $a > 0$, and all real values of m and n, $(a^m)^n$ is equivalent to a^{mn}.

power of a product property For $a > 0$, $b > 0$, and all real values of m, $(ab)^m$ is equivalent to $a^m b^m$.

power of a quotient property For $a > 0$, $b > 0$, and all real values of n, $\left(\dfrac{a}{b}\right)^n$ is equivalent to $\dfrac{a^n}{b^n}$.

power property of equality For all real values of a, b, and n, if $a = b$, then $a^n = b^n$.

power property of logarithms For $a > 0$, $x > 0$, and $n > 0$, $\log_a x^n$ can be rewritten $n \log_a x$.

prediction interval An interval that predicts the distribution of individual points. For example, a 90% prediction interval means that there is a 90% chance that the next measurement will fall in this interval.

prediction line A line of fit for a data set that can be used to predict the expected value of y for any given x-value. The variable $\hat{y}$ is used in place of y to indicate that a line is a prediction line.

principal The initial monetary balance of a loan, debt, or account.

principal value The one solution to an inverse trigonometric function that is within the range for which the function is defined.

probability distribution A function that will give the probability of an event of a discrete random variable or a function in which the probability of a continuous random variable in the interval (a, b) is the area under the function.

probability distribution function (PDF) The function (formula) that gives the probability that a random variable is equal to a value.

product property of exponents For $a > 0$ and $b > 0$, and all real values of m and n, the product $a^m \cdot a^n$ is equivalent to a^{m+n}.

product property of logarithms For $a > 0$, $x > 0$, and $y > 0$, $\log_a xy$ is equivalent to $\log_a x + \log_a y$.

projectile motion The motion of an object that rises or falls under the influence of gravity.

quadratic formula If a quadratic equation is written in the form $ax^2 + bx + c = 0$, the solutions of the equation are given by the quadratic formula, $x = \dfrac{-b \pm \sqrt{b^2 - 4ac}}{2a}$.

quadratic function A polynomial function of degree 2. Quadratic functions are in the family with parent function $y = x^2$.

quotient property of exponents For $a > 0$ and $b > 0$, and all real values of m and n, the quotient $\dfrac{a^m}{a^n}$ is equivalent to a^{m-n}.

quotient property of logarithms For $a > 0$, $x > 0$, and $y > 0$, the expression $\log_a y$ can be rewritten as $\log_a x - \log_a y$.

radian An angle measure in which one full rotation is 2π radians. One radian is the measure of an arc, or the measure of the central angle that intercepts that arc, such that the arc's length is the same as the circle's radius.

radical A square root symbol.

radius See **circle**.

raised to the power A term used to connect the base and the exponent in an exponential expression. For example, in the expression b^x, the base, b, is raised to the power x.

randomization A characteristic of good data collection in which measurements are taken, subjects are chosen, and/or treatments are applied in a random order.

random number A number that is as likely to occur as any other number within a given set.

random process A process in which no individual outcome is predictable.

random sample A sample in which not only is each person (or thing) equally likely, but all groups of persons (or things) are also equally likely.

random variable A variable that takes on numerical values governed by a chance experiment.

range (of a data set) A measure of spread for a one-variable data set that is the difference between the maximum and the minimum.

range (of a relation) The set of output values of a relation.

rational Describes a number or an expression that can be expressed as a fraction or ratio.

rational exponent An exponent that can be written as a fraction. The expression $a^{m/n}$ can be rewritten as $(\sqrt[n]{a})^m$ or $\sqrt[n]{a^m}$, for $a > 0$.

rational function A function that can be written as a quotient, $f(x) = \dfrac{p(x)}{q(x)}$, where $p(x)$ and $q(x)$ are polynomial expressions and $q(x)$ is of degree 1 or higher.

Rational Root Theorem If the polynomial equation $P(x) = 0$ has rational roots, they are in the form $\dfrac{p}{q}$, where p is a factor of the constant term and q is a factor of the leading coefficient.

real axis See **complex plane**.

rectangular form (vector) A way to represent a vector in which the x- and y-coordinates of the vector head are used to name the vector (assuming the vector has been translated to the origin). For example, <2 , 5> describes a vector with its tail at (0, 0) and its head at (2, 5).

recursion Applying a procedure repeatedly, starting with a number or geometric figure, to produce a sequence of numbers or figures. Each term or stage builds on the previous term or stage.

recursive formula A starting value and a recursive rule for generating a sequence.

recursive rule Defines the nth term of a sequence in relation to the previous term(s).

reduced row-echelon form A matrix form in which each row is reduced to a 1 along the diagonal. All entries above and below the diagonal are 0's.

reference angle The acute angle between the terminal side of an angle in standard position and the x-axis.

reference triangle A right triangle that is drawn connecting the terminal side of an angle in standard position to the x-axis. A reference triangle can be used to determine the trigonometric ratios of an angle.

reflection A transformation that flips a graph across a line, creating a mirror image.

regression analysis The process of finding a model with which to make predictions about one variable based on values of another variable.

relation Any relationship between two variables.

relative frequency The count of values divided by the total count. This proportion (or percent) is used to create graphs and experimental probabilities.

relative frequency histogram A histogram in which the height of each bin shows proportions instead of frequencies.

Remainder Theorem If $P(r) = 0$, then r is a zero and $(x - r)$ is a factor of the polynomial function $y = P(x)$. This theorem is used to confirm that a number is a zero of a function.

replication A characteristic of good experimental design in which repeated measurements are made under the same conditions (treatments).

residual For a two-variable data set, the difference between the y-value of a data point and the y-value predicted by the equation of fit.

response variable In statistics, the outcome (dependent) variable that is being studied.

resultant (vector) The sum of two vectors.

rigid transformation A transformation that produces an image that is congruent to the original figure. Translations, reflections, and rotations are rigid transformations.

root mean square error (s) A measure of spread for a two-variable data set, similar to standard deviation for a one-variable data set. It is calculated by the formula $s = \sqrt{\dfrac{\sum_{i=1}^{n}(y_i - \hat{y})^2}{n-2}}$.

roots The solutions of an equation in the form $f(x) = 0$.

row reduction method A method that transforms a matrix into reduced row-echelon form.

sample A part of a population selected to represent the entire population. Sampling is the process of selecting and studying a sample from a population in order to make conjectures about the whole population.

sample space The range of values of a random variable

scalar A real number, as opposed to a matrix or vector.

scalar multiplication The process of multiplying a matrix by a scalar. To multiply a scalar by a matrix, you multiply the scalar by each value in the matrix.

scale factor A number that determines the amount by which a graph is dilated, either horizontally or vertically.

secant The reciprocal of the cosine ratio. If A is an acute angle in a right triangle, the secant of angle A is the ratio of the length of the hypotenuse to the length of the adjacent leg, or $\sec A = \dfrac{hyp}{adj}$. See **trigonometric function**.

sequence An ordered list of numbers.

series The indicated sum of terms of a sequence.

shape (of a data set) Describes how the data are distributed relative to the position of a measure of central tendency.

shifted geometric sequence A geometric sequence that includes an added term in the recursive rule.

sigma notation The mathematical notation for summation, as represented by the Greek letter Σ (sigma). For example, $\sum_{n=1}^{4} \dfrac{1}{n}$ means $\dfrac{1}{1} + \dfrac{1}{2} + \dfrac{1}{3} + \dfrac{1}{4}$.

significant figures (significant digits) The digits in a measurement that give information about its accuracy. For example, 3.00 has three significant digits, whereas 3 has one significant digit.

similar Geometric figures that are dilations of one another by the same horizonal and vertical scale factors. Two figures are similar if and only if all corresponding angles are congruent and lengths of all corresponding sides, edges, or other one-dimensional measures are proportional.

simple event An event consisting of just one outcome.

simple interest A percentage paid on the principal over a period of time.

simple random sample See **random sample**.

simulation A procedure that uses a chance model to imitate a real situation.

sine If A is an acute angle in a right triangle, then the sine of angle A is the ratio of the length of the opposite leg to the length of the hypotenuse, or $\sin A = \dfrac{opp}{hyp}$. See **trigonometric function**.

sine wave A graph of a sinusoidal function. See **sinusoid**.

sinusoid A function or graph for which $y = \sin x$ or $y = \cos x$ is the parent function.

skewed (data) Data that are spread out more on one side of the center than on the other side.

slant asymptote A linear asymptote that is not parallel to either the x- or y-axis.

slope The steepness of a line or the rate of change of a linear relationship. If (x_1, y_1) and (x_2, y_2) are two points on a line, then the slope of the line is $\dfrac{y_2 - y_1}{x_2 - x_1}$, where $x_2 \neq x_1$.

spread The variability in numerical data.

square root function The function that undoes squaring, giving only the positive square root (that is, the positive number that, when multiplied by itself, gives the input). The square root function is written $y = \sqrt{x}$.

standard deviation (s) A measure of spread for a one-variable data set that uses squaring to eliminate the effect of the different signs of the individual deviations. It is the square root of the variance, or $s = \sqrt{\dfrac{\sum_{i=1}^{n}(x_i - \bar{x})^2}{n-1}}$. See also **population standard deviation.**

standard form (of a conic section) The form of an equation for a conic section that shows the transformations of the parent equation.

standard form (of a linear equation) The form $ax + by = c$ of a linear equation.

standard normal distribution A normal distribution with mean 0 and standard deviation 1.

standard position An angle positioned with one side on the positive x-axis.

standardizing the variable The process of converting data values (x-values) to their images (z-values) when a normal distribution is transformed into the standard normal distribution.

statistic A numerical measure of a data set or sample.

statistics A collection of numerical measures, or the mathematical study of data collection and analysis.

stem-and-leaf plot A one-variable data display in which the left digit(s) of the data values, called the stems, are listed in a column on the left side of the plot, and the remaining digits, called the leaves, are listed in order to the right of the corresponding stem.

step function A function whose graph consists of a series of horizontal line segments.

substitution A method of solving a system of equations that involves solving one of the equations for one variable and substituting the resulting expression into the other equation.

survey A type of study in which a researcher collects information about treatment(s) and results from the subjects.

symmetric (data) Data that are balanced, or nearly so, about the center.

synthetic division An abbreviated form of dividing a polynomial by a linear factor.

system of equations A set of two or more equations with the same variables that are solved or studied simultaneously.

tail The end of a vector that does not have an arrowhead or in statistics, the extreme values in a distribution.

tangent If A is an acute angle in a right triangle, then the tangent of angle A is the ratio of the length of the opposite leg to the length of the adjacent leg, or $\tan A = \dfrac{opp}{adj}$. See **trigonometric function.**

term (algebraic) An algebraic expression that represents only multiplication and division between variables and constants.

term (of a sequence) Each number in a sequence.

terminal side The side of an angle in standard position that is not on the positive x-axis.

theoretical probability A probability calculated by analyzing a situation, rather than by performing an experiment, given by the ratio of the number of different ways an event can occur to the total number of equally likely outcomes possible.

third quartile (Q_3) The median of the values greater than the median of a data set.

transcendental number An irrational number that, when represented as a decimal, has infinitely many digits with no pattern, such as π or e, and is not the solution of a polynomial equation with integer coefficients.

transformation A change in the size or position of a figure or graph.

transition diagram A diagram that shows how something changes from one time to the next.

transition matrix A matrix whose entries are transition probabilities.

translation A transformation that slides a figure or graph to a new position.

tree diagram A diagram whose branches show the possible outcomes of an event, and sometimes probabilities.

triangle inequality theorem For any triangle, the length of a side must be less than or equal to the sum of the other two sides, but greater than or equal to the difference between the two sides.

trigonometric function A periodic function that uses one of the trigonometric ratios to assign values to angles with any measure.

trigonometric ratios The ratios of lengths of sides in a right triangle. The three primary trigonometric ratios are sine, cosine, and tangent.

trigonometry The study of the relationships between the lengths of sides and the measures of angles in triangles.

trinomial A polynomial with three terms.

two-way table A summary of counts of value for two categorical variables.

unbiased estimate A simple random sample will produce unbiased estimates because the data are collected from a sample in which every member of the population is equally likely to be selected.

union The set of elements that contains all elements of the sets involved. The word *or* is often associated with the union.

unit circle A circle with radius of one unit. The equation of a unit circle with center (0, 0) is $x^2 + y^2 = 1$.

unit hyperbola The parent equation for a hyperbola, $x^2 - y^2 = 1$ or $y^2 - x^2 = 1$.

variance (s^2) A measure of spread for a one-variable data set that uses squaring to eliminate the effect of the different signs of the individual deviations. It is the sum of the squares of the deviations divided by one less than the number of values, or $s^2 = \dfrac{\sum_{i=1}^{n}(x_i - \bar{x})^2}{n - 1}$.

vector A quantity with both magnitude and direction.

velocity A measure of speed and direction. Velocity can be either positive or negative.

Venn diagram A diagram of overlapping circles that shows the relationships among members of different sets.

vertex (of a conic section) The point or points where a conic section intersects the axis of symmetry that contains the focus or foci.

vertex (of a feasible region) A corner of a feasible region in a linear programming problem.

vertex-edge graph A diagram comprised of a set of points, called vertices, along with segments or arcs, called edges, connecting some or all of the points. The essential information in the graph is shown by the way the vertices are connected by the edges.

vertex form The form $y = a(x - h)^2 + k$ of a quadratic function, where $a \neq 0$. The point (h, k) is the vertex of the parabola, and a is the vertical scale factor.

vertical dilation A transformation that increases or decreases the length of a figure or graph. A vertical dilation by a factor of b multiplies the y-coordinate of every point on a graph by b.

vertical reflection A reflection of a function across the x-axis.

zero exponent For all values of a except 0, $a^0 = 1$.

zero-product property If the product of two or more factors equals zero, then at least one of the factors must equal zero. A property used to find the zeros of a function without graphing.

zeros (of a function) The values of the independent variable (x-values) that make the corresponding values of the function ($f(x)$-values) equal to zero. Real zeros correspond to x-intercepts of the graph of a function. See **roots**.

z-value or **z-score** The number of standard deviations that a given x-value lies from the mean.

Index

absolute-value function, 180–184
addition
    arithmetic sequences and, 33
    of complex numbers, 285, 286
    of rational expressions, 364–366
    See also sum
addition property of equality, 110
addition rules
    for mutually exclusive events, 466–468
    of probability, 465–470
Alcuin of York, 5
algebra, 14
al-Kashi, Jamshid Masud, 322
al-Khwarizm, Muhammad ibn M−us−a, 7
amplitude, 418
analytic geometry, 4
anecdotal studies, 484
angle(s)
    central, 399, 412
    in circular functions, 401–403
    coordinate plane and measurement of, 393–395
    coterminal, 403
    radian measure of, 408–409
    reference, 394, 401
angular speed, 411
applications
    agriculture and horticulture, 66, 132, 266
    appreciation and depreciation, 38, 164, 202–203
    archaeology and anthropology, 44, 107, 173, 241, 407
    architecture, 17, 28, 30, 45, 72, 98, 188, 282, 305, 307, 315, 351
    art, 158, 163, 200, 282
    astronomy, 155, 157, 223, 229, 405, 425, 435, 436
    biology, human, 87, 257, 333, 452, 498
    biology, nonhuman, 131, 224, 228, 341, 496
    botany, 46, 266, 350
    business, 20, 31–32, 36, 50, 65, 73, 80, 82–83, 106, 107, 125, 127–128, 131, 139, 147, 156, 198, 266, 275, 303, 350, 480
    Celsius and Fahrenheit conversion, 235
    chemistry, 186, 223, 441, 497
    computers and Internet, 62, 163, 187
    construction and maintenance, 6, 48–49, 73, 273, 312
    consumer awareness, 12, 16, 22, 61, 67, 113, 139, 186, 208, 235, 257, 377, 382, 463, 487, 514
    cooking, 13, 311, 462
    design and engineering, 13, 72, 166, 302, 312, 351, 379, 391–392, 478
    diet and nutrition, 23, 129–130, 144
    distance calculations, 22, 65, 88, 104, 171, 178, 180, 185, 231–232, 248, 297–300, 387–388, 390–391, 406, 414–415, 416, 420–422
    economics, 32
    education, 81, 120, 122, 257, 513
    employment, 18, 35, 37, 42
    entertainment, 99, 254, 435, 449, 453
    environment, 48, 76–77, 156, 206, 243, 276, 378, 512–513
    gambling, 447–448, 449
    genealogy, 43
    geometry, 155
    government, 88
    income, 63–64, 67, 72, 153
    insurance, 447
    interest, 40–41, 42, 43, 239–240, 240, 441
    investments, 42, 50, 225
    law enforcement, 178, 298
    life expectancy, 93, 496
    loans, credit, and mortgages, 40, 148
    manufacturing, 131, 132, 378, 463, 471, 497
    medicine and health, 47–48, 51, 452
    meteorology and climatology, 54, 115, 249, 517
    music, 44, 125, 165, 221, 242, 286, 514
    oceanography, 55, 399
    pets, 19, 131, 273, 481, 516
    physics, 72, 73, 98, 113, 114, 126, 155, 183–184, 226, 227, 325, 377, 426, 434, 455–456, 471
    polls and surveys, 472
    population, human, 44, 45, 104, 205
    population, nonhuman, 52, 125, 224
    radioactivity, 201, 215, 229
    recycling, 156
    resource consumption and conservation, 22, 70–71, 132, 235, 243, 311, 504
    safety, 93
    seismology, 71–72, 200
    sound, 220
    sports, 46, 73, 78, 94, 158, 167, 170, 223, 272, 377, 414, 463, 472, 502–503, 516, 517
    technology, 241
    telecommunications, 194, 301, 305, 398, 434
    tides, 399, 429–430, 434, 440, 444
    transportation, 22, 23, 126, 195, 241, 504
    velocity and speed calculations, 14–15, 45, 66, 68, 132, 320
arc(s)
    defined, 407
    length of, 407–412
    measure of, 407–408, 467–470
area
    of a parallelogram, 445
    polynomial multiplication and division with, 267
    probability and, 451–453, 492, 499–500
    of sectors, 411–412
area model, 451
Aristotle, 293
arithmetic sequences, 31–32
    applications of, 46–49
    as basic sequence, 33
    common difference of, 31, 318
    defined, 31
    explicit formulas for, 61–64
    graphs of, 54–55
    slope and, 62
arithmetic series, 243–246
    formula for partial sum of, 244–245
    partial sum of, 243, 246
association, 485
asymptotes
    defined, 225
    horizontal, 225–226
    of rational functions, 352–353, 358–359
    slant, 366

Babbage, Charles, 322
base, 209
    of exponents, 209
    of logarithms, 238–239

Bell, Alexander Graham, 220
bell curve. *See* normal distributions
bias in sampling, 485, 486–487
binomial distributions, 491–492. *See also* normal distributions
binomials
    defined, 318
    difference of two squares, 267, 347
    factored, 267
bivariant independence, 473–476
bivariate data, 506–510
bivariate sampling, 506
Boyle, Robert, 223
Braille, Louis, 464
building-block function, 165

Cartesian graphs, 4. *See also* coordinate graphs
causation, 485, 509–510
Celsius, Anders, 234
Celsius and Fahrenheit conversion, 235
central angle, 399, 407, 412
Chu Shih-chieh, 243
circle(s)
    arcs of. *See* arc(s)
    central angle, 399, 407, 412
    circumference, 407
    distance formula to describe points on, 299
    great, 415
    radian measurement of angles, 407–412
    sectors, area of, 411–412
    semicircles, 190
    transformations and, 188–189
    unit. *See* unit circle
    See also circular functions
circular functions, 399–403
    cosine, 399–403
    cyclical patterns modeled by, 399
    notation and terminology for, 399, 401, 403
    as periodic, 400–401
    sine, 399–403
    See also sinusoids (sine waves)
circumference, 407, 409
closed half-plane, 121
Coleman, Ornette, 165
combined variation, 377
common base, 238
common base property of equality, 210
common difference, 31
common logarithm, 238
common ratio, defined, 33
complements, 468–470
completing the square, 268–272

complex arithmetic, 285
complex conjugates, 283
complex numbers, 283–285
    complex conjugates, 283
    conjugate pairs. *See* conjugate pairs
    defined, 284
    imaginary unit (*i*), 283
    modeling using, 285
    operations with, 285
composition of functions, of inverse and its function, 232–233
compound, 450
compound event, 450, 457
compound interest, 40–41
conditional probability, 460
confidence interval, 499–503
    definition, 501
confidence intervals, 501–502
conjugate pairs
    defined, 283–284
    zeros as, 331, 337–338
Connections
    architecture, 98
    business, 275
    careers, 15, 32, 379, 447
    consumers, 22, 38, 236
    cultural, 7, 43, 243, 254, 322, 386, 406, 407, 436, 449
    economics, 241
    engineering, 72, 166, 312
    environment, 48, 78, 243, 276, 311, 378, 512
    health, 144, 486
    history, 7, 10, 34, 88, 234, 243, 245, 282, 283, 298, 322, 407, 415, 436, 449, 464, 489, 491, 495, 498
    language, 156, 160, 386
    mathematics, 4, 32
    music, 165
    recreation, 73, 388, 396, 414
    science, 23, 56, 66, 76, 108, 113, 126, 173, 183, 194, 204, 220, 223, 325, 391, 399, 405, 425, 429, 471, 498
    technology, 398
    See also applications
consistent systems of equations, 111
constant of variation, 14
constraints, 121, 122, 123
continuous graphs, 144, 150
coordinate graphs
    history of, 3
    measuring angles on, 393–395
    problem solving with, 3
correlation, 506–510
    causation distinguished from, 509–510
    defined, 506
correlation coefficient, 506–510

cosecant (csc), 436
cosine (cos), 386
    circular functions and, 399–403
    for real angle values, 393–395
    reciprocal of (secant, sec), 436
    See also sinusoids; trigonometry
cotangent, 436
coterminal angles, 403
coterminal position, 403
cubic functions, 327
    defined, 327
    factored form of, 327–328
    graphs of, 334

data
    bivariate, 506
    extrapolation, 78
    finite differences method and, 320, 322
    interpolation, 78
    See also graphs; samples; statistics; variables
da Vinci, Leonardo, 66
decay, 38–41. *See also* growth and decay
degree of polynomials
    defined, 318
    finite differences method of finding, 319–322
    shape of graph and finding, 336
degrees, conversion to radians, 409–410
de Moivre, Abraham, 491
dependent events, 460
dependent systems of equations, 111
dependent variable, 69
    of functions, 149, 153
    of inverse functions, 230, 232
depreciation, 38, 164, 202–203
Descartes, René, 4
dice, 449
difference of two squares, 267, 347
differences. *See* subtraction
dilations
    absolute-value function, 180–183
    circles and, 188–189
    defined, 184
    horizontal, 181
    as nonrigid transformation, 181
    of rational functions, 354–355
    scale factor and, 180
    summary of, 191
    of trigonometric functions, 419
    vertical, 181
dimensional analysis, 410
directrix, 305
direct variation, 14

discrete graphs, 54
    of functions, 149
    interpretation of, 143
    of sequences, 54
discriminant, 287, 289
distance formula, using, 297–300
distributions. *See* normal
    distributions
division
    inequalities and, 120
    of rational expressions, 368–371
    synthetic, 346–347
domain, 70, 401
double roots, 337
doubling time, 203
dynamic systems, 48

e, 491
elimination, in systems of equations, 110–112
elimination method, 110
ellipse(s), transformations of circles and, 189–190
end behavior, 334
    of function, 203
equality, properties of, 110
equally likely outcomes, 450
equations
    exponential. *See* exponential curves; exponential equations
    linear. *See* linear equations
    normal distribution, 490
    of a parabola, 308
    power, 225–227
    quadratic. *See* quadratic equations
    radian measure, 409–410
    roots of. *See* roots of an equation
    systems of. *See* systems of equations
    unit circle, 188
    See also equations, solving; functions
equations, solving
    exponential equations, with logarithms, 239–240
    exponent properties and, 210–211, 225–227, 237–238
    power equations, 225–227
    quadratic. *See* quadratic equations, solving
    systems of. *See* systems of equations, solving
    undoing order of operations, 213, 225
equivalent quadratic forms, 260–263
    general form of, 260
Euler, Leonhard, 283
even functions, 174

events
    complements, 468–470
    compound, 450
    defined, 450
    dependent, 460
    independent, 460, 468
    simple, 450
expanded form of exponents, 209
experimental design, 484–487
    anecdotal, 484
    association and, 485
    bias in, 485–487
    causation and, 485
    experimental, 484–487
    observational, 484–487
    surveys, 484–487
experimental probability, 450, 452
experimental studies, 484–485, 486
explanatory variables, 508
explicit formulas, 61
exponential curves
    point-ratio form of equation for, 217–220, 225–227
    solving with systems of equations, 258
exponential equations
    applications of, 225–229
    solving, with logarithms, 237–240
exponential functions, 201–204
    applications of, 225–227
    doubling time, 203
    general form of, 203, 211
    graphing, 237
    growth and decay modeled with, 201–204, 225
    properties of, 209–212
    solving, 202–203
    transformations of, 204
exponents
    base of, 209
    expanded form, 209
    negative, 210
    notation for, 209
    positive bases defined for properties of, 212
    properties of, 209–211
    rational. *See* rational exponents
expressions
    operations with rational, 364–366
    See also equations; polynomials; terms
extending trigonometry, 393–395
extrapolation, 78
extreme values, 334

factored form of equations, 267, 327–331, 342–343, 349

factored form of quadratic
    functions, 263
    conversion to general form, 262
    roots corresponding to factors, 381
    vertex found with, 269
factored form of rational functions, 361
factoring polynomials, 327–331
Factor Theorem, 344
Fahrenheit and Celsius conversion, 235
fair coin, 448
family of functions, 165
family trees, 43
feasible region, 121, 123, 127–130
Fibonacci, Leonardo, 12, 37
Fibonacci sequence, 37
finding solutions, 342–347
finite, 243
    differences method, 319–322
fit, 82–86
    line of, 75
focus, 305
formulas
    explicit, 61
    recursive, 29–31
Foucault, Jean Bernard Leon, 155
Foucault pendulum, 155
Fractals, Sierpiński triangle, 32–33
frequency, of cyclic functions, 431
Friedman, Irving, 173
functional response curve, 378
functions
    absolute-value, 180–184
    with arithmetic, 152
    building-block, 165
    building inverses of_, 230–233
    circular. *See* circular functions
    composition of. *See* composition of functions
    defined, 149
    dilations of, 180–184. *See also* dilations
    end behavior of, 203
    even, 174, 196
    exponential. *See* exponential functions
    family of, 165
    inverse. *See* inverse functions
    linear. *See* linear functions
    logarithms. *See* logarithmic functions
    long run behavior of, 203
    notation for, 149–153
    odd, 196
    one-to-one, 230–233
    parent, 165
    periodic, 400
    polynomial, defined, 318
    power, 209–212

projectile motion, 268
quadratic, 165–166
rational. *See* rational functions
reflections of. *See* reflections
square root, 172–175
transformations of, 191
translation of, 158–161, 160
vertical line test to determine, 149
zeros of. *See* zeros of a function
See also equations

Galileo Galilei, 183, 320
Galton, Francis, 498
Gauss, Carl Friedrich, 245
general addition rule, 468
general form, defined, 318
general form of quadratic equations. *See* quadratic equations
general form of quadratic functions, 260, 263
    converting to vertex form, 268–272
    factored form converting to, 262
generalizing patterns, 14–17
general term, 29
geometric probability, 451
geometric sequences, 32–33
    as basic sequence, 33
    common ratio of, 33
    defined, 33
    graphs of, 54–56
    shifted, 48–49
geometric series, partial sums of, 250–252
geometry
    applications, 155
    diagonals of a polygon, 319–320
    language of, 156
    See also Improving Your Geometry Skills
geosynchronous orbits, 194
golden ratio, 282
golden rectangle, 282
graphing
    of absolute-value functions, 180
    of arithmetic sequences, 62
    of exponential functions, 237
    of line of fit, 75
    of normal distributions, 490, 493
    of parabolas, 165–168
    of polynomial functions, 328–331, 342–343
    of quadratic functions, 261
    of rational functions, 352–353, 357–361
    of real angle values, 393–395
    of sequences, 53–56, 62
    of square root functions, 172–175

of systems of equations, 102–103
of systems of inequalities, 120–123
of transformations. *See* transformations
of trigonometric functions. *See* sinusoids (sine waves)
x-intercepts, finding with, 277, 349
graphs
    analyzing, 56
    continuous, 144, 149
    discrete, 54
    of functions and relations, 150
    general shape of, 54
    holes, 357
    interpretation of, 143–145
    linear, 54
    problem solving with, 3
    of rational functions, 356–361
    transformations of. *See* transformations
    vertical line test of, 149
great circle, 415
greatest integer function, 105, 107
growth and decay, 38–41
    asymptotes and, 225–227
    definitions, 40
    doubling time, 203
    exponential functions modeling, 201, 225
    half-life, 204
    recursion modeling, 38–41, 54–56
guess-and-check, 21

Haley, Alex, 43
half-life, 203–204
half-plane, 121
Hendrix, Jimi, 44
HH, 90–89
higher-degree polynomials, 334–337
Hipparchus of Rhodes, 407
holes, 357
Hollings, C. S., 378
horizontal asymptote, 225–227
horizontal dilation, 181, 184
horizontal reflection, 173
How to Solve It (Pólya), 10
hyperbola, 352

i, 283
ibn Musa Al-Khwarizmi, Muhammad, 7
identity, 436
identity, defined, 436
image, 160

image of original point, 160
imaginary number, 283
imaginary unit (i), 283
Improving Your Geometry Skills
    Lines in Motion Revisited, 187
Improving Your Reasoning Skills
    Breakfast Is Served, 208
    Cartoon Watching Causes Small Feet, 115
    The Dipper, 157
    Fair Share, 13
    Fibonacci and the Rabbits, 37
    Sequential Slopes, 67
Improving Your Visual Thinking Skills
    Acorns, 498
    An Equation Is Worth a Thousand Words, 441
    4-in-1, 195
    Miniature Golf, 304
    Sums and Differences, 326
    Think Pink, 52
inconsistent systems of equations, 111
independent events, 460, 468
independent systems of equations, 111
independent variable, 69
    of functions, 149
    of inverse functions, 230, 232
inequalities, 120
    defined, 120
    operations on, 120
    See also systems of inequalities
inflection points, 494
intercept form, 68–70
    defined, 68
    and point-slope, graphical relation of, 158–161
interest, 40–41, 43, 50, 240, 441
interpolation, 78
intersection, 466
inverse functions, 230–233
    one-to-one function, 232
    trigonometric, 387
inverse of function, 230–233
inverse relations, 232
inverse variation, 352. *See also* rational functions
Investigations
    Addition Rule, 467
    Arithmetic Series Formula, 244
    Balloon Blastoff, 69–70
    To Be or Not to Be (a Function), 151
    A Bouncing Spring, 431
    The Box Factory, 329
    The Breaking Point, 351–352
    Bucket Race, 297–298
    Complete the Square, 269–270
    Complex Arithmetic, 285

Doses of Medicine, 47–48
Exponents and Logarithms, 237–238
Flip a Coin, 448
Fold a Parabola, 309
Free Fall, 321
Geometric Series Formula, 251
Getting to the Root, 216
The Inverse, 230–231
The Largest Triangle, 335
Looking for Connections, 506–507
Looking for the Rebound, 39
Make My Graph, 166
Match Point, 63
Match Them Up, 53
Maximizing Profit, 127–128
Monitoring Inventory, 31–32
Movin' Around, 158–159
The Multiplication Rule, 458–459
Paddle Wheel, 400
Paying for College, 120–121
The Pendulum, 183
The Pendulum II, 420
Polar Bear Crossing the Arctic, 3
Population Trends, 104
Predicting Asymptotes and Holes, 358
Problems, Problems, Problems, 10
Properties of Exponents, 209
Pythagorean Identities, 437–438
A Radian Protractor, 408
Radioactive Decay, 201
Rolling Along, 263
Spring Experiment, 85
Steep Steps, 389
Take a Moment to Reflect, 172–173
The Wave, 78
What's Your System?, 111–112
When Is a Circle Not a Circle?, 190
Who Owns the Zebra?, 17

joint variation, 377

Kepler, Johannes, 223
Kit–ab al-jabr wa'al-muq–abalah (al-Khw–arizm), 7, 386

larger systems, 133–135
least common denominator, 364
least squares line, 90–92

length of arc, 407
Leonardo da Vinci, 66
Liber Abaci (Book of Calculations) (Fibonacci), 12, 37
Lindbergh, Charles, 415
line(s)
   as functions, 158
   of reflection, 173, 198
   See also line of fit
linear correlation. *See* correlation coefficient
linear equations, 61–64
   arithmetic sequences and, 61–64
   forms of, 158
   formula, 61
   intercept form, 68, 158
   point-slope form. *See* point-slope form
   systems of. *See* systems of equations
linear functions, 158
   intercept form, 158
   point-slope form, 158
   translation of, 158–161
linear graphs, 54
linear modeling, 75–78, 158
   arithmetic sequence, 31, 61–64
   decay, 38–41
   equations, 61–64
   graphing sequences, 53–56
   growth, 38–41
   intercept form, 68–70
   least squares line, 90–92
   prediction line, 75–78
   recursion, 28–33
   residuals, 82–86
   sequences, 28–33
      applications of, 46–49
   slope, 61, 68–70
linear programming, 127–130
linear systems, 102–105
   of equations, 116–117
line of fit, 75
   correlation coefficient and, 506–510
   defined, 75
   estimation of, 75
   point-slope form and, 76–78
   residuals and, 82–86
   root mean square error, 193
line of reflection, 173, 198
line of symmetry, 165, 196
Li Shun-Fêng, 322
local maximum, 334
local minimum, 334
locus (loci), 299
logarithmic functions, 237–240
   common base, 238
   common logarithm, 238
   definition of, 239

   exponents and, 237–238
   solving exponential equations with, 238–240
long division, 345, 346
long run behavior, 334, 335
   of function, 203
long-run value, 48
lurking variables, 510

magnitude, defined, 180
marginal frequencies, 473
margin of error, 501
matural logarithm, 239
maximum
   local, 334
   of quadratic function, 268
mean
   margin of error, 501
   and normal distributions, 490, 492–495
   notation for, 490
   for population, 490
   z-values and confidence intervals and, 499–503
   See also standard deviation
measure of arc, 407
microgravity, 126
midline, 418
minimum
   local, 334
   of quadratic function, 268
model, defined, 54, 75–78
modeling and tools, in mathematics, 2–4
monomials, 318
motion lines, 158–161
multiplication
   of complex numbers, 285
   geometric sequences and, 33
   inequalities and, 120
   of rational expressions, 368–371
multiplication property of equality, 110
multiplication rule for independent events, 460
multiplication rules
   for independent events, 460
   of probability, 457–460
mutually exclusive, 465

natural logarithm, 239
negative exponents, 210
negative numbers, square root of. *See* complex numbers
Nightingale, Florence, 495
nonlinear inequalities, 123, 141

nonlinear programming, 141
non-linear systems of equations, 116–117
nonrigid transformations, 181
normal curve, 490
normal distributions, 490–495
    binomial distribution and, 491–492
    confidence intervals, 501–503
    data types of, 490–491
    defined, 490
    equation for, 493
    general equation for, 490
    graphs of, 490, 493–494
    inflection points, 494
    margin of error, 501
    notation for, 493
    standard, 491
    standardizing the variable, 500–501
    transformations and, 490, 492
    z-values, 499–503
numbers
    complex. *See* complex numbers
    rounding of, 388

observational studies, 484–486
odd function, 174
one-to-one functions, 230–233, 232
open half-plane, 121
order of operations, undoing, to solve equations, 224
outcomes, defined, 450

parabola(s), 305–309
    defined, 305
    directrix of, 306
    equation of, 308
    focus of, 305, 306
    graphing of, 165–168
    line of symmetry of, 165
    transformations of, 307–308
    vertex of, 165
    See also quadratic equations
paraboloid, 305
parameters, 499
parent functions, 165
partial sums
    of arithmetic series, 243–246
    of geometric series, 250–252
patterns, 28
    problem solving with, 21
    See also recursion
Pearson, Karl, 507
perfect-square trinomials, 267
period, 400, 430–431
periodic functions, 400

Péter, Rózsa, 34
phase shift, 418
point(s)
    inflection, 494
    locus of, 299–300
point(s) estimate, 501
point-ratio form, 218
    applications using, 225
point-slope form
    defined, 76
    and intercept form, graphical relation of, 158–161
    line of fit using, 75–76
    translation direction and, 160
point-slope form, for linear equation, 75–76
Pólya, George, 10
polynomials, 318–322
    addition of, 318
    binomials, 267, 318
    defined, 318
    degree of. *See* degree of polynomials
    end behavior, 334
    expressions, generally, 318
    factored form of, 267, 327–331, 342–343, 349
    as functions, defined, 318
    general form of, defined, 318
    higher-degree, 334–337
    local minimums and maximums (extreme values), 334
    monomials, 318
    trinomials, 318
    zeros, finding, 342–347
    See also quadratic functions
population, 499
    mean of, 490
    standard deviation of, 490
    See also samples
population mean, 490
population standard deviation, 490
power equations, solving, 225–226
power functions
    applications of, 225–227
    general form of, 211
    properties of, 209–212
    rational function as, 217
    solving, 212
power of a power property of exponents, 210, 217
power of a product property of exponents, 210
power of a quotient property of exponents, 210
power property of equality, 210
prediction line, 75–78. *See also* line of fit
principal, 40
probability
    addition rule for, 466–490
    area model, 451

    bivariant independence, 473–476
    conditional probability, 460
    events. *See* events
    experimental probability, 450, 451
    geometric probability, 451
    graphs, 450–451
    multiplication rules of, 457–460
    origins of, 447
    outcomes, 450
    of a path, 460, 465
    random processes and, 447–451
    simulations, 450
    theoretical probability, 450–451
    tree diagrams and, 465
    See also experimental design; statistics
Problems for the Quickening of the Mind (Alcuin of York), 5
problem solving
    acting out the problem, 2
    coordinate graphs and, 3
    diagrams and, 2–3
    group effort and, 3
    strategies for, 2–4, 21
product property of exponents, 210
projectile motion, 268
properties
    addition property of equality, 110
    common base property of equality, 210
    of exponents, 210
    logarithm change-of-base property, 239
    multiplication property of equality, 110
    negative exponents defined, 210
    power of a power property of exponents, 210, 217
    power of a product property of exponents, 210
    power of a quotient property of exponents, 210
    power property of equality, 210
    product property of exponents, 210
    quotient property of exponents, 210
    substitution, 110
    zero exponents, 210
    zero-product, 261–262
puzzles. *See* Improving Your Geometry Skills; Improving Your Reasoning Skills; Improving Your Visual Thinking Skills
Pythagorean identities, 436–438, 437–438, 438
Pythagorean Theorem
    equation of the unit circle and, 188
    See also trigonometry

quadratic equations
　discriminant of, 287
　general form, quadratic formula and, 278–280
　solving, 288–290
　　completing the square, 268–272
　　graphing, 261
　　quadratic formula, 278–280
quadratic formula, 277–280, 278–280
quadratic functions, 165–166, 260
　factored form, 262. *See also* factored form of quadratic functions
　fitting to data, 181–183
　form of, choosing, 263–264
　general form of. *See* general form of quadratic functions
　translations of, 165–168, 181, 260
　vertex form of. *See* vertex form of quadratic functions
　zeros of. *See* zeros of a function
quotient property of exponents, 210

radian, defined, 43
radian measure, 407–412
radical, 177
　equations, solving, 293–295
radioactivity, 201, 215, 229
randomness and probability, 447–451
random numbers, 449
random processes, 447–451
random sample. *See* samples
range, 70
ratio, 216–220
rational equations, solving, 373–376
rational exponents, 216–220
　defined, 217
　point-ratio form of equation, 217–218, 225–226
　as power function, 217
　as roots, 216
　transformations and, 217
rational expressions
　adding and subtracting, 364–366
　multiplying and dividing, 368–371
rational functions, 351–354
　asymptotes of, 352–353
　defined, 352
　factored form of, 361
　graphing of, 352–353, 357–361

　holes of, 357, 358
　as power function, 217
　transformations of, 218, 352–354
Rational Root Theorem, 345
reasoning, 7–11
reciprocal identities, 436–437
rectangle diagrams, 267, 346
recursion, 28–33, 95
　defined, 28, 34
　formula for, 29–31
　growth and decay modeled with, 38–41, 54–56
　partial sum of a series, 244, 250
　rule for, 29
　sequences. *See* sequences
recursive definition, 28–29
recursive formula, 29–31
Recursive Functions in Computer Theory (Péter), 34
recursive rule, 29
recycling, 156
reference angle, 394, 401
reference triangle, 394, 401
reflections, 191
　defined, 173
　definition of, 172
　of functions, 173
　horizontal, 173
　line of, 173
　as rigid transformation, 181
　of square root family, 172–175, 181
　and square root function, 172–175
　summary of, 191
　vertical, 173
Refreshing Your Skills, The Distance Formula, 297–300
regression, linear, 90–92
relation
　defined, 149
　inverse of, 232
remainder theorem, 344
reries, 243–256
residuals, 82–86
　defined, 82
　line of fit and, 82–84
response variables, 508
Richter scale, 200
right triangle trigonometry, 385–390
rigid transformations, 181
Robert of Chester, 386
root mean square error, 195
roots of an equation
　defined, 261
　degree of polynomials and number of, 318
　double, 337
　factored form of polynomials and, 330

　quadratic formula to find, 277–280
　See also zeros of a function
rounding
　of numbers, 388
　of trigonometric ratios, 388

samples
　bias in, 485–487
　correlation coefficient and, 506–510
　notation for mean and standard deviation, 490
　z-values and confidence intervals for, 499–503
　See also data; probability; statistics
scale factor, 180
scientific notation
secant (sec), 436
sectors, area of, 411
semicircles, 190
sequences, 28–33
　applications of, 46–49
　arithmetic, 31, 61–64
　defined, 29
　Fibonacci, 37
　geometric, 33
　graph of, 53–56
　graphs of, 53–56, 62
　shifted, 48–49
　summation of terms in. *See* series
　See also terms
series, 243–256
　arithmetic, 243–246
　defined, 243
　finite number of terms, 243
　geometric, 250–252
　partial sum of, 243–246
　See also sequences
shifted sequences, 48–49
short run behavior, 327
shrinks and stretches. *See* dilations
Sierpiński triangle, 32–33
Sierpiński, Wacław, 32
simple event, 450, 457
simple interest, defined, 40
simulations, 450
sine (sin), 386
　circular functions and, 399–403
　for real angle values, 393–395
　reciprocal of (cosecant, csc), 436
　See also sinusoids; trigonometry
sinusoids (sine waves)
　amplitude of, 418
　cosine and, 418, 430
　defined, 417
　frequency, 431
　modeling with, 429–433

INDEX **567**

period of, 418
phase shift of, 418
transformations of, 417–423
68 - 95 - 99.7 rule, 494
slant asymptotes, 366
slope, 61, 68–70
   arithmetic sequences and, 62
   choice of points to determine, 69–70, 75
   See also point-slope form
Smith, Robert L., 173
solving problems. *See* problem solving
solving quadratic equations, 288–290
solving radical equations, 293–295
solving rational equations, 373–376
speed, angular, 411
sphere(s), great arc, 415
spreadsheets, 31
square, completing the, 268–269, 268–272
square root functions
   fitting to data, 181
   graphing of, 172–175
squares, difference of two, 267, 347
standard deviation
   confidence intervals, 501–502
   inflection points and, 494
   normal distribution and, 490, 492–493
   for population, 490
   z-values, 499–503
standardizing the variable, 90, 500–501
standard normal distribution, 491
standard position of angle, 401
statistics, 499
   confidence intervals, 501–502
   correlation coefficient, 506–510
   least squares line, 90–92
   margin of error, 501
   predictions with. *See* probability
   samples. *See* samples
   z-values, 499–503, 507
   See also data; experimental design; normal distributions
step functions, 105
stretches and shrinks. *See* dilations
substitution, 105
   and elimination, 109–112
   in systems of equations, 104, 109–110
subtraction
   of complex numbers, 285
   of rational expressions, 364–366
sum
   partial sum of series, 243–246
   of the residuals, 82, 84
   See also addition
surveys (experimental design), 484–487

symmetry, line of, 165, 196
systems of equations, 102
   consistent, 111
   dependent, 111
   inconsistent, 111
   independent, 111
systems of equations, solving
   elimination and, 110–112
   exponential curves, 258
   graphing, 102–103
   greatest integer function, 105
   nonlinear, 141
   substitution and, 105, 109–110
systems of inequalities, 120–123
   constraints in, 121–123
   feasible region in, 121–123, 127–130
   graphing solutions for, 120–123
   linear programming and, 127–130
   nonlinear, 123, 141
   operations with, 120

T

Take Another Look
   area, 445
   curves, 383
   exponential curves, 258
   gradians, 445
   loans, 140
   normal distributions, 518
   polynomials, 315, 383
   problem solving, 25
   rational functions, 258, 383
   recursion, 99
   regression analysis, 518
   series, 258
   solving nonlinear systems of equations, 141
   transformations, 198
   trigonometry, 445
tangent, 386
   circle trigonometry and, 422
   identities with, 436–438
   for real angle values, 393–395
   reciprocal of (cotangent, cot), 436
   See also trigonometry
term, 29, 318
terminal side, 401
terms
   general, 29
   graphs of, 54
   of polynomials, 318
   of sequences, 29, 38
   starting, choice of, 38
   See also series
theorems
   Factor, 344
   Pythagorean, 188
   Rational Root, 345

theoretical probability, 450–451. *See also* probability
time, as independent variable, 69
transformations, 165, 188–191
   of circles, 188–191
   defined, 165
   dilations. *See* dilations
   of exponential functions, 203
   nonrigid, 181
   of normal distributions, 490, 492
   of rational functions, 217, 352–354
   reflections. *See* reflections
   rigid, 181
   scale factor and, 180
   summary of, 191
   translations. *See* translations
   of trigonometric functions, 417–423
translations, 158, 191
   of functions, 160
   image, 160
   of linear functions, 158–161
   and quadratic family, 165–168
   of quadratic functions, 165–168, 181, 260
   of rational functions, 354
   as rigid transformation, 181
   summary of, 191
   of trigonometric functions, 414–420, 422
treatments, 484–485
tree diagram, 457
triangle(s)
   area of, 445
   reference, 395, 401
   right. *See* Pythagorean Theorem; trigonometry
trigonometric equations, 429–433
trigonometric functions, 417–423
   cyclical motion and, 399–403
   extending, 393–395
   graphs of. *See* sinusoids (sine waves)
   inverses of, 387–388
   for real angle values, 395
   for real angle values on the coordinate plane, 393–395
   right triangle, 385–390
   transformations of, 417–423
trigonometry, 385
   coterminal angles, 403
   cyclical motion and, 399–401
   defined, 385
   ratios of, 385–390
   reference angle, 401
   rounding of values, 388
   standard position of angle, 401
   terminal side of angle, 401
trinomials
   defined, 318
   factored, 267
   perfect-square, 267

union, 466
unit circle, 188–189
    defined, 188
    equation of, 188
    reference angle on, 401
    standard position of angle in, 401
    terminal side of angle in, 401
    transformation of, 188–189
units of measure, dimensional analysis, 410

variables
    correlation of, 506–510
    dependent. *See* dependent variable
    explanatory, 508
    independent. *See* independent variable
    lurking, 510
    response, 508
variation
    combined, 377
    constant of, 14
    direct, 14
    inverse, 352
    joint, 377
velocity, defined, 66
Venn diagrams, 286
Venn, John, 286
vertex, 121, 165, 166
    of a parabola, 165
    transformations and, 181
vertex form of quadratic functions
    completing the square to find, 268–272
    equation, 260–262
    factored form and, 269, 388
    formulas for h and k to find, 272
    quadratic formula and, 277–278
vertical asymptotes, 352–353, 359
vertical dilation, 181, 184
vertical line test, 149
vertical reflection, 173
volume
    of a cube, 327
    cubic polynomials and, 334

x-intercepts, finding
    factored form, 328
    graphing, 277, 347
    zero-product, 261

y-intercepts, of linear equation, 62

zero exponents, 210
zero-product property, 261–262
zeros of a function
    complex conjugates, 331, 337
    defined, 261
    factored form of polynomial and, 328–331
    Factor Theorem to confirm, 343
    of higher-degree polynomials, 343–347
    quadratic formula to find, 277–280
    Rational Root Theorem to find, 345–347
    zero-product property to find, 261–262
    See also roots of an equation
z-score. *See* z-values
z-values, 499–503, 507

Photo Credits

Abbreviations: top (*t*), middle (*m*), bottom (*b*), left (*l*), right (*r*)

Cover
John Brueske/Shutterstock.com

Front Matter
iii (*t*): NASA; (*m*): Imagentle/Shutterstock.com; **iv** (*t*): National Museum of American Indian, Smithsonian Institution; (*m*): Zurijeta/Shutterstock.com; **v** (*t*): Rich Lindie/Shutterstock.com; (*m*): Josemaria Toscano/Shutterstock.com; **vi** (*t*): Jose Angel Astor Rocha/Shutterstock.com; (*m*): Vadim Petrakov/Shutterstock.com; **vii** (*t*): Erin Cadigan/Shutterstock.com; (*m*): Joseph Sohm/Shutterstock.com

Chapter 0
1 (*l*): NASA; **1** (*tr*): NASA; **1** (*br*): NASA; **4**: Georgios Kollidas/Shutterstock.com; **6**: Alfonso de Tomas/Shutterstock.com; **7** (*t*): Rawpixel.com/Shutterstock.com; **7** (*m*): Yuriy Boyko/Shutterstock.com; **7** (*b*): MSPhotographic/Shutterstock.com; **9**: Minerva Studio/Shutterstock.com; **11**: megainarmy/Shutterstock.com; **12**: JOAT/Shutterstock.com; **13**: Jiri Vaclavek/Shutterstock.com; **14**: Dainis Derics/Shutterstock.com; **15** (*t*): Digital Storm/Shutterstock.com; **15** (*bl*): Carollux/Shutterstock.com; **15** (*br*): Chris Kontoravdis/Shutterstock.com; **16**: Ken Karp Photography; **17** (*t*): Jan Schneckenhaus/Shutterstock.com; **17** (*b*): Daniel Castor; **19** (*t*): GryT/Shutterstock.com; **19** (*b*): Eric Isselee/Shutterstock.com; **20**: Cheryl Fenton; **21** (*tl*): Jan Schneckenhaus/Shutterstock.com; **21** (*mr*): Digital Storm/Shutterstock.com; **23**: NASA

Chapter 1
27 (*l*): Imagentle/Shutterstock.com; **27** (*tr*): mountainpix/Shutterstock.com; **27** (*br*): mezzotint/Shutterstock.com; **28** (*ml*): Photoseeker/Shutterstock.com; **28** (*mr*): angellodeco/Shutterstock.com; **30**: observe.co/Shutterstock.com; **32** (*t*): Rawpixel.com/Shutterstock.com; **32** (*b*): tristan tan/Shutterstock.com; **35**: Javier Brosch/Shutterstock.com; **36**: Ken Karp Photography; **37**: bikeriderlondon/Shutterstock.com; **38**: Fotoluminate LLC/Shutterstock.com; **39**: Ken Karp Photography; **41**: thinz/Shutterstock.com; **43**: Robert Pernell/Shutterstock.com; **44** (*t*): Microgen/Shutterstock.com; **44** (*b*): Anton_Ivanov/Shutterstock.com; **45**: mikecphoto/Shutterstock.com; **46** (*l*): bikeriderlondon/Shutterstock.com; **46** (*r*): P.Burghardt/Shutterstock.com; **47**: Ken Karp Photography; **48**: Kenneth Sponsler/Shutterstock.com; **49**: Ernest R. Prim/Shutterstock.com; **51**: Zigzag Mountain Art/Shutterstock.com; **52**: Cheryl Fenton; **54** (*tl*): Cheryl Fenton; **54** (*tr*): Cheryl Fenton; **54** (*b*): NOAA; **56**: Jeremy Brown/Shutterstock.com; **58**: PandaWild/Shutterstock.com; **59**: Lepas/Shutterstock.com; **60**: Copyright ©1994 by the Estate of Janice Stiefel. Reprinted by permission; **61**: Lissandra Melo/Shutterstock.com; **62**: RF; **64**: wavebreakmedia/Shutterstock.com; **65**: Sean Pavone/Shutterstock.com; **66**: Denis Vrublevski/Shutterstock.com; **68**: View Apart/Shutterstock.com; **69**: Ken Karp Photography; **72**: drserg/Shutterstock.com; **73**: melissaf84/Shutterstock.com; **76**: Ana Phelps/Shutterstock.com; **78** (*tl*): CVADRAT/Shutterstock.com; **78** (*mr*): Eric Broder Van Dyke/Shutterstock.com; **81**: Ken Karp Photography; **88** (*mr*): Cheryl Fenton; **88** (*br*): Everett Historical/Shutterstock.com; **90**: mnoa357/Shutterstock.com; **94**: John Panella/Shutterstock.com; **95** (*tl*): PandaWild/Shutterstock.com; **95** (*mr*): NOAA; **98**: marchesini62/Shutterstock.com

Chapter 2
101 (*l*): Joseph Sohm/Shutterstock.com; **101** (*tr*): Aspen Photo/Shutterstock.com; **101** (*br*): PHB.cz (Richard Semik)/Shutterstock.com; **102**: ZWD/Shutterstock.com; **107**: decade3d/Shutterstock.com; **108**: Microgen/Shutterstock.com; **109**: National Museum of American Indian, Smithsonian Institution; **109**: National Museum of American Indian, Smithsonian Institution; **113** (*tr*): Ken Karp Photography; **113** (*br*): ©Doug Martin/Science Source; **115**: Josemaria Toscano/Shutterstock.com; **116**: ARENA Creative/Shutterstock.com; **122**: Claudio Stocco/Shutterstock.com; **125**: Phovoir/Shutterstock.com; **126**: NASA; **127**: ImageFlow/Shutterstock.com; **132** (*tr*): Rafael Ramirez Lee/Shutterstock.com; **132** (*mr*): Plus69/Shutterstock.com; **133**: Vladitto/Shutterstock.com; **136**: bikeriderlondon/Shutterstock.com; **138** (*tl*): PHB.cz (Richard Semik); **138** (*mr*): ImageFlow/Shutterstock.com; **139**: Elena Dijour/Shutterstock.com; **140**: Orhan Cam/Shutterstock.com

Chapter 3
142 (*l*): NASA; **142** (*tr*): Bennyartist/Shutterstock.com; **142** (*br*): Zurijeta/Shutterstock.com; **144**: GWImages/Shutterstock.com; **147**: Stefano Garau/Shutterstock.com; **154**: Anne Kitzman/Shutterstock.com; **155**: Ella Hanochi/Shutterstock.com; **155**: Ella Hanochi/Shutterstock.com; **156** (*tr*): Peter Hermes Furian/Shutterstock.com; **156** (*mr*): wavebreakmedia/Shutterstock.com; **157**: dedek/Shutterstock.com; **158** (*m*): amophoto.net/Shutterstock.com; **158** (*b*): Ken Karp Photography; **160**: Osugi/Shutterstock.com; **161**: Ivan_Sabo/Shutterstock.com; **163**: Faiz Zaki/Shutterstock.com; **165**: raindrop74/Shutterstock.com; **166**: Neil Weaver Photography/Shutterstock.com; **170**: Jody Dingle/Shutterstock.com; **172**: Ken Karp Photography; **173**: Jennifer White Maxwell/Shutterstock.com; **178** (*tr*): Sopotnicki/Shutterstock.com; **178** (*mr*): stockernumber2/Shutterstock.com; **183**: ©Stefano Bianchetti/Getty Images; **187**: Everett Historical/Shutterstock.com; **188**: Laborant/Shutterstock.com; **190**: huyangshu/Shutterstock.com; **196** (*tl*): Ivan_Sabo/Shutterstock.com; **196** (*mr*): Neil Weaver Photography/Shutterstock.com

Chapter 4
200 (*l*): Cylonphoto/Shutterstock.com; **200** (*tr*): Rich Lindie/Shutterstock.com; **200** (*br*): Zurijeta/Shutterstock.com; **204**: Sergey Kamshylin/Shutterstock.com; **206**: Jerry Whaley/Shutterstock.com; **207**: Trong Nguyen/Shutterstock.com; **213**: Eric Kamischke; **216**: Tupungato/Shutterstock.com; **219**: Ken Wolter/Shutterstock.com; **220**: Photographee.eu/Shutterstock.com; **221**: melis/Shutterstock.com; **223** (*tr*): cigdem/Shutterstock.com; **223** (*b*): Rich Carey/Shutterstock.com; **224**: Maciej Es/Shutterstock.com; **228** (*mr*): Ken Karp Photography; **228** (*br*): Catmando/Shutterstock.com; **229**: NASA; **235**: Galyna Andrushko/Shutterstock.com; **241** (*tr*): Sean Lema/Shutterstock.com; **241** (*m*): dreamerve/Shutterstock.com; **242**: otnaydur/Shutterstock.com; **243**: Wojtek Chmielewski/Shutterstock.com; **245**: Nicku/Shutterstock.com; **247**: kojoku/Shutterstock.com; **248**: Eunika Sopotnicka/Shutterstock.com; **249**: swa182/Shutterstock.com; **250**: Ken Karp Photography; **255** (*tl*): Rich Carey/Shutterstock.com; **255** (*mr*): Ken Wolter/Shutterstock.com

Chapter 5
259 (*l*): Andrew Zarivny/Shutterstock.com; **259** (*tr*): Josemaria Toscano/Shutterstock.com; **259** (*br*): siriratt/Shutterstock.com; **260**: Andrew Zarivny/Shutterstock.com; **263**: Ken Karp Photography; **264**: Ken Karp Photography; **266**: EastVillage Images/Shutterstock.com; **268**: EastVillage Images/Shutterstock.com; **269**: Ken Karp Photography; **271**: Elisabeta Stan/Shutterstock.com; **272**: Jan de Wild/Shutterstock.com; **275** (*tr*): Ken Karp Photography; **275** (*ml*): IgorGolovniov/Shutterstock.com; **275** (*mr*): GongTo/Shutterstock.com; **276** (*tr*): win Verin/Shutterstock.com; **276** (*ml*): StevenRussellSmithPhotos/Shutterstock.com; **276** (*mr*): javarman/Shutterstock.com; **279**: fotandy/Shutterstock.com; **282**: abadesign/Shutterstock.com; **283**: Georgios Kollidas/Shutterstock.com; **285**: Paul Fleet/Shutterstock.com; **292**: vladimir salman/Shutterstock.com; **293**: MidoSemsem/Shutterstock.com; **296**: jiawangkun/Shutterstock.com; **297**: Anton Brand/Shutterstock.com; **298**: Venturelli Luca/Shutterstock.com; **301**: gor Kolos/Shutterstock.com; **302**: daksun/Shutterstock.com; **304**: amolson7/Shutterstock.com; **305**: VladFree/Shutterstock.com; **307**: Supannee Hickman/

Shutterstock.com; **311:** Rosliak Oleksandr/Shutterstock.com; **312 (t):** PavelSh/Shutterstock.com; **312 (b):** Lyudmyla Kharlamova/Shutterstock.com; **313 (tl):** Mariia Golovianko/Shutterstock.com; **313 (mr):** Lyudmyla Kharlamova/Shutterstock.com; **315:** Piotr Sikora/Shutterstock.com

Chapter 6

317 (l): Jose Angel Astor Rocha/Shutterstock.com; **317 (tr):** ©Science Museum/Science & Society Picture Library; **317 (br):** Julien Hautcoeur/Shutterstock.com; **319:** CHEN WS/Shutterstock.com; **321:** Ken Karp Photography; **322:** CHEN WS/Shutterstock.com; **325:** Stuart Monk/Shutterstock.com; **328:** Heather Shimmin/Shutterstock.com; **329:** Ken Karp Photography; **332:** udra11/Shutterstock.com; **341:** powell'sPoint/Shutterstock.com; **342:** angellodeco/Shutterstock.com; **351:** photo.ua/Shutterstock.com; **375:** Prawat/Shutterstock.com; **378:** Troutnut/Shutterstock.com; **379 (m):** ideation90/Shutterstock.com; **379 (b):** Ben Schonewille/Shutterstock.com; **380:** meunierd/Shutterstock.com; **381 (tl):** Mariia Golovianko/Shutterstock.com; **381 (mr):** ideation90/Shutterstock.com

Chapter 7

384 (l): Felix Lipov/Shutterstock.com; **384 (tr):** Vadim Petrakov/Shutterstock.com; **384 (br):** Ihor Pasternak/Shutterstock.com; **385:** ©John Elk/Getty Images; **387:** lzf/Shutterstock.com; **388:** ChristianChan/Shutterstock.com; **396:** ©Ian Walton/Staff/Getty Images; **398:** Christian Delbert/Shutterstock.com; **399:** Jose Lledo/Shutterstock.com; **405:** flowgraph/Shutterstock.com; **406:** Steve Bower/Shutterstock.com; **407:** RYGER/Shutterstock.com; **411:** Brian King/Shutterstock.com; **413:** Sarine Arslanian/Shutterstock.com; **414:** Michel Piccaya/Shutterstock.com; **417:** Tawin Mukdharakosa/Shutterstock.com; **420:** Ken Karp Photography; **425:** Bertl123/Shutterstock.com; **429 (ml):** EpicStockMedia/Shutterstock.com; **429 (mr):** Ethan Daniels/Shutterstock.com; **435:** ©Can Stock Photo Inc./danabeth555; **440 (t):** Josef Hanus/Shutterstock.com; **440 (b):** Verena Matthew/Shutterstock.com; **442 (tl):** Brian King/Shutterstock.com; **442 (mr):** EpicStockMedia/Shutterstock.com

Chapter 8

446 (l): deenphoto/Shutterstock.com; **446 (tr):** Erin Cadigan/Shutterstock.com; **446 (br):** Ihor mj007/Shutterstock.com; **447:** Tootles/Shutterstock.com; **449:** Federica Milella/Shutterstock.com; **451:** evantravels/Shutterstock.com; **452:** vadim kozlovsky/Shutterstock.com; **453:** rook76/Shutterstock.com; **461:** Ken Karp Photography; **464:** zlikovec/Shutterstock.com; **466:** Faraways/Shutterstock.com; **469:** Ken Karp Photography; **473:** Monkey Business Images/Shutterstock.com; **475:** Georgejmclittle/Shutterstock.com; **476:** Marcel Jancovic/Shutterstock.com; **477 (t):** Steve Design/Shutterstock.com; **477 (b):** Herrndorff/Shutterstock.com; **478:** Everett Historical/Shutterstock.com; **479 (tl):** zlikovec/Shutterstock.com; **479 (mr):** Georgejmclittle/Shutterstock.com; **480:** Stuart Monk/Shutterstock.com

Chapter 9

483 (l): Joseph Sohm/Shutterstock.com; **483 (tr):** Katherine Welles/Shutterstock.com; **483 (br):** Gil C/Shutterstock.com; **484:** CandyBox Images/Shutterstock.com; **485:** antoniodiaz/Shutterstock.com; **486:** ©Yves Forestier/Contributor/Getty Images; **488:** Studio KIWI/Shutterstock.com; **489:** ©Underwood Archives/Contributor/Getty Images; **495:** Everett Historical/Shutterstock.com; **496:** Willrow Hood/Shutterstock.com; **498 (t):** Andy Piatt/Shutterstock.com; **498 (b):** Dionisvera/Shutterstock.com; **499:** Rawpixel.com/Shutterstock.com; **502:** Pete Niesen/Shutterstock.com; **504 (t):** photo.ua/Shutterstock.com; **504 (b):** Aleksei Semykin/Shutterstock.com; **506:** Oscity/Shutterstock.com; **510:** Marcio Jose Bastos Silva/Shutterstock.com; **512:** Frontpage/Shutterstock.com; **512:** Frontpage/Shutterstock.com; **514 (l):** Josh Withers/Shutterstock.com; **514 (r):** Mat Hayward/Shutterstock.com; **515 (tl):** Studio KIWI/Shutterstock.com; **515 (mr):** Pete Niesen/Shutterstock.com; **516:** Drpixel/Shutterstock.com; **517:** Jeff Kinsey/Shutterstock.com

Common Core State Standards and *Discovering Advanced Algebra*

Number and Quantity

The Real Number System
Extend the properties of exponents to rational exponents.

N.RN.1   Explain how the definition of the meaning of rational exponents follows from extending the properties of integer exponents to those values, allowing for a notation for radicals in terms of rational exponents.

N.RN.2   Rewrite expressions involving radicals and rational exponents using the properties of exponents.

Quantities*
Reason quantitatively and use units to solve problems.

N.Q.2   Define appropriate quantities for the purpose of descriptive modeling.

The Complex Number System
Perform arithmetic operations with complex numbers.

N.CN.1   Know there is a complex number i such that $i^2 = -1$, and every complex number has the form $a + bi$ with a and b real.

N.CN.2   Use the relation $i^2 = -1$ and the commutative, associative, and distributive properties to add, subtract, and multiply complex numbers.

N.CN.3   (+) — Find the conjugate of a complex number; use conjugates to find moduli and quotients of complex numbers.

Use complex numbers in polynomial identities and equations.

N.CN.7   Solve quadratic equations with real coefficients that have complex solutions.

Algebra

Seeing Structure in Expressions
Interpret the structure of expressions.

A.SSE.1   Interpret expressions that represent a quantity in terms of its context.*

     A.SSE.1a   Interpret parts of an expression, such as terms, factors, and coefficients.

     A.SSE.1b   Interpret complicated expressions by viewing one or more of their parts as a single entity.

A.SSE.2   Use the structure of an expression to identify ways to rewrite it.

Write expressions in equivalent forms to solve problems.

A.SSE.3   Choose and produce an equivalent form of an expression to reveal and explain properties of the quantity represented by the expression.*

     A.SSE.3a   Factor a quadratic expression to reveal the zeros of the function it defines.

     A.SSE.3b   Complete the square in a quadratic expression to reveal the maximum or minimum value of the function it defines.

     A.SSE.3c   Use the properties of exponents to transform expressions for exponential functions.

A.SSE.4   Derive the formula for the sum of a finite geometric series (when the common ratio is not 1), and use the formula to solve problems.

Arithmetic with Polynomials and Rational Expressions
Perform arithmetic operations on polynomials.

A.APR.1   Understand that polynomials form a system analogous to the integers, namely, they are closed under the operations of addition, subtraction, and multiplication; add, subtract, and multiply polynomials.

Understand the relationship between zeros and factors of polynomials.

A.APR.2   Know and apply the Remainder Theorem: For a polynomial $p(x)$ and a number a, the remainder on division by $x - a$ is $p(a)$, so $p(a) = 0$ if and only if $(x - a)$ is a factor of $p(x)$.

A.APR.3   Identify zeros of polynomials when suitable factorizations are available, and use the zeros to construct a rough graph of the function defined by the polynomial.

Use polynomial identities to solve problems.

A.APR.4   Prove polynomial identities and use them to describe numerical relationships.

Rewrite rational expressions.

A.APR.6   Rewrite simple rational expressions in different forms; write $\frac{a(x)}{b(x)}$ in the form $q(x) + \frac{r(x)}{b(x)}$, where $a(x)$, $b(x)$, $q(x)$, and $r(x)$ are polynomials with the degree of $r(x)$ less than the degree of $b(x)$, using inspection, long division, or, for the more complicated examples, a computer algebra system.

A.APR.7 (+) Understand that rational expressions form a system analogous to the rational numbers, closed under addition, subtraction, multiplication, and division by a nonzero rational expression; add, subtract, multiply, and divide rational expressions.

Creating Equations*
Create equations that describe numbers or relationships.

A.CED.1 Create equations and inequalities in one variable and use them to solve problems. *Include equations arising from linear and quadratic functions, and simple rational and exponential functions.*

A.CED.2 Create equations in two or more variables to represent relationships between quantities; graph equations on coordinate axes with labels and scales.

A.CED.3 Represent constraints by equations or inequalities, and by systems of equations and/or inequalities, and interpret solutions as viable or nonviable options in a modeling context.

Reasoning with Equations and Inequalities
Understand solving equations as a process of reasoning and explain the reasoning.

A.REI.1 Explain each step in solving a simple equation as following from the equality of numbers asserted at the previous step, starting from the assumption that the original equation has a solution. Construct a viable argument to justify a solution method.

A.REI.2 Solve simple rational and radical equations in one variable, and give examples showing how extraneous solutions may arise.

Solve equations and inequalities in one variable.

A.REI.4 Solve quadratic equations in one variable.

A.REI.4a Use the method of completing the square to transform any quadratic equation in x into an equation of the form $(x - p)^2 = q$ that has the same solutions. Derive the quadratic formula from this form.

A.REI.4b Solve quadratic equations by inspection (e.g., for $x^2 = 49$), taking square roots, completing the square, the quadratic formula and factoring, as appropriate to the initial form of the equation. Recognize when the quadratic formula gives complex solutions and write them as $a \pm bi$ for real numbers a and b.

Solve systems of equations.

A.REI.5 Prove that, given a system of two equations in two variables, replacing one equation by the sum of that equation and a multiple of the other produces a system with the same solutions.

A.REI.6 Solve systems of linear equations exactly and approximately (e.g., with graphs), focusing on pairs of linear equations in two variables.

A.REI.7 Solve a simple system consisting of a linear equation and a quadratic equation in two variables algebraically and graphically.

Represent and solve equations and inequalities graphically.

A.REI.10 Understand that the graph of an equation in two variables is the set of all its solutions plotted in the coordinate plane, often forming a curve (which could be a line).

A.REI.11 Explain why the x-coordinates of the points where the graphs of the equations $y = f(x)$ and $y = g(x)$ intersect are the solutions of the equation $f(x) = g(x)$; find the solutions approximately, e.g., using technology to graph the functions, make tables of values, or find successive approximations. Include cases where $f(x)$ and/or $g(x)$ are linear, polynomial, rational, absolute value, exponential, and logarithmic functions.*

A.REI.12 Graph the solutions to a linear inequality in two variables as a half-plane (excluding the boundary in the case of a strict inequality), and graph the solution set to a system of linear inequalities in two variables as the intersection of the corresponding half-planes.

Functions

Interpreting Functions
Understand the concept of a function and use function notation.

F.IF.1 Understand that a function from one set (called the domain) to another set (called the range) assigns to each element of the domain exactly one element of the range. If f is a function and x is an element of its domain, then $f(x)$ denotes the output of f corresponding to the input x. The graph of f is the graph of the equation $y = f(x)$.

F.IF.2 Use function notation, evaluate functions for inputs in their domains, and interpret statements that use function notation in terms of a context.

F.IF.3 Recognize that sequences are functions, sometimes defined recursively, whose domain is a subset of the integers.

Interpret functions that arise in applications in terms of the context.

F.IF.4 For a function that models a relationship between two quantities, interpret key features of graphs and tables in terms of the quantities, and sketch graphs showing key features given a verbal description of the relationship.*

F.IF.5 Relate the domain of a function to its graph and, where applicable, to the quantitative relationship it describes.*

F.IF.6 Calculate and interpret the average rate of change of a function (presented symbolically or as a table) over a specified interval. Estimate the rate of change from a graph.*

Analyze functions using different representations.

F.IF.7 Graph functions expressed symbolically and show key features of the graph, by hand in simple cases and using technology for more complicated cases.*

F.IF.7a Graph linear and quadratic functions and show intercepts, maxima, and minima.

F.IF.7b Graph square root, cube root, and piecewisedefined functions, including step functions and absolute value functions.

F.IF.7c Graph polynomial functions, identifying zeros when suitable factorizations are available, and showing end behavior.

F.IF.7d (+) Graph rational functions, identifying zeros and asymptotes when suitable factorizations are available, and showing end behavior.

F.IF.7e Graph exponential and logarithmic functions, showing intercepts and end behavior, and trigonometric functions, showing period, midline, and amplitude.

F.IF.8 Write a function defined by an expression in different but equivalent forms to reveal and explain different properties of the function.

F.IF.8a Use the process of factoring and completing the square in a quadratic function to show zeros, extreme values, and symmetry of the graph, and interpret these in terms of a context.

F.IF.8b Use the properties of exponents to interpret expressions for exponential functions.

F.IF.9 Compare properties of two functions each represented in a different way (algebraically, graphically, numerically in tables, or by verbal descriptions).

Building Functions
Build a function that models a relationship between two quantities.

F.BF.1 Write a function that describes a relationship between two quantities.*

F.BF.1a Determine an explicit expression, a recursive process, or steps for calculation from a context.

F.BF.1b Combine standard function types using arithmetic operations.

F.BF.2 Write arithmetic and geometric sequences both recursively and with an explicit formula, use them to model situations, and translate between the two forms.*

Build new functions from existing functions.

F.BF.3 Identify the effect on the graph of replacing $f(x)$ by $f(x) + k$, $k f(x)$, $f(kx)$, and $f(x + k)$ for specific values of k (both positive and negative); find the value of k given the graphs. Experiment with cases and illustrate an explanation of the effects on the graph using technology. Include recognizing even and odd functions from their graphs and algebraic expressions for them.

F.BF.4 Find inverse functions.

F.BF.4a Solve an equation of the form $f(x) = c$ for a simple function f that has an inverse and write an expression for the inverse.

F.BF.5 (+) Understand the inverse relationship between exponents and logarithms and use this relationship to solve problems involving logarithms and exponents.

Linear, Quadratic, and Exponential Models*
Construct and compare linear, quadratic, and exponential models and solve problems.

F.LE.1 Distinguish between situations that can be modeled with linear functions and with exponential functions.

F.LE.1a Prove that linear functions grow by equal differences over equal intervals, and that exponential functions grow by equal factors over equal intervals.

F.LE.1b Recognize situations in which one quantity changes at a constant rate per unit interval relative to another.

F.LE.1c Recognize situations in which a quantity grows or decays by a constant percent rate per unit interval relative to another.

F.LE.2 Construct linear and exponential functions, including arithmetic and geometric sequences, given a graph, a description of a relationship, or two input-output pairs (include reading these from a table).

F.LE.3 Observe using graphs and tables that a quantity increasing exponentially eventually exceeds a quantity increasing linearly, quadratically, or (more generally) as a polynomial function.

F.LE.4 For exponential models, express as a logarithm the solution to $ab^{ct} = d$ where a, c, and d are numbers and the base b is 2, 10, or e; evaluate the logarithm using technology.

Interpret expressions for functions in terms of the situation they model.

F.LE.5 Interpret the parameters in a linear or exponential function in terms of a context.

Trigonometric Functions
Extend the domain of trigonometric functions using the unit circle.

F.TF.1 Understand radian measure of an angle as the length of the arc on the unit circle subtended by the angle.

F.TF.2 Explain how the unit circle in the coordinate plane enables the extension of trigonometric functions to all real numbers, interpreted as radian measures of angles traversed counterclockwise around the unit circle.

F.TF.3 (+) Use special triangles to determine geometrically the values of sine, cosine, tangent for $\frac{\pi}{3}$, $\frac{\pi}{4}$, and $\frac{\pi}{6}$ and and use the unit circle to express the values of sine, cosine, and tangent for $\pi - x$, $\pi + x$, and $2\pi - x$, in terms of their values for x, where x is any real number.

F.TF.4 (+) Use the unit circle to explain symmetry (odd and even) and periodicity of trigonometric functions.

Model periodic phenomena with trigonometric functions.

F.TF.5 Choose trigonometric functions to model periodic phenomena with specified amplitude, frequency, and midline.*

Prove and apply trigonometric identities.

F.TF.8 Prove the Pythagorean identity $\sin^2(\theta) + \cos^2(\theta) = 1$ and use it to calculate trigonometric ratios.

Geometry
Similarity, Right Triangles, and Trigonometry
Define trigonometric ratios and solve problems involving right triangles.

G.SRT.6 Understand that by similarity, side ratios in right triangles are properties of the angles in the triangle, leading to definitions of trigonometric ratios for acute angles.

G.SRT.7 Explain and use the relationship between the sine and cosine of complementary angles.

G.SRT.8 Use trigonometric ratios and the Pythagorean Theorem to solve right triangles in applied problems.*

Circles
Find arc lengths and areas of sectors of circles.

G.C.5 Derive using similarity the fact that the length of the arc intercepted by an angle is proportional to the radius, and define the radian measure of the angle as the constant of proportionality; derive the formula for the area of a sector.

Expressing Geometric Properties with Equations
Translate between the geometric description and the equation for a conic section.

G.GPE.1 Derive the equation of a circle of given center and radius using the Pythagorean Theorem; complete the square to find the center and radius of a circle given by an equation.

G.GPE.2 Derive the equation of a parabola given a focus and directrix.

Statistics and Probability
Interpreting Categorical and Quantitative Data
Summarize, represent, and interpret data on a single count or measurement variable.

S.ID 4 Use the mean and standard deviation of a data set to fit it to a normal distribution and to estimate population percentages. Recognize that there are data sets for which such a procedure is not appropriate. Use calculators, spreadsheets and tables to estimate areas under the normal curve.

Summarize, represent, and interpret data on two categorical and quantitative variables.

S.ID 5 Summarize categorical data for two categories in two-way frequency tables. Interpret relative frequencies in the context of the data (including joint, marginal and conditional relative frequencies). Recognize possible associations and trends in the data.

S.ID 6 Represent data on two quantitative variables on a scatter plot and describe how the variables are related.

S.ID 6a Fit a function to data; use functions fitted to the data to solve problems in the context of the data. Use given model functions or choose a function suggested by the context. Emphasize linear, quadratic, and exponential models.

S.ID 6b Informally assess the fit of a model function by plotting and analyzing residuals.

S.ID 6c Fit a linear function for a scatter plot that suggests a linear association.

Interpret linear models.

S.ID 7 Interpret the slope (rate of change) and the intercept (constant term) of a linear model in the context of the data.

S.ID 8 Compute (using technology) and interpret the correlation coefficient of a linear fit.

S.ID 9 Distinguish between correlation and causation.

Making Inferences and Justifying Conclusions
Understand and evaluate random processes underlying statistical experiments.

S.IC.1 Understand statistics as a process for making inferences about population parameters based on a random sample from that population.

S.IC.2 Decide if a specified model is consistent with results from a given data-generating process, e.g., using simulation.

Make inferences and justify conclusions from sample surveys, experiments, and observational studies.

S.IC.3 Recognize the purposes of and differences among sample surveys, experiments and observational studies; explain how randomization relates to each.

S.IC.4 Use data from a sample survey to estimate a population mean or proportion; develop a margin of error through the use of simulation models for random sampling.

S.IC.5 Use data from a randomized experiment to compare two treatments; justify significant differences between parameters through the use of simulation models for random assignment.

S.IC.6 Evaluate reports based on data.

Conditional Probability and the Rules of Probability
Understand independence and conditional probability and use them to interpret data.

S.CP.1 Describe events as subsets of a sample space (the set of outcomes) using characteristics (or categories) of the outcomes, or as unions, intersections, or complements of other events ("or," "and," "not").

S.CP.2 Understand that two events A and B are independent if the probability of A and B occurring together is the product of their probabilities, and use this characterization to determine if they are independent.

S.CP.3 Understand the conditional probability of A given B as P(A and B)/P(B), and interpret independence of A and B as saying that the conditional probability of A given B is the same as the probability of A, and the conditional probability of B given A is the same as the probability of B.

S.CP.4 Construct and interpret two-way frequency tables of data when two categories are associated with each object being classified. Use the two-way table as a sample space to decide if events are independent and to approximate conditional probabilities.

S.CP.5 Recognize and explain the concepts of conditional probability and independence in everyday language and everyday situations.

Use the rules of probability to compute probabilities of compound events in a uniform probability model.

S.CP.6 Find the conditional probability of A given B as the fraction of B's outcomes that also belong to A and interpret the answer in terms of the model.

S.CP.7 Apply the Addition Rule, P(A or B) = P(A) + P(B) − P(A and B), and interpret the answer in terms of the model.

S.CP.8 (+) Apply the general Multiplication Rule in a uniform probability model, P(A and B) = P(A)P(B|A) = P(B)P(A|B), and interpret the answer in terms of the model.

Using Probability to Make Decisions
Use probability to evaluate outcomes of decisions.

S.MD 6 (+) Use probabilities to make fair decisions (e.g., drawing by lots, using a random number generator).